Mathematical Physics

Third Edition

for MSc (Physics), BSc (Hons/Pass), TDC, BTech, GATE, NET/SLET (SET)

Mathematical Physics

Third Edition

for **MSc (Physics), BSc (Hons/Pass), TDC, BTech, GATE, NET/SLET (SET)**

SL Kakani
MSc (Physics), PhD

Former Executive Director
Institute of Technology and Management
(Affiliated to Rajasthan Technical University, Kota)
Bhilwara 311001, Rajasthan, India

C Hemrajani
MSc (Physics), PhD

Former Head
Department of Physics
MLV Government PG College
Bhilwara 311001, Rajasthan, India

CBS

CBS Publishers & Distributors Pvt Ltd

New Delhi • Bengaluru • Chennai • Kochi • Kolkata • Mumbai
Hyderabad • Jharkhand • Nagpur • Patna • Pune • Uttarakhand

Mathematical Physics

Third Edition

for MSc (Physics), BSc (Hons/Pass), TDC, BTech, GATE, NET/SLET (SET)

ISBN: 978-93-86478-23-8

Copyright © Authors and Publisher

Third Edition: 2018
Second Edition: 2004
First Edition: 1994

Published by Satish Kumar Jain and produced by Varun Jain for
CBS Publishers & Distributors Pvt Ltd
4819/XI Prahlad Street, 24 Ansari Road, Daryaganj, New Delhi 110 002, India.
Ph: 23289259, 23266861, 23266867   Website: www.cbspd.com
Fax: 011-23243014   e-mail: delhi@cbspd.com; cbspubs@airtelmail.in.
Corporate Office: 204 FIE, Industrial Area, Patparganj, Delhi 110 092
Ph: 4934 4934   Fax: 4934 4935   e-mail: publishing@cbspd.com; publicity@cbspd.com

Branches

- **Bengaluru:** Seema House 2975, 17th Cross, K.R. Road,
  Banasankari 2nd Stage, Bengaluru 560 070, Karnataka
  Ph: +91-80-26771678/79   Fax: +91-80-26771680   e-mail: bangalore@cbspd.com
- **Chennai:** 7, Subbaraya Street, Shenoy Nagar, Chennai 600 030, Tamil Nadu
  Ph: +91-44-26680620, 26681266   Fax: +91-44-42032115   e-mail: chennai@cbspd.com
- **Kochi:** Ashana House, 39/1904, AM Thomas Road, Valanjambalam,
  Ernakulam 682 018, Kochi, Kerala
  Ph: +91-484-4059061-62-64-65   Fax: +91-484-4059065   e-mail: kochi@cbspd.com
- **Kolkata:** 6/B, Ground Floor, Rameswar Shaw Road, Kolkata-700 014, West Bengal
  Ph: +91-33-22891126, 22891127, 22891128   e-mail: kolkata@cbspd.com
- **Mumbai:** 83-C, Dr E Moses Road, Worli, Mumbai-400018, Maharashtra
  Ph: +91-22-24902340/41   Fax: +91-22-24902342   e-mail: mumbai@cbspd.com

Representatives

• Hyderabad	0-9885175004	• Jharkhand	0-9811541605	• Nagpur	0-9021734563
• Patna	0-9334159340	• Pune	0-9623451994	• Uttarakhand	0-9716462459

Printed at: Mudrak, Delhi, India

Preface to the Third Edition

We feel a great pleasure in bringing out the third thoroughly revised and enlarged edition of the book *Mathematical Physics*. To make the book more useful for MSc, BSc (Pass/Hons), BTech and TDC students:

- Four new chapters "linear vector spaces; group theory; numerical methods; probability and statistics" have been included
- Green's functions are discussed in more detail
- New solved problems are added into each chapter
- New MCQs with answers are added at the end of each chapter
- Appendix with GATE (2011–2017) questions and solutions with detailed explanation

The book at a glance

- Simple and cogent style of presentation
- 13 chapters covering the latest revised syllabus of UGC and various universities
- Text supplemented with detailed illustrations
- Fortified with a good number of examples
- Solved problems and exercises with answers at the end of each chapter
- MCQs with answers at the end of each chapter to make the book helpful for GATE, UGC, TDC, NET/SLET (SET) and other all India and state level entrance examinations

We hope that this comprehensive and updated edition will be very handy and user-friendly.

We are extremely thankful and grateful to Mr Satish Kumar Jain, CMD, CBS Publishers & Distributors, New Delhi, for his continued support. We are thankful to Mr YN Arjuna, Senior Vice President Publishing, Editorial and Publicity, and his team comprising Mrs Ritu Chawla, AGM Production; Ms Sanjubala, copy editor; Mr Parmod Kumar, DTP operator and Mr Manish Raj, graphic designer, for bringing out the book in the present form.

Suggestions and constructive criticism are always welcome for effective presentation.

SL Kakani

C Hemrajani

Preface to the Second Edition

The main aim of this book is to provide a suitable text on mathematical physics which meets the requirements of students preparing for honours and postgraduate studies in physics, engineering, advanced mathematics, and AMIE of all Indian and foreign universities.

For better understanding, the language of the book is kept simple and lucid. Each chapter is interspersed with a number of solved problems and a large number of exercise problems of applied nature particularly to emphasize self-learning.

The most striking feature of this text is the inclusion of a good number of objective and short answer questions at the end of each chapter. This will make the book useful for CSIR-UGC: NET and SLET and other entrance examinations.

We are thankful to CBS Publishers & Distributors, New Delhi, for getting the book published in time.

We hope that the text will be appreciated by the readers. Suggestions for improvement of the book shall be gratefully acknowledged.

SL Kakani
C Hemrajani

Contents

7. DIFFERENTIAL EQUATIONS AND POLYNOMIALS   493–622

8. INTEGRAL TRANSFORMS   623–670

Vector Analysis

1.1 INTRODUCTION

A physical quantity possessing both magnitude and direction and can be added according to the law of parallelogram of vectors is known as a vector quantity. An arrow is used to represent a vector quantity. The direction of the quantity is indicated by the head of the arrow, and the magnitude is characterised by the length of the arrow.

Analytically, a vector is represented by a letter with an arrow over it as $\vec{A}$ and its magnitude is denoted by $|A|$ or A. Bold faced type letter, such as A is also used to indicate the vector while $|A|$ or A indicates its magnitude.

Diagrammatically, a vector is represented as shown in Fig. 1.1, where length AB represents the magnitude of vector P and arrow gives its direction. Two vectors are said to be equal when they have the same magnitude as well as direction. Thus, all vectors obtained by parallel displacement or translation of AB are equal. A vector as such remains unchanged by pure displacement. Thus, vector is not localised. The parallel vectors so obtained are called *collinear vectors*. Hence,

$$P = AB = P_1 = -P_2$$

where, P_2 has same scalar magnitude as P_1 but their directions are opposite to each other. The equal vectors that can be defined in this way are called *free vectors*.

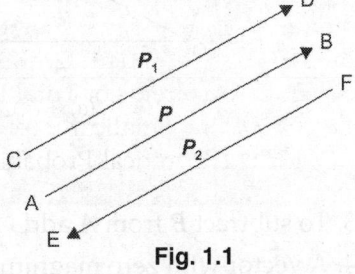

Fig. 1.1

On the other hand, there are vectors such as the forces acting on a rigid body which are restricted to lie on a given line; such vectors are not free, because they can only be shifted along a fixed line as other translations, will alter their dynamic effect. Such vectors are called *line vectors*.

The vector quantities in which the vector is drawn in the direction of the quantity concerned called *polar vectors*. Such as force, displacement, velocity, etc.

In vector quantities, where rotatory action takes place about the axis, the vector is drawn parallel to the axis about which the quantity acts, the length of the vector gives the magnitude of the quantity. Such vectors are called *axial vectors*. The

1

direction of such vectors is fixed by the right-handed screw rule, which says that, in case of clockwise rotation the vector direction is in the direction of linear motion of the screw.

The algebra of numbers (the operation of addition, subtraction, and multiplication) with suitable definition can be extended to an algebra of vectors.

Note:

1. The vectors A and B are said to be equal ($A = B$), if they have the same length and direction. The vector $-A$ is equal in magnitude to vector A but opposite in direction. The multiplication of a vector A by a scalar m, is a vector with magnitude m times the magnitude of A and direction the same as or opposite to that of A according as m is positive or negative.

2. The sum or resultant of the vectors A and B is obtained by placing the origin (initial point) of B at the terminal point of A and then joining initial point of A to the terminal point of B. Analytically, the sum is given by Fig. 1.2:

$$C = A + B \tag{1.1}$$

To add several vectors A, B and C, first find the sum of A and B and then the sum of $A + B$ and C.

From the nature of definition of vector addition, it is apparent that

$$B + A = A + B, \tag{1.2}$$

i.e. vector addition is *commutative*. Similarly

$$A + (B + C) = (A + B) + C, \tag{1.3}$$

i.e. vector addition is *associative*.

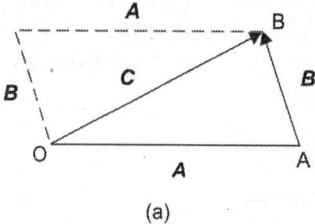

(a)   (b)

Fig. 1.2

3. To subtract B from A add $-B$ to A (Fig. 1.3).

4. A vector with zero magnitude and no specific direction is called *null vector* or *zero vector* and is represented by symbol O.

5. A vector having unit magnitude is called *unit vector*. If A is a vector with magnitude $A \neq 0$ then

$$\frac{A}{|A|} = a, \tag{1.4}$$

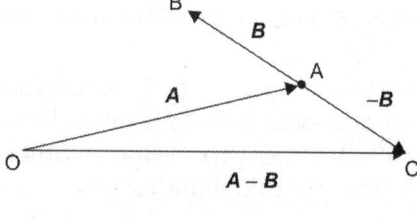

Fig. 1.3

is a unit vector, having the same direction as A but magnitude unity. The most common unit vectors are those which have the direction of right-handed cartesian coordinate system as shown in Fig. 1.4. The vector i is a unit vector having the x-direction of the coordinate system. The unit vector j and the unit vector k have the y-direction and z-direction respectively.

6. The position vector r of an object located at P(x, y, z) is shown in Fig. 1.5.

$$OP = r = OA + AB + BP$$

$$= xi + yj + zk \qquad (1.5)$$

The vectors xi, yj, zk are the three components of r. They are the vector representations of the projections of r on three coordinate axes, respectively.

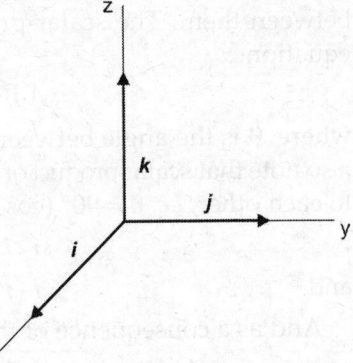

Fig. 1.4

The quantities x, y, and z are the magnitudes of the vector components in the three respective directions. From Fig. 1.5, we find that

$$|OP|^2 = |r|^2 = |OA|^2 + |AB|^2 + |BP|^2$$

$$= x^2 + y^2 + z^2 \qquad (1.6)$$

where $|r|$ is the magnitude of vector r.

7. If the projections of an arbitrary vector A along the three axes of cartesian coordinate system are A_x, A_y, A_z, then the vector A in terms of these three components may be written as

$$A = A_x i + A_y j + A_z k \qquad (1.7)$$

The magnitude of A is given by

$$|A| = (A_x^2 + A_y^2 + A_z^2)^{1/2} \qquad (1.8)$$

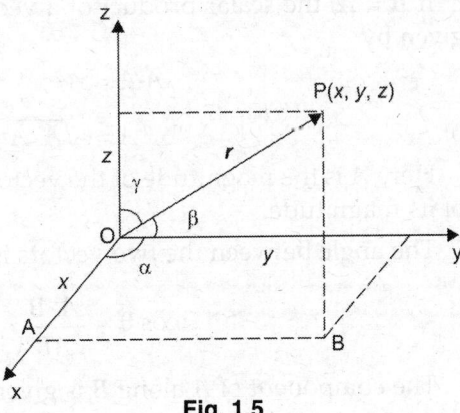

Fig. 1.5

In terms of components, the quantity $A = B$ means

$$A_x = B_x, A_y = B_y, A_z = B_z \qquad (1.9)$$

8. The angles α, β and γ are the angles that OP makes with the three coordinate axes. Here, we have

$$x = rl, y = rm \text{ and } z = rn \qquad (1.10)$$

where, $l = \cos \alpha$, $m = \cos \beta$, and $n = \cos \gamma$. The quantities l, m, n are called direction cosines. We observe that

$$l^2 + m^2 + n^2 = 1 \qquad (1.11)$$

9. For an arbitrary vector A, we may write

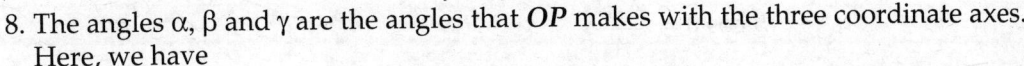

$$a = \frac{A}{|A|} = i \cos \alpha + j \cos \beta + k \cos \gamma \qquad (1.12)$$

Clearly, the quantity $a = \dfrac{A}{|A|}$ is a unit vector since its magnitude is unity.

1.2 SCALAR OR DOT PRODUCT

The scalar or dot product of two vectors is a scalar quantity equal in magnitude to the product of the magnitudes of two given vectors and the cosine of the angle between them. The scalar product of the two vectors A and B is thus given by the equation:

$$A \cdot B = |A| \, |B| \cos \theta \tag{1.13}$$

where, θ is the angle between the directions of two vectors. From Eq. (1.13), we may also note that scalar product of two nonzero vectors will be zero if they are perpendicular to each other, *i.e.* $\theta = 90°$ ($\cos 90° = 0$). Using Eq. (1.13), we get

$$i \cdot i = j \cdot j = k \cdot k = 1 \tag{1.14}$$

and

$$i \cdot j = j \cdot k = k \cdot i = 0 \tag{1.15}$$

And as a consequence of these results, we obtain

$$A \cdot B = (iA_x + jA_y + kA_z) \cdot (iB_x + jB_y + kB_z)$$
$$= A_x B_x + A_y B_y + A_z B_z \tag{1.16}$$

If $B = A$, the scalar product of a vector with itself is called its self product, and is given by

$$A \cdot A = A^2$$

or

$$A = \sqrt{A \cdot A} \tag{1.17}$$

Here A is the magnitude of the vector A. Thus, self product of a vector gives square of its magnitude.

The angle between the two vectors is given by

$$\cos \theta = \frac{A \cdot B}{AB}$$

The component of A along B is given by

$$A \cos \theta = \frac{A \cdot B}{B}$$

The component of B along A is given by

$$B \cos \theta = \frac{A \cdot B}{A}$$

It is clear from the definition of dot product that the commutative law of multiplication holds, that is

$$A \cdot B = B \cdot A \tag{1.18}$$

The fact that the distributive law of multiplication holds can be seen with the help of Fig. 1.6.

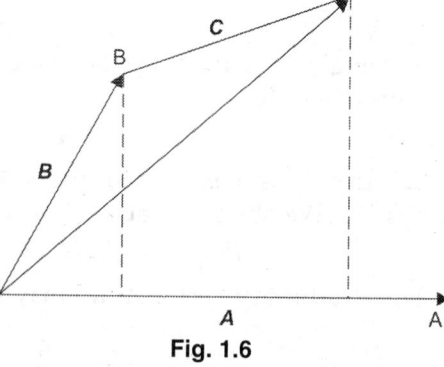

Fig. 1.6

We have,

$$B \cdot A + C \cdot A = (OB) \cdot A + (BC) \cdot A$$
$$= (OC) \cdot A$$
$$= (B + C) \cdot A \tag{1.19}$$

1.3 VECTOR OR CROSS PRODUCT

The vector or cross product of two vectors A and B is denoted as $A \times B$ or $[A\,B]$, and is given by

$$A \times B = |A||B| \sin \theta\, n \tag{1.20}$$

where, θ the angle between the directions of two vectors A and B and n is a unit vector perpendicular to the plane containing vectors A and B, is in the direction of advancement of right hand screw rotated from A to B.

The meaning of this definition is made clear in Fig. 1.7. It is clear from the definition that the commutative law does not hold for this type of multiplication. Instead, we have

$$A \times B = -B \times A \tag{1.21}$$

From Eq. (1.20), we note that the vector product of two vectors is a vector. We also note that it is zero, if both vectors are parallel ($\theta = 0°$, $\sin 0° = 0$). Equation (1.20) also gives

$$i \times j = k,\, j \times k = i,\, k \times i = j \tag{1.21a}$$
$$j \times i = -k,\, k \times j = -i,\, i \times k = -j \tag{1.21b}$$
$$i \times i = j \times j = k \times k = 0 \tag{1.21c}$$

$A \times B$ can also be written as

$$A \times B = (A_x i + A_y j + A_z k) \times (B_x i + B_y j + B_z k) \tag{1.22}$$

Using Eqs (1.21), and (1.22) may be reduced to

$$A \times B = (A_y B_z - A_z B_y) i + (A_z B_x - A_x B_z) j + (A_x B_y - A_y B_x) k \tag{1.22a}$$

The can also be written as

$$A \times B = \begin{vmatrix} i & j & k \\ A_x & A_y & A_z \\ B_x & B_y & B_z \end{vmatrix} \tag{1.22b}$$

The change in sign of the vector product with reversal of the order of the factors manifests itself here as a well-known law states that the interchange of two rows changes the sign of the determinant.

The specification of the product vector $C = A \times B$ gives the following information concerning the parallelogram formed by the two vectors A and B:

i. Its position in space (the plane perpendicular to C).

ii. The relative position of the side A and B with respect to each other.

iii. The area numerically equals to the length of C.

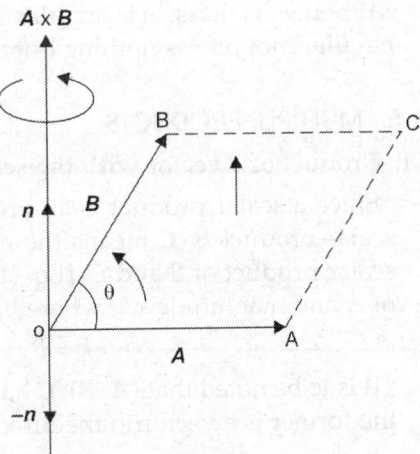

Fig. 1.7

The form of the parallelogram is not given by C. Assuming the form as being unimportant, we are recognizing the fact that every portion of a plane may be represented by a vector whose direction and magnitude are uniquely determined. The extension of this idea leads us to the notion of vector area. Thus, a plane area such as *OACBO* shown in Fig. 1.7 can be regarded as a vector whose magnitude is the geometrical area of parallelogram *OACBO* and direction is normal to the plane of area. The vector areas can be resolved or added just like vector quantities. The total vector area for a closed polygon is zero.

The vector product is *distributive*, i.e.

$$A \times (B + C) = A \times (C + B) \tag{1.23}$$

but the order of the factors must be strictly maintained.

1.4 EXAMPLES OF SCALAR AND VECTOR PRODUCTS IN PHYSICS

i. Work done by a force in displacing a body is defined as the scalar product of displacement vector *r* and force *F*, i.e.

$$W = F \cdot r \tag{1.23a}$$

ii. Moment of a force (*F*) or torque is defined in terms of vector product of position vector *r* and force *F*, i.e.

Torque:

$$\tau = r \times F \tag{1.24}$$

iii. Angular momentum is defined as moment of linear momentum, i.e. vector product of position vector *r* and linear momentum *P*, i.e.

Angular momentum:

$$L = r \times P \tag{1.25}$$

iv. Force on a charged particle of charge *q* moving with velocity *v* in magnetic field *B* is given by (Lorentz force):

$$F = q(v \times B). \tag{1.26}$$

v. Centripetal force = $m(\omega \times r) \times \omega$ \hfill (1.27)

Where *m* is mass, ω is angular velocity and *r* is position vector of the particle (for circular motion *r* is nothing but radius vector).

1.5 MULTIPLE PRODUCTS

1. **Product of a vector with the scalar product of two other vectors:**

Since a scalar product is an ordinary number, the product of vector *A* with the scalar product *B · C* means the multiplication of *A* with the number given by the scalar product of *B* and *C* [Eq. (1.16)], i.e. $A \cdot (B \cdot C)$ is a vector having the direction of *A* and magnitude *kA*, where *k* is a scalar given by:

$$k = B_x C_x + B_y C_y + B_z C_z \tag{1.28}$$

It is to be noted that $(A \cdot B) \cdot C$ has an entirely different meaning from $A \cdot (B \cdot C)$, *viz.* the former is a vector in the direction of *C*.

2. **Scalar product of a vector with the vector product of two other vectors:**

If $P = A \times B$, then the product $C \cdot P = C \cdot (A \times B)$ is the volume of the parallelepiped having the three vectors *A*, *B* and *C* as adjacent edges, as the magnitude of *P* is equal to the area of the parallelogram formed by *A* and *B* as adjacent sides (Fig. 1.8).

Thus, $\quad\quad\quad\quad\quad C \cdot P = CP \cos{(CP)}$, $\quad\quad\quad\quad\quad$ (1.29)

is the volume of the parallelepiped. Because $C \cos{(CP)}$ is equal to its altitude. One might equally well consider the base of the parallelepiped to be formed by the vectors C and A and then form the scalar product of this quantity with B. This means that the volume can also be represented by the product $C \times A \cdot B$. Since the sign of the vector product changes with the change in order of the factors, as seen from Fig. 1.8, i.e. $B \times A$ is a vector directed downwards and hence $\cos{(CP)}$ becomes negative. We may set up the following rule for the sign of the triple scalar product. The product $(A \times B) \cdot C$ is positive if the three vectors are relatively arranged like the axis of a right-handed coordinate system. Thus.

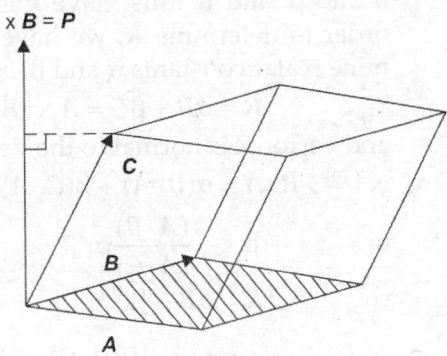

Plane of **A** and **B**

Fig. 1.8

$$A \cdot (B \times C) = B \cdot (C \times A) = C \cdot (A \times B)$$
$$= -A \cdot (C \times B)$$
$$= -B \cdot (A \times C)$$
$$= -C \cdot (B \times A) \quad\quad (1.30)$$

If the components of the three vectors are given, then

$$A \cdot (B \times C) = \{A_x i + A_y j + A_z k\} \cdot \{(B_y C_z - B_z C_y)i + (B_z C_x - B_x C_z)j + (B_x C_y - B_y C_x)k\}$$
$$(1.31)$$

This may be written as determinant:

$$A \cdot (B \times C) = \begin{vmatrix} A_x & A_y & A_z \\ B_x & B_y & B_z \\ C_x & C_y & C_z \end{vmatrix} \quad\quad (1.32)$$

Equation (1.30), thus expresses the well known property of determinants according to which the interchange of two rows causes the change in algebraic sign.

The above discussion of scalar triple product gives the following important results:

i. The cyclic permutation of three vectors does not change the value of scalar triple product while an anticyclic permutation changes the sign leaving the magnitude unchanged.

ii. When the three vectors lie in a plane, the volume of the parallelepiped is zero. *Hence for the three vectors to be coplanar, their scalar triple product vanishes.*

3. **Vector product of a vector with the vector product of two other vectors:**

If $P = B \times C$, then the vector product

$$R = A \times P = A \times (B \times C), \quad\quad (1.33)$$

signifies a vector lying in the plane of B and C, for $P = B \times C$ is perpendicular to this plane, but $R = A \times P$ is in turn perpendicular to P, and so falls in the plane of B and C (Fig. 1.9).

Since any vector *V* lying in the plane of *B* and *C* may be written in the form

$$V = \alpha B + \beta C \qquad (1.34)$$

by suitably choosing the scalar multiplier α and β, and *R* must have this form. In order to determine *R*, we have to determine scalar constants α and β.

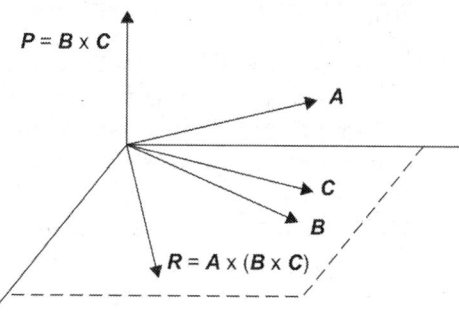

Fig. 1.9

$$R = \alpha B + \beta C = A \times (B \times C)$$

and vector *R* is normal to the vector *A*.

∴ $\qquad R \cdot A = \alpha(B \cdot A) + \beta(C \cdot A) = 0$

or $\qquad \beta = -\dfrac{(A \cdot B)}{(A \cdot C)}\alpha$

so $\qquad R = \alpha B + \beta C$

$$= \frac{\alpha}{A \cdot C}[B(A \cdot C) - (A \cdot B)C]$$

In order to determine the value of $\alpha/A \cdot C$, let us introduce a coordinate system in such a way that the x-axis is in the direction of *C*, y-axis is in the plane of *B* and *C* and perpendicular to *C* and z-axis along $(B \times C)$. This may be done without restricting the generality of discussion. Then, we have

$$\left.\begin{array}{l} A = A_x i + A_y j + A_z k \\ B = B_x i + B_y j \\ C = C_x i \end{array}\right\} \qquad (1.35)$$

The vector product $B \times C$, then assumes the simple form

$$B \times C = -B_y C_x k, \qquad (1.36)$$

and $A \times (B \times C)$ becomes

$$A \times (B \times C) = -A_y B_y C_x i + A_x B_y C_x j \qquad (1.37)$$

If the vector $A_x B_x C_x i$ be added and subtracted, then one obtains

$$A \times (B \times C) = A_x C_x (B_x i + B_y j) - (A_x B_x + A_y B_y) C_x i$$
$$= B(A \cdot C) - C(A \cdot B)$$

or $\qquad A \times (B \times C) = B(C \cdot A) - C(A \cdot B) \qquad (1.38)$

It is apparent that the coefficients α and β are thus the scalar products for $(A \cdot C)$ and $-(A \cdot B)$ respectively, i.e. $\alpha = A \cdot C$ and $\beta = -(A \cdot B)$ yields $\dfrac{\alpha}{A \cdot C} = 1$

Thus, we get

$$(A \times B) \times C \neq A \times (B \times C)$$

and hence the vector triple multiplication is not associative, i.e. the value of a vector triple product is entirely different if the order of the factors is interchanged.

Example 1. *Show that the vectors*

$$A = 3i - 2j + k, \; B = i - 3j + 5k, \; C = 2i + j - 4k$$

form a right angle triangle.

Solution: We have to show first that the vectors form a triangle.

From Figs 1.10a and b, it seems that the vectors will form a triangle if:

(a) one of the vectors, say *A*, is the resultant or sum of *B* and *C*.

(b) the sum or resultant of the vectors $A + B + C$ is zero.

Thus,

$$A = 3i - 2j + k$$
$$= B + C$$
$$= i - 3j + 5k + 2i + j - 4k$$
$$= 3i - 2j + k$$

So the vectors do form a triangle.

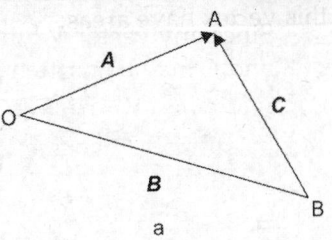

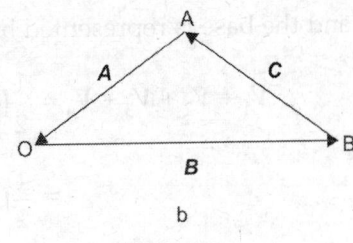

Since
$$A \cdot C = (3)(1) + (-2)(-3) + (1)(5) = 14$$
$$B \cdot C = (1)(2) + (-3)(1) + (5)(-4) = -21$$

and
$$A \cdot C = (3)(2) + (-2)(1) + (1)(-4) = 0$$

Fig. 1.10

it follow that *A* and *C* are perpendicular and the triangle is a right angle triangle.

Example 2. Prove that the angle inscribed in a semicircle is a right angle.

Solution: Let *ACB* be a semicircle (Fig. 1.11) with *AB* as bounding diameter and *C* be any point on the circumference. Let *O* be its centre and *r* the radius.

Let $\quad AO = a = OB$

and $\quad OC = b$

Now $\quad AC = AO + OC = a + b$

and $\quad CB = CO + OB = -b + a$

$\therefore \quad AC \cdot CB = (a + b) \cdot (a - b) = a^2 - b^2$
$$= AO^2 - OC^2 = r^2 - r^2 = 0.$$

i.e. *AC* and *CB* are perpendicular to each other

Hence, angle *ABC* is a right-angle or an angle inscribed in the semicircle is a right angle.

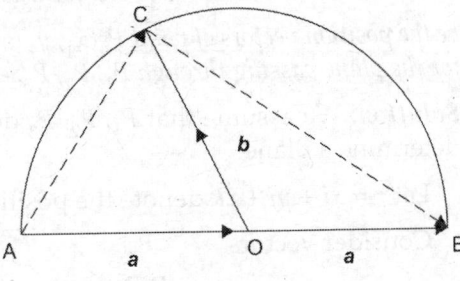

Fig. 1.11

Example 3. *Consider a tetrahedral with faces F_1, F_2, F_3, F_4. Let V_1, V_2, V_3, V_4 be vectors whose magnitudes are equal to the areas of F_1, F_2, F_3, F_4 respectively and whose directions are perpendicular to these faces in the outward direction. Show that*

$$V_1 + V_2 + V_3 + V_4 = 0.$$

Solution: The area of a triangular face determined by vectors *A* and *B* is $\frac{1}{2}|A \times B|$.

If we denote the three vectors emanating from the vertex *O* by *A*, *B* and *C*, then three faces passing through

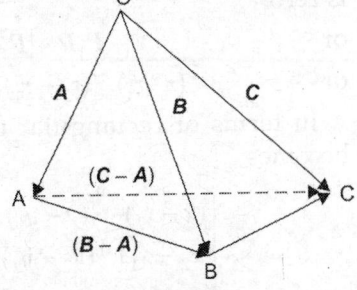

Fig. 1.12

this vertex have areas:

$$V_1 = \frac{1}{2} A \times B$$

$$V_2 = \frac{1}{2} B \times C$$

$$V_3 = \frac{1}{2} C \times A$$

and the base is represented by vectors $(C - A)$ and $(B - A)$, hence its area is

$$V_1 + V_2 + V_3 + V_4 = \frac{1}{2}[(A \times B + B \times C + C \times A) + (C - A) \times (B - A)]$$

$$= \frac{1}{2}[A \times B + B \times C + C \times A + C \times B - C \times A - A \times B + A \times A]$$

$$= 0 \text{ [since } A \times A = 0 \text{ and } C \times B = -B \times C]$$

This result can be generalised to closed polyhedron and in the limiting case to any closed surface.

Example 4. *If*
$$r_1 = x_1 i + y_1 j + z_1 k$$
$$r_2 = x_2 i + y_2 j + z_2 k$$

and
$$r_3 = x_3 i + y_3 j + z_3 k$$

be the position vectors of point $P_1(x_1, y_1, z_1)$, $P_2(x_2, y_2, z_2)$ and $P_3(x_3, y_3, z_3)$. Find an equation for the plane passing through P_1, P_2, P_3.

Solution: We assume that P_1, P_2, P_3 do not lie in the same straight line. Hence, they determine a plane.

Let $r = xi + yj + zk$ denote the position vector of any point $P(x, y, z)$ in the plane.

Consider vectors
$$P_1 P_2 = r_2 - r_1$$
$$P_1 P_3 = r_3 - r_1$$

and
$$P_1 P = r - r_1,$$

which lie in the plane and hence the volume of the parallelepiped formed by them is zero,

or
$$P_1 P \cdot \{P_1 P_2 \times P_1 P_3\} = 0$$

or
$$(r - r_1) \cdot \{(r_2 - r_1) \times (r_3 - r_1)\} = 0$$

In terms of rectangular coordinates, it becomes

$$\begin{vmatrix} (x - x_1) & (y - y_1) & (z - z_1) \\ (x_2 - x_1) & (y_2 - y_1) & (z_2 - z_1) \\ (x_3 - x_1) & (y_3 - y_1) & (z_3 - z_1) \end{vmatrix} = 0$$

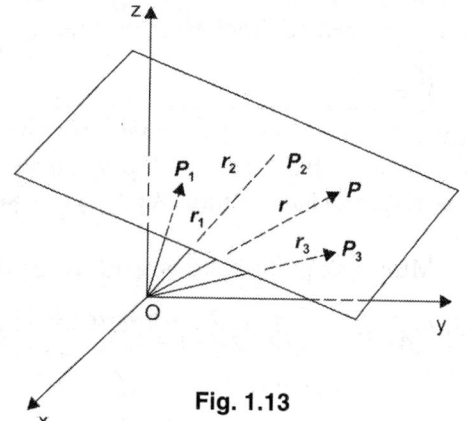

Fig. 1.13

Example 5. *Prove that* $(A \times B) \times (C \times D) = B(A \cdot C \times D) - A(B \cdot C \times D)$
$$= C(A \cdot B \times D) - D(A \cdot B \times C)$$

Solution: Let $A \times B = X$ and $C \times D = Y$

$\therefore$ $X \times (C \times D) = C(X \cdot D) - D(X \cdot C)$

or $(A \times B) \times (C \times D) = C(A \times B \cdot D) - D(A \times B \cdot C)$ [*using* Eq. (1.38)]

Now putting
$$Y = (C \times D)$$

We get
$$(A \times B) \times Y = -Y \times (A \times B) \qquad \text{[\textit{using} Eq. (1.38)]}$$
$$= -A(Y \cdot B) + B(Y \cdot A)$$
$$= B(A \cdot C \times D) - A(B \cdot C \times D)$$

Hence, $(A \times B) \times (C \times D) = B(A \cdot C \times D) - A(B \cdot C \times D)$
$$= C(A \times B \cdot D) - D(A \times B \cdot C)$$

But $A \times B \cdot D = A \cdot B \times D$ and $A \times B \cdot C = A \cdot B \times C$

or $(A \times B) \times (C \times D) = B(A \cdot C \times D) - A(B \cdot C \times D)$
$$= C(A \cdot B \times D) - D(A \cdot B \times C)$$

Example 6. *If A, B, C and a, b, c are any vectors, prove that*
$$(A \cdot B \times C)(a \cdot b \times c) = \begin{vmatrix} A \cdot a & A \cdot b & A \cdot c \\ B \cdot a & B \cdot b & B \cdot c \\ C \cdot a & C \cdot b & C \cdot c \end{vmatrix}$$

Solution: Let $P = B \times C$

Using result from example 5, we have
$$(a \cdot b \times c) P = (P \cdot b \times c) a - (P \cdot a \times c) b + (P \cdot a \times b) c$$

But, we know for any four vectors P, Q, R, S
$$(P \times Q) \cdot (R \times S) = \begin{vmatrix} P \cdot R & P \cdot S \\ Q \cdot R & Q \cdot S \end{vmatrix}$$

We have
$$(a \cdot b \times c)(B \times C) = (B \times C) \cdot (b \times c) a - (B \times C) \cdot (a \times c) b + (B \times C) \cdot (a \times b) c$$

$$= \begin{vmatrix} B \cdot b & B \cdot c \\ C \cdot b & C \cdot c \end{vmatrix} a - \begin{vmatrix} B \cdot a & B \cdot c \\ C \cdot a & C \cdot c \end{vmatrix} b + \begin{vmatrix} B \cdot a & B \cdot b \\ C \cdot a & C \cdot b \end{vmatrix} c$$

Multiplying both sides scalarly with A, we have

$$(A \cdot B \times C)(a \cdot b \times c) = \begin{vmatrix} B \cdot b & B \cdot c \\ C \cdot b & C \cdot c \end{vmatrix} A \cdot a - \begin{vmatrix} B \cdot a & B \cdot c \\ C \cdot a & C \cdot c \end{vmatrix} A \cdot b + \begin{vmatrix} B \cdot a & B \cdot b \\ C \cdot a & C \cdot b \end{vmatrix} A \cdot c$$

or $(A \cdot B \times C)(a \cdot b \times c) = \begin{vmatrix} A \cdot a & A \cdot b & A \cdot c \\ B \cdot a & B \cdot b & B \cdot c \\ C \cdot a & C \cdot b & C \cdot c \end{vmatrix}$ Hence proved.

Example 7. *Prove that* $(b \times c) \cdot (a \times d) + (c \times a) \cdot (b \times d) + (a \times b) \cdot (c \times d) = 0$ *and deduce*

that $\sin(A + B) \sin(A - B) = \sin^2 A - \sin^2 B = \dfrac{1}{2}(\cos 2B \cdot \cos 2A).$

Solution: Here

First part: $(b \times c) \cdot (a \times d) + (c \times a) \cdot (b \times d) + (a \times b) \cdot (c \times d)$

$$= \begin{vmatrix} b \cdot a & b \cdot d \\ c \cdot a & c \cdot d \end{vmatrix} + \begin{vmatrix} c \cdot b & c \cdot d \\ a \cdot b & a \cdot d \end{vmatrix} + \begin{vmatrix} a \cdot c & a \cdot d \\ b \cdot c & b \cdot d \end{vmatrix}$$

$$= (b \times a)(c \cdot d) - (c \cdot a)(b \cdot d) + (c \cdot b)(a \cdot d)$$
$$- (a \cdot b)(c \cdot d) + (a \cdot c)(b \cdot d) - (b \cdot c)(a \cdot d)$$
$$= 0$$

Second part: Let a, b, c, d be four coplanar vectors and n be a unit vector in the direction perpendicular to the plane containing a, b, c, d. Let the angles between the directions of a and b, b and c, c and d be $\theta_1, \theta_2, \theta_3$ respectively.

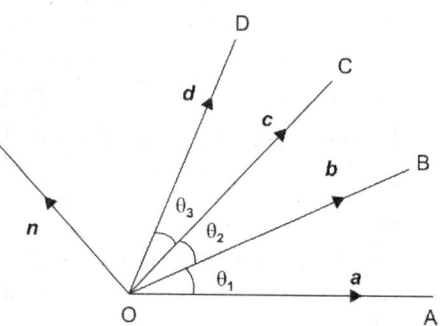

Since, $b \times c = bc \sin \theta_2 n$ and, etc. a, b, c, d being moduli of a, b, c, d respectively.

Fig. 1.14

$\therefore$ $\qquad (b \times c) \cdot (a \times d) + (c \times d) \cdot (b \times d) + (a \times b) \cdot (c \times d) = 0$

$\qquad (bc \sin \theta_2 n) \cdot \{ad \sin(\theta_1 + \theta_2 + \theta_3) n\} + (-) \{ca \sin(\theta_1 + \theta_2) n\}.$

$\qquad \{(bd \sin \theta_2 + \theta_3) n\} + \{ab \sin \theta n\} \times \{cd \sin \theta_3 n\} = 0.$

or $\quad \sin \theta_2 \sin(\theta_1 + \theta_2 + \theta_3) - \sin(\theta_1 + \theta_2) \sin(\theta_2 + \theta_3) + \sin \theta_1 \sin \theta_3 = 0$

Putting $\qquad\qquad \theta_1 = B, \theta_2 = A$ and $\theta_3 = -B$, we have

$$\sin^2 A - \sin(A + B) \sin(A - B) - \sin^2 B = 0$$

or $\qquad \sin(A + B) \sin(A - B) = \sin^2 A - \sin^2 B$

$$= \frac{1}{2}(1 - \cos 2A) - \frac{1}{2}(1 - \cos 2B)$$

$$= \frac{1}{2}(\cos 2B - \cos 2A)$$

Example 8. *Prove that unit normal vector to the plane of two vectors* $2i - j + k$ *and* $3i + j - k$

is $\dfrac{-3i + 5j + 11k}{\sqrt{155}}$ *and the sine of angle between them is* $\sqrt{155/156}$.

Solution: Let $\quad a = 2i - j + k$ and $b = 3i + 4j - k$, we have

$$a \times b = (2i - j + k) \times (3i + 4j - k)$$

$$= \begin{vmatrix} i & j & k \\ 2 & -1 & 1 \\ 3 & 4 & -1 \end{vmatrix} = -3i + 5j + 11k$$

$\therefore$ Unit vector n perpendicular to vectors a and b is

$$n = \frac{a \times b}{|a \times b|} = \frac{-3i + 5j + 11k}{\sqrt{(-3)^2 + (5)^2 + (11)^2}} = \frac{-3i + 5j + 11k}{\sqrt{155}}$$

Further, we have $|a \times b| = ab \sin \theta$

$\therefore \qquad \sin \theta = \dfrac{|a \times b|}{ab} = \dfrac{|-3i + 5j + 11k|}{|2i - j + k| \cdot |3i + 4j - k|}$

$$= \frac{\sqrt{155}}{\sqrt{\{2^2 + (-1)^2 + (1)^2\}} \times \sqrt{\{3^2 + 4^2 + (-1)^2\}}}$$

$$= \sqrt{\frac{155}{\sqrt{6}\sqrt{26}}} = \sqrt{\frac{155}{156}}.$$

1.6 VECTOR DERIVATIVES

In physics, normally we come across differentiation of a vector with respect to time, i.e. scalar and that with respect to space coordinates x, y and z. Here, we shall discuss both types of derivatives:

(a) Differentiation of a Vector w.r.t Scalar

Let a vector A be a continuous function of a continuous scalar variable t, i.e.

$$A = A(t)$$

If the variable t is increased by Δt, the vector will change by an amount

$$\Delta A = A(t + \Delta t) - A(t)$$

and in complete analogy with scalar functions, we define the derivative $\dfrac{dA}{dt}$ as the limit

$$A' = \frac{dA}{dt} = \lim_{\Delta t \to 0} \frac{A(t + \Delta t) - A(t)}{\Delta t} \qquad (1.39)$$

Since, division by a scalar does not alter the vectorial properties, the derivative of a vector with respect to a scalar variable is itself a vector.

Derivatives of a higher order are defined similarly, *e.g.*

$$A'' = \frac{d\left(\dfrac{dA}{dt}\right)}{dt} = \frac{d^2A}{dt^2} = \lim_{\Delta t \to 0} \frac{A'(t + \Delta t) - A'(t)}{\Delta t} \qquad (1.40)$$

As an example of differentiation of a vector w.r.t. a scalar, let us consider vector A to be a position vector r joining the origin O of a coordinate system at any point (x, y, z), then

$$r(t) = x(t)i + y(t)j + z(t)k$$

and specification of the vector function $r(t)$ defines x, y and z functions of t. As t changes, the terminal point of r describes a space curve having parametric equations.

$$x = x(t) \qquad y = y(t) \qquad z = z(t)$$

Then $\qquad \dfrac{\Delta r}{\Delta t} = \dfrac{r(t + \Delta t) - r(t)}{\Delta t},$

is a vector in the direction of Δr (Fig. 1.15). If

$$\lim_{\Delta t \to 0} \frac{\Delta r}{\Delta t} = \frac{dr}{dt} \qquad (1.41)$$

exists, the limit will be a vector in the direction of the tangent to the space curve at (x, y, z) and is given by

$$\frac{dr}{dt} = \frac{dx}{dt}i + \frac{dy}{dt}j + \frac{dz}{dt}k \quad (1.42)$$

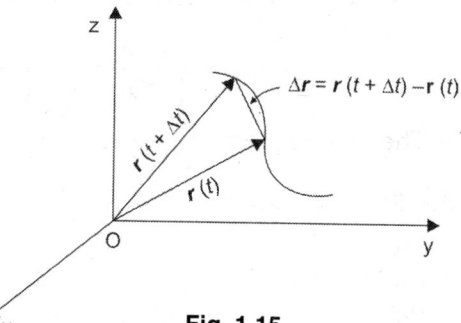

Fig. 1.15

If t is the time, $\dfrac{dr}{dt}$ represents the velocity v with which the terminal point of r describes the curve. Similarly,

$$\frac{dv}{dt} = \frac{d^2 r}{dt^2},$$

represents its acceleration a along the curve.

If vector r does not change in length, since $r \cdot r = r^2 = \text{const.}$

$$\frac{d(r^2)}{dt} = 2r \cdot \frac{dr}{dt} = 0, \qquad (1.43)$$

since neither $\dfrac{dr}{dt}$ nor r is to vanish, this means that the derivative of a vector of constant length is perpendicular to the vector. This is also evident geometrically, if the length is constant, the end point of the vector is restricted to move on a sphere. If the increment is infinitesimal, it is the tangent to the sphere, and hence it is perpendicular to the vector itself.

Since the derivatives of vectors with respect to a scalar variable are deduced by a limiting process from subtraction of vectors and division by scalars, which are operations subject to the rules of ordinary algebra, it follows the rules of the differential calculus and can be extended at once to the differentiation of the sum of vectors:

$$\frac{d}{dt}(A + B) = \frac{dA}{dt} + \frac{dB}{dt} \qquad (1.44)$$

or the product of a scalar and a vector

$$\frac{d}{dt}(mA) = A\frac{dm}{dt} + m\frac{dA}{dt} \qquad (1.45)$$

and, similarly

$$\frac{d}{dt}(A \cdot B) = B \cdot \frac{dA}{dt} + A \cdot \frac{dB}{dt} \qquad (1.45a)$$

$$\frac{d}{dt}(A \times B) = \frac{dA}{dt} \times B + A \times \frac{dB}{dt} \qquad (1.46)$$

$$\frac{d}{dt}(A \cdot B \times C) = A \cdot B \frac{dC}{dt} + A \cdot \frac{dB}{dt} \times C + \frac{dA}{dt} \cdot B \times C \qquad (1.47)$$

$$\frac{d}{dt}\{A \times (B \times C)\} = A \times \left(B \times \frac{dC}{dt}\right) + A \times \left(\frac{dB}{dt} \times C\right) + \frac{dA}{dt} \times (B \times C) \qquad (1.48)$$

The order of vectors in these products is important.

(b) Partial Differentiation

These simple properties of differentiation as applied to vectors can be extended to partial derivatives when a vector is a function of more than one scalar independent variable. The most useful case is that of a vector V which is a function of the cartesian coordinates x, y, z of a point in space. If y and z remain constant while x increases, the partial derivative $\dfrac{\partial V}{\partial x}$ denoting the rate of increase of V with respect to x is defined as

$$\frac{\partial V}{\partial x} = \lim_{\Delta x \to 0} \frac{V(x + \Delta x, y, z) - V(x, y, z)}{\Delta x}, \qquad (1.48a)$$

if this limit exists. Likewise changing y and z alone gives the partial derivatives

$$\frac{\partial V}{\partial y} = \lim_{\Delta y \to 0} \frac{V(x, y + \Delta y, z) - V(x, y, z)}{\Delta y} \qquad (1.48b)$$

and

$$\frac{\partial V}{\partial z} = \lim_{\Delta z \to 0} \frac{V(x, y, z + \Delta z) - V(x, y, z)}{\Delta z} \qquad (1.48c)$$

denoting the rates of increase with respect to y and z respectively. If now x, y and z change simultaneously by differential increments dx, dy, dz then the total change or total differential of V will be

$$dV = \frac{\partial V}{\partial x} dx + \frac{\partial V}{\partial y} dy + \frac{\partial V}{\partial z} dz, \qquad (1.49)$$

which is of frequent occurrence in physical applications of vector analysis. If
$$r = xi + yj + zk$$
is the radius vector from the origin, then its differential increment
$$dr = dxi + dyj + dzk$$
Equation (1.48) can be symbolically written as

$$dV = \left[\frac{\partial}{\partial x} dx + \frac{\partial}{\partial y} dy + \frac{\partial y}{\partial z} dz\right] V$$

We now define the operator ∇ by

$$\nabla = i \frac{\partial}{\partial x} + j \frac{\partial}{\partial y} + k \frac{\partial}{\partial z} \qquad (1.50)$$

It is easy to verify that the scalar product of ∇, regarded as a kind of vector, with dr gives the operator in square brackets. Thus,
$$dV = (\nabla \cdot dr) V \qquad (1.51)$$

The operator ∇ is of immense importance in physical applications of vector analysis, wherein it appears in association with both scalar and vector operands. This operator is called *del operator*.

(c) Vector Integration

For a vector $A = A(t)$, where t is a scalar variable then we define the vector integration as

$$\int A(t)dt = \int \{A_x(t)i + A_y(t)j + A_z(t)k\}dt$$

$$= i\int A_x(t)dt + j\int A_y(t)dt + k\int A_z(t)dt$$

The most important integrals encountered in vector analysis are the line integral, the surface integral and the volume integral.

i. The Line Integral

The integration of a vector along a curve is called *line integral*. Let us consider a vector field defined by

$$A = A(x, y, z)$$

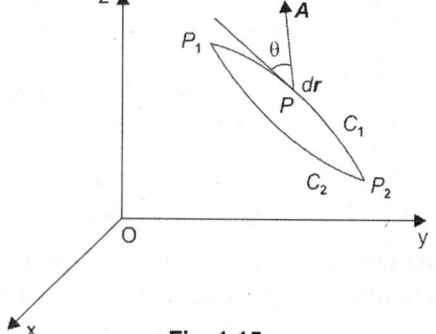

Fig. 1.15a

In this field consider a curve C_1 with end points P_1 and P_2 and at any point P on it, the value of vector be A. At P let us consider a small element of length dr [Fig. 1.15a]. Then

$$A \cdot dr = A\cos\theta\, dr$$

If the vector A varies in magnitude and direction from point to point along the curve, then the intergral

$$\int_{P_1}^{P_2} A \cdot dr = \int_{P_1}^{P_2} A\cos\theta\, dr = \int_{C_1} A\cos\theta\, dr,$$

is defined as the line integral of A along the curve C_1.

If $\int_{C_1} A\cos\theta\, dr$ is independent of the path then the field $A(x, y, z)$ is said to be *conservative*, or *noncurl* or *lamellar vector*. In this case, the line integral depends only on the end value of the curve. Thus,

$$\int_{C_1} A \cdot dr = \int_{P_1}^{P_2} A \cdot dr = \int_{C_2} A \cdot dr = -\int_{P_2}^{P_1} A \cdot dr$$

where, C_2 is another curve having the same end points P_1 and P_2 as those of C_1. Now the integral of the closed path is

$$\oint A \cdot dr = \int_{C_1} A \cdot dr + \int_{C_2} A \cdot dr = \int_{P_1}^{P_2} A \cdot dr + \int_{P_2}^{P_1} A \cdot dr = 0$$

where, the sign $\oint$ represents integral round a closed curve. The vector fields which satisfy the condition $\oint A \cdot dr = 0$ can be written as $A = \nabla\phi$, where ϕ is some scalar field.

The line integrals frequently occur in physical problems in physics. For example, if A is force and dr the displacement element of the path of a particle along any curve, then the line integral denotes the work done in displacing the particle from one end point to the other. Again if A is the electric field strength then line integral expresses the potential difference between the end points.

ii. The Surface Integral

Let us consider a vector field $A = A(x, y, z)$ and draw a surface S in this field.

Let us now take an element of area ds on the surface. Draw a positive normal n of unit length on this element. This positive normal to the closed surface is drawn outward from the enclosed volume. While for open surface this is in the same direction as that of the vector concerned. If θ is the angle between vectors n and A, then the flux of vector A through the element ds of area is $A \cdot n ds$ [Fig. 1.15b]. Then the surface integral of A through the whole surface is defined as

$$\iint_S A \cdot n ds = \iint_S A \cos\theta \, ds = \iint_S (A_x ds_x + A_y ds_y + A_z ds_z)$$

The notation $\oiint_S$ is used to denote the integration over the closed surface.

iii. The Volume Integral

For a vector field A, the volume integral over a closed surface in space enclosing a volume is defined as

$$\iiint A d\tau$$

where, $d\tau$ is the volume element over which the field A is assumed to be uniform and the integral:

$$\oiiint_V A d\tau$$

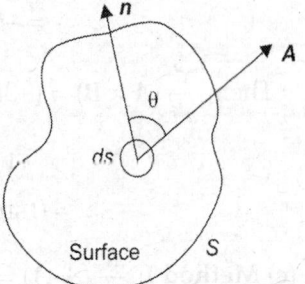

Fig. 1.15b

denotes the volume integral over the closed surface having volume V.

Example 9. If $A = i5t^2 + jt - kt^3$ and $B = i \sin t - j \cos t$, find

(a) $\dfrac{d}{dt}(A \cdot B)$ (b) $\dfrac{d}{dt}(A \times B)$ (c) $\dfrac{d}{dt}(A \cdot A)$

Solution:

(a) **Method I:** $\dfrac{d}{dt}(A \cdot B) = A \cdot \dfrac{dB}{dt} + \dfrac{dA}{dt} \cdot B$

$\qquad = (i\,5t^2 + jt - kt^3) \cdot (i \cos t + j \sin t) + (i\,10t + j - k\,3t^2) \cdot (i \sin t - j \cos t)$

$\qquad = 5t^2 \cos t + t \sin t + 10t \sin t - \cos t$

$\qquad = (5t^2 - 1) \cos t + 11t \sin t.$

Method II: $A \cdot B = 5t^2 \sin t - t \cos t$, then

$\qquad \dfrac{d}{dt}(A \cdot B) = \dfrac{d}{dt}(5t^2 \sin t - t \cos t)$

$\qquad\qquad = 10t \sin t + 5t^2 \cos t - \cos t + t \sin t$

$\qquad\qquad = (5t^2 - 1) \cos t + 11t \sin t.$

(b) **Method I:** $\dfrac{d}{dt}(A \times B) = A \times \dfrac{dB}{dt} + \dfrac{dA}{dt} \times B$

$$= \begin{vmatrix} i & j & k \\ 5t^2 & t & -t^3 \\ \cos t & \sin t & 0 \end{vmatrix} + \begin{vmatrix} i & j & k \\ 10t & 1 & -3t^2 \\ \sin t & -\cos t & 0 \end{vmatrix}$$

$= i(t^3 \sin t - 3t^2 \cos t) - j(t^3 \cos t + 3t^2 \sin t) + k(5t^2 \sin t - t \cos t - 10t \cos t - \sin t)$

$= (t \sin t - 3 \cos t)t^2 \, i - (t \cos t + 3 \sin t) \, t^2 j + (5t^2 \sin t - 11t \cos t \sin t) \, k$

Method II: $A \times B = \begin{vmatrix} i & j & k \\ 5t^2 & t & -t^3 \\ \sin t & -\cos t & 0 \end{vmatrix}$

$= i(-t^3 \cos t) - j(t^2 \sin t) + k(-5t^2 \cos t - t \sin t)$

$= -t^3 \cos t i - t^2 \sin t j - (5t^2 \cos t + t \sin t) k$

Then, $\dfrac{d}{dt}(A \times B) = (-3t^2 \cos t + t^3 \sin t)i - (3t^2 \sin t + t^3 \cos t)j$

$-(10t \cos t - 5t^2 \sin t + \sin t + t \cos t) k$

$= (t \sin t - 3 \cos t)t^2 i - (t \cos t + 3 \sin t)t^2 j + (5t^2 \sin t - 11t \cos t - \sin t)k$

(c) **Method I:** $\dfrac{d}{dt}(A \cdot A) = A \cdot \dfrac{dA}{dt} + \dfrac{dA}{dt} \cdot A = 2A \cdot \dfrac{dA}{dt} = 2(5t^2 i + tj - t^3 k) \cdot (10ti + j - 3t^2 k)$

or $\qquad \dfrac{d}{dt}(A \cdot A) = 2A \cdot \dfrac{dA}{dt} = 100t^3 + 2t + 6t^5$

Method II: $\qquad A \cdot A = (5t^2)^2 + (t)^2 + (t^3)^2 = 25t^4 + t^2 + t^6$

$\therefore \qquad \dfrac{d}{dt}(A \cdot A) = 100t^3 + 2t + 6t^5$

Example 10. *Evaluate* $\dfrac{d}{dt}\left(V \cdot \dfrac{dV}{dt} \times \dfrac{d^2V}{dt^2} \right)$

Solution: $\dfrac{d}{dt}\left(V \cdot \dfrac{dV}{dt} \times \dfrac{d^2V}{dt^2} \right) = \dfrac{dV}{dt} \cdot \dfrac{dV}{dt} \times \dfrac{d^2V}{dt^2} + V \cdot \dfrac{d^2V}{dt^2} \times \dfrac{d^2V}{dt^2} + V \cdot \dfrac{dV}{dt} \times \dfrac{d^3V}{dt^3}$

But $\qquad \dfrac{d^2V}{dt^2} \times \dfrac{d^2V}{dt^2} = 0$

and $\qquad \dfrac{dV}{dt} \cdot \dfrac{dV}{dt} \times \dfrac{d^2V}{dt^2} = 0$

Hence $\qquad \dfrac{d}{dt}\left(V \cdot \dfrac{dV}{dt} \times \dfrac{d^2V}{dt^2} \right) = \dfrac{dV}{dt} \times \dfrac{d^3V}{dt^3}$

Example 11. *A particle moves so that its position vector is given by* $r = \cos \omega t\, i + \sin \omega t\, j$ *where,* ω *is a constant. Show that:*

(a) *the velocity* v *of the particle is perpendicular to* r,

(b) *the acceleration* a *is directed towards the origin and has magnitude proportional to the distance from the origin,*

(c) $r \times v$ *is equal to a constant vector.*

Solution:

(a) $v = \dfrac{dr}{dt} = -\omega \sin \omega t\, i + \omega \cos \omega t\, j$

Then $r \cdot v = (\cos \omega t\, i + \sin \omega t\, j) \cdot (-\omega \sin \omega t\, i + \omega \cos \omega t\, j)$

$\qquad = -\omega \sin \omega t \cos \omega t + \omega \sin \omega t \cos \omega t = 0$

Hence r and v are perpendicular to each other.

(b) $\dfrac{d^2 r}{dt^2} = \dfrac{dv}{dt} = a = -\omega^2 \cos \omega t\, i - \omega^2 \sin \omega t\, j$

$\qquad = -\omega^2 (\cos \omega t\, i + \sin \omega t\, j) = -\omega^2 r$

Thus, acceleration is opposite to the direction of r, i.e. it is directed towards the origin and its magnitude is proportional to $|r|$ which is the distance from the origin.

(c) $r \times v = (\cos \omega t\, i + \sin \omega t\, j) \times (-\omega \sin \omega t\, i + \omega \cos \omega t)\, j$

or $r \times v = \begin{vmatrix} i & j & k \\ \cos \omega t & \sin \omega t & 0 \\ -\omega \sin \omega t & \omega \cos \omega t & 0 \end{vmatrix} = \omega(\cos^2 \omega t + \sin^2 \omega t)k = \omega k$

a constant vector. Physically, it is the motion of a particle moving on the circumference of a circle with constant angular speed ω. The acceleration, directed towards the centre of the circle is the centripetal acceleration.

Example 12. *If* $F = (5xy - x^2)i + (2y - 4x)j$. *Find* $\displaystyle\int_C F \cdot dr$ *along the curve* C *given by* $y = x^3$ *in* $x - y$ *plane from point* (1, 1) *to* (2, 8).

Solution: Along the curve $y = x^3$, $dy = 3x^2\, dx$

$$F = (5x^4 - x^2)i + (2x^3 - 4x)j$$

and

$$r = xi + yj + zk$$

$\therefore$

$$dr = dxi + dyj + dzk = dxi + 3x^2 dxj + dzk$$

$\therefore$

$$F \cdot dr = \left\{(5x^4 - x^2)i + (2x^3 - 4x)j\right\} \cdot \left\{dxi + 3x^2 dxj + dzk\right\}$$

$$= (5x^4 - x^2)dx + (6x^5 - 12x^3)dx$$

$$= (6x^5 + 5x^4 - 12x^3 - x^2)dx$$

$$\therefore \quad \int_C F \cdot dr = \int (6x^5 + 5x^4 - 12x^3 - x^2)dx$$

$$= \left[x^6 + x^5 - 3x^4 - \frac{x^3}{3} \right]_1^2$$

$$= \left\{ (2)^6 + (2)^5 - 3(2)^4 - \frac{(2)^3}{3} \right\} - \left\{ (1)^6 + (1)^5 - 3(1)^4 - \frac{1}{3} \right\}$$

$$= 64 + 32 - 48 - \frac{8}{3} - \left(2 + \frac{10}{3} \right)$$

$$= 48 - 2 - 6 = 40.$$

Example 13. Evaluate $\iint_S r \cdot n\, ds$, where S is the surface of the upper hemisphere of radius a with centre at $(0, 0, 0)$ and r is the position vector. **[Delhi]**

Solution: We have the surface of hemisphere with centre at origin $(0, 0, 0)$ is normal to $r = xi + yj + zk$.

At the surface $r = a$

$$\therefore \qquad\qquad n = \frac{r}{r} = \frac{r}{a}$$

Now, $$\iint_S r \cdot n\, ds = \iint_S r \cdot \frac{r}{a} ds = \iint_S \frac{r^2}{a} ds$$

$$\therefore \qquad \iint_S r \cdot n\, ds = \iint_S a\, ds = a \iint_S ds$$

We have for hemispherical surface

$$\iint_S ds = 2\pi a^2$$

$$\therefore \qquad \iint_S r \cdot n\, ds = a \cdot 2\pi a^2 = 2\pi a^3$$

1.7 SCALAR AND VECTOR FIELDS

A physical quantity can be expressed as a continuous function of the position of a point in a region of space. Such a function is called a *point function* and the region in which it specifies the physical quantity is known as *field*. Fields are of two types, scalar and vector, according to the nature of the quantity concerned.

A typical scalar field, such as the distribution of temperature, density, electric potential or of any other nondirected quantity is represented by a continuous scalar function giving the value of quantity at each point. Such a function does not undergo any abrupt change of magnitude in passing from any point to another close to it, a condition satisfied in all practical cases. The field can be mapped graphically by a series of surface such as isothermal, equidensity or equipotential surfaces upon each of which the scalar has a definite constant value. Such surfaces called equal or level surfaces, are conveniently chosen so that in passing from one to the next, a constant arbitrary difference is made between the scalars which characterise them. It is evident that the level surfaces must lie one within the other and cannot cut, and if two such surfaces could intersect, the scalar values corresponding to both must hold along their common line which is contrary to our definition. Hence scalar point functions are single valued or uniform at every point.

A scalar field which is independent of time is called a *stationary* or *steady-state scalar field*. A typical vector field, such as the distribution of velocity in a fluid or of electric or magnetic field strength is represented at every point by a continuous vector function. At any given point the function is specified by a vector of definite magnitude and direction both of which change continuously from point to point throughout the field region. Starting at any arbitrary place, proceed an infinitesimal distance in the direction of the vector at that place; proceed an infinitesimal distance in the direction of the vector at that place; arriving at a closely-neighbouring point. Proceeding in a similar way, we shall trace out a curved line, the tangent to which at any point gives the direction of the vector there at that point, such a curve is a line of flow or flux line. To represent the magnitude of the vector, at any point on a flux line, draw a very small surface, perpendicular thereto and choose number of points per unit area upon this surface numerically equal to the magnitude of the vector. Through each of point, flux lines can be drawn. The field is then mapped out by flux lines. The direction of the line is that of the vector function, their density, represented by the number of their crossing per unit area perpendicular to their direction, is a measure of the magnitude of the vector. It is clear that lines of flow cannot intersect, since this would involve the indefinite direction of vector at the point where they cut; vector point functions must also be single valued at every point.

A vector field which is independent of time is called a stationary or steady-state vector field.

1.8 THE GRADIENT OF SCALAR FIELD

Let $\phi(x, y, z)$ be a scalar function of position in space, i.e. of coordinates x, y, z. If the coordinates x, y, z are increased by dx, dy, dz respectively then, we have

$$d\phi = \frac{\partial \phi}{\partial x}dx + \frac{\partial \phi}{\partial y}dy + \frac{\partial \phi}{\partial z}dz \qquad (1.52)$$

If R is a position vector of point $P(x, y, z)$ then, we have

$$R = xi + yj + zk$$

and
$$dR = dxi + dyj + dzk \qquad (1.53)$$

Here dR denotes the vector representing the displacement specified by components dx, dy, dz.

Using Eqs (1.52) and (1.53), we have

$$d\phi = \frac{\partial \phi}{\partial x}dx + \frac{\partial \phi}{\partial y}dy + \frac{\partial \phi}{\partial z}dz$$

$$= \left(i\frac{\partial \phi}{\partial x} + j\frac{\partial \phi}{\partial y} + k\frac{\partial \phi}{\partial z} \right) \cdot (dxi + dyj + dzk)$$

$$= (\nabla \phi) \cdot dR \qquad (1.54)$$

The vector $\nabla \phi$ is called the gradient of ϕ and is often written as grad ϕ. Thus,

$$\text{grad } \phi = \nabla \phi = i\frac{\partial \phi}{\partial x} + j\frac{\partial \phi}{\partial y} + k\frac{\partial \phi}{\partial z}$$

Physical Interpretation

The equation $\phi(x, y, z)$ = constant represents a certain surface and as we change the value of the constant, we obtain a family of surfaces. Consider two such surfaces very close together and examine a small portion of them in the neighbourhood of a given point A on the surface characterised by the constant value of the scalar function ϕ; the second surface is specified by a constant value $\phi + d\phi$. This is shown in Fig. 1.16. If R is the position vector from origin to the point A, any point such as B in the second surface is given by $R + dR$. The least distance between the surfaces will be AC, in the direction of the unit normal vector n at A and of length dn.

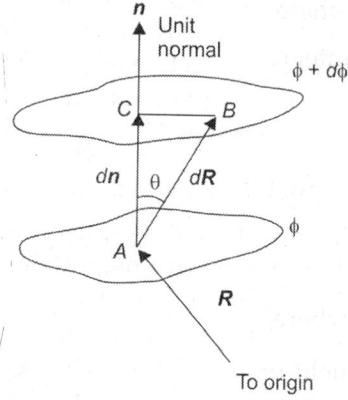

If dR be the length of AB then the magnitude of the rate of increase at A of ϕ in the direction of AB will be $\dfrac{\partial \phi}{\partial R}$ when the two surfaces are vanishingly close together. This rate of increase becomes greatest in the direction of the unit normal n along AC where it has the value $\dfrac{\partial \phi}{\partial n}$, we have

Fig. 1.16

$$dn = n \cdot dR$$

and

$$d\phi = \frac{\partial \phi}{\partial n} dn = \frac{\partial \phi}{\partial n} n \cdot dR \tag{1.55}$$

Here the vector $n \dfrac{\partial \phi}{\partial n}$ gives the greatest rate of increase of ϕ at the point A in magnitude and direction. Comparing Eqs (1.54) and (1.55), we have

$$\text{grad } \phi = \nabla \phi = n \frac{\partial \phi}{\partial n} \tag{1.56}$$

Thus, the gradient of a scalar field is a vector field, the vector at any point having a magnitude equal to the most rapid rate of increase of ϕ at the point and in the direction of this fastest rate of increase, i.e. perpendicular to the level surface at the point.

Let a vector field be obtained from a scalar field by taking gradient of the latter, i.e.

$$V = \text{grad } \phi = \nabla \phi$$

then its line intergral may be calculated as follows.

In Fig. 1.17 on the path C_1, between P_1 and P_2, the vector V makes an angle θ with the element dl of the path. The line integral of vector V for line element dl is defined as

$$V \cdot dl = V \cos \theta \, dl.$$

The path is traced by the extremity of a radius vector R from the origin O. So,

$$dl = AB = R + dR - R = dR$$

$$V \cdot dl = V \cdot dR$$

But

$$V = \text{grad } \phi$$

$\therefore$

$$V \cdot dl = (\text{grad } \phi) \cdot dR = d\phi$$

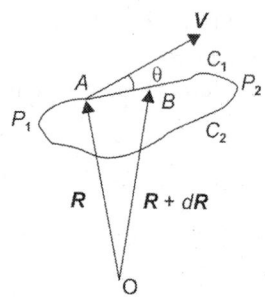

Fig. 1.17

[*from* Eq. (1.54)]

is the line integral of vector V for line element dl. Hence line integral of vector V from P_1 to P_2.

$$\int_{P_1}^{P_2} V \cdot dl = \int_{P_1}^{P_2} d\phi = \phi_{P_1} - \phi_{P_2}$$

where, ϕ_{P_2} and ϕ_{P_1} are the values of the scalars at the extremities of the path. The line integral of V along the path C_2 is

$$\int_{P_2}^{P_1} V \cdot dl = \int_{P_2}^{P_1} d\phi = \phi_{P_1} - \phi_{P_2}$$

So, the line integral of V round a closed path as shown in Fig. 1.17 is zero or

$$\oint (\text{grad}\,\phi) \cdot dl = 0$$

where, $\oint$ shows integral along a closed path. Thus, if a vector field is represented as a field of the gradient of a scalar point function, the value of the line integral of the vector, taken between two points in a region is independent of the path, depends only on the values of scalar at extremities and the line integral over a closed path vanishes. A vector field, derived from the scalar field by taking gradient of the latter, is called *scalar potential field* or the *lamellar vector field*.

Example 14. *Find* $\nabla\phi$ *if:*

(a) $\phi = \log |r|$ 　　　(b) $\phi = \dfrac{1}{r}$ 　　　(c) $\phi = r^n$

where, r is position vector of the point (x, y, z) from origin.

Solution: (a) $r = xi + yj + zk$

Then 　　　$|r| = \sqrt{x^2 + y^2 + z^2}$

and 　　　$\log |r| = \dfrac{1}{2}\log(x^2 + y^2 + z^2)$

$$\nabla\phi = \dfrac{1}{2}\nabla\log(x^2 + y^2 + z^2)$$

$$= \dfrac{1}{2}\left\{ i\dfrac{\partial}{\partial x}\log(x^2 + y^2 + z^2)\right\} + \dfrac{1}{2}\left\{ j\dfrac{\partial}{\partial y}\log(x^2 + y^2 + z^2)\right\}$$

$$+ \dfrac{1}{2}\left\{ k\dfrac{\partial}{\partial z}\log(x^2 + y^2 + z^2)\right\}$$

$$= \dfrac{1}{2}\left\{ i\dfrac{2x}{x^2 + y^2 + z^2} + j\dfrac{2x}{x^2 + y^2 + z^2} + k\dfrac{2x}{x^2 + y^2 + z^2}\right\}$$

$$= \dfrac{xi + yj + zk}{(x^2 + y^2 + z^2)} = \dfrac{r}{r^2}$$

(b) $\quad \nabla\left(\dfrac{1}{r}\right) = \nabla\left(\dfrac{1}{\sqrt{x^2+y^2+z^2}}\right) = \nabla(x^2+y^2+z^2)^{-1/2}$

$\qquad = i\dfrac{\partial}{\partial x}(x^2+y^2+z^2)^{-1/2} + j\dfrac{\partial}{\partial y}(x^2+y^2+z^2)^{-1/2} + k\dfrac{\partial}{\partial z}(x^2+y^2+z^2)^{-1/2}$

$\qquad = i\left\{-\dfrac{1}{2}(x^2+y^2+z^2)^{-3/2}\cdot 2x\right\} + j\left\{-\dfrac{1}{2}(x^2+y^2+z^2)^{-3/2}\cdot 2y\right\}$

$\qquad\qquad\qquad +k\left\{-\dfrac{1}{2}(x^2+y^2+z^2)^{-3/2}\cdot 2z\right\}$

$\qquad = -\dfrac{xi+yj+zk}{(x^2+y^2+z^2)^{3/2}} = -\dfrac{r}{r^3}$

(c) $\quad \nabla(r^n) = \nabla\sqrt{(x^2+y^2+z^2)} = \nabla(x^2+y^2+z^2)^{n/2}$

$\qquad = i\dfrac{\partial}{\partial x}(x^2+y^2+z^2)^{n/2} + j\dfrac{\partial}{\partial y}(x^2+y^2+z^2)^{n/2} + k\dfrac{\partial}{\partial z}(x^2+y^2+z^2)^{n/2}$

$\qquad = i\left\{\dfrac{n}{2}(x^2+y^2+z^2)^{\frac{n}{2}-1}2x\right\} + j\left\{\dfrac{n}{2}(x^2+y^2+z^2)^{\frac{n}{2}-1}2y\right\}$

$\qquad\qquad\qquad +k\left\{\dfrac{n}{2}(x^2+y^2+z^2)^{\frac{n}{2}-1}2z\right\}$

$\qquad = n(x^2+y^2+z^2)^{\frac{n}{2}-2}(ix+jy+kz)$

$\qquad = n\,r\,r^{n-2}$

Example 15. *Let $\phi(x, y, z)$ and $\phi(x + \Delta x, y + \Delta y, z + \Delta z)$ be the temperatures at two neighbouring points $A(x, y, z)$ and $B(x + \Delta x, y + \Delta y, z + \Delta z)$ on a certain region.*

(a) *Interpret physically, the quantity*

$$\dfrac{\Delta\phi}{\Delta s} = \dfrac{\phi(x+\Delta x, y+\Delta y, z+\Delta z) - \phi(x,y,z)}{\Delta s}$$

where Δs is the distance between points A and B.

(b) $\displaystyle\lim_{\Delta s\to 0}\dfrac{\Delta\phi}{\Delta s} = \dfrac{d\phi}{ds}$, *interpret physically.*

(c) *Show that $\dfrac{d\phi}{ds} = \nabla\phi\cdot\dfrac{dr}{ds}$.*

Solution:

(a) Since $\Delta\phi$ is the change in temperature between points A and B and Δs is the distance

between these points, $\dfrac{\Delta\phi}{\Delta s}$ represents the average rate of change in temperature per unit distance between these points, in the direction from A to B.

(b) From the calculus

$$\Delta\phi = \frac{\partial\phi}{\partial x}\Delta x + \frac{\partial\phi}{\partial y}\Delta y + \frac{\partial\phi}{\partial z}\Delta z$$

Then

$$\lim_{\Delta s \to 0}\frac{\Delta\phi}{\Delta s} = \lim_{\Delta s \to 0}\left\{\frac{\partial\phi}{\partial x}\frac{\Delta x}{\Delta s} + \frac{\partial\phi}{\partial y}\frac{\Delta y}{\Delta s} + \frac{\partial\phi}{\partial z}\frac{\Delta z}{\Delta s}\right\}$$

or

$$\frac{d\phi}{ds} = \frac{\partial\phi}{\partial x}\cdot\frac{dx}{ds} + \frac{\partial\phi}{\partial y}\cdot\frac{dy}{ds} + \frac{\partial\phi}{\partial z}\cdot\frac{dz}{ds}$$

$\frac{d\phi}{ds}$ represents the rate of change of temperature with respect to distance at point A in the direction towards B. The is also called the directional derivative of ϕ.

(c) $\quad \dfrac{d\phi}{ds} = \dfrac{\partial\phi}{\partial x}\dfrac{dx}{ds} + \dfrac{\partial\phi}{\partial y}\dfrac{dy}{ds} + \dfrac{\partial\phi}{\partial z}\dfrac{dz}{ds}$

$$= \left(i\frac{\partial\phi}{\partial x} + j\frac{\partial\phi}{\partial y} + k\frac{\partial\phi}{\partial y}\right)\cdot\left(\frac{dx}{ds}i + \frac{dy}{ds}j + \frac{dz}{ds}k\right)$$

$$= \nabla\phi\cdot\frac{dr}{ds}$$

Note: Since $\dfrac{dr}{ds}$ is a unit vector, $\nabla\phi\cdot\dfrac{dr}{ds}$ is the component of $\nabla\phi$ in the direction of this unit vector.

Example 16. *Find the directional derivative of* $\phi = x^2yz + 4xz^2$ *at* $(1, -2, 1)$ *in the direction* $2i - j - 2k$.

Solution: $\nabla\phi = \nabla(x^2yz + 4xz^2) = 2xyzi + 4z^2i + x^2zj + x^2yk + 8xzk$

$\qquad = (2xyz + 4z^2)i + x^2zj + (x^2y + 8xz)k$

$\qquad = 8i - j - 10k$ at $(1, -2, -1)$

The unit vector in the direction of $2i - j - 2k$ is

$$S = \frac{2i - j - 2k}{\sqrt{(2)^2 + (1)^2 + (2)^2}}$$

$$= \frac{2}{3}i - \frac{1}{3}j - \frac{2}{3}k$$

Then, the required directional derivative is

$$\nabla\phi\cdot S = (8i - j - 10k)\cdot\left(\frac{2}{3}i - \frac{1}{3}j - \frac{2}{3}k\right)$$

$$= \frac{16}{3} + \frac{1}{3} + \frac{20}{3} = \frac{37}{3}$$

Since the value is positive, ϕ is increasing in this direction.

1.9 THE CONCEPT OF DIVERGENCE AND GAUSS'S THEOREM

Mathematically, the divergence of a vector A is defined as the scalar product of the vector operator ∇ and the vector A, i.e.

$$\text{div } A = \nabla \cdot A = \left(i\frac{\partial}{\partial x} + j\frac{\partial}{\partial y} + k\frac{\partial}{\partial z} \right) \cdot (iA_x + jA_y + kA_z)$$

$$= \frac{\partial A_x}{\partial x} + \frac{\partial A_y}{\partial y} + \frac{\partial A_z}{\partial z} \tag{1.57}$$

We see that div A is not a vector but is a scalar.

Physical Interpretation

Let us consider flow of a fluid of density $\rho(x, y, z, t)$ and velocity $v = v(x, y, z, t)$ and let A represent the mass of fluid flowing through a unit area normal to the direction of v per unit time. The Y-component of A through the area $ABOC$ (indicated in Fig. 1.18) of infinitesimal element of volume with sides dx, dy, and dz parallel to axes x, y, z in vector field, per unit time is given by $A_y dxdz$.

The flow through the area $DEFG$ per unit time may be represented by the following Taylor's series:

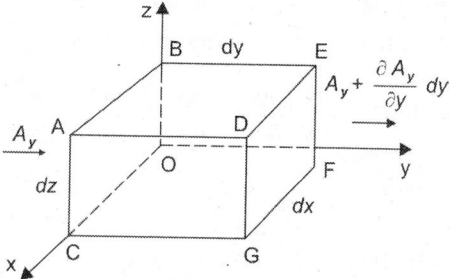

Fig. 1.18

$$A_y(y + dy)\, dx\, dz = \left\{ A_y(y) + \frac{\partial A_y}{\partial y} dy + \cdots \right\} dx\, dz$$

We will neglect higher-order terms in the above expansion. The net increase in the mass of the fluid inside the volume element per unit time due to the flow through the two faces is

$$A_y dx\, dz - \left(A_y + \frac{\partial A_y}{\partial y} dy \right) dx\, dz = -\frac{\partial A_y}{\partial y} dx\, dy\, dz = -\frac{\partial A_y}{\partial y} d\tau$$

Similarly, the net increase in the mass of fluid per unit time due to the flow through $BEFO$ and $ADGC$ is

$$-\frac{\partial A_x}{\partial x} d\tau$$

and that through $FGCO$ and $EDAB$ is

$$-\frac{\partial A_z}{\partial z} d\tau$$

The total increase in mass of fluid per unit volume per unit time due to the excess of inward flow over the outward flow is

$$\frac{\left(-\dfrac{\partial A_x}{\partial x} - \dfrac{\partial A_y}{\partial y} - \dfrac{\partial A_z}{\partial z} \right) d\tau}{d\tau} = -\nabla \cdot A$$

which is just the rate of increase of the density of the fluid inside the volume element $d\tau$. That is to say

$$\frac{\partial \rho}{\partial t} = -\nabla \cdot A \tag{1.58}$$

The above equation is called the *continuity equation.* Using Eq. (1.57), we have

$$\nabla \cdot A = \operatorname{div} A = -\frac{\partial \rho}{\partial t}, \tag{1.59}$$

i.e. the amount of flux per unit volume is defined as the divergence of vector A; and since divergence is the amount of flux, it is essentially a scalar.

If the divergence exists at a point in a fluid, whether liquid or gas, is positive, it expresses the rate at which fluid is flowing away from the point per unit volume. Hence, either the fluid is expanding and its density at the point is falling with time or the point is a source at which fluid is entering the field. When the divergence is negative, it gives the rate at which fluid is flowing towards the point per unit volume. In this case either the fluid is contracting and its density rising at the point or the point is a negative source, a sink, at which fluid is leaving the field. Since most practical liquids are almost incompressible, the existence of divergence means the presence of volume distribution of sources or sinks rather than changes in density.

In case of nonmaterial fluxes, such as those of the thermal, electric or magnetic fields, the existence of divergence means the presence of a source or sink of flux at the point. For example, in the electric field positive divergence means there is positive electric charge at the point, in the thermal field the point is either a source of heat or a place where the temperature is falling.

When the divergence is zero, everywhere the flux entering any element of space is exactly balanced by that leaving it and we may write

$$\operatorname{div} A = 0 \tag{1.60}$$

which is true in many practical problems. In a fluid this means there can be no sources or sinks in the field, nor can its density be changing, i.e. the fluid is incompressible. If the fluxes entering and leaving an element are equal, none can have been generated within it; the lines of flow of the vector A must either form closed curves (i.e. the magnetic field of a current) or terminate up on bounding surfaces (i.e. the electric field in a condenser) or extend to infinity. A vector which satisfies this condition is called *solenoidal vector.*

Gauss's Theorem

The Gauss divergence theorem enables us to transform volume integral into surface integral.

The statement of Gauss divergence theorem is as follows: If A is a continuously differentiable vector point function and S is a closed surface enclosing volume V, then

$$\int_S A \cdot n \, ds = \iiint_V \nabla \cdot A \, dV \tag{1.61}$$

where n is a unit outward normal* or briefly:

The normal surface integral of a vector function A over the boundary of a closed region is equal to the space integral of the divergence of A taken throughout the enclosed space.

---

* The sign convention for the unit vector:
  (a) For a closed surface (a surface which encloses a volume), the outward normal is called positive.
  (b) For an open surface, the right hand screw rule is used; the direction of rotation is the same as that in which the periphery is traversed.

To prove the theorem, we first expand the right-hand side of Eq. (1.61) and obtain

$$\iiint_V \nabla \cdot A \, dV = \iiint_V \left[\frac{\partial A_x}{\partial x} + \frac{\partial A_y}{\partial y} + \frac{\partial A_z}{\partial z} \right] dx \, dy \, dz \qquad (1.62)$$

Although the theorem is valid for an arbitrary shaped closed surface, we choose the volume in (Fig. 1.19) for convenience.

In (Fig. 1.19) we have

$$\left.\begin{aligned}
ds_1 &= -k\,dxdy \,(\text{area } DCHE) \\
ds_2 &= i\,dydz \,(\text{area } AFED) \\
ds_3 &= -j\,dxdz \,(\text{area } ABCD) \\
ds_4 &= -i\,dydz \,(\text{area } BGHC) \\
ds_5 &= k\,dxdy \,(\text{area } ABGF) \\
ds_6 &= j\,dxdz \,(\text{area } EFGH)
\end{aligned}\right\} \qquad (1.63)$$

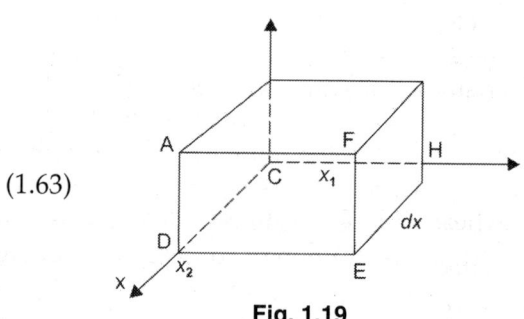

Fig. 1.19

Integrating the first term on the right hand side of Eq. (1.62) with respect to x from x_1 to x_2, we obtain

$$\iint \int_{x_1}^{x_2} \frac{\partial A_x}{\partial x} dx \, dy \, dz = \iint_S \left[A_x(x_2 yz) - A_x(x_1 yz) \right] dy \, dz$$

$$= \iint_{S_6} A_x(x_2 yz) \, dy \, dz - \iint_{S_3} A_x(x_1 yz) \, dy \, dz \qquad (1.64)$$

Note that

and
$$\left.\begin{aligned}
n_4 \cdot i \, dS_4 &= -dy\,dz \\
n_2 \cdot i \, dS_2 &= dy\,dz
\end{aligned}\right\} \qquad (1.65)$$

Using Eqs (1.64) and (1.65), we can write;

$$\iint \int_{x_1}^{x_2} \frac{dA_x}{\partial x} dx \, dy \, dz = \iint_{S_2} A_x(x,y,z) n_2 \cdot i \, dS_2 + \iint_{S_4} A_x(x,y,z) n_4 \cdot i \, dS_4$$

$$= \oint_S A_x n \cdot i \, dS \qquad (1.66)$$

where
$$\left.\begin{aligned}
\iint\limits_{\text{sides}}^{S} (\cdots) = 0
\end{aligned}\right\}$$

since i is perpendicular to dS_1, dS_3, dS_5 and dS_6. Similarly, we can show that

$$\iiint \frac{\partial A_z}{\partial y} dy \, dx \, dz = \oint_S A_y n \cdot j \, dS \qquad (1.67)$$

and
$$\iiint \frac{\partial A_z}{\partial y} dy \, dx \, dz = \oint_S A_z n \cdot j \, dS \qquad (1.68)$$

Combining Eqs (1.66), (1.67) and (1.68), we obtain

$$\iiint_V \nabla \cdot A \, dV = \oint_S A \cdot dS$$

Hence the theorem.

1.10 GREEN'S THEOREM IN THE PLANE

Green's theorem in the plane is a special case of the Stokes theorem. Also, it is of interest to note that Gauss's divergence theorem is a generalization of Green's theorem in the plane where (plane) region R and its closed boundary curve C are replaced by a (space) region V and its closed boundary (surface) S. For this reason, the divergence theorem is often called *Green's theorem in space*.

Green's theorem states that if R is a closed region of the xy plane bounded by a simple closed curve C and if A and B are continuous functions of x and y having continuous derivative in R, then

$$\oint_C (Adx + Bdy) = \iint_R \left(\frac{\partial B}{\partial x} - \frac{\partial A}{\partial x} \right) dx\,dy \tag{1.69}$$

where C is traversed in the positive (counterclockwise) direction (Until and unless stated otherwise, we shall always assume $\oint_S$ to mean that integral is described in the positive sense).

Let the equations of the curves AEB and AFB be $y = y_1(x)$ and $y = y_2(x)$ respectively (Fig. 1.20). If R is the region bounded by C, we have

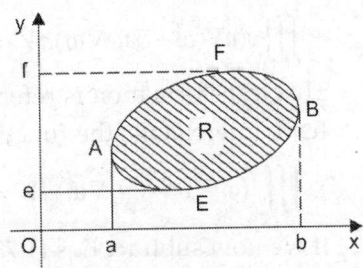

Fig. 1.20

$$\iint_R \frac{\partial A}{\partial y} dx\,dy - \int_{x=a}^{b} \left[\int_{y_1(x)}^{y_2(x)} \frac{\partial A}{\partial y} dy \right] dx$$

$$= \int_{x=a}^{x=b} A(x,y) \Big|_{y=y_1(x)}^{y_2(x)} dx$$

$$= -\int_a^b [A(x,y_1)dx - \int_b^a A(x,y_2)]dx$$

$$= -\oint_C Adx$$

or

$$\oint_C Adx = \iint_R \frac{\partial A}{\partial y} dx\,dy \tag{1.70}$$

Similarly, let the equations of curves EAF and EBF be $x = x_1(y)$ and $x = x_2(y)$ respectively. Then

$$\iint_R \frac{\partial B}{\partial x} dx\,dy = \int_{y=e}^{f} \left[\int_{x=x_1(y)}^{x_2(y)} \frac{\partial B}{\partial x} dx \right] dy$$

$$= \int_e^f \{ B(x_2,y) - B(x_1,y) \} dy$$

$$= \int_f^e B(x_1,y)dy + \int_e^f B(x_2,y)dy = \oint_C Bdy$$

Then

$$\oint_C Bdy = \iint_R \frac{\partial B}{\partial x} \cdot dx\,dy \tag{1.71}$$

Adding Eqs (1.70) and (1.71), we have

$$\oint_C (Adx + Bdy) = \iint_R \left(\frac{\partial B}{\partial x} - \frac{\partial A}{\partial x} \right) dx\,dy \tag{1.72}$$

Green's Identity

Applying Gauss's theorem, we are able to make some important transformations. Consider

$$A = u\nabla\omega$$

i.e., let the vector field A be the product of a scalar function u and the gradient of another scalar function ω. Consider

$$\nabla \cdot A = \frac{\partial A_x}{\partial x} + \frac{\partial A_y}{\partial y} + \frac{\partial A_z}{\partial z}$$

$$= \frac{\partial}{\partial x}\left(u\frac{\partial \omega}{\partial x}\right) + \frac{\partial}{\partial y}\left(u\frac{\partial \omega}{\partial y}\right) + \frac{\partial}{\partial z}\left(u\frac{\partial \omega}{\partial z}\right)$$

$$= u\left(\frac{\partial^2 \omega}{\partial x^2} + \frac{\partial^2 \omega}{\partial y^2} + \frac{\partial^2 \omega}{\partial z^2}\right) + \frac{\partial u}{\partial x}\frac{\partial \omega}{\partial x} + \frac{\partial u}{\partial y}\frac{\partial \omega}{\partial y} + \frac{\partial u}{\partial z}\frac{\partial \omega}{\partial z}$$

$$= u\nabla^2\omega + \nabla u \cdot \nabla\omega.$$

If we place this value of $\nabla \cdot A$ in the left hand side of Gauss's theorem Eq. (1.62), we obtain

$$\iiint_V (u\nabla^2\omega + \nabla u \cdot \nabla\omega)dV = \iint_S (u\nabla\omega) \cdot dS \tag{1.73}$$

This transformation is referred to as the first form of Green's identity or theorem. If we interchange the function u and w in Eq. (1.73), we obtain

$$\iiint_V (\omega\nabla^2 u + \nabla u \cdot \nabla\omega)dV = \iint_S (\omega\nabla u) \cdot dS \tag{1.74}$$

If we now subtract Eqs (1.73) from (1.74), we obtain

$$\iiint_V (u\nabla^2\omega - \omega\nabla^2 u)dV = \iint_S (u\nabla\omega - \omega\nabla u) \cdot dS \tag{1.75}$$

This transformation is referred to as second form of Green's identity or theorem.

Example 1. *(a) Interpret the symbol $A \cdot \nabla$ (b) Give a possible meaning to $(A \cdot \nabla)\,B$ (c) Is it possible to write this as $A \cdot \nabla\,B$ without ambiguity?*

Solution: (a) Let $A = A_x i + A_y j + A_z k$. Then,

$$A \cdot \nabla = A_x\frac{\partial}{\partial x} + A_y\frac{\partial}{\partial y} + A_z\frac{\partial}{\partial z}$$

is an operator. For example

$$(A \cdot \nabla)\phi = \left(A_x\frac{\partial}{\partial x} + A_y\frac{\partial}{\partial y} + A_z\frac{\partial}{\partial z}\right)\phi = A_x\frac{\partial\phi}{\partial x} + A_y\frac{\partial\phi}{\partial y} + A_z\frac{\partial\phi}{\partial z}$$

Note that this is the same as $A \cdot \nabla\phi$.

(b) Using part (a) with ϕ replaced, we obtain

$$B = B_x i + B_y j + B_z k$$

$$(A \cdot \nabla)B = \left(A_x\frac{\partial}{\partial x} + A_y\frac{\partial}{\partial y} + A_z\frac{\partial}{\partial z}\right)B$$

$$= A_x \frac{\partial B}{\partial x} + A_y \frac{\partial B}{\partial y} + A_z \frac{\partial B}{\partial z}$$

$$= \left(A_x \frac{\partial B_x}{\partial x} + A_y \frac{\partial B_x}{\partial y} + A_z \frac{\partial B_x}{\partial z} \right) i + \left(A_x \frac{\partial B_y}{\partial x} + A_y \frac{\partial B_y}{\partial y} + A_z \frac{\partial B_y}{\partial z} \right) j$$

$$+ \left(A_x \frac{\partial B_z}{\partial x} + A_y \frac{\partial B_z}{\partial y} + A_z \frac{\partial B_z}{\partial z} \right) k$$

(c) Using the interpretation of ∇B, the following can be established:

$$A \cdot \nabla B = (A_x i + A_y j + A_z k) \cdot \nabla B$$

$$= A_x i \cdot \nabla B + A_y j \cdot \nabla B + A_z k \cdot \nabla B$$

But grad B or ∇B is defined as

$$\nabla B = \left(\frac{\partial}{\partial x} i + \frac{\partial}{\partial y} j + \frac{\partial}{\partial z} k \right) (B_x i + B_y j + B_z k)$$

$$= \left[\frac{\partial B_x}{\partial x} ii + \frac{\partial B_y}{\partial x} ij + \frac{\partial B_z}{\partial x} ik + \frac{\partial B_x}{\partial y} ji + \frac{\partial B_y}{\partial y} jj + \frac{\partial B_z}{\partial y} jk + \frac{\partial B_x}{\partial z} ki + \frac{\partial B_y}{\partial z} kj + \frac{\partial B_z}{\partial z} kk \right]$$

$$A \cdot \nabla B = (A_x i + A_y j + A_z k) \cdot (\nabla B)$$

$$= A_x \left(\frac{\partial B_x}{\partial x} i + \frac{\partial B_y}{\partial x} j + \frac{\partial B_z}{\partial x} k \right) + A_y \left(\frac{\partial B_x}{\partial y} i + \frac{\partial B_y}{\partial y} j + \frac{\partial B_z}{\partial y} k \right) + A_z \left(\frac{\partial B_x}{\partial z} i + \frac{\partial B_y}{\partial z} j + \frac{\partial B_z}{\partial z} k \right)$$

$$= \left(A_x \frac{\partial B_x}{\partial x} + A_y \frac{\partial B_x}{\partial y} + A_z \frac{\partial B_x}{\partial z} \right) i + j \left(A_x \frac{\partial B_y}{\partial x} + A_y \frac{\partial B_y}{\partial y} + A_z \frac{\partial B_y}{\partial z} \right)$$

$$+ \left(A_x \frac{\partial B_z}{\partial x} + A_y \frac{\partial B_z}{\partial y} + A_z \frac{\partial B_z}{\partial z} \right) k$$

which gives the same result as that given in part (*b*). It follows that $(A \cdot \nabla B) = (A \cdot \nabla)B$ without ambiguity provided the concept of dyadics is introduced with properties as indicated[*].

---

[*] The quantities $i\,i,\ j\,j$, etc. are called unit diads (note that for example $i\,j$ is not the same as $j\,i$). A quantity of the form $a_{11}\,i\,i + a_{12}\,i\,j + a_{13}\,i\,k + a_{21}\,j\,i + a_{22}\,j\,j + a_{23}\,j\,k + a_{31}\,k\,i + a_{32}\,k\,j + a_{33}\,k\,k$ is called a dyadic and the coefficients $a_{11}, a_{12}...$ are its components. Any array of these nine components in the form

$$\begin{bmatrix} a_{11} & a_{12} & a_{13} \\ a_{21} & a_{22} & a_{23} \\ a_{31} & a_{32} & a_{33} \end{bmatrix}$$

is called a 3 by 3 matrix. A dyadic is the generalisation of a vector. Still further generalisation leads to triadics which are quantities consisting of 27 terms of the form $a_{111}\,i\,i\,i + a_{211}\,j\,i\,i + ...$ A study of how the components of a dyadic or triadic transform from one system of coordinates to another leads to the subject of tensor analysis which we shall discuss in later chapters.

Example 2. *Verify divergence theorem for* $F = 2x^2y\,i - y^2j + 4xz^2\,k$ *taken over the region in the first octant bounded by* $y^2 + z^2 = 9$ *and* $x = 2$.

Solution. By divergence theorem, we have

$$\iiint_V \nabla \cdot F\,dV = \iint F \cdot n\,dS$$

To verify this we shall show that the volume integral on the left is equal to the surface integral on the right. We have

$$\nabla \cdot F = \left(i\frac{\partial}{\partial x} + j\frac{\partial}{\partial y} + k\frac{\partial}{\partial z} \right) \cdot (2x^2yi - y^2j + 4xz^2k)$$

$$= \frac{\partial}{\partial x}(2x^2y) + \frac{\partial}{\partial y}(-y^2) + \frac{\partial}{\partial z}(4xz^2)$$

$$= 4xy - 2y + 8xz$$

The required integral over whole volume is given by

$$I_1 = \iiint_V (4xy - 2y + 8xy)\,dx\,dy\,dz$$

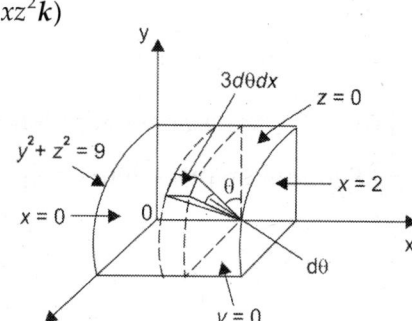

Fig. 1.21

To cover the whole volume the limits of x are from 0 to 2

y are from 0 to $\sqrt{9-z^2}$ $(\because\ y^2 + z^2 = 9)$

z are from 0 to 3 $(\because\ z^2 + 0 = 9)$

and $I_1 = \int_{x=0}^{2} \int_{y=0}^{\sqrt{9-z^2}} \int_{z=0}^{3} [4xy - 2y + 8xz]\,dx\,dy\,dz$

$$= \int_{x=0}^{2} \int_{z=0}^{3} \left[4x(y^2/2) - 2(y^2/2) + 8xyz \right]_{y=0}^{\sqrt{9-2z^2}} dx\,dz$$

$$= \int_{x=0}^{2} \int_{z=0}^{3} \left[(2x-1)(9-z^2) + 8xz\sqrt{9-2z^2} \right] dx\,dz$$

$$= \int_{x=0}^{2} \int_{z=0}^{3} (2x-1)(9-z^2)\,dx\,dz + \int_{x=0}^{2} \int_{z=0}^{3} 8xz\sqrt{9-z^2}\,dx\,dz$$

Putting $z = 3\sin\theta$ in the second integral, we have

$$I_1 = \int_{x=0}^{2} 18(2x-1)\,dx + \int_{x=0}^{2} 72x\,dx = 18 \times 2 + 72 \times 2 = 36 + 144 = 180$$

The surface integral is the contribution due to five parts of the given cylinder, i.e. the surface integral is given by

$$\iint_S F\cdot n\,ds = \iint_{\substack{S_1-\text{Curved}\\ \text{surface}}} F\cdot n\,ds + \iint_{\substack{S_2-\text{Plane}\\ x=0}} F\cdot n\,ds + \iint_{\substack{S_3-\text{Plane}\\ x=0}} F\cdot n\,ds + \iint_{\substack{S_4-\text{Plane}\\ y=0}} F\cdot n\,ds + \iint_{\substack{S_5-\text{Plane}\\ z=0}} F\cdot n\,ds$$

Now, a normal vector to the surface $y^2 + z^2 = 9$ is

$$\nabla\phi = \left(i\frac{\partial}{\partial x} + j\frac{\partial}{\partial y} + k\frac{\partial}{\partial z} \right)(y^2 + z^2) = 2yj + 2zk$$

Unit normal vector **n** along $2yj + 2zk$

$$= \frac{2yj + 2zk}{2\sqrt{y^2 + z^2}} = \frac{yj + zk}{3}$$

$\therefore$

$$\mathbf{F} \cdot \mathbf{n} = (2x^2 yi - y^2 j + 4xz^2 k) \cdot \frac{(yj + zk)}{3}$$

$$= \frac{1}{3}(-y^3 + 4xz^3)$$

Therefore,

$$\iint_{S_1} \mathbf{F} \cdot \mathbf{n} = \iint_{S_1} \frac{1}{3}(-y^3 + 4xz^3) ds$$

[To cover the whole curved surface, we have taken an element $3d\theta \cdot dx$, so that $y = 3\cos\theta$, $z = 3\sin\theta$, and θ varies from 0 to $\dfrac{\pi}{2}$ and x varies from 0 to 2].

or

$$\iint_{S_1} \mathbf{F} \cdot \mathbf{n}\, ds = \int_{x=0}^{2} \int_{\theta=0}^{\pi/2} \left[\frac{1}{3}(-27\cos^3\theta + 4x \times 27 \sin^3\theta) 3 d\theta\, dx \right]$$

$$= 27 \int_{x=0}^{2} \int_{\theta=0}^{\pi/2} (-\cos^3\theta + 4x\sin^3\theta) d\theta\, dx$$

$$= 18 \int_{x=0}^{2} (-1 + 4x) dx = 108$$

For S_2 $\qquad x = 0,\ \mathbf{n} = -i$

and

$$\iint_{S_2} \mathbf{F} \cdot \mathbf{n}\, ds = \iint_{S} (-2x^2 yi - y^2 j + 4xz^2 k) \cdot (-i) ds$$

$$= \iint_{S} -2x^2 y\, ds = 0 \qquad\qquad (\because x = 0)$$

For S_3 $\qquad x = 2,\ \mathbf{n} = i$

$$\iint_{S_3} \mathbf{F} \cdot \mathbf{n}\, ds = \iint 8y\, dy\, dz = \int_{z=0}^{3} \int_{y-0}^{\sqrt{9-z^2}} 8y\, dy\, dz$$

$$= \int_{z=0}^{3} \left(\frac{8y^2}{2} \right)^{\sqrt{9-z^2}} dz = \int_{z=0}^{3} 4(9 - z^2) dz = 72$$

For S_4 $\qquad y = 0,\ \mathbf{n} = -j$

and

$$\iint_{S_4} \mathbf{F} \cdot \mathbf{n}\, ds = \iint_{S_4} (-y^2) ds = 0 \qquad\qquad (\because y = 0)$$

For S_5 $\qquad z = 0,\ \mathbf{n} = -k$

$$\iint_{S_5} \mathbf{F} \cdot \mathbf{n}\, ds = \iint_{S_5} (-4xz^2) ds = 0 \qquad\qquad (\because z = 0)$$

$\therefore$

$$\iint_{S} \mathbf{F} \cdot x\, ds = 108 + 0 + 72 + 0 + 0 = 180$$

Thus, the volume integral = the surface integral

Hence the Gauss's theorem has been verified.

Example 3. *Let **r** be the position vector of any point relative to an origin O, ϕ has continuous derivatives of order two and S be a closed surface enclosing a volume V, then*

$$\int_S \left[\frac{1}{r}\nabla\phi - \phi\nabla\left(\frac{1}{r}\right) \right] ds - \int_V \frac{1}{r}\nabla^2\phi\, dV = 0 \text{ or } 4\pi\phi(P),$$

according as P lies outside or inside S **(Green's formula).**

Solution: Here we apply Green's theorem in symmetrical form to the particular case when $u = \phi$ and $\omega = 1/r$, where r is the distance from P to the points of S. There are two cases.

Case I. *When P lies inside the surface S.* Since $\omega = 1/r$ becomes infinite at $P (r = 0)$, therefore to remove this difficulty enclose P within a small sphere of radius ϵ having its centre at P. Denote the surface of this sphere by S_1. Now in region bounded by S and S_1, (i.e. $V - V_1$ where V_1 is the volume enclosed by S_1) has continuous derivatives up to the order two.

Also we have

$$\nabla\omega = \nabla\left(\frac{1}{r}\right) = -\frac{1}{r^2}r$$

and

$$\nabla^2\omega = \nabla\left(\frac{-r}{r^2}\right) = 0$$

Now applying the Green's second identity Eq. (1.75) for the region bounded by S and S_1, we have

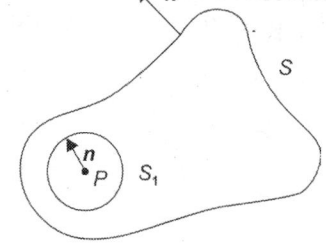

Fig. 1.22

$$\int_{V-V_1}\left[0 - \frac{1}{r}\nabla^2\phi\right]dV = \iint_S\left[\phi\frac{\partial}{\partial n}\left(\frac{1}{r}\right) - \frac{1}{r}\frac{\partial\phi}{\partial n}\right]dS + \iint_{S_1}\left[\phi\frac{\partial}{\partial n}\left(\frac{1}{r}\right) - \frac{1}{r}\frac{\partial\phi}{\partial n}\right]dS_1$$

where unit normal at any point of sphere S_1 is directed towards P. Rearranging the terms, we have

$$\iint_S\left[\phi\frac{\partial}{\partial n}\left(\frac{1}{r}\right) - \frac{1}{r}\frac{\partial\phi}{\partial n}\right]dS + \int_V\frac{\nabla^2\phi}{r}dV = \iint_{S_1}\left[\frac{1}{r}\frac{\partial\phi}{\partial n} - \phi\frac{\partial}{\partial r}\left(\frac{1}{r}\right)\right]dS + \int_V\frac{\nabla^2\phi}{r}dV \quad (1)$$

Consider now what happens as $\epsilon \to 0$. We know that

$$\delta V = \epsilon^2\delta\Omega\,\delta R$$

where Ω is the solid angle and

$$\int_{V_1}\frac{\nabla^2\phi}{r}dV = \int_0^\epsilon\frac{dR}{r}\int_{S_1}\nabla^2\phi\cdot d\Omega$$

If $\nabla^2\phi$ is not infinite at $r = 0$, the above integral tends to zero as $\epsilon \to 0$. On S_1, $n = -r$, $\epsilon^2\delta\Omega$, $r = \epsilon$, and then

$$\int_{S_1}\frac{1}{r}\frac{\partial\phi}{\partial n}ds = \frac{1}{\epsilon}\int\epsilon^2\delta\Omega\frac{\partial\phi}{\partial n} = \epsilon\int\frac{\partial\phi}{\partial n}d\Omega \to 0 \text{ as } \epsilon \to 0$$

and

$$\int_S\phi\frac{\partial}{\partial n}\left(\frac{1}{r}\right)ds = \int_{S_1}\phi\,d\Omega$$

which is independent of ϵ and making $\epsilon \to 0$, it becomes $4\pi\phi(r)$

Hence

$$\iint_S\left[\phi\frac{\partial}{\partial n}\left(\frac{1}{r}\right) - \frac{1}{r}\frac{\partial\phi}{\partial n}\right]ds + \int_V\frac{\nabla^2\phi}{r}dV = -4\pi\phi(r)$$

Case II. *When P lies outside S.*

If P lies outside S, then S_1 and V_1 do not exist and the right hand side of Eq. (1) is zero. Hence

$$\iint_S \left[\phi \frac{\partial}{\partial n}\left(\frac{1}{r}\right) - \frac{1}{r}\frac{\partial \phi}{\partial n} \right] ds + \int_V \frac{\nabla^2 \phi}{r} dv = 0$$

Hence the result.

Example 4. *Verify divergence theorem for*
$$F = x^2 i + y^2 j + z^2 k$$
taken over the cube $0 \le x, y, z, \le 1$.

Solution: Consider the volume ($ABCDEFGH$) as shown in Fig. 1.23. Let us take the adjacent edges EX, EY and EZ as axes of reference with E as origin.

Given $\qquad F = x^2 i + y^2 j + z^2 k$

$$\text{div } F = \nabla \cdot F = \left(i\frac{\partial}{\partial x} + j\frac{\partial}{\partial y} + k\frac{\partial}{\partial z} \right) \cdot (x^2 i + y^2 j + z^2 k)$$

$$= 2(x + y + z).$$

The volume integral is given by

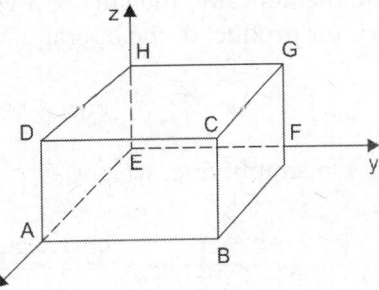

Fig. 1.23

$$\iiint_V \nabla \cdot F \, dV = \iiint 2(x + y + z)\,dx\,dy\,dz$$

$$= \int_{x=0}^1 \int_{y=0}^1 \int_{z=0}^1 2(x + y + z)\,dx\,dy\,dz$$

$$= 2\left[\frac{1}{2} + \frac{1}{2} + \frac{1}{2} \right] = 3$$

For the surface of the cube, we have

$$\iint F \cdot ds = \underset{ABCD}{\iint} F \cdot i\,dy\,dz + \underset{BFGC}{\iint} F \cdot j\,dx\,dz$$

$$+ \underset{FGHE}{\iint} F \cdot (-i)\,dy\,dz + \underset{EHDA}{\iint} F \cdot (-j)\,dx\,dz$$

$$+ \underset{CDHG}{\iint} F \cdot k\,dy\,dx + \underset{ABFE}{\iint} F \cdot (-k)\,dy\,dx$$

$$\underset{ABCD}{\iint} F \cdot i\,dy\,dz = \int_{y=0}^1 \int_{z=0}^1 x^2\,dy\,dz = \int_{y=0}^1 \int_{z=0}^1 dy\,dz = 1$$

(since $x = 1$ for surface $ABCE$)

$$\underset{BFGC}{\iint} F \cdot j\,dx\,dz = \int_{x=0}^1 \int_{z=0}^1 y^2\,dx\,dz = \int_{x=0}^1 \int_{z=0}^1 dx\,dz = 1$$

(since $y = 1$ for surface $BFGC$)

$$\underset{FGHE}{\iint} F \cdot (-i)\,dy\,dz = \int_{y=0}^1 \int_{z=0}^1 (-x)^2\,dy\,dz = 0$$

(since $x = 0$ for surface $FGHE$)

$$\iint_{EHDA} F \cdot (-j) dx\, dz = \int_{x=0}^{1} \int_{z=0}^{1} (-y^2) dx\, dz = 0$$

(since $y = 0$ for surface $EHDA$)

$$\iint_{CDHG} F \cdot k\, dy\, dx = \int_{y=0}^{1} \int_{z=0}^{1} z^2\, dy\, dx = \int_{y=0}^{1} \int_{z=0}^{1} dy\, dx = 1$$

(since $z = 1$ for surface $CDHG$)

$$\iint_{ABFE} F \cdot (-k) dy\, dx = \int_{y=0}^{1} \int_{z=0}^{1} (-z^2) dy\, dx = 0$$

(since $z = 0$ for surface $ABFE$)

$$\therefore \qquad \iint_{S} F \cdot ds = 1 + 1 + 0 + 0 + 1 + 0 = 3$$

Thus, the volume integral = The surface integral.

This verifies the divergence theorem.

1.11. THE CURL OF A VECTOR FIELD AND STOKE'S THEOREM

Mathematically, the curl of a vector A is defined as a vector obtained by taking the vector product of the operator ∇ and the vector A, i.e.

$$\text{curl } A = \nabla \times A = \left(\frac{\partial}{\partial x} i + \frac{\partial}{\partial y} j + \frac{\partial}{\partial z} k \right) \times (A_x i + A_y j + A_z k)$$

On simplifying, we get

$$\text{curl } A = i \left(\frac{\partial A_z}{\partial y} - \frac{\partial A_y}{\partial z} \right) + j \left(\frac{\partial A_x}{\partial z} - \frac{\partial A_z}{\partial x} \right) + k \left(\frac{\partial A_y}{\partial x} - \frac{\partial A_x}{\partial y} \right) \quad (1.76)$$

This may be written conveniently in the following form

$$\text{curl } A = \begin{vmatrix} i & j & k \\ \dfrac{\partial}{\partial x} & \dfrac{\partial}{\partial y} & \dfrac{\partial}{\partial z} \\ A_x & A_y & A_z \end{vmatrix} \qquad (1.76a)$$

curl of A is also called as rotation of A and is denoted as rot A.

Physical significance: It has been referred in Section 1.8 that when a vector field is the gradient of a scalar field, the line integral of the vector taken around any closed path is zero; this result is true no matter what size or shape the path may have.

Many vector fields occur in physical problems, in which the closed path line integral is not zero and which cannot, therefore, be expressed as the gradient of a scalar point function, the magnitude of line integral in such cases measures an important property of the field, especially when the integral is taken for infinite small surface elements, and referred to unit area. Here too, it may be shown that the limiting value is independent of the particular form of the boundary, and again an estimation of the order of magnitude shows that the limit has a meaning when A is a continuous differentiable point function. The value of the integral, however, depends on the orientation of the surface element, as may be shown by a simple example.

Consider a very small region of such a vector field, several lines of flow of which are shown in (Fig. 1.24), the portion is chosen small enough for the lines to be regarded as nearly straight and parallel. Place a small plane area in this field, shown for convenience as rectangular. If the surface element is placed so that its normal is in the direction of A, then A is perpendicular to ds along the entire circuit and the integral vanishes. On the other hand, if the normal is perpendicular to A, then the net contribution of the edges perpendicular to A is zero, but that of the edges parallel to A is not zero since the magnitude of A along the upper edge is assumed to be different from that along the lower edge, the line integral round the boundary has a finite value. The value of the limit

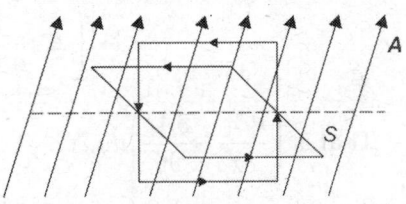

Fig. 1.24

$$L_n = \lim_{\nabla S \to 1} \oint A \cdot dl \tag{1.77}$$

is thus a function of the direction of the normal to the surface element.

In general, if we put a small vector area of any shape at any point in a vector field and compute the line integral of the vector A around its bounding edge there will be an orientation of the area for which the line integral is the greatest. The amount of this maximum line integral expressed per unit area is called the curl of the vector field at the point. The curl of a vector field is a vector quantity directed along the normal to the exploring area which is in the position giving the greatest integral.

To calculate the curl in terms of its cartesian components, take three infinitesimal rectangular areas intersecting mutually at right angles at a point where the vector field A has components of magnitudes A_x, A_y, A_z as in Fig. 1.25. Taking the positive normal to the areas along the positive directions of the x-, y- and z-axes respectively, the circular arrows indicate the positive senses in which their boundaries must be traversed to accord with the right-hand screw rule for vector areas.

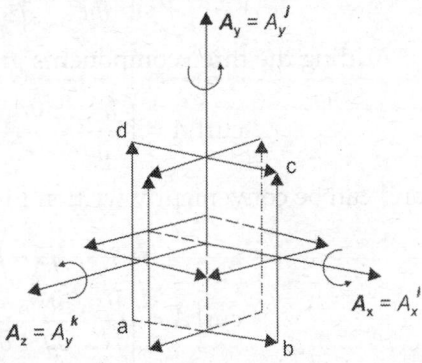

Fig. 1.25

Take one of these areas such as $abcd$ in (Fig. 1.25) with sides dx, dy, its normal being along the axis of z. Since the rectangle is very small, the numerical value of the component of A at the middle of any side may reasonably be taken as the average value for that side; the arrows show the directions in which the components act. Since A_x, A_y, A_z are functions of the coordinates (x, y, z) of the middle of the rectangle, the average value along the four sides ab, bc, dc, ad will be respectively:

$$A_x - \frac{1}{2}\frac{\partial A_x}{\partial y}dy,\ A_y + \frac{1}{2}\frac{\partial A_y}{\partial x}dx,\ A_x + \frac{1}{2}\frac{\partial A_x}{\partial y}dy,$$

and

$$A_y - \frac{1}{2}\frac{\partial A_y}{\partial x}dx$$

Around the contour *abcd* the line integral is, therefore,

$$\left[\left(A_x - \frac{1}{2}\frac{\partial A_x}{\partial y}dy\right) - \left(A_x + \frac{1}{2}\frac{\partial A_x}{\partial y}dy\right)\right]dx$$

$$+\left[\left(A_y + \frac{1}{2}\frac{\partial A_y}{\partial x}dx\right) - \left(A_y - \frac{1}{2}\frac{\partial A_y}{\partial x}dx\right)\right]dy$$

That is $\left(\dfrac{\partial A_y}{\partial x} - \dfrac{\partial A_x}{\partial y}\right)dy\,dx$.

Since the area of the element is $dxdy$, the bracketed term is the magnitude of the component of curl of the vector field taken about the z-axis; we may write

$$\text{curl}_z\, A = \left(\frac{\partial A_y}{\partial x} - \frac{\partial A_x}{\partial y}\right)k$$

By following a precisely similar method with the remaining rectangles, we find the component curls about the y and x to be

$$\text{curl}_y\, A = \left(\frac{\partial A_x}{\partial z} - \frac{\partial A_z}{\partial x}\right)j$$

and $$\text{curl}_x\, A = \left(\frac{\partial A_z}{\partial y} - \frac{\partial A_y}{\partial z}\right)i$$

Adding the three components gives

$$\text{curl}\, A = \left(\frac{\partial A_z}{\partial y} - \frac{\partial A_y}{\partial z}\right)i + \left(\frac{\partial A_x}{\partial z} - \frac{\partial A_z}{\partial x}\right)j + \left(\frac{\partial A_y}{\partial x} - \frac{\partial A_x}{\partial y}\right)k$$

and can be conveniently written in the form of a determinant as

$$\text{curl}\, A = \begin{vmatrix} i & j & k \\ \dfrac{\partial}{\partial x} & \dfrac{\partial}{\partial y} & \dfrac{\partial}{\partial z} \\ A_x & A_y & A_z \end{vmatrix}$$

$\therefore$ $\text{curl}\, A = \nabla \times A$

Since ∇ is invariant, as such the curl is also invariant. Curl is associated with the rotation of fluid in the study of hydrodynamics. This is some times written as rot.

Stoke's Theorem: The line integral of the tangential component of a vector A taken around a simple closed curve C is equal to the surface integral of the normal component of the curl of A taken over any surface S having c as its boundary. Mathematically

$$\oint_C A \cdot dr = \iint_S (\nabla \times A) \cdot n\, ds \tag{1.78}$$

In rectangular coordinates

$$A = A_x i + A_y j + A_z k$$

and $$n = \cos\alpha\, i + \cos\beta\, j + \cos\gamma\, k$$

where $\cos\alpha$, $\cos\beta$, $\cos\gamma$ are the direction cosines of n. Then

$$\nabla \times A = \begin{vmatrix} i & j & k \\ \dfrac{\partial}{\partial x} & \dfrac{\partial}{\partial y} & \dfrac{\partial}{\partial z} \\ A_x & A_y & A_z \end{vmatrix}$$

$$= \left(\frac{\partial A_z}{\partial y} - \frac{\partial A_y}{\partial z}\right)i + \left(\frac{\partial A_x}{\partial z} - \frac{\partial A_z}{\partial x}\right)j + \left(\frac{\partial A_y}{\partial x} - \frac{\partial A_x}{\partial y}\right)k$$

$$(\nabla \times A) \cdot n = \left(\frac{\partial A_z}{\partial y} - \frac{\partial A_y}{\partial z}\right)\cos\alpha + \left(\frac{\partial A_x}{\partial z} - \frac{\partial A_z}{\partial x}\right)\cos\beta + \left(\frac{\partial A_y}{\partial x} - \frac{\partial A_x}{\partial y}\right)\cos\gamma$$

$$A \cdot dr = A_x\,dx + A_y\,dy + A_z\,dz$$

and Stoke's theorem becomes:

$$\iint_S \left[\left(\frac{\partial A_z}{\partial y} - \frac{\partial A_y}{\partial z}\right)\cos\alpha + \left(\frac{\partial A_x}{\partial z} - \frac{\partial A_z}{\partial x}\right)\cos\beta + \left(\frac{\partial A_y}{\partial x} - \frac{\partial A_x}{\partial y}\right)\cos\gamma\right]ds$$

$$= \oint_C (A_x\,dx + A_y\,dy + A_z\,dz)$$

Let S be the surface which is such that its projections on the xy, yz and zx planes are regions bounded by simple closed curves, as indicated in (Fig. 1.26). Assume S to have representation $z = f(x, y)$, $x = g(y, z)$, and $y = h(x, z)$, where f, g, h are single valued, continuous and differentiable functions.

Consider first $\iint_S [\nabla \times (A_x i)] \cdot n\,ds$

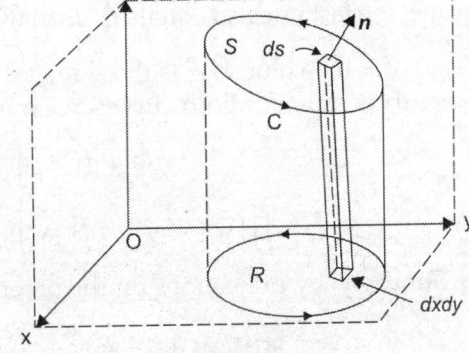

Since
$$\nabla \times (A_x i) = \begin{vmatrix} i & j & k \\ \dfrac{\partial}{\partial x} & \dfrac{\partial}{\partial y} & \dfrac{\partial}{\partial z} \\ A_x & 0 & 0 \end{vmatrix}$$

$$= \frac{\partial A_x}{\partial z}j - \frac{\partial A_x}{\partial y}k$$

and $[\nabla \times (A_x i) \cdot n\,ds]$

Fig. 1.26

$$= \left(\frac{\partial A_x}{\partial z}n \cdot j - \frac{\partial A_x}{\partial y}n \cdot k\right)ds \qquad\qquad (1.78a)$$

If $z = f(x, y)$ is taken as the equation of S, then position vector to any point S is

$$r = xi + yj + zk = xi + yj + f(x, y)k$$

so that
$$\frac{\partial r}{\partial y} = j + \frac{\partial f}{\partial y} \cdot k$$

But $\dfrac{\partial r}{\partial y}$ is a vector tangent to S and thus perpendicular to n.

Thus,
$$n \cdot \frac{\partial r}{\partial y} = n \cdot j + \frac{\partial f}{\partial y} n \cdot k = 0$$

or
$$n \cdot j = \frac{\partial z}{\partial y} n \cdot k \qquad \text{[since } f(x, y) = z]$$

Substituting in Eq. (1.78a), we obtain

$$\left(\frac{\partial A_x}{\partial z} n \cdot j - \frac{\partial A_x}{\partial y} n \cdot k \right) ds = \left(-\frac{\partial A_x}{\partial z} \frac{\partial z}{\partial y} n \cdot k - \frac{\partial A_x}{\partial y} n \cdot k \right) ds$$

or
$$[\nabla \times (A_x i) \cdot n \, ds] = -\left(\frac{\partial A_x}{\partial z} \frac{\partial z}{\partial y} + \frac{\partial A_x}{\partial y} \right) n \cdot k \, ds \qquad (1.78b)$$

Now on S, $A_x (x, y, z) = A_x \{x, y, f(x, y)\} = F(x, y)$
Hence

$$\frac{\partial A_x}{\partial y} + \frac{\partial A_x}{\partial z} \cdot \frac{\partial z}{\partial y} = \frac{\partial F}{\partial y}$$

Substituting in Eq. (1.78b), we obtain

$$[\nabla \times (A_x i) \cdot n \, ds] = -\frac{\partial F}{\partial y} n \cdot k \, ds = -\frac{\partial F}{\partial y} dx \, dy$$

Then
$$\iint_S [\nabla \times (A_x i)] \cdot n \, ds = \iint_R -\frac{\partial F}{\partial y} dx \, dy$$

where R is the projection of S on the xy plane. By Green's theorem, Eq. (1.70) for the plane, the last integral equals $\oint_\Gamma F \, dx$, where Γ is the boundary of R. Since at each point (x, y) of Γ the value of F is the same as the value of A_x at each point (x, y, z) of C and since dx is same for both curves, we must have

$$\oint_\Gamma F \, dx = \oint_C A_x \, dx$$

or
$$\iint_S [\nabla \times (A_x i)] \cdot n \, ds = \oint_C A_x \, dx$$

Similarly, by projections on the other coordinate planes

$$\iint_S [\nabla \times (A_y j)] \cdot n \, ds = \oint_C A_y \, dy$$

and
$$\iint_S [\nabla \times (A_z k)] \cdot n \, ds = \oint_C A_z \, dz$$

Then by adding, we get

$$\iint_S [\nabla \times A] \cdot n \, ds = \oint_C (A_x \, dx + A_y \, dy + A_z \, dz) = \oint_C A \cdot dr$$

The theorem is also valid for surface S which may not satisfy the restrictions imposed above. Let us assume that S can be subdivided into surfaces $S_1, S_2, ..., S_k$ with boundaries $c_1, c_2, c_3, ..., c_k$ which do satisfy the restrictions. Then Stoke's theorem holds for each such surface. Adding these surface integrals, the total surface integral over S

is obtained. Adding the corresponding line integrals over $c_1, c_2, c_3, ..., c_k$, the line integral over c is obtained.

The necessary and sufficient condition, that $\oint_C A \cdot dr = 0$ for every closed curve C is $\nabla \times A = 0$, identically.

Now

Sufficiency: Suppose $\nabla \times A = 0$, then by Stoke's theorem

$$\oint_C A \cdot dr = \iint (\nabla \times A) \cdot n \, ds = 0$$

Necessity: Suppose $\oint_C A \cdot dr = 0$ around every closed path C and assume

$$\nabla \times A \neq 0$$

at some point P. Then assuming $\nabla \times A$ is continuous there will be region with P as an interior point, where

$$\nabla \times A \neq 0$$

Let S be a surface contained in this region whose normal n at each point has the same direction as $\nabla \times A$, i.e. $\nabla \times A = \alpha n$, where α is a positive constant. Let C be the boundary of S. Then by Stoke's theorem

$$\oint_C A \cdot dr = \iint_S (\nabla \times A) \cdot n \, ds = \alpha \iint_S n \ n \, ds > 0$$

which contradicts the hypothesis that $\oint_C A \cdot dr = 0$ and shows that $\nabla \times A = 0$

It follows that $\nabla \times A = 0$ is also a necessary and sufficient condition for a line integral $\int_{P_1}^{P_2} A \cdot dr$ to be independent of the path joining points P_1 and P_2.

Example 1. *Verify Stoke's theorem for $A = (2x - y)i - yz^2 j - y^2 zk$, where S is the upper half surface of the sphere $x^2 + y^2 + z^2 = 1$ and C is its boundary.*

Solution: The boundary C of S is a circle in the xy plane of radius 1 unit and centre at the origin. Let

$$x = \cos t, y = \sin t, z = 0 \text{ and } 0 \leq t < 2\pi$$

be parametric equations of C. Then

$$\oint_C A \cdot dr = \oint_C \left\{ (2x - y)dx - yz^2 dy - y^2 z dz \right\}$$

$$= \int_0^{2\pi} (2\cos t - \sin t)(-\sin t)dt = \pi$$

Also

$$\nabla \times A = \begin{vmatrix} i & j & k \\ \dfrac{\partial}{\partial x} & \dfrac{\partial}{\partial y} & \dfrac{\partial}{\partial z} \\ 2x - y & -yz^2 & -y^2 z \end{vmatrix} = k$$

Then

$$\iint_S (\nabla \times A) \cdot n \, ds = \iint_S k \cdot n \, ds = \iint_S dx \, dy$$

Since $n \cdot k \, ds = dx \, dy$ and R is the projection of S on the xy plane. This last integral equals

$$\int_{x=-1}^{1} \int_{y=-\sqrt{1-x^2}}^{\sqrt{1-x^2}} dy \, dx = 4\int_{0}^{1}\int_{0}^{\sqrt{1-x^2}} dy \, dx = 4\int_{0}^{1}\sqrt{1-x^2}\, dx = \pi$$

Thus, Stoke's theorem is verified.

Example 2. *If R is the projection on the xy plane of the surface S, then*

$$\iint A \cdot ds = \iint A \cdot n \frac{dx \, dy}{|n \cdot k|}$$

where n is unit normal on S and k is unit normal along z-axis.

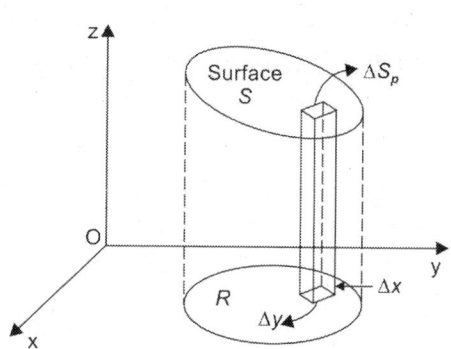

Fig. 1.27

Solution: Let us consider a surface S and take an infinitesimal small element ΔS_p of this surface.

Let this surface be inclined to xy-plane at angle θ. Then

$$\iint_{S} A \cdot ds \text{ is the limit of the sum}$$

$$\sum_{n=1}^{N} A_p \cdot n\Delta S_p \text{ as } N \to \infty \text{ and } \Delta S_p \to 0$$

where A_p is the value of the vector at the middle point of the small element ΔS_p.

The projection of ΔS_p on xy-plane is $\Delta x_p \, \Delta y_p$, so that

$$\Delta x_p \Delta y_p = \Delta S_p \cos \theta = \Delta S_p \, n \cdot k$$

where n is the unit normal for the surface element $\Delta x_p \Delta y_p$.

$\therefore$
$$\Delta S_p = \frac{\Delta x_p \Delta y_p}{|n \cdot k|}$$

Hence
$$\iint A \cdot ds = \lim_{\substack{N \to \infty \\ \Delta S_p \to 0}} \sum_{n=1}^{N} A_p \cdot n\Delta S_p$$

$$= \lim_{\substack{N \to \infty \\ \Delta S_p \to 0}} \sum_{n=1}^{N} A_p \cdot n\frac{\Delta x_p \, \Delta y_p}{|n \cdot k|}$$

$$= \lim_{\substack{N \to \infty \\ \Delta x_p \to 0 \\ \Delta y_p \to 0}} \sum_{n=1}^{N} A_p \cdot n\frac{\Delta x_p \, \Delta y_p}{|n \cdot k|}$$

$$= \iint_{R} A \cdot n\frac{dx \, dy}{|n \cdot k|}$$

Example 3. *Evaluate* $\iint_S A \cdot n \, ds$, *where* $A = zi + xj - 3y^2 zk$ *and* S *is the surface of the cylinder* $x^2 + y^2 = 16$ *included in the first octant between* $Z = 0$ *and* $Z = 5$.

Solution: Project S on the plane xz as in (Fig. 1.28) and call the projection R. Note that projection of S on the xy-plane can not be used here. Then

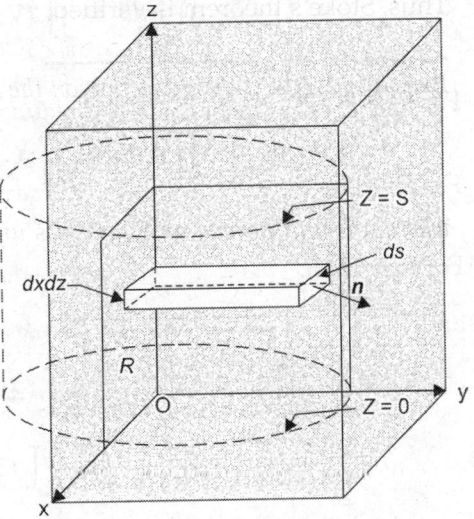

$$\iint_S A \cdot n \, ds = \iint_S A \cdot n \frac{dx \, dz}{|n \cdot j|}$$

A normal to $x^2 + y^2 = 16$ is

$$\nabla(x^2 + y^2) = 2(xi + yj).$$

Thus, the unit normal to S as shown in adjoining (Fig. 1.28) is

$$n = \frac{2xi + 2yj}{\sqrt{(2x)^2 + (2y)^2}} = \frac{xi + yj}{4}$$

Since $x^2 + y^2 = 16$ on S

$$A \cdot n = (zi + xj - 3y^2 zk) \cdot \frac{(xi + yj)}{4}$$

$$= \frac{1}{4}(xz + xy)$$

$$n \cdot j = \frac{xi + yj}{4} \cdot j = \frac{y}{4}$$

Fig. 1.28

Then the surface integral equals

$$\iint_R \frac{xz + xy}{y} \, dx \, dz = \int_{z=0}^{5} \int_{x=0}^{4} \left\{ \frac{xz}{\sqrt{16 - x^2}} + x \right\} dx \, dz$$

$$= \int_{z=0}^{5} (4z + 8) \, dz = 90$$

Example 4. *Prove Stoke's theorem for the vector*
$$A = (x + y)i + (2x - z)j + (y + z)k$$
taken over the triangle ABC cut from the plane $3x + 2y + z = 6$ *by the coordinate plane.*

Solution: The given plane is

$$3x + 2y + z = 6$$

i.e.

$$\frac{x}{2} + \frac{y}{3} + \frac{z}{6} = 1$$

intercepts on axes are 2, 3, 6 respectively

$\therefore$ for A, $y = z = 0, x = 2$

for B, $z = x = 0, y = 3$

for C, $x = y = 0, z = 6$

The given vector is

$$A = (x + y)i + (2x - z)j + (y + z)k$$

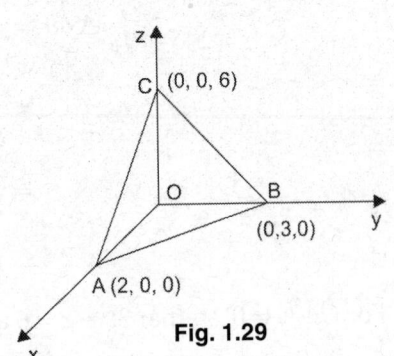

Fig. 1.29

$\therefore$ $\text{curl } A = \nabla \times A = \begin{vmatrix} i & j & k \\ \dfrac{\partial}{\partial x} & \dfrac{\partial}{\partial y} & \dfrac{\partial}{\partial z} \\ (x+y) & (2x-z) & (y+z) \end{vmatrix} = 2i + k$

$\therefore$ $\iint_S (\nabla \times A) \cdot n \, dS = \iint_{BOC} (\nabla \times A) \cdot n \, dS + \iint_{CAO} (\nabla \times A) \cdot n \, dS + \iint_{OAB} (\nabla \times A) \cdot n \, dS$

For face *BOC*: $dS = \dfrac{1}{2} dy \, dz, \; n = i$

For face *CAO*: $dS = \dfrac{1}{2} dz \, dx, \; n = j$

For face *OAB*: $dS = \dfrac{1}{2} dx \, dy, \; n = k$

and $\nabla \times A = 2i + k$

Hence $\iint_S (\nabla \times A) \cdot n \, ds = \int_y \int_z (2i + k) \cdot i \dfrac{1}{2} dy \, dz + \int_z \int_x (2i + k) \cdot j \dfrac{1}{2} dz \, dx$

$$+ \int_x \int_y (2i + k) \cdot k \dfrac{1}{2} dx \, dy$$

or $\iint_S (\nabla \times A) \cdot n \, ds = \int_{y=0}^{3} \int_{z=0}^{6} dy \, dz + \int_z \int_x 0 . dz \, dx + \int_{x=0}^{2} \int_{y=0}^{3} \dfrac{1}{2} dx \, dy$

$$= 18 + 0 + 3 = 21$$

$$\oint_C A \cdot dr = \int_B^C A \cdot dl + \int_C^A A \cdot dl + \int_A^B A \cdot dl$$

But for *BC*, $x = 0$, so that $2y + z = 6$

and $A = yi - zj + (y + z)k$

$\therefore$ $\int_B^C A \cdot dl = \int_B^C \{yi - zj + (y + z)k\} \cdot (dyj + dzk)$

Since in *yz* plane an element of length

$$dl = dy \, j + dz \, k$$

$\therefore$ $\int_B^C A \cdot dl = \int \{-z \, dy + (y + z) \, dz\}$

$$= \int_3^0 \{(2y - 6) \, dy + (6 - y)(-2 \, dy)\}, \qquad \text{(since } dz = -2 \, dy)$$

$$= \int_0^3 \{6 - 2y + 12 - 2y\} \, dy$$

$$= \int_0^3 \{18 - 4y\} \, dy = 36$$

For *CA*, $y = 0$, so that $3x + z = 6$ and

$$A = xi + (2x - z) \, j + zk$$

$$\therefore \qquad \int_C^A A \cdot dl = \int_C^A \{xi + (x-z)j + zk\} \cdot \{dxi + dzk\}$$

$$= \int_C^A (xdx + zdz) = \int_0^2 x\,dx + \int_6^0 z\,dz = -16$$

For *AB*,
$$z = 0, \text{ so that } 3x + 2y = 6 \text{ and}$$
$$A = (x+y)i + 2xj + yk$$

$$\therefore \qquad \int_A^B A \cdot dl = \int_A^B \{(x+y)i + 2xj + yk\} \cdot \{dxi + dyj\}$$

$$= \int_A^B \{(x+y)dx + 2x\,dy\}$$

$$= \int_{x=2}^0 \left\{ \frac{6-x}{2}dx + 2x\left(-\frac{3}{2}dx\right)\right\}$$

$$= \int_0^2 \left(\frac{7}{2}x - 3\right)dx = 1$$

Hence
$$\oint_C A \cdot dl = 36 - 16 + 1 = 21$$

$$\therefore \qquad \iint_S (\nabla \times A) \cdot n\,ds = \oint_C A \cdot dr$$

which is the Stoke's theorem.

Example 5. *If $v = \omega \times r$, prove $\omega = \dfrac{1}{2}$ curl v where ω is a constant vector.*

Solution: Curl $v = \nabla \times v = \nabla \times (\omega \times r)$

$$= \nabla \times \begin{vmatrix} i & j & k \\ \omega_x & \omega_y & \omega_z \\ x & y & z \end{vmatrix}$$

where
$$\omega = \omega_x i + \omega_y j + \omega_z k$$
and
$$r = xi + yj + zk$$

or
$$\text{curl } v = \nabla \times \left[(\omega_y z - \omega_z y)i + (\omega_z x - \omega_x z)j + (\omega_x y - \omega_y x)k\right]$$

$$= \begin{vmatrix} i & j & k \\ \dfrac{\partial}{\partial x} & \dfrac{\partial}{\partial y} & \dfrac{\partial}{\partial z} \\ (\omega_y z - \omega_z y) & (\omega_z x - \omega_x z) & (\omega_x y - \omega_y x) \end{vmatrix}$$

$$= i\left\{\frac{\partial}{\partial y}(\omega_x y - \omega_y x) - \frac{\partial}{\partial z}(\omega_z x - \omega_x z)\right\} + j\left\{\frac{\partial}{\partial z}(\omega_y z - \omega_z y) - \frac{\partial}{\partial x}(\omega_x y - \omega_y x)\right\}$$

$$+ k\left\{\frac{\partial}{\partial x}(\omega_z x - \omega_x y) - \frac{\partial}{\partial y}(\omega_y z - \omega_z y)\right\}$$

$$= i2\omega_x + j2\omega_y + k2\omega_z = 2\omega$$

Then

$$\omega = \frac{1}{2}\nabla \times v = \frac{1}{2} \text{ curl } v.$$

This problem indicates that curl of a vector field has something to do with rotational properties of the field. If the field **F** is due to a moving fluid, for example, then a paddle wheel placed at various points in the field would tend to rotate in regions where curl **F** ≠ 0. While if curl **F** = 0 in the regions there would be no rotation and the field **F** is then called *irrotational*. A field which is not irrotational is sometimes called *vortex field*.

Example 6. *Find constants a, b, c so that vector A*

$$= (x + 2y + az)\,i + (bx - 3y - z)\,j + (4x + cy + 2z)\,k$$

is irrotational. Show also that A can be expressed as gradient for a scalar function.

Solution: A vector A is said to be irrotational if curl $A = 0$.

$$\text{curl } A = \nabla \times A = \begin{vmatrix} i & j & k \\ \dfrac{\partial}{\partial x} & \dfrac{\partial}{\partial y} & \dfrac{\partial}{\partial z} \\ x + 2y + az & bx - 3y - z & 4x + cy + 2z \end{vmatrix}$$

$$= i(c + i) + j(a - 4) + k(b - z)$$

This equals zero, when $a = 4$, $b = 2$ and $c = -1$ and

$$A = (x + 2y + 4z)\,i + (2x - 3y - z)\,j + (4x - y + 2z)\,k$$

Assuming

$$A = \nabla\phi = \frac{\partial\phi}{\partial x}i + \frac{\partial\phi}{\partial y}j + \frac{\partial\phi}{\partial z}k$$

then

$$\frac{\partial\phi}{\partial x} = x + 2y + 4z$$

$$\frac{\partial\phi}{\partial y} = 2x - 3y - z$$

and

$$\frac{\partial\phi}{\partial z} = 4x - y + 2z$$

Integrating $\dfrac{\partial\phi}{\partial x}$ partially keeping y and z constant.

$$\phi = \frac{x^2}{2} + 2xy + 4xz + f(y, z) \tag{1}$$

where $f(x, z)$ is an arbitrary function of y and z.

Similarly, integrating $\dfrac{\partial \phi}{\partial y}$ and $\dfrac{\partial \phi}{\partial z}$ partially keeping z and x, x and y constant respectively.

$$\phi = 2xy - \frac{3}{2}y^2 - yz + f(x,z) \tag{2}$$

$$\phi = 4xy - yz + z^2 + f(x,y) \tag{3}$$

Comparing Eqs (1), (2) and (3), it is evident that there will be a common value of ϕ, if we choose

$$f(y,z) = -\frac{3}{2}y^2 + z^2 - yz$$

$$f(x,z) = \frac{x^2}{2} + z^2 + 4xz$$

$$f(x,y) = -\frac{x^2}{2} - \frac{3y^2}{2} + 2xy$$

So that

$$\phi = \frac{x^2}{2} - \frac{3y^2}{2} + z^2 + 2xy + 4xz - yz$$

Note that we can also add any constant to ϕ. In general if $\nabla \times A = 0$, then we can find ϕ so that $A = \nabla\phi$. A vector field A which can be derived from a scalar field ϕ so that $A = \nabla\phi$ is called a *conservative field* and ϕ is called the *scalar potential*. Note that conversely if $A = \nabla\phi$, then $\nabla \times A = 0$.

Example 7. *The equation of motion of a particle P of mass m is given by*

$$m\frac{d^2 r}{dt^2} = f(r)r_1$$

where r is the position vector of P measured from an origin O, r_1 is a unit vector in the direction r, and f(r) is a function of th]e distance of P from O.

(a) *Show that* $r \times \dfrac{dr}{dt} = c$ *where c is a constant vector.*

(b) *Interpret physically the cases f(r) < 0 and f(r) > 0.*

(c) *Interpret the result in (a) geometrically.*

(d) *Describe how the results obtained relate to the motion of the planets in our solar system.*

Solution: (a) Multiply both sides of $m\dfrac{d^2 r}{dt^2} = f(r)r_1$ by r. Then

$$mr \times \frac{d^2 r}{dt^2} = f(r)\, r \times r_1 = 0$$

Since r and r_1 are collinear and so $r \times r_1 = 0$. Thus,

$$r \times \frac{d^2 r}{dt^2} = 0$$

or

$$\frac{d}{dt}\left(r \times \frac{dr}{dt}\right) = 0$$

Integrating, we have

$$r \times \frac{dr}{dt} = c$$

where c is a constant vector.

(b) If $f(r) < 0$, the acceleration has direction opposite to r_1, hence the force is directed towards O and the particle is always attracted towards O.

If $f(r) > 0$, the force is directed away from O and the particle is under the influence of a repulsive force at O.

A force directed towards or away from a fixed point O and having magnitude depending only on the distance r from O is called central force.

(c) In time ΔT, the particle moves from M to N (Fig. 1.30). The area swept out by the position vector in this time is approximately half the area of parallelogram with sides r and ∇r or $\frac{1}{2} r \times \Delta r$. Then the approximate area swept out by the radius vector per unit time is $\frac{1}{2} r \times \frac{\Delta r}{\Delta t}$, hence the instantaneous time of rate of change in area is

$$\lim_{\Delta t \to 0} \frac{1}{2} r \times \frac{\Delta r}{\Delta t} = \frac{1}{2} r \times \frac{dr}{dt} = \frac{1}{2} r \times v$$

where v is the instantaneous velocity of the particle. The quantity

$$H = \frac{1}{2} r \times \frac{dr}{dt} = \frac{1}{2} r \times v$$

is called the areal velocity. From part (*a*), areal velocity

$$H = \frac{1}{2} r \times \frac{dr}{dt} = \text{constant.}$$

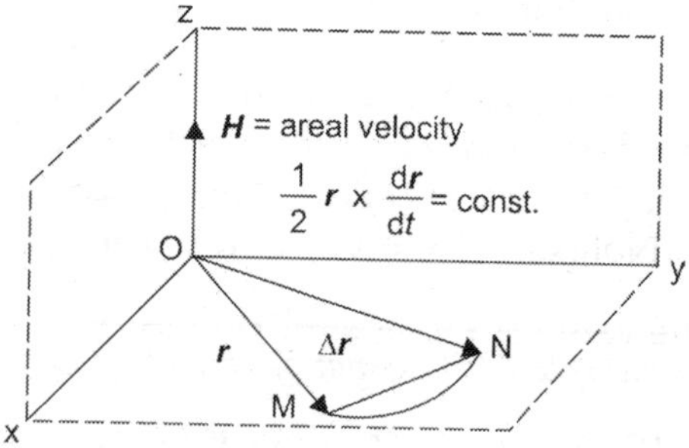

H = areal velocity

$\frac{1}{2} r \times \frac{dr}{dt} = \text{const.}$

Fig. 1.30

Since $r \cdot H = 0$, the motion takes place in plane, which we take as the xy plane as shown in (Fig. 1.30).

(d) A planet (such as the Earth) is attracted towards the Sun according to Newton's universal law of gravitation, which states that any two objects of mass m and M respectively are attracted towards each other with a force of magnitude $F = \dfrac{GMm}{r^2}$, where r is the distance between objects and G is a universal constant. Let m and M be the masses of the planet and sun respectively and choose a set of coordinate axis with the origin O at the sun. Then the equation of motion of the planet is

$$m\frac{d^2 r}{dt^2} = -\frac{GMm}{r^2} r_1 \qquad \text{or} \qquad \frac{d^2 r}{dt^2} = -\frac{GM}{r^2} r_1$$

assuming the influence of the other objects to be negligible.

According to part (c), a planet moves around the sun, so that its position vector sweeps out equal area in equal time. This result is one of Kepler's famous three laws which he deduced empirically from volumes of data complied by, Tycho Brahe (an astronomer). These laws enabled Newton to formulate his universal law of gravitation.

Example 8. *If* $\mathbf{F} = 3xy\mathbf{i} - y^2\mathbf{j}$, *evaluate* $\int_C \mathbf{F} \cdot d\mathbf{r}$, *where C is the curve in the xy plane, y* $= 2x^2$ *from* $(0, 0)$ *to* $(1, 2)$.

Solution: Since the integration is performed in the xy plane ($z = 0$), we can take

$$\mathbf{r} = x\mathbf{i} + y\mathbf{j}$$

Then

$$\int_C \mathbf{F} \cdot d\mathbf{r} = \int_C (3xy\mathbf{i} - y^2\mathbf{j}) \cdot (dx\mathbf{i} + dy\mathbf{j}) = \int_C (3xy\, dx - y^2 dy)$$

Method I: Let $x = t$ in $y = 2x^2$. Then parametric equations of C are $x = t$, $y = 2t^2$, points $(0, 0)$ and $(1, 2)$ correspond to $t = 0$ and $t = 1$ respectively. Then

$$\int_C \mathbf{F} \cdot d\mathbf{r} = \int_{t=0}^1 \left\{ 3(t)(2t^2)dt - (2t^2)^2 d(2t^2) \right\}$$

$$= \int_{t=0}^1 (6t^3 - 16t^5)dt = -\frac{7}{6}$$

Method II: Substitute $y = 2x^2$ directly where x goes from 0 to 1. Then

$$\int_C \mathbf{F} \cdot d\mathbf{r} = \int_{x=0}^1 \{3x(2x)^2 dx - (2x^2)^2 d(2x^2)\}$$

$$= \int_{x=0}^1 (6x^3 - 16x^5)dx = -\frac{7}{6}$$

Note that if the curve were traversed in the opposite sense, i.e. from $(1, 2)$ to $(0, 0)$, the value of the integral would have been $\dfrac{7}{6}$ instead of $-\dfrac{7}{6}$.

1.12. SUCCESSIVE APPLICATIONS OF THE OPERATOR ∇

It frequently happens in various applications of vector analysis that we must operate successively with operator ∇. For example, if S is a scalar function of position in space then grad S, i.e. ∇S is a vector function having the magnitude and direction of the greatest rate of increase of S; this vector is the gradient of S and the flux lines of grad S cut normally through the level surfaces of S. Since grad S is a vector it can have a divergence; using Eqs (1.61) and (1.62), this scalar quantity is

$$\text{div grad } S = \nabla \cdot (\nabla S) = \nabla \cdot \nabla S = \nabla^2 S \qquad (1.79)$$

and using Eq. (1.56), we have

$$\text{div grad } S = \left(\mathbf{i}\frac{\partial}{\partial x} + \mathbf{j}\frac{\partial}{\partial y} + \mathbf{k}\frac{\partial}{\partial z} \right) \cdot \left(\frac{\partial S}{\partial x}\mathbf{i} + \frac{\partial S}{\partial y}\mathbf{j} + \frac{\partial S}{\partial z}\mathbf{k} \right)$$

$$= \frac{\partial^2 S}{\partial x^2} + \frac{\partial^2 S}{\partial y^2} + \frac{\partial^2 S}{\partial z^2} = \left(\frac{\partial^2}{\partial x^2} + \frac{\partial^2}{\partial y^2} + \frac{\partial^2}{\partial z^2} \right) S \qquad (1.79a)$$

Comparing Eqs (1.79) and (1.79a), the operator

$$\text{div grad} = \frac{\partial^2}{\partial x^2} + \frac{\partial^2}{\partial y^2} + \frac{\partial^2}{\partial z^2} = \nabla^2 = \nabla \cdot \nabla \qquad (1.80)$$

and is called as *Laplacian operator*.

Similarly, it is possible to calculate curl of grad S, and thus interpret the operation *curl grad* applied to a scalar point-function. Since components of the vector for which the curl is to be found are $\dfrac{\partial S}{\partial x}, \dfrac{\partial S}{\partial y}, \dfrac{\partial S}{\partial z}$ then

$$\text{curl grad } S = \nabla \times \nabla S = \begin{vmatrix} i & j & k \\ \dfrac{\partial}{dx} & \dfrac{\partial}{dy} & \dfrac{\partial}{dz} \\ \dfrac{\partial S}{dx} & \dfrac{\partial S}{dy} & \dfrac{\partial S}{dz} \end{vmatrix} \tag{1.81}$$

This result follows at once from physical considerations. In Section 1.8, it has been shown that

$$V = \text{grad } S = \nabla S$$

is a lamellar vector, of which $\oint$ is the scalar potential. The characteristic feature of a lamellar or irrotational field is that the line integral of V around any closed path is zero. Since, the curl of a vector field is a particular kind of closed path line integral, curl V = curl grad S is necessarily zero as V is lamellar. A typical noncurl field of this class is electric force field due to static electric charges.

The Operator ∇^2 with Vector Operand

A vector point-function A may be expressed in terms of rectangular components in the usual way.

$$A = A_x i + A_y j + A_z k$$

Since A_x, A_y and A_z are scalar functions of position (x, y, z) respectively, then

$$\nabla^2 A = \nabla^2 A_x i + \nabla^2 A_y j + \nabla^2 A_z k \tag{1.82}$$

which is a result of great importance in electromagnetic theory and hydrodynamics.

The Operator Grad Div

If A is a vector field and div A is a scalar field, which therefore has a gradient. This new vector of which div A is the potential, is necessarily lamellar because curl grad on scalar operand is zero, it may be written as

$$\text{grad div } A = \nabla(\nabla \cdot A)$$

$$= \left(i\frac{\partial}{\partial x} + j\frac{\partial}{\partial y} + k\frac{\partial}{\partial z} \right)\left(\frac{\partial A_x}{\partial x} + \frac{\partial A_y}{\partial y} + \frac{\partial A_z}{\partial z} \right)$$

or

$$\text{grad div } A = \left(\frac{\partial^2 A_x}{\partial x^2} + \frac{\partial^2 A_y}{\partial x \partial y} + \frac{\partial^2 A_z}{\partial x \partial z} \right)i + \left(\frac{\partial^2 A_x}{\partial x \partial y} + \frac{\partial^2 A_y}{\partial y^2} + \frac{\partial^2 A_z}{\partial y \partial z} \right)j$$

$$+ \left(\frac{\partial^2 A_x}{\partial x \partial z} + \frac{\partial^2 A_y}{\partial y \partial z} + \frac{\partial^2 A_z}{\partial z^2} \right)k \tag{1.83}$$

For example, if A is fluid velocity then div A gives the scalar rate at which the density at a point is changing per second, grad div A gives the magnitude and direction in space of the greatest rate of increase of the density with space. In an electric field, A is the electric force, div A, the density of space charge and grad div A, the greatest rate of increase of charge density at any point.

The Operator Curl

The y and z components of curl A are

$$\left(\frac{\partial A_x}{\partial z} - \frac{\partial A_z}{\partial x}\right) j \text{ and } \left(\frac{\partial A_y}{\partial x} - \frac{\partial A_x}{\partial y}\right) k$$

and hence, the x component of curl curl A will be

$$\text{curl}_x \text{ curl } A = \left[\frac{\partial}{\partial y}\left(\frac{\partial A_y}{\partial x} - \frac{\partial A_x}{\partial y}\right) - \frac{\partial}{\partial z}\left(\frac{\partial A_x}{\partial z} - \frac{\partial A_z}{\partial x}\right)\right] i$$

Expanding the factors, adding and subtracting $\dfrac{\partial^2 A_x}{\partial x^2}$ gives

$$\text{curl}_x \text{ curl } A = \left[\frac{\partial^2 A_x}{\partial x^2} + \frac{\partial^2 A_y}{\partial x \partial y} + \frac{\partial^2 A_z}{\partial x \partial z}\right] i - \left[\frac{\partial^2}{\partial x^2} + \frac{\partial^2}{\partial y^2} + \frac{\partial^2}{\partial z^2}\right] A_x i$$

which from Eqs (1.82) and (1.83) may be identified as the x components of grad div A and $\nabla^2 A$ respectively. Permuting i, j, k, x, y, z and A_x, A_y, A_z in cyclic order gives curl_y curl A and curl_z curl A respectively; adding the three components, we have

$$\text{curl curl } A = \nabla \times (\nabla \times A)$$
$$= \text{grad div } A - \nabla^2 A \tag{1.84}$$

So that the operator,

$$\text{curl curl} = \text{grad div} - \nabla^2 \tag{1.85}$$

a theorem which is much used in various problems.

The Operator Div Curl

If A is any vector field, curl A is also a vector field, for which we may calculate the divergence. We have

$$\text{div curl } A = \nabla \cdot (\nabla \times A)$$

$$= \frac{\partial}{\partial x}\left(\frac{\partial A_z}{\partial y} - \frac{\partial A_y}{\partial z}\right) + \frac{\partial}{\partial y}\left(\frac{\partial A_x}{\partial z} - \frac{\partial A_z}{\partial x}\right) + \frac{\partial}{\partial z}\left(\frac{\partial A_y}{\partial x} - \frac{\partial A_x}{\partial y}\right) = 0 \tag{1.86}$$

Hence the curl of any vector field is itself a solenoidal field, i.e. one in which the lines of flow form a closed curve and hence it has no divergence (Section 1.9)

In the same manner the following vector identities may be established by expanding ∇ and the other vectors concerned in terms of their components.

1. $\nabla \cdot (uA) = u(\nabla \cdot A) + A \cdot (\nabla u)$
   div $(uA) = u$ div $A + A \cdot$ grad u $\tag{1.87}$
2. $\nabla \times (uA) = u(\nabla \times A) + (\nabla u) \times A$
   curl $(uA) = u$ curl $A + ($grad $u) \times A$ $\tag{1.87a}$
   Here u is a scalar function of coordinates x, y, z.
3. $\nabla \cdot (A \times B) = B \cdot (\nabla \times A) - A \cdot (\nabla \times B)$
   div $(A \times B) = B \cdot ($curl $A) - A \cdot ($curl $B)$ $\tag{1.87b}$

4. $\nabla \times (A \times B) = A \cdot (\nabla \cdot B) - B \cdot (\nabla \cdot A) + (B \cdot \nabla) A - (A \cdot \nabla) B$

    curl $(A \times B) = A$ div $B - B$ div $A + (B$ grad$) A - (A$ grad$) B$ (1.87c)

In the above set of equations the symbol $(A \cdot \nabla) B$, stands for the vector

$$(A \cdot \nabla) B = i \left(A_x \frac{\partial B_x}{\partial x} + A_y \frac{\partial B_x}{\partial y} + A_z \frac{\partial B_x}{\partial z} \right) + j \left(A_x \frac{\partial B_y}{\partial x} + A_y \frac{\partial B_y}{\partial y} + A_z \frac{\partial B_y}{\partial z} \right)$$

$$+ k \left(A_x \frac{\partial B_z}{\partial x} + A_y \frac{\partial B_z}{\partial y} + A_z \frac{\partial B_z}{\partial z} \right) \tag{1.88}$$

Here if vector B is position vector r, i.e.

$$B = xi + yj + zk \tag{1.89}$$

Then $(A \cdot \nabla) r = i \left\{ A_x \frac{\partial x}{\partial x} \right\} + j \left\{ A_y \frac{\partial y}{\partial y} \right\} + k \left\{ A_z \frac{\partial z}{\partial z} \right\}$

$$= A_x i + A_y j + A_z k = A$$

i.e. $(A \times \nabla) r = A$ (1.90)

5. $\nabla(A \cdot B) = (A\nabla)B + (B\nabla)A + \{A \times (\nabla \times B)\} + \{B \times (\nabla \times A)\}$

    grad $(A \cdot B) = (A$ grad$) B + (B$ grad$) A + (A \times$ curl $B) + (B \times$ curl $A)$ (1.91)

The above formulas are very important in the applications of vector analysis to various branches of engineering and physics.

Example 1. *Prove the following:*

  i. *div* $(\phi$ *grad* $S) = \phi \nabla^2 S + ($*grad* $\phi) \cdot ($*grad* $S)$, *where* ϕ *and* S *are scalar functions.*

  ii. *curl* $(\phi$ *grad* $S) = \nabla\phi \times \nabla S = -$ *curl* $(S$ *grad* $\phi)$

 iii. $\nabla \cdot (\phi \nabla \times A) = (\nabla \times A) \cdot (\nabla \cdot \phi)$

 iv. *curl* $(\nabla\phi \times \nabla S) = \nabla\phi (\nabla \cdot \nabla S) - \nabla S (\nabla \cdot \nabla\phi) + (\nabla S\nabla) \nabla\phi - (\nabla\phi \cdot \nabla) \nabla S.$

Solution:

  i.   div $(\phi$ grad $S) =$ div (ϕA) where $A =$ grad $S = \nabla S$

              $= \nabla \cdot (\phi A) = \phi(\nabla \cdot A) + A \cdot (\nabla\phi)$ *[using* identity (1.87a)]

    ∴ div $(\phi$ grad $S) = \phi (\nabla \cdot \nabla S) + ($grad $S) \cdot ($grad $\phi)$

                  $= \phi \nabla S^2 + ($grad $S) \cdot ($grad $\phi)$

  ii.   curl $(\phi$ grad $S) =$ curl $(\phi A) = \nabla \times (\phi A)$ and using identity Eq. (1.87b)

       $\nabla \times (\phi A) = \phi (\nabla \times A) + (\nabla\phi) \times A$

                $= \phi($curl grad $S) + ($grad $\phi) \times ($grad $S)$ *[using* Eq. (1.86)]

    But curl grad $S = 0$.

    curl $(\phi$ grad $S) = ($grad $\phi) \times ($grad $S)$

                $= (\nabla\phi) \times (\nabla S)$

                $= -(\nabla S) \times (\nabla\phi)$

                $= -\nabla \times (S\nabla\phi)$

                $= -$ curl $(S$ grad $\phi)$.

 iii. $\nabla \cdot (\phi \nabla \times A) = \nabla \cdot (\phi P)$, where $P = \nabla \times A =$ curl A and using identity (1.87a)

       $\nabla \cdot (\phi P) = \phi(\nabla \cdot P) + P \cdot (\nabla\phi)$

or $\quad \nabla \cdot (\phi \nabla \cdot A) = \phi \text{ div (curl } A) + (\text{grad } \phi) \cdot (\text{curl } A)$

But div curl $A = 0$, Hence $\qquad\qquad\qquad\qquad$ [*using* Eq. (1.91)

$$\nabla \times (\phi \nabla \times A) = (\nabla \times A) \cdot (\nabla \phi)$$

iv. Let $\nabla \phi = A$ and $\nabla S = B$

$$\text{curl } (\nabla \phi \times \nabla S) = \text{curl } (A \times B)$$

$$= \nabla \times (A \times B)$$

Using identity Eq. (1.87e)

$$\nabla \times (A \times B) = (B \cdot \nabla) A - (A \cdot \nabla) B - B (\nabla \cdot A) + A (\nabla \cdot B)$$

$\therefore \quad \nabla \times (\nabla \phi \times \nabla S) = (\nabla S \times \nabla)(\nabla \phi) - (\nabla \phi \cdot \nabla)(\nabla S) - \nabla S (\nabla \cdot \nabla \phi) + \nabla \phi (\nabla \cdot \nabla S)$

Example 2. *Find div grad r^m and verify that $\nabla \cdot \nabla r^m = 0$, where $r = xi + yj + zk$.*

Solution: Here

$$\text{grad } r^m = \nabla r^m = \left(i\frac{\partial}{\partial x} + j\frac{\partial}{\partial y} + k\frac{\partial}{\partial z} \right) r^m$$

$$= \left(i\frac{\partial}{\partial x} + j\frac{\partial}{\partial y} + k\frac{\partial}{\partial z} \right)(x^2 + y^2 + z^2)^{m/2} \quad \left(\because |r| = (x^2 + y^2 + z^2)^{1/2} \right)$$

$\therefore \qquad \text{grad } r^m = m(xi + yj + zk)(x^2 + y^2 + z^2)^{(m/2)-1}$

$$= mr^{m-2}(xi + yj + zk)$$

$\therefore \qquad \text{div grad } r^m = \nabla \cdot \nabla r^m$

$$= \left(i\frac{\partial}{\partial x} + j\frac{\partial}{\partial y} + k\frac{\partial}{\partial z} \right) \cdot mr^{m-2}(xi + yj + zk)$$

$$= \frac{\partial}{\partial x}(mxr^{m-2}) + \frac{\partial}{\partial y}(myr^{m-2}) + \frac{\partial}{\partial z}(mzr^{m-2})$$

Now, $\quad \dfrac{\partial}{\partial x}(mxr^{m-2}) = mr^{m-2} + m(m-2)r^{m-3}\dfrac{\partial r}{\partial x}x$

$$= mr^{m-2} + m(m-2)r^{m-4}x^2 \qquad\qquad \left[\because \frac{\partial r}{\partial x} = \frac{x}{r} \right]$$

Similarly,

$$\frac{\partial}{\partial y}(myr^{m-2}) = mr^{m-2} + m(m-2)r^{m-4}y^2$$

and $\quad \dfrac{\partial}{\partial z}(mzr^{m-2}) = mr^{m-2} + m(m-2)r^{m-4}z^2$

Hence, $\quad \text{div grad } r^m = 3mr^{m-2} + m(m-2)r^{m-4}(x^2 + y^2 + z^2)$

$$= 3mr^{m-2} + m(m-2)r^{m-2}$$

$$= m(m+1)\, r^{m-2}$$

Again,
$$\nabla \times \nabla r^m = \left(i\frac{\partial}{\partial x} + j\frac{\partial}{\partial y} + k\frac{\partial}{\partial z}\right) \times \left(i\frac{\partial}{\partial x} + j\frac{\partial}{\partial y} + k\frac{\partial}{\partial z}\right) r^m$$

$$= \left(i\frac{\partial}{\partial x} + j\frac{\partial}{\partial y} + k\frac{\partial}{\partial z}\right) \times (mr^{m-2}x\boldsymbol{i} + mr^{m-2}y\boldsymbol{j} + mr^{m-2}z\boldsymbol{k})$$

$$= \begin{vmatrix} \boldsymbol{i} & \boldsymbol{j} & \boldsymbol{k} \\ \dfrac{\partial}{\partial x} & \dfrac{\partial}{\partial y} & \dfrac{\partial}{\partial z} \\ mr^{m-2}x & mr^{m-2}y & mr^{m-2}z \end{vmatrix}$$

or
$$\nabla \times \nabla r^m = \left\{\frac{\partial}{\partial y}(mr^{m-2}z) - \frac{\partial}{\partial z}(mr^{m-2}y)\right\}\boldsymbol{i} + \left\{\frac{\partial}{\partial z}(mr^{m-2}x) - \frac{\partial}{\partial x}(mr^{m-2}z)\right\}\boldsymbol{j}$$

$$+ \left\{\frac{\partial}{\partial x}(mr^{m-2}y) - \frac{\partial}{\partial y}(mr^{m-2}x)\right\}\boldsymbol{k}$$

$$= \left\{m(m-2)r^{m-3}\frac{yz}{r} - m(m-2)r^{m-3}\frac{yz}{r}\right\}\boldsymbol{i} + \left\{m(m-2)r^{m-3}\frac{zx}{r} - m(m-2)r^{m-3}\frac{zx}{r}\right\}\boldsymbol{j}$$

$$+ \left\{m(m-2)r^{m-3}\frac{xy}{r} - m(m-2)r^{m-3}\frac{xy}{r}\right\}\boldsymbol{k}$$

$$= 0 + 0 + 0 = 0$$

Example 3. *Show that*

$$\operatorname{curl}\frac{\boldsymbol{a}\times\boldsymbol{r}}{r^3} = -\frac{\boldsymbol{a}}{r^3} + \frac{3\boldsymbol{r}}{r^5}(\boldsymbol{a}\cdot\boldsymbol{r})$$

where $\boldsymbol{a}$ *is a constant vector and*

$$\boldsymbol{r} = x\boldsymbol{i} + y\boldsymbol{j} + z\boldsymbol{k}.$$

Solution: Here

$$\operatorname{curl}\frac{\boldsymbol{a}\times\boldsymbol{r}}{r^3} = \nabla\times\frac{(\boldsymbol{a}\times\boldsymbol{r})}{r^3} = \left(i\frac{\partial}{\partial x} + j\frac{\partial}{\partial y} + k\frac{\partial}{\partial z}\right)\times\frac{(\boldsymbol{a}\times\boldsymbol{r})}{r^3}$$

$$= \boldsymbol{i}\times\frac{\partial}{\partial x}\left(\frac{\boldsymbol{a}\times\boldsymbol{r}}{r^3}\right) + \boldsymbol{j}\times\frac{\partial}{\partial y}\times\frac{(\boldsymbol{a}\times\boldsymbol{r})}{r^3} + \boldsymbol{k}\times\frac{\partial}{\partial z}\left(\frac{\boldsymbol{a}\times\boldsymbol{r}}{r^3}\right)$$

Now,
$$\frac{\partial}{\partial x}\left(\frac{\boldsymbol{a}\times\boldsymbol{r}}{r^3}\right) = -\frac{3}{r^4}\frac{\partial r}{\partial x}\boldsymbol{a}\times\boldsymbol{r} + \frac{1}{r^3}\left(\boldsymbol{a}\times\frac{\partial\boldsymbol{r}}{\partial x}\right)$$

$$= -\frac{3x}{r^5}\boldsymbol{a}\times\boldsymbol{r} + \frac{\boldsymbol{a}}{r^3}\times\boldsymbol{i}$$

Since
$$\boldsymbol{r} = x\boldsymbol{i} + y\boldsymbol{j} + z\boldsymbol{k} \text{ gives } \frac{d\boldsymbol{r}}{dx} = \boldsymbol{i}$$

and
$$r^2 = x^2 + y^2 + z^2 \text{ gives } \frac{\partial r}{\partial x} = \frac{x}{r}$$

So that
$$i \times \frac{\partial}{\partial x}\left(\frac{a \times r}{r^3}\right) = -\frac{3x}{r^5}[i \times (a \times r)] + \frac{1}{r^3}i \times (a \times i)$$

$$= -\frac{3x}{r^5}[(i \cdot r)a - (i \cdot a)r] + \frac{1}{r^3}[(i \cdot i)a - (i \cdot a)i]$$

$$= -\frac{3x}{r^5}[xa - (i \cdot a)r] + \frac{1}{r^3}[a - (i \cdot a)i]$$

Similarly,
$$j \times \frac{\partial}{\partial y}\left(\frac{a \times r}{r^3}\right) = -\frac{3y}{r^5}[ya - (j \cdot a)r] + \frac{1}{r^3}[a - (j \cdot a)j]$$

and
$$k \times \frac{\partial}{\partial z}\left(\frac{a \times r}{r^3}\right) = -\frac{3z}{r^5}[za - (k \cdot a)r] + \frac{1}{r^3}[a - (k \cdot a)k]$$

∴
$$i \times \frac{\partial}{\partial x}\left(\frac{a \times r}{r^3}\right) + j \times \frac{\partial}{\partial y}\left(\frac{a \times r}{r^3}\right) + k \times \frac{\partial}{\partial z}\left(\frac{a \times r}{r^3}\right)$$

$$= -\frac{3(x^2 + y^2 + z^2)}{r^5}a + \frac{3r}{r^5}(xi + yj + zk) \cdot a + \frac{3a}{r^3} - \frac{1}{r^3}\{(i \cdot a)i + (j \cdot a)j + (k \cdot a)k\}$$

$$= -\frac{3a}{r^3} + \frac{3r}{r^5}(r \cdot a) + \frac{3a}{r^3} - \frac{a}{r^3}$$

or
$$\text{curl}\left(\frac{a \times r}{r^3}\right) = i \times \frac{\partial}{\partial x}\left(\frac{a \times r}{r^3}\right) + j \times \left(\frac{\partial}{\partial y}\right)\left(\frac{a \times r}{r^3}\right) + k \times \frac{\partial}{\partial z}\left(\frac{a \times r}{r^3}\right) = -\frac{a}{r^3} + \frac{3r}{r^5}(r \cdot a)$$

Example 4. (a) *Prove that* $\nabla \cdot (\nabla\phi_1 \times \nabla\phi_2) = 0$, *where* ϕ_1 *and* ϕ_2 *are arbitrary functions of* (x, y, z). (b) *If* **F** *and* f *are point functions, prove that the normal and tangential components of the former to the level surface* f = 0 *are*
$$\frac{(F \cdot \nabla f)\nabla f}{(\nabla f)^2} \quad and \quad \frac{\nabla f \times (F \times \nabla f)}{(\nabla f)^2}$$

Solution: (a) We have
$$\nabla\phi_1 = \left(i\frac{\partial\phi_1}{\partial x} + j\frac{\partial\phi_1}{\partial y} + k\frac{\partial\phi_1}{\partial z}\right)$$

and
$$\nabla\phi_2 = \left(i\frac{\partial\phi_2}{\partial x} + j\frac{\partial\phi_2}{\partial y} + k\frac{\partial\phi_2}{\partial z}\right)$$

∴
$$\nabla\phi_1 \times \nabla\phi_2 = \left(i\frac{\partial\phi_1}{\partial x} + j\frac{\partial\phi_1}{\partial y} + k\frac{\partial\phi_1}{\partial z}\right) \times \left(i\frac{\partial\phi_2}{\partial x} + j\frac{\partial\phi_2}{\partial y} + k\frac{\partial\phi_2}{\partial z}\right)$$

$$= \left(\frac{\partial\phi_1}{\partial y}\frac{\partial\phi_2}{\partial z} - \frac{\partial\phi_1}{\partial z}\frac{\partial\phi_2}{\partial y}\right)i + \left(\frac{\partial\phi_1}{\partial z}\frac{\partial\phi_2}{\partial x} - \frac{\partial\phi_1}{\partial x}\cdot\frac{\partial\phi_2}{\partial z}\right)j$$

$$+ \left(\frac{\partial\phi_1}{\partial x}\frac{\partial\phi_2}{\partial y} - \frac{\partial\phi_1}{\partial y}\frac{\partial\phi_2}{\partial x}\right)k$$

and
$$\nabla \cdot (\nabla\phi_1 \times \nabla\phi_2) = \frac{\partial}{\partial x}\left[\frac{\partial\phi_1}{\partial y}\frac{\partial\phi_2}{\partial z} - \frac{\partial\phi_1}{\partial z}\frac{\partial\phi_2}{\partial y}\right] + \frac{\partial}{\partial y}\left[\frac{\partial\phi_1}{\partial z}\frac{\partial\phi_2}{\partial x} - \frac{\partial\phi_1}{\partial x}\frac{\partial\phi_2}{\partial z}\right]$$

$$+ \frac{\partial}{\partial z}\left[\frac{\partial\phi_1}{\partial x}\frac{\partial\phi_2}{\partial y} - \frac{\partial\phi_1}{\partial y}\frac{\partial\phi_2}{\partial x}\right]$$

$$= \frac{\partial^2\phi_1}{\partial x \partial y}\frac{\partial\phi_2}{\partial z} + \frac{\partial\phi_1}{\partial y}\frac{\partial^2\phi_2}{\partial x \partial z} - \frac{\partial^2\phi_1}{\partial x \partial z}\frac{\partial\phi_2}{\partial y} - \frac{\partial\phi_1}{\partial z}\frac{\partial^2\phi_2}{\partial x \partial y}$$

$$+ \frac{\partial^2\phi_1}{\partial y \partial z}\frac{\partial\phi_2}{\partial x} + \frac{\partial\phi_1}{\partial z}\frac{\partial^2\phi_2}{\partial y \partial x} - \frac{\partial^2\phi_1}{\partial y \partial x}\frac{\partial\phi_2}{\partial z} - \frac{\partial\phi_1}{\partial x}\frac{\partial^2\phi_2}{\partial y \partial z}$$

$$+ \frac{\partial^2\phi_1}{\partial z \partial x}\frac{\partial\phi_2}{\partial y} + \frac{\partial\phi_1}{\partial x}\frac{\partial^2\phi_2}{\partial z \partial y} - \frac{\partial^2\phi_1}{\partial z \partial y}\frac{\partial\phi_2}{\partial x} - \frac{\partial\phi_1}{\partial y}\frac{\partial^2\phi_2}{\partial z \partial x}$$

$$= \left(\frac{\partial\phi_2}{\partial z}\frac{\partial^2\phi_1}{\partial x \partial y} - \frac{\partial\phi_2}{\partial z}\frac{\partial^2\phi_1}{\partial y \partial x} \right) + \left(\frac{\partial\phi_1}{\partial y}\frac{\partial^2\phi_2}{\partial x \partial z} - \frac{\partial\phi_1}{\partial y}\frac{\partial^2\phi_2}{\partial x \partial z} \right)$$

$$+ \left(\frac{\partial\phi_2}{\partial x}\frac{\partial^2\phi_1}{\partial y \partial z} - \frac{\partial\phi_2}{\partial x}\frac{\partial^2\phi_1}{\partial z \partial y} \right) + \left(\frac{\partial\phi_1}{\partial z}\frac{\partial^2\phi_2}{\partial y \partial x} - \frac{\partial\phi_1}{\partial z}\frac{\partial^2\phi_2}{\partial x \partial y} \right)$$

$$+ \left(\frac{\partial\phi_2}{\partial y}\frac{\partial^2\phi_1}{\partial z \partial x} - \frac{\partial\phi_2}{\partial y}\frac{\partial^2\phi_1}{\partial x \partial z} \right) + \left(\frac{\partial\phi_1}{\partial x}\frac{\partial^2\phi_2}{\partial z \partial y} - \frac{\partial\phi_1}{\partial x}\frac{\partial^2\phi_2}{\partial y \partial z} \right)$$

or $\qquad \nabla \cdot (\nabla\phi_1 \times \nabla\phi_2) = 0$

(b) Let n be the unit vector along the outward, drawn normal at the point P of the level surface $f = 0$. Then normal component of F

$$= (F \cdot n) \, n \qquad (1)$$

Also $\qquad \nabla f = \frac{\partial f}{\partial n} n \qquad (2)$

$\therefore \qquad n = \frac{\nabla f}{\partial f / \partial n} \qquad (3)$

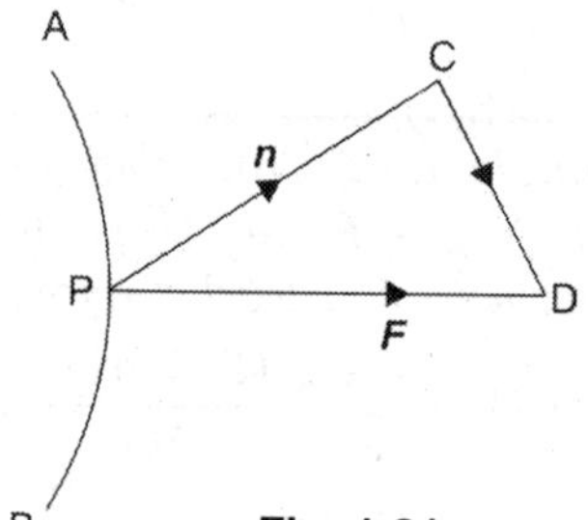

On squaring Eq. (2), we get

$$(\nabla f)^2 = \left(\frac{\partial f}{\partial n} \right)^2 \qquad (4)$$

Fig. 1.31

Making use of Eqs (3) and (4) in Eq. (1), we have:

Normal component of F

$$= \frac{(F \cdot \nabla f)\nabla f}{(\partial f / \partial n)^2} = \frac{(F \cdot \nabla f)\nabla f}{(\nabla f)^2} \qquad \text{[since } (n)^2 = 1]$$

Also tangential component of $F = CD$

$$= F - \text{normal component of } F$$

$$= F - \frac{(F \cdot \nabla f)\nabla f}{(\nabla f)^2}$$

$$= \frac{(\nabla f \cdot \nabla f)F - (F \cdot \nabla f)\nabla f}{(\nabla f)^2}$$

$$= \frac{\nabla f \times (F \times \nabla f)}{(\nabla f)^2}$$

Example 5. *Show that*

(a) *Curl* $[r\,f(r)] = 0$ *where* $f(r)$ *is differentiable.*

(b) $r^n\,r$ *is an irrotational vector for any value of n but is solenoidal only if* $n = -3$ (*r is the position vector of a point*).

(c) $v \cdot \nabla v = \dfrac{1}{2}\nabla v^2 - v \times \text{curl } v.$

Solution: (a) We have

$$\text{curl } [r\,f(r)] = \nabla \times [r\,f(r)] = \nabla \times [f(r)\,xi + f(r)\,yj + f(r)\,zk]$$

$$= \left[z\frac{\partial}{\partial y}f(r) - y\frac{\partial f(r)}{\partial z}\right]i + \left[x\frac{\partial f(r)}{\partial z} - z\frac{\partial f(r)}{\partial x}\right]j + \left[y\frac{\partial f(r)}{\partial x} - x\frac{\partial f(r)}{\partial y}\right]k$$

But
$$\frac{\partial f(r)}{\partial x} = \frac{\partial f}{\partial r}\frac{\partial r}{\partial x} = \frac{\partial f}{\partial r}\frac{\partial}{\partial x}(x^2 + y^2 + z^2)^{1/2}$$

$$= \frac{x}{(x^2 + y^2 + z^2)}f'(r)$$

$$= \frac{xf'(r)}{r}$$

Thus, we have

$$\text{curl } [r\,f(r) = \left[\frac{zyf'(r)}{r} - \frac{yzf'(r)}{r}\right]i + \left[\frac{zxf'(r)}{r} - \frac{xzf'(r)}{r}\right]j + \left[\frac{xyf'(r)}{r} - \frac{yyf'(r)}{r}\right]k = 0$$

(b) curl $r^n\,r = r^n$ curl $r - r \times$ grad r^n [*using* Eq. (1.87b)]

$$= -r \times (nr^{n-2}\,r) \qquad\qquad\qquad \text{(since, curl } r = 0)$$

$$= -r \times r\,(nr^{n-2}) = 0$$

$\therefore$ $r^n\,r$ is an irrotational vector for any value of n.

Again div $r^n\,r = r^n$ div. $r + r$ grad. r^n.   [*using* Eq. (1.87a)]

$$= 3r^n + r\,(nr^{n-2}\,r)$$

$$= (n + 3)\,r^n,$$

which is zero, if $n = -3$. This means $r^n\,r$ is a solenoidal vector when $n = -3$.

(c) Using Eq. (1.91), we have

$$\text{grad } (u \times v) = u \times \text{curl } v + v,\ \text{curl } u + (u \cdot \nabla) \times v + (v \cdot \nabla)\,u.$$

Putting $u = v$ in the above result, we get

$$\text{grad } (v^2) = v \times \text{curl } v + v \times \text{curl } v + (v \cdot \nabla)v + (v \cdot \nabla)\,v.$$

$$= 2v \times \text{curl } v + 2\,(v \cdot \nabla)\,v.$$

or $$(v \cdot \nabla)v = \frac{1}{2}\nabla(v^2) - v \times \text{curl } v.$$

$$= \frac{1}{2}\nabla v^2 - v \times \text{curl } v.$$

Example 6. *A particle moves such that its acceleration at any instant is equal to* $\mu r \times \dfrac{v}{r^3}$, *show that it moves with constant speed and*

$$r^2 = \alpha t^2 + 2\beta t + \gamma,$$

where α, β, γ are constants. Show also that the angular momentum vector consists of two components—one constant in magnitude and direction and other constants in magnitude and in the direction of **r**.

Solution: The equation of motion is given by

$$\frac{d^2 r}{dt^2} = \mu \frac{r \times v}{r^3} = \mu \frac{r}{r^3} \times \frac{dr}{dt} \tag{1}$$

$$\left(\because v = \frac{dr}{dt} \right).$$

Multiplying both sides scalarly by $\dfrac{dr}{dt}$, we have

$$\frac{dr}{dt} \cdot \frac{d^2 r}{dt^2} = \frac{\mu}{r^3} r \times \frac{dr}{dt} \cdot \frac{dr}{dt} = 0$$

Since

$$r \times \frac{dr}{dt} \cdot \frac{dr}{dt} = r \cdot \frac{dr}{dt} \times \frac{dr}{dt} = 0.$$

Integration yields

$$\left(\frac{dr}{dt} \right)^2 = \alpha \ (\text{a constant}) \tag{2}$$

It follows that the speed is constant.

Again multiplying Eq. (1) scalarly by **r**, we get

$$r \cdot \frac{d^2 r}{dt^2} = \frac{\mu}{r^3} r \cdot r \times \frac{dr}{dt} = 0 \tag{3}$$

Since

$$r \cdot r \times \frac{dr}{dt} = \frac{dr}{dt} \cdot r \times r = 0.$$

Adding Eqs (2) and (3), we have

$$r \cdot \frac{d^2 r}{dt^2} + \left(\frac{dr}{dt} \right)^2 = \alpha$$

or

$$\frac{d}{dt} \left(r \cdot \frac{dr}{dt} \right) = \alpha$$

its integration gives

$$r \cdot \frac{dr}{dt} = \alpha t + \beta$$

or

$$\frac{1}{2} \frac{d(r)^2}{dt} = \alpha t + \beta$$

or

$$\frac{d(r)^2}{dt} = 2\alpha t + 2\beta$$

Integrating, we have

$$r^2 = \alpha t^2 + 2\beta t + \gamma$$

which is the required result. Here α, β, and γ are constants.

Now multiplying Eq. (1) vectorially by r, we have

$$r \times \frac{d^2 r}{dt^2} = \frac{\mu}{r^3} r \times \left(r \times \frac{dr}{dt} \right)$$

$$= \frac{\mu}{r^3} \left[\left(r \cdot \frac{dr}{dt} \right) r - (r \cdot r) \frac{dr}{dt} \right]$$

$$= \frac{\mu}{r^3} \left[\frac{1}{2} \frac{d(r^2)}{dt} r - r^2 \frac{dr}{dt} \right] = -\mu \left[\frac{1}{r} \frac{dr}{dt} - \frac{r}{r^2} \frac{dr}{dt} \right]$$

Since

$$\frac{1}{2} \frac{d(r^2)}{dt} = \frac{1}{2} 2r \frac{dr}{dt} = r \frac{dr}{dt}.$$

$$\therefore \qquad r \times \frac{d^2 r}{dt^2} = -\mu \frac{d}{dt} \left(\frac{r}{r} \right)$$

Integrating, we have

$$r \times \frac{dr}{dt} = -\mu \frac{r}{r} + a$$

where a is a constant vector.

It follows from this result that the angular momentum $r \times m \dfrac{dr}{dt}$ consists of two components, one ma which is constant in magnitude and direction and the other $-m\mu \dfrac{r}{r}$ whose magnitude $|m\mu|$ is constant and direction is that of r.

1.13 THE CLASSIFICATION OF VECTOR FIELDS

If to each point (x, y, z) of a region R in space there corresponds a vector $A(x, y, z)$, then A is called a vector function of position, and we say that a vector field A has been defined in R. In practice there are four kinds of vector fields having different associations of curl and divergence. They may be classified as follows:

i. Curl $A = 0$ and div $A = 0$. Such a field is irrotational (lamellar) as well as solenoidal. Hence A can be expressed as $A = \text{grad } \phi$, and second condition gives $A = \text{div grad } \phi = \nabla^2 \phi = 0$ which is Laplace equation. Typical fields of this sort are the electric force in free space due to static charges on boundary or the irrotational motion of an incompressible fluid. This is shown in Fig. 1.32a.

ii. Curl $A = 0$ but div $A \neq 0$. The field is irrotational but not solenoidal. Hence $\nabla^2 \phi \neq 0$, i.e. it is poison's equation and determines ϕ. Examples are the electric field of a volume distribution of charges, e.g. electrons in a thermionic tube or the gravitational force inside a mass. This is shown in Fig. 1.32b.

iii. Curl $A \neq 0$ but div $A = 0$. The field is rotational but can not have a scalar potential but it is solenoidal. Hence curl A can be expressed as curl $A = -\nabla^2 P \neq 0$. Here P is a vector termed as vector potential and determines A. Important typical cases are the magnetic field within a conductor carrying steady current and the rotational motion of an incompressible liquid. The field in shown in Fig. 1.32c.

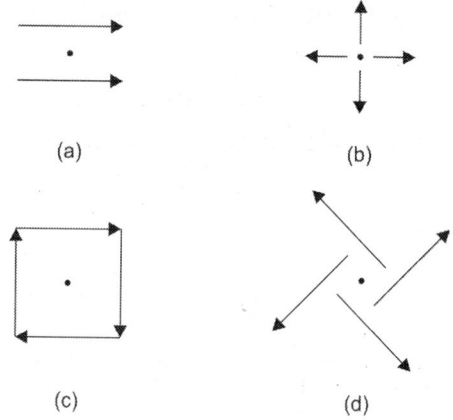

Fig. 1.32

iv. Curl $A \neq 0$ and div $A \neq 0$. This is the most general type of vector field possible found in the rotational motion of compressible fluid and in some branches of electromagnetism. The field can be expressed as the sum of two fields, the first is lamellar vector which has no curl but divergence and the second is a solenoidal vector which has no divergence but can have curl, e.g.

$$A = \text{grad } \phi + \text{curl } P \tag{1.92}$$

Thus, a general vector field may be decomposed into two fields, one being lamellar with scalar potential ϕ and the other solenoidal with vector potential P. This is sometimes known as Helmholtz theorem. Figure 1.32 illustrates diagrammatically the characteristics of vector fields near a point.

Example 1. *Prove that when **B** is solenoidal, a vector **A** exists such that* $B = \nabla \times A$.

Solution: Suppose $B = iB_x + jB_y + kB_z$ and unknown vector is $A = iA_x + jA_y + kA_z$ then we have

$$B_x = \frac{\partial A_z}{\partial y} - \frac{\partial A_y}{\partial z} \tag{1}$$

$$B_y = \frac{\partial A_x}{\partial z} - \frac{\partial A_z}{\partial x} \tag{2}$$

$$B_z = \frac{\partial A_y}{\partial x} - \frac{\partial A_x}{\partial y} \tag{3}$$

Let us choose A in yz plane, i.e. $A_x = 0$

Then $\qquad B_y = -\dfrac{\partial A_z}{\partial x}$ and $B_z = \dfrac{\partial A_y}{\partial x}$

On integration, we have

$$A_y = \int B_z dx + f_1(y,z) \tag{4}$$

and $\qquad A_z = -\int B_y dx + f_2(y,z) \tag{5}$

where f_1 and f_2 are arbitrary functions of y and z but are not of x. Then Eq. (1) gives.

$$\frac{\partial A_z}{\partial y} - \frac{\partial A_y}{\partial z} = -\int_{x_0}^{x}\left(\frac{\partial B_y}{\partial y} + \frac{\partial B_z}{\partial z}\right)dx + \frac{\partial f_2}{\partial y} - \frac{\partial f_1}{\partial z}$$

$$\therefore \qquad \nabla \cdot \boldsymbol{B} = 0, \text{ hence } \frac{\partial B_y}{\partial y} + \frac{\partial B_z}{\partial z} = -\frac{\partial B_x}{\partial x}$$

$$\therefore \qquad \frac{\partial A_z}{\partial y} - \frac{\partial A_y}{\partial z} = \int_{x_0}^{x}\frac{\partial B_x}{\partial x}dx + \frac{\partial f_2}{\partial y} - \frac{\partial f_1}{\partial z}$$

If we choose $f_1 = 0$ and $f_2 = \int_{y_0}^{y} B_x(x,y,z)dy$

Then $\qquad \dfrac{\partial A_z}{\partial y} - \dfrac{\partial A_y}{\partial z} = \displaystyle\int_{x_0}^{x}\dfrac{\partial B_x}{\partial x}dx + \int_{y_0}^{y} B_x(x,y,z)dy = B(x,y,z)$

This equation is in agreement with Eq. (1). Using this choice of f_1 and f_2, one can construct A as follows

$$A = j\int_{x_0}^{x} B_z(x,y,z)dx + k\left[\int_{y_0}^{y} B_x(x,y,z)dy - \int_{x_0}^{x} B_y(x,y,z)dx\right]$$

and hence A exists.

1.14 CURVILINEAR COORDINATES

Many calculations in physics can be simplified by choosing another kind of system that takes advantage of the relation of symmetry involved in the particular problem under consideration instead of a cartesian coordinate system. The purpose of this article is to show how the components of vectors or vector operators may be formulated in a system of curvilinear coordinates, the latter being of so general in nature that it is an easy matter to transform them to any one of the several kinds of special coordinate systems which have been found useful in physical problems.

In cartesian coordinates, the position of a point $P(x, y, z)$ is determined by the intersection of three mutually perpendicular planes $x = $ const, $y = $ const, $z = $ const. Where x, y, and z are related to three new quantities by the following relations.

$$\left.\begin{array}{l} x = x(q_1, q_2, q_3) \\ y = y(q_1, q_2, q_3) \\ z = z(q_1, q_2, q_3) \end{array}\right\} \qquad (1.93)$$

with inverse

$$\left.\begin{array}{l} q_1' = q_1(x,y,z) \\ q_2' = q_2(x,y,z) \\ q_3' = q_3(x,y,z) \end{array}\right\} \qquad (1.94)$$

A given point may be described by specifying either x, y, z or q_1, q_2, q_3, because each relation of Eq. (1.94) represents a surface and the intersection of three such surfaces locates the point. The surface $q_1 = $ const., $q_2 = $ const., $q_3 = $ const. and are called the coordinate surfaces and each pair of these surfaces intersect in curves called coordinate curves or lines (Fig. 1.33). The coordinate axes are determined by the tangents to the coordinates lines at the intersection of these surfaces. They are not in general fixed directions in space, as is true for simple cartesian coordinate. If the coordinate surfaces intersect at right angles, the curvilinear coordinates system is called orthogonal. The quantities (q_1, q_2, q_3) are the curvilinear coordinates of a point $P(x, y, z)$.

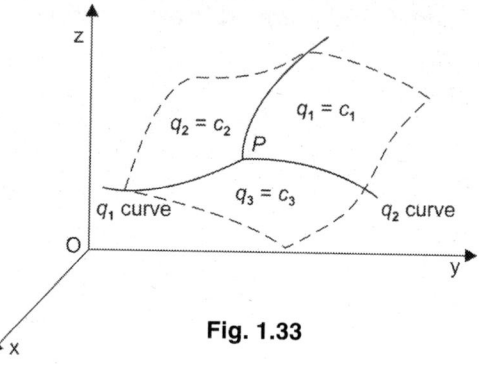

Fig. 1.33

Unit Vectors in Curvilinear Systems

Let $r = xi + yj + zk$, be the position vector of a point P. Then Eq. (1.93) can be written as $r = r(q_1, q_2, q_3)$. A tangent vector to q_1 curve at P (for which q_2 and q_3 are constants) is $\dfrac{dr}{dq_1}$. Then a unit vector in this direction is

$$e_1 = \frac{dr}{dq_1} \bigg/ \left| \frac{dr}{dq_1} \right|$$

So that, $\dfrac{dr}{dq_1} = h_1 e_1$ where, $h_1 = \left| \dfrac{dr}{dq_1} \right|$.

Similarly, if e_2 and e_3 are unit tangent vectors to q_2 and q_3 curves at P respectively, then

$$\frac{dr}{dq_2} = h_2 e_2 \quad \text{and} \quad \frac{dr}{dq_3} = h_3 \cdot e_3$$

where

$$h_2 = \left| \frac{dr}{dq_2} \right| \quad \text{and} \quad h_3 = \left| \frac{dr}{dq_3} \right|.$$

The quantities h_1, h_2, h_3 are called scale factors. The unit vectors e_1, e_2, e_3 are in the direction of increasing q_1, q_2, q_3 respectively.

Since ∇q_1 is a vector at P, normal to the surface $q_1 = c_1$, a unit vector in this direction is given by

$$E_1 = \nabla q_1 / \left| \nabla q_1 \right|$$

and similarly the unit vectors

$$E_2 = \nabla q_2 / \left| \nabla q_2 \right| \quad \text{and } E_3 = \nabla q_3 / \left| \nabla q_3 \right|.$$

at P are normal to the surfaces $q_2 = c_2$ and $q_3 = c_3$ respectively.

Thus at each point P of a curvilinear system there exists, in general, two sets of unit vectors e_1, e_2, e_3 tangent to the coordinate curves and E_1, E_2, E_3 normal to the coordinates surfaces (Fig. 1.34). The sets become identical if and only if the curvilinear coordinate system is orthogonal. Both sets are analogous to i, j, k unit vectors but are unlike them, in that they may change directions from point to point. It can be shown (see solved problems) that the sets

$$\frac{\partial r}{\partial q_1}, \frac{\partial r}{\partial q_2}, \frac{\partial r}{\partial q_3},$$

and

$$\nabla q_1, \nabla q_2, \nabla q_3$$

constitute reciprocal system of vectors.

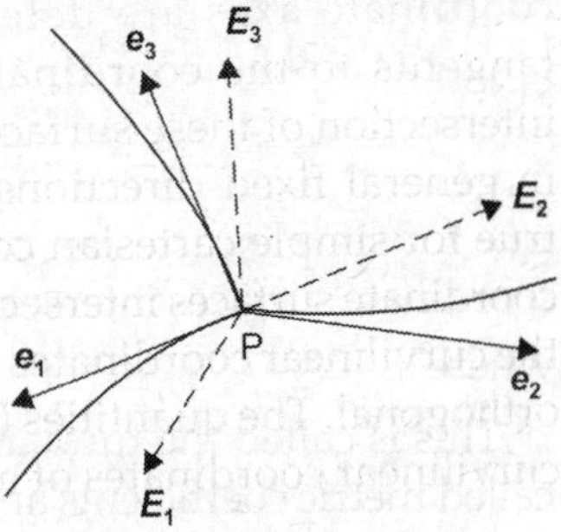

A vector A can be represented in terms of the unit base vectors $e_1, e_2, e_3,$ or E_1, E_2, E_3 in the form

$$A = A_1 e_1 + A_2 e_2 + A_3 e_3$$
$$= \alpha_1 E_1 + \alpha_2 E_2 + \alpha_3 E_3$$

Fig. 1.34

where A_1, A_2, A_3 and $\alpha_1, \alpha_2, \alpha_3$ are respectively components of A in each system.

We can also represent A in terms of the base vectors

$$\frac{\partial r}{\partial q_1}, \frac{\partial r}{\partial q_2}, \frac{\partial r}{\partial q_3} \text{ or } \nabla q_1, \nabla q_2, \nabla q_3,$$

which are called *unitary base vectors* but are not unit vectors in general. In this case

$$A = C_1 \frac{\partial r}{\partial q_1} + C_2 \frac{\partial r}{\partial q_2} + C_3 \frac{\partial r}{\partial q_3}$$
$$= C_1 \alpha_1 + C_2 \alpha_2 + C_3 \alpha_3$$

and

$$A = C_1' \nabla q_1 + C_2' \nabla q_2 + C_3' \nabla q_3$$
$$= C_1' \beta_1 + C_2' \beta_2 + C_3' \beta_3.$$

where C_1, C_2, C_3 are called the contravariant components of A and C_1', C_2', C_3' are called the covariant components of A.

Note that

$$\alpha_P = \frac{\partial r}{\partial q_P} \text{ and } \beta_P = \nabla q_P, \quad q = 1, 2, 3$$

Arc Length and Volume Element

We have,

$$dr = \frac{\partial r}{\partial q_1} dq_1 + \frac{\partial r}{\partial q_2} dq_2 + \frac{\partial r}{\partial q_3} dq_3$$

$$dr = h_1 dq_1 e_1 + h_2 dq_2 e_2 + h_3 dq_3 e_3$$
$$r = r(q_1, q_2, q_3)$$

Where, the differential of arc length ds is determined form

$$ds^2 = dr \cdot dr$$

For orthogonal systems

$$e_1 \cdot e_2 = e_2 \cdot e_3 = e_3 \cdot e_1 = 0$$
$$ds^2 = h_1^2 dq_1^2 + h_2^2 dq_2^2 + h_3^2 dq_3^2$$

For nonorthogonal or general curvilinear system

$$dr = \frac{\partial r}{\partial q_1}dq_1 + \frac{\partial r}{\partial q_2}dq_2 + \frac{\partial r}{\partial q_3}dq_3$$

$$dr = \alpha_1 dq_1 + \alpha_2 dq_2 + \alpha_3 dq_3$$

$$ds = dr \cdot dr = \sum_{i=1}^{3}\sum_{j=1}^{3}\delta_{ij}\, dq_i\, dq_j$$

where

$$\delta_{ij} = \alpha_i \cdot \alpha_j$$

This is called the fundamental quadratic form or metric form. The quantities δ_{ij} are called metric coefficients and are symmetric, i.e. $\delta_{ij} = \delta_{ji}$. If $\delta_{ij} = 0$ and $i \neq j$, then coordinate system is orthogonal. In this case

$$\delta_{11} = h_1^2, \delta_{22} = h_2^2, \delta_{33} = h_3^2$$

The metric form extended to higher dimensional space is of fundamental importance in the theory of relatively.

Along q_1 curve, q_2 and q_3 are constants, so that $dr = h_1 dq_1 e_1$. Then the differential of arc length ds_1 along q_1 at P is $h_1 dq_1$. Similarly, the differential are lengths along q_2 and q_3 at P are $ds_2 = h_2 dq_2$ and $ds_3 = h_3 dq_3$.

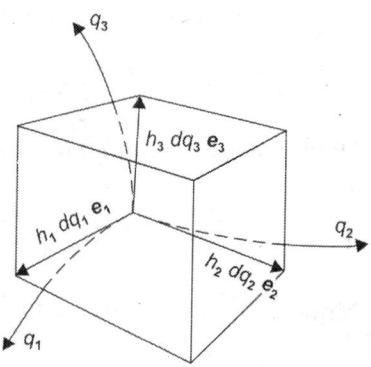

Fig. 1.35

Referring to Fig. 1.35, the volume element for an orthogonal curvilinear coordinate system is given by

$$dv = (h_1 dq_1 e_1) \cdot (h_2 dq_2 e_2) \times (h_3 dq_3 e_3)$$

$$= h_1 h_2 h_3\, dq_1 dq_2 dq_3$$

Since $e_1 \cdot e_2 \times e_3 = 1$.

Gradient, Divergence and Curl

If ϕ is a scalar point function, $\nabla\phi$ must be the same in all coordinate systems, $\nabla\phi$ is a vector whose magnitude and direction give the maximum space rate of change of ϕ. A component of $\nabla\phi$ is its directional derivative in the given direction, thus the component perpendicular to the surface q_i = constant and hence in the direction $ds_i = h_i dq_i$ is

$$\frac{d\phi}{ds_i} = \frac{1}{h_i}\frac{\partial\phi}{\partial q_i}$$

Since it is also possible to regard ∇ as a vector operator, it may be written in terms of unit vectors e_1, e_2, e_3 along the curvilinear coordinate axes. Thus

$$\nabla = \frac{e_1}{h_1}\frac{\partial}{\partial q_1} + \frac{e_2}{h_2}\frac{\partial}{\partial q_2} + \frac{e_3}{h_3}\frac{\partial}{\partial q_3} \tag{1.95}$$

So that

$$\nabla\phi = \frac{e_1}{h_1}\frac{\partial\phi}{\partial q_1} + \frac{e_2}{h_2}\frac{\partial\phi}{\partial q_2} + \frac{e_3}{h_3}\frac{\partial\phi}{\partial q_3} \tag{1.96}$$

Any vector A may be written in terms of curvilinear components A_1, A_2, A_3

$\therefore$
$$A = e_1 A_1 + e_2 A_2 + e_3 A_3$$

But in order to find $\nabla \cdot A$ in curvilinear coordinates, we must know the relation between e_1, e_2, e_3 and x, y, z. We proceed by evaluating $\nabla \cdot e_i$, starting with $\nabla \cdot e_i$, since this is needed to obtain $\nabla \cdot e_i$.

Remembering $\dfrac{e_1}{h_1}$ as the product of a scalar and a vector, we may write using Eq. (1.95)

$$\nabla \times \left(\frac{e_1}{h_1}\right) = \nabla\left(\frac{1}{h_1}\right) \times e_1 + \frac{1}{h_1}(\nabla \times e_1)$$

$$= -e_1 \times \nabla\left(\frac{1}{h_1}\right) + \frac{1}{h_1}(\nabla \times e_1)$$

the change in sign is from the change in order of the vector products. From Eq. (1.95), we note that

$$\frac{e_1}{h_1} = \nabla q_1$$

and from Eq. (1.81) $\quad \nabla \times \nabla q_1 = 0$

Hence $$\nabla \times \frac{e_1}{h_1} = \nabla \times \nabla q_1 = 0$$

and $$e_1 \times \nabla\left(\frac{1}{h_1}\right) = \frac{1}{h_1}(\nabla \times e_1) \tag{1.96a}$$

Now using Eq. (1.95) and performing differentiation, we have

$$\nabla\left(\frac{1}{h_1}\right) = \frac{e_1}{h_1}\frac{\partial h_1}{\partial q_1} - \frac{e_2}{h_1 h_2}\frac{\partial h_1}{\partial q_2} - \frac{e_3}{h_1 h_3}\frac{\partial h_1}{\partial q_3} \tag{1.96b}$$

For orthogonal curvilinear system we have

$$e_i \times e_i = 0, \; e_i \times e_j = e_k$$

Substituting Eqs (1.96a) in (1.96b), we obtain

$$\nabla \times e_1 = h_1\left\{e_1 \times \nabla\left(\frac{1}{h_1}\right)\right\} = h_1\left\{\frac{e_2}{h_1^2 h_3}\frac{\partial h_1}{\partial q_3} - \frac{e_3}{h_1^2 h_2}\cdot\frac{\partial h_1}{\partial q_2}\right\}$$

$$\nabla \times e_1 = \frac{e_2}{h_1 h_3}\frac{\partial h_1}{\partial q_3} - \frac{e_3}{h_1 h_2}\cdot\frac{\partial h_1}{\partial q_2} \tag{1.96c}$$

The scalar product of ∇ and a unit vector may be written as

$$\nabla \cdot e_1 = \nabla \cdot (e_2 \times e_3) = e_3 \cdot (\nabla \times e_2) - e_2 \cdot (\nabla \times e_3)$$

Using Eq. (1.96c), we have

$$\nabla \cdot e_1 = \frac{1}{h_1 h_2 h_3}\frac{\partial(h_2 h_3)}{\partial q_1} \tag{1.96d}$$

In order to determine $\nabla \cdot A$ in curvilinear coordinates, we write

$$\nabla \cdot A = \nabla \cdot (e_1 A_1) + \nabla \cdot (e_2 A_2) + \nabla \cdot (e_3 A_3)$$

Now $$\nabla \cdot (e_1 A_1) = A_1 \nabla \cdot e_1 + e_1 \cdot \nabla A_1$$

Equation (1.96d) gives $\nabla \times e_1$ and (1.96) gives ∇A. Hence,

$$\nabla \cdot A = \frac{1}{h_1 h_2 h_3} \left\{ \frac{\partial}{\partial q_1}(A_1 h_2 h_3) + \frac{\partial}{\partial q_2}(A_2 h_3 h_1) + \frac{\partial}{\partial q_3}(A_3 h_1 h_2) \right\} \tag{1.97}$$

If $A = \nabla \phi$

$$\nabla \cdot \nabla \phi = \nabla^2 \phi = \frac{1}{h_1 h_2 h_3} \left[\frac{\partial}{\partial q_1}\left(\frac{h_2 h_3}{h_1} \frac{\partial \phi}{\partial q_1} \right) + \frac{\partial}{\partial q_2}\left(\frac{h_1 h_3}{h_2} \frac{\partial \phi}{\partial q_2} \right) + \frac{\partial}{\partial q_3}\left(\frac{h_1 h_2}{h_3} \frac{\partial \phi}{\partial q_3} \right) \right] \tag{1.97a}$$

Since, the components of $\nabla \phi$ are $A_i = \left(\dfrac{\partial \phi}{\partial q_i} \right) \dfrac{1}{h_i}$

The curl of a vector in terms of the unit curvilinear vectors becomes

$$\nabla \times A = \nabla \times (e_1 A_1) + \nabla \times (e_2 A_2) + \nabla \times (e_3 A_3)$$

which can be expanded to given terms like

$$\nabla \times (e_i A_i) = A_i (\nabla \times e_i) - e_i \times (\nabla A_i)$$

When three similar equations are added together, the result in determinant form is

$$\nabla \times A = \frac{1}{h_1 h_2 h_3} \begin{vmatrix} h_1 e_1 & h_2 e_2 & h_3 e_3 \\ \partial/\partial q_1 & \partial/\partial q_2 & \partial/\partial q_3 \\ A_1 h_1 & A_2 h_2 & A_3 h_3 \end{vmatrix} \tag{1.98}$$

If $h_1 = h_2 = h_3$ and e_1, e_2, e_3 are replaced by i, j, k these reduce to the usual expressions in rectangular coordinates, where (q_1, q_2, q_3) is replaced by (x, y, z).

Extension of the above results are achieved by a more general theory of curvilinear systems using the method of tensor analysis which is considered in latter chapters.

Special Orthogonal Coordinate Systems

Spherical Polar Coordinates

The coordinate surfaces are families of (i) concentric spheres about the origin ($r = $ const) (ii) right circular cones with apex at the origin and axis along $z(\theta = $ const) (iii) half planes from the z-axis ($\phi = $ const). A point $P(x, y, z)$ is located by specifying the radius r of the sphere on which lies its colatitude θ, and its longitude or azimuth ϕ on the sphere.

From Fig. 1.36, we can derive

$x = r \sin \theta \cos \phi$

$y = r \sin \theta \sin \phi$

$z = r \cos \theta.$

$\therefore \quad dx = dr \sin \theta \cos \phi + r \cos \theta \cdot d\theta \cos \phi - r \sin \theta \sin \phi \, d\phi$

$dy = dr \sin \theta \sin \phi + r \cos \theta \, d\theta \sin \phi + r \sin \theta \cos \phi \, d\phi$

$dz = dr \cos \theta - r \, d\theta \sin \theta$

$\therefore \quad ds^2 = dx^2 + dy^2 + dz^2 = (dr)^2 + r^2 (d\theta)^2 + r^2 \sin^2 \theta \, (d\phi)^2$

On comparing, we have

$h_1 = 1, h_2 = r$ and $h_3 = r \sin \theta \tag{1.99}$

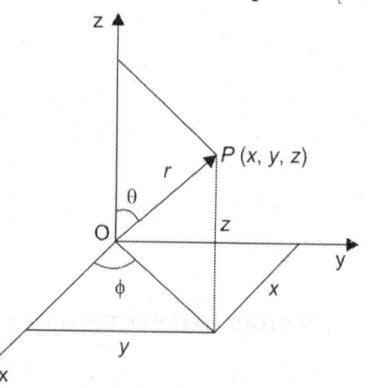

Fig. 1.36

Cylindrical Coordinates

The coordinate surfaces are (i) right circular cylinders which form families of concentric circles about the origin in the xy-plane ($\rho = $ const) (ii) half-planes from the z-axis ($\phi = $ const); (iii) planes parallel to the xy-plane $= z$-const). A point $P(x, y, z)$ is located by giving the distance ρ in the xy-plane from the origin to the cylinder on which the point lies, the angle ϕ in the xy-plane and the distance on the z-axis from this plane to the point. From Fig. 1.37.

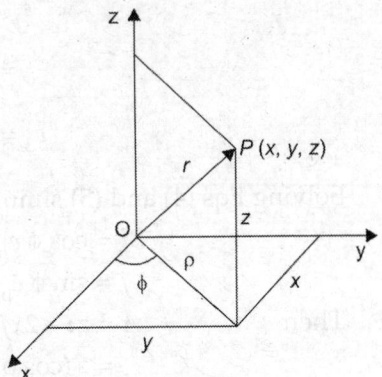

Fig. 1.37

$$x = \rho \cos \phi$$
$$y = \rho \sin \phi$$
$$z = z$$
$$\therefore \quad dx = d\rho \cos \phi - \rho \sin \phi \, d\phi$$
$$dy = d\rho \sin \phi + \rho \cos \phi \, d\phi$$
$$dz = dz$$
$$\therefore \quad ds^2 = (d\rho)^2 + \rho^2 (d\phi)^2 + dz^2$$

Comparing with $ds^2 = h_1^2 dq_1^2 + h_2^2 dq_2^2 + h_3^2 dq_3^2$, we have

$$h_1 = 1, h_2 = \rho \text{ and } h_3 = 1$$

Table 1.1 gives the comparison of the three types (cartesian, cylindrical and spherical) of orthogonal coordinate systems.

Table 1.1

	Cartesian	Cylindrical	Spherical
q_1	x	ρ	r
h_1	1	1	1
q_2	y	ϕ	θ
h_2	1	ρ	r
q_3	z	z	ϕ
h_3	1	1	$r \sin \theta$

Example 1. *Represent the vector $A = zi - 2xj + yk$ in cylindrical coordinates. Thus, determine A_ρ, A_ϕ and A_z.*

Solution: The position vector of any point in cylindrical coordinates is

$$r = xi + yj + zk = \rho \cos \phi i + \rho \sin \phi j + zk.$$

The tangent vector to the ρ, ϕ and z curves are given respectively by

$$\frac{dr}{d\rho}, \frac{\partial r}{\partial \phi} \text{ and } \frac{\partial r}{\partial z}, \text{ where}$$

$$\frac{dr}{d\rho} = \cos \phi i + \sin \phi j \tag{1}$$

$$\frac{dr}{d\phi} = -\rho \sin \phi i + \rho \cos \phi j \tag{2}$$

$$\frac{\partial r}{\partial z} = k \tag{3}$$

Solving simultaneously, the unit vectors in these directions are

$$e_1 = e_\rho = \frac{\partial r / \partial \rho}{|\partial r / \partial \rho|} = \frac{\cos\phi i + \sin\phi j}{\sqrt{\sin^2\phi + \cos^2\phi}}$$

$$= \cos\phi i + \sin\phi j \qquad (4)$$

$$e_2 = e_\phi = \frac{\partial r / \partial \phi}{|\partial r / \partial \phi|} = \frac{-\rho\sin\phi i + \rho\cos\phi j}{\rho\sqrt{\sin^2\phi + \cos^2\phi}}$$

$$= -\sin\phi i + \cos\phi j \qquad (5)$$

$$e_3 = e_z = \frac{\partial r / \partial z}{|\partial r / \partial z|} = k \qquad (6)$$

Solving Eqs (4) and (5) simultaneously, we get

$$i = \cos\phi\, e_\rho - \sin\phi\, e_\phi$$

$$j = \sin\phi\, e_\rho + \cos\phi\, e_\phi.$$

Then $A = zi - 2xj + yk$

$$= z(\cos\phi\, e_\rho - \sin\phi\, e_\phi) - 2 \cdot \rho\cos\phi\,(\sin\phi\, e_\rho + \cos\phi\, e_\phi) + \rho\sin\phi\, e_z$$

$$= (z\cos\phi - 2\rho\cos\phi\sin\phi)\, e_\rho - (z\sin\phi + 2\rho\cos^2\phi)\, e_\phi + \rho\sin\phi\, e_z$$

Thus, $A_\rho = z\cos\phi - 2\rho\cos\phi\sin\phi,$

$A_\phi = -z\sin\phi - 2\rho\cos^2\phi$

$A_z = \rho\sin\phi$

Example 2. *Express the velocity v and acceleration a of a particle in cylindrical coordinates.*

Solution: In rectangular coordinates, the position vector is

$$r = xi + yj + zk,$$

and the velocity and acceleration are

$$v = \frac{dr}{dt} = \dot{x}i + \dot{y}j + \dot{z}k$$

and

$$a = \frac{d^2 r}{dt^2} = \ddot{x}i + \ddot{y}j + \ddot{z}k$$

In cylindrical coordinates (*using* Example 1)

$$r = (\rho\cos\phi)(\cos\phi e_\rho - \sin\phi e_\phi)$$

$$+ (\rho\sin\phi)(\sin\phi e_\rho + \cos\phi e_\phi) + ze_z$$

$$= \rho e_\rho + ze_z.$$

Then $$v = \frac{dr}{dt} = \frac{d\rho}{dt}e_\rho + \rho\frac{de_\rho}{dt} + \frac{dz}{dt}e_z \qquad (1)$$

Using Example 1:

$$e_\rho = \cos\phi i + \sin\phi j, \; e_\phi = -\sin\phi i + \cos\phi j$$

or

$$\frac{de_\rho}{dt} = -(\sin\phi)\dot{\phi}i + (\cos\phi)\dot{\phi}j = (-\sin\phi i + \cos\phi j)\dot{\phi} = \dot{\phi}e_\phi$$

and
$$\frac{de_\phi}{dt} = -(\cos\phi)\dot\phi i - (\sin\phi)\dot\phi j$$
$$= (-\cos\phi i - \sin\phi j)\dot\phi = -\dot\phi e_\rho$$

Substituting in Eq. (1), we have
$$v = \dot\rho e_\rho + \rho\dot\phi e_\phi + \dot z e_z$$

Differentiating again, we get
$$a = \frac{d^2 r}{\partial t^2} = \frac{d}{dt}\{\dot\rho e_\rho + \rho\dot\phi e_\phi + \dot z e_z\}$$
$$= \dot\rho\frac{de_\rho}{dt} + \ddot\rho e_\rho + \rho\dot\phi\frac{de_\phi}{dt} + \rho\frac{d\dot\phi}{\partial t}e_\phi + \frac{d\rho}{\partial t}\dot\phi e_\phi + \ddot z e_z + \dot z\frac{de_z}{dt}$$
$$= \dot\rho\dot\phi e_\phi + \ddot\rho e_\rho + \rho\dot\phi(-\dot\phi e_\rho) + \rho\ddot\phi e_\phi + \dot\rho\dot\phi e_\phi + \ddot z e_z$$
$$= (\ddot\rho - \rho\dot\phi^2)e_\rho + (\rho\ddot\phi + 2\dot\rho\dot\phi)e_\phi + \ddot z e_z.$$

Example 3. *If q_1, q_2, q_3 are general coordinates, show that $\dfrac{\partial r}{\partial q_1}, \dfrac{\partial r}{\partial q_2}, \dfrac{\partial r}{\partial q_3}$ and $\nabla q_1, \nabla q_2, \nabla q_3,$ are reciprocal systems of vectors.*

Solution: We must show that
$$\frac{\partial r}{\partial q_i}\cdot\nabla q_j = \begin{cases} 1 \text{ for } i = j \\ 0 \text{ for } i \neq j \end{cases}$$

where i and j can have any values 1, 2, 3. We have
$$dr = \frac{\partial r}{\partial q_1}dq_1 + \frac{\partial r}{\partial q_2}dq_2 + \frac{\partial r}{\partial q_3}dq_3$$

Multiplying scalarly by ∇q_1, we get
$$\nabla q_1\cdot dr = \left(\nabla q_1\cdot\frac{dr}{dq_1}\right)dq_1 + \left(\nabla q_1\cdot\frac{dr}{dq_2}\right)dq_2 + \left(\nabla q_1\cdot\frac{dr}{dq_3}\right)dq_3$$

or
$$\nabla q_1\cdot\frac{dr}{dq_1} = 1$$

Since
$$\nabla q_1\cdot\frac{\partial r}{dq_2} = \nabla q_1\cdot\frac{\partial r}{dq_3} = 0.$$

Similarly upon multiplying by ∇q_2 and ∇q_3, the remaining relations are proved.

Example 4. *Express in cylindrical coordinates the quantities (a) ∇S (b) $\nabla\cdot A$ (c) $\nabla\times A$ (d) $\nabla^2 S$.*

Solution: For cylindrical coordinates (ρ, ϕ, z)
$$q_1 = \rho, q_2 = \phi, q_3 = z,$$
$$e_1 = e_\rho, e_2 = e_\phi, e_3 = e_z$$

and
$$h_1 = h_\rho = 1, h_2 = h_\phi = \rho, h_3 = h_z = 1 \qquad (see \text{ Table 1.1})$$

(a) $\nabla S = \dfrac{1}{h_1}\dfrac{\partial S}{\partial q_1}e_1 + \dfrac{1}{h_2}\dfrac{\partial S}{\partial q_2}e_2 + \dfrac{1}{h_3}\dfrac{\partial S}{\partial q_2}e_3$ (*using* Eq. 1.96)

$$= \dfrac{1}{1}\dfrac{\partial S}{\partial \rho}e_\rho + \dfrac{1}{\rho}\dfrac{\partial S}{\partial \phi}e_\phi + \dfrac{1}{1}\dfrac{\partial S}{\partial z}e_z$$

$$= \dfrac{\partial S}{\partial \rho}e_\rho + \dfrac{1}{\rho}\dfrac{\partial S}{\partial \phi}e_\phi + \dfrac{\partial S}{\partial z}e_z$$

(b) $\nabla \cdot A = \dfrac{1}{h_1 h_2 h_3}\left[\dfrac{\partial}{\partial q_1^{*}}(h_2 h_3 A_1) + \dfrac{\partial}{\partial q_2}(h_3 h_1 A_2) + \dfrac{\partial}{\partial q_3}(h_1 h_2 A_3)\right]$

$$= \dfrac{1}{(1)(\rho)(1)}\left[\dfrac{\partial}{\partial \rho}\{(\rho)(1)(A_\rho)\} + \dfrac{\partial}{\partial \phi}\{(1)(1)A_\phi\} + \dfrac{\partial}{\partial z}\{(1)(\rho)(A_z)\}\right]$$

$$= \dfrac{1}{\rho}\left[\dfrac{\partial(A_\rho \rho)}{\partial \rho} + \dfrac{\partial A_\phi}{\partial \phi} + \dfrac{\partial(\rho A_z)}{\partial z}\right]$$

$$= \dfrac{A_\rho}{\rho} + \dfrac{\partial A_\rho}{\partial \rho} + \dfrac{1}{\rho}\dfrac{A_\phi}{\partial \phi} + \dfrac{\partial A_z}{\partial z}$$

where $A = A_\rho e_\rho + A_\phi e_\phi + A_z e_z$

(c) $\nabla \times A = \dfrac{1}{h_1 h_2 h_3}\begin{vmatrix} h_1 e_1 & h_2 e_2 & h_3 e_3 \\ \dfrac{\partial}{\partial q_1} & \dfrac{\partial}{\partial q_2} & \dfrac{\partial}{\partial q_3} \\ h_1 A_1 & h_2 A_2 & h_3 A_3 \end{vmatrix}$

$$= \dfrac{1}{\rho}\begin{vmatrix} e_\rho & \rho e_\phi & e_z \\ \dfrac{\partial}{\partial \rho} & \dfrac{\partial}{\partial \phi} & \dfrac{\partial}{\partial z} \\ A_\rho & \rho A_\phi & A_z \end{vmatrix}$$

or $\nabla \times A = \dfrac{1}{\rho}\left[\left\{\dfrac{\partial A_z}{\partial \phi} - \dfrac{\partial(\rho A_\phi)}{\partial z}\right\}e_\rho + \left\{\rho\dfrac{\partial A_\rho}{\partial z} - \rho\dfrac{\partial(A_z)}{\partial \rho}\right\}e_\phi + \left\{\dfrac{\partial(\rho A_\phi)}{\partial \rho} - \dfrac{\partial A_\rho}{\partial \phi}\right\}e_z\right]$

$$= \left(\dfrac{1}{\rho}\dfrac{\partial A_z}{\partial \phi} - \dfrac{\partial A_\phi}{\partial z}\right)e_\rho + \left(\dfrac{\partial A_\rho}{\partial z} - \dfrac{\partial A_z}{\partial \rho}\right)e_\phi + \left\{\dfrac{\partial A_\phi}{\partial \rho} + \dfrac{A_\phi}{\rho} - \dfrac{1}{\rho}\dfrac{\partial A_\rho}{\partial \phi}\right\}e_z$$

(d) $\nabla^2 S = \dfrac{1}{h_1 h_2 h_3}\left[\dfrac{\partial}{\partial q_1}\left(\dfrac{h_2 h_3}{h_1}\dfrac{\partial S}{\partial q_1}\right) + \dfrac{\partial}{\partial q_2}\left(\dfrac{h_3 h_1}{h_2}\dfrac{\partial S}{\partial q_2}\right) + \dfrac{\partial}{\partial q_3}\left(\dfrac{h_1 h_2}{h_3}\dfrac{\partial S}{\partial q_3}\right)\right]$

$$= \dfrac{1}{(1)\rho(1)}\left[\dfrac{\partial}{\partial \rho}\left(\dfrac{(\rho)(1)}{(1)}\dfrac{\partial S}{\partial \rho}\right) + \dfrac{\partial}{\partial \phi}\left(\dfrac{(1)(1)}{\rho}\dfrac{\partial S}{\partial \phi}\right) + \dfrac{\partial}{\partial z}\left(\dfrac{(1)(\rho)}{(1)}\dfrac{\partial S}{\partial z}\right)\right]$$

$$= \dfrac{1}{\rho}\dfrac{\partial}{\partial \rho}\left(\rho\dfrac{\partial S}{\partial \rho}\right) + \dfrac{1}{\rho^2}\dfrac{\partial^2 S}{\partial \phi^2} + \dfrac{\partial^2 S}{\partial z^2} = \dfrac{\partial^2 S}{\partial \rho^2} + \dfrac{1}{\rho}\dfrac{\partial S}{\partial \rho} + \dfrac{1}{\rho^2}\dfrac{\partial^2 S}{\partial \phi^2} + \dfrac{\partial^2 S}{\partial z^2}$$

Example 5. *Express in spherical coordinates, the quantities*

(a) ∇S (b) $\nabla \cdot A$ (c) $\nabla \times A$ (d) $\nabla^2 S$

Solution: Here $q_1 = r, q_2 = \theta, q_3 = \phi$

$$e_1 = e_r, e_2, = e_\theta, e_3 = e_\phi$$

$$h_1 = h_r = 1, h_2 = h_\theta = r,$$

$$h_3 = h_\phi = r \sin \theta$$

(a) $\nabla S = \dfrac{e_1}{h_1}\dfrac{\partial S}{\partial q_1} + \dfrac{e_2}{\partial q_2}\dfrac{\partial S}{\partial q_2} + \dfrac{e_3}{h_3}\dfrac{\partial S}{\partial q_3}$

$$= e_r \frac{\partial S}{\partial r} + \frac{e_\theta}{r}\frac{\partial S}{\partial \theta} + \frac{e_\theta}{r\sin\theta}\frac{\partial S}{\partial \phi}$$

(b) $\nabla \cdot A = \dfrac{1}{h_1 h_2 h_3}\left[\dfrac{\partial}{\partial q_1}(A_1 h_2 h_3) + \dfrac{\partial}{\partial q_2}(A_2 h_3 h_1) + \dfrac{\partial}{\partial q_3}(A_3 h_1 h_2)\right]$

$$= \frac{1}{(1)(r)(r\sin\theta)}\left\{\frac{\partial}{\partial r}(A_r r^2 \sin\theta) + \frac{\partial}{\partial \theta}(A_\theta r\sin\theta) + \frac{\partial}{\partial \phi}(A_\phi r)\right\}$$

$$- \frac{1}{r^2 \sin\theta}\left\{\frac{\partial}{\partial r}(A_r r^2 \sin\theta) + \frac{\partial}{\partial \theta}(A_\theta r\sin\theta) + \frac{\partial}{\partial \phi}(A_\phi r)\right\}$$

$$= \frac{1}{r^2}\frac{\partial}{\partial r}(A_r r^2) + \frac{1}{r\sin\theta}\frac{\partial}{\partial \theta}(A_\theta \sin\theta) + \frac{1}{r\sin\theta}\frac{\partial}{\partial \phi}(A_\phi)$$

(c) $\nabla \times A = \dfrac{1}{h_1 h_2 h_3}\begin{vmatrix} h_1 e_1 & h_2 e_2 & h_3 e_3 \\ \dfrac{\partial}{\partial q_1} & \dfrac{\partial}{\partial q_2} & \dfrac{\partial}{\partial q_3} \\ A_r & rA_\theta & r\sin\theta\, A_\phi \end{vmatrix}$

$$= \frac{1}{(1)(r)(r\sin\theta)}\begin{vmatrix} e_r & re_\theta & r\sin\theta e_\phi \\ \dfrac{\partial}{\partial r} & \dfrac{\partial}{\partial \theta} & \dfrac{\partial}{\partial \phi} \\ A_r & rA_\theta & r\sin\theta A_\phi \end{vmatrix}$$

$$\nabla \times A = \left[\frac{1}{r^2 \sin\theta}\left\{\frac{\partial}{\partial \theta}(r\sin\theta A_\phi) - \frac{\partial}{\partial \phi}(rA_\theta)\right\}e_r\right.$$

$$\left. + \left\{\frac{\partial A_r}{\partial \phi} - \frac{\partial}{\partial r}(r\sin\theta A_\phi)\right\}re_\theta + \left\{\frac{\partial}{\partial r}(rA_\theta) - \frac{\partial A_r}{\partial \theta}\right\}r\sin\theta e_\phi\right]$$

$$= \frac{1}{r\sin\theta}\left\{\frac{\partial}{\partial \theta}(\sin\theta A_\phi) - \frac{\partial A_\theta}{\partial \phi}\right\}e_r + \left\{\frac{1}{r\sin\theta}\frac{\partial A_r}{\partial \phi} - \frac{1}{r}\frac{\partial}{\partial r}(rA_\phi)\right\}e_r$$

$$+ \left\{\frac{1}{r}\frac{\partial}{\partial r}(rA_\theta) - \frac{1}{r}\frac{\partial A_r}{\partial \theta}\right\}e_\phi$$

(d) $\nabla^2 S = \dfrac{1}{h_1 h_2 h_3}\left[\dfrac{\partial}{\partial q_1}\left(\dfrac{h_2 h_3}{h_1}\dfrac{\partial S}{\partial q_1}\right) + \dfrac{\partial}{\partial q_2}\left(\dfrac{h_1 h_3}{h_2}\dfrac{\partial S}{\partial q_2}\right) + \dfrac{\partial}{\partial q_3}\left(\dfrac{h_1 h_2}{h_3}\dfrac{\partial S}{\partial q_3}\right)\right]$

$= \dfrac{1}{(1)(r)(r\sin\theta)}\left[\dfrac{\partial}{\partial r}\left\{\dfrac{(r)(r\sin\theta)}{(1)}\dfrac{\partial S}{\partial r}\right\} + \dfrac{\partial}{\partial \theta}\left\{\dfrac{(1)(r\sin\theta)}{r}\dfrac{\partial S}{\partial \theta}\right\} + \dfrac{\partial}{\partial \phi}\left\{\dfrac{(1)(r)}{r\sin\theta}\dfrac{\partial S}{\partial \phi}\right\}\right]$

$= \dfrac{1}{r^2 \sin\theta}\left[\dfrac{\partial}{\partial r}\left(r^2 \sin\theta\cdot\dfrac{\partial S}{\partial r}\right) + \dfrac{\partial}{\partial \theta}\left(\sin\theta\cdot\dfrac{\partial S}{\partial \theta}\right) + \dfrac{\partial}{\partial \phi}\left(\dfrac{1}{\sin\theta}\dfrac{\partial S}{\partial \phi}\right)\right]$

$= \dfrac{1}{r^2}\dfrac{\partial}{\partial r}\left(r^2\dfrac{\partial S}{\partial r}\right) + \dfrac{1}{r^2 \sin\theta}\dfrac{\partial}{\partial \theta}\left(\sin\theta\dfrac{\partial S}{\partial \theta}\right) + \dfrac{1}{r^2 \sin^2\theta}\dfrac{\partial^2 S}{\partial \phi^2}$

1.15 APPLICATIONS TO HYDRODYNAMICS

The branch of mathematics which deals with the motion of fluids is called fluid mechanics and consists of three branches:

1. *Hydrostatics* deals with the mechanics of fluid at rest.
2. *Kinetatics* deals with velocities and streamlines irrespective of the force and energy.
3. *Hydrodynamics* deals with the relations between velocities and accelerations and forces exerted by or upon fluid in motion considering a fluid particle as a geometrical point.

The Problems in hydrodynamics are generally treated by two methods:

   i. Lagrangian method   ii. Eulerian method

   i. In *Lagrangian method,* a particle of fluid is selected and persued throughout its course making observations of changes in velocity, density and pressure at each instant and at each point.

Consider a region of space containing a fluid of density $\rho(x, y, z, t)$. Let V be the volume inside an arbitrary closed surface S located in this region. Let $Q(t)$ be the mass of fluid inside volume V at any instant. Then

$$Q(t) = \iiint_V \rho dV \tag{1.100}$$

If q denotes the velocity of a typical particle of the fluid, the rate at which the mass of fluid inside V is increasing is given by

$$\frac{dQ}{dt} = -\iint_S (\rho q)\cdot dS \tag{1.100a}$$

Differentiating Eq. (1.100a) w.r.t time and equating the result to Eq. (1.100b), we have

$$\frac{dQ}{dt} = \iiint_V \left(\frac{\partial \rho}{\partial t}\right)dV = -\iint_S (\rho q)\cdot dS \tag{1.101}$$

But by Gauss's theorem, we have

$$\iint_S (\rho q)\cdot dS = \iiint_V \nabla\cdot(\rho q)dV$$

Substituting the value in Eq. (1.101), we have

$$\iiint_V \left[\frac{\partial \rho}{\partial t} + \nabla \cdot (\rho q) \right] dV = 0$$

Now since the integrand is continuous and volume V is arbitrary, we can conclude that

$$\frac{\partial \rho}{\partial t} + \nabla \cdot (\rho q) = 0 \tag{1.102}$$

This is the basic equation of hydrodynamics and is known as the *equation of continuity*.

ρ is constant, if the fluid is incompressible and we have

$$\frac{\partial \rho}{\partial t} = 0 = \rho \nabla \cdot q \tag{1.102a}$$

If the flow is irrotational, then we have

$$\nabla \times q = 0$$

and we know that there exists a scalar function ϕ such that

$$q = \text{grad } \phi = \nabla \phi \tag{1.103}$$

If the fluid is incompressible then from Eq. (1.102a), we have

$$\nabla \cdot q = 0$$

and hence ϕ, called the velocity potential satisfies the equation

$$\nabla \cdot (\nabla \phi) = \nabla^2 \phi = 0$$

On a fixed boundary of the fluid, the velocity has no normal component, and if $\dfrac{\partial}{\partial n}$, denotes differentiation with respect to the normal direction of the boundary, we must have

$$\frac{\partial \phi}{\partial n} = 0$$

as a consequence of Eq. (1.103).

Let us denote any general property of a particle of the fluid, such as its pressure, density, etc., by the function $H(z, y, z, t)$. Then $\dfrac{\partial H}{\partial t}$ denotes the variation of H at a particular point in space as a function of the time. If we take the total differential of $H(x, y, z, t)$, we obtain

$$dH = \frac{\partial H}{\partial x} dx + \frac{\partial H}{\partial y} dy + \frac{\partial H}{\partial z} dz + \frac{\partial H}{\partial t} dt$$

$$\frac{dH}{dt} = \frac{\partial H}{\partial x} \frac{dx}{dt} + \frac{\partial H}{\partial y} \frac{dy}{dt} + \frac{\partial H}{\partial z} \frac{dz}{dt} + \frac{\partial H}{\partial t}$$

The quantity $\dfrac{dH}{dt}$ is the variation of H when we fix our attention on the same particle of fluid.

The velocity of the particle is given by

$$q = i \frac{dx}{dt} + j \frac{dy}{dt} + k \frac{dz}{dt}$$

By definition, we have

$$\nabla H = i\frac{\partial H}{\partial x} + j\frac{\partial H}{\partial y} + k\frac{\partial H}{\partial z}$$

Hence

$$q \cdot \nabla H = \frac{\partial H}{\partial x}\frac{dx}{dt} + \frac{\partial H}{\partial y}\frac{dy}{dt} + \frac{\partial H}{\partial z}\frac{dz}{dt}$$

$$\frac{dH}{dt} = \frac{\partial H}{\partial t} + q \cdot \nabla H \tag{1.104}$$

In a similar way a vector function $A(r)$ may be expressed as

$$\frac{dA}{dt} = \frac{\partial A}{\partial t} + (q \cdot \nabla)A \tag{1.104a}$$

and as already shown (Example 5; Section 1.11)

$$\nabla \times q = \text{curl } q = 2\omega$$

where ω is the angular velocity of an infinitesimal element. This angular velocity is known as *vorticity in hydrodynamics*.

A curve described by a simple particle of fluid during its motion is known as a *path line* and a curve at any point of which the tangent gives the direction of motion of the fluid at that point is known as a *streamline* or *line of flow*.

ii. In *Eulerian method* a point fixed in the space occupied by the fluid is selected and observations of changes in velocity, density and pressure are made as the fluid passes through this point.

To obtain the equation of motion of a frictionless fluid, we consider the forces acting on an element of fluid whose volume is $dx\ dy\ dz$ and whose mass is $\rho\ dx\ dy\ dz$. As a consequence of the pressure of the fluid, P, there will be a force in the x-direction on the element of fluid under consideration of magnitude.

$$P\,dy\,dz - \left(P + \frac{\partial P}{\partial x}dx\right)dy\,dz = -\frac{\partial P}{\partial x}dx\,dy\,dz$$

Let us also consider the action of an external force F acting on unit mass of fluid. The external force acting on the element under consideration in the x-direction is $\dfrac{dq_x}{dt}$. Hence by Newton's second law of motion, we have

$$F_x \rho\,dx\,dy\,dz - \frac{\partial P}{\partial x}dx\,dy\,dz = \frac{dq_x}{dt}\rho\,dx\,dy\,dz$$

By this relation and considering y-and z-directions, we obtain Euler's equations of motion:

$$\frac{dq_x}{dt} = F_x - \frac{1}{\rho}\frac{\partial P}{\partial x} \tag{1.105}$$

$$\frac{dq_y}{dt} = F_y - \frac{1}{\rho}\frac{\partial P}{\partial y} \tag{1.105a}$$

$$\frac{dq_z}{dt} = F_z - \frac{1}{\rho}\frac{\partial P}{\partial z} \tag{1.105b}$$

These three scalar equations may be combined into the single vector equation

$$\frac{dq}{dt} = F - \frac{1}{\rho}\nabla P \qquad (1.106)$$

In Eq. (1.104a), if the vector A is replaced by q, we have

$$\frac{dq}{dt} = \frac{\partial q}{\partial t} + (q \cdot \nabla)q$$

Hence substituting in Eq. (1.106), we have

$$\frac{\partial q}{\partial t} + (q \cdot \nabla)q = F - \frac{1}{\rho}\nabla P \qquad (1.107)$$

which is Euler's equation of motion.

This equation may further be simplified by using Eq. (1.91) which gives

$$(q \cdot \nabla)q = \frac{1}{2}\nabla q^2 - q \operatorname{curl} q$$

Substituting in Eq. (1.107), we get

$$\frac{\partial q}{\partial t} + \frac{1}{2}\nabla q^2 + (\operatorname{curl} q) \times q = F - \frac{1}{\rho}\nabla P \qquad (1.107a)$$

If the external force is conservative, it has a potential S and we have $F = -\nabla S$
If the motion of the fluid is irrotational, $\nabla \times q = 0$ and $q = \nabla\phi$.

In this case Eq. (1.107a) becomes

$$\nabla\left(\frac{\partial\phi}{\partial t}\right) + \frac{1}{2}\nabla q^2 + \nabla S + \frac{1}{\rho}\nabla P = 0 \qquad (1.108)$$

or

$$\nabla\omega = 0 \qquad (1.109)$$

where

$$\omega = \frac{\partial\phi}{\partial t} + \frac{1}{2}q^2 + S + \frac{P}{\rho} \qquad (1.110)$$

Now if dr denotes an arbitrary path in the fluid at any instant, we have

$$dr = idx + jdy + kdz$$

Let us form the scalar product of $\nabla\omega$ and dr, We obtain

$$\nabla\omega \cdot dr = \left(i\frac{\partial\omega}{\partial x} + j\frac{\partial\omega}{\partial y} + k\frac{\partial\omega}{\partial z}\right) \cdot (idx + jdy + kdz)$$

$$= \frac{\partial\omega}{\partial x}dx + \frac{\partial\omega}{\partial y}dy + \frac{\partial\omega}{\partial z}dz = d\omega = 0$$

as a consequence of Eq. (1.109). Hence

$$\omega = \beta(t).$$

where $\beta(t)$ is an arbitrary function of time, since we have integrated along an arbitrary path in the fluid at any instant. That is, we have

$$\omega = \frac{\partial\phi}{\partial t} + \frac{1}{2}q^2 + S + \int\frac{dP}{\rho} = \beta(t) \qquad (1.111)$$

Equation (1.111) is known as Bernoulli's equation.

In the special case where ρ is constant and motion is steady, $\dfrac{\partial \phi}{\partial t} = 0$ and this equation takes the form

$$\frac{q^2}{2} + S + \frac{P}{\rho} = \beta \tag{1.112}$$

where β is now a constant. This equation states that per unit mass of fluid, the sum of the kinetic energy $\dfrac{q^2}{2}$, the potential energy S and the pressure energy $\dfrac{P}{\rho}$ has a constant value β for all points of the fluid.

Usually in aerodynamics the variations in S are so small that they can be neglected, so that we can write

$$\frac{\rho q^2}{2} + P = P_0 \tag{1.113}$$

where P_0 is pressure when the fluid is at rest. It can be seen from this that in this case the pressure diminishes as the velocity increases in the form

$$P = P_0 - \frac{\rho q^2}{2}$$

Aeroplanes are equipped with instruments for measuring P and P_0, so that the velocity of the machine relative to the air can be determined by the equation

$$v = \sqrt{\frac{2(P_0 - P)}{\rho}}$$

1.16 THE EQUATION OF HEAT FLOW IN SOLIDS

Consider a region inside a solid body such as large block of metal. Let a closed surface S be situated inside this region. Let the volume inside the surface S be denoted by V.

If the temperature at any point (x, y, z) of a solid at a time t is $U(x, y, z, t)$ and if K, ρ and c are respectively the thermal conductivity, density and specific heat of solid, then the quantity of heat leaving surface S per unit time is

$$\iint_S (-K\nabla U) \cdot n \, dS$$

Thus, the quantity of heat entering S per unit time (using divergence theorem is)

$$\frac{dQ}{dt} = \iint_S (K\nabla U) \cdot n \, ds = \iiint_V \nabla \cdot (K\nabla U) dV \tag{1.114}$$

In suitable units the amount of heat Q inside the volume V of the body under consideration is given by

$$Q = \iiint_V (Uc\rho) dV$$

Differentiating w.r.t t and equating the result to Eq. (1.14), we obtain

$$\frac{dQ}{dt} = \iiint_V \left(c\rho \frac{\partial U}{\partial t} \right) dV$$

$$= \iiint_V \nabla \cdot (K\nabla U) dV \quad \text{or} \quad \iiint_V \left\{ c\rho \frac{\partial U}{\partial t} - \nabla \cdot (K\nabla U) \right\} dV = 0$$

and since V is arbitrary, the integrand assumed continuous must be idenfically zero, so that

$$c\rho \frac{\partial U}{\partial t} = \nabla \cdot (K \nabla U)$$

or if $c\rho$ and K are constants

$$\frac{\partial U}{\partial t} = \frac{K}{c\rho} \nabla \cdot (\nabla U) = h^2 \nabla^2 U \qquad (1.115)$$

where

$$h^2 = \frac{K}{c\rho}$$

The quantity h^2 is called diffusivity; and the Eq. (1.115) is sometimes called the heat-flow or diffusion equation.

For steady state heat flow, i.e. $\frac{\partial U}{\partial t} = 0$ or U is independent of time, the equation reduces to Laplace's equation

$$\nabla^2 U = 0$$

1.17 THE GRAVITATIONAL POTENTIAL

Let m_1 and m_2 be point masses situated at the point P and Q having position vectors r_1 and r_2 respectively (Fig. 1.38). Then

$$PQ = r = r_2 - r_1$$

The force of attraction between them, according to Newton's law of gravitation in magnitude is given by

$$F = G \frac{m_1 m_2}{r^2}$$

where G is a constant depending upon the units.

If we choose the unit of mass as that of a particle which is placed at a unit distance from one of equal mass, attracts it with unit force, then, the value of G becomes 1 and

$$F = \frac{m_1 m_2}{r^2}$$

or in vector notations

$$F = -\frac{m_1 m_2}{r^3} r$$

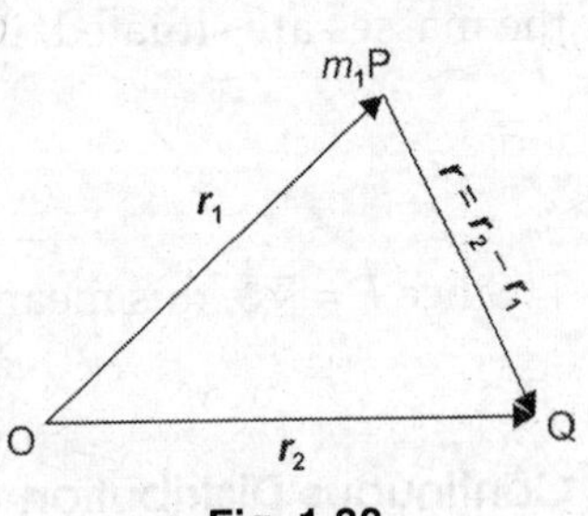

The force per unit mass at Q due to the attracting particle at P can be expressed as

$$F = -\frac{mr}{r^3} = \nabla \left(\frac{m}{r} \right)$$

Fig. 1.38

Here F is called the intensity of force or the gravitational field of force at point Q and we note that it may be expressed as the gradient of the scalar m/r. It follows, therefore, that F satisfies the equation

$$\nabla \times F = \nabla \times \nabla \left(\frac{m}{r} \right) = 0$$

and is therefore, a conservative field of force.

Let us suppose that the particle m is stationary at P and another particle of unit mass moves under the attraction of the former from infinity up to Q along any path. The work done by the force of attraction during an infinitesimal displacement dr of unit mass is $F \cdot dr$. The total work done by the force while the unit particle moves from infinity up to Q is

$$\int_{\infty}^{r} F \cdot dr = \int_{\infty}^{r} \nabla\left(\frac{m}{r}\right) \cdot dr = \frac{m}{r}$$

This is independent of path by which the particle moves up to Q and is called the potential at Q due to the particle of mass m at P. Let us denote it by ϕ. We than have

$$\phi = \frac{m}{r}$$

While the intensity of force at Q due to it is

$$F = \nabla\left(\frac{m}{r}\right) = \nabla\phi$$

That is the intensity at any point is equal to the gradient of the potential.

If we now suppose that there are n particles of masses $m_1, m_2, m_3,...,m_n$, relative to which Q has position vectors $r_1, r_2, r_3,...,r_n$, respectively, then the force of attraction per unit mass at Q due to the system is the vector sum of the intensities due to each, that is

$$F = \nabla\left(\frac{m_1}{r_1}\right) + \nabla\left(\frac{m_2}{r_2}\right) + \cdots + \nabla\left(\frac{m_n}{r_n}\right) = \nabla\sum_{s=1}^{n}\frac{m_s}{r_s}$$

Now by the same relation as before, the work done by the attracting force on a particle of unit mass while it moves from infinity up to Q is

$$\int_{\infty}^{Q} F \cdot dr = \sum_{s=1}^{n} \cdot \frac{m_s}{r_s} = \phi$$

Therefore, the potential at Q due to a system of particles is the sum of the potentials due to each. The potential ϕ is scalar function of position except at the point P, where the masses are situated, it satisfies

$$\nabla^2\phi = \nabla^2\sum_{s=1}^{n}\frac{m_s}{r_s} = \sum_{s=1}^{n}\nabla^2\left(\frac{m_s}{r_s}\right) = 0$$

Since $F = \nabla\phi$, this means that at points excluding matter, we must have

$$\nabla \cdot F = 0$$

Continuous Distribution of Matter

Let ρ be the distribution of mass (density), i.e. the mass per unit volume. We divide the body into infinite number of elements of volume dV, there arise two cases (Fig. 1.39).

1. *When Q is outside the body:* The potential $d\phi$ at point Q due to mass ρdV of element is

$$d\phi = \frac{\rho dV}{r}, \text{ where } r = |PQ|$$

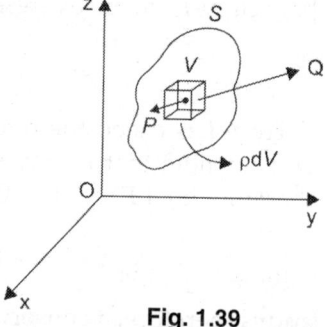

Fig. 1.39

For the entire volume, the potential at Q is

$$\phi = \iiint_V \frac{\rho dV}{r} \tag{1.115a}$$

Similarly, the intensity F is

$$F = \iiint_V \nabla_Q \left(\frac{\rho dV}{r}\right) = \nabla_Q \iiint_V \left(\frac{\rho dV}{r}\right) = \nabla\phi$$

where ∇_Q stands for differentiation w.r.t coordinates at Q.

2. *When Q is inside the body:* When Q is inside the body the integral Eq. (1.115a) become infinite (Fig. 1.40). So the definition of potential as given above is not applicable here. In order to definite potential ϕ at such points, we proceed as follows. Enclose the point Q by a small volume V_1 with boundary S_1 and remove the mass inside volume. We then determine the potential ϕ' at Q due to the remaining mass V_2. If the potential ϕ' approaches a limiting value when the volume of the cavity shrinks to zero the potential ϕ is defined as

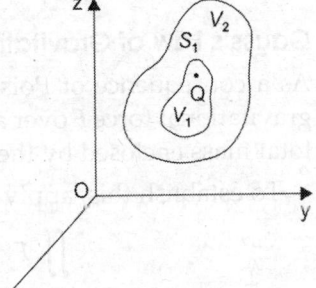

$$\phi = \lim_{\substack{S_1 \to 0 \\ V_1 \to 0}} \phi' = \lim_{\substack{S_1 \to 0 \\ V_1 \to 0}} \int_{V_1} \frac{\rho dV}{r}$$

Similarly, the intensity is defined as

$$F = \lim_{\substack{S_1 \to 0 \\ V_1 \to 0}} \cdot F' = \lim_{\substack{S_1 \to 0 \\ V_1 \to 0}} \int_{V_2} \rho \nabla_Q \left(\frac{1}{r}\right) dV$$

Fig. 1.40

Note that the limits, if exist, are independent of the shape of the cavity. For the sake of convenience, in computation, the cavity usually is taken to be spherical.

We have seen that the potential function satisfies Laplace's equation in the region outside the matter. We now consider the equation satisfied by potential in a region inside the matter.

Enclose the point Q by a small sphere S_1 of radius ϵ, every point of which is in the region (Fig. 1.41). Now the potential function is continuously differentiable at all points of the region V_2 enclosed between S and S_1.

The gravitational intensity at a point Q is the vector sum of the intensities, F_1 produced by the matter outside S_1 and F_2 produced by the matter inside S_1, i.e.

$$F = F_1 + F_2$$
$$\therefore \qquad \nabla \cdot F = \nabla \cdot F_1 + \nabla \cdot F_2$$

But $\nabla \cdot F_1 = 0$, since it is produced by the matter outside S_1, i.e.

$$\nabla \cdot F = \nabla \cdot F_2$$

The intensity at the surface of this sphere

$$F_2 = -\frac{m}{\epsilon^2} = \frac{\frac{4}{3}\pi\epsilon^3\rho}{\epsilon^2} = -\frac{4}{3}\pi\rho\epsilon$$

Fig. 1.41

in magnitude and in the direction of the inwardly drawn normal to the surface, where the mass of the sphere of radius ϵ is $\frac{4}{3}\pi\epsilon^3\rho$.

By Gauss's divergence theorem, we have

$$\lim_{\varepsilon \to 0} \iiint_V (\nabla \cdot F_2) dV = \lim_{\varepsilon \to 0} (\nabla \cdot F_2)\frac{4}{3}\pi\varepsilon^3 = \lim_{\varepsilon \to 0} \iint_S F_2 \cdot dS = \lim_{\varepsilon \to 0}\left(-\frac{4}{3}\pi\rho\varepsilon \cdot 4\pi\varepsilon^2\right)$$

Hence $\nabla \cdot F = \nabla \cdot F_2 = -4\pi\rho$

But $F = \nabla\phi,$

Hence the equation satisfied by the potential in a region containing matter is

$$\nabla^2\phi = -4\pi\rho = \nabla \cdot F \tag{1.116}$$

This relation is known as Poisson's equation. With respect to Eq. (1.115a), we see that we may take

$$\phi = \iiint \frac{\rho dv}{r}$$

as a solution of Poisson's equation.

Gauss's Law of Gravitation

As a consequence of Poisson's equation, we may prove that the surface integral of gravitational force F over a closed surface S drawn in the field is equal to -4π times the total mass enclosed by the surface.

To establish this, apply Gauss's divergence theorem to Eq. (1.116), we have

$$\iint_S F \cdot dS = \iiint_V \nabla \cdot F dV = -4\pi \iiint_V \rho dV$$

But $\iiint_V \rho dV$ is the total mass enclosed by the surface S. This theorem has many applications in potential theory and in the theory of electrostatics.

Example 1. *A vessel filled up to the height h with the fluid of density ρ has a horizontal orifice in the base of vessel. The vessel is constantly full of the fluid up to height h. The cross-section a of the orifice, is much smaller in comparison to the area of cross-section A of the free surface. Show that the velocity of flow of the fluid through orifice is proportional to the square root of the height of free surface above the orifice.*

Solution: Suppose V and v are the velocities of the free surface and the orifice. Bernoulli's equation for fluid of constant density and under steady motion is Eq. (1.112).

$$\frac{q^2}{2} + S + \frac{P}{\rho} = \beta \tag{1}$$

Measuring z downwards, we have

$$S = -gz$$

Eq. (1) becomes

$$\frac{q^2}{2} + \frac{P}{\rho} - gz = \beta \tag{2}$$

If P_0 is atmospheric pressure then for free surface ($z = 0$), and Eq. (2) becomes

$$\frac{P_0}{\rho} + \frac{1}{2}V^2 = \beta \tag{3}$$

and for orifice ($z = h$)

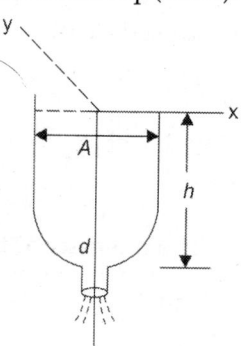

Fig. 1.42

$$\frac{P_0}{\rho} + \frac{1}{2}v^2 - gh = \beta \tag{4}$$

Thus,
$$v^2 - V^2 = 2gh \tag{5}$$

The condition for continuity requires
$$AV = av$$

or
$$V = \frac{a}{A}v$$

Substituting in Eq. (5), we get

$$v^2 = 2gh\left\{1 - \left(\frac{a}{A}\right)^2\right\}^{-1} = 2gh$$

neglecting square and higher powers of a/A.

or
$$v = \sqrt{2gh}$$

or
$$v \propto \sqrt{h}$$

i.e., the velocity of flow of the fluid through orifice is proportional to the square root of the height of free surface above the orifice. This is Torricelli's theorem.

Example 2. (*a*) *In a steady motion of an incompressible homogenous fluid under no force, the velocity at any point is*

$$ax\,i + ay\,j - 2az\,k.$$

Find the surface of equal pressure.

(*b*) *Show that for fluid in equilibrium* $F \cdot \text{curl } F = 0$.

Solution: (a) Euler's equation of motion [Eq. (1.107)] is

$$\frac{dq}{dt} = F - \frac{1}{\rho}\nabla P$$

But
$$\frac{dq}{dt} = \frac{\partial q}{\partial t} + (q \cdot \nabla)q$$

$\therefore$
$$\frac{\partial q}{\partial t} + (q \cdot \nabla)q = F - \frac{1}{\rho}\nabla P$$

For steady motion $\dfrac{\partial q}{\partial t} = 0$; for homogenous fluid $\rho = $ constant and for motion under no forces $F = 0$. Therefore, the equation of motion takes the form

$$q \cdot \nabla q = -\frac{1}{\rho}\nabla P$$

But
$$q \cdot \nabla q = \left(ax\frac{\partial}{\partial x} + ay\frac{\partial}{\partial y} - 2az\frac{\partial}{\partial z}\right)(ax\,i + ay\,j - 2az\,k)$$

$$= a^2x\,i + a^2y\,j + 4a^2z\,k$$

or
$$\nabla P = -\rho q \cdot \nabla q$$

$$= -a^2\rho\,(x\,i + y\,j + 4z\,k)$$

$$\therefore \qquad P = -a^2\rho\left(\frac{x^2}{2} + \frac{y^2}{2} + 2z^2\right) = \text{const.}$$

Therefore, the surfaces of equal pressure (at which pressures are the same) are given by

$$x^2 + y^2 + 4z^2 = \text{const.}$$

which represents spheroids.

(b) The equation for motion is Eq. (1.107)

$$\frac{dq}{dt} = F - \frac{1}{\rho}\nabla P$$

For equilibrium $\dfrac{dq}{dt} = 0$, so that

$$\rho F = \nabla P$$

Taking curl of both sides, we get

$$\nabla \times (\rho F) = \nabla \times (\nabla P) = 0$$

$$\therefore \qquad \rho\nabla \times F + \nabla\rho \times F = 0$$

Multiplying the above equation scalarly by F, we get

$$(\rho\nabla \times F)\cdot F + (\nabla\rho \times F)\cdot F = 0$$

The second term is zero

$$\therefore \qquad F\cdot(\nabla \times F) = 0$$

Example 3. *Steam is rushing from a boiler through a conical pipe, the diameters of the ends of which are D and d; if V and v be corresponding velocities of the steam, and if the motion is supposed to be that of divergence from the vertex of the cone, prove that*

$$\frac{v}{V} = \frac{D^2}{d^2}e^{(v^2-V^2)/2K}$$

where K is the pressure divided by density, and supposed to be constant.

Solution: Let $ABCD$ be a conical pipe whose ends are AB and CD. Let σ_1 and σ_2 be the densities of steam at ends AB and CD.

If at a point the pressure and density of the fluid particles be P and ρ respectively, then

$$K = \frac{P}{\rho}, \text{ i.e. } P = K\rho$$

Now according to equation of continuity, the amount of steam entering at one end must be equal to that emerging from the other end, i.e.

Flux across section AB = Flux across section CD.

or

$$\pi\left(\frac{D}{2}\right)^2 \sigma_1 V = \pi\left(\frac{d}{2}\right)^2 \sigma_2 v$$

$$\frac{\sigma_1}{\sigma_2} = \frac{d^2}{D^2}\frac{v}{V}$$

We have the Bernoulli's Eq. (1.112) (assuming that there exists no external force)

$$\int \frac{dP}{\rho} + \frac{1}{2}q^2 = \beta$$

or

$$K\int \frac{dP}{P} + \frac{1}{2}q^2 = \beta$$

or

$$K \log P + \frac{1}{2}q^2 = \beta \quad (1)$$

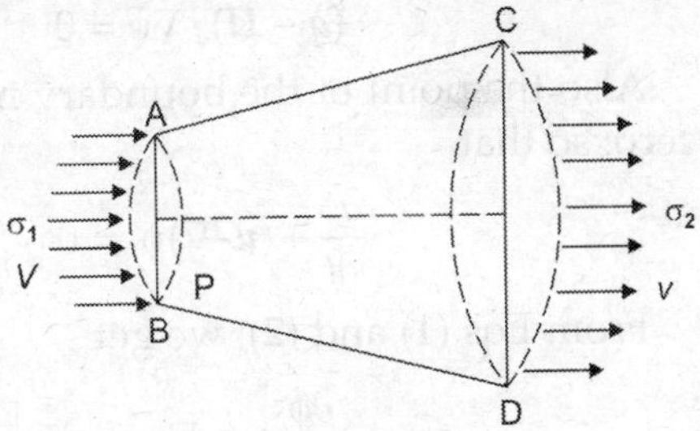

Fig. 1.43

When $\rho = \sigma_1,$ $P = K\sigma_1,$ $q = V$

or

$$K \log K\sigma_1 + \frac{1}{2}V^2 = \beta \tag{2}$$

Again when $\rho = \sigma_2,$ $P = K\sigma_2,$ $q = v$

or

$$K \log K\sigma_2 + \frac{1}{2}v^2 = \beta \tag{3}$$

Subtracting Eqs (3) and (4), we get

$$K \log \frac{K\sigma_1}{K\sigma_2} + \frac{1}{2}(V^2 - v^2) = 0$$

or

$$\frac{\sigma_1}{\sigma_2} = e^{(v^2 - V^2)/2K}$$

But

$$\frac{\sigma_1}{\sigma_2} = \frac{d^2}{D^2}\frac{v}{V}$$

$$\therefore \quad \frac{v}{V} = \frac{D^2}{d^2} e^{(v^2 - V^2)/2K}.$$

Example 4. *(i) If a moving fluid is bounded by a fixed surface $\phi(r) = 0$, then boundary condition is $q \cdot \nabla \phi = 0$ and if the boundary surface is itself moving and given by the equation $\phi(r, t) = 0$, then the boundary condition is*

$$\frac{\partial \phi}{\partial t} + q \cdot \nabla \phi = 0.$$

(ii) Show that the kinetic energy of a body of homogenous liquid with closed surface S moving irrotationally is

$$\frac{1}{2}\rho \int_S \phi \nabla \phi \cdot n \, dS$$

ϕ being the velocity potential.

Solution: (i) When the moving fluid is bounded by a fixed surface the normal component of velocity must be zero. The normal at any point to the surface $\phi(r) = 0$ is parallel to the vector $\nabla \phi$. Therefore, the required condition is $q \cdot \nabla \phi = 0$.

When the moving fluid bounded by moving surface, the normal component of the velocity of the particle relative to the boundary is zero. If U be the velocity of a point of the boundary, then the velocity of fluid particle relative to the boundary is $(q - U)$.

Therefore, the required condition is

$$(q - U) \cdot \nabla\phi = 0 \tag{1}$$

Also the point of the boundary moves so that the individual rate of change of ϕ is zero; so that

$$\frac{\partial\phi}{\partial t} + (u \cdot \nabla)\phi = 0 \tag{2}$$

From Eqs (1) and (2), we get

$$\frac{\partial\phi}{\partial t} + q \cdot \nabla\phi = 0$$

which is the required condition.

(ii) By divergence theorem, we have

$$\frac{1}{2}\rho\int_S \phi\nabla\phi \cdot n\, dS = \frac{1}{2}\rho\int_V \nabla(\phi\nabla\phi)dV$$

$$= \frac{1}{2}\rho\int_V \{\nabla\phi \cdot \nabla\phi + \phi(\nabla \cdot \nabla\phi)\}dV$$

$$= \frac{1}{2}\rho\int_V (\nabla\phi)^2 dV + \frac{1}{2}\rho\int_V \phi\nabla^2\phi\, dV = \frac{1}{2}\int_V \rho q^2 dV$$

$$= \text{kinetic energy}$$

Since $\nabla^2\phi = 0$ and $q = -\nabla\phi$.

Example 5. *A stream in a horizontal pipe after passing a contraction in the pipe at which its sectional area is A, is delivered at the atmospheric pressure at a place where the sectional area is B. Show that if a side tube is connected with the pipe at the former place, water will be sucked up through it into the pipe from the reservoir at a depth* $\dfrac{S^2}{2g}\left(\dfrac{1}{A^2} - \dfrac{1}{B^2}\right)$ *below the pipe, S being the delivery per second.*

Solution: Let P, v and P', v' be the pressures and velocities respectively at A and B (Fig. 1.44). Then, Bernoulli's theorem

$$\frac{P}{\rho} + \frac{1}{2}q^2 = \beta$$

gives

$$\frac{P}{\rho} + \frac{1}{2}v^2 = \frac{P'}{\rho} + \frac{1}{2}v' \tag{1}$$

By equation of continuity, flux across section A and B is equal

i.e. $$Av = Bv' = S \text{ (delivery per second)}$$

$$v = \frac{S}{A}, \quad v' = \frac{S}{B}$$

Substituting in Eq. (1), we have

$$\frac{P}{\rho} + \frac{1}{2}\frac{S^2}{A^2} = \frac{P'}{\rho} + \frac{1}{2}\frac{S^2}{B^2}$$

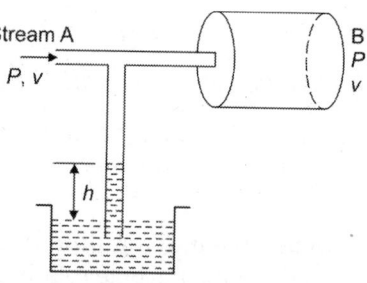

Fig. 1.44

or
$$\frac{S^2}{2}\left(\frac{1}{A^2} - \frac{1}{B^2}\right) = \frac{1}{\rho}(P' - P)$$

Let h be the depth of the reservoir below the pipe. Then pressure difference
$$P' - P = \rho gh$$

$\therefore$ We have
$$\frac{S^2}{2}\left(\frac{1}{A^2} - \frac{1}{B^2}\right) = \frac{\rho gh}{\rho} = gh$$

or
$$h = \frac{S^2}{2g}\left(\frac{1}{A^2} - \frac{1}{B^2}\right)$$

Example 6. *If the components of velocity of a fluid particle are $u = -\dfrac{c^2 y}{r^2}, v = \dfrac{c^2 x}{r^2}, w = 0$, where*

r is the distance of fluid particle from the z-axis which is vertical, find the lines of flow and velocity potential. Also determine the equation of the surface.

Solution: Here
$$u = -\frac{c^2 y}{r^2}, v = \frac{c^2 x}{r^2}, w = 0$$

These values clearly satisfy the equation of continuity
$$\frac{\partial u}{\partial x} + \frac{\partial v}{\partial y} + \frac{\partial w}{\partial z} = 0$$

and therefore the motion represented by these components is possible and therefore, lines of flow are given by
$$\frac{dx}{u} = \frac{dy}{v} = \frac{dz}{w}$$

or
$$\frac{dx}{-c^2 y} r^2 = \frac{dy}{c^2 x} r^2 = \frac{dz}{0}$$

or
$$\frac{dx}{-y} = \frac{dy}{x} = \frac{dz}{0}$$

only first two fractions give
$$xdx + ydy = 0$$
and integration gives
$$\frac{x^2}{2} + \frac{y^2}{2} = \text{const}, \quad \text{i.e.} \quad x^2 + y^2 = \text{const}$$

also
$$dz = 0 \quad \text{or} \quad z = \text{const}$$

Thus, the lines of flow or the streamlines are given by
$$x^2 + y^2 = \text{const} \quad \text{and} \quad z = \text{const}$$

Now
$$\frac{\partial u}{\partial y} = \frac{\partial}{\partial y}\left\{\frac{-c^2 y}{x^2 + y^2}\right\} = -c^2 \frac{(x^2 + y^2) - 2y^2}{(x^2 + y^2)^2}$$

$$= \frac{c^2(y^2 - x^2)}{(x^2 + y^2)^2} = \frac{\partial v}{\partial x}$$

which is the condition for the existence of a surface which cut the streamlines orthogonally and for the expression $udx + vdy + wdz$ to be exact differential.

Thus, $udx + vdy + wdz = -\dfrac{c^2}{x^2 + y^2}(ydx - xdy)$

$$= c^2 \frac{xdy - ydx}{x^2 + y^2}$$

$$= c^2 \frac{1}{1 + \dfrac{y^2}{x^2}} \left\{ \frac{xdy - ydx}{x^2} \right\}$$

$$= c^2 d\left(\tan^{-1} \frac{y}{x} \right)$$

i.e. $-d\phi = d\left\{ c^2 \tan^{-1} \dfrac{y}{x} \right\}$

Now if ϕ is the velocity potential then

$$-d\phi = udx + vdy + wdz$$

Integrating, we get $\phi = -c^2 \tan^{-1}\dfrac{y}{x}$, which is the required velocity potential.

To find the equation of surface, we use the equation

$$S + \frac{P}{\rho} + \frac{1}{2}q^2 = \beta \text{ (const), } S \text{ is the force potential.}$$

Here $q^2 = u^2 + v^2 + w^2 = \dfrac{c^4}{r^4}(x^2 + y^2) = \dfrac{c^4}{r^2}$

and gravity being the only external force $S = gz$.

Substituting the value of S, we have

$$gz + \frac{P}{\rho} + \frac{1}{2}\frac{c^4}{r^2} = \beta \tag{1}$$

taking pressure at the surface to be constant and assuming $z = a$ when $r \to \infty$, we have

$$\beta = \frac{P}{\rho} = ga \tag{2}$$

Eliminating β from Eqs (1) and (2), we have

$$\frac{1}{2}\frac{c^4}{r^2} + gz = ga$$

$$c^4 = 2g(a - z)r^2 = 2g(a - z)(x^2 + y^2)$$

which is the required surface.

SHORT ANSWER QUESTIONS

1. A vector A is drawn from point $(0, -1, 3)$ to $(5, 1, -2)$. Find a unit vector in the direction of A.

Ans. Displacement vector $r = r_2 - r_1 = (5i + j - 2k) - (-j + 3k)$

$$A = 5i + 2j - 5k$$

∴ Hence the unit vector is $A = \left[\dfrac{5i + 2j - 5k}{\sqrt{54}} \right]$

2. Which of the following are the polar vectors and which are axial vectors? Area, velocity, gravitational field intensity, electric field, $(E \times B)$ and torque $(r \times F)$.

Ans. **Axial vectors:** Torque $(r \times F)$

Polar vectors: Area, velocity, gravitational field intensity, E, $E \times B$ (vector along direction of electro-magnetic wave propagation).

3. Find the volume of the parallelopiped whose three conterminous edges are represented by $a = 2i - 5j$, $b = 3j$, $c = 5j + 6k$.

Ans. $V = a \cdot (b \times c) = \begin{vmatrix} a_x & a_y & a_z \\ b_x & b_y & b_z \\ c_x & c_y & c_z \end{vmatrix} = \begin{vmatrix} 2 & -5 & 0 \\ 0 & 3 & 0 \\ 0 & 5 & 6 \end{vmatrix}$

$$= 2(18 - 0) + 5(-0) + 0(0 - 0) = 36 \text{ cubic unit.}$$

4. Three vectors are: $A = i - j - 2k$; $B = 3i + 5j + 6k$ and $C = -i + 4j + mk$. Find the value of m for which these vectors are coplanar.

Ans. Vectors A, B and C will be coplanar, if their scalar triple product vanishes. Thus,

$$A \cdot (B \times C) = 0$$

or $\qquad \begin{vmatrix} 1 & -1 & -2 \\ 3 & 5 & 6 \\ -1 & 4 & m \end{vmatrix} = 0$

or $\quad 1(5m - 24) + 1(3m - 6) - 2(12 + 5) = 0$

or $\quad 8m - 52 = 0$

or $\quad m = \dfrac{52}{8} = 6.5.$

5. In what directions from the point $(1, 3, 2)$ is the directional derivative $\phi = 2xz - y^2$ is maximum? Find the magnitude of this maximum.

Ans. Directional derivative of $\phi = 2xz - y^2$ is obtained by

$$\nabla\phi = \left(i\frac{\partial}{\partial x} + j\frac{\partial}{\partial y} + k\frac{\partial}{\partial z} \right)(2xz - y^2)$$

$$= i(2z) - j(2y) + k(2x)$$

$$= 2zi - 2yj + 2xk$$

At point (1, 2, 3)
$$(\nabla\phi)_{(1, 3, 2)} = 2 \times 2i - 2 \times 3j + 2 \times 1k$$
$$= 4i - 6j + 2k$$

Magnitude of $\qquad \nabla\phi = \sqrt{(4)^2 + (-6)^2 + (2)^2} = 56.$

6. For the position vector $r = ix + jy + kz$, show that $\nabla \cdot r = 3$.

Ans. $\nabla \cdot r = \left(i\dfrac{\partial}{\partial x} + j\dfrac{\partial}{\partial y} + k\dfrac{\partial}{\partial z} \right) \cdot (ix + jy + kz)$

$\qquad = \dfrac{\partial x}{\partial x} + \dfrac{\partial y}{\partial y} + \dfrac{\partial z}{\partial z} = 1 + 1 + 1 = 3$

7. For the position vector $r = ix + jy + kz$, show that $\nabla \cdot (r^n n) = (3 + n) r^n$.

Ans. $\nabla \cdot (r^n r) = \left(i\dfrac{\partial}{\partial x} + j\dfrac{\partial}{\partial y} + k\dfrac{\partial}{\partial z} \right) \cdot [r^n(ix + jy + kz)]$

$\qquad = \dfrac{\partial}{\partial x}(xr^n) + \dfrac{\partial}{\partial y}(yr^n) + \dfrac{\partial}{\partial z}(zr^n)$

$\qquad = r^n + xnr^{n-1}\dfrac{\partial r}{\partial x} + r^n + ynr^{n-1}\dfrac{\partial r}{\partial y} + r^n + znr^{n-1}\dfrac{\partial r}{\partial z}$

$\qquad = 3r^n + nr^{n-1}\left(x\dfrac{\partial r}{\partial x} + y\dfrac{\partial r}{\partial y} + z\dfrac{\partial r}{\partial z} \right)$

$\qquad = 3r^n + nr^n = (3+n)r^n.$

8. For a position vector $r = (ix + jy + kz)$, show that $\nabla \times r = 0$

Ans. $\nabla \times r = \left(i\dfrac{\partial}{\partial x} + j\dfrac{\partial}{\partial y} + k\dfrac{\partial}{\partial z} \right) \times (ix + jy + kz)$

$$= \begin{vmatrix} i & j & k \\ \dfrac{\partial}{\partial x} & \dfrac{\partial}{\partial y} & \dfrac{\partial}{\partial z} \\ x & y & z \end{vmatrix} = 0$$

9. Show that $F = (2xy + z^3)i + x^2j + 3xz^2 k$ is a conservative force field.

Ans. We have, the curl of a conservative force field is always zero. Thus,

$$\nabla \times F = \begin{vmatrix} i & j & k \\ \dfrac{\partial}{\partial x} & \dfrac{\partial}{\partial y} & \dfrac{\partial}{\partial z} \\ (2xy + z^3) & x^2 & 3xz^2 \end{vmatrix}$$

$$= i\left[\frac{\partial}{\partial y}(3xz^2) - \frac{\partial}{\partial z}(x^2)\right] - j\left[\frac{\partial}{\partial x}(3xz^2) - \frac{\partial}{\partial z}(2xy + z^3)\right]$$

$$+ k\left[\frac{\partial}{\partial x}(x^2) - \frac{\partial}{\partial y}(2xy + z^3)\right]$$

$$= i(0 - 0) - j(3z^2 - 3z^2) + k(2x - 2x) = 0$$

Clearly, $\nabla \times F = 0$. This means F is a conservative force field.

10. Show that, $\nabla \cdot (\nabla u \times \nabla v) = 0$

Ans. We have $\nabla \cdot (A \times B) = B \cdot (\nabla \times A) - A \cdot (\nabla \times B)$

Now, substituting $A = \nabla u$ and $B = \nabla v$, one obtains

$$\nabla \cdot (\nabla u \times \nabla v) = \nabla v \cdot \nabla \times (\nabla u) - \nabla u \cdot \nabla \times (\nabla v)$$

But $\qquad \nabla \times (\nabla u) = \nabla \times (\nabla v) = 0$

Hence $\qquad \nabla \cdot (\nabla u \times \nabla v) = 0.$

11. Evaluate $\iiint_V (x^2 + y^2 + z^2)\,dx\,dy\,dz$, where V is the sphere having centre at the origin.

Ans. We have the equation of a sphere of given radius a as

$$x^2 + y^2 + z^2 = a^2$$

$$\therefore \quad \iiint_V (x^2 + y^2 + z^2)\,dx\,dy\,dz = \iiint_V a^2\,dx\,dy\,dz$$

$$= a^2 \iiint_V dx\,dy\,dz = a^2 \times \frac{4}{3}\pi a^3 = \frac{4}{3}\pi a^5$$

12. What do you understand by linear vector space?

Ans. The set $L = (u, v, w, ...)$ in which elements can be multiplied by a complex number α or added to one another to give the number of same set is said to form a linear vector space L.

We may note that the elements of the vector space are vectors. If u and v are in L, then αu and $u + v$, are also in L.

The operations of multiplication and addition must satisfy the conditions below:

   i. $u + v = v + u$ [Commutative law of addition]

  ii. $(u + v) + w = u + (v + w)$ [Associative law of addition]

 iii. $(\alpha + \beta) u = \alpha u + \beta u$ ⎫

 iv. $\alpha(u + v) = \alpha u + \alpha v$ ⎭ [Distributive law of multiplication]

  v. $\alpha\beta(u) = \alpha(\beta u)$ [Associative law of multiplication]

 vi. $1u = u$

The space L will contain a zero (null) vector such that $u + 0 = u$, for all u.

Clearly, linear vector space L forms an *abelian group* under the addition operations and its elements can be multiplied by complex numbers. The linear vector space is also called as *linear space* or *vector space*.

13. Show that vectors $(u + v)$, and $(u - 2v + w)$ are linearly independent provided (u, v, w) are linearly independent.

Ans. Since $(u + v)$, $(u - v)$ and $(u - 2v + w)$ are linearly independent vectors and hence

$$\alpha_1(u + v) + \alpha_2(u - v) + \alpha_3(u - 2v + w) = 0 \qquad\qquad (1)$$

This is possible only, if

$$\alpha_1 = \alpha_2 = \alpha_3 = 0$$

From Eq. (1) we have

$$(\alpha_1 + \alpha_2 + \alpha_3)\, u + (\alpha_1 - \alpha_2 - 2\alpha_3)\, v + \alpha_3 w = 0$$

As u, v, and w are linearly independent vectors and hence we have

$$\alpha_1 + \alpha_2 + \alpha_3 = 0,$$
$$\alpha_1 - \alpha_2 - 2\alpha_3 = 0,\ \alpha_3 = 0$$

Solving these equations, one obtains

$$\alpha_1 = \alpha_2 = \alpha_3 = 0$$

obviously, if (u, v, w) are linearly independent vectors, then $(u + v)$, $(u - v)$ and $(u - 2v + w)$ are also linearly independent vectors.

PROBLEMS

1. The point of application of a force $F = (5, 10, 15)$ is displaced from the point $(1, 0, 3)$ to the point $(4, -1, -6)$. Find the work done by the force.

2. Two sides of a triangle are formed by the vectors $A = 3i + 6j - 2k$ and $B = 4i - j + 3k$. Determine the angles of the triangle.

   [*Ans.* arc cos $\dfrac{7}{\sqrt{75}}$, arc cos $\dfrac{\sqrt{26}}{\sqrt{75}}$, $90°$ or $36°\,4'$, $53°\,36'$, $90°$]

3. Find the scalar product of two diagonals of a unit cube. What is the angle between them?

4. Let a be the position vector of a given point (x_1, y_1, z_1) and r the position vector of any point (x, y, z). Describe the locus of r, if (a) $|r - a| = 3$ (b) $(r - a) \cdot a = 0$ (c) $(r - a) \cdot r = 0$.

   [*Ans.* (a) Sphere, centre (x_1, y_1, z_1) and radius 3 (b) Plane perpendicular to a and passing through its terminal point (c) Sphere with centre at $\left(\dfrac{x_1}{2}, \dfrac{y_1}{2}, \dfrac{z_1}{2}\right)$ and radius $\dfrac{1}{2}\sqrt{x_1^2 + y_1^2 + z_1^2}$ or a sphere with a as diameter]

5. Prove that (a) $(A \times B) \cdot (B \times C) \times (C \cdot A) = 2\,(A \cdot B \times C)^2$
   (b) $(A \times B) \cdot (C \times A) = (A \cdot C)\,(B \cdot A) - (B \cdot C)\,(A \cdot A)$

6. Prove that $(i \cdot m \times n)\,(a \times b) = \begin{vmatrix} i \cdot a & i \cdot b & i \\ m \cdot a & m \cdot b & m \\ n \cdot a & n \cdot b & n \end{vmatrix}$

7. If A, B, C and D are four vectors, prove that
   (a) $A \times \{B \times (C \times D)\} = (B \cdot D)\,(A \times C) - (B \cdot C)\,(A \times D)$ and hence expand $A \times [B \times \{C \times (D \times E)\}]$
   (b) $(B \times C) \cdot (A \times D) + (C \times A) \cdot (B \times D) + (A \times B) \cdot (C \times D) = 0$
   (c) $(A + B) \cdot (B + C) \times (C + A) = 2\,(A \cdot B \times C)$

8. Show that if $A \neq 0$ and both of the conditions (a) $A \cdot B = A \cdot C$ and (b) $A \times B = A \times C$ hold simultaneously then $B = C$, but if only one of these conditions holds, $B \neq C$ necessarily.

9. If $A = x_1 a + y_1 b + z_1 c$, $B = x_2 a + y_2 b + z_2 c$ and $C = x_3 a + y_3 b + z_3 c$, prove that

$$A \times B \times C = \begin{bmatrix} x_1 & y_1 & z_1 \\ x_2 & y_2 & z_2 \\ x_3 & y_3 & z_3 \end{bmatrix} (a \cdot b \times c)$$

10. Prove that the necessary and sufficient condition for
$(A \times (B \times C) = (A \times B) \times C$ is $(A \times C) \times B = 0$. Discuss the cases where $A \cdot B = 0$ or $B \times C = 0$

11. The angular velocity of a rotating rigid body about an axis of rotation is given by $\omega = 4i + j - 2k$. Find the linear velocity of a point P on the body whose position vector relative to a point on the axis of rotation is $2i - 3j + k$
     [*Ans.* $-5i - 8j - 14k$]

12. Given $R = \sin ti + \cos tj + tk$, find

    (a) $\dfrac{dR}{dt}$　　(b) $\dfrac{d^2 R}{dt^2}$　　(c) $\left|\dfrac{dR}{dt}\right|$　　(d) $\left|\dfrac{d^2 R}{dt^2}\right|$

    [*Ans.* (a) $\cos ti - \sin tj + k$ (b) $-\sin ti - \cos tj$ (c) $\sqrt{2}$ (d) 1]

13. A particle moves along the curve $x = 2t^2$, $y = t^2 - 4t$, $z = 3t - 5$, where t is the time. Find the components of its velocity and acceleration at time $t = 1$ in the direction

    $i - 3j + 2k$.
    $\left[Ans. \ \dfrac{8\sqrt{14}}{7}, \dfrac{\sqrt{14}}{7} \right]$

14. (a) Find the unit tangent vector to any point on the curve
    $$x = t^2 + 1, y = 4t - 3, z = 2t^2 - 6t$$
    (b) Determine the unit tangent at the point, where $t = 2$.

    $\left[Ans. \ (a) \ \dfrac{2ti + 4j + (4t - 6)k}{\sqrt{(2t)^2 + (4)^2 + (4t-6)^2}} \ (b) \ \dfrac{2}{3}i + \dfrac{2}{3}j + \dfrac{1}{3}k \right]$

15. If $A = (2x^2 y - x^4) i + (e^{xy} - y \sin x) j + (x^2 \cos y) k$

    Find $\dfrac{\partial A}{\partial x}, \dfrac{\partial A}{\partial y}, \dfrac{\partial^2 A}{\partial x^2}, \dfrac{\partial^2 A}{\partial y^2}, \dfrac{\partial^2 A}{\partial x \partial y}, \dfrac{\partial^2 A}{\partial y \partial x}$.

16. Find the angle between the surfaces $x^2 + y^2 + z^2 = 9$ and $z = x^2 + y^2 - 3$ at the point $(2, -1, 2)$.
     [*Ans.* $54° \, 25'$]

17. Find the directional derivative of $\phi = x^2 yz + 4 xz^2$ at $(1, -2, -1)$ in the direction
    $(2i - j - 2k)$.
    $\left[Ans. \ \dfrac{37}{3} \right]$

18. If $A = 2yzi - x^2 yj + xz^2 k$, $B = x^2 i + yzj - xyk$ and $\phi = 2x^2 z^3$, find
    (a) $(A \cdot \nabla)\phi$, (b) $A \cdot \nabla\phi$, (c) $(B \cdot \nabla)A$, (d) $(A \times \nabla)\phi$, (e) $A \times \nabla\phi$.

    [*Ans.* (a) $8xy^2 z^4 - 2x^4 yz^3 + 6x^3 yz^4$
    (b) $8xy^2 z^4 - 2x^4 yz^3 + 6x^3 yz^4$
    (c) $(2yz^2 - 2xy^2) i - (2x^3 y + x^2 yz) j + (x^2 z^2 - 2x^2 yz)k$
    (d) $-(6yx^4 y^2 z^2 + 2x^3 z^5) i + (4x^2 yz^5 - 12x^2 y^2 z^3)j + (4x^2 yz^4 + 4x^3 y^2 z^3)k$
    (e) $-(6x^4 y^2 z^2 + 2x^3 z^5) i + (4x^2 yz^5 - 12x^2 y^2 z^3)j$
    $+ (4x^2 yz^4 + 4x^3 y^2 z^3) k$, i.e. $(A \times \nabla) \phi = A \times \nabla\phi$]

19. Two rectangular x, y, z and x', y', z' coordinate systems having the same origin are rotated with respect to each other. Derive the transformation equations between the coordinates of a point in the two systems.

20. If $A = 2x^2 i - 3yzj + xz^2 k$ and $\phi = 2z - x^3 y$, find $A \cdot \nabla\phi$ and $A \times \nabla\phi$ at the point $(1, -1, 1)$.
    [*Ans.* $5, 7i - j - 11k$]

21. If $\nabla\phi = (y^2 - 2xyz^3) i + (3 + 2xy - x^2 z^3) j + (6z^3 - 3x^2 yz^2) k$, find ϕ.

$$\left[Ans.\ \phi = xy^2 - x^2 yz^3 + 3y + \left(\frac{3}{2}\right) z^4 = const \right]$$

22. Prove $\nabla\left(\dfrac{F}{G}\right) = \dfrac{G\nabla F - F\nabla G}{G^2}$, if $G \neq 0$

23. Find the unit outward drawn normal to the surface.

    $(x - 1)^2 + y^2 + (z + 2)^2 = 9$ at the point $(3, 1, -4)$.　　$\left[Ans.\ \left(\dfrac{2i - j - 2k}{3}\right) \right]$

24. Prove the vector $A = 3y^4 z^2 i + 4x^3 z^2 j - 3x^2 y^2 k$ is solenoidal.

25. Show that the divergence of an inverse square force is zero. What is the divergence of a gradient?

26. Show that $A = (2x^2 + 8xy^2 z) i + (3x^2 y - 3xy) j - (4y^2 z^2 + 2x^3 z)k$ is not solenoidal but $B = xyz^2 A$ is solenoidal.

27. If $A = 2xz^2 i - yz^3 j + 3xz^3 k$ and $\phi = x^2 yz$. Find

    (a) $\nabla \times A$ (b) curl (ϕA) (c) $\nabla \times (\nabla \times A)$ (d) $\nabla[A \cdot \text{curl } A]$ (e) curl grad (ϕA) at the point $(1, 1, 1)$.　　[*Ans.* (a) $i + j$ (b) $5i - 3j - 4k$ (c) $5i + 3k$ (d) $-2i + j + 8k$ (e) 0]

28. If $F = x^2 yz$, $G = xy - 3z^2$. Find

    (a) $\nabla[(\nabla F) \cdot (\nabla G)]$　　　(b) $\nabla \times [(\nabla F) \times (\nabla G)]$　　　(c) $\nabla \times [(\nabla F) \times (\nabla G)]$
    [*Ans.* (a) $(2y^2 z + 3x^2 z - 12xyz) i + (4xyz - 6x^2 z) j + (2xy^2 + x^3 - 6x^2 y)k$
    (b) 0
    (c) $(x^2 z - 24xyz) i - (12x^2 z + 2xyz) j + (2xy^2 + 12yz^2 + x^3)k]$

29. If $A = yz^2 i - 3xz^2 j + 2xyzk$, $B = 3xi + 4zj - xyk$ and $\phi = xyz$, find
    (a) $A \times (\nabla\phi)$,　　　　　(b) $(A \times \nabla)\phi$
    (c) $(\nabla \times A) \times B$　　　(d) $B \cdot \nabla A$

    [*Ans.* (a) $-5x^2 yz^2 i + xy^2 z^2 j + 4xyz^3 k$
    (b) $-5x^2 yz^2 i + xy^2 z^2 j + 4xyz^3 k$
    (c) $16z^3 i + (8x^2 yz - 12xz^2) j + 32\ xz^2 k$
    (d) $24x^2 z + 4xyz^2].$

30. Prove $\nabla \cdot (A \times B) = B \cdot (\nabla \times A) - A \cdot (\nabla \times B)$.

31. Prove $\nabla \times (A \times B) = (B \cdot \nabla) A - B (\nabla \cdot A) - (A \cdot \nabla) B + A (\nabla \times B)$

32. Prove $\nabla (A \cdot B) = (B \cdot \nabla) A + (A \cdot \nabla) B + B \times (\nabla \times A) + A \times (\nabla \times B)$.

33. If $F = 3xyi - y^2 j$, evaluate $\displaystyle\int_C F \cdot dr$, where C is the curve in xy plane, $y = 2x^2$ from $(0, 0)$ to $(1, 2)$.　　　　$\left[Ans.\ -\dfrac{7}{6} \right]$

34. Compute the integral $\int_C (xy\,dx - y\,dy - dz)$ over the following paths.

    (a) Straight line joining $(0, 0, 0)$ and $(1, 1, 1)$ $\left[Ans.\ \dfrac{5}{6} \right]$

    (b) Straight line joining $(0, 0, 1)$ and $(0, 1, 1)$ $\left[Ans.\ -\dfrac{1}{2} \right]$

    (c) Straight line joining $(0, 0, 0)$ and $(1, 2, 3)$ $\left[Ans.\ \dfrac{5}{3} \right]$

35. (a) If $F = \nabla\phi$, where ϕ is single-valued and has continuous partial derivatives, show that the work done in moving a particle from one point $P_1 = (x_1, y_1, z_1)$ in this field to another point $P_2 = (x_2, y_2, z_2)$ is independent of the path joining the two points.

    (b) Conversely, if $\int_C F \cdot dr$ is independent of the path C joining any two points, show that there exists function ϕ such that $F = \nabla\phi$.

36. (a) If F is conservative field, prove that curl $F = \nabla \times F = 0$ (i.e. F is irrotational).

    (b) Conversely if $\nabla \times F = 0$ (i.e. F is irrotational), prove that F is conservative.

37. Find the areal velocity of a particle which moves along the path $r = a \cos \omega t\, i + b \sin \omega t\, j$ where a, b, ω are constants and t is time. $\left[Ans.\ \dfrac{1}{2} ab\omega k \right]$

38. Evaluate $\iint_S A \cdot n\, dS$ over the entire surface of the region above xy plane bounded by cone $z^2 = x^2 + y^2$ and the plane $z = 4$, if $A = 4xzi + xyz^2 j + 3zk$.   [Ans. $320\,\pi$]

39. Verify theorem in the plane for $\iint_C [3x^2 - 8y^2)dx + (4y - 6xy)dy]$ where C is the boundary of the region defined by

    (a) $y = \sqrt{x}, y = x^2$ (b) $x = 0, y = 0, x + y = 1$ $\Big[Ans.\ \text{(a)Common value} = \dfrac{3}{2}$

                                                (b)Common value $= \dfrac{5}{2} \Big]$

40. Find the area bounded by the hypocyloid $x^{2/3} + y^{2/3} = a^{2/3}, a > 0$.

    [Hint: Parametric equations are $x = a \cos^3 \theta, y = a \sin^3 \theta$] $\left[Ans.\ \dfrac{3\pi a^2}{8} \right]$

41. Evaluate $\iint_S F \cdot n\, ds$ where $F = 2xy\, i + yz^2 j + xy\, k$ and S is

    (a) the surface of a parallelopiped bounded by $x = 0, y = 0, z = 0, x = 2; y = 1$ and $z = 3$.   [Ans. 30]

    (b) the surface of the region bounded by $x = 0, y = 0, y = 3, z = 0$ and $x + 2z = 6$. $\left[Ans.\ \dfrac{351}{2} \right]$

42. Verify the divergence theorem for $A = 2x^2yi - y^2j + 4xz^2k$ taken over the region in the first octant bounded by $y^2 + z^2 = 9$ and $x = 2$.   [*Ans.* 180]

43. Verify stoke's theorem for $A = (y - z + 2)i + (yz + 4)j - xzk$; where S is the surface of the cube at $x = 0$, $y = 0$, $x = 2$, $y = 2$, $z = 2$ above xy plane.

                 [*Ans.* Common value = −4]

44. Evaluate $\iint_S (\nabla \times A) \cdot n\,dS$, where $A = (x^2 + y - 4)i + 3xyj + (2xz + z^2)k$ and S is the surface of

    (a) the hemisphere $x^2 + y^2 + z^2 = 16$ above xy plane.

    (b) the paraboloid $z = 4 - (x^2 + y^2)$ above xy plane.   [*Ans.* (a) − 16π (b) -4π]

45. Let r be the position vector of any point relative to an origin O. Suppose ϕ has continuous derivatives of order two, at least, and let S be a closed surface bounding a volume V. Denote ϕ at O by ϕ_0, show that

$$\iint_S \left[\frac{1}{r}\nabla\phi - \phi\nabla\left(\frac{1}{r}\right) \right] \cdot dS = \iiint_S \frac{\nabla^2\phi}{r}\,dV + \alpha$$

where $\alpha = 0$ or $4\pi\phi_0$ according as O is outside or inside S.

46. Prove the following:

    (a) $\nabla \times (a \times r) = 2a$, where a is a constnat vector

    (b) $\nabla \times (e \times r) = 0$, $\nabla \times [(e \times r) \times e] = 0$, where e is a unit vector

    (c) $\nabla \times [(r \times a) \times b] = (b \times a)$

    (d) Curl curl curl curl $F = \nabla^2\nabla^2 = \nabla^4F - F$ being solenoidal vector.

47. Prove the following:

    (a) $\nabla\phi$ is a vector perpendicular to the surface $\phi(x, y, z) = C$, where C is a constant.

    (b) if $\rho F = \nabla p$, where ρ, P F are point functions. $F \times$ curl $F = 0$.

    (c) $\quad a \cdot \nabla\left[b \cdot \nabla\left(\frac{1}{r}\right) \right] = \dfrac{3(a \cdot r)(b \cdot r)}{r^5} - \dfrac{a \cdot b}{r^3}$

    where a and b are constant vectors.

48. Prove the following:

    (a) $a \cdot \{\nabla(V \cdot a) - \nabla \times (V \cdot a)\} = a^2$ div V, where a is a constant vector.

    (b) $a \cdot \{\nabla(V \cdot a) - \nabla \times (V \cdot a)\} = 3$ div V, where a is a constant vector having each component equal to unity.

    (c) curl grad $r^m = 0$.

49. The vector r satisfies the vector equation

$$m\left(\frac{d^2r}{dt^2} \right) = eE + \left(\frac{e}{c} \right)\left(\frac{dr}{dt} \right) \times H$$

where $E = (0, E, 0)$, $H = (0, 0, H)$ and e, m c, E, H are constants. Write the equation in the component form and show by solving the equation that

$$x = \left(\frac{eEt}{H} \right) - \left(\frac{mc^2E}{eH^2} \right)\sin\left(\frac{eHt}{mc} \right)$$

$$y = \left(\frac{mc^2 E}{eH^2} \right) \left\{ 1 - \cos\left(\frac{eHt}{mc} \right) \right\}, z = 0]$$

is the solution that satisfies the condition $r = 0$, $\frac{dr}{dt} = 0$ at $t = 0$.

50. Prove $\dot{e}_r = \dot{\theta} e_\theta = \sin\theta \, \dot{\phi} e_\phi$, $\dot{e}_\theta = -\dot{\theta} e_r + \cos\theta \, \dot{\phi} e_\phi$

   $\dot{e}_\phi = -\sin\theta \dot{\phi} e_r - \cos\phi \dot{e}_\theta$

51. Represent the vector $A = 2yi - zj + 3xk$ in spherical coordinates and determine A_r, A_θ and A_ϕ.

52. Prove that the necessary and sufficient condition for curvilinear coordinate system to be orthogonal is $g_{pq} = 0$ for $p \neq q$.

53. Prove that in any orthogonal curvilinear coordinate system, div curl $A = 0$ and curl grad $\phi = 0$.

54. Let (x, y, z) and (u_1, u_2, u_3) be the rectangular and curvilinear coordinates of a point respectively.
   (a) if $x = 3u_1 + u_2 - u_3$, $y = u_1 + 2u_2 + 2u_3$, $z = 2u_1 - u_2 - u_3$ is the system with u_1, u_2, u_3 orthogonal?
   (b) find ds^2 and g for the system.

55. Prove that the motion of an incompressible perfect fluid is possible when the velocity q is given by
   (a) $q = x(y - z)i + y(z - x)j + z(x - y)k$
   (b) $q = (bz - cy)i + (cx - az)j + (ay - bx)k$.
   Examine if the motion in either case is irrotational or not. Show also, that in case (b) the fluid moves like a rigid body.

56. If the velocity of an incompressible fluid at the point (x, y, z) is given by
   $$\frac{3xy}{r^5}, \frac{3yz}{r^5}, \frac{3z^2 - r^2}{r^5}$$
   prove that the fluid motion is possible and that the velocity potential is $\frac{z}{r^3}$. Find also the streamlines.

57. Air obeying Boyle's law is in motion in a uniform tube of small section; prove that if ρ be the density and v the velocity at a distance x from a fixed point at time t, then
   $$\frac{\partial^2 \rho}{\partial t^2} = \left(\frac{\partial^2}{\partial x^2} \right) \{(v^2 + k)\rho\}$$

58. A quantity of liquid occupies a length $2l$ of a straight tube of uniform small bore under the action of a force to a point in the tube varying as the distance from the point. Determine the motion and the pressure.

59. Show that $\frac{x^2}{a^2} \tan^2 t + \frac{y^2}{b^2} \cot^2 t = 1$ is a possible form for the bounding surface of a liquid.

60. (a) For an orthogonal set of curvilinear coordinates u_1, u_2, u_3, prove that $|\nabla u_i| = |h_i|^{-1}$, where h_i are scalar factors and $i = 1, 2, 3$.
   (b) Prove that in any orthogonal curvilinear coordinate system curl grad $\phi = 0$

61. Obtain an expression for curl A in general orthogonal curvilinear coordinates and hence find all the components of curl A in cylindrical coordinate system.

62. Verify Stoke's theorem for vector $A = 3yi - xzj + yz^2k$, where S is the surface of the paraboloid $2z = x^2 + y^2$ bounded by $z = 2$ and C is the boundary.

63. Apply Stoke's theorem to find the value of $\int_C ydx + zdy + xdz$, where C is the curve

    of intersection $x^2 + y^2 + z^2$ and $x + z = a$ **[Ans. $-Na^2/\sqrt{2}$]**

64. Show that (a) div $\dfrac{a \times r}{r^2} = 0$ (b) $\dfrac{a \times r}{r^2} = -\dfrac{a}{r^2} + \dfrac{3r(a \cdot r)}{r^5}$

65. Derive an expression for $\nabla\phi$ in curvilinear coordinates. Express $\nabla\phi$ in spherical coordinates.

66. Express the position and velocity vectors of a particle in cylindrical coordinates.

67. Prove that $\iiint_V (\phi\nabla^2\psi - \psi\nabla^2\phi)dV = \iint_S (\phi\nabla\psi - \phi\nabla\phi) \cdot dS$

MULTIPLE CHOICE QUESTIONS

1. When two vectors A and B of magnitude a and b are added, the magnitude of the resultant vector is always
   (a) equal to $(a + b)$ (b) less than $(a + b)$
   (c) greater than $(a + b)$ (d) never greater than $(a + b)$

2. If the magnitude of vectors A, B and C are 12, 5 and 13 units and $A + B = C$, the angle between A and B is
   (a) 0   (b) π
   (c) $\pi/2$ (d) $\pi/4$

3. If two vectors A and B (of same magnitude P) have the resultant having magnitude (P), the angle between the vectors is
   (a) 0   (b) $\pi/4$
   (c) $\pi/3$ (d) $2\pi/3$

4. If $A = 5i + 7j - 3k$ and $B = 2i + 2j - ck$ are perpendicular to each other, the value of C is
   (a) -2 (b) 8
   (c) -7 (d) -8

5. A vector perpendicular to $(4i - 3j)$ is
   (a) $4i - 3j$ (b) $7k$
   (c) $6i$ (d) $3i - 4j$

6. Position vector of a particle is $r = (a \cos \omega t) i + (a \sin \omega t)j$. The velocity of the particle is
   (a) parallel to position vector   (b) perpendicular to position vector
   (c) directed towards origin   (d) directed away from origin

7. Three vectors A, B, C satisfy the relation $A \cdot B = 0$ and $A \cdot C = 0$, the vector A is parallel to
   (a) B
   (b) C
   (c) $B \cdot C$
   (d) $B \times C$

8. The value of triple product $A \times (B \times C)$ is
   (a) $B(C \cdot A) - C(A \cdot B)$
   (b) $B(C \cdot A) + C(A \cdot B)$
   (c) $A \times B - A \times C$
   (d) $A \times B + A \times C$

9. If $A + B + C = 0$, then correct relation is
   (a) $A \cdot B = B \cdot C = C \cdot A$
   (b) $A \times B = B \times C = C \times A$
   (c) $A \times B \times C \neq 0$
   (d) $A \times B = B \times C = C \times A = 0$

10. If vector $A + B = C$, if θ is the angle between A and B then $\cos \theta$ has the value
    (a) $\dfrac{C^2 - A^2 - B^2}{2AB}$
    (b) $\dfrac{A^2 - B^2 - C^2}{2BC}$
    (c) $\dfrac{B^2 - C^2 - A^2}{2AC}$
    (d) $\dfrac{A^2 - B^2 - C^2}{2AB}$

11. If r is position vector with magnitude r then the value of ∇r^n is
    (a) $nr^{n-2}r$
    (b) $nr^{n-1}r$
    (c) nr^{n-1}
    (d) $nr^{n-2}r$

12. Two vectors A and B are such that $A + B = A - B$, then
    (a) $A + B = 0$
    (b) $A = 0$
    (c) $B = 0$
    (d) A and $B = 0$
    or, are perpendicular to each other.

13. Forces 5 N, 12 N and 13 N are in equilibrium, the angle between 5 N and 13 N forces is
    (a) $13°$
    (b) $90°$
    (c) $113°$
    (d) $45°$

14. The vector sum of N coplanar forces each of magnitude F, when each force is making an angle $2\pi/N$ with the preceding one is
    (a) NF
    (b) $\dfrac{NF}{2}$
    (c) $\dfrac{F}{2}$
    (d) zero

15. For vectors P, Q and R to be coplanar
    (a) $(P \times Q) \cdot R = 0$
    (b) $(P \times Q) \times R = 0$
    (c) $P \times Q \times R = 0$
    (d) $(P \cdot Q) \cdot R = 0$

16. $i \times (j \times k)$ is
    (a) $i + j + k$
    (b) $i - j + k$
    (c) zero vector
    (d) unit vector

17. A region is specified by the potential function $V = (5x + 4y - 6z)$ then the potential gradient is
    (a) $5i + 4j - 6k$
    (b) $5i + 4j + 6k$
    (c) $-5i - 4j + 6k$
    (d) $-5i - 4j - 6k$

18. For any two vectors A and B if $A \cdot B = |A \times B|$, the magnitude of $C = A + B$ is equal to

    (a) $\sqrt{A^2 + B^2}$

    (b) $A + B$

    (c) $\sqrt{A^2 + B^2 + \left(\dfrac{AB}{\sqrt{2}}\right)}$

    (d) $\sqrt{A^2 + B^2 + \sqrt{2}\,AB}$

19. Vector $0.4i + 0.8j + Ck$ represents a unit vector, when C is

    (a) -0.2

    (b) $\sqrt{0.2}$

    (c) $\sqrt{0.8}$

    (d) 0

20. A boat is moving with velocity $3i + 4j$ w.r.t ground. The water in the river is moving with a velocity $-3i - 4j$ w.r.t ground. The relative velocity of the boat w.r.t water is

    (a) $8i$

    (b) $-6i - 8j$

    (c) $6i + 8j$

    (d) $5\sqrt{2}$

21. The equation of continuity in electromagnetism is

    (a) $\operatorname{div} j + \dfrac{\partial \rho}{\partial t} = 0$

    (b) $\operatorname{div} e - \dfrac{\partial j}{\partial t} = 0$

    (c) $\operatorname{curl} j + \dfrac{\partial \rho}{\partial t} = 0$

    (d) $\operatorname{div} j - \dfrac{\partial \rho}{\partial t} = 0$

22. If S is a closed surface enclosing volume V, n is the unit normal vector to the surface and r is the position vector, then the value of the following integral $\int n \cdot dS$ is

    (a) V

    (b) $2V$

    (c) 0

    (d) $3V$ **[GATE]**

    [*Hint:* using Gauss's theorem $\displaystyle\iint_S n \cdot dS = \iiint_V (\nabla \cdot n)\, dV = 3\iiint_V dV = 3V$

    $$\left(\because \nabla \cdot n = \frac{dx}{dx} + \frac{dy}{dy} + \frac{dz}{dz} = 3 \right)]$$

23. Consider the set of vectors $\dfrac{1}{\sqrt{2}}(1,1,0)$, $\dfrac{1}{\sqrt{2}}(0,1,1)$ and $\dfrac{1}{\sqrt{2}}(1,0,1)$.

    (a) three vectors are orthonormal.

    (b) three vectors are linearly independent.

    (c) three vectors cannot form a basis in a three dimensional real vector space.

    (d) $\dfrac{1}{\sqrt{2}}(1,1,0)$ can be written as a linear combination of $\dfrac{1}{\sqrt{2}}(0,1,1)$ and $\dfrac{1}{\sqrt{2}}(1,0,1)$.

    [GATE]

    [*Hint:* Option (c) and (d) can be simultaneously correct or simultaneously incorrect and hence one can drop these options. Now, each of the three given vectors is a normalized vector. However, their multiplication does not give zero. This means that they are not orthogonal. Therefore, the remaining option (b) is correct. We can also prove this in the following manner.

Let the given vectors are represented by A, B and C.

Now, if $aA + bB + cC = 0$ for $a, b, c \in R$. For A, B and C to be linearly independent we have $a = b = c = 0$

$\therefore$ $$\frac{a}{\sqrt{2}}(i+j) + \frac{b}{\sqrt{2}}(j+k) + \frac{c}{\sqrt{2}}(k+i) = 0$$

or $$\frac{i}{\sqrt{2}}(a+c) + \frac{j}{\sqrt{2}}(b+a) + \frac{k}{2}(b+c) = 0$$

This leads to $a + c = 0$, or $b + a = 0$ or $b + c = 0$, means $a = b = c = 0$. Hence option (b) is correct, i.e. three vectors are linearly independent].

24. If $A = xe_x + ye_y + ze_z$, then $\nabla^2 A$ equals

(a) 1

(b) 3

(c) 0

(d) –3   [GATE]

[**Hint:** $\nabla^2 A = \left(i\dfrac{\partial^2}{\partial x^2} + j\dfrac{\partial^2}{\partial y^2} + k\dfrac{\partial^2}{\partial z^2} \right) A = 0 + 0 + 0 = 0$].

25. Which of the following vectors is orthogonal to the vector $(ai + bj)$, where a and b $(a \neq b)$ are constants, and i and j are unit orthogonal vectors?   [GATE]

(a) $-bi + aj$

(b) $-ai + bj$

(c) $-ai - bj$

(d) $-bi - aj$

[**Hint:** Two vectors A and B are said to be orthogonal if $A \cdot B = 0$.

Here $(ai + bj) \cdot (-bi + aj) = 0$.

26. The unit vector normal to the surface $3x^2 + 4y + z$ at the point $(1, 1, 7)$ is

(a) $(6i + 4j + k)/\sqrt{53}$

(b) $(4i + 6j - k)/\sqrt{53}$

(c) $(6i + 4j - k)/\sqrt{53}$

(d) $(4i + 6j + k)/\sqrt{53}$ [GATE]

[**Hint:** Here $f(x) = 3x^2 + 4y + z$ and $\nabla f = 6xi + 4j - k$

$\therefore$ Normalized $\nabla f = \dfrac{6i + 4j - k}{\sqrt{53}}$].

27. Two vectors $p = i$, $q = (i + j)/\sqrt{2}$ one

(a) related by a rotation

(b) related by reflection through the xy plane

(c) related by an inversion

(d) not linearly independent   [GATE]

[**Hint:** $OP = OQ$, obviously, they are related by a rotation]

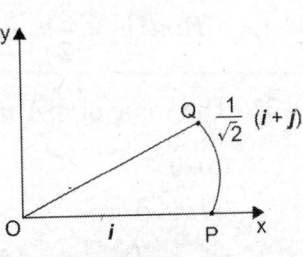

Fig. 1.45

28. The curl of a vector $A = zi + xj + yk$ is given by
    (a) $i + j + k$
    (b) $i - j + k$
    (c) $i + j - k$
    (d) $-i - j - k$ [GATE]

[*Hint:* $\nabla \times A = \begin{vmatrix} i & j & k \\ \frac{\partial}{\partial x} & \frac{\partial}{\partial y} & \frac{\partial}{\partial z} \\ z & x & y \end{vmatrix} = i(1 - 0) - j(0 - 1) + k(1 - 0) = i + j + k$]

29. The unit normal to the curve $x^3y^2 + xy = 17$ at the point $(2, 0)$ is
    (a) $(i + j)/\sqrt{2}$
    (b) $-i$
    (c) $-j$
    (d) j [GATE]

[*Hint:* $f(x) = x^3y^2 + xy = 17$
            $\nabla f = (3x^2y^2 + y)i + (2x^3y + x)j$
    $\therefore$ $\nabla f_{(2, 0)} = 2j$].

30. If a vector field is $F = xi + 2yj + 3zk$, then $\nabla \times (\nabla \times F)$ is
    (a) 0
    (b) i
    (c) $2j$
    (d) $3k$ [GATE]

[*Hint:* $\Delta \times F = \begin{vmatrix} i & j & k \\ \frac{\partial}{\partial x} & \frac{\partial}{\partial y} & \frac{\partial}{\partial z} \\ x & 2y & 3z \end{vmatrix} = i(0 - 0) + j(0 - 0) + k(0 - 0) = 0$

    $\therefore$ $\nabla \times (\nabla \times F) = 0$]

31. Given the four vectors $u_1 = \begin{pmatrix} 1 \\ 2 \\ 1 \end{pmatrix}$, $u_2 \begin{pmatrix} 3 \\ -5 \\ 1 \end{pmatrix}$, $u_3 \begin{pmatrix} 2 \\ 4 \\ -8 \end{pmatrix}$, $u_4 \begin{pmatrix} 3 \\ 6 \\ -12 \end{pmatrix}$, the linearly dependent

    pair is
    (a) u_1, u_2
    (b) u_1, u_3
    (c) u_1, u_4
    (d) u_3, u_4 [GATE]

[*Hint:* $u_4 = \frac{3}{2}u_3$ and hence u_3 and u_4 are linearly dependent pairs].

32. The value of $\oint A \cdot dl$ along a square loop of side L in a uniform field A is
    (a) 0
    (b) $2LA$
    (c) $4LA$
    (d) L^2A [GATE]

[*Hint:* $\oint A \cdot dl = \iint_S (\nabla \times A) \cdot dS = \iint_S 0 \cdot dS = 0$].

33. The value of $\oint \dfrac{r \cdot dS}{r^3}$, where r is the position vector and S is the closed surface enclosing the origin is

(a) 0

(b) π

(c) 4π

(d) 8π **[GATE]**

[**Hint:** $\oint \dfrac{r \cdot dS}{r^3} = \oint \dfrac{r}{r^3} \cdot dS = \iiint_V \nabla\left(\dfrac{r}{r^3}\right) dV$

Now, $\quad \nabla\left(\dfrac{r}{r^3}\right) = (\nabla \cdot r)\dfrac{1}{r^3} + r \cdot \nabla\left(\dfrac{1}{r^3}\right) \quad \left(\because \nabla \cdot r = 3 \text{ and } \nabla\left(\dfrac{1}{r^3}\right) = \dfrac{-3r}{r^3}\right)$

$\quad\quad\quad\quad\quad = \dfrac{3}{r^3} - r \cdot \dfrac{3r}{r^4} = \dfrac{3}{r^3} - \dfrac{3}{r^3} = 0$

or $\quad\quad \oint \dfrac{r \cdot dS}{r^3} = \iiint_V 0 \cdot dV = 0$].

34. A vector field is defined everywhere as $F = \dfrac{y^2}{z}i + zk$. The net flux of F associated with a cube of side L with one vertex at the origin and sides along the positive X, Y, and Z axes is

(a) $2L^3$

(b) $4L^3$

(c) $8L^3$

(d) $10L^3$ **[GATE]**

[**Hint:** $\nabla \cdot F = 0 + 1 = 1 \oiint_S F \cdot dS = \iiint_V (\nabla \cdot F)dV = \iiint 1 \cdot dV = L^3$, obviously all given options are wrong].

35. If $r = xi + yj$, then

(a) $\nabla \cdot r = 0$ and $\nabla(r) = r$

(b) $\nabla \cdot r = 2$ and $\nabla(r) = r$

(c) $\nabla \cdot r = 2$ and $\nabla(r) = r/r$

(d) $\nabla \cdot r = 3$ and $\nabla(r) = r/r$ **[GATE]**

[**Hint:** $\nabla \cdot r = \dfrac{\partial x}{\partial x} + \dfrac{\partial y}{\partial y} = 2$

Also $\quad |r| = (x^2 + y^2)^{1/2}$

Thus, $\nabla|r| = \dfrac{xi}{\sqrt{x^2 + y^2}} + \dfrac{yj}{\sqrt{x^2 + y^2}} = \dfrac{r}{r}$].

36. Consider a vector $p = 2i + 3j + 2k$ in the coordinate system (i, j, k). The axes are rotated anti-clockwise about y-axis by an angle of $60°$. The vector p in the rotated coordinate system (i, j, k) is

(a) $(1-\sqrt{3})i' + 3j' + (1+\sqrt{3})k'$

(b) $(1+\sqrt{3})i' + 3j' + (1-\sqrt{3})k'$

(c) $(1-\sqrt{3})i' + (3+\sqrt{3})j' + 2k'$

(d) $(1-\sqrt{3})i' + (3-\sqrt{3})j' + 2k'$ **[GATE]**

[**Hint:** We have (about y-axis)

$\begin{pmatrix} \cos\theta & \sin\theta \\ -\sin\theta & \cos\theta \end{pmatrix} \begin{pmatrix} z \\ x \end{pmatrix} = \begin{pmatrix} z' \\ x' \end{pmatrix}$ Given, $\theta = 60°$

$$\therefore \qquad z' = \frac{1}{2}z + \frac{\sqrt{3}}{2}x$$

$$x' = \frac{\sqrt{3}}{2}z + \frac{1}{2}x$$

$$= (\sqrt{3}k' + i') + 3j' + (k' - \sqrt{3}i')$$

$$= (1 - \sqrt{3})i' + 3j' + (1 + \sqrt{3})k'$$

and $\qquad\qquad y' = y$

One can see that new coordinated system can be rotated by '60°' to get the original axis.

$$\begin{pmatrix} \cos(-60°) & \sin(-60°) \\ -\sin(-60°) & \cos(-60°) \end{pmatrix} \begin{pmatrix} z' \\ x' \end{pmatrix} = \begin{pmatrix} z \\ x \end{pmatrix}$$

This leads to $\qquad Z = \frac{1}{2}z' - \frac{\sqrt{3}}{2}x'$

$$X = \frac{\sqrt{3}}{2}z' + \frac{1}{2}x'$$

$$P' = 2\left(\frac{\sqrt{3}}{2}k, +\frac{1}{2}i'\right) + 2\left(\frac{1}{2}k - \frac{\sqrt{3}}{2}i\right).$$

37. The curl of a vector field F is $2x$. Identify the appropriate vector field F from the options given below:
    (a) $F = 2zx + 3zy + 5yz$
    (b) $F = 3zy + 5yz$
    (c) $F = 3xy + 5yz$
    (d) $F = 2x + 5yz$ [GATE]

    [*Hint:* We have $\nabla \times F = 2i + 0j + 0k$ Hence $\frac{\partial}{\partial x}F_y - \frac{\partial}{\partial y}F_x = 0;$ $\frac{\partial}{\partial y}F_z - \frac{\partial}{\partial z}F_y = 0$

    and $\frac{\partial}{z}F_x - \frac{\partial}{\partial x}F_z = 0$. Obviously, option (b) is correct].

38. Consider the set of vectors in three dimensional real vector space
    $$\mathbb{R}^3, S = \{(1, 1, 1), (1, -1, 1), (1, 1, -1)\}.$$
    Which one of the following statements is true?
    (a) S is not a linearly independent set
    (b) S is a basis for $\mathbb{R}^3$
    (c) The vectors in S are orthogonal
    (d) An orthonormal set of vectors cannot be generated from S.
    [*Hint:* There cannot exist real numbers x and y such that
    $$i + j + k = x(i - j + k) + y(i + j - k)$$
    But $\mathbb{R}^3$ space requires three independent spaces. Now
    $$(i - j + k) \cdot (i + j - k) \neq 0$$
    clearly, they are not orthogonal. We can think of orthogonal set of vectors in $3D$ space. Thus, option (3) is correct].

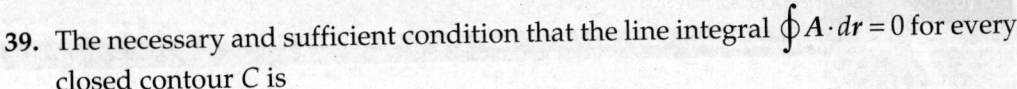

39. The necessary and sufficient condition that the line integral $\oint A \cdot dr = 0$ for every closed contour C is

    (a) $\oint \text{div}\, A = 0$

    (b) $\oint \text{curl}\, A = 0$

    (c) $\oint \text{div}\, A \neq 0$

    (d) $\oint \text{curl}\, A \neq 0$

40. In cylindrical coordinate system, the values of scale factors h_1, h_2 and h_3 are

    (a) $h_1 = 1, h_2 = r, h_3 = 1$

    (b) $h_1 = 1, h_2 = r, h_3 = r \sin \theta$

    (c) $h_1 = h_2 = h_3 = 1$

    (d) $h_1 = 1, h_2 = r \sin \theta, h_3 = r \cos \theta$

41. In spherical polar coordinates, the values of scalar factors h_1, h_2 and h_3 are

    (a) $h_1 = 1, h_2 = r, h_3 = 1$

    (b) $h_1 = 1, h_2 = r, h_3 = r \sin \theta$

    (c) $h_1 = h_2 = h_3 = 1$

    (d) $h_1 = 1, h_2 = r \sin \theta, h_3 = r \cos \theta$

42. The gradient of a scalar field is always

    (a) a scalar

    (b) a vector

    (c) a numeric

    (d) sometimes a scalar and sometimes a vector

43. Div $(A + B)$ is

    (a) div A + div B

    (b) div A + A grad B

    (c) $B \cdot$ curl A + $A \cdot$ curl B

    (d) $A \cdot$ div B + $B \cdot$ div A

ANSWERS

1. (d)	2. (c)	3. (d)	4. (d)	5. (b)	6. (b)	7. (d)	8. (a)
9. (b)	10. (a)	11. (a)	12. (d)	13. (c)	14. (d)	15. (a)	16. (c)
17. (a)	18. (d)	19. (b)	20. (c)	21. (a)	22. (d)	23. (b)	24. (c)
25. (a)	26. (c)	27. (a)	28. (a)	29. (d)	30. (a)	31. (d)	32. (a)
33. (a)	34. (c)	35. (c)	36. (a)	37. (b)	38. (c)	39. (b)	40. (a)
41. (b)	42. (b)	43. (a)					

2

Matrices

2.1 INTRODUCTION

A matrix is a rectangular array of numbers. Such arrays occur in various branches of applied mathematics. In many cases, they form the coefficients of linear transformation equations or systems of linear equations arising, for instance, from electrical networks, frameworks in mechanics, curve fittings in statistics and transportation problems. Matrices are useful because they enable us to consider an array of many numbers as a single object, denote it by one symbol and perform calculations with these symbols in a very compact form. The mathematical shorthand thus obtained is very elegant and powerful, and is suitable for various practical problems.

Matrices occur in physics mainly in two ways: First in the solution of systems of linear equations and second in the solutions of eigen value problems in classical and quantum mechanics. Both these types of problems arise, in turn, from transformations of vectors in vector spaces and the operation of linear operators on vector spaces. The methods of vector algebra are not only useful but essential in handling such problems. In this chapter, we shall discuss various operations with matrices and different situations in which they can be applied.

2.2 BASIC CONCEPTS

A rectangular array of numbers (real or complex) is called a matrix. The array is usually enclosed within a square or curved bracket such as:

$$\begin{bmatrix} a_{11} & a_{12} & a_{13} & \cdots & a_{1n} \\ a_{21} & a_{22} & a_{23} & \cdots & a_{2n} \\ a_{31} & a_{32} & a_{33} & \cdots & a_{3n} \\ \vdots & \vdots & \vdots & & \\ a_{m1} & a_{m2} & a_{m3} & \cdots & a_{mn} \end{bmatrix}$$

or

$$\begin{bmatrix} a_{11} & a_{21} & a_{31} & \cdots & a_{n1} \\ a_{12} & a_{22} & a_{32} & \cdots & a_{n2} \\ a_{13} & a_{23} & a_{33} & \cdots & a_{n3} \\ \vdots & \vdots & \vdots & & \\ a_{1m} & a_{2m} & a_{3m} & \cdots & a_{nm} \end{bmatrix}$$

104

The number $a_{11},..., a_{mn}$ are called the *elements* of the matrix. The horizontal lines are called rows or *row vectors* and the vertical lines are called *columns* or *column vectors* of the matrix. A matrix with m rows and n columns is called $m \times n$ matrix (read "*m* by *n*" matrix).

Generally matrix is denoted by capital bold-faced letters A, B, etc. or by writing the general elements of matrix enclosed in brackets such as $A = \{a_{ij}\}, B = [b_{ij}]$.

In double subscript notation for elements, the first subscript always denotes the row and the second subscript the column containing the given element.

A matrix of only one row is called a *row matrix* or *row vector* and a matrix having only one column is called a *column matrix* or *column vector*. Row and column matrices are denoted generally by small bold-faced letters such as:

$$a = \begin{bmatrix} a_1 & a_2 & \cdots & a_n \end{bmatrix}, b = \begin{bmatrix} b_1 \\ b_2 \\ \vdots \\ b_n \end{bmatrix}$$

Although, we have defined a matrix here with reference to numbers, we can easily extend the definition to a matrix whose elements are functions such as:

$$\begin{bmatrix} F_{11}(x) & F_{12}(x) & \cdots & F_{1n}(x) \\ F_{21}(x) & F_{22}(x) & \cdots & F_{2n}(x) \\ \vdots & & & \\ F_{m1}(x) & F_{m2}(x) & \cdots & F_{mn}(x) \end{bmatrix}$$

A matrix having the same number of rows and columns is called a *square matrix* and the number of rows is called its *order*.

Two matrices $A = [a_{ij}]$ and $B = [b_{ij}]$ are *equal* if and only if A and B have the same number of rows and same number of columns and corresponding elements are equal, i.e.

$$a_{ij} = b_{ij} \text{ for all occurring } i \text{ and } j$$

Then we write

$$A = B.$$

Any matrix obtained by some rows and columns from a given $m \times n$ matrix A is called a *submatrix* of A.

For example, if $A = \begin{bmatrix} 1 & 2 & 3 \\ 0 & 4 & 5 \\ 2 & 3 & 1 \end{bmatrix}$ then matrix $\begin{bmatrix} 4 & 5 \\ 3 & 1 \end{bmatrix}$ or $\begin{bmatrix} 1 & 2 \\ 0 & 4 \end{bmatrix}$, etc. are called sub-matrices of A.

A matrix A of arbitrary order is said to be *zero matrix* or *null matrix*, if and only if every element of A equals zero. A zero matrix is also denoted by 0. If it is necessary to specify the order of a zero matrix, we may write it as $0_{m \times n}$, e.g.

$$0 = \begin{bmatrix} 0 & 0 & 0 \\ 0 & 0 & 0 \\ 0 & 0 & 0 \end{bmatrix}.$$

This matrix is called 3×3 null matrix and is written as $0_{3 \times 3}$.

A square matrix which has nonvanishing elements only along the principal diagonals and zeros elsewhere is called a *diagonal matrix*. Thus, a diagonal matrix would has the form

$$A = \begin{bmatrix} a_{11} & 0 & 0 & \cdots & 0 \\ 0 & a_{22} & 0 & \cdots & 0 \\ \vdots & & & & \\ 0 & 0 & 0 & \cdots & a_{mn} \end{bmatrix} = \text{diag } [a_{ij}]$$

from which we have

$$(A)_{ij} = a_{ij}\, \delta_{ij}^{*}$$

If all the nonvanishing elements of a diagonal matrix happen to be equal to each other, it is said to be *constant matrix*. The elements of constant matrix are thus given by

$$(A)_{ij} = a\delta_{ij}$$

where a is a scalar. Constant matrix is also called *scalar matrix*.

A diagonal matrix with nonvanishing elements

$$A = \begin{bmatrix} \lambda & 0 & 0 & \cdots & 0 \\ 0 & \lambda & 0 & \cdots & 0 \\ \vdots & & & & \\ 0 & 0 & 0 & \cdots & \lambda \end{bmatrix} = \text{diag } [\lambda ... \lambda ... \lambda]$$

If every nonvanishing element of a constant matrix equals to unity, i.e. $a = 1$, then we have *unit matrix*. A unit matrix is generally denoted by the symbol I. In order to specify the order of unit matrix it is written as I_n for unit matrix of order n.

The unit matrices of different orders are specified as:

$$I_2 = \begin{bmatrix} 1 & 0 \\ 0 & 1 \end{bmatrix}, \quad I_3 = \begin{bmatrix} 1 & 0 & 0 \\ 0 & 1 & 0 \\ 0 & 0 & 1 \end{bmatrix}$$

and

$$I_n = \begin{bmatrix} 1 & 0 & \cdots & \cdots & \cdots & 0 \\ 0 & 1 & \cdots & \cdots & \cdots & 0 \\ 0 & 0 & 1 & \cdots & \cdots & 0 \\ \vdots & \cdots & \cdots & \cdots & \cdots & \cdots \\ 0 & 0 & 0 & 0 & \cdots & 1 \end{bmatrix}$$

Here I_2 is unit matrix of order 2, I_3 of order 3, I_n of order n, etc.

$^*(A)_{ij}$ means the ijth element of A is a_{ij} and δ_{ij} is Kroneker delta $= 1$ for $i = j = 0$ for $i \neq j$.

The *transpose* A^T of an $m \times n$ matrix $A = [a_{ij}]$ is the $n \times m$ matrix obtained by interchanging the rows and columns in A that is the ith row of A becomes the ith column of A^T.

$$A^T = [a_{ji}] = \begin{bmatrix} a_{11} & a_{21} & \cdots & a_{m1} \\ a_{12} & a_{22} & \cdots & a_{m2} \\ \vdots & & & \\ a_{1n} & a_{2n} & \cdots & a_{mn} \end{bmatrix}$$

and

$$A = [a_{ij}] = \begin{bmatrix} a_{11} & a_{12} & \cdots & a_{1n} \\ a_{21} & a_{22} & \cdots & a_{2n} \\ \vdots & & & \\ a_{m1} & a_{m2} & \cdots & a_{mn} \end{bmatrix}$$

It is easily seen that the transpose of A^T is A itself, i.e. $(A^T)^T = A$.

A square matrix A such that $A^T = A$ is called *symmetric*. Thus, a matrix $A = [a_{ij}]$ is symmetric provided $a_{ij} = a_{ji}$ for all values of i and j.

For example

$$A - \begin{bmatrix} a & h & g \\ h & b & f \\ g & f & c \end{bmatrix}$$

is symmetric matrix because its transpose

$$A^T = \begin{bmatrix} a & h & g \\ h & b & f \\ g & f & c \end{bmatrix}$$

has all the elements same as those of A.

A square matrix A such that $A^T = -A$ is called *skew-symmetric*. Thus, a square matrix A is a skew-symmetric provided $a_{ij} = -a_{ji}$ for all values of i and j. Clearly the diagonal elements are zero.

For example

$$A = \begin{bmatrix} 0 & h & g \\ -h & 0 & f \\ -g & -f & 0 \end{bmatrix}$$

is skew-symmetric.

When A is a matrix having complex numbers as elements, the matrix obtained from A by replacing each element by its conjugate is called the *conjugate* of A and is denoted by $\bar{A}$.

For example, if

$$A = \begin{bmatrix} 2+i & 5-3i & i \\ 0 & 3 & 6+i \end{bmatrix}$$

then

$$\bar{A} = \begin{bmatrix} 2-i & 5+3i & -i \\ 0 & 3 & 6-i \end{bmatrix}$$

The conjugate of the transpose of a matrix A is called the conjugate transpose of A and is denoted by A^+ or A^θ. Thus,

$$A^\theta = A^+ = \overline{A^T}$$

If
$$A = \begin{bmatrix} 2i+3 & 0 & i \\ 2 & 1-i & 4 \end{bmatrix} \text{ then } A^T = \begin{bmatrix} 2i+3 & 2 \\ 0 & 1-i \\ i & 4 \end{bmatrix}$$

and
$$A^\theta = \overline{A^T} = \begin{bmatrix} 2i-3 & 2 \\ 0 & 1+i \\ -i & 4 \end{bmatrix}$$

Note: It is clear that

$$(\overline{\overline{A}}) = A \text{ and } (A^\theta)^\theta = A.$$

Hermitian and Skew-Hermitian Matrices

A square matrix, conjugate transpose of which coincides with the matrix itself is called *Hermitian matrix*. Thus, a matrix $A = [a_{ij}]$ is a Hermitian matrix if

$$A^+ = A$$

Taking transpose of this equation, we have

$$(A^+)^T = A^T \text{ or } \left((\overline{A})^T\right)^T = A^T \text{ or } \overline{A} = A^T$$

Hence transpose of a Hermitian matrix coincides with its conjugate. This means matrix $A = (a_{ij})$.

$$a_{ji} = \overline{a}_{ij}$$

This means (ij)th element of the Hermitian matrix must coincide with the conjugate of its (ji)th element. Setting $i = j$ gives

$$a_{ii} = \overline{a}_{ii}$$

which means that every diagonal element of Hermitian matrix is equal to its conjugate and hence every diagonal element of Hermitian matrix must be a real number, e.g.

$$A = \begin{bmatrix} 2 & 0 & 4+i \\ 0 & 1 & -i \\ 4-i & i & 3 \end{bmatrix}$$

is a Hermitian matrix. It is obvious that the real Hermitian matrix is a symmetric matrix because for real matrix $A^+ = A^T = A$. Which is the condition for symmetric matrix.

A square matrix, the conjugate transpose of which is the matrix itself with negative sign is called skew-Hermitian or anti-Hermitian matrix. Hence for a square matrix $A = [a_{ij}]$ to be *skew-Hermitian*,

$$A^+ = -A \text{ or } \overline{A^T} = -\overline{A} \text{ or } a_{ij} = -\overline{a}_{ji}$$

i.e. (ij)th element is negative conjugate of (ji)th element. Setting $i = j$, we get

$$a_{ii} = \overline{a}_{ii}$$

which means every diagonal element of skew-Hermitian matrix is either zero or purely imaginary number. Thus the matrix

$$A = \begin{bmatrix} 2i & 0 & 4+i \\ 0 & 0 & -1 \\ -4+i & -i & 3i \end{bmatrix}$$

is a skew-Hermitian matrix. It is obvious that a real skew-Hermitian matrix is anti-symmetric.

Determinant of a Square Matrix

Determinant having the same array as that of matrix A is called *determinant of matrix A* and is denoted by $|A|$. Thus if $A = [a_{ij}]$ then $|A| = |a_{ij}|$. A square matrix whose deter-minant is equal to zero is called *singular matrix*. Thus, a square matrix A is singular if

$$|A| = |a_{ij}| = 0$$

In case $|A| \neq 0$ the *matrix is nonsingular*.

If the row and column of the element a_{ij} of a given $n \times n$ square matrix A are crossed, then determinant of remaining $(n-1) \times (n-1)$ square matrix is called *minor of the element a_{ij}* and written as M_{ij}. Thus for a square matrix

$$A = \begin{bmatrix} a & b & c \\ e & f & g \\ h & i & j \end{bmatrix}$$

M_{ij} of (a_{11}), i.e. $M_{11} = \begin{vmatrix} f & g \\ i & j \end{vmatrix}$

and cofactor of a_{ij} is $(-1)^{i+j}$ times its minor and it is denoted as

$$A^{ij} = (-1)^{i+j} M_{ij}$$

Related to this definition of cofactor, we have two theorems

i. $\displaystyle\sum_{\text{row}} a_{ij} A^{ik} = \sum a_{ij} A^{ik} \begin{cases} |A| \text{ if } j=k \\ 0 \quad \text{if } j \neq k \end{cases}$

ii. $\displaystyle\sum_i a_{ij} A^{kj} = \begin{cases} |A| \text{ if } j=k \\ 0 \quad \text{if } j \neq k \end{cases}$

Proof: If $A = [a_{ij}]$, $n \times n$ matrix, its determinant is given by

$$|A| = (-1)^{i+1} a_{i1} M_{i1} + (-1)^{i+2} a_{i2} M_{i2} + \cdots + (-1)^{i+a_{ij}} M_{ij} + \cdots = \sum_j a_{ij} (-1)^{i+j} M_{ij} = \sum_j a_{ij} A^{ij}$$

where, $\displaystyle\sum_j a_{ij} A^{kj}$ is a determinant whose two rows are identical, and hence is zero. Thus we have

$$\sum_j a_{ij} A^{kj} = |A| \delta_{ik}$$

Similarly, chosing a column and summing over rows, we have $\displaystyle\sum_i a_{ij} A^{ij} = |A|$ and $\displaystyle\sum_i a_{ij} A^{ik}$ is determinant with two identical columns. Hence $\displaystyle\sum_i a_{ij} A^{ik} = |A| \delta_{jk}$.

Determinant of Hermitian Matrix is Real

If A be a Hermitian matrix then $A^+ = A$

or $$A^T = \bar{A}$$

$\therefore$ $$|A^T| = |\bar{A}| = |A|$$

which is possible only when $|A|$ is real.

Similarly if A is skew-Hermitian matrix, then it is either zero or purely imaginary.

Adjugate Matrix or Adjoint of a Matrix

Adjoint or adjugate of a matrix is the transpose of a matrix whose elements are cofactors of corresponding elements of the given matrix. Thus, if A is a square matrix $[a_{ij}]$ is called, the *adjoint of A*, where A^{ij} is the cofactor of a^{ij} in $|A|$. It is denoted by *adj A* or A. Thus if matrix

$$A = \begin{bmatrix} a_1 & b_1 & c_1 \\ a_2 & b_2 & c_2 \\ a_3 & b_3 & c_3 \end{bmatrix}$$

Then cofactor of $a_1 = \begin{vmatrix} b_2 & c_2 \\ b_3 & c_3 \end{vmatrix} = b_2 c_3 - b_3 c_2$

Cofactor of $\quad b_1 = a_3 c_2 - a_2 c_3$

Cofactor of $\quad c_1 = a_2 b_3 - b_2 a_3$ etc.

Hence $\quad$ adj $A = \begin{bmatrix} b_2 c_3 - b_3 c_2 & a_3 c_2 - a_2 c_3 & a_2 b_3 - b_2 a_3 \\ b_3 c_1 - b_1 c_3 & a_1 c_3 - a_3 c_1 & a_3 b_1 - b_3 a_1 \\ b_1 c_2 - b_2 c_1 & a_2 c_1 - a_1 c_2 & a_1 b_2 - b_1 a_2 \end{bmatrix}$

A square matrix A is called nonsingular, if $|A| \neq 0$. Otherwise A is called singular.

2.3 ADDITION AND MULTIPLICATION OF MATRICES

For doing calculations with matrices, we shall introduce algebraic operations for matrices, such as addition, multiplication etc.

Addition of matrices is defined only for those having the same number of rows and the same number of columns. The sum of two $m \times n$ matrices $A = [a_{ij}]$ and $B = [b_{ij}]$ is the $m \times n$ matrix $c = [c_{ij}]$ such that

$$c_{ij} = a_{ij} + b_{ij}$$

and is written as $\quad C = A + B$

i.e. $(A + B)$ is obtained by adding corresponding elements of A and B. We see from the definition that matrix addition enjoys properties which are quite similar to those of the addition of real numbers.

   i. $A + B = B + A$ $\qquad$ (commutative law)
   ii. $(A + B) + C = A + (B + C)$ $\quad$ (associative law)
   iii. $A + 0 = A$
   iv. $A + (-A) = 0$

where $-A = [-a_{ij}]$ is the $m \times n$ matrix obtained by multiplying every element of A by -1 and is called the *negative of A*.

Instead of $A + (-B)$ we simply write $A - B$ and call this matrix the difference of A and B. Obviously its elements are obtained by subtracting corresponding elements of A and B.

Multiplication of Matrices by Scalars (Numbers)

The product of an $m \times n$ matrix A by a scalar K (a number) is denoted by KA or AK and is the $m \times n$ matrix obtained by multiplying every element of A by K, that is, if

$$A = \begin{bmatrix} a_{11} & a_{12} & a_{13} \\ a_{21} & a_{22} & a_{23} \end{bmatrix}$$

then

$$KA = \begin{bmatrix} Ka_{11} & Ka_{12} & Ka_{13} \\ Ka_{21} & Ka_{22} & Ka_{23} \end{bmatrix}$$

From the definition, we see that for $m \times n$ matrices (with fixed m and n) and for any numbers, we have the following relations:

i. $K(A + B) = KA + KB$

ii. $(K + I) A = KA + IA$

iii. $K(IA) - (KI) A$

iv. $IA = A$

Multiplication of Matrices

Let $A = [a_{ij}]$ be an $m \times n$ matrix and $B = [b_{ij}]$ an $r \times p$ matrix. Then the product AB (in this order) is defined only when $r = n$, i.e. the number of rows of second matrix are equal to the number of columns of first matrix and is an $m \times p$ matrix $C = [c_{ij}]$ whose elements are

$$c_{ij} = a_{i1} b_{i1} + a_{i2} b_{2i} + \cdots + a_{in} b_{nj} = \sum_{l=1}^{n} a_{ij} b_{ij}$$

We see that c_{ij} is the inner product (dot product) of the ith row vector of the first matrix A and the jth column vector of the second matrix B. The process of matrix multiplication is, therefore, conveniently referred to the multiplication of rows and columns.

Examples of Matrix Multiplication

i. If $A = \begin{bmatrix} a_{11} & a_{12} \\ a_{21} & a_{22} \end{bmatrix}$ and $B = \begin{bmatrix} b_{11} & b_{12} \\ b_{21} & b_{22} \end{bmatrix}$

Then $AB = \begin{bmatrix} a_{11}b_{11} + a_{12}b_{21} & a_{11}b_{12} + a_{12}b_{22} \\ a_{21}b_{11} + a_{22}b_{21} & a_{21}b_{12} + a_{22}b_{22} \end{bmatrix}$

ii. If $A = \begin{bmatrix} 2 & -2 & 1 \\ 0 & 1 & 3 \end{bmatrix}$ and $B = \begin{bmatrix} 3 & 0 & -4 \\ -2 & 1 & 0 \\ 1 & 2 & 4 \end{bmatrix}$

Then $AB = \begin{bmatrix} 6+4+1 & 0-2+2 & -8+0+4 \\ 0-2+3 & 0+1+6 & 0+0+12 \end{bmatrix} = \begin{bmatrix} 11 & 0 & -4 \\ 1 & 7 & 12 \end{bmatrix}$

Note: In example (ii) the product BA is not possible as the number of columns in $B(3)$ is not equal to the number of rows (2) in A.

Properties of Matrix Multiplication

Matrix multiplication has some properties which are quite unusual because they have no counterparts in the usual multiplication of numbers. We illustrate these properties in (b) and (c).

(a) Matrix multiplication is associative and distributive with respect to addition of matrix, that is

$$(KA)B = K(AB) = A(KB) \text{ written } KAB \text{ or } AKB$$
$$(AB)C = A(BC) \qquad \text{written } ABC$$
$$(A + B)C = AC + BC$$
$$C(A + B) = CA + CB.$$

provided A, B, and C are such that the expression of the left are defined (K is any number).

(b) Matrix multiplication is not commutative, that is if A and B are matrices such that both AB and BA are defined, then

$$AB \neq BA, \text{ in general.}$$

(c) The cancellation law is not true, in general, that is

$$AB = 0 \text{ does not necessarily imply } A = 0 \text{ or } B = 0.$$

For the examples of property (a) it is left as an exercise for readers and for b.

If $\qquad\qquad A = \begin{bmatrix} 1 & 0 \\ 0 & 0 \end{bmatrix}$ and $B = \begin{bmatrix} 0 & 1 \\ 1 & 0 \end{bmatrix}$

then $\qquad\qquad AB = \begin{bmatrix} 0 & 1 \\ 0 & 0 \end{bmatrix}$ and $BA = \begin{bmatrix} 0 & 0 \\ 1 & 0 \end{bmatrix}$, i.e. $AB \neq BA$.

attribute (c) can be seen from the following example:

$$\begin{bmatrix} 1 & 1 \\ 2 & 2 \end{bmatrix} \begin{bmatrix} -1 & 1 \\ 1 & -1 \end{bmatrix} = \begin{bmatrix} 0 & 0 \\ 0 & 0 \end{bmatrix}$$

But there are some matrices which commute, e.g.

If $\qquad\qquad A = \begin{bmatrix} 2 & 3 \\ 1 & 7 \end{bmatrix}$ and $B = \begin{bmatrix} 2 & 0 \\ 0 & 2 \end{bmatrix}$

then $\qquad\qquad AB = BA = \begin{bmatrix} 4 & 6 \\ 2 & 14 \end{bmatrix}.$

There are some matrices which satisfy the equation

$$XY = -YX$$

The matrices X and Y are said to anticommute, e.g.

If $\qquad\qquad \sigma_x = \begin{bmatrix} 0 & 1 \\ 1 & 0 \end{bmatrix}, \sigma_y = \begin{bmatrix} 0 & -i \\ i & 0 \end{bmatrix}$ and $\sigma_z = \begin{bmatrix} 1 & 0 \\ 0 & -1 \end{bmatrix}$

Then we have
$$\sigma_x\sigma_y = -\sigma_y\sigma_x = \begin{bmatrix} i & 0 \\ 0 & -i \end{bmatrix}$$

$$\sigma_y\sigma_z = -\sigma_z\sigma_y = \begin{bmatrix} 0 & i \\ i & 0 \end{bmatrix}$$

and
$$\sigma_z\sigma_x = -\sigma_x\sigma_z = \begin{bmatrix} 0 & 1 \\ -1 & 0 \end{bmatrix}$$

i.e. σ_x, σ_y, σ_z anticommute with each other.

Theorem: Matrices A and Adj A are commutative and that their product is a scalar matrix each diagonal element of which is $|A|$, i.e. A (adj A) = (adj A) $A = |A|I$.

Proof: Let us write (ij)th element of the product of A and $Adj\ A$.

$$(A\ \text{adj}\ A)_{ij} = \sum_k A_{ik}(\text{adj}\,A)_{kj}$$

where A^{jk} is cofactor of a_{jk} and is given by $(-1)^{j+k}$ times its minor. It is denoted as
$$A^{jk} = (-1)^{j+k}\,M_{jk}$$

where M_{jk} is minor and is the determinant left after row and column of a_{jk} is crossed.

$$(A\ \text{adj}\ A)_{ij} = |A|\delta_{ij}^* = \left[|A|\delta_{ij}\right]\ \text{for every } i \text{ and } j.$$

Similarly
$$\{(A\,\text{adj})A\}_{ij} = \sum_k A^{ki}a_{kj} = \sum_k a_{kj}A^{ki} = |A|\delta_{ij}$$

or
$$(\text{Adj}\ A) = |A|I$$

From this it is obvious that if A is a singular matrix then (Adj A) A = (Adj A) = 0 (null matrix). The following results may be proved directly

i. $(\text{Adj}\ A^T) = (\text{Adj}\ A)^T$
ii. $\text{Adj}\ A^\theta = (\text{Adj}\ A)^\theta$
iii. Adjoint of a symmetric matrix is itself symmetric
iv. Adjoint of a Hermitian matrix is also Hermitian

Adjoint of a Product

If A and B are n-square matrices then
$$\text{Adj}\ AB = \text{Adj}\ B \cdot \text{Adj}\ A$$

Proof: We have
$$A\ \text{adj}\ A = |A|I = (\text{Adj}\ AB)\ AB$$
or
$$AB\ \text{adj}\ AB = |AB|I = (\text{Adj}\ AB)\ AB$$

---

* If $A = [a_{ji}]\ n \times n$ is a square matrix, its determinant is given by
$$\det A = |A| = (-1)^{i+1}\,a_{i1}\,M_{i1} + (-1)^{i+2}\,a_{i2}\,M_{i2} + \cdots (-1)^{i+j}\,a_{ij}\,M_{ij} + \cdots$$

$$= \sum_i a_{ij}(-1)^{i+j}\,M_{ij} = \sum_j a_{ij}A^{ji} = |A|\delta_{ik}$$

while $\sum_i a_{ij}A^{ik}$ is a determinant whose two rows are identical and hence 0.

But $\qquad AB \text{ Adj } B \text{ adj } A = A (B \text{ Adj } B) \text{ adj } A$
$$= A (|B|I) \text{ Adj } A$$
$$= |B| (A \text{ Adj } A)$$
$$= |B| \ |A|I = |AB|I$$

and $\qquad (\text{Adj } B \text{ Adj } A) AB = \text{Adj } B \{(\text{Adj } A) A\}B$
$$= \text{Adj } B \ |A|IB = |AB|I$$

Hence Adj AB = Adj B Adj A

2.4 SPECIAL MATRICES

Inverse or Reciprocal Matrix

A matrix A is said to be nonsingular if there exists a matrix B such that
$$AB = I_n \text{ and } BA = I_m \tag{2.1}$$
where I_n and I_m are unit matrices, not necessarily of the same order. Then A is said to be the inverse of B and *vice versa*. For a given matrix A if there exists no matrix B which satisfies Eq. (2.1), then A has no inverse and is said to be singular. The inverse matrix is also denoted by symbol A^{-1}.

We have already proved that
$$A \text{ adj } A = |A|I \tag{2.2}$$
or
$$A \frac{\text{adj} A}{|A|} = I$$

and from Eq. (2.1) we have $A^{-1} = A \dfrac{\text{adj} A}{|A|}$ $\qquad$ (2.3)

i. If A has an inverse, the inverse is unique

Indeed, if both B and C are inverse of A, then $AB = I$ and $CA = I$, so that we obtain uniqueness from $B = IB = (CA) B = C(AB) = CI = C$.

Note: Every matrix commutes with its inverse.

ii. Inverse of the product of two square matrices of the same order is the product of their inverse in the reverse order, i.e.
$$[AB]^{-1} = B^{-1} A^{-1}$$

Consider the product
$$AB[B^{-1}A^{-1}] = A[BB^{-1}]A^{-1} = AIA^{-1} = AA^{-1} = I$$
and $\qquad [B^{-1} A^{-1}] AB = B^{-1} [A^{-1} A] B = B^{-1}IB = B^{-1} B = I$
or $\qquad AB [B^{-1} A^{-1}] = [B^{-1} A^{-1}] AB = I$
or $\qquad [AB]^{-1} = B^{-1} A^{-1}$

iii. Inverse of the transpose of a square matrix is the transpose of its inverse, i.e. the operations of transposing and inversing are commutative, i.e.
$$[A^T]^{-1} = [A^{-1}]^T$$

To prove this let us consider the fact
$$AA^{-1} = I = A^{-1}A$$

On transposing this relation, we have
$$[AA^{-1}]^T = [A^{-1}]^T[A^T] = I^T = A^T[A^{-1}]^T$$

which means $[A^{-1}]^T = [A^T]^{-1}$.

Similarly we can prove

$$[A^{-1}]^+ = [A^+]^{-1}.$$

Reversal law given above for the inverse of the product of two matrices can be generalised for any finite number of nonsingular matrices of the same order.

Let $A_1, A_2, ..., A_n$ be n such matrices then

$$[A_1 A_2, ..., A_n]^{-1} = A_n^{-1} A_{n-1}^{-1}, ..., A_2^{-1} A_1^{-1}$$

If we put $A_1 = A_2 = A_3 =, ..., = A_n = A$, then we have

$$[A\ A\ A - n \text{ times}]^{-1} = A^{-1}\ A^{-1}, ..., n \text{ times}$$

or $$[A^n]^{-1} = [A^{-1}]^n = A^{-n}.$$

Orthogonal Matrix

$$AA^T = I_n \text{ and } A^T A = I_m \tag{2.4}$$

is called an orthogonal matrix. It is apparent that if A satisfies Eq. (2.3) then A must be a square matrix.

Let $$|A| = d$$

Taking determinants of both sides of Eq. (2.4) we have

$$d^2 = 1 \text{ or } d = \pm 1.$$

This shows that the determinant of orthogonal matrix can only have values $+1$ or -1. At the same time it shows that A is nonsingular so that A^{-1} exists. Now

$$(AA^T)\,A^{-1} = I_n A^{-1}$$

or $$A^T = A^{-1}$$

This is an alternate way of defining orthogonal matrix, i.e. an orthogonal matrix is a real matrix for which the inverse equals its transpose.

If $|A| = +1$, then matrix A is said to be *proper orthogonal matrix* and if $|A| = -1$ it is called *improper orthogonal matrix*.

If $$A = \begin{bmatrix} \cos\theta & \sin\theta \\ -\sin\theta & \cos\theta \end{bmatrix}$$

then $$A^T = \begin{bmatrix} \cos\theta & -\sin\theta \\ \sin\theta & \cos\theta \end{bmatrix}$$

or $$AA^T = A^T A = \begin{bmatrix} 1 & 0 \\ 0 & 1 \end{bmatrix} = I$$

Here A is a orthogonal matrix.

Note the products of orthogonal matrices are orthogonal matrices.

Equating the ijth elements of both sides of Eq. (2.4), we have

$$\left. \begin{aligned} \sum_{k=1}^{n} a_{ik}\, a_{jk} = \delta_{ij} \quad 1 \le i, j \le n \\[2mm] \sum_{k=1}^{n} a_{ki}\, a_{kj} = \delta_{kj} \quad 1 \le i, j \le n \end{aligned} \right\} \tag{2.4a}$$

and

where n is the order of the matrix A and a_{ik} is the ik^{th} element of A. These are the conditions satisfied by the elements of an orthogonal matrix, i.e. the row vectors and also the column vectors of an orthogonal matrix $A = [a_{ij}]$ form a system of orthogonal unit vectors, i.e. the sum of the products of the corresponding elements of two distinct rows of an orthogonal matrix is zero while the sum of the squares of the elements of any rows is equal to unity.

Unitary Matrix

A square matrix A satisfying the relation.

$$AA^\theta = A^\theta A = I \tag{2.5}$$

is called a unitary matrix, i.e. a unitary matrix when multiplied with its transpose conjugate gives unit matrix. The elements of unitary matrix may be complex. In fact it is evident from Eq. (2.5) that a real unitary matrix is orthogonal.

If $|A| = d$ then $|A^\theta| = d^*$ where d^* is complex conjugate of d. Then we have

$$|AA^\theta| = |A^\theta A| = dd^* = I \text{ or magnitude } d = 1.$$

This shows that the determinant of a unitary matrix can be a complex number of unit magnitude, i.e. a number of the form $e^{i\theta}$ where, θ is a real number. It also shows that a unitary matrix is nonsingular and possesses an inverse.

Equating the ijth element of both sides of Eq. (2.5), we have

$$\sum_{k=1}^{n} A_{ik} A_{jk}^* = \delta_{ij}, \quad \sum_{k=1}^{n} A_{ki} A_{kj} = \delta_{ij} \quad 1 \le i, j \le n$$

where A_{ik} is the ikth element of A and n its order.

If a unitary matrix is also Hermitian, then

$$A^+ = A = A^{-1} \quad \text{or} \quad AA^+ = I = A^2.$$

For example, if

$$A = \begin{bmatrix} \cos\theta & i\sin\theta \\ i\sin\theta & \cos\theta \end{bmatrix} \quad \text{then } A^+ = \begin{bmatrix} \cos\theta & -i\sin\theta \\ -i\sin\theta & \cos\theta \end{bmatrix}$$

and

$$AA^+ = \begin{bmatrix} 1 & 0 \\ 0 & 1 \end{bmatrix} = I$$

which shows that A is a unitary matrix.

Similarly, matrix

$$A = \begin{bmatrix} \dfrac{1}{\sqrt{2}} & \dfrac{i}{2} \\ -\dfrac{1}{\sqrt{2}} & -\dfrac{1}{\sqrt{2}} \end{bmatrix}$$

can be proved that it is unitary.

Generalising the result we can prove that every multiplication of two unitary matrices is a unitary matrix. For this to prove, consider two unitary $n \times n$ matrices A and B then

$$A^+ = A \quad B = B^+$$

and

$$[AB]^+ [AB] = B^+ A^+ AB = B^+ B = I.$$

Similarly

$$[AB] [AB]^+ = I$$

which shows that AB is a unitary matrix. In this way it can be shown that BA is also a unitary matrix.

From the definition we can also say that *a real unitary matrix is an orthogonal matrix.*

Periodic Matrix

A matrix A for which $A^{k+1} = A$, where k is a positive integer, is called periodic. If k is the least positive integer for which $A^{k+1} = A$, then A is said to be of period k.

If $k = 1$, so that $A^2 = A$, then A is said *idempotent*.

A matrix A for which $A^p = 0$ where, p is a positive integer is called *nilpotent*. If p is the least positive integer for which $A^p = 0$, then A is said to be *nilpotent of index p*.

A matrix A for which $A^2 = I$ is called *involutory matrix.*

Trace of a Square Matrix

The sum of the diagonal elements of a square matrix A is called the trace of A, i.e. if
$$A = [a_{ij}]\ 1 \le i \le n, 1 \le j \le n.$$

then trace of $\quad A = (T_r A) = \sum_{i=1}^{n} a_{ii}$

It can be shown that the trace of the product of a finite number of matrices is invariant under any cyclic permutation of the matrices, i.e.
$$T_r(A_1 A_2 A_3, ..., A_k) = T_r(A_{r+1}, ..., A_k A_1, ..., A_r), i \le r \le k-1.$$
To prove this, let us take two matrices A and B so that

$$T_r[AB] = \sum_{i}[AB]_{ii} = \sum_{i}\left\{\sum_{j} a_{ij} b_{ji}\right\}$$

$$= \sum_{j}\sum_{i} a_{ij} b_{ji} = \sum_{j}\sum_{i} b_{ji} a_{ij} = \sum_{j}[BA]_{jj} = T_r[BA]$$

This holds even if $AB \ne BA$.

The result can be generalised to show
$$T_r[A_1 A_2 A_3, ..., A_k] = T_r[A_{r+1}, ..., A_k A_1, ..., A_r]$$
In this way it can be proved that the trace of the product of a symmetric and an antisymmetric matrix is zero.

Let matrix A be symmetric, i.e. $A = A^T$ and B an antisymmetric matrix, i.e. $B = -B^T$

Then $\qquad T_r[AB] = \sum_{i}[AB]_{ii} = \sum_{i}\sum_{j} a_{ij} b_{ji} = \sum_{i}\sum_{j}(-b_i) a_{ji}$

$$= -\sum_{j}\sum_{i} b_{ij} a_{ji} = -\sum_{i}(BA)_{ii} = -T_r[BA]$$

$$= -T_r[AB] \quad \text{or} \quad T_r[AB] = 0$$

It is obvious from the definition that trace of an antisymmetric matrix is zero and trace of Hermitian matrix is real while the trace of skew-Hermitian matrix is either purely imaginary or zero.

Partitioning of Matrices

It is often convenient to divide matrices into blocks or to partition them, for the purpose of addition and multiplication. These blocks containing the elements of the given matrix are defined as the *submatrices* of the given matrix. This process of matrix subdivision helps us to obtain smaller and more manageable portions. This is largely a matter of individual preference. As we shall see below, this procedure also facilitates in obtaining the inverse of a large matrix.

The square submatrix of a square matrix is known as *principal submatrix*. It is obtained by deleting the corresponding rows and columns of a square matrix.

The principal square submatrix is obtained by deleting only some of the last rows and the corresponding columns of a square matrix then it is known as leading sub-matrix.

Thus a partitioned matrix

$$A = \begin{bmatrix} 1 & 2 & 3 \\ 4 & 5 & 6 \\ \hline 7 & 8 & 9 \end{bmatrix}$$

can be written as

$$A = \begin{bmatrix} A_{11} & A_{12} \\ A_{21} & A_{22} \end{bmatrix}$$

where

$$A_{11s} = \begin{bmatrix} 1 & 2 \\ 4 & 5 \end{bmatrix} \quad A_{12} = \begin{bmatrix} 3 \\ 6 \end{bmatrix} \quad A_{21} = \begin{bmatrix} 7 & 8 \end{bmatrix} \quad A_{22} = [9]$$

The matrix A can be partitioned as follows:

$$A = \begin{bmatrix} A_1 & A_2 & A_3 \end{bmatrix}$$

where

$$A_1 = \begin{bmatrix} 1 \\ 4 \\ 7 \end{bmatrix} \quad A_2 = \begin{bmatrix} 2 \\ 5 \\ 8 \end{bmatrix} \quad A_3 = \begin{bmatrix} 3 \\ 6 \\ 9 \end{bmatrix}$$

Two matrices A and B of the same order are said to be *identically partitioned* if the corresponding blocks have the same order. Thus if

$$A = \begin{bmatrix} 1 & 2 & 3 \\ 4 & 5 & 6 \\ \hline 7 & 8 & 9 \end{bmatrix}, \ B = \begin{bmatrix} 2 & 4 & 6 \\ 7 & 8 & 9 \\ \hline 1 & 0 & 3 \end{bmatrix}$$

Then the above partition is identical, since we can write

$$A = \begin{bmatrix} A_{11} & A_{12} \\ A_{21} & A_{22} \end{bmatrix} \text{ and } B = \begin{bmatrix} B_{11} & B_{12} \\ B_{21} & B_{22} \end{bmatrix}$$

where

$$A_{11} = \begin{bmatrix} 1 \\ 4 \end{bmatrix} \quad A_{12} = \begin{bmatrix} 2 & 3 \\ 5 & 6 \end{bmatrix}$$

$$A_{21} = [7] \text{ and } A_{22} = [8 \quad 9]$$

and
$$B_{11} = \begin{bmatrix} 2 \\ 7 \end{bmatrix} \quad B_{12} = \begin{bmatrix} 4 & 6 \\ 8 & 9 \end{bmatrix}$$

$$B_{21} = [1] \text{ and } B_{22} = [0 \quad 3]$$

It can be easily seen that

$$A + B = \begin{bmatrix} A_{11} + B_{11} & A_{12} + B_{12} \\ A_{21} + B_{21} & A_{22} + B_{22} \end{bmatrix}$$

In order to partition matrices conformably for multiplication let us consider $A = [a_{ij}]$ and $B = [b_{ij}]$, two $m \times n$ and $n \times p$ matrices so that their product AB exists. The matrix A is in effect partitioned into m matrices of order $1 \times P$ and B into n matrices of order $P \times 1$. Other partitions may be used in such a way that the columns of A and the rows of B are partitioned in exactly the same way. For example

$$A = \begin{bmatrix} (m_1 \times P_1) & (m_1 \times P_2) & (m_1 \times P_3) \\ (m_2 \times P_1) & (m_2 \times P_2) & (m_2 \times P_3) \end{bmatrix}$$

$$B = \begin{bmatrix} (P_1 \times n_1) & (P_1 \times n_2) \\ (P_2 \times n_1) & (P_2 \times n_2) \\ (P_3 \times n_1) & (P_3 \times n_2) \end{bmatrix}$$

or
$$A = \begin{bmatrix} A_{11} & A_{12} & A_{13} \\ A_{21} & A_{22} & A_{23} \end{bmatrix}, B = \begin{bmatrix} B_{11} & B_{12} \\ B_{21} & B_{22} \\ B_{31} & B_{33} \end{bmatrix}$$

However, m_1, m_2, n_1, n_2 may be any nonnegative (including 0) integers such that $m_1 + m_2 = m$ and $n_1 + n_2 = n$. Then

$$AB = \begin{bmatrix} A_{11}B_{11} + A_{12}B_{21} + A_{13}B_{31} & A_{11}B_{12} + A_{12}B_{22} + A_{13}B_{33} \\ A_{21}B_{11} + A_{22}B_{21} + A_{23}B_{31} & A_{21}B_{12} + A_{22}B_{22} + A_{23}B_{33} \end{bmatrix}$$

$$= \begin{bmatrix} C_{11} & C_{12} \\ C_{21} & C_{22} \end{bmatrix} = C.$$

Upper and Lower Triangular Matrices

A square matrix of order $n \times n$ is called *upper triangular matrix* if its elements $a_{ij} = 0$ for $i > j$ for all values ranging from 1 to n. Thus an upper triangular matrix has nonzero elements only in the upper triangle (above the principal diagonal), whereas all the elements in the lower triangle are zero. Thus upper triangular matrix is

$$\begin{bmatrix} a_{11} & a_{12} & \cdots & \cdots & a_{1n} \\ 0 & a_{22} & \cdots & \cdots & a_{2n} \\ 0 & 0 & a_{33} & \cdots & a_{3n} \\ \cdots & \cdots & \cdots & \cdots & \cdots \\ \cdots & \cdots & \cdots & \cdots & \cdots \\ 0 & 0 & 0 & \cdots & a_{2n} \end{bmatrix}$$

Similarly, if all the elements in the upper triangle of a square matrix are zero it is known as lower triangular matrix. Thus for a lower triangular matrix $a_{ij} = 0$ for $i < j$ for all i and j ranging from 1 to n. It has nonzero elements only in the lower triangle (below the principal diagonal). It is given by

$$\begin{bmatrix} a_{11} & 0 & 0 & 0 & \cdots & 0 \\ a_{21} & a_{22} & 0 & 0 & \cdots & 0 \\ a_{31} & 0 & a_{33} & 0 & \cdots & 0 \\ \vdots & & & & & \\ a_{41} & a_{42} & a_{43} & \cdots & \cdots & a_{44} \end{bmatrix}$$

Matrix Division

Let us consider the matrix equations

$$B = AX \text{ and } B = YA$$

where A and B are square matrices of the same order and X, Y, are unknown matrices. There are three possibilities:

i. Suppose A be a nonsingular matrix. Premultiplying the first equation with A^{-1} and postmultiplying second equation with A^{-1}, we have

$$A^{-1}B = A^{-1}AX = X \text{ and } BA^{-1} = YAA^{-1} = Y$$

as such X and Y exist uniquely.

ii. Suppose A is singular, i.e. $|A| = 0$ then we have

$$|B| = |AX| = |A||X| = 0$$

So that B cannot exist unless $|B| = 0$, i.e. B is also singular. Similarly Y can not exist unless B is also singular.

iii. Suppose A and B both are singular then X and Y may or may not exist.

Hence we can conclude that X as well as Y exist uniquely if and only if A is nonsingular. If A is nonsingular then

$X = A^{-1}B$ and $Y = BA^{-1}$ are called the left and right quotients of B by A respectively.

2.5 MINORS AND RANK OF MATRICES

Minors of Matrix

Let A be any $(n \times n)$ square matrix or rectangular matrix $(m \times n)$. The determinant of any $p \times p$ $(p \leq m, p \leq n)$ square submatrix of the matrix A is said to be the *p-square minor* (or *minor of order p*) of matrix A. A single element of A may be considered to be a minor of order 1. The square submatrix so chosen is called principal or leading, submatrix and its determinant is called the principal or leading minor. For example, if

$$A = \begin{bmatrix} a_{11} & a_{12} & a_{13} & a_{14} \\ a_{21} & a_{22} & a_{23} & a_{24} \\ a_{31} & a_{32} & a_{33} & a_{34} \end{bmatrix}$$

then
$$\begin{bmatrix} a_{11} & a_{12} & a_{13} \\ a_{21} & a_{22} & a_{23} \\ a_{31} & a_{32} & a_{33} \end{bmatrix}, \begin{bmatrix} a_{12} & a_{13} & a_{14} \\ a_{22} & a_{23} & a_{24} \\ a_{32} & a_{33} & a_{34} \end{bmatrix}$$

are 3-square minors (or minors of order 3) of A.

and
$$\begin{bmatrix} a_{11} & a_{12} \\ a_{21} & a_{22} \end{bmatrix}, \begin{bmatrix} a_{13} & a_{14} \\ a_{23} & a_{24} \end{bmatrix}$$

etc. are 2-square minors of A while $a_{11}, a_{12}, \dots$ are minors of order 1.

Rank of Matrix

A natural number r is said to be the rank of a matrix A if it has the following two properties:

i. There is at least one nonzero minor of matrix A of order r.

ii. Every minor of A of order $(r + 1)$, if any, vanishes.

As every minor of order $(r + 2)$ can be expanded as the sum of multiples of minor of order $(r + 1)$, then it is implied that every minor of order $(r + 2)$ vanishes if minor of order $(r + 1)$ vanishes. In fact property (ii) implies that every minor of order greater than r will vanish.

Thus briefly we can say that the rank of a matrix is the largest order of any nonvanishing minor of the matrix.

Usually the rank of the matrix A is denoted by the symbol $\rho(A)$.

From the above definition of the rank of the matrix we have the following useful results for determining the rank of a matrix:

i. The rank of every zero (null) matrix is zero.

ii. The rank of every nonzero matrix is ≥ 1.

iii. If every $(r + 1)$th row minor of a matrix vanishes or if matrix does not possess any $(r + 1)$th row minor then the rank of the matrix $\leq r$.

iv. If there is at least one nonzero minor of order r of a matrix then the rank of the matrix is $\geq r$.

v. The rank of every n-square nonsingular matrix is n.

vi. If every $(r + 1)$th row minor of a matrix is zero, then every higher order minor is automatically zero.

vii. The rank of any $m \times n$ matrix is $\leq m$ if $m \leq n$ and less than n if $n \leq m$.

If it is a unit matrix of order n then
$$|I_n| = 1 \neq 0 \text{ so that } \rho(I_n) = n$$
and if A is any diagonal matrix, with nonzero diagonal elements then
$$|A| \neq 0 \text{ so that } \rho(A) = n.$$

The rank of a matrix remains invariant under elementary operations, i.e.

(a) equivalent matrices have the same rank.

(b) the interchange of any two rows does not alter the rank of a matrix.

(c) the multiplication of the elements of a row nonzero number does not alter the rank of the matrix.

2.6 ELEMENTARY TRANSFORMATIONS

The following operations called transformations on a matrix do not change either its order or its rank:

i. The interchange of the ith and jth rows, is denoted by H_{ij}, and the interchange of the ith and jth columns is denoted by k_{ij}.

ii. The multiplication of every element of the ith row by a nonzero scalar p, is denoted by $H_i(p)$ and the multiplication of every element of the ith column by a nonzero scalar p, is denoted by $K_i(p)$.

iii. The addition to the elements of the ith row of p times the corresponding elements of the jth row is denoted by $H_{ij}(p)$ and the addition to the elements of the ith column p times the corresponding elements of the jth column, is denoted by $K_{ij}(p)$. The transformations H are called elementary row transformations and the transformations k are called elementary column transformations.

The inverse of an elementary transformation: The inverse of an elementary transformation is an operation which undergoes the effect of the elementary transformation; that is after A has been subjected to one of the elementary transformations and then the resulting matrix has been subjected to the inverse of that elementary transformation; the final result is matrix A, i.e. if A is a matrix and P is an elementary transformation such that $PA = B$ and if there exists an elementary transformation Q such that $QB = A$ then Q is said to be an inverse of P and denoted by $Q = P^{-1}$. The inverse elementary transformations are:

i. $H_{ij}^{-1} = H_{ij}$ $\qquad$ $K_{ij}^{-1} = K_{ij}$

ii. $H_i^{-1}(p) = H_i(1/p)$ $\qquad$ $K_i^{-1}(p) = K_i(1/p)$

iii. $H_{ij}^{-1}(p) = H_{ij}(-p)$ $\qquad$ $K_{ij}^{-1}(p) = K_{ij}(-p)$

Therefore, inverse of an elementary transformation is an elementary transformation of the same type.

Elementary Matrices: The matrix which results when an elementary row (column) transformation is applied to the identity matrix I_n is called an elementary matrix. It is denoted as E-matrix or by the symbol introduced to denote the elementary transformation which produces the matrix.

For example:

$$\begin{bmatrix} 1 & 0 & 0 \\ 0 & 1 & 0 \\ 0 & 0 & 1 \end{bmatrix} \xrightarrow{H_2} \begin{bmatrix} 0 & 1 & 0 \\ 1 & 0 & 0 \\ 0 & 0 & 1 \end{bmatrix} = K_{12}$$

$$\begin{bmatrix} 1 & 0 & 0 \\ 0 & 1 & 0 \\ 0 & 0 & 1 \end{bmatrix} H_3(p) = \begin{bmatrix} 1 & 0 & 0 \\ 0 & 1 & 0 \\ 0 & 0 & p \end{bmatrix} = K_3(p)$$

$$H_{23}(p) = \begin{bmatrix} 1 & 0 & 0 \\ 0 & 0 & p \\ 0 & 1 & 0 \end{bmatrix} = K_{32}(p)$$

are elementary matrices because they are obtained from the matrix

$$I_3 = \begin{bmatrix} 1 & 0 & 0 \\ 0 & 1 & 0 \\ 0 & 0 & 1 \end{bmatrix}$$

The elementary matrices are represented by bold letters. Thus in general there are three row elementary matrices H_{ij}, $H_i(p)$, $H_{ij}(p)$ and three column elementary matrices K_{ij}, $K_i(p)$, $K_{ij}(p)$ corresponding to six elementary transformations.

Now let $A = \begin{bmatrix} a_{11} & a_{12} & a_{13} \\ a_{21} & a_{22} & a_{23} \\ a_{31} & a_{32} & a_{33} \end{bmatrix}$ then $\xrightarrow{H_{13}}$ $\begin{bmatrix} a_{31} & a_{32} & a_{33} \\ a_{21} & a_{22} & a_{23} \\ a_{11} & a_{12} & a_{13} \end{bmatrix}$

and the elementary matrix

$$H_{13} = \begin{bmatrix} 0 & 0 & 1 \\ 0 & 1 & 0 \\ 1 & 0 & 0 \end{bmatrix}$$

So $H_{13}A = \begin{bmatrix} 0 & 0 & 1 \\ 0 & 1 & 0 \\ 1 & 0 & 0 \end{bmatrix}\begin{bmatrix} a_{11} & a_{12} & a_{13} \\ a_{21} & a_{22} & a_{23} \\ a_{31} & a_{32} & a_{33} \end{bmatrix} = \begin{bmatrix} a_{31} & a_{32} & a_{33} \\ a_{21} & a_{22} & a_{23} \\ a_{11} & a_{12} & a_{13} \end{bmatrix}$

Thus $A \xrightarrow{H_{13}} H_{13}A$ or in general $A \xrightarrow{H_{ij}} H_{ij}A$.

Similarly $A \xrightarrow{H_i(p)} H_i(p)A$ and $A \xrightarrow{H_{ij}(p)} H_{ij}(p)A$.

In this way, every elementary row transformation of a matrix can be brought about by premultiplying (multiplying from left) with the corresponding elementary matrix.

Similarly, column transformations can be obtained by postmultiplying (multiplying from the right) with corresponding elementary matrix, i.e.

$$A \xrightarrow{K_{ij}} AK_{ij} \quad A \xrightarrow{K_i(p)} AK_i(p) \text{ and } A \xrightarrow{K_{ij}(p)} AK_{ij}(p)$$

Equivalent matrices: Two matrices A and B are called equivalent ($A \sim B$) if one can be obtained from the other by a sequence of elementary transformation. Equivalent matrices have the same order and the same rank.

If matrix B is obtained by simply row operations on A, then B is said to be row-equivalent of A and on the other hand if matrix B is obtained by simply column operations on A, then B is said to be column equivalent of A.

If matrix B is equivalent to matrix A, then by definition B can be obtained by performing certain finite elementary row and column operations on A. Let these transformations be designated as $H_1, H_2,...,H_i, K_1, K_2,...,K_i$ where, H_1 is first row transformation, H_2 second..., K_1 is the first column transformation, K_2 the second..., then

$$H_i,..., H_2 \cdot H_1 \cdot A\ K_1 \cdot K_2,...,K_i = PAQ = B$$

where $P = H_i,...,H_2 \cdot H_1$ and $Q = K_1 \cdot K_2,...,K_i$

Thus, we can say that the two matrices A and B are equivalent if and only if there exists two nonsingular matrices P and Q such that

$$B = PAQ$$

Properties of Equivalence of Matrices

(a) **Reflexivity:** Every matrix is equivalent to itself, i.e. $A \sim A$ for
$$A = IAI \text{ so that } P = I \text{ and } Q = I.$$

(b) **Symmetry:** If $B \sim A$ then $A \sim B$.

$B \sim A$ implies that there exists two nonsingular matrices P and Q such that
$$B = PAQ$$
This implies, $\qquad A = P^{-1} BQ^{-1}$

As P^{-1} and Q^{-1} being inverse of P and Q, are nonsingular, therefore
$$A \sim B.$$

(c) **Transitivity:** If $A \sim B$, $B \sim C$, then $A \sim C$, $A \sim B$ and $B \sim C$ imply that there exists nonsingular matrices such that
$$A = PBQ, \quad B = RCS$$
Then $\qquad A = PRCSQ = (PR)(SQ)$

As PR and SQ being the products of nonsingular matrices, therefore it implies that
$$A \sim C$$

Canonical Matrix

A matrix of rank r is said to be canonical when it possesses the following properties:

i. One or more elements of each of the first r rows is nonzero while all other rows have only zero elements.

ii. In the ith row ($i = 1, 2,..., r$) the first nonzero element is 1, let the column in which this element stands be numbered j_i.

iii. $j_1 < j_2 < j_3,..., \le j_r$.

where j_i ($i = 1, 2, 3,...,r$) is the 1st nonzero column number in ith row.

iv. The only nonzero element in the column numbered j_i ($i = 1, 2,...,r$) is the element 1 of the ith row.

For example
$$A = \begin{bmatrix} 1 & 5 & 0 & 3 \\ 0 & 0 & 1 & 2 \\ 0 & 0 & 0 & 0 \end{bmatrix}$$

is a canonical matrix. The rank of this matrix is 2. The first two rows have one and 2 zeroes respectively, and all the elements of 3rd row are zero. The first two has 1st nonzero element 1 which lies in 1st column, i.e. $j_1 = 1$. Similarly $j_2 = 3$ or $j_1 < j_2$ and the only nonzero element in 1st and 3rd columns is 1 satisfying the fourth property. Similarly
$$B = \begin{bmatrix} 1 & 0 & 0 & 2 \\ 0 & 1 & 0 & 4 \\ 0 & 0 & 1 & 3 \\ 0 & 0 & 0 & 0 \end{bmatrix}$$

is also a canonical matrix.

The Normal Form of the Matrix

By means of elementary transformations, any matrix A of rank $r > 0$ can be reduced to one of the forms

$$I_{r'} \begin{bmatrix} I_r & 0 \\ 0 & 0 \end{bmatrix}, [I_r \quad 0], \begin{bmatrix} I_r \\ 0 \end{bmatrix}$$

called its *normal form*. Here I_r is unit matrix of order r and 0 is zero matrix. A zero matrix is its own normal form.

Since both row and column transformations may be used here, the element 1 of the 1st row obtained in the section above can be moved into the first column. Then both the first row and first column can be cleared of other nonzero elements. Similarly, the element 1 of the second row can be brought into the second column and so on. For example, the sequences $H_{21}(-2)$, $H_{31}(1)$, $H_2(1/5)$, $H_{12}(1)$, $H_{32}(-5)$, $K_{41}(-17/5)$, $K_{43}(3/5)$, $K_{21}(-2)$ and K_{23} applied to matrix

$$A = \begin{bmatrix} 1 & 2 & -1 & 4 \\ 2 & 4 & 3 & 5 \\ -1 & -2 & 6 & -7 \end{bmatrix}$$

yields
$$\begin{bmatrix} 1 & 0 & 0 & 0 \\ 0 & 1 & 0 & 0 \\ 0 & 0 & 0 & 0 \end{bmatrix} = \begin{bmatrix} I_2 & 0 \\ 0 & 0 \end{bmatrix}$$

It is essential to note that the normal form of the matrix depends only on its order and rank, i.e. if the order and rank of a matrix is known, its normal form can be written without going into the details of the elementary transformations. For example, the normal form of 3 × 4 matrix of rank 2 will always be

$$\begin{bmatrix} 1 & 0 & 0 & 0 \\ 0 & 1 & 0 & 0 \\ 0 & 0 & 0 & 0 \end{bmatrix} = \begin{bmatrix} I_2 & 0 \\ 0 & 0 \end{bmatrix}$$

and of 3 × 4 matrix of rank 3 will always be

$$\begin{bmatrix} 1 & 0 & 0 & 0 \\ 0 & 1 & 0 & 0 \\ 0 & 0 & 1 & 0 \end{bmatrix} = [I_3 \quad 0]$$

2.7 METHODS OF COMPUTING THE INVERSE OF A GENERAL NONSINGULAR MATRIX

As already defined that if A and B are n square matrices such that $AB = BA = I$, B is called the inverse of A ($B = A^{-1}$) and A is called the inverse of B ($A = B^{-1}$).

An n-square matrix A has an unique inverse if and only if it is nonsingular.

Here are some methods of obtaining the inverse of a general nonsingular matrix:

i. By Elementary Transformations

A nonsingular n-square matrix A can be reduced to its normal form I by elementary row transformations. If in order to bring matrix A into its normal form I, S elementary row transformations are essential and the elementary matrices related to these elementary row transformations be H_1, H_2,...,H_3, then

$$I = H_S,...,H_2 H_1 A.$$

Multiplying both sides with A^{-1}, we have

$$IA^{-1} = H_S,...,H_2 \cdot H_1 \cdot AA^{-1} = H_S,...,H_2 \cdot H_1 \cdot I = A^{-1}$$

We observe that the order in which the elementary row transformation transforms matrix A into identity matrix I, in the same order of elementary row transformations of IA^{-1} is obtain. Hence in order to obtain inverse of n-ordered matrix A, it is written along with identity matrix I_n, $[AI_n]$. Then, simultaneous elementary row transformations are carried out till the matrix A is transformed into an identity matrix, then matrix I_n is converted into A^{-1}. For example, to obtain inverse of matrix

$$A = \begin{bmatrix} 1 & 2 & 3 \\ 1 & 3 & 4 \\ 1 & 4 & 3 \end{bmatrix}$$

We write $[AI_n] = \begin{bmatrix} 1 & 2 & 3 & 1 & 0 & 0 \\ 1 & 3 & 4 & 0 & 1 & 0 \\ 1 & 4 & 3 & 0 & 0 & 1 \end{bmatrix}$

$$\sim \begin{bmatrix} 1 & 2 & 3 & 1 & 0 & 0 \\ 0 & 1 & 1 & -1 & 1 & 0 \\ 0 & 2 & 0 & -1 & 0 & 1 \end{bmatrix} \quad \text{[By } H_{21}(-1) \, H_{31}(-1) \text{ transformations]}$$

$$\sim \begin{bmatrix} 1 & 2 & 3 & 1 & 0 & 0 \\ 0 & 2 & 0 & -1 & 0 & 1 \\ 0 & 1 & 1 & -1 & 1 & 0 \end{bmatrix} \quad \text{[By } H_{32} \text{ transformation]}$$

$$\sim \begin{bmatrix} 1 & 2 & 3 & 1 & 0 & 0 \\ 0 & 1 & 0 & -1/2 & 0 & 1/2 \\ 0 & 0 & 1 & -1 & 1 & 0 \end{bmatrix} \quad \text{[By } H_2(1/2) \text{ transformation]}$$

$$\sim \begin{bmatrix} 1 & 2 & 3 & 1 & 0 & 0 \\ 0 & 1 & 0 & -1/2 & 0 & 1/2 \\ 0 & 0 & 1 & -1/2 & 1 & -1/2 \end{bmatrix} \quad \text{[By } H_{32}(-1) \text{ transformation]}$$

$$\sim \begin{bmatrix} 1 & 0 & 0 & 7/2 & -3 & 1/2 \\ 0 & 1 & 0 & -1/2 & 0 & 1/2 \\ 0 & 0 & 1 & -1/2 & 1 & -1/2 \end{bmatrix} \quad \text{[By } H_{12}(-2) \, H_{13}(-3) \text{ transformations]}$$

Hence $\quad A^{-1} = \begin{bmatrix} 7/2 & -3 & 1/2 \\ -1/2 & 0 & 1/2 \\ -1/2 & 1 & -1/2 \end{bmatrix}$

ii. By Inverse from the Adjoint

We have from Eq. (2.3a)

$$A^{-1} = \frac{\text{adj}\,A}{|A|}$$

Hence A^{-1} can be obtained from adj A and $|A|$.

Hence to obtain the A^{-1} of matrix A of previous example, we have

$$A = \begin{bmatrix} 1 & 2 & 3 \\ 1 & 3 & 4 \\ 1 & 4 & 3 \end{bmatrix}$$

$$|A| = 1(9-16) - 2(3-4) + 3(4-3) = -2$$

$$\text{Adj}\,A = \begin{bmatrix} -7 & 6 & -1 \\ 1 & 0 & -1 \\ 1 & -2 & 1 \end{bmatrix}$$

Hence,

$$A^{-1} = -\frac{1}{2}\begin{bmatrix} -7 & 6 & -1 \\ 1 & 0 & -1 \\ 1 & -2 & 1 \end{bmatrix}$$

$$= \begin{bmatrix} 7/2 & -3 & 1/2 \\ -1/2 & 0 & 1/2 \\ -1/2 & 1 & -1/2 \end{bmatrix}$$

iii. By Solving Algebraic Equations

In general, this method is very difficult but it is useful to discuss theoretically as it helps in solving the equations when the inverse matrix is known by some other method.

Let us find the inverse of n ordered nonsingular matrix.

$$A = \begin{bmatrix} a_{11} & a_{12} & \cdots & a_{1n} \\ a_{21} & a_{22} & \cdots & a_{2n} \\ \cdots & \cdots & \cdots & \cdots \\ a_{n1} & a_{n2} & \cdots & a_{nm} \end{bmatrix}$$

Let $n \times 1$ ordered two matrices

$$X = \begin{bmatrix} x_1 \\ x_2 \\ \vdots \\ x_n \end{bmatrix} \text{ and } Y = \begin{bmatrix} y_1 \\ y_2 \\ \vdots \\ y_n \end{bmatrix}$$

are related as follows.

$$AX = Y$$

i.e.
$$\begin{bmatrix} a_{11} & a_{12} & \cdots & a_{1n} \\ a_{21} & a_{22} & \cdots & a_{2n} \\ \vdots & \cdots & \cdots & \cdots \\ a_{n1} & a_{2n} & \cdots & a_{nn} \end{bmatrix} \begin{bmatrix} x_1 \\ x_2 \\ \vdots \\ x_n \end{bmatrix} = \begin{bmatrix} y_1 \\ y_2 \\ \vdots \\ y_n \end{bmatrix}$$

or
$$\begin{bmatrix} a_{11}x_1 + a_{12}x_2 + a_{13}x_3 + \cdots + a_{1n}x_n \\ a_{21}x_1 + a_{22}x_2 + \cdots + a_{2n}x_n \\ \cdots\cdots\cdots\cdots\cdots\cdots\cdots\cdots \\ a_{n1}x_1 + a_{2n}x_2 + \cdots + a_{nn}x_n \end{bmatrix} = \begin{bmatrix} y_1 \\ y_2 \\ \vdots \\ y_n \end{bmatrix}$$

Since the matrices on both sides are equal, hence equating corresponding elements, we have

$$a_{11}x_1 + a_{12}x_2 + a_{13}x_3 + \cdots + a_{1n}x_n = y_1$$
$$a_{21}x_1 + a_{22}x_2 + a_{23}x_3 + \cdots + a_{2n}x = y_2$$
$$\cdots\cdots\cdots\cdots\cdots\cdots\cdots\cdots$$
$$a_{n1}x_1 + a_{n2}x_2 + a_{n3}x_3 + \cdots + a_{nn}x_n = y_n$$

On solving for $x_1, x_2,..., x_n$ in terms of $y_1, y_2,..., y_n$ let us say we have

$$x_1 = b_{11}y_1 + b_{12}y_2 + \cdots + b_{1n}y_n$$
$$x_2 = b_{21}y_1 + b_{22}y_2 + \cdots + b_{2n}y_n$$
$$\cdots\cdots\cdots\cdots\cdots\cdots\cdots$$
$$x_n = b_{n1}y_1 + b_{n2}y_2 + \cdots + b_{nn}y_n$$

or
$$\begin{bmatrix} x_1 \\ x_2 \\ \vdots \\ x_n \end{bmatrix} = \begin{bmatrix} b_{11} & b_{12} & \cdots & b_{1n} \\ b_{21} & b_{22} & \cdots & b_{2n} \\ \vdots & \cdots & \cdots & \cdots \\ b_{n1} & b_{2n} & \cdots & b_{nn} \end{bmatrix} \begin{bmatrix} y_1 \\ y_2 \\ \vdots \\ y_n \end{bmatrix}$$

i.e.
$$X = BY$$

where
$$B = \begin{bmatrix} b_{11} & b_{12} & \cdots & b_{1n} \\ b_{21} & b_{22} & \cdots & b_{2n} \\ \vdots & \cdots & \cdots & \cdots \\ b_{n1} & b_{2n} & \cdots & b_{nn} \end{bmatrix}$$

or $\quad\quad\quad A^{-1} = AX = A^{-1}Y$

or $\quad\quad\quad X = A^{-1}Y$

i.e. $\quad\quad\quad A^{-1} = B$

Example 1. *Find the inverse of the following by algebraic method of matrix;*

$$A = \begin{bmatrix} 1 & 2 & 3 \\ 1 & 3 & 4 \\ 1 & 4 & 3 \end{bmatrix}$$

Solution: We write $AX = Y$

or
$$\begin{bmatrix} 1 & 2 & 3 \\ 1 & 3 & 4 \\ 1 & 4 & 3 \end{bmatrix} \begin{bmatrix} x_1 \\ x_2 \\ x_3 \end{bmatrix} = \begin{bmatrix} y_1 \\ y_2 \\ y_3 \end{bmatrix}$$

or
$$x_1 + 2x_2 + 3x_3 = y_1$$
$$x_1 + 3x_2 + 4x_3 = y_2$$
$$x_1 + 4x_2 + 3x_3 = y_3$$

On solving for x_1, x_2 and x_3, we have

$$x_1 = \frac{7}{2}y_1 - 3y_2 + \frac{1}{2}y_3$$

$$x_2 = -\frac{1}{2}y_1 + 0y_2 + \frac{1}{2}y_3$$

$$x_3 = \frac{1}{2}y_1 + y_2 - \frac{1}{2}y_3$$

i.e.
$$\begin{bmatrix} x_1 \\ x_2 \\ x_3 \end{bmatrix} = \begin{bmatrix} \frac{7}{2} & -3 & \frac{1}{2} \\ -\frac{1}{2} & 0 & \frac{1}{2} \\ \frac{1}{2} & 1 & -\frac{1}{2} \end{bmatrix} \begin{bmatrix} y_1 \\ y_2 \\ y_3 \end{bmatrix}$$

or
$$A^{-1} = \begin{bmatrix} \frac{7}{2} & -3 & \frac{1}{2} \\ -\frac{1}{2} & 0 & \frac{1}{2} \\ \frac{1}{2} & 1 & -\frac{1}{2} \end{bmatrix}$$

iv. By Partitioning of the Matrix

A given square matrix M of order $(m \times n)$ be partitioned as

$$M = \left[\begin{array}{c|c} A & B \\ \hline C & D \end{array} \right]$$

where A is a matrix of order $m \times m$, B of order $m \times n$, C of order $n \times m$ and D of order $n \times n$. Let the required inverse matrix is

$$Z = \left[\begin{array}{c|c} P & Q \\ \hline R & S \end{array} \right]$$

which is also a square matrix of order $m \times n$ and has been identically partitioned as the given matrix. Then we must have

$$\left[\begin{array}{c|c} A & B \\ \hline C & D \end{array} \right] \left[\begin{array}{c|c} P & Q \\ \hline R & S \end{array} \right] = \left[\begin{array}{c|c} I & 0 \\ \hline 0 & I \end{array} \right]$$

or
$$\left[\begin{array}{c|c} AP+BR & AQ+BS \\ \hline CP+DR & CQ+DS \end{array}\right] = \left[\begin{array}{c|c} I & 0 \\ \hline 0 & I \end{array}\right]$$

which yields the four matrix equations

$$AP + BR = I_m \tag{2.6}$$
$$CP + DR = 0 \tag{2.7}$$
$$AQ + BS = 0 \tag{2.8}$$
$$CQ + DS = I_n \tag{2.9}$$

In solving these equations for the unknown matrices P, Q, R and S, we must remember that B and C are rectangular matrices and hence their inverse is not defined. From Eq. (2.7), we have

$$R = -D^{-1} CP \tag{2.10}$$

using (2.10) in (2.6), we have

$$AP = I_m - BR = I_m + BD^{-1} CP$$

or
$$(A - BD^{-1}C)^{-1} P = I_m$$

so that
$$P = (A - BD^{-1}C)^{-1} \tag{2.11}$$

Substituting the value of P in Eq. (2.10), we have

$$R = -D^{-1}C(A - BD^{-1}C)^{-1} \tag{2.12}$$

Similarly from Eqs (2.8) and (2.9), we have on solving

$$S = (D - CA^{-1}B)^{-1} \tag{2.13}$$
$$Q = -A^{-1}B(D - CA^{-1}B)^{-1} \tag{2.14}$$

Using these in matrix Z, the desired inverse is obtained. Thus, we see that the inverse of a large matrix can be obtained from the inverse of a smaller matrices. As an example let us find inverse of matrix

$$M = \left[\begin{array}{cc|cc} \cos\alpha & \sin\alpha & \cos\beta & \sin\beta \\ -\sin\alpha & \cos\alpha & -\sin\beta & \cos\beta \\ \hline 0 & 0 & \cos\gamma & \sin\gamma \\ -0 & 0 & -\sin\gamma & \cos\beta \end{array}\right]$$

Here
$$A = \begin{bmatrix} \cos\alpha & \sin\alpha \\ -\sin\alpha & \cos\alpha \end{bmatrix} \quad B = \begin{bmatrix} \cos\beta & \sin\beta \\ -\sin\beta & \cos\beta \end{bmatrix}$$

$$C = \begin{bmatrix} 0 & 0 \\ 0 & 0 \end{bmatrix} \quad D = \begin{bmatrix} \cos\gamma & \sin\gamma \\ -\sin\gamma & \cos\gamma \end{bmatrix}$$

and Eq. (2.11) gives
$$P = A^{-1} \begin{bmatrix} \cos\alpha & -\sin\alpha \\ \sin\alpha & \cos\alpha \end{bmatrix}$$

Eq. (2.12) gives
$$R = 0$$

Eq. (2.13) gives
$$S = D^{-1} = \begin{bmatrix} \cos\gamma & -\sin\gamma \\ \sin\gamma & \cos\gamma \end{bmatrix}$$

and Eq. (2.14) gives $Q = -A^{-1}BD^{-1}$

$$= - \begin{bmatrix} \cos\alpha & -\sin\alpha \\ \sin\alpha & \cos\alpha \end{bmatrix} \begin{bmatrix} \cos\beta & \sin\beta \\ -\sin\beta & \cos\beta \end{bmatrix} \begin{bmatrix} \cos\gamma & -\sin\gamma \\ \sin\gamma & \cos\gamma \end{bmatrix}$$

$$= - \begin{bmatrix} \cos(\alpha-\beta+\gamma) & -\sin(\alpha-\beta+\gamma) \\ \sin(\alpha-\beta+\gamma) & \cos(\alpha-\beta+\gamma) \end{bmatrix}$$

Hence

$$M^{-1} = \begin{bmatrix} P & Q \\ R & S \end{bmatrix}$$

$$= \begin{bmatrix} \cos\alpha & -\sin\alpha & -\cos(\alpha-\beta+\gamma) & \sin(\alpha-\beta+\gamma) \\ \sin\alpha & \cos\alpha & -\sin(\alpha-\beta+\gamma) & -\cos(\alpha-\beta+\gamma) \\ 0 & 0 & \cos\gamma & -\sin\gamma \\ 0 & 0 & \sin\gamma & \cos\gamma \end{bmatrix}$$

v. Inverse of a Diagonal Matrix

The inverse of a nonsingular diagonal matrix [diag. $(p_1, p_2,...,p_n)$] is the diagonal matrix

$$\left[\text{diag} \left(\frac{1}{p_1}, \frac{1}{p_2}, ..., \frac{1}{p_n} \right) \right]$$

If $A_1, A_2,..., A_s$ are nonsingular matrices, then inverse of the direct sum diag $(A_1, A_2,..., A_s)$ is

$$\text{diag}(A_1^{-1}, A_2^{-1}, ..., A_s^{-1})$$

For example, let us consider the matrix

$$M = \begin{bmatrix} 1 & 2 & 3 & \vdots & 0 & 0 \\ 1 & 3 & 4 & \vdots & 0 & 0 \\ 1 & 4 & 3 & \vdots & 0 & 0 \\ ... & ... & ... & ... & ... & ... \\ 0 & 0 & 0 & \vdots & 1 & 2 \\ 0 & 0 & 0 & \vdots & 0 & 1 \end{bmatrix}$$

which can be written as

$$M = \begin{bmatrix} A_1 & 0 \\ 0 & A_2 \end{bmatrix}$$

where

$$A_1 = \begin{bmatrix} 1 & 2 & 3 \\ 1 & 3 & 4 \\ 1 & 4 & 3 \end{bmatrix}, \quad A_2 = \begin{bmatrix} 1 & 2 \\ 0 & 1 \end{bmatrix}$$

and its inverse

$$M^{-1} = \begin{bmatrix} A_1^{-1} & 0 \\ 0 & A_2^{-1} \end{bmatrix}$$

But

$$A_1^{-1} = \begin{bmatrix} \dfrac{7}{2} & -3 & \dfrac{1}{2} \\[2mm] -\dfrac{1}{2} & 0 & \dfrac{1}{2} \\[2mm] \dfrac{1}{2} & 1 & -\dfrac{1}{2} \end{bmatrix} \quad \text{and} \quad A_2^{-1} = \begin{bmatrix} 1 & -2 \\ 0 & 1 \end{bmatrix}$$

Hence

$$M^{-1} = \begin{bmatrix} \dfrac{7}{2} & -3 & +\dfrac{1}{2} & 0 & 0 \\[2mm] -\dfrac{1}{2} & 0 & \dfrac{1}{2} & 0 & 0 \\[2mm] \dfrac{1}{2} & 1 & -\dfrac{1}{2} & 0 & 0 \\[2mm] 0 & 0 & 0 & 1 & -2 \\[2mm] 0 & 0 & 0 & 0 & 1 \end{bmatrix}$$

vi. Choleski's Method

This is a very important method of finding inverse matrix and is highly useful for symmetric matrices. This method is also known as square root method.

In this method the symmetric matrix M is written as a product of upper triangular matrix L and its transpose matrix L^T, i.e.

$$M = L^T L.$$

Hence

$$M^{-1} = (L^T L)^{-1} = L^{-1}(T)^{-1} = L^{-1}(L^{-1})^T$$

as

$$(L^T)^{-1} = (L^{-1})^T$$

The use of this is explained in the following example:

Let

$$M = \begin{bmatrix} 1 & 2 & 3 \\ 2 & 3 & 4 \\ 3 & -4 & 4 \end{bmatrix} = L^T L$$

$$= \begin{bmatrix} a & 0 & 0 \\ b & d & 0 \\ c & e & f \end{bmatrix} \begin{bmatrix} a & b & c \\ 0 & d & e \\ 0 & 0 & f \end{bmatrix}$$

or

$$\begin{bmatrix} 1 & 2 & 3 \\ 2 & 3 & 4 \\ 3 & -4 & 4 \end{bmatrix} = \begin{bmatrix} a^2 & ab & ac \\ ab & b^2 + d^2 & bc + de \\ ac & bc + de & c^2 + e^2 + f^2 \end{bmatrix}$$

i.e.

$$a = 1,\ b = 2,\ c = 3,\ d = \sqrt{-1} = i,\ e = 2\sqrt{-1} = 2i$$

and

$$f = \sqrt{-1} = i.$$

or

$$L = \begin{bmatrix} 1 & 2 & 3 \\ 0 & i & 2i \\ 0 & 0 & i \end{bmatrix}$$

Since the inverse of an upper triangular matrix is also an upper triangular matrix, hence

$$L^{-1} = \begin{bmatrix} P & Q & R \\ 0 & S & T \\ 0 & 0 & U \end{bmatrix}$$

Hence

$$\begin{bmatrix} 1 & 2 & 3 \\ 0 & i & 2i \\ 0 & 0 & i \end{bmatrix}\begin{bmatrix} P & Q & R \\ 0 & S & T \\ 0 & 0 & U \end{bmatrix} = \begin{bmatrix} 1 & 0 & 0 \\ 0 & 1 & 0 \\ 0 & 0 & 1 \end{bmatrix}$$

or

$$\begin{bmatrix} P & Q+2S & R+2T+3U \\ 0 & iS & i(T+2U) \\ 0 & 0 & iU \end{bmatrix} = \begin{bmatrix} 1 & 0 & 0 \\ 0 & 1 & 0 \\ 0 & 0 & 1 \end{bmatrix}$$

i.e. $P = 1, S = -i, Q = 2i, U = -i, T = 2i, R = -i$

Hence $M^{-1} = L^{-1}(L^{-1})^T$

$$= \begin{bmatrix} 1 & 2i & -i \\ 0 & -i & 2i \\ 0 & 0 & i \end{bmatrix}\begin{bmatrix} 1 & 0 & 0 \\ 2i & -i & 0 \\ -i & 2i & i \end{bmatrix}$$

$$= \begin{bmatrix} -4 & 4 & -1 \\ 4 & -5 & 2 \\ -1 & 2 & -1 \end{bmatrix}$$

vii. Right Inverse and Left Inverse

If the product of A (may or may not be the square matrix) with B

$$AB = I$$

Then B is said to be right inverse of A

Similarly if $CA = I$, then C is said to be left inverse of A.

Here I is the identity matrix and definitely a square matrix.

Example 2. *If A and B are idempotent matrices, then A + B will be idempotent if and only if AB = BA = 0.*

Solution: As A and B are idempotent matrices, we must have

$$A^2 = A \text{ and } B^2 = B$$

If $A + B$ is idempotent, then

$$(A + B)^2 = A + B$$

We have $(A + B)^2 = (A + B)(A + B)$

$$= A^2 + AB + BA + B^2$$

$$= A + AB + BA + B$$

$$= A + B + (AB + BA)$$

We see that $(A + B)^2$ will be equal to $A + B$, i.e. $A + B$ will be idempotent if and only if

$$AB + BA = 0$$

or
$$AB = BA = 0$$

Example 3. *If A^T and B^T be the transposes of matrices A and B respectively, then*

(a) $(A^T)^T = A$

(b) $(\lambda A)^T = \lambda A^T$, λ *being any scalar (real or complex)*

(c) $(A + B)^T = A^T + B^T$, *A and B being conformable for addition*

(d) $(AB)^T = B^T A^T$, *A and B being conformable for multiplication*

Solution: (a) Let A be a $m \times n$ matrix, i.e.

$$A = [a_{ij}]_{m \times n}$$

Then
$$A^T = [a_{ji}]_{n \times m}$$

Therefore
$$(A^T)^T = [a_{ij}]_{m \times n} = A$$

(b) As λ is a scalar, λA is also, $m \times n$ matrix given by

$$\lambda A = [\lambda a_{ij}]_{m \times n}$$

Then
$$[\lambda A]^T = [\lambda a_{ji}]_{n \times m}$$

Again
$$A^T = [a_{ji}]_{n \times m}$$

So
$$\lambda A^T = [\lambda a_{ji}]_{n \times m} = [\lambda A]^T$$

(c) Let
$$A = [a_{ij}]_{n \times m}$$

Then
$$A^T = [a_{ji}]_{n \times m}$$

and
$$B^T = [b_{ji}]_{n \times m}$$

also
$$A + B = [a_{ij} + b_{ij}]_{n \times m}$$

Therefore
$$(A + B)^T = [a_{ji}]_{n \times m} + [b_{ji}]_{n \times m}$$
$$= A^T + B^T$$

(d) Let
$$A = [a_{ij}]_{m \times n}$$

and
$$B = [b_{ij}]_{n \times p}$$

$$A^T = [a_{ji}]_{n \times m}$$

and
$$B^T = [b_{ji}]_{p \times n}$$

Then $B^T A^T$ is a matrix of order $p \times m$

It is obvious that $C = AB = [c_{ij}]$ is a matrix of order $m \times p$

Then
$$(AB)^T = [c_{ji}]_{p \times m}$$

Thus the matrices $(AB)^T$ and $B^T A^T$ are conformable because each being a matrix of order $p \times m$.

Now $(j, i)^{th}$ element of $(AB)^T$

$$= (i, j)\text{th element of } AB$$

$$= \sum_{k=1}^{n} a_{ik} b_{kj} \text{ for all values of } i \text{ and } j$$

The elements of jth row of B^T are $(b_{1j}, b_{2j}, ..., b_{nj})$ and the elements of the ith column of A^T are $(a_{i1}, a_{i2}, ..., a_{in})$, then the (j, i)th element of $B^T A^T$ is

$$\sum_{k=1}^{n} b_{kj} a_{ik} = \sum_{k=1}^{n} a_{ik} b_{kj} \text{ for all values of } i \text{ and } j.$$

Thus the matrices $(AB)^T$ and $B^T A^T$ are conformable and their corresponding elements are equal. Hence

$$(AB)^T = B^T A^T.$$

This result is known as the *reversal law of transpose*.

Example 4. If $\bar{A}$ and $\bar{B}$ are the conjugates of matrices A and B respectively, then

(a) $(\bar{\bar{A}}) = A$

(b) $(\overline{A+B}) = \bar{A} + \bar{B}$, *A and B being conformable for addition*

(c) $(\overline{\lambda A}) = \bar{\lambda}\,\bar{A}$, *where λ being any complex number and $\bar{\lambda}$ its conjugate*

(d) $(\overline{AB}) = \bar{A}\bar{B}$ *and B are conformable for multiplication*

Solution:

(a) Let $A = [a_{ij}]_{m \times n}$ then $\bar{A} = [\bar{a}_{ij}]_{mxn}$, where $\bar{a}_{ij}$ is the complex conjugate of a_{ij}.

Now $(\bar{\bar{A}})$ = complex conjugate of $[\bar{a}_{ij}]_{mxn} = [a_{ij}]_{mxn} = A.$

(b) Let $\qquad A = [a_{ij}]_{m \times n}$

and $\qquad\qquad B = [b_{ij}]_{m \times n}$

Then $\qquad A + B = [a_{ij} + b_{ij}]_{m \times n}$

So, $\qquad (\overline{A+B})$ = complex conjugate of $[a_{ij} + b_{ij}]_{m \times n}$

$\qquad\qquad\qquad$ = complex conjugate of $[a_{ij}]_{m \times n}$ + complex conjugate of $[b_{ij}]_{m \times n}$

$\qquad\qquad\qquad$ = $[\bar{a}_{ij}]_{mxn} + [\bar{b}_{ij}]_{mxn} = \bar{A} + \bar{B}.$

(c) Let $\qquad A = [a_{ij}]_{m \times n}$

then $\qquad\qquad \bar{A} = [\bar{a}_{ij}]_{mxn}$

and $\qquad\qquad (\overline{\lambda A}) = [(\overline{\lambda a_{ij}})]_{mxn}$

But $\qquad\qquad \overline{\lambda a_{ij}} = \bar{\lambda}\,\bar{a}_{ij}$

$\qquad\qquad\qquad\quad = \bar{\lambda}[\bar{a}_{ij}]_{mxn} = \bar{\lambda}\,\bar{A}$

(d) Let $\quad\quad A = [a_{ij}]_{m\times n}$

and $\quad\quad\quad B = [b_{ij}]_{n\times p}$

Then $\quad\quad\quad \overline{A} = [\overline{a}_{ij}]_{m\times n}$

and $\quad\quad\quad \overline{B} = [\overline{b}_{ij}]_{n\times p}$

It is obvious that $(\overline{AB})$ is a matrix of order $(m \times p)$ and $\overline{A}\,\overline{B}$ is also a $(m \times p)$ matrix.

Hence $(\overline{AB})$ and $\overline{A}\,\overline{B}$ are conformable.

(i, j)th element of $\overline{A}\,\overline{B} = \displaystyle\sum_{k=1}^{n} \overline{a}_{ik}\,\overline{b}_{kj}$ $\quad\quad\quad\quad\quad\quad$ (1)

(i, j)th element of $(AB) = \displaystyle\sum_{k=1}^{n} a_{ik}\,b_{kj}$

Hence (i, j)th element of

$$(\overline{AB}) = \sum_{k=1}^{n} \overline{a}_{ik}\,\overline{b}_{kj} \quad\quad\quad\quad\quad\quad (2)$$

Since Eqs (1) and (2) hold for all values of i and j, we have

$$(\overline{AB})\,\overline{A}\,\overline{B},$$

i.e. the conjugate of the product of two matrices is the product in the same order of their conjugates.

Example 5. *If A^+ and B^+ are the transposed conjugates of matrices A and B respectively, then*

(a) $(A^+)^+ = A$

(b) $(A + B)^+ = A^+ + B^+$, *A and B being conformable for addition*

(c) $(\lambda A)^+ = \overline{\lambda}A^+$, λ *being any complex number and $\overline{\lambda}$ its conjugate*

(d) $(AB)^+ = A^+ B^+$, *A and B being conformable for multiplication*

Solution: (a) Let $\quad A = [a_{ij}]_{m\times n}$

$$A^+ = [\overline{A}^T] = [a_{ji}]_{m\times n}$$

Therefore $(A^+)^+ = $ tranposed conjugate of A^+

$= $ transposed conjugate of $[\overline{a}_{ji}]_{n\times m} = [a_{ij}]_{m\times n} = A$

(b) Let $\quad\quad\quad A = [a_{ij}]_{m\times n}$

and $\quad\quad\quad\quad B = [b_{ij}]_{m\times n}$

then $\quad\quad\quad\quad A^+ = [\overline{a}_{ji}]_{n\times m}$

and $\quad\quad\quad\quad B^+ = [\overline{b}_{ji}]_{n\times m}$

Also $\quad\quad A + B = [a_{ij} + b_{ij}]_{m\times n}$

Hence $\quad (A + B)^+ = [\overline{a_{ji} + b_{ji}}]_{n\times m} = [\overline{a}_{ji} + \overline{b}_{ji}]_{n\times m}$

$\quad\quad\quad\quad\quad = \left([\overline{a}_{ji}]_{m\times n} + [\overline{b}_{ji}]_{n\times m}\right) = A^+ + B^+$

(c) Let $\qquad A = [a_{ij}]_{m \times n}$

then $\qquad (\lambda A)^+ = [\overline{(\lambda a_{ji})}]_{n \times m}$

$$= \overline{\lambda}[\overline{a}_{ji}]_{n \times m} = \overline{\lambda} A^+$$

(d) Let $\qquad A = [a_{ij}]_{m \times n}$

and $\qquad B = [b_{ij}]_{n \times p}$

It is obvious that

$$C = AB = [c_{ij}]_{m \times p}$$

Then $\qquad [AB]^+ = [\overline{c}_{ji}]_{p \times m}$

$$= \text{a matrix of order } p \times m$$

Also it is obvious that $B^+ A^+$ is a matrix of order $p \times m$

Thus matrices $(AB)^+$ and $B^+ A^+$ are conformable because each being a matrix of order $p \times m$.

Now (j, i)th element of $(AB)^+$

$$= (i, j)\text{th element of } (\overline{A} \overline{B})$$

$$= \left(\overline{\sum_{k=1}^{n} (a_{ik} b_{kj})} \right)$$

$$= \sum_{k=1}^{n} \overline{a}_{ik} \overline{b}_{kj} \text{ for all values of } i \text{ and } j.$$

The elements of jth row of B^+ are

$$\overline{b}_{1j}, \overline{b}_{2j}, \overline{b}_{3j}, ..., \overline{b}_{nj}$$

and the elements of ith column of A^+ are $\overline{a}_{1j}, \overline{a}_{2j}, \overline{a}_{3j}, ..., \overline{a}_{ij}$, then (j, i)th element of $B^+ A^+$

$$= \sum_{k=1}^{n} \overline{b}_{kj} \overline{a}_{ik}$$

$$= \sum_{k=1}^{n} \overline{a}_{ik} \overline{b}_{kj} \text{ for all values of } i \text{ and } j.$$

Thus the matrices $(AB)^+$ and $B^+ A^+$ are conformable and their corresponding elements are equal. Hence

$$(AB)^+ = B^+ A^+$$

Example 6. *Show that every square matrix can be uniquely expressed as the sum of a symmetric- and a skew-symmetric matrix.*

Solution: Let $A = [a_{ij}]$ be a square matrix.

Then evidently the matrix A may be expressed as

$$A = \frac{1}{2}(A + A^T) + \frac{1}{2}(A - A^T)$$

where A^T is the transpose of A.

Substituting $P = \dfrac{1}{2}(A + A^T)$ and $Q = \dfrac{1}{2}(A - A^T)$ we have

$$A = P + Q$$

Now transpose of P

$$= P^T = \left\{\frac{1}{2}(A + A^T)\right\}^T = \frac{1}{2}(A^T + A) = P$$

and transpose of $\qquad Q = Q^T = \left\{\frac{1}{2}(A - A^T)\right\}^T$

$$= \frac{1}{2}(A^T - A) = -Q$$

This proves that P is symmetric and Q is skew-symmetric.

Thus the matrix A may be expressed as a sum of a symmetric and a skew-symmetric matrix.

In order to prove that $A = P + Q$ is unique, lets assume

$$A = R + S$$

be another possible representation of A, where R is symmetric and S is skew-symmetric so that

$$R^T = R \text{ and } S^T = -S$$

Then representation $\qquad A = R + S$ gives

$$A^T = (R + S)^T = R^T + S^T = R - S.$$

Therefore $\qquad A + A^T = R + S + R - S = 2R$

and $\qquad A - A^T = R + S - R + S = 2S$

Hence $\qquad R = \dfrac{1}{2}(A + A^T) = P$

and $\qquad S = \dfrac{1}{2}(A - A^T) = Q$

This proves the representation

$$A = P + Q = \frac{1}{2}(A + A^T) + \frac{1}{2}(A - A^T) \text{ is unique.}$$

Example 7. *If A and B are symmetric matrices, then show that AB is symmetric if and only if A and B commute.*

Solution: Let $A = [a_{ij}]_{n \times n}$ and $B = [b_{ij}]_{n \times n}$ be given symmetric matrices, then

$$A^T = A \text{ and } B^T = B \qquad \text{(1)}$$

where A^T and B^T are transposes of A and B respectively. For AB to be symmetric, we have

$$(AB)^T = AB \qquad \text{(2)}$$

From the properties of transposes of matrices we have

$$(AB)^T = B^T A^T = BA \qquad \text{from Eq. (1)}$$

Comparing it with (*ii*), we have

$$(AB)^T = BA = AB$$

i.e. if and only if A and B commute, then only AB is symmetric.

Example 8. Show that:

(a) *Every square matrix can be uniquely expressed as the sum of a Hermitian and a skew-Hermitian matrix.*

(b) *If H is Hermitian matrix, then iH is skew-Hermitian.*

(c) *If A is Hermitian matrix, then B⁺AB is Hermitian for every matrix B.*

[Lucknow]

Solution: (a) Let A be a square matrix.

Then evidently the matrix A can be expressed as

$$A = \frac{1}{2}(A+A^+)+\frac{1}{2}(A-A^+) \tag{1}$$

where A^+ is transpose conjugate of A.

Now substituting

$$P = \frac{1}{2}(A+A^+)$$

and

$$Q = \frac{1}{2}(A-A^+)$$

From Eq. (1), we have

$$A = P + Q \tag{2}$$

We have

$$P^+ = \left\{\frac{1}{2}(A+A^+)^+\right\}$$

$$= \frac{1}{2}\left\{(A^+)+(A^+)^+\right\}$$

$$= \frac{1}{2}\{A^+ + A\} = P$$

and

$$Q^+ = \left\{\frac{1}{2}(A-A^+)\right\}^+$$

$$= \frac{1}{2}\left\{(A^+)-(A^+)^+\right\}$$

$$= \frac{1}{2}\{A^+ - A\} = -Q.$$

This proves that P is Hermitian and Q is skew-Hermitian.

Thus Eq. (2) shows that the matrix A can be expressed as the sum of a Hermitian and a skew-Hermitian matrix. We have now to prove that Eq. (2) is unique.

Let $A = R + S$ be another possible representation of A, where R is Hermitian and S is skew-Hermitian matrix, so that

$$R^+ = R \text{ and } S^+ = -S$$

and

$$A^+ = (R+S)^+ = R^+ + S^+ = R - S$$

Hence

$$A + A^+ = R + S + R - S = 2R$$

or
$$R = \frac{1}{2}(A + A^+) = P$$

and
$$A - A^+ = R + S - R + S = 2S$$

or
$$S = \frac{1}{2}(A - A^+) = Q$$

This proves that $A = P + Q = \frac{1}{2}(A - A^+)$ is unique.

(b) If H is Hermitian matrix, then
$$H^+ = H$$

If iH is skew-Hermitian, we must have
$$i(H)^+ = -iH$$

But
$$(iH)^+ = iH^+ = -iH$$

Thus iH is skew-Hermitian matrix.

(c) If A is Hermitian matrix, then
$$A^+ = A$$

If $B^+ AB$ is Hermitian, matrix, then
$$(B^+ AB)^+ = B^+ AB$$

We have
$$(B^+ AB)^+ = B^+A^+(B^+)^+ = B^+A^+B = B^+AB$$

Since
$$A^+ = A$$

Hence B^+AB is Hermitian.

Example 9. *If A and B are two n-square, nonsingular matrices, then AB, A^T, A^+ are all invertible and prove that*

(a) $(AB)^{-1} = B^{-1}A^{-1}$

(b) $(A^T)^{-1} = (A^{-1})^T$

(c) $(A^+)^{-1} = (A^{-1})$

Solution: As A and B are nonsingular matrices, therefore
$$|A| \neq 0 \text{ and } |B| \neq 0$$

so that
$$|AB| = |A||B| \neq 0$$

This implies AB is a nonsingular matrix and hence invertible.

As $|A^T| = |A| \neq 0$, therefore A^T is invertible.

(a) Let A^{-1} and B^{-1} be the inverse of matrices A and B respectively.

∴
$$AA^{-1} = A^{-1} A = 1$$

and
$$BB^{-1} = B^{-1} B = I$$

Now consider a matrix C given by
$$C = B^{-1}A^{-1}$$

Then
$$C(AB) = (B^{-1}A^{-1})(AB)$$
$$= B^{-1}(A^{-1}A) B = B^{-1} IB$$
$$= B^{-1} B = I$$

i.e.
$$(B^{-1}A^{-1})(AB) = I.$$

Similarly, it can be shown that

$$(AB)(B^{-1}A^{-1}) = I$$

Proving that $B^{-1}A^{-1}$ is inverse of AB, i.e.

$$(AB)^{-1} = B^{-1}A^{-1}$$

Note: This result may be extended to any number of square matrices which are conformable for multiplication. For example:

For three matrices A, B and C, we have

$$(ABC)^{-1} = C^{-1}B^{-1}A^{-1}$$

Here writing $\qquad X = BC$, we have

$$(AX)^{-1} = X^{-1}A^{-1}$$

But $\qquad\qquad X^{-1} = (BC)^{-1} = C^{-1}B^{-1}$

or $\qquad\qquad (ABC)^{-1} = C^{-1}B^{-1}A^{-1}$

Hence for any number of matrices $ABC,...,GH$ we get

$$[ABC,...,GH]^{-1} = H^{-1}G^{-1},...,C^{-1}B^{-1}A^{-1}$$

(b) We have

$$A A^{-1} = I = A^{-1}A$$

Taking transpose we have

$$(A A^{-1})^T = (A^{-1}A)^T = I^T = I.$$

or $\qquad (A^{-1})^T A^T = A^T(A^{-1})^T = I$

From this it follows that $(A^{-1})^T$ is the inverse of A^T, i.e.

$$(A^T)^{-1} = (A^{-1})^T$$

(c) Again taking the conjugate transpose of both sides of

$$A A^{-1} = A^{-1}A = I$$

we have $\qquad (A A^{-1})^+ = (A^{-1}A)^+ = I^+$

or $\qquad (A^{-1})^+ A^+ = A^+(A^{-1})^+ = I$

Since $\qquad\qquad I^+ = I$

which shows that $(A^{-1})^+$ is the inverse of A^+, i.e.

$$(A^+)^{-1} = (A^{-1})^+$$

Example 10. (a) *If A is a real skew-symmetric matrix and $A^2 + I = 0$, then show that A is orthogonal.* (b) *If H is a Hermitian matrix what kind of matrix is e^{iB}?*

Solution:

(a) Let A be the real skew-symmetric matrix

Then $\qquad\qquad A^T = -A$

and $\qquad\qquad A^T A = -A A = -A^2 = I$

Since $\qquad\qquad A^2 + I = 0$

Hence A is orthogonal

(b) Let H be a Hermitian matrix. Then

$$H^+ = H$$

For any matrix M, we have

$$(e^M)^+ = e^{M^+}$$

Let

$$e^{iH} = A$$

Then

$$A^+ A = (e^{iH})^+ e^{iH} = e^{(iH)^+} e^{iH} = e^{-iH^+} e^{iH}$$

$$= e^{-iH} e^{iH}$$

i.e.

$$A^+ A = I,$$

which is the condition for the matrix $A = e^{iH}$ to be unitary. Hence the matrix e^{iH} is unitary matrix.

Example 11. *A matrix $A = [a_{ij}]$ of order n and its inverse $B = [b_{ij}]$ are partitioned into submatrices of indicated orders*

$$\left[\begin{array}{c|c} A_{11} & A_{12} \\ (p \times p) & (p \times q) \\ \hline A_{21} & A_{22} \\ (q \times p) & (q \times p) \end{array}\right] \quad and \quad \left[\begin{array}{c|c} B_{11} & B_{12} \\ (p \times p) & (p \times q) \\ \hline B_{21} & B_{22} \\ (q \times p) & (q \times p) \end{array}\right] \quad where \ p + q = n$$

Find the value of B_{11}, B_{12}, B_{21}, and B_{22} in terms of A_{11}, A_{12}, A_{21} and A_{22}.

Solution: Since B is inverse of A, hence

$$AB = BA = I_n$$

or

$$A_{11} B_{11} + A_{12} B_{21} = I_p \tag{1}$$

$$A_{11} B_{12} + A_{12} B_{22} = 0 \tag{2}$$

$$B_{21} A_{11} + B_{22} A_{21} = 0 \tag{3}$$

$$B_{21} A_{12} + B_{22} A_{22} = I_q \tag{4}$$

Now, let

$$M^{-1} = B_{22}$$

From, Eq. (2), we have

$$A_{11} B_{12} = -A_{12} B_{22}$$

or

$$(A_{11}^{-1} A_{11}) B_{12} = -A_{11}^{-1}(A_{12} B_{22})$$

or

$$B_{12} = -(A_{11}^{-1} A_{12}) M^{-1}$$

From Eq. (3) we have

$$B_{21} = -B_{22} A_{21} A_{11}^{-1} = -M^{-1}(A_{21} A_{11}^{-1})$$

and from Eq. (1)

$$B_{11} = A_{11}^{-1} - A_{11}^{-1} A_{12} B_{21} = A_{11}^{-1} + (A_{11}^{-1} A_{12}) M^{-1}(A_{21}^{-1} A_{11})$$

and finally substituting the value of I_q in Eq. (4)

$$-M^{-1}(A_{21} A_{11}^{-1}) A_{12} + M^{-1} A_{22} = I_q$$

or

$$M^{-1}\left\{ A_{22} - (A_{21} A_{11}^{-1}) A_{12} \right\} = I_q$$

or

$$M = A_{22} - (A_{21} A_{11}^{-1}) A_{12}$$

Example 12. *Compute inverse of the symmetric matrix*

$$A = \begin{bmatrix} 2 & 1 & -1 & 2 \\ 1 & 3 & 2 & -3 \\ -1 & 2 & 1 & -1 \\ 2 & -3 & -1 & 4 \end{bmatrix}$$

Solution: Consider the submatrix $X = \begin{bmatrix} 2 & 1 & -1 \\ 1 & 3 & 2 \\ -1 & 2 & 1 \end{bmatrix}$ partitioned, so that

$$A_{11} = \begin{bmatrix} 2 & 1 \\ 1 & 3 \end{bmatrix}, A_{12} = \begin{bmatrix} -1 \\ 2 \end{bmatrix}, A_{21} = [-1 \ \ 2], A_{22} = [1]$$

Now,
$$A_{11}^{-1} = \begin{bmatrix} \dfrac{3}{5} & -\dfrac{1}{5} \\ -\dfrac{1}{5} & \dfrac{2}{5} \end{bmatrix}$$

and
$$A_{11}^{-1}A_{12} = \begin{bmatrix} \dfrac{3}{5} & -\dfrac{1}{5} \\ -\dfrac{1}{5} & \dfrac{2}{5} \end{bmatrix}\begin{bmatrix} -1 \\ 2 \end{bmatrix} = \begin{bmatrix} -1 \\ 1 \end{bmatrix}$$

Hence
$$M = A_{22} - A_{21}(A_{11}^{-1}A_{12}) = [1] - [-1 \ \ 2]\begin{bmatrix} -1 \\ 1 \end{bmatrix}$$

$$= [-2] \text{ and } M^{-1} = \left[-\dfrac{1}{2}\right]$$

$$\therefore \quad B_{11} = A_{11}^{-1} - A_{11}^{-1}A_{12}B_{21} = A_{11}^{-1} + A_{11}^{-1}A_{12}M^{-1}(A_{21}A_{11}^{-1})$$

$$= \begin{bmatrix} \dfrac{3}{5} & -\dfrac{1}{5} \\ -\dfrac{1}{5} & \dfrac{2}{5} \end{bmatrix} + \begin{bmatrix} -1 \\ 1 \end{bmatrix}\left[-\dfrac{1}{2}\right][-1 \ \ 2]\begin{bmatrix} \dfrac{3}{5} & -\dfrac{1}{5} \\ -\dfrac{1}{5} & \dfrac{2}{5} \end{bmatrix}$$

$$= \begin{bmatrix} \dfrac{3}{5} & -\dfrac{1}{5} \\ -\dfrac{1}{5} & \dfrac{2}{5} \end{bmatrix} + \begin{bmatrix} -1 \\ 1 \end{bmatrix}\left[-\dfrac{1}{2}\right][-1 \ \ 1]$$

$$= \begin{bmatrix} \dfrac{3}{5} & -\dfrac{1}{5} \\ -\dfrac{1}{5} & \dfrac{2}{5} \end{bmatrix} + \begin{bmatrix} -1 \\ 1 \end{bmatrix}\begin{bmatrix} \dfrac{1}{2} & -\dfrac{1}{2} \end{bmatrix}$$

$$= \begin{bmatrix} \dfrac{3}{5} & -\dfrac{1}{5} \\ -\dfrac{1}{5} & \dfrac{2}{5} \end{bmatrix} + \begin{bmatrix} -\dfrac{1}{2} & \dfrac{1}{2} \\ \dfrac{1}{2} & -\dfrac{1}{2} \end{bmatrix} = \dfrac{1}{10}\begin{bmatrix} 1 & 3 \\ 3 & -1 \end{bmatrix}$$

$$B_{12} = -\begin{bmatrix} -1 \\ 1 \end{bmatrix}\begin{bmatrix} -\dfrac{1}{2} \end{bmatrix} = \begin{bmatrix} -\dfrac{1}{2} \\ \dfrac{1}{2} \end{bmatrix}$$

$$B_{21} = -\begin{bmatrix} -\dfrac{1}{2} \end{bmatrix}[-1 \quad 1] = \begin{bmatrix} -\dfrac{1}{2} & \dfrac{1}{2} \end{bmatrix}$$

$$B_{22} = M^{-1} = \begin{bmatrix} -\dfrac{1}{2} \end{bmatrix}$$

Hence

$$X^{-1} = \dfrac{1}{10}\begin{bmatrix} 1 & 3 & -5 \\ 3 & -1 & 5 \\ -5 & 5 & -5 \end{bmatrix}$$

Consider now the matrix A partitioned, so that

$$A_{11} = \begin{bmatrix} 2 & 1 & -1 \\ 1 & 3 & 2 \\ -1 & 2 & 1 \end{bmatrix} A_{12} = \begin{bmatrix} 2 \\ -3 \\ -1 \end{bmatrix}$$

$$A_{21} = [2 \quad -3 \quad -1], \ A_{22} = [4]$$

$$A_{11}^{-1} = X^{-1} = \dfrac{1}{10}\begin{bmatrix} 1 & 3 & -5 \\ 3 & -1 & 5 \\ -5 & 5 & -5 \end{bmatrix}$$

$$A_{11}^{-1}A_{12} = \dfrac{1}{10}\begin{bmatrix} 1 & 3 & -5 \\ 3 & -1 & 5 \\ -5 & 5 & -5 \end{bmatrix}\begin{bmatrix} 2 \\ -3 \\ -1 \end{bmatrix} = \begin{bmatrix} -\dfrac{1}{5} \\ \dfrac{2}{5} \\ -2 \end{bmatrix}$$

and

$$M = A_{22} - A_{21}(A_{11}^{-1}A_{12})$$

$$= [4] - [2 \quad -3 \quad -1]\begin{bmatrix} -\dfrac{1}{5} \\ \dfrac{2}{5} \\ -2 \end{bmatrix} = [4] - \begin{bmatrix} \dfrac{2}{5} \end{bmatrix} = \begin{bmatrix} \dfrac{18}{5} \end{bmatrix}$$

and

$$M^{-1} = \begin{bmatrix} \dfrac{5}{18} \end{bmatrix}$$

and $\qquad B_{21} = -M^{-1}\left(A_{21}\,A_{11}^{-1}\right)$

$$= -\frac{1}{10}\left[\frac{5}{18}\right][2 \quad -3 \quad -1]\begin{bmatrix} 1 & 3 & -5 \\ 3 & -1 & 5 \\ -5 & 5 & -5 \end{bmatrix}$$

$$= -\frac{1}{10}\left[\frac{5}{18}\right][-2 \quad +4 \quad +20]$$

$$= \frac{1}{18}[1 \quad -2 \quad 10]$$

$$B_{22} = M^{-1} = \left[\frac{5}{18}\right]$$

Now $\qquad B_{12} = -(A_{11}^{-1}A_{12})M^{-1}$

$$= -\begin{bmatrix} \dfrac{1}{5} \\[4pt] \dfrac{2}{5} \\[4pt] -2 \end{bmatrix}\left[\frac{5}{18}\right] = \left[\frac{1}{18}\right]\begin{bmatrix} 1 \\ -2 \\ 10 \end{bmatrix}$$

and $\qquad B_{11} = A_{11}^{-1} - A_{11}^{-1}A_{12}A_{21}$

$$= \frac{1}{10}\begin{bmatrix} 1 & 3 & -5 \\ 3 & -1 & 5 \\ -5 & 5 & -5 \end{bmatrix} - \begin{bmatrix} -\dfrac{1}{5} \\[4pt] \dfrac{2}{5} \\[4pt] -2 \end{bmatrix}\frac{1}{18}[1 \quad -2 \quad 10]$$

$$= \frac{1}{10}\begin{bmatrix} 1 & 3 & -5 \\ 3 & -1 & 5 \\ -5 & 5 & -5 \end{bmatrix} - \frac{1}{18}\begin{bmatrix} -\dfrac{1}{5} & \dfrac{2}{5} & -2 \\[4pt] \dfrac{2}{5} & -\dfrac{4}{5} & 4 \\[4pt] -2 & +4 & -20 \end{bmatrix}$$

$$= \frac{1}{90}\begin{bmatrix} 9 & 27 & -45 \\ 27 & -9 & 45 \\ -45 & 45 & -45 \end{bmatrix} - \frac{1}{90}\begin{bmatrix} -1 & 2 & -10 \\ 2 & -4 & 20 \\ -10 & +20 & -100 \end{bmatrix}$$

$$= \frac{1}{90}\begin{bmatrix} 10 & 25 & -35 \\ 25 & -5 & 25 \\ -35 & 25 & +55 \end{bmatrix} = \frac{1}{18}\begin{bmatrix} 2 & 5 & -7 \\ 5 & -1 & 5 \\ -7 & 5 & 11 \end{bmatrix}$$

and $\qquad A^{-1} = \dfrac{1}{18}\begin{bmatrix} 2 & 5 & -7 & 1 \\ 5 & -1 & 5 & -2 \\ -7 & 5 & 11 & 10 \\ 1 & -2 & 10 & 5 \end{bmatrix}$

2.8 SOLUTION OF LINEAR EQUATIONS

Definitions

The matrix methods are useful in solving a set of linear equations. An equation of the form

$$a_1 x_1 + a_2 x_2 + a_3 x_3 + \cdots + a_n x_n = 0$$

is called homogeneous linear equation in variables $x_1, x_2, x_3, \ldots, x_n$ while an equation of the form

$$a_1 x_1 + a_2 x_2 + a_3 x_3 + \cdots + a_n x_n = Y$$

is called inhomogeneous or nonhomogeneous linear equation. A set of linear equations is said to be consistent if the set has either one solution or an infinite number of solutions. Similarly, a set of linear equations is said to be inconsistent if the set has no solution.

Consider a system of linear equations in n unknown variables

$$x_1, x_2, x_3, \ldots, x_n.$$

$$\left.\begin{aligned}
a_{11} x_1 + a_{12} x_2 + a_{13} x_3 + \cdots + a_{1n} x_n &= b_1 \\
a_{21} x_1 + a_{22} x_2 + a_{23} x_3 + \cdots + a_{2n} x_n &= b_2 \\
\vdots \qquad \vdots \qquad\qquad \vdots \qquad \vdots \\
\vdots \qquad \vdots \qquad\qquad \vdots \qquad \vdots \\
a_{m1} x_1 + a_{m2} x_2 + a_{m3} x_3 + \cdots + a_{mn} x_n &= b_m
\end{aligned}\right\} \tag{2.15}$$

This set of equations can be written in the from of matrix equation

$$AX = B \tag{2.15a}$$

where A, X, and B are matrices such as

$$A = [a_{ij}] = \begin{bmatrix} a_{11} a_{12} \ldots a_{1n} \\ a_{21} a_{22} \ldots a_{2n} \\ \vdots \\ \vdots \\ a_{m1} a_{m2} \ldots a_{mn} \end{bmatrix}$$

and is called coefficient matrix of the given system of equations and

$$X = \begin{bmatrix} x_1 \\ x_2 \\ \vdots \\ \vdots \\ x_n \end{bmatrix} \quad \text{and} \quad B = \begin{bmatrix} b_1 \\ b_2 \\ \vdots \\ \vdots \\ b_n \end{bmatrix}$$

The matrix $[AB]$ formed by adding one more column that of B (column matrix) to A is called augmented matrix, i.e.

$$[AB] = \begin{bmatrix} a_{11} a_{12} \ldots a_{1n} \, b_1 \\ a_{21} a_{22} \ldots a_{2n} \, b_2 \\ \vdots \\ \vdots \\ a_{m1} a_{m2} \ldots a_{mn} b_n \end{bmatrix}$$

A homogeneous linear equation in matrix form will be written as

$$AX = 0 \tag{2.16}$$

It is obvious that $X = 0$, i.e. $x_1 = 0$, $x_2 = 0$, $x_n = 0$ is a solution of Eq. (2.16) and this solution is called trivial solution.

Every other solution of Eq. (2.16) is called nontrivial solution.

Fundamental Theorems for the Solution of Linear Equations

i. Homogeneous Linear Equations

Theorem 1. If A be a $m \times n$ matrix of rank r, then the number of linearly independent solutions of the system of equations

$$AX = 0 \text{ is } n - r.$$

Proof: The system of equations is

$$AX = 0 \tag{2.17}$$

where A is a $m \times n$ matrix of rank r and X is $n \times 1$ column matrix.

We know that if A be a $m \times n$ matrix of rank r, there exists nonsingular matrices P and Q such that

$$PAQ = \begin{bmatrix} I_r & 0 \\ 0 & 0 \end{bmatrix} \tag{2.18}$$

Now premultiplying Eq. (2.17) by P, we get

$$PAX = 0$$

and taking $X = QY$, we get

$$PAQY = 0$$

and using Eq. (2.18), we have

$$\begin{bmatrix} I_r & 0 \\ 0 & 0 \end{bmatrix} Y = 0$$

or

$$\begin{bmatrix} I_r & 0 \\ 0 & 0 \end{bmatrix} \begin{bmatrix} Y_1 \\ Y_2 \\ \vdots \\ Y_r \end{bmatrix} = 0, \quad \text{i.e.} \quad \begin{bmatrix} Y_1 \\ Y_2 \\ \vdots \\ Y_i \\ \vdots \\ Y_r \end{bmatrix} = 0$$

which implies that $Y_1 = 0$, $Y_2 = 0,..., Y_r = 0$

and

$$Y = \begin{bmatrix} 0 \\ 0 \\ \vdots \\ 0 \\ Y_{r+1} \\ Y_n \end{bmatrix}$$

Now let
$$Q = \begin{bmatrix} q_{11} & q_{12} & \cdots & q_{1n} \\ q_{21} & q_{22} & \cdots & q_{2n} \\ \vdots & \vdots & & \\ & & & \\ q_{n1} & q_{n2} & \cdots & q_{nn} \end{bmatrix} = [Q_1 Q_2 \cdots Q_n]_{n \times 1}$$

where Q_r is r^{th} column of $Q = \begin{bmatrix} q_{1r} \\ q_{2r} \\ \vdots \\ \vdots \\ q_{nr} \end{bmatrix}$

$\therefore$
$$X = QY = [Q_1 Q_2 \cdots Q_n] \begin{bmatrix} 0 \\ 0 \\ \vdots \\ 0 \\ Y_{r+1} \\ Y_{r+2} \\ \vdots \\ Y_n \end{bmatrix}$$

i.e.
$$X = Q_{r+1} Y_{r+1} + Q_{r+2} Y_{r+2}, ..., Q_n Y_n \qquad (2.19)$$

Thus $x_1, x_2, x_3, ..., x_n$ can be expressed in terms of $(n-r)$th parameters, i.e. $Y_{r+1}, Y_{r+2}, ..., Y_n$.

Now let us consider a particular case of these parameters where $Y_{r+1} = 1$, $Y_{r+2} = 0$, $Y_{r+3} = 0, ..., Y_n = 0$.

Then Eq. (2.19) is reduced to
$$X = Q_{r+1}$$

This indicates that $(r + 1)$th column of Q is a solution of Eq. (2.17). Let this solution be X_1. Then
$$X_1 = Q_{r+1}$$

Similarly, we may consider a particular case where
$$Y_{r+1} = 0, \ Y_{r+2} = 1, \ Y_{r+3} = 0, ..., Y_n = 0.$$

giving
$$X = Q_{r+2} = X_2 \text{ (say).}$$

Repeating the process, we see that
$$X_3 = Q_{r+3}, \ X_4 = Q_{r+4}, ..., X_n = Q_n$$

This shows that each of the last $(n - r)$th columns of Q is a solution of Eq. (2.17). The given system of Eq. (2.17) possess $(n - r)$th linearly independent solutions.

If X is any general solution of Eq. (2.17), then it may be expressed as
$$X = K_{r+1} X_{r+1} + K_{r+2} X_{r+2} + \cdots + K_n X_n$$

where $K_{r+1}, K_{r+2} ... K_n$ are scalars.

Corollary: Equation (2.17) is equivalent to a set of m homogeneous equations in n unknowns $x_1, x_2,...,x_n$.

By above theorem, the number of linearly independent solutions of Eq. (2.17) is $(n-r)$.

Case 1. When $r = n$, i.e. $\rho(A) = n$, then the number of linearly independent solutions of Eq. (2.17) is $n - r = n - n = 0$. This implies that there is no linearly independent solution of Eq. (2.17). In this case $X = 0$ is the only solution of Eq. (2.17).

Case 2. When $r < n$, i.e. $\rho(A) < n$, then the number of linearly independent solutions of Eq. (2.17) is $(n - r)$. Every general solution of Eq. (2.17) can be expressed as a linear combination of $(n - r)$ solutions.

Thus we conclude,

A system of m homogeneous equations in n unknowns has a solution other than the trivial one if and only if the rank of the matrix A of coefficients is less than n.

ii. Nonhomogeneous Equations

Theorem 2. The system $AX = Y$ is consistent (i.e. possesses a solution), if and only if the matrices A and AY have the same rank.

Proof: The given system of equation

$$AX = Y \tag{2.20}$$

is equivalent to n a nonhomogeneous equations, i.e. Eq. (2.15) in n unknown variables, $x_1, x_2, x_3, x_4,...,x_n$. Let $c_1, c_2, c_3,...,c_n$ denote the column vectors of the matrix A. Then the given system of equations may be written as:

$$[c_1, c_2, c_3,...,c_n]\begin{bmatrix} x_1 \\ x_2 \\ \vdots \\ x_n \end{bmatrix} = \begin{bmatrix} y_1 \\ y_2 \\ \vdots \\ y_n \end{bmatrix} \tag{2.21}$$

i.e. $\quad x_1 c_1 + x_2 c_2 + \cdots + x_n c_n = Y \tag{2.21a}$

As the rank of the matrix A is r, i.e. $\rho(A) = r$.

There exist r linearly independent columns of A. Without the loss of generality we can suppose that the first r columns, i.e. $C_1, C_2,...,C_r$ of the matrix A form linearly independent set so that each of the last $(n - r)$ columns i.e. $C_{r+1}, C_{r+2},...,C_n$ of the matrix A can be expressed as a linear combination of the first r-columns.

Let the set of Eq. (2.20) be consistent, i.e. possess a solution. Let this solution be represented by

$$X = \begin{bmatrix} k_1 \\ k_2 \\ \vdots \\ k_n \end{bmatrix} \tag{2.22}$$

Then Eq. (2.21) becomes

$$K_1 C_1 + K_2 C_2 + \cdots + K_n C_n = Y \tag{2.23}$$

We have seen that each of the last $(n - r)$ columns of A, i.e. $C_{r+1}, C_{r+2},...,C_n$ can be expressed as a linear combination of first r columns, i.e. $C_1, C_2, C_3,...,C_r$.

Now Eq. (2.23) implies that Y can be expressed as a linear combination of $C_1, C_2,...,C_r$.

This proves that the maximum number of linearly independent columns of $[A^{-1}]$, the augmented matrix is r. Hence the necessary condition for the set of Eq. (2.20) to be consistent in that rank of $[AY] = \rho[AY] = \rho(A)$.

Now we shall show that system of Eq. (2.20) to be consistent, $\rho(A) = \rho(AY) = r$, is the sufficient condition also,

Let $$\rho(A) = \rho(AY) = r$$

Then there exist a set of r linearly independent columns, i.e. $C_1, C_2,...,C_r$ of $[AY]$.

This means that every other column of $[AY]$ can be expressed as linear combination of r column vectors $C_1, C_2,...,C_r$.

As Y is a column of $[AY]$ hence Y may be expressed as a linear combination of r column vectors $C_1, C_2,...,C_r$ so that there exist r scalars $K_1, K_2,...,K_r$ which are not all zero.

Then
$$Y = K_1 C_1 + K_2 C_2 + K_3 C_3 + \cdots + K_r C_r$$
$$= K_1 C_1 + K_2 C_2 + K_3 C_3 + \cdots + K_r C_r + OC_{r+1} + OC_{r+2} + \cdots + OC_n.$$

or
$$Y = [C_1, C_2, C_3,..., C_n] \begin{bmatrix} k_1 \\ k_2 \\ \vdots \\ k_r \\ 0 \\ 0 \\ 0 \\ \vdots \\ 0 \end{bmatrix}_{n \times 1}$$

Comparing with Eq. (2.21) we get

$$X = \begin{bmatrix} K_1 \\ K_2 \\ \vdots \\ K_r \\ 0 \\ 0 \\ \vdots \\ 0 \end{bmatrix}_{n \times 1} = [K_1, K_2,..., K_r, 0, 0, 0,..., 0]^T$$

is a solution of Eq. (2.20). Hence the system of Eq. (2.21) is consistent.

Thus the necessary and sufficient condition for the system of equation $AX = Y$ to be consistent is that the matrices A and AY are of the same rank.

General Solution

Let $AX = Y$ be a system of consistent nonhomogeneous equation. Then, we have $\rho(AY) = \rho(A) = r$.

Let X_0 be any particular solution of Eq. (2.20) and X_g any general solution. Then

$$AX_0 = Y \text{ and } AX_g = Y.$$

These equations give

$$A(X_g - X_0) = 0$$

This equation shows that $X_g - X_0$ is a solution of the auxiliary equation

$$AX = 0 \tag{2.24}$$

As $\rho(A) = r$, the total number of linearly independent solutions of Eq. (2.24) are $(n - r)$.

Let these $(n - r)$ solutions be $X_1, X_2,...,X_{n-r}$. Then any other solution of Eq. (2.24) is a linear combination of these $(n - r)$ solutions so that there exist $(n - r)$ scalars $K_1, K_2,...,K_{n-r}$ not all zero such that

$$X_g - X_0 = K_1 X_1 + K_2 X_2 + K_3 X_3 +\cdots+ K_{n-r} X_{n-r},$$

i.e.

$$X_g = X_0 + K_1 X_1 + K_2 X_2 +\cdots+ K_{n-r} X_{n-r} \tag{2.25}$$

iii. Cramer's Theorem (Solution of Linear Equations by Determinants)

If the determinants $D = |A|$ of a system of n linear equations $AX = Y$ is not zero, the system has precisely one solution. This solution is given by the formula:

$$x_1 = \frac{D_1}{D}, x_2 = \frac{D_2}{D},..., x_n = \frac{D_n}{D}$$

where D_k is the determinant obtained from D by replacing the Kth column with elements $y_1, y_2,...,y_n$. Here, if the equation is homogeneous and $D \neq 0$, it has only the trivial solution $x_1 = 0, x_2 = 0, x_n = 0$.

Proof: The given system of n equations in n unknown variables $x_1, x_2,...,x_n$ is

$$\left.\begin{array}{c} a_{11}x_1 + a_{12}x_2 + \cdots + a_{1n}x_n = y_1 \\ a_{21}x_1 + a_{22}x_2 + \cdots + a_{2n}x_n = y_2 \\ \cdots\cdots\cdots\cdots\cdots\cdots\cdots\cdots\cdots \\ a_{n1}x_1 + a_{n2}x_2 + \cdots + a_{nn}x_n = y_n \end{array}\right\} \tag{2.26}$$

Here $D = |A| \neq 0$ implies that $\rho(A) = n$ and hence the system has unique solution.

Now let the cofactor of a_{ij} in determinant $|A|$ be A_{ij}. Then multiplying the given Eq. (2.26) by $A_{11}, A_{21},...,A_{n1}$ respectively and adding, we have

$$\sum_{i=1}^{n} a_{i1}A_{i1}x_1 + \sum_{i=1}^{n} a_{i2}A_{i1}x_{n2} +\cdots+ \sum_{i=1}^{n} a_{in}A_{i1}x_n = \sum_{i=1}^{n} y_i A_{i1} \tag{2.27}$$

We observe that all but first term on the left hand side of Eq. (2.27) are zero and the first term on left becomes

$$|A| x_1 = Dx_1$$

and the right term becomes

$$\begin{vmatrix} y_1 & a_{12}...a_{1n} \\ y_2 & a_{22}...a_{2n} \\ \vdots & \\ y_n & a_{n2}...a_{nn} \end{vmatrix} = D_1,$$

i.e. its determinant is obtained by replacing 1st column in $|A|$ by the column of Y which we have named as D_1.

As such we have

$$x_1 = \frac{D_1}{D}.$$

Similarly, by multiplying Eq. (2.26) respectively by $A_{12}, A_{22},...,A_{n2}$ and adding we have

$$x_2 = \frac{D_2}{D}$$

In this way we come to the conclusion that

$$x_k = \frac{D_k}{D} \tag{2.28}$$

Note: Equation Eq. (2.28) tells us that the value of $x_1, x_2,...,x_r$ can be determined by this method only when $D = |A| \neq 0$, i.e. the rank of square matrix A is equal to its order.

***

Example 1. *Solve the following system of equations using Cramer's rule*

$$2x_1 + x_2 + 5x_3 + x_4 = 5$$
$$x_1 + x_2 - 3x_3 - 4x_4 = -1$$
$$3x_1 + 6x_2 - 2x_3 + x_4 = 8$$
$$2x_1 + 2x_2 + 2x_3 - 3x_4 = 2.$$

Solution: We find $D = |A| = \begin{vmatrix} 2 & 1 & 5 & 1 \\ 1 & 1 & -3 & -4 \\ 3 & 6 & -2 & 1 \\ 2 & 2 & 2 & -3 \end{vmatrix} = -120$

$$D_1 = |A| = \begin{vmatrix} 5 & 1 & 5 & 1 \\ -1 & 1 & -3 & -4 \\ 8 & 6 & -2 & 1 \\ 2 & 2 & 2 & -3 \end{vmatrix} = -240$$

and $$x_1 = \frac{D_1}{D} = \frac{-240}{-120} = 2$$

$$D_2 = \begin{vmatrix} 2 & 5 & 5 & 1 \\ 1 & -1 & -3 & -4 \\ 3 & 8 & -2 & 1 \\ 2 & 2 & 2 & -3 \end{vmatrix} = -24$$

and $$x_2 = \frac{D_2}{D} = \frac{-24}{-120} = \frac{1}{5}$$

$$D_3 = \begin{vmatrix} 2 & 1 & 5 & 1 \\ 1 & 1 & -1 & -4 \\ 3 & 6 & 8 & 1 \\ 2 & 2 & 2 & -3 \end{vmatrix} = 0$$

and
$$x_3 = \frac{D_3}{D} = \frac{0}{-120} = 0$$

$$D_4 = \begin{vmatrix} 2 & 1 & 5 & 5 \\ 1 & 1 & -3 & -1 \\ 3 & 6 & -2 & 8 \\ 2 & 2 & 2 & 2 \end{vmatrix} = -96$$

and
$$x_4 = \frac{D_4}{D} = \frac{-96}{-120} = \frac{4}{5}$$

Example 2. *Solve*

$$x_1 + x_2 - 2x_3 + x_4 + 3x_5 = 1$$
$$2x_1 - x_2 + 2x_3 + 2x_4 + 6x_5 = 2$$
$$3x_1 + 2x_2 - 4x_3 - 3x_4 - 9x_5 = 3$$

Solution: The augmented matrix

$$[A \quad B] = \begin{bmatrix} 1 & 1 & -2 & 1 & 3 & 1 \\ 2 & -1 & 2 & 2 & 6 & 2 \\ 3 & 2 & -4 & -3 & -9 & 3 \end{bmatrix}$$

$$\sim \begin{bmatrix} 1 & 1 & -2 & 1 & 3 & 1 \\ 0 & -3 & 6 & 0 & 0 & 0 \\ 0 & -1 & 2 & -6 & -18 & 0 \end{bmatrix}$$

$$\sim \begin{bmatrix} 1 & 1 & -2 & 1 & 3 & 1 \\ 0 & -1 & 2 & 0 & 0 & 0 \\ 0 & -1 & 2 & -6 & -18 & 0 \end{bmatrix}$$

$$\sim \begin{bmatrix} 1 & 1 & 0 & 1 & 3 & 1 \\ 0 & +1 & -2 & 0 & 0 & 0 \\ 0 & -1 & 2 & -6 & -18 & 0 \end{bmatrix}$$

$$\sim \begin{bmatrix} 1 & 0 & 0 & 1 & 3 & 1 \\ 0 & 1 & -2 & 0 & 0 & 0 \\ 0 & 0 & 0 & -6 & -18 & 0 \end{bmatrix}$$

$$\sim \begin{bmatrix} 1 & 0 & 0 & 1 & 3 & 1 \\ 0 & 1 & -2 & 0 & 0 & 0 \\ 0 & 0 & 0 & +1 & 3 & 0 \end{bmatrix}$$

$$\sim \begin{bmatrix} 1 & 0 & 0 & 0 & 0 & 1 \\ 0 & 1 & -2 & 0 & 0 & 0 \\ 0 & 0 & 0 & 1 & 3 & 0 \end{bmatrix}$$

Writing in matrix form, we have

$$\begin{bmatrix} 1 & 0 & 0 & 0 & 0 \\ 0 & 1 & -2 & 0 & 0 \\ 0 & 0 & 0 & 1 & 3 \end{bmatrix} \begin{bmatrix} x_1 \\ x_2 \\ x_3 \\ x_4 \\ x_5 \end{bmatrix} = \begin{bmatrix} 1 \\ 0 \\ 0 \end{bmatrix}$$

i.e. $x_1 = 1,\ x_2 - 2x_3 = 0,\ x_4 + 3x_5 = 0.$

Take $x_3 = a,\ x_5 = b$, where a and b are arbitrary.

The complete solution is

$$x_1 = 1,\ x_2 = 2a,\ x_3 = a,\ x_4 = -3b,\ x_5 = b.$$

or $$X = [1\ \ 2a\ a - 3b\ b]^T.$$

Example 3. *Is the following system of equations consistent?*

$$x + y + z + 3 = 0$$
$$3x + y - 2z + 2 = 0$$
$$2x + 4y + 7z + 7 = 0$$

In case of consistency, find the complete solution.

Solution: The given system of equations is equivalent to the single matrix equation

$$\begin{bmatrix} 1 & 1 & 1 \\ 3 & 1 & -2 \\ 2 & 4 & 7 \end{bmatrix} \begin{bmatrix} x \\ y \\ z \end{bmatrix} = \begin{bmatrix} -3 \\ -2 \\ -7 \end{bmatrix} \tag{1}$$

Applying row operations $H_{21}\,(-3),\ H_{31}\,(-2)$ to the coefficient matrix and also to the matrix on the right hand side of Eq. (1), we have

$$\begin{bmatrix} 1 & 1 & 1 \\ 0 & -2 & -5 \\ 0 & 2 & 5 \end{bmatrix} \begin{bmatrix} x \\ y \\ z \end{bmatrix} = \begin{bmatrix} -3 \\ 7 \\ -1 \end{bmatrix}$$

Applying row operation $H_{32}\,(1)$ we get

$$\begin{bmatrix} 1 & 1 & 1 \\ 0 & -2 & -5 \\ 0 & 0 & 0 \end{bmatrix} \begin{bmatrix} x \\ y \\ z \end{bmatrix} = \begin{bmatrix} -3 \\ 7 \\ 6 \end{bmatrix}$$

i.e.
$$\left. \begin{array}{r} x + y + z = -3 \\ -2y - 5z = 7 \\ 0 = 6 \end{array} \right\} \tag{2}$$

The last segment Eq. (2) being obviously false implies that the system of Eq. (2) is inconsistent. Consequently the given system of equations is inconsistent.

Note: If at any stage of process of reduction of the coefficient matrix to triangular form it is found that a row of the coefficient matrix consists entirely of zeros, and the element in the corresponding row on the right hand side is not zero, the system of equations will be incosistent.

Example 4. *For what value of* μ *the equations*

$$x + y + z = 1$$
$$x + 2y + 4z = \mu$$
$$x + 4y + 10z = \mu^2$$

have a solution and solve them completely.

Solution: The matix equation of the given system is

$$\begin{bmatrix} 1 & 1 & 1 \\ 1 & 2 & 4 \\ 1 & 4 & 10 \end{bmatrix} \begin{bmatrix} x \\ y \\ z \end{bmatrix} = \begin{bmatrix} 1 \\ \mu \\ \mu^2 \end{bmatrix} \tag{1}$$

Applying row operations $H_{21}(-1)$ and $H_{31}(-1)$ to the coefficient matrix and also to the matrix on RHS of Eq. (1), we get

$$\begin{bmatrix} 1 & 1 & 1 \\ 0 & 1 & 3 \\ 0 & 3 & 9 \end{bmatrix} \begin{bmatrix} x \\ y \\ z \end{bmatrix} = \begin{bmatrix} 1 \\ \mu - 1 \\ \mu^2 - 1 \end{bmatrix}$$

Again applying row operation $H_{32}(-3)$ we get

$$\begin{bmatrix} 1 & 1 & 1 \\ 0 & 1 & 3 \\ 0 & 0 & 0 \end{bmatrix} \begin{bmatrix} x \\ y \\ z \end{bmatrix} = \begin{bmatrix} 1 \\ \mu - 1 \\ \mu^2 - 3\mu + 2 \end{bmatrix}$$

This implies that

$$\left.\begin{aligned} x + y + z &= 1 \\ y + 3z &= \mu - 1 \\ 0 &= \mu^2 - 3\mu + 2 \end{aligned}\right\} \tag{2}$$

The last segment of Eq. (2) represents the condition for the given system of equations to be consistent we have

either $\mu = 1$ or $\mu = 2$

Case 1. When $\qquad\qquad \mu = 1$

We have $\qquad\qquad x + y + z = 1 \quad y + 3z = 0$

i.e. $\qquad\qquad\qquad\qquad y = -3z$

and $\qquad\qquad\qquad\qquad x = 1 + 2z$

If $z = a$, where a is any arbitrary number then

$$x = 1 + 2a,\ y = -3a,\ z = a.$$

Case 2. When $\qquad\qquad \mu = 2$

$$x + y + z = 1 \quad y + 3z = 1$$

i.e. $\qquad\qquad\qquad\qquad y = 1 - 3z$

and $\qquad\qquad\qquad\qquad x = 2z$

If $z = b$, where b is any arbitrary number, then

$$x = 2b,\ y = 1 - 3b,\ z = b.$$

As such we have an infinite number of solutions.

Example 5. If $A = \begin{bmatrix} 1 & 2 \\ -2 & 1 \end{bmatrix}$, obtain A^2, Find scalars a and b such that $I + aA + bA^2 = 0$, where

I is the unit matrix and O is the null matrix, both of order two.

Solution: Given $A = \begin{bmatrix} 1 & 2 \\ -2 & 1 \end{bmatrix}$

$$A^2 = AA = \begin{bmatrix} 1 & 2 \\ -2 & 1 \end{bmatrix}\begin{bmatrix} 1 & 2 \\ -2 & 1 \end{bmatrix}$$

$$= \begin{bmatrix} -3 & 4 \\ -4 & -3 \end{bmatrix}$$

Since I and O are unit and null matrices respectively and each of order 2 we have

$$I = \begin{bmatrix} 1 & 0 \\ 0 & 1 \end{bmatrix} \text{ and } O = \begin{bmatrix} 0 & 0 \\ 0 & 0 \end{bmatrix}$$

Therefore $I + aA + bA^2 = 0$ can be expressed as

$$\begin{bmatrix} 1 & 0 \\ 0 & 1 \end{bmatrix} + a\begin{bmatrix} 1 & 2 \\ -2 & 1 \end{bmatrix} + b\begin{bmatrix} -3 & 4 \\ -4 & -3 \end{bmatrix} = \begin{bmatrix} 0 & 0 \\ 0 & 0 \end{bmatrix}$$

or

$$\begin{bmatrix} 1+a-3b & 2a+4b \\ -2a-4b & 1+a-3b \end{bmatrix} = \begin{bmatrix} 0 & 0 \\ 0 & 0 \end{bmatrix}$$

or $1 + a - 3b = 0$ and $2a + 4b = 0$

or $a + 2b = 0$

Solving for a and b we have

$$a = -\frac{2}{5} \text{ and } b = \frac{1}{5}$$

which are the required values of the scalars.

Example 6. *Solve*

$$x_1 - 2x_2 + x_3 - x_4 = 0$$
$$x_1 + x_2 - 2x_3 + 3x_4 = 0$$
$$4x_1 + x_2 - 5x_3 + 8x_4 = 0$$
$$5x_1 - 7x_2 + 2x_3 - x_4 = 0$$

Solution: The coefficient matrix

$$A = \begin{bmatrix} 1 & -2 & 1 & -1 \\ 1 & 1 & -2 & 3 \\ 4 & 1 & -5 & 8 \\ 5 & -7 & 2 & -1 \end{bmatrix} \sim \begin{bmatrix} 1 & 0 & -1 & \frac{5}{3} \\ 0 & 1 & -1 & \frac{4}{3} \\ 0 & 0 & 0 & 0 \\ 0 & 0 & 0 & 0 \end{bmatrix}$$

By applying elementary row transformations $H_{21}(-1)$, $H_{41}(-5)$, $H_{42}(-1)$, $H_{32}(-3)$, $H_2\left(\dfrac{1}{3}\right)$, $H_{12}(2)$.

Hence, we have

$$x_1 - x_3 + \frac{5}{3}x_4 = 0$$

$$x_2 - x_3 + \frac{4}{3}x_4 = 0$$

It is clear that there are 4 unknowns and rank is 2. Hence the value of unknown $4 - 2 = 2$ will be arbitrary. Let $x_3 = a$ and $x_4 = 6$.

Hence $\qquad x_1 = a - \dfrac{5}{3}b, \quad x_2 = a - \dfrac{4}{3}b, \quad x_3 = a, \quad x_4 = b,$

Note: We will have different solutions for different values of a and b. But here $n - r = 2$, hence only two solutions will be linearly independent and others can be expressed as linear combinations of these two. In order to obtain linearly independent solutions, we take one of the unknowns as 0 and other 1 in one solution and in other solution 1st unknown is taken as 1 and the other zero, i.e. for 1st solution.

$$a = 0, \ b = 1$$

Then $\qquad x_1 = -\dfrac{5}{3}, \ x_2 = -\dfrac{4}{3}, \ x_3 = 0, \ x_4 = 1$

$$X = \begin{bmatrix} x_1 \\ x_2 \\ x_3 \\ x_4 \end{bmatrix} = \begin{bmatrix} -\dfrac{5}{3} \\ -\dfrac{4}{3} \\ 0 \\ 1 \end{bmatrix} = X_1 \ \text{(say)}$$

and for 2nd solution

$$a = 1, \ b = 0$$

$$x_2 = 1, \ x_2 = 1, \ x_3 = 1, \ x_4 = 0.$$

or $\qquad X = \begin{bmatrix} x_1 \\ x_2 \\ x_3 \\ x_4 \end{bmatrix} = \begin{bmatrix} 1 \\ 1 \\ 1 \\ 0 \end{bmatrix} = X_2 \ \text{(say)}$

Hence the complete solution is

$$X = C_1 X_1 + C_2 X_2.$$

or $\qquad \begin{bmatrix} x_1 \\ x_2 \\ x_3 \\ x_4 \end{bmatrix} = \begin{bmatrix} C_2 - \dfrac{5}{3} \, C_1 \\ C_2 - \dfrac{4}{3} \, C_1 \\ C_2 \\ C_1 \end{bmatrix}$

Example 7. *Find the general solution of*

$$x_1 + x_2 - 2x_3 + x_4 + 3x_5 = 1$$
$$2x_1 - x_2 + 2x_3 + 2x_4 + 6x_5 = 2$$
$$3x_1 + 2x_2 - 4x_3 - 3x_4 - 9x_5 = 3$$
$$4x_1 - 2x_2 + 4x_3 + 4x_4 + 12x_5 = 4$$

Solution: Particular solution for X_0 is obtained by taking augmented matrix

$$\begin{bmatrix} 1 & 1 & -2 & 1 & 3 & 1 \\ 2 & -1 & 2 & 2 & 6 & 2 \\ 3 & 2 & -4 & -3 & -9 & 3 \\ 4 & -2 & 4 & 4 & 12 & 4 \end{bmatrix}$$

which is transformed to matrix

$$\begin{bmatrix} 1 & 0 & 0 & 0 & 0 & 1 \\ 0 & 1 & -2 & 0 & 0 & 0 \\ 0 & 0 & 0 & 1 & 3 & 0 \\ 0 & 0 & 0 & 0 & 0 & 0 \end{bmatrix}$$

by elementary row transformations $H_{42}(-2)$, $H_{21}(-2)$, $H_2\left(-\dfrac{1}{3}\right)$, H_{32} (+1), $H_3\left(\dfrac{1}{3}\right)$, H_{13} (+1), $H_1\left(\dfrac{1}{2}\right)$, $H_{12}(-1)$, $H_{31}(-1)$, $H_{32}(-1)$.

It is clear that the rank of coefficient matrix and augmented matrix is 3, hence solution is possible. Since unknowns are 5 and rank is 3, thus independent solutions will be $n - r = 3$, hence the values of 2 unknowns will be arbitrary. From the equivalent augmented matrix, we have

$$x_1 = 1, \quad x_2 - 2x_3 = 0, \quad x_4 + 3x_5 = 0$$

If $x_3 = a$, and $x_5 = b$ we have

$$x_1 = 1, \ x_2 = 2a, \ x_3 = a, \ x_4 = -3b, \ x_5 = b.$$

i. Since $x_1 = 1$ as such x_1 can not have arbitrary value.

ii. *a* and *b* can take infinite values, hence there will be infinite possible solutions. In order to find X_0, we have to solve the equation

$$AX = 0$$

Here coefficient matrix

$$A = \begin{bmatrix} 1 & 1 & -2 & 1 & 3 \\ 2 & -1 & 2 & 2 & 6 \\ 3 & 2 & -4 & -3 & -9 \\ 4 & -2 & 4 & 4 & 12 \end{bmatrix}$$

again by elementary row transformations the matrix is

$$\begin{bmatrix} 1 & 0 & 0 & 0 & 0 \\ 0 & 1 & -2 & 0 & 0 \\ 0 & 0 & 0 & 1 & 3 \\ 0 & 0 & 0 & 0 & 0 \end{bmatrix}$$

Hence $x_1 = 0,\quad x_2 - 2x_3 = 0,\quad x_4 + 3x_5 = 0$

taking $x_3 = c$ and $x_5 = d$, we have

$x_1 = 0, x_2 = 2c,\ x_3 = c,\ x_4 = -3d,\ x_5 = d.$

and the complete solution is

$$X = X_0 + X_p$$

$$= \begin{bmatrix} 0 \\ 2c \\ c \\ -3d \\ d \end{bmatrix} + \begin{bmatrix} 1 \\ 2a \\ a \\ -3b \\ b \end{bmatrix}$$

$$= \begin{bmatrix} 1 \\ 2(c+a) \\ (c+a) \\ -3(b+d) \\ b+d \end{bmatrix} = \begin{bmatrix} 1 \\ 2k_1 \\ k_1 \\ -3k_2 \\ k_2 \end{bmatrix}$$

where $k_1 - a + c$ and $k_2 = b + d$.

Since a, b, c and d are arbitrary hence, k_1 and k_2 are also arbitrary.

Note: Here we observe that complete or general solution and particular solution are same, hence particular solution may be called complete solution. This equality is universal. Hence determination of a particular solution is sufficient and it is useless to solve $AX = 0$ and add to the particular solution.

2.9 VECTOR SPACES

A vector in three dimensional space is completely described by three components. If the three components along three mutually perpendicular axes x, y, z be a_1, a_2, a_3, then vector can be expressed as

$$a_1 i + a_2 j + a_3 k.$$

This representation is equivalent to writing $[a_1, a_2, a_3]$ and this ordered set of 3 numbers is called a 3-vector.

In a similar way, in ordered set of n-numbers (complex or real), it is called n-vector or n-dimensional vector. All the n components can be written in the form of row matrix or column matrix, i.e.

$$A = [a_1, a_2, a_3,...,a_n]$$

or $$A = \begin{bmatrix} a_1 \\ a_2 \\ \vdots \\ a_n \end{bmatrix} = [a_1, a_2, a_3,...,a_n]$$

Here the elements $a_1, a_2, a_3,....$ are called the first, second, third... nth components of A respectively.

If each element is zero, then the vector is said to be zero vector.

Algebraic Operations on Vectors

Let A and B be two vectors whose components are $a_1, a_2, a_3,...,a_n$ and $b_1, b_2, b_3,...,b_n$ respectively. Then we have

i. $A + B = [a_1, a_2, a_3,...,a_n] + [b_1, b_2, b_3,...,b_n]$

       $= [a_1 + b_1, a_2 + b_2, a_3 + b_3,...,a_n + b_n]$

ii. If k is a scalar, then

        $KA = k[a_1, a_2, a_3,...,a_n]$

            $= [ka_1, ka_2, ka_3,..., ka_n]$

iii. $A \cdot B = \bar{a}_1 b_1 + \bar{a}_2 b_2 + \cdots + \bar{a}_n b_n$

where $A \cdot B$ is known as inner product of vectors and $\bar{a}_r$ is complex conjugate of a_r. Hence,

$$A \cdot A = \bar{a}_1 a_1 + \bar{a}_2 a_2 + \cdots + \bar{a}_n a_n$$

$$= \|a_1\|^2 + \|a_2\|^2 + \cdots + \|a_n\|^2 = \|A\|^2.$$

Here $\|A\|$ is the magnitude of A and $\|a_r\|$ is magnitude of a_r.

iv. A vector is unit vector $\|A\| = 1$.

v. The vectors A and B will be orthogonal if $A \cdot B$, i.e. their inner product is zero.

Linearly Dependent and Linearly Independent Vectors

A set of mn vectors $X_1, X_2,...,X_n$ are said to be linearly dependent if there exists a set of m nonzero scalars $k_1, k_2,...,k_m$ such that

$$k_1 X_1 + k_2 X_2 + \cdots + k_m X_m = 0 \qquad (2.29)$$

If the Eq. (2.29) is true when $k_1 = k_2 = k_3 = ... = k_m = 0$, the vectors are said to be linear independent.

We can easily prove the following facts:

    i. If on adding a vector X_{m+1} to a set of m linearly independent vectors $X_1, X_2,...,X_m$ makes the whole set linearly dependent then vector X_{m+1} can be expressed as the linear combination of vectors $X_1, X_2, X_3,...,X_m$.

    ii. If in a set of m vectors, $X_1, X_2,...,X_m$, r vectors $(m \geq r > 0)$ are linearly dependent, then whole set is linearly dependent.

    iii. If a set of m vectors is linearly independent then its any subset will be linearly independent.

    iv. A set of mn-vectors can be expressed in the form of a $n \times m$ matrix. If the rank of this matrix is r, then out of m vectors, n vectors are linearly independent and rest $(m - r)$ vectors can be expressed as the combination of r-vectors.

Vector space: Any set of n vectors over a field F which is closed with respect to both addition and scalar multiplication is called a vector space over field F. This vector space is usually denoted by $V_n(F)$ or simply by V_n. The elements of F relative to $V_n(F)$ are called scalars.

If $A_1, A_2,...,A_n$ be a set of n vectors, then the set of all linear combinations $k_1 A_1 + k_2 A_2 + ... + k_r A_r$ is a vector space over F, where $k_1, k_2,...,k_r$ are scalars and belong to field F. This vector space is said to be spanned by the vectors $A_1, A_2,...,A_n$.

Any subset S of a vector space $V_n(F)$ is called a subspace of $V_n(F)$ if S is closed with respect to vector addition and scalar multiplication.

For example, a set of all 3-vectors is a vector space V_3. In this a set of vectors $A_i = [a_i, 0, 0]^T$ is also closed with respect to vector addition and scalar multiplication, as such this set of vectors is a vector subspace. The number of vectors in vector space or vector subspace may be finite or infinite. Vector $[a_i, 0, 0]^T$ is a vector subspace containing only one vector.

A vector space V is said to be *spanned* by set of vectors $X_1, X_2,..., X_m$, when X_i ($i = 1, 2,..., m$) are members of vector space V and each vector of vector space V can be expressed as the linear combination $X_1, X_2,..., X_m$. (It is not necessary that spanning set $X_1, X_2, X_3,..., X_m$ is linearly independent).

The *dimension* of a vector space is defined as the maximum number of linearly independent vectors in the space. It is also defined as the minimum number of linearly independent vectors required to span vector space V. It is easy to see that the two definitions are equivalent. A n-vector space having dimension r is also represented by V_n^r.

A set of n-linearly independent vectors in n-dimensional vector space V_n is called the *basis* for the vector space. Obviously, the basis is not unique and may be chosen in an infinite number of ways.

If the basis set of vector space V_n^r is a set of mutually perpendicular vectors, the basis is said to be *orthogonal basis*. If these mutually orthogonal r vectors happen to be unit vectors the basis is called *orthogonal basis*.

Each of the m-rows of a matrix A of order $m \times n$ consists of n elements and therefore is an n-vector. Similarly, each of n-columns, consisting of m elements, is a m vector.

A *row space* of A means a subspace of V_n spanned by m rows and the *column space* means a subspace V_n spanned by n-columns. The dimensions of row and column spaces are, known as row rank and column rank of the matrix respectively.

Two subspaces $_1V_n^r$ and $_2V_n^r$ are said to be *identical subspaces* if each vector of $_1V_n^r$

is a vector of $_2V_n^r$ and each vector of $_2V_n^r$ is a vector of $_1V_n^r$.

The subspace generated by the vector X such that $XA = 0$, is said to be *row null space* of matrix A and the dimension of the subspace is called *row-nullity* of matrix A. Again the subspace generated by the vector Y such that $AY = 0$ is said to be *column null space* of matrix A and its dimension is called the *column nullity* of the matrix A. The nullity of matrix A is represented by N_A.

i. If the rank of $m \times n$ matrix A is r, then $N_A = n - r$.

ii. Rank of A + column nullity of A = number of columns in A.

iii. Rank of A + row nullity of A = number of rows in A.

iv. In a square matrix row nullity and column nullity are equal.

Elementary vectors: The following n vectors are called elementary vectors.

$$E_1 = [1, 0, 0...0]^T$$
$$E_2 = [0, 1, 0...0]^T$$
$$E_3 = [0, 0, 1...0]^T \qquad (2.30)$$
$$\vdots \quad \vdots \quad \vdots \quad \vdots$$
$$E_n = [0, 0, 0...1]^T$$

The vector E_j whose jth element is 1 and all the other elements are zero is called jth elementary vector. All these elementary vectors are mutually perpendicular to each other and linearly independent and all have unit magnitude. Any vector of vector space V_n can be expressed as the linear combination of these n-vectors. Hence a set of these vectors is a very important basis set. We will call it here as E-base. E-base is an example of orthogonal basis.

Any vector $X = [x_1, x_2,...,x_n]^T$ can be written with the help of E-base.

$$X = \sum_{i=1}^{n} x_i E_i = x_1 E_1 + x_2 E_2 + \cdots + x_n E_n \qquad (2.31)$$

Elements $x_1, x_2,...,x_n$ are called the coordinates of X in E-base.

If $z_1, z_2, z_3,...,z_n$ be another basis of V_n, there exist unique scalar $a_1, a_2,...,a_n$ in F such that

$$X = \sum_{i=1}^{n} a_i z_i = a_1 z_1 + a_2 z_2 + \cdots + a_n z_n$$

These scalars $a_1, a_2, a_3,...,a_n$ are called the coordinates of X relative to the z-basis. Writing,

$$X_z = [a_1, a_2,..., a_n]^T$$

We have $\qquad X = [z_1, z_2,...,z_n], \ X_z = zX_z$

or $\qquad X_z = z^{-1}X \qquad (2.32)$

where z is the matrix whose columns are the basis vectors $z_1, z_2,...,z_n$.

2.10 LINEAR TRANSFORMATION

Let $X = [x_1, x_2,...,x_n]^T$ and $Y = [y_1, y_2,...,y_n]^T$ be two vectors of V_n, their coordinates being relative to the same basis of the space. Suppose that the coordinates of X and Y are related by

$$y_1 = a_{11}x_1 + a_{12}x_2 + \cdots + a_{1n}x_n$$
$$y_2 = a_{21}x_1 + a_{22}x_2 + \cdots + a_{2n}a_n$$
$$\cdots\cdots\cdots\cdots\cdots\cdots\cdots\cdots\cdots \qquad (2.33)$$
$$y_n = a_{n1}x_1 + a_{n2}x_2 + \cdots + a_{nn}x_n$$

In matrix form this equation can be written as

$$AX = Y$$

where $\qquad A = [a_{ij}] \qquad (2.33a)$

The Eqs (2.33) and (2.33a) can be considered as transformation equation which transforms any vector X of V_n into another vector Y of the same space. Vector Y is called the *image* of vector A. This transformation is called *linear* because if it transform X_1 into Y_1, then

i. it carries KX_1 into KY_1, for every scalar K, and

ii. it carries $ax_1 + bx_2$ into $ay_1 + by_2$, for every pair of scalars a and b.

If $|A| \neq 0$, the transformation is called *nonsingular* and if $|A| = 0$, the transformation is called *singular*.

The rank of transformation matrix A is said to be the rank of transformation.

A nonsingular linear transformation carries linearly independent vectors into linearly independent vectors.

Under a nonsingular transformation the image of a vector space $V_n^r(F)$ is a vector $V_n^r(F)$, that is the dimension of the vector space is preserved. In particular the transformation is a mapping of $V_n(F)$ onto itself.

The elementary vectors E_i of V_n may be transformed into any set of n linearly independent n-vectors by a nonsingular linear transformation and conversely.

If transformation $Y = AX$ transforms a vector X into Y and if $Z = BY$ carries Y into Z. then $Z = BY = (BA)X$ carries the vector X into Z.

When any two sets of n linearly independent n-vectors are given, there exists a nonsingular linear transformation which carries the vectors of one set into the vectors of the other.

Change of base: Relative to z-base let $Y_z = AX_z$ be a linear transformation of V_n, and let X_w and Y_w be the coordinates of X_z and Y_z relative to base w. Then using Eq. (2.11) there exists a nonsingular matrix P such that $X_w = PX_z$ and $Y_w = PY_z$, or setting $P^{-1} = Q$, such that

$$X_z = QX_w \text{ and } Y_z = QY_w$$

then
$$Y_w = Q^{-1}Y_z = Q^{-1}AX_z = Q^{-1}AQX_w = BX_w$$

where
$$B = Q^{-1}AQ.$$

Two matrices A and B, such that there exists a nonsingular matrix Q for which $B = R^{-1}AQ$ are called *singular*, i.e. if $Y_z = AX_z$ is a linear transformation of V_n relative to a given basis Z and $Y_w = BX_w$ is the same linear transformation relative to another basis w, then A and B are similar. This transformation of matrix A into B is known as *similarity transformation*.

Here we may write:

$$QBQ^{-1} = QQ^{-1} AQQ^{-1} = IAI = A$$

A matrix equation retains its form under similarity transformation.

For example, if $\qquad AB = C$

then $\qquad\qquad Q^{-1}(AB)Q = Q^{-1}CQ$

i.e. $\qquad\qquad Q^{-1}AQQ^{-1}BQ = Q^{-1}CQ$

or $\qquad\qquad (Q^{-1}AQ)(Q^{-1}BQ) = Q^{-1}CQ$

i.e. $\qquad\qquad\qquad A'B' = C'$

where A', B' and C' are related to corresponding matrices A, B and C by similarity transformation through Q.

2.11 UNITARY AND ORTHOGONAL TRANSFORMATIONS

Consider a linear transformation

$$Y = AX$$

such that X and Y are column vectors of order $n \times 1$ and A is a square transformation matrix of order $n \times n$. If the matrix A of transformation is unitary, the linear transformation is said to be a unitary transformation. From the definition of unitary matrix $A^+ A = I$. Hence we have

$$Y^+Y = (AX)^+ (AX) = X^+ A^+ AX = X^+X$$

or
$$\sum_{i=1}^{n} y_i^* y_i = \sum_{i=1}^{n} x_i^* x_i$$

This shows that the magnitude (norm) of the vector remains invariant under a unitary transformation.

As an inverse case consider a linear transformation matrix A which leaves the magnitude of every vector in vector space unchanged that is

$$\|Y\| = \|X\| \quad \text{or} \quad \|AX\| = \|X\|$$

or
$$X^+X = (AX)^+ (AX) = X^+A^+AX.$$

Obviously $A^+A = I$, showing A a unitary matrix. Thus, a linear transformation acting on a vector space is unitary if and only if it leaves the magnitude of every vector in vector space unchanged.

If the matrix of linear transformation is further restricted to be real, it is called a *real orthogonal transformation*. In this case we have

$$Y^+Y = X^TA^TAX = X^TX \qquad (2.34)$$

i.e.
$$\sum_{i=1}^{n} y_i^2 = \sum_{i=1}^{n} x_i^2 \qquad (2.34a)$$

Obviously the condition of transformation $Y = AX$ to be orthogonal is that the transformation matrix A satisfies the condition $A^+A = AA^+ = I$ and transformed vectors satisfy the condition of both Eqs (2.13) and (2.13a).

Example 1. *A certain linear transformation $Y = AX$ carries linearly independent vectors $X_1 = [1, 0, 1]^T$, $X_2 = [1, -1, 1]^T$ and $X_3 = [1, 2, -1]^T$ into $Y_1 = [2, 3, -1]^T$, $Y_2 = [3, 0, -2]^T$, $Y_3 = [-2, 7, -1]^T$, respectively. Find the images of elementary vectors E_1, E_2, E_3 and matrix A.*

Solution: The elementary vectors E_1, E_2 and E_3 can be expressed as linear combination of linearly independent vectors X_1, X_2 and X_3. Let

$$E_1 = a_1X_1 + a_2X_2 + a_3X_3$$

or
$$E_1 = \begin{bmatrix} 1 \\ 0 \\ 0 \end{bmatrix} = a_1 \begin{bmatrix} 1 \\ 0 \\ 1 \end{bmatrix} + a_2 \begin{bmatrix} 1 \\ -1 \\ 1 \end{bmatrix} + a_3 \begin{bmatrix} 1 \\ 2 \\ -1 \end{bmatrix} = \begin{bmatrix} a_1 + a_2 + a_3 \\ -a_2 + 2a_3 \\ a_1 + a_2 - a_3 \end{bmatrix}$$

or
$$a_1 = -\frac{1}{2}, a_2 = 1 \text{ and } a_3 = \frac{1}{2}$$

i.e.
$$E_1 = -\frac{1}{2}X_1 + X_2 + \frac{1}{2}X_3$$

Hence image of
$$E_1 = AE_1 = A\left[-\frac{1}{2}X_1 + X_2 + \frac{1}{2}X_3 \right]$$

$$= -\frac{1}{2}Y_1 + Y_2 + \frac{1}{2}Y_3 = [1, 2, -2]^T$$

E_2 and E_3 can be similarly determined and their images AE_2 and AE_3 obtained. Image of $E_2 = AE_2 = [-1, 3, 1]^T$ and image $E_3 = AE_3[1, 1, 1]^T$

To obtain matrix A let $A = [a_{ij}]$. Hence

$$AE_1 = \begin{bmatrix} a_{11} & a_{12} & a_{13} \\ a_{21} & a_{22} & a_{23} \\ a_{31} & a_{32} & a_{33} \end{bmatrix} \begin{bmatrix} 1 \\ 0 \\ 0 \end{bmatrix} = \begin{bmatrix} a_{11} \\ a_{21} \\ a_{31} \end{bmatrix} = \begin{bmatrix} 1 \\ 2 \\ -2 \end{bmatrix}$$

Similarly $\qquad AE_2 = [a_{12}, a_{22}, a_{32}]^T = [-1, 3, 1]^T$

and $\qquad AE_3 = [a_{13}, a_{23}, a_{33}]^T = [1, 1, 1]^T$

Hence $\qquad A = \begin{bmatrix} 1 & -1 & 1 \\ 2 & 3 & 1 \\ -2 & 1 & 1 \end{bmatrix}$

Example 2. *The vector* $X = [2, 3, 4]^T$ *is relative to basis* $Z_1 = [1, 1, 0]^T$, $Z_2 = [1, 0, 1]^T$, $Z_3 = [1, 1, 1]^T$. *Find its components relative to basis* $w_1 = [1, 1, 2]^T$, $w_2 = [2, 2, 1]^T$, $w_3 = [1, 2, 2]^T$ *If relative to z-basis the linear transformation is given by*

$$Y_z - AX_z - \begin{bmatrix} 1 & 1 & 2 \\ 2 & 2 & 1 \\ 3 & 1 & 2 \end{bmatrix} X_z$$

then find matrix which satisfies the linear transformation.

$$Y_w = BX_w \text{ relative to w-basis.}$$

Solution: We have

$$X_w = w^{-1} ZX_z.$$

where X_w is a vector relative to w-basis.

$$w\text{-basis vector} = \begin{bmatrix} 1 & 2 & 1 \\ 1 & 2 & 2 \\ 2 & 1 & 2 \end{bmatrix} \qquad \therefore \ w^{-1} = \frac{1}{3} \begin{bmatrix} 2 & -3 & 2 \\ 2 & 0 & -1 \\ -3 & 3 & 0 \end{bmatrix}$$

and $z\text{-basis vector} = \begin{bmatrix} 1 & 1 & 1 \\ 1 & 0 & 1 \\ 0 & 1 & 1 \end{bmatrix}$

$$\therefore \qquad X_w = \frac{1}{3} \begin{bmatrix} 2 & -3 & 2 \\ 2 & 0 & 1 \\ -3 & 3 & 0 \end{bmatrix} \begin{bmatrix} 1 & 1 & 1 \\ 1 & 0 & 1 \\ 0 & 1 & 1 \end{bmatrix} \begin{bmatrix} 2 \\ 3 \\ 4 \end{bmatrix} = \begin{bmatrix} 14/3 \\ 11/3 \\ -3 \end{bmatrix}$$

If we have $\qquad Y_z = AX_z$ and $Y_w = BX_w$

Then $\qquad B = QAQ^{-1}$, where $Q = w^{-1}Z.$

$$\therefore \qquad Q = \frac{1}{3} \begin{bmatrix} 2 & -3 & 2 \\ 2 & 0 & -1 \\ -3 & 3 & 0 \end{bmatrix} \begin{bmatrix} 1 & 1 & 1 \\ 1 & 0 & 1 \\ 0 & 1 & 1 \end{bmatrix} = \frac{1}{3} \begin{bmatrix} -1 & 4 & 1 \\ 2 & 1 & 1 \\ 0 & -3 & 0 \end{bmatrix}$$

and $\quad Q^{-1} = \begin{bmatrix} -1 & 1 & -1 \\ 0 & 0 & -1 \\ 2 & 1 & 3 \end{bmatrix}$

Hence $\quad B = QAQ^{-1} = \dfrac{1}{3}\begin{bmatrix} -1 & 4 & 1 \\ 2 & 1 & 1 \\ 6 & -3 & 0 \end{bmatrix}\begin{bmatrix} 1 & 1 & 2 \\ 2 & 2 & 1 \\ 3 & 1 & 2 \end{bmatrix} = \begin{bmatrix} -1 & 1 & -1 \\ 0 & 0 & 1 \\ 2 & 1 & 3 \end{bmatrix}$

$$= \dfrac{1}{3}\begin{bmatrix} -2 & 14 & -6 \\ 7 & 14 & 9 \\ 0 & -9 & 3 \end{bmatrix} = \begin{bmatrix} -2/3 & 14/3 & -2 \\ 7/3 & 14/3 & 3 \\ 0 & -3 & 1 \end{bmatrix}$$

Example 3. *Relative to basis* $Z = \begin{bmatrix} 1 & 0 \\ 0 & 1 \end{bmatrix}$, *the vectors X, Y and Pauli's spin matrices* $\sigma_{x'}$ σ_y *and* σ_z *are*

$$X = \dfrac{1}{\sqrt{2}}\begin{bmatrix} 1 & 1 \\ 1 & -1 \end{bmatrix}, Y = \dfrac{1}{\sqrt{2}}\begin{bmatrix} 1 & 0 \\ i & -i \end{bmatrix},$$

$$\sigma_x = \sigma_x = \begin{bmatrix} 0 & 1 \\ 1 & 0 \end{bmatrix},$$

$$\sigma_y = \begin{bmatrix} 0 & -i \\ i & 0 \end{bmatrix}, \sigma_z = \begin{bmatrix} 1 & 0 \\ 0 & -1 \end{bmatrix}$$

Find the vectors **Z, X, Y** *and matrices* $\sigma_{x'}$ $\sigma_{y'}$ σ_z *relative to base X.* **[Udaipur]**

Solution: For transformation from Z-base to X-base the transformation matrix is $X^{-1}Z$.

Here X is a unitary matrix, hence

$$X^{-1} = X^+ = \dfrac{1}{\sqrt{2}}\begin{bmatrix} 1 & 1 \\ 1 & -1 \end{bmatrix} = X$$

∴ $\qquad X_z^{-1} = \dfrac{1}{\sqrt{2}}\begin{bmatrix} 1 & 1 \\ 1 & -1 \end{bmatrix}\begin{bmatrix} 1 & 0 \\ 0 & 1 \end{bmatrix} = \dfrac{1}{\sqrt{2}}\begin{bmatrix} 1 & 1 \\ 1 & -1 \end{bmatrix} = U$ (say)

Let the vectors **Z, X, Y,** and $\sigma_{x'}$ σ_y and σ_z relative to X be **Z′, X′ Y′,** $\sigma'_{x'}$ σ'_y and σ'_z. Then we have

$$Z' = UZ = \dfrac{1}{\sqrt{2}}\begin{bmatrix} 1 & 1 \\ 1 & -1 \end{bmatrix}\begin{bmatrix} 1 & 0 \\ 0 & 1 \end{bmatrix} = \dfrac{1}{\sqrt{2}}\begin{bmatrix} 1 & 1 \\ 1 & -1 \end{bmatrix}$$

$$X' = UX = \dfrac{1}{\sqrt{2}}\begin{bmatrix} 1 & 1 \\ 1 & -1 \end{bmatrix}\dfrac{1}{\sqrt{2}}\begin{bmatrix} 1 & 1 \\ 1 & -1 \end{bmatrix}$$

$$= \dfrac{1}{2}\begin{bmatrix} 2 & 0 \\ 0 & 2 \end{bmatrix} = \begin{bmatrix} 1 & 0 \\ 0 & 1 \end{bmatrix}$$

$$Y' = UY = \frac{1}{\sqrt{2}}\begin{bmatrix} 1 & 1 \\ 1 & -1 \end{bmatrix}\frac{1}{\sqrt{2}}\begin{bmatrix} 1 & 1 \\ i & -i \end{bmatrix}$$

$$= \frac{1}{2}\begin{bmatrix} 1+i & 1-i \\ 1-i & 1+i \end{bmatrix}$$

Again $\sigma'_x = U\sigma_x U^{-1}$

Now $U^{-1} = U^+ = \frac{1}{\sqrt{2}}\begin{bmatrix} 1 & 1 \\ 1 & -1 \end{bmatrix} = U$ (again)

∴ $\sigma'_x = \frac{1}{\sqrt{2}}\begin{bmatrix} 1 & 1 \\ 1 & -1 \end{bmatrix}\begin{bmatrix} 0 & 1 \\ 1 & 0 \end{bmatrix}\frac{1}{\sqrt{2}}\begin{bmatrix} 1 & 1 \\ 1 & -1 \end{bmatrix}$

$$= \frac{1}{2}\begin{bmatrix} 1 & 1 \\ -1 & 1 \end{bmatrix}\begin{bmatrix} 1 & 1 \\ 1 & -1 \end{bmatrix} = \frac{1}{2}\begin{bmatrix} 2 & 0 \\ 0 & -2 \end{bmatrix} = \begin{bmatrix} 1 & 0 \\ 0 & 1 \end{bmatrix}$$

Similarly $\sigma'_y = U\sigma_y U^{-1} = \begin{bmatrix} 0 & i \\ -1 & 0 \end{bmatrix}$

and $\sigma'_z = U\sigma_z U^{-1} = \begin{bmatrix} 0 & 1 \\ 1 & 0 \end{bmatrix}$

Example 4. *Prove that the trace of a matrix remains invariant under similarity transformation.*

Solution: We have $Q^{-1}AQ = B$ for similarity transformation. The trace of a matrix is the sum of its diagonal terms, i.e.

$$\text{Trace } B = \sum_i B_{ii} = \text{Trace } (Q^{-1}AQ) = \sum_i (Q^{-1}AQ)_{ii}$$

$$= \sum_i \sum_{jk} Q^{-1}_{ij} A_{jk} Q_{ki}$$

$$= \sum_i \sum_j \sum_k (Q^{+1}_{ki} Q^{-1}_{ij}) A_{jk}$$

$$= \sum_j \sum_k (QQ^{-1})_{kj} A_{jk}$$

$$= \sum_k (QQ^{-1})_{kk} A_{kk}$$

$$= \text{Trace } A$$

2.12 EIGEN VALUES, EIGEN VECTORS, CHARACTERISTIC EQUATION OF MATRIX

Let $A = [a_{ij}]$ be a square matrix of order n. Suppose that there is an n-dimensional nonzero column vector X such that the matrix product AX gives a vector which is just a multiple of X, that is

$$AX = \lambda X \tag{2.35}$$

where λ is a scalar. The transformation represented by Eq. (2.35) just multiplies the vector X by a scalar λ.

A vector X defined by Eq. (2.35) is called *invariant vector* under the linear transformation. Equation (2.35) may be written as

$$(A - \lambda I) X = 0 \tag{2.35a}$$

where I is unit matrix.

Any value of λ for which Eqs (2.35) or (2.35a) has a nonzero solution (i.e. $X \neq 0$) is called an *eigen value or characteristic root* or *latent root* of the matrix A and the corresponding nonzero solution X is called *eigen vector or characteristic vector* or *latent vector* of A corresponding to that value of λ.

The matrix $[A - \lambda I]$ is called the *characteristic matrix* of A. The determinant $D(\lambda) = |A - \lambda I|$ is called the *characteristic polynomial of A*.

The system of homogeneous Eqs (2.35) or (2.35a) have *nontrivial solutions* if and only if matrix $[A - \lambda I]$ is *singular, i.e.*

$$D(\lambda) = |A - \lambda I| = 0 \tag{2.36}$$

or its equivalent

$$D(\lambda) = a_0 + a_1\lambda + a_2\lambda^2 + \cdots + a_n\lambda^n = 0 \tag{2.37}$$

where a's are the functions of the elements of A.

Equations (2.36) and (2.37) are called the *characteristic equation or secular equation* of A. From Eq. (2.36), it follows that *every characteristic root λ of matrix A is a root of its characteristic equation*. The n-roots $\lambda_1, \lambda_2, ..., \lambda_n$ of the characteristic equation are not necessarily all different.

Cayley–Hamilton Theorem

Every square matrix satisfies its own characteristic equation, i.e. if for a square matrix A of order n, the characteristic polynomial is

$$|A - I\lambda| = a_0 + a_1\lambda + a_2\lambda^2 + \cdots + a_n\lambda^n$$

then matrix equation

$$a_0I + a_1X + a_2X^2 + \cdots + a_nX^n = 0$$

is satisfied by $X = A$.

Proof: The characteristic polynomial is

$$|A - \lambda I| = a_0 + a_1\lambda + a_2\lambda^2 + \cdots + a_n\lambda^n = 0 \tag{2.38}$$

The characteristic equation of A is

$$|A - \lambda I| = a_0 + a_1\lambda + a_2\lambda^2 + \cdots + a_n\lambda^n = 0 \tag{2.39}$$

The matrix equation is

$$a_0I + a_1X + a_2X^2 + \cdots + a_nX^n = 0 \tag{2.40}$$

If this equation is satisfied by A then we have to show that

$$a_0I + a_1A + a_2A^2 + \cdots + a_nA^n = 0 \tag{2.41}$$

Since each element of characteristic matrix $(A - \lambda I)$ is an ordinary polynomial of degree n (at most) therefore the cofactor of every element of $|A - \lambda I|$ is an ordinary polynomial of degree $n - 1$ (at most). Consequently each element of

$$B = \text{adj}\,(A - \lambda I)$$

is an ordinary polynomial of degree $n - 1$ (at most).

Therefore, we can write

$$B = \text{adj}(A - \lambda I) = B_0 + B_1\lambda + B_2\lambda_2 + \cdots + B_{n-1}\lambda^{n-1} \tag{2.42}$$

where $B_0, B_1, B_2, ..., B_{n-1}$ are all matrices (square) of some order n whose elements are polynomials in the elements of A. But we know that, A and adj A are commutative, thus

$$A(\text{adj}\,A) = (\text{Adj}\,A)\,A = |A|\,I.$$

Hence we have

$$(A - \lambda I)\,\text{Adj}(A - \lambda I) = |A - \lambda I|\,I.$$

Using Eqs (2.42) and (2.38), we have

$$|A - \lambda I|\,[B_0 + B_1\lambda + B_2\lambda^2 + \cdots + B_{n-1}\lambda^{n-1}] = [a_0 + a_1\lambda + a_2\lambda^2 + \cdots + a_n\lambda^n]I \tag{2.43}$$

Comparing the coefficients of like powers of λ, we have

$$AB_0 = a_0 I$$
$$AB_1 - B_1 = a_2 I$$
$$AB_2 - B_0 = a_1 I$$

$$\dotfill$$

$$AB_{n-1} - B_{n-2} = a_n I$$

Now premultiplying these equations by $I, A, A^2, A^3, ..., A^n$ respectively and then on adding we get

$$0 = a_0 I + a_1 A + a_2 A^2 + \cdots + a_n A^n$$

which is Eq. (2.41). Thus, the theorem is proved.

Corollary: The inverse of a nonsingular matrix

Let A be a nonsingular matrix of order n so that $|A| \neq 0$, then characteristic equation of A is

$$|A - \lambda I| = a_0 + a_1\lambda + a_2\lambda^2 + \cdots + a_n\lambda^n = 0 \tag{2.44}$$

And as per Cayley–Hamilton theorem

$$a_0 I + a_1 A + a_2 A^2 + \cdots + a_n A^n = 0 \tag{2.45}$$

The characteristic polynomial is

$$|A - \lambda I| = a_0 + a_1\lambda + a_2\lambda^2 + \cdots + a_n\lambda^n \tag{2.46}$$

Equation (2.46) holds for all values of λ and hence in particular for $\lambda = 0$; as such for $\lambda = 0$ Eq. (2.46) becomes

$$|A| = a_0$$

as

$$|A| \neq 0 \text{ therefore } a_0 \neq 0$$

Now dividing Eq. (2.45) by a_0 we have

$$I = -\left(\frac{a_1}{a_0}A + \frac{a_2}{a_0}A^2 + \cdots + \frac{a_n}{a_0}A^n\right)$$

Premultiplying by A^{-1} we get

$$A^{-1} = -\left(\frac{a_1}{a_0}I + \frac{a_2}{a_0}A + \cdots + \frac{a_n}{a_0}A^{n-1}\right)$$

Hence Cayley–Hamilton theorem gives a direct method of evaluating A^{-1}.

This also shows that every characteristic root λ of A is a root of characteristic Eq. (2.35) conversely if λ is a root of the characteristic Eq. (2.35) then the matrix Eq. (2.35a) becomes

$$[A - \lambda I]X = 0$$

necessarily possesses a nonzero solution X and hence there exists nonzero X such that

$$AX = \lambda IX = \lambda X$$

Therefore, every root of the characteristic equation of A is its characteristic root.

Some Important Theorems on Eigen Values and Eigen Vectors

1. The eigen values of a Hermitian matrix are real

For a Hermitian matrix A

$$A^+ = A \tag{2.47}$$

Now let X be any eigen vector of A corresponding to the eigen value λ, then

$$AX = \lambda X \tag{2.48}$$

Premultiplying Eq. (2.48) by X^+, we have

$$X^+AX = X^+ \lambda X = \lambda X^+X \tag{2.49}$$

Taking transpose conjugate of Eq. (2.49), we have

$$(X^+AX)^+ = (\lambda X^+X)^+$$

or $\qquad\qquad X^+A^+(X^+)^+ = \bar{\lambda}X^+(X^+)^+ \cdot \bar{\lambda}$ is the complex conjugate of λ

or $\qquad\qquad X^+A^+X = \bar{\lambda}X^+X$

or $\qquad\qquad X^+AX = \bar{\lambda}X^+X \qquad\qquad$ [*using* Eq. (2.47)]

or $\qquad\qquad X^+\lambda X = \bar{\lambda}X^+X \qquad\qquad$ [*using* Eq. (2.48)]

$$(\lambda - \bar{\lambda}) = X^+X = 0$$

As X is an eigen vector, $X \neq 0$ therefore $X^+X \neq 0$

Hence $\qquad\qquad \lambda - \bar{\lambda} = 0 \quad$ or $\lambda = \bar{\lambda}$

This means the conjugate of λ is equal to itself. This is only possible when λ is real. Thus the eigen-values of a Hermitian matrix are all real.

2. The eigen values (characteristic roots) of a real symmetric matrix are all real.

For a real symmetric matrix A, we have

$$\bar{A} = A \tag{2.49a}$$
$$A^T = A \tag{2.50}$$

Taking complex conjugate of Eq. (2.50) and keeping in mind Eq. (2.49), we get

$$(\bar{A}^T) = A \quad \text{i.e. } A^+ = A$$

Thus a real symmetric matrix is a Hermitian matrix. According to theorem (1) the eigen values of a Hermitian matrix are all real. Consequently, the eigen values of a real symmetric matrix are all real.

3. The eigen values (characteristic roots) of skew-Hermitian matrix are either zero or purely imaginary.

If A is a skew-Hermitian matrix, then we have

$$A^+ = -A \tag{2.51}$$

Let X be an eigen vector of A corresponding to eigen value λ, then

$$AX = \lambda X \tag{2.52}$$

Taking transpose conjugate on both sides, we have

$$X^+ A^+ = \bar{\lambda} X^+ \quad \text{or} \quad -X^+ A = \bar{\lambda} X^+ \tag{2.53}$$

Multiplying Eq. (2.52) by X^+ on left and Eq. (2.53) by X on the right, we get

$$X^+ AX = \lambda X^+ X \quad \text{and} \quad X^+ AX = -\bar{\lambda} X^+ X$$

Comparing these equations, we get

$$\lambda = -\bar{\lambda}$$

which gives $\lambda = 0$ or a purely imaginary number.

4. The eigen values of a real skew-symmetric matrix are either zero or purely imaginary

For a real skew-symmetric matrix A, we have

$$\bar{A} = A \tag{2.54}$$
$$A^T = -A \tag{2.55}$$

Taking complex conjugate of Eq. (2.55), we get

i.e. $$A^+ = -A \qquad \qquad [using \text{ Eq. (2.54)}]$$

This implies that the matrix A is skew-Hermitian.

Hence, according to Theorem (3), the eigen values of a real skew-symmetric matrix are either zero or purely imaginary.

5. The eigen values (characteristic roots) of a unitary matrix are of unit modulus

For a unitary matrix A we have

$$A^T A = I \tag{2.56}$$

Let X be an eigen vector of A corresponding to eigen value λ. Then eigen value equation is

$$AX = \lambda X \tag{2.57}$$

Taking transpose conjugate of Eq. (2.57), we get

$$(AX)^+ = (\lambda X)^+$$

$$X^+ A^+ = \bar{\lambda} X^+ \tag{2.58}$$

Postmultiplying Eqs (2.58) by (2.57), we get

$$(X^+ A^+)(AX) = (\bar{\lambda} X^+)(\lambda X)$$

$$X^+ (A^+ A) X = \lambda \bar{\lambda} X^+ X$$

or $$X^+ I X = \lambda \bar{\lambda} X^+ X \qquad \qquad [using \text{ Theorem Eq. (2.56)}]$$

i.e. $$X^+ X = \lambda \bar{\lambda} X^+ X \quad (\text{since } IX = X)$$

i.e. $$X^+ X (1 - \lambda \bar{\lambda}) = 0$$

As $X \neq 0$, therefore $X^+ X \neq 0$

Hence $1 - \lambda \bar{\lambda} = 0$, i.e. $\lambda \bar{\lambda} = 1$

or
$$\|\lambda^2\| = 1$$

$$\|\lambda\| = 1$$

Thus eigen values of unitary matrix are of unit modulus.

6. The eigen values of an orthogonal matrix are unimodular (i.e. of unit modulus).

For an orthogonal real matrix A, we have

$$\bar{A} = A \tag{2.59}$$
$$A^T A = I \tag{2.60}$$

Taking complex conjugate of Eq. (2.60), we get

$$(A^T A)^* = I^{*\#}$$

i.e.
$$(A^T) = A^* = I, \text{ since } I^* = I$$
$$A^T A = I$$

which shows that matrix A is unitary matrix. But we have proved in Theorem (4) that the eigen values of a unitary matrix are of unit magnitude. Hence the eigen values of a real orthogonal matrix are unimodular.

7. The eigen values of diagonal matrix are precisely the elements in the diagonal

Let $A = \text{diag} [a_{11}, a_{22},...,a_{nn}]$

Then $(A - \lambda I) = [a_{11} - \lambda, a_{22} - \lambda,..., a_{nn} - \lambda]$

So the characteristic polynomial

$$|A - \lambda I| = (a_{11} - \lambda)(a_{21} - \lambda)(a_{22} - \lambda)...(a_{nn} - \lambda)$$

The characteristic equation $|A - \lambda I| = 0$ gives

$$(a_{11} - \lambda)(a_{22} - \lambda)...(a_{nn} - \lambda) = 0$$

i.e.
$$\lambda = a_{11}, a_{22},...,a_{nn}.$$

As $a_{11}, a_{22},...,a_{nn}$ are the diagonal elements of A therefore the eigen value λ of a diagonal matrix are the elements in the diagonal.

8. Any two eigen vectors corresponding to distinct eigen values of a Hermitian matrix are orthogonal

Let X_1, X_2 be two eigen vectors corresponding to two distinct eigen values λ_1, λ_2 of a Hermitian matrix A. Then,

$$A^+ = A \tag{2.61}$$
$$AX_1 = \lambda_1 X_1 \tag{2.62}$$
$$AX_2 = \lambda_2 X_2 \tag{2.63}$$

But we know from Theorem (2) that eigen values of a Hermitian matrix are real. Hence λ_1 and λ_2 are real.

i.e.
$$\lambda_1^* = \lambda_1 \text{ and } \lambda_2^* = \lambda_2 \tag{2.64}$$

#Complex conjugate

Premultiply Eqs (2.62) and (2.63) by X_2^+ and X_1^+, respectively we get

$$X_2^+ A X_1 = \lambda_1 X_2^+ X_1 \tag{2.65}$$
$$X_1^+ A X_2 = \lambda_2 X_1^+ X_2 \tag{2.66}$$

Taking transpose conjugate of (2.65), we have

$$(X_2^+ A X_1)^+ = (\lambda_1 X_2^+ X_1)^+$$
$$X_1^+ A^+ (X_2^+)^+ = \lambda_1^+ X_1^+ (X_2^+)^+$$

or $\qquad\qquad X_1^+ A^+ X_2 = \lambda_1 X_1^+ X_2$

or $\qquad\qquad X_1^+ A X_2 = \lambda_1 X_1^+ X_2 \qquad$ [*using* Eq. (2.61)] $\tag{2.67}$

Comparing Eqs (2.66) and (2.67), we have

$$\lambda_1 X_1^+ X_2 = \lambda_2 X_1^+ X_2$$

or $\qquad\qquad (\lambda_1 - \lambda_2) X_1^+ X_2 = 0$

As the eigen values are different, i.e. $\lambda_1 \neq \lambda_2$ or $\lambda_1 - \lambda_2 \neq 0$, therefore, we have

$$X_1^+ X_2 = 0$$

which is the condition for eigen vectors X_1, X_2 to be orthogonal. Hence any two eigen vectors corresponding to two distinct eigen values of a Hermitian matrix are real.

9. Any two eigen vectors of a real symmetric matrix are orthogonal, provided the corresponding eigen values are different

Let A be a real symmetric matrix. Then

$$A^+ = A \tag{2.68}$$
$$A^T = A \tag{2.69}$$

Taking conjugate of Eq. (2.69), we get

$$(A^T)^* = A^*$$

i.e. $\qquad\qquad A^+ = A \qquad\qquad$ [*using* Eq. (2.68)]

which is the condition for matrix A to be Hermitian. Hence real symmetric matrix A is Hermitian.

But we know that any two eigen vectors corresponding to two distinct eigen values of a Hermitian matrix are orthogonal, hence any two eigen vectors corresponding to two distinct eigen values of a real symmetric matrix are orthogonal.

10. Any two eigen vectors corresponding to distinct eigen values of a unitary matrix are orthogonal

Let A be a unitary matrix. Then

$$A^+ A = I \tag{2.70}$$

Let X_1, X_2 be two eigen vectors corresponding to two distinct eigen values λ_1, λ_2 of unitary matrix A. Then

$$A X_1 = \lambda_1 X_1 \tag{2.71}$$
$$A X_2 = \lambda_2 X_2 \tag{2.72}$$

Taking transpose conjugate of Eq. (2.71), we have

$$(A X_1)^+ = (\lambda_1 X_1)^+$$

i.e. $\qquad\qquad X_1^+ A^+ = \lambda_1^* X_1^+ \tag{2.73}$

Postmultiplying Eqs (2.73) by (2.72), we have

$$(X_1^+ A^+)(A X_2) = (\lambda_1^* X_1^+)(\lambda_2 X_2)$$
$$X_1^+ (A^+ A) X_2 = \lambda_1^* \lambda_2 X_1^+ X_2$$

i.e.
$$X_1^+ X_2 = \lambda_1^* \lambda_2 X_1^+ X_2$$

Since
$$A^+ A = I \qquad\qquad\qquad [using \text{ Eq. (2.70)}]$$

or
$$(1 - \lambda_1^* \lambda_2) X_1^+ X_2 = 0$$

But we know that the eigen values of a unitary matrix are of unit modulus, hence

$$\lambda_1^* \lambda_1 = 1$$

Now $(1 - \lambda_1^* \lambda_2) = (\lambda_1^* \lambda_1 - \lambda_1^* \lambda_2) = \lambda_1^* (\lambda_1 - \lambda_2) \neq 0$ since $\lambda_1 \neq \lambda_2$.
Hence it follows that

$$X_1^+ X_2 = 0$$

which implies that X_1 and X_2 are orthogonal.

11. Eigen values of a matrix are invariant under a similarity transformation

Let us consider two similar matrices A and B, related through similarity transformation such that

$$B = Q^{-1} A Q$$

The characteristic equation of matrix B is $|B - \lambda I| = 0$ where λ is an eigen value of B.

i.e.
$$|Q^{-1} A Q - \lambda I| = 0$$

or
$$|Q^{-1} A Q - Q^{-1} \lambda I Q| = 0$$

or
$$|Q^{-1}(A - \lambda I) Q| = 0$$

$$|Q^{-1}||A - \lambda I||Q| = 0$$

or
$$|Q^{-1} Q||A - \lambda I| = 0$$

or
$$|A - \lambda I| = 0$$

This shows that λ is an eigen value of A also which proves the theorem.

12. The eigen vectors corresponding to distinct eigen values of a matrix are linearly independent

Let $X_1, X_2, \ldots, X_n$ be the eigen vectors corresponding to distinct eigen values $\lambda_1, \lambda_2, \ldots, \lambda_n$ of matrix A.

$$A = [a_{ij}]$$

Then we have to prove that only solution of

$$a_1 X_1 + a_2 X_2 + \cdots + a_n X_n = 0$$

is
$$a_1 = a_2 = a_3 = \ldots = a_n = 0.$$

Let us assume that

$$a_1 X_1 + a_2 X_2 + \cdots + a_n X_n = 0 \qquad\qquad (2.74)$$

where $a_1, a_2, a_3, \ldots, a_n$ are not all zero.

Pre-multiplying Eq. (2.74) by matrix A we have

$$a_1 A X_1 + a_2 A X_2 + \cdots + a_n A X_n = 0$$

or
$$a_1 \lambda_1 X_1 + a_2 \lambda_2 X_2 + \cdots + a_n \lambda_n X_n = 0$$

Multiplying again by A, we have

$$a_1\lambda_1^2 X_1 + a_2\lambda_2^2 X_2 + \cdots + a_n\lambda_n^2 X_n = 0 \qquad (2.75)$$

In this way by multiplying n times, we have

$$a_1\lambda_1^3 X_1 + a_2\lambda_2^3 X_2 + \cdots + a_n\lambda_n^3 X_n = 0$$

$$\cdots\cdots\cdots\cdots\cdots\cdots\cdots\cdots\cdots\cdots\cdots$$

$$a_1\lambda_1^n X_1 + a_2\lambda_2^n X_2 + \cdots + a_n\lambda_n^n X_n = 0$$

In matrix form Eqs (2.74) and (2.75) can be written as

$$P = \begin{bmatrix} 1 & 1 & 1 & 1 \\ \lambda_1 & \lambda_2 & \cdots & \lambda_n \\ \vdots & \vdots & & \\ \lambda_1^n & \lambda_2^n & \cdots & \lambda_n^n \end{bmatrix} \begin{bmatrix} a_1 X_1 \\ a_2 X_2 \\ \vdots \\ a_n X_n \end{bmatrix} = \begin{bmatrix} 0 \\ 0 \\ 0 \\ \vdots \\ 0 \end{bmatrix}$$

or $\qquad\qquad PY = 0 \qquad\qquad\qquad\qquad (2.76)$

where $\qquad P = \begin{bmatrix} 1 & 1 & 1 & \cdots & 1 \\ \lambda_1 & \lambda_2 & \lambda_3 & \cdots & \lambda_n \\ \lambda_1^2 & \lambda_2^2 & \lambda_3^2 & \cdots & \lambda_n^2 \\ \vdots & & & & \\ \lambda_1^n & \lambda_2^n & \lambda_3^n & \cdots & \lambda_n^n \end{bmatrix}, \; Y = \begin{bmatrix} a_1 X_1 \\ a_2 X_2 \\ \vdots \\ \vdots \\ a_n X_n \end{bmatrix}$

The determinant

$$|P| = \begin{vmatrix} 1 & 1 & 1 & \cdots & 1 \\ \lambda_1 & \lambda_2 & \lambda_3 & \cdots & \lambda_n \\ \lambda_1^2 & \lambda_2^2 & \lambda_3^2 & \cdots & \lambda_n^2 \\ \vdots & & & & \\ \lambda_1^n & \lambda_2^n & \lambda_3^n & \cdots & \lambda_n^n \end{vmatrix}$$

$$= (\lambda_n - \lambda_{n-1})(\lambda_{n-1} - \lambda_{n-1})\ldots(\lambda_3 - \lambda_2)(\lambda_2 - \lambda_1)(\lambda_n - \lambda_1) \neq 0$$

Since $\lambda_1, \lambda_2, \ldots, \lambda_n$ are all distinct. Thus the matrix P is nonsingular or P^{-1} exists. Multiplying Eq. (2.76) with P^{-1} from left, we have

$$P^{-1}PY = Y = 0$$

or $\qquad\qquad a_1 X_1 = a_2 X_2 = \ldots = a_n X_n = 0$

But $X_1, X_2, \ldots, X_n$ being eigen vectors can not be zero. So,

$$a_1 = a_2 = \ldots = a_n = 0$$

i.e. $\qquad a_1 X_1 + a_2 X_2 + \cdots + a_n X_n = 0$

or $\qquad\qquad X_1, X_2, X_3, \ldots, X_n$

are linearly independent eigen vectors.

2.13. DIAGONALIZATION OF MATRICES

The process of reducing a matrix to diagonal form, by orthogonal similarity transformation or by a unitary transformation is known as diagonalisation of matrix. A matrix A which is similar to a diagonal matrix is called *diagonable matrix*.

The characteristic roots of a diagonal matrix are simply the diagonal elements. Thus the problem of reducing any matrix to the diagonal form by means of similarity transformation is closely related to the problem of finding its eigen values. For the reduction of matrices to diagonal form we discuss following two cases:

The eigen values are different

Let $X_1, X_2,...,X_n$ be the n eigen vectors corresponding to n distinct eigen values $\lambda_1, \lambda_2,...,\lambda_n$ of a matrix A. These eigen vectors form a linearly independent set as such let us write these eigen vectors as column matrices

$$X_i = \begin{bmatrix} x_{1i} \\ x_{2i} \\ \vdots \\ x_{ni} \end{bmatrix} \qquad (2.77)$$

Then a matrix P constructed by these columns is a nonsingular matrix, i.e. for the matrix

$$P = [X_1, X_2,...,X_n] = \begin{bmatrix} X_{11} & X_{12} & \cdots & X_{1n} \\ X_{21} & X_{22} & \cdots & X_{2n} \\ \cdots & \cdots & \cdots & \cdots \\ X_{n1} & X_{n2} & \cdots & X_{nn} \end{bmatrix},$$

$\therefore P^{-1}$ exists.

Suppose that D is a diagonal matrix

$$D = \text{diag}\,(\lambda_1, \lambda_2,...,\lambda_n) = \begin{bmatrix} \lambda_1 & 0 & 0 & \cdots & 0 \\ 0 & \lambda_2 & 0 & \cdots & 0 \\ \vdots & & & & \\ 0 & \cdots & \cdots & \cdots & \lambda_n \end{bmatrix}$$

With eigen values $\lambda_1, \lambda_2,...,\lambda_n$. Then we have

$$PD = \begin{bmatrix} \lambda_1 X_{11} & \lambda_2 X_{12} & \cdots & \lambda_n X_{1n} \\ \lambda_1 X_{21} & \lambda_2 X_{22} & \cdots & \lambda_n X_{2n} \\ \vdots & & & \\ \lambda_1 X_{n1} & \lambda_2 X_{n2} & \cdots & \lambda_n X_{nn} \end{bmatrix}$$

$$= [\lambda_1 X_1, \lambda_2 X_2,...,\lambda_n X_n]$$
$$= [AX_1, AX_2, AX_3,...,AX_n]$$
$$= [A]\,[X_1, X_2,...,X_n] = AP \qquad (2.78)$$

If P is nonsingular (i.e in the case of all eigen values being distinct), then P^{-1} exists and Eq. (2.78) gives

$$P^{-1}PD = D = P^{-1}AP \qquad (2.79)$$

i.e. A is similar to D. This exhibits the diagonalisation of A. For diagonalisation of A the essential requirement is the existence P^{-1} which is possible when and only when all eigen vectors X_i of A are linearly independent. We may, therefore, have the following theorem.

The necessary and sufficient condition for $n \times n$ matrix A to be similar to a diagonal matrix is that the set of eigen vectors of A includes a set of n linearly independent vectors.

Let us first prove that the condition is necessary, i.e. there exists a nonsingular matrix P such that

$$P^{-1} AP = D \tag{2.80}$$

i.e.

$$P^{-1} AP = D = \text{diag} (\lambda_1, \lambda_2,...,\lambda_n)$$

where $\lambda_1, \lambda_2,...,\lambda_n$ are eigen values of A.

Pre-multiplying this by P, we have

$$AP = PD$$

Let us designate the columns of the matrix P as $X_1, X_2,...,X_n$, i.e.

$$P = [X_1, X_2,...,X_n]$$

Since P is nonsingular, all these column vectors are linearly independent. Substituting the value of P in Eq. (2.80), we get

$$A[X_1, X_2,...,X_n] = [X_1, X_2,...,X_n]D$$
$$= [X_1, X_2,...,X_n] \text{ diag } [\lambda_1, \lambda_2,...,\lambda_n]$$
$$= [\lambda_1 X_1, \lambda_2 X_2,...,\lambda_n X_n]$$

i.e.

$$AX_1 = \lambda_1 X_1, AX_2 = \lambda_2 X_2,...,AX_n = \lambda_n X_n.$$

Thus $X_1, X_2,...,X_n$ are eigen vectors of A corresponding to eigen values $\lambda_1, \lambda_2,...,\lambda_n$. Hence the necessary condition for the matirx A to be equivalent to a diagonal matrix is that all its eigen vectors form linearly independent set.

In order to prove the sufficient condition, let us assume that vectors $X_1, X_2,...,X_n$ form linearly independent set of all n eigen vectors of matrix A. Let $\lambda_1,...,\lambda_n$ be the corresponding eigen values. Then by definition

$$AX_1 = \lambda_1 X_1$$
$$AX_2 = \lambda_2 X_2$$
$$...............$$
$$AX_n = \lambda_n X_n$$

or

$$A[X_1 X_2...X_n] = [\lambda_1 X_1, \lambda_2 X_2,...,\lambda_n X_n]$$

or

$$AP = [X_1, X_2,...,X_n] \text{ diag } [\lambda_1, \lambda_2,...,\lambda_n]$$
$$= PD \tag{2.81}$$

where P is the matrix with columns identical to vectors $X_1, X_2,...,X_n$, i.e.

$$P = [X_1, X_2...X_n]$$

All its columns being linearly independent, the matrix P is nonsingular and hence P^{-1} exists. Therefore, Eq. (2.81) can be written as

$$P^{-1}AP = D = \text{diag} (\lambda_1, \lambda_2,...,\lambda_n)$$

which diagonalises A and therefore linear independence of the eigen vectors of a matrix A is the sufficient condition for its diagonalisation.

The eigen values are not all different

When two or more of the eigen values of a matrix A are equal to each other, its reduction to true diagonal form is not always possible. Suppose λ_1 is an eigen value of A repeated r_1 times. We find an eigen vector X_1 such that

$$AX_1 = \lambda_1 X_1$$

If the vector is the first column of a square matrix X, the first column of AX will be $\lambda_1 X_1$ and the first column of $X^{-1}AX$ will consist of λ, followed by $(n-1)$ zeros. Let this matrix be B. Then.

$$B = X^{-1}AX = \begin{bmatrix} \lambda_1 & B_j \\ 0 & B_{ij} \end{bmatrix}, (ij = 2,3,...,n) \tag{2.82}$$

where B_j is a row matrix with $(n-1)$ elements and B_{ij} is a square matrix of order $(n-1)$. Since B is the transform of A, it also has the eigen value λ_1 repeated r_1 times but B_{ij} contains that eigen value $(r_1 - 1)$ times. Thus the matrix is subjected to the same procedure as A, we find an eigen vector and form a new matrix Y whose first column is that eigen vector. Matrix Y has obviously $(n-1)$ rows and columns.

and

$$Y^{-1}B_{ij} B = \begin{bmatrix} \lambda_1 & c_k \\ 0 & c_{kl} \end{bmatrix}, k,l = 3,4,...,n \tag{2.83}$$

Therefore, the matrix $\begin{bmatrix} 1 & 0 \\ 0 & Y \end{bmatrix}$ will transform the matrix B into the form

$$\begin{bmatrix} \lambda_1 & B_j & Y \\ 0 & \lambda_1 & c_k \\ 0 & 0 & k_{kl} \end{bmatrix}$$

Repeated application of similar transformations will eventually result in a single matrix Z such that

$$Z^{-1}AZ = \begin{bmatrix} A_1 & F \\ 0 & G \end{bmatrix} \tag{2.84}$$

where

$$A_1 = \begin{bmatrix} \lambda_1 & H_{12} & H_{13} & \cdots & H_{1r_1} \\ 0 & \lambda_1 & H_{23} & \cdots & H_{2r_1} \\ 0 & 0 & \lambda_1 & \cdots & H_{3r_1} \\ \vdots & & & & \\ 0 & 0 & 0 & \cdots & \lambda_1 \end{bmatrix}$$

Matrix F is a rectangular matrix of order $r_1 \times (n - r_1)$, G is a square matrix of order $(n-1)$. If a matrix Z_1 is composed from the first r_1 columns of Z, then we may remove the unwanted matrix F from Eq. (2.84). Since,

$$Z_1^{-1}AZ_1 = A$$

then the matrix G can be treated in the similar way until it is reduced to the form of A_1 with its eigen value λ_2 along the diagonal. We then continue with each remaining matrix until every eigen value has been used. Finally if we join together all of the rectangular matrices Z_i to form a square matrix W, we will have

$$W^{-1}AW = \text{diag}(A_1, A_2,...,A_r)$$

where r is the number of distinct eigen values of A.

Here the matrix diag $(A_1, A_2,...,A_r)$ is not purely a diagonal matrix but it is in diagonal block form such that its elements are grouped into blocks. All the blocks which do not lie along principal diagonal are the blocks of zero elements, and the blocks A_1, $A_2,...,A_r$ of nonzero elements are arranged along the principal diagonal.

Nature of Diagonalizing Matrices for Special Matrices

1. The diagonalising matrix of a real symmetric matrix is orthogonal

Let A be a real symmetric matrix and the matrix P be its diagonalizing matrix. Then

$$P^{-1}AP = \text{diag}(\lambda_1 \lambda_2,...,\lambda_n)$$

where $\lambda_1, \lambda_2,...,\lambda_n$ are the eigen values of the matrix A.

Taking the transpose of both sides we have

$$(P^{-1}AP)^T = [\text{diag}(\lambda_1, \lambda_2,...,\lambda_n)]^T = \text{diag}(\lambda_1, \lambda_2,...,\lambda_n)$$

or $\qquad P^TA(P^{-1})^T = \text{diag}(\lambda_1, \lambda_2,...,\lambda_n) = P^{-1}AP$

or $\qquad\qquad P^T = P^{-1}$

or P is an orthogonal matrix.

2. Diagonalizing matrix of a Hermitian matrix is unitary

Let H be a Hermitian matrix and R be its diagonalizing matrix, then

$$R^{-1}HR = \text{diag}(\lambda_1, \lambda_2,...,\lambda_n)$$

where $\lambda_1, \lambda_2,...,\lambda_n$ are the eigen values of H and are real.

Taking conjugate transpose on both sides we have

$$(R^{-1}HR)^+ = \{\text{diag}(\lambda_1, \lambda_2,...,\lambda_n)\}^+$$

or $\qquad R^+H(R^{-1})^+ = \text{diag}(\lambda_1\lambda_2,...,\lambda_n) = R^{-1}HR$

or $\qquad\qquad R^+ = R^{-1}$

or $\qquad\qquad RR^+ = I$

or R is unitary matrix.

3. If Y is an eigen vector of $B = R^{-1}AR$ corresponding to eigen value λ, then $U = RY$ is an eigen vector of A corresponding to the same eigen value

Since Y is an eigen vector of B corresponding to the eigen value λ. We have

$$BY = \lambda Y$$

or $\qquad\qquad RBY = R\lambda Y$

or $\qquad\qquad ARY = \lambda RY \quad$ [since $RB = AR$]

or $\qquad\qquad A(RY) = \lambda(RY)$

or (RY) is an eigen vector of A corresponding to eigen value λ.

4. The invariant (eigen) vectors of a diagonal matrix are unit vectors

Let $\qquad E_1 = [1 \ 0 \ 0 \ \cdots \ 0]^T, \quad E_2 = [0 \ 1 \ 0 \ \cdots \ 0]^T$

$\qquad\qquad E_3 = [0 \ 0 \ 1 \ \cdots \ 0]^T, \quad E_n = [0 \ 0 \ 0 \ 0 \ 0 \ \cdots \ 1]^T$

and a diagonal matrix

$$D = \text{diag } (\lambda_1, \lambda_2, \dots, \lambda_n)$$

Then $D_1 = \text{diag } (\lambda_1, \lambda_2, \dots, \lambda_n),\ E_1 = [\lambda_1, 00\dots0]^T = \lambda_1 E_1.$

So E_1 is an eigen vector of D corresponding to eigen value λ_1. Similarly $E_2, E_3, \dots, E_n$ are its eigen vectors corresponding to eigen values $\lambda_2, \lambda_3, \dots, \lambda_n$.

5. Any matrix similar to a diagonal matrix has a linearly independent invariant vectors

Let A be the given matrix with the diagonalizing matrix R, i.e.

$$R^{-1}AR = \text{diag } (\lambda_1, \lambda_2, \dots, \lambda_n) = D.$$

The invariant eigen vector of D are unit matrices $E_1, E_2, E_3, \dots, E_n$ which are linearly independent.

From Eq. (2.79) we have $RE_1, RE_2, \dots, RE_n$ as invariant eigen vector of A. These vectors are also linearly independent since $E_1, E_2, \dots, E_n$ are independent vectors.

6. If a *n*-rowed square matrix *A* has *n* linearly independent invariant vectors then it is similar to *Q* diagonal matrix

Let $X_1, X_2, \dots, X_n$ be the linearly independent invariant eigen vectors of the matrix A corresponding to eigen values $\lambda_1, \lambda_2, \dots, \lambda_n$ then

$$AX_i = \lambda_i A$$

Let us consider a square matrix R with its column as vectors X_i, i.e.

$$R = [X_1, X_2, \dots, X_n]$$

Then $AR = A[X_1, X_2, \dots, X_n] = [AX_1, AX_2, \dots, AX_n]$

$$= [\lambda_1 X_1, \lambda_2 X_2 \dots \lambda_n X_n]$$

$$= [X_1, X_2, \dots, X_n] \begin{bmatrix} \lambda_1 & 0 & 0 & \cdots & 0 \\ 0 & \lambda_2 & 0 & \cdots & 0 \\ \vdots & & & & \\ 0 & 0 & \cdots & \cdots & \lambda_n \end{bmatrix}$$

$$= R \text{ diag } (\lambda_1 \lambda_2, \dots, \lambda_n)$$

Multiplying both sides with R^{-1} from left we have

$$R^{-1}AR = \text{diag } (\lambda_1, \lambda_2, \dots, \lambda_n)$$

or A is similar to diagonal matrix.

Example 1. *Find the eigen values and normalising eigen vectors of the matrix*

$$\begin{bmatrix} 1 & 0 & 0 \\ 0 & 1 & 1 \\ 0 & 1 & 1 \end{bmatrix}$$

Solution: Let $A = \begin{bmatrix} 1 & 0 & 0 \\ 0 & 1 & 1 \\ 0 & 1 & 1 \end{bmatrix}$

We have, $A - \lambda I = \begin{bmatrix} 1 & 0 & 0 \\ 0 & 1 & 1 \\ 0 & 1 & 1 \end{bmatrix} - \lambda \begin{bmatrix} 1 & 0 & 0 \\ 0 & 1 & 0 \\ 0 & 0 & 1 \end{bmatrix} = \begin{bmatrix} 1-\lambda & 0 & 0 \\ 0 & 1-\lambda & 1 \\ 0 & 1 & 1-\lambda \end{bmatrix}$

The characteristic equation of A is

$$|A - \lambda I| = \begin{vmatrix} 1-\lambda & 0 & 0 \\ 0 & 1-\lambda & 1 \\ 0 & 1 & 1-\lambda \end{vmatrix} = 0$$

or $(1 - \lambda)[(1 - \lambda)^2 - 1] = 0$

or $\lambda(1 - \lambda)(\lambda - 2) = 0$

i.e. $\lambda = 0, 1, 2.$

The eigen values of matrix A are $0, 1, 2$. Eigen value equation is

$$(A - \lambda I) X = 0$$

For $\lambda = 0$ we have

$$AX = 0$$

or

$$\begin{bmatrix} 1 & 0 & 0 \\ 0 & 1 & 1 \\ 0 & 1 & 1 \end{bmatrix} \begin{bmatrix} x_1 \\ x_2 \\ x_3 \end{bmatrix} = \begin{bmatrix} 0 \\ 0 \\ 0 \end{bmatrix}$$

which is equivalent to

$$x_1 = 0, \qquad x_2 + x_3 = 0$$

or $x_2 = -3x_3.$

Within an arbitrary scale factor and an arbitrary sign (or phase factor) eigen vector corresponding to $\lambda = 0$ is

$$X_1 = \begin{bmatrix} x_1 \\ x_2 \\ x_3 \end{bmatrix} = [0, 1 - 1]^T$$

If the eigen vectors be normalised to unity, then

$$\|x_1\| = 1.$$

so that the normalised eigen vector of matrix A corresponding to eigen value $\lambda = 0$ is given by

$$X_1 = \left[0, \frac{1}{\sqrt{2}}, \frac{-1}{\sqrt{2}} \right]^T$$

For $\lambda = 1$ we have

$$\begin{bmatrix} 0 & 0 & 0 \\ 0 & 0 & 1 \\ 0 & 1 & 0 \end{bmatrix} \begin{bmatrix} x_1 \\ x_2 \\ x_3 \end{bmatrix} = \begin{bmatrix} 0 \\ 0 \\ 0 \end{bmatrix}$$

or $x_2 = 0, \quad x_3 = 0$

so that $X_2 = [x_1 \, x_2 \, x_3]^T = [1, 0, 0]^T$

is the suitable eigen vector. Thus, the normalised eigen vector of matrix A, corresponding to eigen value $\lambda = 1$ is given by

$$X_2 = [1, 0, 0]^T$$

For $\lambda = 2$ we have

$$(A - \lambda I)X = \begin{bmatrix} -1 & 0 & 0 \\ 0 & -1 & 1 \\ 0 & 1 & -1 \end{bmatrix} \begin{bmatrix} x_1 \\ x_2 \\ x_3 \end{bmatrix} = \begin{bmatrix} 0 \\ 0 \\ 0 \end{bmatrix}$$

$-x_1 = 0$, i.e. $x_1 = 0$

$x_2 - x_3 = 0$, i.e. $x_2 = x_3$.

Hence the eigen vector corresponding to $\lambda = 2$ within the arbitrary scale is

$$X_3 = [0, 1, 1]^T.$$

The normalised eigen vector of matrix A corresponding to eigen value $\lambda = 2$ is

$$X_3 = \left[0, \frac{1}{\sqrt{2}}, \frac{1}{\sqrt{2}} \right]^T$$

Thus, the normalised eigen vector of the given matrix A corresponding to the eigen values 0, 1, 2 are

$$X_1 = \left[0, \frac{1}{\sqrt{2}}, \frac{1}{\sqrt{2}} \right]^T, \ X_2 = [1, 0, 0]^T$$

$$X_3 = \left[0, \frac{1}{\sqrt{2}}, \frac{1}{\sqrt{2}} \right]^T \quad \text{respectively.}$$

Example 2. Find the characteristic equation of the matrix A

$$A = \begin{bmatrix} 2 & -1 & 1 \\ -1 & 2 & -1 \\ 1 & -1 & 2 \end{bmatrix}$$

and verify that it is satisfied by A. Hence find the inverse of A.

Solution: We have

$$A - \lambda I = \begin{bmatrix} 2 & -1 & 1 \\ -1 & 2 & -1 \\ 1 & -1 & 2 \end{bmatrix} - \lambda \begin{bmatrix} 1 & 0 & 0 \\ 0 & 1 & 0 \\ 0 & 0 & 1 \end{bmatrix} = \begin{bmatrix} 2-\lambda & -1 & 1 \\ -1 & 2-\lambda & -1 \\ 1 & -1 & 2-\lambda \end{bmatrix}$$

The characteristic equation of A is

$$|A - \lambda I| = \begin{bmatrix} 2-\lambda & -1 & 1 \\ -1 & 2-\lambda & -1 \\ 1 & -1 & 2-\lambda \end{bmatrix} = 0$$

i.e. $-\lambda^3 + 6\lambda^2 + 9\lambda + 4 = 0$

i.e. $\lambda^3 - 6\lambda^2 + 9\lambda - 4 = 0$ (1)

If this characteristic equation is satisfied by A itself, then we must have

$$A^3 - 6A^2 + 9A - 4I = 0$$ (2)

$$A^2 = \begin{bmatrix} 2 & -1 & 1 \\ -1 & 2 & -1 \\ 1 & -1 & 2 \end{bmatrix}\begin{bmatrix} 2 & -1 & 1 \\ -1 & 2 & -1 \\ 1 & -1 & 2 \end{bmatrix} = \begin{bmatrix} 6 & -5 & 5 \\ -5 & 6 & -5 \\ 5 & -5 & 6 \end{bmatrix}$$

and

$$A^3 = A^2 \cdot A \begin{bmatrix} 6 & -5 & 5 \\ -5 & 6 & -5 \\ 5 & -5 & 6 \end{bmatrix}\begin{bmatrix} 2 & -1 & 1 \\ -1 & 2 & -1 \\ 1 & -1 & 2 \end{bmatrix} = \begin{bmatrix} 22 & -21 & 21 \\ -21 & -22 & -21 \\ 21 & -21 & 22 \end{bmatrix}$$

So that $A^3 - 6A^2 + 9A - 4I$

$$= \begin{bmatrix} 22 & -21 & 21 \\ -21 & -22 & -21 \\ 21 & -21 & 22 \end{bmatrix} - 6\begin{bmatrix} 6 & -5 & 5 \\ -5 & 6 & -5 \\ 5 & -5 & 6 \end{bmatrix} + \begin{bmatrix} 2 & -1 & 1 \\ -1 & 2 & -1 \\ 1 & -1 & 2 \end{bmatrix} - 4\begin{bmatrix} 1 & 0 & 0 \\ 0 & 1 & 0 \\ 0 & 0 & 1 \end{bmatrix}$$

$$= \begin{bmatrix} 0 & 0 & 0 \\ 0 & 0 & 0 \\ 0 & 0 & 0 \end{bmatrix}$$

i.e. $A^3 - 6A^2 + 9A - 4I = 0.$

Hence characteristic equation of A is satisfied by A itself.

From Eq. (2), we have

$$I = \frac{1}{4}(A^3 - 6A^2 + 9A)$$

Premultiplying by A^{-1} we have

$$A^{-1} = \frac{1}{4}(A^2 - 6A + 9I) = \frac{1}{4}\begin{bmatrix} 3 & 1 & -1 \\ 1 & 3 & 1 \\ -1 & 1 & 3 \end{bmatrix}$$

Example 3. *For Hermitian matrix*

$$H = \begin{bmatrix} 7 & -2 & 1 \\ -2 & 10 & -2 \\ -1 & -2 & 7 \end{bmatrix}$$

Find the matrix U such that $U^{-1}HU$ is a diagonal matrix.

Solution: Characteristic equation is

$$|H - \lambda I| = \begin{vmatrix} 7-\lambda & -2 & 1 \\ -2 & 10-\lambda & -2 \\ 1 & -2 & 7-\lambda \end{vmatrix} = -\lambda^3 + 24\lambda^2 - 180\lambda + 432 = 0$$

or $(\lambda - 6)^2 (12 - \lambda) = 0.$

Hence characteristic roots are $\lambda_1 = 12$, $\lambda_2 = 6$, and $\lambda_3 = 6$. Let X_1 be the eigen vector corresponding to eigen value 12. Then we have

$$(H - 12I) X_1 = 0$$

i.e.
$$\begin{bmatrix} -5 & -2 & 1 \\ -2 & -2 & -2 \\ 1 & -2 & -5 \end{bmatrix}\begin{bmatrix} X_{11} \\ X_{21} \\ X_{31} \end{bmatrix} = 0$$

or
$$-5X_{11} - 2X_{21} + X_{31} = 0$$
$$2X_{11} + 2X_{21} + 2X_{31} = 0$$
$$X_{11} - 2X_{21} - 5X_{31} = 0.$$

From these equations, we get the following solution

$$X_1 = \begin{bmatrix} a \\ -2a \\ a \end{bmatrix}$$

where a is some arbitrary constant. For normal form of X_1 we take $a = \dfrac{1}{\sqrt{6}}$.

Hence normalised X_1 is

$$X_1 = \begin{bmatrix} \dfrac{1}{\sqrt{6}} \\ -\dfrac{2}{\sqrt{6}} \\ \dfrac{1}{\sqrt{6}} \end{bmatrix}$$

Corresponding to root 6 we have the equation

$$\begin{bmatrix} 1 & -2 & 1 \\ -2 & 4 & -2 \\ 1 & -2 & 1 \end{bmatrix}\begin{bmatrix} x_{11} \\ x_{12} \\ x_{13} \end{bmatrix} = 0$$

we have
$$x_{11} - 2x_{12} + x_{13} = 0$$
$$-2x_{11} + 4x_{12} - 2x_{13} = 0$$
$$x_{11} - 2x_{12} + x_{13} = 0$$

Its two independent solutions are

$$X_2 = \begin{bmatrix} b \\ c \\ 2c - b \end{bmatrix}$$

Here b and c are arbitrary constants.

In order to find two linearly independent solutions we will choose any arbitrary values of b and c for one solutions X_2 and for the other solution X_3, the values of b and c are so chosen that X_2 and X_3 are orthogonal.

For X_2 let $b = 0$ and $c = p$. We have

$$X_2 = \begin{bmatrix} 0 \\ 0 \\ 2p \end{bmatrix}$$

and for normalised X_2, we have $p = \dfrac{1}{\sqrt{5}}$. Hence normalised

$$X_2 = \begin{bmatrix} 0 \\ \dfrac{1}{\sqrt{5}} \\ \dfrac{2}{\sqrt{5}} \end{bmatrix}$$

and for X_3 the values of b and c are such that $X_2 X_3 = 0$, i.e.

$$\begin{bmatrix} 0 & \dfrac{1}{\sqrt{5}} & \dfrac{2}{\sqrt{5}} \end{bmatrix} \begin{bmatrix} b \\ c \\ 2c - b \end{bmatrix} = \dfrac{1}{\sqrt{5}}[c + 4c - 2b] = 0$$

i.e.

$$b = \dfrac{5}{2}c$$

Hence

$$X_3 = \begin{bmatrix} \dfrac{5}{\sqrt{30}} \\ \dfrac{2}{\sqrt{30}} \\ \dfrac{-1}{\sqrt{30}} \end{bmatrix}$$

Hence unitary matrix

$$U = [X_1 \ X_2 \ X_3]$$

$$= \begin{bmatrix} \dfrac{1}{6} & 0 & \dfrac{5}{\sqrt{30}} \\ \dfrac{-2}{\sqrt{6}} & \dfrac{1}{\sqrt{5}} & \dfrac{2}{\sqrt{30}} \\ \dfrac{1}{\sqrt{6}} & \dfrac{2}{\sqrt{5}} & \dfrac{-1}{\sqrt{30}} \end{bmatrix}$$

It is clear that

$$U^{-1} H U = U^{+} H U = \begin{bmatrix} 12 & 0 & 0 \\ 0 & 6 & 0 \\ 0 & 0 & 6 \end{bmatrix} = \text{diag } (12, 6, 6)$$

Note: We have chosen the value of $b = 0$ and $c = p$ for obtaining X_2. The values of b and c can be anything and correspondingly the values of X_1 will be different as such X_3 will also be different. Hence the unitary matrix U is not unique.

Example 4. *Prove that it is possible to find a common set of eigen vectors for two commuting matrices.*[*]

Solution: Let A and B be two square matrices each of order n which commute with each other, so that

$$AB - BA = [A, B] = 0 \tag{1}$$

First, let λ be an eigen value of A with multiplicity 1, corresponding to the eigen vector X. Hence we have

$$AX = \lambda X \tag{2}$$

and premultiplying by B we have

$$BAX = A(BX) = \lambda(BX) \tag{3}$$

Equation (3) shows that BX is also an eigen vector of A with the same eigen value λ. But since X is a nondegenerate eigen vector of A, any other vector which is eigen vector A with same eigen value as that of X must be a multiple of X, i.e.

$$BX = \mu X$$

where μ is a scalar. This shows that X is also an eigen vector of B with the eigen value m. *Thus for two commuting matrices, every nondegenerate eigen vector of one is also an eigen vector of the other and vice versa.*

Next let λ be an eigen value of A with multiplicity K. This means that A has K linearly independent eigen vectors say $X_1, X_2, ..., X_k$ each corresponding to the eigen value λ or

$$AX_i = \lambda X_i \quad 1 \leq i \leq k \tag{4}$$

Again premultiplying Eq. (4) by B we obtain

$$(BAX_i) = A(BX_i) = \lambda(BX_i) \tag{5}$$

This shows that BX is also an eigen vector of A with the same eigen value λ. Now the most general eigen vector of A having the eigen value λ must be a linear combination of the degenerate eigen vectors $X_1, X_2, ..., X_k$ so that

$$BX_i = \sum_{j=1}^{k} C_{ij} X_j \quad 1 \leq i \leq k \tag{6}$$

where C_{ij} are certain scalar coefficients. Let us define a matrix $C \equiv [c_{ij}]$ of order K and let the matrix C be diagonalised in the form

$$DCD^{-1} = \begin{bmatrix} b_1 & & & 0 \\ & b_2 & & \\ & & \ddots & \\ 0 & & & b_k \end{bmatrix} \equiv Q$$

i.e.

$$DC = QD \tag{7}$$

which defines Q as the diagonal matrix with diagonal elements μ_i.

Let ij-th element of D be d_{ij}, then equating the ij-th elements of both sides of Eq. (7) we have

$$(DC)_{ij} = (QD)_{ij}$$

[*] This theorem also holds good in quantum mechanics only replacing the word 'matrix' by 'operator'. The proof also holds good with suitable modification.

or
$$\sum_{j=1}^{k} d_{ij} c_{ij} = \sum_{l=1}^{k} \mu_i Q_{il} d_{li} = \mu_i d_{ij} \qquad (8)$$

Let us define new vectors $Y_i's$ by

$$Y_i = \sum_{l=1}^{k} d_{il} BX_l \quad 1 \le i \le k \qquad (9)$$

which are linear combinations of the $X_i's$. Since the matrix D is nonsingular, the k vectors Y_i will be linearly independent. Moreover, since they are linear combinations of the degenerate eigen vectors, each Y_i will also be an eigen vector of A with same eigen value λ. Now consider the operation of B on Y_i

$$BY_i = \sum_{l=1}^{k} d_{il} BX_l = \sum_{lj=1}^{k} d_{il} c_{ij} X_j \qquad [using \ Eq. \ (6)]$$

$$= \sum_{j=1}^{k} \mu d_{ij} X_i \qquad [using \ Eq. \ (7)]$$

$$= \mu_i Y_i \qquad [using \ Eq. \ (9)]$$

This show that $Y_i (l \le i \le K)$ are also eigen vectors of B associated respectively with the eigen values μ_i. The k vectors Y_i are therefore, common eigen vectors of A and B.

We have now proved that if A has k degenerate eigen vectors X_i then it is possible to find a set of k independent liner combinations of $X_i's$ which are also eigen vectors of B.

Continuing this process for every distinct eigen value of A, we see that finally it would be possible to find a common set of eigen vectors for the two commuting matrices, thus proving the theorem.

Note: The theorem can obviously be extended to any number of mutually commuting matrices, that is a set of matrices every one of which commutes with other member of the set. We therefore, have the generalised theorem that it is possible to find a common set of eigen vectors for any number of mutually commuting matrices. Alternately, we can say that mutually commuting matrices can be simultaneously diagonalized.

Example 5. *Find a common set of eigen vectors for the two matrices*

$$A = \begin{bmatrix} -1 & \sqrt{6} & \sqrt{2} \\ \sqrt{6} & 0 & \sqrt{3} \\ \sqrt{2} & \sqrt{3} & -2 \end{bmatrix}, B = \begin{bmatrix} 10 & \sqrt{6} & -\sqrt{2} \\ \sqrt{6} & 9 & \sqrt{3} \\ -\sqrt{2} & \sqrt{3} & 11 \end{bmatrix}$$

Solution: It is easily verified that A and B commute, in fact

$$AB = BA = \begin{bmatrix} -6 & 9\sqrt{6} & 15\sqrt{2} \\ 9\sqrt{6} & 9 & 9\sqrt{3} \\ 15\sqrt{2} & 9\sqrt{3} & -11 \end{bmatrix}$$

The eigen values of A are found to be 3, –3, –3. The eigen value 3 has multiplicity 1 while the eigen value –3 has multiplicity 2. The eigen vectors corresponding to $\lambda = 3$ is found to be

$$X_1 = \{\sqrt{2}, \sqrt{3}, 1\}^T$$

Since X_1 is a nondegenerate eigen vector of A, it must also be an eigen vector of B, as can be readily verified.

$$\begin{bmatrix} 10 & \sqrt{6} & -\sqrt{2} \\ \sqrt{6} & 9 & \sqrt{3} \\ -\sqrt{2} & \sqrt{3} & 11 \end{bmatrix} \begin{bmatrix} \sqrt{2} \\ \sqrt{3} \\ 1 \end{bmatrix} = \begin{bmatrix} 12\sqrt{2} \\ 12\sqrt{3} \\ 12 \end{bmatrix} = 12 \begin{bmatrix} \sqrt{2} \\ \sqrt{3} \\ 1 \end{bmatrix}$$

Thus 12 is one of the eigen values of B associated with the eigen vector X_1.

For the double degenerate eigen value of $\lambda = -3$, we shall proceed in the following manner. The most general eigen vector of A having the eigen value $\lambda = -3$ is found to be

$$X = [a, b, -\sqrt{2}a - \sqrt{3}b]^T \tag{1}$$

which involves two arbitrary constants. We wish to obtain two linearly independent eigen vectors of B. We will, therefore, find the conditions on a and b for which X is an eigen vector of B. We have

$$BX = \begin{bmatrix} 10 & \sqrt{6} & -\sqrt{2} \\ \sqrt{6} & 9 & \sqrt{3} \\ -\sqrt{2} & \sqrt{3} & 11 \end{bmatrix} \begin{bmatrix} a \\ b \\ -\sqrt{2}a - \sqrt{3}b \end{bmatrix}$$

$$= \begin{bmatrix} 12a + 2\sqrt{6}b \\ 6b \\ 12\sqrt{2}a - 10\sqrt{3}b \end{bmatrix}$$

For X to be eigen vector of B we must, therefore, have

$$\left. \begin{array}{c} 12a + 2\sqrt{6}b = \mu a \\ 6b = \mu b \\ 12\sqrt{2}a + 10\sqrt{3}b = \mu(\sqrt{2}a + \sqrt{3}b) \end{array} \right\} \tag{2}$$

Second part of Eq. (2) gives either $\mu = 6$ or $b = 0$

Taking $b = 0$ we find first and third part of Eq. (2) is satisfied for $\mu = 12$. The corresponding eigen vector with $a = 1$ is

$$X_2 = [1, 0, -\sqrt{2}]^T \tag{3}$$

Taking $\mu = 6$, the first and third part of Eq. (2) are satisfied if $b = -\sqrt{3}a/\sqrt{2}$. Choosing $a = \sqrt{2}$ the corresponding eigen vector is found to be

$$X_3 = [\sqrt{2}, -\sqrt{3}, 1]^T \tag{4}$$

The three eigen vectors X_1, X_2 and X_3 given respectively by Eqs (1), (3) and (4) are a common set of eigen vectors for A and B. Their values with respect to A are respectively 3, –3, –3 while those with respect to B respectively 12, 12, 6.

Note that any linear combination of X_2 and X_3 will be an eigen vector of B but not of A.

Example 6. If $A = \begin{bmatrix} 1 & 0 & 0 \\ 1 & 0 & 1 \\ 0 & 1 & 0 \end{bmatrix}$ then find A^{50} and A^{-1}.

Solution: The characteristic equation of matrix A is

$$|A - \lambda| = \begin{vmatrix} 1-\lambda & 0 & 0 \\ 1 & -\lambda & 1 \\ 0 & 1 & -\lambda \end{vmatrix} = 0$$

or $\qquad \lambda^3 - \lambda^2 - \lambda + 1 = 0 \qquad\qquad (1)$

Hence according to Hamilton–Cayley theorem, the Eq. (1) must be satisfied by matrix A, i.e.

$$A^3 - A^2 - A + I = 0$$

or $\qquad\qquad A^3 = A^2 + A - I \qquad\qquad (2)$

Multiplying Eq. (2) by $A, A^2, A^3, ..., A^{n-3}$ respectively, we get

$$\left. \begin{aligned} A^4 &= A^3 + A^2 - A \\ A^5 &= A^4 + A^3 - A^2 \\ &\cdots\cdots\cdots\cdots\cdots\cdots \\ A^{n-1} &= A^{n-2} + A^{n-3} - A^{n-4} \\ A^n &= A^{n-1} + A^{n-2} - A^{n-3} \end{aligned} \right\} \qquad (3)$$

Adding all parts of Eqs (3) and (2) we have

$$A^n = A^{n-2} + A^2 - I \ (n \geq 3) \qquad\qquad (4)$$
$$A^{50} = A^{48} + A^2 - I$$
$$A^{48} = A^{46} + A^2 - I$$
$$\cdots\cdots\cdots\cdots\cdots\cdots$$
$$A^4 = A^2 + A^2 - I$$

Adding all the above equations we have

$$A^{50} = A^2 + 24A^2 - 24I = 25A^2 - 24I$$

But $\qquad A^2 = \begin{bmatrix} 1 & 0 & 0 \\ 1 & 0 & 1 \\ 0 & 1 & 0 \end{bmatrix}\begin{bmatrix} 1 & 0 & 0 \\ 1 & 0 & 1 \\ 0 & 1 & 0 \end{bmatrix} = \begin{bmatrix} 1 & 0 & 0 \\ 1 & 1 & 0 \\ 1 & 0 & 1 \end{bmatrix}$

Hence $\qquad A^{50} = 25\begin{bmatrix} 1 & 0 & 0 \\ 1 & 1 & 0 \\ 1 & 0 & 1 \end{bmatrix} - 24\begin{bmatrix} 1 & 0 & 0 \\ 0 & 1 & 0 \\ 0 & 0 & 1 \end{bmatrix}$

$$= \begin{bmatrix} 1 & 0 & 0 \\ 25 & 1 & 0 \\ 25 & 0 & 1 \end{bmatrix} \qquad\qquad \textit{[using Eq. (2) gives]}$$

$$I = -A^3 + A^2 + A$$

Multiplying with A^{-1} we have

$$A^{-1} = -A^2 + A + I$$

$$= -\begin{bmatrix} 1 & 0 & 0 \\ 1 & 1 & 0 \\ 1 & 0 & 1 \end{bmatrix} + \begin{bmatrix} 1 & 0 & 0 \\ 1 & 0 & 1 \\ 0 & 1 & 0 \end{bmatrix} + \begin{bmatrix} 1 & 0 & 0 \\ 0 & 1 & 0 \\ 0 & 0 & 1 \end{bmatrix}$$

$$= \begin{bmatrix} 1 & 0 & 0 \\ 0 & 0 & 0 \\ -1 & 1 & 0 \end{bmatrix}$$

Example 7. Given $A = \begin{bmatrix} 1 & 1 & 2 \\ 3 & 1 & 1 \\ 2 & 3 & 1 \end{bmatrix}$, use the fact that A satisfies its characteristic equation, to

compute A^3 and A^4, also, since A is nonsingular to compute A^{-1} and A^{-2}.

Solution: The characteristic equation of matrix A is

$$|A - \lambda I| = \begin{vmatrix} 1-\lambda & 1 & 2 \\ 3 & 1-\lambda & 1 \\ 2 & 3 & 1-\lambda \end{vmatrix} = 11 + 7\lambda + 3\lambda^2 - \lambda^3 = 0$$

or $\qquad \lambda^3 - 3\lambda^2 - 7\lambda - 11 = 0$ $\qquad\qquad$ (1)

Hence according to Hamilton–Gayley theorem, the Eq. (1) must be satisfied by matrix A, i.e.

$$A^3 = 3A^2 + 7A + 11I = 0$$

or $\qquad 11I = A^3 + 3A^2 - 7A$ $\qquad\qquad$ (2)

or

$$A^3 = 3\begin{bmatrix} 1 & 1 & 2 \\ 3 & 1 & 1 \\ 2 & 3 & 1 \end{bmatrix}\begin{bmatrix} 1 & 1 & 2 \\ 3 & 1 & 1 \\ 2 & 3 & 1 \end{bmatrix} + 7\begin{bmatrix} 1 & 1 & 2 \\ 3 & 1 & 1 \\ 2 & 3 & 1 \end{bmatrix} + 11\begin{bmatrix} 1 & 0 & 0 \\ 0 & 1 & 0 \\ 0 & 0 & 1 \end{bmatrix}$$

$$= \begin{bmatrix} 42 & 31 & 29 \\ 45 & 39 & 31 \\ 43 & 45 & 42 \end{bmatrix}$$

and $\qquad A^4 = 3A^3 + 7A^2 + 11A = \begin{bmatrix} 42 & 31 & 29 \\ 45 & 39 & 31 \\ 43 & 45 & 42 \end{bmatrix} + 7\begin{bmatrix} 8 & 8 & 5 \\ 8 & 7 & 8 \\ 13 & 8 & 8 \end{bmatrix} + 11\begin{bmatrix} 1 & 1 & 2 \\ 3 & 1 & 1 \\ 2 & 3 & 1 \end{bmatrix}$

$$= \begin{bmatrix} 193 & 160 & 144 \\ 224 & 177 & 160 \\ 272 & 224 & 193 \end{bmatrix}$$

From Eq. (2) we have

$$11I = A^3 - 3A^2 - 7A$$

or

$$A^{-1} = \frac{1}{11}\{A^2 - 3A - 7I\}$$

$$= \frac{1}{11}\begin{bmatrix} 8 & 8 & 5 \\ 8 & 7 & 8 \\ 13 & 8 & 8 \end{bmatrix} - 3\begin{bmatrix} 1 & 1 & 2 \\ 3 & 1 & 1 \\ 2 & 3 & 1 \end{bmatrix} - 7\begin{bmatrix} 1 & 0 & 0 \\ 0 & 1 & 0 \\ 0 & 0 & 1 \end{bmatrix}$$

$$= \frac{1}{11}\begin{bmatrix} -2 & 5 & -1 \\ -1 & -3 & 5 \\ 7 & -1 & -2 \end{bmatrix}$$

$$A^{-2} = \frac{1}{11}[A - 3I - 7A^{-1}]$$

$$= \frac{1}{11}\begin{bmatrix} 1 & 1 & 2 \\ 3 & 1 & 1 \\ 2 & 3 & 1 \end{bmatrix} - \frac{3}{11}\begin{bmatrix} 1 & 0 & 0 \\ 0 & 1 & 0 \\ 0 & 0 & 1 \end{bmatrix} - \frac{7}{11}\cdot\frac{1}{11}\begin{bmatrix} -2 & 5 & -1 \\ -1 & -3 & 5 \\ 7 & -1 & -2 \end{bmatrix}$$

$$= \frac{1}{121}\begin{bmatrix} -8 & -24 & 29 \\ 40 & -1 & -24 \\ -27 & 40 & -1 \end{bmatrix}$$

2.14. FUNCTIONS OF MATRIX

As we define and study various functions of a variable in algebra, it is possible to define and evaluate functions of a matrix such as integral powers (positive and negative), fractional powers (rots), exponential, logarithm, trigonometric and hyperbolic functions.

There are two methods by which a function of matrix can be evaluated. The first is rather a straight forward method based on the diagonalization of a matrix and is, therefore, applicable to diagonalizable matrices only. The second method, i.e. based on the existence of a minimal polynomial and can be used to evaluate functions of any matrix.

Functions of Diagonalizable Matrix

If A be a diagonalizable matrix and let P a diagonalizing matrix for A, so that

$$P^{-1} AP = D, \quad A = PDP^{-1} \tag{2.85}$$

where D is diagonal matrix containing the eigen value of A.

Now if f is any function of a matrix, then we have

$$f(A) = Pf(D)P^{-1} \tag{2.86}$$

Thus, if we can define a function of a diagonal matrix, we can define and evaluate the function of any diagonalizable matrix. The discussion of this section evidently applies to square matrices only.

Power of a Matrix

We define the square of a matrix by $A^2 = AA$. The cube by $A^3 = AAA$, etc. In general, if k is a positive integer, we define the kth power of A as a matrix obtained by multiplying A with itself k times, that is

$$A^k = AAA...A \ (k \text{ times}) \tag{2.87}$$

If A is nonsingular, we have defined its inverse A^{-1} as a matrix whose product with A gives the unit matrix. The negative powers of A are then similarly defined. If m is a negative integer, let $k = -m$, so that

$$A^m = (A^{-1}) = A^{-1} A^{-1} A^{-1}...A^{-1} \ (k \text{ times}) \tag{2.88}$$

Finally, in analogy with the functions of a variable we have

$$A^0 = I \tag{2.89}$$

Although all the integral powers of A have thus been defined in a straightforward manner, the actual evaluation may be tedious for large values of K. The calculation is considerably simplified by using the diagonalizationality of A. For taking the Kth power of A and using Eqs (2.85) and (2.86), we have

$$A^k = (PDP^{-1}) (PDP^{-1})...(PDP^{-1}) \ (k \text{ times})$$
$$= PD^k P^{-1} \tag{2.90}$$

Similarly if $m = -k$ is a negative integer and A is nonsingular then

$$A^m = PD^m P^{-1} = P(D^{-1})^k P^{-1} \tag{2.91}$$

In the same manner if q is any fraction, we have

$$Aq = PDq \ P^{-1} \tag{2.92}$$

Exponential of a Matrix

The exponential series is defined as

$$e^\lambda = \sum_{k=0}^{\infty} \frac{\lambda^k}{k!}$$

Similarly, we shall define the exponential of matrix A by

$$e^A = \sum_{k=0}^{\infty} \frac{A^k}{k!}$$

Which will be a matrix of the same order as A and will exist provided all the elements (and hence eigen values) of A are finite.

To begin with let us obtain the exponential of a diagonal matrix. Let D be a diagonal matrix with elements $(D)_{ij} = \lambda_i \delta_{ij}$. Then

$$e^D = \sum_{k=0}^{\infty} \frac{D^k}{k!}$$

The ijth element of e^D will be given by

$$(e^D)_{ij} = \sum_{k=0}^{\infty} \frac{(D^k)_{ij}}{k!} = \sum_{k=0}^{\infty} \lambda_i^k \delta_{ij} / k! = e^{\lambda_i} \delta_{ij}$$

It is, therefore, evident that if

$$D = \begin{bmatrix} \lambda_1 & & & & 0 \\ & \lambda_2 & & & \\ & & \lambda_3 & & \\ & & & \ddots & \\ 0 & & & & \lambda_n \end{bmatrix} \quad \text{then} \quad e^D = \begin{bmatrix} e^{\lambda_1} & & & & 0 \\ & e^{\lambda_2} & & & \\ & & e^{\lambda_3} & & \\ & & & \ddots & \\ 0 & & & & e^{\lambda_n} \end{bmatrix}$$

Now consider the series

$$e^A = I + A + \frac{A^2}{2!} + \frac{A^3}{3!} + \cdots + \frac{A^k}{k!} + \cdots$$

Let P be a matrix which brings A to a diagonal form D. Then we have

$$P^{-1} e^A P = I + D \frac{D^2}{2!} + \frac{D^3}{3!} + \cdots + \frac{D^k}{k!} + \cdots$$

It follows immediately that

$$e^A = P(e^D) P^{-1} \tag{2.93}$$

Having defined the exponential function, it is possible to define the matrix exponent of any number. We have

$$a^x = e^{x \log a}$$

Hence we define

$$a^A = e^{A \log a} = P e^{D \log a} P^{-1} \tag{2.94}$$

where

$$e^{D \log a} = a^D = \begin{bmatrix} a^{\lambda_1} & & & & 0 \\ & a^{\lambda_2} & & & \\ & & \ddots & & \\ 0 & & & & a^{\lambda_n} \end{bmatrix}$$

Logarithm of a Matrix

We say x is the natural logarithm of y if $e^x = y$ and write it as $\log y$. Similarly, given a matrix A, we shall say that a matrix B is the natural logarithm of A if $e^B = A$.

Therefore, by definition

$$B = \log A, \text{ i.e. } e^B = A$$

We shall first find the logarithm of a diagonal matrix. If $D = [\lambda_i \, \delta_{ij}]$ is a diagonal matrix, its natural logarithm is diagonal matrix of the same order given by

$$\log D = \begin{bmatrix} \log \lambda_1 & & & 0 \\ & \log \lambda_2 & & \\ & & \ddots & \\ 0 & & & \log \lambda_n \end{bmatrix} = X \text{ (say)}$$

To prove this, consider e^x, and remembering that $e^{\log x} = x$, we have

$$e^x = e^{\log D} = D \tag{2.95}$$

Therefore, by definition, it follows that x is the natural logarithm of D. Let A be diagonalizable matrix. We now state that

$$B = \log A = P(\log D)P^{-1} = PxP^{-1}$$

and to prove that we must show that $e^B = A$. We have

$$e^B = I + B + \frac{B^2}{2!} + \cdots + \frac{B^k}{k!} + \cdots$$

$$= I + PxP^{-1} + \frac{(PxP^{-1})^2}{2!} + \cdots + \frac{(PxP^{-1})^k}{k!} + \cdots$$

$$= P\left[I + X + \frac{X^2}{2!} + \cdots + \frac{X^k}{k!} + \cdots \right]P^{-1}$$

$$= Pe^x P^{-1} = PDP^{-1} = A$$

Evaluation of Functions using Cayley–Hamilton Theorem

The method discussed so far, for evaluating the various functions of a matrix is based on the principle that if $A = PDP^{-1}$, where D is diagonal matrix, then $f(A) = P f(D)P^{-1}$. It has however, the major drawback that it is applicable to diagonalizable matrices only. There is an alternative method for evaluating the functions of a matrix which is based on the use of the Cayley–Hamilton theorem and which is applicable to any matrix.

We have already shown that polynomial of any degree in a matrix is equal to a polynomial of degree $\leq m - 1$, where m is the degree of the minimal polynomial. The result, infact holds good not only for polynomials but also for any arbitrary function of matrix A provided the function is sufficiently differentiable. Thus, if the degree of the minimal polynomial of a matrix A is m, any function $f(A)$ can be expressed as a linear combination of the linearly independent matrices $I, A, A^2, ..., A^{m-1}$, i.e.

$$f(A) = r(A) = \alpha_{m-1} A^{m-1} + \alpha_{m-2} A^{m-2} + \cdots + \alpha_0 I$$

The scalars α_i are determined as follows. If λ_i is a k-fold degenerate eigen value of A, the algebraic function $f(\lambda)$ and $r(\lambda)$ satisfy the equations

$$f(\lambda_i) = r(\lambda_i)$$

$$\frac{df(\lambda_i)}{d\lambda} = \frac{dr(\lambda_i)}{d\lambda}$$

$$\frac{d^2 f(\lambda_i)}{d^2\lambda} = \frac{d^2 r(\lambda_i)}{d\lambda^2}$$

$$\cdots\cdots\cdots\cdots\cdots\cdots$$

$$\frac{d^{k-1} f(\lambda_i)}{d\lambda^{k-1}} = \frac{d^{k-1} r(\lambda_i)}{d\lambda^{k-1}}$$

Here the notation $\dfrac{d^l f(\lambda_i)}{d\lambda^l}$ denotes the lth derivative of $f(\lambda)$ evaluated at $\lambda = \lambda_i$.

Differentiation and Integration of Matrices

Consider a matrix $A(t)$ whose elements a_{ij} are functions of scalar variable t, then the matrix $A(t)$ is called matrix function of variable t. Then the derivative of this matrix with respect to t, if exists is defined by the relation

$$\frac{d}{dt}[A(t)] = \left[\frac{d}{dt}a_{ij}(t)\right]$$

Obviously, the elements of the derivative matrix $[A(t)]'$ are the derivatives of the corresponding elements of the matrix $A(t)$.

The integration of the matrix $A(t)$ is defined by the relation

$$\int A(t)\,dt = \left[\int a_{ij}(t)\,dt\right]$$

Obviously, the integral of the matrix exists, provided that the integral of each element of the matrix $A(t)$ exists and is represented by a matrix whose elements are obtained by integrating the corresponding elements of the matrix $A(t)$.

Example 1. Find A^k, where k is any integer, positive or negative and

$$A - \begin{bmatrix} \dfrac{4}{3} & \dfrac{\sqrt{3}}{3} \\ \dfrac{\sqrt{2}}{3} & \dfrac{5}{3} \end{bmatrix}$$

Solution: Eigen values and the eigen vectors of A are found to be

i. $1,[\sqrt{2} \ -1]^T$, ii. $2,[1 \ \sqrt{2}]^T$.

We, therefore have

$$P = \begin{bmatrix} \sqrt{2} & 1 \\ -1 & \sqrt{2} \end{bmatrix}, P^{-1}AP^{-1}\begin{bmatrix} 1 & 0 \\ 0 & 2 \end{bmatrix} = D.$$

The matrix A is seen to be nonsingular. For any integral k, therefore, we have

$$A^k = PD^k P^{-1}$$

$$= \begin{bmatrix} \sqrt{2} & 1 \\ -1 & \sqrt{2} \end{bmatrix}\begin{bmatrix} 1 & 0 \\ 0 & 2^k \end{bmatrix}\begin{bmatrix} \dfrac{\sqrt{2}}{3} & -\dfrac{1}{3} \\ \dfrac{1}{3} & \dfrac{\sqrt{2}}{3} \end{bmatrix}$$

Now. If $k = 50$

$$A^{50} = \frac{1}{3}\begin{bmatrix} 2^{50}+2 & (2^{50}-1)\sqrt{2} \\ (2^{50}-1)\sqrt{2} & (2^{51}+1) \end{bmatrix}$$

and for $k = -10$

$$A^{-10} = \frac{1}{3}\begin{bmatrix} 2^{-10}+2 & (2^{-10}-1)\sqrt{2} \\ (2^{-10}-1)\sqrt{2} & (2^{-9}+1) \end{bmatrix}$$

Similarly $$A^{3/7} = \frac{1}{3}\begin{bmatrix} 2^{3/7}+2 & (2^{3/7}-1)\sqrt{2} \\ (2^{-37}-1)\sqrt{2} & (2^{10/7}+1)\sqrt{2} \end{bmatrix}$$

Example 2. *Find the square roots of the matrix*

$$A = \begin{bmatrix} \dfrac{3}{2} & \dfrac{1}{2} \\ \dfrac{1}{2} & \dfrac{3}{2} \end{bmatrix}$$

Solution: The eigen values and eigen vectors of the given matrix are found to be

i. $2, [1, \; 1]$
ii. $1, [1, \; -1]^T$.

We have then

$$P = \begin{bmatrix} 1 & 1 \\ 1 & -1 \end{bmatrix}, P^{-1} = \frac{1}{2}\begin{bmatrix} 1 & 1 \\ 1 & -1 \end{bmatrix}, \; D = \begin{bmatrix} 2 & 0 \\ 0 & 1 \end{bmatrix}$$

The square roots of D are

$$A^{1/2} = \begin{bmatrix} \pm\sqrt{2} & 0 \\ 0 & \pm 1 \end{bmatrix}$$

We can choose any of the four matrices to obtain $A^{1/2}$ by using the relation
$$A^{1/2} = PD^{1/2}P^{-1}.$$

We observe that A has four square roots given by $\pm B$ and $\pm C$, where

$$B = \begin{bmatrix} \left(\dfrac{\sqrt{2}+1}{2}\right) & \left(\dfrac{\sqrt{2}-1}{2}\right) \\ \left(\dfrac{\sqrt{2}-1}{2}\right) & \left(\dfrac{\sqrt{2}+1}{2}\right) \end{bmatrix} = \frac{1}{2}\begin{bmatrix} (\sqrt{2}+1) & \sqrt{2}-1 \\ \sqrt{2}-1 & \sqrt{2}+1 \end{bmatrix}$$

$$C = \begin{bmatrix} \left(\dfrac{\sqrt{2}-1}{2}\right) & \left(\dfrac{\sqrt{2}+1}{2}\right) \\ \left(\dfrac{\sqrt{2}+1}{2}\right) & \left(\dfrac{\sqrt{2}-1}{2}\right) \end{bmatrix} = \frac{1}{2}\begin{bmatrix} (\sqrt{2}-1) & \sqrt{2}+1 \\ \sqrt{2}+1 & \sqrt{2}-1 \end{bmatrix}$$

Example 3. *Evaluate E^A and 4^A if A is the matrix*

$$A = \begin{bmatrix} 1 & 0 \\ 0 & 2 \end{bmatrix}$$

Solution: The eigen values of given diagonal matrix are
$$\lambda_1 = 1 \quad \text{and} \quad \lambda_2 = 2$$

Therefore

$$e^A = \begin{bmatrix} e^{\lambda_1} & 0 \\ 0 & e^{\lambda_2} \end{bmatrix} = \begin{bmatrix} e^1 & 0 \\ 0 & e^2 \end{bmatrix} = \begin{bmatrix} e & 0 \\ 0 & e^2 \end{bmatrix}$$

and

$$4^A = \begin{bmatrix} 4^1 & 0 \\ 0 & 4^2 \end{bmatrix} = \begin{bmatrix} 4 & 0 \\ 0 & 16 \end{bmatrix}$$

Example 4. *Evaluate e^A and 6^A if A is a matrix and*

$$A = \begin{bmatrix} \dfrac{3}{2} & \dfrac{1}{2} \\ \dfrac{1}{2} & \dfrac{3}{2} \end{bmatrix}$$

Solution: The eigen values and eigen vectors of the given matrix are

$$2, [1, \ 1]^T, \quad 1, [1 \ -1]^T,$$

∴   We have $P = \begin{bmatrix} 1 & 1 \\ 1 & -1 \end{bmatrix}$

$$P^{-1} = \frac{1}{2}\begin{bmatrix} 1 & 1 \\ 1 & -1 \end{bmatrix}, \ D = \begin{bmatrix} 2 & 0 \\ 0 & 1 \end{bmatrix}$$

∴ $e^A = Pe^D P^{-1}$

$$= \begin{bmatrix} 1 & 1 \\ 1 & -1 \end{bmatrix}\begin{bmatrix} e^2 & 0 \\ 0 & e^1 \end{bmatrix}\frac{1}{2}\begin{bmatrix} 1 & 1 \\ 1 & -1 \end{bmatrix}$$

$$= \frac{1}{2}\begin{bmatrix} e^2+e & e^2-e \\ e^2-e & e^2+e \end{bmatrix}$$

$$6^A = P6^D P^{-1}$$

$$= \begin{bmatrix} 1 & 1 \\ 1 & -1 \end{bmatrix}\begin{bmatrix} 6^2 & 0 \\ 0 & 6 \end{bmatrix}\frac{1}{2}\begin{bmatrix} 1 & 1 \\ 1 & -1 \end{bmatrix}$$

$$= \frac{1}{2}\begin{bmatrix} 36+6 & 36-6 \\ 36-6 & 36+6 \end{bmatrix}$$

$$= \frac{1}{2}\begin{bmatrix} 42 & 30 \\ 30 & 42 \end{bmatrix} = \begin{bmatrix} 21 & 15 \\ 15 & 21 \end{bmatrix} = 3\begin{bmatrix} 7 & 5 \\ 5 & 7 \end{bmatrix}$$

Example 5. *Find the logarithm of the matrix*

$$A = \begin{bmatrix} 39 & -50 & -20 \\ 15 & -16 & -10 \\ 30 & -50 & -11 \end{bmatrix}$$

Solution: For the given matrix A, the matrices, P, D and P^{-1} are found to be

$$P = \begin{bmatrix} 3 & 1 & 2 \\ 1 & 1 & 1 \\ 2 & -1 & 2 \end{bmatrix}, \ D = \begin{bmatrix} 9 & 0 & 0 \\ 0 & 9 & 0 \\ 0 & 0 & -6 \end{bmatrix}$$

$$P^{-1} = \frac{1}{3}\begin{bmatrix} 3 & -4 & -1 \\ 0 & 2 & -2 \\ -3 & 5 & 2 \end{bmatrix}$$

We have

$$\log D = \begin{bmatrix} \log 9 & 0 & 0 \\ 0 & \log 9 & 0 \\ 0 & 0 & \log 6 + i\pi \end{bmatrix}$$

Therefore

$$\log A = P(\log D)P^{-1} = \begin{bmatrix} 3 & 1 & 2 \\ 1 & 1 & 1 \\ 2 & -1 & 2 \end{bmatrix}\begin{bmatrix} \log 9 & 0 & 0 \\ 0 & \log 9 & 0 \\ 0 & 0 & \log 6 + i\pi \end{bmatrix} \times \frac{1}{3}\begin{bmatrix} 3 & -4 & -1 \\ 0 & 2 & -2 \\ -3 & 5 & 2 \end{bmatrix}$$

$$= \frac{1}{3}\begin{bmatrix} 9a - 6b & 10(b-a) & 4(b-a) \\ 3(a-b) & 5b - 2a & 2(b-a) \\ 6(a-b) & 10(b-a) & 4b - a \end{bmatrix}$$

where $a = \log 9$, $b = \log(-6) = \log 6 + i\pi$.

Example 6. *Find $\sin \pi A$ and $\cos \pi A$ where*

$$A = \begin{bmatrix} -\dfrac{47}{2} & 53 & 30 \\ -12 & \dfrac{53}{2} & 15 \\ 2 & -\dfrac{7}{2} & -2 \end{bmatrix}$$

Solution: The matrices P, D, and P^{-1} associated with the given matrix A are found to be

$$P = \begin{bmatrix} 5 & 8 & 1 \\ 0 & 2 & 1 \\ 4 & 3 & -1 \end{bmatrix}, \quad D = \begin{bmatrix} \dfrac{1}{2} & 0 & 0 \\ 0 & 1 & 0 \\ 0 & 0 & -\dfrac{1}{2} \end{bmatrix}$$

$$P^{-1} = \begin{bmatrix} 5 & -11 & -6 \\ -4 & 9 & 5 \\ 8 & -17 & -10 \end{bmatrix}$$

We have

$$\sin \pi A = \frac{1}{2i}(e^{i\pi A} - e^{-i\pi A})$$

$$= \sum_{k=0}^{\infty} \frac{(-1)^k (\pi A)^{2k+1}}{(2k+1)!}$$

and
$$\cos \pi A = \frac{1}{2}(e^{i\pi A} + e^{-i\pi A})$$

$$= \sum_{k=0}^{\infty} \frac{(-1)^k A^{2k}}{(2k)!}$$

Hence
$$\sin \pi D = \sum_{k=0}^{\infty} \frac{(-1)^k (\pi D)^{2k+1}}{(2k+1)!}$$

$$= \begin{bmatrix} \sin\dfrac{\pi}{2} & 0 & 0 \\ 0 & \sin\pi & 0 \\ 0 & 0 & -\sin\dfrac{\pi}{2} \end{bmatrix} = \begin{bmatrix} 1 & 0 & 0 \\ 0 & 0 & 0 \\ 0 & 0 & -1 \end{bmatrix}$$

Therefore
$$\sin \pi A = P \sin (pD) P^{-1}$$

$$= \begin{bmatrix} 5 & 8 & 1 \\ 0 & 2 & 1 \\ 4 & 3 & -1 \end{bmatrix} \begin{bmatrix} 1 & 0 & 0 \\ 0 & 0 & 0 \\ 0 & 0 & -1 \end{bmatrix} \begin{bmatrix} 5 & -11 & -6 \\ -4 & 9 & 5 \\ 8 & -17 & -10 \end{bmatrix}$$

$$= \begin{bmatrix} 17 & -38 & -20 \\ -8 & 17 & 10 \\ 28 & -61 & -34 \end{bmatrix}$$

Similarly
$$\cos \pi D = \begin{bmatrix} 0 & 0 & 0 \\ 0 & -1 & 0 \\ 0 & 0 & 0 \end{bmatrix}$$

so that
$$\cos \pi A = P \cos (pD) P^{-1}$$

$$= \begin{bmatrix} 32 & -72 & -40 \\ 8 & -18 & -10 \\ 12 & -27 & -15 \end{bmatrix}$$

Example 7. *Find A^P, where P is any number and A is a matrix where*

$$A = \begin{bmatrix} \dfrac{4}{3} & \dfrac{\sqrt{2}}{3} \\ \dfrac{\sqrt{2}}{3} & \dfrac{5}{3} \end{bmatrix}$$

Solution: The matrix A is of order 2 and has distinct eigen values $\lambda_1 = 1$, $\lambda_2 = 2$. The degree of minimal polynomial is therefore $m = 2$, we have $f(A) = A^P$, so that $f(\lambda) = \lambda^P$.

Let
$$r(A) = \alpha_1 A + \alpha_0 I$$
so that
$$r(\lambda) = \alpha_{11} + \alpha_0$$

Since both the eigen values are nondegenerate, we have the two conditions

$$f(\lambda_1) = \gamma(\lambda_1), f(\lambda_2) = \gamma(\lambda_2)$$

which gives

$$f(\lambda_1) = I^P = \alpha_1 + \alpha_0 = 1$$

$$f(\lambda_2) = 2^P = 2\alpha_1 + \alpha_0 = 2^P$$

or

$$\alpha_1 = 2^P - I \quad \text{and} \quad \alpha_0 = 2 - 2^P.$$

∴

$$f(A) = \gamma(A) = \alpha_1 A + \alpha_0 I$$

$$= (2^P - 1)\begin{bmatrix} \dfrac{4}{3} & \dfrac{\sqrt{2}}{3} \\ \dfrac{\sqrt{2}}{3} & \dfrac{5}{3} \end{bmatrix} + (2 - 2^P)\begin{bmatrix} 1 & 0 \\ 0 & 1 \end{bmatrix}$$

or

$$A^P = \frac{1}{3}\begin{bmatrix} 2^P + 2 & (2^P - 1)\sqrt{2} \\ (2^P - 1)\sqrt{2} & (2^{P+1} + 1) \end{bmatrix}$$

which agress with Example 1. If P is nonintegral this will not be the only matrix equal to A^P.

Example 8. *Determine 4A and log A where A is a matrix, and*

$$A = \begin{bmatrix} 3 & 1 \\ 2 & 2 \\ 1 & 3 \\ 2 & 2 \end{bmatrix}$$

Solution: We have the eigen values of matrix A as

$$\lambda_1 = 2, \lambda_2 = 1$$

The matrix is of order 2 and has distinct eigen values. The degree of the minimal polynomial is, therefore, $m = 2$.

i. Here $f(A) = 4^A$ so that $f(\lambda) = 4^\lambda$.

Let

$$r(A) = \alpha_1 A + A_0 I \quad \text{so that} \quad r(\lambda) = \alpha_1 \lambda + \alpha_0$$

so that

$$4^2 = 2\alpha_1 + \alpha_0 = 16$$

$$4^1 = \alpha_1 + \alpha_0 = 4$$

∴

$$\alpha_1 = 12 \quad \text{and} \quad a_0 = -8$$

Here

$$4^A = 12\begin{bmatrix} \dfrac{3}{2} & \dfrac{1}{2} \\ \dfrac{1}{2} & \dfrac{3}{2} \end{bmatrix} - 8\begin{bmatrix} 1 & 0 \\ 0 & 1 \end{bmatrix} = \begin{bmatrix} 10 & 6 \\ 6 & 10 \end{bmatrix}$$

ii. Here $f(A) = \log A$ and $f(\lambda) = \log \lambda$

Let

$$r(A) = \alpha_1 A + \alpha_0 I \quad \text{so that} \quad r(\lambda) = \alpha_1 \lambda + \alpha_0$$

so that

$$2\alpha_1 + \alpha_0 = \log 2$$

$$\alpha_1 + \alpha_0 = \log 1 = 0.$$

∴

$$\alpha_1 = \log 2 \quad \text{and} \quad \alpha_0 = -\log 2$$

$$\therefore \qquad \log A = \log 2 \begin{bmatrix} \dfrac{3}{2} & \dfrac{1}{2} \\ \dfrac{1}{2} & \dfrac{3}{2} \end{bmatrix} - \log 2 \begin{bmatrix} 1 & 0 \\ 0 & 1 \end{bmatrix}$$

$$= \frac{1}{2} \log 2 \begin{bmatrix} 1 & 1 \\ 1 & 1 \end{bmatrix}$$

Example 9. Find e^{3A}, e^{-A} and cosh (At) where A is a matrix,

$$A = \begin{bmatrix} 3 & 1 & -1 \\ 2 & 2 & -1 \\ 2 & 2 & 0 \end{bmatrix}$$

Solution: The given matrix has eigen values $\lambda_1 = 1$ with multiplicity 1 and $\lambda_2 = 2$ with multiplicity 2. It would be convenient to find the function $f(A) = e^{AT}$, where t is a parameter. We, therefore, take $f(\lambda) = e^{\lambda t}$. Also let

$$r(A) = \alpha_2 A^2 + \alpha_1 A + \alpha_0 I$$

and

$$r(\lambda) = \alpha_2 \lambda^2 + \alpha_1 \lambda + \alpha_0.$$

The coefficients α_0, α_1, α_2 are determined by the conditions

$$f(1) - r(1), f(2) = r(2), f'(\lambda)|_{\lambda=2} - r'(\lambda)|_{\lambda=2}$$

These give

$$e^t = \alpha_2 + \alpha_1 + \alpha_0$$
$$e^{2t} = 4\alpha_2 + 2\alpha_1 + \alpha_0$$
$$te^{2t} = 4\alpha_2 + \alpha_1$$

which have the solution

$$\alpha_2 = (t-1) e^{2t} + e^t$$
$$\alpha_1 = (4-3t) e^{2t} - 4e^t$$
$$\alpha_0 = (2t-3) e^{2t} + 4e^t$$

The matrix e^{At} is, therefore, given by

$$e^{At} = \alpha_2 \begin{bmatrix} 9 & 3 & -4 \\ 8 & 4 & -4 \\ 10 & 6 & -4 \end{bmatrix} + \alpha_1 \begin{bmatrix} 3 & 1 & -1 \\ 2 & 2 & -1 \\ 2 & 2 & 0 \end{bmatrix} + \alpha_0 \begin{bmatrix} 1 & 0 & 0 \\ 0 & 1 & 0 \\ 0 & 0 & 1 \end{bmatrix}$$

$$= \begin{bmatrix} 2te^{2t} + e^t & e^{2t} - e^t & -te^{2t} \\ 2te^{2t} & e^{2t} & -te^{2t} \\ (4t-2)e^{2t} & 2e^{2t} - 2e^t & (1-2t)e^{2t} \end{bmatrix}$$

From this we have

$$e^{3A} = \begin{bmatrix} 6e^6 + e^3 & e^6 - e^3 & -3e^6 \\ 6e^6 & e^6 & -3e^6 \\ 10e^6 + 2e^3 & 2e^6 - 2e^3 & -5e^6 \end{bmatrix}$$

and

$$e^{-A} = \begin{bmatrix} e^{-1} - 2e^{-2} & e^{-2} - e^{-1} & e^{-2} \\ -2e^{-2} & e^{-2} & e^{-2} \\ -6e^{-2} + 2e^{-1} & 2e^{-2} - 2e^{-1} & 3e^{-2} \end{bmatrix}$$

The function cosh At can be found out either directly by taking $f(\lambda) = \cosh(\lambda t)$ and following the same procedure, or by finding e^{-At} [replacing t by $-t$ in e^{At}] and using $\cosh At = (e^{At} + e^{-At})/2$. The result is

$$\cosh At = \begin{bmatrix} 2r\sin 2t \cosh t & \cosh 2t - \cosh t & -t\sinh 2t \\ 2t\sinh 2t & \cosh 2t & -\sinh 2t \\ 2t\sinh 2t + 2\cosh t - 2\cosh 2t & 2\cosh 2t - 2\cosh t & \cosh 2t - 2t\sinh 2t \end{bmatrix}$$

Example 10. *Find* $\log A$, *where*

$$A = \begin{bmatrix} 2 & -2 & 0 \\ 0 & 2 & 0 \\ 1 & -3 & 2 \end{bmatrix}$$

Solution: The given matrix has only one distinct eigen value $\lambda = 2$ with multiplicity 3, we have

$$f(A) = \log A, f(\lambda) = \log \lambda$$
$$r(A) = \alpha_2 A^2 + \alpha_1 A + A_0 I$$

and

$$r(\lambda) = \alpha_2\lambda^2 + \alpha_1\lambda + \alpha_0$$

The coefficients are determined by the conditions

$$f(2) = r(2), f'(2) = r'(2), f''(2) = r''(2)$$

which gives

$$\log 2 = 4\alpha_2 + 2\alpha_1 + \alpha_0$$

$$\frac{1}{2} = 4\alpha_2 + \alpha_1$$

$$-\frac{1}{4} = 2\alpha_2$$

or

$$\alpha_0 = \log 2 - \frac{3}{2}, \quad \alpha_1 = 1 \quad \text{and} \quad \alpha_2 = -\frac{1}{8}.$$

Therefore

$$\log A = -\frac{1}{8}\begin{bmatrix} 4 & -8 & 0 \\ 0 & 4 & 0 \\ 4 & -14 & 4 \end{bmatrix} + \begin{bmatrix} 2 & -2 & 0 \\ 0 & 2 & 0 \\ 1 & -3 & 2 \end{bmatrix} + \left(\log 2 - \frac{3}{2}\right)\begin{bmatrix} 1 & 0 & 0 \\ 0 & 1 & 0 \\ 0 & 0 & 1 \end{bmatrix}$$

$$= \begin{bmatrix} \log 2 & -1 & 0 \\ 0 & \log 2 & 0 \\ \frac{1}{2} & -\frac{5}{4} & \log 2 \end{bmatrix}.$$

Example 11. *For square matrix A, show that*

$$\det(e^A) = e^{(Tr\cdot A)} = |e^A|$$

Solution: We shall prove this result by two methods. The first in applicable to diagonalizable matrices only while the second is general.

i. If A is a diagonalizable matrix, let

$$P^{-1}AP = D, \quad A = PDP^{-1}$$

where D is a diagonal matrix containing the eigen values λ_i of A, then we have

$$e^A = Pe^D P^{-1}$$

Therefore $\det (e^A) = \det (P) \det (e^D) \det (P^{-1}) = \det (e^D)$.

Now e^D is a diagonal matrix with diagonal elements e^{λ_i} hence it is evident that

$$\det (e^A) = e^{\lambda_1} e^{\lambda_2} ... e^{\lambda_n}$$

$$= e^{(\lambda_2 + \lambda_2 + \cdots + \lambda_n)} = e^{Tr \cdot (A)}$$

ii. Now we shall prove the theorem for a square matrix in general. We know that the square matrix can be brought to a triangular form by a similarity transformation. Let therefore,

$$P^{-1}AP = T, \quad A = PTP^{-1}$$

where T is a triangular matrix whose diagonal elements are $(T)_{ii} = \lambda_i$ which are the eigen values of A. Hence, we have

$$e^A = \sum_{K=0}^{\infty} \frac{A^K}{K!} = P\left[\sum_{K=0}^{\infty} \frac{T^K}{K!}\right] P^{-1} = Pe^T P^{-1} \tag{1}$$

Since T is a triangular matrix, the diagonal elements of the Kth power of T are λ_i^K, where K is a positive integer. Therefore, it is evident from the exponential series that the diagonal elements are $(T)_{ii} = \lambda_i$. Since the determinant of a triangular matrix equals the product of its diagonal elements we have

$$\det (e^T) = e^{\lambda_1 + \lambda_2 + \cdots + \lambda_n} = e^{(Tr \cdot A)}$$

From Eq. (1), we have

$$\det (e^T) = \det (P) \det (e^T) \det (P^{-1}) = \det (e^T)$$

$$= e^{Tr \cdot A}.$$

SHORT ANSWER QUESTIONS

1. Prove that the inverse of a matrix is unique.
2. What do you mean by characteristic polynomial?
3. Define indentity and orthogonal matrix.
4. Show that eigen values of Hermitian matrices are real and orthogonal to each other.
5. What do on mean by eigen values of a matrix?
6. What are symmetric matrices?
7. If A and B are two nonsingular matrices of the same order then $(AB)^{-1} = \bar{B} A^{-1}$
8. Define singular and nonsingular matrices.
9. What is the characteristic equation of a matrix $A = \begin{bmatrix} a_{11} & a_{12} \\ a_{21} & a_{22} \end{bmatrix}$?
10. Show that eigen values of a Hermitian matrix are equal.
11. Define Hermitian matrix. What is the condition to be satisfied by its elements?

12. If A and B are two matrices and form the product AB, then show that $(AB)^T = B^T A^T$.

13. Define inverse of a matrix.

14. What do you mean by eigen values and eigen vectors of a matrix.

15. Show that the trace of a matrix remains changed under similar trans-formations.

16. Name the matrix in which the sum of elements of each row is 1. Give one example.

17. Define trace of a matrix.

18. What do you mean by transpose of a matrix?

19. What do you mean by the rank of a matrix?

20. What do you mean by transformation?

21. What do you mean by orthogonal transformation?

22. What do you mena by similarity transformation?

PROBLEMS

1. For three pauli's spin matrices

$$\sigma_1 = \begin{bmatrix} 0 & 1 \\ 1 & 0 \end{bmatrix}, \sigma_2 = \begin{bmatrix} 0 & -i \\ i & 0 \end{bmatrix}, \sigma_3 = \begin{bmatrix} 1 & 0 \\ 0 & -1 \end{bmatrix}$$

prove that

i. $\sigma_1^2 = I$

ii. $\sigma_i \sigma_j = i \sigma_k$, where i, j, k are cyclic permutations of indices.

iii. $\sigma_i \sigma_j + \sigma_j \sigma_k = 2\delta_{ii} I$.

2. If matrices M_x, M_y and M_z are given by

$$M_x = \frac{1}{\sqrt{2}} \begin{bmatrix} 0 & 1 & 0 \\ 1 & 0 & 1 \\ 0 & 1 & 0 \end{bmatrix}, M_y = \frac{1}{\sqrt{2}} \begin{bmatrix} 0 & -i & 0 \\ i & 0 & -i \\ 0 & i & 0 \end{bmatrix}$$

and

$$M_z = \frac{1}{\sqrt{2}} \begin{bmatrix} 1 & 0 & 0 \\ 0 & 0 & 0 \\ 0 & 0 & -1 \end{bmatrix}$$

prove that

i. $[M_x, M_y] = iM_z$ etc. in cyclic order

ii. $M_x^2 + M_y^2 + M_z^2 = 2I = M^2$

iii. $[M^2, M_i] = 0$

iv. $[M_z, L_+] = L_+$ and $[L_+, L_-] = 2 M_z$.
    when $L_+ = M_x + iM_y$ and $L_- = M_x - iM_y$.

3. With {X} an N-dimensional column vector and [Y] an N-dimensional row vector, show that

$$\text{trace} (\{X\} [Y]) = [Y] = \{X\}$$

4. Prove that

$$\begin{bmatrix} \cos\theta & -\sin\theta \\ \sin\theta & \cos\theta \end{bmatrix} = \begin{bmatrix} 1 & -\tan\dfrac{1}{2}\theta \\ \tan\dfrac{1}{2}\theta & 1 \end{bmatrix}\begin{bmatrix} 1 & \tan\dfrac{1}{2}\theta \\ -\tan\dfrac{1}{2}\theta & 1 \end{bmatrix}$$

5. Show that the matrix $\begin{bmatrix} 1 & 2 & 3 \\ 3 & -2 & 1 \\ 4 & 3 & 1 \end{bmatrix}$ satisfies the equation

$$A^3 - 23A - 40I = 0.$$

6. If A is the matrix

$$A = \begin{bmatrix} 0 & 1 & 0 \\ 0 & 0 & 1 \\ p & q & r \end{bmatrix}$$

and I is the unit-matrix of order 3, show that

$$A^3 = PI + qA + rA^2.$$

7. i. Define identity matrix, transposed matrix and singular matrix.

   ii. If A, B, C are three matrices such that

$$A = [X, \quad Y, \quad Z]; \ B = \begin{bmatrix} a & h & g \\ h & b & f \\ g & f & c \end{bmatrix}; \ C = 3\begin{bmatrix} X \\ Y \\ Z \end{bmatrix} \text{ evaluate } ABC.$$

   iii. Show that the matrix $\begin{bmatrix} 13 & 16 & 19 \\ 14 & 17 & 20 \\ 15 & 18 & 21 \end{bmatrix}$ is singular.

8. i. Define Hermitian and unitary matrices. Prove that the eigen values of a Hermitain matrix are all real.

   iii Find the inverse of the matrix, $A = \begin{bmatrix} 1 & 3 \\ 2 & 1 \end{bmatrix}$

9. If two matrices are both reducible to diagonal form, show that the necessary and sufficient condition that they shall be reducible by the same transformation is that they shall commute.

10. A matrix A satisfying $A^2 = A$, is neither identity nor the null matrix. Show that the characteristic equation of A is of the form $\lambda^P(1 - \lambda)^q = 0$ and that λ is neither orthogonal nor skew-symmetric.

11. i. Prove that any two eigen vectors of a real symmetric matrix are orthogonal provided corresponding eigen values are different.

ii. Show that the matrix

$$A = \begin{bmatrix} \dfrac{1}{\sqrt{3}} & \dfrac{1}{\sqrt{6}} & \dfrac{1}{\sqrt{2}} \\[2mm] \dfrac{1}{\sqrt{3}} & -\dfrac{\sqrt{2}}{3} & 0 \\[2mm] \dfrac{1}{\sqrt{2}} & \dfrac{1}{\sqrt{6}} & -\dfrac{1}{\sqrt{2}} \end{bmatrix}$$

is an orthogonal matrix and find its inverse.

12. i. Show that the eigen values of a Hermitian matrix are all real and its eigen vectors corresponding to two distinct eigen-values are orthogonal.

ii. What do you mean by diagonalization of a matrix? Show that the necessary and sufficient condition for the reduction of two matrices to the diagonal form by the same transformation is that they commute.

13. i. Determine the eigen values and eigen vector of the matrix $A = \begin{bmatrix} 5 & 4 \\ 1 & 2 \end{bmatrix}$

ii. Evaluate e^A when the matrix A is given by $A = \begin{bmatrix} 1 & 0 \\ 0 & 2 \end{bmatrix}$

iii. Show that the following matrices are orthogonal.

(a) $\begin{bmatrix} \cos\theta & -\sin\theta \\ \sin\theta & \cos\theta \end{bmatrix}$ 　　(b) $\begin{bmatrix} 0 & 1 & 0 \\ 1 & 0 & 0 \\ 0 & 0 & 1 \end{bmatrix}$

14. If A is a square matrix of order n and X and Y are n-vectors then show that

$$XAY = AXY$$

15. Represent each of transformations

$$\begin{aligned} y_1 &= x_1 + 4x_2 & \text{and} \quad x_1 &= w_1 - w_2 \\ y_2 &= 4x_1 + 2x_2 & x_2 &= w_1 + w_2 \end{aligned}$$

by the use of matrices find the composite transformation which represents y_1, y_3 in terms of ω_1 and ω_2.

16. Show that the linear transformation $Y = AX$ with matrix

$$A = \begin{bmatrix} \cos\theta & -\sin\theta \\ \sin\theta & \cos\theta \end{bmatrix} \quad \text{and} \quad \begin{bmatrix} x_1 \\ x_2 \end{bmatrix} Y = \begin{bmatrix} y_1 \\ y_2 \end{bmatrix}$$

is with counterclockwise rotation of the cartesian $x_1\, x_2$ coordinate system in the plane about the origin.

17. Show that

$$A^n = \begin{bmatrix} \cos n\theta & -\sin n\theta \\ \sin n\theta & \cos n\theta \end{bmatrix}, \quad \text{where } A = \begin{bmatrix} \cos\theta & -\sin\theta \\ \sin\theta & \cos\theta \end{bmatrix}$$

what does this result mean geometrically.

18. Let $AX = 0$ be a system of n homogeneous equations in n unknowns and suppose A has rank $r = n - 1$. Show that any nonzero vector of cofactors $[\alpha_{i1}, \alpha_{i2}, ..., \alpha_{in}]^T$ of a row of A is solution of $AX = 0$.

19. Solve linear equations through matrices or through Crammer's rule.

    i. $x + y + 3z = 6$ ii. $x_1 - 2x_2 + 3x_3 = 0$

       $2x + 3y - 4z = 6$ $2x_1 + 5x_2 + 6x_3 = 0$

       $3x + 2y + 7z = 0$

   iii. $2x_1 + 3x_2 - x_3 = 0$ iv. $4x + 9y + z = -8$

       $3x_1 - 4x_2 + 2x_3 = 0$ $3x - y + 2z = 4$

                                               $-x - y - z = 2.$

20. Define Hermitian and orthogonal matrices. Give one example of each type.

Find the matrices C and C^{-1} required to reduce the matrix $A = \begin{bmatrix} 1 & -1 \\ -1 & 1 \end{bmatrix}$ to the diagonal form by transformation $C^{-1}AC$.

21. If $A = \begin{bmatrix} 2 & 5 \\ 4 & 1 \end{bmatrix}$, then

    i. Find whether A is singular or not

   ii. Find eigen values of A

  iii. Find trace of A

   iv. Find AA^T

    v. Verify Cayley–Hamilton theorem for A. Hence or otherwise find A^{-1}.

22. i. If A be an n-rowed square matrix, prove that
$$A(adj\ A) = |A| I_n,$$
where I_n a is n-rowed unit matrix.

   ii. Find the inverse of the matrices

    (a) $\begin{bmatrix} 1 & -1 & 3 \\ -1 & 1 & 2 \\ 3 & 2 & -1 \end{bmatrix}$

    (b) $\begin{bmatrix} \cos\theta & -\sin\alpha \\ \sin\alpha & \cos\alpha \end{bmatrix}$

23. i. If A and B are Hermitian matrices, show that $AB + BA$ is also Hermitian.

   ii. If A and B are orthogonal matrices, show that AB is also orthogonal.

  iii. If $\lambda_1, \lambda_2, \lambda_3$ are the eigen values of a matrix M and x_1, x_2, x_3 are the corresponding eigen vectors, find the eigen values and eigen vectors of product MM.

24. What is similarity transformation? If T is the trace of a matrix, find its trace after a similarity transformation has been performed on it. Show that the eigen-values and eigen functions will be affected by the similarity transformation.

25. If λ is a nonzero characteristic root of a nonsingular matrix A, show that λ^{-1} is a characteristic root of A^{-1}.

26. Show that every orthogonal matrix A can be expressed as
$$A = (I + S)(I - S)^{-1}$$
by a suitable choice of a real skew-symmetric matrix S such that -1 is not a characteristic root of A.

27. If H is a Hermitian matrix, show that
$$(I - iH)(I + iH)^{-1} = (I + iH)^{-1}(I - iH) = U$$
where U is a unitary matrix, and that if λ is an eigen value of H, then $(1 - i\lambda)(1 + i\lambda)$ is an eigen value of U.

28. If A and B are two symmetric matrices such that $|A - \lambda B| = 0$ the roots of the equation are all distinct, then show that there exists a matrix, such that $P^{-1}AP$ and $P^{-1}BP$ are both diagonal matrices.

29. Show that every unitary matrix A can, by a suitable choice of anti-Hermitian matrix S, be expressed as
$$A = (I + S)(I - S)^{-1}$$
provided that -1 is not a characteristic root of A.

30. Obtain $\log (I + A)$, where A is the matrix and
$$A = \begin{bmatrix} \dfrac{3}{8} & \dfrac{1}{8} & \dfrac{1}{8} \\ 0 & \dfrac{1}{3} & 0 \\ \dfrac{1}{8} & \dfrac{1}{24} & \dfrac{3}{8} \end{bmatrix}$$
by any of the methods. Evaluate the expression
$$A - \frac{A^2}{2} + \frac{A^3}{3} - \frac{A^4}{4} \dots$$
and compare this with $\log (I + A)$.

31. Find A^k, e^A, and $\log A$, where
$$A = \begin{bmatrix} 1 & 1 & 1 \\ 0 & 1 & 1 \\ 0 & 0 & 1 \end{bmatrix}$$

Sum the series $\displaystyle\sum_{K=20}^{\infty} A^k / k!$ exactly and verify that the result agrees with the direct calculations of e^A.

32. If A and B are any square matrices of the same order and λ is a parameter, show that
$$e^{\lambda A} b e^{-\lambda A} = B\lambda[A, B] + \frac{\lambda^2}{2!}[A,[A, B]]$$
$$+ \frac{\lambda^3}{3!}\left[A,[A,[A,B]]\right] + \cdots$$

33. By finding the inverse of the matrix of coefficients, inverse the transformation

$$u = 19X + 2Y - 9Z$$
$$v = -4X - Y + 2Z$$
$$w = -4X + Z.$$

34. Solve the equation

(a) $AX = 2X$ (b) $AX = -X$ (c) $AX = 3X$, where

$$A = \begin{bmatrix} 0 & 1 & 1 \\ 1 & 0 & 1 \\ 1 & 1 & 0 \end{bmatrix}, \quad X = \begin{bmatrix} x_1 \\ x_2 \\ x_3 \end{bmatrix}$$

35. Show that the matrix $A = \begin{bmatrix} a & h \\ h & b \end{bmatrix}$ is transformed to diagonal matrix $B = T_\theta A\, T_\theta$, where matrix

$$T_\theta = \begin{bmatrix} \cos\theta & \sin\theta \\ -\sin\theta & \cos\theta \end{bmatrix} \quad \text{and} \quad \tan 2\theta = \frac{2h}{a-b}.$$

36. A certain rigid body may be represented by three pointmasses $m_1 = 1$ at $(1, 1, -2)$, $m_2 = 2$ at $(-1, 0, 0)$ and $m_3 = 1$ at $(1, 1, 2)$. (a) Find the inertia matrix and (b) diagonalise the inertia matrix.

37. For any matrix A, prove that

  i. $\sin A = A - \dfrac{A^3}{3!} + \dfrac{A^5}{5!} + \cdots$ ii. $\cos A = I - \dfrac{A^2}{2!} + \dfrac{A^4}{4!} + \cdots$

  iii. $\sinh A = A + \dfrac{A^3}{3!} + \dfrac{A^5}{5!} + \cdots$ iv. $\cosh A = I + \dfrac{A^2}{2!} + \dfrac{A^4}{4!} + \cdots$

MULTIPLE CHOICE QUESTIONS

1. If matrix $A = \begin{bmatrix} 1 & 1 \\ 0 & 0 \end{bmatrix}$ and $B = \begin{bmatrix} 1 & 0 \\ -1 & 0 \end{bmatrix}$ then

(a) $AB = 0$ (b) $BA = 0$

(c) $AB = BA \neq 0$ (d) $AB \neq BA \neq 0$

2. Out of the following matrices the null matrix is

(a) $\begin{bmatrix} 1 & 0 \\ 0 & 1 \end{bmatrix}$ (b) $\begin{bmatrix} 1 & 1 \\ 0 & 0 \end{bmatrix}$

(c) $\begin{bmatrix} 0 & 1 \\ 0 & 1 \end{bmatrix}$ (d) $\begin{bmatrix} 0 & 0 \\ 0 & 0 \end{bmatrix}$

3. The elements of a scalar matrix are given by

(a) a_{ij} (b) $a_{ij} = \lambda_{ij}\, \delta_{ij}$

(c) $a_{ij} = \lambda\, \delta_{ij}$ (d) $[a_{ij}]^T$

4. Determinant of a Hermitian matrix is
   - (a) zero of purely imaginary
   - (b) is real
   - (c) purely imaginary but not zero
   - (d) may be real may be imaginary

5. A *Adj A* is the adjoint of matrix *A*, then
   - (a) $A^{-1} |A| Adj A$
   - (b) $A^{-1} \dfrac{|A|}{Adj A}$
   - (c) $A^{-1} = \dfrac{Adj A}{|A|}$
   - (d) none of the above

6. If *A* is a square matrix, then *A* will be a unitary matrix if (*I* is a unit matrix)
   - (a) $AA^+ = I$
   - (b) $(AA)^+ = I$
   - (c) $AA^T = I$
   - (d) $(AA)^T = I$

7. If the product of two square matrices is a null matrix and one of them is a singular matrix, then other is a
   - (a) Hermitian matrix
   - (b) unitary matrix
   - (c) diagonal matrix
   - (d) null matrix

8. The product of a singular matrix with its adjoint is a
   - (a) null matrix
   - (b) unitary matrix
   - (c) diagonal matrix
   - (d) inverse matrix

9. If a matrix *A* commutes with a diagonal matrix *D*, and all the diagonal elements of which are distinct, then *A* is
   - (a) a diagonal matrix with all elements equal
   - (b) a diagonal matrix with all elements distinct
   - (c) a matrix with elements $a_{ij} = 0$ for $i = j$
   - (d) a matrix with elements $a_{ij} = 0$ for $i \neq j$

10. The rank of matrix $A = \begin{bmatrix} 1 & 2 & 0 \\ 3 & 7 & 1 \\ 5 & 9 & 3 \end{bmatrix}$ is
    - (a) 1
    - (b) 2
    - (c) 3
    - (d) 0

11. For three pauli's spin matrices $\sigma_1 = \begin{bmatrix} 0 & 1 \\ 1 & 0 \end{bmatrix}$, $\sigma_2 = \begin{bmatrix} 0 & -i \\ i & 0 \end{bmatrix}$, $\sigma_3 = \begin{bmatrix} 1 & 0 \\ 0 & -1 \end{bmatrix}$, the value of $\sigma_i \sigma_j + \sigma_j \sigma_i$ is
    - (a) $\delta_{ij} I$
    - (b) $\dfrac{1}{2} \delta_{ij} I$
    - (c) $2 \delta_{ij} I$
    - (d) none of the above

12. If *Y* is an eigen vector of $B = R^{-1}AR$. Corresponding to an eigen value λ, then an eigen vector of *A* corresponding to the same eigen value is
    - (a) $R^{-1}BR$
    - (b) RB
    - (c) RY
    - (d) $R^{-1}YR$

13. If A and B are two square matrices and A is nonsingular then the matrices having same eigen values are
    (a) $A^{-1}B, B^{-1}A$
    (b) $A^{-1}B, BA^{-1}$
    (c) $A^{-1}B^{-1}, BA$
    (d) $(AB)^{-1}, (AB)$

14. If A is a diagonal matrix $\begin{bmatrix} 1 & 0 & 0 \\ 0 & 2 & 0 \\ 0 & 0 & 3 \end{bmatrix}$ then its eigen values are
    (a) $6, 0, 0$
    (b) $4, 0, 0$
    (c) $1, 2, 3$
    (d) $3, 0, 0$

15. Choose the incorrect statement.
    (a) the modulus of each eigen value of a unitary matrix is unity
    (b) any two eigen vectors corresponding to two distinct eigen values of unitary matrix are orthogonal
    (c) the eigen vectors corresponding to distinct eigen values of a matrix are linearly dependent.
    (d) the modulus of each eigen value of an orthogonal matrix is unity.

16. Rank of matrix
    (a) does not change by premultiplying with a nonsingular matrix.
    (b) does not change by postmultiplying with a nonsingular matrix.
    (c) does not change by premultiplying or postmultiplying with nonsingular matrix.
    (d) does not change under all the above cases.

17. If A is a nonsingular square matrix, B its inverse and I unit matrices, then
    (a) $AI = B$
    (b) $IB = A$
    (c) $AB = I$
    (d) there is no such relation between them.

18. If $A = [a_{ij}]$ is a square matrix then its trace is
    (a) $T_r A = \sum_i \sum_j a_{ij}$
    (b) $T_r A = \sum_j a_{ij}$
    (c) $T_r A = \sum_i a_{ij}$
    (d) $T_r A = \sum_i a_{ii}$

19. If A is a singular matrix then product $A \cdot adj\, A$ is a
    (a) unitary matrix
    (b) null matrix
    (c) diagonal matrix with all elements same
    (d) diagonal matrix with all different elements

20. If λ is an eigen value of A, then eigen values of KA are (K is nonzero scalar)
    (a) $k\lambda$
    (b) $\dfrac{k}{\lambda}$
    (c) $\dfrac{\lambda}{k}$
    (d) λ

21. If A and B any symmetric matrices then the matrices AB is symmetric only if
    (a) $AB = BA$
    (b) $AB = -BA$
    (c) $AB = 1$
    (d) either $A = 0$ or $B = 0$

22. If A and B are idempotent matrices then matrix $A + B$ will be idempotent if and only if
    (a) $AB = BA \neq 0$
    (b) $AB = BA = 0$
    (c) $AB = -BA$
    (d) $AB = 0$ but $BA \neq 0$

23. Every square matrix can be uniquely expressed as a
    (a) sum of two symmetric matrices
    (b) symmetric two antisymmetric matrices
    (c) symmitric and antisymmetric matrix
    (d) linear combination of any two matrices.

24. A matrix A $[a_{ij}]$ is Hermitain if
    (a) $a_{ij} = a_{ji}$
    (b) $A^+ = A$
    (c) $a_{ij} = 0$ for $i \neq j$
    (d) its all diagonal elements are zero

25. A square matrix is of the order of $n \times n$ then what is the value of $|KA|^n$?
    (a) $K|A|$
    (b) $K^n|A|$
    (c) $\dfrac{|A|}{K^2}$
    (d) $K|A|^n$

26. A square matrix A necessarily possesses an inverse if
    (a) all its diagonal elements are zero (b) det $A = 0$
    (c) del $A \neq 0$
    (d) it is a diagonal matrix

27. If A and B are orthogonal matrices then the product AB is
    (a) symmetric
    (b) Q symmetric
    (c) orthogonal
    (d) unitary

28. The product of two unitary matrices A and B is
    (a) unitary
    (b) symmetric
    (c) antisymmetric
    (d) orthogonal

29. Trace of matrix $\begin{bmatrix} a_{11} & a_{12} & a_{13} \\ a_{21} & a_{22} & a_{23} \\ a_{31} & a_{32} & a_{33} \end{bmatrix}$ is

    (a) $a_{11} + a_{22} + a_{33}$
    (b) $a_{22} a_{33} - a_{32} a_{23}$
    (c) $a_{11} a_{33} - a_{13} a_{31}$
    (d) $a_{11} + a_{13} + a_{31} + a_{33}$

30. The matrix $\begin{bmatrix} 1 & 0 \\ 0 & -1 \end{bmatrix}$ is

    (a) unit matrix
    (b) orthogonal matrix
    (c) both (d) and (b)
    (d) neither (a) nor (b)

31. The matrix $\begin{bmatrix} \dfrac{1}{\sqrt{2}} & \dfrac{1}{\sqrt{2}} \\ -\dfrac{1}{\sqrt{2}} & -\dfrac{1}{\sqrt{2}} \end{bmatrix}$ is

    (a) only Hermitian
    (b) only unitary
    (c) Hermitan and unitary both
    (d) neither Hermitian nor unitary

32. The eigen vectors of a Hermitan matrix are
    (a) real
    (b) imaginary
    (c) complex
    (d) ± 1

33. The product of eigen values of the matrix $\begin{bmatrix} \alpha & 1 & 0 \\ 0 & \beta & 1 \\ 0 & 0 & r \end{bmatrix}$ is

    (a) $\alpha\beta\gamma$

    (b) $\dfrac{1}{\alpha\beta\gamma}$

    (c) $\dfrac{\alpha\beta\gamma}{\sqrt{\alpha^2 + \beta^2 + \gamma^2}}$

    (d) $\dfrac{1}{\alpha\beta\gamma}(\alpha\beta + \beta\alpha + \gamma\alpha)$

34. The eigen values of matrix $\begin{bmatrix} \cos\theta & -\sin\theta \\ \sin\theta & \cos\theta \end{bmatrix}$ are

    (a) $\pm e^{i\theta}$
    (b) $e^{\pm i\theta}$
    (c) $\cos \pm \sin\theta$
    (d) $\pm \tan\theta$

35. The eigen values of an antisymmetric matrix are
    (a) zero $\pm$
    (b) $\pm i$
    (c) real or zero
    (d) zero or imaginary

36. The eigen values of an orthogonal matrix are
    (a) zero
    (b) imaginary
    (c) real
    (d) of unit modulus

37. If A is a skew-symmetric matrix of odd order then the determinant of A is
    (a) -1
    (b) 0
    (c) 1
    (d) a real number

38. If A and B are two matrices and form the product AB, then the value $(AB)^T$ is (when T is a transposed matrix)
    (a) $A^T B^T$
    (b) $B^T A^T$
    (c) $\dfrac{A^T}{B^T}$
    (d) $A^T + B^T$.

39. A square matrix A possesses an inverse if
    (a) A is singular
    (b) A is nonsingular
    (c) A is imaginary
    (d) A is real

40. *A* square matrix is said to be orthogonal if
    (a) *A* is singular
    (b) *A* is nonsingular
    (c) $A^T A = 1$
    (d) $A = -A^T$.

41. A nonsingular matrix possesses
    (a) unique inverse
    (b) two inverse
    (c) no inverse
    (d) none of these

42. The trace of a square matrix is
    (a) sum of its diagonal terms
    (b) sum of its nondiagonal terms
    (c) product of its diagonal terms
    (d) product of its nondiagonal terms

43. The matrix $\begin{bmatrix} 1 & 2+3i & 3+i \\ 2-3i & 2 & 1-2i \\ 3-i & 1+2i & 3 \end{bmatrix}$ is
    (a) Hermitian
    (b) skew-Hermitian
    (c) idempotent
    (d) orthogonal

44. Inverse of matrix *A*, i.e A^{-1}
    (a) $|A|\, Adj\, A$
    (b) $\dfrac{Adj\, A}{|A|}$
    (c) $\dfrac{Adj\, A}{|A|^2}$
    (d) $|A|^2\, Adj\, A$.

45. If $A = \begin{bmatrix} 2 & 1 \\ 2 & 3 \end{bmatrix}$, $B = \begin{bmatrix} -3 & 1 \\ 2 & 0 \end{bmatrix}$ then the matrix *AB* is
    (a) $\begin{bmatrix} -4 & 2 \\ 2 & 0 \end{bmatrix}$
    (b) $\begin{bmatrix} -4 & 2 \\ 0 & 2 \end{bmatrix}$
    (c) $\begin{bmatrix} 4 & 2 \\ 0 & -1 \end{bmatrix}$
    (d) $\begin{bmatrix} 0 & 2 \\ 0 & 2 \end{bmatrix}$

46. Eigen values of matrices $\begin{bmatrix} 1 & -1 \\ -1 & 1 \end{bmatrix}$ are
    (a) ± 1
    (b) 0 and + 1
    (c) 0 and + 2
    (d) 1 and 2

47. If $AY = PY$ then $Y =$
    (a) PYA
    (b) PYA^{-1}
    (c) $A^{-1}PY$
    (d) PYP^{-1}

48. For two matrices *A* and *B*
    (a) $AB = BA$
    (b) $AB \neq BA$
    (c) $AB = 1$
    (d) $AB = 0$

49. For two matrices A and B, $(A + B)^2$ is equal to
    (a) $A^2 + B^2 + 2AB$
    (b) $A^2 + B^2 + AB$
    (c) $A^2 + B^2 + AB + BA$
    (d) $A^2 + B^2$

50. The inverse of this matrix is $\begin{bmatrix} 1 & -1 \\ 1 & 1 \end{bmatrix}$

    (a) $\begin{bmatrix} 1 & -1 \\ 1 & 1 \end{bmatrix}$
    (b) $\dfrac{1}{2}\begin{bmatrix} 1 & 1 \\ -1 & 1 \end{bmatrix}$

    (c) $\dfrac{1}{\sqrt{2}}\begin{bmatrix} 1 & 1 \\ 1 & -1 \end{bmatrix}$
    (d) does not exist

51. For arbitrary matrices EFG and H if $EFFE = 0$ the trace $(EFGH)$ is equal to
    (a) trace $(FGHE)$
    (b) trace (E) trace (F) trace (G) trace (H)
    (c) trace $(FGEH)$
    (d) trace $(EGHF)$ **[GATE]**

52. An arbitrary matrix $\begin{bmatrix} ae^{i\alpha} & ce^{i\beta} \\ b & d \end{bmatrix}$ is given when $a, b, c\, d, \alpha$ and β real. The inverse

    of the matrix is

    (a) $\begin{bmatrix} ae^{i\alpha} & -ce^{i\beta} \\ b & d \end{bmatrix}$
    (b) $\begin{bmatrix} ae^{i\alpha} & ce^{i\beta} \\ b & d \end{bmatrix}$

    (c) $\begin{bmatrix} ae^{-i\alpha} & b \\ ce^{-i\beta} & d \end{bmatrix}$
    (d) $\begin{bmatrix} ae^{-i\alpha} & ce^{-i\beta} \\ b & d \end{bmatrix}$ **[GATE]**

53. Eigen values of matrix $\begin{bmatrix} 0 & i \\ c & 0 \end{bmatrix}$ are

    (a) real and distinct
    (b) complex and distinct
    (c) complex and coinciding
    (d) complex and coinciding **[GATE]**

54. If $\sigma(i = 1, 2, 3)$ represents the pauli's spin matrix, which one of the following is not true?
    (a) $\sigma_i\sigma_j + \sigma_j\sigma_i = 2\delta_{ij}$
    (b) $T_r(\sigma_i) = 0$
    (c) the eigen values of σ_i and ± 1
    (d) del $(\sigma_i) = 1$ **[GATE]**

55. The eigen values of matrix $\begin{bmatrix} 2 & 3 & 0 \\ 3 & 2 & 0 \\ 0 & 0 & 1 \end{bmatrix}$ are

    (a) $5, 2, -2$
    (b) $-5, -1, -1$
    (c) $5, 1, -1$
    (d) $-5, 1, 1$ **[GATE]**

56. The eigen values of matrix $\begin{bmatrix} \cos\theta & -\sin\theta \\ \sin\theta & \cos\theta \end{bmatrix}$ are

    (a) $\dfrac{1}{2}(\sqrt{3} \pm i)$ when $\theta = 45°$
    (b) $\dfrac{1}{2}(\sqrt{3} \pm i)$ when $\theta = 30°$

    (c) ± 1 sin 4 matrix is unitary
    (d) $\dfrac{1}{\sqrt{2}}(1 \pm i)$ when $\theta = 30°$ **[GATE]**

ANSWERS

1. (a)	2. (d)	3. (c)	4. (b)	5. (c)	6. (a)	7. (d)	8. (a)
9. (b)	10. (c)	11. (c)	12. (c)	13. (b)	14. (c)	15. (c)	16. (d)
17. (c)	18. (d)	19. (b)	20. (a)	21. (a)	22. (b)	23. (c)	24. (b)
25. (b)	26. (c)	27. (c)	28. (a)	29. (a)	30. (b)	31. (c)	32. (a)
33. (a)	34. (b)	35. (d)	36. (d)	37. (b)	38. (b)	39. (b)	40. (c)
41. (a)	42. (a)	43. (a)	44. (b)	45. (b)	46. (c)	47. (c)	48. (b)
49. (c)	50. (b)	51. (a)	52. (d)	53. (b)	54. (d)	55. (c)	56. (b)

3

Tensor Analysis

3.1. INTRODUCTION

It is a fundamental postulate of physics that the laws of nature can be expressed by equations that are valid for all *reference frames,* i.e. coordinate axes. This means that the laws of nature are *covariant,* i.e. they have the same form in all coordinate systems. A systematic method of investigating the behavior of quantities that undergo a coordinate transformation is the subject matter of tensor analysis.

The main aim of this chapter is the development, in an introductory manner of the essentials of tensor algebra and calculus tensor that are needed for an understanding of a tensorial presentation of mechanics, electromagnetic theory and special, and general theory of relativity. We must note that tensor analysis is indispensable in a mathematical treatment of general relativity.

A *tensor* consists of a set of quantities called *component*, whose properties are independent of the coordinate system used to describe them. The components of a tensor in two different coordinate systems are related by the characteristic transformation laws. The cartesian coordinate systems are related to one another through *linear orthogonal transformations*. The tensors in this case are called *cartesian tensors*. This is only a special class of the *general tensors* and are concerned only with linear orthogonal transformation.

3.2. NOTATIONS

A collection of indices (subscripts and/or superscripts) is used to make the mathematical development of tensor analysis compact. The superscripts, contravariant indices are used to denote the contravariant components of a tensor $A^{\mu\nu}$. The subscripts, covariant indices are used to represent the covariant components of a tensor $A_{\mu\nu}$. The component of a mixed tensor are specified by indicating both subscripts and superscripts A^{ij}_{lmn}. We must note that the suffix μ in the coordinates x^{μ} do not have the character of power indices. Usually powers are indicated by the use of brackets, i.e. $(x^{\mu})^2$.

3.3. THE RANK AND NUMBER OF COMPONENTS OF A TENSOR

The *rank* (order) of a tensor is the number (without counting an index which appears once as a subscript and once as a superscript) of indices in the symbol representing a tensor (or the components of tensor).

For example,

A is a tensor of rank zero (scalar)

B^μ is a contravariant tensor of rank one (vector)

C_v is a covariant tensor of rank one (vector)

$D_{\mu v}$ is a covariant tensor of rank two

$E^{\mu v}$ is contravariant tensor of rank two

T^{ij}_{jkl} is a mixed tensor of rank three

In an n-dimensional space, the number of components of a tensor of rank r is n^r.

3.4 *n*-DIMENSIONAL SPACE

In a 3-dimensional space, a point is determined by a set of three coordinates, e.g. (x, y, z) in the rectangular coordinate system. By analogy, if a point is represented by a set of n real variables $(x_1, x_2, x_3,...,x_n)$ or more conveniently $(x^1, x^2, x^3,...,x^n)$, then all the points corresponding to all values of coordinates (variables) are said to form an n-dimensional space. Let it be denoted by S_n.

A *curve* in n-dimensional space (V_n) is defined as the collection of points which satisfy the equation

$$x^\mu = x^\mu(u), \qquad\qquad (\mu = 1, 2, 3,...,n)$$

where u is a parameter and $x^\mu(u)$ are n functions of u, which satisfy certain continuity equations.

A *subspace*, say $V_m (m < n)$ of V_n is defined as the collection of points which satisfy n equations

$$x^\mu = x^\mu(u^1, u^2,...,u^m) \qquad\qquad (\mu = 1, 2,...,n)$$

where $u^1, u^2, u^3,...,u^m$ are m parameters and $x^\mu(u^1, u^2,...,u^m)$ are m functions of $u^1, u^2,..., u^m$ which satisfy certain continuity conditions.

3.5. TRANSFORMATION OF COORDINATES IN LINEAR SPACES

Consider an ordered set of n mutually independent real variables, $x^1, x^2,...,x^n = \{x^i\}$ called the coordinates of a point. The collection of all such points corresponding to all the sets of values $\{x^i\}$ forms an n-dimensional linear space which we specify by V_n. The set of n equations

$$\bar{x}^i = \phi^i(x^1, x^2, x^3,...,x^n)\,(i = 1,2,...,n) \tag{3.1}$$

defines a new coordinate system specified by the mutually independent variables $\bar{x}^1, \bar{x}^2,...,\bar{x}^n$. The ϕ^i are assumed to be single-valued, real functions of the coordinates and possess continuous partial derivatives.

On differentiation Eq. (3.1) w.r.t. x^j, one obtains the following representation for an infinitesimal displacement in the original coordinate system x^j, in terms of the new coordinate system $\bar{x}^j$.

$$d\bar{x}^i = \sum_{j=1}^{n} \frac{\partial \bar{x}^i}{\partial x^j} dx^j \tag{3.1a}$$

$$= \sum_{j=1}^{n} \frac{\partial \bar{x}^i}{\partial x^j} dx^j = \frac{\partial \bar{x}^i}{\partial x^j} dx^j \tag{3.2}$$

3.6. INDICIAL AND SUMMATION CONVENTIONS

The following two conventions are usually in practice.

i. **Indicial Convention:** Any index used either as subscript or superscript takes all values from 1 to n unless the contrary is specified. One can rewrite Eq. (3.1) as

$$\bar{x}^i = \phi^i(x^\alpha) \tag{3.3}$$

This reveals that there are n equations with $i = 1, 2,...,n$ and ϕ^i are the functions of n coordinates x^α with $\alpha = 1, 2,...,n$.

ii. **Einstein's Summation Convention:** Any index which occurs twice in the same term is to be summed, then a summation with respect to that index over the range $1, 2, 3,..., n$ is implied. According to Einstein's summation convention, instead of expression $\sum_{\mu=1}^{n} a_\mu x^\mu$ one can merely write $a_\mu x^\mu$.

In accordance with the above two conventions, one can express Eq. (3.1a) is:

$$d\bar{x}^i = \frac{\partial \bar{x}^i}{\partial x^j} dx^j \tag{3.4}$$

Clearly, the convention implies the sum of the term for the index appearing twice in that term over defined range.

3.7. DUMMY AND REAL INDICES

Any index which is related in a given term, so that the summation convention applies, is called *dummy index*, and one can replace it freely by any other index not already used in that term. Obviously, α is a dummy index in $a_\alpha^\mu x^\alpha$ and also α is a dummy index in

$$d\bar{x}^\mu = \frac{\partial \bar{x}^\mu}{\partial x^\alpha} dx^\alpha$$

Thus, one can write the above equation as

$$d\bar{x}^\mu = \frac{\partial \bar{x}^\mu}{\partial x^\beta} dx^\beta = \frac{\partial \bar{x}^\mu}{\partial x^\gamma} dx^\gamma \tag{3.5}$$

One can interchange two or more dummy indices. However, one must not use the same index more than twice in any single term. For example $\left(\sum_{\mu} a_\mu x^\mu\right)^2$ cannot be expressed as $a_\mu x^\mu a_\mu x^\mu$ but we can express it as $a_\mu a_\nu x^\mu x^\nu$.

In a given term, if any index is not repeated then the index is called a real index, e.g. μ is a real index in $a_\alpha^\mu x^\alpha$. One cannot replace a real index by another real index, i.e.

$$a_\alpha^\mu \bar{x}^\alpha \neq a_\alpha^\nu \bar{x}^\alpha$$

3.8. KNONECKER DELTA

It is defined as

$$\partial_\nu^\mu = \begin{cases} 1 \text{ if } \mu = \nu \\ 0 \text{ if } \mu \neq \nu \end{cases} \tag{3.6}$$

A few important properties of this symbol are:

i. If $x^1, x^2, x^3, ..., x^n$ are independent variables, then

$$\frac{\partial x^\mu}{\partial x^\nu} = \partial_\nu^\mu \tag{3.7}$$

ii.
$$\partial_\nu^\mu A^\mu = A^\nu \tag{3.8}$$

iii. If one is dealing with n-dimensions, then

$$\delta_\mu^\mu = \delta_\nu^\nu = n \tag{3.9}$$

i.e.
$$\delta_\mu^\mu = \delta_1^1 + \delta_2^2 + \delta_3^3 + \cdots + \delta_n^n$$
$$= 1 + 1 + 1 + \cdots + 1 = n \qquad \text{(by } summation\ convention\text{)}$$

iv.
$$\delta_\nu^\mu \delta_\sigma^\nu = \delta_\sigma^\mu \tag{3.10}$$

We have from summation convention,

$$\delta_\nu^\mu \delta_\sigma^\nu = \delta_1^\mu \delta_\sigma^1 + \delta_2^\mu \delta_\sigma^2 + \delta_3^\mu \delta_\sigma^3 + \cdots + \delta_\mu^\mu \delta_\sigma^\mu + \cdots + \delta_n^\mu \delta_\sigma^n$$
$$= 0 + 0 + 0 + \cdots + 1\delta_\sigma^\mu + \cdots + 0 = \delta_\sigma^\mu$$

v.
$$\frac{\delta x^\sigma}{\delta \overline{x}^\nu} \frac{\delta x^\mu}{\delta \overline{x}^\sigma} = \frac{\delta \overline{x}^\mu}{\delta x^\sigma} = \delta_\sigma^\mu \tag{3.11}$$

Generalised Kronecker delta

It is represented as

$$\delta_{\nu_1 \nu_2 \nu_3 ... \nu_m}^{\mu_1 \mu_2 \mu_3 ... \mu_m}$$

It is defined as

i. The subscripts and superscripts can have any value ranging from 1 to n.

ii. If either at least two superscripts or any two subscripts have the same value or the subscripts are not the same set as superscripts, then the generalised Kronecker delta is zero, i.e.

$$\delta_{\alpha\beta\gamma}^{\mu\nu\nu} = \delta_{\alpha\beta\beta}^{\mu\nu\sigma} = \delta_{\alpha\beta\gamma}^{\mu\nu\sigma} = \delta_{\sigma\alpha\beta}^{\mu\nu\sigma} = 0$$

iii. If all the subscripts are separately different and are the same set of numbers as the superscripts, then the generalized Kronecker delta has the value ±1 accordingly as whether it requires an even or odd number of permutations to arrange the superscripts in the same order as the subscripts, i.e.

$$\delta_{123}^{123} = \delta_{231}^{123} = \delta_{312}^{132} = +1$$

and
$$\delta_{213}^{123} = \delta_{232}^{123} = \delta_{312}^{132} = -1$$

one must note that

$$\delta_{123...n}^{\mu_1 \mu_2 \mu_3 ... \mu_n} = \delta_{\nu_1 \nu_2 \nu_3 ... \nu_n}^{123...n} = \delta_{\nu_1 \nu_2 \nu_3 ... \nu_n}^{\mu_1 \mu_2 \mu_3 ... \mu_n}$$

3.9. SCALARS (INVARIANTS), CONTRAVARIANT AND COVARIANT VECTORS

Scalars or Invariants or Tensors of Zero Order

If a function has value ϕ in a system of variables of x_μ and $\overline{\phi}$ in another system of variables $\overline{x}_\mu$, and if

$$\phi = \overline{\phi}$$

then the function ϕ is said to be a scalar or invariant or a tensor of rank zero. We must note that the quantity $\delta_\mu^\mu = \delta_1^1 + \delta_2^2 + \delta_3^3 + \cdots + \delta_n^n = n$ is a scalar or an invariant or a tensor of order zero.

Contravariant Vectors

Let us consider a set of n quantities, say $A^1, A^2, A^3,...,A^n$ in a system of variables x^μ and let these quantities have values $\bar{A}^1, \bar{A}^2, \bar{A}^3,..., \bar{A}^n$ in another system of variables $\bar{x}_\mu$. If these quantities obey the following transformation relation

$$\bar{A}^\mu = \frac{\partial \bar{x}_\mu}{\partial x^\alpha} A^\alpha \tag{3.12}$$

then the quantities A^α are termed the components of a *contravariant vector* or a *contravariant tensor of first rank*.

One can choose any n functions as the components of contravariant vector in a system of variables $\bar{x}_\mu$.

Multiplying Eq. (3.12) by $\dfrac{\partial x^\beta}{\partial \bar{x}^\mu}$ and taking the sum over the index μ from 1 to n, one obtains

$$\frac{\partial x^\beta}{\partial \bar{x}^\mu} \bar{A}^\mu = \frac{\partial x^\beta}{\partial \bar{x}^\mu} \frac{\partial \bar{x}^\mu}{\partial x^\alpha} A^\alpha = \frac{\partial x^\beta}{\partial x^\alpha} A^\alpha$$

$$= \delta_\alpha^\beta A^\alpha = A^\beta = \frac{\partial x^\beta}{\partial \bar{x}^\mu} \bar{A}^\mu \tag{3.13}$$

Obviously, Eq. (3.13) represents the solution of Eq. (3.12).

One can write the transformation of differentials dx^μ and $d\bar{x}^\mu$ in terms of variables x^μ and $\bar{x}^\mu$ respectively as

$$d\bar{x}^\mu = \frac{\partial \bar{x}^\mu}{\partial x^\alpha} dx^\alpha \tag{3.14}$$

We note that Eqs (3.12) and (3.14) are similar transformations. This reveals that the differentials dx^μ form the components of contravariant vector, whose components in any other system are the differentials $d\bar{x}^\mu$ of that system, concluding one can say that the contravariant tensor of rank one. Thus, the tensors whose components transform like coordinate differentials are called *contravariant tensors*.

Let us consider a further change of variables from $\bar{x}^\mu$ to x'^μ. The new components A'^μ will be given by

$$A'^\mu = \frac{\partial x'^\mu}{\partial \bar{x}^\mu} \bar{A}^\alpha = \frac{\partial x'^\mu}{\partial \bar{x}^\alpha} \frac{\partial \bar{x}^\alpha}{\delta x^\lambda} A^\lambda \qquad \textit{[using Eq. (3.12)]}$$

$$= \frac{\partial x'^\mu}{\partial x^\lambda} A^\lambda \tag{3.15}$$

We note that Eq. (3.15) has the same form as Eq. (3.12). This reveals that the *transformations of contravariant vectors form a group.*

We must remember that a single superscript is always used to indicate a contravariant vectors unless the contrary is explicitly expressed.

Covariant Vectors

Let us consider a set of n quantities $A_1, A_2, A_3,...,A_n$ in a system of variables x^μ and let these quantities have values $\bar{A}_1, \bar{A}_2, \bar{A}_3,..., \bar{A}_n$ in another system of variables $\bar{x}^\mu$. If these quantities follow the following transformation equations

$$\bar{A}_\mu = \frac{\partial x^\alpha}{\partial \bar{x}^\alpha} A_\alpha \tag{3.16}$$

then the quantities A_α are called the *component of a covariant vector* or a *covariant tensor of rank one.*

One can choose any n functions as the components of a covariant vector in a system of variables x^μ and Eq. (3.16) determines the n components in the new system of variables $\bar{x}^\mu$. Multiplying Eq. (3.16) by $\frac{\partial \bar{x}^\mu}{\partial x^\beta}$ and taking the sum over the index μ varying from 1 to n, one obtains

$$\frac{\partial \bar{x}^\mu}{\partial x^\beta} \bar{A}_\mu = \frac{\partial \bar{x}^\mu}{\partial x^\beta} \frac{\partial x^\alpha}{\partial \bar{x}^\mu} A_\alpha = \frac{\partial x^\alpha}{\partial x^\beta} A_\alpha = A_\beta$$

Clearly
$$A_\beta = \frac{\partial \bar{x}^\mu}{\partial x^\beta} \bar{A}_\mu \tag{3.17}$$

Obviously, Eq. (3.17) represents solution of Eq. (3.16).

We now consider a further change of variables from $\bar{x}^\mu$ to x'^μ. The new components A'_μ must be given by

$$A'_\mu = \frac{\partial \bar{x}^\alpha}{\partial x'^\mu} \bar{A}_\alpha = \frac{\partial \bar{x}^\alpha}{\partial x'^\mu} \frac{\partial x^\lambda}{\partial \bar{x}^\alpha} A_\lambda$$

$$= \frac{\partial x^\lambda}{\partial x'^\mu} A_\lambda \tag{3.18}$$

Equation (3.18) has the same form as Eq. (3.16). This reveals that the *transformations of covariant vectors form a group.*

$$\frac{\partial \psi}{\partial \bar{x}^\mu} = \frac{\partial \psi}{\partial x^\alpha} \frac{\partial x^\alpha}{\partial \bar{x}^\mu} = \frac{\partial x^\alpha}{\partial \bar{x}^\mu} \frac{\partial \psi}{\partial x^\alpha}$$

From Eq. (3.16), we note that $\frac{\partial \psi}{\partial x^\alpha}$ form the components of a covariant vector whose components in any other system are the corresponding partial derivatives $\frac{\partial \psi}{\partial \bar{x}^\mu}$. This covariant vector is termed grad ψ. We must note that a single subscript is always used to indicate a covariant vector unless contrary is explicitly stated. However, the exception occurs in the notation of coordinates.

3.10. TENSORS OF HIGHER RANKS

Contravarient Tensors of Second Rank

Consider n^2 quantities $A^{\mu\nu}$, where μ and ν take the values from 1 to n independently in a system of variables x^{μ}. Let these quantities have values $\bar{A}^{\mu\nu}$ in another system of variables $\bar{x}^{\mu}$. If these quantities follow the following transformation equation

$$\bar{A}^{\mu\nu} = \frac{\partial \bar{x}^{\mu}}{\partial x^{\alpha}} \frac{\partial \bar{x}^{\nu}}{\partial x^{\beta}} A^{\alpha\beta}, \tag{3.19}$$

then the quantities $A^{\mu\nu}$ are termed the components of a *contravariant tensor of second rank*.

Equation (3.19) [transformation law] is the generalisation of Eq. (3.12) [transformation law]. One can choose any set of $(n)^2$ quantities as the components of a contravariant tensor of second rank in a system of variables x^{μ} and then with the help of Eq. (3.19), one can find $(n)^2$ components in any other system of variables $\bar{x}^{\mu}$.

Covariant Tensor of Second Rank

If $(n)^2$ quantities $A^{\mu\nu}$ in a system of variables x^{μ} are related to other $(n)^2$ quantities in another system of variables $\bar{x}^{\mu}$ through the following transformation equation

$$A_{\mu\nu} = \frac{\partial x^{\alpha}}{\partial \bar{x}^{\mu}} \frac{\partial x^{\beta}}{\partial \bar{x}^{\nu}} A_{\alpha\beta}, \tag{3.20}$$

then the quantities $A_{\mu\nu}$ are termed the components of a *covariant tensor of second rank*. Obviously, Eq. (3.20) is a generalisation of Eq. (3.16). One can choose any set of $(n)^2$ quantities as the components of a covanient tensor of second rank in a system of variable x^{μ} and then with the help of Eq. (3.20), one can determine $(n)^2$ components in any system of variable $\bar{x}^{\mu}$.

Mixed Tensor of Second Rank

If $(n)^2$ quantities in a system of variables x^{μ} are related to the other $(n)^2$ quantities $\bar{A}^{\mu}_{\nu}$ in another system of variables through the following transformation equation

$$\bar{A}^{\mu}_{\nu} = \frac{\partial \bar{x}^{\mu}}{\partial x^{\alpha}} \frac{\partial x^{\beta}}{\partial \bar{x}^{\nu}} A^{\alpha}_{\beta}, \tag{3.21}$$

then the quantities A^{μ}_{ν} are said to be the components of a *mixed tensor of second rank*. Kronecker delta δ^{μ}_{ν} is a familiar example of the mixed tensor of second rank.

One must remember that in contravariant tensors, the indices are placed as superscripts. The mixed tensor A^{μ}_{ν} transforms like a contravariant vector with respect to index μ and like a covariant vector with respect to index ν.

Rank of a Tensor and Tensors of Higher Ranks

The rank of a tensor only indicates the number of indices attached to it per component.

For example, $A^{\mu\nu\sigma}_{\lambda}$ are the components of a mixed tensor of rank 4 (contravariant of rank 3 and covariant of rank 1), if they transform according to the following relation

$$\bar{A}^{\mu\nu\sigma}_{\lambda} = \frac{\partial \bar{x}^{\mu}}{\partial x^{\alpha}} \frac{\partial \bar{x}^{\nu}}{\partial x^{\beta}} \frac{\partial \bar{x}^{\sigma}}{\partial x^{\gamma}} \frac{\partial x^{\delta}}{\partial \bar{x}_{\lambda}} \bar{A}^{\alpha\beta\gamma}_{\delta} \tag{3.22}$$

The rank of a tensor, when raised as power to the number of dimensions gives the number of components of the tensor, e.g. a tensor of rank r in the n-dimensional space has $(n)^r$ components. Obviously, the rank of a tensor gives the number of mode of changes of a physical quantity when passing from one system to the other which is in rotation relative to the first. This means that a quantity which remains unchanged when axes are rotated is tensor of zero rank, which is scalar or invariant. We must remember that the *tensors of rank one are vectors.*

3.11. ALGEBRAIC OPERATIONS ON TENSORS

(i) Addition

The sum of two tensors can only be defined in the case of tensors of *same rank and same type*. Same type means the same number of contravariant and covariant indices. We must note that the sum of two or more tensors of the same rank and type is also a tensor of the same rank and type. For example, $A_\sigma^{\mu\nu}$ and $B_\sigma^{\mu\nu}$ are two tensors, then

$$A_\sigma^{\mu\nu} + B_\sigma^{\mu\nu} = C_\sigma^{\mu\nu} \quad \text{(defined)}$$

where $C_\sigma^{\mu\nu}$ is a tensor of the same rank and type as the given tensors. The transformation laws for the given tensors are

$$\bar{A}_\sigma^{\mu\nu} = \frac{\partial \bar{x}^\mu}{\partial x^\alpha} \frac{\partial \bar{x}^\nu}{\partial x^\beta} \frac{\partial x^\gamma}{\partial \bar{x}^\sigma} A_\gamma^{\alpha\beta} \tag{3.23}$$

$$\bar{B}_\sigma^{\mu\nu} = \frac{\partial \bar{x}^\mu}{\partial x^\alpha} \frac{\partial \bar{x}^\nu}{\partial x^\beta} \frac{\partial x^\gamma}{\partial \bar{x}^\sigma} B_\gamma^{\alpha\beta} \tag{3.24}$$

on addition, Eqs (3.23) and (3.24), will give

$$\bar{C}_\sigma^{\mu\nu} = \frac{\partial \bar{x}^\mu}{\partial x^\alpha} \frac{\partial \bar{x}^\nu}{\partial x^\beta} \frac{\partial x^\gamma}{\partial \bar{x}^\sigma} C_\gamma^{\alpha\beta} \tag{3.25}$$

which is a transformation law for the sum of tensors represented by Eqs (3.23) and (3.24). We must note that $C_\sigma^{\mu\nu}$, the sum in itself is a tensor.

The addition of tensors is *commutative* and *associative.*

(ii) Subtraction

The difference of two tensors of the same rank and type is also a tensor of the same rank and type. Let us consider tensors $A_\sigma^{\mu\nu}$ and $B_\sigma^{\mu\nu}$.

$$A_\sigma^{\mu\nu} - B_\sigma^{\mu\nu} = D_\sigma^{\mu\nu} \quad \text{(say)} \tag{3.26}$$

Taking the difference of Eqs (3.23) and (3.24), one obtains

$$\bar{B}_\sigma^{\mu\nu} - \bar{A}_\sigma^{\mu\nu} = \frac{\partial \bar{x}^\mu}{\partial x^\alpha} \frac{\partial \bar{x}^\nu}{\partial x^\beta} \frac{\partial x^\gamma}{\partial \bar{x}^\sigma} \left(A_\gamma^{\alpha\beta} - B_\gamma^{\alpha\beta} \right) = \frac{\partial \bar{x}^\mu}{\partial x^\alpha} \frac{\partial \bar{x}^\nu}{\partial x^\beta} \frac{\partial x^\gamma}{\partial \bar{x}^\sigma} D_\gamma^{\alpha\beta}$$

which is again a tensor transformation law for subtraction. Obviously $D_\gamma^{\alpha\beta}$ is a tensor of the same rank and same type as the given tensors.

(iii) Equality of Tensors

Two tensors of the same rank and type are said to be equal if their components are (one to one) equal, i.e. if

$$\bar{A}_{\sigma\lambda}^{\mu\nu} = \bar{B}_{\sigma\lambda}^{\mu\nu} \quad \text{(for all values of indices)} \tag{3.27}$$

If two tensors are equal in one coordinate system, they will be also equal in any other coordinate system, i.e. equality remains invariant under coordinate transformations.

(iv) Tensor Multiplication

In the case of general tensors one can define following two types of products.

a. Outer product

b. Inner product

a. **Outer product:** The outer product of two tensors is a tensor whose rank is the sum of the ranks of given tensors. If 'n' and 'm' are the ranks of the given tensors, their outer product will be a tensor of rank $(n + m)$, e.g.

$$A_{\mu\nu} B_{\sigma} = D_{\mu\nu\sigma}$$

$$A_{\mu\nu} B^{\sigma r} = D^{\sigma r}_{\mu\nu}$$

$$A^{\mu\nu} B^{\sigma r} = D^{\mu\nu\sigma r}$$

and $\qquad A^{\mu\nu}_{\sigma} B^{\gamma}_{r} = D^{\mu\nu\gamma}_{\sigma r}$ $\hfill$ (3.28)

To illustrate the above results, we consider the last relation, viz.

$$A^{\mu\nu}_{\sigma} B^{\gamma}_{r} = D^{\mu\nu\gamma}_{\sigma r}$$

We start from the law of transformation of the given tensors $A^{\mu\nu}_{\sigma}$ and B^{γ}_{r} of ranks 3 and 2 respectively.

Since $\qquad \overline{A}^{\mu\nu}_{\gamma} = \dfrac{\partial \overline{x}_{\alpha}}{\partial x_{\mu}} \dfrac{\partial \overline{x}_{\beta}}{\partial x_{\nu}} \dfrac{\partial x_{\sigma}}{\partial \overline{x}_{\gamma}} A^{\mu\nu}_{\sigma}$

and $\qquad \overline{B}^{\delta}_{\varepsilon} = \dfrac{\partial \overline{x}_{\delta}}{\partial x_{\xi}} \dfrac{\partial x_{r}}{\partial \overline{x}_{\varepsilon}} B^{\xi}_{r}$

$\therefore \qquad \overline{A}^{\alpha\beta}_{\gamma} \overline{B}^{\delta}_{\varepsilon} = \dfrac{\partial \overline{x}_{\alpha}}{\partial x_{\mu}} \dfrac{\partial \overline{x}_{\beta}}{\partial x_{\nu}} \dfrac{\partial x_{\sigma}}{\partial \overline{x}_{\gamma}} A^{\mu\nu}_{\sigma} \dfrac{\partial \overline{x}_{\delta}}{\partial x_{\xi}} \dfrac{\partial x_{r}}{\partial \overline{x}_{\varepsilon}} B^{\varepsilon}_{r}$

$\qquad\qquad = \dfrac{\partial \overline{x}_{\alpha}}{\partial x_{\mu}} \dfrac{\partial \overline{x}_{\beta}}{\partial x_{\nu}} \dfrac{\partial x_{\sigma}}{\partial \overline{x}_{\gamma}} \dfrac{\partial \overline{x}_{\delta}}{\partial x_{\xi}} \dfrac{\partial x_{r}}{\partial \overline{x}_{\varepsilon}} A^{\mu\nu}_{\sigma} B^{\xi}_{r}$

or $\qquad T^{\alpha\beta\delta}_{\gamma\varepsilon} = \dfrac{\partial \overline{x}_{\alpha}}{\partial x_{\mu}} \dfrac{\partial \overline{x}_{\beta}}{\partial x_{\nu}} \dfrac{\partial x_{\sigma}}{\partial \overline{x}_{\gamma}} \dfrac{\partial \overline{x}_{\delta}}{\partial x_{\xi}} \dfrac{\partial x_{r}}{\partial \overline{x}_{\varepsilon}} T^{\mu\nu\xi}_{\sigma r}$ $\hfill$ (3.29)

Equation (3.29) is the law of transformation for a tensor of rank 5, i.e. $T^{\alpha\beta\delta}_{\gamma\varepsilon}$ is tensor of rank $3 + 2 = 5$.

The outer product of tensors is *commutative* and *associative*. This is also called the direct product.

Contraction of tensors: The *algebraic operation through which the rank of a mixed tensor is lowered by 2 is known as contraction of a mixed tensor.* In this process, one contravariant index and one covariant index of a mixed tensor are set equal and the repeated index summed over, the result is a tensor of rank lower by two than the original tensor. To illustrate the contraction process, we consider a mixed tensor $A^{\mu\nu\sigma}_{\lambda\rho}$ of rank 5 with contravariant indices μ, ν, σ and covariant indices λ, ρ. One can write the transformation law of the given tensor as

$$\bar{A}^{\mu\nu\sigma}_{\lambda\rho} = \frac{\partial \bar{x}^{\mu}}{\partial x^{\alpha}} \frac{\partial \bar{x}^{\nu}}{\partial x^{\beta}} \frac{\partial \bar{x}^{\sigma}}{\partial x^{\gamma}} \frac{\partial x^{\delta}}{\partial \bar{x}^{\lambda}} \frac{\partial x^{r}}{\partial \bar{x}_{\rho}} A^{\alpha\beta\gamma}_{\delta r} \tag{3.30}$$

Putting $\rho = \sigma$ for applying the contraction, one obtains

$$\bar{A}^{\mu\nu\sigma}_{\lambda\rho} = \frac{\partial \bar{x}^{\mu}}{\partial x^{\alpha}} \frac{\partial \bar{x}^{\nu}}{\partial x^{\beta}} \frac{\partial \bar{x}^{\sigma}}{\partial x^{\gamma}} \cdot \frac{\partial x^{\delta}}{\partial \bar{x}^{\lambda}} \frac{\partial x^{r}}{\partial \bar{x}^{\sigma}} A^{\alpha\beta\gamma}_{\delta r}$$

$$= \frac{\partial \bar{x}^{\mu}}{\partial x^{\alpha}} \frac{\partial \bar{x}^{\nu}}{\partial x^{\beta}} \frac{\partial x^{\delta}}{\partial \bar{x}^{\lambda}} \cdot \frac{\partial x^{r}}{\partial \bar{x}^{\sigma}} \frac{\partial \bar{x}^{\sigma}}{\partial x^{\gamma}} A^{\alpha\beta\gamma}_{\delta r}$$

$$= \frac{\partial \bar{x}^{\mu}}{\partial x^{\alpha}} \frac{\partial \bar{x}^{\nu}}{\partial x^{\beta}} \frac{\partial x^{\delta}}{\partial \bar{x}^{\lambda}} \delta^{r}_{\gamma} A^{\alpha\beta\gamma}_{\delta r} \qquad \left(\because \frac{\partial x^{r}}{\partial \bar{x}^{\sigma}} \frac{\partial \bar{x}^{\sigma}}{\partial x^{\gamma}} = \frac{\partial x^{r}}{\partial x^{\gamma}} = \delta^{r}_{\gamma} \right)$$

$$= \frac{\partial \bar{x}^{\mu}}{\partial x^{\alpha}} \frac{\partial \bar{x}^{\nu}}{\partial x^{\beta}} \frac{\partial x^{\delta}}{\partial \bar{x}^{\lambda}} A^{\alpha\beta\gamma}_{\delta\gamma} \tag{3.31}$$

Obviously, Eq. (3.31) is a transformation law for a mixed tensor of rank 3, i.e. $A^{\alpha\beta\gamma}_{\delta\gamma}$ is a mixed tensor of rank 3 and may be denoted as $A^{\alpha\beta}_{\delta}$. One can further apply the contraction process and obtain the contravariant vector $A^{\mu\sigma\nu}_{\nu\sigma}$ or A^{μ}. Obviously, the process of contraction enables one to obtain a tensor of rank $(n - 2)$ from a fixed tensor of rank n.

Let us consider one more example of mixed tensor of rank 2, whose transformation law is

$$\bar{A}^{\mu}_{\nu} = \frac{\partial \bar{x}^{\mu}}{\partial x^{\alpha}} \frac{\partial x^{\nu}}{\partial \bar{x}^{\beta}} A^{\alpha}_{\beta}$$

Now, let $\nu = \mu$ to apply the contraction process, one obtains

$$\bar{A}^{\mu}_{\nu} = \frac{\partial \bar{x}^{\mu}}{\partial x^{\alpha}} \frac{\partial x^{\nu}}{\partial \bar{x}^{\beta}} A^{\alpha}_{\beta}$$

$$= \delta^{\beta}_{\alpha} A^{\alpha}_{\beta} = A^{\alpha}_{\alpha}$$

Obviously, we get a scalar or tensor of rank zero.

b. ***Inner product:*** *The outer product of two tensors followed by a contraction is called the inner product or inner multiplication of two tensors.* To illustrate it we consider the following examples.

Example 1. Let us consider two tensors $A^{\mu\nu}_{\sigma}$ and B^{λ}_{ρ}. The outer product of these tensor is

$$A^{\mu\nu}_{\sigma} B^{\lambda}_{\rho} = C^{\mu\nu\lambda}_{\sigma\rho} \text{ (say).}$$

Now, let $\rho = \mu$ to apply the contraction process, we obtain

$$A^{\mu\nu}_{\sigma} B^{\lambda}_{\rho} = C^{\rho\nu\lambda}_{\sigma\rho} = D^{\nu\lambda}_{\sigma} \quad \text{(a new tensor)}$$

Obviously, $D^{\nu\lambda}_{\sigma}$ is the inner product of given two tensors $A^{\mu\nu}_{\sigma}$ and B^{λ}_{ρ}.

Example 2. Let us consider two tensors A^μ and B_ν, each of rank 1. The outer product of these two tensors is

$$A^\mu B_\nu = C_\nu^\mu \text{ (say)}$$

Now, let $\mu = \nu$ to apply the contraction, one obtains $A^\mu B_\mu = C_\mu^\mu$, a scalar or a tensor of rank zero. Obviously, the inner product of two tensors of each of rank 1 is a tensor of rank zero, i.e. a invariant.

3.12. THE QUOTIENT LAW

This law helps one to ascertain whether a particular entity in tensor analysis is a tensor or not. According to this law, *an entity, which on inner multiplication with any arbitrary tensor (contravariant or covariant) always gives a tensor, is itself a tensor.* To make it more clear, we consider the following example.

Let us consider an entity $A(\mu, \nu, \sigma)$, we want to know whether it is a tensor or not. For this purpose, take an arbitrary tensor $B_\sigma^{\nu r}$ and suppose that the inner product of $B_\sigma^{\nu r}$ and $A(\mu, \nu, \sigma)$ is a tensor, i.e.

$$A(\mu,\nu,\sigma)B_\sigma^{\nu r} = C_\mu^r \text{ (say)} \tag{3.32}$$

i.e. C_μ^r is a tensor. In the transformed system, one obtains

$$\bar{A}(\mu,\nu,\sigma)\bar{B}_\gamma^{\beta\delta} = \bar{C}_\alpha^\delta \tag{3.33}$$

$$\therefore \quad \bar{A}(\alpha,\beta,\gamma)\frac{\partial\bar{x}^\beta}{\partial x^\nu}\frac{\partial\bar{x}^\delta}{\partial x^r}\frac{\partial x^\sigma}{\partial\bar{x}^\gamma}B_\sigma^{\nu r} = \frac{\partial\bar{x}^\delta}{\partial x^r}\frac{\partial x^\mu}{\partial\bar{x}^\alpha}C_\mu^r = \frac{\partial\bar{x}^\delta}{\partial x^r}\frac{\partial x^\mu}{\partial\bar{x}^\alpha}A(\mu,\nu,\sigma)B_\sigma^{\nu r} \tag{3.34}$$

or $\quad \dfrac{\partial\bar{x}^\delta}{\partial x^r}\left[\bar{A}(\alpha,\beta,\gamma)\dfrac{\partial\bar{x}^\beta}{\partial x^\nu}\dfrac{\partial x^\sigma}{\partial\bar{x}^\gamma} - \dfrac{\partial x^\mu}{\partial\bar{x}^\alpha}\bar{A}(\mu,\nu,\sigma)\right]B_\sigma^{\nu r} = 0$ $\tag{3.34a}$

On inner multiplication with $\dfrac{\partial x^\sigma}{\partial\bar{x}^\delta}$, one obtains from Eq. (3.34a)

$$\frac{\partial x^\sigma}{\partial\bar{x}^\delta}\frac{\partial\bar{x}^\delta}{\partial x^r}\left[\bar{A}(\alpha,\beta,\gamma)\frac{\partial\bar{x}^\beta}{\partial x^\nu}\frac{\partial x^\sigma}{\partial\bar{x}^\gamma} - \frac{\partial x^\mu}{\partial\bar{x}^\alpha}\bar{A}(\mu,\nu,\sigma)\right]B_\sigma^{\nu r} = 0$$

or $\quad \delta_r^\rho\left[\bar{A}(\alpha,\beta,\gamma)\dfrac{\partial\bar{x}^\beta}{\partial x^\nu}\dfrac{\partial x^\sigma}{\partial\bar{x}^\gamma} - \dfrac{\partial x^{\mu'}}{\partial\bar{x}^\alpha}A(\mu,\nu,\sigma)\right]B_\sigma^{\nu r} = 0$

or $\quad \left[\bar{A}(\alpha,\beta,\gamma)\dfrac{\partial\bar{x}^\beta}{\partial x^\nu}\dfrac{\partial x^\sigma}{\partial\bar{x}^\gamma} - \dfrac{\partial x^\mu}{\partial\bar{x}^\alpha}A(\mu,\nu,\sigma)\right]B_\sigma^{\nu\rho} = 0$

But $B_\sigma^{\nu\rho}$ is an arbitrary tensor and it is not equal to zero, hence

$$\bar{A}(\alpha,\beta,\gamma)\frac{\partial\bar{x}^\beta}{\partial x^\nu}\frac{\partial x^\sigma}{\partial\bar{x}^\gamma} - \frac{\partial x^\mu}{\partial\bar{x}^\alpha}A(\mu,\nu,\sigma) = 0$$

Again on inner multiplication with $\dfrac{\partial x^\nu}{\partial\bar{x}^\delta}\dfrac{\partial\bar{x}^\rho}{\partial x^\sigma}$, one obtains

$$\frac{\partial x^\nu}{\partial\bar{x}^\delta}\frac{\partial\bar{x}^\rho}{\partial x^\sigma}\frac{\partial\bar{x}^\beta}{\partial x^\nu}\frac{\partial x^\sigma}{\partial\bar{x}^\gamma}\bar{A}(\alpha,\beta,\gamma) - \frac{\partial x^\mu}{\partial\bar{x}^\alpha}\frac{\partial\bar{x}^\nu}{\partial x^\delta}\frac{\partial\bar{x}^\rho}{\partial x^\sigma}A(\mu,\nu,\sigma) = 0$$

On regrouping, one obtains

$$\frac{\partial x^\nu}{\partial \overline{x}^\delta} \frac{\partial \overline{x}^\beta}{\partial x^\nu} \frac{\partial \overline{x}^\rho}{\partial x^\sigma} \frac{\partial x^\sigma}{\partial \overline{x}^\gamma} \overline{A}(\alpha,\beta,\gamma) - \frac{\partial x^\mu}{\partial \overline{x}^\alpha} \frac{\partial x^\nu}{\partial \overline{x}^\delta} \frac{\partial \overline{x}^\rho}{\partial x^\sigma} A(\mu,\nu,\sigma) = 0$$

or

$$\delta_\delta^\beta \delta_\gamma^\rho \overline{A}(\alpha,\beta,\gamma) - \frac{\partial x^\mu}{\partial \overline{x}^\alpha} \frac{\partial x^\nu}{\partial \overline{x}^\delta} \frac{\partial \overline{x}^\rho}{\partial x^\sigma} A(\mu,\nu,\sigma) = 0$$

or

$$\overline{A}(\alpha,\delta,\rho) = \frac{\partial x^\mu}{\partial \overline{x}^\alpha} \frac{\partial x^\nu}{\partial \overline{x}^\delta} \frac{\partial \overline{x}^\rho}{\partial x^\sigma} A(\mu,\nu,\sigma) \qquad (3.35)$$

Obviously, Eq. (3.34) is the transformation of law for the given entity $A(\mu, \nu, \sigma)$, which is clearly a tensor law of transformation. This shows that $A(\mu, \nu, \sigma)$ is a tensor. We must note that Eq. (3.35) does not tell us only that $A(\mu, \nu, \sigma)$ is a tensor but also tells its type (the number of contravariant and covariant indices). On inspection of RHS of Eq. (3.34), we note that it is once contravariant and on L.H.S $B_\sigma^{\nu r}$ is twice contravariant and once covariant. Obviously, $A(\mu, \nu, \sigma)$ should be twice covariant and once contravariant, which is evident from Eq. (3.35).

3.13. EXTENSION OF RANK OF A TENSOR

One can extend the rank of a tensor by differentiating each of its components with respect to variables x^μ. As an example, we consider a simple case in which the original tensor of rank zero, i.e. a scalar $S(x^\mu)$ whose derivatives relative to variables x^μ are $\frac{\partial S(x^\mu)}{\partial x^\mu}$. In the other system of variables $\overline{x}^\mu$, the scalar is $\overline{S}(\overline{x}^\mu)$, such that

$$\frac{\partial \overline{S}}{\partial \overline{x}^\mu} = \frac{\partial S}{\partial x^\alpha} \frac{\partial x^\alpha}{\partial \overline{x}^\mu} = \frac{\partial x^\alpha}{\partial \overline{x}^\mu} \frac{\partial S}{\partial x^\alpha}$$

Obviously, $\frac{\partial S}{\partial x^\alpha}$ transforms like the components of a tensor of rank one. This shows that the differentials of a tensor of rank zero give a tensor of rank one. Concluding, one can say that the *differentiation of tensor with respect to variables x^m yields a new tensor of rank one greater than the original tensor.*

One can also extend the rank of a tensor when a tensor depends upon another tensor and the differentiation w.r.t. that tensor is performed. For example, consider a tensor S of rank zero, i.e. a scalar, depending upon another tensor $A_{\mu\nu}$, then

$$\frac{\partial S}{\partial A_{\mu\nu}} = B_{\mu\nu} \text{ a tensor of rank 2} \qquad (3.36)$$

This reveals that the rank of the tensor of rank zero has been extended by 2.

3.14. SYMMETRIC AND ANTISYMMETRIC TENSORS

Symmetric Tensors

A tensor is said to be a symmetric tensor when two contravariant or covariant indices can be interchanged without altering the tensor, i.e.

or

$$\left.\begin{array}{l} A^{\mu\nu} = A^{\nu\mu} \\ A_{\mu\nu} = A_{\nu\mu} \end{array}\right\} \qquad (3.37)$$

Obviously, the contravariant tensor of second rank $A^{\mu\nu}$ or covariant tensor of second rank $A_{\mu\nu}$ is said to be symmetric with respect to two indices μ and ν. This also applies for a tensor of higher rank, i.e.

$$A_\lambda^{\mu\nu\sigma} = A_\lambda^{\nu\mu\sigma}$$

Obviously, $A_\lambda^{\mu\nu\sigma}$ is symmetric w.r.t. indices μ and ν.

The symmetric property of a tensor is independent of coordinate system used, i.e. a tensor is symmetric with respect to two indices in any coordinate system. One can easily verify it. Let $A_\lambda^{\mu\nu\sigma}$ is symmetric with respect to first two indices μ and ν. We have

$$A_\lambda^{\mu\nu\sigma} = A_\lambda^{\nu\mu\sigma} \tag{3.38}$$

Using tensor transformation law, one obtains

$$\bar{A}_\lambda^{\mu\nu\sigma} = \frac{\partial \bar{x}^\mu}{\partial x^\alpha} \frac{\partial \bar{x}^\nu}{\partial x^\beta} \frac{\partial \bar{x}^\sigma}{\partial x^\gamma} \frac{\partial x^\delta}{\partial \bar{x}^\lambda} A_\delta^{\alpha\beta\gamma}$$

$$= \frac{\partial \bar{x}^\mu}{\partial x^\alpha} \frac{\partial \bar{x}^\nu}{\partial x^\beta} \frac{\partial \bar{x}^\sigma}{\partial x^\gamma} \frac{\partial x^\delta}{\partial \bar{x}^\lambda} A_\delta^{\beta\alpha\gamma}$$

Interchanging the dummy indices α and β, one obtains

$$\bar{A}_\lambda^{\mu\nu\sigma} = \frac{\partial \bar{x}^\mu}{\partial x^\beta} \frac{\partial \bar{x}^\nu}{\partial x^\alpha} \frac{\partial \bar{x}^\sigma}{\partial x^\gamma} \frac{\partial x^\delta}{\partial \bar{x}^\lambda} A_\delta^{\alpha\beta\gamma}$$

$$= \frac{\partial \bar{x}^\nu}{\partial x^\alpha} \frac{\partial \bar{x}^\mu}{\partial x^\beta} \frac{\partial \bar{x}^\sigma}{\partial x^\gamma} \frac{\partial x^\delta}{\partial \bar{x}^\lambda} A_\delta^{\alpha\beta\gamma} = \bar{A}_\lambda^{\nu\mu\sigma}$$

This shows that the given tensor is again symmetric w.r.t. first two indices in the new coordinate system. One can easily prove this result for covariant indices. Obviously, the symmetry property of a tensor is independent of coordinate system.

One cannot usually define the symmetry of property w.r.t. two indices, of which one is contravariant and the other is covariant, because this symmetry property cannot be preserved after a coordinate transformation. One can easily prove it.

Let $A_\lambda^{\mu\nu\sigma}$ be a symmetric tensor w.r.t. indices, one contravariant μ and the other covariant λ, then we have

$$A_\lambda^{\mu\nu\sigma} = A_\mu^{\lambda\nu\sigma} \tag{3.39}$$

Using tensor transformation law, one obtains

$$\bar{A}_\lambda^{\mu\nu\sigma} = \frac{\partial \bar{x}^\mu}{\partial x^\alpha} \frac{\partial \bar{x}^\nu}{\partial x^\beta} \frac{\partial \bar{x}^\sigma}{\partial x^\gamma} \frac{\partial x^\delta}{\partial \bar{x}^\lambda} A_\delta^{\alpha\beta\gamma}$$

Using Eq. (3.39), one obtains

$$\bar{A}_\lambda^{\mu\nu\sigma} = \frac{\partial \bar{x}^\mu}{\partial x^\alpha} \frac{\partial \bar{x}^\nu}{\partial x^\beta} \frac{\partial \bar{x}^\sigma}{\partial x^\gamma} \frac{\partial x^\delta}{\partial \bar{x}^\lambda} A_\alpha^{\delta\beta\gamma}$$

Now, interchanging dummy indices α and δ, one obtains

$$\bar{A}_\lambda^{\mu\nu\sigma} = \frac{\partial \bar{x}^\mu}{\partial x^\delta} \frac{\partial \bar{x}^\nu}{\partial x^\beta} \frac{\partial \bar{x}^\sigma}{\partial x^\lambda} \frac{\partial x^\alpha}{\partial \bar{x}^\lambda} A_\alpha^{\delta\beta\gamma} \tag{3.40}$$

However, according to tensor transformation law

$$\bar{A}_\mu^{\lambda\mu\sigma} = \frac{\partial \bar{x}^\lambda}{\partial x^\alpha}\frac{\partial \bar{x}^\nu}{\partial x^\beta}\frac{\partial \bar{x}^\sigma}{\partial x^\gamma}\frac{\partial x^\delta}{\partial \bar{x}^\mu}A_\delta^{\alpha\beta\gamma} \tag{3.41}$$

On comparing Eqs (3.40) and (3.41), one obtains

$$\bar{A}_\lambda^{\mu\nu\sigma} \neq \bar{A}_\mu^{\lambda\nu\sigma} \tag{3.42}$$

Obviously, symmetry property is not preserved after a change of coordinate system if we consider the symmetry w.r.t. one contravariant and one covariant indices. However, we must remember that Kronecker delta which is a mixed tensor is symmetric w.r.t. its indices.

We must remember that when all the indices of either a contravariant or a covariant tensor can be interchanged without altering the tensor, the tensor is said to be symmetric. A symmetric tensor of rank 2 in n-dimensional space has at most $\dfrac{n(n+1)}{2}$ independent components. One can easily show it as follows.

The total number of components in the array are n^2. Out of these n^2, all the n-diagonal terms are different and rest $(n^2 - n)$ are equal in pairs. Obviously, the number of pairs will be $\left(\dfrac{n^2-n}{2}\right)$. Therefore, the total number of independent components

$$= n + \frac{n^2-n}{2} = \frac{n(n+1)}{2}$$

Antisymmetric Tensor

A tensor, whose each component changes in sign but remains unaltered in magnitude, when two contravariant or covariant indices are interchanged, is said to be antisymmetric or skew-symmetric w.r.t. these two indices. For example.

$$\left.\begin{array}{l} A^{\mu\nu} = -A^{\nu\mu} \\ A_{\mu\nu} = -A_{\nu\mu} \end{array}\right\} \tag{3.43}$$

Obviously, covariant tensor $A^{\mu\nu}$ and contravariant tensor $A_{\mu\nu}$ of second rank following the property [Eq. (3.43)] are antisymmetric. This property is also true for any higher rank tensor. For example, if

$$A_\lambda^{\mu\nu\sigma} = -A_\lambda^{\mu\sigma\nu}$$

then $A_\lambda^{\mu\sigma\nu}$ is antisymmetric w.r.t. to indices ν and σ.

The antisymmetric property of a tensor is also independent of the choice of the coordinate system, i.e. if a tensor is antisymmetric w.r.t. to any two indices in any coordinate system then it remains antisymmetric w.r.t. these two indices in any other coordinate system. One can easily show it as follows.

Let a tensor $A_\lambda^{\mu\sigma\nu}$ is antisymmetric w.r.t. first two indices μ and ν, then we have

$$A_\lambda^{\mu\nu\sigma} = -A_\lambda^{\mu\nu\sigma} \tag{3.44}$$

Using tensor transformation law, one obtains

$$\bar{A}_\lambda^{\mu\nu\sigma} = \frac{\partial \bar{x}^\mu}{\partial x^\alpha}\frac{\partial \bar{x}^\nu}{\partial x^\beta}\frac{\partial \bar{x}^\sigma}{\partial x^\gamma}\frac{\partial x^\delta}{\partial \bar{x}^\lambda}A_\delta^{\alpha\beta\gamma}$$

$$= \frac{\partial \bar{x}^\mu}{\partial x^\alpha} \frac{\partial \bar{x}^\nu}{\partial x^\beta} \frac{\partial \bar{x}^\sigma}{\partial x^\gamma} \frac{\partial x^\delta}{\partial \bar{x}^\lambda} A_\delta^{\beta\alpha\gamma} \qquad\qquad [using \text{ Eq. (3.44)}]$$

On interchanging dummy indices α and β, one obtains

$$\bar{A}_\lambda^{\mu\nu\sigma} = -\frac{\partial \bar{x}^\mu}{\partial x^\beta} \frac{\partial \bar{x}^\nu}{\partial x^\alpha} \frac{\partial \bar{x}^\sigma}{\partial x^\gamma} \frac{\partial x^\delta}{\partial \bar{x}^\lambda} A_\delta^{\alpha\beta\gamma} = -\bar{A}_\lambda^{\nu\mu\sigma}$$

Obviously, the given tensor is again antisymmetric w.r.t. first two indices in new coordinate system and this antisymmetric property is preserved under coordinate transformation.

We must note that antisymmetric property, like symmetric property, cannot be defined w.r.t. two indices of which one is contravariant and the other is covariant.

When all the indices of a contravariant or covariant tensor can be interchanged so that its components change sign at each interchange of a pair of indices, then the tensor is said to be antisymmetric. For example, if $A^{\mu\nu\sigma}$ is an antisymmetric tensor, then

$$A^{\mu\nu\sigma} = -A^{\nu\mu\sigma} = +A^{\nu\sigma\mu}$$

Obviously, a contravariant or covariant tensor is antisymmetric if its components change sign under an odd permutation of its indices and do not change sign under an even permutation of its indices. An antisymmetric tensor of rank 2 in n-dimensional space has $\frac{n(n-1)}{2}$ independent components. One may easily show it as follows.

The total number of components in the array is $(n)^2$. Out of these all diagonal terms of the array are zero since all the quantities $A^{\mu\nu}$ (no summation) are zero. The rest are equal in pairs, i.e. the number of pairs $= \frac{n^2-n}{2}$. Therefore, the total number of independent components $= \frac{n^2-n}{2} = \frac{n(n-1)}{2}$. In general a skew-symmetric tensor of rank r in n-dimensional space will have at most $^nC_r = \frac{n!}{r!(n-r)!}$ independent components. If $n = r$, then $^nC_r = {}^nC_n = 1$, i.e. any antisymmetric tensor of rank equal to the number of dimensions of space has only one independent component. Anti-symmetric tensor of rank higher than the number of dimensions of the space are identically zero.

We must note that any tensor having either two contravariant or two covariant indices can be expressed as a sum of two parts. One symmetric and the other antisymmetric, i.e.

$$A^{\mu\nu} = \frac{1}{2}(A^{\mu\nu} + A^{\nu\mu}) + \frac{1}{2}(A^{\mu\nu} - A^{\nu\mu}) \qquad\qquad (3.45)$$

The first term on R.H.S of Eq. (3.45) is symmetric part and the second term is antisymmetric part. The symmetric part is a symmetric tensor and vice versa.

The above process of expressing a tensor as a sum of symmetric and antisymmetric tensors is quite general, e.g. a tensor $A_{\sigma\lambda}^{\mu\nu}$ can be expressed as

$$A_{\sigma\lambda}^{\mu\nu} = \frac{1}{2}\left[A_{\sigma\lambda}^{\mu\nu} + A_{\sigma\lambda}^{\mu\nu}\right] + \frac{1}{2}\left[A_{\sigma\lambda}^{\mu\nu} - A_{\sigma\lambda}^{\nu\mu}\right]$$

$$\text{(symmetric part)} \qquad \text{(antisymmetric part)}$$

$$= \frac{1}{2}\left[A^{\mu\nu}_{\sigma\lambda} + A^{\mu\nu}_{\lambda\sigma}\right] + \frac{1}{2}\left[A^{\mu\nu}_{\sigma\lambda} - A^{\mu\nu}_{\lambda\sigma}\right]$$

(symmetric part)   (antisymmetric part)

3.15. INVARIANT TENSORS

The tensors which have the same components in all coordinate systems are said to be invariant tensors. Kronecker delta symbol $\left(\delta^{\mu}_{\nu}\right)$ and Levi–Civita symbols (alternate tensor) or epsilon tensor are two important examples of the invariant tensors.

Kronecker delta: It is an invariant mixed tensor of rank two. It is defined as

$$\delta^{\mu}_{\nu} = \begin{matrix} 0, \text{ if } \mu \neq \nu \\ 1, \text{ if } \mu = \nu \end{matrix}$$

To prove that δ^{μ}_{ν} is a mixed tensor of rank two, we apply *quotient law*. Let A^{ν} be any arbitrary contravariant vector, then $\delta^{\mu}_{\nu}A^{\nu} = A^{\mu}$ = contravariant tensor of rank one.

The inner product of δ^{μ}_{ν} with arbitrary vector (tensor of rank one) is a tensor. Therefore, by quotient law, δ^{μ}_{ν} is a tensor. Since δ^{μ}_{ν} has two distinct indices one of which is contravariant and the other is covariant. Obviously δ^{μ}_{ν} is a mixed tensor of rank two. Since, Kronecker delta is a mixed tensor of rank two and hence it must transform according to the rule.

$$\bar{\delta}^{\mu}_{\nu} = \frac{\partial \bar{x}^{\mu}}{\partial x^{\alpha}} \frac{\partial x^{\beta}}{\partial \bar{x}^{\nu}} \delta^{\alpha}_{\beta}$$

$$= \frac{\partial \bar{x}^{\mu}}{\partial x^{\alpha}} \frac{\partial x^{\beta}}{\partial \bar{x}^{\nu}} \frac{\partial x^{\alpha}}{\partial x^{\beta}} = \frac{\partial \bar{x}^{\mu}}{\partial x^{\alpha}} \frac{\partial x^{\alpha}}{\partial \bar{x}^{\nu}} \tag{3.46}$$

We note that new variables x^{μ} are the functions of old variables x^{μ} which in turn are the functions of new variables $\bar{x}^{\nu}$. One obtains by chain rule

$$\frac{\partial \bar{x}^{\mu}}{\partial \bar{x}^{\nu}} = \frac{\partial \bar{x}^{\mu}}{\partial \bar{x}^{\alpha}} \frac{\partial \bar{x}^{\alpha}}{\partial \bar{x}^{\nu}} \tag{3.47}$$

This gives the change $\delta \bar{x}^{\mu}$ consequent upon a change $\delta \bar{x}^{\nu}$. However, $\bar{x}^{\mu}$ and $\bar{x}^{\nu}$ are the coordinates of the same system, hence their variations are independent of each other unless $\mu = \nu$ where $\delta \bar{x}^{\mu} = \delta \bar{x}^{\nu}$. This gives

$$\frac{\partial \bar{x}^{\mu}}{\partial \bar{x}^{\nu}} = \begin{cases} 1 \text{ for } \mu = \nu \\ 0 \text{ for } \mu \neq \nu \end{cases} = \delta^{\mu}_{\nu} \tag{3.48}$$

From Eqs (3.46) to (3.48), one obtains

$$\bar{\delta}^{\mu}_{\nu} = \frac{\partial \bar{x}^{\mu}}{\partial \bar{x}^{\nu}} = \delta^{\mu}_{\nu} \tag{3.49}$$

This shows that Kronecker delta has the same components in every coordinate system. Clearly, Kronecker delta is an invariant mixed tensor or rank two.

Levi–Civita Tensor (or alternate tensor, or permutation tensor or Epsilon tensor):
The Levi–Civita symbol ε_{ijk} is defined as follows.

$$\begin{cases} \varepsilon_{ijk} = 0, \text{ if any two indices are equal} \\ \quad = +1, \text{ if } ijk \text{ is an even permutation of } 1, 2, 3 \\ \quad = -1, \text{ if } ijk \text{ is an odd permutation of } 1, 2, 3 \end{cases} \qquad (3.50)$$

Since ε_{ijk} is antisymmetric in every pair of indices, therefore,

$$\varepsilon_{ijk} = -\varepsilon_{jik} = +\varepsilon_{jki} = -\varepsilon_{kji} = +\varepsilon_{kij}$$

One can easily show that,

$\overline{\varepsilon}_{ijk} = \varepsilon_{ijk}$, for all values of indices (i, j, k), i.e. Levi–Civita tensor is an invariant.

An important result involving the Levi–Civita symbol is

$$A_{il} A_{jm} A_{kn} \varepsilon_{lmn} = \varepsilon_{ijk} \det A \qquad (3.51)$$

where, A_{ij} is any 3×3 matrix. One can easily prove it as follows.

Since, ε_{lmn} is zero, if any two of the indices l, m, n are equal, the L.H.S of Eq. (3.51)

$$= A_{i1} A_{j2} A_{k3} \varepsilon_{123} + A_{i1} A_{j3} A_{k2} \varepsilon_{132} + A_{i2} A_{j3} A_{k1} \varepsilon_{231}$$

$$+ A_{i2} A_{j1} A_{k3} \varepsilon_{213} + A_{i3} A_{j1} A_{k2} \varepsilon_{312} + A_{i3} A_{j2} A_{k1} \varepsilon_{321}$$

$$= A_{i1} A_{j2} A_{k3} - A_{i1} A_{j3} A_{k2} + A_{i2} A_{j3} A_{k1} - A_{i2} A_{j1} A_{k3} + A_{i3} A_{j1} A_{k2} - A_{i3} A_{j2} A_{k1}$$

It is now easy to verify by assigning values 1, 2, 3 for each of the indices i, j, k, that the above expression

$$= + \det A, \text{ if } i, j, k \text{ is an even permutation of } 1, 2, 3$$
$$= - \det A, \text{ if } i, j, k \text{ is an odd permutation of } 1, 2, 3$$
$$= 0 \text{ if any two of the indices } i, j, k \text{ are equal.}$$

Using the definition of ε_{ijk} one can then easily write the L.H.S of Eq. (3.51) as $\varepsilon_{ijk} \det A$.

We must note the Levi–Civita tensor is a tensor of rank 3 in three-dimensional space. However, Levi–Civita tensor is a *pseudo-tensor* of rank three and also an *isotropic* tensor.

If ϕ is a scalar, the quantities $\phi\varepsilon_{iklm}$ are called *pseudoscalars*, since they have only one component.

From every antisymmetric tensor A_{ik} of the second rank, we can obtain a pseudo-tensor A_{ik} of the same rank by multiplying former with pseudotensor of rank 4, i.e.

$$\overline{A}_{ik} = \frac{1}{2} \sum_{lm=0}^{3} \epsilon^{iklm} A_{ik}$$

Thus, the product of a tensor with a pseudo-tensor is a pseudo-tensor. It is called dual of a given tensor.

Epsilon tensor can be used to write the cross-product of two vectors A and B.

Let $\qquad\qquad C = A \times B$

Then $\qquad\qquad C_1 = A_2 B_3 - A_3 B_2 = \varepsilon_{123} A_2 B_3 + \epsilon_{132} A_3 B_3 = \varepsilon_{1jk} A_j B_k$

Similarly $\qquad C_2 = \varepsilon_{2jk} A_j B_k$ and $C_3 = \varepsilon_{3jk} A_j B_k$ and in general

$$C_i = \varepsilon_{ijk} A_j B_k$$

Evaluating various possible combinations, we can prove

$$\varepsilon^{ijk} \varepsilon_{klm} = \delta_{lm}^{ij} = \delta_l^i \delta_m^j - \delta_m^i \delta_l^j$$

Then for $r = D \cdot (A \times B) = D \cdot C$, where $C = A \times B$.

We have $r = D_i C^i = D_i \, \varepsilon^{ijk} A_j B_k$

Similarly, vector triple product of three vectors can also be written as

$$E = D \times (A \times B) = D \times C$$

So that $E^i = \varepsilon^{ijk} D_j C_k = \varepsilon_{ijk} \, D_j \left(\varepsilon_{kln} A^l B^n \right)$

$$= \varepsilon^{ijk} \varepsilon_{kln} D_j A_l B^n = \delta^{ij}_{ln} D_j A^l B^n = \left(\delta^i_l \delta^j_n - \delta^i_n \delta^j_l \right) D_j A^l B^n$$

Since $\delta^i_l A^l = A^l$, etc. then we have

$$E^i = \delta^j_n A^i D_j B^n - \delta^i_l B^i D_j A^l = A^i (D_n B^n) - B^i (D_l A^l)$$

Hence $E = A(B \cdot D) - B(D \cdot A)$

3.16. RECIPROCAL OR CONJUGATE TENSORS

Consider a symmetric covariant tensor of second rank, $A_{\mu\nu}$, whose determinant $|A_{\mu\nu}| = A$ (say) $\neq 0$. Let us define

$$A_{\mu\nu} = \frac{\text{cofactor of } A_{\mu\nu} \text{ in } A}{A} = \frac{a_{\mu\nu}}{A} \qquad (3.52)$$

Here $a_{\mu\nu}$ is the cofactor of $A_{\mu\nu}$ in A. Since $A_{\mu\nu}$ is symmetric, A is symmetric which implies that $a_{\mu\nu}$ is symmetric and so $A^{\mu\nu}$ is symmetric.

Let B^μ is an arbitrary vector. Using quotient law

$$B_\nu = A_{\mu\nu} B^\mu \qquad (3.53)$$

is an arbitrary covariant vector. Multiplying Eq. (3.53) by $A^{\nu\sigma}$, one obtains

$$B_\nu A^{\nu\sigma} = A_{\mu\nu} B^\mu A^{\nu\sigma}$$

$$= A_{\mu\nu} B^\mu \frac{a_{\nu\sigma}}{A} \qquad \text{[using (3.52)]}$$

$$= \frac{A_{\mu\nu} a_{\nu\sigma}}{A} B^\mu$$

$$= \delta^\sigma_\mu B^\mu \qquad \text{(from the theory of determinant)}$$

$$\therefore \qquad B_\nu A^{\nu\sigma} = B^\sigma$$

Since B^σ is an arbitrary vector, hence from quotient law $A^{\nu\sigma}$, (i.e. $A_{\mu\nu}$) defined by Eq. (3.52) is a *contravariant* tensor of rank 2.

The symmetric covariant tensor of rank two, $A^{\mu\nu}$ defined by Eq. (3.52) is called reciprocal for conjugate tensor of $A_{\mu\nu}$.

3.17. RELATIVE AND ABSOLUTE TENSORS

If the components of tensor $A^{\mu_1\mu_2\cdots}_{\nu_1\nu_2\cdots}$ are transformed according to the following equation.

$$A^{\mu_1\mu_2\cdots}_{\nu_1\nu_2\cdots} = \left| \frac{\partial x}{\partial \overline{x}} \right|^w \left(\frac{\partial \overline{x}^{\mu_1}}{\partial x^{\alpha_1}} \frac{\partial \overline{x}^{\mu_2}}{\partial x^{\alpha_2}} \cdots \right) \left(\frac{\partial x^{\beta_1}}{\partial \overline{x}^{\nu_1}} \frac{\partial x^{\beta_2}}{\partial \overline{x}^{\nu_2}} \right) A^{\alpha_1\alpha_2\cdots}_{\beta_1\beta_2\cdots} \qquad (3.54)$$

where, $\left| \dfrac{\partial x}{\partial \overline{x}} \right|^w$ is the Jacobian of transformation, then the tensor $A^{\alpha_1\alpha_2\cdots}_{\beta_1\beta_2\cdots}$ is termed a relative tensor of weight W. If the weight W is zero, then the relative tensor is called the absolute tensor. If $W = 1$, then the relative tensor is called the tensor density.

We must remember that the algebraic operations of relative tensors are similar to those of absolute tensors.

3.18. METRIC OR FUNDAMENTAL TENSOR

The expression which expresses the distance between two adjacent points is called a *line element* or *metric*. In three-dimensional space, the distance between two adjacent points (x, y, z) and $(x + dx, y + dy, z + dz)$, i.e. a line element in cartesian coordinates is given by

$$dS^2 = dx^2 + dy^2 + dz^2$$

The line element in terms of general curvilinear coordinates becomes

$$dS^2 = \sum_{\mu=1}^{3} \sum_{\nu=1}^{3} g_{\mu\nu} dq_\mu dq_\nu = g_{\mu\nu} dq_\mu dq_\nu$$

[*from* summation convention]

Riemann generalized the above idea to n-dimensional space.

The distance between two neighbouring points having coordinate x^μ and $x^\mu + dx^\mu$ is given by

$$dS^2 = \sum_{\mu-1}^{n} \sum_{\nu-1}^{n} g_{\mu\nu} dx^\mu dx^\nu = g_{\mu\nu} dx^\mu dx^\nu \qquad (3.55)$$

[*from* summation convention]

Here the coefficients $g_{\mu\nu}$ are the functions of the coordinates x^μ, subject to the restriction $g = $ determinant of g_μ, i.e. $|g_{\mu\nu}| \neq 0$.

The quadratic differential form $g_{\mu\nu} dx^\mu dx^\nu$ is independent of the coordinate system and called as the *Riemannian metric for n-dimensional space*. The space characterised by Riemannian metric is termed *Reimannian space*. We must note that here $g_{\mu\nu}$ are components of a covariant symmetric tensor of rank two, called the *fundamental* or *metric* tensor.

When the metric is represented by $(dx^1)^2 + (dx^2)^2 + (dx^3)^2 + \cdots + (dx^n)^2$ or by $(dx^\mu)^2$, then the space is called n-dimensional *Euclidean space*. The Euclidean spaces are the particular cases of Riemannian spaces. We had defined $dS^2 = g_{\mu\nu} dx^\mu dx^\nu$ in general Riemannina space. It is invariant, i.e. independent of the coordinate system.

3.19. FUNDAMENTAL TENSORS

Covariant Fundamental Tensor

We have seen that the line element in Riemannian space is given by

$$dS^2 = g_{\mu\nu} dx^\mu dx^\nu$$

where, $dx^\mu dx^\nu$ are contravariant vectors and dS^2 is invariant for arbitrary choice of vectors dx^μ and dx^ν. From quotient law, one can easily show that $g_{\mu\nu}$ is a covariant tensor of second rank. One can also prove it with the help of transformation law of tensors as follows.

$$dS^2 = g_{\mu\nu} dx^\mu dx^\nu \text{ in system of variables } x^\mu$$

$$= \bar{g}_{\mu\nu} d\bar{x}^\mu d\bar{x}^\nu \text{ in system of variables } \bar{x}^\mu$$

This means $g_{\mu\nu} dx^\mu dx^\nu = \bar{g}_{\alpha\beta} d\bar{x}^\alpha d\bar{x}^\beta \qquad (3.56)$

Applying inverse transformation law to dx^α and dx^β, one obtains

$$\overline{g}_{\mu\nu} d\overline{x}^\mu d\overline{x}^\nu = g_{\alpha\beta} \frac{\partial x^\alpha}{\partial \overline{x}^\mu} \frac{\partial x^\beta}{\partial \overline{x}^\nu} d\overline{x}^\nu d\overline{x}^\mu$$

$$= g_{\alpha\beta} \frac{\partial x^\alpha}{\partial \overline{x}^\mu} \frac{\partial x^\beta}{\partial \overline{x}^\nu} d\overline{x}^\mu d\overline{x}^\nu$$

or $\quad\left(\overline{g}_{\mu\nu} - g_{\alpha\beta} \dfrac{\partial x^\alpha}{\partial \overline{x}^\mu} \dfrac{\partial x^\beta}{\partial \overline{x}^\nu} \right) d\overline{x}^\mu d\overline{x}^\nu = 0$

Since $\partial \overline{x}^\mu$ and $\partial \overline{x}^\nu$ are arbitrary contravariant vectors, one must have

$$\overline{g}_{\mu\nu} - g_{\alpha\beta} \frac{\partial x^\alpha}{\partial \overline{x}^\mu} \frac{\partial x^\beta}{\partial \overline{x}^\nu} = 0$$

or $\qquad\qquad\qquad \overline{g}_{\mu\nu} = \dfrac{\partial x^\alpha}{\partial \overline{x}^\mu} \dfrac{\partial x^\beta}{\partial \overline{x}^\nu} g_{\alpha\beta}$ $\qquad\qquad$ (3.57)

Obviously, Eq. (3.57) is the transformation law for the second order covariant tensor. Thus, $g_{\mu\nu}$ is a covariant tensor of rank 2. One can express $g_{\mu\nu}$ as

$$g_{\mu\nu} = \frac{1}{2}(g_{\mu\nu} + g_{\nu\mu}) + \frac{1}{2}(g_{\mu\nu} - g_{\nu\mu})$$

$$= A_{\mu\nu} + B_{\mu\nu} \qquad\qquad (3.58)$$

where, $A_{\mu\nu}\left[= \dfrac{1}{2}(g_{\mu\nu} + g_{\nu\mu}) \right]$ is a symmetric tensor and $B_{\mu\nu}\left[= \dfrac{1}{2}(g_{\mu\nu} - g_{\nu\mu}) \right]$ is an antisymmetric tensor. Using the above results, one can write the line element as

$$dS^2 = g_{\mu\nu} \, dx^\mu \, dx^\nu = (A_{\mu\nu} + B_{\mu\nu}) \, dx^\mu \, dx^\nu$$

Now we will prove that $g_{\mu\nu}$ is a covariant symmetric tensor of rank 2. We have

$$B_{\mu\nu} \, dx^\alpha \, dx^\nu = B_{\mu\nu} \, dx^\nu \, dx^\mu$$

$$= -B_{\mu\nu} \, dx^\mu \, dx^\nu \qquad\qquad [\because B_{\mu\nu} \text{ is antisymmetric}]$$

$\therefore \qquad\qquad 2B_{\mu\nu} \, dx^\mu \, dx^\nu = 0$

Since dx^μ and dx^ν are arbitrary vectors, we have

$$B_{\mu\nu} = 0,$$

i.e. $\qquad\qquad \dfrac{1}{2}(g_{\mu\nu} - g_{\nu\mu}) = 0$

thus $\quad g_{\mu\nu}$ is symmetric, we can write

$$g_{\mu\nu} = \frac{1}{2}(g_{\mu\nu} + g_{\nu\mu}) \qquad\qquad (3.59)$$

This shows that matrix tensor $g_{\mu\nu}$ is the covariant symmetric tensor of rank 2. Obviously, $g_{\mu\nu}$ is the covariant fundamental tensor of rank 2.

Contravariant Fundamental Tensor, $g^{\mu\nu}$

In order to define $g^{\mu\nu}$, let us consider the determinant $g = |g_{\mu\nu}|$

i.e.
$$g = g^{\mu\nu} = \begin{vmatrix} g_{11} & g_{12} & g_{13} & \cdots & g_{1n} \\ g_{21} & g_{22} & \cdots & \cdots & g_{2n} \\ \vdots & \vdots & \cdots & \cdots & \vdots \\ g_{n1} & g_{n2} & \cdots & \cdots & g_{nm} \end{vmatrix} \qquad (3.60)$$

If $G_{\mu\nu}$ is the cofactor of $g_{\mu\nu}$ in g, i.e. it is $(-1)^{\mu+\nu}$ times the determinant obtained by deleting μ^{th} row and ν^{th} column from g, then from the theory of determinants, we have

$$g_{\mu\nu} G_{\mu\nu} = g$$

and
$$g_{\mu\nu} G^{\sigma\nu} = 0$$

These two results when combined give

$$g_{\mu\nu} \frac{G^{\sigma\nu}}{g} = \delta^\sigma_\mu$$

Let us now define quantity $g^{\sigma\nu}$ by the equation

$$g^{\sigma\nu} = \frac{G^{\sigma\nu}}{g}$$

Then we have

$$g_{\mu\nu} g^{\sigma\nu} = \delta^\sigma_\mu$$

Hence using the quotient law, it can be seen that $g^{\sigma\nu}$ is a contravariant tensor of rank two. This tensor is called *contravarient metric tensor*.

Since $g_{\mu\nu}$ and g are symmetric and hence cofactor of $g_{\mu\nu}$ in g is symmetric. It is then obvious that $g^{\mu\nu}$ is also a symmetric tensor.

Let A^μ be an arbitrary contravariant vector. From the quotient law, we have

$$A_\nu = g_{\mu\nu} A^\mu \qquad (3.61)$$

i.e. A_ν is an arbitrary covariant vector. Multiplying Eq. (3.61) by $g^{\nu\sigma}$, one obtains

$$g^{\nu\sigma} A_\nu = g_{\mu\nu} g^{\nu\sigma} A^\mu \qquad (3.62)$$

But
$$g_{\mu\nu} g^{\nu\sigma} = \frac{\text{cofactor of } g_{\nu\sigma} \text{ in } g}{g}$$

$$= \delta^\sigma_\mu \qquad (\textit{from the theory of determinants}) \qquad (3.63)$$

Thus, we obtain from Eq. (3.62)

$$g^{\nu\sigma} A_\nu = \delta^\sigma_\mu A^\mu = A^\sigma \qquad (3.64)$$

Obviously, the inner product of $g^{\nu\sigma}$ with an arbitrary covariant vector A_ν yields a contravariant vector. Clearly $g^{\nu\sigma}$ (or $g^{\mu\nu}$) is contravariant tensor of rank 2.

Concluding, we can say that $g^{\mu\nu}$ is a symmetric contravariant tensor of rank 2. $g^{\nu\sigma}$ is reciprocal of $g_{\mu\nu}$ and called the *conjugate metric tensor* or *covariant fundamental tensor* of rank 2.

Mixed Fundamental Tensor

g^μ_ν or δ^μ_ν. From Eq. (3.63), we have

$$g_{\mu\nu} g^{\nu\sigma} = \delta^\sigma_\mu \qquad (3.65)$$

Since $g_{\mu\nu}$ and $g^{\nu\sigma}$ are covariant and contravariant tensors of rank 2, respectively and hence δ^σ_μ is also a tensor of rank 2. Obviously, δ^σ_μ is a mixed tensor, contravariant in index σ and covariant in index μ. This is called as *mixed fundamental tensor*. This is an invariant tensor, i.e. its component have the same value in all coordinate systems.

The above mentioned fundamental tensors are of basic importance and are widely used in the theory of relativity.

3.20. ASSOCIATED TENSORS

Raising and Lowering of Indices

We have read that by raising a suffix, a covariant index is changed to a contravariant index and by lowering a suffix, a contravariant index is changed to a covariant index. We must note that the operations of raising and lowering the indices merely depend on the inner multiplication of a given tensor with the contravariant or covariant fundamental tensors $g^{\mu\nu}$ and $g_{\mu\nu}$, respectively. One can define the raising of a suffix of a first rank covariant tensor A_ν as

$$g^{\mu\nu} A_\nu = A^\mu \tag{3.66}$$

Similarly, one can also define the operation of raising the suffix for a tensor of higher rank as

$$g^{\mu\nu} A^\sigma_{\lambda\nu} = A^{\sigma\mu}_\lambda \tag{3.67}$$

Now, we can define the lowering of a suffix of a first rank contravariant tensor A^ν as

$$g_{\mu\nu} A^\nu = A_\mu \tag{3.68}$$

The operation of lowering of a suffix for a tensor of higher rank as

$$\left. \begin{array}{l} g_{\mu\nu} A^{\sigma\nu}_{...\lambda} = A^{\sigma...}_{\mu\lambda} \\ g_{\mu\sigma} A^{\sigma\nu}_{...\lambda} = A^{.\nu.}_{\mu\lambda} \end{array} \right\} \tag{3.69}$$

Here, we have introduced the *dot notation* to indicate which indices has been raised or lowered. Where there is no possibility of confusion, we shall omit the dots.

We now show that the entities obtained as a result of the raising and lowering operations with fundamental tensors are themselves tensors. To illustrate it, we consider Eq. (3.66). Rewriting Eq. (3.66) as

$$\bar{g}^{\mu\nu} \bar{A}_\nu = \bar{A}^\mu$$

i.e.
$$\bar{A}^\mu = \left(\frac{\partial \bar{x}^\mu}{\partial x^\alpha} \frac{\partial \bar{x}^\nu}{\partial x^\beta} g^{\alpha\beta} \right) \left(\frac{\partial x^\gamma}{\partial \bar{x}^\nu} A_\gamma \right)$$

$$= \frac{\partial \bar{x}^\mu}{\partial x^\alpha} \frac{\partial \bar{x}^\nu}{\partial x^\beta} \frac{\partial x^\gamma}{\partial \bar{x}^\nu} g^{\alpha\beta} A_\gamma$$

$$= \frac{\partial \bar{x}^\mu}{\partial x^\alpha} \delta^\gamma_\beta g^{\alpha\beta} A_\gamma \qquad \left(\because \frac{\partial \bar{x}^\nu}{\partial x^\beta} \frac{\partial x^\gamma}{\partial \bar{x}^\nu} = \frac{\partial x^\gamma}{\partial \bar{x}^\nu} \frac{\partial \bar{x}^\nu}{\partial x^\beta} = \delta^\gamma_\beta \right)$$

$$= \frac{\partial \bar{x}^\mu}{\partial x^\alpha} g^{\alpha\beta} A_\beta = \frac{\partial \bar{x}^\mu}{\partial x^\alpha} A^\alpha$$

Obviously, A^μ is a contravariant tensor of rank one.

Associated Tensors

The tensors obtained as a result of raising and lowering operations with fundamental tensors are termed associated tensors.

We must note that the two processes of raising and lowering the indices are reciprocal to each other, i.e. if the index is first raised and then lowered or vice versa, the original tensor is obtained.

To illustrate it, we consider the original tensor, $A^{\sigma}_{\mu\nu}$. First raising the index ν by taking its inner product with contravariant fundamental tensor $g^{\lambda\nu}$, one obtains

$$g^{\lambda\nu} A^{\sigma}_{\mu\nu} = A^{\sigma\lambda}_{\mu} \qquad (3.70)$$

Now for lowering the index λ by taking its inner product with covariant fundamental tensor $g_{\rho\lambda}$, one obtains

$$g_{\rho\lambda} A^{\sigma\lambda}_{\mu} = g_{\rho\lambda} g^{\lambda\nu} A^{\sigma}_{\mu\nu}$$

$$= \delta^{\nu}_{\rho} A^{\sigma}_{\mu\nu} \qquad (\because \ g_{\rho\lambda} g^{\lambda\nu} = \delta^{\nu}_{\rho})$$

$$= A^{\sigma}_{\mu\rho}$$

Obviously, R.H.S. is the same as the given tensor. This proves that the processes of raising and lowering the indices are reciprocal to each other.

The salient features of the process of obtaining associated tensors from a given tensor by the raising and lowering operations are:

i. Inner multiplication with contravariant fundamental tensor $g^{\mu\nu}$ gives substitution with raising,

ii. Inner multiplication with covariant fundamental tensor $g_{\mu\nu}$ gives substitution with lowering,

iii. Inner multiplication with fundamental tensor g^{μ}_{ν} or δ^{μ}_{ν} gives substitution without raising or lowering.

It is obvious that all the metric tensors $g_{\mu\nu}$, $g^{\mu\nu}$ and g^{ν}_{μ} are associated tensors.

3.21. VECTOR ANALYSIS USING TENSOR NOTATION

The results in vector analysis can be conveniently obtained by using tensor notation. In this section, a study of vector analysis using tensor notation is presented.

i. **Length of a vector:** In vector analysis, the square of magnitude of a vector is its scalar product with itself, i.e. $A^2 = A \cdot A$. Similarly, the length L, i.e. the magnitude of a vector (i.e. tensor of rank one) in *Riemannian space* is defined by the scalar product $A^{\mu} A_{\mu}$ as

$$L^2 = A^{\mu} A_{\mu}$$
$$= A_{\mu} A^{\mu} = g_{\mu\nu} A^{\mu} A^{\nu} = g_{\mu\nu} A_{\mu} A_{\nu} \qquad (3.71)$$

ii. **Orthogonality of two vectors:** In vector analysis, the scalar product of two vectors A and B is defined as

$$A \cdot B = |A| \, |B| \, \cos\theta$$

where, θ is the angle between the two vectors A and B, i.e.

$$\cos\theta = \frac{A \cdot B}{|A| \, |B|}$$

In a similar manner, the angle between two vectors A^μ and B_μ in Riemannian space is given by

$$\cos\theta = \frac{A^\mu B_\mu}{\sqrt{(A^\mu A_\mu)(B^\nu B_\nu)}}$$

$$= \frac{g_{\mu\nu} A^\mu B^\nu}{\sqrt{(g_{\mu\nu} A^\mu A^\nu)(g_{\sigma\lambda} B^\sigma B^\lambda)}} \tag{3.72}$$

If the vectors A^μ and B_μ are orthogonal, then Eq. (3.72) yields

$$A^\mu B_\mu - g^{\mu\nu} A^\mu B^\nu = 0 \tag{3.73}$$

Equation (3.73) represents the condition of orthogonality of two vectors A^μ and B^μ in Riemannian space.

We shall discuss tensor form of gradient, divergence, Laplacian and curl in subsequent sections.

3.22. CHRISTOFFEL'S 3-INDEX SYMBOLS

In this section, we introduce two expressions (not tensors) from the fundamental tensors known as Christoffel's 3-index symbols of first and second kind, namely:

Christoffel's Symbols of First Kind

It is written as:

$$\Gamma^\lambda_{\mu\sigma} \text{ or } \left\{ \begin{matrix} \lambda \\ \mu\sigma \end{matrix} \right\} = \frac{1}{2}\left[\frac{\partial g_{\mu\nu}}{\partial x^\sigma} + \frac{\partial g_{\nu\sigma}}{\partial x^\mu} - \frac{\partial g_{\mu\sigma}}{\partial x^\nu} \right] \tag{3.74}$$

Christoffel's Symbols of Second Kind

It is written as:

$$\Gamma^\lambda_{\mu\sigma} \text{ or } \left\{ \begin{matrix} \lambda \\ \mu\sigma \end{matrix} \right\} = \frac{1}{2} g^{\nu\lambda}\left[\frac{\partial g_{\mu\nu}}{\partial x^\sigma} + \frac{\partial g_{\nu\sigma}}{\partial x^\mu} - \frac{\partial g_{\mu\sigma}}{\partial x^\nu} \right] \tag{3.75}$$

From these equations, one can easily verify that Christoffel's symbols are symmetrical w.r.t. the indices μ and σ,

i.e. $\qquad \Gamma_{\mu\sigma, \nu} = \Gamma_{\sigma\mu, \nu} \quad \text{or} \quad [\mu\sigma, \nu] = [\sigma\mu, \nu]$

and $\qquad \Gamma^\lambda_{\mu\sigma} = \Gamma^\lambda_{\sigma\mu} \quad \text{or} \quad \left\{ \begin{matrix} \lambda \\ \mu\sigma \end{matrix} \right\} = \left\{ \begin{matrix} \lambda \\ \sigma\mu \end{matrix} \right\} \tag{3.76}$

Relation between First and Second Kinds of Christoffel's Symbols

Form Eq. (3.74), we have

$$\Gamma_{\mu\sigma, \nu} = \frac{1}{2}\left[\frac{\partial g_{\mu\nu}}{\partial x^\sigma} + \frac{\partial g_{\nu\sigma}}{\partial x^\mu} - \frac{\partial g_{\mu\sigma}}{\partial x^\nu} \right]$$

$$\therefore \qquad \Gamma_{\nu\sigma, \mu} = \frac{1}{2}\left[\frac{\partial g_{\nu\mu}}{\partial x^\sigma} + \frac{\partial g_{\mu\sigma}}{\partial x^\nu} - \frac{\partial g_{\nu\sigma}}{\partial x_\mu} \right]$$

On adding the above equations, one obtains

$$\Gamma_{\mu\sigma,\,v} + \Gamma_{v\sigma,\,\mu} = \frac{\partial g_{\mu v}}{\partial x^{\sigma}} \qquad (\text{since } g_{\mu v} = g_{v\mu}) \tag{3.77}$$

From Eqs (3.74) and (3.75), one can see that

$$\Gamma^{\lambda}_{\mu\sigma} = g^{\mu v \lambda}\,\Gamma_{\mu\sigma,\,v}$$

and since

$$g_{\mu v}\, g^{v\lambda} = 1 \qquad (\text{when } \mu = \lambda) \tag{3.78}$$

$$\therefore \qquad\qquad \Gamma_{\mu\sigma,\,v} = g_{v\lambda}\Gamma^{\lambda}_{\mu\sigma} \tag{3.79}$$

One can easily see that the process of transformation from Christoffel's symbol of first kind to second kind and vice versa is the same process as used in raising and lowering of an index.

3.23. TRANSFORMATION LAWS FOR CHRISTOFFEL SYMBOLS

Now, we derive the laws of transformation for the Christoffel symbols and from these laws it will be seen that those are not tensors.

Transformation Laws for the Christoffel Symbols of the First Kind

We start from the law of transformation for the covariant fundamental tensor, which is as follows:

$$\overline{g}_{\mu v} = \frac{\partial x^{\alpha}}{\partial \overline{x}^{\mu}}\frac{\partial x^{\beta}}{\partial \overline{x}^{v}} g_{\alpha\beta} \tag{3.80}$$

Differentiating Eq. (3.80) w.r.t. to $\overline{x}^{\sigma}$, one obtains

$$\frac{\partial \overline{g}_{\mu v}}{dx^{\sigma}} = \frac{\partial x^{\alpha}}{\partial \overline{x}^{\mu}}\frac{\partial x^{\beta}}{\partial \overline{x}^{v}}\frac{\partial g_{\alpha\beta}}{\partial x^{\gamma}}\frac{\partial x^{\gamma}}{\partial \overline{x}^{\sigma}} + \frac{\partial^2 x^{\alpha}}{\partial \overline{x}^{\sigma}\partial \overline{x}^{\mu}}\frac{\partial x^{\beta}}{\partial \overline{x}^{v}}g_{\alpha\beta} + \frac{\partial^2 x^{\beta}}{\partial \overline{x}^{\sigma}\partial \overline{x}^{v}}\frac{\partial x^{\alpha}}{\partial \overline{x}^{\mu}}g_{\alpha\beta}$$

Interchanging the dummy suffixes α and β in the third term on the RHS and using $g_{\alpha\beta} = g_{\beta\alpha}$, one obtains

$$\frac{\partial \overline{g}_{\mu v}}{\partial \overline{x}^{\sigma}} = \frac{\partial x^{\alpha}}{\partial \overline{x}^{\mu}}\frac{\partial x^{\beta}}{\partial \overline{x}^{v}}\frac{\partial g_{\alpha\beta}}{\partial x^{\gamma}}\frac{\partial x^{\gamma}}{\partial \overline{x}^{\sigma}} + g_{\alpha\beta}\left[\frac{\partial^2 x^{\alpha}}{\partial \overline{x}^{\sigma}\partial \overline{x}^{\mu}}\frac{\partial x^{\beta}}{\partial \overline{x}^{v}} + \frac{\partial^2 x^{\alpha}}{\partial \overline{x}^{\sigma}\partial \overline{x}^{v}}\frac{\partial x^{\beta}}{\partial \overline{x}^{\mu}}\right] \tag{3.81}$$

Similarly, through cyclic interchange of μ, v and σ, one obtains

$$\frac{\partial \overline{g}_{v\sigma}}{\partial \overline{x}^{\mu}} = \frac{\partial x^{\alpha}}{\partial \overline{x}^{v}}\frac{\partial x^{\beta}}{\partial \overline{x}^{\sigma}}\frac{\partial g_{\alpha\beta}}{\partial x^{\gamma}}\frac{\partial x^{\gamma}}{\partial \overline{x}^{\mu}} + g_{\alpha\beta}\left[\frac{\partial^2 x^{\alpha}}{\partial \overline{x}^{\mu}\partial \overline{x}^{v}}\frac{\partial x^{\beta}}{\partial \overline{x}^{\sigma}} + \frac{\partial^2 x^{\alpha}}{\partial \overline{x}^{\mu}\partial \overline{x}^{\sigma}}\frac{\partial x^{\beta}}{\partial \overline{x}^{v}}\right] \tag{3.82}$$

and

$$\frac{\partial \overline{g}_{\sigma\mu}}{\partial \overline{x}^{v}} = \frac{\partial x^{\alpha}}{\partial \overline{x}^{\sigma}}\frac{\partial x^{\beta}}{\partial \overline{x}^{\sigma}}\frac{\partial g_{\alpha\beta}}{\partial x^{\gamma}}\frac{\partial x^{\gamma}}{\partial \overline{x}^{v}} + g_{\alpha\beta}\left[\frac{\partial^2 x^{\alpha}}{\partial \overline{x}^{v}\partial \overline{x}^{\sigma}}\frac{\partial x^{\beta}}{\partial \overline{x}^{\mu}} + \frac{\partial^2 x^{\alpha}}{\partial \overline{x}^{v}\partial \overline{x}^{\mu}}\frac{\partial x^{\beta}}{\partial \overline{x}^{\sigma}}\right] \tag{3.83}$$

Now changing the dummy suffixes α, β, and γ in the first term of the RHS of Eq. (3.82), i.e. writing β for α, γ for β and α for γ, we obtain

$$\frac{\partial \overline{g}_{v\sigma}}{\partial \overline{x}_{\mu}} = \frac{\partial x^{\beta}}{\partial \overline{x}^{v}}\frac{\partial x^{\gamma}}{\partial \overline{x}^{\sigma}}\frac{\partial x^{\alpha}}{\partial \overline{x}^{\mu}}\frac{\partial g_{\beta\gamma}}{\partial x^{\alpha}} + g_{\alpha\beta}\left[\frac{\partial^2 x^{\alpha}}{\partial \overline{x}^{\mu}\partial \overline{x}^{v}}\frac{\partial x^{\beta}}{\partial \overline{x}^{\sigma}} + \frac{\partial^2 x^{\alpha}}{\partial \overline{x}^{\mu}\partial \overline{x}^{\sigma}}\frac{\partial x^{\beta}}{\partial \overline{x}^{v}}\right] \tag{3.84}$$

Now changing the dummy suffixes α, β, and γ in the first term on the RHS of Eq. (3.83), i.e. writing γ for α, β for γ and α for β, one obtains

$$\frac{\partial \overline{g}_{\sigma\mu}}{\partial \overline{x}^{\nu}} = \frac{\partial x^{\gamma}}{\partial \overline{x}^{\sigma}}\frac{\partial x^{\alpha}}{\partial \overline{x}^{\mu}}\frac{\partial x^{\beta}}{\partial \overline{x}^{\nu}}\frac{\partial g_{\gamma\alpha}}{\partial x^{\beta}} + g_{\alpha\beta}\left[\frac{\partial^2 x^{\alpha}}{\partial \overline{x}^{\nu}\partial \overline{x}^{\sigma}}\frac{\partial x^{\beta}}{\partial \overline{x}^{\mu}} + \frac{\partial^2 x^{\alpha}}{\partial \overline{x}^{\nu}\partial \overline{x}^{\mu}}\frac{\partial x^{\beta}}{\partial \overline{x}^{\sigma}}\right] \tag{3.85}$$

Now adding Eqs (3.81) and (3.84), subtracting Eq. (3.85) from the sum and dividing the result throughout by 2, one obtains

$$\frac{1}{2}\left[\frac{\partial \overline{g}_{\mu\nu}}{\partial \overline{x}^{\sigma}} + \frac{\partial \overline{g}_{\nu\sigma}}{\partial \overline{x}^{\mu}} - \frac{\partial \overline{g}_{\sigma\mu}}{\partial \overline{x}^{\nu}}\right] = \frac{1}{2}\frac{\partial x^{\alpha}}{\partial \overline{x}^{\mu}}\frac{\partial x^{\beta}}{\partial \overline{x}^{\nu}}\frac{\partial x^{\gamma}}{\partial \overline{x}^{\sigma}} \times \left[\frac{\partial g_{\alpha\beta}}{\partial x^{\gamma}} + \frac{\partial g_{\beta\gamma}}{\partial x^{\alpha}} - \frac{\partial g_{\gamma\alpha}}{\partial x^{\beta}}\right]$$

$$+ \frac{1}{2}g_{\alpha\beta}\left[\frac{\partial^2 x^{\alpha}}{\partial \overline{x}^{\sigma}\partial \overline{x}^{\mu}}\frac{\partial x^{\beta}}{\partial \overline{x}^{\nu}} + \frac{\partial^2 x^{\alpha}}{\partial \overline{x}^{\sigma}\partial \overline{x}^{\nu}}\frac{\partial x^{\beta}}{\partial \overline{x}^{\mu}} + \frac{\partial^2 x^{\alpha}}{\partial \overline{x}^{\mu}\partial \overline{x}^{\nu}}\frac{\partial x^{\beta}}{\partial \overline{x}^{\sigma}}\right.$$

$$\left. + \frac{\partial^2 x^{\alpha}}{\partial \overline{x}^{\mu}\partial \overline{x}^{\sigma}}\frac{\partial x^{\beta}}{\partial \overline{x}^{\nu}} - \frac{\partial^2 x^{\alpha}}{\partial \overline{x}^{\nu}\partial \overline{x}^{\sigma}}\frac{\partial x^{\beta}}{\partial \overline{x}^{\mu}} + \frac{\partial^2 x^{\alpha}}{\partial \overline{x}^{\nu}\partial \overline{x}^{\mu}}\frac{\partial x^{\beta}}{\partial \overline{x}^{\sigma}}\right] \tag{3.86}$$

Making use of Eqs (3.80) in (3.86), one obtains

$$\overline{\Gamma}_{\mu\sigma,\nu} = \frac{\partial x^{\alpha}}{\partial \overline{x}^{\mu}}\frac{\partial x^{\beta}}{\partial \overline{x}^{\nu}}\frac{\partial x^{\gamma}}{\partial \overline{x}^{\sigma}}\Gamma_{\alpha\gamma,\beta} + g_{\alpha\beta}\frac{\partial^2 x^{\alpha}}{\partial \overline{x}^{\mu}\partial \overline{x}^{\sigma}}\frac{\partial x^{\beta}}{\partial \overline{x}^{\nu}} \tag{3.87}$$

Equation (3.87) is the law of transformation for the Christoffel symbol of first kind. Obviously, this is not a tensor transformation law on account of the presence of the 2nd term of the R.H.S. of Eq. (3.87). However, in the special case of linear transformations connecting cartesian coordinate systems $\dfrac{\partial^2 x^{\alpha}}{\partial \overline{x}^{\mu}\partial \overline{x}^{\sigma}}$ will be zero as $\dfrac{\partial x^{\nu}}{\partial \overline{x}^{\mu}}$, the coefficients will all be constant there and hence this symbol will transform as a tensor.

Transformation Laws for the Christoffel Symbol of the Second Kind

We start from the law of transformation for the contravariant fundamental tensor

$$\overline{g}^{\nu\lambda} = \frac{\partial \overline{x}^{\nu}}{\partial x^{\eta}}\frac{\partial \overline{x}^{\lambda}}{\partial x^{\varepsilon}}g^{\eta\varepsilon} \tag{3.88}$$

Multiplying the corresponding sides of Eqs (3.87) and (3.88), one obtains

$$\overline{g}^{\nu\lambda}\overline{\Gamma}_{\mu\sigma,\nu} = \frac{\partial x^{\alpha}}{\partial \overline{x}^{\mu}}\frac{\partial x^{\beta}}{\partial \overline{x}^{\nu}}\frac{\partial x^{\nu}}{\partial \overline{x}^{\sigma}}\frac{\partial \overline{x}^{\nu}}{\partial x^{\eta}}\frac{\partial \overline{x}^{\lambda}}{\partial x^{\varepsilon}}g^{\eta\varepsilon}\Gamma_{\alpha\gamma,\beta} + g^{\eta\varepsilon}g_{\alpha\beta}\frac{\partial^2 x^{\alpha}}{\partial \overline{x}^{\mu}\partial \overline{x}^{\sigma}}\frac{\partial x^{\beta}}{\partial \overline{x}^{\nu}}\frac{\partial \overline{x}^{\nu}}{\partial x^{\eta}}\frac{\partial \overline{x}^{\lambda}}{\partial x^{\varepsilon}} \tag{3.89}$$

Raising the suffix on L.H.S and rearranging of R.H.S, one obtains

$$\overline{\Gamma}^{\lambda}_{\mu\sigma} = \frac{\partial x^{\alpha}}{\partial \overline{x}^{\mu}}\frac{\partial x^{\gamma}}{\partial \overline{x}^{\sigma}}\frac{\partial \overline{x}^{\lambda}}{\partial x^{\varepsilon}}\frac{\partial x^{\beta}}{\partial \overline{x}^{\nu}}\frac{\partial \overline{x}^{\nu}}{\partial x^{\eta}}g^{\eta\varepsilon}\Gamma_{\alpha\gamma,\beta} + \frac{\partial^2 x^{\alpha}}{\partial \overline{x}^{\mu}\partial \overline{x}^{\sigma}}\frac{\partial x^{\beta}}{\partial \overline{x}^{\nu}}\frac{\partial \overline{x}^{\nu}}{\partial x^{\eta}}g^{\eta\varepsilon}g_{\alpha\beta}\frac{\partial \overline{x}^{\lambda}}{\partial x^{\varepsilon}}$$

$$= \frac{\partial x^{\alpha}}{\partial \overline{x}^{\mu}}\frac{\partial x^{\lambda}}{\partial \overline{x}^{\sigma}}\frac{\partial \overline{x}^{\lambda}}{\partial x^{\varepsilon}}\delta^{\beta}_{\eta}g^{\eta\varepsilon}\Gamma_{\alpha\gamma,\beta} + \frac{\partial^2 x^{\alpha}}{\partial \overline{x}^{\mu}\partial \overline{x}^{\sigma}}\delta^{\beta}_{\eta}g^{\eta\varepsilon}g_{\alpha\beta}\frac{\partial \overline{x}^{\lambda}}{\partial x^{\varepsilon}}$$

where the substitution operation $\dfrac{\partial x^{\beta}}{\partial \overline{x}^{\nu}}\dfrac{\partial \overline{x}^{\nu}}{\partial x^{\eta}}$ has been replaced by δ^{β}_{η}.

or
$$\bar{\Gamma}^{\lambda}_{\mu\sigma} = \frac{\partial x^{\alpha}}{\partial \bar{x}^{\mu}} \frac{\partial x^{\gamma}}{\partial \bar{x}^{\sigma}} \frac{\partial \bar{x}^{\lambda}}{\partial x^{\varepsilon}} g^{\beta\varepsilon} \Gamma_{\alpha\gamma,\beta} + \frac{\partial^2 x^{\alpha}}{\partial \bar{x}^{\mu} \partial \bar{x}^{\sigma}} g^{\beta\varepsilon} g_{\alpha\beta} \frac{\partial \bar{x}^{\lambda}}{\partial x^{\varepsilon}}$$

or
$$\bar{\Gamma}^{\lambda}_{\mu\sigma} = \frac{\partial x^{\alpha}}{\partial \bar{x}^{\mu}} \frac{\partial x^{\gamma}}{\partial \bar{x}^{\sigma}} \frac{\partial \bar{x}^{\lambda}}{\partial x^{\varepsilon}} \Gamma^{\varepsilon}_{\alpha\gamma} + \frac{\partial^2 x^{\alpha}}{\partial \bar{x}^{\mu} \partial \bar{x}^{\sigma}} \delta^{\varepsilon}_{\alpha} \frac{\partial \bar{x}^{\lambda}}{\partial x^{\varepsilon}}$$

or
$$\bar{\Gamma}^{\lambda}_{\mu\sigma} = \frac{\partial x^{\alpha}}{\partial \bar{x}^{\mu}} \frac{\partial x^{\gamma}}{\partial \bar{x}^{\sigma}} \frac{\partial \bar{x}^{\lambda}}{\partial x^{\varepsilon}} \Gamma^{\varepsilon}_{\alpha\gamma} + \frac{\partial^{\varepsilon} x^{\alpha}}{\partial \bar{x}^{\mu} \partial \bar{x}^{\sigma}} \frac{\partial \bar{x}^{\lambda}}{\partial x^{\alpha}} \tag{3.89a}$$

Equation (3.89a) is the law transformation for the Christoffel symbol of the second kind and we must note that it is not a tensor transformation law.

The above results show that the Christoffel's 3-index symbols are not the tensor.

However, in a very special case of linear transformation of coordinates of the type

$$x^{\alpha} = a^{\alpha}_{\mu} + b^{\alpha} \tag{3.90}$$

where a^{α}_{μ} and b^{α} are constants, so that $\dfrac{\partial^2 x^{\alpha}}{\partial \bar{x}^{\mu} \partial \bar{x}^{\nu}} = 0$, then Eqs (3.87) and (3.89a) become

$$\bar{\Gamma}_{\mu\nu,\sigma} = \frac{\partial x^{\alpha}}{\partial \bar{x}^{\mu}} \cdot \frac{\partial x^{\beta}}{\partial \bar{x}^{\nu}} \cdot \frac{\partial x^{\gamma}}{\partial \bar{x}_{\sigma}} \Gamma_{\alpha\beta,\gamma} \tag{3.91}$$

$$\bar{\Gamma}^{\lambda}_{\mu\nu} = \frac{\partial x^{\alpha}}{\partial \bar{x}^{\mu}} \cdot \frac{\partial x^{\beta}}{\partial \bar{x}^{\nu}} \cdot \frac{\partial x^{\lambda}}{\partial \bar{x}^{\gamma}} \Gamma^{\gamma}_{\alpha\beta} \tag{3.92}$$

Equations (3.91) and (3.92) indicate that the Christoffel's 3-index symbol transform like tensors relative to the linear transformation of coordinates of type Eq. (3.90). In such cases Christoffel's 3-index symbols are called *pseudo tensors*.

Note: Equation (3.89a) may be used to express the second partial derivatives of x^p w.r.t. to $\bar{x}^{\mu}$ in terms of first derivatives and Christoffel's symbols of second kind.

Taking inner product of Eq. (3.89a) with $\dfrac{\partial x^p}{\partial \bar{x}^{\lambda}}$, one obtains

$$\frac{\partial x^p}{\partial \bar{x}^{\lambda}} \bar{\Gamma}^{\lambda}_{\alpha\sigma} = \frac{\partial x^{\mu}}{\partial \bar{x}^{\alpha}} \frac{\partial x^{\nu}}{\partial \bar{x}^{\sigma}} \frac{\partial \bar{x}^{\lambda}}{\partial x^{\eta}} \frac{\partial x^p}{\partial \bar{x}^{\lambda}} \Gamma^{\eta}_{\mu\nu} + \frac{\partial^2 x^{\mu}}{\partial \bar{x}^{\alpha} \partial \bar{x}^{\sigma}} \frac{\partial \bar{x}^{\lambda}}{\partial x^{\mu}} \frac{\partial x^p}{\partial \bar{x}^{\lambda}}$$

$$= \frac{\partial x^{\mu}}{\partial \bar{x}^{\alpha}} \frac{\partial x^{\nu}}{\partial \bar{x}^{\sigma}} \delta^p_{\eta} \Gamma^{\eta}_{\mu\nu} + \frac{\partial^2 x^{\mu}}{\partial \bar{x}^{\alpha} \partial \bar{x}^{\sigma}} \delta^p_{\mu}$$

$$= \frac{\partial x^{\mu}}{\partial \bar{x}^{\alpha}} \frac{\partial x^{\nu}}{\partial \bar{x}^{\sigma}} \Gamma^{\eta}_{\mu\nu} + \frac{\partial^2 x^p}{\partial \bar{x}^{\alpha} \partial \bar{x}^{\sigma}}$$

i.e.
$$\frac{\partial^2 x^p}{\partial \bar{x}^{\alpha} \partial \bar{x}^{\sigma}} = \bar{\Gamma}^{\lambda}_{\alpha\sigma} \frac{\partial x^p}{\partial \bar{x}^{\lambda}} - \frac{\partial x^{\mu}}{\partial \bar{x}^{\alpha}} \frac{\partial x^{\nu}}{\partial \bar{x}^{\sigma}} \Gamma^p_{\mu\nu} \tag{3.92a}$$

3.24. GEODESICS

In flat or euclidean three-dimensional space, the path of the shortest distance between two fixed points is a straight line. In curved Riemannian space, if we draw different curves joining two points A and B, the shortest or the longest (extremum) curve or path as shown in Fig. 3.1 will be called a *'geodesic'*. Obviously, geodesic in

Riemannian space is the equivalent of a straight line in Euclidean space. The term 'minimum distance' is correct for straight lines in flat space, but in curved space, one can assert only the property of 'stationary distance'. Thus, geodesic is the path between the two points such that

$$\int_A^B dS \text{ is stationary or } \int_A^B \delta(dS) = 0 \quad (3.93)$$

But
$$\int_A^B \delta(dS) = \frac{1}{2}\int_A^B \frac{\delta}{dS}(dS^2) = 0 \quad (3.94)$$

The interval between the two points in Riemannian space is given by

$$dS^2 = g_{\mu\nu}\, dx^\mu\, dx^\nu \quad (3.95)$$

Keeping the end points A and B fixed, let the path be deformed by giving every intermediate point an arbitrary infinitesimal displacement δx^σ, then Eq. (3.95) yields on differentiation:

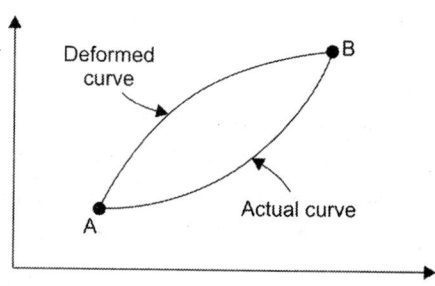

Deformed curve

B

Actual curve

A

Fig. 3.1

$$2\, dS\, \delta\,(dS) = \delta(g_{\mu\nu})dx^\mu dx^\nu + g_{\mu\nu}\delta(dx^\mu)dx^\nu + g_{\mu\nu}dx^\mu\delta(dx^\nu)$$

$$= dx^\mu dx^\nu \frac{\partial g_{\mu\nu}}{\partial x^\sigma}\delta x^\sigma + g_{\mu\nu}\, dx^\nu\delta(dx^\mu) + g_{\mu\nu}\, dx^\mu\delta(dx^\nu)$$

Dividing both sides by $2\,dS$ and making use of the relation

$$\delta\left(\frac{dx^\mu}{dS}\right) = \frac{d}{dS}(\delta x^\mu), \text{ one obtains}$$

$$\delta(dS) = \frac{1}{2}\left[\frac{dx^\mu}{dS}\frac{dx^\nu}{dS}\frac{\partial g_{\mu\nu}}{\partial x^\sigma}\delta x^\sigma + g_{\mu\nu}\frac{dx^\nu}{dS}\frac{d}{dS}(\delta x^\mu) + g_{\mu\nu}\frac{dx^\mu}{dS}\frac{d}{dS}(\delta x^\nu)\right]dS \quad (3.96)$$

From Eqs (3.94) and (3.96), one obtains

$$\frac{1}{2}\int_A^B\left[\frac{\partial x^\mu}{dS}\frac{dx^\nu}{dS}\frac{\partial g_{\mu\nu}}{\partial x^\sigma}\delta x^\sigma + g_{\mu\nu}\frac{dx^\nu}{dS}\frac{d}{dS}(\delta x^\mu) + g_{\mu\nu}\frac{dx^\mu}{dS}\frac{d}{dS}(\delta x^\nu)\right]dS = 0$$

Now changing the dummy indices in the last two terms, i.e. $\mu \to \sigma$ in the first term and $\nu \to \sigma$ in the second term, and rearranging, one obtains

$$\frac{1}{2}\int_A^B\left[\frac{dx^\mu}{dS}\frac{dx^\nu}{dS}\frac{\partial g_{\mu\nu}}{\partial x^\sigma}\delta x^\sigma + \left(g_{\mu\sigma}\frac{dx^\mu}{dS} + g_{\sigma\nu}\frac{dx^\nu}{dS}\right)\frac{d}{dS}(\delta x^\sigma)\right]dS = 0$$

Integrating the second term by parts and keeping in mind that the variation δ is zero at the fixed points A and B, one obtains

$$\frac{1}{2}\int_A^B\left[\frac{dx^\mu}{dS}\frac{dx^\nu}{dS}\frac{\partial g_{\mu\nu}}{\partial x^\sigma} - \frac{d}{dS}\left(g_{\mu\sigma}\frac{dx^\mu}{dS} + g_{\sigma\nu}\frac{dx^\nu}{dS}\right)\right]\delta x^\sigma dS = 0$$

Since δx^σ are arbitrary and hence for the integral to be stationary, the coefficients of δx^σ in the integrand must vanish at all points on the path, i.e.

$$\frac{1}{2}\left[\frac{dx^\mu}{dS}\frac{dx^\nu}{dS}\frac{\partial g_{\mu\nu}}{\partial x^\sigma} - \frac{d}{dS}\left(g_{\mu\sigma}\frac{dx^\mu}{dS} + g_{\sigma\nu}\frac{dx^\nu}{dS}\right)\right] = 0$$

or $\quad\frac{1}{2}\frac{dx^\mu}{dS}\frac{dx^\nu}{dS}\frac{\partial g_{\mu\nu}}{\partial x^\sigma} - \frac{1}{2}\frac{\partial g_{\mu\sigma}}{\partial S}\frac{dx^\mu}{dS} - \frac{1}{2}g_{\mu\sigma}\frac{d^2x^\mu}{dS^2} - \frac{1}{2}\frac{dg_{\sigma\nu}}{dS}\frac{dx^\nu}{dS} - \frac{1}{2}g_{\sigma\nu}\frac{d^2x^\nu}{dS^2} = 0$ (3.97)

But $\qquad\qquad \dfrac{dg_{\mu\sigma}}{dS} = \dfrac{\partial g_{\mu\sigma}}{\partial x^\nu}\dfrac{dx^\nu}{dS}$

and $\qquad\qquad \dfrac{dg_{\sigma\nu}}{dS} = \dfrac{\partial g_{\sigma\nu}}{\partial x^\mu}\dfrac{dx^\mu}{dS}$

Using the above relation, Eq. (3.97) becomes

$$\frac{1}{2}\frac{dx^\mu}{dS}\frac{dx^\nu}{dS}\frac{\partial g_{\mu\nu}}{\partial x^\sigma} - \frac{1}{2}\frac{\partial g_{\mu\sigma}}{\partial x^\nu}\frac{dx^\nu}{dS}\frac{dx^\mu}{dS} - \frac{1}{2}g_{\mu\sigma}\frac{d^2x^\mu}{dS^2} - \frac{1}{2}\frac{\partial g_{\sigma\nu}}{\partial x^\mu}\frac{dx^\mu}{dS}\frac{dx^\nu}{dS} - \frac{1}{2}g_{\sigma\nu}\frac{d^2x^\nu}{dS^2} = 0$$

or $\quad\dfrac{1}{2}\dfrac{dx^\mu}{dS}\dfrac{dx^\nu}{dS}\left(\dfrac{\partial g_{\sigma\mu}}{\partial x^\nu} + \dfrac{\partial g_{\nu\sigma}}{\partial x^\mu} - \dfrac{\partial g_{\mu\nu}}{\partial x^\sigma}\right) + g_{\rho\sigma}\dfrac{d^2x^\rho}{dS^2} = 0$

Using the definition of Christoffel's symbol of first kind, one can write the above as

$$\frac{dx^\mu}{dS}\frac{dx^\nu}{dS}\Gamma_{\sigma,\nu\mu} + g_{\rho\sigma}\frac{d^2x^\rho}{dS^2} = 0$$

Multiplying throughout by $g^{\sigma\lambda}$, one obtains

$$\frac{dx^\mu}{dS}\frac{dx^\nu}{dS}g^{\sigma\lambda}\Gamma_{\sigma,\mu\nu} + g^{\sigma\lambda}g_{\rho\sigma}\frac{d^2x^\rho}{dS^2} = 0$$

$$\frac{\partial x^\mu}{dS}\frac{dx^\nu}{dS}g^{\sigma\lambda}\Gamma_{\sigma,\mu\nu} + \delta_\rho^\lambda\frac{d^2x^\rho}{dS^2} = 0$$

or $\qquad\qquad \dfrac{d^2x^\lambda}{dS^2} + \dfrac{dx^\mu}{dS}\dfrac{dx^\nu}{dS}\Gamma_{\mu\nu}^\lambda = 0$ (3.98)

Equation (3.98) is the equation of a geodesic and also it represents the equation of a curve in Riemannian space. For $\lambda = 1, 2, 3,...,n$, Eq. (3.98) gives n differential equations which determine a geodesic. However, in the cartesian coordinate systems, since $\Gamma_{\mu\nu}^\lambda = 0$, Eq. (3.98) reduces to

$$\frac{d^2x^\lambda}{dS^2} = 0 \quad\text{or}\quad x^\lambda = a^\lambda S + b^\lambda$$ (3.99)

which represents a straight line. Obviously, in cartesian systems a geodesic reduces to a straight line.

3.25. COVARIANT DIFFERENTIATION

Covariant Derivative of a Covariant Vector

The derivative of a covariant vector is not a tensor. Let A_μ be a covariant vector, then from the tensor transformation law, we have

$$\bar{A}_\mu = \frac{\partial x^p}{\partial \bar{x}^\mu} A_p \tag{3.100}$$

Differentiating Eq. (3.100) w.r.t. $\bar{x}^\nu$, one obtains

$$\frac{\partial \bar{A}_\mu}{\partial \bar{x}^\nu} = \frac{\partial}{\partial \bar{x}^\nu}\left(\frac{\partial x^p}{\partial \bar{x}^\mu} A_p\right) = \frac{\partial x^p}{\partial \bar{x}^\mu}\frac{\partial}{\partial \bar{x}^\nu}(A_p) + \frac{\partial}{\partial \bar{x}^\nu}\frac{x^p}{\partial \bar{x}^\mu} A_p$$

$$= \frac{\partial x^p}{\partial \bar{x}^\mu}\frac{\partial A_p}{\partial x^\nu}\frac{\partial x^\nu}{\partial \bar{x}^\nu} + \frac{\partial^2 x^p}{\partial \bar{x}^\nu \partial \bar{x}^\mu} A_p$$

$$= \frac{\partial x^p}{\partial \bar{x}^\mu}\frac{\partial x^q}{\partial \bar{x}^\nu}\frac{\partial A_p}{\partial x^q} + \frac{\partial^2 x^p}{\partial \bar{x}^\mu \partial \bar{x}^\nu} A_p \tag{3.101}$$

The presence of second term in Eq. (3.101) shows that $\dfrac{\partial \bar{A}_\mu}{\partial \bar{x}^\nu}$ do not transform like the components of a tensor.

We are interesting in finding those quantities which are obtained through partial differentiation and transform like tensors. One can do it by diminishing the partial derivative of the second order,

$$\frac{\partial^2 x^p}{\partial \bar{x}^\mu \partial \bar{x}^\nu} = \bar{\Gamma}^\lambda_{\mu\nu}\frac{\partial x^p}{\partial \bar{x}^\lambda} - \frac{\partial x^\alpha}{\partial \bar{x}^\mu}\frac{\partial x^\beta}{\partial \bar{x}^\nu}\Gamma^p_{\alpha,\beta} \tag{3.102}$$

Using Eq. (3.102), one obtains

$$\frac{\partial \bar{A}_\mu}{\partial \bar{x}^\nu} = \frac{\partial x^p}{\partial \bar{x}^\mu}\frac{\partial x^q}{\partial \bar{x}^\nu}\frac{\partial A_p}{\partial x^q} + \left(\bar{\Gamma}^\lambda_{\mu\nu}\frac{\partial x^p}{\partial \bar{x}^\lambda} - \frac{\partial x^\alpha}{\partial \bar{x}^\mu}\frac{\partial x^\beta}{\partial \bar{x}^\nu}\Gamma^p_{\alpha,\beta}\right) A_p$$

$$= \frac{\partial x^p}{\partial \bar{x}^\mu}\frac{\partial x^q}{\partial \bar{x}^\nu}\frac{\partial A_p}{\partial x^q} + \bar{\Gamma}^\lambda_{\mu\nu}\bar{A}_\lambda - \frac{\partial x^\alpha}{\partial \bar{x}^\mu}\frac{\partial x^\beta}{\partial \bar{x}^\nu}\Gamma^p_{\alpha\beta} A_p \qquad [using \; Eq. (3.100)]$$

or $\quad \dfrac{\partial \bar{A}_\mu}{\partial \bar{x}^\nu} - \bar{\Gamma}^\lambda_{\mu\nu}\bar{A}_\lambda = \dfrac{\partial x^p}{\partial \bar{x}^\mu}\dfrac{\partial x^q}{\partial \bar{x}^\nu}\dfrac{\partial A_p}{\partial x^q} - \dfrac{\partial x^\alpha}{\partial \bar{x}^\mu}\dfrac{\partial x^\beta}{\partial \bar{x}^\nu}\Gamma^p_{\alpha\beta} A_p$

By changing dummy indices, one obtains

$$\frac{\partial \bar{A}_\mu}{\partial \bar{x}^\nu} - \bar{\Gamma}^\lambda_{\mu\nu}\bar{A}_\lambda = \frac{\partial x^\alpha}{\partial \bar{x}^\mu}\frac{\partial x^\beta}{\partial \bar{x}^\nu}\left(\frac{\partial A_\alpha}{\partial x^\beta} - \Gamma^\gamma_{\alpha\beta} A_\gamma\right) \tag{3.103}$$

Using the following comma notation

$$A_\mu; \nu = \frac{\partial A_\mu}{\partial x^\nu} - \Gamma^\lambda_{\mu\nu} A_\lambda \tag{3.104}$$

Equation (3.103) takes the following form

$$\bar{A}_\mu; \nu = \frac{\partial x^\alpha}{\partial \bar{x}^\mu}\frac{\partial x^\beta}{\partial \bar{x}^\nu} A_{\alpha;\beta} \tag{3.105}$$

Equation (3.105) shows that $A_{\mu}; v$ defined by Eq. (3.104) is a covariant tensor of rank two and is termed the *covariant derivative* of A_{μ} w.r.t. x^v.

We must note that in the special case of linear transformation of type $x^{\alpha} = a^{\alpha}_{\mu} + b^{\alpha}$, the Christofell's symbol $\Gamma^{\lambda}_{\mu v}$ vanish and hence the covariant derivative of a covariant vector in Riemannian space transforms in the same way as ordinary derivative. The transformation law of partial derivative, Eq. (3.103) takes the following form.

$$\frac{\partial \bar{A}_{\mu}}{\partial \bar{x}^{v}} = \frac{\partial x^{\alpha}}{\partial \bar{x}^{\mu}} \frac{\partial x^{\beta}}{\partial \bar{x}^{v}} \frac{\partial A_{\alpha}}{\partial x^{\beta}} \tag{3.106}$$

Obviously, the ordinary derivatives of a covariant vector is a second rank covariant tensor relative to linear transformation of coordinates.

Covariant Derivative of a Contravariant Vector

Let us consider that A^{μ} be a contravariant vector. Tensor transformation law gives

$$\bar{A}^{\mu} = \frac{\partial \bar{x}^{\mu}}{\partial x^{\alpha}} A^{\alpha} \tag{3.107}$$

Differentiating Eq. (3.107) w.r.t. $\bar{x}^v$, one obtains

$$\frac{\partial \bar{A}^{\mu}}{\partial \bar{x}^{v}} = \frac{\partial}{\partial \bar{x}^{v}}\left(\frac{\partial \bar{x}^{\mu}}{\partial x^{\alpha}} A^{\alpha}\right) = \frac{\partial}{\partial \bar{x}^{\beta}}\left(\frac{\partial \bar{x}^{\mu}}{\partial x^{\alpha}} A^{\alpha}\right)\frac{\partial x^{\beta}}{\partial \bar{x}^{v}}$$

$$= \frac{\partial \bar{x}^{\mu}}{\partial x^{\alpha}}\frac{\partial A^{\alpha}}{\partial x^{\beta}}\frac{\partial x^{\beta}}{\partial \bar{x}^{v}} + \frac{\partial^2 \bar{x}^{\mu}}{\partial x^{\beta}\partial x^{\alpha}}\frac{\partial x^{\beta}}{\partial \bar{x}^{v}} A^{\alpha} \tag{3.108}$$

The presence of last term in the R.H.S. of Eq. (3.108) shows that $\dfrac{\partial A^{\mu}}{\partial x^{v}}$ do not transform like the components of tensor. One is interested to find those quantities which are obtained through partial differentiation and follow tensor transformation law. One can achieve it by eliminating the second order partial derivatives as follows.

In Eq. (3.92a) by interchanging x and $\bar{x}$ coordinates, one obtains

$$\frac{\partial^2 \bar{x}^{\mu}}{\partial x^{\alpha}\partial x^{\beta}} = \Gamma^{\gamma}_{\alpha\beta}\frac{\partial \bar{x}^{\mu}}{\partial x^{\gamma}} - \frac{\partial \bar{x}^{\sigma}}{\partial x^{\alpha}}\frac{\partial \bar{x}^{\lambda}}{\partial x^{\beta}}\Gamma^{\mu}_{\sigma\lambda} \tag{3.109}$$

Using Eq. (3.109), Eq. (3.108) becomes

$$\frac{\partial \bar{A}^{\mu}}{\partial \bar{x}^{v}} = \frac{\partial \bar{x}^{\mu}}{\partial x^{\alpha}}\frac{\partial x^{\beta}}{\partial \bar{x}^{v}}\frac{\partial A^{\alpha}}{\partial x^{\beta}} + \left(\Gamma^{\gamma}_{\alpha\beta}\frac{\partial \bar{x}^{\mu}}{\partial x^{\gamma}} - \frac{\partial \bar{x}^{\sigma}}{\partial x^{\alpha}}\frac{\partial \bar{x}^{\lambda}}{\partial x^{\beta}}\Gamma^{\mu}_{\sigma\lambda}\right)\frac{\partial x^{\beta}}{\partial \bar{x}^{v}}A^{\alpha}$$

$$= \frac{\partial \bar{x}^{\mu}}{\partial x^{\alpha}}\frac{\partial x^{\beta}}{\partial \bar{x}^{v}}\frac{\partial A^{\alpha}}{\partial x^{\beta}} + \Gamma^{\gamma}_{\alpha\beta}\frac{\partial \bar{x}^{\mu}}{\partial x^{\gamma}}\frac{\partial x^{\beta}}{\partial \bar{x}^{v}}A^{\alpha} - \delta^{\lambda}_{v}\frac{\partial \bar{x}^{\sigma}}{\partial x^{\alpha}}\bar{\Gamma}^{\mu}_{\sigma\lambda}A^{\alpha}$$

$$= \frac{\partial \bar{x}^{\mu}}{\partial x^{\alpha}}\frac{\partial x^{\beta}}{\partial \bar{x}^{v}}\frac{\partial A^{\alpha}}{\partial x^{\beta}} + \Gamma^{\gamma}_{\alpha\beta}\frac{\partial \bar{x}^{\mu}}{\partial x^{\gamma}}\frac{\partial x^{\beta}}{\partial \bar{x}^{v}}A^{\alpha} - \bar{\Gamma}^{\mu}_{\sigma v}\frac{\partial \bar{x}^{\sigma}}{\partial x^{\alpha}}A^{\alpha}$$

Using Eq. (3.107), one obtains

$$\frac{\partial \bar{A}^{\mu}}{\partial \bar{x}^{v}} = \frac{\partial \bar{x}^{\mu}}{\partial x^{\alpha}}\frac{\partial x^{\beta}}{\partial \bar{x}^{v}}\frac{\partial A^{\alpha}}{\partial x^{\beta}} + \Gamma^{\gamma}_{\alpha\beta}\frac{\partial \bar{x}^{\mu}}{\partial x^{\gamma}}\frac{\partial x^{\beta}}{\partial \bar{x}^{v}}A^{\alpha} - \bar{\Gamma}^{\mu}_{\sigma v}\bar{A}^{\sigma} \tag{3.110}$$

On interchanging dummy indices α and γ in the second term on R.H.S of Eq. (3.110), one obtains

$$\frac{\partial \bar{A}^\mu}{\partial \bar{x}^\nu} + \bar{\Gamma}^\mu_{\sigma\nu}\bar{A}^\sigma = \frac{\partial \bar{x}^\mu}{\partial x^\alpha}\frac{\partial x^\beta}{\partial \bar{x}^\nu}\left(\frac{\partial A^\alpha}{\partial x^\beta} + \Gamma^\alpha_{\gamma\beta}A^\gamma\right) \tag{3.111}$$

Using the comma notation, one can write

$$A^\mu; \nu = \frac{\partial A^\mu}{\partial x^\nu} + \Gamma^\mu_{\sigma\nu}A^\sigma \tag{3.112}$$

Using Eq. (3.112), Eq. (3.111) can be written as

$$\bar{A}^\mu; \nu = \frac{\partial \bar{x}^\mu}{\partial x^\alpha}\frac{\partial x^\beta}{\partial \bar{x}^\nu}A^\alpha; \beta \tag{3.113}$$

Equation (3.113) shows that $A^\mu; \nu$ defined by Eq. (3.112) is a mixed tensor of rank 2, called the covariant derivative of contravariant vector A^μ w.r.t. x^ν.

We must note that in the special case of linear transformation of the type

$$x^\alpha = \alpha^\alpha_\mu + b^\alpha$$

the Christoffel symbol $\Gamma^\mu_{\sigma\nu}$ vanish and obviously the covariant derivative of a contravariant vector is the same as ordinary derivative, so Eq. (3.111) takes the form

$$\frac{\partial \bar{A}^\mu}{\partial \bar{x}^\nu} = \frac{\partial \bar{x}^\mu}{\partial x^\alpha}\frac{\partial x^\beta}{\partial \bar{x}^\nu}\frac{\partial A^\alpha}{\partial x^\beta}$$

This shows that ordinary partial derivative of a contravariant vector is a second order mixed tensor relative to linear transformation of coordinates.

3.26. COVARIANT DIFFERENTIATION OF TENSORS OF HIGHER RANK

In previous section, we have seen that the covariant derivatives of a covariant and contravariant vector are tensors of second rank. Now, we shall extend the same to the tensors of higher rank. As an example, we consider the mixed tensor of rank 2, A^μ_ν. Tensor transformation law gives

$$\bar{A}^\mu_\nu = \frac{\partial \bar{x}^\mu}{\partial x^\alpha}\frac{\partial x^\beta}{\partial \bar{x}^\nu}A^\alpha_\beta \tag{3.114}$$

The inner product of Eq. (3.114) by $\dfrac{\partial x^\alpha}{\partial \bar{x}^\mu}$ gives

$$\bar{A}^\mu_\nu = \frac{\partial x^\alpha}{\partial \bar{x}^\mu} = \frac{\partial x^\beta}{\partial \bar{x}^\nu}A^\alpha_\beta \tag{3.115}$$

By partial differentiation of Eq. (3.115) w.r.t. $\bar{x}^\sigma$, one obtains

$$\frac{\partial \bar{A}^\mu_\nu}{\partial \bar{x}^\sigma}\frac{\partial x^\alpha}{\partial \bar{x}^\mu} + \frac{\partial^2 x^\alpha}{\partial \bar{x}^\mu \partial \bar{x}^\sigma}\bar{A}^\mu_\nu = \frac{\partial x^\beta}{\partial \bar{x}^\nu}\frac{\partial A^\alpha_\beta}{\partial x^\gamma}\frac{\partial x^\gamma}{\partial \bar{x}^\sigma} + \frac{\partial^2 x^\beta}{\partial \bar{x}^\nu \partial \bar{x}^\sigma}A^\alpha_\beta$$

Using Eq. (3.92a), one obtains

$$\frac{\partial \bar{A}^\mu_\nu}{\partial \bar{x}^\sigma}\frac{\partial x^\alpha}{\partial \bar{x}^\mu} + \left(\bar{\Gamma}^\lambda_{\mu\sigma}\frac{\partial x^\alpha}{\partial \bar{x}^\lambda} - \frac{\partial x^\rho}{\partial \bar{x}^\mu}\frac{\partial x^q}{\partial \bar{x}^\sigma}\Gamma^\alpha_{pq}\right)\bar{A}^\mu_\nu = \frac{\partial x^\beta}{\partial \bar{x}^\nu}\frac{\partial x^\gamma}{\partial \bar{x}^\sigma}\frac{\partial A^\alpha_\beta}{\partial x^\gamma} + \left(\bar{\Gamma}^\rho_{\nu\sigma}\frac{\partial x^\beta}{\partial \bar{x}^\rho} - \frac{\partial x^m}{\partial \bar{x}^\nu}\frac{\partial x^q}{\partial \bar{x}^\sigma}\Gamma^\beta_{mq}\right)A^\alpha_\beta$$

Using Eq. (3.115) and changing appropriate dummy indices, one obtains

$$\left(\frac{\partial \bar{A}_v^\mu}{\partial x^\gamma}+\bar{\Gamma}_{\lambda v}^\mu \bar{A}_\lambda^\mu - \bar{\Gamma}_{v\sigma}^\mu \bar{A}_\lambda^\mu\right)\frac{\partial x^\sigma}{\partial \bar{x}^\mu} = \frac{\partial x^\beta}{\partial \bar{x}^v}\frac{\partial x^\gamma}{\partial \bar{x}^\sigma}\left(\frac{\partial A_\beta^\alpha}{\partial x^\gamma}+\Gamma_{\mu\gamma}^\alpha A_\beta^m - \Gamma_{\beta\gamma}^m A_m^\alpha\right)$$

Inner multiplication with $\dfrac{\partial \bar{x}^\mu}{\partial x^\alpha}$ gives

$$\frac{\partial \bar{A}_v^\mu}{\partial x^\sigma}+\bar{\Gamma}_{\lambda\sigma}^\mu \bar{A}_v^\lambda - \bar{\Gamma}_{v\sigma}^\mu \bar{A}_\lambda^\mu = \frac{\partial \bar{x}^\mu}{\partial x^\alpha}\frac{\partial x^\beta}{\partial \bar{x}^v}\frac{\partial x^\gamma}{\partial \bar{x}^\sigma}\left(\frac{\partial A_\beta^\alpha}{\partial x^\gamma}+\Gamma_{m\gamma}^\alpha A_\beta^m - \Gamma_{\beta\gamma}^\mu A_m^\alpha\right) \tag{3.116}$$

Writing the comma notation as

$$A_v^\mu,\sigma = \frac{\partial A_v^\mu}{\partial x^\sigma}+\Gamma_{\lambda\sigma}^\mu A_v^\lambda - \Gamma_{v\sigma}^\lambda A_\lambda^\mu \tag{3.117}$$

Using Eq. (3.117), Eq. (3.116) may be expressed as

$$A_v^\mu,\sigma = \frac{\partial \bar{x}^\mu}{\partial x^\alpha}\frac{\partial x^\beta}{\partial \bar{x}^v}\frac{\partial x^\gamma}{\partial \bar{x}^\sigma}A_\beta^\alpha;\gamma \tag{3.118}$$

Obviously A_v^μ, σ is a mixed tensor of rank 3 and called the covariant derivative of A_v^μ w.r.t. x^σ.

From Eq. (3.118), we note that the covariant derivative of A_v^μ contains the following three terms (i) partial derivatives (ii) a term with positive sign similar to that contained in the covariant derivative of a contravariant vector and (iii) a term with negative sign similar to that contained in the covariant derivative of a covariant vector.

On the basis of the above results, one can easily conclude that the covariant differentiation is applicable to the tensors of higher ranks. Generalising the above result, one can easily write the covariant derivative of the tensor $A_{v_1 v_2 \ldots v_s}^{\mu_1\mu_2\ldots\mu_p}$ w.r.t. x^σ as

$$A_{v_1 v_2 \ldots v_s,\sigma}^{\mu_1\mu_2\ldots\mu_p} = \frac{\partial A_{v_1 v_2 \ldots v_s}^{\mu_1\mu_2\ldots\mu_p}}{\partial x^\sigma}+\Gamma_{\lambda\sigma}^{\mu_1} A_{v_1 v_2 \ldots v_s}^{\lambda\mu_2\ldots\mu_p} + \Gamma_{\lambda\sigma}^{\mu_2} A_{v_1 v_2 \ldots v_s}^{\mu_1\lambda\mu_3\ldots\mu_p} +\cdots+\Gamma_{\lambda\sigma}^{\mu_p} A_{v_1 v_2 \ldots v_s}^{\mu_1\mu_2\ldots\mu_{p-1}\lambda}$$

$$-\Gamma_{v_1\sigma}^\lambda A_{\lambda v_2 \ldots v_s}^{\mu_1\mu_2\ldots\mu_p} - \Gamma_{v_2\sigma}^\lambda A_{v_1\lambda v_3 \ldots v_s}^{\mu_1\mu_3\ldots\mu_p} - \Gamma_{v_s\sigma}^\lambda A_{v_1 v_2 \ldots v_{s-1}\lambda}^{\mu_1\mu_2\ldots\mu_p} \tag{3.119}$$

3.27. COVARIANT DERIVATIVES OF FUNDAMENTAL TENSORS $g_{\mu v}$, $g^{\mu v}$ AND G_ρ^μ

i. Covariant derivative of $g_{\mu v}$ with x^σ

We have

$$g_{\mu,}v\sigma = \frac{\partial g_{\mu v}}{\partial x^\sigma}-\Gamma_{\mu\sigma}^\lambda g_{\lambda v} - \Gamma_{v\sigma}^\lambda g_{\mu\lambda}$$

$$= \frac{\partial g_{\mu v}}{\partial x^\sigma}-\Gamma_{v,\mu\sigma} - \Gamma_{\mu,v\sigma} \quad \text{(property of lowering an index has been used)}$$

$$= \frac{\partial g_{\mu v}}{\partial x^\sigma}-\frac{1}{2}\left(\frac{\partial g_{v\mu}}{\partial x^\sigma}+\frac{\partial g_{\sigma v}}{\partial x^\mu}-\frac{\partial g_{\mu\sigma}}{\partial x^v}\right) -\frac{1}{2}\left(\frac{\partial g_{\mu v}}{\partial x^\sigma}+\frac{\partial g_{\sigma\mu}}{\partial x^v}-\frac{\partial g_{v\sigma}}{\partial x^\mu}\right) = 0$$

(since $g_{\mu v}$ is symmetric)

$$\therefore \quad g_{\mu v,}\varepsilon = 0 \tag{3.120}$$

ii. Covariant derivative of $g^{\mu\nu}$ with x^σ

We have
$$g_{\mu\nu',\sigma} = \frac{\partial g^{\mu\nu}}{\partial x^\sigma} + \Gamma^\mu_{\lambda\sigma} g^{\lambda\nu} + \Gamma^\nu_{\lambda\sigma} g^{\mu\lambda}$$

We have
$$g_{\mu\nu} g^{\nu\sigma} = g^\sigma_\mu = \begin{cases} 1, \text{ for } \mu = \sigma \\ 0, \text{ for } \mu \neq \sigma \end{cases}$$

Differentiating above w.r.t. x^ρ, one obtains

$$g_{\mu\nu} \frac{\partial g^{\nu\sigma}}{\partial x^\rho} + g^{\nu\sigma} \frac{\partial g_{\mu\nu}}{\partial x^\rho} = 0$$

Multiplying throughout by $g^{\mu\lambda}$, one obtains

$$g^{\mu\lambda} g_{\mu\nu} \frac{\partial g^{\nu\sigma}}{\partial x^\rho} + g^{\mu\lambda} g^{\nu\sigma} \frac{\partial g_{\mu\nu}}{\partial x^\rho} = 0$$

or
$$\delta^\lambda_\nu \frac{\partial g^{\nu\sigma}}{\partial x^\rho} + g^{\mu\lambda} g^{\nu\sigma} \left(\Gamma_{\nu,\mu\rho} + \Gamma_{\mu,\nu\rho} \right) = 0$$

or
$$\frac{\partial g^{\lambda\sigma}}{\partial x^\rho} + g^{\mu\lambda} \Gamma^\sigma_{\mu\rho} + g^{\nu\sigma} \Gamma^\lambda_{\nu\rho} = 0$$

or
$$\frac{\partial g^{\lambda\sigma}}{\partial x^\rho} + \Gamma^\lambda_{\nu\rho} g^{\nu\sigma} + \Gamma^\sigma_{\nu\rho} g^{\nu\lambda} = 0$$

(change of dummy index $\mu \to \nu$ has been made)

or
$$g^{\lambda\sigma},\rho = 0 \tag{3.121}$$

This gives
$$g^{\mu\nu},\sigma = 0$$

iii. Covariant derivative of g^μ_ν with x^σ

We have

$$g^\mu_\nu,\sigma = \frac{\partial g^\mu_\nu}{\partial x^\sigma} + \Gamma^\mu_{\lambda\sigma} g^\lambda_\nu - \Gamma^\lambda_{\nu\sigma} g^\mu_\lambda$$

$$= 0 + \Gamma^\mu_{\mu\sigma} - \Gamma^\mu_{\nu\sigma} = 0$$

or
$$g^\mu_\nu,\sigma = 0 \tag{3.122}$$

The above results clearly reveal that the covariant derivatives of the fundamental tensors $g_{\mu\nu}, g^{\mu\nu}$, and g^μ_ν are all identically zero, i.e. all the three fundamental tensors w.r.t. covariant differentiation may be treated as constants. This is why $g_{\mu\nu}, g^{\mu\nu}$, and g^μ_ν are defined as *covariant constants*.

3.28. LAWS OF COVARIANT DIFFERENTIATION OF TENSORS

i. The covariant derivative of the sum (or difference) of the two tensors is the sum (or difference) of their covariant derivatives.

ii. The covariant derivative of an outer (or inner) product of two tensors is equal to the sum of the two tensors obtained by outer (or inner) multiplication of each tensor with the covariant derivative of the other tensor.

To illustrate it, let us consider two tensors A^μ and $B_{v\sigma}$. The outer product is

$$A^\mu B_{v\sigma} = C^\mu_{\mu\sigma} \qquad (3.123)$$

The covariant derivative of Eq. (3.123) w.r.t. x^ρ can be expressed as

$$(A^\mu B_{\mu\sigma}), \rho = (C^\mu_{\mu\sigma}), \rho$$

$$= \frac{\partial C^\mu_{v\sigma}}{\partial x^\rho} + \Gamma^\mu_{\lambda\rho} C^\lambda_{v\sigma} - \Gamma^\lambda_{v\rho} C^\mu_{\lambda\sigma} - \Gamma^\lambda_{\sigma\rho} C^\mu_{v\lambda}$$

$$= \frac{\partial(A^\mu B_{v\sigma})}{\partial x^\rho} + \Gamma^\mu_{\lambda\rho}(A^\lambda B_{v\sigma}) - \Gamma^\lambda_{v\rho}(A^\mu B_{\lambda\sigma}) - \Gamma^\lambda_{\sigma\rho}(A^\mu B_{v\lambda})$$

$$= B_{v\sigma}\left(\frac{\partial A^\mu}{\partial x^\rho} + \Gamma^\lambda_{v\rho} A^\mu\right) + A^\mu\left(\frac{\partial B_{v\sigma}}{\partial x^\rho} - \Gamma^\lambda_{v\rho} B_{\lambda\sigma} - \Gamma^\lambda_{\sigma\rho} B_{v\lambda}\right)$$

or $\qquad (A^\mu B_{v\sigma}), \rho = B_{v\sigma} A^\mu; \rho + A^\mu B_{v\sigma}; \rho \qquad (3.123a)$

The above proof is a general one and can be applied to any outer product of two tensors.

The contraction of μ and σ gives

$$(A^\sigma B_{v\sigma}), \rho = B_{v\sigma} A^\sigma, \rho + A^\sigma B_{v\sigma}; \rho$$

This shows that the above rule [Eq. (3.123)] is applicable to inner products also. Here we must remember that the inner product of two tensors is a tensor formed by outer product and contraction.

iii. The fundamental tensors $g_{\mu v}$, $g^{\mu v}$ and g^μ_v or δ^μ_v are constants w.r.t. covariant differentiation, e.g.

$$(g^{\mu v} A_{\mu\sigma}); \rho = g^{\mu v}; \rho A_{\mu\sigma} + g^{\mu v} A_{\mu\sigma}; \rho$$

Obviously, the covariant differentiation for products, sums and differences obeys the same laws as in the case of ordinary differentiation.

3.29. COVARIANT DERIVATIVE OF AN INVARIANT

It is same as its ordinary derivative. One can easily prove it as follows.

Let ϕ be an invariant and A_μ be a covariant vector. Obviously ϕA_μ is a covariant vector.

The covariant derivative of the product (ϕA_μ) w.r.t. x^v is given by

$$(\phi A_\mu); v = \frac{\partial(\phi A_\mu)}{\partial x^v} - \Gamma^\lambda_{\mu v}(\phi A_\lambda) = A_\mu \frac{\partial\phi}{\partial x^v} + \phi \frac{\partial A_\mu}{\partial x^v} - \phi A_\lambda \Gamma^\lambda_{\mu v}$$

or $\qquad (\phi A_\mu); v = \phi\left(\frac{\partial A_\mu}{\partial x^v} - \Gamma^\lambda_{\mu v} A_\lambda\right) + A_\mu \frac{\partial\phi}{\partial x^v} = \phi A_\mu; v + A_\mu \frac{\partial\phi}{\partial x^v} \qquad (3.124)$

Since the covariant differentiation of products obeys the same laws as in the case of ordinary differentiation, one obtains

$$(\phi A_\mu); v = \phi; v A_\mu + \phi A_\mu; v \qquad (3.125)$$

From Eqs (3.124) and (3.125), one obtains

$$\phi; \nu A_\mu + \phi A_{\mu}; \nu = \phi A_\mu; \nu + A_\mu \frac{\partial \phi}{\partial x^\nu}$$

or

$$\left(\phi; \nu - \frac{\partial \phi}{\partial x^\nu} \right) A_\mu = 0$$

Since A_μ is arbitrary and hence we have

$$\phi; \nu = \frac{\partial \phi}{\partial x^\nu} \tag{3.125a}$$

Obviously, the covariant derivative of an invariant or scalar is same as ordinary derivative.

3.30. TENSOR NOTATION AND FORM OF GRADIENT; DIVERGENCE, LAPLACIAN AND CURL

In Chapter 1, we have discussed vector analysis using tensor notations. In this section, we will find tensor notation and form of gradient, divergence, Laplacian and curl.

i. **Gradient of a scalar function:** The gradient of a scalar or an invariant ϕ is defined as

$$\text{grad } \phi = \nabla \phi = \frac{\partial \phi}{\partial x^\mu} = \phi; \mu \tag{3.126}$$

Obviously, Eq. (3.126) is the tensor form of a gradient of a scalar function.

ii. **Divergence of a vector:** Let A^μ is a contravariant vector. Its divergence is defined as the contraction of its covariant derivative w.r.t. x^ν, i.e.

$$\text{div } A^\nu = \text{contraction of } A^\nu; \nu = A^\nu; \nu \tag{3.127}$$

However, from definition

$$A^\mu; \nu = \frac{\partial A^\mu}{\partial x^\nu} + \Gamma^\mu_{\lambda\nu} A^\lambda$$

Thus

$$\text{div } A^\mu = A^\mu, \mu = \frac{\partial A^\mu}{\partial x^\mu} + \Gamma^\mu_{\lambda\mu} A^\lambda$$

$$= \frac{\partial A^\mu}{\partial x^\mu} + \frac{\partial \log \sqrt{g}}{\partial x^\lambda} A^\lambda \qquad (see \text{ Example 20})$$

$$= \frac{\partial A^\mu}{\partial x^\mu} + \frac{1}{\sqrt{g}} \frac{\partial \sqrt{g}}{\partial x^\lambda} A^\lambda = \frac{\partial A^\mu}{\partial x^\mu} + \frac{1}{\sqrt{g}} \frac{\partial \sqrt{g}}{\partial x^\mu} A^\mu$$

$$= \frac{1}{\sqrt{g}} \left(\sqrt{g} \frac{\partial A^\mu}{\partial x^\mu} + \frac{\partial \sqrt{g}}{\partial x^\mu} A^\mu \right)$$

$$= \frac{1}{\sqrt{g}} \frac{\partial (A^\mu \sqrt{g})}{\partial x^\mu} \tag{3.128}$$

Now, we find the divergence of a covariant vector A_μ;

$$\text{div } A_\mu = g^{v\sigma} A_v; \sigma = (g^{v\sigma} A_v); \sigma \qquad (\because g^{v\sigma}; \sigma = 0)$$

$$= A^\sigma; \sigma = \text{div } A_\sigma = \text{div } A^\mu \qquad (3.129)$$

iii. **Laplacian:** $\nabla\phi$ is covariant vector A_μ (say), where ϕ is a scalar, then

$$A_\mu = \phi; \mu = \frac{\partial\phi}{\partial x^\mu} \qquad (3.130)$$

The contravariant vector associated with ϕ; μ can be expressed as

$$A^\mu = g^{\mu v} A_v = g^{\mu v} \frac{\partial\phi}{\partial x^v} \qquad (3.131)$$

From Eq. (3.128)

$$\text{div } A_\mu = \text{div } A^\mu = \frac{1}{\sqrt{g}} \frac{\partial(A^\mu \sqrt{g})}{\partial x^\mu}$$

or

$$\text{div grad } \phi = \frac{1}{\sqrt{g}} \frac{\partial(A^\mu \sqrt{g})}{\partial x^\mu} \phi$$

or

$$\nabla^2\phi = \frac{1}{\sqrt{g}} \frac{\partial}{\partial x^\mu}\left(g^{\mu v} \frac{\partial\phi}{\partial x^v} \sqrt{g} \right)$$

$$= \frac{1}{\sqrt{g}} \frac{\partial}{\partial x^\mu}\left(\sqrt{g}\, g^{\mu v} \frac{\partial\phi}{\partial x^v} \right) \qquad (3.132)$$

iv. **Curl of a vector**

$$\text{curl } A_\mu = A_\mu; v - A_v; \mu$$

$$= \left(\frac{\partial A_\mu}{\partial x^v} - \Gamma^\lambda_{\mu v} A_\lambda \right) - \left(\frac{\partial A_v}{\partial x^\mu} - \Gamma^\lambda_{v\mu} A_\lambda \right)$$

Using the symmetry property of $\Gamma^\lambda_{\mu v}$ w.r.t. μ, v, one obtains

$$\text{curl } A_\mu = \frac{\partial A_\mu}{\partial x^v} - \frac{\partial A_v}{\partial x^\mu}$$

or

$$\text{curl } A_\mu = A_\mu; v - A_v; \mu = \frac{\partial A_\mu}{\partial x^v} - \frac{\partial A_v}{\partial x^\mu} \qquad (3.133)$$

3.31. DIVERGENCE OF A TENSOR

It is defined as its contracted covariant derivative w.r.t. the index of differentiation of any superscript.

Let us consider a contravariant tensor $A^{\mu v}$ of rank 2. The covariant derivative of A^μ w.r.t. x^σ can be defined as

$$A^{\mu v}; \sigma = \frac{\partial A^{\mu v}}{\partial x^\sigma} + \Gamma^\mu_{\lambda\sigma} A^{\lambda v} + \Gamma^v_{\lambda\sigma} A^{\mu\lambda}$$

or $\quad$ div $A^{\mu\nu} = A^{\mu\nu}; \mu = \dfrac{\partial A^{\mu\nu}}{\partial x^\mu} + \Gamma^\mu_{\lambda\mu} A^{\lambda\nu} + \Gamma^\nu_{\lambda\mu} A^{\mu\lambda}$

$$= \dfrac{\partial A^{\mu\nu}}{\partial x^\mu} + \dfrac{\partial}{\partial x^\lambda}(\log \sqrt{g}) A^{\lambda\nu} + \Gamma^\nu_{\lambda\mu} + A^{\mu\lambda}$$

$$= \dfrac{\partial A^{\mu\nu}}{\partial x^\mu} + \dfrac{1}{\sqrt{g}}\dfrac{\partial \sqrt{g}}{\partial x^\lambda} A^{\lambda\nu} + \Gamma^\nu_{\lambda\mu} A^{\mu\lambda} = \dfrac{1}{\sqrt{g}}\left(\sqrt{g}\dfrac{\partial A^{\mu\nu}}{\partial x^\mu} + \dfrac{\partial \sqrt{g}}{\partial x^\mu} A^{\mu\nu}\right) + \Gamma^\nu_{\mu\lambda} A^{\mu\lambda}$$

where the dummy index in the second term of the above equation has been changed from λ to μ.

$\therefore \qquad\qquad\qquad\qquad$ div $A^{\mu\nu} = A^{\mu\nu}; \mu = \dfrac{1}{\sqrt{g}}\dfrac{\partial(\sqrt{g}A^{\mu\nu})}{\partial x^\mu}$ $\qquad\qquad\qquad$ (3.134)

3.32. THE INTRINSIC DERIVATIVES OF A VECTOR

The intrinsic derivative of a covariant vector w.r.t. the parameter t can be defined as

$$\dfrac{\delta A_\mu}{\delta t} = \dfrac{dA_\mu}{dt} - \Gamma^\sigma_{\mu\nu} A_\sigma \dfrac{dx^\nu}{dt} \qquad\qquad\qquad (3.135)$$

and it can be easily shown that it is a covariant vector. Generalising, one can show that the intrinsic derivative is tensor of the same order and type as the original tensor.

Similarly, the intrinsic derivative of a contravariant vector w.r.t. the parameter 't' is defined as

$$\dfrac{\delta A^\mu}{\delta t} = \dfrac{dA^\mu}{dt} + \Gamma^\alpha_{\beta\gamma} A^\beta \dfrac{dx^\gamma}{dt} \qquad\qquad\qquad (3.136)$$

which is a *contravariant vector*.

The intrinsic derivative of an invariant ϕ is given by

$$\dfrac{\delta\phi}{\delta t} = \phi; \sigma\dfrac{dx^\sigma}{dt} = \dfrac{\partial\phi}{\partial x^\sigma}\dfrac{dx^\sigma}{dt} = \dfrac{d\phi}{dt} \qquad\qquad\qquad (3.137)$$

Obviously, the intrinsic derivative of an invariant coincides with its total derivative.

The intrinsic derivatives of the fundamental tensor $g_{\mu\nu}, g^{\mu\nu}$, and g^μ_ν can be expressed as

$$\dfrac{\partial g_{\mu\nu}}{\partial t} = g_{\mu\nu}; \sigma\dfrac{dx^\sigma}{dt} = 0 \qquad\qquad (\because g_{\mu\nu,\sigma} = 0)$$

$$\dfrac{\partial g^{\mu\nu}}{\partial t} = g^{\mu\nu}; \sigma\dfrac{dx^\sigma}{dt} = 0$$

$$\dfrac{\partial g^\mu_\nu}{\partial t} = g^\mu_\nu; v\sigma\dfrac{dx^\sigma}{dt} = 0$$

Thus $\qquad\qquad\qquad \dfrac{\partial g_{\mu\nu}}{\partial t} = \dfrac{\delta g^{\nu\mu}}{\partial t} = \dfrac{\delta g^\mu_\nu}{\delta t} = 0 \qquad\qquad\qquad$ (3.138)

Obviously, the intrinsic derivatives of the fundamental tensors are all identically equal or zero.

From the definition of intrinsic derivatives, one can easily conclude that they obey the same three laws which apply to covariant derivatives.

3.33. INTRINSIC AND COVARIANT DERIVATIVES OF HIGHER RANK TENSORS

Generalisation of Eqs (3.135) and (3.136) to define the intrinsic derivatives of higher rank tensor is quite straight. The intrinsic derivative of a third rank mixed tensor $A^{\mu}_{\nu\sigma}$ w.r.t. the parameter t is defined by the following equation.

$$\frac{\delta A^{\mu}}{\delta t^{\nu\sigma}} = \frac{dA^{\mu}}{dt^{\nu\sigma}} + \Gamma^{\mu}_{\alpha\beta} A^{\mu}_{\nu\sigma} \frac{dx^{\beta}}{dt} - \Gamma^{\alpha}_{\nu\beta} A^{\nu}_{\beta\sigma} \frac{dx^{\beta}}{dt} - \Gamma^{\alpha}_{\beta\sigma} A^{\mu}_{\nu\sigma} \frac{dx^{\alpha}}{dt} \qquad (3.139)$$

The above is a mixed tensor of the same rank and type as the tensor whose intrinsic derivative has been defined, i.e. $A^{\mu}_{\nu\sigma}$.

The covariant derivative of $A^{\mu}_{\nu\sigma}$ can be defined as

$$A^{\mu}_{\nu\sigma,\beta} = \frac{\partial A^{\mu}_{\nu\sigma}}{\partial x^{\beta}} + \Gamma^{\mu}_{\alpha\beta} A^{\alpha}_{\nu\sigma} - \Gamma^{\alpha}_{\nu\beta} A^{\mu}_{\alpha\sigma} - \Gamma^{\alpha}_{\alpha\beta} A^{\mu}_{\sigma\alpha} \qquad (3.140)$$

which is a tensor of 4th rank, i.e. one rank higher than the tensor whose covariant derivative has been found.

3.34. GEODESIC COORDINATES

A coordinate system x^{μ} is said to be a geodesic coordinate system with pole P_0 if all the Christoffel's symbols are zero at the point P_0. Obviously, for geodesic coordinates, we have

$$\Gamma_{\sigma,\mu\nu} = \Gamma^{\sigma}_{\mu\sigma} = 0 \text{ at pole } P_0 \qquad (3.141)$$

This clearly shows that at the pole of a geodesic coordinate system, the first order covariant derivatives reduce to the corresponding ordinary derivatives. A familiar example is the covariant derivative of A^{μ}_{σ} w.r.t. x^{σ} at the pole P_0 of geodesic coordinate system, i.e.

$$A^{\mu}_{\nu;\sigma} = \frac{\partial A^{\mu}_{\nu}}{\partial x^{\sigma}} + \Gamma^{\mu}_{\lambda\sigma} A^{\lambda}_{\nu} - \Gamma^{\lambda}_{\nu\sigma} A^{\mu}_{\lambda} = \frac{\partial A^{\mu}_{\nu}}{\partial x^{\sigma}}$$

Now we find the necessary and sufficient condition that given coordinate system be geodesic with pole at P_0. We have from Eq. 3.92 (a)

$$\overline{\Gamma}^{\lambda}_{\mu\nu} \frac{\partial x^{\gamma}}{\partial \overline{x}_{\lambda}} = \Gamma^{\gamma}_{\alpha,\beta} \frac{\partial x^{\alpha}}{\partial \overline{x}_{\mu}} \frac{\partial x^{\beta}}{\partial \overline{x}^{\nu}} + \frac{\partial^2 x^{\gamma}}{\partial \overline{x}^{\mu} \partial \overline{x}^{\nu}}$$

On interchanging x and $\overline{x}$, one obtains

$$\Gamma^{\lambda}_{\mu\nu} \frac{\partial \overline{x}^{\gamma}}{\partial x^{\lambda}} = \overline{\Gamma}^{\gamma}_{\alpha,\beta} \frac{\partial \overline{x}^{\alpha}}{\partial x^{\mu}} \frac{\partial \overline{x}^{\beta}}{\partial x^{\nu}} + \frac{\partial^2 \overline{x}^{\gamma}}{\partial x^{\mu} \partial x^{\nu}}$$

$$\frac{\partial^2 \overline{x}^{\gamma}}{\partial x^{\mu} \partial x^{\nu}} - \Gamma^{\lambda}_{\mu\nu} \frac{\partial \overline{x}^{\gamma}}{\partial x^{\lambda}} = \overline{\Gamma}^{\gamma}_{\alpha\beta} \frac{\partial \overline{x}^{\alpha}}{\partial x^{\mu}} \frac{\partial \overline{x}^{\beta}}{\partial x^{\nu}} \qquad (3.142)$$

For a given value of γ, $\bar{x}^\gamma$ is a scalar function of x^μ. Obviously, $\dfrac{\partial x^\gamma}{\partial x^\mu}$ is a covariant vector. Let us say that

$$\frac{\partial \bar{x}^\gamma}{\partial x^\mu} = \bar{x}^\gamma; \mu = A_\mu$$

From Eq. (3.142), one obtains

$$\frac{\partial A_\mu}{\partial x^\nu} - \Gamma^\lambda_{\mu\nu} A_\lambda = -\bar{\Gamma}^\gamma_{\alpha\beta} \frac{\partial \bar{x}^\alpha}{\partial x^\mu} \frac{\partial \bar{x}^\beta}{\partial x^\nu}$$

But

$$\frac{\partial A_\mu}{\partial x^\nu} - \Gamma^\lambda_{\mu\nu} A_{\mu;\nu} = (\bar{x}^\gamma; \mu); \quad \nu = \bar{x}^\gamma; \mu\nu$$

$\therefore$

$$\bar{x}^\gamma; \mu\nu = -\bar{\Gamma}^\gamma_{\alpha\beta} \frac{\partial \bar{x}^\alpha}{\partial x^\mu} \frac{\partial \bar{x}^\beta}{\partial x^\nu} \qquad (3.143)$$

(a) Let $\bar{x}^\gamma$ be a geodesic coordinate system with its pole at P_0, then one obtains

$$\bar{\Gamma}^\gamma_{\alpha\beta} = 0 \text{ at } P_0$$

Therefore, from Eq. (3.143), one obtains

$$\bar{x}^\gamma; \mu\nu = 0 \text{ at } P_0$$

Obviously, this gives the necessary condition for a coordinate system to be geodesic.

(b) Conversely, suppose that

$$\bar{x}^\gamma; \mu\nu = 0 \text{ at pole } P_0$$

From Eq. (3.143), we have

$$\bar{\Gamma}^\gamma_{\alpha\beta} \frac{\partial \bar{x}^\alpha}{\partial x^\mu} \frac{\partial \bar{x}^\beta}{\partial x^\nu} = 0 \text{ at pole } P_0$$

Since $\dfrac{\partial \bar{x}^\alpha}{\partial x^\mu}$ and $\dfrac{\partial \bar{x}^\beta}{\partial x^\nu}$ are arbitrary, one obtains

$$\bar{\Gamma}^\gamma_{\alpha\beta} = 0 \text{ at pole } P_0$$

This shows that $\bar{x}^\gamma$ is a geodesic coordinate system with pole at P_0.

Concluding, one can say that the necessary and sufficient condition that a given coordinate system to be geodesic with the pole at P_0, is that their second covariant derivatives w.r.t. space coordinates must vanish at P_0.

3.35. RIEMANN–CHRISTOFFEL TENSOR

We have seen that the covariant derivatives of fundamental tensors $g_{\mu\nu}$, $g^{\mu\nu}$ and g^μ_ν (or δ^μ_ν) are identically equal to zero. This shows that there is no tensor of rank 3 which can be obtained entirely from fundamental tensors and their covarient derivatives. However, there are two tensors, one of rank two $R_{\mu\nu}$ and other of rank four $R^\lambda_{\mu\nu\sigma}$ which involves only the first partial derivatives of the metric tensor. These tensors are obtained from a repeated process of covariant differentiation.

Let us consider a covariant vector A_μ. Its covariant derivative w.r.t. ν is given by

$$A_{\mu;\nu} = \frac{\partial A_\mu}{\partial x^\nu} - \Gamma^\lambda_{\mu\nu} A_\lambda \tag{3.144}$$

which is covariant tensor of rank two.

A second covariant differentiation of Eq. (3.144) yields

$$(A_{\mu;\nu\sigma}) = A_{\mu;\nu\sigma} = \frac{\partial A_{\mu;\nu}}{\partial x^\sigma} - \Gamma^\lambda_{\mu\nu} A_{\lambda;\nu} - \Gamma^\lambda_{\nu\sigma} A_{\mu;\lambda}$$

or

$$A_{\mu;\nu\sigma} = \frac{\partial}{\partial x^\sigma}\left(\frac{\partial A_\mu}{\partial x^\nu} - \Gamma^\lambda_{\mu\nu} A_\lambda\right) - \Gamma^\lambda_{\mu\sigma}\left(\frac{\partial A_\lambda}{\partial x^\nu} - \Gamma^\alpha_{\lambda\nu} A_\alpha\right) - \Gamma^\lambda_{\nu\sigma}\left(\frac{\partial A_\mu}{\partial x^\lambda} - \Gamma^\alpha_{\mu\lambda} A_\alpha\right)$$

$$= \frac{\partial^2 A_\mu}{\partial x^\sigma \partial x^\mu} - \Gamma^\lambda_{\mu\nu}\frac{\partial A_\lambda}{\partial x^\sigma} - A_\lambda \frac{\partial}{\partial x^\sigma}(\Gamma^\lambda_{\mu\nu}) - \Gamma^\lambda_{\mu\sigma}\frac{\partial A_\lambda}{\partial x^\nu}$$

$$+ A^\lambda_{\mu\sigma} \Gamma^\lambda_{\mu\sigma} A_\alpha - \Gamma^\lambda_{\nu\sigma}\frac{\partial A_\mu}{\partial x^\lambda} + \Gamma^\lambda_{\nu\sigma} \Gamma^\alpha_{\mu\lambda} A_\alpha$$

On rearrangement, one obtains

$$A_{\mu;\nu\sigma} = \frac{\partial^2}{\partial x^\sigma}\frac{A_\mu}{\partial x^\nu} - \Gamma^\lambda_{\mu\nu}\frac{A_\lambda}{\partial x^\sigma} - \Gamma^\lambda_{\mu\sigma}\frac{\partial A_\lambda}{\partial x^\nu} - 1^\lambda_{\nu\sigma}\frac{\partial A_\mu}{\partial x^\lambda} - A_\lambda\frac{\partial}{\partial x^\sigma}(\Gamma^\lambda_{\mu\nu})$$

$$+ A_\alpha\left[\Gamma^\lambda_{\mu\sigma}\Gamma^\alpha_{\lambda\nu} + \Gamma^\lambda_{\nu\sigma}\Gamma^\alpha_{\mu\lambda}\right] \tag{3.145}$$

Now, interchanging ν and σ, one obtains

$$A_{\mu;\nu\sigma} = \frac{\partial^2}{\partial x^\nu}\frac{A_\mu}{\partial x^\sigma} - \Gamma^\lambda_{\mu\sigma}\frac{A_\lambda}{\partial x^\nu} - \Gamma^\lambda_{\mu\nu}\frac{\partial A_\lambda}{\partial x^\sigma} - \Gamma^\lambda_{\sigma\nu}\frac{\partial A_\mu}{\partial x^\lambda} - A_\lambda\frac{\partial}{\partial x^\nu}(\Gamma^\lambda_{\mu\sigma})$$

$$+ A_\alpha\left[\Gamma^\lambda_{\mu\nu}\Gamma^\alpha_{\lambda\sigma} + \Gamma^\lambda_{\sigma\nu}\Gamma^\alpha_{\mu\lambda}\right] \tag{3.146}$$

Subtracting Eqs (3.146) from (3.145) and keeping in mind the symmetry property of Christoffel's symbols $\left(\Gamma^\lambda_{\mu\nu} = \Gamma^\lambda_{\nu\mu}\right)$ and the commutative property of partial derivatives,

i.e. $\dfrac{\partial^2}{\partial x^\mu \partial x^\nu} = \dfrac{\partial^2}{\partial x^\nu \partial x^\mu}$, one obtains

$$A_{\mu;\nu\sigma} - A_{\mu;\sigma\nu} = A_\lambda\left[\frac{\partial}{\partial x^\nu}(\Gamma^\lambda_{\mu\sigma}) - \frac{\partial}{\partial x^\sigma}(\Gamma^\lambda_{\mu\nu})\right] + A_\alpha\left[\Gamma^\lambda_{\mu\sigma}\Gamma^\alpha_{\lambda\nu} - \Gamma^\lambda_{\mu\nu}\Gamma^\alpha_{\lambda\sigma}\right]$$

On interchanging the dummy indices α and λ in the last two terms, one obtains

$$A_{\mu;\nu\sigma} - A_{\mu;\sigma\nu} = \left[\frac{\partial}{\partial x^\nu}\Gamma^\lambda_{\mu\sigma} - \frac{\partial}{\partial x^\sigma}\Gamma^\lambda_{\mu\nu} + \Gamma^\alpha_{\mu\sigma}\Gamma^\lambda_{\alpha\nu} - \Gamma^\alpha_{\mu\nu}\Gamma^\lambda_{\alpha\sigma}\right]A_\lambda \tag{3.147}$$

or $\quad A_{\mu;\nu\sigma} - A_{\mu;\sigma\nu} = R^\lambda_{\mu\nu\sigma} A_\lambda \tag{3.148}$

Here $\quad R^\lambda_{\mu\nu\sigma} = \left[\dfrac{\partial}{\partial x^\nu}\Gamma^\lambda_{\mu\sigma} - \dfrac{\partial}{\partial x^\sigma}\Gamma^\lambda_{\mu\nu} + \Gamma^\alpha_{\mu\sigma}\Gamma^\lambda_{\alpha\nu} - \Gamma^\alpha_{\mu\nu}\Gamma^\lambda_{\alpha\sigma}\right] \tag{3.149}$

L.H.S of Eq. (3.148) is the difference of two covariant tensors A_{μ}; $v\sigma$ and A_{μ}; σv each of rank 3. Obviously L.H.S of Eq. (3.148) is a tensor of rank 3. Since A_{λ} is an arbitrary covariant tensor of rank 1, it follows from the quotient law that $R^{\lambda}_{\mu v \sigma}$ is a tensor of rank 4, contravariant in index λ and covariant in indices μ, v and σ. The tensor $R^{\lambda}_{\mu v \sigma}$ given by Eq. (3.149) is called *Riemann–Christoffel tensor* or sometimes called as the curvature tensor. From Eq. (3.149), it is obvious that Riemann–Christoffel tensor contains only the Christoffel's symbols and their derivatives which in turn contains fundamental tensor and their partial derivatives. Obviously, Riemann–Christoffel tensor also belongs to the class of fundamental tensors.

Mathematical importance of the Riemann–Christoffel tensor is that in Euclidean or flat space $g_{\mu v}$ are constants and hence the Christoffel symbols vanish as is the case with space-time continuum of special relativity. The vanishing of Christoffel symbols also implies the vanishing of $R^{\alpha}_{\mu v \sigma}$. Since $R^{\alpha}_{\mu v \sigma}$ is a tensor and hence if it vanishes in one coordinate system, it will continue to vanish in other coordinate system as well. Thus Euclidean space is characterised by the vanishing of Riemann–Christoffel tensor. Riemann–Christoffel tensor has the following properties.

i. **Antisymmetric property:** On interchanging the indices v and σ in Eq. (3.149), one obtains

$$R^{\lambda}_{\mu v \sigma} = \frac{\partial}{\partial x^{\sigma}} \Gamma^{\lambda}_{\mu v} - \frac{\partial}{\partial x^{v}} \Gamma^{\lambda}_{\mu \sigma} + \Gamma^{\alpha}_{\mu v} \Gamma^{\lambda}_{\alpha \sigma} - \Gamma^{\alpha}_{\mu \sigma} \Gamma^{\lambda}_{\alpha v}$$

$$= -\left[\frac{\partial}{\partial x^{v}} \Gamma^{\lambda}_{\mu \sigma} - \frac{\partial}{\partial x^{\sigma}} \Gamma^{\lambda}_{\mu v} + \Gamma^{\alpha}_{\mu \sigma} \Gamma^{\lambda}_{\alpha v} - \Gamma^{\alpha}_{\mu v} \Gamma^{\lambda}_{\alpha \sigma} \right] \tag{3.150}$$

On comparing Eqs (3.149) and (3.150), one obtains

$$R^{\lambda}_{\mu \sigma v} = -R^{\lambda}_{\mu \sigma v} \tag{3.151}$$

Equation (3.151) shows that $R^{\lambda}_{\mu \sigma v}$ is antisymmetric w.r.t. indices μ and σ.

ii. **Cyclic property:** On permuting the indices μ, v and σ in a cyclic order and adding, one can easily prove that

$$R^{\lambda}_{\mu \sigma v} + R^{\lambda}_{v \sigma \mu} + R^{\lambda}_{\sigma \mu v} = 0 \tag{3.152}$$

3.36. COVARIANT CURVATURE TENSOR

The mixed Riemann–Christoffel tensor is

$$R^{\lambda}_{\mu v \sigma} = \frac{\partial}{\partial x^{v}} \Gamma^{\lambda}_{\mu \sigma} - \frac{\partial}{\partial x^{\sigma}} \Gamma^{\lambda}_{\mu v} + \Gamma^{\lambda}_{\mu \sigma} \Gamma^{\lambda}_{\alpha v} - \Gamma^{\alpha}_{\mu v} \Gamma^{\lambda}_{\alpha \sigma} \tag{3.153}$$

Taking the inner product of both sides Eq. (3.153) with $g_{\rho \lambda}$, one obtains

$$R_{\rho \mu v \sigma} = g_{\rho \lambda} R^{\lambda}_{\mu v \sigma}$$

$$= g_{\rho \lambda} \left[\frac{\partial}{\partial x^{v}} \Gamma^{\lambda}_{\mu \sigma} - \frac{\partial}{\partial x^{\sigma}} \Gamma^{\lambda}_{\mu v} + \Gamma^{\alpha}_{\mu \sigma} \Gamma^{\lambda}_{\alpha v} - \Gamma^{\alpha}_{\mu v} \Gamma^{\lambda}_{\alpha \sigma} \right] \tag{3.154}$$

Since $\quad g_{\rho \lambda} \dfrac{\partial}{\partial x^{v}} \Gamma^{\lambda}_{\mu \sigma} = \dfrac{\partial}{\partial x^{v}} (g_{\rho \lambda} \Gamma^{\lambda}_{\mu \sigma}) - \dfrac{\partial g_{\rho \lambda}}{\partial x^{v}} \Gamma^{\lambda}_{\mu \sigma} = \dfrac{\partial}{\partial x^{v}} \Gamma_{\rho, \mu \sigma} - \dfrac{\partial g_{\rho \lambda}}{\partial x^{v}} \Gamma^{\lambda}_{\mu \sigma} \tag{3.155}$

Using Eq. (3.155), Eq. (3.154) becomes

$$R_{\rho\mu\nu\sigma} = \frac{\partial}{\partial x^{\nu}} \Gamma_{\rho,\mu\sigma} - \frac{\partial g_{\rho\lambda}}{\partial x^{\nu}} \Gamma^{\lambda}_{\mu\sigma} - \frac{\partial}{\partial x^{\sigma}} \Gamma_{\rho,\mu\nu}$$

$$= \frac{\partial g_{\rho\lambda}}{\partial x^{\sigma}} \Gamma^{\lambda}_{\mu\nu} + \Gamma^{\alpha}_{\mu\sigma} \Gamma_{\rho,\alpha\nu} - \Gamma^{\alpha}_{\mu\nu} \Gamma_{\rho,\alpha\sigma}$$

Changing the dummy index α to λ in the last two terms of the above equation, one obtains

$$R_{\rho\mu\nu\sigma} = \frac{\partial}{\partial x^{\nu}} \Gamma_{\rho,\mu\sigma} - \frac{\partial}{\partial x^{\sigma}} \Gamma_{\rho,\mu\nu} + \Gamma^{\lambda}_{\mu\nu}\left(\Gamma_{\rho,\lambda\nu} - \frac{\partial g_{\rho\lambda}}{\partial x^{\nu}} \right) - \Gamma^{\lambda}_{\mu\nu}\left(\Gamma_{\rho,\lambda\sigma} - \frac{\partial g_{\rho\lambda}}{\partial x^{\sigma}} \right) \quad (3.156)$$

We have

$$\frac{\partial g_{\rho\lambda}}{\partial x^{\nu}} = \Gamma_{\lambda,\rho\nu} + \Gamma_{\rho,\lambda\nu}$$

and

$$\frac{\partial g_{\rho\lambda}}{\partial x^{\sigma}} = \Gamma_{\lambda,\rho\sigma} + \Gamma_{\rho,\lambda\sigma} \quad (3.157)$$

Substituting Eq. (3.157) in the last two terms of Eq. (3.156), and the values of $g^S_{\mu\nu}$ in the first two terms of Eq. (3.156), one obtains

$$R_{\rho\mu\nu\sigma} = \frac{1}{2}\left(\frac{\partial^2 g_{\rho\sigma}}{\partial x^{\nu}\partial x^{\mu}} + \frac{\partial^2 g_{\mu\nu}}{\partial x^{\sigma}\partial x^{\mu}} - \frac{\partial^2 g_{\mu\sigma}}{\partial x^{\nu}\partial x^{\rho}} - \frac{\partial^2 g_{\sigma\nu}}{\partial x^{\sigma}\partial x^{\mu}} \right) + g_{\lambda\alpha}\left[\Gamma^{\lambda}_{\mu\nu}\Gamma^{\alpha}_{\rho\sigma} - \Gamma^{\lambda}_{\mu\sigma}\Gamma^{\alpha}_{\rho\nu} \right] \quad (3.158)$$

Equation (3.158) represents the covariant curvature tensor or covariant form of Riemann–Christoffel tensor.

We must note that in the expression for covariant curvature tensor we have expressed the lowered index in the covariant curvature tensor as the first of the four covariant indices. However, there is no universally accepted norm for writing of indices of the covariant curvature tensor. We will follow consistently the notation used in Eq. (3.158).

Properties of Covariant Curvature Tensor

i. **Covariant curvature tensor is antisymmetric**: Interchanging the indices ρ and μ in Eq. (3.158) and comparing the two we note that

$$R_{\rho\mu\nu\sigma} = -R_{\mu\rho\nu\sigma} \quad (3.159)$$

Obviously, covariant curvature tensor $R_{\rho\mu\nu\sigma}$ is antisymmetric in the first two indices ρ and μ.

One can also prove that $R_{\rho\mu\nu\sigma}$ is antisymmetric in the last two indices μ and σ. On interchanging ν and σ in Eq. (3.158) and comparing the two equations, one obtains

$$R_{\rho\mu\nu\sigma} = -R_{\rho\mu\sigma\nu} \quad (3.160)$$

Obviously, $R_{\rho\mu\nu\sigma}$ is antisymmetric in the last two indices ν and σ.

One can further show that $R_{\rho\mu\nu\sigma}$ is antisymmetric in the indices ρ and σ, i.e.

$$R_{\rho\mu\nu\sigma} = - R_{\sigma\mu\nu\rho}$$

or

$$R_{\rho\mu\nu\sigma} + R_{\sigma\mu\nu\rho} = 0 \quad (3.161)$$

ii. Covariant curvature tensor is symmetric in the index pairs: Interchanging first pair of indices ($\rho\mu$) and second pair of indices ($\nu\sigma$) without changing the order of indices in each pair in Eq. (3.158) and comparing the two, one finds that

$$R_{\rho\mu\nu\sigma} = R_{\nu\sigma\rho\mu} \tag{3.162}$$

Obviously, covariant curvature tensor is symmetric w.r.t. an interchange of the first and second pair of indices without changing the order of indices in each pair.

iii. Cyclic property of the covariant curvature tensor: Permuting the last three indices $\mu\nu\sigma$ in Eq. (3.158) in cyclic order and then taking sum of all the three, one obtains

$$R_{\rho\mu\nu\sigma} + R_{\rho\nu\sigma\mu} + R_{\rho\sigma\mu\nu} = 0 \tag{3.163}$$

3.37. INDEPENDENT COMPONENTS OF THE CURVATURE TENSOR $R_{\rho\mu\nu\sigma}$

A general tensor of rank 4 has n^4 components in an n-dimensional space. Thus in 4-dimensional space it has $4^4 = 256$ components. However, in the case of covariant curvature tensor the number of algebraically independent components of the curvature tensor $R_{\rho\mu\nu\sigma}$ is greatly reduced due to its symmetric properties. Because of symmetries and antisymmetries, it can be proved that the curvature tensor has only $\dfrac{n^2(n^2-1)}{12}$ independent components. Thus, in a space of 4-dimensions, the curvature tensor has only $\dfrac{4^2(4^2-1)}{12} = 20$ independent components. The total number of algebraically independent components can be broken up into the following three categories:

i. Independent components of *first category* with two different indices, i.e. the two pairs in indices are identical, $N_I = \dfrac{1}{2}n(n-1)$. In this case the curvature tensor is $R_{\sigma\mu\sigma\mu}$.

ii. Independent components of *second category*, i.e. when the three or four indices are different is given by $N_{II} = \dfrac{1}{2}n(n-1)(n-2)$. In this case the curvature tensor is $R_{\sigma\mu\sigma\rho}$.

iii. Independent components of third category, i.e. when all the four indices are different, given by $N_{III} = \dfrac{1}{12}n(n-1)(n-2)(n-3)$. The total number of algebraically independent components of curvature tensor $R_{\rho\mu\nu\sigma}$ is given by

$$N = N_I + N_{II} + N_{III}$$

$$= \frac{1}{12}n^2(n^2-1)$$

3.38. CONTRACTION OF RIEMANN–CHRISTOFFEL TENSOR

The Riemann–Christoffel tensor $R^\lambda_{\mu\nu\sigma}$ can be contracted in the following three ways:

i. contraction w.r.t. indices λ and σ

ii. contraction w.r.t. indices λ and ν and

iii. contraction w.r.t. indices λ and μ.

Since the Riemann–Christoffel tensor $R^\lambda_{\mu\nu\sigma}$ is antisymmetric w.r.t. indices ν and σ, the processes of contraction (i) and (ii) give only one independent tensor. Obviously, there are only two ways of contracting Riemann–Christoffel tensor. The one way of contraction leads to *Ricci tensor* while the other may lead to a *zero tensor*. We have

$$R^\lambda_{\mu\sigma\nu} = \frac{\partial}{\partial x^\nu}\Gamma^\lambda_{\mu\sigma} - \frac{\partial}{\partial x^\sigma}\Gamma^\lambda_{\mu\nu} + \Gamma^\alpha_{\mu\sigma}\Gamma^\lambda_{\alpha\nu} - \Gamma^\alpha_{\mu\nu}\Gamma^\lambda_{\alpha\sigma} \qquad (3.164)$$

i. **Ricci tensor:** The contraction of Riemann–Christoffel tensor w.r.t. σ gives us the tensor of rank 2. This tensor is called Ricci tensor and denoted by $R_{\mu\nu}$. Ricci tensor is defined as

$$R_{\mu\nu} = R^\lambda_{\mu\nu\lambda} = \frac{\partial}{\partial x_\nu}\Gamma^\lambda_{\mu\nu} - \frac{\partial}{\partial x^\lambda}\Gamma^\lambda_{\mu\nu} + \Gamma^\alpha_{\mu\lambda}\Gamma^\lambda_{\alpha\nu} - \Gamma^\alpha_{\mu\nu}\Gamma^\lambda_{\alpha\nu} \qquad (3.165)$$

Since $\Gamma^\lambda_{\mu\nu} = \frac{\partial}{\partial x^\lambda}(\log\sqrt{g})$, hence Ricci tensor can also be expressed as

$$R_{\mu\nu} = \frac{\partial^2(\log\sqrt{g})}{\partial x^\nu \partial x^\mu} - \frac{\partial}{\partial x^\lambda}\Gamma^\lambda_{\mu\nu} + \Gamma^\alpha_{\mu\lambda}\Gamma^\alpha_{\alpha\nu} - \Gamma^\mu_{\mu\lambda}\Gamma^\lambda_{\alpha\lambda} \qquad (3.166)$$

On interchanging the indices μ and ν in Eq. (3.166) and comparing the resulting equation with Eq. (3.166) one finds that

$$R_{\mu\nu} = R_{\nu\mu} \qquad (3.167)$$

Obviously, Ricci tensor is a symmetric tensor.

ii. **Zero tensor:** On contracting Riemann–Christoffel tensor w.r.t. indices λ and μ, one obtains

$$R^\lambda_{\lambda\nu\sigma} = \frac{\partial}{\partial x^\nu}\Gamma^\lambda_{\lambda\sigma} - \frac{\partial}{\partial x^\sigma}\Gamma^\lambda_{\lambda\nu} + \Gamma^\alpha_{\lambda\sigma}\Gamma^\lambda_{\alpha\nu} - \Gamma^\alpha_{\lambda\nu}\Gamma^\lambda_{\alpha\sigma}$$

$$= \frac{\partial^2(\log\sqrt{g})}{\partial x^\nu\partial x^\sigma} - \frac{\partial^2(\log\sqrt{g})}{\partial x^\sigma\partial x^\nu} + \Gamma^\alpha_{\lambda\sigma}\Gamma^\lambda_{\alpha\nu} - \Gamma^\alpha_{\lambda\nu}\Gamma^\lambda_{\alpha\sigma}$$

Now, interchanging the dummy indices λ and α in third term on RHS of the above, one obtains

$$R^\lambda_{\lambda\nu\sigma} = 0 \qquad (3.168)$$

The Scalar Curvature

If Ricci tensor is further contracted, one obtains an invariant R, called the *scalar curvature*. It is given by

$$R = g^{\mu\nu}R_{\mu\nu} \qquad (3.169)$$

3.39. BIANCHI IDENTITIES

The mixed curvature tensor at the pole P_0 of geodesic coordinate system is given by

$$R^\lambda_{\mu\nu\sigma} = \frac{\partial}{\partial x^\nu}\Gamma^\lambda_{\mu\sigma} - \frac{\partial}{\partial x^\sigma}\Gamma^\lambda_{\mu\nu}$$

Now, taking covariant derivative w.r.t. ρ at the pole of geodesic coordinate system, one obtains

$$R^\lambda_{\mu\nu\sigma;\rho} = \frac{\partial^2}{\partial x^\rho\partial x^\nu}\Gamma^\lambda_{\mu\sigma} - \frac{\partial^2}{\partial x^\rho\partial x^\sigma}\Gamma^\lambda_{\mu\nu} \qquad (3.170)$$

On permuting the indices ν, σ, ρ in a cyclic order, one obtains two more equations from Eq. (3.170), i.e.

$$R^{\lambda}_{\mu\sigma\rho;\nu} = \frac{\partial^2}{\partial x^{\sigma}\partial x^{\nu}}\Gamma^{\lambda}_{\mu\rho} - \frac{\partial^2}{\partial x^{\nu}\partial x^{\rho}}\Gamma^{\lambda}_{\mu\sigma} \tag{3.171}$$

and

$$R^{\lambda}_{\mu\rho\nu;\sigma} = \frac{\partial^2}{\partial x^{\sigma}\partial x^{\rho}}\Gamma^{\lambda}_{\mu\nu} - \frac{\partial^2}{\partial x^{\sigma}\partial x^{\nu}}\Gamma^{\lambda}_{\mu\rho} \tag{3.172}$$

Adding Eqs (3.170) and (3.173), one obtains at pole P_0

$$R^{\lambda}_{\mu\nu\sigma;\rho} + R^{\lambda}_{\mu\sigma\rho;\nu} + R^{\lambda}_{\mu\rho\nu;\sigma} = 0 \tag{3.173}$$

Although we have shown that the above relation is valid in a local or geodesic coordinate system, but since it is a tensor relation, it must hold good in any other coordinate system whatsoever. Further, the chosen point is an arbitrary one and hence the result is valid at all points of the space. The relations expressed by Eq. (3.173) are called *Bianchi identities*. The covariant form of Bianchi identities can be obtained by taking inner product of Eq. (3.173) with $g_{\alpha\lambda}$, i.e.

$$R^{\lambda}_{\alpha\mu\nu\sigma;\rho} + R^{\lambda}_{\alpha\mu\sigma\rho;\nu} + R^{\lambda}_{\alpha\mu\rho\nu;\sigma} = 0 \tag{3.174}$$

3.40. THE EINSTEIN'S TENSOR: CONTRACTION OF BIANCHI IDENTITIES

The Bianchi identities are given by Eq. (3.173). Applying antisymmetric property in the second term of (3.173), one obtains

$$R^{\lambda}_{\mu\nu\sigma;\rho} - R^{\lambda}_{\mu\rho\sigma;\nu} + R^{\lambda}_{\mu\rho\nu;\sigma} = 0 \tag{3.175}$$

Now, contracting it w.r.t. σ and λ, i.e. setting $\lambda = \sigma$, one obtains

$$R^{\lambda}_{\mu\nu\lambda;\rho} - R^{\lambda}_{\mu\rho\lambda;\nu} + R^{\lambda}_{\mu\rho\nu;\lambda} = 0 \tag{3.176}$$

Using the definition of Ricci tensor, one obtains

$$R^{\lambda}_{\mu\nu\lambda} = R_{\mu\nu} \text{ and } R^{\lambda}_{\mu\rho\lambda} = R_{\mu\rho}$$

Therefore, Eq. (3.176) yields

$$R_{\mu\nu;\rho} - R_{\mu\rho;\nu} + R^{\lambda}_{\mu\rho\nu;\lambda} = 0 \tag{3.177}$$

We know that derivatives of fundamental tensors are zero. Obviously, one can express Eq. (3.177) as

$$(g^{\mu\rho}R_{\mu\nu})_{;\rho} - (g^{\mu\rho}R_{\mu\rho}) + (g^{\mu\rho}R^{\lambda}_{\mu\rho\nu})_{;\lambda} = 0 \tag{3.178}$$

Using the result

$$R_{;\nu} = \frac{\partial R}{\partial x^{\nu}} = \frac{\partial}{\partial x^{\mu}}(\delta^{\mu}_{\nu}R) = (\delta^{\mu}_{\nu}R)_{;\mu}$$

Equation (3.178) becomes

$$2R^{\mu}_{\nu;\mu} - (\delta^{\mu}_{\nu}R)_{;\mu} = 0$$

or

$$\left(R^{\mu}_{\nu} - \frac{1}{2}\delta^{\mu}_{\nu}R\right)_{;\mu} = 0 \tag{3.179}$$

or $\qquad\qquad G^{\mu}_{v;\mu} = 0$ $\qquad\qquad$ (3.180)

where the tensor

$$G^{\mu}_{v} = R^{\mu}_{v} - \frac{1}{2}\delta^{\mu}_{v}R \qquad\qquad (3.181)$$

G_{v}; μ, given by Eq. (3.180) is called Einstein's tensor. This tensor plays a very fundamental role in general theory of relativity.

Covariant from of Einstein's Tensor

$$G_{v\lambda} = g_{\lambda\mu}G^{\mu}_{v} = g_{\lambda\mu}\left(R^{\mu}_{v} - \frac{1}{2}\delta^{\mu}_{v}R\right) = R_{\lambda v} - \frac{1}{2}g_{\lambda v}R$$

or $\qquad\qquad G_{v\lambda} = R_{\lambda v} - \frac{1}{2}g_{\lambda v}R \qquad\qquad$ (3.182)

Divergence of Einstein's Tensor

We have

$$\text{div.}(G^{\mu}_{v}) = G^{\mu}_{v;\mu}$$

∴ Using Eq. (3.180), one obtains

$$\text{div.}G^{\mu}_{v} = 0 \qquad\qquad (3.183)$$

Obviously, the divergence of Einstein's tensor vanishes.

3.41. APPLICATIONS OF TENSORS IN MECHANICS

In this Chapter, we will consider a few simple applications of tensor analysis to the mechanics of a particle.

Dynamics of a Particle

Consider a particle of mass m moving in space. Let the position of the particle be represented by a set of curvilinear coordinates x^{μ}. As the time t varies, the particle describes a curve in space called the trajectory of the particle, whose equation is

$$x^{\mu} = x^{\mu}(t), \ \mu = 1, 2, 3, \qquad\qquad (3.184)$$

If the curvilinear coordinate system x^{μ} is transformed to another curvilinear coordinate system x^{μ}, according to transformation

$$\bar{x}^{\mu} = \bar{x}^{\mu}(x^1, x^2, x^3) \qquad\qquad (3.185)$$

then the trajectory of the particle in terms of new set of curvilinear coordinates can be obtained by substituting Eqs (3.184) in (3.185).

The velocity of the particle in terms of old set of curvilinear coordinates x^{μ} is given by

$$V^{\mu} = \frac{dx^{\mu}}{dt} \qquad\qquad (3.186)$$

In new coordinate system $\bar{x}^{\mu}$, the velocity of the particle can be expressed as

$$\bar{V}^{\mu} = \frac{d\bar{x}^{\mu}}{dt} = \frac{\partial \bar{x}^{\mu}}{\partial x^{v}}\frac{dx^{v}}{dt} = \frac{\partial \bar{x}^{\mu}}{\partial x^{v}}V^{v} \qquad\qquad (3.187)$$

Obviously, the velocity of the particle transforms like a contravariant vector. The *intrinsic derivative* of this velocity vector w.r.t. the parameter t is given by

$$a^\mu = \frac{\partial \bar{v}^\mu}{\partial t} = \frac{d^2 x^\mu}{dt^2} + \Gamma^\mu_{\alpha\beta} \frac{\partial x^\alpha}{dt} \frac{dx^\beta}{dt} \qquad (3.188)$$

where, $\Gamma^\mu_{\alpha\beta}$ are Christoffel symbols of system x^μ. Eq. (3.188) gives a quantity called the *acceleration vector* of the particle in curvilinear coordinates x^μ. This is in cartesian coordinates. We know that Christoffel symbols are identically zero in rectangular cartesian coordinate system, and hence Eq. (3.188) in cartesian coordinates reduces to

$$a^\mu = \frac{d^2 x^\mu}{dt^2} \qquad (3.189)$$

If the mass of the particle is constant then Eq. (3.188) represents a quantity which is independent of any coordinate system and also independent of time t, i.e. it is an invariant.

The force vector acting on the particle in curvilinear coordinate systems may be expressed in accordance with Newton's second law as

$$F^\mu = ma^\mu \qquad (3.190)$$

where the contravariant vector F^μ is the *force vector*. The force vector completely specifies the magnitude and direction of the force acting on the particle.

From Eqs (3.189) and (3.190), one obtains the force vector in cartesian coordinates as

$$F^\mu = m \frac{d^2 x^\mu}{dt^2}, \qquad \mu = 1, 2, 3 \qquad (3.191)$$

Substituting a^μ from Eqs (3.188) in (3.190), one obtains

$$F^\mu = m \frac{d^2 x^\mu}{dt^2} + m \Gamma^\mu_{\alpha\beta} \frac{dx^\alpha}{dt} \frac{dx^\beta}{dt} \qquad (3.192)$$

Work or Energy

Let us consider that the new system of coordinates $\bar{x}^\mu$ to be a rectangular cartesian. Let us consider a force $\bar{F}^\mu$ in this system. Let the point of application of the force moves through a distance $\delta \bar{x}^\mu$, then the work done by the force $\bar{F}^\mu$ is given by

$$\delta W = \bar{F}^\mu \delta x^\mu \qquad (3.193)$$

Keeping in mind the transformation relation of $x^{-\mu}$ and x^μ, one can write the work done as

$$\delta W = F_\mu \, \delta x^\mu \qquad (3.194)$$

Since δW is an invariant, i.e. a scalar and δx^μ is a contravariant vector, it follows that F_μ are the components of a covariant vector in the coordinates x^μ. The components are related to the following relations.

$$\text{and} \qquad \left. \begin{array}{l} F_\mu = g_{\mu\nu} F^\nu \\ F^\mu = g^{\mu\nu} F_\nu \end{array} \right\} \qquad (3.195)$$

If the work done $\delta W = F_\mu \, \delta x^\mu$ is a perfect differential, then the force F_μ is said to be conservative. Then we may define a potential function V such that

$$V = -\int F_\mu dx^\mu \qquad (3.196)$$

Sometimes V is also called as the *force potential*.

The kinetic energy of the particle (T) is given by

$$T = \frac{1}{2}mv^2 = \frac{1}{2}mg_{\mu\nu}v^\mu v^\nu$$

$$= \frac{1}{2}mg_{\mu\nu}\dot{x}^\mu\dot{x}^\nu \tag{3.197}$$

where, $g_{\mu\nu}$ is the metric tensor of curvilinear coordinates x^μ.

Lagrange's Equations

Differentiating Eq. (3.197) w.r.t. $\dot{x}^\lambda$ partially, one obtains

$$\frac{\partial T}{\partial \dot{x}^\lambda} = \frac{1}{2}mg_{\mu\nu}\left[\frac{\partial \dot{x}^\mu}{\partial \dot{x}^\lambda}\dot{x}^\nu + \dot{x}^\mu\frac{\partial \dot{x}^\nu}{\partial \dot{x}^\lambda}\right]$$

$$= \frac{1}{2}m\left[g_{\mu\nu}\delta^\mu_\lambda\dot{x}^\nu + g_{\mu\nu}\delta^\nu_\lambda\dot{x}^\mu\right]$$

$$= \frac{1}{2}m\left[g_{\mu\nu}\dot{x}^\nu + g_{\mu\lambda}\dot{x}^\mu\right]$$

Replacing dummy indices μ and ν in the last term of the above and keeping in mind $g_{\mu\nu} = g_{\nu\mu}$, one obtains

$$\frac{\partial T}{\partial \dot{x}^\lambda} = mg_{\lambda\nu}\dot{x}^\nu \tag{3.198}$$

$$\therefore \qquad \frac{d}{dt}\left(\frac{\partial T}{\partial \dot{x}^\lambda}\right) = \frac{d}{dt}(mg_{\lambda\nu}\dot{x}^\nu)$$

$$= m\frac{d}{dt}(g_{\lambda\nu}\dot{x}^\nu)$$

$$= m\left[\frac{\partial g_{\lambda\nu}}{\partial x^\sigma}\frac{\partial x^\sigma}{\partial t}\dot{x}^\nu + g_{\lambda\nu}\ddot{x}^\nu\right]$$

$$= m\left[g_{\lambda\nu}\ddot{x}^\nu + \frac{\partial g_{\lambda\nu}}{\partial x^\sigma}\dot{x}^\sigma\dot{x}^\nu\right] \tag{3.199}$$

Similarly
$$\frac{\partial T}{\partial x^\lambda} = \frac{\partial}{\partial x^\lambda}\left(\frac{1}{2}mg_{\mu\nu}\dot{x}^\mu\dot{x}^\nu\right)$$

$$= \frac{1}{2}m\frac{\partial g_{\mu\nu}}{\partial x^\lambda}\dot{x}^\mu\dot{x}^\nu$$

On replacing dummy index μ by σ, one obtains

$$\frac{\partial T}{\partial x^\lambda} = \frac{1}{2}m\frac{\partial g_{\sigma\nu}}{\partial x^\lambda}\dot{x}^\sigma\dot{x}^\nu \tag{3.200}$$

Using Eqs (3.199) and (3.200), one may write

$$\frac{d}{dt}\left(\frac{\partial T}{\partial \dot{x}^\lambda}\right)-\frac{\partial T}{\partial x^\lambda}=m\left[g_{\lambda v}\ddot{x}^v+\frac{\partial g_{\lambda v}}{\partial x^\sigma}\dot{x}^\sigma\dot{x}^v\right]-\frac{1}{2}m\frac{\partial g_{\mu v}}{\partial x^\lambda}\dot{x}^\sigma\dot{x}^v$$

$$=m\left[g_{\lambda v}\ddot{x}^v+\frac{1}{2}\frac{\partial g_{\lambda v}}{\partial x^\sigma}\dot{x}^\sigma\dot{x}^v+\frac{1}{2}\frac{\partial g_{\lambda v}}{\partial x^\sigma}\dot{x}^\sigma\dot{x}^v\right]-\frac{1}{2}m\frac{\partial g_{\mu v}}{\partial x^\lambda}\dot{x}^\sigma\dot{x}^v$$

Interchanging dummy indices v and σ in the last term within the bracket, one obtains

$$\frac{d}{dt}\left(\frac{\partial T}{\partial \dot{x}^\lambda}\right)-\frac{\partial T}{\partial x^\lambda}=m\left[g_{\lambda v}\ddot{x}^v+\frac{1}{2}\left(\frac{\partial g_{\lambda v}}{\partial x^\sigma}+\frac{\partial g_{\lambda\sigma}}{\partial x^v}-\frac{\partial g_{\mu v}}{\partial x^\lambda}\right)\dot{x}^\sigma\dot{x}^v\right]$$

$$=m\left[g_{\lambda v}\ddot{x}^v+\Gamma_{v\sigma,\lambda}\dot{x}^v\dot{x}^\sigma\right] \tag{3.201}$$

where, $\Gamma_{v\sigma,\lambda}=\dfrac{1}{2}\left(\dfrac{\partial g_{\lambda v}}{\partial x^\sigma}+\dfrac{\partial g_{\lambda\sigma}}{\partial x^v}-\dfrac{\partial g_{\mu v}}{\partial x^\lambda}\right)$ are Christoffel symbols of first kind.

We may write Eq. (3.201) as

$$\frac{d}{dt}\left(\frac{\partial T}{\partial \dot{x}^\lambda}\right)-\frac{\partial T}{\partial x^\lambda}=mg_{\lambda\mu}\left(\ddot{x}^\mu+\Gamma^\mu_{v\sigma}\dot{x}^v\dot{x}^\sigma\right) \tag{3.202}$$

Using Eq. (3.187), one finds

$$\ddot{x}^\mu+\Gamma^\mu_{v\sigma}\dot{x}^v\dot{x}^\sigma=a^\mu$$

and hence Eq. (3.202) takes the form

$$\frac{d}{dt}\left(\frac{\partial T}{\partial \dot{x}^\lambda}\right)-\frac{\partial T}{\partial x^\lambda}=mg_{\lambda\mu}a^\mu$$

or

$$\frac{d}{dt}\left(\frac{\partial T}{\partial \dot{x}^\lambda}\right)-\frac{\partial T}{\partial x^\lambda}=ma_\lambda=F_\lambda \tag{3.203}$$

Equation (3.203) represent *Lagrange's equations*. If the force system is conservative, i.e. $F_\lambda=-\dfrac{\partial V}{\partial x^\lambda}$, then Eq. (3.203) takes the form

$$\frac{d}{dt}\left(\frac{\partial T}{\partial \dot{x}^\lambda}\right)-\frac{\partial T}{\partial x^\lambda}+\frac{\partial V}{\partial x^\lambda}=0 \tag{3.204}$$

Elasticity

When an elastic body is subjected to an external force or stress, the body is deformed or strained. In the formulation of the equations governing the equilibrium of an elastic solid, the cartesian tensor of second rank are used. The study of elasticity in terms of tensors falls into three sections (i) a description of the strain or deformation (ii) a description of the force or stress and (iii) a generalization of Hooke's law.

Stress

It is defined as the restoring force per unit area of a deformed body, i.e. the stress at a given point in a solid depends on the orientation of an element of area considered at that point. Obviously, for complete specification of the stress at a point in the solid, one must specify the stresses on three mutually perpendicular elements of area at that point.

Let us consider that P as a point in the solid at which the stress is to be specified. Consider a rectangular cartesian system within the body with P as origin as shown in Fig. 3.2. Let us consider three elements of area da_x, da_y, and da_z in the yz, zx and xy planes respectively such that each plane encloses the point P as shown in Fig. 3.2. The net stress at may be specified by three force F_1, F_2 and F_3 acting on elementary areas da_x, da_y and da_z respectively. One may resolve each of these three forces into its three cartesian components, e.g. F_{1x}, F_{1y} and F_{1z} are components of force F_1 (acting on area da_x) along x, y and z axes respectively. Stress is defined as the force per unit area and hence one can represent the corresponding stress components as

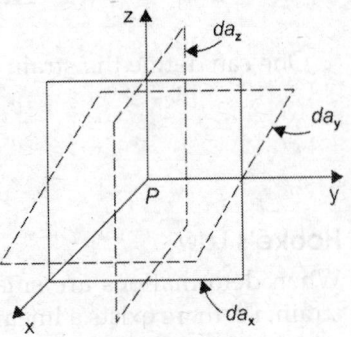

Fig. 3.2

$$x_x = \frac{F_{1x}}{da_x}, \quad y_x = \frac{F_{1y}}{da_x}$$

and
$$z_x = \frac{F_{1z}}{da_x} \qquad (3.205)$$

The stress components represented by Eq. (3.205) and acting on elementary area da_x are shown in Fig. 3.3. The capital letter indicates the direction of force and small letter as subscript indicates normal to the plane to which the stress has been applied. Similarly, one may also introduce stress components (x_y, y_y, z_y) and (x_z, y_z, z_z) which define the force per unit area acting on elementary area da_y and da_z respectively. These nine components are called the cartesian components of the *stress tensor* of rank two.

Let us consider that $x_{\mu\nu}$, where $\mu, \nu = 1, 2, 3$; denote that the cartesian components of stress tensor, writing 1, 2, 3 for x, y, z respectively, one finds that x_{11} stands for x_x, x_{12} for x_y and x_{21} for y_x and so on.

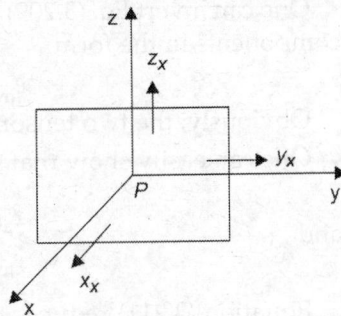

Fig. 3.3

Strain

When a body is subjected to some external force, deformation of the body takes place. As the result of this force, the configuration of the body changes and one can say that the body is in a *strained state*. The strain in a body can be specified in terms of a strain tensor $\rho_{\mu\nu}$ which is defined below.

Let us imagine three orthogonal vectors x, y, z of unit length at point P in the unstrained body. After a small deformation of the body, the axes are distorted in orientation and length. One may write the new axes x', y', z' in terms of old ones in the following form.

$$\left.\begin{array}{l} x' = (1+\varepsilon_{xx})x + \varepsilon_{xy}y + \varepsilon_{xz}z \\ y' = \varepsilon_{yz}x + (1+\varepsilon_{yy})y + \varepsilon_{yz}z \\ z' = \varepsilon_{zx}x + \varepsilon_{zy}y + (1+\varepsilon_{zz})z \end{array}\right\} \qquad (3.206)$$

The coefficient $\varepsilon_{\mu\nu}$ appearing in Eq. (3.206) define the deformation and are dimensionless. When strain is small, i.e. $\varepsilon_{\mu\nu} \leq 1$

Let us define $x' - \hat{x} = \Delta x_1$, $y' - \hat{y} = \Delta x_2$ and $z' - \hat{z} = \Delta x_3$ and denoting $\hat{x}, \hat{y}$ and $\hat{z}$ by x_1, x_2 and x_3 respectively, one can express Eq. (3.206) as

$$\Delta x_\mu = \sum_{v=1}^{3} \varepsilon_{\mu v} x_v \qquad (\mu = 1, 2, 3) \qquad (3.207)$$

One can define the strain tensor of rank two by relations

$$e_{xx} = \varepsilon_{xx}; \quad e_{yy} = \varepsilon_{yy}; \quad e_{zz} = \varepsilon_{zz}$$
$$e_{xy} = e_{yx} = \varepsilon_{xy} + \varepsilon_{yx}; \quad e_{yz} = e_{zy} = \varepsilon_{yz} + \varepsilon_{zy};$$
$$e_{zx} = e_{xz} = \varepsilon_{zx} + \varepsilon_{xz} \qquad\qquad (3.208)$$

Hooke's Law

When deformations are small, Hooke's law states that stress is proportional to the strain, i.e. there exists a linear relationship between stress and strain components. One may express this relationship in tensor form as,

$$x_{\mu v} = E_{\mu v \sigma \rho} e_{\sigma \rho} \qquad (3.209)$$

where the summation over the repeated indices is implied. The coefficients $E_{\mu v \sigma \rho}$ are called elastic stiffness constants or moduli of elasticity. Obviously, these components are components of a fourth rank tensor and evidently their number are $3^4 = 81$ in number.

One can invert Eq. (3.209) to express strain components as a linear function of stress components in the form

$$e_{\mu v} = S_{\mu v \sigma \rho} x_{\sigma \rho} \qquad (3.210)$$

Obviously, the two tensors $E_{\mu v \sigma \rho}$ and $S_{\sigma \rho \lambda \delta}$ are reciprocal of each other.

One can easily show that stress and strain tensors are symmetric. So that

and
$$\left. \begin{array}{l} x_{\mu v} = x_{v \mu} \\ e_{\mu v} = e_{v \mu} \end{array} \right\} \qquad (3.211)$$

Equation (3.211) reduces the number of independent components of either tensor from 9 to 6 and the number of independent components of elastic moduli tensor $E_{\mu v \sigma \rho}$ or elastic compliance tensor from 81 to 6^2 (= 36), one can also show that these four-rank tensors are symmetric under the interchange of the first two and the last two indices, i.e.

and
$$\left. \begin{array}{l} E_{\mu v \sigma \rho} = E_{\sigma \rho \mu v} \\ S_{\mu v \sigma \rho} = S_{\sigma \rho \mu v} \end{array} \right\} \qquad (3.212)$$

Equation (3.212) further the number of independent elastic constants of any solid to $21 = \dfrac{6(6+1)}{2}$. If the solid has internal symmetries then the number of independent elastic constants further get reduced. For example, a cubic crystal has only three independent elastic constants which may be chosen as

$$E_{11} = E_{xxxx}, \quad E_{12} = E_{xxyy}, \quad E_{44} = E_{yzyz}$$

The remaining components either vanishes or each becomes equal to one of the above three components.

Moment of Inertia

Let us consider a rigid body consisting of a number of point masses undergo rotatory motion with angular velocity ω. Let m_i be the mass of ith particle and r_i be its position vector relative to some arbitrary origin O. Let V_i is the linear velocity of ith particle, then the total angular momentum of the system relative to origin O is given by

$$J = \sum_i r_i \times p_i = \sum_i r \times m_i v_i$$

$$= \sum_i r_i \times m_i (\omega_i \times r_i) = \sum_i m_i r_i \times (\omega_i \times r_i)$$

$$= \sum_i m_i \left[(r_i \cdot r_i)\omega_i - (r_i \cdot \omega_i) r_i \right] = \sum_i m_i \left(r_i^2 \omega - [(r_i \cdot \omega) r_i] \right) \qquad (3.213)$$

Let (J_x, J_y, J_z), $(\omega_x, \omega_y, \omega_z)$ and (x_i, y_i, z_i) represent the cartesian components of vector J, ω and r_i respectively, then Eq. (3.213) in component form may be expressed as

$$\left. \begin{aligned} J_x &= I_{xx}\omega_x + I_{xy}\omega_y + I_{xz}\omega_z \\ J_y &= I_{yx}\omega_x + I_{yy}\omega_y + I_{yz}\omega_z \\ J_z &= I_{zx}\omega_x + I_{zy}\omega_y + I_{zz}\omega_z \end{aligned} \right\} \qquad (3.214)$$

where
$$I_{xx} = \sum_i m_i (r_i^2 - x_i^2), \ I_{yy} = \sum_i m_i (r_i^2 - y_i^2),$$

and
$$I_{zz} = \sum_i m_i (r_i^2 - z_i^2),$$

$$\left. \begin{aligned} I_{yx} &= I_{yx} = -\sum_i m_i x_i y_i \\ I_{yz} &= I_{zy} = -\sum_i m_i y_i z_i \\ I_{zx} &= I_{xz} = -\sum_i m_i z_i x_i \end{aligned} \right\} \qquad (3.215)$$

and

Equation (3.214) shows that angular momentum J is not necessarily parallel to angular velocity ω. One may express Eq. (3.214) as

$$J_\mu = I_{\mu\nu}\omega_\nu \qquad (3.216)$$

where J_μ and ω_ν are vectors. Using quotient law, one can say that $I_{\mu\nu}$ is a tensor of rank two. $I_{\mu\nu}$ is a symmetric tensor and it is called as the moment of inertia tensor of a rigid body.

Example 1. If $a_{\alpha\beta} x^\alpha x^\beta = 0$ for all values of variables $x^1, x^2, x^3,..., x^n$, then prove that $a_{\mu\nu} + a_{\nu\mu} = 0$.

Solution: Given $a_{\alpha\beta} x^\alpha x^\beta = 0$

Differentiating it w.r.t. x^μ, one obtains

$$a_{\alpha\beta} x^\alpha \frac{\partial x^\beta}{\partial x^\mu} + a_{\alpha\beta} x^\beta \frac{\partial x^\sigma}{\partial x^\mu} = 0$$

or
$$a_{\alpha\beta}x^\alpha\delta_\mu^\beta + a_{\alpha\beta}x^\beta\delta_\mu^\alpha = 0$$

or
$$a_{\alpha\mu}x^\alpha + a_{\mu\beta}x^\beta = 0$$

Differentiating the above w.r.t. x^ν, one obtains

$$a_{\alpha\mu}\frac{\partial x^\alpha}{\partial x^\nu} + a_{\mu\beta}\frac{\partial x^\beta}{\partial x^\nu} = 0$$

or
$$a_{\alpha\mu}\delta_\nu^\beta + a_{\mu\beta}\delta_\nu^\beta = 0$$

or
$$a_{\nu\mu} + a_{\mu\nu} = 0.$$

Example 2. *A covariant tensor has components xy, 2y – z², xz in rectangular coordinates. Find its covariant components in spherical coordinates.*

Solution: Here $x^1 = x$, $x^2 = y$, $x^3 = z$

and
$$\bar{x}^1 = r, \ \bar{x}^2 = \theta, \ \bar{x}^3 = \phi$$

The covariant components A_1 in rectangular coordinates are
$$A_1 = xy = x^1 x^2$$
$$A_2 = 2y - z^2 = 2x^2 - (x^3)^2$$
$$A_3 = xz = x^1 x^3$$

If $\bar{A}_\mu$ denote the covariant components in spherical coordinates, then

$$\bar{A}_\mu = \frac{\partial x^\alpha}{\partial \bar{x}^\mu} A_\alpha \qquad (\alpha = 1, 2, 3) \qquad (1)$$

The transformation equations between the coordinate systems can be expressed as
$$x^1 = \bar{x}^1 \sin \bar{x}^2 \cos \bar{x}^3$$
$$x^2 = \bar{x}^1 \sin \bar{x}^2 \sin \bar{x}^3$$
$$x^3 = \bar{x}^1 \cos \bar{x}^2 \qquad (2)$$

One can write the covariant components from Eq. (1) as
$$\bar{A}_1 = \frac{\partial x^1}{\partial \bar{x}^1}\bar{A}_1 + \frac{\partial x^2}{\partial \bar{x}^1}\bar{A}_2 + \frac{\partial x^3}{\partial \bar{x}^1}A_2$$

$$= (\sin\bar{x}^2 \cos\bar{x}^3)(x^1 x^2) + (\sin\bar{x}^2 \sin\bar{x}^3)[2x^2 - (x^3)^2] + (\cos\bar{x}^2)(x^1 x^3)$$
$$= (\sin\theta \cos\theta)(r^2 \sin^2\theta \sin\phi \cos\phi) + (\sin\theta \sin\phi)(2r\sin\theta \sin\phi - r^2 \cos^2\theta)$$
$$+ (\cos\theta)(r^2 \sin\theta \cos\theta \cos\phi)$$

$$\bar{A}_2 = \frac{\partial x^1}{\partial \bar{x}^2}A_1 + \frac{\partial x^2}{\partial \bar{x}^2}A_2 + \frac{\partial x^3}{\partial \bar{x}^2}A_3$$

$$= (\bar{x}^1 \cos\bar{x}^2 \cos\bar{x}^3)(x^1 x^2) + (\bar{x}^1 \cos\bar{x}^2 \sin\bar{x}^3)[2x^2 - (x^3)^2] + (-\bar{x}^1 \sin\bar{x}^2)(x^1 x^3)$$
$$= (r \cos\theta \cos\phi)(r^2 \sin^2\theta \sin\phi \cos\phi) + (r \cos\theta \sin\phi)(2r\sin\theta \sin\phi - r^2 \cos^2\theta)$$
$$+ (-r \sin\theta)(r^2 \sin\theta)(\cos\theta \cos\phi)$$

$$\overline{A}_3 \ = \ \frac{\partial x^1}{\partial \overline{x}^3} A_1 + \frac{\partial x^2}{\partial \overline{x}^3} A_2 + \frac{\partial x^3}{\partial \overline{x}^3} A_3$$

$$= (-\overline{x}^1 \sin \overline{x}^2 \sin \overline{x}^3)(x^1 x^2) + (\overline{x}^1 \sin \overline{x}^2 \cos \overline{x}^3)[2x^2 - (x^3)^2] + 0$$

$$= (-r \sin \theta \sin \phi)(r^2 \sin^2 \theta \sin \phi \cos \phi) + (r \sin \theta \cos \phi)(2r \sin \theta \sin \phi - r^2 \cos^2 \phi)$$

Example 3. *Show that in cartesian coordinate system, the contravariant and covariant components of a vector are identical.*

Solution: Let us consider a point P whose coordinates relative to orthogonal cartesian coordinate systems S and S^1 be (x, y, z) and $(\overline{x}, \overline{y}, \overline{z})$ respectively. Let direction cosines of axes x, y, z be (l_1, m_1, n_1), (l_2, m_2, n_2) and (l_3, m_3, n_3), then we have

$$\left. \begin{array}{l} \overline{x} = l_1 x + m_1 y + n_1 z \\ \overline{y} = l_2 x + m_2 y + n_2 z \\ \overline{z} = l_3 x + m_3 y + n_3 z \end{array} \right] \tag{1}$$

From Eq. (1), one obtains

$$\left. \begin{array}{l} x = l_1 \overline{x} + l_2 \overline{y} + l_3 \overline{z} \\ y = m_1 \overline{x} + m_2 \overline{y} + m_3 \overline{z} \\ z = n_1 \overline{x} + n_2 \overline{y} + n_3 \overline{z} \end{array} \right] \tag{2}$$

Writing $x^1 = x$, $x^2 = y$ and $x^3 = z$, one can express Eqs (1) and (2) as follows.

$$\left. \begin{array}{l} \overline{x}^1 = l_1 x^1 + m_1 x^2 + n_1 x^3 \\ \overline{x}^2 = l_2 x^1 + m_2 x^2 + n_2 x^3 \\ \overline{x}^3 = l_3 x^1 + m_3 x^2 + n_3 x^3 \end{array} \right] \tag{3}$$

and

$$x^1 = l_1 \overline{x}^1 + l_2 \overline{x}^2 + l_3 \overline{x}^3 \tag{4}$$

$$x^2 = m_1 \overline{x}^1 + m_2 \overline{x}^2 + m_3 \overline{x}^3$$

$$x^3 = n_1 \overline{x}^1 + n_2 \overline{x}^2 + n_3 \overline{x}^3$$

Using the transformation law for contravariant vector A^μ, i.e.

$$\overline{A}^\mu \ = \ \frac{\partial \overline{x}^\mu}{\partial x^\alpha} A^\alpha = \frac{\partial \overline{x}^\mu}{\partial x^1} A^1 + \frac{\partial \overline{x}^\mu}{\partial x^2} A^2 + \frac{\partial \overline{x}^\mu}{\partial x^3} A^3, \text{ one can write}$$

$$\left. \begin{array}{l} \overline{A}^1 = \dfrac{\partial \overline{x}^1}{\partial x^1} A^1 + \dfrac{\partial \overline{x}^1}{\partial x^2} A^2 + \dfrac{\partial \overline{x}^1}{\partial x^3} A^3 \\[2mm] \overline{A}^2 = \dfrac{\partial \overline{x}^2}{\partial x^1} A^1 + \dfrac{\partial \overline{x}^2}{\partial x^2} A^2 + \dfrac{\partial \overline{x}^2}{\partial x^3} A^3 \\[2mm] \overline{A}^3 = \dfrac{\partial \overline{x}^3}{\partial x^1} A^1 + \dfrac{\partial \overline{x}^3}{\partial x^2} A^2 + \dfrac{\partial \overline{x}^3}{\partial x^3} A^3 \end{array} \right] \tag{5}$$

From Eq. (3), one can write Eq. (5) as

$$\overline{A}^1 = l_1 A^1 + m_1 A^2 + n_1 A^3$$
$$\overline{A}^2 = l_2 A^1 + m_2 A^2 + n_2 A^3$$
$$\overline{A}^3 = l_3 A^1 + m_3 A^2 + n_3 A^3$$

(6)

The transformation equation for covariant vector A_μ is

$$\overline{A}_\mu = \frac{\partial x^\alpha}{\partial \overline{x}^\mu} A_\alpha = \frac{\partial x^1}{\partial \overline{x}^\mu} A^1 + \frac{\partial x^2}{\partial \overline{x}^\mu} A^2 + \frac{\partial x^3}{\partial \overline{x}^\mu} A_3$$

one can write

$$\overline{A}_1 = \frac{\partial x^1}{\partial \overline{x}^1} A_1 + \frac{\partial x^2}{\partial \overline{x}^1} A_2 + \frac{\partial x^3}{\partial \overline{x}^3} A_3$$
$$\overline{A}_2 = \frac{\partial x^1}{\partial \overline{x}^2} A_1 + \frac{\partial x^2}{\partial \overline{x}^2} A_2 + \frac{\partial x^3}{\partial \overline{x}^3} A_3$$
$$\overline{A}_3 = \frac{\partial x^1}{\partial \overline{x}^3} A_1 + \frac{\partial x^2}{\partial \overline{x}^3} A_2 + \frac{\partial x^3}{\partial \overline{x}^3} A_3$$

(7)

With the help of Eq. (4), one can express Eq. (7) as

$$\overline{A}_1 = l_1 A_1 + m_1 A_2 + n_1 A_3$$
$$\overline{A}_2 = l_2 A_1 + m_2 A_2 + n_2 A_3$$
$$\overline{A}_3 = l_3 A_1 + m_3 A_2 + n_3 A_3$$

(8)

Equations (6) and (8) show that there is no distinction between contravariant and covariant vectors in contravariant coordinate system.

Example 4. *If A^μ and B_μ are any two vectors, one contravariant, and the other covariant then show that sum $A^\mu B_\mu$ is invariant.*

Solution:

$$\because \qquad A^\mu = \frac{\partial \overline{x}^\mu}{\partial x^\alpha} A^\alpha$$

and

$$B_\mu = \frac{\partial x^\beta}{\partial \overline{x}^\mu} B_\beta$$

We have

$$A^\mu B_\mu = \frac{\partial \overline{x}^\mu}{\partial x^\alpha} \frac{\partial x^\beta}{\partial \overline{x}^\mu} A^\alpha B_\beta = \frac{\partial x^\beta}{\partial x^\alpha} A^\alpha B_\beta$$

$$= \delta_\alpha^\beta A^\alpha B_\beta$$

Obviously

$$A^\mu B_\mu = A^\mu B_\mu$$

Example 5. *If A^μ and B_ν are the components of a contravariant and covariant tensors of rank one, show that $C_\nu^\mu = A^\mu B_\nu$ are the components of a mixed tensor of rank two.*

Solution: We have the tensor transformation law

$$A^\mu = \frac{\partial \overline{x}^\mu}{\partial x^\alpha} A^\alpha$$

$$B_\nu = \frac{\partial x^\beta}{\partial \bar{x}^\nu} B_\beta$$

$$\therefore \qquad G_\nu^\mu = \bar{A}^\mu \bar{B}_\nu = \frac{\partial \bar{x}^\mu}{\partial x^\alpha} A^\alpha \frac{\partial x^\beta}{\partial \bar{x}^\nu} B_\beta$$

$$= \frac{\partial \bar{x}^\mu}{\partial x^\alpha} \frac{\partial x^\beta}{\partial \bar{x}^\nu} A^\alpha B_\beta$$

$$= \frac{\partial \bar{x}^\mu}{\partial x^\alpha} \frac{\partial x^\beta}{\partial \bar{x}^\nu} C_\beta^\alpha$$

Obviously, the above is a transformation equation for a mixed tensor of rank two.

Example 6. *If A^μ is an arbitrary contravariant vector and $C_{\mu\nu} A^\mu A^\nu$ is an invariant, then show that $(C_{\mu\nu} + C_{\nu\mu})$ is a covariant tensor of second order.*

Solution: Given that $C_{\mu\nu} A_\mu A^\nu$ is an invariant. This means

$$\bar{C}_{\mu\nu} \bar{A}^\mu \bar{A}^\nu = C_{\mu\nu} A^\mu A^\nu \tag{1}$$

Using tensor transformation law, one obtains

$$\bar{C}_{\mu\nu} \bar{A}^\mu \bar{A}^\nu = \bar{C}_{\mu\nu} \frac{\partial \bar{x}^\mu}{\partial x^\alpha} A^\alpha \frac{\partial \bar{x}^\nu}{\partial x^\beta} A^\beta$$

$$= \bar{C}_{\mu\nu} \frac{\partial \bar{x}^\mu}{\partial x^\alpha} \frac{\partial \bar{x}^\nu}{\partial x^\beta} A^\alpha A^\beta \tag{2}$$

Interchanging the dummy indices μ and ν, one obtains

$$\bar{C}_{\nu\mu} \bar{A}^\nu \bar{A}^\mu = \bar{C}_{\nu\mu} \frac{\partial \bar{x}^\nu}{\partial x^\alpha} \frac{\partial \bar{x}^\mu}{\partial x^\beta} A^\alpha A^\beta$$

Now, interchanging the dummy indices α and β, one obtains

$$\bar{C}_{\nu\mu} \bar{A}^\mu \bar{A}^\nu = \bar{C}_{\nu\mu} \frac{\partial \bar{x}^\mu}{\partial x^\alpha} \frac{\partial \bar{x}^\nu}{\partial x^\beta} A^\alpha A^\beta \tag{3}$$

On adding Eqs (2) and (3), one obtains

$$(\bar{C}_{\mu\nu} + \bar{C}_{\nu\mu}) \bar{A}^\mu \bar{A}^\nu = (\bar{C}_{\mu\nu} + \bar{C}_{\nu\mu}) \frac{\partial \bar{x}^\mu}{\partial x^\alpha} \frac{\partial \bar{x}^\nu}{\partial x^\beta} A^\alpha A^\beta$$

From Eq. (1), we have

$$\bar{C}_{\mu\nu} \bar{A}^\nu \bar{A}^\mu = C_{\nu\mu} A^\mu A^\nu = C_{\alpha\beta} A^\alpha A^\beta \tag{4}$$

and $\qquad \bar{C}_{\mu\nu} \bar{A}^\nu \bar{A}^\mu = C_{\nu\mu} A^\nu A^\mu = C_{\beta\alpha} A^\beta A^\alpha \tag{5}$

Equations (5) and (4) can be expressed as

$$(C_{\alpha\beta} + C_{\beta\alpha}) A^\alpha A^\beta = (\bar{C}_{\mu\nu} + \bar{C}_{\nu\mu}) \frac{\partial \bar{x}^\mu}{\partial x^\alpha} \frac{\partial \bar{x}^\nu}{\partial x^\beta} A^\alpha A^\beta$$

or
$$\left[(\bar{C}_{\mu\nu}+\bar{C}_{\nu\mu})\frac{\partial\bar{x}^{\mu}}{\partial x^{\alpha}}\frac{\partial\bar{x}^{\nu}}{\partial x^{\beta}}-(C_{\alpha\beta}+C_{\beta\alpha})\right]A^{\alpha}A^{\beta}=0$$

Since A^{α} and A^{β} are arbitrary vectors hence the expression within brackets must vanish.

$$(\bar{C}_{\mu\nu}+\bar{C}_{\nu\mu})\frac{\partial\bar{x}^{\mu}}{\partial x^{\alpha}}\frac{\partial\bar{x}^{\nu}}{\partial x^{\beta}}-(C_{\alpha\beta}+C_{\beta\alpha})=0$$

Multiplying above with $\dfrac{\partial x^{\alpha}}{\partial\bar{x}^{\mu}}\dfrac{\partial x^{\beta}}{\partial\bar{x}^{\nu}}$, one obtains

$$(\bar{C}_{\mu\nu}+\bar{C}_{\nu\mu})\frac{\partial\bar{x}^{\mu}}{\partial x^{\alpha}}\frac{\partial x^{\alpha}}{\partial\bar{x}^{\mu}}\frac{\partial\bar{x}^{\nu}}{\partial x^{\beta}}\frac{\partial x^{\beta}}{\partial\bar{x}^{\nu}}-(C_{\alpha\beta}+C_{\beta\alpha})\frac{\partial x^{\alpha}}{\partial\bar{x}^{\mu}}\frac{\partial x^{\beta}}{\partial\bar{x}^{\nu}}=0$$

or
$$(\bar{C}_{\mu\nu}+\bar{C}_{\nu\mu})\delta^{\alpha}_{\mu}\delta^{\nu}_{\beta}=(C_{\alpha\beta}+C_{\beta\alpha})\frac{\partial x^{\alpha}}{\partial\bar{x}^{\mu}}\frac{\partial x^{\beta}}{\partial\bar{x}^{\nu}}$$

or
$$(\bar{C}_{\mu\nu}+\bar{C}_{\nu\mu})=\frac{\partial x^{\alpha}}{\partial\bar{x}^{\mu}}\frac{\partial x^{\beta}}{\partial\bar{x}^{\nu}}(C_{\alpha\beta}+C_{\beta\alpha})$$

Obviously, the above is transformation law for the covariant tensor of second rank. This shows that $(C_{\mu\nu}+C_{\nu\mu})$ is a covariant tensor of the second rank.

Example 7. If $A=\begin{pmatrix}-xy & y^{2}\\ x^{2} & xy\end{pmatrix}$ is a tensor, then $B=\begin{pmatrix}-xy & -y^{2}\\ x^{2} & -xy\end{pmatrix}$ is not a tensor.

Solution: Given that A is a tensor. Obviously, its components must obey transformation laws. Components of tensor A are
$$A^{11}=-xy,\ A^{12}=y^{2},\ A_{21}=x^{2}\ \text{and}\ A_{22}=xy$$

Let us consider two-dimensional axes x and y of a cartesian coordinate system. Let the coordinates (x, y) are rotated counterclockwise through an angle θ. Keeping r fixed, one obtains the following relationship between components resolved in original (unprimed) frame and those resolved in new rotated (primed) frame as,

$$\left.\begin{array}{l}x'^{1}=x\cos\theta+y\sin\theta\\ y'^{1}=-x\sin\theta+y\sin\theta\end{array}\right\}\qquad(1)$$

Equation (1) yields

$$\frac{\partial x'}{\partial x}=\cos\theta,\qquad\frac{\partial x'}{\partial y}=\sin\theta$$

$$\frac{\partial y'}{\partial x}=-\sin\theta,\qquad\frac{\partial y'}{\partial y}=\cos\theta\qquad(2)$$

In the primed coordinate frame $(A^{11})'$ component must be $-x'\,y'$.

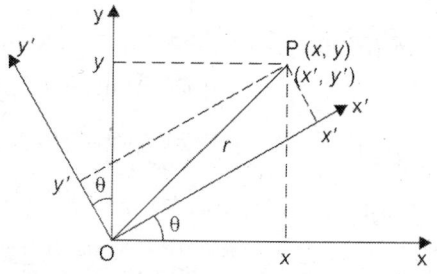

Fig. 3.4

Hence from tensor transformation law

$$(A^{11})' = x'y' = \sum_{\mu v} \frac{\partial x'_1}{\partial x^\mu} \frac{\partial x'_1}{\partial x^v} A^{\mu v} \qquad \left[\begin{array}{c} \text{Here } x_1 = x, x_2 = y \\ \mu = 1,2; v = 1,2 \end{array} \right]$$

$$= \frac{\partial x'}{\partial x} \frac{\partial x'}{\partial x} A^{11} + \frac{\partial x'}{\partial x} \frac{\partial x'}{\partial y} A^{12} + \frac{\partial x'}{\partial y} \frac{\partial x'}{\partial x} A^{21} + \frac{\partial x'}{\partial y} \frac{\partial x'}{\partial y} A^{22}$$

$$= \cos^2 \theta A^{11} + \cos\theta \sin\theta A^{12} + \sin\theta \cos\theta A^{21} + \sin^2 \theta A^{22}$$

Expressing the L.H.S of the above in the unprimed coordinate system from Eq. (1) and substituting the values of components $A_{11}, A_{12}, A_{21}, A_{22}$ is R.H.S one obtains

$$- (x \cos \theta + y \sin \theta)(-x \sin \theta + y \cos \theta)$$

$$= -\cos^2 \theta xy + \cos\theta \sin\theta y^2 + \sin\theta \cos\theta x^2 + \sin^2 \theta xy \quad [\text{from Eq. (2)}]$$

or

$$x^2 \sin\theta \cos\theta - xy \cos^2 \theta + xy \sin^2 \theta + y^2 \sin\theta \cos\theta$$

$$= -\cos^2 \theta xy + \cos\theta \sin\theta y^2 + \sin\theta \cos\theta x^2 + \sin^2 \theta xy$$

The above equation is an identity. This shows that tensor transformation law is obeyed by the element $A^{11} = -xy$. Similarly, one can show that other components also transform according to tensor transformation law. Obviously, one may conclude that matrix A is a tensor. Now, we apply the tensor transformation law to the matrix

$$B = \begin{pmatrix} -xy & -y^2 \\ x^2 & -xy \end{pmatrix}$$

It has components $B^{11} = -xy, B^{12} = -y^2,$
$$B^{21} = x^2 \text{ and } B^{12} = -xy,$$

We must remember that in primed coordinate system the components

$$(B^{11})' = -x'y',$$

or

$$= \sum_{\mu v} \frac{\partial' x_1}{\partial x^\mu} \frac{\partial' x_1}{\partial x^v} B^{\mu v} \quad (\text{Here } x_1 = x, \; x_2 = y, \; \mu, \; v = 1, 2)$$

$$= \frac{\partial x'}{\partial x} \frac{\partial x'}{\partial x} B^{11} + \frac{\partial x'}{\partial x} \frac{\partial x'}{\partial y} B^{12} + \frac{\partial x'}{\partial y} \frac{\partial x'}{\partial x} B^{21} + \frac{\partial x'}{\partial y} \frac{\partial x'}{\partial y} B^{22}$$

$$= \cos^2 \theta B^{11} + \cos\theta \sin\theta B^{12} + \sin\theta \cos\theta B^{21} + \sin^2 \theta B^{22}$$

Expressing the L.H.S in the unprimed coordinates form Eq. (1) and putting the values of B^{11}, B^{12}, B^{21}, and B^{22} in R.H.S, one obtains

$$- (x \cos \theta + y \sin \theta)(-x \sin \theta + y \cos \theta)$$

$$= \cos^2 \theta \, xy + \cos\theta \sin\theta (-y^2) + \sin\theta \cos\theta x^2 \sin^2 \theta(-xy)$$

$$= -xy \cos^2 \theta - y^2 \sin\theta \cos\theta + x^2 \sin\theta \cos\theta - xy \sin^2 \theta$$

$$= -xy(\sin^2 \theta + \cos^2 \theta) + \sin\theta \cos\theta(x^2 + y^2)$$

$$= -xy + \sin\theta \cos\theta(x^2 - y^2)$$

We note that the above equation is not satisfied because R.H.S and L.H.S are distinct. Obviously, the components of B do not satisfy the tensor transformation laws. Thus matrix represented by B is not tensor.

Example 8. Show the $A^\mu_\nu B^\mu C^\nu$ is an invariant if B^μ and C^ν are contravariant vectors and $A_{\mu\nu}$ is a covariant tensor.

Solution: Using the following transformation laws of tensors, one obtains

$$\bar{A}_{\mu\nu} = \frac{\partial x^\alpha}{\partial \bar{x}^\mu} \frac{\partial x^\beta}{\partial \bar{x}^\nu} A_{\alpha\beta}$$

$$\bar{B}^\mu = \frac{\partial \bar{x}^\mu}{\partial x^\alpha} B^\alpha$$

$$\bar{C}^\nu = \frac{\partial \bar{x}^\nu}{\partial x^\beta} C^\beta$$

$$\bar{A}_{\mu\nu} \bar{B}^\mu \bar{C}^\nu = \frac{\partial x^\alpha}{\partial \bar{x}^\mu} \frac{\partial x^\beta}{\partial \bar{x}^\nu} A_{\alpha\beta} \frac{\partial \bar{x}^\mu}{\partial x^\alpha} B^\alpha \frac{\partial \bar{x}^\nu}{\partial x^\beta} C^\beta$$

$$= \frac{\partial x^\alpha}{\partial \bar{x}^\mu} \frac{\partial \bar{x}^\mu}{\partial x^\alpha} \frac{\partial x^\beta}{\partial \bar{x}^\nu} \frac{\partial x^\nu}{\partial x^\beta} A_{\alpha\beta} B^\alpha C^\beta$$

$$= \delta^\alpha_\mu \delta^\nu_\beta A_{\alpha\beta} B_\alpha C^\beta$$

$$\therefore \qquad \bar{A}_{\mu\nu} \bar{B}^\mu \bar{C}^\nu = A_{\mu\nu} B^\mu C^\nu$$

This shows that $A_{\mu\nu} B^\nu C^\nu$ is an invariant tensor.

Example 9. Show that any tensor of rank 2 can be expressed as a sum of a symmetric and an antisymmetric tensor, both of rank 2.

Solution: One can express any tensor $A^{\mu\nu}$ of rank 2 as

$$A^{\mu\nu} = \frac{1}{2}(A^{\mu\nu} + A^{\nu\mu}) + \frac{1}{2}(A^{\mu\nu} - A^{\nu\mu})$$

$$= B^{\mu\nu} + C^{\mu\nu}$$

where

$$B^{\mu\nu} = \left(\frac{1}{2} A^{\mu\nu} + A^{\nu\mu}\right) \text{ and } C^{\mu\nu} = \frac{1}{2}(A^{\mu\nu} - A^{\nu\mu})$$

Addition and subtraction laws of tensors show that $B^{\mu\nu}$ and $C^{\mu\nu}$ are tensors of rank 2. Now, interchanging indices in $B^{\mu\nu}$ and $C^{\mu\nu}$, one obtains

$$B^{\nu\mu} = \frac{1}{2}(A^{\nu\mu} + A^{\mu\nu}) = \frac{1}{2}(A^{\mu\nu} + A^{\nu\mu}) = B^{\mu\nu}$$

$$C^{\nu\mu} = \frac{1}{2}(A^{\nu\mu} - A^{\mu\nu}) = -\frac{1}{2}(A^{\mu\nu} - A^{\nu\mu}) = -C^{\mu\nu}$$

This shows that $B^{\mu\nu}$ is symmetric, whereas $C^{\mu\nu}$, is antisymmetric and both are tensors of rank 2.

Example 10. *If A_{ij} is an antisymmetric tensor of second order and u^i is a tensor of rank are, then show that $A_{ij} u^i u^j = 0$ (summation over repeated indices is assumed).*

Solution: Interchanging dummy indices i and j, in $A_{ij} u^i u^j$, one obtains

$$A_{ij} u^i u^j = A_{ji} u^j u^i \tag{1}$$

Since A_{ij} is an antisymmetric tensor, we have

$$A_{ij} = -A_{ji}$$

or
$$A_{ij} u^i u^j = -A_{ji} u^i u^j \tag{2}$$

On adding Eqs (1) and (2), one obtains

$$A_{ij} u^i u^j = 0$$

Example 11. *Show that an antisymmetric tensor of the second order can be associated with a vector in three-dimensions. Obtain the corresponding result in four dimensions.*

Solution. The number of independent components of an antisymmetric tensor of rank r in n-dimensional space is given by

$$^nC_r = \frac{n!}{r!(n-r)!} \tag{1}$$

∴ The number of independent components of an antisymmetric tensor of rank 2 in three-dimensional space is

$$^3C_2 = \frac{3!}{2!1!} = 3$$

Since the independent components of a vector in 3-dimensional space are 3 in number and hence one may conclude that an anti-symmetric tensor of second order can be associated with a vector in 3-dimensions.

Now, the number of independent components of an anti-symmetric tensor of second order 4-dimensional space

$$^4C_2 = \frac{4!}{2!2!} = 6$$

However, the number of independent components of a vector in 4-dimensional space is 4. This shows that an antisymmetric tensor of second order cannot be associated with a vector in 4-dimensional space. However, a 4-dimensional antisymmetric tensor of second order can be associated with two 3-dimensional vectors.

Example 12. *The components of a tensor are zero in one coordinate system. Prove that its components in all coordinate systems will be zero.*

Solution. Let us consider the components of a tensor of weight W in n-dimensional space in the form $S^{\mu_1\mu_2\cdots}_{\nu_1\nu_2\cdots}$ where the indices $\mu_1, \mu_2 \dots \nu_1, \nu_2 \dots$ run through the integer $1, 2, \dots, n$.

Transformation law of tensors gives

$$\bar{S}^{\mu_1\mu_2\cdots}_{\nu_1\nu_2\cdots} = \left|\frac{\partial x}{\partial \bar{x}}\right|^W \left(\frac{\partial \bar{x}^{\mu_1}\partial \bar{x}^{\mu_2}}{\partial x^{\sigma_1}\partial x^{\sigma_2}}\cdots\right)\left(\frac{\partial x^{\beta_1}\partial x^{\beta_2}}{\partial \bar{x}^{\nu_1}\partial \bar{x}^{\nu_2}}\right) S^{\alpha_1\alpha_2\cdots}_{\beta_1\beta_2\cdots}$$

Here $S^{\alpha_1\alpha_2\cdots}_{\beta_1\beta_2\cdots}$ are the components of tensor S in system of variables x^μ whereas $\overline{S}^{\mu_1\mu_2\cdots}_{\nu_1\nu_2\cdots}$ as those in system of variable $\overline{x}^\mu$. From the above equation, it is clear that if $S^{\alpha_1\alpha_2\cdots}_{\beta_1\beta_2\cdots}$ are zero, then $\overline{S}^{\mu_1\mu_2\cdots}_{\nu_1\nu_2\cdots}$ are also zero.

Obviously, if the components of a tensor are zero in one coordinate system, they are also zero in all other coordinate system.

Example 13. *Prove that*

   i. $dg_{\alpha\beta} = -g_{\mu\alpha}\, g_{\nu\beta}\, dg^{\mu\nu}$

   ii. $A^{\alpha\beta}dg_{\alpha\beta} = -A_{\alpha\beta}\, dg^{\alpha\beta}$

Solution. i. Since $g_{\mu\alpha}\, g^{\mu\nu} = \delta^\nu_\alpha$

Taking differentials of both the sides, one obtains

$$dg_{\mu\alpha}\, g^{\mu\nu} + g_{\mu\alpha}\, dg^{\mu\nu} = 0 \qquad\qquad (\because \delta^\nu_\alpha \text{ is a constant})$$

$$g^{\mu\nu}dg_{\mu\alpha} = -g_{\mu\alpha}\, dg^{\mu\nu}$$

Multiplying the above by $g_{\nu\beta}$, one obtains

$$g_{\nu\beta}\, g^{\mu\nu}dg_{\mu\alpha} = -g_{\nu\beta}\, g_{\mu\alpha}\, dg^{\mu\nu}$$

or $\qquad\qquad \delta^\mu_\beta dg_{\mu\alpha} = -g_{\mu\alpha}\, g_{\nu\beta}\, dg^{\mu\nu}$

or $\qquad\qquad dg_{\beta\alpha} = -g_{\mu\alpha}\, g_{\nu\beta}\, dg^{\mu\nu}$

or $\qquad\qquad dg_{\alpha\beta} = -g_{\mu\alpha}\, g_{\nu\beta}\, dg^{\mu\nu} \qquad\qquad (1)$

$$(\because g_{\alpha\beta} \text{ is symmetric})$$

ii. Multiplying Eq. (1) by $A^{\alpha\beta}$, one obtains

$$A^{\alpha\beta}dg_{\alpha\beta} = -g_{\mu\alpha}\, g_{\nu\beta}\, A^{\alpha\beta}dg^{\mu\nu}$$

$$= -A_{\mu\nu}\, dg^{\mu\nu} \qquad \text{(by lowering both indices } \alpha \text{ and } \beta\text{)}$$

$$= -A_{\alpha\beta}\, dg^{\alpha,\beta}$$

where we have changed indices μ and ν by α and β respectively.

Example 14. *Show that $L^2 = g^{\mu\nu}\, A_\mu A_\nu$ is an invariant.*

Solution: Let A^μ and A_ν are the components of a contravariant and covariant vectors respectively and $g^{\mu\nu}$ is a fundamental contravariant tensor. Using the raising operation, one gets

$$g^{\mu\nu}\, A_\nu = A^\mu$$

$$\therefore \qquad L^2\, g^{\mu\nu}\, A_\mu A_\nu = A^\mu A_\mu$$

Using transformation law of tensors, one can write

$$\overline{L}^2 = \overline{A}^\mu \overline{A}_\mu = \frac{\partial \overline{x}^\mu}{\partial x^\alpha} A^\alpha \frac{\partial x^\beta}{\partial \overline{x}^\mu} A_\beta$$

$$= \frac{\partial x^\beta}{\partial x^\mu} \frac{\partial \overline{x}^\mu}{\partial x^\alpha} A^\alpha A_\beta = \delta^\beta_\alpha A^\alpha A_\beta = A^\alpha A_\alpha = L^2.$$

Example 15. Prove that $\dot{x}^J = \dfrac{1}{m} g^{Jn}\left(\dfrac{\partial T}{\partial \dot{x}^n}\right)$, where g_{mn} being the metric tensor. Given that

$T = \dfrac{1}{2} M g_{mn}\, \dot{x}^m\, \dot{x}^n.$

Solution: Given $T = \dfrac{1}{2} M g_{mn}\, \dot{x}^m\, \dot{x}^n$ \hfill (1)

Differentiating Eq. (1) partially w.r.t. $\dot{x}^r$, one obtains

$$\frac{\partial T}{\partial \dot{x}^r} = \frac{1}{2} M g_{mn}\left(\dot{x}^m \frac{\partial \dot{x}^n}{\partial \dot{x}^r} + \dot{x}^n \frac{\partial \dot{x}^m}{\partial \dot{x}^r}\right)$$

$(g_{mn}$ is independent of velocity components $\dot{x}^r)$

$$= \frac{1}{2} M g_{mn}\, \dot{x}^m \frac{\partial \dot{x}^n}{\partial \dot{x}^r} + \frac{1}{2} M g_{mn}\, \dot{x}^n \frac{\partial \dot{x}^m}{\partial \dot{x}^r}$$

$$= \frac{1}{2} M g_{mn}\, \dot{x}^m \delta^n_r + \frac{1}{2} M g_{mn}\, \dot{x}^n \delta^m_r$$

Replacing indices m and n by l and knowing that g_{mn} is symmetric, one obtains

$$\frac{\partial T}{\partial \dot{x}^r} = M g_{lr}\, \dot{x}^l \tag{2}$$

Multiplying Eq. (2) by g^{rj}, one obtains

$$g^{rj} \frac{\partial T}{\partial \dot{x}^r} = M g^{rj} g_{lr}\, \dot{x}^l$$

$$= M \delta^j_l\, \dot{x}^l \qquad \left(\because g^{rj} g_{lr} = \delta^j_l\right)$$

$$= M \dot{x}^J$$

Thus $\qquad \dot{x}^J = \dfrac{1}{M} g^{rj} \dfrac{\partial T}{\partial \dot{x}^r} = \dfrac{1}{M} g^{nj} \dfrac{\partial T}{\partial \dot{x}^n}$

Example 16. Obtain the transformation $dS^2 = dx^2 + dy^2 + dz^2$ in spherical and cylindrical coordinates.

Solution: Given $\qquad dS^2 = dx^2 + dy^2 + dz^2$ \hfill (1)

In general, we have $\qquad dS^2 = g_{\mu\nu}\, dx^\mu\, dx^\nu$ \hfill (2)

$(\mu, \nu = 1, 2, 3)$

From Eqs (1) and (2), one obtains

$$g_{11} = g_{22} = g_{33} = 1 \text{ and } G_{\mu\nu} = 0 \text{ for } \mu \neq \nu$$

Since $g_{\mu\nu}$ is a covariant tensor of rank 2, and hence according to tensor transformation law

$$\bar{g}_{\mu\nu} = \frac{\partial x^\alpha}{\partial \bar{x}^\mu} \frac{\partial x^\beta}{\partial \bar{x}^\nu} g_{\alpha\beta}$$

In the given problem, we have

$$\bar{g}_{\mu\nu} = \sum_{\alpha=1}^{3} \frac{\partial x^\alpha}{\partial \bar{x}^\mu} \frac{\partial x^\alpha}{\partial \bar{x}^\nu} g_{\alpha\alpha} \qquad\qquad (\because g_{\mu\nu} = 0 \text{ for } \mu \neq \nu)$$

$$= \sum_{\alpha=1}^{3} \frac{\partial x^\alpha}{\partial \bar{x}^\mu} \frac{\partial x^\alpha}{\partial \bar{x}^\nu} \qquad\qquad (3)$$

$$(\because g_{\alpha\alpha} = 1 \text{ as there is no summation involved})$$

(a) Transformation into spherical coordinates (r, θ, ϕ)

Cartesian to spherical coordinates transformation equations can be expressed as

$$\left. \begin{array}{l} x = r\sin\theta\cos\phi \\ y = r\sin\theta\sin\phi \\ z = r\cos\phi \end{array} \right\} \qquad\qquad (4)$$

Let

$$\left. \begin{array}{l} x^1 = x, x^2 = y, x^3 = z \\ \bar{x}^1 = r, \bar{x}^2 = \theta, \bar{x}^3 = \phi \end{array} \right\} \qquad\qquad (5)$$

From Eq. (3), we have

$$\bar{g}_{11} = \frac{\partial x^1}{\partial \bar{x}^1} \frac{\partial x^1}{\partial \bar{x}^1} + \frac{\partial x^2}{\partial \bar{x}^1} \frac{\partial x^2}{\partial \bar{x}^1} + \frac{\partial x^3}{\partial \bar{x}^1} \frac{\partial x^3}{\partial \bar{x}^1}$$

Using Eq. (5), one obtains

$$\bar{g}_{11} = \left(\frac{\partial x}{\partial r}\right)^2 + \left(\frac{\partial y}{\partial r}\right)^2 + \left(\frac{\partial z}{\partial r}\right)^2$$

Using Eq. (4), one obtains

$$\bar{g}_{11} = (\sin\theta\cos\phi)^2 + (\sin\theta\sin\phi)^2 + (\cos\theta)^2 = 1$$

From Eq. (3), we have

$$\bar{g}_{22} = \frac{\partial x^1}{\partial \bar{x}^2} \frac{\partial x^1}{\partial \bar{x}^2} + \frac{\partial x^2}{\partial \bar{x}^2} \frac{\partial x^2}{\partial \bar{x}^2} + \frac{\partial x^3}{\partial \bar{x}^2} \frac{\partial x^3}{\partial \bar{x}^2}$$

Using Eq. (5), one obtains

$$\bar{g}_{22} = \left(\frac{\partial x}{\partial \theta}\right)^2 + \left(\frac{\partial y}{\partial \theta}\right)^2 + \left(\frac{\partial z}{\partial \theta}\right)^2$$

or

$$\bar{g}_{22} = (r\cos\theta\cos\phi)^2 + (r\cos\theta\sin\phi)^2 + (-r\sin\theta)^2 = r^2$$

From Eq. (3), we have

$$\bar{g}_{33} = \frac{\partial x^1}{\partial \bar{x}^3} \frac{\partial x^1}{\partial \bar{x}^3} + \frac{\partial x^2}{\partial \bar{x}^3} \frac{\partial x^2}{\partial \bar{x}^3} + \frac{\partial x^3}{\partial \bar{x}^3} \frac{\partial x^3}{\partial \bar{x}^3}$$

Using Eq. (5), one obtains

$$\bar{g}_{33} = \left(\frac{\partial x}{\partial \phi}\right)^2 + \left(\frac{\partial y}{\partial \phi}\right)^2 + \left(\frac{\partial z}{\partial \phi}\right)^2$$

Using Eq. (4), one obtains

$$\bar{g}_{33} = (-r\sin\theta\sin\phi)^2 + (r\sin\theta\cos\phi)^2 + 0 = r^2 \sin^2\theta.$$

From Eq. (3), we have

$$\bar{g}_{12} = \frac{\partial x^1}{\partial \bar{x}^1}\frac{\partial x^1}{\partial \bar{x}^2} + \frac{\partial x^2}{\partial \bar{x}^1}\frac{\partial x^2}{\partial \bar{x}^2} + \frac{\partial x^3}{\partial \bar{x}^1}\frac{\partial x^3}{\partial \bar{x}^2}$$

$$= \left(\frac{\partial x}{\partial r}\right)\left(\frac{\partial x}{\partial \theta}\right) + \frac{\partial y}{\partial r}\frac{\partial y}{\partial \theta} + \frac{\partial z}{\partial r}\frac{\partial z}{\partial \theta}$$

$$= (\sin\theta\,\cos\phi)(r\cos\theta\cos\phi) + (\sin\theta\sin\phi)$$
$$+ (r\cos\theta\sin\phi) + (\cos\theta)(-r\sin\theta)$$

$$= r\sin\theta\cos\theta(\cos^2\phi + \sin^2\phi - 1) = 0$$

or $\quad\quad \bar{g}_{12} = 0.$

Similarly, one obtains

$$\bar{g}_{13} = \bar{g}_{31} = 0 \quad\text{and}\quad \bar{g}_{23} = \bar{g}_{32} = 0$$

or $\quad\quad \bar{g}_{\mu\nu} = 0 \;\text{ for }\; \mu \neq \nu$

$\therefore\quad\quad dS^2 = \bar{g}_{\mu\nu}\,d\bar{x}^\mu d\bar{x}^\nu$

$$= \sum_{\mu=1}^{n}\bar{g}_{\mu\mu}d\bar{x}^\mu d\bar{x}^\mu \quad\quad\quad (\text{since } \bar{g}_{\mu\nu}=0 \text{ for } \mu \neq \nu)$$

$$= \bar{g}_{11}(d\bar{x}^1)^2 + \bar{g}_{22}(d\bar{x}^2)^2 + \bar{g}_{33}(d\bar{x}^3)^2$$

$$= dr^2 + r^2 d\theta^2 + r^2\sin^2\theta\, d\phi^2 \quad\quad\quad\quad (6)$$

Equation (6) gives the desired transformation in spherical coordinates.

(b) Transformation in cylindrical coordinates

Cartesian to cylindrical coordinates transformation equations can be expressed as

$$x = r\cos\theta, \; y = r\sin\theta \;\text{ and }\; z = z \quad\quad\quad (7)$$

Let
$$\left.\begin{array}{l} x^1 = x, x^2 = y, x^3 = z \;\text{ and} \\ \bar{x}^1 = r, \bar{x}^2 = \theta, \;\text{ and }\; \bar{x}^3 = z \end{array}\right\} \quad\quad (8)$$

From Eq. (3), one obtains

$$\bar{g}_{11} = \frac{\partial x^1}{\partial \bar{x}^1}\frac{\partial x^1}{\partial \bar{x}^1} + \frac{\partial x^2}{\partial \bar{x}^1}\frac{\partial x^2}{\partial \bar{x}^1} + \frac{\partial x^3}{\partial \bar{x}^1}\frac{\partial x^3}{\partial \bar{x}^1}$$

Using Eq. (8), one obtains

$$\bar{g}_{11} = \left(\frac{\partial x}{\partial r}\right)^2 + \left(\frac{\partial y}{\partial r}\right)^2 + \left(\frac{\partial z}{\partial r}\right)^2$$

Using Eq. (7), one obtains

$$\bar{g}_{11} = (\cos\theta)^2 + (\sin\theta)^2 + 0 = 1.$$

Also from Eq. (3), one obtains

$$\bar{g}_{22} = \frac{\partial x^1}{\partial \bar{x}^2}\frac{\partial x^1}{\partial \bar{x}^2} + \frac{\partial x^2}{\partial \bar{x}^2}\frac{\partial x^2}{\partial \bar{x}^2} + \frac{\partial x^3}{\partial \bar{x}^2}\frac{\partial x^3}{\partial \bar{x}^2}$$

Using Eq. (8), one obtains

$$\bar{g}_{22} = \left(\frac{\partial x}{\partial \theta}\right)^2 + \left(\frac{\partial y}{\partial \theta}\right)^2 + \left(\frac{\partial z}{\partial \theta}\right)^2$$

Using Eq. (7), one obtains

$$\bar{g}_{22} = (-r\sin\theta)^2 + (r\cos\theta)^2 + 0 = r^2.$$

Also from Eq. (3), one obtains

$$\bar{g}_{33} = \frac{\partial x^1}{\partial \bar{x}^3}\frac{\partial x^1}{\partial \bar{x}^3} + \frac{\partial x^2}{\partial \bar{x}^3}\frac{\partial x^2}{\partial \bar{x}^3} + \frac{\partial x^3}{\partial \bar{x}^3}\frac{\partial x^3}{\partial \bar{x}^3}$$

Using Eq. (8), one obtains

$$\bar{g}_{33} = \left(\frac{\partial x}{\partial z}\right)^2 + \left(\frac{\partial y}{\partial z}\right)^2 + \left(\frac{\partial z}{\partial z}\right)^2$$

Using Eq. (7), one obtains

$$\bar{g}_{33} = 0 + 0 + 1 = 1.$$

Again from Eq. (3), one obtains

$$\bar{g}_{12} = \frac{\partial x^1}{\partial \bar{x}^1}\frac{\partial x^1}{\partial \bar{x}^2} + \frac{\partial x^2}{\partial \bar{x}^1}\frac{\partial x^2}{\partial \bar{x}^2} + \frac{\partial x^3}{\partial \bar{x}^1}\frac{\partial x^3}{\partial \bar{x}^2}$$

$$= \left(\frac{\partial x}{\partial r}\right)\left(\frac{\partial x}{\partial \theta}\right) + \left(\frac{\partial y}{\partial r}\right)\left(\frac{\partial y}{\partial \theta}\right) + \left(\frac{\partial z}{\partial r}\right)\left(\frac{\partial z}{\partial \theta}\right)$$

$$= (\cos\theta)(-r\sin\theta) + (\sin\theta)(r\cos\theta) + 0 = 0$$

Obviously, $\bar{g}_{12} = \bar{g}_{21} = 0$

Similarly, one can easily show that

$$\bar{g}_{23} = \bar{g}_{32} = 0 \quad \text{and} \quad \bar{g}_{31} = \bar{g}_{13} = 0$$

i.e., $\bar{g}_{\mu\nu} = 0 \quad \text{for } \mu \neq \nu$

∴ $dS^2 = \bar{g}_{\mu\nu} d\bar{x}^\mu d\bar{x}^\nu$

$$= \sum_{\mu=1}^{3} \bar{g}_{\mu\nu} d\bar{x}^\mu d\bar{x}^\nu \qquad \text{(since } \bar{g}_{\mu\nu} = 0 \text{ for } \mu \neq \nu\text{)}$$

$$= \bar{g}_{11}(d\bar{x}^1)^2 + \bar{g}_{22}(d\bar{x}^2)^2 + \bar{g}_{33}(d\bar{x}^3)^2$$

$$= 1dr^2 + r^2 d\theta^2 + 1dz^2$$

$$= dr^2 + r^2 d\theta^3 + dz^2 \tag{9}$$

Example 17. Obtain expressions for metric tensor in (a) spherical coordinates and (b) cylindrical coordinates.

Solution: (a) Metric tensor in spherical coordinates

The metric tensor $g_{\mu\nu}$ can be expressed as

$$dS^2 = g_{\mu\nu} dx^\mu dx^\nu \tag{1}$$

In spherical coordinates, we have
$$dS^2 = dr^2 + r^2\, d\theta^2 + r^2 \sin^2\theta d\phi^2$$
Let $x^1 = r$, $x^2 = \theta$ and $x^3 = \phi$, then we have
$$g_{11} = 1,\ g_{22} = r^2,\ g_{33} = r^2 \sin^2\theta$$
and
$$g_{12} = g_{21} = 0,\ g_{23} = g_{32} = 0,\ g_{31} = g_{13} = 0$$
$g_{\mu\nu}$ in matrix form can be expressed as

$$g_{mn} = \begin{bmatrix} g_{11} & g_{12} & g_{13} \\ g_{21} & g_{22} & g_{23} \\ g_{31} & g_{32} & g_{33} \end{bmatrix}$$

$$= \begin{bmatrix} 1 & 0 & 0 \\ 0 & r^2 & 0 \\ 0 & 0 & r^2 \sin^2\theta \end{bmatrix}$$

(b) Matric tensor in cylindrical coordinates

We have, $dS^2 = dr^2 + r^2\,(d\theta)^2 + dz^2$ (cylindrical coordinates)
Let $x^1 = r$, $x^2 = \theta$ and $x^3 = z$, then
$$g_{11} = 1, g_{22} = r^2, g_{33} = 1$$
and
$$g_{12} = g_{21} = 0,\ g_{23} = g_{32} = 0,\ g_{31} = g_{13} = 0$$
One can express the metric tensor $g_{\mu\nu}$ in cylindrical coordinates as

$$g_{\mu\nu} = \begin{bmatrix} g_{11} & g_{12} & g_{13} \\ g_{21} & g_{22} & g_{23} \\ g_{31} & g_{32} & g_{33} \end{bmatrix}$$

$$= \begin{bmatrix} 1 & 0 & 0 \\ 0 & r^2 & 0 \\ 0 & 0 & 1 \end{bmatrix}.$$

Example 18. Show that

$$[\mu\nu, \sigma] + [\sigma\nu, \mu] = \frac{\partial g_{\sigma\mu}}{\partial x^\nu}$$

Solution: Using the definition of Christoffel's symbols of first kind, one can write

$$[\mu\nu, \sigma] = \Gamma_{\alpha,\mu\nu} = \frac{1}{2}\left(\frac{\partial g_{\sigma\mu}}{\partial x^\nu} + \frac{\partial g_{\nu\sigma}}{\partial x^\mu} - \frac{\partial g_{\mu\nu}}{\partial x^\sigma} \right) \qquad (1)$$

and

$$[\sigma\nu, \mu] = \Gamma_{\mu,\sigma\nu} = \frac{1}{2}\left(\frac{\partial g_{\mu\sigma}}{\partial x^\nu} + \frac{\partial g_{\nu\mu}}{\partial x^\sigma} - \frac{\partial g_{\sigma\nu}}{\partial g^\mu} \right) \qquad (2)$$

Adding Eqs (1) and (2) and using symmetry property of $g_{\mu\nu}$, one obtains

$$[\mu\nu, \sigma] + [\sigma\nu, \mu] = \frac{\partial g_{\sigma\mu}}{\partial x^\nu} \qquad (3)$$

Example 19. *Show that*

i. $\left\{\begin{matrix}\mu\\\mu\nu\end{matrix}\right\} = \dfrac{\partial}{\partial x^\nu}\log\sqrt{g}$

and   ii. $\dfrac{\partial g^{pq}}{\partial x^m} = g^{p\alpha}\left\{\begin{matrix}q\\\alpha m\end{matrix}\right\} + g^{q\alpha}\left\{\begin{matrix}p\\\alpha m\end{matrix}\right\}$

Solution: i. Differentiating the determinant $g = \{g_{\mu\nu}\}$ and keeping in mind that $g^{\sigma\mu} g$ is the of cofactor $g_{\sigma\mu}$. In this determinant, one obtains

$$\frac{\partial g}{\partial x^\nu} = g^{\sigma\mu} g\frac{dg_{\sigma\mu}}{dx^\nu}$$

$$= gg^{\sigma\mu}\left(\Gamma_{\sigma,\mu\nu} + \Gamma_{\mu,\sigma\nu}\right) \quad \text{(from the results of \textbf{problem 18})}$$

$$= g\left(g^{\mu\sigma}\Gamma_{\sigma,\mu\nu} + g^{\sigma\mu}\Gamma_{\mu,\sigma\nu}\right) \qquad \left(\because g^{\mu\sigma} = g^{\sigma\mu}\right)$$

$$= g\left(\Gamma^\mu_{\mu\nu} + \Gamma^\sigma_{\sigma\nu}\right) = 2g\,\Gamma^\mu_{\mu\nu}$$

$$\therefore \qquad \Gamma^\mu_{\mu\nu} = \frac{1}{2g}\frac{\partial g}{\partial x^\nu}$$

or $\qquad \left\{\begin{matrix}\mu\\\mu\nu\end{matrix}\right\} = \dfrac{\partial}{\partial x^\nu}(\log\sqrt{g})$

If g is negative, then the above will take the form

$$\left\{\begin{matrix}\mu\\\mu\nu\end{matrix}\right\} = \Gamma^\mu_{\mu\nu} = \frac{\partial}{\partial x^\nu}\log\sqrt{-g}$$

ii. We have from the properties of fundamental tensors

$$g_{\mu\nu} g^{\nu\sigma} = \delta^\sigma_\mu \tag{1}$$

Differentiating Eq. (1) w.r.t. x^m, one obtains

$$g_{\mu\nu}\frac{\partial g^{\nu\sigma}}{\partial x^m} + g^{\nu\sigma}\frac{\partial g_{\mu\nu}}{\partial x^m} = 0 \tag{2}$$

$$(\because \delta^\sigma_\mu \text{ is a constant})$$

Taking inner product of Eq. (2) with $g^{\mu\rho}$, one obtains

$$g_{\mu\nu}g^{\mu\rho}\frac{\partial g^{\nu\sigma}}{\partial x^m} + g^{\nu\sigma}g^{\mu\rho}\frac{\partial g_{\mu\nu}}{\partial x^m} = 0$$

or $\qquad \delta^\rho_\nu\dfrac{\partial g^{\nu\sigma}}{\partial x^m} + g^{\nu\sigma}g^{\mu\rho}\left(\Gamma_{\nu,\mu m} + \Gamma_{\mu,vm}\right)$

or $\qquad \dfrac{\partial g^{\rho\sigma}}{\partial x^m} + g^{\mu\rho}g^{\sigma\nu}\Gamma_{\nu,\mu m} + g^{\nu\sigma}g^{\rho\mu}\Gamma_{\mu,vm} = 0 \qquad (\because g^{\mu\nu} \text{ is a symmetric tensor})$

or $\qquad \dfrac{\partial g^{\rho\sigma}}{\partial x^m} + g^{\mu\rho}\Gamma^\sigma_{\mu m} + g^{\nu\sigma}\Gamma^\rho_{vm} = 0$

or $\qquad \dfrac{\partial g^{\rho\sigma}}{\partial x^m} = -g^{\mu\rho}\Gamma^\sigma_{\mu m} - g^{\nu\sigma}\Gamma^\rho_{vm}$

Replacing σ by q, μ and ν by α, one obtains

$$\frac{\partial g^{pq}}{\partial x^m} = -g^{\alpha p}\Gamma^q_{\alpha m} - g^{\alpha q}\Gamma^p_{\alpha m}$$

Since $g_{\mu\nu}$ is a symmetric tensor, one can express the above equation as

$$\frac{\partial g^{pq}}{\partial x^m} = -g^{pq}\left\{ \begin{matrix} q \\ \alpha m \end{matrix} \right\} - g^{q\alpha}\left\{ \begin{matrix} p \\ \alpha m \end{matrix} \right\}.$$

Example 20. *The length dS of a line element in a two-dimensional surface θ – φ is given by*

$$dS^2 = R^2\, d\theta^2 + R^2 \sin^2\theta\, d\phi^2$$

where, R is a constant. Find all components of the metric tensor $g_{\mu\nu}$ and the Christoffel's symbols of first kind for this surface.

Solution: Given $\quad\quad dS^2 = R^2\, d\theta^2 + R^2 \sin^2\theta\, d\phi^2 \quad\quad\quad$ (1)

The general line element is given by

$$dS^2 = g_{\mu\nu}\, dx^\mu\, dx^\nu \quad\quad\quad (2)$$

On comparing Eqs (1) and (2), one obtains

$$x^1 = \theta,\; x^2 = \phi$$
$$g^{11} = R^2,\; g_{22} = R^2 \sin^2\theta$$

and $\quad\quad\quad g_{12} = g_{21} = 0$

∴ The components of the metric tensor $g_{\mu\nu}$ are

$$g_{11} = R^2,\; g_{22} = R^2 \sin^2\theta,\; g_{12} = g_{21} = 0$$

The Christoffel's symbols of first kind are given by

i. $\Gamma_{\mu,\nu\sigma} = 0$ ii. $\Gamma_{\mu,\mu\mu} = \dfrac{1}{2}\dfrac{\partial g_{\mu\mu}}{\partial x^\mu}$ iii. $\Gamma_{\mu,\mu\nu} = -\Gamma_{\nu,\mu\nu} = \dfrac{1}{2}\dfrac{\partial g_{\mu\mu}}{\partial x^\nu}$

∴ The Christoffel's symbols of first kind in the this case are given by

$$\Gamma_{1,11} = \frac{1}{2}\frac{\partial g_{11}}{\partial x^1} = \frac{1}{2}\frac{\partial R^2}{\partial\theta} = 0;$$

$$\Gamma_{2,22} = \frac{1}{2}\frac{\partial g_{22}}{\partial x^2} = \frac{1}{2}\frac{\partial(R^2\sin^2\theta)}{\partial\phi} = 0;$$

$$\Gamma_{1,12} = \frac{1}{2}\frac{\partial g_{11}}{\partial x^2} = \frac{1}{2}\frac{\partial R^2}{\partial\phi} = 0;$$

$$\Gamma_{1,21} = \Gamma_{1,12} = 0$$

$$\left.\begin{array}{l} \Gamma_{1,22} = \dfrac{1}{2}\dfrac{\partial g_{22}}{\partial x^1} = -\dfrac{1}{2}\dfrac{\partial(R^2\sin^2\theta)}{\partial\theta} = -R^2\sin\theta\cos\theta \\[4mm] \Gamma_{2,21} = \dfrac{1}{2}\dfrac{\partial g_{22}}{\partial x^1} = \dfrac{1}{2}\dfrac{\partial(R^2\sin^2\theta)}{\partial\theta} = R^2\sin\theta\cos\theta \end{array}\right\} \quad (3)$$

$$\Gamma_{2,21} = \Gamma_{2,21} = R^2\sin\theta\sin\theta$$

Obviously, the nonvanishing Christoffel's symbols of first kind for the given surface are given by Eq. (3).

Example 21. *If* $g_{\mu\nu} = 0$ *for* $\mu \neq \nu$ *and if* μ, ν, σ *are unequal indices, then show that*

i. $\Gamma_{\mu,\nu\sigma} = 0$ ii. $\Gamma_{\mu,\mu\nu} = -\Gamma_{\nu,\nu\mu} = \dfrac{1}{2}\dfrac{\partial g_{\mu\mu}}{\partial x^\nu}$ iii. $\Gamma^\mu_{\nu\sigma} = 0$

iv. $\Gamma^\nu_{\mu\mu} = -\dfrac{1}{2g_{\nu\nu}}\dfrac{\partial g_{\mu\mu}}{\partial x^\nu}$ v. $\Gamma^\mu_{\mu\nu} = \dfrac{1}{2}\dfrac{\partial(\log g_{\mu\mu})}{\partial x^\nu}$ vi. $\Gamma_{\mu,\mu\mu} = \dfrac{1}{2}\dfrac{\partial g_{\mu\mu}}{\partial x^\mu}$

vii. $\Gamma^\mu_{\mu\mu} = \dfrac{1}{2}\left(\dfrac{\partial \log g_{\mu\mu}}{\partial x^\mu}\right)$

Solution: Let us consider that μ, ν and σ be unequal indices.

Given $g_{\mu\nu} = 0$ for $\mu \neq \nu$.

$\therefore$ $g_{\mu\nu} = g_{\nu\sigma} = g_{\sigma\mu} = 0$ (1)

$g_{\mu\nu} = 0$ for $\mu \neq \nu$ implies that $g^{\mu\nu} = 0$ for $\mu \neq \nu$

$\therefore$ $g^{\mu\nu} = g^{\nu\sigma} = g^{\sigma\mu} = 0$ (2)

This gives $g^{\mu\mu}g_{\mu\mu} = \delta^\mu_\mu$ (no summation) $= 1$

or $g_{\mu\mu} = \dfrac{1}{g^{\mu\mu}}$ for every value of μ (3)

i. $\Gamma_{\mu,\nu\sigma} = \dfrac{1}{2}\left(\dfrac{\partial g_{\mu\nu}}{\partial x^\sigma} + \dfrac{\partial g_{\sigma\mu}}{\partial x^{\mu\nu}} - \dfrac{\partial g_{\nu\sigma}}{\partial x^\mu}\right)$

            $= \dfrac{1}{2}(0 + 0 - 0) = 0$ (4)

ii. $\Gamma_{\mu,\mu\nu} = \dfrac{1}{2}\left(\dfrac{\partial g_{\mu\mu}}{\partial x^\nu} + \dfrac{\partial g_{\nu\mu}}{\partial x^\mu} - \dfrac{\partial g_{\mu\nu}}{\partial x^\mu}\right) = \dfrac{1}{2}\dfrac{g_{\mu\mu}}{\partial x^\nu}$ (5)

and $\Gamma_{\nu,\mu\mu} = \dfrac{1}{2}\left(\dfrac{\partial g_{\nu\mu}}{\partial x^\mu} + \dfrac{\partial g_{\mu\nu}}{\partial x^\mu} - \dfrac{\partial g_{\mu\mu}}{\partial x^\mu}\right) = -\dfrac{1}{2}\dfrac{\partial g_{\mu\mu}}{\partial x^\nu}$ (6)

From Eqs (5) and (6), one obtains

            $\Gamma_{\mu,\mu\nu} = -\Gamma_{\nu,\mu\mu} = \dfrac{1}{2}\dfrac{\partial g_{\mu\mu}}{\partial x^\nu}$ (7)

iii. $\Gamma^\mu_{\mu\sigma} = g^{\mu\lambda}\Gamma_{\lambda,\nu\sigma} = g^{\mu\mu}\Gamma_{\mu,\nu\sigma}$ $(\because g^{\mu\lambda} = 0$ for $\mu \neq \lambda)$ (8)

On using Eq. (4), one obtains

            $\Gamma^\mu_{\nu\sigma} = 0$

iv. Using Eqs (3) and (6), one can write

            $\Gamma^\mu_{\mu\mu} = g^{\nu\lambda}\Gamma_{\lambda,\mu\mu} = g^{\nu\nu} = \dfrac{1}{g^{\nu\nu}}\left(-\dfrac{1}{2}\dfrac{\partial g_{\mu\mu}}{\partial x^\nu}\right)$ (9)

v. Using Eqs (3) and (5), we have

$$\Gamma^{\mu}_{\mu\nu} = g^{\mu\mu}\Gamma_{\lambda,\mu\nu} = g^{\mu\mu}\Gamma_{\mu,\mu\nu}$$

$$= \frac{1}{g_{\mu\mu}}\left(\frac{1}{2}\frac{\partial g_{\mu\mu}}{\partial x^{\nu}}\right)$$

or $\qquad \Gamma^{\mu}_{\mu\nu} = \frac{1}{2}\frac{\partial(\log g_{\mu\mu})}{\partial x^{\nu}}$ (10)

vi. $\qquad \Gamma_{\mu,\mu\mu} = \frac{1}{2}\left(\frac{\partial g_{\mu\mu}}{\partial x^{\mu}} + \frac{\partial g_{\mu\mu}}{\partial x^{\mu}} - \frac{\partial g_{\mu\mu}}{\partial x^{\mu}}\right)$

$$= \frac{1}{2}\frac{\partial g_{\mu\mu}}{\partial x^{\mu}}$$ (11)

vii. Using Eqs (3) and (11), one obtains

$$\Gamma^{\mu}_{\mu\mu} = g^{\mu\lambda}\Gamma_{\lambda,\mu\mu} = g^{\mu\mu}\Gamma_{\mu,\mu\mu} = \frac{1}{g_{\mu\mu}}\left(\frac{1}{2}\frac{\partial g_{\mu\mu}}{\partial x^{\mu}}\right)$$

$\therefore \qquad \Gamma^{\mu}_{\mu\mu} = \frac{1}{2}\frac{\partial(\log g_{\mu\mu})}{\partial x^{\mu}}$ (12)

Example 22. *If $g_{\mu\nu}$ and $a_{\mu\nu}$ are the components of two symmetric covariant tensors and $\left(\Gamma^{\mu}_{\nu\nu}\right)_{g}$ and $\left(\Gamma^{\mu}_{\nu\sigma}\right)_{a}$ are the corresponding Christoffel's symbols of second kind, then prove that the quantities*

$$\left(\Gamma^{\mu}_{\nu\sigma}\right)_{g} - \left(\Gamma^{\mu}_{\nu\sigma}\right)_{a}$$

are the components of a mixed tensor, μ being an index of contravariance and ν, σ being the indices of covariance.

Solution: From Eq. (3.92a), one obtains

$$\frac{\partial^2 x^p}{\partial \bar{x}^{\mu}\partial \bar{x}^{\nu}} = \bar{\Gamma}^{\lambda}_{\mu\nu}\frac{\partial x^p}{\partial \bar{x}^{\lambda}} - \frac{\partial x^{\alpha}}{\partial \bar{x}^{\mu}}\frac{\partial x^{\beta}}{\partial \bar{x}^{\nu}}\Gamma^{p}_{\alpha\beta}$$

or $\qquad \bar{\Gamma}^{\lambda}_{\mu\nu} = \left[\frac{\partial^2 x^p}{\partial \bar{x}^{\mu}\partial \bar{x}^{\nu}} + \frac{\partial x^{\alpha}}{\partial \bar{x}^{\mu}}\frac{\partial x^{\beta}}{\partial \bar{x}^{\sigma}}\Gamma^{p}_{\alpha\beta}\right]\frac{\partial \bar{x}^{\lambda}}{\partial x^p}$

Using the above, one can write

$$\left(\bar{\Gamma}^{\mu}_{\nu\sigma}\right)_{g} = \left[\frac{\partial^2 x^{\alpha}}{\partial \bar{x}^{\nu}\partial \bar{x}^{\sigma}} + \frac{\partial x^{\beta}}{\partial \bar{x}^{\nu}}\frac{\partial x^{\gamma}}{\partial \bar{x}^{\sigma}}\left(\Gamma^{\alpha}_{\beta\gamma}\right)_{g}\right]\frac{\partial \bar{x}^{\mu}}{\partial x^{\alpha}}$$ (1)

and $\qquad \left(\bar{\Gamma}^{\mu}_{\nu\sigma}\right)_{a} = \left[\frac{\partial^2 x^{\alpha}}{\partial \bar{x}^{\nu}\partial \bar{x}^{\sigma}} + \frac{\partial x^{\beta}}{\partial \bar{x}^{\nu}}\frac{\partial x^{\gamma}}{\partial \bar{x}^{\sigma}}\left(\Gamma^{\alpha}_{\beta\gamma}\right)_{a}\right]\frac{\partial \bar{x}^{\mu}}{\partial x^{\alpha}}$ (2)

Subtracting Eq. (2) from Eq. (1), one obtains

$$\left[\left(\bar{\Gamma}^\mu_{v\sigma}\right)_g - \left(\bar{\Gamma}^\mu_{v\sigma}\right)_\alpha\right] = \frac{\partial \bar{x}^\mu}{\partial x^\alpha}\frac{\partial x^\beta}{\partial \bar{x}^v}\frac{\partial x^v}{\partial \bar{x}^\sigma}\left[\left(\Gamma^\alpha_{\beta\gamma}\right)_g - \left(\Gamma^\alpha_{\beta\gamma}\right)_\alpha\right] \qquad (3)$$

From Eq. (3), it is evident that $\left(\Gamma^\mu_{v\sigma}\right)_g - \left(\Gamma^\mu_{v\sigma}\right)_\alpha$ obey the transformation law of mixed tensor. Hence are the components of a mixed tensor.

Example 23. *Find the nonvanishing 3-index symbols for the metric relation*

$$dS^2 = f(x)\,dx^2 + dy^2 + dz^2 + \frac{1}{f(x)}dt^2$$

and hence obtain the differential equations of the geodesics for this space.
Solution: Given

$$dS^2 = f(x)\,dx^2 + dy^2 + dz^2 + \frac{1}{f(x)}dt^2 \qquad (1)$$

We have proved that $dS^2 = g_{\mu v}\,dx^v\,dx^v$ $\qquad (2)$
Comparing Eqs (1) and (2), we obtain

$$x^1 = x,\ x^2 = y,\ x^3 = z,\ x^4 = t$$

and $\qquad\qquad g_{11} = f(x),\ g_{22} = 1,\ g_{44} = \dfrac{1}{f(x)}$ $\qquad (3)$

This shows that $g^{\mu v} = 0$ for $\mu \ne v$

Further $\qquad\qquad g^{\mu\mu} = \delta^\mu_\mu$ $\qquad\qquad$ (no summation = 1)

i.e. $\qquad\qquad g^{\mu\mu} = \dfrac{1}{g_{\mu\mu}}$ $\qquad\qquad$ (for all values of μ)

In Example 20, we have proved that

i. $\Gamma_{\mu, v\sigma} = 0$ $\qquad$ ii. $\Gamma_{\mu,\mu\mu} = \dfrac{1}{2}\dfrac{\partial g_{\mu\mu}}{\partial x^\mu}$ $\qquad$ iii. $\Gamma_{\mu,\mu v} = -\Gamma_{v,\mu\mu} = \dfrac{1}{2}\dfrac{\partial g_{\mu\mu}}{\partial x^v}$

In Example 21, we have proved that

and $\quad$ i. $\Gamma^\mu_{v\sigma} = 0$ $\qquad\qquad\qquad\qquad$ ii. $\Gamma^\mu_{\mu\mu} = \dfrac{1}{2}\dfrac{(\partial \log g_{\mu\mu})}{\partial x^\mu}$

$\quad$ iii. $\Gamma^v_{\mu\mu} = \dfrac{1}{2g_{vv}}\dfrac{\partial g_{\mu\mu}}{\partial x^v}$ $\qquad\qquad$ iv. $\Gamma^\mu_{\mu v} = \dfrac{1}{2}\dfrac{\partial(\log g_{\mu\mu})}{\partial x^v}$

$\therefore$ The nonvanishing Christoffel's symbols of second kind can be expressed as

$$\Gamma_{1,11} = \frac{1}{2}\frac{\partial g_{11}}{\partial x^1} = \frac{1}{2}\frac{\partial f}{\partial x}$$

$$\Gamma_{4,41} = -\Gamma_{1,44} = \frac{1}{2}\frac{\partial g_{44}}{\partial x^1} = \frac{\partial\left(\frac{1}{f}\right)}{\partial x}$$

$$= -\frac{1}{2f^2}\frac{\partial f}{\partial x}$$

and
$$\Gamma^1_{11} = \frac{1}{2}\frac{\partial(\log g_{11})}{\partial x^1} = \frac{1}{2}\frac{\partial(\log f)}{\partial x} = \frac{1}{2f}\frac{\partial f}{\partial x}$$

$$\Gamma^1_{44} = -\frac{1}{2g_{11}}\frac{\partial g_{44}}{\partial x^1} = -\frac{1}{2f}\frac{\partial\left(\frac{1}{f}\right)}{\partial x} = \frac{1}{2f^3}\frac{\partial f}{\partial x}$$

$$\Gamma^4_{41} = \frac{1}{2}\frac{\partial(\log g_{44})}{\partial x^1} = \frac{1}{2}\frac{\partial\log\left(\frac{1}{f}\right)}{\partial x} = -\frac{1}{2f}\frac{\partial f}{\partial x}$$

Obviously, the nonvanishing Christoffel's 3-index symbols are

$$\Gamma_{1,11} = \frac{1}{2}\frac{\partial f}{\partial x}, \Gamma_{4,41} = -\frac{1}{2f^2}\frac{\partial f}{\partial x}$$

$$\Gamma_{1,44} = \frac{1}{2f^2}\frac{\partial f}{\partial x}$$

$$\Gamma^1_{11} = \frac{1}{2f}\frac{\partial f}{\partial x}, \Gamma^1_{44} = \frac{1}{2f^3}\frac{\partial f}{\partial x}, \Gamma^4_{41} = -\frac{1}{2f}\frac{\partial f}{\partial x}$$

Now, the equations of geodesics are

$$\frac{d^2x^\lambda}{\partial S^2} + \Gamma^\lambda_{\mu\nu}\frac{\partial x^\mu}{dS}\frac{\partial x^\nu}{dS} = 0; \ \lambda = 1, 2, 3, 4$$

For $\lambda = 1$, one obtains

$$\frac{d^2x^1}{dS^2} + \Gamma^1_{\mu,\nu}\frac{dx^\mu}{dS}\frac{dx^\nu}{dS} = 0$$

or
$$\frac{d^2x^1}{dS^2} + \Gamma^1_{11}\frac{dx^1}{dS}\frac{dx^2}{dS} + \Gamma^1_{44}\frac{dx^4}{dS}\frac{dx^4}{dS} = 0$$

or
$$\frac{d^2x}{\partial S^2} + \frac{1}{2f}\frac{\partial f}{\partial x}\left(\frac{dx}{dS}\right) + \frac{1}{2f^2}\frac{\partial f}{\partial x}\left(\frac{dt}{dS}\right)^2 = 0$$

i.e.
$$\frac{d^2x}{dS^2} + \frac{1}{2f}\frac{\partial f}{\partial x}\left[\left(\frac{dx}{dS}\right) + \frac{1}{f^2}\left(\frac{dt}{dS}\right)^2\right] = 0 \tag{1}$$

$$\lambda = 2$$

$$\frac{d^2x^2}{dS^2} + \Gamma^2_{\mu\nu}\frac{dx^\mu}{dS}\frac{dx^\nu}{dS} = 0$$

or
$$\frac{d^2y}{dS^2} = 0 \tag{2}$$

$$\lambda = 3$$

$$\frac{d^2x^3}{dS^2} + \Gamma^3_{\mu\nu}\frac{dx^\mu}{dS}\frac{dx^\nu}{dS} = 0$$

or
$$\frac{d^2z}{dS^2} = 0 \qquad (3)$$

For
$$\lambda = 4$$

$$\frac{d^2x^4}{dS^2} + \Gamma^4_{\mu\nu}\frac{dx^\mu}{dS}\frac{dx^\nu}{dS} = 0$$

or
$$\frac{d^2x^4}{dS^2} + \Gamma^4_{41}\frac{dx^4}{dS}\frac{dx^1}{dS} + \Gamma^4_{14}\frac{dx^1}{dS}\frac{dx^4}{dS} = 0$$

or
$$\frac{d^2t}{dS^2} - \frac{1}{f}\frac{\partial f}{\partial x}\frac{dx}{dS}\frac{dt}{dS} = 0 \qquad \because \ \Gamma^4_{41} = \Gamma^4_{14}$$

or
$$\frac{d^2t/dS^2}{dt/dS} = \frac{1}{f}\frac{\partial f}{\partial x}\frac{\partial x}{\partial S}[\log f(x)]\frac{dx}{dS}$$

or
$$\int \frac{d^2t/dS^2}{dt/dS}dS = \int \frac{d}{dx}[\log f(x)]\frac{dx}{dS}dS + \log K$$

where log K is a constant of integration.

or
$$\log \frac{dt}{dS} = \log f(x) + \log K$$

or
$$\frac{dt}{dS} = f(x) + C \qquad (4)$$

Equations (1) to (4) are the desired equations of geodesic.

***

Example 24. *If $A_{\mu\nu}$ is an antisymmetric tensor of second rank, then show that*
$$\frac{\partial A_{\mu\nu}}{\partial x^\sigma} + \frac{\partial A_{\nu\sigma}}{\partial x^\mu} + \frac{\partial A_{\sigma\mu}}{\partial x^\nu}$$
is a tensor, or otherwise show that
$$A_{\mu\nu};\sigma + A_{\nu\sigma};\mu + A_{\sigma\mu};\nu = \frac{\partial A_{\mu\nu}}{\partial x^\sigma} + \frac{\partial A_{\nu\sigma}}{\partial x^\mu} + \frac{\partial A_{\sigma\mu}}{\partial x^\nu}$$

Solution: We have

$$A_{\mu\nu};\sigma + A_{\nu\sigma};\mu + A_{\sigma\mu};\nu$$

$$= \left(\frac{\partial A_{\mu\nu}}{\partial x^\sigma} - \Gamma^\lambda_{\mu\sigma}A_{\lambda\nu} - \Gamma^\lambda_{\nu\sigma}A_{\mu\lambda}\right) + \left(\frac{\partial A_{\nu\sigma}}{\partial x^\mu} - \Gamma^\lambda_{\nu\mu}A_{\lambda\sigma} - \Gamma^\lambda_{\sigma\mu}A_{\nu\lambda}\right)$$

$$+ \left(\frac{\partial A_{\sigma\mu}}{\partial x^\nu} - \Gamma^\lambda_{\sigma\nu}A_{\lambda\mu} - \Gamma^\lambda_{\mu\nu}A_{\sigma\lambda}\right)$$

$$= \frac{\partial A_{\mu\nu}}{\partial x^\sigma} + \frac{\partial A_{\nu\sigma}}{\partial x^\mu} + \frac{\partial A_{\sigma\mu}}{\partial x^\nu} - (A_{\lambda\sigma} + A_{\sigma\lambda})\Gamma^\lambda_{\mu\nu} - (A_{\mu\lambda} + A_{\lambda\mu})\Gamma^\lambda_{\nu\sigma} - (A_{\lambda\nu} + A_{\nu\lambda})\Gamma^\lambda_{\sigma\mu} \qquad (1)$$

Since $A_{\mu\nu}$ is an antisymmetric tensor, i.e. $A_{\mu\nu} = -A_{\nu\lambda}$, one gets

$$A_{\lambda\sigma} = -A_{\sigma\lambda}, \ A_{\mu\lambda} = -A_{\lambda\mu}, \ A_{\lambda\nu} = -A_{\nu\lambda}$$

Eq. (1) gives

$$A_{\mu\nu};\sigma + A_{\nu\sigma};\mu + A_{\sigma\mu};\nu$$

$$= \frac{\partial A_{\mu\nu}}{\partial x^{\sigma}} + \frac{\partial A_{\nu\sigma}}{\partial x^{\mu}} + \frac{\partial A_{\sigma\mu}}{\partial x^{\nu}} \qquad (2)$$

L.H.S of Eq. (2) contains the sum of three covariant tensors each of rank 3. Obviously, R.H.S of Eq. (2) is a covariant tensor of rank 2.

Example 25. *If A_{μ} is a vector, then prove that, in general $\dfrac{\partial A_{\mu}}{\partial A^{\nu}}$ is not a tensor that but $\dfrac{\partial A_{\mu}}{\partial A^{\nu}} - \dfrac{\partial A_{\nu}}{\partial A^{\mu}}$ is a tensor.*

Solution: Let A_{μ} be a tensor of rank 2, then by transformation law, we have

$$\bar{A}_{\mu} = \frac{\partial x^{\alpha}}{\partial \bar{x}^{\mu}} A_{\alpha} \qquad (1)$$

Differentiating Eq. (1) partially w.r.t. $\bar{x}^{\nu}$, one obtains

$$\frac{\partial \bar{A}_{\mu}}{\partial \bar{x}^{\nu}} = \frac{\partial x^{\alpha}}{\partial \bar{x}^{\mu}} \frac{\partial A_{\mu}}{\partial \bar{x}^{\nu}} + \frac{\partial^2 x^{\alpha}}{\partial \bar{x}^{\nu} \partial \bar{x}^{\mu}} A_{\alpha}$$

$$= \frac{\partial x^{\alpha}}{\partial \bar{x}^{\mu}} \frac{\partial A_{\alpha}}{\partial x^{\beta}} \frac{\partial x^{\beta}}{\partial \bar{x}^{\nu}} + \frac{\partial^2 x^{\alpha}}{\partial \bar{x}^{\mu} \partial \bar{x}^{\nu}} A_{\alpha}$$

$$= \frac{\partial x^{\alpha}}{\partial \bar{x}^{\mu}} \frac{\partial x^{\beta}}{\partial \bar{x}^{\nu}} \frac{\partial A_{\alpha}}{\partial \bar{x}^{\beta}} + \frac{\partial^2 x^{\alpha}}{\partial \bar{x}^{\mu} \partial \bar{x}^{\nu}} A_{\alpha} \qquad (2)$$

Tensor transformation law shows that $\dfrac{\partial A_{\mu}}{\partial x^{\nu}}$ is a tensor only when the second term

of Eq. (2) is zero. This shows that $\dfrac{\partial A_{\mu}}{\partial x^{\nu}}$ is in general, not a tensor.

Now,

$$\frac{\partial A_{\mu}}{\partial x^{\nu}} - \frac{\partial A_{\nu}}{\partial x^{\mu}} = \frac{\partial A_{\mu}}{\partial x^{\nu}} - \Gamma_{\mu\nu}^{\lambda} A_{\lambda} + \Gamma_{\mu\nu}^{\lambda} A_{\lambda} - \frac{\partial A_{\nu}}{\partial x^{\mu}}$$

$$= \left(\frac{\partial A_{\mu}}{\partial x^{\nu}} - \Gamma_{\mu\nu}^{\lambda} A_{\lambda} \right) - \left(\frac{\partial A_{\nu}}{\partial x^{\mu}} - \Gamma_{\mu\nu}^{\lambda} A_{\lambda} \right)$$

$$= \left(\frac{\partial A_{\mu}}{\partial x^{\nu}} - \Gamma_{\mu\nu}^{\lambda} A_{\lambda} \right) - \left(\frac{\partial A_{\nu}}{\partial x^{\mu}} - \Gamma_{\mu\nu}^{\lambda} A_{\lambda} \right)$$

or

$$\frac{\partial A_{\mu}}{\partial x^{\nu}} - \frac{\partial A_{\nu}}{\partial x^{\mu}} = A_{\mu};\nu - A_{\nu};\mu \qquad (3)$$

Obviously, R.H.S of Eq. (3) being the difference of two covariant tensors of rank 2,

i.e. a covariant tensor. Clearly L.H.S of Eq. (3), i.e. $\dfrac{\partial A_{\mu}}{\partial x^{\nu}} - \dfrac{\partial A_{\nu}}{\partial x^{\mu}}$ is also a tensor of rank 2.

Example 26. *If the second order derivatives of coordinates $\bar{x}^\mu$ w.r.t. space coordinates are zero, then*

$$\bar{x}^\mu = x^\mu - x_0^\mu + \frac{1}{2}\left(\Gamma^\mu_{pq}\right)_0 \left(x^p - x_0^p\right)\left(x^q - x_0^q\right)$$

are geodesic coordinates with the pole at P_0.

Solution: Let us consider a general coordinate system x^μ whose values at particular point p_0 are x_0^μ. Let $\bar{x}^\mu$ be another coordinate system defined as

$$\bar{x}^\mu = x^\mu - x_0^\mu + \frac{1}{2}\left(\Gamma^\mu_{pq}\right)_0 \left(x^p - x_0^p\right)\left(x^q - x_0^q\right) \tag{1}$$

Here the index '0' attached to any entity simply indicates its value at P_0. Differentiating Eq. (1) w.r.t. x^ν, one obtains

$$\frac{\partial \bar{x}^\mu}{\partial x^\nu} = \frac{\partial x^\mu}{\partial x^\nu} + \frac{1}{2}\left(\Gamma^\mu_{pq}\right)_0 \frac{\partial x^p}{\partial x^\nu}\left(x^q - x_0^q\right) + \frac{1}{2}\left(\Gamma^\mu_{pq}\right)_0 \frac{\partial x^q}{\partial x^\nu}\left(x^p - x_0^p\right)$$

$$= \delta^\mu_\nu + \frac{1}{2}\left(\Gamma^\mu_{pq}\right)_0 \delta^p_\nu\left(x^q - x_0^q\right) + \frac{1}{2}\left(\Gamma^\mu_{pq}\right)_0 \delta^q_\nu\left(x^p - x_0^p\right)$$

$$= \delta^\mu_\nu + \frac{1}{2}\left(\Gamma^\mu_{vq}\right)_0 \left(x^q - x_0^q\right) + \frac{1}{2}\left(\Gamma^\mu_{pv}\right)_0 \left(x^p - x_0^p\right)$$

Changing index q to p, one obtains

$$= \delta^\mu_\nu + \left(\Gamma^\mu_{vp}\right)_0 \left(x^p - x_0^p\right) \tag{2}$$

$$\therefore \quad\quad \left(\frac{\partial \bar{x}^\mu}{\partial x^\nu}\right)_0 = \delta^\mu_\nu \quad \text{(because } x^p = x_0^p \text{ at pole } p_0\text{)} \tag{3}$$

Differentiating Eq. (2) w.r.t. x^σ, one obtains

$$\frac{\partial^2 \bar{x}^\mu}{\partial x^\sigma \partial x^\nu} = \left(\Gamma^\mu_{vp}\right)_0 \frac{\partial x^p}{\partial x^\sigma} = \left(\Gamma^\mu_{vp}\right)_0 \delta^p_\sigma$$

Obviously, $$\left(\frac{\partial^2 \bar{x}^\mu}{\partial x^\sigma \partial x^\nu}\right)_0 = \left(\Gamma^\mu_{v\sigma}\right)_0 \tag{4}$$

The second order derivative of $\bar{x}^\mu$ is given by

$$\left(\bar{x}^\mu; v\sigma\right)_0 = \frac{\partial^2 \bar{x}^\mu}{\partial x^\sigma \partial x^\nu} - \Gamma^\lambda_{v\sigma}\frac{\partial \bar{x}^\mu}{\partial x^\lambda} \tag{5}$$

Using Eqs (3) and (4), one obtains at pole P_0

$$\left(\bar{x}^\mu; v\sigma\right)_0 = \left(\Gamma^\mu_{v\sigma}\right)_0 - \left(\Gamma^\lambda_{v\sigma}\right)_0 \delta^\mu_\lambda$$

$$= \left(\Gamma^\mu_{v\sigma}\right) - \left(\Gamma^\mu_{v\sigma}\right) = 0$$

Obviously, $\bar{x}^\mu; v\sigma = 0$ at pole P_0

This shows that the second order derivatives of coordinates $\bar{x}^\mu$ w.r.t. space coordinates are zero. This means that the coordinates $\bar{x}^\mu$, defined by Eq. (1) are geodesic coordinates with the pole at P_0.

SHORT ANSWER QUESTIONS

1. Express KE of a particle in tensor form.

2. What do you mean by fundamental tensors $g^{\mu\nu}$, $g_{\mu\nu}$ and g_{μ}^{ν}? What is the value of $g^{\mu\nu} A_{\mu}$?

Ans. A^{ν}

3. If τ_{ij} is a tensor of rank 2, then prove that

   (a) τ_i^j is a tensor of rank 2 and $\tau_{ij} \neq \tau_i^j$.

   (b) $\dfrac{\partial \tau_{ij}}{\partial x^k}$ is not a tensor.

4. Show that the outer product of two tensors is a tensor of rank equal to the sum of the ranks of two given tensors.

5. Define a tensor giving examples.

6. What is Einstein's tensors? Show that divergence of Einstein's tensor is zero?

7. What do you mean by dyadic? Is it a tensor?

8. Define outer and inner products of two tensors.

9. Show that the sum of two tensors of same rank is a tensor of the same rank as well.

10. Show that ds^{μ} is a contravariant vector.

11. Show that the contraction of A_q^p is a scalar.

12. What does contraction mean in tensors? What is obtained if a second order tensor is contracted?

13. Explain quotient law.

14. If S is a scalar show that $\dfrac{\partial S}{\partial \overline{x}^a}$ is a covariant tensor of rank 1.

15. Define symmetric and antisymmetric tensors.

16. What is the total number of algebraically independent components of curvature tensor in n-dimensional space? Hence, find the number of these components in 4-dimensional space.

17. What do you mean by dummy and real indices in tensor analysis?

18. Explain Einstein's summation convention with examples.

19. Show that every contraction of a mixed tensor reduces its rank by 2.

20. Express tensor $A^{\mu\nu}$ as a sum of symmetric and an antisymmetric tensors.

21. What is covariant curvature tensor? Explain its symmetric, antisymmetric and cyclic properties.

22. If A_{ij} is an antisymmetric tensor, what is the value of A_{11}?

Ans. 0

23. Show that Kronecker delta is a mixed tensor of rank 2.

24. What is Riemann–Christoffel tensor? Give its antisymmetric and cyclic properties.

25. What do you mean by reciprocal tensor?

26. If $a_{\alpha\beta} x^\alpha x^\beta = 0$ for all variables x^μ then show that $a_{\mu\nu} + a_{\nu\mu} = 0$.

27. Write tensor form of divergence and curl of vector form.

28. What do you mean by a geodesic? Write differential equation of a geodesic and show that in cartesian coordinate system, a geodesic reduces to a straight line.

29. Show that in linear transformation of coordinates, the Christoffel's symbols transform like tensors.

30. What do you mean by Christoffel's 3-index symbols? Are they tensors?

31. What are the possible values of epsilon tensor?

Ans. 0, 1, –1

32. What is the number of independent components of a

   i. symmetric tensor

   ii. skew-symmetric tensor of rank 2 in n-dimensional space?

33. Give transformation equation for a tensor A^{ijk}_{lm} and write its rank and order.

34. Show that if $A^\mu = \dfrac{\partial \bar{x}^\mu}{\partial x^\alpha} A^\alpha$, then $A^\alpha = \dfrac{\partial x^\alpha}{\partial \bar{x}^\mu} \bar{A}^\mu$.

PROBLEMS

1. Show that if the components of a tensor are zero in one reference system, its components are also zero in all other reference systems.

2. Define contravariant, covariant and mixed tensors. Give examples for each kind.

3. (a) If A^{pq} and B^{pq} are tensors, then prove that their sum and difference are also tensors.

   (b) Show that by contraction, the rank of a tensor is reduced by two.

4. Show that every tensor can be expressed as sum of two tensors, one of which is symmetric and the other is skew-symmetric in a pair of covariant or contravariant indices.

5. State and prove quotient law of tensors.

6. Define Kronecker delta δ^m_n and find its tensor character.

7. Given a symmetric covariant tensor of rank two, define the reciprocal tensor and prove its tensor character.

8. Give transformation equation for the tensor A^{ijk}_{lm} and write its rank and order.

9. Show that the dummy suffixes have a certain freedom of movement between the tensor factors of an expression and prove that

   i. $A_{\mu\nu} B^{\sigma\nu} = A^\nu_\mu B^\sigma_\nu$ ii. $A_{\mu\nu} B^{\mu\nu} = A^{\mu\nu} B_{\mu\nu}$

10. If T_{ij} is a second rank tensor, prove whether or not

   i. T_i^j equal to T_j^i ii. $\dfrac{\partial T_{ij}}{\partial x^k}$ is a tensor

   [*Ans. (i)* yes *(ii)* no]

11. Prove that an entity which on inner multiplication with an arbitrary tensor (contravariant or covariant) always gives a tensor, is itself a tensor.

   [*Hint. see* quotient law]

12. Show that the covariant derivative of an invariant is the same as its ordinary derivative.

13. Find the covariant and contravariant components of acceleration vector in cylindrical and spherical coordinates.

14. Show that any inner product of the tensors A_r^p and B_t^{gs} is a tensor of rank 3.

15. Show that $\dfrac{\partial A_p}{\partial x^q}$ is not a tensor even A_p is a covariant tensor of rank one.

16. If a contravariant tensor is symmetric in one coordinate system, show that it is symmetric in every other coordinate system.

17. If A_{ij} is an antisymmetric tensor and S_{ij} is a symmetric tensor, find out whether or not any of the following tensors is antisymmetric or symmetric.

   i. $A_{ij} A_{jk}$ ii. $A_{ij} S_{jk}$ iii. $S_{ij} A_{jk}$

   iv. $S_{ij} S_{jk}$ v. $A_{ij} S_{jk} + S_{ij} A_{jk}$ vi. $A_{ij} S_{jk} - S_{ij} A_{jk}$

   vii. $A_{im} A_{mn} A_{nk}$ viii. $A_{im} S_{mn} A_{nk}$

18. If S_{ij} is a symmetric tensor and A_{ij} is antisymmetric tensor, show that the product $A_{ij} S_{ij}$ vanishes.

19. Given that $dS^2 = g_{\mu\nu}\, dx^\mu\, dx^\nu$ is invariant, show that $g_{\mu\nu}$ is a covariant tensor of second rank.

20. Show that the multiplication of tensors results in addition of their ranks and contraction reduces the rank by two. Given

$$T = \frac{1}{2} M g_{mn} \dot{x}^m \dot{x}^n$$

   show that $x^j = \dfrac{1}{m} g^{jn} \dfrac{\partial T}{\partial x^\mu}$

   g_{mn} being the metric tensor.

21. Show that

   i. $dg_{\alpha\beta} = -g_{\mu\alpha} g_{\nu\beta}\, dg^{\mu\nu}$

   ii. $A^{\alpha\beta} dg_{\alpha\beta} = -A_{\alpha\beta} dg^{\alpha\beta}$

   iii. $\dfrac{dg}{g} = -g_{\mu\nu}\, dg^{\mu\nu}$

22. If $A^{\mu\nu}$ and $A_{\nu\mu}$ are reciprocal symmetrical tensors and if u_ν are the components of a covariant tensor of rank one, show that $A_{\mu\nu} u^\mu u^\nu = A^{\mu\nu} u_\mu u_\nu$, where $u^\mu = A^{\mu\alpha} u_\alpha$.

23. Define Christoffel's symbols of first and second kinds. Derive their transformation laws and show that they are not the components of a tensor, but when linear transformations are assumed, they transform like the components of a tensor.

24. Show that the ordinary derivative of a vector is not a tensor, but a tensor derivative can be defined with the help of Christoffel's symbols and the vector.

25. Determine the Christoffel's symbols corresponding to
    i. $dS^2 = a^2 \, d\theta^2 + a^2 \sin^2 \theta \, d\phi^2$
    ii. $dS^2 = dr^2 + r^2 \, d\theta^2 + r^2 \sin^2 \theta \, d\phi^2$

26. Surface of a sphere can be regarded as a two-dimensional Riemanian space. Obtain its fundamental metric tensor.

27. Define covariant derivative of a tensor and that it is a covariant tensor of rank one greater than the given tensor.

28. Separate the following tensor

$$\begin{pmatrix} -xy & x^2 \\ -y^2 & xy \end{pmatrix}$$

into symmetric and antisymmetric parts.

29. Show that $L^2 = g_{\mu\nu} A^\mu B^\nu$ is an invariant.

30. Show that Kronecker delta is a mixed tensor of rank 2.

31.   i. Show that the covariant differentiation of products, sum and difference obey the same rule as in the case of ordinary differentiation.
      ii. Define covariant derivative vector and show that it is a tensor.

32.   i. Show that

$$\frac{\partial A^l}{\partial x^i} + A^k T_{kj}^l \text{ transform like a tensor.}$$

      ii. Show that grad ϕ is a vector.
      iii. Show that acceleration is a contravariant vector.
      iv. Show that a vector can always be associated with an antisymmetric tensor of rank two in three-dimensional space.

33. Obtain the covariant derivative of a contravariant vector A.

34. Define a tensor and distinguish between symmetric and skew symmetric tensors. Obtain the tensor form of gradient, divergence and curl.

35. Show that divergence of Einstein's tensors vanishes identically.

36. Discuss the application of tensor analysis to the mechanics of a particle.

37. Show that the curvature tensor may be contracted in two ways: one of these leads to zero tensor and other method leads to Ricci tensor.

38. Define intrinsic and covariant derivatives of a contravariant vector. Use the expression for the intrinsic derivative to obtain the components of acceleration vector for the metric $dS^2 = dr^2 + r^2\, d\theta^2$.

39. Prove that covariant differentiation in Euclidean space is commutative. Thus prove that the Riemann–Christoffel tensor and curvature tensor are zero in a Euclidean space.

40. i. Explain Bianchi identities;

$$R^{\alpha}_{ijk,\,l} + R^{\alpha}_{ikl,\,j} + R^{\alpha}_{ilj,\,k} = 0$$

   ii. Show that the number of associate covariant curvature tensor does not exceed 20 in space–time continuum.

41. Show that in two-dimensions, a skew symmetric tensor of second rank is a pseudo scalar.

42. Show that covariant curvature tensor R_{hijk} is (i) skew symmetric in first two indices (ii) skew symmetric in last two indices (ii) symmetric in two pairs of indices.

43. $B_{v\sigma}$ is an arbitrary covariant tensor and if $A(\mu, v)\, B_{\mu v} = C_{\mu\sigma}$, where $C_{\mu\sigma}$ is a tensor, prove that $A(\mu, v)$ is a mixed tensor.

44. If $A_{\mu v}$ is an antisymmetric tensor, then prove that

$$\left(B^{\mu}_{v}\, B^{\alpha}_{\beta} + B^{\mu}_{\beta}\, B^{\alpha}_{v} \right) A_{\mu\alpha} = 0$$

45. i. Show that when the Riemann–Christoffel tensor vanishes, the differential equations

$$A_{\mu v} = \frac{\partial A_{\mu}}{\partial x^{v}} - \Gamma^{\alpha}_{\mu v}\, A_{\alpha} = 0; \text{ are integrable.}$$

   ii. Prove that the divergence of $R^{\mu}_{v} - g^{\mu}_{v} R$ is identically zero.

46. What is meant by a metric tensor? The surface of a sphere is a two-dimensional Riemannian space. Find its fundamental metric tensor.

47. Show that finite rotation in space is tensor of rank two.

48. Define a tensor with one physical example. Show that Kronecker delta δ^{μ}_{v} is a mixed tensor and has the same components in every coordinate system.

49. Write the Christoffel symbols of the second kind for the matrix

$$dS^2 = (dx^1)^2 + [(x^2)^2 - (x^1)^2](dx^2)^2$$

and the corresponding geodesics.

[*Hint:* $\Gamma^1_{22} = x^1, \Gamma^2_{12} = \Gamma^2_{21} = \dfrac{x^1}{(x^1)^2 - (x^2)^2}$

$$\Gamma_{22}^2 = \frac{x^2}{(x^2)^2 - (x^1)^2} \text{ and all these are zero.}$$

Geodesics: $\dfrac{d^2 x^1}{dS^2} + x^1 \left(\dfrac{dx^2}{dS} \right) = 0$

$$\frac{d^2 x^2}{dS^2} + \frac{2x^1}{(x^1)^2 - (x^2)^2} \frac{dx^1}{dS} \frac{dx^2}{dS} + \frac{x^2}{(x^2)^2 - (x^1)^2} \left(\frac{dx^2}{dS} \right)^2 = 0]$$

50. i. Prove that: $\dfrac{\partial g_{\mu\nu}}{\partial x^\sigma} = [\mu\sigma, \alpha] + [\nu\sigma, \mu]$

   ii. Prove that $[\mu\nu, \sigma] = [\nu\mu, \sigma]$

51. Define geodesic. Obtain differential equations of a geodesic in a given space.

52. Define metric tensor and write it in spherical coordinates.

53. A quantity $A\,(j, k, l, m)$ which is a function of coordinates x^i transforms to another coordinate system according to the rule

$$\bar{A}(p,q,r,s) = \frac{\partial x^j}{\partial \bar{x}^p} \frac{\partial \bar{x}^q}{\partial x^k} \frac{\partial \bar{x}^r}{\partial x^l} \frac{\partial x^s}{\partial x^m} A(j,k,l,m)$$

   i. Is the quantity of a tensor?

   ii. If so, write the tensor in a suitable notation, and

   iii. Give the contravariant and covariant rank.

54. Prove that the contraction of the tensor A_q^p is a scalar or invariant.

55. i. Prove that S_q^p is a mixed tensor of second rank.

   ii. Define metric tensor. Determine the metric tensor in cylindrical coordinates.

56. Taking $\begin{bmatrix} \eta_{11} & \eta_{12} & \eta_{13} \\ \eta_{21} & \eta_{22} & \eta_{23} \\ \eta_{31} & \eta_{32} & \eta_{33} \end{bmatrix} = \begin{bmatrix} \eta/3 & 0 & 0 \\ 0 & \eta/3 & 0 \\ 0 & 0 & \eta/3 \end{bmatrix}$

$$+ \begin{bmatrix} \eta_{11} - \eta/3 & \eta_{12} & \eta_{13} \\ \eta_{21} & \eta_{22} - \eta/3 & \eta_{23} \\ \eta_{31} & \eta_{32} & \eta_{33} - \eta/3 \end{bmatrix}$$

where, η is contracted to η_{ij}. Show that the tensor $S_{ij} = \eta_{ij} - \dfrac{1}{2} \eta \delta_{ij}$ describes a change

(shear) with no change in volume to first order and the tensor $V_{ij} = \eta_{ij} - \dfrac{1}{3} \eta \delta_{ij}$ describes a change in volume and no change of shape.

57. Show that $\Gamma^{\alpha}_{\mu\sigma} = \dfrac{\partial}{\partial x^{\mu}}\log(-g)$, where $g = g_{\mu\nu}$. Prove that in Cartesian coordinates, there is no distinction between covariant and contravariant vectors.

58. Show that

     i. $dg^{\alpha\beta} = -g^{\mu\alpha}g^{\nu\beta}dg_{\mu\nu}$ ii. $dg_{\alpha\beta} = -g_{\mu\alpha}g_{\nu\beta}dg^{\mu\nu}$

MULTIPLE CHOICE QUESTIONS

1. If A^i and B_j are components of contravariant and covariant tensors of rank one, then $A^i B_j$ are the components of
   (a) contravariant tensor of rank 2   (b) covariant tensor of rank 2
   (c) mixed tensor of rank 1   (d) mixed tensor of rank 2

2. If A^{ji} and A_{ij} are reciprocal symmetric tensors and if $u_i\left(= A^{i\alpha}u_\alpha\right)$ are components of a covariant tensor of rank 1, then the correct relation is
   (a) $A_{ij}u^iu^j = A^{ij}u_iu_j$ (b) $A_{ij}u^iu^j = A_{ij}u^iu^j$
   (c) $A_{ij}u^iu^j - A_{ij}u^{ij}$ (d) $A_{ij}u^iu^j = A^{ij}u_{ij}$

3. If A^i are the components of an absolute contravariant tensor of rank one, then $\dfrac{\partial A_i}{\partial x^j}$ are the components of
   (a) covariant tensor of rank 1   (b) contravariant tensor of rank 1
   (c) mixed tensor of rank 1   (d) covariant tensor of rank 2.

4. If array $A = \begin{bmatrix} -xy & -y^2 \\ x^2 & xy \end{bmatrix}$ is a tensor of rank 2, then the array $B = \begin{bmatrix} -xy & -y^2 \\ x^2 & -xy \end{bmatrix}$ is
   (a) a tensor of rank 2   (b) a tensor of rank 1
   (c) is not a tensor   (d) none of the above

5. If K is a second rank tensor, A and B are vectors then
   (a) $\mathbf{B} \cdot \mathbf{K} \cdot \mathbf{A} = \mathbf{A} \cdot \mathbf{K} \cdot \mathbf{B}$;   if K is symmetric
   (b) $\mathbf{B} \cdot \mathbf{K} \cdot \mathbf{A} = -\mathbf{A} \cdot \mathbf{K} \cdot \mathbf{B}$;   if K is symmetric
   (c) $\mathbf{B} \cdot \mathbf{K} \cdot \mathbf{A} = \mathbf{A} \cdot \mathbf{K} \cdot \mathbf{B}$;   if K is antisymmetric
   (d) $\mathbf{B} \cdot \mathbf{K} \cdot \mathbf{A} = \mathbf{A} \cdot \mathbf{B} \cdot \mathbf{K}$;   if K is antisymmetric

6. Isotropic tensors of rank three are
   (a) scalar multiples of ε_{ikm} (b) vector multiples of ε_{ikm}
   (c) scalar multiples of $\delta^{\mu}_i \delta^{m}_k$ (d) vector multiples of $\delta^{\mu}_i \delta^{m}_k$

7. The equation $K_{ij} A_{jk} = B_{ik}$ holds for all orientations of the coordinate systems. If A and B are second rank tensors then the rank of K is
   (a) 1   (b) 2
   (c) 3   (d) 4

8. T_{iklm} is antisymmetric with respect to all pairs of indices, then the no of independent components in 3-dimensional space are

(a) 1   (b) 4

(c) 256   (d) zero

9. If u is an antisymmetric dyadic and v is a vector then

(a) $u \cdot v \cdot u = 0$ (b) $v \cdot u \cdot v = 0$

(c) $v \cdot v \cdot u = 0$ (d) $u \cdot u \cdot v = 0$

10. We know that dummy suffixes have a certain freedom of movement between the tensor factors of an expression, then correct relation is

(a) $A_{\mu\alpha} B^{\nu\alpha} = A_\mu^\alpha B_\alpha^\nu$ (b) $A_{\mu\alpha} B^{\nu\alpha} = A_\alpha^\mu B_\alpha^\nu$

11. In case of covariant differentiation of product, the correct relation is

(a) $(B_\mu C_\nu)_{;\sigma} = B_{\mu;\sigma} C_\nu + B_\mu C_{\nu;\sigma}$ (b) $(B_\mu C_\nu)_{;\sigma} = B_{\mu;\nu} C_\sigma + B_\sigma C_{\mu;\sigma}$

(c) $(B_\mu C_\nu)_{;\sigma} = B_{\nu;\mu} C_\sigma + B_\mu C_{\nu;\sigma}$ (d) $(B_\mu C_\nu)_{;\sigma} = B_{\mu;\sigma} C_\nu + B_\mu C_{\sigma;\nu}$

12. In case of covariant curvature tensor $R_{\alpha\nu\mu\rho}$, the value of $R_{\alpha\nu\mu\rho} + R_{\alpha\mu\rho\nu} + R_{\alpha\rho\nu\mu}$ is

(a) $R_{\alpha\nu\mu\rho}$ (b) $-R_{\alpha\nu\mu\rho}$

(c) $R_{\alpha\nu\mu\rho}$ (d) zero

13. Covariant curvature tensor is antisymmetric in indices

(a) ν, μ and α, ρ (b) ν, ρ and α, μ

(c) α, ρ and α, μ (d) ν, μ and $\alpha\mu$

14. For Levi–Civita tensor (epsilon tensor), we have

(a) $\varepsilon_{ijk} = \varepsilon_{jik}$ (b) $\varepsilon_{ijk} = -\varepsilon_{jki}$

(c) $\varepsilon_{ijk} = -\varepsilon_{kij}$ (d) $\varepsilon_{ijk} = -\varepsilon_{jik}$

15. If $g_{\mu\nu}$ is a covariant fundamental tensor of rank 2 and $g^{\mu\sigma}$ is contravariant fundamental tensor of rank 2 then their product $g_{\mu\nu} g^{\nu\sigma}$ is

(a) contravariant tensor of rank 4   (b) covariant tensor of rank 4

(c) zero   (d) δ_μ^σ

16. Moment of inertia is a

(a) scalar   (b) vector

(c) tensor of rank 2   (d) a tensor of higher rank

17. The number of independent components of a symmetric tensor of rank 2 in n-dimensional space is

(a) $2n$ (b) n^2

(c) $\dfrac{n(n+1)}{2}$ (d) $\dfrac{n(n-1)}{2}$

18. If the multiplication of a dyadic with an arbitrary vector is commutative then dyadic is

(a) symmetric   (b) antisymmetric

(c) diagonal   (d) scalar

19. If $A^{\mu\nu} = A^{\nu\mu}$, then $A^{\mu\nu}$ is

    (a) 0

    (b) a symmetric tensor

    (c) an antisymmetric tensor

    (d) a scalar

20. Kronecker delta δ_l^k is a

    (a) scalar

    (b) vector

    (c) tensor of rank 1

    (d) tensor of rank 2

21. The value of δ_k^k in n-dimensional space is

    (a) 0

    (b) 1

    (c) k

    (3) n

22. Christoffel's 3-index symbols are

    (a) in variant

    (b) vectors

    (c) tensors

    (d) not a tensor

23. The number of independent components of an antisymmetric tensor of rank 2 in n-dimensional space is

    (a) $2n$

    (b) n^2

    (c) $\dfrac{n(n+1)}{2}$

    (d) $\dfrac{n(n-1)}{2}$

24. If A^μ and B_μ are the components of both contravariant and covariant vectors, then the product $A^\mu B_\mu$ is a

    (a) scalar

    (b) contravariant tensor of rank 2

    (c) covariant tensor of rank 2

    (d) mixed tensor

25. The components of a quantity A follow the transformation law in two coordinate system $\bar{A}^\mu = \dfrac{\partial \bar{x}^\mu}{\partial x^\mu} A^\mu$, then the quantity A is a

    (a) scalar

    (b) contravariant vector

    (c) covariant vector

    (d) tensor of rank 2

26. The components of quantity A follow the transformation law $\bar{A}_\mu = \dfrac{\partial x_\mu}{\partial \bar{x}_\mu} A_\mu$, then the quantity A is a

    (a) scalar

    (b) contravariant vector

    (c) covariant vector

    (d) tensor of rank 2

27. If A^μ and B_ν are the components of contravariant and covariant vectors, then the product $A^\mu B_\nu$ is a

    (a) scalar

    (b) covariant tensor of rank 2

    (c) contravariant tensor of rank 2

    (d) mixed tensor

28. A tensor of rank 2 is n-dimensional space has components

    (a) n

    (2) $2n$

    (c) n^2

    (d) 2^n

29. Differentiation of a scalar with respect to variable x^μ results
    - (a) a scalar
    - (b) a covariant vector
    - (c) a contravariant vector
    - (d) a tensor of rank 2

30. The number of independent components of antisymmetric tensor of rank n in n-dimensional space is
    - (a) n
    - (b) $n(n-1)$
    - (c) 3
    - (d) 1

31. A tensor $A^{\mu\nu}$ of rank 2 may be expressed as
    - (a) the sum of two symmetric tensors each of rank 1
    - (b) the sum of two antisymmetric tensors each of rank 1
    - (c) the sum of symmetric and antisymmetric tensors each of rank 2
    - (d) the product of a symmetric and an antisymmetric tensor

32. If A_{ij} is an antisymmetric tensor, then the value of A_{11} is
    - (a) 0
    - (b) 1
    - (c) $\dfrac{n}{2}$
    - (d) n

33. Inner product of two mixed tensors A^μ_ν and B^m_n each of rank 2 is
    - (a) a scalar
    - (b) a tensor of rank 2
    - (c) a tensor of rank 4
    - (d) does not exist

34. The outer product of two mixed tensors A^μ_ν and B^m_n each of rank 2 is
    - (a) a scalar
    - (b) a tensor of rank 2
    - (c) a tensor of rank 4
    - (d) does not exist

35. The expression for distance ds between two adjacent points is given by $ds^2 = g_{\mu\nu}\, dx^\mu\, dx^\nu$, the quantity $g_{\mu\nu}$ is a
    - (a) scalar
    - (b) covariant fundamental tensor
    - (c) contravariant fundamental tensor
    - (d) mixed fundamental tensor

36. The velocity of a particle is a
    - (a) scalar
    - (b) contravariant vector
    - (c) covariant vector
    - (d) tensor of rank 2

37. The transformation law of tensor $A^{\mu\nu}_\sigma$ of rank 3 is
    - (a) $\dfrac{\partial \bar{x}^\mu}{\partial x^\alpha} \dfrac{\partial \bar{x}^\nu}{\partial x^\beta} \dfrac{\partial \bar{x}^\sigma}{\partial x^\gamma}$
    - (b) $\dfrac{\partial \bar{x}^\mu}{\partial x^\alpha} \dfrac{\partial \bar{x}^\nu}{\partial x^\beta} \dfrac{\partial \bar{x}^\gamma}{\partial x^\sigma}$
    - (c) $\dfrac{\partial \bar{x}^\alpha}{\partial x^\mu} \dfrac{\partial \bar{x}^\beta}{\partial x^\nu} \dfrac{\partial \bar{x}^\gamma}{\partial x^\sigma}$
    - (d) $\dfrac{\partial x^\alpha}{\partial \bar{x}^\mu} \dfrac{\partial x^\beta}{\partial \bar{x}^\nu} \dfrac{\partial x^\nu}{\partial \bar{x}^\sigma}$

38. If $A_{\mu\nu}$ is an antisymmetric tensor of rank 2 and B^μ is a tensor of rank 1, then the quantity $A_{\mu\nu} B^\mu B^\nu$ (summation index applicable) is
    - (a) a scalar
    - (b) a tensor of rank 4
    - (c) n
    - (d) 0

39. The total number of algebraically independent components of a covariant curvature tensor in four-dimensional space is
(a) 256
(b) 264
(c) 20
(d) 16

40. If $g_{\mu\nu}$ and $a_{\mu\nu}$ are two symmetric covariant tensors and $\left(\Gamma^{\mu}_{\nu\sigma}\right)_g$ and $\left(\Gamma^{\mu}_{\nu\sigma}\right)_a$ are the corresponding Christoffel's symbols then the quantity $\left(\Gamma^{\mu}_{\nu\sigma}\right)_g - \left(\Gamma^{\mu}_{\nu\sigma}\right)_a$ is
(a) zero
(b) a scalar
(c) a covariant tensor of rank 2
(d) a mixed tensor of rank 3

41. If $S_{\mu\nu}$ is a symmetric tensor and $A_{\mu\nu}$ is an antisymmetric tensor then the product $S_{\mu\nu} A_{\mu\nu}$ is
(a) a symmetric tensor of rank 4
(b) a scalar
(c) an antisymmetric tensor of rank 4
(d) zero

42. Curvature tensor $R^{\lambda}_{\mu\nu\sigma}$ is
(a) symmetric with respect to indices μ, ν
(b) symmetric with respect to indices μ, σ
(c) antisymmetric with respect to indices ν, μ
(d) antisymmetric with respect to indices ν, σ

43. The divergence of Einstein's tensor G^{μ}_{ν} is
(a) zero
(b) a scalar
(c) a vector
(d) a tensor

44. Which of the following does not obey tensor transformation laws in nonlinear space?
(a) covariant derivative of contravariant vector
(b) covariant derivative of a tensor
(c) Christoffel's 3-index symbols
(d) all of the above

45. The intrinsic derivative of a fundamental tensor $g_{\mu\nu}$ is
(a) a constant
(b) zero
(c) a scalar
(d) a tensor

46. Which of the following is nonvanishing?
(a) covariant derivative of $g_{\mu\nu}$
(b) covariant derivative of a vector
(c) divergence of Einstein's tensor
(d) Epsilon tensor $\varepsilon_{\mu\nu}$

47. The value of epsilon tensor $\varepsilon_{\mu\mu\nu}$ is
(a) 1
(b) –1
(c) 0
(d) infinity

48. If $g_{\mu\nu}$ is a fundamental covariant tensor A^{μ} is a contravariant tensor and $g_{\mu\nu} A^{\mu} B^{\nu} = 0$, then angle between A and B is
(a) 0
(b) $\pi/2$
(c) π
(d) between 0 and $\pi/2$

49. In the equation $d\bar{x}^\mu = \dfrac{\partial \bar{x}^\mu}{\partial x^\alpha} dx^\alpha$

(a) μ and α are dummy indices

(b) μ and α are real indices

(c) μ is real and α dummy index

(d) μ is dummy and α is real index

50. A tensor $A^\sigma_{\mu\nu}$ is said to be symmetric with respect to

(a) μ, ν

(b) μ, σ

(c) ν, σ

(d) all indices μ, ν, σ

51. If $g_{\mu\nu} = 0$ for $\mu \neq \nu$ and μ, ν, σ are unequal indices then which of the following vanishes

(a) $\Gamma^\mu_{\mu;\sigma}$

(b) $\Gamma_{\mu;\mu\nu}$

(c) $\Gamma^\nu_{\mu\mu}$

(d) $\Gamma^\mu_{\mu\mu}$

52. If $R_{\mu\nu}$ is Ricci tensor and $g^{\mu\nu}$ is covariant fundamental tensor, then quantity of $g^{\mu\nu}$ is $R_{\mu\nu}$ is

(a) a tensor of rank 4

(b) a tensor of rank 2

(c) a pseudo tensor

(d) scalar curvature

53. Given that A is a covariant vector then $\dfrac{\partial A_i}{\partial x^j}$ is

(a) a covariant tensor

(b) a contravariant tensor

(c) a mixed tensor

(d) not a tensor

54. If $B^{\lambda\sigma}_{\mu\nu}$ and $C^{\lambda\sigma}_\sigma$ are tensors and $A(\mu,\nu,\sigma)B^{\lambda\sigma}_{\mu\nu} = C^{\lambda\sigma}_\sigma$ then $A(\mu, \nu, \sigma)$ is

(a) a vector

(b) not a vector

(c) covariant tensor

(d) a mixed tensor

55. The gradient of a scalar function is

(a) scalar

(b) covariant vector

(c) contravarient vector

(d) covariant tensor

56. If A_μ is a vector, then $\dfrac{\partial A_\mu}{\partial x^\nu}$ is

(a) a scalar

(b) a vector

(c) always a tensor

(d) not a tensor

57. If A_μ is a vector then $\dfrac{\partial A_\mu}{\partial x^\nu} - \dfrac{\partial A_\nu}{\partial x^\mu}$ is

(a) zero

(b) a scalar

(c) a vector

(d) always a tensor

58. Consider an antisymmetric tensor p_{ij} with the indices running from 1 to 5. The number of independent components of the tensor is

(a) 1

(b) 10

(c) 9

(d) 6   **[GATE]**

ANSWERS

1. (d)	2. (a)	3. (c)	4. (d)	5. (a)	6. (a)	7. (b)	8. (a)
9. (b)	10. (a)	11. (a)	12. (d)	13. (a)	14. (d)	15. (d)	16. (c)
17. (c)	18. (a)	19. (b)	20. (d)	21. (d)	22. (d)	23. (d)	24. (a)
25. (b)	26. (c)	27. (d)	28. (c)	29. (c)	30. (d)	31. (c)	32. (a)
33. (b)	34. (c)	35. (b)	36. (b)	37. (b)	38. (d)	39. (c)	40. (d)
41. (d)	42. (d)	43. (a)	44. (c)	45. (b)	46. (b)	47. (c)	48. (b)
49. (c)	50. (a)	51. (a)	52. (d)	53. (a)	54. (d)	55. (b)	56. (d)
57. (d)	58. (b)						

4

Factorial and Related Functions

4.1 INTRODUCTION

We will discuss here a few of the more common special functions which arise in context other than the Forbenius series solution of some second order differential equations. These functions appear occasionally in physical and mathematical problems such as normalisation of Coulomb wave functions and computation of probabilities in statistical mechanics. These functions known as *beta functions* and *gamma functions*, have their usefulness in developing other functions, such as exponential integrals, elliptic function, periodic functions etc. which have direct physical applications.

4.2. FACTORIAL FUNCTION

We define factorial function by

$$z! = \int_0^\infty u^z e^{-u} du \tag{4.1}$$

where, $R(z) > -1$ (here $R(z)$ means real part of z). Convergence is uniform in any region $R(z) \geq -1 + \delta$, where $\delta > 0$

It follows immediately by integration by parts of Eq. (4.1) that

$$(z + 1)! = (z + 1) z!$$

and hence z is a positive integer, since

$$0! = 1$$

$$z! = 1.2.3.4...z \tag{4.2}$$

Also z ! is an analytical function because it has a derivative

$$\frac{dz!}{dz} = \int_0^\infty u^z \log u e^{-u} du \tag{4.3}$$

which converges uniformly in any closed region of z for $R(z) > -1$.

4.3. GAMMA FUNCTIONS

The gamma function can be viewed as a function of the complex variable and can be defined by three different convenient ways. The first definition is given in terms of

Euler's definite integral which describes it only as a function of real argument of z. It is

$$\Gamma(z) = \int_0^\infty e^{-u} u^{z-1} du \qquad (4.4)$$

The integral converges for $R(z) > 0$. For $R(z) \leq 0$, the integral diverges and cannot be used to define $\Gamma(z)$. We shall show later, how to define $\Gamma(z)$ for $n \leq 0$. This definition of gamma functions is used to establish relationship of gamma function with well known factorial function and Gauss's pi function.

For positive integral values of $z = n$, we have

$$\Gamma(n + 1) = \int_0^\infty e^{-u} u^n du$$

integrating by parts, we have

$$\Gamma(n + 1) = n\Gamma(n) \qquad (4.5)$$

The recursion relation Eq. (4.5) gives

$$\Gamma(n + 1) = 1.2.3.4...n = n\ ! \qquad (4.5a)$$

[The factorial function $z!$ is usually denoted by letter $\Gamma(z + 1)$, when z is not a positive integer].

Equation (4.5) is called recursion relation for gamma functions. If the value of $\Gamma(n)$ is known for n between any two successive positive integers, then the value of $\Gamma(n)$ for any positive n may be found by successive application of this equation. For example, gamma function is usually tabulated for n between 1 and 2. Now using Eq. (4.5), we can find $\Gamma(n)$ for n between 2 and 3 and so on. This equation gives

$$\Gamma(1.5) = 0.5\ \Gamma(0.5)$$

Now, if $\Gamma(0.5)$ is known from tabulated values then $\Gamma(1.5)$ can be determined.

Equation Eq. (4.4) yields

$$\Gamma(1) = \int_0^\infty e^{-u} du = 1$$

and Eq. (4.5) gives

$$\Gamma(n) = \frac{\Gamma(n+1)}{n} \to \infty \text{ as } n \to 0.$$

The Eqs (4.4) and (4.5) are incapable of defining $\Gamma(n)$ for $n = 0, -1, -2,...,$ etc.

For n either real or complex number, $\Gamma(n)$ is also defined in terms of Euler's infinite limit.

$$\Gamma(n) = \lim_{m \to \infty} \frac{1.2.3...m}{n(n+1)...(n+m)} m^n \qquad (4.6)$$

The definition is equivalent of the definition given by Eq. (4.4) and to show the equivalence, we take

$$F(n, m) = \int_0^m \left(1 - \frac{t}{m}\right)^m t^{n-1} dt \quad R(n) > 0 \qquad (4.7)$$

where m is a positive integer.

Here we have $\lim\limits_{m\to\infty}\left(1-\dfrac{t}{m}\right)^m = e^{-t}$

and
$$\lim_{m\to\infty} F(n,m) = F(n,\infty) = \int_0^m e^{-t}t^{n-1}dt = \Gamma(n) \qquad (4.8)$$

Again substituting $u = \dfrac{t}{m}$ in Eq. (4.7), we have

$$F(n, m) = m^n \int_0^1 (1-u)^m u^{n-1} du$$

Integrating by parts, we have

$$F(n, m) = \frac{m(m-1)...1}{n(n+1)...(n+m)}m^n \qquad (4.9)$$

Combining Eqs (4.8) and (4.9), we have

$$\Gamma(n) = \lim_{m\to\infty} F(n,m) = \lim_{m\to\infty} \frac{1.2.3...m}{(n+1)(n+2)...(n+m)}m^n$$

From this definition also we have

$$\overline{(n+1)} = \lim_{m\to\infty} \frac{1.2.3...m}{(n+1)(n+2)...(n+m+1)}m^{n+1}$$

$$= \lim_{m\to\infty}\left(\frac{mn}{n+m+1}\right)\lim_{m\to\infty}\frac{1.2.3...m}{n(n+1)...(n+m)}m^n$$

$$= \lim_{m\to\infty}\frac{n}{1+\dfrac{n+1}{m}}\overline{(n)} = n\overline{(n)}$$

This equation is identical to Eq. (4.5).

It is also evident that $\overline{(n)} = \infty$ for $n = 0, -1, -2,..., \infty$ and

$$\overline{1} = \lim_{m\to\infty}\frac{1.2.3...m}{1.2.3...(m+1)}m = \lim_{m\to\infty}\frac{m}{m+1} = 1$$

$\therefore \qquad \overline{1} = 1$

The third definition of gamma functions is given in terms of Weierstrass' infinite product

$$\frac{1}{\Gamma(n)} = ne^{n\gamma}\prod_{m=1}^{\infty}\left(1+\frac{n}{m}\right)e^{-n/m} \qquad (4.10)$$

where $\prod\limits_{n=1}^{\infty}$ stands for continuous product and the constant γ can be determined from the relation

$$e^{-\gamma} = \prod_{m=1}^{\infty}\left(1+\frac{1}{m}\right)e^{-1/m}$$

$$= \lim_{m \to \infty} \left(\sum_{\rho=1}^{m} \rho^{-1} - \log m \right) \tag{4.11}$$

and is called the Euler–Mascheroni constant. A numerical evaluation of γ yields,

$$\gamma = 0.57721566...$$

To arrive at this form of γ-function, we rewrite Eq. (4.6) as

$$\Gamma(n) = \lim_{m \to \infty} \frac{1}{n} \left(\frac{1}{n+1} \right) \left(\frac{2}{n+2} \right) \left(\frac{3}{n+3} \right) \cdots \left(\frac{m}{n+m} \right) m^n$$

$$= \lim_{m \to \infty} \frac{1}{n} (1+n)^{-1} \left(1 + \frac{n}{2} \right)^{-1} \left(1 + \frac{n}{3} \right)^{-1} \cdots \left(1 + \frac{n}{m} \right)^{-1} m^n$$

$$= \lim_{m \to \infty} \frac{1}{n} \prod_{\rho=1}^{m} \left\{ \left(1 + \frac{n}{\rho} \right)^{-1} \right\} m^n$$

or

$$\frac{1}{\Gamma(n)} = \lim_{m \to \infty} nm^{-n} \prod_{\rho=1}^{m} \left(1 + \frac{n}{\rho} \right)^{+1}$$

$$= n \lim_{m \to \infty} e^{-(\log m)^n} \prod_{\rho=1}^{m} \left(1 + \frac{n}{\rho} \right)^{+1} \qquad \because \quad m^{-n} = e^{(-\log m)^n}$$

or

$$\frac{1}{\Gamma(n)} = n \lim_{m \to \infty} e^{-(\log m)^n} \prod_{\rho=1}^{m} \left(1 + \frac{n}{\rho} \right)^{+1} \prod_{\rho=1}^{m} e^{n/\rho} \Big/ \prod_{\rho=1}^{m} e^{n/\rho}$$

Dividing and multiplying by

$$\exp\left[\left(1 + \frac{1}{2} + \frac{1}{3} + \cdots + \frac{1}{m} \right)^n \right] = \prod_{\rho=1}^{m} e^{n/\rho} \text{ we get}$$

$$\frac{1}{\Gamma(n)} = n \lim_{m \to \infty} \left[e^{\left\{ 1 + \frac{1}{2} + \frac{1}{3} + \cdots + \frac{1}{m} - \log m \right\}^n} \prod_{\rho=1}^{m} \left(1 + \frac{n}{\rho} \right) e^{-n/\rho} \right]$$

$$\frac{1}{\Gamma(n)} = \lim_{m \to \infty} \cdot n e^{n\gamma} \prod_{\rho=1}^{m} \left(1 + \frac{n}{\rho} \right) e^{-n/\rho}$$

which is Eq. (4.10). Thus, the three definitions are equivalent.

From Eq. (4.5), we have

$$\overline{(n+1)} = n\overline{n} = n(n-1)\overline{n-1}...$$

$$= n(n-1)(n-2)...2.1\overline{(1)} = n!$$

It should be remembered that

$$\overline{(0)} = (-1)! = \infty, \ \overline{(1)} = (0)! = 1, \overline{(-n)} = \infty$$

For n a positive integer and noninteger negative values of n gamma function has finite values but for negative integral values of n, the value is infinite. Gauss's pi function is also defined as

$$\prod(n) = n!$$

and hence

$$\prod(n) = \overline{(n+1)}$$

4.4. OTHER FORMS OF GAMMA FUNCTION

i. Putting $u = \lambda v$ in Eq. (4.4) and for $z = n$, we have

$$\Gamma(n) = \int_0^\infty e^{-\lambda v} \lambda^{n-1} v^{n-1} \lambda \, dv$$

$$= \lambda^n \int_0^\infty e^{-\lambda v} v^{n-1} dv$$

or

$$\frac{\Gamma(n)}{\lambda^n} = \int_0^\infty e^{-\lambda v} v^{n-1} dv \qquad (4.12)$$

ii. Again let $e^{-u} = v$　　　　　　　$\because \ u = \log_e \dfrac{1}{v}$

and

$$du = -\frac{dv}{v}$$

Equation (4.4) becomes

$$\Gamma(n) = \int_0^\infty e^{-u} u^{n-1} du = -\int_1^0 \left(\log \frac{1}{v}\right)^{n-1} dv$$

$$= \int_0^1 \left(\log \frac{1}{y}\right)^{n-1} dy \qquad (4.13)$$

iii. Now substituting $u^n = v$　$\therefore \ u = v^{1/n}$ and $du = \dfrac{1}{n} v^{(1-n)/n} dv$ in Eq. (4.4), we have

$$\Gamma(n) = \int_0^\infty e^{-u} u^{n-1} du$$

$$= \int_0^\infty e^{-v^{1/n}} \cdot v^{(n-1)/n} \cdot \frac{1}{n} \cdot v^{(1-n)/n} dv$$

$$= \frac{1}{n} \int_0^\infty e^{-v^{1/n}} dv \qquad (4.14)$$

Equations (4.12)–(4.14) also represent the gamma functions. Here from Eq. (4.14), we have

$$n\Gamma(n) = \int_0^\infty e^{-v^{1/n}} du$$

Put $n = \dfrac{1}{2}$. Then, we have

$$\frac{1}{2}\Gamma\!\left(\frac{1}{2}\right) = \int_0^\infty e^{-v^2}\,dv = \frac{\sqrt{\pi}^{\,*}}{2}$$

$\therefore$

$$\Gamma(n) = \frac{1}{n}\Gamma(n+1)$$

$$\Gamma\!\left(-\frac{1}{2}\right) = \frac{1}{-\frac{1}{2}}\Gamma\!\left(\frac{1}{2}\right) = -2\sqrt{\pi}$$

Similarly

$$\Gamma\!\left(-\frac{3}{2}\right) = \frac{\Gamma\!\left(-\frac{1}{2}\right)}{-\frac{3}{2}} = -\frac{2}{3}\cdot(-2\sqrt{\pi}) = \frac{4}{3}\sqrt{\pi} \quad \text{and so on.}$$

4.5 THE BETA (β) FUNCTION

β (beta) function in terms of definite Eulerian integral is defined as

$$\beta(m,\,n) = \int_0^1 t^{m-1}(1-t)^{n-1}\,dt$$

where m and n are positive, i.e. $m > 0$, $n > 0$. For m and n complex numbers, this definition is valid for Real $m > 0$ and Real $n > 0$. For m and n negative real numbers and complex numbers with negative values of Real m and Real n this integral does not converge.

For $R(m)$, $R(n) > -1$, we have from Eq. (4.1)

$$n!\,m! = \int_0^\infty e^{-x}x^m\,dx\int_0^\infty e^{-y}y^n\,dy = \lim_{A\to\infty}\int_0^A\int_0^A e^{-(x+y)}x^m y^n\,dx\,dy \qquad (4.15)$$

Put $x + y = z$ and eliminate y then

$$n!\,m! = \lim_{A\to\infty}\int_0^A dz\int_0^z e^{-z}x^m(z-x)^n\,dx$$

$${}^{*}I = \int_0^\infty e^{-x^2}\,dx \qquad \text{Let } x = ay \quad \therefore \quad dx = a\,dy$$

$$= \int_0^\infty e^{-a^2 y^2}\cdot a\,dy$$

or

$$I = \int_0^\infty e^{-a^2}\,da = \int_0^\infty e^{-a^2 y^2}a\,dy$$

or

$$I^2 = \int_0^\infty\int_0^\infty e^{-a^2(1+y^2)}a\,dy\,da$$

$$= \int_0^\infty \frac{dy}{2(1+y^2)} = \frac{1}{2}\left[\tan^{-1}y\right]_0^\infty$$

or

$$I^2 = \frac{\pi}{4} \quad \text{or} \quad I = \frac{\sqrt{\pi}}{2}$$

and putting $x = tz$, we have

$$n!\,m! = \lim_{A\to\infty}\int_0^A dz\int_0^1 z^{m+n+1}e^{-z}t^m(1-t)^n\,dt$$

$$= \int_0^1 t^m(1-t)^n\,dt \cdot \int_0^\infty z^{m+n+1}e^{-z}\,dz$$

$$= (m+n+1)!\int_0^1 t^m(1-t)^n\,dt$$

or

$$\frac{n!\,m!}{(m+n+1)!} = \int_0^1 t^m(1-t)^n\,dt \tag{4.16}$$

The integral of Eq. (4.16) is usually denoted by $\beta(m+1, n+1)$ and is known as beta function. Replacing m by $(m-1)$ and n by $(n-1)$, we have

$$\beta(m, n) = \int_0^1 t^{m-1}(1-t)^{n-1}\,dt \tag{4.17}$$

It can be shown that β-function is symmetric in m and n. For this let us put

$$t = 1 - p \qquad \because\ dt = -dp \quad \text{in Eq. (4.17).}$$

$$\therefore \qquad \beta(m, n) = \int_1^0 (1-p)^{(m-1)}\{1-1+p\}^{(n-1)}(-dp)$$

$$= \int_0^1 p^{(n-1)}(1-p)^{(m-1)}\,dp \tag{4.17a}$$

The definition of beta function as given by Eqs (4.17) and (4.17a) are identical. This definition in definite integral form is useful in establishing integral representations of Bessel's functions and hypergeometric functions discussed in later chapters.

4.6. OTHER FORMS OF BETA FUNCTION

i. In Eq. (4.16), let us put

$$t = \frac{u}{1+u}$$

Then

$$\beta(m+1, n+1) = \frac{n!\,m!}{(m+n+1)!} = \int_0^\infty \frac{u^m\,du}{(1+u)^{m+n+2}} \tag{4.18}$$

Replacing m by $(m-1)$ and n by $(n-1)$, we have

$$\beta(m, n) = \int_0^\infty \frac{u^{m-1}}{(1+u)^{m+n}} \tag{4.18a}$$

From Eq. (4.18a), we have

$$\beta(m, n) = \int_0^\infty \frac{u^{m-1}\,du}{(1+u)^{m+n}} = \int_0^1 \frac{u^{m-1}\,du}{(1+u)^{m+n}} + \int_1^\infty \frac{u^{m-1}}{(1+u)^{m+n}}\,du^* = \int_0^1 \frac{u^{m-1}+u^{n-1}}{(1+u)^{m+n}}\,du \tag{4.19}$$

---

*Put $u = \dfrac{1}{x}$

$$\int_1^\infty \frac{u^{m-1}\,du}{(1+u)^{m+n}} = \int_1^0 \frac{(1/x)^{m-1}}{(1+1/x)^{m+n}}\left(-\frac{1}{x^2}\right)dx = \int_0^1 \frac{x^{n-1}}{(1+x)^{m+n}}\,dx$$

ii. Putting $u = \dfrac{a}{b}x$ in Eq. (4.18a), we have

$$\beta(m, n) = \int_0^\infty \frac{u^{m-1} du}{(1+u)^{m+n}}$$

$$= a^m b^n \int_0^\infty \frac{x^{m-1}}{(ax+b)^{m+n}} dx \qquad (4.19a)$$

iii. In Eq. (4.17), put $t = \sin^2 \theta$; then

$$\beta(m, n) = \int_0^{\pi/2} \sin^{2m-2}\theta \, \cos^{2n-2}\theta \, d\theta \, \sin\theta \cos\theta$$

$$= 2 \int_0^{\pi/2} \sin^{2m-1}\theta \, \cos^{2n-1}\theta \, d\theta \qquad (4.20)$$

iv. Substituting $t = \dfrac{u}{a}$ in Eq. (4.17), we have

$$\beta(m, n) = \frac{1}{a^{m+n-1}} \int_0^a u^{m-1}(a-u)^{n-1} du \qquad (4.21)$$

Reduction of Certain Classes of Integrals to Gamma Functions

(a) Equations (4.18) and (4.18a) give

$$\int_0^\infty \frac{u^{m-1}}{(1+u)^{m+n}} du = \beta(m,n) = \frac{(n-1)!(m-1)!}{(m+n-1)!} = \frac{\lfloor n \rfloor m}{\lfloor m+n \rfloor} \qquad (4.21a)$$

(b) Equation (4.19a) gives

$$a^m b^n \int_0^\infty \frac{x^{m-1}}{(ax+b)^{m+n}} dx = \beta(m,n) = \frac{\lfloor n \rfloor m}{\lfloor m+n \rfloor}$$

Now put $x = \tan^2 \theta$, we get

$$a^m b^n \int_0^{\pi/2} \frac{(\sin\theta)^{2m-1}(\cos\theta)^{2n-1}}{(a\sin^2\theta + b\cos^2\theta)^{m+n}} d\theta = \beta(m,n) = \frac{\lfloor n \rfloor m}{\lfloor m+n \rfloor} \qquad (4.21b)$$

(c) Equation (4.19) gives

$$\int_0^1 \frac{u^{m-1} + u^{n-1}}{(1+u)^{m+n}} du = \beta(m,n) = \frac{\lfloor n \rfloor m}{\lfloor m+n \rfloor} \qquad (4.21c)$$

(d) Equation (4.20) gives

$$2 \int_0^{\pi/2} (\sin\theta)^{2m-1}(\cos\theta)^{2n-1} d\theta = \beta(m,n) = \frac{\lfloor n \rfloor m}{\lfloor m+n \rfloor} \qquad (4.21d)$$

or

$$\int_0^{\pi/2} (\sin\theta)^{2m-1}(\cos\theta)^{2n-1} d\theta = \frac{1}{2}\beta(m,n) = \frac{1}{2}\frac{\lfloor n \rfloor m}{\lfloor m+n \rfloor}$$

Now put $2m - 1 = p$ or $m = \dfrac{p+1}{2}$

and $2n - 1 = q$ or $n = \dfrac{q+1}{2}$

Then
$$\int_0^{\pi/2} (\sin\theta)^p (\cos\theta)^q \, d\theta = \frac{\left\lfloor \dfrac{p+1}{2} \right. \left\lfloor \dfrac{q+1}{2} \right.}{2 \left\lfloor \dfrac{p+q}{2} + 1 \right.} \quad p, q > -1 \tag{4.21e}$$

This integral is used in integral calculus for evaluating a large variety of integrals involving sine and cosine functions. In special case if $p = 0$, then

$$\int_0^{\pi/2} (\cos\theta)^q \, d\theta = \frac{\left\lfloor \dfrac{1}{2} \right. \left\lfloor \dfrac{q+1}{2} \right.}{2 \left\lfloor \dfrac{p+q}{2} + 1 \right.} = \frac{\left\lfloor \dfrac{q+1}{2} \right. \sqrt{\pi}}{2 \left\lfloor \dfrac{q}{2} + 1 \right.} \tag{4.21f}$$

Similarly if $q = 0$, then

$$\int_0^{\pi/2} (\sin\theta)^p \, d\theta = \frac{\left\lfloor \dfrac{p+1}{2} \right. \sqrt{\pi}}{2 \left\lfloor \dfrac{p}{2} + 1 \right.} \tag{4.21g}$$

Using these results, we can derive an important formula known as *duplication formula*.

If we put $2n = 1$ in (d) we have

$$\int_0^{\pi/2} (\sin\theta)^{2m-1} \, d\theta = \frac{\left\lfloor \dfrac{1}{2} \right. \left\lfloor m \right.}{2 \left\lfloor m + \dfrac{1}{2} \right.} = \frac{\left\lfloor m \right. \sqrt{\pi}}{2 \left\lfloor m + \dfrac{1}{2} \right.} \tag{4.21h}$$

and on putting $m = n$, we have

$$\int_0^{\pi/2} (\sin\theta)^{2m-1} (\cos\theta)^{2m-1} \, d\theta = \frac{\{\left\lfloor m \right.\}^2}{2 \left\lfloor 2m \right.}$$

or
$$\frac{\{\left\lfloor m \right.\}^2}{2 \left\lfloor 2m \right.} = \frac{1}{2^{2m-1}} \int_0^{\pi/2} (\sin 2\theta)^{2m-1} \, d\theta$$

$$= \frac{1}{2^{2m}} \int_0^{\pi} (\sin\phi)^{2m-1} \, d\phi \quad \text{(putting } 2\theta = \phi)$$

$$= \frac{1}{2^{2m-1}} \int_0^{\pi/2} (\sin\phi)^{2m-1} \, d\phi = \frac{1}{2^{2m-1}} \int_0^{\pi/2} (\sin\theta)^{2m-1} \, d\theta$$

Thus
$$\int_0^{\pi/2} (\sin\theta)^{2m-1} \, d\theta = \frac{2^{2m-2} \{\left\lfloor m \right.\}^2}{\left\lfloor 2m \right.} \tag{4.21i}$$

From Eqs (4.21h) and (4.21i), we get

$$\frac{\sqrt{\pi}\,\overline{|m}}{2\cdot\overline{\left|m+\dfrac{1}{2}\right.}} = 2^{2m-2}\,\frac{\{\overline{|m}\}^2}{\overline{|2m}} \qquad (4.21j)$$

or

$$\sqrt{\pi}\,\overline{|2m} = 2^{2m-1}\,\overline{|m}\;\overline{\left|m+\dfrac{1}{2}\right.} \qquad (4.22)$$

This is called *Legendre's duplication formula*.

(e) Using Eq. (4.17), we have

$$\beta(m,\,n) = \int_0^1 t^{m-1}(1-t)^{n-1}\,dt = \frac{\overline{|m}\;\overline{|n}}{\overline{|m+n}}$$

Now putting $m = n$, we have

$$\int_0^1 t^{m-1}(1-t)^{m-1} = \frac{\overline{|m}\;\overline{|m}}{\overline{|2m}}$$

Now put $t = (1+x)/2$ in the integral. We have

$$\frac{\overline{|m}\;\overline{|m}}{\overline{|2m}} = \frac{1}{2^{2m-1}}\int_{-1}^{+1}(1+x)^{m-1}(1-x)^{m-1}\,dx$$

$$= \frac{1}{2^{2m-1}}\int_{-1}^{+1}(1-x^2)^{m-1}\,dx$$

$$= \frac{2}{2^{2m-1}}\int_0^1(1-x^2)^{m-1}\,dx$$

Thus

$$\int_0^1(1-x^2)^{m-1}\,dx = 2^{2m-2}\cdot\frac{\overline{|m}\;\overline{|m}}{\overline{|2m}} \qquad (4.22a)$$

Now putting $x^2 = y$, we get

$$\frac{1}{2}\int_0^1(1-y)^{m-1}y^{-1/2}\,dy = \frac{1}{2}\beta\!\left(m,\,\frac{1}{2}\right) \quad \text{by definition.}$$

$$= \frac{\overline{|m}\;\overline{\left|\dfrac{1}{2}\right.}}{2\,\overline{\left|m+\dfrac{1}{2}\right.}}$$

$$\therefore \qquad \frac{\overline{|m}\;\overline{|m}}{\overline{|2m}}\,2^{m-2} = \frac{\overline{|m}\;\overline{\left|\dfrac{1}{2}\right.}}{2\,\overline{\left|m+\dfrac{1}{2}\right.}}$$

$$\therefore \qquad \sqrt{\pi}\,\overline{|2m} = 2^{2m-1}\,\overline{|m}\;\overline{\left|m+\dfrac{1}{2}\right.} \qquad (4.23)$$

(f) If we put $m + n = 1$ in Eq. (4.23a), we have

$$\int_0^\infty \frac{u^{m-1}}{1+u}du = \beta(m, 1-m) = \frac{\overline{|m|1-m}}{\overline{|1}} \quad \text{and} \quad \overline{|1} = 1$$

$\therefore$

$$\int_0^\infty \frac{u^{m-1}}{1+u}du = \overline{|m|1-m} \tag{4.23a}$$

This integral is solved using contour integration method and value is $\dfrac{\pi}{\sin m\pi}$.

$\therefore$

$$\overline{|m|1-m} = \frac{\pi}{\sin m\pi} \tag{4.23b}$$

(g) Putting $m = m + 1$ and $n = n + 1$ in Eq. (4.23a), we get

$$\int_0^\infty \frac{u^m du}{(1+u)^{m+n+2}} = \frac{\overline{|m+1|n+1}}{\overline{|m+n+2}}$$

Now putting $m = -n$, we get

$$\int_0^\infty \frac{u^m du}{(1+u)^2} = \frac{\overline{|m+1|1-m}}{\overline{|2}}$$

But

$$\overline{|m+1|1-m} = m\overline{|m|1-m} = \frac{m\pi}{\sin m\pi} \qquad [using \text{ Eq. (4.23b)}]$$

$\therefore$

$$\int_0^\infty \frac{u^m du}{(1+u)^2} = \frac{m\pi}{\overline{|2}\sin m\pi} = \frac{m\pi}{\sin m\pi} \qquad \because \overline{|2} = 1$$

This relation may be written as

$$\int_0^\infty \frac{u^m du}{(1+u)^2} = (m)!(-m)! = \frac{m\pi}{\sin m\pi} \tag{4.23c}$$

4.7. DERIVATIVES OF GAMMA FUNCTIONS (DIGAMMA AND POLYGAMMA FUNCTIONS)

The gamma functions as defined by Eq. (4.4) cannot be differentiated conveniently. As such the natural logarithm of the function is taken which converts the product to sum and then differentiated. Now to deal with the derivatives of gamma functions let us take Eq. (4.10) which is

$$\Gamma(n) = n^{-1}e^{-\gamma n}\prod_{m=1}^{\infty}\left(1+\frac{n}{m}\right)^{-1}e^{n/m}$$

On taking natural logarithm, this equation reduces to

$$\log\Gamma(n) = -\log n - \gamma n + \sum_{m=1}^{\infty}\{-\log(m+n) + \log m + n/m\}$$

Differentiating w.r.t. n, we get

$$\frac{d\log\Gamma(n)}{dn} = -\frac{1}{n} - \gamma + \sum_{m=1}^{\infty}\left\{\frac{1}{m} - \frac{1}{m+n}\right\}$$

This logarithmic derivative of gamma function $F_1(n)$ is called as *digamma function*. So that

$$F_1(n) = \frac{d \log \Gamma n}{dn} = \frac{\Gamma(n)'}{\Gamma(n)} = -\frac{1}{n} - \gamma + \sum_{m=1}^{\infty} \left\{ \frac{1}{m} - \frac{1}{m+n} \right\} \tag{4.24}$$

When n is an integer N, then

$$F_1(N) = -\gamma - \frac{1}{N} + \sum_{m=1}^{\infty} \left\{ \frac{1}{m} - \frac{1}{m+N} \right\} = -\gamma + \sum_{m=1}^{\infty} \frac{1}{m} - \sum_{m=0}^{\infty} \frac{1}{m+N} \tag{4.25}$$

and it is obvious that

$$F_1(1) = -\gamma \tag{4.26}$$

The derivative of $F_1(n)$ is derived from the Eq. (4.24)

$$F_2(n) = \frac{d^2 \log \Gamma(n)}{dn^2} = \frac{dF_1(n)}{dn} = \frac{1}{n^2} + \sum_{m=1}^{\infty} \frac{1}{(m+n)^2} = \sum_{m=0}^{\infty} \frac{1}{(m+n)^2} \tag{4.27}$$

and

$$F_3(n) = \frac{dF_2(n)}{dn} = (-1)^3 2! \sum_{m=0}^{\infty} \frac{1}{(m+n)^3}$$

Thus in general

$$F_p(n) = \frac{dF_{p-1}(n)}{dx} = (-1)^p \Gamma(p) \sum_{m=0}^{\infty} \frac{1}{(m+n)^p} \tag{4.28}$$

These derivatives of $\log \Gamma(n)$ of order two or higher than two are defined as the polygamma functions.

$$F_p(n+1) = (-1)^p \Gamma(p) \sum_{m=0}^{\infty} \frac{1}{(m+n+1)^p}$$

$$= (-1)^p \Gamma(p) \sum_{m=1}^{\infty} \frac{1}{(m+n)^p} = (-1)^p \Gamma(p) \left\{ \sum_{m=0}^{\infty} \frac{1}{(m+n)^p} - \frac{1}{n^p} \right\}$$

$$F_p(n+1) = F_p(n) - (-1)^p \Gamma(p)/np \tag{4.29}$$

Again in order to represent $F_1(n)$ as as definite integral, we write the Eq. (4.24) in the form

$$F_1(n) = -\gamma + \sum_{m=1}^{\infty} \frac{1}{m} - \sum_{m=0}^{\infty} \frac{1}{m+n}$$

$$= -\gamma + \lim_{m \to \infty} \left(1 + \frac{1}{2} + \frac{1}{3} + \cdots + \frac{1}{m} \right) - \lim_{m \to \infty} \left(\frac{1}{n} + \frac{1}{n+1} + \cdots + \frac{1}{n+m} \right)$$

$$= \int_0^{\infty} e^{-t} \log t \, dt + \lim_{m \to \infty} \sum_{p=1}^{m} \int_0^{\infty} e^{-pt} \, dt - \lim_{m \to \infty} \sum_{p=0}^{m} \int_0^{\infty} e^{-(n+p)t} \, dt$$

where
$$-\gamma = \left.\frac{|dz!}{dz}\right|_{z=0} = \int_0^\infty e^{-t} \log t\, dt \qquad \text{[using Eqs (4.3) and (4.11)]}$$

$\therefore$
$$F_1(n) = \int_0^\infty e^{-t} \log t\, dt + \int_0^\infty \frac{e^{-t}}{1-e^{-t}} dt - \int_0^\infty \frac{e^{-nt}}{1-e^{-t}} dt$$

$$= \int_0^\infty e^{-t} \log t\, dt + \int_0^\infty \frac{e^{-t} - e^{-nt}}{1-e^{-t}} dt \qquad (4.30)$$

Equation (4.30) represents integral form of digamma function. To obtain integral formula for polygamma functions, we differentiate Eq. (4.30) w.r.t.. n. Thus, we get

$$F_p(n) = (-1)^p \int_0^\infty \frac{t^{p-1} e^{-nt}}{1-e^{-t}} dt \qquad (4.31)$$

Polygamma functions are used for the ready evaluation of sums, each term of which is a rational function of summation index. For example the sum;

$$S = \sum_{P=0}^\infty \frac{1}{(P+1)^2 (P+a)^2}$$

can be written as

$$S = \frac{1}{(a-1)^2} \sum_{P=0}^\infty \left\{ \left[\frac{1}{(P+1)^2} + \frac{1}{(P+a)^2} \right] - \frac{2}{a-1} \left[\frac{1}{P+1} + \frac{1}{P+a} \right] \right\}$$

$$= \frac{1}{(a-1)^2} \left\{ [F_2(1) + F_2(a)] - \frac{2}{a-1}[F_1(1) + F_1(a)] \right\}$$

Another important use of polygamma functions is made in the evaluation of Riemann zeta functions of different orders. A general Riemann zeta function is defined as

$$\xi(m) = 1 + \frac{1}{2^m} + \frac{1}{3^m} + \cdots + = \sum_{P=1}^\infty \frac{1}{P^m}$$

If we put $n = 1$ in Eq. (4.28), we have

$$F_p(1) = (-)^P \Gamma(P) \sum_{m=0}^\infty \frac{1}{(m+1)^P}$$

$$= (-1)^P \Gamma(P) \sum_{m=1}^\infty \frac{1}{m^P} = (-)^P \Gamma(P)\xi(P)$$

Thus
$$\xi(P) = (-)^P \frac{1}{\Gamma(P)} F_p(1)$$

This equation determines $\xi(p)$ in terms of polygamma functions.

4.8. STIRLING'S FORMULA FOR LARGE n

It is very difficult to simplify algeberiacally or to differentiate formula involving $n!$ or $\Gamma(n + 1)$. Stirling derived an approximate formula for gamma functions and factorial functions involving large n. Stirling formula can be derived as follows.

$$\Gamma(n + 1) = n! = \int_0^\infty t^n e^{-t} dt \qquad\qquad [\,using \text{ Eq. (4.1)}]$$

$$= \int_0^\infty e^{\log t^n} e^{-t} dt = \int_0^\infty e^{(\log t^n - 1)} dt$$

Let $t = n + y\sqrt{n}$. Then

$$n! = \int_{-\sqrt{n}}^\infty e^{n \log(n + y\sqrt{n}) - (n + y\sqrt{n})} \cdot \sqrt{n}\, dy$$

For large n, the logarithm can be expanded in the following series.

$$\log(n + y\sqrt{n}) = \log n + \log\left(1 + \frac{y}{\sqrt{n}}\right)$$

$$= \log n + \frac{y}{\sqrt{n}} - \frac{y^2}{2n} + \cdots$$

Neglecting higher powers, we have approximately

$$\log(n + y\sqrt{n}) \approx \log n + \frac{y}{\sqrt{n}} - \frac{y^2}{2n}.$$

$$\therefore \qquad n! = \int_{-\sqrt{n}}^\infty e^{\{n \log n + y\sqrt{n} - (y^2/2) - n - y\sqrt{n}\}} \sqrt{n}\, dy$$

$$= \sqrt{n}\, e^{(n \log n - n)} \int_{-\sqrt{n}}^\infty e^{-y^2/2} dy$$

$$= \sqrt{n}\, n^n e^{-n} \left\{ \int_{-\infty}^{+\infty} e^{-y^2/2} dy - \int_{-\infty}^{-\sqrt{n}} e^{-y^2/2} dy \right\}$$

Here the second integral tends to zero as $n \to \infty$ and we have

$$n! = \sqrt{n}\, n^n e^{-n} \int_{-\infty}^{+\infty} e^{-y^2/2} dy$$

$$= \sqrt{n}\, n^n e^{-n} 2 \int_0^{+\infty} e^{-y^2/2} dy$$

$$= \sqrt{n}\, n^n e^{-n} 2 \frac{\sqrt{2\pi}}{2}$$

$$\therefore \qquad n! = \sqrt{2n\pi}\, n^n e^{-n} \qquad\qquad \dots(4.32)$$

The most immediate application of Stirling's approximate formula is the derivation of Wallis, historical formula for π. Writing Eq. (4.32) as

$$\left(\frac{\pi}{2}\right)^{1/2} = \lim_{n \to \infty} n!(4n)^{-1/2} n^{-n} e^n$$

$$= \lim_{n \to \infty} n! 2^{-1} n^{-(n+1/2)} e^n$$

Taking logarithm on both sides, we have

$$\frac{1}{2}\log\frac{\pi}{2} = \lim_{n\to\infty}[\log n! - (n+1/2)\log n + n - \log 2]$$

or

$$\log\frac{\pi}{2} = 2\lim_{n\to\infty}\left[\log n! - \left(n+\frac{1}{2}\right)\log n + n - \log 2\right] \tag{4.33}$$

Equation (4.33) is known as *Wallis formula* for π. This formula was derived by Wallis by a complicated method using limiting inequalities of definite integrals.

4.9. INCOMPLETE BETA AND GAMMA FUNCTIONS

With variable upper limit in definition of beta and gamma functions are known as incomplete beta and incomplete gamma functions. Thus, incomplete beta function is defined as

$$B_x(m, n) = \int_0^x t^{m-1}(1-t)^{n-1}dt \tag{4.34}$$

for $0 \le x \le 1$, $m \ge 0$, $n \ge 0$. Its series expansion is given by

$$B_x(m, n) = x^m\left\{\frac{1}{m} + \frac{1-n}{m+1}x + \frac{(1-n)(2-n)}{2!(m+2)}x^2 + \cdots + \frac{(1-n)(2-n)\cdots(\rho-n)}{\rho!(m+\rho)}x^\rho + \cdots\right\} \tag{4.35}$$

For $x = 1$, $B_1(m, n) = B(m, n)$ we get the regular beta function.

$$\therefore \qquad B(m, n) = \int_0^1 t^{m-1}(1-t)^{n-1}dt$$

$$= \left\{\frac{1}{m} + \frac{1-n}{m+1} + \frac{(1-n)(2-n)}{2!(m+2)} + \cdots + \frac{(1-n)(2-n)\cdots(\rho-n)}{\rho!(m+\rho)} + \cdots\right\}$$

The above relation gives the series expansion for regular beta functions.

We have already given Euler's definition for gamma functions, [Eq. (4.4)]. In a similar manner to definite incomplete beta functions, the incomplete gamma functions can be defined by using variable limit integrals. Here two incomplete generalised gamma functions are defined as follows.

$$\gamma(n, x) = \int_0^x e^{-t}t^{n-1}dt \tag{4.36}$$

and

$$\Gamma(n, x) = \int_x^\infty e^{-t}t^{n-1}dt \tag{4.37}$$

where real $n > 0$, if n is complex, and $n > 0$, if n is real. Obviously

$$\gamma(n, x) + \Gamma(n, x) = \Gamma(n) \tag{4.38}$$

For n positive integer, the integrals in Eqs (4.36) and (4.37) are easily solved completely by parts (repeatedly). Thus, we have

$$\gamma(n, x) = (n-1)!\left(1 - e^{-x}\sum_{\rho=0}^{n-1}\frac{x^\rho}{\rho!}\right) \tag{4.39}$$

and

$$\Gamma(n, x) = (n-1)!\,e^{-x}\sum_{\rho=0}^{n-1}\frac{x^\rho}{\rho!} \tag{4.40}$$

If n is noninteger, then Eq. (4.36) can be written as power series expansion by putting $e^{-t} = \sum\limits_{\rho=0}^{\infty} \dfrac{(-1)^\rho t^\rho}{\rho!}$

Then

$$\gamma(n, x) = \int_0^x \sum_{\rho=0}^{\infty} (-1)^\rho \cdot \frac{t^{\rho+n-1}}{\rho!} dt$$

$$= \sum_{\rho=0}^{\infty} (-1)^\rho \cdot \frac{x^{\rho+n}}{\rho!(\rho+n)} \qquad (4.41)$$

Similarly for nonintegral n, the function $\Gamma(n, x)$ can be expressed as series expansion by integrating right hand side by parts. Thus

$$\Gamma(n, x) = e^{-x} x^{n-1} + \int_x^\infty (n-1)e^{-t} t^{n-2} dt$$

$$= e^{-x} x^{n-1} - (1-n)e^{-x} x^{n-2} + (1-n)(2n-n)e^{-x} x^{n-3}$$

$$\cdots + (-1)^m (1-n)(2-n)\cdots(m-n)\int_x^\infty e^{-t} t^{n-m-1} dt$$

Here the series diverges for nonintegral n and hence few last integral cannot be solved. But we have

$$\Gamma(n, x) - S_m(n, x) = (-1)^{m+1} \frac{(m-n+1)!}{(-n)!} \int_x^\infty e^{-t} t^{n-m-2} dt \qquad (4.42)$$

where,

$$S_m(n, x) = \sum_{\rho=0}^{m} (-1)^\rho e^{-x} \frac{(\rho-n)!}{(-n)!} x^{n-\rho-1} \qquad (4.43)$$

In absolute value

$$|\Gamma(n,x) - S_m(n,x)| \le \left| \frac{(m-n+1)!}{(-n)!} \int_x^\infty e^{-t} t^{n-m-2} dt (-1)^{m+1} \right|$$

$$\le \frac{(m-n+1)!}{(-n)!} \int_x^\infty t^{n-m-2} dt$$

$$\le \frac{(m-n)!}{(-n)!} x^{n-m-1}$$

$$\le \frac{(m-n)!}{(-n)!} \frac{1}{x^{m+1-n}}$$

$$\le 0$$

If x is sufficiently large and hence if we take x large enough, the partial sum S_m is a arbitrary good approximation to the function $\Gamma(n, x)$. Thus for large x

$$\Gamma(n, x) = \lim_{m\to\infty} S_{(m)}(n,x) = \sum_{\rho=0}^{\infty} (-1)^\rho e^{-x} \frac{(\rho-n)!}{(-n)!} x^{n-\rho-1}$$

or
$$\overline{(n,x)} = x^{n-1}e^{-x}\sum_{\rho=0}^{\infty}(-1)^{\rho}\frac{(\rho-n)!}{(-n)!}\cdot\frac{1}{x^{\rho}} \tag{4.43a}$$

4.10. EXPONENTIAL AND RELATED INTEGRALS

Exponential Integrals

These are related to a case of the incomplete gamma function which we shall denote as

$$e_1(z) = \int_z^{\infty}\frac{e^{-t}}{t}dt \tag{4.44}$$

Comparing with Eq. (4.37), we observe that
$$\Gamma(0,z) = e_1(z) \tag{4.45}$$

Now
$$\int_z^{\infty}\frac{e^{-t}}{t}dt = \lim_{P\to\infty}\int_z^P\frac{e^{-t}}{t}dt = \lim_{P\to\infty}\int_z^P\frac{dt}{t} - \lim_{P\to\infty}\int_z^P\frac{1-e^{-t}}{t}dt$$

But by expansion and integration, we have

$$\int_z^P\frac{e^{-t}}{t}dt = \int_z^P\frac{dt}{t} - \int_z^P\frac{1-e^{-t}}{t}dt = \left[\log t - \left\{t - \frac{t^2}{2.2!} + \frac{t^3}{3.3!}\cdots\right\}\right]_z^P$$

The left hand side tends to definite limit as $p \to \infty$: hence the right hand side does the same. Suppose the definite limit is c. Then

$$\int_z^{\infty}\frac{e^{-t}}{t}dt = c - \log z + \sum_{P=1}^{\infty}(-1)^P\frac{z^P}{p.p!} \tag{4.46}$$

To identify constant c, we have
$$\overline{(n+1)} = \int_0^{\infty}e^{-u}u^n du$$

or
$$\frac{d\Gamma(n+1)}{dn} = \frac{d}{dn}\int_0^{\infty}e^{-u}u^n du = \int_0^{\infty}e^{-u}\log u\, u^n du$$

[using Eqs (4.24) and (4.25) by replacing n by $(n + 1)$]

$$F_1(n + 1) = \frac{d\log\overline{n+1}}{dn} = \int_0^{\infty}e^{-u}\log u\, u^n du$$

Putting $n = 0$ and using Eq. (4.26), we have

$$F_1(1) = -\gamma = \int_0^{\infty}e^{-t}\log t\, dt \tag{4.47}$$

Now
$$\int_z^{\infty}e^{-t}\log t\, dt = \left[-e^{-t}\log t\right]_z^{\infty} + \int_z^{\infty}\frac{e^{-t}}{t}dt$$

$$= e^{-z}\log z + e_1(z)$$

$$= c - (1-e^{-z})\log z + \sum_{p=1}^{\infty}(-1)^P\frac{z^P}{p.p!}$$

and again

$$\int_z^\infty e^{-t} \log t\, dt = \int_0^\infty e^{-t} \log t\, dt - \int_0^z e^{-t} \log t\, dt$$

$$= -\gamma - \int_0^z e^{-t} \log t\, dt$$

or $\quad c - (1 - e^{-z}) \log z + \sum_{p=1}^\infty (-1)^p \dfrac{z^p}{p.p!} = -\gamma - \int_0^z e^{-t} \log t\, dt$

Putting $z = 0$ yields

$$c = -\gamma$$

Substitution of the value of c in Eq. (4.46) we get power series expansion for exponential function for small z.

$$e_1(z) = -\gamma - \log z + \sum_{p=1}^\infty (-1)^p \frac{z^p}{p.p!} \tag{4.48}$$

An asymptotic expansion for large value of z may be derived directly from Eq. (4.43a) by substituting $n = 0$ since

$$e_1(z) = \Gamma(0, z)$$

Thus, the asymptotic expansion for $e_1(z)$ is

$$e_1(z) \sim z^{-1} e^{-z} \sum_{p=0}^\infty (-1)^p \frac{p!}{z^p} \tag{4.49}$$

The power series for sine and cosine integrals is derived if we put $z = iy$ where, y is real and positive.

$$e_1(iy) = \int_{iy}^\infty \frac{e^{-t}}{t} dt = \int_y^\infty \frac{e^{-iu}}{u} du = \int_y^\infty \frac{\cos u}{u} du - i \int_y^\infty \frac{\sin u}{u} du \tag{4.49a}$$

$$= -\gamma - \log y - \frac{1}{2}\pi i + i\left(y - \frac{y^3}{3.3!} + \frac{y^5}{5.5!} + \cdots \right) + \left(\frac{y^2}{2.2!} - \frac{y^4}{4.4!} + \cdots \right)$$

Hence $\quad \displaystyle\int_y^\infty \frac{\cos u}{u} du = -\gamma - \log y + \frac{y^2}{2.2!} - \frac{y^4}{4.4!} + \cdots \tag{4.50}$

and $\quad \displaystyle\int_y^\infty \frac{\sin u}{u} du = \frac{\pi}{2} - y + \frac{y^3}{3.3!} - \frac{y^5}{5.5!} + \cdots \tag{4.51}$

The asymptotic expansions for these integrals follow from Eq. (4.49) by putting $z = iy$ which gives

$$e_1(iy) = e^{-iy}\left\{ -\frac{i}{y} + \frac{1}{y^2} - \frac{i.2!}{y^3} - \frac{3!}{y^4} + \cdots \right\}$$

Using Eq. (4.49a), we have

$$e_1(iy) = \int_y^\infty \frac{\cos u}{u} du - i\int_y^\infty \frac{\sin u}{u} du$$

$$= (\cos y - i\sin y)\left\{ -\frac{i}{y} + \frac{1}{y^2} - \frac{i.2!}{y^3} - \frac{3!}{y^4} + \cdots \right\}$$

Now equating real and imaginary parts we have

$$\int_y^\infty \frac{\cos u}{y}\,du \sim \cos y\left\{\frac{1}{y^2} - \frac{3!}{y^4} + \cdots\right\} - \sin y\left(\frac{1}{y} - \frac{2!}{y^3} + \frac{4!}{y^5}\cdots\right) \tag{4.52}$$

and

$$\int_y^\infty \frac{\sin u}{y}\,du \sim \cos y\left\{\frac{1}{y} - \frac{2!}{y^3} + \frac{4!}{y^5}\cdots\right\} + \sin y\left(\frac{1!}{y^2} - \frac{3!}{y^4}\cdots\right) \tag{4.53}$$

The special forms related to exponential integrals are sine, cosine and logarithmic integrals, defined by

$$S_i(y) = \int_y^\infty \frac{\sin u}{u}\,du \tag{4.54}$$

$$C_i(y) = -\int_y^\infty \frac{\cos u}{u}\,du \tag{4.55}$$

$$l_i(y) = \int_0^y \frac{du}{\log u} \tag{4.56}$$

Hence using Eqs (4.50) and (4.51), we have

$$C_i(y) = y + \log y - \frac{y^2}{2.2!} + \frac{y^4}{4.4!} + \cdots \tag{4.56a}$$

$$S_i(y) = -\frac{\pi}{2} + y - \frac{y^3}{3.3!} + \frac{y^5}{5.5!} + \cdots \tag{4.56b}$$

The asymptotic expansion for these integrals is given in Eqs (4.52) and (4.53). If we put $z = -\log y$ and $t = -\log u$ in Eq. (4.44) we get

$$e_1(-\log y) = -\int_0^y \frac{du}{\log u}$$

Hence the logarithmic integral of Eq. (4.56) takes the form

$$l_i(y) = -e_1(-\log y) \tag{4.56c}$$

The principal value being taken if y is real and greater than 1.

Error Integral

The error functions are frequently used in probability theory and statistical mechanics. The power series of incomplete gamma functions are used to define error integrals. The error integrals are defined as

$$e_r f(x) = \frac{2}{\sqrt{\pi}}\int_0^x e^{-t^2}\,dt \tag{4.57}$$

and the complementary error function $e_r f_c(x)$ is defined as

$$e_r f_c(x) = \frac{2}{\sqrt{\pi}}\int_x^\infty e^{-r^2}\,dt \tag{4.58}$$

In order to write them as incomplete gamma functions, let us substitute $t^2 = y$ in Eq. (4.57). Then

$$e_r f(x) = \frac{2}{\sqrt{\pi}} \int_0^{x^2} \frac{1}{2} e^{-y} \cdot y^{-1/2} dy$$

$$= \frac{1}{\sqrt{\pi}} \int_0^{x^2} e^{-y} \cdot y^{-1/2} dy = \frac{1}{\sqrt{\pi}} \gamma\left(\frac{1}{2}, x^2\right) \tag{4.59}$$

The power series expansion $e_r f(x)$ for small x follows directly from Eq. (4.41) by putting $n = \frac{1}{2}$ and replacing x by x^2. Thus

$$e_r f(x) = \frac{2}{\sqrt{\pi}} \sum_{P=0}^{\infty} \frac{(-1)^P x^{2P+1}}{p!(2p+1)}$$

$$= \frac{2}{\sqrt{\pi}} \left(x - \frac{x^3}{3.1!} + \frac{x^5}{5.2!} \cdots\right) \tag{4.60}$$

Similarly substituting $t^2 = y$ in Eq. (4.57), we have

$$e_r f_c(x) = \frac{1}{\sqrt{\pi}} \Gamma\left(\frac{1}{2}, x^2\right) \tag{4.61}$$

Now

$$e_r f(x) = \frac{2}{\sqrt{\pi}} \int_0^x e^{-t^2} dt$$

$$= \frac{2}{\sqrt{\pi}} \left[\int_0^{\infty} e^{-t^2} dt - \int_x^{\infty} e^{-t^2} \cdot dt\right]$$

$$= \frac{2}{\sqrt{\pi}} \left[\frac{\sqrt{\pi}}{2} - \int_x^{\infty} e^{-t^2} dt\right]$$

$$= 1 - \frac{2}{\sqrt{\pi}} \int_x^{\infty} e^{-t^2} dt = 1 - e_r f_c(x) \tag{4.62}$$

Now using Eqs (4.62) and (4.60), the power series for $e_r f_c(x)$ may be written directly. The asymptotic expansion of $e_r f(x)$ is directly obtained using Eq. (4.62). Thus we have

$$e_r f(x) = 1 - \frac{2}{\sqrt{\pi}} \int_x^{\infty} e^{-t^2} dt$$

Here

$$\int_x^{\infty} e^{-t^2} dt = \int_x^{\infty} \frac{1}{t} \frac{d}{dt}\left(-\frac{1}{2} e^{-t^2}\right) dt$$

$$= \frac{1}{t}\left(-\frac{1}{2} e^{-t^2}\right) - \int_x^{\infty} \left(-\frac{1}{2} e^{-t^2}\right)\left(-\frac{1}{t^2}\right) dt$$

$$= \frac{1}{2x} e^{-x^2} - \frac{1}{2} \int_x^{\infty} \frac{1}{t^2} e^{-t^2} dt \qquad \text{(integrating by parts)}$$

Similarly
$$\int_x^\infty \frac{1}{t^2} e^{-t^2} dt = \frac{1}{2x^3} e^{-x^2} - \frac{3}{2} \int_x^\infty \frac{1}{t^4} e^{-t^2} dt$$

Continuing this process, we obtain

$$e_r f(x) \sim 1 - \frac{1}{\sqrt{\pi} x} e^{-x^2} \left\{ 1 - \frac{1}{2x^2} + \frac{1.3}{4x^4} - \frac{1.3.5}{8x^6} + \cdots \right\} \tag{4.63}$$

which is the required asymptotic expansion for $e_r f(x)$. The asymptotic expression for $e_r f(x)$ directly follows from the Eqs (4.63) and (4.62).

From the definition of $e_r f(x)$, we have
$$e_r f(-x) = - e_r f(x)$$
$$e_r f(0) = 0$$

$$\left.\begin{array}{l} e_r f(\infty) = \dfrac{2}{\sqrt{\pi}} \int_0^\infty e^{-t^2} dt = \dfrac{2}{\sqrt{\pi}} \dfrac{\sqrt{\pi}}{2} = 1 \\[3mm] e_r f(iy) = \dfrac{2i}{\sqrt{\pi}} \int_0^y e^{-t^2} dt = \dfrac{2i}{\sqrt{\pi}} \cdot \dfrac{\sqrt{\pi}}{2} = i = \sqrt{-1} \end{array}\right] \tag{4.64}$$

Elliptic Integrals

The elliptic integrals are very useful because they frequently come in applied problems. Though these integrals are not related to gamma or beta functions still it is worthwhile to give them here in brief.

Legendre form of the elliptic integrals of first and second kind are given by

$$F(K, \theta) = \int_0^\theta \frac{d\phi}{(1 - K^2 \sin^2 \phi)^{1/2}} \quad (0 < K < 1) \tag{4.65}$$

and
$$E(K, \theta) = \int_0^\theta \frac{d\phi}{(1 - K^2 \sin^2 \phi)^{-1/2}} = \int_0^\phi (1 - K^2 \sin^2)^{1/2} d\phi \quad (0 < K < 1) \tag{4.65a}$$

where K is called the modulus and θ the amplitude of the elliptic integrals. These integrals are represented in terms of hypergeometric functions (Chapter 7). Values of these integrals for $\theta = \dfrac{\pi}{2}$ are defined as *complete elliptical integrals*. These are given by

$$K(k) = F(k, \pi/2) = \int_0^{\pi/2} \frac{d\phi}{(1 - K^2 \sin^2 \phi)^{1/2}} \tag{4.65b}$$

and
$$E(k) = E(k, \pi/2) = \int_0^{\pi/2} (1 - K^2 \sin^2 \phi)^{1/2} d\phi \tag{4.66}$$

From these equations it is obvious that
$$F(k, n\pi \pm \theta) = 2nK \pm F(K, \theta) \tag{4.67}$$
$$E(k, n\pi \pm \theta) = 2nE(K) \pm (K, \theta) \tag{4.67a}$$

Expanding the integrands in Eqs (4.65) and (4.66), using binomial theorem, we get convergent infinite series since $K^2 \sin^2 \phi < 1$ (for $K < 1$).

Integrals of functions $(1 - K^2 \sin^2 \phi)^{1/2}$ and $(1 - K^2 \sin^2 \phi)^{-1/2}$ with any limits can be represented in terms of elliptical integrals. For example,

$$\int_{\theta_1}^{\theta_2} \frac{d\phi}{(1 - K^2 \sin \phi)^{1/2}} = \int_0^{\theta_2} \frac{d\phi}{(1 - K^2 \sin^2 \phi)^{1/2}} - \int_0^{\theta_1} \frac{d\phi}{(1 - K^2 \sin^2 \phi)^{1/2}}$$

$$= F(K, \theta_2) - F(K, \theta_1) \tag{4.68}$$

Similarly

$$\int_{\theta_1}^{\theta_2} \frac{d\phi}{(1 - K^2 \sin^2 \phi)^{-1/2}} = E(K, \theta_2) - E(K, \theta_1) \tag{4.68a}$$

Elliptical integrals with negative amplitude are

$$F(K, -\theta) = \int_0^{-\theta} \frac{d\phi}{(1 - K^2 \sin^2 \phi)^{1/2}} = \int_0^{\theta} \frac{d\phi}{(1 - K^2 \sin^2 \phi)^{1/2}} = -F(K, \theta) \tag{4.69}$$

and $E(K, -\theta) = -E(K, \theta)$ \hfill (4.69a)

Jacobi forms of elliptic integrals are derived by putting $\sin \phi = x$ in Eqs (4.65) and (4.66). Thus

$$F(K, \theta) = \int_0^x \frac{dx}{\sqrt{(1 - x^2)(1 - k^2 x^2)}} \tag{4.70}$$

$$E(K, \theta) = \int_0^x \sqrt{\frac{1 - k^2 x^2}{1 - x^2}} \, dx \tag{4.70a}$$

Similarly

$$K(k) = F(k, \pi/2) = \int_0^1 \frac{dx}{\sqrt{(1 - x^2)(1 - k^2 x^2)}} \tag{4.71}$$

and

$$E(k) = E(k, \pi/2) = \int_0^1 \sqrt{\frac{1 - k^2 x^2}{1 - x^2}} \, dx \tag{4.71a}$$

A number of integrals can be reduced to one or a combination of these forms.

Example 1. *Evaluate* $\Gamma\left(\dfrac{1}{n}\right)\Gamma\left(\dfrac{2}{n}\right)\Gamma\left(\dfrac{3}{n}\right)\cdots\Gamma\left(\dfrac{n-1}{n}\right)$ *where, n is a positive integer.*

Solution: Let $\quad P = \Gamma\left(\dfrac{1}{n}\right)\Gamma\left(\dfrac{2}{n}\right)\Gamma\left(\dfrac{3}{n}\right)\cdots\Gamma\left(\dfrac{n-1}{n}\right)$ \hfill (1)

and by reversing the order we have

$$P = \Gamma\left(\frac{n-1}{n}\right)\Gamma\left(\frac{n-2}{n}\right)\Gamma\left(\frac{n-3}{n}\right)\cdots\Gamma\left(\frac{2}{n}\right)\Gamma\left(\frac{1}{n}\right) \tag{2}$$

Multiplying Eqs (1) and (2), we have

$$P^2 = \left\{\Gamma\left(\frac{1}{n}\right)\Gamma\left(\frac{n-1}{n}\right)\right\}\left\{\Gamma\left(\frac{2}{n}\right)\Gamma\left(\frac{n-2}{n}\right)\right\}\left\{\Gamma\left(\frac{3}{n}\right)\Gamma\left(\frac{n-3}{n}\right)\right\}\cdots\left\{\Gamma\left(\frac{n-1}{n}\right)\Gamma\left(\frac{1}{n}\right)\right\}$$

$$= \left\{\Gamma\left(\frac{1}{n}\right)\Gamma\left(1 - \frac{1}{n}\right)\right\}\left\{\Gamma\left(\frac{2}{n}\right)\Gamma\left(1 - \frac{2}{n}\right)\right\}\left\{\Gamma\left(\frac{3}{n}\right)\Gamma\left(1 - \frac{3}{n}\right)\right\}\cdots\left\{\Gamma\left(\frac{n-1}{n}\right)\Gamma\left(1 - \frac{n-1}{n}\right)\right\}$$

From the definition of beta functions in Eq. (4.18), we have

$$\beta(m, n) = \frac{(n-1)!(m-1)!}{(m+n-1)!} = \frac{\Gamma(n)\Gamma(m)}{\Gamma(m+n)} = \int_0^\infty \frac{u^{n-1}}{(1+u)^{m+n}} du$$

Putting $m + n = 1$, we have

$$\Gamma(n)\,\Gamma(1-n) = \int_0^\infty \frac{u^{n-1}}{1+u} du = \frac{\pi}{\sin n\pi} \qquad \text{[since } \Gamma(1) = 1\text{]}$$

$$\therefore \qquad P^2 = \frac{\pi}{\sin\dfrac{\pi}{n}} \cdot \frac{\pi}{\sin\dfrac{2\pi}{n}} \cdot \frac{\pi}{\sin\dfrac{3\pi}{n}} \cdots \frac{\pi}{\sin\dfrac{(n-1)\pi}{n}}$$

$$= \frac{\pi^{n-1}}{\sin\dfrac{\pi}{n} \cdot \sin\dfrac{2\pi}{n} \cdot \sin\dfrac{3\pi}{n} \cdots \sin\dfrac{(n-1)\pi}{n}} \qquad (3)$$

But we know that

$$\frac{\sin n\theta}{\sin \theta} = 2^{n-1} \sin\left(\theta + \frac{\pi}{n}\right) \sin\left(\theta + \frac{2\pi}{n}\right) \cdots \sin\left\{\theta + \frac{(n-1)\pi}{n}\right\}$$

Taking the limit as $\theta \to 0$

$$\lim_{\theta \to \infty} \frac{\sin n\theta}{\sin \theta} = \lim_{\theta \to \infty}\left\{\frac{n\sin n\theta}{n\theta} \cdot \frac{\theta}{\sin\theta}\right\} = n$$

$$= 2^{n-1} \sin\frac{\pi}{n} \cdot \sin\frac{2\pi}{n} \cdots \sin\frac{n-1}{n}\pi$$

Substitution in Eq. (3) gives

$$P^2 = \frac{\pi^{n-1} 2^{n-1}}{n}$$

$$P = \frac{(2\pi)^{\frac{n-1}{2}}}{\sqrt{n}}$$

Example 2. *(i) Express period for 180° swings (back and fourth from –90° to + 90°) of a simple pendulum in terms of beta functions. (ii) Express the time period of a simple pendulum with large swing, in terms of elliptical integral.*

Solution: The differential equation of motion of simple pendulum is given by

$$\ddot{\theta} = -\frac{g}{l}\sin\theta \qquad (1)$$

where θ is the angle of swing and l the length of pendulum. Multiplying with $\dot{\theta}$ and then integrating, we get

$$\frac{1}{2}\dot{\theta}^2 = \frac{g}{l}\cos\theta + \text{constant} \qquad (2)$$

i. For 180° swing, i.e. when $\theta = \pm 90°$, $\dot{\theta} = 0$ and hence constant of Eq. (2) is zero. Then

$$\frac{d\theta}{dt} = \sqrt{\frac{2g}{l}} \cdot \sqrt{\cos\theta}$$

or
$$\frac{d\theta}{\sqrt{\cos\theta}} = \sqrt{\frac{2g}{l}} dt$$

In one quarter of a period the pendulum swings from $\theta = 0$ to $\theta = \dfrac{\pi}{2}$ and hence

$$\int_0^{\pi/2} \frac{d\theta}{\sqrt{\cos\theta}} = \sqrt{\frac{2g}{l}} \int_0^{T/4} dt = \sqrt{\frac{2g}{l}} \cdot T/4$$

or
$$\sqrt{\frac{2g}{l}} \cdot T/4 = \int_0^{\pi/2} \sin\theta \cdot \cos^{-1/2}\theta \, d\theta = \frac{1}{2}\beta\left(\frac{1}{2}, \frac{1}{4}\right)$$

Using $\beta(m,n) = 2\displaystyle\int_0^{\pi/2} \sin^{2m-1}\theta \cdot \cos^{2n-1}\theta \cdot d\theta$

$$\therefore \qquad T = 2\sqrt{\frac{l}{?g}} \beta\left(\frac{1}{2}, \frac{1}{4}\right) \qquad (3)$$

which represents the period in terms of β (beta) functions.

And using the relation $\beta(m,n) = \dfrac{\overline{|m} \, \overline{|n}}{\overline{|m+n}}$, we have

$$T = 2\sqrt{\frac{l}{2g}} \cdot \frac{\Gamma\left(\frac{1}{2}\right)\Gamma\left(\frac{1}{4}\right)}{\Gamma\left(\frac{3}{4}\right)}$$

ii. In Eq. (2) let us consider the swings of amplitude α. Then $\dot{\theta} = 0$ at $\theta = \pm \alpha$. Hence Eq. (2) becomes

$$\dot{\theta}^2 = \frac{2g}{l}(\cos\theta - \cos\alpha)$$

or
$$\frac{d\theta}{dt} = \sqrt{\frac{2g}{l}} \sqrt{\cos\theta - \cos\alpha}$$

and integrating this we have

$$\int_0^\alpha \frac{d\theta}{\sqrt{\cos\theta - \cos\alpha}} = \sqrt{\frac{2g}{l}} \cdot \frac{T_\alpha}{4} \qquad (4)$$

where T_α is the period for swings from $-\alpha$ to $+\alpha$ and back.
We have

$$\cos\theta - \cos\alpha = 2\sin^2\alpha/2\left(1 - \frac{\sin^2\theta/2}{\sin^2\alpha/2}\right)$$

and Let
$$x = \frac{\sin\theta/2}{\sin\alpha/2} \qquad \because \quad dx = \frac{\cos\theta/2 \cdot d\theta/2}{\sin\alpha/2}.$$

or
$$d\theta = \frac{2 \cdot dx \sin\alpha/2}{\cos\theta/2} = \frac{2 \cdot \sin\alpha/2 \cdot dx}{\sqrt{(1-x^2 \sin^2\alpha/2)}}$$

Substituting in Eq. (4), we get

$$\sqrt{\frac{2g}{l}} \frac{T_\alpha}{4} = \int_0^1 \frac{2\sin\alpha/2}{\sqrt{2}\sin\alpha/2} \frac{dx}{\sqrt{(1-x^2\sin^2\alpha/2)}\sqrt{(1-x^2)}}$$

or
$$\sqrt{\frac{2g}{l}} \frac{T_\alpha}{4} = \sqrt{2}\int_0^1 \frac{dx}{\sqrt{(1-x^2)}\sqrt{(1-x^2\sin^2\alpha/2)}} = \sqrt{2}K(\sin\alpha/2)$$

Using Jacobi form of elliptic integral as

$$K(k) = F(k, \pi/2) = \int_0^1 \frac{dx}{\sqrt{(1-x^2)}\sqrt{(1-k^2x^2)}}$$

Hence
$$T_\alpha = 4\sqrt{l/g}\, K(\sin\alpha/2)$$

which is the required result.

Example 3. *Show that* $\displaystyle\int_0^{\pi/2} \sin^p\theta \cdot \cos^q\theta \cdot d\theta$

$$= \frac{\Gamma\left(\dfrac{p+1}{2}\right) \cdot \Gamma\left(\dfrac{q+1}{2}\right)}{2\left|\dfrac{p+q+2}{2}\right.}$$

Hence evaluate $\displaystyle\int_0^{\pi/2} \sin^p\theta \cdot d\theta$ *and* $\displaystyle\int_0^{\pi/2}\cos^q\theta \cdot d\theta$

Solution: We have from the definition of beta functions

$$\beta(m, n) = \int_0^1 x^{m-1}(1-x)^{n-1}dx$$

Substituting $x = \sin^2\theta$, we have

$$\beta(m, n) = 2\int_0^{\pi/2}\sin^{2m-2}\theta \cdot \cos^{2n-2}\theta \cdot 2\sin\theta\cos\theta \cdot d\theta$$

$$= 2\int_0^{\pi/2}\sin^{2m-1}\theta \cdot \cos^{2n-1}\theta \cdot d\theta = \frac{\overline{|m}\,\overline{|n}}{\overline{|m+n}}$$

Replacing $2m - 1 = p$ and $2n - 1 = q$, we have

$$\int_0^{\pi/2}\sin^p\theta \cdot \cos^q\theta \cdot d\theta = \frac{\Gamma\left(\dfrac{p+1}{2}\right)\Gamma\left(\dfrac{q+1}{2}\right)}{2\Gamma\left(\dfrac{p+q+2}{2}\right)} \qquad (1)$$

Substituting $q = 0$ we get

$$\int_0^{\pi/2} \sin^p \theta \cdot d\theta = \frac{\Gamma\left(\dfrac{p+1}{2}\right)\Gamma\left(\dfrac{1}{2}\right)}{2\Gamma\left(\dfrac{p+2}{2}\right)} = \frac{\left|\dfrac{(p+1)}{2}\right.\sqrt{\pi}}{2\Gamma\left(\dfrac{p}{2}+1\right)}$$

Again substituting $p = 0$ in Eq. (1), we get

$$\int_0^{\pi/2} \cos^q \theta \cdot dx = \frac{\sqrt{\pi}\,\Gamma\left(\dfrac{q+1}{2}\right)}{2\Gamma\left(\dfrac{q}{2}+1\right)}$$

Example 4. *Evaluate* (i) $\displaystyle\int_0^1 \frac{dx}{\sqrt{1-x^n}}$ (ii) $\displaystyle\int_0^1 \frac{dx}{(1-x^n)^{1/n}}$

Solution: Substituting $x^n = \sin^2 \theta$, gives

$$x = \sin^{2/n} \theta \text{ and } dx = \frac{2}{n}\sin^2 \theta^{\frac{2}{n}-1} \theta \cdot \cos\theta \cdot d\theta$$

i.

$$\int_0^1 \frac{dx}{\sqrt{1-x^n}} = \int_0^{\pi/2} \frac{\dfrac{2}{n}\sin^{\frac{2}{n}-1}\theta \cdot \cos\theta \cdot d\theta}{\cos\theta}$$

$$= \int_0^{\pi/2} 2/n \sin^{\frac{2}{n}-1}\theta \cdot d\theta$$

$$= \frac{2}{n}\frac{\left|1/n\right|1/2}{2\Gamma\left(\dfrac{1}{n}+\dfrac{1}{2}\right)} = \frac{\sqrt{\pi}\,\Gamma(1/n)}{n\Gamma\left(1/n+\dfrac{1}{2}\right)}$$

ii.

$$\int_0^1 \frac{dx}{(1-x^n)^{1/n}} = \int_0^{\pi/2} \frac{2/n\sin^{\left(\frac{2}{n}-1\right)}\theta \cdot \cos\theta \cdot d\theta}{\cos^{2/n}\theta}$$

$$= \int_0^{\pi/2} \frac{2}{n}\cdot\sin^{\frac{2}{n}-1}\theta \cdot \cos^{1-\frac{2}{n}}\theta \cdot d\theta$$

$$= \frac{2}{n}\cdot\frac{\Gamma\dfrac{1}{n}\Gamma\left(1-\dfrac{1}{n}\right)}{2\Gamma 1} = \frac{2}{n}\frac{\pi}{2\cdot\sin\pi/n}$$

$$= \frac{\pi}{n\sin\pi/n} \qquad \left[\text{since } \Gamma\frac{1}{n}\Gamma\left(1-\frac{1}{n}\right) = \frac{\pi}{\sin\pi/n}\right]$$

Example 5. *Show that*

$$\int_0^1 \frac{x^{m-1}(1-x)^{n-1}}{(a+x)^{m+n}}\,dx \;=\; \frac{\Gamma(m)\Gamma(n)}{a^n(1+a)^m\Gamma(m+n)}$$

Solution: Put $\dfrac{x(1+a)}{a+x} = y$

$$\therefore \qquad \frac{a(1+a)}{(a+x)^2}\,dx = dy$$

and $$x = \frac{ay}{1+a-y}$$

and $$a+x = \frac{a(1+a)}{1+a-y} \quad \text{and} \quad 1-x = \frac{(1+a)(1-y)}{1+a-y}$$

and $$dx = \frac{(a+x)^2}{a(1+a)}\cdot dy$$

$$= \frac{a(1+a)}{(1+a-y)^2}\cdot dy$$

Thus $$\int_0^1 \frac{x^{m-1}(1-x)^{n-1}}{(a+x)^{m+n}}\,dx = \int_0^1 \frac{a^{m-1}y^{m-1}(1+a)^{n-1}(1-y)^{n-1}\cdot(1+a-y)^{m+n}\,a(1+a)dy}{(1+a-y)^{m-1}(1+a-y)^{n-1}\,a^{m+n}(1+a)^{m+n}(1+a-y)^2}$$

$$= \frac{1}{a^n(1+a)^m}\int_0^1 y^{m-1}(1-y)^{n-1}dy$$

$$= \frac{1}{a^n(1+a)^m}\beta(m,n) = \frac{1}{a^n(1+a)^m}\frac{\Gamma(m)\Gamma(n)}{\Gamma(m+n)}$$

Example 6. *Evaluate the integrals*

(i) $\displaystyle\int_0^\infty e^{-ax}x^{m-1}\cos bx\,dx$ (ii) $\displaystyle\int_0^\infty e^{-ax}x^{m-1}\sin bx\,dx$

Solution: From the definition of gamma functions, we have

$$\Gamma(n) = \int_0^\infty e^{-t}t^{n-1}dt$$

$$\therefore \qquad \int_0^\infty e^{-mt}t^{n-1}dt = \frac{\Gamma(n)}{m^n}$$

$$\therefore \qquad \int_0^\infty e^{-(a+ib)x}x^{m-1}dx = \frac{\Gamma(m)}{(a+ib)^m} = \frac{\Gamma(m)(a-ib)^m}{(a^2+b^2)^m}$$

Now let $a = r\cos\theta$ and $b = r\sin\theta$

$$\int_0^\infty e^{-(a+ib)x}x^{m-1}dx = \frac{\Gamma(m)r^m\cdot(\cos\theta-i\sin\theta)^m}{r^{2m}}$$

or $$\int_0^\infty e^{-ax}(\cos bx - i\sin bx)x^{m-1}dx = \frac{\Gamma(m)(\cos m\theta - i\sin m\theta)}{r^m}$$

Equating real and imaginary parts we have

$$\int_0^\infty e^{-ax}\cos bx\, x^{m-1}dx = \frac{\Gamma(m)\cos m\theta}{r^m}$$

and $$\int_0^\infty e^{-ax}\sin bx\, x^{m-1}dx = \frac{\Gamma(m)\sin m\theta}{r^m}$$

where $$r^2 = (a^2+b^2)^{1/2} \text{ and } \theta = \tan^{-1}b/a.$$

Example 7. Evaluate the following integrals.

i. $\displaystyle\int_{-\infty}^{+\infty} e^{-\alpha|x|+iKa}dx \ \ \alpha > 0 \text{ and } |x| \text{ is the absolute value of } x$

ii. $\displaystyle\frac{1}{\beta}\int_0^\infty (x^\beta - 1)e^{-\alpha x}dx$

iii. In second integral take the limit $\beta \to 0$ and hence evaluate the integral

$$\int_0^\infty \log x e^{-\alpha x}dx.$$

Using approximations $\alpha^\beta = 1 + \beta\tan\alpha$ and $\sqrt{1+\beta} = 1+\gamma\beta$, where $\gamma = -0.57721$ for small β.

Solution: i. $$I = \int_{-\infty}^{+\infty} e^{-\alpha|x|+iKx}dx = \int_{-\infty}^{+\infty} e^{-\alpha|x|}\{\cos Kx + i\sin Kx\}dx$$

$$= \int_{-\infty}^{+\infty} e^{-\alpha|x|}\cos Kx\, dx + i\int_{-\infty}^{+\infty} e^{-\alpha|x|}\sin Kx\, dx$$

$$I_1 = \int_{-\infty}^{+\infty} e^{-\alpha|x|}\cos Kx\, dx = i2\int_0^{+\infty} e^{-a|x|}\cos Kx\, dx$$

[since integrand is an even function of x]

$$= 2\left[-\frac{e^{-\alpha x}}{\alpha}\cos Kx\right]_0^\infty - \frac{K}{\alpha}\int_0^\infty e^{-\alpha|x|}\sin Kx\, dx$$

$$= \frac{2}{\alpha} - \frac{2K}{\alpha}\left[-\frac{e^{-\alpha x}}{\alpha}\sin Kx\int_0^\infty + \frac{K}{\alpha}\int_0^\infty e^{-\alpha x}\cdot\cos Kx\, dx\right]$$

$$= \frac{2}{\alpha} - \frac{2K^2}{\alpha^2}\frac{I_1}{2}$$

or $$I_1\left(1+\frac{K^2}{\alpha^2}\right) = \frac{2}{\alpha}$$

or $$I_1 = \frac{2\cdot\alpha}{\alpha^2+K^2}$$

$$I_2 = \int_{-\infty}^{+\infty} e^{-\alpha|x|} \sin Kx\,dx$$

$$= \int_{-\infty}^{0} e^{-\alpha|x|} \sin Kx\,dx + \int_{0}^{+\infty} e^{-\alpha|x|} \sin Kx\,dx = 0$$

Changing x to $-x$ in first integral, we have

$$I_2 = -\int_{0}^{\infty} e^{-\alpha|x|} \sin Kx\,dx + \int_{0}^{\infty} e^{-\alpha|x|} \sin Kx\,dx = 0$$

Hence $\qquad I = I_1 + iI_2 = \dfrac{2\alpha}{\alpha^2 + K^2}$

ii.

$$I = \frac{1}{\beta}\int_{0}^{\infty} (x^{\beta} - 1)e^{-\alpha x}\,dx = \frac{1}{\beta}\int_{0}^{\infty} e^{-\alpha x} x^{\beta}\,dx - \frac{1}{\beta}\int_{0}^{\infty} e^{-\alpha x}\,dx$$

$$= \frac{1}{\beta}\int_{0}^{\infty} e^{-\alpha x} \cdot x^{(\beta+1)-1}\,dx - \frac{1}{\beta}\int_{0}^{\infty} e^{-\alpha x}\,dx$$

$$= \frac{1}{\beta}\left[\frac{\Gamma(\beta+1)}{\alpha^{\beta+1}} - \frac{1}{\alpha} \right], \quad \text{since} \quad \int_{0}^{\infty} e^{-\alpha x} x^{\beta-1} = \frac{\Gamma\beta}{\alpha^{\beta}}$$

$$= \frac{1}{\alpha\beta}\left[\frac{|\beta+1}{\alpha^{\beta}} - 1 \right]$$

iii.

$$\frac{1}{\beta}\int_{0}^{\infty} (x^{\beta} - 1)e^{-\alpha x}\,dx = \frac{1}{\alpha}\left[\frac{\Gamma(\beta+1)}{\beta\alpha^{\beta}} - \frac{1}{\beta} \right]$$

$$= \frac{1}{\alpha}\left[\frac{\Gamma(\beta+1) - \alpha^{\beta}}{\beta\alpha^{\beta}} \right]$$

Taking the limit as $\beta \to 0$, we have

$$\lim_{\beta\to\infty} \frac{\displaystyle\int_{0}^{\infty} \frac{d}{d\beta}(x^{\beta}-1)e^{-\alpha x}\,dx}{\dfrac{d\beta}{d\beta}} = \lim_{\beta\to\infty}\left[\frac{(1+\gamma\beta)-(1+\beta\tan\alpha)}{\beta(1+\beta\tan\alpha)} \right] \cdot \frac{1}{\alpha}$$

or $\qquad \displaystyle\lim_{\beta\to\infty}\int_{0}^{\infty} e^{-\alpha x} \cdot x^{\beta} \log x \cdot dx = \lim_{\beta\to\infty}\left[\frac{\gamma - \tan\alpha}{1+\beta\tan\alpha} \right] \cdot \frac{1}{\alpha}$

or $\qquad \displaystyle\int e^{-\alpha x} \log x \cdot dx = \frac{\gamma - \tan\alpha}{\alpha}$

Example 8. *Prove that*

$$\int_{0}^{1} \frac{x^2\,dx}{(1-x^4)^{1/2}} \times \int_{0}^{1} \frac{dx}{(1+x^4)^{1/2}} = \frac{\pi}{4\sqrt{2}}$$

Solution: Let $\qquad I_1 = \displaystyle\int_{0}^{1} \frac{x^2\,dx}{(1-x^4)^{1/2}} \quad$ and $\quad I_2 = \displaystyle\int_{0}^{1} \frac{dx}{(1+x^4)^{1/2}}$

In I_1, let us put $x^2 = \sin\theta$

$\therefore$ $\qquad\qquad 2x\,dx = \cos\theta\,d\theta$

and I_1 becomes

$$I_1 = \frac{1}{2}\int_0^{\pi/2}\frac{\sqrt{\sin\theta}\cdot\cos\theta\cdot d\theta}{\cos\theta} = \frac{1}{2}\int_0^{\pi/2}\sin^{1/2}\theta\cdot d\theta$$

$$= \frac{1}{2}\cdot\frac{\left\lfloor\dfrac{3}{4}\right.\left\lfloor\dfrac{1}{2}\right.}{2\left\lfloor\dfrac{5}{4}\right.} = \frac{1}{4}\cdot\frac{\left\lfloor\dfrac{3}{4}\right.\left\lfloor\dfrac{1}{2}\right.}{\left\lfloor\dfrac{1}{4}\right.\left\lfloor\dfrac{1}{4}\right.} = \frac{\left\lfloor\dfrac{3}{4}\right.\left\lfloor\dfrac{1}{2}\right.}{\left\lfloor\dfrac{1}{4}\right.}$$

In I_2, put $x^2 = \tan\phi$ $\because$ $2x\,dx = \sec^2\phi\,d\phi$ and I_2 becomes

$$I_2 = \frac{1}{2}\int_0^{\pi/4}\frac{\sec^2\phi\cdot d\phi}{\sqrt{\tan\phi\cdot\sec\phi}} = \frac{1}{2}\int_0^{\pi/4}\frac{\sqrt{2}\cdot d\phi}{\sqrt{2\sin\phi\cdot\cos\phi}}$$

$$= \frac{1}{\sqrt{2}}\int_0^{\pi/4}\frac{d\phi}{\sqrt{\sin 2\phi}}$$

Now putting $2\phi = \alpha$, $d\phi = \dfrac{1}{2}d\alpha$

$$I_2 = \frac{1}{2\sqrt{2}}\int_0^{\pi/2}\sin^{-1/2}\alpha\,d\alpha = \frac{1}{2\sqrt{2}}\frac{\left\lfloor\dfrac{1}{4}\right.\left\lfloor\dfrac{1}{2}\right.}{2\left\lfloor\dfrac{3}{4}\right.} = \frac{\left\lfloor\dfrac{1}{4}\right.\left\lfloor\dfrac{1}{2}\right.}{4\sqrt{2}\left\lfloor\dfrac{3}{4}\right.}$$

$\therefore$

$$I_1\times I_2 = I = \frac{\left\lfloor\dfrac{3}{4}\right.\left\lfloor\dfrac{1}{2}\right.}{\left\lfloor\dfrac{1}{4}\right.}\cdot\frac{\left\lfloor\dfrac{1}{4}\right.\left\lfloor\dfrac{1}{2}\right.}{4\sqrt{2}\left\lfloor\dfrac{3}{4}\right.} = \frac{\pi}{4\sqrt{2}}\qquad\left(\because\left\lfloor\dfrac{1}{2}\right. = \sqrt{\pi}\right)$$

Example 9. *The equation of motion of a particle moving from rest under central force of attraction is $\dfrac{d^2x}{dt^2} + \dfrac{K}{x} = 0$ (where K is constant). If at t = 0, the particle is at x = a then show that the particle reaches the centre of attraction in time $T = a\sqrt{\dfrac{\pi}{2K}}$*

Solution: Equation of motion is $\dfrac{d^2x}{dt^2} + \dfrac{K}{x} = 0$

or $\qquad\qquad \dfrac{d}{dt}\left(\dfrac{dx}{dt}\right) = -\dfrac{K}{x}$

or $\qquad\qquad \dfrac{dv}{dt} = -\dfrac{K}{x}$ $\qquad\qquad\left[\text{where } v = \dfrac{dx}{dt}\right]$

$\therefore$ $\qquad\qquad \dfrac{dv}{dx}\cdot\dfrac{dx}{dt} = -K/x$

or $\qquad\qquad v\,dv = -\dfrac{K}{x}dx$

Integration gives

$$\frac{v^2}{2} = -K \log x + c, \text{ where } c \text{ is constant.}$$

But $v = 0$ when $x = a$

$\therefore$

$$c = K \log a$$

and

$$v^2 = 2K \log \frac{a}{x}$$

or

$$\left(\frac{dx}{dt}\right)^2 = 2K \log \frac{a}{x}$$

or

$$\frac{dx}{dt} = \sqrt{2K \cdot \log a / x}$$

or

$$dt = \frac{dx}{\sqrt{2K}} (\log a / x)^{-1/2}$$

$\therefore$

$$\int_0^T dt = \frac{1}{\sqrt{2K}} \int_0^a [\log(a/x)]^{-1/2} dx$$

Put

$$\log a / x = p$$
$$dx = -x dp = -ae^{-p} dp$$

and

$$\int_0^T dt = \frac{1}{\sqrt{2K}} \int_0^a \left[\log \frac{a}{x}\right]^{-1/2} dx$$

or

$$T = -\frac{a}{\sqrt{2K}} \int_\infty^0 (p)^{-1/2} e^{-p} \cdot dp$$

Then

$$T = \frac{a}{\sqrt{2K}} \int_0^\infty (p)^{(1/2-1)} \cdot e^{-p} dp = \frac{a}{\sqrt{2K}} \left|\frac{1}{2}\right.$$

$$= \frac{a}{\sqrt{2K}} \sqrt{\pi} = a\sqrt{\frac{\pi}{2K}}$$

Then the particle will reach the centre of attraction at

$$T = a\sqrt{\frac{\pi}{2K}}$$

Example 10. *Using Legendre duplication formula, prove that*

$$(m + 1/2)! = \pi^{1/2}(2m+1)!! / 2^{m+1}$$

where $(2m + 1)!! = 1.3.5... (2m - 1) (2m + 1)$

Solution: Legendre duplication formula is

$$2^{2m-1} \overline{(m)} \left| \left(m + \frac{1}{2}\right)\right. = \sqrt{\pi} \overline{(2m)}$$

Multiplying both sides by $m\,(m+1/2)$, we have

$$2^{2m-1}\,m\overline{|(m)}\frac{(2m+1)}{2}\left|m+\frac{1}{2}\right. = \sqrt{\pi}\,m\left(\frac{2m+1}{4}\right)\overline{|(2m)!}$$

or $\qquad 2^{2m-1}(m)!(m+1/2)! = \sqrt{\pi}\left(\frac{2m+1}{4}\right)(2m)!$

$$\therefore \qquad \left(m+\frac{1}{2}\right)! = \frac{\sqrt{\pi}\,(2m+1)!}{(m)!\,2^{2m+1}} = \frac{\sqrt{\pi}\,(2m+1)!}{2.4.6\cdots 2m(2^{m+1})}$$

$$= \sqrt{\pi}\,\frac{1.3.5\cdots(2m-1)(2m+1)}{2^{m+1}} = \frac{\sqrt{\pi}}{2^{m+1}}(2m+1)!!$$

which is the required result.

Example 11. *Derive Legendre duplication formula from Weierstrass infinite limit product for gamma function.*

Solution: The ratio of $\overline{|n}$ and $\overline{|2n}$ from Weierstrass infinite limit product is

$$\frac{\overline{|n}}{\overline{|2n}} = 2e^{\gamma n}\frac{\displaystyle\prod_{m=1}^{\infty}\left[1+\frac{2n}{m}\right]e^{-2n/m}}{\displaystyle\prod_{m=1}^{\infty}\left[1+\frac{n}{m}\right]e^{-n/m}}$$

or $\qquad \dfrac{\overline{|n}}{\overline{|2n}} = 2e^{\gamma n}(1+2n)e^{-2n}\dfrac{\displaystyle\prod_{p=1}^{\infty}[1+(n+1/2)/p]e^{-(n+1/2)/p}}{\displaystyle\prod_{p=1}^{\infty}\left(1+\frac{1}{2}pe^{-\frac{1}{2p}}\right)}$

$$= 2e^{\gamma n}(1+2n)e^{-2n}\prod_{p=1}^{\infty}\left[1+\frac{1}{2}p\right]e^{-\frac{1}{2p}}\prod_{p=1}^{\infty}e^{n\left[\frac{1}{p}-\frac{1}{p+1/2}\right]}$$

$$= \frac{2\left|\overline{\frac{1}{2}}\right|e^{-2n}}{\left|\overline{\left(n+\frac{1}{2}\right)}\right|}\exp\left[n\sum_{p=1}^{\infty}\left(\frac{1}{p}-\frac{1}{p+\frac{1}{2}}\right)\right]$$

If we put $n = 1$, we have

$$\frac{\overline{|1}}{\overline{|2}} = 1 = \frac{2\left|\overline{\frac{1}{2}}\right|}{\left|\overline{\frac{3}{2}}\right|}e^{-2}\exp\left[\sum_{p=1}^{\infty}\left(\frac{1}{p}-\frac{1}{p+\frac{1}{2}}\right)\right]$$

or $\qquad \displaystyle\exp\sum_{p=1}^{\infty}\left(\frac{1}{p}-\frac{1}{p+1/2}\right) = \frac{e^2}{4} = \frac{e^2}{2^2}$

Hence $\qquad \displaystyle\exp\sum_{p=1}^{\infty}n\left(\frac{1}{p}-\frac{1}{p+1/2}\right) = \frac{e^{2n}}{2^{2n}}$

Substituting the value of exponential, we have

$$\frac{\lceil n}{\lceil 2n} = \frac{2\left\lceil\dfrac{1}{2}\right.}{\left\lceil n+\dfrac{1}{2}\right.} \cdot \frac{1}{2^{2n}} = \frac{\sqrt{\pi}}{2^{2n-1}\left\lceil n+\dfrac{1}{2}\right.}$$

or
$$2^{2n-1}\lceil n\left\lceil n+\dfrac{1}{2}\right. = \sqrt{\pi}\lceil 2n$$

which is the Legendre's duplication formula.

SHORT ANSWER QUESTIONS

1. Define β function. Give integral formula for β functions.

2. Prove that $\lceil n+1 = n\lceil n$.

3. What is the value of erf (∞)?

4. Find the value of β(1, 2) and β(2, 4).

Ans. [0.5 and 0.05]

5. Define gamma functions and write its integral formula.

6. Show that $\left\lceil\dfrac{1}{2}\right. = \sqrt{\pi}$

7. Show the relation between β and γ functions.

8. Use the integral $\displaystyle\int_0^{\pi/2} \cos^m\theta \cdot \sin^m\theta\, d\theta = \dfrac{\left\lceil\dfrac{m+1}{2}\right. \cdot \left\lceil\dfrac{n+1}{2}\right.}{2\left\lceil\dfrac{m+n+2}{2}\right.}$ to find the value of

$\displaystyle\int_0^{\pi/2}\sqrt{\tan\theta}\, d\theta$.

Ans. $\dfrac{\left\lceil\dfrac{1}{4}\right.\left\lceil\dfrac{3}{4}\right.}{2}$

9. Define error function and hence find the value of erf (∞)

Ans. 1

10. Show that β(m, n) = β(n, m).

PROBLEMS

1. Prove that

   i. $\displaystyle\int_0^\infty e^{-r}\log r\, dr = -\gamma$

   ii. $\displaystyle\int_0^\infty re^{-r}\log r\, dr = 1-\gamma$

   iii. $\displaystyle\int_0^\infty r^n e^{-r}\log r\, dr = \Gamma(n) + n\int_0^\infty r^{n-1}e^{-r}\log r\, dr, n = 1,2,3,\cdots$

   Here γ is Euler–Mascheroni constant.

2. Prove that

$$|\Gamma(\alpha+i\beta)| = \frac{\Gamma(\alpha+1)}{\alpha}\prod_{n=0}^{\infty}\left[1+\frac{\beta^2}{(\alpha+n)^2}\right]^{-1/2}$$

3. Verify the following beta function identities.

    i. $\beta_{(m,n)} = \beta(m+1,n)+\beta(m,n+1)$

    ii. $\beta_{(m,n)} = \dfrac{m+n}{n}\beta(m,n+1)$

    iii. $\beta_{(m,n)} = \dfrac{n-1}{m}(m+1,n-1)$

    iv. $\beta_{(m,n)}\,\beta_{(m+n,\rho)} = \beta(n,\rho)\beta(m,n+\rho)$

4. Evaluate the following integrals in terms of beta functions

$$\int_{-1}^{+1}(1+x)^n(1-x)^n\,dx$$

5. When x is positive, show that

$$\frac{\Gamma(x)\Gamma\dfrac{1}{2}}{\Gamma\left(x+\dfrac{1}{2}\right)} = \sum_{n=0}^{\infty}\frac{2n!}{2^{2n}\cdot n!}\frac{1}{(x+n)}$$

6. Show that $\displaystyle\int_0^1\frac{x^n}{\sqrt{1-x^2}}\,dx = \dfrac{1.3.5\cdots(n-1)}{2.4.6\cdots n}\cdot\dfrac{\pi}{2}$ if n is even

$$= \frac{2.4.6\cdots(n-1)}{1.3.5\cdots n}$$ if n is odd.

7. Show that $\displaystyle\int_0^{+1}\frac{x^{m-1}(1-x)^{n-1}}{(a+x)^{m+n}}\,dx = \dfrac{\beta(m,n)}{a^n(1+a)^m}$.

8. Derive Legendre duplication formula

$$2^{2n-1}\Gamma(n)\Gamma\left(n+\frac{1}{2}\right) = \sqrt{\pi}\,\Gamma(2n)$$

    from Weierstrass' limit product for gamma functions.

9. Prove that

    i. $\displaystyle\int_0^{\pi}\frac{\sqrt{\sin x}}{(5+3\cos x)^{3/2}}\,dx = \dfrac{\left\{\Gamma\left(\dfrac{3}{4}\right)\right\}^2}{2\sqrt{\pi}}$ ii. $\displaystyle\int_0^{\pi/2}(\cos\theta)^{1/2}\,d\theta = \dfrac{(2\pi)^{3/2}}{16\left\{\Gamma\left(\dfrac{5}{4}\right)\right\}^2}$

    iii. $\displaystyle\int_2^{\pi/2}\cos^n\theta\cdot d\theta = \dfrac{\sqrt{\pi}\,\Gamma\left(n+\dfrac{1}{2}\right)}{2\Gamma\left(\dfrac{1}{2}n+1\right)}$ iv. $\displaystyle\int_0^1(x^2)^p(1-x^2)^q\,dx = \dfrac{\Gamma\left(p+\dfrac{1}{2}\right)\Gamma(q+1)}{\Gamma\left(p+q+\dfrac{3}{2}\right)}$

10. Prove the following properties of incomplete gamma functions.

   i. $\gamma(n,x) = e^{-x} \sum\limits_{m=0}^{\infty} \dfrac{(n-1)!}{(n+m)!} x^{n+m}$

   ii. $\dfrac{d^m}{dx^m}[x^m \gamma(n,x)] = (-1)^m x^{-n-m} \gamma(n+m,x)$

   iii. $\dfrac{d^m}{dx^m}[e^x \gamma(n,x)] = \dfrac{e^x \Gamma(n)}{\Gamma(n-m)} \gamma(n-m,x)$

   iv. $\gamma(n+1,x) = n\gamma(n,x) - x^n e^{-n}$

   v. $\Gamma(n+1,x) = n\Gamma(n,x) + x^n e^{-x}$

11. Show that

   i. $\displaystyle\int_0^1 \dfrac{dx}{(1-x^6)^{1/6}} = \dfrac{\pi}{3}$

   ii. $\displaystyle\int_0^{\infty} \dfrac{x^8(1-x^6)}{(1+x)^{24}} dx = 0$

   iii. $\displaystyle\int_0^1 x^m (1-x^n)^{\rho} dx = \dfrac{1}{n} \dfrac{\Gamma\left(\dfrac{m+1}{n}\right)\Gamma(\rho+1)}{\Gamma\left(\dfrac{m+1}{n}+\rho+1\right)}$

   iv. $\displaystyle\int_0^p x^m (p^q - x^q)^n dx = \dfrac{p^{2n+m+1}}{q} B\left((n+1),\dfrac{m+1}{q}\right)$

   if $p > 0.\, q > 0,\, m+1 > 0,\, n+1 > 0$.

12. The equation of motion of a particle moving from rest towards the centre of attraction point, situated at distance 'a' from it is given by

$$\dfrac{d^2 x}{dt^2} + \dfrac{K}{x} = 0, \text{ where } K \text{ is constant.}$$

Apply your knowledge of gamma functions, to show that it will reach the centre of attraction point in time given by

$$T = a\sqrt{\dfrac{\pi}{2K}}$$

13. Prove that if $R(z) > -1$

$$\lim_{n \to \infty}\left\{\log n - \int_0^{\infty} e^{-tz}\dfrac{e^{-t} - e^{-(n+1)t}}{1-e^{-t}} dt\right\} = \int_0^{\infty}\left\{\dfrac{e^{-t}}{t} - \dfrac{e^{-(n+1)t}}{1-e^{-t}}\right\} dt$$

14. Prove that

$$\int_0^{\infty} \dfrac{u^m}{(Au^2 + B)^{\frac{1}{2(m+n+2)}}} du = \dfrac{\left\{\dfrac{1}{2}(m-1)\right\}!\left\{\dfrac{1}{2}(n-1)\right\}!}{\left\{\dfrac{1}{2}(m+n)\right\}!} \cdot \dfrac{1}{2A^{\frac{1}{2}(m+1)}B^{\frac{1}{2}(n+1)}} \quad [m > -1, n > -1]$$

15. Prove that

i. $\displaystyle\int_0^\infty x^{-n}e^{-k/x^2}\,dx = \dfrac{\left(\dfrac{1}{2}n-\dfrac{3}{2}\right)!}{2\cdot k^{\left(\frac{1}{2n}-\frac{1}{2}\right)}}\quad (n>1)$

ii. Prove that for positive real n

$$\dfrac{1}{n}-\log\dfrac{n}{n-1}<0,\quad \dfrac{1}{n}-\log\dfrac{n+1}{n}>0$$

and hence show that the limit defining Euler's constant exists.

16. If $\displaystyle f(x)=\int_0^x \dfrac{u(y)}{(x-y)^n}\cdot dy$

where $f(0)=0,\ 0<n<1$ and $f'(x)$ is continuous, prove that

$$u(x) = \dfrac{\sin n\pi}{\pi}\dfrac{d}{dx}\int_0^x \dfrac{f(P)dP}{(x-P)^{1-n}}$$

17. Prove that

$$\beta\left(\dfrac{1}{4},\dfrac{1}{2}\right) = 2\sqrt{2}\left(\dfrac{1}{\sqrt{2}},\dfrac{\pi}{2}\right)=2\sqrt{2}k\left(\dfrac{1}{\sqrt{2}}\right)\ \text{and hence}$$

$$k=\left(\dfrac{1}{\sqrt{2}}\right)=\dfrac{1}{4\sqrt{\pi}}\left[\left(\dfrac{1}{4}\right)\right]^2$$

18. If $C(x)$ and $S(x)$ are Fresnel's integrals defined for real x by

$$C(x)+iS(x) = \int_0^x e^{it^2}\,dt$$

prove that for large positive x.

$$C(x) = \dfrac{1}{2}\sqrt{\dfrac{\pi}{2}}-P\cos x^2 + Q\sin x^2$$

$$S(x) = \dfrac{1}{2}\sqrt{\dfrac{\pi}{2}}-P\sin x^2 - Q\cos x^2$$

where $\quad P(x)\sim \dfrac{1}{2}\left(\dfrac{1}{2x^2}-\dfrac{1.3.5}{2^3 x^7}+\cdots\right)\quad Q(x)\sim \dfrac{1}{2}\left(\dfrac{1}{x}-\dfrac{1.3}{2^2 x^5}+\dfrac{1.3.5}{2^4 x^9}+\cdots\right)$

19. Express the following integrals as beta functions.

i. $\displaystyle\int_0^1 \dfrac{x^4}{\sqrt{1-x^2}}\,dx$

ii. $\displaystyle\int_0^{\pi/2} \sin^{3/2}x\cos^{1/2}x\,dx$

iii. $\displaystyle\int_0^1 \dfrac{dx}{(1-x^3)^{1/2}}$

iv. $\displaystyle\int_0^1 x^2(1-x^2)^{3/2}\,dx$

v. $\displaystyle\int_0^\infty \dfrac{x^3}{(1+x)^5}\,dx$

vi. $\displaystyle\int_0^\infty \dfrac{y}{(1+y^3)^2}\,dy$

20. Prove that

$$\int_0^\infty \dfrac{\cos x}{x^P}\,dx = \dfrac{\pi}{2\Gamma(P)\cos\left(\dfrac{P\pi}{2}\right)}\quad 0<P<1$$

21. Prove that

$$\int_0^\infty xe^{-x^3}dx \int_0^\infty x^2 e^{-x^4}dx = \frac{\pi}{16\sqrt{2}}$$

22. Show that

$$\frac{\left\lfloor (x)\left\lfloor\left(\frac{1}{2}\right)\right\rfloor\right\rfloor}{\left\lfloor\left(x+\frac{1}{2}\right)\right\rfloor} = \sum_{n=0}^\infty \left\{\frac{(2n)!}{2^{2n}n!}\right\}\left\{\frac{1}{x+n}\right\}$$

23. Show that

$$\left\lfloor\left(\frac{1}{2}-n\right)\right\rfloor\left\lfloor\left(\frac{1}{2}+n\right)\right\rfloor = (-1)^n\,\pi. \quad \text{where } n \text{ is an integer.}$$

24. Use di- and poly-gamma functions to sum the series

i. $\displaystyle\sum_{n=1}^\infty \frac{1}{n(n+1)}$ ii. $\displaystyle\sum_{n=2}^\infty \frac{1}{n^2-1}$

25. Using Stirling formula, and evaluate

$$\lim_{m\to\infty} \frac{(2m)!\sqrt{(m)}}{2^{2m}(m!)^2}.$$

MULTIPLE CHOICE QUESTIONS

1. The value of $\left\lfloor\frac{1}{2}\cdot\left\lfloor\frac{3}{4}\right.\right.$ is

(a) $\sqrt{\pi}$ (b) $2\sqrt{\pi}$

(c) $\sqrt{2\pi}$ (d) $\sqrt{2\pi}$

2. The correct relation for beta function is

(a) $\beta(m,n)=\dfrac{\lfloor n \lfloor m}{\lfloor m+n}$ (b) $\beta(m,n)=\dfrac{\lfloor n+m}{\lfloor n \lfloor m}$

(c) $\beta(m,n)=\dfrac{\lfloor n}{\lfloor n \lfloor m+n}$ (d) $\beta(m,n)=\dfrac{\lfloor m}{\lfloor n \lfloor m+n}$

3. The value of integral $I=\displaystyle\int_0^{\pi/2}(\tan\theta)^{1/2}d\theta$ is

(a) $\dfrac{\left\lfloor\frac{3}{4}\right.\left\lfloor\frac{1}{4}\right.}{2}$ (b) $\dfrac{\left\lfloor\frac{1}{2}\right.\left\lfloor\frac{1}{4}\right.}{2}$

(b) $\dfrac{\left\lfloor\frac{3}{4}\right.\left\lfloor\frac{1}{4}\right.}{2}$ (d) $\dfrac{2\left\lfloor\frac{3}{4}\right.}{\left\lfloor\frac{1}{4}\right.}$

4. Using the definition of beta functions the value of integral $I = \int_0^\infty \frac{x^3}{(1+x)^5} dx$ is

(a) $\dfrac{1}{2}$

(b) $\dfrac{1}{3}$

(c) $\dfrac{1}{4}$

(d) ∞

5. The correct from of Legendre's duplication formula is

(a) $\dfrac{\overline{|2n}}{\overline{|n}} = \dfrac{2^{2n-1}\sqrt{\pi}}{\left| n+\dfrac{1}{2}\right.}$

(b) $\dfrac{\overline{|2n}}{\overline{|n}} = \dfrac{2^{2n-1}\sqrt{n+\dfrac{1}{2}}}{\overline{|\pi}}$

(c) $\dfrac{\overline{|2n}}{\overline{|n}} = \dfrac{\left| n+\dfrac{1}{2}\right.}{2^{2n-1}\sqrt{\pi}}$

(d) $\dfrac{\overline{|2n}}{\overline{|n}} = \dfrac{\sqrt{\left| \pi\, \overline{\left(\pi\, \dfrac{1}{2}\right)}\right.}}{2^{2n-1}}$

6. The value $\sqrt{-\dfrac{3}{2}}$ is

(a) ∞

(b) 0

(c) $\dfrac{3}{4}\sqrt{\pi}$

(d) $\dfrac{4}{3}\sqrt{\pi}$

7. The value $\beta(3, 2)$ is

(a) $\dfrac{1}{12}$

(b) $\dfrac{1}{6}$

(c) $\dfrac{1}{3}$

(d) $\dfrac{1}{2}$

8. The value of $\beta\left(\dfrac{1}{2}, \dfrac{3}{2}\right)$ is

(a) 2π

(b) $\sqrt{2\pi}$

(c) $\sqrt{\dfrac{\pi}{2}}$

(d) $\dfrac{\pi}{2}$

9. Stirling formula in correct form is

(a) $n! = \sqrt{(2\pi n)}n^n e^{-n}$

(b) $n! = 2\pi n^{n+1/2} e^{-n}$

(c) $n! = \sqrt{2\pi}\, n^{n+1} e^{-n}$

(d) $n! = 2\pi n^{n+1} e^{-n}$

10. If n is an integer then the value of $\sqrt{\left| \left(\dfrac{1}{2}-n\right)\right| \left(\dfrac{1}{2}+n\right)}$ is

(a) π

(b) $-\pi$

(c) $(-1)^n \pi$

(d) $(-1)^n \sqrt{\pi}$

11. If m and n are positive integers, then the value of $\beta(m, n)$ is

(a) $\dfrac{n!m!}{(m+n)!}$

(b) $\dfrac{(m-1)!(n-1)!}{(m+n-1)!}$

(c) $\dfrac{(n+1)!(m+1)!}{(m+n+2)!}$

(d) $\dfrac{(m-1)!(n-1)!}{(m+n-2)!}$

12. If n is a positive integer then the value of $\sqrt{n}$ is

(a) $n!$

(b) $(n-1)!$

(c) $(n-2)!$

(d) $\dfrac{(n-1)!}{2^n}$

13. The value of $\left|\dfrac{1}{2}\right|$ is

(a) 0

(b) $\dfrac{\pi}{2}$

(c) π

(d) $\sqrt{\pi}$

14. The value of $\beta(z, 1)$ is

(a) $\dfrac{1}{z}$

(b) $\dfrac{1}{z+1}$

(c) $\dfrac{1}{z(z+1)}$

(d) $\dfrac{z}{z+1}$

15. The value of $\beta(2, z)$ is

(a) $\dfrac{1}{z}$

(b) $\dfrac{1}{z+1}$

(c) $\dfrac{1}{z(z+1)}$

(d) $\dfrac{z-1}{z+1}$

16. The relation between β and γ functions is

(a) $\beta(m,n) = \dfrac{\overline{|m}\,\overline{|n}}{2\overline{|(m+n)}}$

(b) $\beta(m,n) = \dfrac{\overline{|m}\,\overline{|n}}{\overline{|(m+n)}}$

(c) $\beta(m,n) = \dfrac{\overline{|(m-1)}\,\overline{|(n-1)}}{\overline{|(m+n-2)}}$

(d) $\beta(m,n) = \dfrac{\overline{\left|\dfrac{(m+1)}{2}\right.}\,\overline{\left|\dfrac{(n+1)}{2}\right.}}{2\overline{\left|\dfrac{(m+n+2)}{2}\right.}}$

17. The value of $\beta(a, b)\,\beta[(a + b), c]$ is

(a) $\overline{|a}\,\overline{|b}\,\overline{|c}$

(b) $\dfrac{\overline{|a}\,\overline{|b}\,\overline{|c}}{\overline{|(a+b+c)}}$

(c) $\overline{|b}\,\overline{|(a+b)}$

(c) $\overline{|c}\,\overline{|(a+b)}\,\overline{|a}$

18. The value of $\int_0^{2\pi} \sin^p \theta \cdot \sin^q \theta \cdot d\theta$ is

(a) $\dfrac{\lfloor p \rfloor q}{\lfloor (p+Q)}$

(b) $\dfrac{\lfloor p+1 \cdot \lfloor q+1}{\lfloor (p+q+2)}$

(c) $\dfrac{\lfloor p \rfloor q}{2 \lfloor p+q+2}$

(d) $\dfrac{\dfrac{\lfloor (p+1)}{2} \cdot \dfrac{\lfloor (q+1)}{2}}{2 \dfrac{\lfloor (p+q+2)}{2}}$

19. The value $\lfloor m \lfloor (1-m)$ is

(a) $\sin m \sin (1 - m)$

(b) $\dfrac{\pi}{\sin m\pi}$

(c) $\dfrac{\pi}{\sin(m-1)\pi}$

(d) $\dfrac{\sin(m-1)\pi}{\sin(m+1)\pi}$

20. The value of $\int_0^\infty \sqrt{\dfrac{k}{y}} e^{-ky} dy$ is

(a) $\dfrac{1}{2}$

(b) $\dfrac{3}{2}$

(c) $\dfrac{\pi}{2}$

(d) $\dfrac{k}{2}$

21. Given $\lfloor 3 \dfrac{5}{2} = C \lfloor 6$, the value of C is

(a) $\sqrt{\pi}$

(b) $\dfrac{\sqrt{\pi}}{2}$

(c) $\dfrac{\sqrt{\pi}}{2^3}$

(d) $\dfrac{\sqrt{\pi}}{2^5}$

22. Given $1.3.5 \cdots (2n-1) = \dfrac{2^n}{\sqrt{\pi}} \lfloor P$, then the value of P is

(a) $n + \dfrac{1}{2}$

(b) $n - \dfrac{1}{2}$

(c) $\left(n + \dfrac{1}{3} \right)$

(d) $n + \dfrac{2}{3}$

23. The value of $\lfloor (1+m) \cdot \lfloor (1-m)$ is

(a) $\dfrac{m\pi}{\sin m\pi}$

(b) $\dfrac{\pi}{\sin m\pi}$

(c) $\dfrac{\pi}{m\sin m\pi}$

(d) $\dfrac{\sin(1-m)\pi}{\sin(1+m)\pi}$

24. What is the ratio $\dfrac{\left|\dfrac{-\dfrac{3}{2}}{}\right.}{\left|\dfrac{3}{2}\right.}$

(a) 1

(b) $\dfrac{3}{8}$

(c) $\dfrac{8}{3}$

(d) $\dfrac{2}{3}$

25. The value of error function erf (α) is

(a) 0

(b) 1

(c) α

(d) $\dfrac{\pi}{2}$

26. The diagram function is defined as

(a) diagram representation of gamma function
(b) logarithmic gamma functions
(c) logarithmatic derivative of gamma functions
(d) none of the above

27. The Legendre's duplication formula is

(a) $\overline{n}\left|n+\dfrac{1}{2}\right. = \dfrac{\sqrt{2\pi}\,\sqrt{2n}}{2^{2n-1}}$

(b) $\overline{n}\left|n+\dfrac{1}{2}\right. = \dfrac{\pi\overline{2n}}{2n-1}$

(c) $\overline{n}\left|n+\dfrac{1}{2}\right. = \dfrac{\sqrt{\pi}\,\sqrt{2n}}{2^{2n-1}}$

(d) $\overline{n}|n+1 = \dfrac{\sqrt{\pi}\,\overline{2n}^2}{2^{2n+1}}$

28. What is the ratio $\dfrac{\left|\dfrac{-\dfrac{1}{2}}{}\right.}{\left|\dfrac{1}{2}\right.}$

(a) 2

(b) –2

(c) $\dfrac{1}{2}$

(d) $-\dfrac{1}{2}$

29. Error function erf (x) is

(a) $2\pi \displaystyle\int_0^x e^{-x^2}\,dx$

(b) $\dfrac{2}{\sqrt{\pi}} \displaystyle\int_0^x e^{-x^2}\,dx$

(c) $\dfrac{1}{2\pi} \displaystyle\int_0^x e^{-x^2}\,dx$

(d) $\displaystyle\int_0^{\pi/2} e^{x^2}\,dx$

ANSWERS

1. (c)	2. (a)	3. (a)	4. (c)	5. (b)	6. (d)	7. (a)	8. (d)
9. (a)	10. (c)	11. (b)	12. (b)	13. (d)	14. (a)	15. (c)	16. (b)
17. (b)	18. (d)	19. (b)	20. (a)	21. (d)	22. (a)	23. (a)	24. (c)
25. (b)	26. (c)	27. (c)	28. (b)	29. (b)			

5

Dirac Delta Function and Green's Functions

5.1 DIRAC DELTA FUNCTION

It is a function used to get a precise notation for dealing with quantities involving a certain kind of infinity, i.e. the quantities having infinitely high and infinitely narrow peak (e.g. in the description of an impulsive force, charge density for a point charge, short range forces such as nuclear force etc.). One can define one-dimensional Dirac delta function, $\delta(x)$, as a singular function which vanishes everywhere except at $x = 0$. At the point $x = 0$, Dirac delta function is so large that the integral of the function over an interval containing the point $x = 0$ is equal to unity, i.e.

$$\int_{-\infty}^{\infty} \delta(x)dx = 1 \tag{5.1}$$

and
$$\delta(x) = 0 \qquad \text{for } x \neq 0 \tag{5.2}$$

To understand it properly, let us consider a function of real variable x which vanishes everywhere except within a small domain of length ε surrounding the point $x = 0$ and which is so large within this domain that its integral over the domain is unity. This function in the limit $\varepsilon \to 0$ will go over into $\delta(x)$, i.e. Dirac delta function.

$\delta(x)$ is frequently called as Dirac delta function for historical reasons while it is not a function in the usual mathematical sense because a mathematical function must have values for points in its domain. It is usually referred to as *improper function*. Obviously, $\delta(x)$ cannot be used in mathematical analysis like an ordinary function. $\delta(x)$ is defined not by giving its values at different points, but it is defined by assigning a rule for integrating its product with a continuous function $f(x)$, i.e.

$$\int_{-\infty}^{\infty} f(x)\delta(x)dx = 0 \tag{5.3}$$

The validity of Eq. (5.3) is clear from the above relation of the function $\delta(x)$. LHS of Eq. (5.3) can depend only on the values of $f(x)$ very close to the origin, so that one can replace $f(x)$ by its value at the origin, $f(0)$. By shifting the origin, one can transform Eq. (5.3) to the following form.

$$\int_{-\infty}^{\infty} f(x)\delta(x-a)dx = f(a) \tag{5.4}$$

Equation (5.4) is valid for any continuous function independent of whether it is a scalar, vector or tensor. Equation (5.4) reveals that the process of multiplying a function of x by $\delta(x - a)$ and integrating over all x, is equivalent to the process of substituting a for x.

Dirac delta function $\delta(x)$ can also be defined as

$$\delta(x) = 0 \qquad \text{for} \quad x \neq 0 \tag{5.5}$$

and
$$\int_{-a}^{+b} \delta(x)\delta(x)dx = 1 \qquad a, b, > 0 \tag{5.6}$$

Eqs (5.5) and (5.6) imply

$$\int f(y)\delta(x-y)dy = f(x) \tag{5.7}$$

for any function $f(x)$, provided the range of integration includes the point x.

It is worthwhile to mention that the range of integration in Eqs (5.3) and (5.4) is not necessarily to be $-\infty$ to $+\infty$. It can be over any domain surrounding the point at which $\delta(x)$ function is not zero, as is obvious from Eq. (5.6). Obviously, limits in these integrations need not be mentioned and one may understand that the domain of the integration is a subtle one.

Sometimes it is useful to employ an explicit expression for the δ-function as the limit of a sequence of analytic functions, e.g.

$$\delta(x) = \lim_{L \to \infty} \frac{\sin xL}{\pi x} \tag{5.8}$$

At $x = 0$, the limiting value of $\dfrac{\sin xL}{\pi x}$ is equal to $\dfrac{L}{\pi}$ and its value oscillates with a period $\dfrac{2\pi}{L}$ when x increases. Its integral within the limit $-\infty$ to ∞ is unity and independent of the value of L. Obviously, $\lim\limits_{L \to \infty} \dfrac{\sin xL}{\pi x}$ has all the properties of a delta function.

An integral representation of the δ function is

$$\delta(x) = \frac{1}{2\pi}\int_{-\infty}^{\infty} e^{ikx}dk \tag{5.9}$$

The proof of this is as follows. One can write RHS of Eq. (5.9) as

$$\lim_{L \to \infty}\int_{-L}^{L} e^{ikx}dk = \lim_{L \to \infty}\frac{e^{ixL} - e^{-ixL}}{ix}$$

$$= 2\pi \lim_{L \to \infty}\frac{e^{iLx} - e^{-iLx}}{2i(\pi x)} = 2\pi \lim_{L \to \infty}\frac{\sin Lx}{\pi x}$$

$$= 2\pi\delta(x).$$

or
$$\delta(x) = \frac{1}{2\pi}\int_{-\infty}^{\infty} e^{ikx}dk \tag{5.10}$$

Separating the real and imaginary parts, one obtains

$$\frac{1}{2\pi}\int_{-\infty}^{\infty} \cos kx\, dx = \delta(x) \tag{5.11}$$

and
$$\int_{-\infty}^{\infty} \sin kx\, dx = 0 \tag{5.12}$$

Equation (5.11) is most commonly used as an explicit expression for delta function $\delta(x)$.

One can also define the δ-function as the differential coefficient $\varepsilon'(x)$ of the function $\varepsilon(x)$ given by

$$\varepsilon(x) = 0 \text{ for } x < 0$$
$$= 1 \text{ for } x > 0 \tag{5.13}$$

To verify the above definition of delta function, let us substitute $\varepsilon'(x)$ for $\delta(x)$ in Eq. (5.3) and integrate by parts. One obtains for a and b positive numbers.

$$\int_{-b}^{a} f(x)\varepsilon'(x)\,dx = f(x)\varepsilon(x)\big|_{-b}^{a} - \int_{-b}^{a} f'(x)\varepsilon(x)\,dx$$

$$= f(a) - \int_{0}^{a} f'(x)\,dx \qquad \text{[using Eq. (5.13)]}$$

$$= f(0)$$

Obviously, the result agrees with Eq. (5.3).

The delta function satisfies the following properties.

$$\delta(-x) = \delta(x) \tag{5.14}$$
$$x\delta(x) = 0 \tag{5.15}$$

$$\delta(ax) = \frac{1}{|a|}\delta(x) \tag{5.16}$$

$$f(x)\,\delta(x-a) = \delta(a)\,\delta(x-a) \tag{5.17}$$

$$\int \delta(x-a)\delta(x-b)\,dx = \delta(a-b) \tag{5.18}$$

$$\delta(x^2 - a^2) = \frac{\delta(x-a) + \delta(x+a)}{2|a|} \tag{5.19}$$

The property Eq. (5.14) is trivial, since $\delta(x)$ is an even function of x. Property Eq. (5.15) is also obvious since for any continuous function $f(x)$, we have

$$\int x\delta(x)\,dx = 0 \qquad \text{[using Eq. (5.3)]}$$

This shows that $x\delta(x)$ is a factor in an integrand and is equivalent to zero.

Similarly, one can verify Eqs (5.16) and (5.19). In order to verify Eq. (5.18), we consider a continuous function $f(a)$ of a, then

$$\int f(a)\,da \int \delta(a-x)\,dx\delta(x-b) = \int \delta(x-b)\,dx \int f(a)\,da\delta(a-x)$$

$$= \int \delta(x-b)\,dx f(x) = f(b)$$

and
$$f(b) = \int f(a)\,da\,\delta(a-b)$$

$$\therefore \quad \int f(a)\,da \int \delta(x-a)\,dx\,\delta(x-b) = \int f(a)\,da\,\delta(a-b)$$

Obviously, $\int \delta(a-x)\delta(x-b)\,dx$ and $\delta(a-b)$ are equivalent. Equation (5.17) directly follows from Eq. (5.18).

5.2. DERIVATIVE OF δ-FUNCTION

One can always define the derivative $\delta'(x)$ of δ-function with respect to x as

$$\delta'(x) = \frac{1}{\pi}\lim_{L\to\infty}\left[\frac{L\cos Lx}{x} - \frac{\sin Lx}{x^2}\right] \tag{5.20}$$

Now solving integral I involving $\delta'(x)$ by parts, we get

$$I = \int_a^b \delta'(x)f(x)\,dx = f(x)\delta(x)\Big|_a^b - \int_a^b \delta(x)\cdot f'(x)\,dx = -f'(0) \tag{5.21}$$

$$\because \qquad \delta(x) = 0 \text{ for } x = a \text{ and } x = b$$

Now solving $\int x\delta'(x)\,dx$ by parts, we get

$$I = \int x\delta'(x)\,dx = x\delta(x) - \int \delta(x)\,dx = -1 \tag{5.21a}$$

Since $x\delta(x) = 0$
Equation (5.21a) gives

$$\int x\delta'(x)\,dx = -\int \delta(x)\,dx$$

or $\qquad x\delta'(x) = -\delta(x)$

where $\delta'(x)$, i.e. the derivative of $\delta(x)$ satisfies

$$x\delta'(x) = -\delta(x) \tag{5.22}$$

Then it follows from Eq. (5.22), since $\delta(x)$ is an even function of x, its derivative $\delta'(x)$ is an odd function.

5.3. THREE-DIMENSIONAL DELTA-FUNCTION, δ(r)

It is defined as

$$\delta'(r) = \delta(x)\,\delta(y)\,\delta(z) \tag{5.23}$$

Obviously, we have

$$\delta(r) = 0 \quad \text{for } r \neq 0 \tag{5.24}$$

and

$$\int \delta(r)f(r)\,d^3r = f(0) \tag{5.25}$$

where $\qquad d^3r = dx\,dy\,dz$

Also

$$\int \delta(r)\,d^3r = \iiint_{x\,y\,z} \delta(x)\delta(y)\delta(z)\,dx\,dy\,dz = 1 \tag{5.26}$$

and in spherical coordinates.

$$\iiint_{\theta\,\phi\,r} \delta(r)r^2 dr \sin\theta\, d\theta\, d\phi = 1 \tag{5.27}$$

which corresponds to a singularity at the origin. Equation (5.26) can be generalized to

$$\iiint_{x\,y\,z} \delta(r_2 - r_1)r_2^2 dr_2 \sin\theta_2\, d\theta_2\, d\phi_2 = 1 \tag{5.28}$$

where $\delta(r_2 - r_1) = \delta(r_1 - r_2)$

and $\delta(r_2 - r_1) = 0,$ for $r_1 \neq r_2$ (5.29)

Using Eqs (5.9) and (5.22), one obtains

$$\delta(r) = (2\pi)^{-3} \int e^{i(k_x x + k_y y + k_z z)} dk_x dk_y dk_z$$

$$= (2\pi)^{-3} \int e^{-(k\cdot r)} d^3 k \tag{5.30}$$

Such transform pairs are of interest in quantum mechanics. If $f(x)$ is the wave function of a particle, the Fourier transform $\phi(k)$ is the so-called 'wave function in momentum space' and $|f(x)|^2$ and $|\phi(k)|^2$ are the probability distributions for the position and momentum respectively.

One useful relation is

$$\delta(r) = \frac{\delta(r)}{2\pi r^2}\delta(r' - r) = \frac{2}{r^2}\delta(n' - n)\delta(r' - r)$$

Here n' and n are unit vectors along r' and r respectively.

5.4. A FEW SIMPLE ILLUSTRATIONS OF DELTA FUNCTION

i. Equation (5.1) shows that if we divide both sides by a variable x which can take on the value zero, one should add on to one side an arbitrary multiple of $\delta(x)$, i.e. form an equation

$$a = b$$

one cannot conclude that

$$\frac{a}{x} = \frac{b}{x}$$

thus, one can draw only inference

$$\frac{a}{x} = \frac{b}{x} = c\delta(x) \tag{5.31}$$

where c is unknown.

As an illustration of above expression, we consider the differentiation of log x for which the usual formula

$$\frac{d}{dx}(\log x) = \frac{1}{x} \tag{5.31a}$$

requires the consideration for the neighborhood of $x = 0$. In order to make this function well defined in the neighborhood of $x = 0$, one must impose an extra condition such that its integral from $-\varepsilon$ to ε vanishes, where ε is an infinitesimally small positive number. This condition makes RHS of Eq. (5.31a) vanishes but LHS equal to log (–1). Obviously, Eq. (5.31a) is not a correct expression. One can correct it by keeping in mind that if we take principal values, log x has a pure imaginary term $i\pi$ for negative values of x. As x passes through the value zero, above imaginary term gives $-i\pi\delta(x)$ on differentiation. Obviously, Eq. (5.31a) must be corrected to

$$\frac{d}{dx}(\log x) = \frac{1}{x} - i\pi\delta(x) \tag{5.32}$$

Similarly, one can evaluate the integrals involving singular function, e.g.

$$\xi(x) = \lim_{t \to \infty} \frac{1 - e^{ixt}}{ix} \tag{5.33}$$

by keeping in mind that

$$\xi(x) = \pi\delta(x) - iP\left(\frac{1}{x}\right) \tag{5.34}$$

where P denotes the principal value.

ii. δ-function representation is used mostly in terms of a complete other normal system of functions. For the function $\phi_n(y)$ corresponding to a discrete spectrum, the orthonormality condition is described as

$$\delta(x - x') = \sum_{n=1}^{\infty} \phi_n^*(x)\phi_n(x') \tag{3.35}$$

For the function ϕ corresponding to continuous spectrum, the orthonormality is described as

$$\delta(x - x') = \int \phi_F^*(x)\phi_F(x')dx \tag{5.36}$$

Here * (star) functions are complex conjugate of – * (nonstar) functions.

5.5. LAPLACE AND FOURIER TRANSFORMS OF DELTA FUNCTION

One can write Laplace transform of delta function $\delta(t - a)$ by definition as

$$L\{\delta(t - a)\} = \int_0^\infty \delta(t - a)e^{-st}dt \tag{5.37}$$

We can evaluate the integral appearing on the right hand side of Eq. (5.37) by using Eq. (5.4), one obtains

$$\int_0^\infty \delta(t - a)e^{-st}dt = e^{-sa}, \quad a > 0 \tag{5.38}$$

and hence

$$L\{\delta(t - a)\} = e^{-sa} \tag{5.39}$$

Thus Laplace transform of $\delta(t)$ is

$$L\{\delta(t)\} = \int_0^\infty \delta(t)e^{-st}dt = 1$$

Fourier Transform of δ-Function

Using definition of the Fourier transforms, we can obtain Fourier transforms of Dirac delta functions.

Thus
$$F\{\delta(t-a)\} = \int_{-\infty}^{+\infty} \delta(t-a)e^{-ist}dt = e^{-isa} \tag{5.40}$$

Thus
$$F\{\delta(t)\} = 1 \qquad \because \ a = 0$$

Similarly, Fourier inverse transform of $\delta(s)$ is defined as

$$\frac{1}{2\pi}\int_{-\infty}^{+\infty} \delta(s)e^{ist}ds = \frac{1}{2\pi} \tag{5.40a}$$

i.e. the inverse Fourier transform of $2\pi\delta(s) = 1$

Inverse Fourier transform of the function $2\pi\delta(s-a)$ is given by

$$\frac{1}{2\pi}\int_{-\infty}^{+\infty} 2\pi\delta(s-a)e^{ist}ds = \int_{-\infty}^{+\infty} \delta(s-a)e^{ist}ds = e^{iat} \tag{5.40b}$$

i.e.
$$F\{e^{iat}\} = 2\pi\delta\,(s-a)$$

and
$$F\{e^{-iat}\} = 2\pi\delta\,(s+a)$$

Using these relations, we have

$$F\{\cos at\} = \frac{1}{2}F[e^{iat} + e^{-iat}] = \pi\{\delta(s-a) + (s+a)\}$$

and
$$F\{\sin at\} = i\pi\{\delta(s+a) - \delta(s-a)\}$$

Example 1. *Prove that* $\dfrac{d}{dx}\theta(x) = \delta(x)$, *where* $\theta(x)$ *is called the unit step function defined as*

$$\theta(x) = \begin{bmatrix} 1 \text{ for } x > 0 \\ 0 \text{ for } x < 0 \end{bmatrix}$$

Solution: $\delta(x)$ satisfies the following relation

$$\int_{-\infty}^{\infty} f(x)\delta(x)dx = f(0)$$

Substituting $\delta(x)$ for two positive numbers a and b and integrating by parts, one obtains

$$\int_{-b}^{a} f(x)\frac{d}{dx}\theta(x)dx = \left[f(x)\theta(x)\right]_{-b}^{a} - \int_{-b}^{a} f'(x)\theta(x)dx$$

$$= f(a) - \int_0^a f'(x)dx = f(0)$$

Now taking a and b as infinite, one obtains

$$\frac{d}{dx}\theta(x) = \delta(x).$$

Example 2. Show that delta function obeys the following properties.

(a) $\delta(x) = \delta(-x)$ (b) $\delta'(x) = -\delta'(-x)$

(c) $x\delta(x) = 0$ (d) $x\delta'(x) = -\delta(x)$

(e) $\delta(ax) = \dfrac{1}{a}\delta(x)$ for $a > 0$ (f) $\delta(x^2 - a^2) = \dfrac{1}{2a}[\delta(x-a) + \delta(x+a)]$ for $a > 0$.

Solution: Let us choose that $f(x)$ is a differentiable function, then

(a) We know $I = \displaystyle\int_{-\infty}^{+\infty} f(x)\delta(x)dx = f(0)$

Let us consider integral

$$I = \int_{-\infty}^{+\infty} f(x)\delta(-x)dx$$

Put $x = -u$ $\therefore$ $dx = -du$

$\therefore$ $I = \displaystyle\int_{+\infty}^{-\infty} f(-u)\delta(u)(-dx) = \int_{-\infty}^{+\infty} f(-u)\delta(u)dx = f(0)$

One obtains $\delta(x) = \delta(-x)$

(b) Let us consider the integral

$$I = \int_{-\infty}^{+\infty} \delta'(x)f(x)dx$$

Integrating by parts, one obtains

$$I = \left[\delta(x)f(x)\right]_{-\infty}^{+\infty} - \int_{-\infty}^{+\infty} \delta(x)f'(x)dx$$

$$= -\int_{-\infty}^{\infty} \delta(x)f'(x)dx$$

$\therefore$ $I = -f'(0)$

Now consider the integral

$$I = \int_{-\infty}^{\infty} \delta'(-x)f(x)dx = \int_{-\infty}^{\infty} \delta'(u)f(-u)du = f'(0) \text{ (By putting } -x = u)$$

Comparing the two, one obtains

$$\delta'(x) = -\delta'(-x)$$

(c) Consider the integral

$$I = \int_{-\infty}^{+\infty} f(x)\delta(x)dx = f(0)$$

Now if $f(x) = x$

Then $\displaystyle\int_{-\infty}^{+\infty} x\delta(x)dx = 0$

or $x\delta(x) = 0$

(d) Consider the integral $\int_{-\infty}^{+\infty} x\delta'(x)dx$ and by integration by parts, we have

$$\int_{-\infty}^{+\infty} x\delta'(x)dx = x\delta(x) - \int_{-\infty}^{+\infty} \delta(x)dx$$

Since $\qquad x\delta(x) = 0$

$\therefore \qquad \int_{-\infty}^{+\infty} x\delta'(x)dx = -\int_{-\infty}^{+\infty} \delta(x)dx = \int_{-\infty}^{+\infty} (-1)\delta(x)dx$

$\therefore \qquad x\delta'(x) = -\delta(x)$

(e) We have

$$\int_{-\infty}^{+\infty} \delta(ax)f(x)dx = \frac{1}{a}\int_{-\infty}^{+\infty} \delta(u)f\left(\frac{u}{a}\right)du = \frac{1}{a}f(0)$$

by putting $ax = u$

Again $\int_{-\infty}^{+\infty} \frac{1}{a}\delta(x)f(x)dx = \frac{1}{a}f(0)$

Comparing the two, we obtain

$$\delta(ax) = \frac{1}{a}\delta(x)$$

(f) Let us consider the integral

$$I = \int_{-\infty}^{+\infty} f(x)\delta(x^2 - a^2)dx = \int_{-\infty}^{0} f(x)\delta(x^2 - a^2)dx + \int_{0}^{\infty} f(x)\delta(x^2 - a^2)dx \qquad (1)$$

Let $x^2 - a^2 = u$, one obtains

$$x = \pm\sqrt{u+a^2} \text{ and } dx = \pm\frac{1}{2}\frac{du}{\sqrt{u+a^2}}$$

Taking negative sign in the first term and positive sign in the second term, one obtains

$$I = \int_{-\infty}^{+\infty} f(x)\delta(x^2 - a^2)dx = \int_{-a^2}^{+\infty} \frac{f\{-\sqrt{u+a^2}\}}{2\sqrt{u+a^2}}\delta(u)du + \int_{-a^2}^{\infty} \frac{f\{\sqrt{u+a^2}\}}{2\sqrt{u+a}}\delta(u)du$$

$$= \frac{f(-a)}{2a} + \frac{f(a)}{2a} = \frac{1}{2a}[f(a) + f(-a)] \qquad (2)$$

We know that

$$\frac{1}{2a}\int_{-\infty}^{\infty} f(x)[\delta(x-a) + \delta(x+a)]dx = \frac{1}{2a}[f(a) + f(-a)] \qquad (3)$$

From Eqs (2) and (3), one obtains

$$\delta(x^2 - a^2) = \frac{1}{2a}[\delta(x-a) + \delta(x+a)]$$

Example 3. *Show that the delta function obeys the following properties.*

(i) $\int \delta(a-x)\delta(x-b)dx = \delta(a-b)$

(ii) $f(x)\delta(x-a) = f(a)\delta(x-a)da$

Solution: (i) $\int \delta(a-x)\delta(x-b)dx = \delta(a-b)$

Multiplying both sides by $f(a)$ and integrating over da, one obtains

$$\text{LHS} = \iint f(a)\delta(a-x)\delta(x-b)dx\,da = \int f(x)\delta(x-b)dx = f(b)$$

$$\text{RHS} = \int f(a)\delta(a-b)da = f(b)$$

This gives $\int \delta(a-x)\delta(x-b)dx = \delta(a-b)$

(ii) $f(x)\,\delta(x-a) = f(a)\,\delta(x-a)$

Integrating both sides over x, one obtains

$$\text{LHS} = \int f(x)\delta(x-a)dx = f(a) \quad \text{and} \quad \text{RHS} = \int f(a)\delta(x-a)dx$$

$$f(a) = \int \delta(x-a)dx = f(a)$$

$\because$ $\int \delta(x-a)dx = 1$ for $x = a$ and zero for $x \neq a$

Hence $f(x)\,\delta(x-a) = f(a)\,\delta(x-a)$

Example 4. *Discuss the behaviour of $\delta(x)$ for small and large x.*

Solution: We have

$$\delta(x) = \lim_{a\to\infty} \frac{1}{2\pi} \int_{-a}^{+a} e^{ikx}dk$$

$$= \frac{1}{2\pi}\lim_{a\to\infty}\left(\frac{e^{iax} - e^{iax}}{ix}\right) = \lim_{a\to\infty}\frac{\sin ax}{\pi x} \tag{1}$$

Here a is positive and real. Let us examine the behaviour of this function for small x, i.e. the limit as x goes to zero;

$$\lim_{x\to 0}\frac{\sin ax}{\pi x} = \frac{a}{\pi}\lim_{a\to 0}\frac{\sin ax}{ax} = \frac{a}{\pi}$$

Thus $\delta(0) = \lim\dfrac{a}{\pi} \to \infty$ as $a \to \infty$ or the amplitude becomes infinite at the singularity.

For large $|x|$, $\dfrac{\sin ax}{\pi x}$ oscillates with period $\dfrac{2\pi}{a}$, and its amplitude falls as $1/|x|$.

But in the limit as $a \to \infty$, the period gets infinitesimally narrow so that function approaches zero everywhere except for the infinite spike of infinitesimally width at the singularity. The integral [Eq. (1)] over all space is unity, i.e.

$$\int_{-\infty}^{\infty} \lim_{a\to\infty}\frac{\sin ax}{\pi x}dx = \lim_{a\to\infty}\frac{2}{\pi}\int_0^{\infty}\frac{\sin ax}{x}dx = \frac{2}{\pi}\cdot\frac{\pi}{2} = 1$$

Example 5. *Prove that*

$$\delta[f(x)] = \sum_i \frac{1}{\left|\frac{\partial f}{\partial x}(x_i)\right|} \delta(x - x_i)$$

where f(x) is assumed to have only simple zeros located at x = x_1.

Solution: Since $\delta(x) = 0$ for $x \neq 0$, then contribution to the integral

$$I = \int_{-\infty}^{\infty} \delta[f(x)]g(x)dx$$

will come only from the points $x_1, x_2,..., x_n$ which are the roots of equation $f(x) = 0$.

or

$$f(x) = (x - x_1)(x - x_2)(x - x_3)\cdots(x - x_n) = 0$$

Hence, we can write the integral as

$$\int_{-\infty}^{\infty} \delta[f(x)]g(x)dx = \int_{x_1-\varepsilon}^{x_1+\varepsilon} \delta[f(x)]g(x)dx + \int_{x_2-\varepsilon}^{x_2+\varepsilon} \delta[f(x)]g(x)dx + \cdots + \int_{x_n-\varepsilon}^{x_n+\varepsilon} \delta[f(x)]g(x)dx$$

Let us consider the term corresponding to the root x_i in the right hand side

$$\int_{x_i-\varepsilon}^{x_i+\varepsilon} \delta[(x - x_1)(x - x_2)\cdots(x - x_{i-1})(x - x_i)(x_i - x_{i+1})\cdots(x_i - x_n)]g(x)dx$$

But $(x_i - x_1)(x_i - x_2)\cdots(x_i - x_{i-1})(x_i - x_{i+1})\cdots(x_i - x_n) = \frac{df}{dx}(x_i)$

$$\therefore \quad \int_{x_i-\varepsilon}^{x_i+\varepsilon} \delta[f(x)]g(x)dx = \int_{x_i-\varepsilon}^{x_i+\varepsilon} \delta\left[\left(\frac{df}{dx}(x_i)\right)(x - x_i)\right]g(x)dx$$

$$= \frac{1}{\left|\frac{df}{dx}(x_i)\right|} \int_{x_i-\varepsilon}^{x_i+\varepsilon} \delta(x - x_i)g(x)dx = \frac{g(x_i)}{\left|\frac{df}{dx}(x_i)\right|}$$

Here we have used the result $\delta(ax) = \frac{1}{|a|}\delta(x)$

But

$$g(x_i) = \int_{-\infty}^{+\infty} \delta(x - x_i)g(x)dx$$

and

$$g(x_i) = \left|\frac{df}{dx}(x_i)\right| \int_{x_i-\varepsilon}^{x_i+\varepsilon} \delta[f(x)]g(x)dx$$

$$\therefore \quad \int_{x_i-\varepsilon}^{x_i+\varepsilon} \delta[f(x)]g(x)dx = \frac{1}{\left|\frac{df}{dx}(x_i)\right|} \int_{-\infty}^{\infty} \delta(x - x_i)g(x)dx$$

$$\therefore \quad \int_{-\infty}^{\infty} \delta[f(x)]g(x)dx = \sum_{i=1}^{n} \frac{1}{\left|\frac{df}{dx}(x_i)\right|} \int_{-\infty}^{\infty} \delta(x - x_i)g(x)dx$$

or

$$\delta[f(x)] = \sum_{i=1}^{n} \frac{1}{\left|\frac{df}{dx}(x_i)\right|} \delta(x - x_i)$$

5.6. GREEN'S FUNCTIONS

Green's functions provide a general method for solving differential equations satisfying certain boundary conditions in the form of an integral equation.

Let us consider the inhomogeneous equation

$$LU(x) - \lambda U(x) = f(x) \tag{5.41}$$

over a domain Ω, with L a Hermitian operator, with $u(x)$ subject to the usual type of (homogeneous) boundary conditions and λ is given constant. To solve Eq. (5.41) expand $U(x)$ and $f(x)$ in eigen functions of the operator L, one obtains

$$U(x) = \sum_n C_n U_n(x), \quad f(x) = \sum_n d_n U_n(x) \tag{5.42}$$

Using Eq. (5.42), Eq. (5.41) becomes

$$\sum_n C_n(\lambda_n - \lambda)U_n(x) = \sum_n d_n U_n(x)$$

Since the functions $U_n(x)$ are linearly independent, therefore, one obtains

$$C_n = \frac{d_n}{\lambda_n - \lambda} = \frac{U_n \cdot f}{\lambda_n - \lambda} \quad (\because d_n = U_n \cdot f) \tag{5.43}$$

Using Eq. (5.43) and Eq. (5.42) becomes

$$U(x) = \sum_n \frac{U_n U_n \cdot f}{\lambda_n - \lambda} = \sum_n \frac{U_n(x)}{\lambda_n - \lambda} \int_\Omega U_n^*(x')f(x')d^3x'$$

or

$$u(X) = \int_\Omega G(X, X')f(X')(dX')^3 \tag{5.44}$$

where

$$G(X, X') = \sum_n \frac{U_n(X) \cdot U_n(X')}{\lambda_n - \lambda} \tag{5.45}$$

$G(X, X')$ given by Eq. (5.45) is called *Green's function*. We must note that the Green's function is determined by a differential operator (nearly always a Hermitian) for a particular region, and suitable boundary conditions. Sometimes one writes Green's function as $G(x, x'; \lambda)$ to emphasize the dependence of G on λ as well as on x and x'.

Example 6. *A string of length l is vibrating with frequency ω. The equation and boundary conditions are*

$$\frac{d^2u}{dx^2} + k^2u = 0, \quad u(0) = 0 = u(l), \quad k = \frac{\omega}{c}$$

Determine the Green's function.

Solution: Suppose

$$k^2 = -\lambda \text{ so that } \frac{d^2u}{dx^2} = \lambda u$$

Eigen values

$$\lambda_n = -\left(\frac{n\pi}{l}\right)^2$$

$$n = 1, 2, 3$$

Normalized eigen function

$$u_n = \sqrt{\frac{2}{x}} \sin \frac{n\pi x}{l}$$

Green's function

$$G(x, x') = \sum_n \frac{u_n(x) \cdot \bar{u}_n(x')}{\lambda_n - \lambda}$$

$$\therefore \qquad G(x, x') = \sum_n \frac{\sin\left(\frac{n\pi x}{l}\right) \sin\left(\frac{n\pi x'}{l}\right)}{k^2 - \left(\frac{n\pi}{l}\right)^2}$$

These results reveal an important general symmetry relation for Green's functions

$$G(x, x') = [G(x, x')]^* \qquad (5.46)$$

The physical significance of Eq. (5.46), for real Green's functions, is the reciprocity relation, that the response at x to a unit point disturbance at x' is the same as the response at x' to a unit point disturbance at x.

Example 7. *Determine the Green's function when the equation and boundary conditions in the case of circular drum are given as*

$$\nabla^2 u + k^2 u = 0, \qquad u = 0, \quad \text{when } r = R$$

Solution: It is evident from physical considerations that $G(X, X')$ can depend only on r, r' and the angle between X' and, X, i.e. θ.

Now, we consider the equation

$$\nabla^2 G + k^2 G = \delta(X - X') \qquad (1)$$

Obviously, G is the solution of Eq. (5.47) where $\delta(X - X')$ is a two-dimensional delta function such that $\int \delta(X - X')(dx)^2 = 1$ for any area of integration which surrounds X'.

When $X \neq X'$, one obtains $\nabla^2 G + k^2 G = 0$. The solution of this equation with the given boundary conditions can be put as

$$G = \sum A_m J_m(kr)\cos m\theta \quad \text{for } r < r'$$

$$G = \sum B_m [J_m(kr)Y_m(kR) - Y_m(kr)J_m(kR)\cos m\theta] \quad \text{for } r > r' \qquad (2)$$

where $J_m(kr)$ and $Y_m(kR)$ are Bessel's functions of 1st and 2nd kind respectively. Values of constants A_m and B_m can be determined as follows.

Let us substitute our solutions given by Eq. (5.48) together along the circle $r = r'$. Now integrating the differential equation $\nabla^2 G + k^2 G = \delta(X - X')$ over an area which surrounds the point $X = X'$, one obtains

$$\int \nabla^2 G \, dx^2 = \int (\nabla G)_n \, dl = 1 \quad \text{(using Gauss's theorem } \int_V \nabla \cdot u (dx)^3 = \int_S u \cdot ds \text{)}$$

Therefore, for an area enclosing the point X', one has

$$\int_{r'+\varepsilon} \frac{\partial G}{\partial r} \, dl - \int_{r'-\varepsilon} \frac{\partial G}{\partial r} \, dl = 1$$

But $l = r'\theta;$ $\therefore$ $dl = r'\, d\theta$

$$\therefore \quad \int_{r'+\varepsilon} \frac{\partial G}{\partial r} r'\, d\theta - \int_{r'-\varepsilon} \frac{\partial G}{\partial r} r'\, d\theta = 1$$

The leads to

$$\int_{r'+\varepsilon} \frac{\partial G}{\partial r} d\theta - \int_{r'-\varepsilon} \frac{\partial G}{\partial r} d\theta = \frac{1}{r'} = \left. \frac{\partial G}{\partial r} \right|_{r'+\varepsilon} - \left. \frac{\partial G}{\partial r} \right|_{r'-\varepsilon} = \frac{1}{r'}\delta(\theta)$$

Suppose $\left. \dfrac{\partial G}{\partial r} \right|_{r'+\varepsilon} - \left. \dfrac{\partial G}{\partial r} \right|_{r'-\varepsilon} = \displaystyle\sum C_m \cos m\theta$

Multiplying both sides of the above relation by $\cos m'\theta$ and integrating w.r.t. θ from $-\pi$ to π, one obtains

$$\therefore \qquad \frac{1}{r'} = \pi C_{m'}\, \varepsilon_{m'} \quad \text{where} \quad \begin{cases} 1 \text{ if } m' > 0 \\ 2 \text{ if } m' = 0 \end{cases}$$

$$\therefore \qquad \left. \frac{\partial G}{\partial r} \right|_{r'+\varepsilon} - \left. \frac{\partial G}{\partial r} \right|_{r'-\varepsilon} = \frac{1}{\pi r'} \sum \frac{1}{\varepsilon_m} \cos m\theta \tag{3}$$

From the condition of the continuity of G at $r = r'$, one can obtain

$$A_m J_m(kr') = B_m[J_m(kr')Y_m(kR) - Y_m(kr')J_m(kR)]$$

From Eq. (5.49) one obtains

$$B_m[J'_m(kr')Y_m(kR) - Y'_m(kr')J_m(kR)] - A_m J'_m(kr') = \frac{1}{4\pi\varepsilon_m kr'}$$

Finally one obtains

$$A_m = \frac{J_m(kR)Y_m(kr') - J_m(kr')Y_m(kR)}{2\varepsilon_m J_m(kR)} \tag{4}$$

and

$$B_m = \frac{-J_m(kr')}{2\varepsilon_m J_m(kR)} \tag{5}$$

[we have used $J_m(x)Y'_m - J'_m(x)Y_m(x) = (2/\pi x)$]

Using Eqs (5.50), (5.51) and (5.48) one can obtain the desired Green's function.

5.7. ALTERNATIVE SOLUTION OF GREEN'S FUNCTION EQUATION

We again consider the original Green's function equation

$$\nabla^2 G + k^2 G = \delta(X - X') \tag{5.47}$$

where $\delta(X - X')$ is the two-dimensional delta function located at the 'source point X'.

There are two difficulties in obtaining the solution: (i) to ensure the proper singularity at the source point X' and the other is to satisfy the boundary conditions. Sometimes it is convenient to separate the two difficulties by finding a solution of the form

$$G = U(X, X') + V(X, X') \tag{5.48}$$

where U has the singularity at X' and remains unaffected over under boundary conditions while V is smooth at X' but is such so as to make $G(X, X')$ satisfy the boundary conditions.

The singularity at X' can be determined by integrating Eq. (5.52) over an elementary area surrounding X' and finding U in the form $U(\rho)$, where

$$\rho = |X - X'|$$

i.e.,

$$\int_0^\rho \nabla^2 G \, d\rho + k^2 \int_0^\rho G \, d\rho = \int_a^\rho \delta(X - X') d\rho$$

Using Gauss's theorem, one obtains

$$\left[\int_0^\rho \nabla^2 G 2\pi\rho \, d\rho = 2\pi\rho \frac{\partial G}{\partial \rho} \right]$$

or

$$2\pi\rho \frac{\partial G}{\partial \rho} + k^2 \int_0^\rho G 2\pi\rho \, d\rho = 1$$

and

$$G(\rho) \rightarrow \frac{1}{2\pi} \ln\rho + \text{constant} \qquad [\text{as } \rho \rightarrow 0] \tag{5.49}$$

$U|X, X'|$ should then have this behaviour near $\rho = 0$. The singular solution of $\nabla^2 G + k^2 G = 0$, $Y_0(k\rho)$, has the behaviour for small ρ.

$$Y_0(k\rho) \rightarrow \frac{2}{\pi} \ln\rho + \text{constant}$$

Thus

$$G = \frac{1}{4} Y_0(k\rho) + V|X, X'| \tag{5.50}$$

Writing $V = \sum_n A_n J_n(kr)\cos n\theta$ with A_n chosen so that G satisfies the boundary conditions. That is

$$G(r = R) = 0 = \frac{1}{4} Y_0(k\rho_R) + \sum_n A_n J_n(kr)\cos n\theta$$

$$A_n = -\frac{1}{4\pi J_n(kR)\varepsilon_n} \int_0^{2\pi} Y_0(k\rho_R)\cos n\theta \, d\theta$$

where

$$\rho_R^2 = R^2 + r'^2 - 2Rr' \cos\theta$$

The result is

$$G(x, x') = \frac{1}{4} Y_0(k\rho) - \sum_{n=0} \frac{J_n(kr)\cos n\theta}{2\pi J_n(kr)\varepsilon_n} \int_0^\pi Y(k\rho_R)\cos n\theta \, d\theta \tag{5.51}$$

5.8. GREEN'S FUNCTIONS IN ELECTRODYNAMICS

In this section, we shall restrict ourself to the brief discussion of two Green's functions which are important in electrodynamics.

i. First, we consider Laplace's equation

$$\nabla^2 \phi = 0$$

in an infinite region. The boundary condition is that $\phi \rightarrow 0$ as $r \rightarrow \infty$. To find the Green's function, one must solve

$$\nabla^2 \phi(x) = \delta(x - x') \tag{5.52}$$

Obviously, ϕ can depend only on the scalar quantity $r = |x - x'|$. Thus, we choose our region of (spherical) coordinates to be at the point x'.

- For all $x \neq x'$, $\nabla^2\phi(x) = 0$. Therefore,

$$\phi \sim \left\{ \begin{matrix} r \\ r^{-(l+1)} \end{matrix} \right\} P_l^m(\cos\theta) e^{\pm im\phi}$$

Because of the spherical symmetry and the boundary condition at infinity, the only possibility is

$$\phi = \frac{A}{r}$$

To find A, we integrate the Eq. (5.57) over the volume of a sphere of radius r surrounding the origin. This gives

$$\int ds \cdot \nabla\phi = 1 = 4\pi r^2 \left(\frac{-A}{r^2} \right) \Rightarrow A = -\frac{1}{4\pi}$$

The Green's function which gives the potential of a point charge or coulomb potential is given by

$$G(x, x') = \frac{-1}{4\pi(x - x')} \tag{5.53}$$

This is *fundamental solution* of Eq. (5.57).

ii. As a second example, we consider the wave equation in an infinite region

$$\nabla^2\phi(x, t) - \frac{1}{c^2}\frac{\partial^2\phi(x,t)}{\partial t^2} = 0 \tag{5.54}$$

We want to solve the equation

$$\nabla^2\phi - \frac{1}{c^2}\frac{\partial^2\phi}{\partial t^2} = \delta(x - x')\,\delta(t - t') \tag{5.55}$$

It is evident that the solution depends only on $x - x'$ and $t - t'$; in other words, the wave Eq. (5.59) possesses translational invariance in x and t. Therefore, one can, without loss of generality, set $x' = 0$, $t' = 0$. Now, we introduce the Fourier transform of ϕ as

$$\phi(k, \omega) = \int \frac{d^3k}{(2\pi)^3} \int \frac{d\omega}{2\pi} \phi(k, \omega) e^{-i(k \cdot x - \omega t)} \tag{5.56}$$

$$\phi(x, t) = \int d^3x \int dt\, \phi(x, t) e^{-i(k \cdot x - \omega t)} \tag{5.57}$$

The Fourier transform of differential Eq. (5.60) is

$$\left(-k^2 + \frac{\omega^2}{c^2} \right)\phi = 1$$

$$\therefore \qquad \phi(k, \omega) = \frac{c^2}{\omega^2 - k^2 c^2} \tag{5.58}$$

and

$$\phi(x, t) = \int \frac{d^3k}{(2\pi)^3} \int \frac{d\omega}{2\pi} c^2 \frac{e^{i(k \cdot x - \omega t)}}{\omega^2 - k^2 c^2} \tag{5.59}$$

If we choose the axis of polar coordinates in k-space along the vector x, the angular integrations are straight forward, and one obtains

$$\phi(x, t) = \frac{1}{(2\pi)^3} \frac{c^2}{ir} \int_{-\infty}^{\infty} k\, dk \int_{-\infty}^{\infty} d\omega \frac{e^{i(k \cdot x - \omega t)}}{\omega^2 - k^2 c^2} \quad (r = 1 \times 1) \tag{5.60}$$

From Eq. (5.65), it is clear that there are two poles on the path of integration

(i) $\omega = kc$ and (ii) $\omega = -kc$

Further Eq. (5.60) involves as disturbance localised at $t = 0$ and $x = 0$. Taking the initial boundary condition as $\phi = 0$ for $t = 0$, one must make the contour to go over both the poles.

Thus, is when $t < 0$, one should complete the contour by an upper semicircular which includes no poles; when $t > 0$ one should complete the contour by a lower semicircle.

$$\phi(x, t) = \frac{1}{(2\pi)^3} \frac{c^2}{ir} \int_{-\infty}^{\infty} k\, dk\, e^{ikr} \int_{-\infty}^{\infty} d\omega \frac{e^{-i\omega t}}{\omega^2 - k^2 c^2}$$

$$= \frac{1}{(2\pi)^3} \frac{c}{2} \frac{1}{ir} \int_{-\infty}^{\infty} dk\, e^{ikr} \int_{-\infty}^{\infty} d\omega\, e^{-i\omega t} \left(\frac{1}{\omega - kc} - \frac{1}{\omega + kc} \right)$$

$$= -\frac{c}{8\pi^2 r} \int_{-\infty}^{\infty} dk\, e^{-ikr} (e^{-ikct} - e^{ikct})$$

$$= -\frac{c}{4\pi r} [\delta(r - ct) - \delta(r + ct)]$$

The second delta function does not contribute, because r and t are both positive. Therefore the desired Green's function is

$$G(x - x', t - t') = \left\{ -\frac{c}{4\pi(x - x')} \delta[(n - n') - c(t - t')] \right\} \quad \text{if } t > t' \tag{5.61}$$

Example 8. *A circuit consisting of an inductance L_0 and capacitance C_0 in series has no current initially at $t = 0$. An emf of very large value is applied for a very short time so that it may be expressed as $E_0 \delta(t)$. Determine the subsequent current in the circuit.*

Solution: The differential equation for current in the given circuit is

$$L_0 \frac{di}{dt} + \frac{q}{C_0} = E_0 \delta(t)$$

where

$$i = \frac{dq}{dt}$$

Taking Laplace transforms on both sides, we get

$$L \left\{ L_0 \frac{di}{dt} \right\} + L \left\{ \frac{q}{C_0} \right\} = L \{ E_0 \delta(t) \}$$

$$L_0 sL\{i\} + \frac{1}{C_0} L\{q\} = E_0$$

$$L_0 s L\{i\} + \frac{1}{C_0} \frac{L\{i\}}{s} = E_0$$

$$\left[L_0 s + \frac{1}{C_0 s} \right] L\{i\} = E_0$$

$$L\{i\} = \frac{E_0 s}{L_0 \left(s^2 + \frac{1}{L_0 C_0} \right)} = \frac{E_0}{L_0} \frac{s}{(s^2 + \omega^2)}$$

where $\omega^2 = \dfrac{1}{L_0 C_0}$. Using inverse Laplace transform, we have

$$i = \frac{E_0}{L_0} \cos \omega t = \frac{E_0}{L_0} \cos \frac{t}{\sqrt{L_0 C_0}}$$

Example 9. *Find the effect produced when a particle of mass m situated at the origin is acted upon by an impulsive force F_0 applied to the mass at t = 0.*

Solution: If the impulse acts in x-direction, then the equation of motion is

$$m \frac{d^2 x}{dt^2} = F_0 \delta(t)$$

where the mass is at $x = 0$ when $t = 0$ and has no velocity

Taking Laplace transforms, we have

$$mL \left[\frac{d^2 x}{dt^2} \right] = F_0 L\{\delta(t)\} = F_0$$

or
$$ms^2 L\{x\} = F_0$$

or
$$L\{x\} = \frac{F_0}{ms^2}$$

and the inverse Laplace transform gives

$$x = L^{-1} \left\{ \frac{F_0}{ms^2} \right\} = \frac{F_0 t}{m}$$

i.e., the particle will move with constant velocity.

SHORT ANSWER QUESTIONS

1. Define Dirac delta function.
2. Find Fourier transform of delta function.
3. Express completeness condition in terms of delta function.
4. Show that the derivative of unit step function is the Dirac delta function.
5. Find Laplace transform of delta function.
6. State two properties of delta function.
7. Express green's function for the Poison's equation.
8. Find the derivative of Dirac delta function at the origin $x = 0$.

9. Show that $\delta[c(x-a)] = \dfrac{1}{c}\delta(x-a)$.

10. Show that $\displaystyle\int_{-\infty}^{+\infty} f(x)\delta'(x)dx = -f'(0)$.

11. Show that $x\delta(x) = 0$, where $\delta(x)$ is Dirac delta function.

12. What is Green's function, state its symmetry properties.

13. Show that $\delta(a^2 - x^2) = \dfrac{1}{2a}[\delta(x-a) + \delta(x+a)]$.

14. Given Dirac delta function $\delta(x) = \dfrac{1}{2\pi}\displaystyle\int_{-\infty}^{+\infty} e^{i\omega}d\omega$. Find $\delta(x^2 - a^2)$.

15. Prove that $x\delta'(x) = -\delta(x)$.

16. Show that $\displaystyle\int_{-\infty}^{+\infty}\delta(x-a)\delta(x-b)dx = \delta(a-b)$.

PROBLEMS

1. Show that $x\dfrac{d}{dx}\delta(x) = -\delta(x)$

2. If $\delta_n(x) = \begin{cases} 0 & x < -\dfrac{1}{2n} \\ n & -\dfrac{1}{2n} < x < \dfrac{1}{2n}, \\ 0 & \dfrac{1}{2n} < x \end{cases}$

   show that $\displaystyle\lim_{n\to\infty}\int_{-\infty}^{\infty} f(x)\delta_n(x)dx = f(0)$, where $f(x)$ is continues at $x = 0$.

3. Show that $\delta[f(x)] = \left|\dfrac{df(x)}{dx}\right|^{-1}\delta(x - x_0)$, where x_0 is chosen so that $f(x_0) = 0$.

4. If $\delta_n(x) = \dfrac{n}{2\cosh^2 nx}$, show that $\displaystyle\int_{-\infty}^{\infty}\delta_n(x)dx = 1$ independent of n.

5. Obtain the solution of Poisson's equations using Green's function.

6. Prove that the Green's function satisfying the equation

$$(\nabla^2 + k^2)G(r_1, r_2) = -\delta(r_1 - r_2) \text{ is}$$

$$G(r_1, r_2) = \dfrac{e^{ik|r_1 - r_2|}}{4\pi|r_1 - r_2|}$$

7. Using Fourier transforms, show that Green's functions satisfying the nonhomogeneous Helmholtz equation.

$$(\nabla^2 + k_0^2)G(r_1, r_2) = -\delta(r_1 - r_2) \text{ is}$$

$$G(r_1, r_2) = \dfrac{1}{(2\pi)^3}\int\dfrac{e^{ik\cdot(r_1 - r_2)}}{k^2 - k_0^2}d^3k$$

8. Construct the dimensional Green's function for the modified Helmholtz equation

$$(\nabla^2 - K^2)\psi(x) = f(x)$$

where the boundary conditions are that the Green's functions must vanish for $x \to \infty$ and for $x \to -\infty$.

Ans. $\left[G(x_1 x_2) = \dfrac{1}{2k} \exp\left(-\dfrac{k}{x_1 - x_2} \right) \right]$

9. Show that $G(r_1, r_2) = G(r_2, r_1)$
10. Define Dirac delta function and state its properties.
11. Establish the following properties of the delta function.

   i. $\delta(x) = \lim\limits_{g \to \infty} \dfrac{\sin gx}{\pi x}$ ii. $\delta(k) = \dfrac{1}{2\pi} \int_{-\infty}^{\infty} e^{ikx} dx$

   iii. $\delta[f(x)] = \sum\limits_i \delta(x - x_i) \Big/ \left| \dfrac{df}{dx} \right|$ iv. $\delta(x^2 - a^2) = \dfrac{1}{2a}[\delta(x - a) + \delta(x + a)]$

MULTIPLE CHOICE QUESTIONS

1. If $\dfrac{d}{dx}[\delta(x)f(x)] = f(x)$, then $f(x)$ must be

  (a) 1   (b) e^x

  (c) $\log x$ (d) unit step function $u(x)$

2. Which of the following does not represent delta function?

  (a) $\lim\limits_{\sigma \to 0} R_\sigma(x)$ for $R_\sigma(x) = \dfrac{1}{2\sigma}$ for $-\sigma < x < \sigma$ and 0 for $|x| > 0$

  (b) $\lim\limits_{\sigma \to 0} \dfrac{1}{\sigma \sqrt{2\pi}} e^{-x^2/2\sigma^2}$ for $\sigma > 0$

  (c) $\lim\limits_{g \to \infty} \dfrac{\sin \pi x}{\pi x}$

  (d) $\lim\limits_{\sigma \to 0} e^{-x^2/\sigma}$

3. For an arbitrary well behaved function, the value of $\int_{-\infty}^{+\infty} f(x)\delta(x - a)dx$ is

  (a) 0   (b) $f(x)$

  (c) $f(a)$ (d) a

4. What is the value of $\delta[c(x - a)]$?

  (a) a (b) $\dfrac{1}{c}\delta(x - a)$

  (c) $c\delta(x - a)$ (d) ca

5. The value of $f(x)\,\delta(x - a)$ is

  (a) 0   (a) a

  (c) $f(x)$ (d) $f(x)\,\delta(x - a)$

6. The completeness condition in terms of delta function is equal to

(a) $\sum_n \phi_n^x(x')\phi_n(x) = \delta(x-x')$

(b) $\int \phi_n^x(x')\phi_n(x)dx = \delta(x-x')$

(c) $\sum \phi_n^*(x')\phi_n(x) = \dfrac{\delta(x)}{\delta(x')}$

(d) $\int \phi_n^*(x')\phi_n(x)dx = \delta(x) - \delta(x')$

7. The Laplace transform of $\delta(t)$ is

(a) 1

(b) 0

(c) $\sqrt{2\pi}$

(d) $\dfrac{1}{\sqrt{2\pi}}$

8. The fourier transform of $\delta(t)$ is

(a) 1

(b) 0

(c) $\sqrt{2\pi}$

(d) $\dfrac{1}{\sqrt{2\pi}}$

9. The derivative of heavy side step function $u(x-a) = \begin{bmatrix} 0 \text{ for } x < a \\ 1 \text{ for } x > a \end{bmatrix}$ is equal to

(a) $\delta(1)$

(b) $\delta(0)$

(c) $\delta(a)$

(d) $\delta(x-u)$

10. The solution of Poison's equation $\nabla^2\phi = -4\pi\,\delta(r)$ is

(a) $\dfrac{1}{r^2}$

(b) $\dfrac{1}{r}$

(c) $\dfrac{e^{ikr}}{r}$

(d) $\dfrac{1}{4\pi r}e^{ikr}$

11. If a function is discontinuous at $x = a$ and has discontinuity of c, then its derivative $\dfrac{d}{dx}[cf(x-a)]_{x=a}$ is equal to

(a) $c\delta(x-a)$

(b) $c\delta(x-a)$

(c) $\dfrac{\delta(x-a)}{c}$

(d) $\dfrac{\delta(x+a)}{c}$

12. Green's function for Poison's equation $G(r_1, r_2)$ is equal to

(a) $\dfrac{|r_2 - r_1|}{4\sqrt{\pi}}$

(b) $\dfrac{1}{4\pi|r_2 - r_1|}$

(c) $\dfrac{|r_2 - r_1|}{|r_2 + r_1|}$

(d) $\dfrac{r_2 - r_1}{|r_2 - r_1|}$

13. If $\delta(r_1, r_2)$ is Dirac delta function and $G(r_1, r_2)$ is Green's function, then which of the following is correct?

(a) $\delta(r_1, r_2) = \delta(r_2, r_1)$ but $G(r_1, r_2) \neq G(r_2, r_1)$

(b) $G(r_1, r_2) = G(r_2, r_1)$ but $\delta(r_1, r_2) \neq \delta(r_2, r_1)$

(c) $\delta(r_1, r_2) = \delta(r_2, r_1)$ and $G(r_1, r_2) = G(r_2, r_1)$

(d) $\delta(r_1, r_2) \neq \delta(r_2, r_1)$ and $G(r_1, r_2) \neq G(r_2, r_1)$

14. The value of $\int_{-\infty}^{+\infty} xf(x-z)dx$

(a) 0   (b) 2

(c) e^2 (d) x^2

15. The derivative of heavy side step function, i.e. $\dfrac{d}{dx}[\varepsilon(x)]$ is

(a) 0   (b) $\delta(x)$

(c) $\delta(1)$ (d) $\delta(0)$

ANSWERS

1. (d)	2. (d)	3. (c)	4. (b)	5. (d)	6. (a)	7. (a)	8. (d)
9. (d)	10. (b)	11. (a)	12. (b)	13. (c)	14. (b)	15. (b)	

6

Functions of Complex Variables

6.1. INTRODUCTION

It is fairly safe to state that the subject of complex variables is fundamental to the subject of applied mechanics. For example, they are used in (i) potential theory (ii) response theory (iii) power series solution of certain types of differential equations (Forhenius–Fuchs method) (iv) evaluating certain definite integrals (v) quantum theory, and (vi) investigations of stability conditions (dispersion relation). As such a solid background in this area is essential for the application of many useful methods of analysis to obtain solution to a wide range of practical problems.

The number system solely based on the real numbers is not sufficient for all mathematical needs, e.g. there is no real number, rational or irrational, which represents the root of equation $x^2 + 1 = 0$. In fact, we are not justified in assuming that there is any number whose square is -1, satisfying the equation $x^2 + 1 = 0$, until we have constructed a field having an element with this property. To make up for the deficiency, Euler for the first time, introduced a symbol i for the square root of -1, with the property that $i^2 = -1$. Bauss introduced a number of the form $\alpha + ib$, which satisfies every algebraic equation with real coefficients. The real root is a particular case of such numbers obtained by putting $\beta = 0$.

The square roots of $-$ve numbers are defined as imaginary numbers. Initially it was thought that these numbers do not have any practical use but now as we see they are of great importance in variety of applied fields. This chapter is intended as a review and brief summary of many of the useful results of complex function theory. It has been assumed by the authors that the reader has some familiarity with complex numbers[*].

6.2. SOME DEFINITIONS

A complex variable $z = x + iy, (i = \sqrt{-1})$, is an ordered pair of real variables x and y which satisfies certain laws of operation. In general, $x + iy \neq y + ix$; hence the term *ordered pair* is important. The real and imaginary parts of z are Re $z = x$ and

---

[*] Readers who would like a brief review of complex numbers can refer to *Applied Mathematics for Engineers and Physicists*, Pipes and Harvill, McGraw Hill, 3rd edition (1970).

369

$\operatorname{Im} z = y$ respectively. Note also that $z = 0$ implies that $x = y = 0$; and that $z_1 = z_2$ means that $x_1 = x_2, y_1 = y_2$.

Complex Plane or Argand Diagram

The graphical representation of the complex variable $z = x + iy$ is called Argand diagram. Hence x, the real part of z is plotted as abscissa (on x-axis) and the imaginary part of z as the ordinate. The complex plane or Argand diagram is shown in Fig. 6.1. Any point $P(x, y)$ in this plane is labelled as

$$z = x + iy$$

where z is called the complex coordinate of the point P. x-axis of this plane is called *real axis* and y-axis is called *imaginary axis*. It is obvious that the point (x, y) does not coincide with the point (y, x) except under special case.

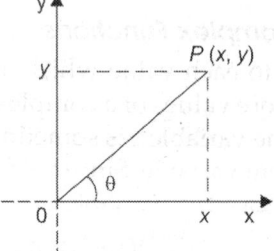

Fig. 6.1

Set of points: The set of points in Argand diagram is a collection of points finite or infinite in number.

Neighbourhood

A neighbourhood of a point z_0 is the set of all points z for which

$$|z - z_0| < \varepsilon$$

where ε is some positive constant. Thus, a neighbourhood consists of all points of a disc, or circular region including the centre z_0 but excluding the points on the bounding circle. The term neighbourhood will be used consistently in this sense. A deleted neighbourhood of z_0 is one in which the point z_0 is omitted, i.e. $0 < |z - z_0| < \varepsilon$.

Limit Point

A point z_0 is a limit point for a set of points in the z-plane if every neighbourhood of z_0 contains points, other than z_0 of the set. Thus each point on the circle $|z| < c$ is a limit point for the set $|z| < c$, and these limit points do not belong to the set. Each point of the set $|z| < c$ is also a limit point of the set. As an example $z = \frac{1}{n}(n = 1,2,3,\cdots)$ has the limit point $z = 0$. Note that the limit point (z_0) may or may not belong to the set.

Interior, Exterior and Boundary Points

A point z_0 is called an interior point of a set S if we find c-neighbourhood of z_0 all of whose points belong to S. If every c-neighbourhood of z_0 contains points belonging to S and also points not belonging to S, then z_0 is called a *boundary point*. In particular, then, every limit point that does not belong to the set is a boundary point. The origin $z = 0$, as well as each-point on the unit circle $|z| = 1$, is a boundary point for either of two sets

$$0 < |z| < 1 \quad \text{or} \quad 0 < |z| \le 1$$

If a point is not interior or boundary point of a set S, it is an exterior point of S.

Open and Closed Regions

An open set is a set which consists only of interior points. For example, the set of points z such that $|z| \le 1$ is an open set.

An open set S is said to be connected if any two points of the set can be joined by a path consisting of straight line segments (i.e. a polygonal path) all points of which are in S.

An open connected set is called an open region or *domain*.

If to set S we add all the limit points of S, the new set is called closed set. The closure of an open region or domain is called a closed region. In this book, whenever we use the word 'region' without qualifying it, we shall mean open region or domain.

Complex Functions

If to each value which a complex variable, z, can assume, there corresponds one or more values of a complex variable w, we say that w is a function of z and write $w = f(z)$. The variable z is sometimes called an independent variable, while w is called a dependent variable. Since $z = x + iy$, $f(z)$ will be of the form $u + iv$, where u and v are functions of two real variables x and y. We may then write

$$w = f(z) = u(x, y) + iv(x, y)$$

w is said to be single valued or multivalued function of z according as for a given value of z there corresponds one or more than one values of w.

Transformations and Mapping

If $w = f(z)$, then corresponding to each point (x, y) in the z-plane, in a domain of definition of f, there is a point (u, v) in the w-plane, where $w = u + iv$. The correspondence between points in the two planes is called a mapping or transformation of points from the z-plane into points of the w-plane by the function f. Corresponding points or corresponding sets of points are called *images* of each other.

Example. If $w = z^2$ then $u + iv = (x + iy)^2 = x^2 - y^2 + 2iyx$ and the transformation is

$$u = x^2 - y^2$$
$$v = 2xy$$

The image of a point $(1, 2)$ in the z-plane is the point $(-3, 4)$ in w-plane (Fig. 6.2). If $f(z)$ is a multivalued function a point (or curve) in z-plane is mapped in general into more than one point (or curve) in the w-plane.

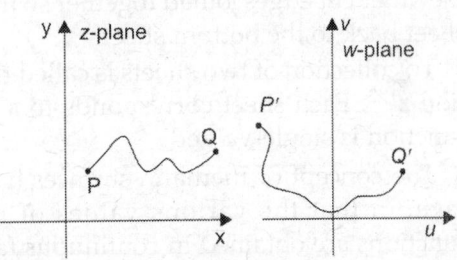

Fig. 6.2

Branch Points and Branch Lines

A point of the complex plane having the property that, after the completion of any cycle around it, a given function is not restored to its initial value (or, more precisely, to the value we have assigned to it initially) is called a branch point of the function.

Example. Let us consider a multivalued function $w = z^{1/2}$ and allow z to make a complete circuit (counterclockwise) around the origin starting from point A (Fig. 6.3). We have $z = re^{i\theta}$.

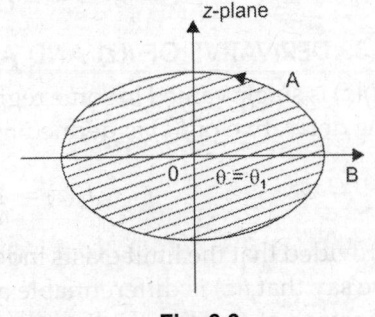

Fig. 6.3

$$w = \sqrt{r}\, e^{i\theta/2}, \text{ so that}$$

At A, $\theta = \theta_1$ and $w = \sqrt{r}\, e^{i\theta_1/2}$

After a complete circuit back to A

$$\theta = \theta_1 + 2\pi \quad \text{and } w = \sqrt{r}\, e^{(i\theta_1/2+\pi i)}$$

or

$$w = -\sqrt{r}\, e^{i\theta/2}$$

Thus, we have not achieved the same value of w with which we started. However, by making a second complete circuit back to A.

i.e. $$\theta = \theta_1 + 4\pi \qquad w = \sqrt{r}\, e^{i(\theta_1+4\pi/2)} = \sqrt{r}\, e^{i\theta_1/2}$$

and we obtain the same value of w with which we started. We can describe the above by stating that if $0 \le \theta < 2\pi$, we are on one branch of the multivalued function $z^{1/2}$ while if $2\pi \le \theta < 4\pi$, we are on the other branch of the function.

It is clear that each branch of the function is single-valued. We set up an artificial barrier such as OB, where B is at infinity (although any other line from O can be used) which we agree not to cross. This barrier is called a branch line or branch cut, and point O is called a branch point. It should be noted that a circuit around any point other than $r = 0$ does not lead to different values, thus $z = 0$ is the only finite branch point.

Riemann Surfaces

There is another way to achieve the purpose of the branch line described above. To see this, we imagine that the z-plane consists of two sheets superimposed on each other (Fig. 6.4). We now cut the sheets along OB and imagine that the lower edge of the bottom sheet is joined to the upper edge of the top sheet. Then starting in the bottom sheet and making one complete circuit we arrive in the top sheet. We must now imagine the other cut edges joined together so that by continuing the circuit we go from the top sheet back to the bottom sheet.

To collection of two sheets is called the Riemann surface corresponding to the function $z^{1/2}$. Each sheet corresponds to a branch of the function and on each sheet the function is single-valued.

The concept of Riemann surfaces has an advantage, in that the various values of multivalued functions are obtained in continuous fashion.

The ideas are easily extended. For example, for the function $z^{1/2}$ the Riemann surface has 2 sheets, for $\log z$, the Riemann surface has infinitely many sheets.

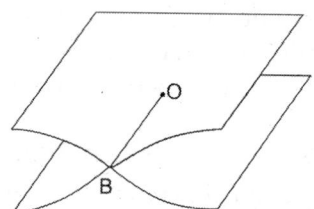

Fig. 6.4: The Riemann surface of the function $f(z) = z^{1/2}$

6.3. DERIVATIVE OF f(z) AND ANALYTICITY

If $f(z)$ is single valued in some region R of the z-plane, the derivative of $f(z)$ is defined as

$$f'(z) = \lim_{\Delta z \to 0} \frac{f(z+\Delta z) - f(z)}{\Delta z}, \tag{6.1}$$

provided that the limit exists independent of the manner in which $\Delta z \to 0$. In such case we say that $f(z)$ is differentiable at z. In the definition (Section 6.1), we sometimes use h instead of Δz. Here the limit $\Delta z \to 0$ is a actually a double limit in as much as both the real part Δx and the imaginary part Δy of Δz each tend separately to zero. Since there are an infinite number of ways by which Δz can tend to zero, there are in general an infinite number of possible values that the limits can assume. If the limit exists, it is finite and does not depend on the manner in which $\Delta z \rightarrow 0$, we say that a function of complex argument z is differentiable at a given point $z = z_0$.

It should be remembered that in Eq. (6.1) if the limit depends on amplitude Δz, the function $f(z)$ is not differentiable.

If function $f(z)$ is differentiable at $z = z_0$ it is necessarily continuous at this point for

$$|f(z_0 + \Delta z) - f(z_0)| \rightarrow 0 \quad \text{as} \quad \Delta z \rightarrow 0$$

if $\quad \lim\limits_{\Delta z \rightarrow 0} \dfrac{f(z_0 + \Delta z) - f(z_0)}{\Delta z}$

exists. But continuity does not imply differentiability as may be seen from the following example. Function $f(z) = |z|^2 = x^2 + y^2$ is continuous everywhere but

$$f'(z_0) = \lim_{\Delta z \rightarrow 0} \frac{(z_0 + \Delta z)(\bar{z}_0 + \Delta \bar{z}) - z_0 \bar{z}_0}{\Delta z}$$

$$= \lim_{\Delta z \rightarrow 0} \left(\bar{z}_0 + \Delta \bar{z} + z_0 \frac{\Delta \bar{z}}{\Delta z} \right) = 0 \text{ at point } z_0 = 0$$

and $\quad \dfrac{\Delta \bar{z}}{\Delta z} = \cos 2\phi - \sin 2\phi \text{ at } z_0 \neq 0, \text{ where } \phi = \arg (\Delta z)$

Thus, $\lim\limits_{\Delta z \rightarrow 0} \dfrac{\Delta \bar{z}}{\Delta z}$ does not exist at any other point different from zero and therefore function $|z|^2$ which is continuous everywhere is differentiable at $z = 0$ only and nowhere else.

On the other hand if $f(z)$ is differentiable in a domain except for a finite number of points, such points are called singular points of functions $f(z)$.

The definition of derivative is too broad and in the theory of functions of a complex variable it is restricted by demanding that the function $f(z) = u(x, y) + iv (x, y)$ shall have a definite derivative for a given value of z. However, it will now be shown that the uniqueness of the derivative requires the functions u and v to satisfy certain condition.

Analytic Function: If the derivative $f'(z)$ exists at all points z of a region R, then $f(z)$ is said to be analytic in R and is referred to as an analytic function in R. The term regular and holomorphic are sometimes used as synonyms for analytic.

A function $f(z)$ is said to be analytic at a point z_0, if there exists a neighbourhood $|z - z_0| < \varepsilon$ at all points of which $f'(z)$ exists.

Necessary and Sufficient Conditions for $f(z)$ to be Analytic: $f(z)$ in order to be analytic, the limit

$$f'(z) = \lim_{\Delta z \rightarrow 0} \frac{f(z + \Delta z) - f(z)}{\Delta z}$$

$$= \lim_{\substack{\Delta x \rightarrow 0 \\ \Delta y \rightarrow 0}} \frac{[u(x + \Delta x, y + \Delta y) + iv(x + \Delta x, y + \Delta y) - \{u(x,y) + iv(x,y)\}}{\Delta x + i\Delta y} \tag{6.2}$$

must exist independent of the manner in which $\Delta z(\Delta x \text{ and } \Delta y)$ will approach zero. We consider two possible approaches.

Case 1: $\Delta y = 0$, $\Delta x \to 0$

In this case Eq. (6.2) becomes

$$\lim_{\Delta x \to 0} \left\{ \frac{u(x+\Delta x, y) - u(x,y)}{\Delta x} + i\frac{v(x+\Delta x) - v(x,y)}{\Delta x} \right\} = \frac{\partial u}{\partial x} + i\frac{\partial v}{\partial x}.$$

provided the derivative exists.

Case 2: $\Delta x = 0$, $\Delta y \to 0$

In this case Eq. (6.2) becomes

$$\lim_{\Delta y \to 0} \left\{ \frac{u(x, y+\Delta y) - u(x,y)}{i\Delta y} + \frac{v(x, y+\Delta y) - v(x,y)}{\Delta y} \right\} = -i\frac{\partial u}{\partial y} + \frac{\partial v}{\partial y}$$

Now $f(z)$ cannot possibly be analytic unless these two limits are identical. Thus a necessary condition that $f(z)$ be analytic is

$$\frac{\partial u}{\partial x} + i\frac{\partial v}{\partial x} = -i\frac{\partial u}{\partial y} + \frac{\partial v}{\partial y}$$

Equating real and imaginary quantities, we have

$$\frac{\partial u}{\partial x} = \frac{\partial v}{\partial y} \quad \text{and} \quad \frac{\partial v}{\partial x} = -\frac{\partial u}{\partial y} \tag{6.3}$$

Equation (6.3) is known as *Cauchy–Riemann equations* and is the necessary condition for $f(z)$ to be analytic.

Sufficiency: For the existence of derivative of a function, the sufficient requirement is continuity of these partial derivatives satisfying Cauchy–Riemann conditions [Eq. (6.3)].

Since $\partial u/\partial x$ and $\partial u/\partial y$ are supposed to be continuous, we have

$$\Delta u = u(x + \Delta x, y + \Delta y) - u(x, y)$$

$$= \{u(x+\Delta x, y+\Delta y) - u(x, y+\Delta y) + u(x, y+\Delta y) - u(x, y)\}$$

$$= \left(\frac{\partial u}{\partial x} + \varepsilon_1 \right) \Delta x + \left(\frac{\partial u}{\partial y} + \eta_1 \right) \Delta y$$

where $\varepsilon_1 \to 0$ and $\eta_1 \to 0$ as $\Delta x \to 0$, $\Delta y \to 0$

Similarly, since $\partial u/\partial x$ and $\partial u/\partial y$ are supposed to be continuous, we have

$$\Delta v = \left(\frac{\partial v}{\partial x} + \varepsilon_2 \right) \Delta x + \left(\frac{\partial v}{\partial y} + \eta_2 \right) \Delta y$$

$$= \frac{\partial v}{\partial x} \Delta x + \frac{\partial v}{\partial y} \Delta y + \varepsilon_2 \Delta x + \eta_2 \Delta y$$

where $\varepsilon_2 \to 0$, $\eta_2 \to 0$ as $\Delta x \to 0$ and $\Delta y \to 0$

Since $w = f(z) = u + iv$

$$\Delta w = \Delta u + i\Delta v$$

or

$$\Delta w = \left(\frac{\partial u}{\partial x} + i\frac{\partial v}{\partial x} \right) \Delta x + \left(\frac{\partial u}{\partial y} + i\frac{\partial v}{\partial y} \right) \Delta y + \varepsilon \Delta x + \eta \Delta y$$

where $\varepsilon = \varepsilon_1 + i\varepsilon_2 \to 0$ and $\eta = \eta_1 + i\eta_2 \to 0$ as $\Delta x \to 0$ and $\Delta y \to 0$.

Using Cauchy–Riemann equations, we have

$$\Delta w = \left(\frac{\partial u}{\partial x} + i \frac{\partial v}{\partial x} \right) \Delta x + \left(-\frac{\partial v}{\partial x} + i \frac{\partial u}{\partial x} \right) \Delta y + \varepsilon \Delta x + \eta \Delta y$$

$$= \left(\frac{\partial u}{\partial x} + i \frac{\partial v}{\partial x} \right) (\Delta x + i \Delta y) + \varepsilon \Delta x + \eta \Delta y$$

Then on dividing by $\Delta z = \Delta x + i \Delta y$ and taking the limit as $\Delta z \to 0$, we have

$$\frac{dw}{dz} = f'(z) = \lim_{\Delta z \to 0} \frac{\Delta w}{\Delta z} = \frac{\partial u}{\partial x} + i \frac{\partial v}{\partial x} = \frac{\partial f}{\partial x}$$

so that the derivative exists and is unique, i.e. derivative of $f(z)$ w.r.t z is independent of the direction of approach in the complex plane as long as derivatives are continuous.

Harmonic Functions

Let the functions $f = u + iv$ be analytic in some domain of the z-plane, then at every point of the domain

$$\frac{\partial u}{\partial x} = \frac{\partial v}{\partial y}, \frac{\partial u}{\partial y} = -\frac{\partial v}{\partial x}$$

If the second partial derivatives of u and v w.r.t. x and y exist (we shall show in later chapters that when f is analytic, the partial derivatives of u and v of all orders exist and are continuous functions of x and y) and are continuous in a region R, then we find that

$$\frac{\partial^2 u}{\partial x^2} + \frac{\partial^2 u}{\partial y^2} = 0, \quad \frac{\partial^2 v}{\partial x^2} + \frac{\partial^2 v}{\partial y^2} = 0 \tag{6.4}$$

It follows that under these conditions, the real and imaginary parts of an analytic function satisfy Laplace's equations denoted by

$$\frac{\partial^2 \psi}{\partial x^2} + \frac{\partial^2 \psi}{\partial y^2} = 0 \qquad \text{or} \quad \nabla^2 \psi = 0$$

The functions such as $u(x, y)$ and $v(x, y)$ which satisfy Laplace's equation in a region R are called harmonic functions. Here u and v are called conjugate harmonic functions. Given one of two conjugate harmonic functions, the Cauchy–Riemann equations can be used to find the other (*see* solved Problem No. 2).

Polar Form of Cauchy–Riemann Equations

In polar coordinates (r, θ)

$$x = r \cos \theta, \quad y = r \sin \theta$$

or

$$r = \sqrt{x^2 + y^2}, \quad \theta = \tan^{-1} \frac{y}{x}$$

Then

$$\frac{\partial u}{\partial x} = \frac{\partial u}{\partial r} \cdot \frac{\partial r}{\partial x} + \frac{\partial u}{\partial \theta} \cdot \frac{\partial \theta}{\partial x} = \frac{\partial u}{\partial r} \left(\frac{x}{\sqrt{x^2 + y^2}} \right) + \frac{\partial u}{\partial \theta} \left(\frac{-y}{x^2 + y^2} \right)$$

$$= \frac{\partial u}{\partial r} \cos \theta - \frac{1}{r} \frac{\partial u}{\partial \theta} \sin \theta \tag{6.4a}$$

and

$$\frac{\partial u}{\partial y} = \frac{\partial u}{\partial r} \cdot \frac{\partial r}{\partial y} + \frac{\partial u}{\partial \theta} \cdot \frac{\partial \theta}{\partial y} = \frac{\partial u}{\partial r}\left(\frac{y}{\sqrt{y^2 + x^2}}\right) + \frac{\partial u}{\partial \theta}\left(\frac{x}{x^2 + y^2}\right)$$

$$= \frac{\partial u}{\partial r}\sin\theta + \frac{1}{r}\frac{\partial u}{\partial \theta}\cos\theta \qquad (6.4b)$$

Similarly,

$$\frac{\partial v}{\partial x} = \frac{\partial v}{\partial r} \cdot \frac{\partial r}{\partial x} + \frac{\partial v}{\partial \theta} \cdot \frac{\partial \theta}{\partial x}$$

$$= \frac{\partial v}{\partial r}\cos\theta - \frac{1}{r}\frac{\partial v}{\partial \theta}\sin\theta \qquad (6.4c)$$

$$\frac{\partial v}{\partial y} = \frac{\partial v}{\partial r} \cdot \frac{\partial r}{\partial y} + \frac{\partial v}{\partial \theta} \cdot \frac{\partial \theta}{\partial r}$$

$$= \frac{\partial v}{\partial r}\sin\theta + \frac{1}{r}\frac{\partial v}{\partial \theta}\cos\theta \qquad (6.4d)$$

From Cauchy–Riemann equation $\dfrac{\partial u}{\partial x} = \dfrac{\partial v}{\partial y}$, we have using Eqs (6.4a) and (6.4d)

$$\left(\frac{\partial u}{\partial r} - \frac{1}{r}\frac{\partial v}{\partial \theta}\right)\cos\theta - \left(\frac{\partial v}{\partial r} + \frac{1}{r}\frac{\partial u}{\partial \theta}\right)\sin\theta = 0 \qquad (6.4e)$$

From Cauchy–Riemann equation $\dfrac{\partial u}{\partial y} = -\dfrac{\partial v}{\partial x}$, we have using Eqs (6.4b) and (6.4c)

$$\left(\frac{\partial u}{\partial r} - \frac{1}{r}\frac{\partial v}{\partial \theta}\right)\sin\theta + \left(\frac{\partial v}{\partial r} + \frac{1}{r}\frac{\partial u}{\partial \theta}\right)\cos\theta = 0 \qquad (6.4f)$$

Multiplying Eq. (6.4e) by $\cos\theta$ and Eq. (6.4f) by $\sin\theta$ and adding yields

$$\frac{\partial v}{\partial r} + \frac{1}{r}\frac{\partial u}{\partial \theta} = 0 \qquad \text{or} \qquad \frac{\partial u}{\partial r} = -\frac{1}{r}\frac{\partial v}{\partial \theta} \qquad (6.5)$$

Multiplying Eq. (6.4e) by $-\sin\theta$ and Eq. (6.4f) by $\cos\theta$ and on adding yields

$$\frac{\partial v}{\partial r} + \frac{1}{r}\frac{\partial u}{\partial \theta} = 0 \qquad \text{or} \qquad \frac{\partial v}{\partial r} = -\frac{1}{r}\frac{\partial u}{\partial \theta} \qquad (6.5a)$$

Equations (6.5a) and (6.5b) represent Cauchy–Riemann equations in polar coordinates.

The derivative of the function $w = f(z)$ in polar form is

$$\frac{dw}{dz} = \frac{dw}{dx} = \frac{\partial w}{\partial r} \cdot \frac{\partial r}{\partial x} + \frac{\partial w}{\partial \theta} \cdot \frac{\partial \theta}{\partial x}$$

$$= \frac{\partial w}{\partial r}\cos\theta - \frac{\sin\theta}{r} \cdot \frac{\partial w}{\partial \theta}$$

$$= \frac{\partial w}{\partial r}\cos\theta - \frac{\sin\theta}{r} \cdot \left\{\frac{\partial u}{\partial \theta} + i\frac{\partial v}{\partial \theta}\right\}$$

Using Cauchy–Riemann equation in polar form

$$\frac{dw}{dz} = \frac{\partial w}{\partial r}\cos\theta - \frac{\sin\theta}{r}\left\{-r\frac{\partial v}{\partial r} + ir\frac{\partial u}{\partial r}\right\}$$

$$= \left\{\frac{\partial u}{\partial r} + i\frac{\partial v}{\partial r}\right\}\cos\theta - i\sin\theta\left\{\frac{\partial u}{\partial r} + i\frac{\partial v}{\partial r}\right\}$$

$$= (\cos\theta - i\sin\theta)\frac{\partial w}{\partial r} \tag{6.6}$$

Similarly, it can be shown by taking the change in z as wholly imaginary that

$$\frac{dw}{dz} = -\frac{1}{r}\frac{\partial w}{\partial\theta}(\sin\theta + i\cos\theta) \tag{6.6a}$$

Example 1. *Draw the Riemann surfaces for the following complex functions.*

i. $f(z) = z^{1/2}$ ii. $f(z) = (z^2 - 1)^{1/2}$ iii. $f(z) = \log z$

Solution: (i) Setting $z = re^{i\theta}$; $f(z) = z^{1/2} = r^{1/2}\,e^{i\theta/2}$

We easily see that the point $r = 0$ and $z = \infty$ are branch points of $f(z) = z^{1/2}$. For example, a cycle around $z = 0$ changes in θ by 2π results in changing the sign of the function at a given point.

$$f(r, \theta) = -f(r, \theta + 2\pi)$$

The branch cut of $f(z) = z^{1/2}$ can be chosen to connect the branch points $z = 0$ and $z = \infty$ along the positive real axis.

We define the first Riemann sheet[*] by fixing the value of $f(z)$ on the upper tip of the cut

$$f(z) = f(r, 0) = r^{1/2} \text{ for Re } z > 0 \qquad\qquad [\text{Im } z = + 0]$$

Then on the lower tip of the cut we have

$$f(z) = f(r, 2\pi) = -r^{1/2} \text{ for Re } z > 0 \qquad\qquad [\text{Im } z = - 0]$$

Crossing the cut, we move to the next sheet, where the values of $f(z)$ on the upper tip of the cut are same as the values of $f(z)$ on the lower tip on the first sheet, while the values of $f(z)$ on the lower tip are obtained by adding 2π to θ.

$$f(z) = (r, 4\pi) = r^{1/2} \text{ for Re } z > 0 \qquad\qquad [\text{Im } z = - 0]$$

on the second sheet. Hence a second crossing of the cut does not yield a new value for $f(z)$, since

$$f(r, 0) = f(r, 4\pi).$$

Therefore, the Riemann surface of $f(z) = z^{1/2}$, constructed so as to make this function continuous everywhere, has two sheets that are connected along the cut, the lower tip of the second sheet must be reconnected to the upper tip of the 1st sheet, i.e. if the cut is crossed from the second sheet, the function returns to the first sheet. In other words the Riemann surface is closed surface (Fig. 6.5).

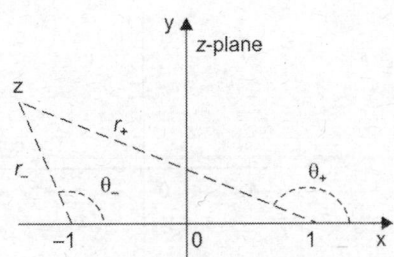

Fig. 6.5: Definition of $r_\pm$ and $\theta_\pm$

---

[*] Obviously, when there are no external conditions imposed, e.g. physical conditions, we are free to define the first sheet at will. The values of the function on the other sheets will then follow.

In general, the Riemann surface of the function

$$f(z) = z^{1/n} \ (n = 2, 3, 4...)$$

is a closed n-sheet surface, the nth sheet being reconnected to the first sheet.

(ii) $f(z) = (z^2 - 1)^{1/2}$

Writing

$$z + 1 = r\_ e^{i\theta\_}$$

and

$$z - 1 = r_+ e^{i\theta_+}$$

It is easy to see that $f(z)$ has two branch points $z = \pm 1$.

We take the cut along the segment of the real axis $-1 \le x \le 1$ as in (Fig. 6.6a). Then, in order to remain on the same Riemann sheet $\theta\_$ should vary, for example from $-\pi$ to $+\pi$. This choice of the range of variation of angles $\theta_\pm$ defines the first Riemann sheet.

1. $f(z) = i\sqrt{1-x^2}$ for $-1 \le x \le 1$, $y = +0$

   Since these values of x and y correspond to $\theta\_ = 0$ and $\theta_+ = \pi$.

2. $f(z) = -i\sqrt{1-x^2}$ for $-1 \le x \le 1, y = -0$

   Since $\theta\_ = 0$ and $\theta_+ = -\pi$

3. $f(z) = -\sqrt{x^2 - 1}$ for $x \le -1$, $y = \pm 0$

   Since $\theta\_ = \theta_+ = \pi$

4. $f(z) = +\sqrt{x^2 - 1}$ for $x \ge +1$, $y = \pm 0$

   Since $\theta\_ = \theta_+ = 0$

Comparing conditions (1) and (2), it is seen that $f(z)$ is discontinuous across the cut, as it should be. On the other hand, it is not difficult to see that a second crossing of the cut restores $f(z)$ to its initial value. Hence, the Riemann surface is a closed two sheeted surface constructed in such a manner that $f(z)$ changes sheets after any cycle that surrounds one of its initial branch points only, whereas it restores $f(z)$ to its initial value after the completion of any cycle that surrounds both branch points.

The cut in the complex plane equally will be chosen as in Fig. 6.6b. In a sense, this corresponds to joining the points $z = \pm 1$ by a path going through the point at infinity. In that case, a sheet is defined by letting θ_+ vary from 0 to 2π and $\theta\_$ from $-\pi$ to $+\pi$. We

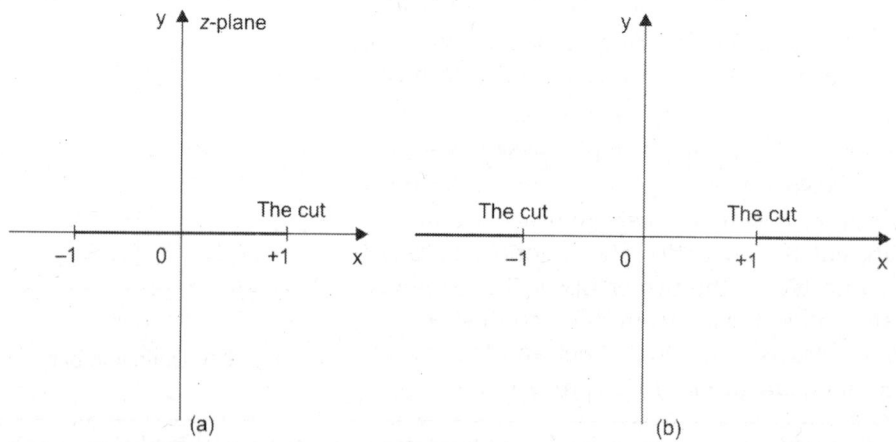

(a)   (b)

Fig. 6.6: Two possible ways of choosing cuts in the z-plane for the function $f(z) = (z^2 - 1)^{1/2}$

leave it to the reader to find the values that $f(z)$ takes near the cuts of this Riemann sheet, as we have done for the other choice of the cut.

(iii) $f(z) = \log z = \log |z| + i \arg z$

$$\arg z = \theta + 2\pi n \qquad (n = 0, \pm 1, \pm 2, \pm 3...)$$

The point $z = 0$ and the point at infinity are the branch points of the function

$$f(z) = \log z = \log |z| + i(\theta + 2\pi n)$$
$$= \log |z| + i(\theta + 2\pi n)$$

A curve joining the two points of $\log z$ which in our case is a line starting from the origin and extending out to infinity in an arbitrary direction is the branch line of the function $f(z)$.

For definiteness, we shall assume that the z-plane is cut along the negative half of the real axis. One can then define the single-valued function.

$$f_0(z) = f_0(r, \theta) = \log r + i\theta \qquad (-\pi < \theta < \pi)\ 0 < r$$

one can also define single-valued function

$$f_1(z) = f_1(r, \theta) = \log r + i(\theta + 2\pi) \qquad (-\pi < \theta < \pi)\ 0 < r$$

and $\qquad f_{-1}(z) = f_{-1}(r, \theta) = \log r + i(\theta - 2\pi) \qquad (-\pi < \theta < \pi)\ 0 < r$

In their range of definition $f_1(z)$ and $f_{-1}(z)$ take on the same values and the logarithm takes in the polar angle range $\pi < \theta < 3\pi$ and $-3\pi < \theta < -\pi$ respectively. In general we can define an infinite series of single-valued functions.

$$f_0(z), f_{\pm 1}(z), f_{\pm 2}(z),...., f_{\pm n}(z)...$$

where $\qquad f_n(z) = f_n(r, \theta) = \log r + i(\theta + 2\pi n) \qquad (-\pi < \theta < \pi,\ 0 < r)$

so that $f_n(z)$ takes on the same values for $-\pi < \theta < \pi$ that the logarithm takes in the polar angle range

$$(2n - 1)\pi < \theta < (2n + 1)\pi$$

We have now replaced the multivalued logarithmic function by a series of different functions that are analytic in cut z-plane. This suggests the following geometrical construction. We superpose an infinite series of cut complex planes one on top of the other, each plane corresponding to a different value of n ($= 0, \pm 1, \pm 2,...$). The adjacent planes are connected along the cut, the upper tip of the cut in the nth plane is connected to the lower tip of the cut in the $(n + 1)$st plane, the branch points are common to all the planes. Hence a crossing of the cut is equivalent to going to one off the two adjacent complex planes (Fig. 6.7).

The geometrical surface obtained from this helix-like superposition of planes is called a Riemann surface.

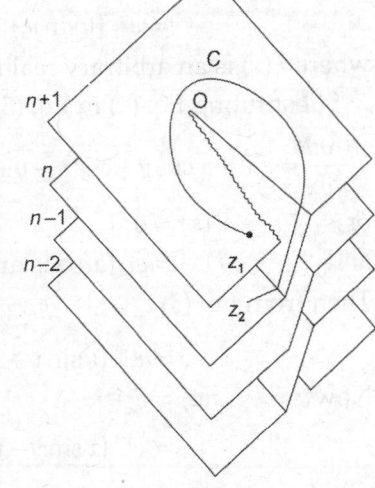

Fig. 6.7

Example 2. Prove that $u = e^{-x}(x \sin y - y \cos y)$ is a harmonic function. Find v such that $f(z) = u + iv$ is analytic. Find also $f(z)$.

Solution: $\dfrac{\partial u}{\partial x} = e^{-x} \sin y + (-e^{-x})(x \sin y - y \cos y)$

$$= e^{-x}\{\sin y - x \sin + y \cos y\}$$

$$\frac{\partial^2 u}{\partial x^2} = e^{-x}(-\sin y) + (-e^{-x})(\sin y - x\sin y + \cos y)$$

$$= e^{-x}\{-2\sin y + x\sin y - y\cos y\} \tag{1}$$

$$\frac{du}{dy} = e^{-x}\{x\cos y - \cos y + y\sin y\}$$

$$\frac{\partial^2 u}{\partial y^2} = e^{-x}\{-x\sin y + \sin y + \sin y + y\cos y\}$$

$$= e^{-x}\{-x\sin y + 2\sin y + y\cos y\} \tag{2}$$

Adding Eqs (1) and (2) yields

$$\frac{\partial^2 u}{\partial x^2} + \frac{\partial^2 u}{\partial y^2} = 0, \text{ hence } u \text{ is a harmonic function}$$

From Cauchy–Riemann equations

$$\frac{\partial v}{\partial y} = \frac{\partial u}{\partial x} = e^{-x}\{\sin y - x\sin y + y\cos y\} \tag{3}$$

$$\frac{\partial v}{\partial x} = -\frac{\partial u}{\partial y} = e^{-x}\{-x\cos y + \cos y - y\sin y\} \tag{4}$$

Integrating Eq. (3) w.r.t y keeping x constant we have

$$v = \int e^{-x}\{\sin y - x\sin y + y\cos y\}\, dy$$

$$= e^{-x}\{-\cos y + x\cos y + y\sin y + \cos y\} + F(x)$$

$$= e^{-x}\{y\sin y + x\cos y\} + F(x) \tag{5}$$

where $F(x)$ is an arbitrary real function of x.

Substituting Eq. (4) in Eq. (5), we obtain

$$\frac{\partial u}{\partial x} = e^{-x}\{-x\cos y + \cos y - y\sin y\} = e^{-x}\{-x\cos y + \cos y - y\sin y\} + F'(x)$$

or $\qquad\qquad F'(x) = 0$

and $\qquad\qquad F(x) = c$ (a constant).

Then from Eq. (5)

$$v = e^{-x}\{y\sin y + x\cos y\} + c$$

Now $\qquad\qquad f(z) = u + iv$

$$= e^{-x}\{x\sin y - y\cos y\} + ie^{-x}\{y\sin y + x\cos y\} + ic$$

$$= e^{-x}(x + iy)\{\sin y + i\cos y\} + ic$$

$$= e^{-x}(x + iy)ie^{-iy} + ic$$

$$= i(x + iy)e^{-(x+iy)} + ic$$

$$= ize^{-z} + ic = i\{ze^{-z} + c\}$$

is the required analytic function.

Example 3. *Show that function*

$$f(z) = u + iv = \frac{x^3(1+i) - y^3(1-i)}{x^2 + y^2} \qquad [(z \neq 0, \ f(0) = 0)]$$

is continuous and that the Cauchy–Riemann equations are satisfied at the origin, yet $f'(0)$ does not exist.

Solution: Here $u = \dfrac{x^3 - y^3}{x^2 + y^2}, \quad v = \dfrac{x^3 + y^3}{x^2 + y^2}$

When $z \neq 0$, u and v are rational functions of x and y with nonzero denominators. It follows that they are continuous. To test them for continuity at $z = 0$, we get, on changing to polar coordinates,

$$u = r \, (\cos^3 \theta - \sin^3 \theta) \qquad v = r \, (\cos^3 \theta + \sin^3 \theta)$$

each of which tends to zero as $r \to 0$ whatever the value θ may have.

Now the actual value of u and v at the origin are zero, since $f(0) = 0$. Since the actual and limiting values of u and v are equal at the origin, they are continuous there. Hence $f(z)$ is a continuous function for all values of z. Now at the origins

$$\frac{\partial u}{\partial x} = \lim_{x \to 0} \frac{u(x,0) - u(0,0)}{x} = \lim_{x \to 0} \frac{x - 0}{x} = 1$$

Similarly

$$\frac{\partial u}{\partial y} = \lim_{y \to 0} \frac{u(y,0) - u(0,0)}{y} = \lim_{y \to 0} \frac{-y - 0}{y} = -1$$

$$\frac{dv}{dx} = \lim_{x \to 0} \frac{v(x,0) - v(0,0)}{x} = \lim_{x \to 0} \frac{x - 0}{x} = 1$$

and

$$\frac{\partial v}{\partial y} = \lim_{y \to 0} \frac{v(0,y) - v(0,0)}{y} = \lim_{y \to 0} \frac{y}{y} = 1$$

Hence

$$\frac{\partial u}{\partial x} = \frac{\partial v}{\partial y} \quad \text{and} \quad \frac{\partial u}{\partial y} = -\frac{\partial v}{\partial x}$$

The Cauchy–Riemann equations are therefore satisfied.

Again $$f'(z) = \lim_{z \to 0} \frac{f(z) - f(0)}{z}$$

$$= \lim_{z \to 0} \frac{(x^3 - y^3) + i(x^3 + y^3)}{x^2 + y^2} \cdot \frac{1}{x + iy}$$

Let $z \to 0$ along the path $y = mx$.

$$f'(0) = \lim_{x \to 0} \frac{x^3(1+i) - m^3 x^3(1-i)}{(x^2 + m^2 x^2)(x + imx)}$$

$$= \frac{(1+i) - m^3(1-i)}{(1+m^2)(1+im)}$$

which is not unique as it depends on the argument of $z \to 0$. Hence $f(z)$ is continuous at $z = 0$ but $f'(z)$ does not exist at $z = 0$, i.e. at origin.

Example 4. If $f(z) = u + iv$ is an analytic function of $z = x + iy$ and ϕ any function of x and y with differential coefficient of the first and second order, then

i. $\left(\dfrac{\partial\phi}{\partial x}\right)^2 + \left(\dfrac{\partial\phi}{\partial y}\right)^2 = \left\{\left(\dfrac{\partial\phi}{\partial u}\right)^2 + \left(\dfrac{\partial\phi}{\partial v}\right)^2\right\}\,|\,f'(z)\,|^2$

ii. $\dfrac{\partial^2\phi}{\partial x^2} + \dfrac{\partial^2\phi}{\partial y^2} = \left(\dfrac{\partial^2\phi}{\partial u^2} + \dfrac{\partial^2\phi}{\partial v^2}\right)|\,f'(z)\,|^2$ and

iii. $\left(\dfrac{\partial^2}{\partial x^2} + \dfrac{\partial^2}{\partial y^2}\right)|\,f(z)\,|^2 = 4\,|\,f'(z)\,|^2$

Solution: i.

$$\frac{\partial\phi}{\partial x} = \frac{\partial\phi}{\partial u}\cdot\frac{\partial u}{\partial x} + \frac{\partial\phi}{\partial v}\frac{\partial v}{\partial x} \qquad (1)$$

$$\frac{\partial\phi}{\partial y} = \frac{\partial\phi}{\partial u}\frac{\partial u}{\partial y} + \frac{\partial\phi}{\partial v}\cdot\frac{\partial v}{\partial y}$$

or

$$\frac{\partial\phi}{\partial y} = -\frac{\partial\phi}{\partial u}\frac{\partial v}{\partial x} + \frac{\partial\phi}{\partial v}\cdot\frac{\partial u}{\partial x} \qquad (2)$$

by using Cauchy–Riemann equations.

Squaring and adding Eqs (1) and (2), we have

$$\left(\frac{\partial\phi}{\partial x}\right)^2 + \left(\frac{\partial\phi}{\partial y}\right)^2 = \left\{\left(\frac{\partial\phi}{\partial u}\right)^2 + \left(\frac{\partial\phi}{\partial v}\right)^2\right\}\left\{\left(\frac{\partial u}{\partial x}\right)^2 + \left(\frac{\partial v}{\partial x}\right)^2\right\} = \left\{\left(\frac{\partial\phi}{\partial u}\right)^2 + \left(\frac{\partial\phi}{\partial v}\right)^2\right\}f'\,|\,z\,|^2$$

Since

$$f'(z) = \frac{\partial u}{\partial x} + i\frac{\partial v}{\partial x}$$

ii. Since $w = u + iv$, $\bar{w} = u - iv$

or

$$u = \frac{1}{2}(w + \bar{w}), \; v = \frac{1}{2i}(w - \bar{w})$$

So that

$$\frac{\partial}{\partial w} = \frac{\partial}{\partial u}\frac{\partial u}{\partial w} + \frac{\partial}{\partial v}\frac{\partial v}{\partial w} = \frac{1}{2}\left(\frac{\partial}{\partial u} - i\frac{\partial}{\partial v}\right)$$

$$\frac{\partial}{\partial\bar{w}} = \frac{\partial}{\partial u}\frac{\partial u}{\partial\bar{w}} + \frac{\partial}{\partial v}\frac{\partial v}{\partial\bar{w}} = \frac{1}{2}\left(\frac{\partial}{\partial u} + i\frac{\partial}{\partial v}\right)$$

$\therefore$

$$\frac{\partial}{\partial w}\frac{\partial}{\partial\bar{w}} = \frac{1}{4}\left(\frac{\partial}{\partial u} - i\frac{\partial}{\partial v}\right)\left(\frac{\partial}{\partial u} + i\frac{\partial}{\partial v}\right)$$

where u and v are treated as functions of two independent variables, w and $\bar{w}$.

$\therefore$

$$\frac{\partial}{\partial w}\frac{\partial}{\partial\bar{w}} = \frac{1}{4}\left(\frac{\partial^2}{\partial u^2} + \frac{\partial^2}{\partial v^2}\right)$$

Hence

$$\frac{\partial^2\phi}{\partial u^2} + \frac{\partial^2\phi}{\partial v^2} = 4\frac{\partial^2\phi}{\partial w\,\partial\bar{w}} \qquad (3)$$

Similarly $\qquad \dfrac{\partial^2 \phi}{\partial x^2} + \dfrac{\partial^2 \phi}{\partial y^2} = 4\dfrac{\partial^2 \phi}{\partial z\, \partial \bar{z}}$ $\qquad\qquad$ (4)

Also $\qquad\qquad w = f(z), \quad \bar{w} = f(\bar{z})$

$\therefore \qquad 4\dfrac{\partial^2}{\partial w\, \partial \bar{w}} = 4\left(\dfrac{\partial}{\partial z}\dfrac{\partial z}{\partial w}\right)\left(\dfrac{\partial}{\partial \bar{z}}\dfrac{\partial \bar{z}}{\partial \bar{w}}\right)$

$\qquad\qquad\qquad = 4\left\{\dfrac{1}{f'(z)}\dfrac{\partial}{\partial z}\right\}\left\{\dfrac{1}{f'(\bar{z})}\dfrac{\partial}{\partial \bar{z}}\right\}$

$\qquad\qquad\qquad = 4\dfrac{1}{f'(z)f'(\bar{z})}\dfrac{\partial^2}{\partial z\, \partial \bar{z}}$ $\qquad\qquad$ (5)

Since z and $\bar{z}$ are independent variables

Hence from Eqs (3)–(5), we have

iii. $\qquad \dfrac{\partial^2 \phi}{\partial u^2} + \dfrac{\partial^2 \phi}{\partial v^2} = 4\dfrac{\partial^2 \phi}{\partial w\, \partial \bar{w}} = \dfrac{4}{f'(z)f'(\bar{z})}\dfrac{\partial^2 \phi}{\partial z\, \partial \bar{z}} = \dfrac{1}{f'(z)f'(\bar{z})}\left\{\dfrac{\partial^2 \phi}{\partial x^2} + \dfrac{\partial^2 \phi}{\partial y^2}\right\}$

$\qquad \dfrac{\partial^2 \phi}{\partial x^2} + \dfrac{\partial^2 \phi}{\partial y^2} = \left(\dfrac{\partial^2 \phi}{\partial u^2} + \dfrac{\partial^2 \phi}{\partial v^2}\right) f'(z)f'(\bar{z})$

$\qquad\qquad\qquad = \left(\dfrac{\partial^2 \phi}{\partial u^2} + \dfrac{\partial^2 \phi}{\partial v^2}\right) |f'(z)|^2$

Equations (3) and (4) give $4\dfrac{\partial^2}{\partial z\, \partial \bar{z}} = \dfrac{\partial^2}{\partial x^2} + \dfrac{\partial^2}{\partial y^2}$

$\left(\dfrac{\partial^2}{\partial x^2} + \dfrac{\partial^2}{\partial y^2}\right)|f(z)|^2 = 4\dfrac{\partial^2}{\partial z\, \partial \bar{z}}[f(z)f(\bar{z})]$

$\qquad\qquad\qquad = 4f'(z)f'(\bar{z}) = 4\,|f'(z)|^2.$

Example 5. *If f(z) = u + iv, an analytic function of z = x + iy, find f(z) in terms of z if*

(a) $u - v = (x - y)(x^2 + 4xy + y^2)$

(b) $u + v = \dfrac{2\sin 2x}{e^{2y} + e^{-2y} - 2\cos 2x}$

(c) $u - v = \dfrac{\cos x + \sin x - e^{-y}}{2\cos x - e^{y} - e^{-y}}$ *subject to the condition* $f(\pi/2) = 0$.

Solution: (a) We have

$\qquad\qquad f(z) = u + iv \quad \text{or} \quad \text{if } (z) = iu - v$

$\therefore \qquad (1 + i)\, f(z) = (u - v) + i(v + u) = U + iV \text{ (say)}$

$\qquad\qquad \dfrac{\partial U}{\partial x} = \dfrac{\partial u}{\partial x} - \dfrac{\partial v}{\partial x} = x^2 + 4xy + y^2 + (x - y)(2x + 4y)$

$\qquad\qquad\qquad = 3x^2 + 6xy - 3y^2 = \phi_1(x, y) \text{ (say)}$

$$\frac{\partial U}{\partial y} = -(x^2 + 4xy + y^2) + (x-y)(4x+2y)$$

$$= 3x^2 - 6xy - 3y^2 = \phi_2(x,y) \text{ (say)}$$

Also $$(1+i)f'(z) = \frac{\partial U}{\partial x} + i\frac{\partial V}{\partial x} = \frac{\partial U}{\partial x} - i\frac{\partial U}{\partial y} = \phi_1 - i\phi_2$$

$\therefore$ $$(1+i)f(z)^* = \int [\phi_1(z,0) - i\phi_2(z,0)] dz + c$$

$$= \int (3z^2 - i3z^2) dz + c = (1-i)z^3 + c$$

$\therefore$ $$f(z) = \frac{1-i}{1+i}z^3 + \frac{c}{1+i} = -iz^3 + A,$$

where A is some other constant.

(b) As above, we have

$$(1+i)f(z) = (u-v) + i(u+v) = U + iV$$

or $$(1+i)f'(z) = \frac{\partial U}{\partial x} + i\frac{\partial V}{\partial x} = \frac{\partial V}{\partial y} - i\frac{\partial U}{\partial y}$$

$$\frac{\partial V}{\partial x} = 4\left[\frac{\cos 2x(e^{2y} + e^{-2y}) - 2}{(e^{2y} + e^{-2y} - 2\cos 2x)^2}\right] = \phi_2(x,y) \qquad \text{(say)}$$

$$\frac{\partial V}{\partial y} = -4\frac{\sin 2x(e^{2y} - e^{-2y})}{(e^{2y} + e^{-2y} - 2\cos 2x)^2} = \phi_1(x,y) \qquad \text{(say)}$$

$\therefore$ $$(1+i)f(z) = \int [\phi_1(z,0) + i\phi_2(z,0)] dz + c$$

$$(1+i)f(z) = \int \left[0 + i\frac{4(2\cos 2z - 2)}{(2 - 2\cos 2z)^2}\right] dz + c$$

$$= 2i\int \frac{dz}{\cos 2z - 1} + c = -\int \text{cosec}^2 z\, dz + c$$

$$= +i \cot z + c$$

$$f(z) = \frac{i}{1+i}\cot z + \frac{c}{1+i}$$

or $$= \frac{1+i}{2}\cot z + A$$

where A is some other constant.

---

*If $\frac{\partial u}{\partial x} = \phi_1(x,y)$ and $\frac{\partial u}{\partial y} = \phi_2(x,y)$

Then $f'(x+iy) = f'(z) = \frac{\partial u}{\partial x} + i\frac{\partial v}{\partial x} = \frac{\partial u}{\partial x} - i\frac{\partial u}{\partial y} = \phi_1(x,y) - i\phi_2(x,y)$

If we put $x = z$ and $y = 0$

Then $f'(x) = \phi_1(z,0) - i\phi_2(z,0)$ $\therefore f(z) = \int \phi_1(z,0) dz - i\int \phi_2(z,0) dz + c$, where c is any arbitrary constant.

(c) Here we are given

$$u - v = \frac{\cos x + \sin x - e^{-y}}{2\cos x - e^{y} - e^{-y}}$$

$$= \frac{1}{2}\left[1 + \frac{2\cos x + 2\sin x - 2e^{-y}}{2\cos x - e^{y} - e^{-y}} - 1\right]$$

$$= \frac{1}{2}\left[1 + \frac{2\sin x + e^{y} - e^{-y}}{2\cos x - e^{y} - e^{-y}}\right]$$

$$= \frac{1}{2}\left[1 + \frac{\sin x + \sinh y}{\cos x - \cosh y}\right]$$

$$\frac{\partial u}{\partial x} - \frac{\partial v}{\partial x} = \frac{1}{2}\left[\frac{\cos x(\cos x - \cosh y) + (\sin x + \sinh y)\sin x}{(\cos x - \cosh y)^2}\right]$$

$$= \frac{1}{2}\left[\frac{1 + \sin x \sinh y - \cos x \cosh y}{(\cos x - \cosh y)^2}\right] \tag{1}$$

$$\frac{\partial u}{\partial y} - \frac{\partial v}{\partial y} = \frac{1}{2}\left[\frac{\cosh y(\cos x - \cosh y) + (\sin x + \sinh y)\sinh y}{(\cos x - \cosh y)^2}\right]$$

$$\frac{\partial u}{\partial y} - \frac{\partial v}{\partial y} = \frac{1}{2}\left[\frac{\cosh y \cos x + \sinh y \sin x - 1}{(\cos x - \cosh y)^2}\right] \tag{2}$$

Using Cauchy–Riemann conditions, we have

$$\frac{\partial u}{\partial y} = -\frac{\partial v}{\partial x} \quad \text{and} \quad \frac{\partial v}{\partial y} = \frac{\partial u}{\partial x}$$

Then (2) becomes

$$-\frac{\partial v}{\partial x} - \frac{\partial u}{\partial x} = \frac{1}{2}\left[\frac{\cosh y \cos x + \sinh y \sin x - 1}{(\cos x - \cosh y)^2}\right] \tag{3}$$

Solving Eqs (1) and (3), we obtain

$$\frac{\partial u}{\partial x} = \frac{1}{2}\left[\frac{1 - \cosh y \cos x}{(\cos x - \cosh y)^2}\right] = \phi_1(x,y) \quad \text{(say)}$$

and

$$\frac{\partial v}{\partial x} = -\frac{1}{2}\left[\frac{\sinh y \sin x}{(\cos x - \cosh y)^2}\right] = \phi_2(x,y) \quad \text{(say)}$$

and

$$f'(z) = \frac{\partial u}{\partial x} + i\frac{\partial v}{\partial x} = \phi_1(x,y) + i\phi_2(x,y)$$

$$f(z) = \int[\phi_1(z,0) + i\phi_2(z,0)]dz + c$$

$$= \frac{1}{2}\int\left[\frac{1 - \cos z}{(\cos z - 1)^2} + 0\right]dz + c$$

$$= \frac{1}{4}\int \mathrm{cosec}^2 \frac{z}{2} dz + c$$

$$= -\frac{1}{2}\cot\frac{z}{2} + c$$

To determine the constant, let us put $z = \pi/2$, then

$$f(z) = 0 = -\frac{1}{2}\cot\frac{\pi}{4} + c$$

or
$$c = \frac{1}{2}$$

∴
$$f(z) = \frac{1}{2}\left(1 - \cot\frac{z}{2}\right).$$

6.4. COMPLEX INTEGRATION

Before defining complex integral we will define some terms.

Length of the curve: Let us consider an arc of the plane curve defined by equations $x = \phi(t)$ and $y = \psi(t)$, where $\alpha \leq t \leq \beta$. Let us divide the interval by points $t_0, t_1, t_2,..., t_n$ and let these points on the curve be denoted by $P_0 P_1, P_2,...,P_n$. Length of polynomial line $P_0 P_1 P_2...P_i...P_n$ will be the sum of the length of lines $P_0 P_1, P_2,..., P_{n-1} P_n$ (Fig. 6.8). If $Z_0, Z_1, Z_2,..., Z_n$ be the points on the arc corresponding to the values $t_0, t_1, t_2,..., t_n$ of t then length of the polynomial arc

$$= \sum_{i=1}^{n} [(x_i - x_{i-1})^2 + (y_i - y_{i-1})^2]^{1/2}$$

If the value of this sum depends upon the mode of subdivision, then sum is called the *length of an inscribed polygon.* If the arc is such that the length of all the inscribed polygons have a finite upperbound S, the curve is known to be *rectifiable* and S is called *length of the curve.*

The necessary and sufficient condition for the arc to be rectifiable are that the function $\phi(t)$, and $\psi(t)$ must be bounded variation in the interval (α, β). In case $\phi'(t)$ and $\psi'(t)$ are continuous, the curve denoted by $x = \phi(t)$, $y = \psi(t)$, where $\alpha \leq t \leq \beta$ is rectifiable and its length is

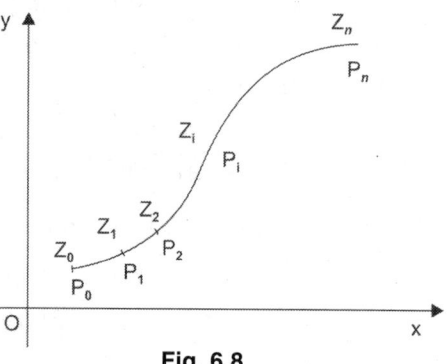

Fig. 6.8

$$S = \int_{\alpha}^{\beta} \left\{[\phi'(t)]^2 + [\psi'(t)]^2\right\}^{1/2} dt$$

In case $\phi(t)$ and $\psi(t)$ are real continuous functions of the real variables t in the interval (α, β), the arc is said to be a continuous arc. A *continuous arc* having no multiple points is known as *Jordan arc.*

Regular arc of a Jordan curve: An arc defined by equation $z = \phi(t) + i\psi(t)$, when $\alpha \leq t \leq \beta$ and if z be expressed as single valued and $\phi(t)$, $\psi(t)$ and $\phi'(t)$, $\psi'(t)$ as well, are continuous in interval $\alpha \leq t \leq \beta$, the arc is then called *regular arc of Jordan curve.*

Contour: A continuous Jordan curve consisting of chain of finite number of regular arcs is known as the contour. Obviously, a contour is rectifiable. Thus contour is a smooth curve having no multiple points. Such a smooth curve is composed of arcs which join on continuously and each arc having continuous tangent. The last requirement eliminates some pathological possibilities such as contour being sufficiently irregular, i.e. of infinite length. A closed curve traversed in a counterclockwise direction with respect to the domain enclosed by the contour is said to be described in positive direction. The clockwise traversed curve is in negative direction.

Integration along the contour C, which consists of a finite number of regular arcs L_r is defined as

$$\int_C f(z)\,dz = \sum_{r=1}^{n} \int_{L_r} f(z)\,dz$$

If the contour is a closed one, the integration along it will be symbolised by $\oint_C$ or simply by $\int_C$.

Definition of integral: The definition of the integral of a function of a complex variable is a straight generalisation of the definition of the Riemann integral of a function of a real variable. Let $f(z)$ be a function of the complex variable $x + iy$, and let us consider a curve C in the complex plane with end points a and b. We shall suppose that the curve C is a regular one, by which it will be meant that it can be described by a parametric equation.

$$z = z(t) \equiv x(t) + iy(t); \quad t_a \leq t \leq t_b$$

where t is a real parameter and $x(t)$ and $y(t)$ are real single-valued functions that have continuous first-order derivatives. Our discussion will also be valid for a piecewise, regular curve C', i.e. for a continuous curve consisting of a finite number of regular arcs.

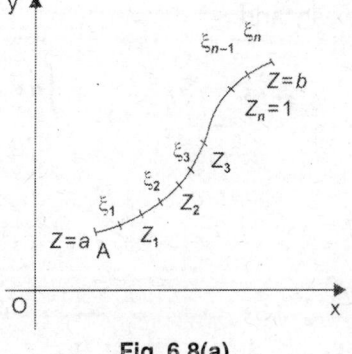

Fig. 6.8(a)

As shown in Fig. (6.8a), we first subdivide the arc ab into n intervals by introducing the $(n + 1)$ points.

$$Z_0,\ Z_1,\ Z_2,...,\ Z_{n-1},\ Z_n$$

On this arc Z_0 and Z_n will be taken to coincide with the end points a and b respectively.

Let $\qquad \Delta Z_r = Z_r - Z_{r-1} \quad [r = 1, 2,..., n]$

represents chord vectors of the curve C. Let ξ_C be a point on the curve C located between Z_{r-1} and Z_r (this point may also coincide with Z_{r-1} or with Z_r). Let the following summation may be performed.

$$S_n = \sum_{r=1}^{n} f(\xi_r)\Delta Z_r \qquad\qquad (6.7)$$

If the curve C is divided into smaller and smaller parts so that $n \to \infty$. $|\Delta Z_r| \to 0$ and if summation tends to a limit, that is independent of the choice of the intermediate points and of the manner in which the division is performed, then the above limit is called contour integral of $f(z)$ along c and is written as

$$I = \lim_{n \to \infty} S_n = \int_c f(z)dz = \int_a^b f(z)dz \qquad (6.8)$$

But it must be remembered that the value of the contour integral depends in general on the path connecting the points a and b.

If u and v denote the components of $f(z)$, when z is on c then

$$f(z) = u + iv$$

and $\qquad\qquad\qquad dz = dx + idy$

The complex line integral, i.e. Eq. (6.8) can be expressed in terms of real line integral as

$$\int f(z)dz = \int_c (u + iv)(dx + idy)$$

$$= \int_c (udx - vdy) + i\int_c (vdx + udy) \qquad (6.9)$$

Equation (6.9) is sometimes taken as definition of complex line integral.

Example 1. *If c is any rectifiable arc joining the points z = a and z = b, prove that*

i. $\displaystyle\int_c dz = b - c$ \qquad ii. $\displaystyle\int_c zdz = \frac{1}{2}(b^2 - a^2)$ \qquad iii. $\displaystyle\int_c |dz| = l$

where l is the length of curve c.

Solution: i. Here the given function $f(z) = 1$ is continuous on c. Therefore, the integral exists and

$$\int dz = \lim_{n \to \infty} \sum_{r=1}^{n} f(\xi_r)(z_r - z_{r-1})$$

$$= \lim_{n \to \infty} \sum_{r=1}^{n} (z_r - z_{r-1}), \text{ since } f(\xi_r) = 1$$

$$= \lim_{n \to \infty} [(z_1 - z_0) + (z_2 - z_1) + \cdots + (z_n - z_{n-1})]$$

$$= z_n - z_0 = b - a.$$

Note 1: Since the result is independent of the particular curve c taken, we may write the result as

$$\int_a^b dz = (b - a)$$

Note 2: If c be a closed curve then clearly $\displaystyle\int_c dz = 0$

ii.

$$I = \int_c zdz = \lim_{n \to \infty} \sum_{r=1}^{n} f(\xi_r)(z_r - z_{r-1})$$

$$= \lim_{n \to \infty} \sum_{r=1}^{n} \xi_r(z_r - z_{r-1})$$

In this case, integral does not depend on the point ξ_r, which we choose in summation to define the integral. Thus taking $\xi_r = z_r$, we obtain

$$I = \int_c z\,dz = \lim_{n \to \infty} \sum_{r=1}^{n} z_r(z_r - z_{r-1})$$

Again if we take $\xi_r = z_{r-1}$, then

$$I = \int_c z\,dz = \lim_{n \to \infty} \sum_{r=1}^{n} z_{r-1}(z_r - z_{r-1})$$

Adding we have

$$2I = \lim_{n \to \infty} \sum_{r=1}^{n} (z_r^2 - z_{r-1}^2)$$

$$= \lim_{n \to \infty} \{(z_1^2 - z_0^2) + (z_2^2 - z_1^2) + \cdots + (z_n^2 - z_{n-1}^2)\}$$

$$= z_n^2 - z_0^2 = b^2 - a^2.$$

Since the result is again independent of the path, we may express the result as

$$\int_a^b z\,dz = \frac{1}{2}(b^2 - a^2)$$

Note 3: If the given curve c be a closed curve, so that $b = a$, then

$$\int_c z\,dz = 0$$

iii. $\displaystyle I = \int_c |dz| = \lim_{n \to \infty} \sum_{r=1}^{n} |z_r - z_{r-1}|$

$$= (|z_1 - z_0| + |z_2 - z_1| + \cdots + |z_n - z_{n-2}|)$$

= length of the chord between z_1 and z_0 + length of the chord between z_2 and z_1 + $\cdots$

= length of the curve $c = l$

where length of every chord $(z_r - z_{r-1})$ is of infinitesimally small length such that it is equal to arc $(z_r - z_{r-1})$.

Example 2. *Evaluate* $\displaystyle \int_c z^2\,dz$ *where c is an arc of the circle* $x = r \cos\theta$, $y = r \sin\theta$ *from* $\theta = \alpha$ *to* $\theta = \beta$.

Solution: Here $f(z) = z^2 = (x + iy)^2$

$$= (x^2 - y^2) + 2ixy$$

∴ $u = x^2 - y^2$, $v = 2xy$

$dx = -r \sin\theta\,d\theta$, $dy = r \cos\theta\,d\theta$

which gives $u\,dx = -r^3(\cos^2\theta - \sin^2\theta)\sin\theta\,d\theta$

$v\,dy = 2r^3 \cos^2\theta \sin\theta \cdot d\theta$

$u\,dy = r^3(\cos^2\theta - \sin^2\theta)\cos\theta\,d\theta$

$v\,dx = -2r^3 \cos\theta \sin^2\theta \cdot d\theta$

∴ $$\int_C f(z)dz = \int_C (udx - vdy) + i\int_C (vdx + udy)$$

$$= r^3 \int_\alpha^\beta (\sin^3\theta - 3\cos^2\theta\sin\theta)d\theta + ir^3 \int_\alpha^\beta (\cos^3\theta - 3\sin^2\theta\cos\theta)d\theta$$

$$= r^3 \int_\alpha^\beta (1 - 4\cos^2\theta)\sin\theta\, d\theta + ir^3 \int_\alpha^\beta (1 - 4\sin^2\theta)\cos\theta\, d\theta$$

$$= r^3 \left[\left(-\cos\theta + \frac{4}{3}\cos^3\theta\right) + i\left(\sin\theta - \frac{4}{3}\sin^3\theta\right)\right]_\alpha^\beta$$

$$= \frac{1}{3}r^3 [(\cos 3\beta - \cos 3\alpha) + i(\sin 3\beta - \sin 3\alpha)].$$

6.5. AN UPPER BOUND FOR A CONTOUR INTEGRAL

Theorem: If a function $f(z)$ is continuous on a contour C of length l, and if M be the upper bound of $|f(z)|$ on C, then

$$\left|\int_C f(z)dz\right| \le Ml$$

In order to consider the upper bounds of certain contour integrals, let us consider the integral (which is supposed to exist)

$$I = \int_C f(z)dz$$

where C is a piecewise regular path in the complex plane. We shall further assume that $|f(z)|$ is bounded on C. As discussed in the preceding section, the integral I is the limit as $n \to \infty$ of the sum

$$I_n = \sum_{r=1}^n (z_r - z_{r-1})f(\xi_r)$$

Denoting M as the maximum modulus of $f(z)$ on C, we find

$$|I_n| \le \sum_{r=1}^n |z_r - z_{r-1}|\,|f(\xi_r)|$$

$$\le M\sum_{r=1}^n |z_r - z_{r-1}|$$

The sum of the R.H.S of inequality is the length of a polygon inscribed in the curve C and is therefore smaller than the arc length l of the curve itself. Hence, for all n

$$|I_n| \le Ml$$

In particular, as $n \to \infty$

$$\left|\int_C f(z)dz\right| \le Ml \qquad\qquad (6.10)$$

Equation (6.10) is called *Darboux's inequality*. It will be frequently used in subsequent sections.

6.6. CAUCHY'S INTEGRAL THEOREM

If C denotes a piecewise, regular closed curve in the complex plane and let $f(z)$ be analytic on C and within the whole region enclosed by C. Then

$$\int_C f(z)dz = 0 \tag{6.11}$$

The theorem in its original form required not only that the derivative of $f(z)$ to exist, but also that it to be continuous. Goursat has shown that the latter requirement is unnecessary. For this reason, this theorem is sometimes called the Cauchy–Goursat theorem.

Elementary Proof

Since $f(z) = u + iv$ is analytic and has a continuous derivative

$$f'(z) = \frac{\partial u}{\partial x} + i\frac{\partial v}{\partial x} = \frac{\partial v}{\partial y} - i\frac{\partial u}{\partial y}$$

It follows that the partial derivatives:

i. $\dfrac{\partial u}{\partial x} = \dfrac{\partial v}{\partial y}$ 　　　　 ii. $\dfrac{\partial u}{\partial y} = -\dfrac{\partial v}{\partial x}$

are continuous inside and on C. Thus, the Green's theorem can be applied and we have

$$\oint_C f(z)dz = \oint_C (u + iv)(dx + idy)$$

$$= \oint_C (udx - vdy) + i\oint_C (udy + vdx)$$

$$= \iint_R \left(-\frac{\partial v}{\partial x} - \frac{\partial u}{\partial y}\right)dx\,dy + i\iint_R \left(\frac{\partial v}{\partial y} - \frac{\partial u}{\partial x}\right)dx\,dy = 0$$

Using Cauchy–Riemann equations.

Goursat's Proof

Before we proceed with the actual proof of the theorem, we prove a lemma called Goursat's lemma.

Lemma: It is always possible to divide up the region inside C into a finite number of squares C_n and partial squares on D_n such that within or upon the boundary of each of these subregions there exists a point z_0 such that as z describes the boundary of the region, we have

$$\left|\frac{f(z) - f(z_0)}{z - z_0} - f'(z_0)\right| < \varepsilon \tag{6.12}$$

where ε is a previously assigned arbitrarily small positive number.

In order to prove this lemma, we apply the method of successive subdivision. Suppose that we start with a network of lines parallel to the real and imaginary axes at constant distance (Fig. 6.9). By this means the given region is subdivided into smaller regions of the subregions so obtained some may satisfy condition of Eq. (6.11) while others do not. Without changing the subregions which satisfy Eq. (6.11), we subdivided the others in the same way. The process may end either after a finite number of steps.

When the condition of Eq. (6.11) is satisfied for every subdivision, or the process may go on indefinitely. In the second case, we obtain a sequence of squares (each contained in the preceding one) whose limit point is z_0 which lies inside C and at which condition Eq. (6.11) is not satisfied.

i.e. $$\left| \frac{f(z) - f(z_0)}{z - z_0} - f'(z_0) \right| \not< \varepsilon$$

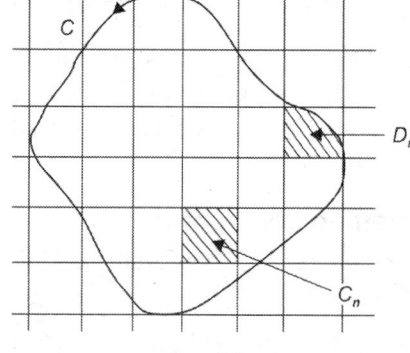

This shows that $f(z)$ is not differentiable at z_0, or in other words $f(z)$ is not analytic at z_0. This contradicts the hypothesis that $f(z)$ is analytic at all the points within and on the contour C. Hence the lemma is true, i.e.

$$\left| \frac{f(z) - f(z_0)}{z - z_0} - f'(z_0) \right| < \varepsilon$$

Fig. 6.9

or $$\frac{f(z) - f(z_0)}{z - z_0} - f'(z_0) = \eta(z), \text{ where } |\eta| < \varepsilon$$

and $$\eta \to 0 \text{ as } z \to z_0.$$

Thus $$f(z) = f(z_0) + (z - z_0) f'(z_0) + \eta (z - z_0) \tag{6.13}$$

This is also known as **L'Hospital's rule**

We now notice that the integral

$$\int_C f(z)\,dz$$

can be replaced by a sum of mesh integrals; where the meshes correspond to the boundaries C_j of the areas A_j;

$$\int_C f(z)\,dz = \sum_{j=1}^{n} \int_{C_j} f(z)\,dz$$

The process is legitimate because the common boundary of the two adjacent subregions gives equal and opposite contribution to mesh integrals in each of the adjacent subregions, provided all the integrals including $\int_C f(z)\,dz$ are performed in a counterclockwise direction.

Using Eq. (6.12), we can write

$$\int_{C_j} f(z)\,dz = \int_{C_j} f(z_0)\,dz + \int_{C_j} f'(z_0)z\,dz - \int_{C_j} z_0 f'(z_0)\,dz + \int_{C_j} \eta(z - z_0)\,dz$$

Remembering Darboux inequality (Eq. 6.10) and

$$\int_{C_j} dz = 0, \quad \int_{C_j} z\,dz = 0,$$

we get $$\left| \int_C f(z)\,dz \right| < \begin{cases} 4\sqrt{2}\,\varepsilon l_j^2 & \text{if } A_j \text{ is an interior square } C_n \\ \sqrt{2}\,\varepsilon l_j(4l_j + S_j) & \text{if } A_j \text{ has a segment} \\ & \text{of } c \text{ as part of its boundary, i.e. } D_n. \end{cases}$$

where S_j is the arc length of C included in the square D_n continuous to C.

Let l be the length of the edge of a square that contains the entire region and S be the length of the contour C, then

$$l_j < l; \; \sum_{j=1}^{n} S_j = S, \; \sum_{j=1}^{n} l_j^2 \le l^2$$

The last inequality simply means that the sum of the areas of all the D_n (and consequently of all the C_n) is not larger than l^2. Finally, we have

$$\left| \int_C f(z)dz \right| < \sum_{j=1}^{n} \left| \int_{C_j} f(z)dz \right| < \varepsilon \left\{ 4\sqrt{2}\,l^2 + \sqrt{2}\,ls \right\}_{\varepsilon \to 0} \to 0$$

This completes the proof of Cauchy's theorem. It should be noted carefully that this proof relies heavily on the fact that the function $f(z)$ has a derivative in the region enclosed by the contour C.

6.7. SIMPLY-CONNECTED AND MULTIPLY-CONNECTED REGIONS

There exists a very simple and important classification of regions in the complex plane. For example, a region could consist of the entire domain contained within a closed curve, or it could consist of the entire domain with exclusion of a number of holes punched out of it. A region is said to be simply connected if it is such that all closed paths within it contain only points that belong to the region (no holes !). Otherwise, the region is said to be multiply connected.

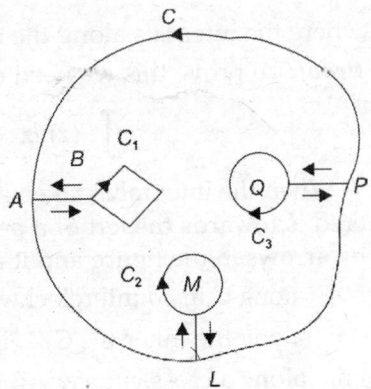

The region enclosed by the curves C_1 or C_2 or C_3 in Fig. 6.10 is simply connected while the region enclosed between C and C_1, C_2, C_3 is multiply-connected. A multiply-connected region can be made simply-connected by introducing additional boundary lines (cross cuts) joining the inner curves to the outer curve (Fig. 6.11).

The boundary consisting of curves C, C_1, C_2, C_3 and the lines AB, LM, PQ form a closed contour and Cauchy's theorem applies to such a region.

Fig. 6.10

6.8. CONSEQUENCES OF CAUCHY'S THEOREM

Theorem 1: If $F(z)$ is analytic in a simply-connected region R, then $\int_a^b F(z)dz$ is independent of the path in R joining any two points a and b in R.

Proof: By Cauchy's theorem

$$\int_{ADBEA} F(z)dz = 0$$

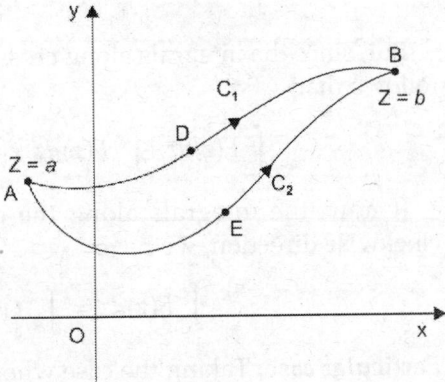

Fig. 6.11

or $\qquad \int\limits_{ADB} F(z)dz + \int\limits_{BEA} F(z)dz = 0$

or $\qquad \int\limits_{ADB} F(z)dz = -\int\limits_{BEA} F(z)dz = \int\limits_{AEB} F(z)dz$

Thus $\qquad \int_{C_1} F(z)dz = \int_{C_2} F(z)dz = \int_a^b F(z)dz$

This result gives us an important remark, that is, if $F(z)$ is analytic in a simply connected region R then the function

$$\phi(z) = \int_a^b F(z)dz$$

is also analytic in the region R.

Theorem 2: Let C be a closed contour and $C_1, C_2, C_3,...$ be a finite number of closed contours inside C, and $F(z)$ be analytic in the region between them. Then

$$\int_C F(z)dz = \int_{C_1} F(z)dz + \int_{C_2} F(z)dz + \int_{C_3} F(z)dz + \cdots$$

where the integrals along the contours are taken in the counter-clockwise directions.
Proof: To prove this we need only to put cross-cuts AB, LM, PQ (Fig. 6.11). Then

$$\int F(z)dz = 0$$

When the integral is taken along the complete boundary so that the region considered is towards the left of a person going round the boundary. The path is indicated by arrows in the figure and it is seen that the integral is taken

i. along C in counterclockwise direction

ii. along the inner $C_1, C_2,...$ in clockwise direction and

iii. along a cross-cut say AB it is taken once along AB and once along BA.

Hence taking the integrals in the directions given above we get

$$\int_C F(z)dz + \int_{C_1} F(z)dz + \int_{AB} F(z)dz + \int_{BA} F(z)dz + \cdots = 0$$

But, since the integrals along cross-cuts being taken in opposite directions cancel, it follows that

$$\int_C F(z)dz + \int_{C_1} F(z)dz + \cdots = 0$$

If now, the integrals along the curves $C_1, C_2,...$ are also taken in the counter-clockwise direction, we have

$$\int_C F(z)dz = \int_{C_1} F(z)dz + \int_{C_2} F(z)dz + \cdots \qquad (6.14)$$

Particular case: Taking the case when there is only one inner curve C_1, we get

$$\int_C F(z)dz = \int_{C_1} F(z)dz \qquad (6.15)$$

6.9. CAUCHY'S INTEGRAL FORMULAS

If $F(z)$ is analytic inside and on the simple closed curve C, and 'a' is any point inside C then

$$F(a) = \frac{1}{2\pi i} \oint_C \frac{F(z)}{z - a} dz \qquad (6.16)$$

where C is transversed in the positive (counterclockwise) sense.

Also the nth derivative of $F(z)$ at $z = a$ is given by

$$F^n(a) = \frac{n!}{2\pi i} \oint_C \frac{F(z)}{(z - a)^{n+1}} dz \qquad (6.17)$$

$$n = 1, 2, 3...$$

Equation (6.16) can be considered a special case of Eq. (6.17) with $n = 0$.

Proof:

Method 1. The function $\dfrac{F(z)}{z - a}$ is analytic inside and on C except at the point $z = a$ (Fig. 6.12).

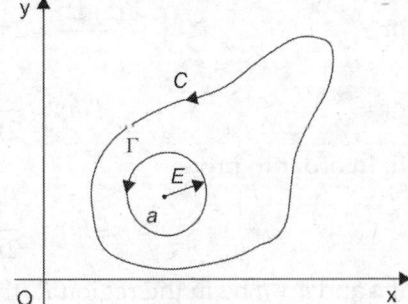

Fig. 6.12

Draw a circle Γ of radius ε about $z = a$ lying entirely within c. In the region between c and Γ,

the function $\dfrac{F(z)}{z - a}$ is analytic and using Eq. (6.15)

we have

$$\int_C \frac{F(z)}{z - a} dz = \int_\Gamma \frac{F(z)}{z - a} dz$$

The equation for Γ is $|z - a| = \varepsilon$ or $z - a = \varepsilon e^{i\theta}$, where $0 \le \theta \le 2\pi$. Substituting $z = a + \varepsilon e^{i\theta}$ and $dz = i\varepsilon e^{i\theta} d\theta$, the integral becomes

$$\int_C \frac{F(z)dz}{z - a} = \int_\Gamma \frac{F(z)}{z - a} dz = \int_0^{2\pi} \frac{F(a + \varepsilon e^{i\theta}) i\varepsilon e^{i\theta} d\theta}{\varepsilon e^{i\theta}}$$

$$= i \int_0^{2\pi} F(a + \varepsilon e^{i\theta}) d\theta$$

Taking the limit as $\varepsilon \to 0$ and making use of the continuity of $F(z)$, we have

$$\int_C \frac{F(z)dz}{z - a} = \lim_{\varepsilon \to 0} i \int_0^{2\pi} F(a + \varepsilon e^{i\theta}) d\theta = 2\pi i F(a)$$

so that we have the required result

$$F(a) = \frac{1}{2\pi i} \int_C \frac{F(z)}{z - a} dz$$

Method 2. We have

$$\int_C \frac{F(z)}{z - a} dz = \int_\Gamma \frac{F(z)}{z - a} dz = \int_\Gamma \frac{F(z) - F(a)}{z - a} dz + \int_\Gamma \frac{F(a)}{z - a} dz$$

$$= \int_\Gamma \frac{F(z) - F(a)}{z - a} dz + 2\pi i F(a)$$

The required result will follow, if we can prove

$$\int_\Gamma \frac{F(z)-F(a)}{z-a}\,dz = 0$$

Using Eq. (6.13), we have

$$\int_\Gamma \frac{F(z)-F(a)}{z-a}\,dz = \int_\Gamma F'(a)\,dz + \int_\Gamma \eta\,dz = \int_\Gamma \eta\,dz$$

Since

$$\int_\Gamma dz = 0$$

Now, choosing Γ so small that for all points on Γ we have $|\eta| < \delta/2\pi$ and we obtain

$$\left| \int_\Gamma \eta\,dz \right| < \left(\frac{\delta}{2\pi} \right)(2\pi\varepsilon) = \delta\varepsilon \to 0 \text{ as } \varepsilon \to 0$$

Thus, we have $\int_\Gamma \eta\,dz = 0$

or

$$\int_C \frac{F(z)\,dz}{z-a} = 2\pi i F(a)$$

or

$$F(a) = \frac{1}{2\pi i}\int_C \frac{F(z)\,dz}{z-a}$$

ii. In order to prove

$$F'(a) = \frac{1}{2\pi i}\int_C \frac{F(z)}{(z-a)^2}\,dz$$

if a and $a+h$ be in the region R, then we have

$$\frac{F(a+h)-F(a)}{h} = \frac{1}{2\pi i}\int_C \frac{1}{h}\left\{ \frac{1}{z-(a+h)} - \frac{1}{z-a} \right\} F(z)\,dz$$

$$= \frac{1}{2\pi i}\int_C \frac{F(z)}{(z-a-h)(z-a)}\,dz$$

$$= \frac{1}{2\pi i}\int_C \frac{F(z)}{(z-a)^2}\,dz + \frac{h}{2\pi i}\int_C \frac{F(z)}{(z-a-h)(z-a)^2}\,dz$$

The result follows on taking the limit as $h \to 0$, if we can show that the last term approaches zero.

To show this we use the fact that if Γ is a circle of radius ε and centre 'a' which lies entirely in R (Fig. 6.13) then

$$\frac{h}{2\pi i}\int_C \frac{F(z)\,dz}{(z-a)^2(z-a-h)} = \frac{h}{2\pi i}\int_\Gamma \frac{F(z)\,dz}{(z-a)^2(z-a-h)}$$

Choosing h, so small in absolute value that $a+h$ lies in Γ and $|h| < \varepsilon/2$, and the fact that Γ has the equation

$$|z-a| = \varepsilon$$

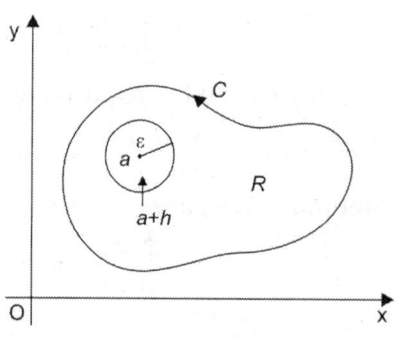

Fig. 6.13

Then $$|z-a-h| \geq |z-a| - |h|$$

$$|z-a-h| > \varepsilon - \frac{\varepsilon}{2} = \frac{\varepsilon}{2}$$

Then since the length of Γ is $2\pi\varepsilon$, we have using Eq. (6.10)

$$\left| \frac{h}{2\pi i} \int_C \frac{F(z)dz}{(z-a)^2(z-a-h)} \right| \leq \frac{|h|}{2\pi} \frac{M2\pi\varepsilon}{\varepsilon/2 \cdot \varepsilon^2} = \frac{2|h|M}{\varepsilon^2}$$

and it follows that

$$\lim_{h \to 0} \frac{h}{2\pi i} \int_C \frac{F(z)dz}{(z-a-h)(z-a)^2} \to 0$$

Hence $$F'(0) = \frac{1}{2\pi i} \int_C \frac{F(z)dz}{(z-a)^2} \qquad (6.18)$$

It is of interest to observe that the result is equivalent to

$$\frac{d}{da}F(a) = \frac{d}{da}\left\{ \frac{1}{2\pi i} \int_C \frac{F(z)dz}{(z-a)} \right\}$$

$$= \frac{1}{2\pi i} \int_C \frac{F(z)dz}{(z-a)^2}$$

which is an extension to contour integrals of Leibnitz's rule for differentiating under the integral sign.

iii. To prove the result

$$F^n(a) = \frac{n!}{2\pi i} \int_C \frac{F(z)}{(z-a)^{n+1}} dz$$

for $n = 2$, we use case (ii) for $n = 1$ where a and $a + h$ lie in R to obtain

$$\frac{F'(a+h) - F'(a)}{h} = \frac{1}{2\pi i} \int_C \frac{1}{h}\left\{ \frac{F(z)}{(z-a-h)^2} - \frac{F(z)}{(z-a)^2} \right\} dz$$

$$= \frac{2!}{2\pi i} \int_C \frac{F(z)}{(z-a)^3} dz + \frac{h}{2\pi i} \int_C \frac{3(z-a) - 2h \cdot F(z)dz}{(z-a-h)^2(z-a)^3}$$

The result follows on taking the limits $h \to 0$, if we can show that the last term approaches zero. The proof is similar to that of case (ii), for using the fact that the integral around C equals the integral around Γ, we have

$$\left| \frac{h}{2\pi i} \int_\Gamma \frac{3(z-a) - 2h}{(z-a-h)^2(z-a)^3} F(z)dz \right| \leq \frac{|h|}{2\pi} \frac{M(2\pi\varepsilon)}{(\varepsilon/2)^2(\varepsilon)^3}$$

$$= \frac{4|h|M}{\varepsilon^4} \to 0 \quad \text{as} \quad h \to 0.$$

Since M exists such that

$$|\{3(z-a) - 2h\}F(z)| < M.$$

In a similar manner, we can establish the result for $n = 3, 4,...$

The result is equivalent to

$$\frac{d^n}{da^n}F(a) = \frac{d^n}{da^n}\left\{\frac{1}{2\pi i}\int_C \frac{F(z)dz}{(z-a)}\right\}$$

$$= \frac{1}{2\pi i}\int_C \frac{d^n}{da^n}\left\{\frac{F(z)dz}{(z-a)}\right\} \qquad (6.19)$$

The results of Eqs (6.16) and (6.17) are called Cauchy's integral formulas and are quite remarkable because they show that if a function $F(z)$ is known on the simple closed curve C then the values of the function and all its derivatives can be found at all points inside C. Thus, if a function of complex variable has a first derivative that is analytic in a simply connected region R, all its higher derivatives exist in R. This is not necessarily true for functions of real variables.

6.10. POISSON'S INTEGRAL FORMULAS FOR A CIRCLE

(a) Let $F(z)$ be analytic inside and on the circle C defined by $|z| = R$ and let $z = re^{i\theta}$ be any point inside C (Fig. 6.14). Prove that

$$F(re^{i\theta}) = \frac{1}{2\pi}\int_0^{2\pi} \frac{R^2-r^2}{R^2-2Rr\cos(\theta-\phi)+r^2}F(Re^{i\phi})d\phi$$

(b) If $u(r, \theta)$ and $v(r, \theta)$ are the real and imaginary parts of $F(re^{i\theta})$ prove that

$$u(r, \theta) = \frac{1}{2\pi}\int_0^{2\pi} \frac{(R^2-r^2)u(R.\phi)d\phi}{R^2-2Rr\cos(\theta-\phi)+r^2}$$

$$v(r, \theta) = \frac{1}{2\pi}\int_0^{2\pi} \frac{(R^2-r^2)v(R.\phi)d\phi}{R^2-2Rr\cos(\theta-\phi)+r^2}$$

The results are called Poisson's integral formulas for the circle.

(c) Since $z = re^{i\theta}$ is any point inside C, we have Cauchy's integral formula

$$F(z) = F(re^{i\theta}) = \int_C \frac{F(w)}{w-z}dw \qquad (6.19a)$$

The inverse of the point z with respect to C lies outside C and is given by $\dfrac{R^2}{\bar{z}}$.

Hence by Cauchy's theorem

$$0 = \frac{1}{2\pi i}\int_C \frac{F(w)}{w-R^2/\bar{z}}dw \qquad (6.19b)$$

Subtracting Eqs (6.19b) from (6.19a), we have

$$F(z) = \frac{1}{2\pi i}\int_C \left\{\frac{1}{w-z}-\frac{1}{w-R^2/\bar{z}}\right\}F(w)dw$$

$$= \frac{1}{2\pi i}\int_C \frac{z-R^2/\bar{z}}{(w-z)(w-R^2/\bar{z})}F(w)dw$$

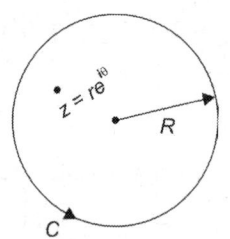

Fig. 6.14

Now let $z = re^{i\theta}$ and $w = Re^{i\phi}$. Then since $\bar{z} = re^{i\theta}$, case (c) yields

$$F(re^{i\theta}) = \frac{1}{2\pi i} \int_0^{2\pi} \frac{\left\{ re^{i\theta} - (R^2/r)e^{i\theta} \right\} F(Re^{i\phi}) iRe^{i\phi} d\phi}{(Re^{i\phi} - re^{i\theta})\left\{ Re^{i\phi} - (R^2/r)e^{i\theta} \right\}}$$

$$= \frac{1}{2\pi i} \int_0^{2\pi} \frac{(r^2 - R^2)e^{i(\theta+\phi)} F(Re^{i\phi})}{(Re^{i\phi} - re^{i\theta})(re^{i\theta} - Re^{i\theta})} d\phi$$

$$= \frac{1}{2\pi} \int_0^{2\pi} \frac{(R^2 - r^2)F(Re^{i\phi}) d\phi}{(Re^{i\phi} - re^{i\theta})(Re^{-i\phi} - re^{i\theta})} = \frac{1}{2\pi} \int_0^{2\pi} \frac{(R^2 - r^2)F(Re^{i\phi}) d\phi}{R^2 - 2Rr\cos(\theta - \phi) + r^2} \qquad (6.20)$$

(d) Since
$$F(re^{i\theta}) = u(r, \theta) + iv(r, \theta)$$

and
$$F(Re^{i\phi}) = u(R, \phi) + iv(R, \phi)$$

We have then from Eq. (6.19)

$$u(r, \theta) + iv(r, \theta) = \frac{1}{2\pi} \int_0^{2\pi} \frac{(R^2 - r^2)\{u(R, \phi) + iv(R, \phi)\} d\phi}{R^2 - 2Rr\cos(\theta - \phi) + r^2}$$

or $u(r, 0) + iv(r, 0) - \dfrac{1}{2\pi} \displaystyle\int_0^{2\pi} \dfrac{(R^2 - r^2)u(R, \phi) d\phi}{R^2 - 2Rr\cos(\theta - \phi) + r^2} + \dfrac{1}{2\pi} \displaystyle\int_0^{2\pi} \dfrac{(R^2 - r^2)v(R, \phi) d\phi}{R^2 - 2Rr\cos(\theta - \phi) + r^2}$

Equating real and imaginary parts, we have

$$u(r, \theta) = \frac{1}{2\pi} \int_0^{2\pi} \frac{(R^2 - r^2)u(R, \phi) d\phi}{R^2 - 2Rr\cos(\theta - \phi) + r^2} \qquad (6.20a)$$

and
$$v(r, \theta) = \frac{1}{2\pi} \int_0^{2\pi} \frac{(R^2 - r^2)v(R, \phi) d\phi}{R^2 - 2Rr\cos(\theta - \phi) + r^2} \qquad (6.21)$$

Morera's Theorem: This theorem is converse of Cauchy's integral theorem. It states that if a function $F(z)$ is continuous in a simply-connected region R and $\displaystyle\int_C F(z) dz = 0$ for every contour C within R then $F(z)$ is analytic throughout R.

Proof: We have seen in case of Cauchy's integral theorem that if closed path integral of $F(z)$ vanishes, the integral of $F(z)$ between points $z_1 = a$ and $z_2 = b$ depends only on these end points. This we have already mentioned as

$$\phi(z) = \int_a^b F(z) dz$$

Thus we can write the result as

$$\phi(b) - \phi(a) = \int_a^b F(z) dz$$

where points a and b lie in the region R. Hence

$$\frac{\phi(b) - \phi(a)}{(b - a)} = \frac{1}{b - a} \int_a^b F(z) dz = \frac{1}{b - a} \int_a^b F(t) dt$$

where t is another complex variable.

We can also write $\qquad F(a) = \dfrac{F(a)}{b-a}\displaystyle\int_a^b dt = \int_a^b \dfrac{F(a)}{b-a}dt$

Thus we have an identity that

$$\frac{\phi(b)-\phi(a)}{b-a} - F(a) = \frac{\displaystyle\int_a^b [F(t)-F(a)]dt}{b-a}$$

Now if we take the limit $b \to a$, then the right hand side of the identity vanishes, i.e.

$$\lim_{b \to a} \frac{1}{(b-a)}\int_a^b [F(t)-F(a)]dt = 0$$

since $F(t)$ is continuous.

Thus, we have

$$\lim_{b \to a} \frac{\phi(b)-\phi(a)}{b-a} = \phi'(z)\big|_{z=a} = F(a)$$

This shows that $\phi'(z)$ at $z = a$ exists and equals to $F(a)$. Since a is any point in R, $\phi(z)$ is analytic in R. Then by Cauchy's integral formula for derivative $\phi'(z)$ is analytic. Hence $F(z) = \phi'(z)$ is also analytic in the region R. This proves Morera theorem.

This theorem serves as a means for the identification of an analytic function and is thus the integral analogue of the differential requirement given by Cauchy–Riemann conditions. The physical interpretation of Morera's theorem as given by electrostatic analogue is an interesting aspect of this theorem.

Example 1. *Evaluate* (i) $\displaystyle\int_C \frac{\sin \pi z^2 + \cos \pi z^2}{(z-1)(z-2)}dz$, *and* (ii) $\displaystyle\int_C \frac{e^{2z}}{(z+1)^4}dz$, *where C is a circle of radius* $|z| = 3$.

Solution: i.
$$\frac{1}{(z-1)(z-2)} = \frac{1}{z-2} - \frac{1}{z-1}$$

$$I = \int_C \frac{\sin \pi z^2 + \cos \pi z^2}{(z-1)(z-2)}dz$$

$$= \int_C \frac{\sin \pi z^2 + \cos \pi z^2}{(z-2)}dz - \int_C \frac{\sin \pi z^2 - \cos \pi z^2}{(z-1)}dz$$

Using Cauchy's integral formula

$$F(a) = \frac{1}{2\pi i}\int_C \frac{F(z)}{z-a}dz, \text{ we have}$$

$$\int_C \frac{\sin \pi(z^2) + \cos \pi z^2}{z-2}dz = 2\pi i \{\sin \pi(2)^2 + \cos \pi(2)^2\} = 2\pi i$$

$$\int_C \frac{\sin \pi z^2 + \cos \pi z^2}{z-1}dz = 2\pi i \{\sin \pi 1^2 + \cos \pi 1^2\} = -2\pi i$$

or $\qquad\qquad\qquad I = 2\pi i - (-2\pi i) = 4\pi i$

ii. Let $f(z) = e^{2z}$ and $a = -1$ Cauchy's integral formula

$$f^n(a) = \frac{n!}{2\pi i} \int_C \frac{f(z)dz}{(z-a)^{n+1}}$$

If

$$n = 3$$
$$f^3(z) = 8e^{2z} \quad \text{and} \quad f^3(-1) = 8e^{-2}$$

or

$$8e^{-2} = \frac{3!}{2\pi i} \int_C \frac{e^{2z}}{(z+1)^4} dz$$

or

$$\int_C \frac{e^{2z}}{(z+1)^4} dz = \frac{8e^{-2} \cdot 2\pi i}{6} = \frac{8\pi i e^{-2}}{3}$$

Example 2. *Using Cauchy's integral formula, evaluate the integral*

$$\int_C \frac{zdz}{(9-z^2)(z+i)}$$

where C is the circle, $|z| = 2$ described in the positive sense.

Solution: Let us choose

$$f(z) = \frac{z}{9-z^2} \quad \text{and} \quad a = -i$$

Since C is a circle, $|z| = 2$, hence $f(z) = \dfrac{z}{9-z^2}$ is analytical at all the points inside and

on the boundary of C.

Since point $z = a$ is within the circle, hence

$$f(-i) = \frac{1}{2\pi i} \int_C \frac{f(z)dz}{z-a} = \frac{1}{2\pi i} \int_C \frac{z/(9-z^2)}{z+i} dz$$

$$= \frac{1}{2\pi i} \int_C \frac{z}{(9-z^2)(z+i)} dz$$

Then

$$\frac{1}{2\pi i} \int_C \frac{z}{(9-z^2)(z+i)} dz = f(-i) = \lim_{z \to -i} \left\{ \frac{z}{9-z^2} \right\} = \frac{-i}{9-i^2} = \frac{-i}{10}$$

or

$$\int_C \frac{z}{(9-z^2)(z+i)} dz = \frac{2\pi}{10} = \frac{\pi}{5}$$

Example 3. *If f(z) is analytic in a region R bounded by two concentric circles C_1 and C_2 and on the boundary, prove that if z_0 is any point in R, then*

$$f(z_0) = \frac{1}{2\pi i} \int_{C_1} \frac{f(z)}{z-z_0} dz - \frac{1}{2\pi i} \int_{C_2} \frac{f(z)}{z-z_0} dz$$

Solution: Construct cross-cut EH connecting circles C_1 and C_2. Then $f(z)$ is analytic in the region bounded by EFGEHJIHE.

Hence by Cauchy's integral formula, we obtain

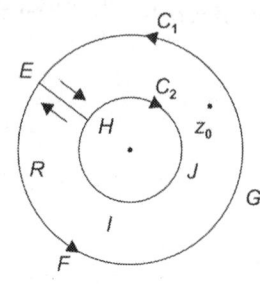

$$f(z_0) = \frac{1}{2\pi i} \int_{EFGEHJIHE} \frac{f(z)}{z - z_0} dz$$

$$= \frac{1}{2\pi i} \int_{EFGE} \frac{f(z)}{z - z_0} dz + \frac{1}{2\pi i} \int_{EH} \frac{f(z)}{z - z_0} dz$$

$$+ \frac{1}{2\pi i} \int_{HJIH} \frac{f(z)}{z - z_0} dz + \frac{1}{2\pi i} \int_{HE} \frac{f(z)}{z - z_0} dz$$

$$= \frac{1}{2\pi i} \int_{C_1} \frac{f(z)}{z - z_0} dz - \frac{1}{2\pi i} \int_{C_2} \frac{f(z)}{z - z_0} dz$$

Fig. 6.15

Since the integrals along EH and HE cancel.

6.11. INFINITE SERIES: TAYLOR'S AND LAURENT'S SERIES

Taylor's Theorem: If $f(z)$ be analytic inside and on a simple closed curve C, and a and $a + h$ be points inside C, then

$$f(a + h) = f(a) + h f'(a) + \frac{h^2}{2!} f''(a) + \cdots + \frac{h^n}{n!} f^{(n)}(a) + \cdots \qquad (6.22)$$

or writing $z = a + h$, $h = (z - a)$

$$f(z) = f(a) + f^{(1)}(a)(z - a) + \frac{f^{(2)}(a)}{2!}(z - a)^2 + \cdots + \frac{f^{(n)}(a)}{n!}(z - a)^n + \cdots \qquad (6.23)$$

i.e.

$$f(z) = \sum_{n=0}^{\infty} \frac{f^n(a)}{n!}(z - a)^n \cdots, \qquad (6.23a)$$

where z is a point in C, such that $|z - a| = r < R$ and $f^{(n)}(a)$ is the nth derivative of $f(z)$ at $z = a$.

This is called Taylor's theorem and the series, i.e. Eqs (6.22) or (6.23) is called Taylor's series or expansion for $f(a + h)$ or $f(z)$.

The region for convergence of series in Eq. (6.23) is given by $|z - a| < R$, where the radius of convergence R is the distance from a to the nearest singularity of the function $f(z)$. On $|z - a| = R$, the series may or may not converge. For $|z - a| > R$ the series diverges for all z.

If the nearest singularity* of $f(z)$ is at infinity, the radius of convergence is infinite, i.e. the series converges for all z.

If $a = 0$ in Eqs (6.22) and (6.23), the resulting series

$$f(z) = f(0) = f'(0)z + \frac{f''(0)}{2!}z^2 + \cdots + \frac{f^{(n)}(0)}{n!}z^n + \cdots \qquad (6.24)$$

is often called Maclaurin series.

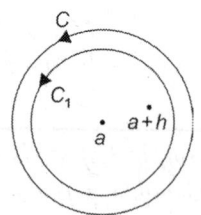

Fig. 6.16

Proof: Let $(a + h)$ be any point inside C. Construct a circle C with centre at a and enclosing $(a + h)$ (Fig. 6.16). Then by Cauchy's integral formula [Eq. (6.16)]

$$f(a + h) = \frac{1}{2\pi i} \int_C \frac{f(z)}{z - a - h} dz$$

---

* A point at which $f(z)$ fails to be analytic is called singular point or singularity of $f(z)$.

We have $\dfrac{1}{z-a-h} = \dfrac{1}{(z-a)} + \dfrac{h}{(z-a)(z-a-h)}$

$$= \frac{1}{z-a} + \frac{h}{z-a}\left\{\frac{1}{z-a} + \frac{h}{(z-a)(z-a-h)}\right\}$$

$$= \frac{1}{z-a} + \frac{h}{(z-a)^2} + \frac{h^2}{(z-a)^2(z-a-h)}$$

$$= \frac{1}{z-a} + \frac{h}{(z-a)^2} + \frac{h^2}{(z-a)^2}\left\{\frac{1}{z-a} + \frac{h}{(z-a)(z-a-h)}\right\}$$

$$= \frac{1}{z-a} + \frac{h}{(z-a)^2} + \frac{h^2}{(z-a)^3} + \cdots + \frac{h^{n-1}}{(z+a)^n} + \frac{h^n}{(z-a)^n(z-a-h)} \qquad (6.25)$$

Multiplying both sides by $f(z)$ and using Cauchy's integral formula, we have

$$f(a+h) = \frac{1}{2\pi i}\int_C \frac{f(z)}{z-a-h}dz$$

$$= \frac{1}{2\pi i}\int_{C_1}\frac{f(z)}{z-a}dz + \frac{h}{2\pi i}\int_{C_1}\frac{f(z)}{(z-a)^2}dz + \cdots + \frac{h^{n-1}}{2\pi i}\int_{C_2}\frac{f(z)}{(z-a)^n}dz + U_n \cdots \quad (6.26)$$

where $U_n = \dfrac{1}{2\pi i}\int_{C_1}\dfrac{h^n f(z)}{(z-a)^n(z-a-h)}dz$

Using Cauchy's integral formula

$$f^n(a) = \frac{n!}{2\pi i}\int_{C_1}\frac{f(z)}{(z-a)^{n+1}}dz, \quad n = 0,1,2,3... \qquad (6.27)$$

Equation (6.26) becomes

$$f(a+h) = f(a) + f'(a)h + \frac{f^2(a)}{2!}h^2 + \cdots + \frac{f^{n-1}(a)}{(n-1)!}h^{n-1} + U_n \qquad (6.28)$$

We have obtained the required result by integrating Eq. (6.25) term by term. Only a uniformly convergent series of analytic function can be integrated term by term along any path lying in the region of convergence. Thus,

$$\lim_{n\to\infty} U_n = 0$$

This leads to

$$\left|\frac{h}{z-a}\right| = \varepsilon < 1$$

where ε is a constant and h is within C_1. Also we have $f(z)$ is analytic on C_1 and within, then $|f(z)| < M$ where M is constant (upper bound limit of analytic function).

$$|z-a-h| \geq |z-a| - |h| = r_1 - |h|$$

where r_1 is the radius of C_1. Hence

$$|U_n| = \frac{1}{2\pi}\left|\int_{C_1}\left(\frac{h}{z-a}\right)^n \cdot \frac{f(z)dz}{z-a-h}\right| \leq \frac{1}{2\pi}\frac{\varepsilon^n M}{r_1-|h|}2\pi r_1 = \frac{\varepsilon^n M r_1}{r_1-|h|}$$

and $\lim_{n\to\infty} U_n = 0$; hence completing the proof.

About the radius of convergence of the Taylor's series, it is apparent that it is obtained immediately by inspecting Cauchy's formula as Eqs (6.16) and (6.17) on which the proof of Taylor's expansion rests. Indeed this formula breaks down when C_1 goes through circles having a singularity of $f(z)$. Therefore, it is concluded that the radius of convergence of the power series cannot be greater than the distance from the point $z = a$ to the nearest singularity of $f(z)$.

Note 1: We have shown that an analytic function can be expanded in Taylor's series at any of its regular points. The converse statement is also true, i.e. a function that can be expanded in power series.

$$f(z) = \sum_{n=0}^{\infty} a_n (z-a)^n \tag{6.29}$$

which is convergent in some neighbourhood of point $z = a$ (for example, for $|z-a| < R$) is necessarily analytic. Indeed the convergence of the infinite sum in Eq. (6.29) for $|z-a| = r < R$ implies the existence of a constant A such that

$$|a_n| r^n < A \tag{6.30}$$

for any n. Therefore, the sum of terms with $M \le n < N$ satisfies.

$$\left| \sum_{n=M}^{N} a_n (z-a)^n \right| \le \sum_{n=M}^{N} |a_n| \, |r-a|^n$$

$$< A \sum_{n=M}^{N} \left\{ \frac{|z-a|}{r} \right\}^n$$

$$= A \left\{ \frac{|z-a|}{r} \right\}^m \frac{1 - \left\{ \frac{|z-a|}{r} \right\}^{N-M}}{1 - \left\{ \frac{|z-a|}{r} \right\}} \tag{6.31}$$

For $|z-a| < r$, the expression on R.H.S of Eq. (6.31), inequality can be made smaller than any arbitrary constant independent of z by choosing M large enough. Hence, the convergence of the power series in Eq. (6.29) is uniform for $|z-a| < r$ and so it can be integrated term by term to give

$$\int_C f(z) dz = \sum_{n=0}^{\infty} a_n \int_C (z-a)^n dz = 0 \tag{6.32}$$

Here C is an arbitrary closed path lying within the circle of radius r and centred at $z = a$. Therefore, by virtue of Morera's theorem[*], $f(z)$ is analytic for $|z-a| < r$. But r can be arbitrarily close to R, and therefore the power series as in Eq. (6.29) is an analytic function of z for $|z-a| < R$.

---

[*] Morera's theorem is a converse of Cauchy's theorem, and follows from the fact that derivative of an analytic function is itself analytic and states that if $\int_C f(z) dz$ of a function which is continuous in some region, vanishes for any closed contour C lying within the region, then $f(z)$ is analytic in that region.

Note 2: If $|f(z)|$ has a maximum of $|z - a| = r < \varepsilon$ then

$$|a_n| \leq \frac{M}{r^n} \quad \text{where} \quad a_n = \frac{f^{(n)}(a)}{n!},$$

for if C is the circle $|z - a| = r$, we have

$$|a_n| = \left| \frac{1}{2\pi i} \int_C \frac{f(z)dz}{(z-a)^{n+1}} \right|$$

$$\leq \frac{1}{2\pi} \frac{M}{r^{n+1}} \int_C |dz|$$

$$\leq \frac{1}{2\pi} \frac{M}{r^{n+1}} 2\pi r = \frac{M}{r^n} \qquad \qquad ...(6.32a)$$

This is called Cauchy's inequality.

Note 3: If $f(z)$ is analytic in the whole complex plane, then $f(z)$ must be a constant provided that $|f(z)| < K$ for all values of z.

Since function is analytic, it can be expanded into Taylor's series with

$$|a_n| \leq \frac{M(r)}{r^n} \quad \text{where} \quad M(r) = |f(z)| \text{ at } z = r$$

But $|f(z)| < K$ for all values of z including $z = \infty$.

Then $M(r) < K$

or $\qquad \qquad |a_n| \leq \frac{K}{r^n} \leq 0 \quad \text{when} \quad r \to \infty \ n > 0$

Therefore, Taylor's expansion has only one nonzero term, i.e. $f(z) = a_0 = f(a) = $ constant. This theorem is known as *Lioville's theorem*.

The Laurent Series

The Laurent series is in fact an extension of Taylor's theorem to the case of a multiply-connected region. If a function $f(z)$ is not analytic throughout the whole interior of a circle (as it was assumed to be in the derivation of the Taylor series) but only throughout the annular region between two concentric circles C_1 and C_2 (shown shaded in Fig. 6.16a). Let $a + h$ be any point in R. Then the values of $f(z)$ at the point $(a + h)$ is given by

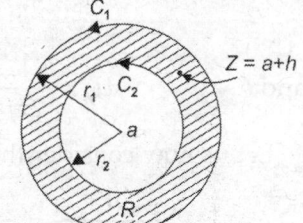

$$f(a + h) = a_0 + a_1 h + a_2 h^2 + \cdots + \frac{b_1}{h} + \frac{b_2}{h^2} + \frac{b_3}{h^3} + \cdots$$

where $\qquad a_n = \frac{1}{2\pi i} \int_{C_1} \frac{f(z)}{(z-a)^{n+1}} dz, \qquad n = 0, 1, 2,...$

and $\qquad b_n = \frac{1}{2\pi i} \int_{C_2} (z-a)^{n-1} f(z) dz, \quad n = 1, 2, 3,...$

Fig. 6.16a

and C_1 and C_2 being traversed in the positive direction with respect to their interiors. We can in the above integration replace C_1 and C_2 by concentric circle C between them, then the coefficients a_n and b_n can be written as

$$a_n = \frac{1}{2\pi i} \int_C \frac{f(z)}{(z-a)^{n+1}} dz, \qquad n = 0, \pm 1, \pm 2, \pm 3...$$

and putting $h + a = z$, the series can be written as

$$f(z) = a_0 + a_1(z-a) + a_2(z-a)^2 + \cdots + \frac{b_1}{z-a} + \frac{b_2}{(z-a)^2} + \cdots$$

This is called Laurent's theorem and series with coefficients a_n and b_n is called Laurent's series or expansion.

The part $a_0 + a_1(z-a) + a_2(z-a)^2 + \cdots$ is called the analytic part of Laurent's series, while the remainder of the series which consists of inverse powers of $(z - a)$ is called principal part. If the principal part is zero, the Laurent's series reduces to Taylor series.

Proof: By Cauchy's integral formula, we have

$$f(a+h) = \frac{1}{2\pi i}\int_{C_1} \frac{f(z)}{(z-a-h)}dz - \frac{1}{2\pi i}\int_{C_2} \frac{f(z)}{(z-a-h)}dz \qquad (6.33)$$

Consider the first integral in Eq. (6.33). We have from Eq. (6.25)

$$\frac{1}{z-a-h} = \frac{1}{z-a} + \frac{h}{(z-a)(z-a-h)}$$

$$= \frac{1}{z-a} + \frac{h}{(z-a)^2} + \cdots + \frac{h^{n-1}}{(z-a)^n} + \frac{h^n}{(z-a)^n}\cdot\frac{1}{z-a-h}$$

So that

$$\frac{1}{2\pi i}\int_{C_1}\frac{f(z)}{z-a-h}dz = \frac{1}{2\pi i}\int_{C_1}\frac{f(z)}{z-a}dz + \frac{h}{2\pi i}\int_{C_1}\frac{f(z)}{(z-a)^2}dz + \cdots + \frac{h^{n-1}}{2\pi i}\int_{C_1}\frac{f(z)}{(z-a)^n}dz + U_n$$

$$= a_0 + a_1 h + a_2 h^2 + \cdots + a_{n-1}h^{h-1} + U_n \qquad (6.33a)$$

where
$$a_0 = \frac{1}{2\pi i}\int_{C_1}\frac{f(z)}{z-a}dz, \qquad a_1 = \int_{C_1}\frac{f(z)}{(z-a)^2}dz$$

$$a_{n-1} = \frac{1}{2\pi i}\int_{C_1}\frac{f(z)}{(z-a)^n}dz$$

and
$$U_n = \frac{1}{2\pi i}\int_{C_1}\frac{h^n f(z)dz}{(z-a)^n(z-a-h)}$$

Let us now consider the second integral in Eq. (6.33), we have

$$\frac{1}{h-(z-a)} = \frac{1}{h\left\{1-\dfrac{z-a}{h}\right\}} = \left\{\frac{1}{h} + \frac{(z-a)/h}{\left(1-\dfrac{z-a}{h}\right)h}\right\}$$

$$= \left[\frac{1}{h} + \frac{(z-a)}{h}\left\{\frac{1}{h} + \frac{(z-a)/h}{\left(1-\dfrac{z-a}{h}\right)h}\right\}\right]$$

$$= \left[\frac{1}{h} + \frac{(z-a)}{h^2} + \cdots + \frac{(z-a)^{n-1}}{h^n} + \frac{(z-a)^n}{h^n}\cdot\frac{1}{h-(z-a)}\right]$$

So that

$$-\frac{1}{2\pi i}\int_{C_2}\frac{f(z)}{(z-a-h)}dz = \frac{1}{2\pi i}\int_{C_2}\frac{f(z)}{h-(z-a)}dz$$

$$= \frac{1}{2\pi i}\int_{C_2}\frac{f(z)}{h}dz + \frac{1}{2\pi i}\int_{C_2}\frac{(z-a)}{h^2}f(z)dz + \cdots$$

$$+ \frac{1}{2\pi i}\int_{C_2}\frac{(z-a)^{n-1}}{h^n}f(z)dz + V_n$$

$$= \frac{b_1}{h} + \frac{b_2}{h^2} + \cdots + \frac{b_n}{h^n} + V_n \tag{6.34}$$

where $$b_n = \frac{1}{2\pi i}\int_{C_2}(z-a)^{n-1}f(z)dz \tag{6.35}$$

and $$V_n = \frac{1}{2\pi i}\int_{C_2}\left(\frac{z-a}{h}\right)^n\frac{f(z)}{h-(z-a)}dz$$

Equations (6.33) and (6.34) give

$$f(a+h) = (a_0 + a_1 h + a_2 h^2 + \cdots + a_{n-1} h^{n-1}) + \left(\frac{b_1}{h} + \frac{b_2}{h^2} + \cdots + \frac{b_n}{h^n}\right) + U_n + V_n \tag{6.36}$$

The required result follows if we can show that:

(a) $\lim\limits_{n\to\infty} U_n = 0$ and (b) $\lim\limits_{n\to\infty} V_n = 0$. The proof of (a) follows from the proof of Taylor's theorem. To prove (b) we first note that since z is on C_2,

$$\left|\frac{z-a}{h}\right| = \eta < 1$$

where η is a constant. Also we have $|f(z)| < M$; where M is a constant and

$$|a+h-z| = |h-(z-a)| \geq |h| - r_2$$

Hence, we have

$$|V_n| = \frac{1}{2\pi}\left|\int_{C_2}\left(\frac{z-a}{h}\right)^n\frac{f(z)}{(a+h-z)}dz\right|$$

$$\leq \frac{1}{2\pi}\frac{\eta^n M}{|h|-r_2}2\pi r_2 = \frac{\eta^n M r_2}{|h|-r_2}$$

Then $\lim\limits_{n\to\infty} V_n = 0$ and the proof is complete.

Thus, we have

$$f(a+h) = \sum_{n=0}^{\infty}a_n h^n + \sum_{m=0}^{\infty}\frac{b_m}{h^m} \tag{6.37}$$

where $$a_n = \frac{1}{2\pi i}\int_{C_1}\frac{f(z)}{(z-a)^{n+1}}dz, \quad n = 0,1,2,\dots$$

and $$b_m = \frac{1}{2\pi i}\int_{C_2}(z-a)^{m-1}f(z)dz, \quad m = 1,2,3,\dots \tag{6.38}$$

Note: The Laurent's series is unique only for a specified annulus. In general, a function $f(z)$ may possess two or more entirely different Laurent's series about a given point valid for different (nonoverlapping) regions. For instance,

$$\frac{1}{z(1-z)} = \frac{1}{z} + 1 + z + z^2 + z^3 + \cdots \quad [0 < |z| < 1]$$

$$\frac{1}{z(1-z)} = -\frac{1}{z^2} - \frac{1}{z^3} - \frac{1}{z^4} - \cdots \quad [1 < |z| < \infty]$$

6.12. COMMON TECHNIQUES FOR THE CONSTRUCTION OF TAYLOR AND LAURENT SERIES

6.12.1. Use of Geometric Series

$$f(z) = \frac{1}{z-a} \qquad (a = \text{nonzero complex constant})$$

$$\frac{1}{1-z} = 1 + z + z^2 + z^3 + \cdots = \sum_{n=0}^{\infty} z^n \qquad (|z| < 1)$$

Therefore $\qquad f(z) = \frac{1}{z-a} = -\left(\frac{1}{a}\right)\frac{1}{1-z/a} = -\frac{1}{a}\sum_{n=0}^{\infty}\left(\frac{z}{a}\right)^n \qquad (|z| < |a|)$

This is the Taylor's series of $f(z)$ about the point $z = 0$. Its radius of convergence is $R = |a|$ because at the distance R from the origin there is a point $z = a$ where $f(z)$ fails to be analytic. Therefore $f(z)$ should possess a Laurent's series about $z = 0$ which will be valid for $(|z| > |a|)$. Write

$$f(z) = \frac{1}{z-a} = \frac{1}{z}\cdot\frac{1}{1-a/z}$$

If $|z| > |a|$, then $\left|\frac{a}{z}\right| < 1$, and we can expand

$$\frac{1}{1-a/z} = \sum_{n=0}^{\infty}\left(\frac{a}{z}\right)^n \qquad (|z| > |a|)$$

Therefore $\qquad f(z) = \frac{1}{z-a} = \frac{1}{z}\sum_{n=0}^{\infty}\left(\frac{a}{z}\right)^n = \sum_{n=0}^{\infty}\frac{a^n}{z^{n+1}} \qquad (|z| > |a|)$

This is desired Laurent's series.

The function $f(z)$ can be expanded by this method about any point $z = b$. In this case write

$$f(z) = \frac{1}{z-a} = \frac{1}{(z-b)-(a-b)} \qquad (b \neq a)$$

Then either $\qquad f(z) = -\frac{1}{a-b}\sum_{n=0}^{\infty}\frac{(z-b)^n}{(a-b)^n} \qquad (|z-b| < |a-b|)$

or $\qquad f(z) = \sum_{n=0}^{\infty}\frac{(a-b)^n}{(z-b)^{n+1}} \qquad (|z-b| > |a-b|)$

6.12.2. Rational Fraction Decomposition

$$f(z) = \frac{1}{z^2 - (2+i)z + 2i}$$

The roots of denominator are $a = i$, $b = 2$, (simple and distinct). Therefore $f(z)$ fails to be analytic only at $z = i$ and $z = 2$ and should possess a Taylor series about $z = 0$ valid for $|z| < (|i| = 1)$ and two Laurent's series about $z = 0$ for $1 < |z| < 2$ and $|z| > 2$. To obtain these three series, we use the identities

$$z^2 - (2 + i)z + 2i = (z - i)(z - 2)$$

and
$$f(z) = \frac{1}{(z-i)(z-2)} = \frac{1}{(2-i)}\left(\frac{1}{z-2} - \frac{1}{z-i}\right)$$

Suppose that Laurent's series valid for $1 < |z| \leq 2$ is desired. The function $\dfrac{1}{z-2}$ should be expanded in Taylor series about $z = 0$ (Example 1). This series is in particular, valid for $1 < |z| < 2$. The function $\dfrac{1}{z-i}$ should be expanded in Laurent's series about $z = 0$ valid for $|z| > 1$ (Example 1). This series is also valid for $1 < |z| < 2$. If these two series are subtracted we may obtain $\left(\text{multiplying by } \dfrac{1}{z-1}\right)$ a series for $f(z)$ valid for $1 < |z| \leq 2$, which is the desired Laurent series.

6.12.3. Differentiation

Because of the double root of the denominator, the method applied in subsection (6.12.2) fails here. Among the alternative methods the simplest one is, perhaps, to observe that

$$\frac{1}{(z-1)^2} = \frac{d}{dz}\left(\frac{1}{1-z}\right)$$

The series

$$\frac{1}{1-z} = 1 + z + z^2 + \cdots = \sum_{n=0}^{\infty} z^n \qquad (|z| < 1|$$

can be differentiated term by term within the circle of convergence. Therefore

$$\frac{1}{(z-1)^2} = 1 + 2z + 3z^2 + \cdots = \sum_{n=0}^{\infty}(n+1)z^n$$

6.12.4. Integration

$$f(z) = \log(1+z) = \log|1+z| + i\arg(1+z)$$

This is the principal branch of the (multivalued) logarithmic function. The branch line extends from $-\infty$ to -1 and $\log(1 + z)$ is analytic within the circle $|z| = 1$.

We know that

$$\frac{d}{dz}\log(1+z) = \frac{1}{1+z}$$

Therefore, we may expand

$$\frac{1}{1+z} = 1 - z + z^2 - z^3 + z^4 + \cdots = \sum_{n=0}^{\infty} (-1)^n z^n \qquad (|z| < 1)$$

and integrate term by term

$$\int \frac{dz}{1+z} = z - \frac{z^2}{2} + \frac{z^3}{3} - \frac{z^4}{4} + \frac{z^5}{5} + \cdots + c \qquad (|z| < 1)$$

where c is constant of integration. Since $\log 1 = 0$, it follows that $c = 0$ and

$$\log(1 + z) = z - \frac{z^2}{2} + \frac{z^3}{3} + \cdots = \sum_{n=0}^{\infty} (-1)^{n+1} \frac{z^n}{n} \qquad (|z| < 1)$$

Other branches of $\log(1 + z)$ will have the same series except for different values of constant c.

6.13. ZEROS AND SINGULARITIES

Point $z = a$ is called a zero (or a root) of the function $f(z)$ if $f(a) = 0$; if $f(z)$ is analytic at $z = a$, then its Taylor series is given by

$$f(z) = \sum_{n=0}^{\infty} c_n (z - a)^n$$

must have $c_0 = 0$. If $c_1 \neq 0$, then point $z = a$ is called a simple zero (or a zero of order one). It may happen that c_1 and perhaps, several other next coefficients vanish. Let c_m be the first nonvanishing coefficients (unless $f(z) = 0$ such coefficients must exist), then zero is said to be of order m. The order of a zero may be evaluated (without the knowledge of Taylor series) by calculating

$$\lim_{z \to 0} \frac{f(z)}{(z - a)^n}$$

for $n = 1, 2, 3,...$ the lowest value of n for which this limit will not vanish is equal to the order of zero.

A singularity of a function is the point at which the function ceases to be analytic.

If $z = a$ is a singularity of $f(z)$, it is said to be an **isolated singularity** if there exists a neighborhood of $z = a$ containing no other singularity.

Consider a function $f(z)$ which is analytic at points of a certain region in the z-plane except at $z = a$. Surround the point $z = a$ by a small circle Γ with 'a' as centre (Fig. 6.17). Then in the ring shaped space between Γ and some larger concentric circle C', $f(z)$ can be expanded by Laurent theorem in the form

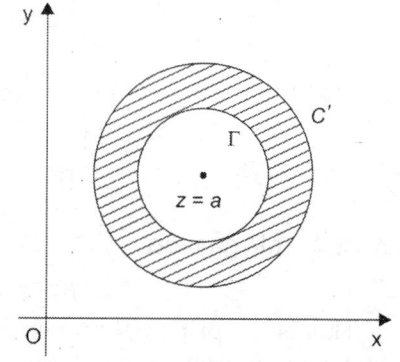

Fig. 6.17

$$f(z) = a_0 + a_1(z - a) + a_2(z - a)^2 + \cdots + a_n(z - a)^n + \cdots + \frac{b_1}{z - a} + \frac{b_2}{(z - a)^2} + \cdots + \frac{b_n}{(z - a)^n} + \cdots$$

The terms with coefficients b' is called the **principal part** of the expansion of the function at singularities. Now three cases are possible.

i. All the coefficients b_n may be zero. The function then is analytic function except at $z = a$. This is the case of **removable singularity** and is of little importance.

ii. The terms in the principal part may be unlimited in number, i.e. the series

$$\frac{b_1}{z-a} + \frac{b_2}{(z-a)^2} + \cdots \text{ may be an infinite series. In this case the point } z = a \text{ is said to be}$$

an **essential singularity** of the function $f(z)$.

iii. The principal part may contain a finite number of terms, i.e.

$$f(z) = a_0 + a_1(z-a) + \cdots + \frac{b_1}{z-a} + \frac{b_2}{(z-a)^2} + \cdots + \frac{b_n}{(z-a)^n}$$

In this case, the function is said to have a **pole of order** n. When $n = 1$, the singularity is called a **simple pole**. The coefficient b_1 is of great importance in some investigations. In all cases it is called the **residue** of $f(z)$ at the singularity $z = a$.

Mathematically a function $f(z)$ is said to have a zero of order n at $z = a$, if

$$f(a) = \left.\frac{df(z)}{dz}\right|_{z=a} = \cdots = \left.\frac{d^{n-1}f(z)}{dz^{n-1}}\right|_{z=a} = 0$$

but

$$\left.\frac{d^n f(z)}{dz^n}\right|_{z=a} \neq 0$$

Then the first n coefficients in Taylor's expansion of $f(z)$ about $z = a$ vanish so that

$$f(z) = a_n(z-a)^n + a_{n+1}(z-a)^{n+1} + \cdots$$

$$= (z-a)^n \sum_{k=0}^{\infty} a_{n+k}(z-a)^k$$

$$= (z-a)^n \sum_{k=0}^{\infty} a_{n+k}\phi(z)$$

where $\phi(z)$ is analytic and vanishing at $z = a$.

If $f(z)$ is analytic in a given region R except at a point $z = a$ and

$$\lim_{z \to a} f(z) = \infty$$

then $f(z)$ is said to have a pole at $z = a$. Further, if

$$\lim_{z \to a}(z-a)^n f(z) = c$$

where c is a nonvanishing constant, the pole at $z = a$ is said to be of nth order and

$$f(z) = \frac{\phi(z)}{(z-a)^n}$$

where $\phi(z)$ is analytic at $z = a$.

If $n = 1$, a is a simple pole.

If no value of n can be found such that

$$\lim_{z \to a}(z-a)^n f(z)$$

is finite then $z = a$ is said to be an essential singularity of $f(z)$.

Example 1. *Expand* $f(z) = \dfrac{1}{(z+1)(z+3)}$ *in a Laurent series valid for*

(a) $1 \le |z| < 3$ (b) $|z| > 3$
(c) $0 < |z+1| < 2$ (d) $|z| < 3$.

Solution: (a) Resolving into partial fractions,

$$\frac{1}{(z+1)(z+3)} = \frac{1}{2}\left(\frac{1}{z+1}\right) - \frac{1}{2}\left(\frac{1}{z+3}\right)$$

If $|z| > 1$

$$\frac{1}{2(z+1)} = \frac{1}{2z(1+1/z)} = \frac{1}{2z}\left(1 + \frac{1}{z}\right)^{-1}$$

$$= \frac{1}{2z}\left\{1 - \frac{1}{z} + \frac{1}{z^2} - \frac{1}{z^3} + \cdots\right\}$$

$$= \frac{1}{2z} - \frac{1}{2z^2} + \frac{1}{2z^3} - \frac{1}{2z^4} + \cdots$$

If $|z| < 3$

$$\frac{1}{2(z+3)} = \frac{1}{6\left(1+\dfrac{z}{3}\right)} = \frac{1}{6}\left(1+\frac{z}{3}\right)^{-1}$$

$$= \frac{1}{6}\left(1 - \frac{z}{3} + \frac{z^2}{9} - \frac{z^3}{27} + \cdots\right)$$

$$= \frac{1}{6} - \frac{z}{18} + \frac{z^2}{54} - \frac{z^3}{162} + \cdots$$

Then the required Laurent expansion valid for both $|z| > 1$ and $|z| < 3$ is

$$\cdots - \frac{1}{2z^4} + \frac{1}{2z^3} - \frac{1}{2z^2} + \frac{1}{2z} - \frac{1}{6} + \frac{z}{18} - \frac{z^2}{54} + \frac{z^3}{162} \cdots$$

The principal part contains an infinite number of terms hence the only singularity in the region $1 < |z| < 3$ is $z = -1$ which is the essential singularity.

(b) If $|z| > 1$, we have as in part (a)

$$\frac{1}{2(z+1)} = \frac{1}{2z} - \frac{1}{2z^2} + \frac{1}{2z^3} - \frac{1}{2z^4} + \cdots$$

If $|z| > 3$

$$\frac{1}{2(z+3)} = \frac{1}{2z\left(1+\dfrac{3}{z}\right)} = \frac{1}{2z}\left(1+\frac{3}{z}\right)^{-1} = \frac{1}{2z}\left(1 - \frac{3}{z} + \frac{3^2}{z^2} - \frac{3^3}{z^3} \cdots\right)$$

Then the required Laurent expansion valid for both $|z| > 1$ and $|z| > 3$, i.e. $|z| > 3$ is by subtracting

$$\frac{1}{z^2} - \frac{4}{z^3} + \frac{13}{z^4} - \frac{40}{z^4} + \cdots$$

In this case the analytic part is absent and the region contains two singularities at $z = -1$ and $z = -3$. The singularities are essential singularities because the principal part contains an infinite number of terms.

(c) Let $z + 1 = u$, then

$$\frac{1}{(z+1)(z+3)} = \frac{1}{u(u+2)} = \frac{1}{2\left(1+\dfrac{u}{2}\right)} = \frac{1}{2u}\left(1+\frac{u}{2}\right)^{-1}$$

$$= \frac{1}{2u}\left\{1 - \frac{u}{2} + \frac{u^2}{2^2} - \frac{u^3}{2^3} + \frac{u^4}{2^4}\cdots\right\} = \frac{1}{2u} - \frac{1}{2^2} + \frac{u}{2^3} - \frac{u^2}{2^4} + \frac{u^3}{2^5}\cdots$$

$$= \frac{1}{2(z+1)} - \frac{1}{2^2} + \frac{(z+1)}{2^3} - \frac{(z+1)^2}{2^4} + \frac{(z+1)^3}{2^5}\cdots$$

Valid for $|u| < 2, u \neq 0$ or $0 < |z+1| < 2$.

The principal part contains only one term, hence the singularity at $z = -1$ is the simple pole.

(d) If $|z| < 1$

$$\frac{1}{2(z+1)} = \frac{1}{2}(1+z)^{-1} = \frac{1}{2}(1 - z + z^2 - z^3 + z^4 - \cdots)$$

If $|z| < 3$, we have by part (a)

$$\frac{1}{2(z+3)} = \frac{1}{6} - \frac{z}{18} + \frac{z^2}{54} - \frac{z^3}{162} + \cdots$$

Then the required Laurent expansion valid for both $|z| < 1$ and $|z| < 3$, i.e. $|z| < 1$ is by subtraction

$$\frac{1}{3} - \frac{4}{9}z + \frac{13}{27}z^2 - \frac{40}{81}z^3 + \cdots$$

All the coefficients of principal part are zero. Hence the singularities $z = -1$ and $z = -3$ are removable singularities.

Example 2. *Find Laurent series about the indicated singularity for each of the following functions. Name the singularity in each case and give the region of convergence for each series.*

(a) $\dfrac{e^{2z}}{(z-1)^3}; z = 1$

(b) $(z-3)\sin\left(\dfrac{1}{z+2}\right); z = -2$

(c) $\dfrac{z - \sin z}{z^3}; z = 0$

(d) $\dfrac{z}{(z+1)(z+2)}; z = -2$

(e) $\dfrac{z}{(z+1)(z+2)}; z = 3$

Solution: (a) Let $z - 1 = u$ then $z = u + 1$

$$\frac{e^{2z}}{(z-1)^3} = \frac{e^{2(u+1)}}{u^3} = \frac{e^2 \cdot e^{2u}}{u^3}$$

$$= \frac{e^2}{u^3}\left\{1 + 2u + \frac{(2u)^2}{2!} + \frac{(2u)^3}{3!} + \frac{(2u)^4}{4!} + \cdots\right\}$$

or

$$\frac{e^{2z}}{(z-1)^3} = \frac{e^2}{(z-1)^3} + \frac{2e^2}{(z-1)^2} + \frac{2^2 e^2}{(z-1)}\frac{1}{2!} + \frac{2^3 e^2}{3!} + \frac{2^4 e^2}{4!}(z-1) + \cdots$$

$z = 1$ is a pole of order 3, or triple pole.

The series converges for all values of $z \neq 1$

(b) $(z-3)\sin\left(\dfrac{1}{z+2}\right)$ $z = -2$

Let $z + 2 = u$ or $z = u - 2$. Then

$$(z-3)\sin\left(\frac{1}{z+2}\right) = (u-5)\sin\frac{1}{u}$$

$$= (u-5)\left\{\frac{1}{u} - \frac{1}{3!u^3} + \frac{1}{5!u^5}\cdots\right\}$$

$$= 1 - \frac{5}{u} - \frac{1}{3!u^2} + \frac{5}{3!}\cdot\frac{1}{u^3} + \frac{1}{5!}\cdot\frac{1}{u^4} - \frac{5}{5!}\frac{1}{u^5} + \cdots$$

$$= 1 - \frac{5}{z+2} - \frac{1}{3!(z+2)^2} + \frac{5}{3!}\frac{1}{(z+2)^2} + \frac{1}{5!}\frac{1}{(z+2)^4}\cdots$$

$z = -2$ is an essential singularity.

The series converges for all values of $z \neq -2$.

(c) $\dfrac{z - \sin z}{z^3}, z = 0$

$$\frac{z - \sin z}{z^3} = \frac{1}{z^3}\left\{z - \left(z - \frac{z^3}{3!} + \frac{z^5}{5!} - \frac{z^7}{7!} + \cdots\right)\right\}$$

$$= \frac{1}{z^3}\left\{\frac{z^3}{3!} - \frac{z^5}{5!} + \frac{z^7}{7!}\cdots\right\}$$

$$= \frac{1}{3!} - \frac{z^2}{5!} + \frac{z^4}{7!}\cdots$$

$z = 0$ is a removable singularity.

The series converges for all values of z.

(d) $\dfrac{z}{(z+1)(z+2)}, z = -2$

Let $z + 2 = u$. Then

$$\frac{z}{(z+1)(z+2)} = \frac{(u-2)}{(u-1)u} = \frac{2-u}{u}\cdot\frac{1}{(1-u)}$$

$$= \frac{2-u}{u}(1 + u + u^2 + u^3 + u^4 + \cdots)$$

$$= \frac{2}{u} - 1 + 2 - u + 2u - u^2 + 2u^2 - u^3 + \cdots$$

$$= \frac{2}{u} + 1 + u + u^2 + \cdots$$

$$= \frac{2}{z+2} + 1 + (z+2) + (z+2)^2 + \cdots$$

$z = -2$ is a pole of order one or simple pole.

The series converges for all values of z such that

$$0 < |z+2| < 1$$

(e) $\dfrac{1}{z^2(z-3)^2} ; z = 3$

Let $z - 3 = u$. Then

$$\frac{1}{z^2(z-3)^2} = \frac{1}{u^2(u+3)^2} = \frac{1}{9u^2\left(1+\dfrac{u}{3}\right)^2}$$

$$\frac{1}{9u^2}(1+u/3)^{-2} = \frac{1}{9u^2}\left\{1 + (-2)(u/3) + \frac{(-2)(-3)}{2!}(u/3)^2 + \frac{(-2)(-3)(-4)}{3!}\left(\frac{u}{3}\right)^3 + \cdots\right\}$$

$$= \frac{1}{9u^2} - \frac{2}{27u} + \frac{1}{27} - \frac{4}{343}u + \cdots$$

or $\dfrac{1}{z^2(z-3)^2} = \dfrac{1}{9(z-3)^2} - \dfrac{2}{27(z-3)} + \dfrac{1}{27} - \dfrac{4}{343}(z-3) + \cdots$

$z = 3$ is a pole of order 2 or double pole.

The series converges for all values of z such that

$$0 < |z-3| < 3$$

Example 3. *By using the integral representation of $f^n(a)$, prove that* $\dfrac{x^n}{n!} = \dfrac{1}{2\pi i}\displaystyle\int_C \dfrac{e^{xz}}{z^{n+1}}dz,$

where C is any closed contour surrounding the origin. Hence prove that

$$\int_0^{2\pi} e^{2x\cos\theta}d\theta = 2\pi \sum_{n=0}^{\infty} \left(\frac{x^4}{n!}\right)^2$$

Solution: We have

$$f^n(a) = \frac{n!}{2\pi i}\int_C \frac{f(z)dz}{(z-a)^{n+1}}$$

$\therefore$ $\qquad f^n(0) = \dfrac{n!}{2\pi i}\displaystyle\int_C \dfrac{f(z)dz}{z^{n+1}}$ \hfill (1)

Now if we take $f(z) = e^{xz}$

Then differentiating n times w.r.t. z, we have

$$f^n(z) = x^n \cdot e^{xz}$$

and
$$f^n(0) = x^n \qquad\qquad (2)$$

Comparing Eqs (1) and (2), we have

$$x^n = \frac{n!}{2\pi i} \int_C \frac{e^{xz}}{z^{n+1}} dz$$

or
$$\frac{x^n}{n!} = \frac{1}{2\pi i} \int_C \frac{e^{xz}}{z^{n+1}} dz$$

Hence the result.

Now multiplying both sides by $\dfrac{x^n}{n!}$, we have

$$\left\{\frac{x^n}{n!}\right\}^2 = \frac{1}{2\pi i} \int_C \frac{x^n}{n!} \frac{e^{xz}}{z^{n+1}} dz$$

or
$$\sum_{n=0}^{\infty} \left(\frac{x^n}{n!}\right)^2 = \frac{1}{2\pi i} \int_C \left\{\frac{e^{xz}}{z} dz \sum_{n=0}^{\infty} \frac{x^n}{n! z^n}\right\}$$

$$= \frac{1}{2\pi i} \int_C \left\{\frac{e^{xz}}{z} dz \sum \frac{(x/n)^n}{n!}\right\}$$

$$= \frac{1}{2\pi i} \int_C \left\{\frac{e^{xz}}{z} dz \cdot e^{x/z}\right\} = \frac{1}{2\pi i} \int_C \frac{e^{x\left(z+\frac{1}{z}\right)}}{z} dz$$

Now let us put $z = e^{i\theta}$ $\quad\therefore\quad \dfrac{1}{z} = e^{-i\theta}$ and $dz = i e^{i\theta}\, d\theta$

$$\therefore\quad \sum_{n=0}^{\infty} \left(\frac{x^n}{n!}\right)^2 = \frac{1}{2\pi i} \int_0^{2\pi} \frac{e^{x(e^{i\theta}+e^{-i\theta})}}{e^{i\theta}} i e^{i\theta} \cdot d\theta$$

$$= \frac{1}{2\pi} \int_0^{2\pi} e^{2x\cos\theta} \cdot d\theta$$

$$\therefore\quad \int_0^{2\pi} e^{2x\cos\theta} d\theta = 2\pi \sum_{n=0}^{\infty} \left\{\frac{x^n}{n!}\right\}^2$$

Example 4. *Prove that* $\cosh\left[\dfrac{1}{z}+z\right] = a_0 + \displaystyle\sum_{n=1}^{\infty} a_n\left(z^n + \dfrac{1}{z^n}\right)$ *where*

$$a_n = \frac{1}{2\pi} \int_0^{2\pi} \cos n\theta \cdot \cosh(2\cos\theta)\, d\theta.$$

Solution: Laurent's expansion gives

$$\cosh\left[\frac{1}{z}+z\right] = a_0 + \sum_{n=1}^{\infty} a_n z^n + \sum_{n=1}^{\infty} b_n z^{-n} \qquad\qquad (1)$$

putting $\dfrac{1}{z}$ for z, we get

$$\cosh\left[\frac{1}{z}+z\right] = a_0 + \sum_{n=1}^{\infty} a_n z^{-n} + \sum_{n=1}^{\infty} b_n z^n \qquad (2)$$

Equations (1) and (2) give $a_n = b_n$

The coefficients of expansion Eqs (1) and (2) are given by

$$a_n = \frac{1}{2\pi i}\int_C \frac{\cosh\left(z+\dfrac{1}{z}\right)}{z^{n+1}}\,dz$$

and

$$b_n = \frac{1}{2\pi i}\int_C \cosh\left(z+\frac{1}{z}\right)dz\cdot z^{n-1}$$

Now putting $z = e^{i\theta}$ by taking a circle of unit radius

$$\frac{1}{z} = e^{-i\theta} \quad\text{and}\quad z+\frac{1}{z}=2\cos\theta; \quad dz = ie^{i\theta}\,d\theta \quad\text{and}\quad \frac{dz}{z} = id\theta$$

$$\therefore \quad a_n = \frac{1}{2\pi i}\int_{C_1} \frac{\cosh(2\cos\theta)id\theta}{e^{in\theta}} = \frac{1}{2\pi}\int_0^{2\pi}\cosh(2\cos\theta)e^{-in\theta}\cdot d\theta$$

and

$$b_n = \frac{1}{2\pi}\int_C \cosh(2\cos\theta)e^{in\theta}\cdot d\theta$$

$$\therefore \quad a_n + b_n = \frac{1}{2\pi}\int_0^{2\pi}\cosh(2\cos\theta)\cdot 2\cos n\theta\, d\theta$$

$$\because \quad e^{in\theta} + e^{-in\theta} = 2\cos n\theta$$

but

$$a_n = b_n = \frac{1}{2\pi}\int_0^{2\pi}\cosh(2\cos\theta)\cdot\cos n\theta\cdot d\theta \qquad (3)$$

$$\therefore \quad \cosh\left[\frac{1}{z}+z\right] = a_0 + \sum_{n=1}^{\infty} a_n\left(z^n + \frac{1}{z^n}\right)$$

where a_n is by Eq. (3). Hence the result.

6.14. THE CALCULUS OF RESIDUE

Previously, we defined the residue of a function $f(z)$ at the pole $z = a$; to be the coefficient of $(z-a)^{-1}$ in the Laurent expansion of $f(z)$.

When $z = a$ is an isolated singular point of $f(z)$, a positive number ρ exists such that the function is analytic at each point z for which $0 < |z-a| < \rho$. In that domain, the function is represented by the Laurent expansion (Eqs 6.37 and 6.38).

$$f(z) = \sum_{n=0}^{\infty} a_n(z-a)^n + \frac{b_1}{z-a} + \frac{b_2}{(z-a)^2} + \cdots$$

where the coefficients are given by

$$a_n = \frac{1}{2\pi i}\int_C \frac{f(z)dz}{(z-a)^{n+1}}$$

and
$$b_n = \frac{1}{2\pi i}\int_C f(z)(z-a)^{n-1}dz$$

where C is any closed contour around 'a' described in positive sense, such that $f(z)$ is analytic on C and interior to C except at the point a itself is

In particular
$$b_1 = \frac{1}{2\pi i}\int_C f(z)dz \tag{6.39}$$

In the above expansion for $f(z)$, b_1 the coefficient of $(z-a)^{-1}$ is called residue of $f(z)$ at isolated singular point $z = a$ and its value is given by Eq. (6.39).

Also when $z = a$ is a simple pole (pole of order one), the coefficients
$$b_2 = b_3 = b_4 = ... = 0$$

Then
$$f(z) = \sum_{n=0}^{\infty} a_n(z-a)^n + \frac{b_1}{z-a}$$

or
$$b_1 = (z-a)f(z) - \sum_{n=0}^{\infty} a_n(z-a)^{n+1} = \lim_{z\to a}(z-a)f(z) \tag{6.40}$$

Since
$$\lim_{\substack{z\to a \\ n\to 0}} \sum^{\infty} a_n(z-a)^{n+1} = 0$$

Obviously (Eq. 6.40) gives the formula for determining the residue when $z = a$ is a simple pole.

It $z = a$ is a pole of order m, the Laurent expansion of $f(z)$ takes the form
$$f(z) = \sum_{n=0}^{\infty} a_n(z-a)^n + \sum_{k=1}^{\infty} \frac{b_k}{(z-a)^k}$$

Multiplying both sides by $(z-a)^m$, we have
$$(z-a)^m f(z) = \sum_{n=0}^{\infty} a_n(z-a)^{n+m} + b_1(z-a)^{m-1} + b_2(z-a)^{m-2} + \cdots + b_m$$

Differentiating $(m-1)$ times, we have
$$\frac{d^{m-1}}{dz^{m-1}}\{(z-a)^m f(z)\} = \sum \text{terms involving } (z-a)+(m-1)!b_1$$

∴
$$b_1 = \lim_{z\to a}\frac{1}{(m-1)!}\frac{d^{m-1}}{dz^{m-1}}\{(z-a)^m f(z)\} \tag{6.41}$$

Thus Eq. (6.41) gives the formula for determining the residue if $z = a$ is a pole of order m.

Now when $m = 1$, this formula for residue of $f(z)$ reduces to Eq. (6.40).
$$b_1 = \lim_{z\to a}(z-a)f(z)$$

Sometimes it becomes difficult to calculate limiting Eq. (6.40), particularly in the case where $f(z)$ having pole at $z = a$ has the form
$$f(z) = \frac{\phi(z)}{\psi(z)} \tag{6.42}$$

where $\phi(a) \neq 0$ and $\psi(a) = 0$, i.e. $\psi(z)$ has a simple pole at $z = a$. In this case

$$\text{res. } f(z)_{z=a} = \frac{\phi(a)}{\psi'(a)} \tag{6.43}$$

To prove this we proceed as follows.

Since $z = a$ is a simple pole of $\psi(z)$, we may have $\psi(z) = (z - a) F(z)$ where $F(a) \neq 0$.

Therefore, residue of $f(z) = \dfrac{\phi(z)}{\psi(z)}$ at $z = a$ is given by

$$\lim_{z \to a}(z-a)f(z) = \lim_{z \to a}\frac{(z-a)\phi(z)}{\psi(z)} = \lim_{z \to a}\frac{\phi(z)}{F(z)}$$

Now since ϕ and $F(z)$ are analytic at $z = a$ such that ϕ and $F(z)$ can be expanded in Taylor's series about $z = a$. Taylor expansion gives

$$\frac{\phi(z)}{F(z)} = \frac{\phi(a)+(z-a)\phi'(a)+(z-a)^2\phi''(a)/2!+\cdots}{F(a)+(z-a)F'(a)+(z-a)^2 F''(a)/2!+\cdots}$$

Therefore, residue of $f(z)$ at $z = a$ is

$$\text{res. } f(z)_{z=a} = \lim_{z \to a}(z-a) f(z)$$

$$= \lim_{z \to a}\frac{\phi(a)+(z-a)\phi'(a)+\cdots}{F(a)+(z-a)F'(a)+\cdots} = \frac{\phi(a)}{F(a)}$$

But we have

$$\psi(z) = (z - a) F(z)$$

and differentiating w.r.t z, we have

$$\psi'(z) = (z - a) F'(a) + F(z)$$

$$\therefore \qquad \psi'(z)_{z=a} = F(a) = \psi'(a)$$

$$\therefore \qquad \text{res. } f(z)_{z=a} = \frac{\phi(a)}{\psi'(a)}$$

Hence

$$b_1 = \frac{\phi(a)}{\psi'(a)}$$

The Point at Infinity

In the theory of the complex variable, it is convenient to regard infinity as a single point. The behaviour of $f(z)$ at infinity is considered by making the substitution

$$z = \frac{1}{t}$$

and expanding $f\left(\dfrac{1}{t}\right)$ at $t = 0$. We then say that $f(z)$ is analytic or has a pole or an essential singularity at infinity according as $f\left(\dfrac{1}{t}\right)$ has the corresponding property at $t = 0$.

The residue of $f(z)$ at infinity is defined as

$$b_1 = \frac{1}{2\pi i} \int_C f(z) dz \tag{6.44}$$

where C is a large circle that encloses all singularities of $f(z)$ except at $z = \infty$. The integration is taken around C in the negative sense, that is negative with respect to the origin, provided that the integral has a definite value.

If we apply the transformation

$$z = \frac{1}{t},$$

Eq. (6.44) becomes

$$b_1 = \frac{1}{2\pi i} \int_C -f\left(\frac{1}{t}\right) \frac{dt}{t^2}$$

where the integration is performed in a positive sense about a small circle whose centre is at the origin. It follows that, if

$$\lim_{t \to 0}\left[-f\left(\frac{1}{t}\right)\frac{1}{t}\right] = \lim_{z \to \infty}[-z\,f(z)]$$

has a definite value, then that value is the residue of $f(z)$ at infinity.

This definition of residue at ∞ gives an interesting result for the residue at infinity. It states that "residue of $f(z)$ at infinity is the negative of the coefficient of $\dfrac{1}{z}$ in the expansion $f(z)$ for the values of z in the neighbourhood of $z = \infty$".

To prove this, let $f(z)$ has a pole of order m at $z = \infty$. Then $f(1/z)$ has a pole of order m at $z = 0$. As such $f(1/z)$ can be expanded in Laurent's series for values of z in the annulus $0 < |z| < \rho$, where ρ is small, that is

$$f(1/z) = \sum_{n=0}^{\infty} a_n z^n + \sum_{n=1}^{\infty} b_n (1/z)^n \tag{6.44a}$$

Replacing z by $1/z$ we get expansion of $f(z)$ in the neighbourhood having $z = \infty$.

$$\therefore \qquad f(z) = \sum_{n=1}^{\infty} b_n (z)^n + \sum_{n=1}^{\infty} a_n z^{-n} \tag{6.44b}$$

so that the residue of $f(z)$ at infinity $= -\dfrac{1}{2\pi i} \int_C f(z) dz$, where C is taken counterclockwise. Thus, the residue at infinity is

$$\frac{1}{2\pi i} \int_C \left[\sum_{n=1}^{\infty} b_n z^n + \sum_{n=1}^{\infty} a_n z^{-n} \right] dz = \frac{1}{2\pi i} \int_{(-C)} \frac{a_1}{z} dz$$

Since other integrals vanish, being of the form

$$\int_C \frac{dz}{z^\rho} \quad (\rho \neq 1)$$

Hence residue is

$$\text{res} = -\frac{1}{2\pi i}\int_C \frac{a_1}{z}dz = -\frac{a_1}{2\pi i}\int_C \frac{dz}{z} = -a_1,$$

i.e. residue = negative of the coefficient of $(1/z)$ in the expansion of $f(z)$ for value of z in neighbourhood of $z = \infty$.

Residue Theorem

If $f(z)$ is analytic on and inside a closed contour C except for a finite number of isolated singularities at $z = a_1, a_2,...,a_n$ which are located inside C, then

$$\int_C f(z)dz = 2\pi i \sum_{k=1}^{n} \text{res}.f(a_k)$$

where integral is taken counterclockwise around C.

With centres at $a_1, a_2, a_3,...$ respectively, construct circles $C_1, C_2, C_3,...$ which lie entirely inside C as shown in Fig. 6.18. This can be done since $a_1, a_2, a_3,...$ are interior points. By Eq. (6.14) (consequences of Cauchy's theorem), we have

$$\int_C f(z)dz = \int_{C_1} f(z)dz + \int_{C_2} f(z)dz + \int_{C_3} f(z)dz + \cdots$$

Using Eq. (6.39), we have

$$\int_{C_1} f(z)dz = 2\pi i \quad (\text{residue of } f(z) \text{ at singularity})$$

$$= 2\pi i R_1$$

where R_1 is residue of $f(z)$ at $z = a_1$

Similarly,

$$\int_{C_2} f(z)dz = 2\pi i R_2$$

$$\int_{C_3} f(z)dz = 2\pi i R_3$$

where $R_2, R_3,...$ are the residues of $f(z)$ at singularities $z = a_2, a_3,...$ Hence, we have

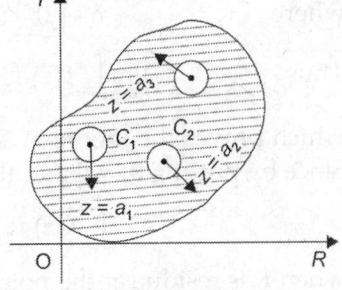

Fig. 6.18

$$\int_C f(z)dz = 2\pi i (R_1 + R_2 + R_3...)$$

$$= 2\pi i \text{ (sum of residues at the singularities)}$$

$$= 2\pi i \sum_{k=1}^{n} \text{res}.f(a_k) \tag{6.44c}$$

The proof given here establishes the residue theorem for simply-connected regions containing a finite number of singularities of $f(z)$. It can be extended to regions with infinitely many isolated singularities and to multiply-connected regions.

Two Useful Theorems Related to Cauchy's Residue Theorem

(a) If C is the arc of a semi-circle $|r-a| = \rho$ between the angles θ_1 and θ_2, $\theta_1 \leq |z-a| \leq \theta_2$ and if $\lim_{z \to \infty}(z-a)f(z) = R$ then

$$\lim_{\rho \to 0}\int_C f(z)dz = iR(\theta_2 - \theta_1)$$

where $f(z)$ is a continuous function.

Proof: Let us consider the function such that $F(z) = (z - a) f(z)$ which is obviously continuous when $f(z)$ is continuous and hence

$$|F(z) - F(a)| < \eta \qquad \text{as } |z - a| < \varepsilon$$

where $\varepsilon \to 0$ as $z \to a$. Let us choose an infinitesimal small $\delta < \eta$ such that

$$|F(z) - F(a)| = \delta \text{ or } F(z) = F(a) \pm \delta$$

But

$$F(a) = \lim_{z \to a} F(z) = \lim_{z \to a}(z - a)f(z) = R$$

and hence

$$F(z) = R \pm \delta$$

The given contour C is an arc of circle of radius ρ and centre at $z = a$. Let this arc be BC as shown in Fig. 6.18a. Then we have

$$\int_{BC} f(z)dz = \int_{BC} \frac{F(z)}{z - a}dz = \int_{BC} \frac{(R \pm \delta)}{z - a}dz \qquad (6.45a)$$

The equation of the arc BC may be taken as

$$z - a = \rho e^{i\theta}, \quad \theta_1 \le \theta \le \theta_2$$

Substituting this value in Eq. (6.45a), we get

$$\int_{BC} f(z)dz = i\int_{\theta_1}^{\theta_2}(R \pm \delta)d\theta = \int_C f(z)dz$$

or

$$\int_C f(z)dz = i(R \pm \delta)(\theta_2 - \theta_1)$$

where

$$\delta \to 0 \qquad \text{as } \varepsilon \to 0 \quad \text{or } \rho \to 0.$$

Thus

$$\lim_{\rho \to 0}\int_C f(z)dz = iR(\theta_2 - \theta_1)$$

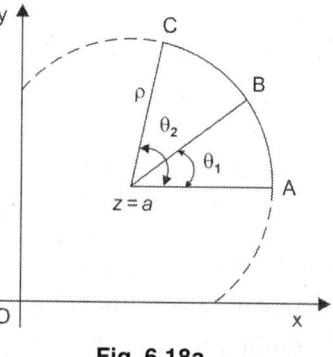

Fig. 6.18a

which proves the theorem. This theorem is general case of Cauchy residue theorem. Since by putting $\theta_1 = 0$ and $\theta_2 = 2\pi$ this reduces to

$$\int_C f(z)dz = 2\pi iR$$

when R is residue at the pole enclosed in closed contour C.

(b) Let C be the arc $\theta_1 \le \arg(z) \le \theta_2$ of the circle $|z| = R$ and if $\lim_{R \to \infty} z f(z) \to b$ uniformly then

$$\lim_{R \to \infty}\int_C f(z)dz = ib(\theta_2 - \theta_1) \qquad (6.45b)$$

Proof: Consider R sufficiently large, we may have

$$|z f(z) - b| < \varepsilon \text{ or } z f(z) = b + \eta$$

where

$$|\eta| < \varepsilon \text{ and } \eta \to 0 \text{ as } R \to \infty$$

Hence

$$\int_C f(z)dz = \int_C \frac{b + \eta}{z}dz = \int_{\theta_1}^{\theta_2}(b + \eta)id\theta$$

Since

$$z = Re^{i\theta} \text{ on arc } C.$$

$\therefore$

$$\int_C f(z)dz = bi(\theta_2 - \theta_1) + \int_{\theta_1}^{\theta_2} i\eta d\theta$$

$$\left| \int_C f(z)dz - bi(\theta_2 - \theta_1) \right| \le \int_{\theta_1}^{\theta_2} |\eta| \, d\theta \le \varepsilon(\theta_2 - \theta_1)$$

$$\lim_{R \to \infty} \left| \int_C f(z)dz - bi(\theta_2 - \theta_1) \right| \le 0$$

or
$$\lim_{R \to \infty} \int_C f(z)dz = ib(\theta_2 - \theta_1)$$

which proves the theorem.

An Extension of Cauchy's Residue Theorem

In Cauchy's residue theorem, we have assumed that all the singularities, residues of which are included in ΣR lie within the closed contour C in such a manner that any small circle drawn around any one of them does not cross or touch the contour C. If there is an isolated first order pole directly on the contour of integration, then we may deform the contour and include or exclude the residue as desired, by including a semicircular detour of infinitesimal radius as shown in Fig. 6.18b.

The integration over the semicircle gives $\pi i b_1 = \pi i R_0$, if the semicircle is transversed counterclockwise at point B or $\pi i b_1 = -\pi i R_0$ if it is transversed clockwise at A where R_0 is the residue at the concerned pole. This contribution appears on left hand side of Eq. (6.44c) and should be added to the right side also. If the detour (deletion) were clockwise, the pole would not be enclosed and there would not be the corresponding term of the right side of Eq. (6.44c). Since in this situation, the concerned pole does not lie in the region enclosed by the deformed contour. If the detour is counterclockwise, the pole is enclosed in this region and therefore a term $\pi i R_0$ should be added to the right hand side of Eq. (6.44c).

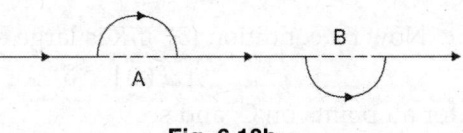

Fig. 6.18b

In this case Eq. (6.44c) becomes

$$\int_C f(z)dz = 2\pi i \sum_k R_k + \pi i R_0$$

If a finite number of simple poles $a_1, ..., a_k$ lie directly on the closed contour C, then this equation becomes

$$\int_C f(z)dz = 2\pi i \sum R + \pi i \sum (R_0) \tag{6.46}$$

Evaluation of Definite Integrals

The evaluation of definite integrals is often achieved by using the residue theorem together with a suitable function $f(z)$ and a suitable closed path or contour C, the choice of which may require ingenuity. The following types are most common in practice.

1. $\int_{-\infty}^{+\infty} F(x)dx, F(x)$ is a *rational function*.

In order to solve such type of integrals, we consider

$$I = \int_C F(z)dz$$

along a contour C consisting of the line along the x-axis from $-R$ to $+R$ and the semi-circle C above x-axis having this line as diameter (Fig. 6.19). $F(z)$ is a function that satisfies the following restrictions.

(a) It is analytic in upper half plane except at finite number of points.

(b) It has no poles on real axis.

(c) $zF(z) \to 0$ uniformly as $|z| \to \infty$ for

$0 \le \arg z \le \pi$.

(d) When x is real, $xF(x) \to 0$ as $x \to \pm \infty$ in such a way that

$$\int_{\infty}^{0} F(z)dx \text{ and } \int_{-\infty}^{0} F(x)dx$$

both converge then

$$\int_{-\infty}^{+\infty} F(x)dx = 2\pi i \sum R_k \qquad (6.46)$$

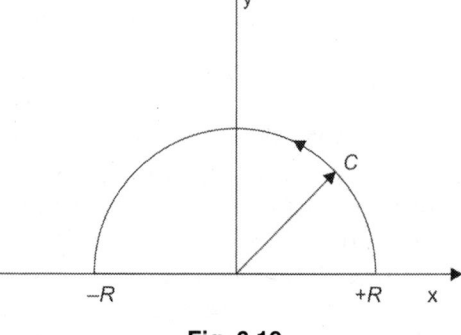

Fig. 6.19

where $\sum R_k$ denotes the sum of the residues of $F(z)$ at its poles in the upper half plane.

To prove this, refer to Fig. 6.19 and using residue theorem, we have

$$\int_{-R}^{+R} F(x)dx + \int_C F(z)dz = 2\pi i \sum R_k \qquad (6.46a)$$

Now by condition (3), if R is large enough, we have

$$|zF(z)| < \delta$$

for all points on C, and so

$$\left| \int_C F(z)dz \right| = \left| \int_0^{\pi} F(Re^{i\theta})iRe^{i\theta} \cdot d\theta \right| \le \delta \int_0^{\pi} d\theta = \delta\pi$$

Hence as $R \to \infty$, the integral around C tends to zero and if condition (4) is satisfied, we have Eq. (6.46) as

$$\int_{-\infty}^{+\infty} F(x)dx = 2\pi i \sum R_k$$

Example 1. Evaluate $I = \int_0^{\infty} \dfrac{dx}{x^6 + 1}$

Solution: (a) Consider

$$I = \oint_C \frac{dz}{z^6 + 1}$$

Here C is the closed contour of Fig. 6.19 consisting of the line from $-R$ to $+R$ and the semicircle C traversed in positive (counterclockwise) sense.

$z^6 + 1 = 0$ when $z = e^{\pi i/6}, e^{3\pi i/6}, e^{5\pi i/6}, e^{7\pi i/6}, e^{9\pi i/6}$ and $e^{11\pi i/6}$. These are simple

poles of $\dfrac{1}{z^6 + 1}$. Only the poles $e^{\pi i/6}, e^{3\pi i/6}$ and $e^{5\pi i/6}$ lie within C.

Then residue at $e^{\pi i/6}$ will be

$$= \lim_{z \to e^{\pi i/6}} \left\{ (z - e^{i\pi/6}) \frac{1}{z^6 + 1} \right\} = \lim_{z \to e^{i\pi/6}} \frac{1}{6z^5} = \frac{1}{6} e^{-5\pi i/6}$$

Similarly residue at $e^{3i\pi/6}$

$$= \lim_{x \to e^{3i\pi/6}} \frac{1}{6z^5} = \frac{1}{6} e^{-5\pi i/2}$$

and residue at $e^{5\pi i/6}$

$$= \lim_{x \to e^{5\pi i/6}} \frac{1}{6z^5} = \frac{1}{6} e^{-25\pi i/6}$$

Thus
$$\oint \frac{dz}{z^6 + 1} = \frac{2\pi i}{6} \left\{ e^{\frac{-5\pi i}{6}} + e^{\frac{-5\pi i}{2}} + e^{\frac{-25\pi i}{6}} \right\}$$

$$= \frac{2\pi}{3}$$

or
$$\int_{-R}^{+R} \frac{dx}{x^6 + 1} + \int \frac{dz}{z^6 + 1} = \frac{2\pi}{3}$$

Taking the limits as $R \to \infty$, we have

$$\int_{-\infty}^{+\infty} \frac{dx}{x^6 + 1} + \lim_{R \to \infty} \int_C \frac{dz}{z^6 + 1} = \frac{2\pi}{3} \tag{1}$$

Since on C
$$z = Re^{i\theta}.$$

$$\left| \frac{1}{z^6 + 1} \right| = \left| \frac{1}{R^6 e^{6i\theta} + 1} \right| \le \frac{1}{|R^6 e^{6i\theta}| - 1} = \frac{1}{R^6 - 1}$$

Using inequality $|z_1 + z_2| \le |z_1| - |z_2|$, with $z_1 = R^6 e^{6i\theta}$ and $z_2 = 1$.

Hence
$$\lim_{R \to \infty} \left| \int_C \frac{dz}{z^6 + 1} \right| \le \lim_{R \to \infty} \int_0^\pi \frac{|Re^{i\theta} id\theta|}{R^6 - 1}$$

$$= \lim_{R \to \infty} \int_0^\pi \frac{Rd\theta}{R^6 - 1} = \lim_{R \to \infty} \int_0^\pi \frac{d\theta}{6R^5} = \lim_{R \to \infty} \cdot \frac{\pi}{R^5} = 0$$

Equation (1) then gives
$$\int_{-\infty}^{+\infty} \frac{dx}{x^6 + 1} = 2 \int_0^{+\infty} \frac{dx}{x^6 + 1} = \frac{2\pi}{3}$$

or
$$\int_0^{+\infty} \frac{dx}{x^6 + 1} = \frac{\pi}{3}.$$

(b) $\int_0^{2\pi} G(\sin\theta, \cos\theta) d\theta$

$G(\sin\theta, \cos\theta)$ is a rational function of $\sin\theta$ and $\cos\theta$ that is finite in the range of integration, may be evaluated by transformation

$$e^{i\theta} = z$$

Then
$$\sin\theta = \frac{z - z^{-1}}{2i}, \quad \cos\theta = \frac{z + z^{-1}}{2} \quad \text{and} \quad dz = ie^{i\theta} \cdot d\theta.$$

or
$$d\theta = \frac{dz}{iz}$$

The given integral is equivalent to

$$I = \int_C F(z)dz = 2\pi i \sum R_k$$

where C is unit circle with centre at origin.

Example 2. *Evaluate*

$$I = \int_0^{2\pi} \frac{\sin^2\theta}{a+b\cos\theta}\,d\theta = \frac{2\pi}{b^2}\left\{a - \sqrt{a^2-b^2}\right\}$$

if $a > b > 0.$

Solution: Let $e^{i\theta} = z$

$$\frac{1}{z} = e^{-i\theta} \quad \text{and} \quad d\theta = \frac{dz}{iz}$$

$\therefore$
$$I = \frac{1}{i}\int_C \frac{i^2\left(z-\dfrac{1}{z}\right)^2}{2z\left\{2a + b\left(z+\dfrac{1}{z}\right)\right\}}\,dz = \frac{i}{2b}\int_C \frac{\left(z-\dfrac{1}{z}\right)^2}{z\left(z^2 + \dfrac{2az}{b} + 1\right)}\,dz$$

then
$$I = \frac{i}{2b}\int_C \frac{(z^2-1)^2\,dz}{z^2(z-p)(z-q)}$$

where
$$p = \frac{-a + \sqrt{a^2-b^2}}{b} \quad \text{and} \quad q = \frac{-a - \sqrt{a^2-b^2}}{b}$$

are the two roots of the quadratic equation

$$z^2 + \frac{2az}{b} + 1 = 0$$

It is seen that $z = p$ is the only simple pole of the integrand inside the unit circle C, and the origin is a pole of order 2. We must now compute the residue of

$$f(z) = \frac{(z^2-1)^2}{z^2(z-p)(z-q)}$$

at the poles $z = p$ and $z = 0$.
The residue at $z = p$ may be evaluated by using Eq. (6.40). We thus obtain

$$\operatorname*{Res}_{z=p} \cdot f(z) = \lim_{z\to p} \frac{(z^2-1)^2}{z^2(z-q)}$$

$$= \frac{(p^2-1)^2}{p^2(p-q)}$$

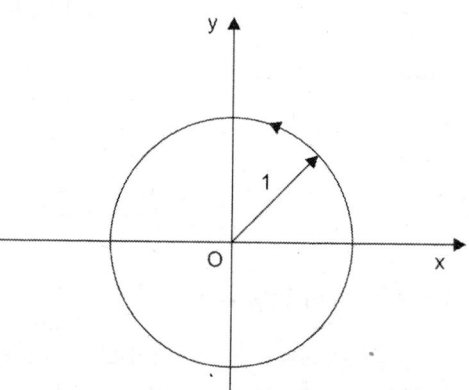

Fig. 6.20

$$= \frac{(p - 1/p)^2}{p - q} = \frac{(p - q)^{2*}}{(p - q)} = p - q = \frac{2\sqrt{a^2 - b^2}}{b}$$

The residue at the double pole $z = 0$ may be evaluated by Eq. (6.41)

$$\text{Res·} f(z) = \lim_{z \to 0} \frac{d}{dz} \left[z^2 f(z) = -\frac{2a}{b} \right]$$

$$I = \frac{i}{2b} \int f(z) \, dz$$

$$= \frac{i}{2b} \cdot 2\pi i \left(-\frac{2a}{b} + \frac{2\sqrt{a^2 - b^2}}{b} \right) = \frac{2\pi}{b^2} \left\{ a - \sqrt{a^2 - b^2} \right\}$$

This proves the result.

In a similar way, we can have

$$\int_0^{2\pi} \frac{d\theta}{(a + b\cos\theta)} = \frac{4\pi}{\sqrt{a^2 - b^2}}$$

and

$$I = \int_0^{2\pi} \frac{\cos^2\theta \, d\theta}{a + b\cos\theta} = \frac{2\pi a}{b^2\sqrt{a^2 - b^2}} \left\{ a - \sqrt{a^2 - b^2} \right\} \text{ for } a > b > 0$$

3. Integrals of the type

$$\int_{-\infty}^{+\infty} F(x) \begin{Bmatrix} \cos mx \\ \sin mx \end{Bmatrix} dx$$

Before solving such type of integrals we will prove a very useful and important theorem, usually known as **Jordan's Lemma.**

Jordan's Lemma

Let $F(z)$ be a function of the complex variable z that satisfies the following conditions

  i. It is analytic in the upper half plane except at a finite number of poles

  ii. $F(z) \to 0$ uniformly as $|z| \to \infty$ for $0 < \arg z < \pi$

  iii. m is a positive number

Then $\qquad \lim_{R \to \infty} \int_c e^{imz} F(z) \, dz = 0 \qquad\qquad$ (6.47)

where c is semicircle with its centre at the origin and radius R.

$$* \ \frac{1}{p} = \frac{b}{-a + \sqrt{a^2 - b^2}} = \frac{b\left(a + \sqrt{a^2 - b^2}\right)}{\left(-a + \sqrt{a^2 - b^2}\right)\left(a + \sqrt{a^2 - b^2}\right)}$$

$$= \frac{b \cdot \left(a + \sqrt{a^2 - b^2}\right)}{-a^2 + (a^2 - b^2)}$$

$$= \frac{b\left(a + \sqrt{a^2 - b^2}\right)}{-b^2} = -\frac{\left(a + \sqrt{a^2 - b^2}\right)}{b} = q$$

Proof: For all points on C we have (Fig. 6.21)

$$z = Re^{i\theta} = R(\cos\theta + i\sin\theta) \quad (1)$$

Now $\quad dz = i\, Re^{i\theta}\, d\theta \quad (2)$

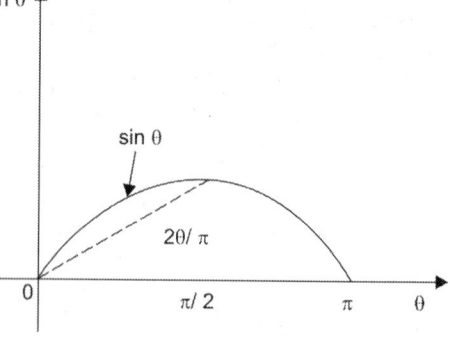

Fig. 6.21

$$\left|e^{imz}\right| = \left|e^{imR(\cos\theta + i\sin\theta)}\right| = \left|e^{-mR\sin\theta}\right|^{*}$$

By Eq. (2), if R is sufficiently large, we have for all points on C

$$|F(z)| < d, \quad \text{where } \delta \to 0 \text{ as } R \to \infty$$

Hence $\quad \left|\displaystyle\int_C F(z)e^{imz}\,dz\right| = \left|\displaystyle\int_0^\pi F(z)e^{imz}Re^{i\theta}\cdot i\,d\theta\right|$

$$\le \delta\int_0^\pi Re^{-mR\sin\theta}\,d\theta$$

$$\le 2\delta R\int_0^{\pi/2} e^{-mR\sin\theta}\,d\theta$$

From Fig. 6.21, it is seen geometrically that

$$\sin\theta \ge \frac{2\theta}{\pi} \quad \text{for } 0 \le \theta \le \frac{\pi}{2}$$

$$\left(\text{since } \frac{\sin\theta}{\theta} \text{ decreases steadily from 1 to } \frac{2}{\pi} \text{ as } \theta \text{ increases from 0 to } \frac{\pi}{2}\right)$$

Therefore $\quad \left|\displaystyle\int_C F(z)e^{imz}\,dz\right| \le 2\delta R\int_0^\pi e^{-2mR\theta/\pi}\,d\theta = \frac{\pi\delta}{m}(1 - e^{-mR})$

or $\quad \displaystyle\lim_{R\to\infty}\left|\int_C F(z)e^{imz}\,dz\right| \le \lim_{R\to\infty}\frac{\pi\delta}{m}(1 - e^{-mR}) = 0$

Example 3. Show that

$$\int_0^\infty \frac{\cos mx\, dx}{x^2 + a^2} = \frac{\pi e^{-ma}}{2a}$$

and

$$\int_{-\infty}^\infty \frac{\sin mx}{x^2 + a^2}\,dx = 0, \quad where \ m > 0.$$

Solution: Consider $I = \displaystyle\oint_C \frac{e^{imz}}{z^2 + a^2}\,dz$

where C is the contour of Fig. 6.19. The integrand has a simple pole at $z = \pm ia$ but only $z = +ia$ lies inside C. Residue at $z = ia$ is

$$\lim_{z\to ia}\left\{(z - ia)\frac{e^{imz}}{(z - ia)(z + ia)}\right\} = \frac{e^{-ma}}{2ia}$$

$*\ \left|e^{imR\cos\theta}\right|^2 = e^{imR\cos\theta}\cdot e^{-imR\cos\theta} = 1$

or $\left|e^{imR\cos\theta}\right| = 1$

Then
$$\oint_C \frac{e^{imz}\,dz}{z^2+a^2} = 2\pi i\left(\frac{e^{-ma}}{2ia}\right) = \frac{\pi e^{-ma}}{a}$$

or
$$\int_{-R}^{+R} \frac{e^{imx}}{x^2+a^2}\,dx + \int_C \frac{e^{imz}}{z^2+a^2}\,dz = \frac{\pi e^{-ma}}{a}$$

Taking Lt as $R \to \infty$, we have using Eq. 6.47.

$$\int_{-\infty}^{+\infty} \frac{(\cos mx + i\sin mx)}{(x^2+a^2)}\,dx = \frac{\pi e^{-ma}}{a}$$

Equating real and imaginary parts, we have

$$\int_{-\infty}^{+\infty} \frac{\sin mx}{x^2+a^2}\,dx = 0,$$

$$\int_{-\infty}^{+\infty} \frac{\cos mx}{x^2+a^2}\,dx = \frac{\pi e^{-ma}}{a}$$

or
$$2\int_{0}^{\infty} \frac{\cos mx}{x^2+a^2}\,dx = \frac{\pi e^{-ma}}{a}$$

$$\int_{0}^{\infty} \frac{\cos mx}{x^2+a^2}\,dx = \frac{\pi e^{-ma}}{2a}$$

Miscellaneous Definite Integrals

By extending the method used in Example 3, the integrals whose integrands have singularity on real axis can be solved.

Example 4. *Prove that*

$$\int_{0}^{\infty} \frac{\sin mx}{x}\,dx = \frac{\pi}{2}, \text{ if } m > 0$$

Solution: To prove this let us consider the integral

$$I = \oint_C \frac{e^{imz}}{z}\,dz$$

Since $z = 0$ is the singularity lying on the real axis and we cannot integrate through singularity as such, thus we modify the path at $z = 0$ as shown in Fig. 6.22 which we call contour C or ABDEFGHJA and since $z = 0$ is outside C, we have

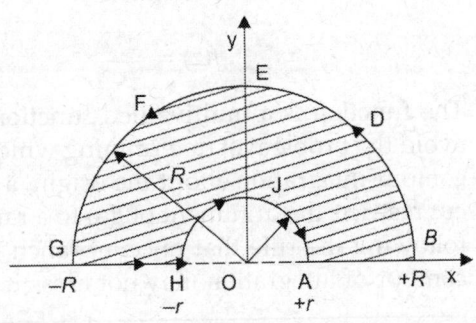

Fig. 6.22

$$\oint_C \frac{e^{imz}}{z}\,dz = 0$$

or
$$\int_{-R}^{-r} \frac{e^{imx}}{x}\,dx + \int_{HJA} \frac{e^{imz}}{z}\,dz + \int_{+r}^{+R} \frac{e^{imx}}{x}\,dx + \int_{BDEFG} \frac{e^{imz}}{z}\,dz = 0$$

Replacing x by $-x$ in the first integral and combining with 3rd integral, we get*

$$\int_r^R \frac{e^{imx} - e^{-imx}}{x} dx = -\int_{HJA} \frac{e^{imz}}{z} dz - \int_{BDEFG} \frac{e^{imz}}{z} dz \qquad (1)$$

Let $r \to 0$ and $R \to \infty$. Using Eq. (6.46), we have

$$\lim_{R \to \infty} \int_{BDEFG} \frac{e^{imz}}{z} dz \to 0$$

for

$$\lim_{r \to 0} \int_{HJA} \frac{e^{imz}}{z} dz,$$

Let $z = re^{i\theta}$ and $dz = ire^{i\theta} \cdot d\theta$

$$\lim_{r \to 0} \int_{HJA} \frac{e^{imz}}{z} dz = \lim_{r \to 0} \int_\pi^0 \frac{\varepsilon^{imre^{i\theta}}}{re^{i\theta}} \cdot ire^{i\theta} \cdot d\theta$$

$$= \lim_{r \to 0} \int_\pi^0 ie^{imre^{i\theta}} d\theta = \int_\pi^0 id\theta = -i\pi$$

Equation (1) then becomes

$$2i \int_0^\infty \frac{\sin mx}{x} dx = +i\pi$$

or

$$\int_0^\infty \frac{\sin mx}{x} dx = \frac{\pi}{2}$$

Example 5. Show that

$$\int_0^\infty \frac{x^{p-1}}{x+1} dx = \frac{\pi}{\sin px}, \quad 0 < p < 1$$

Solution:

Method I: Consider the function

$$f(z) = \frac{z^{p-1}}{z+1}$$

The function is a multivalued function with origin as the branch point. In order to avoid the problem of determining which branch the function is on when performing a contour integration about the origin, a branch cut is introduced into the z-plane. The cut restricts the argument of $f(z)$ to a range of only 2π. This is accomplished by simply following the rule that once a branch cut has been introduced into the z-plane, any contour of integration may not cross it.

---

* 1st integral $= \int_{-R}^{-r} \frac{e^{imx}}{x} dx$; let $x = -x$

∴ $dx = -dx$

$$\int_R^r \frac{e^{-imx}}{(-x)} (-dx) = -\int_r^R \frac{e^{-imx}}{x} dx$$

Method of evaluating integrals of type

$$I = \int_0^\infty x^{n-1} F(x) dx$$

is to a contour consisting of a large circle with centre at the origin and radius R. The plane must be cut along the real axis from 0 to ∞ and the branch point $z = 0$ enclosed by a small circle of radius r as shown in Fig. 6.23; where AB and GH are actually coincident with the x-axis but are shown separated for visual purposes.

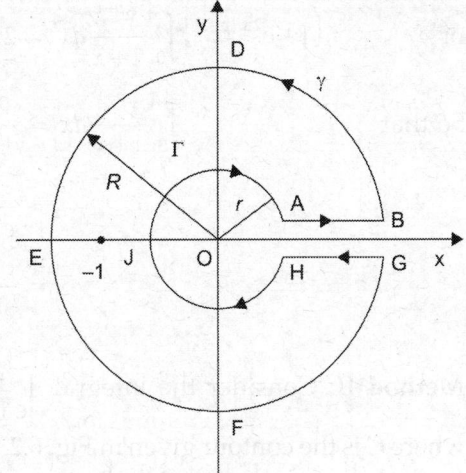

Fig. 6.23

For

$$\oint_C \frac{z^{p-1}}{z+1} dz$$

the integrand has a simple pole at $z = -1$ inside C.

Residue at $z = -1 = e^{\pi i}$ is

$$\lim_{z \to -1} (z+1) \frac{z^{p-1}}{z+1} = (e^{\pi i})^{p-1} = e^{(p-1)i\pi}$$

Then

$$\oint_C \frac{z^{p-1}}{z+1} dz = 2\pi i e^{(p-1)i\pi}$$

or

$$\int_r^R \frac{x^{p-1}}{x+1} dx + \int_0^{2\pi} \frac{(Re^{i\theta})^{p-1} iRe^{i\theta} d\theta}{1+Re^{i\theta}} + \int_R^r \frac{(xe^{2\pi i})^{p-1}}{1+xe^{2\pi i}} dx + \int_{2\pi}^0 \frac{(re^{i\theta})^{p-1} ire^{i\theta} \cdot d\theta}{1+re^{i\theta}} = 2\pi i e^{(p-1)\pi i}$$

where we have used $z = xe^{2\pi i}$ for the integral along GH since the argument of z is increased by 2π in going around the circle BDEFG[*].

Taking the limit as $r \to 0$ and $R \to \infty$ and noting that the second and fourth integrals[**] approach zero, we find

$$\int_0^\infty \frac{x^{p-1}}{1+x} dx + \int_\infty^0 \frac{x^{p-1} e^{2\pi i(p-1)}}{1+x} dx = 2\pi i e^{(p-1)i\pi}$$

---

[*] Here along the path BDEFG, $z = Re^{i\theta}$ and along the path HJA, $z = re^{i\theta}$

[**] Second integral $= \int_0^{2\pi} \frac{R^{p-1} e^{pi\theta} id\theta}{1/R + e^{i\theta}}$ taking limits as $R \to \infty, \frac{1}{R} \to 0$ and $R^{p-1} \to 0$

Since $0 \le P \le 1$, and the

hence $\lim_{R \to \infty} \int_0^{2\pi} \frac{R^{p-1} e^{pi\theta} id\theta}{\frac{1}{R} + e^{i\theta}} \to 0$ fourth integral $= \int_{2\pi}^0 \frac{r^p e^{ip\theta} id\theta}{1+re^{i\theta}}$ taking the limits as $r \to 0$

$\lim_{r \to 0} \int_0^{2\pi} \frac{r^p e^{i\theta p} id\theta}{1+re^{i\theta}} \to 0$

or

$$\left\{1 - e^{2\pi i(p-1)}\right\} \int_0^\infty \frac{x^{p-1}}{1+x}\,dx = 2\pi i e^{(p-1)i\pi}$$

So that

$$\int_0^\infty \frac{x^{p-1}}{1+x}\,dx = \frac{2\pi i e^{(p-1)i\pi}}{1 - e^{2\pi i(p-1)}} = \frac{2\pi i}{e^{-(p-1)i\pi} - e^{(p-1)\pi i}}$$

$$= \frac{2\pi i}{e^{p\pi i} - e^{-p\pi i}} \qquad\qquad (\text{since } e^{\pi i} = -1)$$

$$= \frac{\pi}{\sin P\pi}$$

Method II: Consider the integral $\int_C \frac{z^{p-1}}{1-z}\,dz$,

where C is the contour given in Fig. 6.23a, and the real axis is indented at $z = 0$ and $z = 1$, such that radii of indentation are r_1 and r_2. Now the integrand is regular within and on C, hence by Cauchy's residue theorem

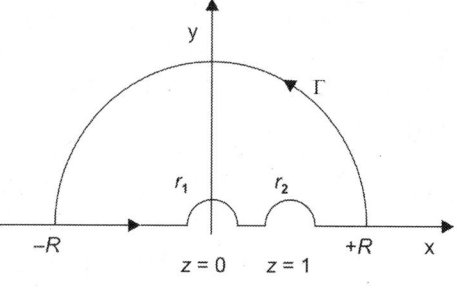

Fig. 6.23a

$$\int_{-R}^{-r_1} \frac{x^{P-1}}{1-x}\,dx + \int_{\Gamma_1} \frac{z^{P-1}}{1-z}\,dz + \int_{r_1}^{1-r_2} \frac{x^{P-1}}{1-x}\,dx + \int_{\Gamma_2} \frac{z^{P-1}}{1-z}\,dz + \int_{1+r_2}^{R} \frac{x^{P-1}}{1-x}\,dx + \int_{\Gamma} \frac{z^{P-1}}{1-z}\,dz = 0$$

or

$$\left| \int_{\Gamma} \frac{z^{P-1}}{1-z}\,dz \right| \leq \int_0^\pi \left| \frac{z^{P-1}}{1-z} \right| |\,dz\,| \leq \int_0^\pi \frac{R^P\,d\theta}{R-1} \leq \frac{R^P \cdot \pi}{R-1} \to 0 \quad \text{as } R \to \infty, \text{ since } P < 1$$

Similarly

$$\left| \int_{\Gamma_1} \frac{z^{P-1}}{1-z} \right| \leq \int_0^\pi \left| \frac{z^{P-1}}{1-z} \right| |\,dz\,| \leq \int_0^\pi \frac{r_1^P\,d\theta}{1-r_1} \leq \frac{\pi r_1^P}{1-r_1} \to 0, \quad \text{as } r_1 \to 0, \text{ since } 0 < P$$

$$\int_{\Gamma_2} \frac{z^{P-1}}{1-z}\,dz = -\int_\pi^0 \frac{(1 + r_2 e^{i\theta})}{r_2 e^{i\theta}} \cdot i r_2 e^{i\theta} \cdot d\theta, \quad \text{where } z - 1 = r_2 e^{i\theta}$$

$$= i\int_0^\pi (1 + r_2 e^{i\theta})^{P-1}\,d\theta = \pi i \quad \text{as } r_2 \to 0$$

Hence taking the limits $R \to \infty$, $r_1 \to 0$ and $r_2 \to 0$, we have

$$\int_{-\infty}^0 \frac{x^{P-1}}{1-x}\,dx + \int_0^1 \frac{x^{P-1}}{1-x}\,dx + \int_1^\infty \frac{x^{P-1}}{1-x}\,dx = -\pi i$$

or

$$\int_{-\infty}^0 \frac{x^{P-1}}{1-x}\,dx + \int_0^\infty \frac{x^{P-1}}{1-x}\,dx = -\pi i \qquad\qquad (1)$$

It is emphasised here that $z = 0$ is the branch point of the integrand $\dfrac{z^{P-1}}{1-z}$ and hence the two integrals of Eq. (1) cannot be combined directly. However, we may replace x by $-x$ in the first integral. Then this integral becomes

$$-\int_{\infty}^{0}(-1)^{P-1}\frac{x^{P-1}}{1+x}dx = -\int_{0}^{\infty}(-1)^{P}\frac{x^{P-1}}{1+x}dx$$

$$= -\int_{0}^{\infty}e^{i\pi P}\frac{x^{P-1}}{1+x}dx, \quad \text{since } e^{i\pi} = -1.$$

Thus Eq. (1) becomes

$$-e^{i\pi P}\int_{0}^{\infty}\frac{x^{P-1}}{1+x}dx + \int_{0}^{\infty}\frac{x^{P-1}}{1-x}dx = -\pi i$$

or $$(\cos\pi P + i\sin\pi P)\int_{0}^{\infty}\frac{x^{P-1}}{1+x}dx - \int_{0}^{\infty}\frac{x^{P-1}}{1-x}dx = +\pi i$$

Equating imaginary quantities, we have

$$\sin\pi P\int_{0}^{\infty}\frac{x^{P-1}}{1+x}dx = \pi$$

or $$\int_{0}^{\infty}\frac{x^{P-1}}{1+x}dx = \frac{\pi}{\sin\pi P}. \text{ Hence the result.}$$

Equating real parts, we have

$$\cos\pi P\int_{0}^{\infty}\frac{x^{P-1}}{1+x}dx - \int_{0}^{\infty}\frac{x^{P-1}}{1-x}dx = 0$$

or $$\int_{0}^{\infty}\frac{x^{P-1}}{1-x}dx = \cos\pi P \cdot \frac{\pi}{\sin P\pi} = \pi\cot P\pi$$

Example 6. *Prove that*

$$\int_{0}^{\infty}\frac{\log(x^2+1)}{x^2+1}dx = \pi\log 2.$$

Solution. Consider $\oint_{C}\dfrac{\log(z+i)}{z^2+1}dz$ around the contour C consisting of the real axis from $-R$ to $+R$ and semicircle C of radius R (Fig. 6.24).

The only pole of $\dfrac{\log(z+i)}{z^2+1}$ inside C is the simple pole at $z = i$, and residue is

$$\lim_{z\to i}\frac{(z-i)\log(z+i)}{(z-i)(z+i)} = \frac{\log(2i)}{2i}$$

Hence by residue theorem

$$\int_{C}\frac{\log(z+i)}{z^2+1}dz = 2\pi i\left\{\frac{\log 2i}{2i}\right\}$$

$$= \pi\log 2i = \pi\log 2 + \frac{1}{2}\pi^2 i$$

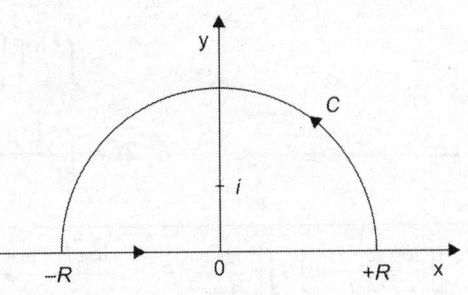

Fig. 6.24

On writing $\qquad \log(2i) = \log 2 + \log i = \log 2 + \log e^{i\pi/2}$

$$= \log 2 + i\pi/2$$

Using principal values of the logarithms, the result can be written as

$$\int_{-R}^{+R} \frac{\log(x+i)}{x^2+1} dx + \int_C \frac{\log(z+i)}{z^2+1} dz = \pi \log 2 + \frac{1}{2}\pi^2 i$$

or $\quad \int_{-R}^{0} \frac{\log(x+i)}{x^2+1} dx + \int_0^R \frac{\log(x+i)}{x^2+1} dx + \int_C \frac{\log(z+i)}{z^2+1} dz = \pi \log 2 + \frac{1}{2}\pi^2 i$

Replacing x by $-x$ in the first integral, this can be written as

$$\int_0^R \frac{\log(i-x)}{x^2+1} dx + \int_0^R \frac{\log(x+i)}{x^2+1} dx + \int_C \frac{\log(z+i)}{z^2+1} dz = \pi \log 2 + \frac{1}{2}\pi^2 i$$

or since $\log(i-x) + \log(i+x) = \log(i^2 - x^2) = \log(x^2+1) + \pi i$

$$\int_0^R \frac{\log(x^2+1)}{x^2+1} dx + \int_0^R \frac{\pi i}{x^2+1} dx + \int_C \frac{\log(z+i)}{z^2+1} dz = \pi \log 2 + \frac{1}{2}\pi^2 i$$

As $R \to \infty$, the integral around C approaches[*] zero. Hence equating real and imaginary parts, we have

$$\lim_{R\to\infty} \int_0^R \frac{\log(x^2+1)}{x^2+1} dx = \int_0^\infty \frac{\log(x^2+1)}{x^2+1} dx = \pi \log 2$$

and

$$\int_0^\infty \frac{dx}{x^2+1} = \frac{\pi}{2}$$

Remark. Let

$$I = \int_0^1 \frac{(\log(x+1/x)}{1+x^2} dx$$

$$= \int_0^\infty \frac{\log(x+1/x)}{1+x^2} dx + \int_\infty^1 \frac{\log(x+1/x)}{1+x^2} dx$$

Putting $x = 1/t$ in second integral, we have

$$I_1 = \int_\infty^1 \frac{\log(x+1/x)}{x^2+1} dx$$

$$= \int_0^1 \frac{\log(t+1/t)}{1+1/t^2}\left(-\frac{1}{t^2}\right) dt = -\int_0^1 \frac{\log(t+1/t)}{1+t^2} dt = -I$$

i.e.

$$2I = \int_0^\infty \frac{\log(x+1/x)}{1+x^2} dx$$

---

[*] $\left| \int \frac{\log(z+i)}{z^2+1} dz \right| \le \int \left| \frac{\log(z+i)}{1+z^2} \right| dz \le \frac{\log R + |\log(1+i/R)|}{R^2-1} R \int_0^\pi d\theta$

$$\le \frac{R\left[\log(R) + \frac{1}{2}\log(1+1/R^2)\right]}{R^1-1} \to 0 \text{ as } R \to \infty$$

Since $R^{-1} \log R \to 0$ as $R \to \infty$.

or
$$I = \frac{1}{2}\int_0^\infty \frac{\log(x^2+1)}{1+x^2}dx - \frac{1}{2}\int_0^\infty \frac{\log x}{1+x^2}dx$$

$$= \frac{\pi\log 2}{2}$$

Since
$$\int_0^\infty \frac{\log x}{1+x^2}dx = 0$$

Example 7. *By integrating* $(\log z)^2/(1+z^2)$ *round a suitable contour, prove that*

i. $\displaystyle\int_0^\infty \frac{(\log x)^2}{1+x^2}dx = \pi^3/8$

ii. $\displaystyle\int_0^\infty \frac{(\log x)}{1+x^2}dx = 0$

Solution: Consider the integral

$$I = \oint_C \frac{(\log z)^2}{z^2+1}dz$$

where C is the contour given in (Fig. 6.25), where origin has been indented by a small semicircle Γ of radius r. Obviously, the only pole lying in the contour is $z = i$ and residue is

$$\lim_{z\to i}\frac{(z-i)(\log z)^2}{(z+i)(z-i)} = \frac{(\log i)^2}{2i}$$

$$= \frac{(i\pi/2)^2}{2i} = -\frac{\pi^2}{8i}$$

Hence by residue theorem

$$I = \oint_C \frac{(\log z)^2}{z^2+1}dz = \int_{-R}^{-r}\frac{(\log x)^2}{x^2+1}dx + \int_\Gamma \frac{(\log z)^2}{z^2+1}dz + \int_r^R \frac{(\log x)^2}{x^2+1}dx$$

$$+\int_\gamma \frac{(\log z)^2}{z^2+1}dz = -\frac{\pi^2}{8i}\cdot 2\pi i = -\frac{\pi^3}{4} \tag{1}$$

But $\displaystyle\left|\int_\Gamma \frac{(\log z)^2}{z^2+1}dz\right| \le \int_\pi^0 \frac{\left|\log(re^{i\theta})\right|^2}{|z^2+1|}|dz|$

$$\le \int_\pi^0 \frac{\left[(\log r)^2+\theta^2\right]}{r^2-1}rd\theta$$

Since $|dz| = |re^{i\theta}id\theta| = rd\theta$

$$\le \int_0^\pi \frac{r\left[(\log r)^2+\theta^2\right]}{1-r^2}d\theta$$

or $\displaystyle\left|\int_\Gamma \frac{(\log z)^2}{z^2+1}dz\right| \le \frac{r\left[(\log(r))^2\pi + \dfrac{\pi^3}{3}\right]}{1-r^2}$

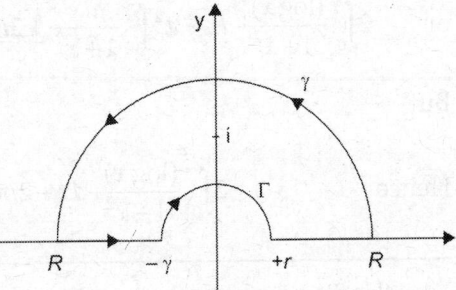

Fig. 6.25

and $\quad\displaystyle\lim_{r\to 0}\left|\int_{\Gamma}\frac{(\log z)^2}{z^2+1}dz\right| \le \lim_{r\to 0}\frac{r\left[(\log (r))^2\pi+\dfrac{\pi^3}{3}\right]}{1-r^2}\to 0$

Since $\quad\displaystyle\lim_{r\to 0}\frac{r(\log r)^2}{1-r^2} = \lim_{r\to 0}\frac{2\cdot\log r}{1r-r}$

$$= 2\lim_{r\to 0}\frac{1}{r}\cdot\frac{1}{\left(-\dfrac{1}{r^2}\right)-1}$$

$$= -2\lim_{r\to 0}\cdot\frac{r}{1+r^2} = 0$$

and $\quad\displaystyle\left|\int_{\gamma}\frac{(\log z)^2}{z^2+1}dz\right| \le \int_0^{\pi}\frac{\left|\log (Re^{i\theta})\right|^2}{|z^2+1|}|dz|$

$$\le \int_0^{\pi}\frac{R\left[(\log R)^2+\theta^2\right]d\theta}{R^2-1}$$

$$\le \frac{R\left[(\log R)^2\pi+\pi^3/3\right]}{R^2-1}$$

or $\quad\displaystyle\lim_{R\to\infty}\left|\int_{\gamma}\frac{(\log z)^2}{z^2+1}dz\right| \le \lim_{R\to\infty}\frac{R\left[(\log R)^2\pi+\pi^3/3\right]}{R^2-1}\to 0$

Since $\quad\displaystyle\lim_{R\to\infty}\frac{R\cdot\left[(\log R)^2\pi+\pi^3/3\right]}{R^2-1} = \lim_{R\to\infty}\left[\frac{R^{-1}\left[(\log R)^2\pi+\pi^3/3\right]}{1-1/R^2}\right]^* = 0$

Hence as $r\to 0$ and $R\to\infty$, we have from Eq. (1)

$$\int_{-\infty}^0\frac{(\log x)^2}{1+x^2}dx+\int_0^{\infty}\frac{(\log x)^2}{1+x^2}dx = -\frac{\pi^3}{4}$$

Let $x = -x = xe^{i\pi}$ for first integral

$$\int_0^{\infty}\frac{[\log xe^{i\pi}]^2}{1+x^2}dx+\int_0^{\infty}\frac{(\log x)^2}{1+x^2}dx = -\pi^3/4$$

or $\quad\displaystyle 2\int_0^{\infty}\frac{(\log x)^2}{1+x^2}dx-\pi^2\int_0^{\infty}\frac{dx}{1+x^2}+2i\pi\int_0^{\infty}\frac{\log x}{1+x^2}dx = -\pi^3/4$

But $\quad\displaystyle\int_0^{\infty}\frac{dx}{1+x^2} = \pi/2$ $\qquad$ (*refer to* Example 6)

Hence $\quad\displaystyle 2\int_0^{\infty}\frac{(\log x)^2}{1+x^2}dx+2i\pi\int_0^{\infty}\frac{\log x}{1+x^2}dx = -\frac{\pi^3}{4}+\frac{\pi^3}{2} = \frac{\pi^3}{4}$

$^*\displaystyle\lim_{R\to\infty}\frac{(\log R)^2}{R} = \lim_{R\to\infty}\frac{2\log R}{R} = \lim_{R\to\infty}\frac{2}{R} = \lim_{R\to\infty}\frac{2}{R} = 0$

Equating real and imaginary parts, we have

$$2\int_0^\infty \frac{(\log x)^2}{1+x^2}dx = \frac{\pi^3}{4}$$

or

$$\int_0^\infty \frac{(\log x)^2}{1+x^2}dx = \frac{\pi^3}{8}$$

and

$$\int_0^\infty \frac{\log x}{1+x^2}dx = 0$$

Example 8. If $-1 < p < 1$ and $-\pi < \lambda < \pi$ show by contour integration that

$$\int_0^\infty \frac{x^{-p}dx}{1+2x\cos\lambda+x^2} = \frac{\pi}{\sin p\pi}\frac{\sin p\lambda}{\sin\lambda}$$

Explain clearly the necessity of conditions on P and λ.

Solution: Let

$$f(z) = \frac{z^{-p}}{1+2z\cos\lambda+z^2}$$

Here $z = 0$ is a branch point and we take double circle contour (*see* Fig. 6.23)

Since

$$f(z) = \frac{z^{-p}}{1+z(e^{i\lambda}+e^{-i\lambda})+z^2}$$

$$= \frac{z^{-p}}{e^{i\lambda}\cdot e^{-i\lambda}+z(e^{i\lambda}+e^{-i\lambda})+z^2}$$

$$= \frac{z^{-p}}{(z+e^{i\lambda})(z+e^{-i\lambda})}$$

Hence the singularities within the contour consist of simple poles at

$$z = -e^{i\lambda}$$

and

$$z = -e^{-i\lambda}$$

Residue at

$$z = -e^{i\lambda} \text{ is}$$

$$\lim_{z\to e^{i\lambda}}\frac{(z+e^{i\lambda})z^{-p}}{(z+e^{-i\lambda})(z+e^{i\lambda})} = \frac{(-1)^{-p}e^{-i\lambda p}}{-e^{+i\lambda}+e^{-i\lambda}}$$

$$= \frac{e^{-P\pi i}e^{-i\lambda P}}{-2i\sin\lambda} = \frac{e^{-iP(\pi+\lambda)}}{2i\sin\lambda}$$

Similarly residue at $z = -e^{-i\lambda}$ is $\dfrac{e^{-iP(\pi-\lambda)}}{2i\sin\lambda}$

Therefore, sum of residues

$$= \frac{e^{-iP\pi}}{2i\sin\lambda}\{-e^{-ip\lambda}+e^{ip\lambda}\}$$

$$= e^{-iP\pi}\frac{\sin p\lambda}{\sin\lambda}$$

Now
$$\int_r^R f(x)dx + \int_\gamma f(z)dz + \int_R^r f(xe^{2\pi i})d(xe^{2\pi i}) - \int_\Gamma f(z)dz = 2\pi i R_k$$

where R_k is the sum of residues at the poles,

but
$$\left|\int_\gamma f(z)dz\right| \le \int_\gamma \frac{|z|^{-P}|dz|}{|z+e^{i\lambda}||z+e^{-i\lambda}|}$$

$$\le \frac{R^{-P}2\pi R}{(R+1)(R+1)} = -\frac{2\pi}{R^{1+P}\left(1+\dfrac{1}{R}\right)^2}$$

∵
$$|z + e^{i\lambda}| \le |z| + |e^{i\lambda}|$$
$$\le |Re^{i\theta}| + |e^{i\lambda}|$$
$$\le R + 1$$

or
$$\lim_{R\to\infty}\left|\int_\gamma f(z)dz\right| \le \lim_{R\to\infty}\frac{2\pi}{R^{1+P}\left(1+\dfrac{1}{R}\right)^2} \to 0$$

Since $-1 < P < 1$

Hence under limit $r \to 0$ and $R \to \infty$, we have

$$\int_0^\infty \frac{x^{-P}}{1+2x\cos\lambda + x^2}dx - \int_0^\infty \frac{(xe^{2\pi i})^{-P}e^{2\pi i}dx}{1+2xe^{2\pi i}\cos\lambda + (xe^{2\pi i})^2} = 2\pi i e^{-P\pi i}\frac{\sin p\lambda}{\sin\lambda}$$

or
$$(1-e^{-2\pi i P})\int_0^\infty \frac{x^{-P}dx}{1+2x\cos\lambda + x^2} = 2\pi i e^{-P\pi i}\frac{\sin p\lambda}{\sin\lambda}$$

or
$$\int_0^\infty \frac{x^{-P}}{1+2x\cos\lambda + x^2}dx = 2\pi i\frac{\sin p\lambda}{\sin\lambda}\frac{e^{-P\pi i}}{1+e^{-2\pi P i}}$$

$$= \frac{\sin p\lambda}{\sin\lambda}\frac{2\pi i}{e^{P\pi i} - e^{-\pi P i}}$$

$$= \frac{\sin p\lambda}{\sin\lambda}\frac{\pi}{\sin P\pi}$$

The singularities within the contour consist of simple poles $z = e^{i(\pi - \lambda)}$ and $e^{i(\pi + \lambda)}$ and amplitudes of these lie between 0 to 2π under the condition $-p < \lambda < \pi$.

Example 9. *Prove that*

$$\int_0^\infty \sin x^2 dx = \int_0^\infty \cos x^2 dx = \frac{1}{2}\sqrt{\pi/2}$$

Solution: Let C be contour as indicated in Fig. 6.26, where AB is the arc of a circle with centre at O and radius R. By Cauchy's theorem $\int_C e^{iz^2}dz = 0$.

or
$$\int_{OA} e^{iz^2}dz + \int_{AB} e^{iz^2}dz + \int_{BO} e^{iz^2}dz = 0$$

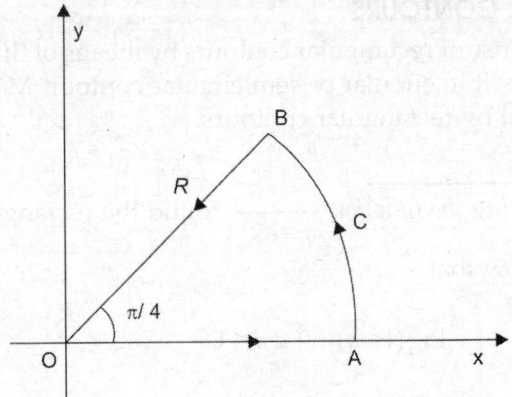

Fig. 6.26

Now on OA, $z = x$ (from $x = 0$ to $x = R$) on AB, $z = Re^{i\theta}$ (from $\theta = 0$ to $\theta = \pi/4$), on BO $z = re^{i\pi/4}$ (from $r = R$ to $r = 0$).

Hence, we have

$$\int_0^R e^{ix^2}dx + \int_0^{\pi/4} e^{iR^2e^{2i\theta}} \cdot iRe^{i\theta}\,d\theta + \int_R^0 e^{ir^2e^{i\pi/2}}e^{i\pi/4}dr = 0$$

$$\int_0^R (\cos x^2 + i\sin x^2)dx = \int_0^R e^{i\pi/4}e^{-r^2}dr - \int_0^{\pi/4}e^{(iR^2\cos 2\theta - R^2\sin 2\theta)}iRe^{i\theta}d\theta$$

Taking the limit as $R \to \infty$, we have

$$e^{i\pi/4}\int_0^\infty e^{-r^2}dr = e^{i\pi/4}\frac{\sqrt{\pi}}{2} = \frac{\sqrt{\pi}}{2}\left\{\cos\frac{\pi}{4} + i\sin\frac{\pi}{4}\right\}$$

$$= \frac{\sqrt{\pi}}{2}\left(\frac{1}{\sqrt{2}} + i\frac{1}{\sqrt{2}}\right)$$

The absolute value of the second integral on R.H.S is

$$\left|\int_0^{\pi/4}e^{(iR^2\cos 2\theta - R^2\sin 2\theta)}iRe^{i\theta}\,d\theta\right| \le \int_0^{\pi/4}e^{-R^2\sin 2\theta}R\,d\theta = \frac{R}{2}\int_0^{\pi/2}e^{-R^2\sin\phi}d\phi$$

where we have used the transformation $2\theta = \phi$.

Taking the limits $R \to \infty$ on R.H.S, approaches to zero (*see* proof of Jordon's Lemma).

Hence, we have

$$\int_0^\infty (\cos x^2 + i\sin x^2)dx = \frac{1}{2}\sqrt{\frac{\pi}{2}} + i\frac{1}{2}\sqrt{\frac{\pi}{2}}$$

Equating real and imaginary parts, we have

$$\int_0^\infty [\cos x^2]dx = \int_0^\infty \sin x^2 dx = \frac{1}{2}\sqrt{\frac{\pi}{2}}$$

6.15. RECTANGULAR CONTOURS

Now we illustrate the use of rectangular contours by means of the following examples. There is no special merit in circular or semicircular contour. Many of the above integrals can be calculated by rectangular contours.

Example 1. *By integrating the function* $\dfrac{z}{a - e^{-iz}}$ *round the rectangle with vertices at the points* $\pm\pi, \pm\pi + iR$ *show that*

$$\int_0^\pi \frac{x \sin x}{1 - 2a\cos x + a^2}\,dx = \begin{bmatrix} \dfrac{\pi}{a}\log(1 + a)\text{ if }0 < a \le 1 \\[2mm] \dfrac{\pi}{a}\log(1 + 1/a)\text{ if }a > 1 \end{bmatrix}$$

Solution: Consider the integral

$$I = \oint_C \frac{z}{a - e^{iz}}\,dz,$$

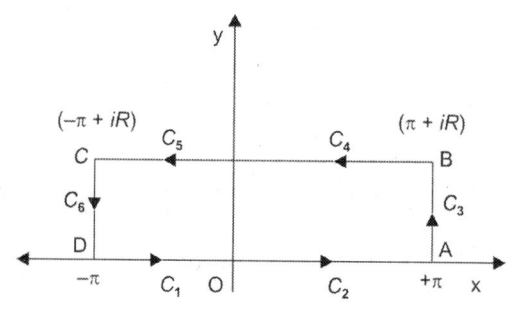

Fig. 6.27

C being the rectangle with vertices $\pm\pi, \pm\pi + iR$ (Fig. 6.27). The poles of the integrand are given by $a - e^{-iz} = 0$

or $\qquad e^{-iz} = a = ae^{2n\pi i}$

where n being any integer 0, 1, 2,...

hence $\qquad -iz = 2n\pi i + \log a$

i.e. $\qquad z = i\log a - 2n\pi.$

The only pole lying inside the contour is $i\log a$ if $a > 1$ and there is no pole inside the contour if $a \le 1$.

$$\text{Residue} = \begin{bmatrix} \dfrac{z}{\dfrac{d}{dz}\{a - e^{iz}\}} \end{bmatrix}_{z = i\log a} = \frac{i\log a}{ia} = \frac{\log a}{a}\ \text{ if } a > 1$$

Hence by Cauchy's residue theorem

$$\oint_C f(z)\,dz = \int_{-\pi}^0 \frac{x}{a - e^{-ix}}\,dx + \int_0^\pi \frac{x}{a - e^{-ix}}\,dx + \int_0^R \frac{\pi + iy}{a - e^{-i(\pi + iy)}}\,idy$$

$$+ \int_\pi^{-\pi} \frac{x + iR}{a - e^{-i(x + iR)}}\,dx + \int_R^0 \frac{-\pi + iy}{a - e^{-i(-\pi + iy)}}\,idy$$

$$= \begin{cases} \dfrac{2\pi i\log a}{a} & \text{if } a > 1 \\[2mm] 0 & \text{if } a < 1 \end{cases}$$

But $\qquad \left| \displaystyle\int_\pi^{-\pi} \frac{x + iR}{a - e^{-i(x + iR)}}\,dx \right| \le \int_\pi^{-\pi} \frac{|x| + |iR|\,dx}{a - e^R\,|e^{-ix}|}$

$$= \int_{-\pi}^\pi \frac{|x| + R}{e^R - a}\,dx \to 0 \text{ as } R \to \infty$$

and $\displaystyle\int_0^R \frac{(\pi + iy)}{a - e^{-i(\pi + iy)}} i\,dy + \int_R^0 \frac{-\pi + iy}{a - e^{-i(-\pi + iy)}} i\,dy = \int_0^{*R} \frac{(\pi + iy)i\,dy}{a + e^y} + \int_0^R \frac{\pi - iy}{a + e^y} i\,dy$

$$= \int_0^R \frac{2\pi i}{a + e^y}\,dy = 2\pi i \int_0^R \frac{e^{-y}}{ae^{-y} + 1}\,dy$$

$$= \frac{2\pi i}{a}[\log(1 + a) - \log(1 + ae^{-R})]$$

and taking the limits as $R \to \infty$, we have

$$\int_0^\infty \frac{2\pi i}{a + e^y}\,dy = \frac{2\pi i}{a}\log(1 + a)$$

Hence, we have

$$\int_{-\pi}^0 \frac{x}{a - e^{-ix}}\,dx + \int_0^\pi \frac{x}{a - e^{-ix}}\,dx = \begin{cases} -\dfrac{2\pi i}{a}\log(1 + a) + \dfrac{2\pi i}{a}\log a & \text{if } a > 1 \\[2mm] = -\dfrac{2\pi i}{a}\log(1 + a) & \text{if } a < 1 \end{cases}$$

Putting $x = -x$ in first integral, we have

$$-\int_0^\pi \frac{x}{a - e^{ix}}\,dx + \int_0^\pi \frac{x}{a - e^{-ix}}\,dx = \begin{cases} -\dfrac{2\pi i}{u}\log\left(1 + \dfrac{1}{a}\right) & \text{if } a > 1 \\[2mm] -\dfrac{2\pi i}{a}\log(1 + a) & \text{if } a < 1 \end{cases}$$

$$\text{L.H.S.} = \int_0^\pi \left\{ \frac{x}{a - e^{ix}} - \frac{x}{a - e^{ix}} \right\} dx$$

$$= \int_0^\pi \frac{-x(e^{ix} - e^{-ix})}{1 - a(e^{ix} + e^{-ix}) + a^2}\,dx = -2i\int_0^\pi \frac{x\sin x\,dx}{1 - 2\cos x + a^2}$$

or $$\int_0^\pi \frac{x\sin x}{1 - 2a\cos x + a^2}\,dx = \begin{cases} \dfrac{\pi}{a}\log\left(1 + \dfrac{1}{a}\right) & \text{if } a > 1 \\[2mm] \dfrac{\pi}{a}\log(1 + a) & \text{if } a < 1 \end{cases}$$

Example 2. *Prove that when* $0 < a < 1$

$$\int_0^\infty \frac{t^{a-1}}{1 + t}\,dt = \int_{-\infty}^{+\infty} \frac{e^{ax}}{1 + e^x}\,dx = \frac{\pi}{\sin a\pi}$$

Solution: We have already evaluated this integral (Example 5, miscellaneous definite integrals). Now we evaluate it by the use of rectangular contour.

Putting $$t = e^x$$

$$\int_0^\infty \frac{t^{a-1}}{1 + t}\,dt = \int_{-\infty}^{+\infty} \frac{(e)^{xa-x}}{1 + e^x}\cdot e^x\,dx = \int_{-\infty}^{+\infty} \frac{e^{ax}\,dx}{1 + e^x}$$

$^*e^{-i\pi} = e^{i\pi} = -1$

Now consider $I = \oint_C \dfrac{e^{az}}{1+e^z} dz$, where C is the rectangle ABCD with vertices at R, $R + 2\pi i$, $-S + 2\pi i$, $-S$. Here R and S are positive (Fig. 6.28). Clearly $f(z) = \dfrac{e^{az}}{1+e^z}$ has simple poles given by

$$e^z = -1 = e^{(2n+1)i\pi}$$

or $\qquad z = (2n+1)i\pi$

The only pole inside the rectangle is $z = \pi i$

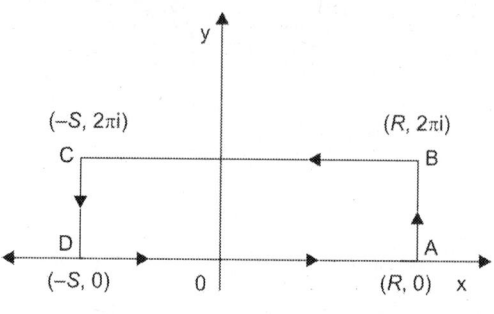

Fig. 6.28

Residue at $z = \pi i$ is $\left[\dfrac{e^{az}}{\dfrac{d}{dz}(1+e^z)} \right]_{z=\pi i} = \dfrac{e^{a\pi i}}{e^{\pi i}} = -e^{a\pi i}$

∴ By residue theorem, we get

$$\oint_C \dfrac{e^{az}}{1+e^z} dz = \int_{-S}^{R} \dfrac{e^{ax}}{1+e^x} dx + \int_0^{2\pi} \dfrac{e^{a(R+iy)}}{1+e^{(R+iy)}} i\, dy + \int_R^{-S} \dfrac{e^{ax} \cdot e^{2\pi ai}}{1+e^{x+2\pi i}} dx$$

$$+ \int_{2\pi}^0 \dfrac{e^{a(-S+iy)}}{1+e^{(-S+iy)}} i\, dy = 2\pi i(-e^{ai\pi})$$

∵ $z = x$ along DA, $z = R + iy$ along AB, $z = x + 2\pi i$ along BC
$z = -S + iy$ along CD. Hence

$$\int_{-S}^{R} \dfrac{e^{ax}(1-e^{2\pi ia})}{1+e^x} dx + \int_0^{2\pi} \dfrac{e^{a(R+iy)}}{1+e^{(R+iy)}} i\, dy - \int_0^{2\pi} \dfrac{e^{a(-S+iy)}}{1+e^{(-S+iy)}} i\, dy = -2\pi i\, e^{a\pi i}$$

Now $\qquad \left| \displaystyle\int_0^{2\pi} \dfrac{e^{a(R+iy)}}{1+e^{(R+iy)}} i\, dy \right| \leq \int_0^{2\pi} \dfrac{|e^{aR}| \, |e^{iay}|}{1+|e^R| \cdot |e^{iy}|} |i| \, dy$

$$\leq \int_0^{2\pi} \dfrac{e^{aR}}{e^R + 1} dy = \dfrac{2\pi e^{aR}}{e^R + 1} \to 0 \qquad \text{as } R \to \infty \text{ (since } 0 < a < 1)$$

and $\qquad \left| \displaystyle\int_0^{2\pi} \dfrac{e^{a(-S+iy)}}{1+e^{(-S+iy)}} i\, dy \right| \leq \int_0^{2\pi} \dfrac{e^{-aS}}{1-e^{-S}} dy$

$$= \dfrac{2\pi e^{-aS}}{1-e^{-S}} \to 0 \text{ as } S \to \infty \qquad\qquad [\text{since } a > 0]$$

Hence as $R \to \infty$ and $S \to \infty$, we get

$$\int_{-\infty}^{+\infty} \dfrac{e^{ax}(1-e^{2\pi ia})}{1+e^x} dx = -2\pi i\, e^{a\pi i}$$

or $\qquad \displaystyle\int_{-\infty}^{+\infty} \dfrac{e^{ax}}{1+e^x} dx = -\dfrac{2\pi i\, e^{a\pi i}}{1-e^{2\pi ia}} = -\dfrac{2\pi i}{e^{-\pi ia} - e^{\pi ia}} = \dfrac{\pi}{\dfrac{e^{\pi/a} - e^{-\pi ia}}{2i}} = \dfrac{\pi}{\sin \pi a}$

Example 3. By integrating $\dfrac{e^{az}}{e^{-2iz}-1}$ round a suitable contour, show that

$$\int_0^\infty \frac{\sin ay}{e^{2y}-1}\,dy = \frac{1}{4}\pi\coth(\pi a/2) - \frac{1}{2a}$$

Solution: Consider the integral $I = \displaystyle\oint_C \frac{e^{az}}{e^{-2iz}-1}\,dz$, C being the rectangle indented at the origin and $z = p$, shown in (Fig. 6.29). By Cauchy's theorem

$$\oint_C \frac{e^{az}}{e^{-2iz}-1}\,dz = \int_{\rho_1}^{\pi-\rho_2}\frac{e^{ax}}{e^{-2ix}-1}\,dx + \int_{\Gamma_2}f(z)dz + \int_{\rho_2}^{R}\frac{e^{a(\pi+iy)}}{e^{-2i(\pi+iy)}-1}\,idy$$

$$+\int_\pi^0 \frac{e^{a(x+iR)}}{e^{-2i(x+iR)}-1}\,dx + \int_R^{\rho_1}\frac{e^{aiy}}{e^{2y}-1}\,idy + \int_{\Gamma_1}f(z)dz = 0$$

(since within the contour C function $f(z) = \dfrac{e^{az}}{e^{-2iz}-1}$ is regular)

or

$$I_1 + I_2 + I_3 + I_4 + I_5 + I_6 = 0$$

But

$$|I_4| = \int_\pi^0 \left|\frac{e^{ax}e^{iuR}}{e^{-2ix}\cdot e^{2R}-1}\right|dx \le \int_\pi^0 \frac{e^{ax}}{e^{2R}-1}\,dx$$

$$= \frac{1-e^{\pi a}}{(e^{2R}-1)a} \to 0 \quad\text{as } R \to \infty$$

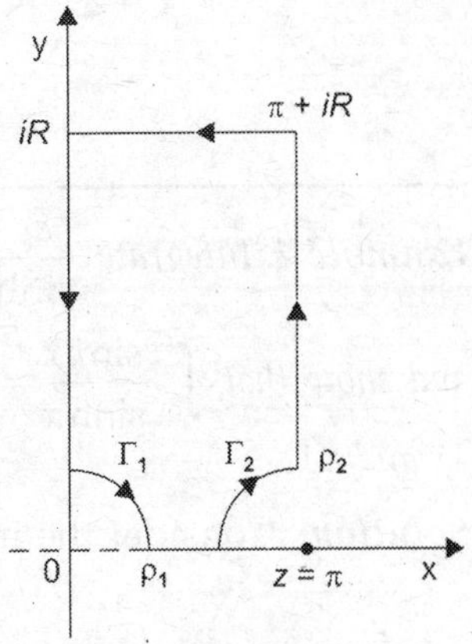

and

$$I_6 = \int_{\Gamma_1}^0 \frac{e^{az}}{e^{-2iz}-1}\,dz = \int_{\pi/2}^0 \frac{e^{a\rho_1 e^{i\theta}}}{e^{-2i\rho_1 e^{i\theta}}-1}\,\rho_1 i e^{i\theta}\cdot d\theta$$

where

$$z = \rho_1 e^{i\theta}$$

$$I_6 = \int_{\pi/2}^0 \frac{e^{a\rho_1 e^{i\theta}}\cdot\rho_1 i e^{i\theta}}{-2i\rho_1 e^{i\theta}+0\left(\rho_1^2\right)}\cdot d\theta^*$$

$$= -\int_{\pi/2}^0 \frac{d\theta}{2} = \frac{\pi}{4}\quad\text{as } r_1 \to 0$$

Fig. 6.29

$$I_2 = \int_{\Gamma_2}\frac{e^{az}}{e^{-2iz}-1}\,dz = \int_\pi^{\pi/2}\frac{e^{a\pi}\cdot e^{a\rho_2 e^{i\theta}}}{e^{-2i(\pi+\rho_2 e^{i\theta})}-1}\,\rho_2 e^{i\theta}\cdot id\theta$$

where

$$z = \pi + \rho_2 e^{i\theta}$$

$$I_2 = \int_\pi^{\pi/2}\frac{e^{a\pi}\left\{1+0\left(\rho_2^2\right)\right\}\rho_2 e^{i\theta}\cdot id\theta}{-2i\rho_2 e^{i\theta}+0\left(\rho_2^2\right)}\quad\text{since } e^{-2i\pi} = +1$$

---

*Using the relation

$e^x - 1 = x + \dfrac{x^2}{2!} + \dfrac{x^3}{3!} + \cdots = x + 0(x^2)$, where $0\,(x)^2$ is some function involving x and higher powers of x.

$$= -\int_{\pi}^{\pi/2} \frac{e^{a\pi}}{2} d\theta \quad \text{as } \rho_2 \to 0$$

$$= e^{a\pi} \cdot \frac{\pi}{4}.$$

Hence, we have

$$\int_0^{\pi} \frac{e^{ax}}{e^{-i2x}-1} dx + \int_0^{\infty} \frac{e^{\pi a} \cdot e^{iay}}{e^{2y}-1} i dy + \int_{\infty}^{0} \frac{e^{aiy}}{e^{2y}-1} i dy = -\frac{\pi}{4}(1+e^{a\pi})$$

or

$$-\int_0^{\infty} \frac{(1-e^{a\pi})e^{iay}}{e^{2y}-1} i dy + \int_0^{\pi^*} \frac{e^{ax}}{e^{-2ix}-1} dx = -\frac{\pi}{4}(1+e^{a\pi})$$

$$-(1-e^{\pi a}) \int_0^{\infty} \frac{(\cos ay + i \sin ay)i dy}{e^{2y}-1} + \int_0^{\pi} \frac{e^{ax}(\cos x + i \sin x)}{(-2i)\sin x} dx = -\frac{\pi}{4}(1+e^{a\pi})$$

Equating real parts, we have

$$(1-e^{a\pi})\int_0^{\infty} \frac{\sin ay}{e^{2y}-1} dy - \frac{1}{2}\frac{(e^{a\pi}-1)}{a} = -\frac{\pi}{4}(1+e^{a\pi})$$

or

$$\int_0^{\infty} \frac{\sin ay}{e^{2y}-1} dy = \frac{\pi}{4}\frac{(e^{a\pi}+1)}{(e^{a\pi}-1)} - \frac{1}{2a}$$

$$= \frac{\pi}{4}\frac{(e^{a\pi/2}+e^{-a\pi/2})}{(e^{a\pi/2}-e^{-a\pi/2})} - \frac{1}{2a} = \frac{\pi}{4}\cosh\frac{a\pi}{2} - \frac{1}{2a}$$

Example 4. *Integrate* $\dfrac{e^{iaz}}{\sinh z}$ *round a suitable indented rectangle with vertices* $\pm R, \pm R + i\pi$

and show that $\displaystyle\int_0^{\infty} \frac{\sin ax}{\sinh x} = \frac{1}{2}\pi\frac{e^{a\pi}-1}{e^{a\pi}+1}$. *Provided that the imaginary part of* α *lies between*

−1 and +1.

Solution: Consider the integral

$$I = \int_C \frac{e^{iaz}}{\sinh z} dz$$

C being the rectangle with vertices $\pm R, \pm R + i\pi$, which has been indented at $z = 0$ and $z = i\pi$ (Fig. 6.30). The poles of

$$f(z) = \frac{e^{iaz}}{\sinh z}$$

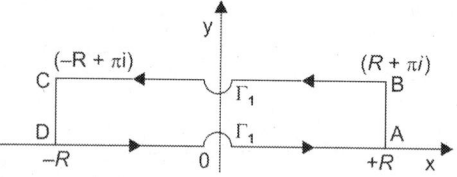

Fig. 6.30

are given by $\sinh z = 0$, i.e. $e^z - e^{-z} = 0$ or $e^{2z} = e^{2\pi ni}$

or $z = n\pi i$ where $n = 0, \pm 1, \pm 2, \pm 3,...$

$$* \quad \frac{1}{e^{-2ix}-1} = \frac{1}{e^{-ix}(e^{-ix}-e^{+ix})} = \frac{e^{ix}}{(-)(e^{ix}-e^{-ix})} = \frac{\cos x + i \sin x}{(-2i)\sin x}$$

Real part of the integral $= -\dfrac{1}{2}\displaystyle\int_0^{\pi} e^{ax} dx = -\dfrac{1}{2}\dfrac{(e^{a\pi}-1)}{a}$

It is quite obvious that none of the poles lie inside the contour, and $z = 0$, πi are indented by small semicircles Γ_1, Γ_2 of radii ρ_1, ρ_2 respectively.

By Cauchy's theorem

$$\oint_C f(z)\,dz = \int_{-R}^{-\rho_1} f(x)\,dx + \int_{\Gamma_1} f(z)\,dz + \int_{\rho_1}^{R} f(x)\,dx + \int_0^\pi \frac{e^{i\alpha(R+iy)}}{\sinh(R+iy)}\,i\,dy + \int_R^{\rho_2} \frac{e^{i\alpha(x+i\pi)}}{\sinh(x+i\pi)}\,dx$$

$$+ \int_{\Gamma_2} f(z)\,dz \int_{-\rho_2}^{-R} \frac{e^{i\alpha(x+i\pi)}}{\sinh(x+i\pi)}\,dx + \int_\pi^0 \frac{e^{i\alpha(-R+iy)}}{\sinh(-R+iy)}\,i\,dy = 0$$

$$\lim_{\rho_1 \to 0} \int_{\Gamma_1} f(z)\,dz = \lim_{\rho_1 \to 0} \int_c \frac{e^{i\alpha z}}{\sinh z}\,dz$$

$$= \lim_{\rho_1 \to 0} \int_\pi^0 \frac{e^{i\alpha(\rho_1 e^{i\theta})}}{\sinh \rho_1 e^{i\theta}}\,\rho_1 i e^{i\theta} \cdot d\theta$$

$$= \lim_{\rho_1 \to 0} \int_\pi^0 \frac{e^{i\alpha(\rho_1, e^{i\theta})}\rho_1, i e^{i\theta}\,d\theta^{*}}{\rho_1, e^{i\theta} + O(\rho_1^3)}$$

$$= i \int_\pi^0 d\theta = -\pi i$$

Similarly

$$\lim_{\rho_2 \to 0} \int_{\Gamma_2} f(z)\,dz = \lim_{\rho_2 \to 0} \int_{\Gamma_2} \frac{e^{i\alpha z}}{\sinh z}\,dz$$

$$= \lim_{\rho_2 \to 0} \int_0^{-\pi} \frac{e^{i\alpha(\pi i + \rho_2 e^{i\theta})}}{\sinh(i\pi + \rho_2 e^{i\theta})}\,\rho_2 i e^{i\theta} \cdot d\theta$$

[here $z = i\pi + \rho_2 e^{i\theta}$]

$$= \lim_{\rho_2 \to 0} \int_0^{-\pi} \frac{e^{i\alpha(\pi i + \rho_2 e^{i\theta})}}{-\sinh(\rho_2 e^{i\theta})}\,\rho_2 i e^{i\theta}\,d\theta^{**}$$

$$= \lim_{\rho_2 \to 0} \int_0^{-\pi} \frac{e^{i\alpha(\pi i + \rho_2 e^{i\theta})}}{\rho_2 e^{i\theta} + O(\rho_2^3)}\,\rho_2 i e^{i\theta}\,d\theta$$

$$= -\int_0^{-\pi} i e^{-\alpha\pi}\,d\theta = \pi i e^{-\pi\alpha}$$

While $$\int_0^\pi \frac{e^{i\alpha(R+iy)}}{\sinh(R+iy)}\,i\,dy = 2i \int_0^\pi \frac{e^{i\alpha R} e^{-\alpha y}}{e^{(R+iy)} - e^{-(R+iy)}}\,dy$$

$${}^{*}\ \sinh x = \frac{e^x - e^{-x}}{2} = x + \frac{x^3}{3!} + \frac{x^4}{5!} + \cdots$$

$$= x + O(x^3)$$

$${}^{**}\ \sinh(i\pi + \rho_2 e^{i\theta}) = \frac{e^{(i\pi + \rho_2 e^{i\theta})} - e^{(-i\pi - \rho_2 e^{i\theta})}}{2} = -\frac{e^{\rho_2 e^{i\theta}} + e^{-\rho_2 e^{i\theta}}}{2} = -\sinh(\rho_2 e^{i\theta})$$

or $\qquad \left| 2i \int_0^\pi \dfrac{e^{\alpha iR} e^{-\alpha y}}{e^{(R+iy)} - e^{-(R+yi)}} dy \right| \le 2 \int_0^\pi \dfrac{e^{-\alpha y}}{e^R - e^{-R}} dy = \dfrac{2}{\alpha} \cdot \dfrac{(1-e^{-\alpha\pi})}{(e^R - e^{-R})} \to 0$ as $R \to \infty$

Similarly, $\quad \left| 2i \int_0^\pi \dfrac{e^{-i\alpha R} e^{-\alpha y}}{e^{-R+iy} - e^{R-yi}} dy \right| \le 2 \int_0^\pi \dfrac{e^{-\alpha y}}{e^{-R} - e^{+R}} dy$

$$= -\dfrac{2}{\alpha} \dfrac{(1-e^{-\pi\alpha})}{(-e^{-R} + e^R)} \to 0 \text{ as } R \to \infty$$

Hence, taking the limits as $R \to \infty$, $\rho_1 \to 0$ and $\rho_2 \to 0$, we have

$$\int_{-\infty}^0 \dfrac{e^{i\alpha x}}{\sinh x} dx - \pi i + \int_0^\infty \dfrac{e^{i\alpha x}}{\sinh x} dx + \int_\infty^0 \dfrac{e^{-\alpha\pi} e^{i x\alpha}}{-\sinh x} dx + \pi i e^{-\alpha\pi} + \int_0^{-\infty} \dfrac{e^{-\alpha\pi} e^{i\alpha x}}{-\sinh x} dx = 0$$

or $\qquad \int_{-\infty}^0 \dfrac{(1+e^{-\alpha\pi}) e^{i\alpha x}}{\sinh x} dx + \int_0^\infty \dfrac{(1+e^{-\alpha\pi}) e^{i\alpha x}}{\sinh x} dx = \pi i (1 - e^{-\alpha\pi})$

Putting $x = -x$ in first integral, we have

$$\int_0^\infty \dfrac{(1+e^{-\alpha\pi})(e^{i\alpha x} - e^{-i\alpha x})}{\sinh x} dx = \pi i (1 - e^{-\alpha\pi})$$

or $\qquad \displaystyle\int_0^\infty \dfrac{2i \sin \alpha x}{\sinh x} dx = \dfrac{\pi i (1-e^{-\alpha\pi})}{(1+e^{-\alpha\pi})}$

or $\qquad \displaystyle\int_0^\infty \dfrac{\sin \alpha x}{\sinh x} dx = \dfrac{\pi (e^{\alpha\pi} - 1)}{2 (e^{\alpha\pi} + 1)}$

Example 5. Prove that $\displaystyle\int_0^\infty e^{-x^2} \cos 2ax \, dx = \dfrac{1}{2}\sqrt{\pi} e^{-a^2}$

Solution: Let us choose the integrand $f(z) = e^{-z^2}$ in a complex plane and integrate it around the rectangular contour OABC (Fig. 6.31). Here we take the contour C as the rectangle vertices of which are $0, R, R + ia$ and ia, such that variables on OA is $z = x$, on AB it is $R + iy$ on BC it is $x + ia$ and CO it is iy. Using Cauchy's integral theorem, we have

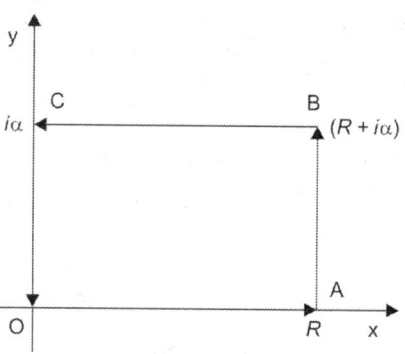

Fig. 6.31

$$\int_C e^{-z^2} dz = \int_0^R e^{-x^2} dx + \int_0^a e^{-(R+iy)^2} i\,dy + \int_R^0 e^{-(x+ia)^2} dx + \int_a^0 e^{-(iy)^2} i\,dy = 2\pi i \Sigma R_K \tag{1}$$

where ΣR_K is the sum of the residues of e^{-z^2} at its poles enclosed in C. But e^{-z^2} does not have any pole enclosed in the contour C and hence $\Sigma R_K = 0$. Eq. (1) gives

$$\int_0^R e^{-x^2} dx + \int_0^a e^{-(R+iy)^2} i\,dy + \int_R^0 e^{-(x+ia)^2} dx + \int_a^0 e^{y^2} i\,dy = 0 \tag{2}$$

where $\qquad \left| \displaystyle\int_0^a e^{-(R+iy)^2} i\,dy \right| \le \int_0^a \left| e^{-(R+iy)^2} \right| dy \le \int_0^a e^{-R^2 + y^2} dy \le e^{-R^2} \int_0^a e^{y^2} dy \le 0$ as $R \to \infty$

Since $e^{-R^2} \to 0$ as $R \to \infty$ and $\int_0^a e^{y^2} dy$ is finite.

Thus $\qquad \lim_{R \to \infty} \int_0^\infty e^{-(R+iy)^2} i \, dy = 0$

Now Eq. (2) becomes

$$\int_0^\infty e^{-x^2} dx + \int_{-\infty}^0 e^{-(x+ia)^2} dx = i \int_0^a e^{y^2} dy$$

or $\displaystyle \int_0^\infty e^{-x^2} dx - e^{a^2} \int_0^\infty e^{-x^2} [\cos 2ax - i \sin 2ax] dx = i \int_0^a e^{y^2} dy$

equating real parts on both sides of this equation, we have

$$\int_0^\infty e^{-x^2} dx - e^{a^2} \int_0^\infty e^{-x^2} \cos 2ax \, dx = 0$$

or $\qquad\qquad\qquad \displaystyle \int_0^\infty e^{-x^2} \cos 2ax \, dx = e^{-a^2} \int_0^\infty e^{-x^2} dx$

But $\qquad\qquad\qquad\qquad \displaystyle \int_0^\infty e^{-x^2} dx = \frac{1}{2}\sqrt{\pi}$

$\therefore$ $\qquad\qquad\qquad \displaystyle \int_0^\infty e^{-x^2} \cos 2ax \, dx = \frac{1}{2}\sqrt{\pi} \, e^{-a^2}$

Integration Round a Sector of a Circle

Method for integrals of type $I = \int_0^\infty x^{n-1} f(x) dx$, we have already discussed. But sometimes the conditions required in the method are not satisfied. For example,

$$\lim_{R \to \infty} \int_\Gamma z^{n-1} f(z) dz \text{ is not zero as required and}$$

$$\lim_{\rho \to 0} \int_\Gamma z^{n-1} f(z) dz \text{ is also not zero}$$

These conditions may be satisfied for parts of circles, i.e. sector of a circle. Then the contour used for evaluation of these integrals is chosen as a sector of circle. In order to explain this method let us consider the following example.

Example 6. *Integrate* $e^{iz} z^{\alpha-1}$ *round a quadrant and prove that*

i. $\displaystyle \int_0^\infty x^{\alpha-1} \cos x \, dx = \overline{|(\alpha)} \cos \frac{\pi\alpha}{2}, \quad 0 < \alpha < 1$

ii. $\displaystyle \int_0^\infty x^{\alpha-1} \sin x \, dx = \overline{|(\alpha)} \sin \frac{\pi\alpha}{2}, \quad 0 < \alpha < 1$

where $\overline{|\alpha}$ *is the gamma function.*

Solution: Integrand has a branch point (a multipoint) at $z = 0$. In order to exclude it from the region of integration, we choose the contour C shown in Fig. 6.32 which consists of a quadrant Γ of large circle of radius R, quadrant γ of small circle of radius ρ to indent the point $z = 0$ and two lines AB and DC along x-and y-axes. Then by Cauchy's integral theorem, we have

$$\int_C e^{iz}z^{\alpha-1}dz = \int_{\Gamma} e^{iz}z^{\alpha-1}dz + \int_{\gamma} e^{iz}z^{\alpha-1}dz + \int_{\rho}^{R} e^{ix}x^{\alpha-1}dx + \int_{R}^{\rho} e^{i(iy)}(iy)^{\alpha-1}idy = 0 \qquad (1)$$

Since there is no pole inside C. In this equation, third and fourth integrals are taken along AB and DC lines respectively.

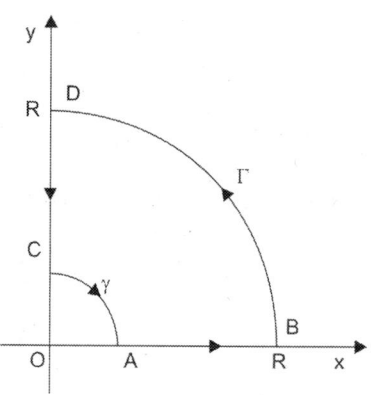

Fig. 6.32

$$\left|\int_{\Gamma} e^{iz}z^{\alpha-1}dz\right| = \left|\int_0^{\pi/2} e^{iRe^{i\theta}}(Re^{i\theta})^{\alpha-1}iRe^{i\theta}d\theta\right|$$

Here $z = Re^{i\theta}$ on the curve Γ, thus we have

$$\left|\int_{\Gamma} e^{iz}z^{\alpha-1}dz\right| \le \int_0^{\pi/2}\left|e^{iRe^{i\theta}}\right|\left|Re^{i\theta}\right|^{\alpha-1}\left|Re^{i\theta}\right|d\theta$$

$$\le \int_0^{\pi/2} e^{-R\sin\theta}R^{\alpha}d\theta \le \frac{\pi R^{\alpha}}{2R}(1-e^{-R})$$

using Jordon's inequality.

$$\le \frac{\pi}{2R^{1-\alpha}}(1-e^{-R}) \le 0 \text{ as } R \to \infty \quad \because \ 0 < \alpha < 1$$

Similarly $\left|\int_{\gamma} z^{\alpha-1}e^{iz}dz\right| \le \frac{\pi}{2}\frac{\rho^{\alpha}}{\rho}(1-e^{-\rho}) \le \frac{\pi}{2}\frac{\rho^{\alpha}}{\rho}\left(\rho - \frac{\rho^2}{2!}+\cdots\right)$

$$\le \frac{\pi}{2}\rho^{\alpha}\left(1-\frac{\rho}{2!}+\cdots\right) \le 0 \text{ as } \rho \to 0$$

or $\qquad \lim_{\rho\to0}\int_{\gamma} z^{\alpha-1}e^{i\alpha}dz = 0,$

Thus Eq. (1) becomes

$$\int_0^{\infty} e^{ix}x^{\alpha-1}dx - \int_0^{\infty} e^{-y}i^{\alpha}y^{\alpha-1}dy = 0$$

or $\qquad \int_0^{\infty} e^{ix}x^{\alpha-1}dx = i^{\alpha}\int_0^{\infty} y^{\alpha-1}e^{-y}dy = (e^{i\pi/2})^{\alpha}\int_0^{\infty} y^{\alpha-1}e^{-y}dy$

$$= e^{i\alpha\pi/2}\overline{|(\alpha)}$$

Equating real and imaginary parts on both sides, we have

$$\int_0^{\infty} x^{\alpha-1}\cos x\,dx = \overline{|(\alpha)}\cos\frac{\alpha\pi}{2}$$

and $\qquad \int_0^{\infty} x^{\alpha-1}\sin x\,dx = \overline{|(\alpha)}\sin\frac{\alpha\pi}{2}$

Example 7. *Integrate* e^{-z^2} *over a suitable contour and prove that*

i. $\displaystyle\int_0^\infty e^{-x^2\cos 2\alpha}\cos(x^2\sin 2\alpha)dx = \frac{\sqrt{\pi}}{2}\cos\alpha$

ii. $\displaystyle\int_0^\infty e^{-x^2\cos 2\alpha}\sin(x^2\sin 2\alpha)dx = \frac{\sqrt{\pi}}{2}\sin\alpha, \quad 0<\alpha<\frac{\pi}{4}$.

Solution: In order to evaluate this integral, we take the contour as shown in (Fig. 6.33), which consists of a sector Γ of a circle of radius R, the line AB along the real axis and the line AC along the radius making an angle α with the real axis. Now using Cauchy's theorem we have (since no pole is enclosed in C).

$$\int_C e^{-z^2}dz = \int_0^R e^{-x^2}dx + \int_\Gamma e^{-z^2}dz + \int_R^0 e^{-(re^{i\alpha})^2}d(re^{i\alpha}) = 0 \tag{1}$$

The first integral is taken along OB and the last integral along CO, any point on which is given by $z = re^{i\alpha}$, where r is the distance of point from origin. Now

$$\left|\int_\Gamma e^{-z^2}dz\right| = \left|\int_0^\alpha e^{-(Re^{i\theta})^2}iRe^{i\theta}d\theta\right| \qquad \because\ z = Re^{i\theta}\ \text{on}\ \Gamma$$

$$\left|\int_\Gamma e^{-z^2}dz\right| \le \int_0^\alpha \left|e^{-(Re^{i\theta})^2}\right|\left|iRe^{i\theta}\right|d\theta$$

$$\le \int_0^\alpha e^{-|R^2 e^{2i\theta}|}R\,d\theta$$

$$\le R\int_0^\alpha e^{-R^2\cos 2\theta}d\theta \le \frac{R}{2}\int_{\pi/2-2\alpha}^{\pi/2} e^{-R^2\sin\phi}d\phi$$

This is obtained by putting $\phi = \dfrac{\pi}{2} - 2\theta$

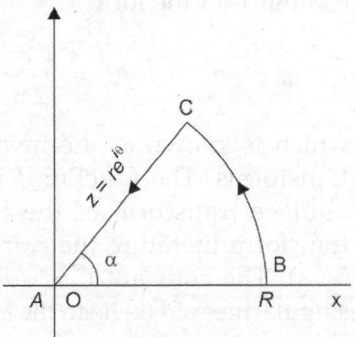

Fig. 6.33

$$\therefore\quad \left|\int_\Gamma e^{-z^2}dz\right| = \frac{R}{2}\int_{\pi/2-2\alpha}^{\pi/2} e^{-2R^2\phi/\pi}d\phi \qquad \text{(using Jordan's inequality)}$$

$$\le \frac{\pi}{4R}(e^{-2R^2(\pi/2-2\alpha)/\pi} - e^{-R^2})\le 0 \quad \text{as } R\to\infty$$

Since $\qquad\qquad 0 < \alpha < \pi/4$

Thus we have $\quad \displaystyle\lim_{R\to\infty}\int_\Gamma e^{-z^2}dz = 0$

Using this result in Eq. (1), we get

$$\int_0^\infty e^{-x^2}dx + \int_\infty^0 e^{-r^2 e^{2i\alpha}}e^{i\alpha}dr = 0$$

or $\qquad\qquad \displaystyle\int_0^\infty e^{-x^2}dx = \int_0^\infty e^{-r^2 e^{2i\alpha}}\cdot e^{i\alpha}dr$

But $\qquad\qquad \displaystyle\int_0^\infty e^{-x^2}dx = \frac{\sqrt{\pi}}{2}$

$\therefore\qquad \displaystyle\int_0^\infty e^{-r^2\cos 2\alpha}e^{-r^2\sin 2\alpha}dr = \frac{1}{2}e^{-i\alpha}\cdot\sqrt{\pi}$

Thus, we have

$$\int_0^\infty e^{-r^2\cos 2\alpha}\left\{\cos(r^2\sin 2\alpha)-i\sin(r^2\sin 2\alpha)\right\}dr = \frac{1}{2}e^{i\alpha}\sqrt{\pi} = \frac{\sqrt{\pi}}{2}(\cos\alpha - i\sin\alpha)$$

Equating real and imaginary quantities, we get

$$\int_0^\infty e^{-r^2\cos 2\alpha}\left\{\cos(r^2\sin 2\alpha)\right\}dr = \frac{\sqrt{\pi}}{2}\cos\alpha$$

and $$\int_0^\infty e^{-r^2\cos 2\alpha}\left\{\sin(r^2\sin 2\alpha)\right\}dr = \frac{\sqrt{\pi}}{2}\sin\alpha$$

6.16. BROMWICH CONTOUR INTEGRALS

A particular type of integral that occurs with great frequency in mathematical analysis is the Bromwich contour integral which is illustrated in (Fig. 6.34). This contour extends from $C - i\infty$ to $C + i\infty$, where $C \geq 0$. A typical integral involving the Bromwich contour is of the form

$$f(t) = \frac{1}{2\pi i}\int_{C-i\infty}^{C+i\infty} F(z)e^{zt}dz \qquad (6.46a)$$

which is known as the inversion formula for Laplace transforms. The function $f(t)$ is said to be the inverse Laplace transform of the function $F(z)$ (in standard transform literature the complex variable z is replaced by s). The constant C is adjusted such that all of the singularities of $F(z)$ lie to the left of the Bromwich contour. The requirement is necessary for the integral to exist.

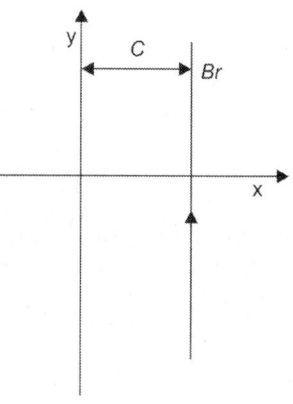

Fig. 6.34

The evaluation of any integral involving the contour indicated by Eq. (6.46a) is accomplished in the same manner as discussed in the previous sections. In most cases, the Bromwich contour is closed by adding a semicircular path Γ on the left, as illustrated in (Fig. 6.35). Jordan's Lemma can be applied in this case to obtain conditions on $F(z)$ such that

$$\lim_{R\to\infty}\int_\Gamma F(z)e^{zt}dz \to 0$$

and hence Cauchy's residue theorem gives

$$\int_{C-i\infty}^{C+i\infty} F(z)e^{zt}dz = 2\pi i\sum \text{Res}\cdot\text{in}(Br+\Gamma) \qquad (6.47)$$

Comparison of this expression with Eq. (6.46a) leads to the result that

$$f(t) = \sum \text{Res}\cdot\text{in}(Br+\Gamma) \qquad (6.48)$$

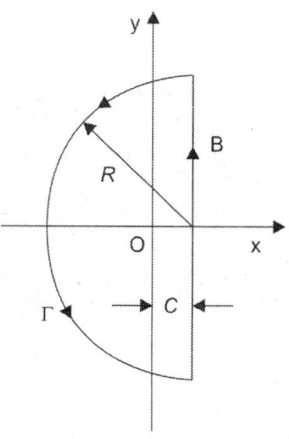

Fig. 6.35

Example 1. *Evaluate*

$$\frac{1}{2\pi i} \int_{C-i\infty}^{C+i\infty} \frac{e^{zt}}{\sqrt{z+1}} dz$$

where C and t are any positive constants.

Solution: The integrand has a branch point at $z = -1$. We shall take as branch line that part of the real axis to the left of $z = -1$. Since we cannot cross this branch line, let us consider

$$\oint_C \frac{e^{zt}}{\sqrt{z+1}} dz$$

where C is the contour ABDEFGHJKA as shown in Fig. 6.36, here EF and HJ actually lie on the real axis but have been shown separated for visual purposes. Also FGH is a circle of radius ε while BDE and JKA represent arcs of a circle of radius R.

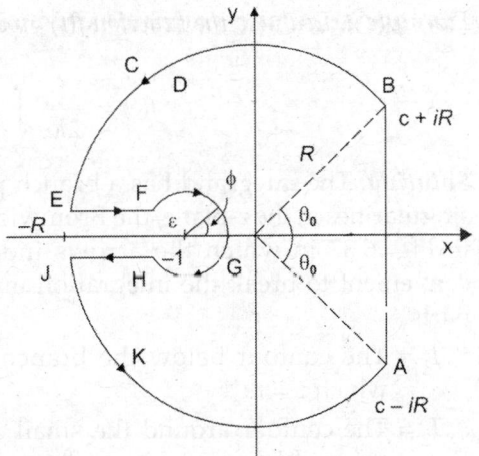

Since $e^{zt}\sqrt{z+1}$ is analytical inside and on C, we have Cauchy's theorem

$$\oint_C \frac{e^{zt}}{\sqrt{z+1}} dz = 0$$

Omitting the integrand, this can be written as

$$\int_{AB} + \int_{BDE} + \int_{EF} + \int_{FGH} + \int_{HI} + \int_{JKA} = 0$$

Fig. 6.36

Now on BDE and JKA, $z = Re^{i\theta}$, where θ goes from θ_0 to π and π to $2\pi - \theta_0$ respectively.

On EF, $z+1 = ue^{\pi i}, \sqrt{z+1} = \sqrt{u}\, e^{i\pi/2} = i\sqrt{u}$

whereas on HJ, $z+1 = ue^{-\pi i}, \sqrt{z+1} = \sqrt{u}\, e^{-\pi i/2} = -i\sqrt{u}$

In both cases $z = -1$, u or $dz = -du$, where u varies from $R - 1$ to ε along EF and ε to $R - 1$ along HJ.

On FGH, $z + 1 = \varepsilon e^{i\phi}$ where ϕ goes from $+\pi$ to $-\pi$.

Thus, we have

$$\int_{C-iR}^{C+iR} \frac{e^{zt}}{\sqrt{z+1}} dz + \int_{\theta_0}^{\pi} \frac{e^{tRe^{i\theta}}}{\sqrt{Re^{i\theta}+1}} i Re^{i\theta} \cdot d\theta + \int_{R-1}^{\varepsilon} \frac{e^{-(u+1)t}(-du)}{i\sqrt{u}} + \int_{\pi}^{-\pi} \frac{e^{(\varepsilon e^{i\phi}-1)t}}{\sqrt{\varepsilon e^{i\phi}}} i\varepsilon e^{i\phi} d\phi$$

$$+ \int_{\varepsilon}^{R-1} \frac{e^{-(u+1)t}}{-i\sqrt{u}}(-du) + \int_{\pi}^{2\pi-\theta_0} \frac{e^{Re^{i\theta}t}}{\sqrt{Re^{i\theta}+1}} i Re^{i\theta} \cdot d\theta = 0$$

taking the limits as $R \to \infty$ and $\varepsilon \to 0$. Second, fourth and sixth integral approach zero.

Hence we have

$$\int_{C-i\infty}^{C+i\infty} \frac{e^{zt}}{\sqrt{z+1}} dz = \lim_{\substack{\varepsilon \to 0 \\ R \to \infty}} 2i \int_{\varepsilon}^{R-1} \frac{e^{-(u+1)t}}{\sqrt{u}} du$$

$$= 2i \int_{0}^{\infty} \frac{e^{(-u+1)t}}{\sqrt{u}} du$$

or

$$\frac{1}{2\pi i} \int_{C-i\infty}^{C+i\infty} \frac{e^{zt}}{\sqrt{z+1}} dz = \frac{1}{\pi} \int_{0}^{\infty} \frac{e^{-(u+1)t}}{\sqrt{u}} du = \frac{e^{-t}}{\pi} \sqrt{\frac{\pi}{t}} = \frac{e^{-t}}{\sqrt{\pi t}}$$

Example 2. *Evaluate the function f(t) given by*

$$f(t) = \frac{1}{2\pi i} \int_{a-i\infty}^{a+i\infty} \frac{e^{-cz^{1/2}}}{z} e^{zt} \cdot dz \qquad\qquad \textbf{[Bangalore 1999]}$$

Solution: The integrand has a branch point at the origin. Because there are no other singularities in the z-plane, the Bromwich contour may be deformed into the one shown in Fig. 6.37 in which the arrows indicate the direction of integration. It will be convenient to break the integration into three parts.

I_1 = The contour below the branch cut on which $z = re^{-i\pi}$

I_2 = The contour around the small circle at the origin on which $z = \varepsilon e^{i\theta}$

I_3 = The contour above the branch cut on which $z = re^{i\pi}$

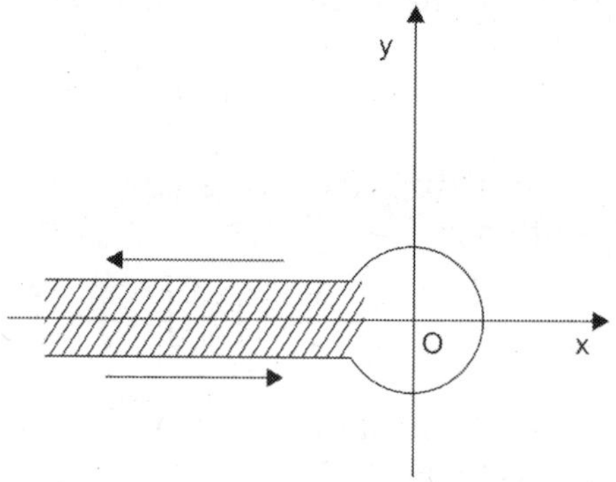

Thus we may write

$$I_1 = \frac{1}{2\pi i} \int_{\infty}^{\varepsilon} e^{icr^{1/2}} e^{-rt} \frac{dr}{r}$$

Fig. 6.37

$$I_2 = \frac{1}{2\pi} \int_{-\pi}^{\pi} e^{-c\varepsilon^{1/2}(\cos\theta/2 + i\sin\theta/2)} \cdot e^{\varepsilon t(\cos\theta + i\sin\theta)} d\theta$$

$$I_3 = \frac{1}{2\pi i} \int_{\varepsilon}^{\infty} e^{-icr^{1/2}} e^{-rt} \frac{dr}{r}$$

$$f(t) = \lim_{\varepsilon \to \infty}(I_1 + I_2 + I_3)$$

Examining I_2, the limit and integral may be interchanged.

$$\int_{0}^{\infty} \frac{e^{-t} \cdot e^{ut}}{\sqrt{u}} du \qquad\qquad\qquad \text{[Put } u = v^2 \text{ or } du = 2v \cdot dv]$$

$$= e^{-t} \int_{0}^{\infty} \frac{e^{-v^2 t}}{v} \cdot 2v \cdot dv = 2e^{-t} \int_{0}^{\infty} e^{-tv^2} dv = 2e^{-t} \cdot \frac{1}{2}\sqrt{\frac{\pi}{t}}$$

Performing this and the indicated operation yields $\lim\limits_{\varepsilon \to 0} I_2 = 1$.

Since I_1 and I_3 involve ε only in limits of integration, ε may be set directly to zero and we have

$$I_1 + I_3 = -\frac{1}{2\pi i} \int_0^\infty \left(e^{icr^{1/2}} - e^{-icr^{1/2}} \right) e^{-rt} \frac{dr}{r}$$

$$= -\frac{1}{\pi} \int_0^\infty \sin cr^{1/2} \cdot e^{-rt} \frac{dr}{r} \tag{1}$$

Expanding the sine function in Taylor's series

$$\sin cr^{1/2} = cr^{1/2} - \frac{1}{3!}(cr^{1/2})^3 + \frac{1}{5!}(cr^{1/2})^5 \cdots$$

and integrating Eq. (1) term by term results in

$$I_1 + I_2 = -\frac{2}{\pi^{1/2}} \left[\frac{x}{2} - \frac{1}{3.1!}\left(\frac{x}{2}\right)^3 + \frac{1}{5.2!}\left(\frac{x}{2}\right)^5 \cdots \right]$$

where $x = c/t^{1/2}$

The series is an expansion of the function known as the error function, that is $erf\,\dfrac{1}{2}x^*$

which permits the series to be expressed in the following form.

$$I_1 + I_2 = -erf\,\frac{c}{2t^{1/2}}$$

Combining all components of the integral, we have

$$f(t) = 1 - erf\,\frac{c}{2t^{1/2}} = erf_c\,\frac{c}{2t^{1/2}}$$

where erf_c denotes the complimentary error function as defined in Chapter 5.

Example 3. *Evaluate the function f(t) given by*

$$f(t) = \frac{1}{2\pi i} \int_{c-i\infty}^{c+i\infty} \frac{e^{zt}}{(z^2 + a^2)^{1/2}} dz$$

Solution: To evaluate this integral, the Bromwich contour will be deformed into the dumbell contour as shown in (Fig. 6.38).

A branch cut introduced between $z = \pm ia$, since the integrand has two branch points one at each of the designated points. The evaluation $f(t)$ is accomplished by breaking the integral into four parts.

$$I_1 = \frac{1}{2\pi i} \int_{-a+\varepsilon}^{a-\varepsilon} \frac{e^{iyt}}{(a^2 - y^2)^{1/2}} i\,dy = \frac{1}{2\pi} \int_{-a+\varepsilon}^{a-\varepsilon} \frac{e^{iyt}}{(a^2 - y^2)^{1/2}} dy$$

---

* The function of $erf \cdot x$ has the following integral representation:

$$erf \cdot x = \frac{2}{\pi^{1/2}} \int_0^x e^{-\alpha^2} d\alpha = \frac{1}{\pi^{1/2}} \int_0^{x^2} e^{-t} t^{-1/2} dt$$

$$I_2 = \frac{\varepsilon}{2\pi} \int_{-\pi/2}^{+\pi/2} \frac{e^{(ia+\varepsilon e^{i\theta})} e^{i\theta} d\theta}{\left[(ia+\varepsilon e^{i\theta})^2 + a^2\right]^{1/2}} \qquad z = ia + \varepsilon e^{i\theta}$$

$$I_3 = -\frac{1}{2\pi} \int_{a-\varepsilon}^{-a+\varepsilon} \frac{e^{iyt} dy}{(a^2-y^2)^{1/2}} \qquad z = iye^{i\pi} = -iy$$

$$I_4 = \frac{\varepsilon}{2\pi} \int_{-3\pi/2}^{\pi/2} \frac{e^{(-ia+\varepsilon e^{i\theta})} e^{i\theta} d\theta}{\left[(-ia+\varepsilon e^{i\theta})^2 + a^2\right]^{1/2}} \qquad z = -ia + \varepsilon e^{i\theta}$$

If we now take the limit as $\varepsilon \to 0$ we find that I_2 and I_4 both vanish. This leaves us with the result

$$f(t) = I_1 + I_2$$

$$= \frac{1}{2\pi} \int_{-a}^{+a} \frac{e^{iyt}}{(a^2-y^2)^{1/2}} dy - \frac{1}{2\pi} \int_{a}^{-a} \frac{e^{-iyt}}{(a^2-y^2)^{1/2}} dy$$

$$= \frac{1}{\pi} \int_{-a}^{+a} \frac{\cos yt}{(a^2-y^2)^{1/2}} dy$$

To evaluate this, put $y = a \sin\theta$, $dy = a\cos\theta \cdot d\theta$.

or $f(t) = \frac{1}{\pi} \int_{-\pi/2}^{\pi/2} \cos(at\sin\theta) d\theta$

This expression is standard form for the representation of Bessel's function of zeroth order that is

$$f(t) = J_0(at)$$

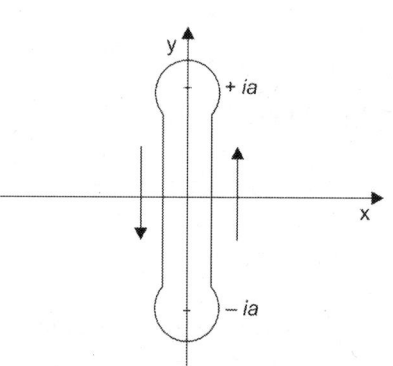

Fig. 6.38

Example 4. Evaluate the function $f(t)$ given by

$$f(t) = \frac{1}{2\pi i} \int_{c-i\infty}^{c+i\infty} \frac{e^{zt}}{(z^2+a^2)^2} dz$$

Solution: By inspection, it is easily seen that $F(z) = \dfrac{1}{(z^2+a^2)^2}$ has two double poles one at $z = ia$ and other at $z = -ia$. Since these poles lie on the imaginary axis, the positive constant c in the Bromwich contour integral must be chosen such that $c > 0$, or else the contour will cross the singularities.

All that remains now is to evaluate the residue of $\dfrac{e^{zt}}{(z^2+a^2)^2}$ at the two poles. To do this, for double pole at $z = ia$, we have

$$\text{Res (at } z = ia) = \frac{1}{1!} \frac{d}{dz}\left[(z-ia)^2 \frac{e^{zt}}{(z^2+a^2)^2}\right]_{z=ia}$$

$$= \frac{d}{dz}\left\{\frac{e^{zt}}{(z+ia)^2}\right\}_{z=ia} = \frac{e^{iat}}{4ia^3} - \frac{te^{iat}}{4a^2} \qquad (1)$$

To evaluate residue at $z = -ia$, simply take the complex conjugate of Eq. (1) that is put $i = -i$, we have

$$\text{Res}\big|_{z=-ia} = -\frac{e^{-iat}}{4ia^3} - \frac{te^{-iat}}{4a^2} \tag{2}$$

Adding the results of Eqs (1) and (2) give the sum of both residues and hence

$$f(t) = \frac{e^{iat} - e^{-iat}}{4ia^3} - \frac{t}{4a^2}(e^{iat} + e^{-iat})$$

$$= \frac{\sin at}{2a^3} - \frac{t}{2a^2}\cos at$$

Example 5. Solve the equation $f(t) = \dfrac{1}{2\pi i} \displaystyle\int_{c-i\infty}^{c+i\infty} \dfrac{e^{zt}}{z+a}\,dz$ to find the explicit expression of $f(t)$.

Solution: Here $F(z) = \dfrac{1}{z+a}$, and hence using Eq. (6.47), we obtain

$$\int_\Gamma \frac{e^{zt}\,dz}{z+a} + \int_{c-iR}^{c+iR} \frac{e^{zt}}{z+a}\,dz = 2\pi i \Sigma \,(\text{Res}) \tag{1}$$

where

$$\int_\Gamma \frac{e^{zt}}{z+a}\,dz = \int_{\pi/2}^{3\pi/2} \frac{e^{(c+Re^{i\theta})t}}{c+a+Re^{i\theta}} \cdot iRe^{i\theta}\cdot d\theta$$

or

$$\lim_{c\to 0}\left|\int_\Gamma \frac{e^{zt}}{z+a}\,dz\right| \le R \int_{\pi/2}^{3\pi/2} \frac{e^{iR\cos\theta}}{R-a}\,d\theta \le \frac{R}{R-a}\int_0^\pi e^{-iR\sin\phi}\,d\phi$$

Substituting $\phi = \dfrac{3\pi}{2} - \theta$

$$\therefore \qquad \lim_{c\to 0}\left|\int_\Gamma \frac{e^{zt}}{z+a}\,dz\right| \le 0 \text{ as } R \to \infty \quad \text{(using Jordan's Lemma)}$$

Thus

$$\lim_{c\to 0}\left|\int_\Gamma \frac{e^{zt}}{z+a}\,dz\right| = 0$$

Substituting in Eq. (1), we have

$$\int_{c-i\infty}^{c+i\infty} \frac{e^{zt}}{z+a}\,dz = 2\pi i \,\Sigma\,(\text{Res})$$

Pole of $F(z) = \dfrac{1}{z+a}$ occurs at $z = -a$. Hence, residue at $z = -a$ is

$$\lim_{z\to -a}\left\{\frac{e^{zt}}{z+a}(z+a)\right\} = e^{-at}$$

and hence we have

$$\int_{c-i\infty}^{c+i\infty} \frac{e^{zt}}{z+a} dz = 2\pi i e^{-at}$$

$$\therefore \qquad f(t) = \frac{1}{2\pi i} \int_{c-i\infty}^{c+i\infty} \frac{e^{zt}}{z+a} dz = e^{-at}$$

6.17. CONFORMAL MAPPING (TRANSFORMATION)

In complex variable theory, we seek a graphical representation of $w = f(z)$, where $z = x + iy$ and $w = v(x, y) + iv(x, y)$. That is to say, each pair of values x and y corresponds to two values u and v. Hence, we need a four dimensional space to plot the real values u, v, x and y. The mathematical subject of quaternions was developed by Hamilton (1805–1865) and Frobenius (1849–1917) to treat such system. Limited applications of quaternions in quantum mechanics have been made in recent years, but it is not a widely discussed subject in mathematical methods of physics courses.

Riemann developed a mode of visualising the relationship $W = F(z)$ which uses to separate complex planes (z-plane and w-plane). Riemann viewed a single-valued analytic function of a complex variable as indicating a specific one-to-one mapping between the points of two different two-dimensional surfaces. The relation $W = f(z)$ established the connection between points of given region Γ_z in z-plane and corresponding region Γ_w in the w-plane.

Suppose that under transformation point (x_0, y_0) of the xy plane is mapped into point (u_0, v_0) of the uv plane (Fig. 6.39). While curves C_1 and C_2 [intersecting at (x_0, y_0) are mapped respectively into curves C_1' and C_2' [intersecting at (u_0, v_0)]. Then if the transformation is such that the angle at (x_0, y_0) between C_1 and C_2 is equal to the angle at (u_0, v_0) between C_1' and C_2' both in magnitude and sense, the transformation or mapping is said to be conformal at (x_0, y_0). A mapping which preserves the magnitude of angles but not necessarily the sense is called *isogonal*.

Theorem: If $f(z)$ is analytic and $f'(z) \neq 0$ in a region R, then the mapping $W = f(z)$ is conformal at all points of R.

For conformal mappings or transformations small figures in the neighbourhood of a point z in the z-plane map into similar figures in the w-plane and are magnified

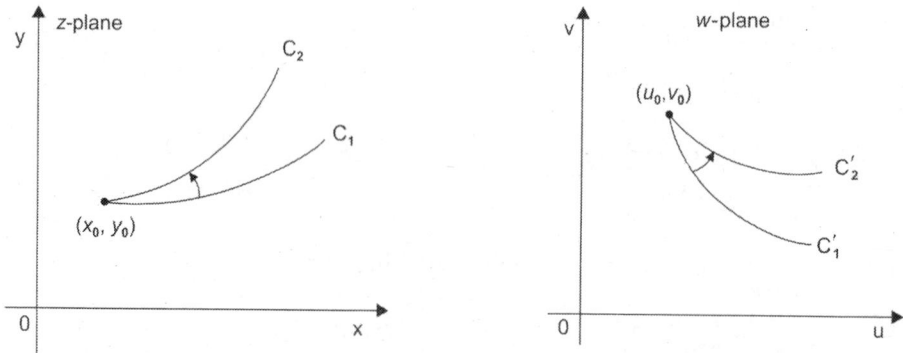

Fig. 6.39

(or reduced) by an amount given approximately by $|f'(z_0)|^2$, called the *area magnification factor* or simply magnification factor. Short distance in the z-plane in the neighbourhood of z_0 are magnified (or reduced) in the w-plane by an amount given approximately by $|f'(z_0)|$, called the *linear magnification factor*. Larger figures in the z-plane usually map into figures in the w-plane which are far from similar.

Some General Transformations

In the following α, β, are given complex constants while, a, θ_0 are real constants.

i. Translation ($w = z + \beta$)

By this translation, figures in the z-plane are displaced or translated in the direction of vector β.

ii. Rotation ($w = e^{i\theta}z$)

By this transformation, figures in the z-plane are rotated through an angle θ_0. If $\theta_0 > 0$, the rotation is counter-clockwise, while if $\theta_0 < 0$ the rotation is clockwise.

iii. Stretching ($w = az$)

By this transformation, figures in the z-plane are stretched (or contracted) in the direction z if $a > 1$ (or $0 < a < 1$). We consider contraction as a special case of stretching.

iv. Inversion ($w = 1/z$)

v. Linear transformation $w = \alpha z + \beta$

where α and β are complex constants. It is observed that general linear transformation is a combination of the transformation of translation, rotation and stretching.

vi. Bilinear fractional transformation $w = \dfrac{\alpha z + \beta}{\gamma z + \delta}$

$$\alpha\delta - \beta\gamma \neq 0$$

This transformation can be considered as combination of the transformation of translation, rotation, stretching and inversion.

Example 1. *Determine the region of the w-plane into which each of the following is mapped.*

i. *Region bounded by $x = 1$, $y = 1$ and $x + y = 1$ by transformation $w = z^2$.*

ii. *Rectangular region in the z-plane bounded by $x = 0$, $y = 0$, $x = 2$, $y = 1$ by transformation*
$$w = \sqrt{2}\,e^{\pi i/4}z + (1 - 2i)$$

Solution: i. Since $w = z^2$ is equivalent to
$$w = (x + iy)^2 = x^2 - y^2 + 2ixy$$

or
$$u = x^2 - y^2 \text{ and } v = 2xy$$

Then line $x = 1$ maps into $u = 1 - y^2$ and $v = 2y$

or
$$u = 1 - \frac{v^2}{4}$$

Line $y = 1$ maps into $u = \dfrac{v^2}{4} - 1$ and line $x + y = 1$ maps into $v = \dfrac{1}{2}(1 - u^2)$. The regions appear shaded as in Fig. 6.40 where points A, B, C map into A′ B′ C′. Note that the angles of triangle ABC are equal respectively to the angles of curvilinear triangle A′ B′ C′. This is the consequence of the fact that the mapping is conformal.

ii. If $w = \sqrt{2}\, e^{i\pi/4} z + (1 - 2i)$

Then $\qquad u + iv = \sqrt{2}\left(\cos\dfrac{\pi}{4} + i\sin\dfrac{\pi}{4}\right)(x + iy) + (1 - 2i)$

or $\qquad\qquad u = x - y + 1$

and $\qquad\qquad v = x + y - 2$

The lines $x = 0$, $y = 0$, $x = 2$ and $y = 1$ are mapped respectively into

$$u + v = -1$$
$$u - v = 3$$
$$u + v = 3 \text{ and } u - v = 1 \text{ (Fig. 6.41)}$$

The mapping accomplishes both rotation and stretching, and subsequent translation.

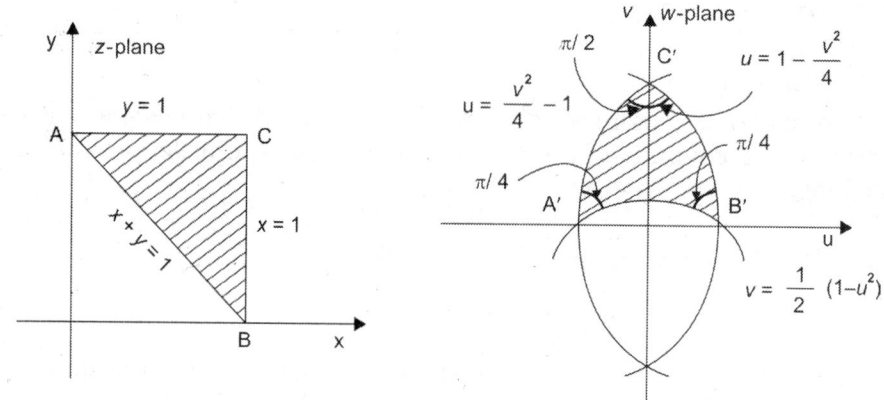

Fig. 6.40

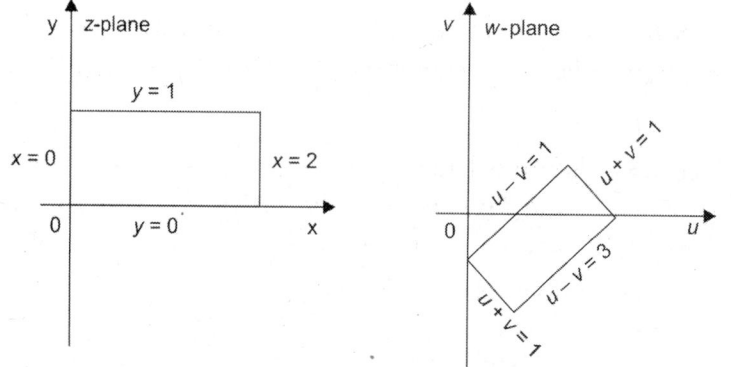

Fig. 6.41

Example 2. *Show that the transformation* $w = z + \dfrac{1}{z}$ *maps (i) the interior of semicircle*

$r = 1$ *in the upper half of the z-plane into lower half of w-plane and the exterior of the semicircle* $r = 1$ *in the upper half of z-plane into the upper half of w-plane.*

(ii) The semi-circular ring between the circle $|z| = 1$. $|z| = K > 1$ *in the upper half of z-plane*

into the semiellipse. $\left(\dfrac{Ku}{K^2 + 1}\right)^2 + \left(\dfrac{Kv}{K^2 - 1}\right)^2 = 1$ *in the upper half of w-plane.*

Solution: (i) Putting $z = re^{i\theta} : w = u + iv$, we have

$$u = \left(r + \frac{1}{r}\right)\cos\theta, \quad v = \left(r - \frac{1}{r}\right)\sin\theta$$

On eliminating θ, we get

$$\frac{u^2}{\left(r + \dfrac{1}{r}\right)^2} + \frac{v^2}{\left(r - \dfrac{1}{r}\right)^2} = 1$$

which is an ellipse. As $r \to 1$, the major-axis $\to 2$, while minor axis $\to 0$. Hence the ellipse reduces to a slit in the w-plane from -2 to $+2$. As $r \to 0$ or $r \to \infty$ both semi-major and semi-minor axis $\to \infty$ and it follows that the inside and outside of the unit circle in the z-plane both correspond to the whole w-plane, cut along the real axis from -2 to $+2$. If $0 \le \theta \le \pi, r < 1$, then $v > 0$. Hence the exterior of the semicircle $r = 1$ in the upper half of z-plane corresponds to the upper half of w-plane.

(ii) The circle $|z| = K, |z| = \dfrac{1}{K}$ both correspond to the ellipse.

$$\frac{u^2}{\left(K + \dfrac{1}{K}\right)^2} + \left(\frac{v^2}{K - \dfrac{1}{K}}\right)^2 = 1$$

or

$$\left(\frac{Ku}{K^2 + 1}\right) + \left(\frac{Kv}{K^2 - 1}\right)^2 = 1$$

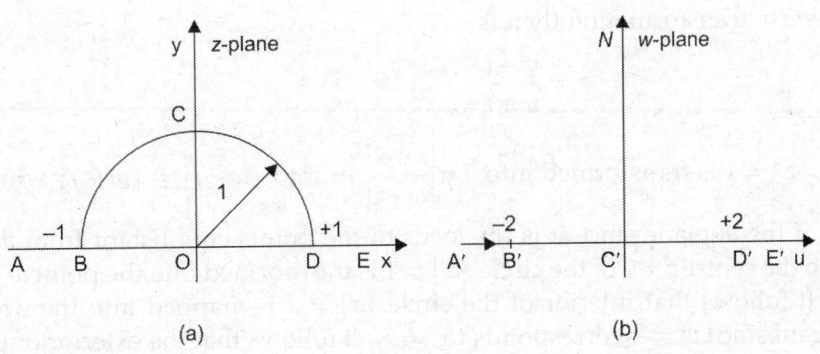

Fig. 6.42

Now the exterior of the semicircle $|z| = 1$ in upper half of z-plane corresponds to the upper half of w-plane and the interior of the semicircle $|z| = K$ ($K > 1$) in the upper half z-plane corresponds to the interior of the semiellipse in the upper half w-plane [$\because$ for $0 \leq \theta \leq \pi$ and $r > 1$, $v > 0$]. Hence semicircular ring between the circle $|z| = 1$ and $|z| = K > 1$ in the upper half of z-plane corresponds to the semiellipse in the upper half of w-plane.

Example 3. *Find a bilinear transformation which transforms the unit circle $|z| = 1$ into the real axis in such a way that the point $z_1 = 1$ is mapped into $w_1 = 0$, the point $z_2 = +1$ is mapped into $w_2 = 1$ and $z_3 = -1$ is mapped into $w_3 = \infty$. Into what regions the interior and exterior of the circle are mapped?*

Solution: Let the transformation be

$$w = \frac{az + b}{cz + d}$$

Substituting $z_1 = 1$, $w_1 = 0$, $z_2 = i$, $w_2 = 1$ and $z_3 = -1$, $w_3 = \infty$, we have

$$a + b = 0 \qquad \text{or } a = -b \qquad\qquad (1)$$
$$ci + d = ai + b \qquad\qquad\qquad\qquad (2)$$

and

$$-c + d = 0 \qquad \text{or } c = \pm d \qquad\qquad (3)$$

Substituting Eqs (3) and (1) in (2), we have

$$c(i + 1) = a(i - 1)$$

or

$$\frac{c}{a} = \frac{i - 1}{i + 1}$$

The transformation then becomes

$$w = \frac{az - a}{cz + c} = \frac{a(z - 1)}{c(z + 1)}$$

$$= \frac{(i + 1)\,(z - 1)}{(i - 1)\,(z + 1)}$$

or

$$w = \frac{i(1 - z)}{1 + z}$$

The inverse transformation then is

$$z = \frac{w - i}{w + i}$$

Hence $|z| = 1$ is transformed into $1 = \left|\dfrac{w - i}{w + i}\right|$ or into $|w - i| = |w + i|$ which is the

real axis of the w-plane since it is the locus of the points equidistant from $w = i$ and $w = -i$ also the centre $z = 0$ of the circle $|z| = 1$ is transformed into the point $w = i$ of the w-plane. It follows that interior of the circle $|z| = 1$ is mapped into the upper half plane. Again since $w = -i$ corresponds to $z = \infty$, it follows that the exterior of the circle $|z| = 1$ is transformed into lower half of the w-plane.

Example 4. *Prove that transformation* $w = \tan^2 \dfrac{\pi\sqrt{z}}{4}$ *transforms the interior of the unit circle* $|w| = 0$ *into the space lying within a parabola in z-plane.*

Solution: For the given transformation

$$\frac{1-w}{1+w} = \frac{1 - \tan^2(\pi\sqrt{z}/4)}{1 + \tan^2(\pi\sqrt{z}/4)} \cos(\pi\sqrt{z}/2)$$

This in polar coordinates becomes

$$\frac{1-w}{1+w} = \cos\left[\frac{\pi}{2}\cdot\sqrt{r}\,e^{i\theta/2}\right] = \cos(a+ib) \tag{1}$$

where
$$a = \frac{\pi\sqrt{r}}{2}\cos\theta/2 \text{ and } b = \frac{\pi}{2}\sqrt{r}\sin\theta/2$$

Let
$$1 + w = Re^{i\phi} \text{ then Eq. (1) becomes}$$

$$\frac{2 - Re^{i\phi}}{Re^{i\phi}} = \cos a \cosh b - i\sin a \sinh b$$

Equating real and imaginary parts we get

$$\cos a \cosh b = \frac{2}{R}\cos\phi - 1 \tag{2}$$

$$\sin a \sinh b = \frac{2}{R}\sin\phi \tag{3}$$

For a = constant, $\dfrac{\pi}{2}\sqrt{r}\cos\theta/2$ = constant or $r\cos^2\theta/2$ = constant, which represents a parabola in z-plane. Equations (2) and (3) give

$$\frac{(2\cos\phi - R)^2}{\cos^2 a} - \frac{4\sin^2\phi}{\sin^2 a} = R^2$$

which gives

$$\frac{2}{R}\cos\phi > 1 \text{ for } a < \frac{\pi}{2} \text{ and } R > 2\cos\phi \text{ for } a > \frac{\pi}{2}$$

Thus the space lying within the parabola $a = \dfrac{\pi}{2}$, i.e. $r\cos^2\theta/2 = 1$ in z-plane is mapped into the interior of circle $R = 2\cos\phi$ in w-plane.

Example 5. *Determine the centre and radius of the circle in the w-plane which corresponds to the real axis in the z-plane, where* $w = \dfrac{ze^\alpha - i}{z - ie^\alpha}$ *(α is a real constant).*

Solution: From the given transformation, we have
$$w(z - ie^\alpha) = ze^\alpha - i$$

or
$$z = \frac{i(we^\alpha - 1)}{w - e^\alpha} \tag{1}$$

Equation of the real axis in z-plane $= z - \bar{z} = 0$

Substituting in Eq. (1), we have

$$\frac{i(we^\alpha - 1)}{w - e^\alpha} + \frac{i(\bar{w}e^\alpha - 1)}{\bar{w} - e^\alpha} = 0$$

or $$w\bar{w} - \frac{w(1 + e^{2\alpha})}{2e^\alpha} - \frac{\bar{w}(1 + e^{2\alpha})}{2e^\alpha} + 1 = 0$$

or $$(w - \cosh \alpha)(\bar{w} - \cosh \alpha) = \cosh^2 \alpha - 1 = \sinh^2 \alpha$$

or $$|w - \cosh \alpha|^2 = \sinh^2 \alpha$$

which describes a circle in w-plane with its centre at $\omega = \cosh \alpha$ and radius equal to $\sinh \alpha$.

6.18. PHYSICAL APPLICATIONS OF CONFORMAL MAPPING

Many problems of science and engineering when formulated mathematically lead to partial differential equations and associated conditions called boundary conditions. The problem of determining solutions to a partial differential equation which satisfy the boundary conditions is called a boundary value problem.

Application to Heat Flow

Consider a solid having a temperature distribution which may be varying. We are often interested in the quantity of heat conducted per unit area per unit time across a surface located in the solid. This quantity, sometimes called heat flux across the surface and is given by

$$Q = - K \text{ grad } \phi$$

where ϕ is the temperature and K, assumed to be constant, is called the thermal conductivity and depends on the material of which the solid is made.

If we restrict ourselves to problem of two-dimensional type, then

$$Q = - K\left(\frac{\partial \phi}{\partial x} + \frac{i\partial \phi}{\partial y}\right) = Q_x + iQ_y \qquad (6.49)$$

where $$Q_x = - K\frac{\partial \phi}{\partial x}, Q_y = -K\frac{\partial \phi}{\partial y}$$

Let C be any simple closed curve in the z-plane (representing the cross-section of cylinder). If Q_t and Q_n are the tangential and normal components of the heat flux and if steady state conditions prevail so that there is no net accumulation of heat inside C, then we have,

$$\left.\begin{aligned}\oint_C Q_n ds = \oint_C (Q_x dy + Q_y dx) = 0\\ \oint_C Q_t ds = \oint_C (Q_x dy - Q_y dx) = 0\end{aligned}\right\} \qquad (6.50)$$

assuming no sources or sink inside C. The first Eq. (6.50) yields

$$\frac{\partial Q_x}{\partial x} + \frac{\partial Q_y}{\partial y} = 0 \qquad (6.51)$$

Which becomes [on using Eq. (6.49)]

$$\frac{\partial^2 \phi}{\partial x^2} + \frac{\partial^2 \phi}{\partial y^2} = 0$$

i.e. ϕ is harmonic. Introducing the harmonic conjugate function ψ, we see that

$$T(z) = \phi(x, y) + i\psi(x, y) \tag{6.52}$$

is analytic. The families of curves

$$\phi(x, y) = \alpha \text{ and } \psi(x, y) = \beta \tag{6.53}$$

are called isothermal lines and flux lines respectively, while $T(z)$ is called complex temperatures.

The analogies with fluid flow and electrostatics are evident. In fluid flow (fluid dynamics, hydrodynamics), the quantities like temperature, and heat flux are replaced by velocity potential and velocity of fluid whereas in electrostatics, these quantities are replaced by electrostatic potential and electric field intensity.

Examples 1. *An infinite wedge-shaped region ABDE of angle $\pi/4$ (shaded in Fig. 6.40a) has one of its sides AB maintained at constant temperature T_1. The other side BDE has part BD (of unit length) insulated while the remaining part DE is maintained at constant temperature T_2. Find the temperature everywhere in the region.*

Solution: By transformation $w' = z^2$, the shaded region of the z-plane (Fig. 6.43a) is mapped into the shaded region as in Fig. 6.40b with the indicated boundary conditions.

By transformation $w' = \sin \dfrac{\pi w}{z}$, the shaded region of w'-plane (Fig. 6.43b) is mapped into the region shaded in Fig. 6.43c with the indicated boundary conditions.

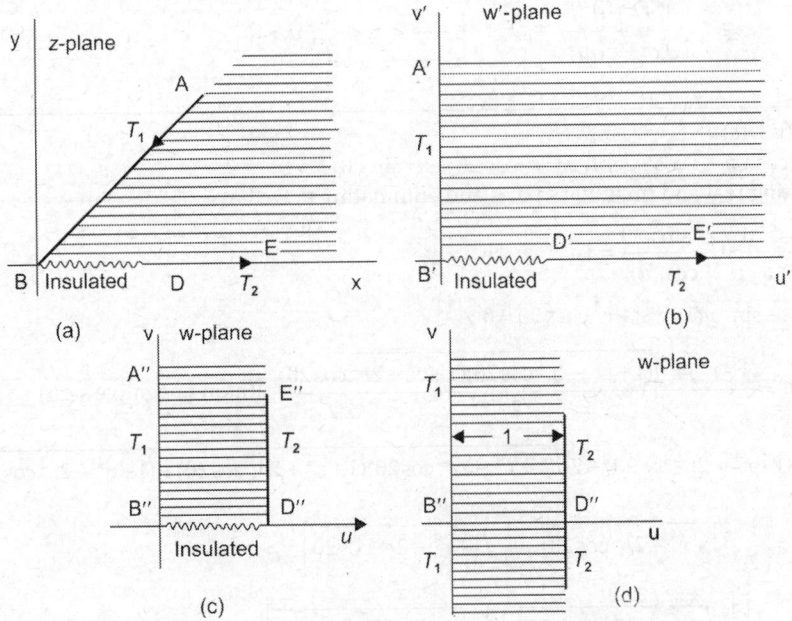

Fig. 6.43

Now the temperature problem represented by (Fig. 6.43c) with $B''D''$ insulated is equivalent to the temperature problem represented by (Fig. 6.43d) by symmetry so no heat transfer can take place across $B''D''$. But this is the problem of determining the temperature between two parallel planes kept at constant temperature T_1 and T_2 respectively. In this case the temperature variation is linear and so must be given by $T_1 + (T_2 - T_1)u$.

From $w' = z^2$ and $w' = \sin\dfrac{\pi w}{2}$, eliminate w', then

$$w = \frac{2}{\pi}\sin^{-1}z^2 \quad \text{or} \quad u = \frac{2}{\pi}\mathrm{Re}(\sin^{-1}z^2)$$

Thus the required temperature is

$$T_1 + \frac{2}{\pi}(T_2 - T_1)\mathrm{Re}\{(\sin^{-1}(z^2)\}$$

In polar coordinates (r, θ) this can be written as*

$$T_1 + \frac{2}{\pi}(T_2 - T_1)\sin^{-1}\left\{\frac{1}{2}\sqrt{r^4 + 2r^2\cos 2\theta + 1} - \frac{1}{2}\sqrt{r^4 - 2r^2\cos 2\theta + 1}\right\}$$

Example 2. *Find the formula for the steady temperature $T(x, y)$ in a semi-infinite slab bounded by the planes $x = \pi/2, x = -\pi/2$ and $y = 0$, when the first two boundaries are kept at temperature zero and the last at temperature $T = 1$.*

Solution: The problem is to find the temperatures in a plate having the form of a semi-infinite (strip, where the faces of the plate are perfectly insulated (Fig. 6.44a). The boundary value problem to be solved here can be written as

$$\frac{\partial^2 T}{\partial x^2} + \frac{\partial^2 T}{\partial y^2} = 0 \quad \left(-\frac{\pi}{2} < x < \frac{\pi}{2}, y > 0\right) \tag{1}$$

---

* $\sin^{-1} z^2 = (u + iv)$

or $\sin u \cosh v + i \cos u \sinh v = r^2 \cos 2\theta + ir^2 \sin 2\theta$

Equating real and imaginary parts and eliminating v, we have

$$\frac{r^4 \cos^2 2\theta}{\sin^2 u} - \frac{r^4 \sin^2 2\theta}{\cos^2 u} = 1$$

or $\sin^4 u - \sin^2 u(1 + r^4) + r^4 \cos^2 2\theta = 0$

or $\sin^2 u = \dfrac{(1 - r^4) \pm \sqrt{(1 + r^4 - 2r^2\cos 2\theta)(1 + r^4 + 2r^2\cos 2\theta)}}{2}$ (taking negative sign)

$$= \frac{1}{4}\left\{(1 + r^4 + 2r^2\cos 2\theta) - 2\sqrt{(1 + r^4 - 2r^2\cos 2\theta)(1 + r^4 + 2r^2\cos 2\theta)} + (1 + r^4 - 2r^2\cos 2\theta)\right\}$$

or $\sin u = \left\{\dfrac{1}{2}\sqrt{(1 + r^4 + 2r^2\cos 2\theta)} - \dfrac{1}{2}\sqrt{1 + r^4 - 2r^2\cos 2\theta}\right\}$

or $u = \sin^{-1}\left\{\dfrac{1}{2}\sqrt{(1 + r^4 + 2r^2\cos 2\theta)} - \dfrac{1}{2}\sqrt{(1 + r^4 - 2r^2\cos 2\theta)}\right\}$

$$T\left(-\frac{\pi}{2},y\right) = T\left(\frac{\pi}{2},y\right) = 0 \quad (y>0) \tag{2}$$

$$T(x, 0) = 1 \quad \left(-\frac{\pi}{2}<x<\frac{\pi}{2}\right) \tag{3}$$

Also $|T(x,0)| < M$, where M is some constant, a condition that could be replaced by the condition that T is to approach zero as y tends to infinity.

In fact the boundary conditions are all of the type $T = c$, a type that is invariant under conformal transformation. The transformation $z' = \sin z$ transforms the strip into the upper half of the z'-plane (Figs 6.44a and b). As indicated, the image of the base of the strip is the segment of the x'-axis between the points $z' = -1$ and $z' = 1$ and the images of the sides are the remaining parts of the x'-axis.

Again this half plane is transformed into the infinite strip between the lines $v = 0$ and $v = \pi$ by the transformation

$$w = \log\frac{z'-1}{z'+1} = \log\frac{r_1}{r_2}+i(\theta_1-\theta_2)$$

$$(0 < \theta_1 < \pi, 0 < \theta_2 < \pi)$$

As indicated in Figs 6.44b and c. The segment of the x' axis between $z' = -1$ and $z' = 1$ maps on to the upper side of the strip; and the rest of that axis on to the lower side.

A harmonic function u and v that is zero on the side $v = 0$ of the strip, unity on the side $v = \pi$, and bounded in the strip is clearly

$$T = \frac{1}{\pi}v \tag{4}$$

for this the imaginary coefficient of the analytic function

$$f(w) = \frac{w}{\pi},$$

changing to the coordinates x' and y' means the transformation

$$w = \log\frac{z'-1}{z'+1} = \log\left|\frac{z'-1}{z'+1}\right|+i\arg\frac{z'-1}{z'+1} \tag{5}$$

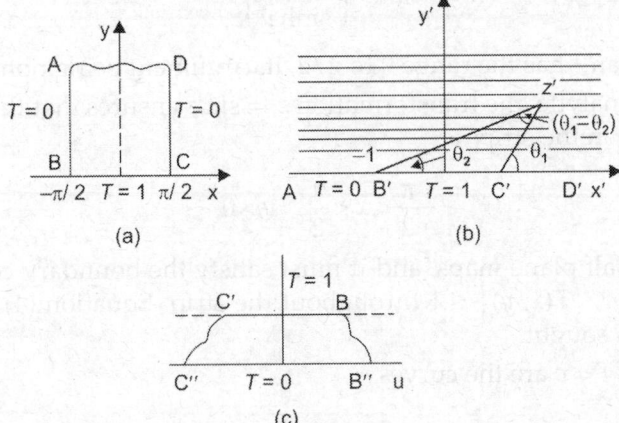

(a)

(b)

(c)

Fig. 6.44

We find that

$$v = \arg\left(\frac{x' - 1 + iy'}{x' + 1 + iy'}\right) = \arg\left[\frac{x'^2 + y'^2 - 1 + 2iy'}{(x' + 1)^2 + y'^2}\right]$$

$$v = \tan^{-1}\frac{2y'}{x'^2 + y'^2 - 1} \qquad (\tan^{-1} \text{ has the range } 0 \text{ to } \pi)$$

The function T given by Eq. (4) therefore becomes

$$T = \frac{1}{\pi}\tan^{-1}\left(\frac{2y'}{x'^2 + y'^2 - 1}\right)$$

The function represents the steady temperatures in the semi-infinite plate $y' \geq 0$ with section $(-1 < x' < 1)$ of its boundary $y' = 0$ kept at the temperature $T = 1$ and the rest at temperature zero. The isotherms $T = c$ $(0 < c < 1)$ are the circles.

$$x'^2 + y'^2 - \frac{2}{\tan \pi c}y' - 1 = 0$$

with their centres on the y'-axis and passing through the points $(\pm 1, 0)$.

In the transformation $z' = \sin z$

$$x' = \sin x \cosh y \text{ and } y' = \cos x \sinh y$$

and

$$T = \frac{1}{\pi}\tan^{-1}\left(\frac{2\cos x \sin y}{\sin^2 x \cosh^2 y + \cos^2 x \sinh^2 y - 1}\right)$$

$$= \frac{1}{\pi}\tan^{-1}\left\{\frac{2\cos x \sinh y}{1 - (\cos x / \sinh y)^2}\right\}$$

$$= \frac{1}{\pi}\tan^{-1}(\tan 2\alpha)$$

where

$$\tan \alpha = \frac{\cos x}{\sinh y}$$

$$T = \frac{2}{\pi}\tan^{-1}\left(\frac{\cos x}{\sinh y}\right) \qquad (6)$$

The function $\tan^{-1}$ has the range 0 to $\pi/2$, its argument being nonnegative.

Since $\sin z$ is analytic, the transformation $z' = \sin z$ ensures that the function Eq. (6) will be harmonic in the strip

$$-\frac{\pi}{2} < x < \frac{\pi}{2}, \ y > 0$$

onto which the half plane maps, and it must satisfy the boundary conditions Eqs (2) and (3). Moreover, $|T(x, y)| \leq 1$ throughout the strip. Equation (4) is therefore temperature formula sought.

The isotherms $T = c$ are the curves

$$\cos x = \tan\frac{\pi c}{2}\sinh y$$

each of which passes through the points $(\pm \pi/2, 0)$. If K is thermal conductivity, the flux of heat into the wall through its base is

$$-K\frac{\partial T}{\partial y}\bigg|_{y=0} = \frac{2K}{\pi \cos x} \qquad \left(-\frac{\pi}{2} < x < \frac{\pi}{2}\right)$$

and the flux outward through the plane $x = \pi/2$ is

$$-K\frac{\partial T}{\partial x}\bigg|_{x=\pi/2} = \frac{2K}{\pi \sinh y} \qquad (y > 0)$$

The product of harmonic function by a constant is also harmonic. The function

$$T = \frac{2T_0}{\pi}\tan^{-1}\left(\frac{\cos x}{\sin y}\right)$$

represents the temperature in the above slab when the base is kept at temperature T_0 and the sides at zero.

Example 3. *Find the potential V in the space between the planes y = 0 and y = p if V = 0 on the part x > 0 of each of those and V = 1 on parts x < 0.*

Solution: The transformation $z' = e^z$ transforms the strip (Fig. 6.45a) into the upper half of the z' plane (Fig. 6.45b). The BC and ED parts of the strip are mapped as B'C' and E'D' with point C' and D' coinciding. The half plane is transformed into the infinite strip between the lines $v = 0$ and $v = \pi$ by the transformation,

$$w = \log\frac{z'-1}{z'+1} = \log\frac{r_1}{r_2} + i(\theta_1 - \theta_2)\begin{cases}0 < \theta_1 < \pi \\ 0 < \theta_2 < \pi\end{cases}$$

As indicated in Figs 6.45b and c the segment of the x' axis between $z' = -1$ and $z' = 1$ maps on the upper side of the strip and the rest of that axis on the lower side.

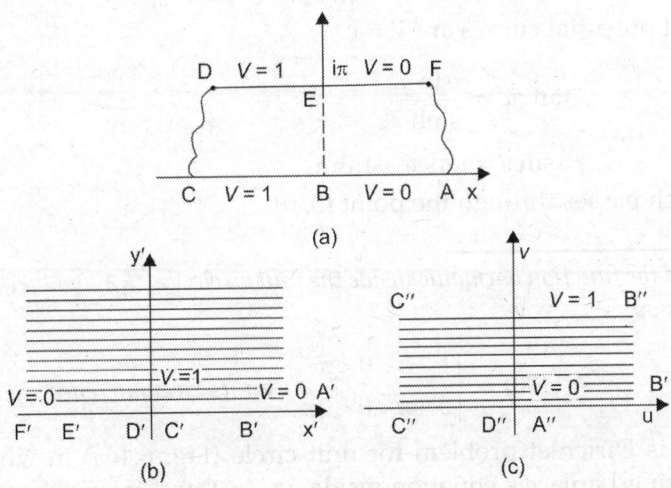

Fig. 6.45

The harmonic function of u and v that is zero on the side $v = 0$ of the strip and unity on the side $v = \pi$ and bounded in the strip is clearly

$$V = \frac{1}{\pi} v$$

This is the imaginary coefficient of the analytic function

$$f(w) = \frac{w}{\pi}$$

Changing to the coordinates x' and y' by means of the transformation

$$w = \log \frac{z' - 1}{z' + 1}$$

$$= \log \left| \frac{z' - 1}{z' + 1} \right| + i \arg \frac{z' - 1}{z' + 1}$$

or

$$v = \arg \left[\frac{x' - 1 + iy'}{x' + 1 + iy'} \right]$$

$$= \arg \left[\frac{x'^2 + y'^2 - 1 + 2iy'}{(x' + 1)^2 + y'^2} \right]$$

or

$$v = \tan^{-1} \frac{2y'}{x'^2 + y'^2 - 1}$$

or

$$V = \frac{1}{\pi} \tan^{-1} \frac{2y'}{x'^2 + y'^2 - 1}$$

Again using the transformation

$$z' = e^z$$

$$x' = \cosh x \cos y + \sinh x \sin y$$

and

$$y' = \cosh x \sin y - \sinh x \cos y$$

∴

$$V = \frac{1}{\pi} \tan^{-1} \left(\frac{\sin y}{\sinh x} \right) \qquad\qquad (0 \le \tan^{-1} \theta \le \pi)$$

The constant potential curves are $V = c$

$$\tan \pi c = \frac{\sin y}{\sinh x}$$

or

$$\sin y = \tan \pi c \cdot \sinh x$$

Each of which passes through the point $(0, 0)$.

Example 4. *Find the function harmonic inside the unit circle $|z| = 1$ and taking the prescribed values of y given by*

$$F(\theta) = \begin{cases} 1 & 0 < \theta < \pi \\ 0 & \pi < \theta < 2\pi \end{cases} \quad \text{on its circumference.}$$

Solution: This is Dirichlet problem for unit circle (Fig. 6.46a) in which we seek a function satisfying Laplace's equation inside $|z| = 1$ and taking the values 0 on arc ABC and 1 on arc CDE.

We map the interior of the circle $|z| = 1$ on to the upper half of the w-plane (Fig. 6.46b) by using the mapping function

$$z = \frac{i-w}{i+w}$$

or

$$w = i\left(\frac{1-z}{1+z}\right)$$

Under this transformation, arcs ABC and CDE are mapped on to the negative and positive real axis A′B′C′ and C′D′E′ respectively of the w-plane; the boundary conditions $\phi = 0$ on arc ABC and $\phi = 1$ on arc CDE become respectively $\phi = 0$ on A′B′C′ and $\phi = 1$ on C′D′E′.

Thus, we have reduced the problem to find a function ϕ harmonic in upper half of w-plane and taking the values 0 for $u < 0$ and 1 for $u > 0$. The solution of the problem is thus

$$\phi = 1 - \frac{1}{\pi}\tan^{-1}\frac{v}{u} \qquad (1)$$

Now for

$$w = i\left(\frac{1-z}{1+z}\right), \text{ we find}$$

$$v = \frac{2y}{(1+x)^2 + y^2}$$

and

$$u = \frac{1-(x^2+y^2)}{(1+x)^2 + y^2}$$

Thus, the required solution is

$$\phi = 1 - \frac{1}{\pi}\tan^{-1}\left\{\frac{2y}{1-(x^2+y^2)}\right\}$$

or in polar coordinates (r, θ), where $x = r\cos\theta$, $y = r\sin\theta$.

$$\phi = 1 - \frac{1}{\pi}\tan^{-1}\left\{\frac{2r\sin\theta}{1-r^2}\right\}$$

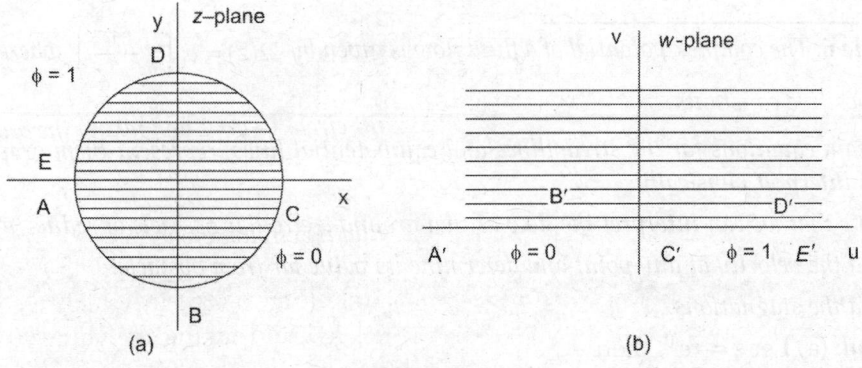

Fig. 6.46

Example 5. *(a) Find the complex potential for a fluid moving with constant speed v_0 in a direction making an angle δ with the positive x-axis.*

(b) Determine the velocity potential and stream functions.

(c) Determine the equations for the streamlines and equipotential lines.

Solution: (a) The x and y components of velocity are

$$v_x = v_0 \cos \delta, \; v_y = v_0 \sin \delta$$

The complex velocity is

$$v = v_x + iv_y$$

$$= v_0 \cos \delta + iv_0 \sin \delta = v_0 e^{i\delta}$$

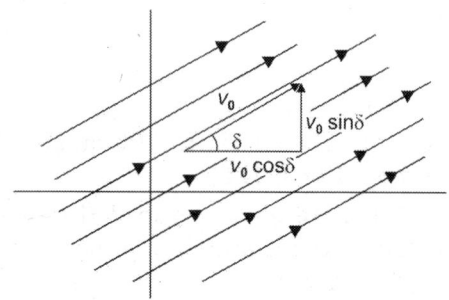

The complex potential $\Omega(z)$ is given by

$$\frac{d\Omega(z)}{dz} = \bar{v} = v_0 e^{-i\delta}$$

and integrating $\Omega(z) = v_0 e^{-i\delta} z$. Leaving the constant of integration.

Fig. 6.47

(b) The velocity potential ϕ and stream function ψ are the real and imaginary parts of the complex potential.

Thus
$$\Omega(z) = \phi + i\psi = v_0 e^{-i\delta} z$$
$$= v_0 (\cos \delta - i \sin \delta)(x + iy)$$
$$= v_0 (x \cos \delta + y \sin \delta) + iv_0 (y \cos \delta - x \sin \delta)$$

and
$$\phi = v_0 (x \cos \delta + y \sin \delta)$$
$$\psi = v_0 (y \cos \delta - x \sin \delta)$$

(c) The streamlines are given by
$$\psi = v_0 (y \cos \delta - x \sin \delta) = \beta$$

for different values of β. Physically, under steady state conditions, a streamline represents the path actually taken by a fluid particle, in this case a straight line path.

The equipotential lines are given by
$$\phi = v_0 (x \cos \delta + y \sin \delta) = \alpha$$

for different values of α. Geometrically they are lines perpendicular to the streamlines, all points on an equipotential lines are at equal potential.

Example 6. *The complex potential of a fluid flow is given by $\Omega(z) = v_0 \left(z + \dfrac{a^2}{z} \right)$ where v_0 and 'a' are positive constants.*

(a) Obtain equations for the streamlines and equipotential lines, represent them graphically and interpret physically.

(b) Show that we can interpret the flow as that around a circular obstacle of radius 'a'.

(c) Find the velocity at any point and determine its value far from obstacle.

(d) Find the stagnations.

Solution: (a) Let $z = re^{i\theta}$, then
$$\Omega(z) = \phi + i\psi$$

$$= v_0\left(re^{i\theta} + \frac{a^2}{r}e^{-i\theta}\right)$$

$$= v_0\left(r + \frac{a^2}{r}\right)\cos\theta + iv_0\left(r - \frac{a^2}{r}\right)\sin\theta$$

from which
$$\phi = v_0\left(r + \frac{a^2}{r}\right)\cos\theta$$

and
$$\psi = v_0\left(r - \frac{a^2}{r}\right)\sin\theta$$

The streamlines are given by $\psi = \text{const} = \beta$, i.e.

$$v_0\left(r - \frac{a^2}{r}\right)\sin\theta = \beta$$

These are given by the curves of Fig. 6.48 and show the actual paths by fluid particles. Note that $\psi = 0$ corresponds to $r = a$ and $\theta = 0$ or π.

The equipotential lines are given by $\phi = \text{constant} = \alpha$, i.e.

$$v_0\left(r + \frac{a^2}{r}\right)\cos\theta = \alpha$$

These are indicated by dashed curves of Fig. 6.48 and are orthogonal to the family of streamlines.

(b) The circle $r = a$ represents a streamline, and since there can not be any flow across a streamline, it can be considered as a circular obstacle of radius a placed in the path of the fluid.

(c) We have

$$\frac{d\Omega(z)}{dz} = \Omega'(z) = v_0\left(1 - \frac{a^2}{z^2}\right)$$

$$= v_0\left(1 - \frac{a^2}{r^2}e^{-2i\theta}\right)$$

$$= v_0\left(1 - \frac{a^2}{r^2}\cos 2\theta\right) + i\frac{v_0 a^2}{r^2}\sin 2\theta$$

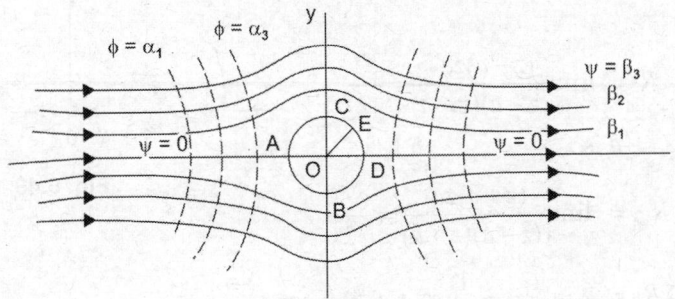

Fig. 6.48

Then complex velocity is

$$v = \Omega'(z)$$

$$= v_0\left(1 - \frac{a^2}{r^2}\cos 2\theta\right) + i\frac{v_0 a^2}{r^2}\sin 2\theta \qquad (1)$$

and its magnitude is

$$v = \sqrt{v_0^2\left(1 - \frac{a^2}{r^2}\cos 2\theta\right)^2 + v_0^2\left(\frac{a^2}{r^2}\sin 2\theta\right)^2}$$

$$= v_0\sqrt{1 - \frac{2a^2}{r^2}\cos 2\theta + \frac{a^4}{r^4}} \qquad (2)$$

Far from the obstacle, From Eq. (1) $v = v_0$ approximately, i.e. the fluid is travelling in the direction of the positive x axis with constant speed v_0.

(d) The stagnation points, i.e. points at which the velocity is zero, are given by $\Omega'(z) = 0$, i.e.

$$v_0\left(1 - \frac{a^2}{z^2}\right) = 0 \quad \text{or} \quad z = a \text{ and } z = -a$$

The stagnation points are therefore at A and D (Fig. 6.48).

Miscellaneous Solved Problems on Contour Integration

Example 1. Integrate $\displaystyle\int_{-\infty}^{+\infty} \frac{x\sin x}{x^2 - a^2}dx$, where a is real and positive.

Solution: Let us choose that integrand $\dfrac{ze^{iz}}{z^2 - a^2}$ which has poles at $z = a$ and $z = -a$ and therefore, we choose the contour as doubly intended semicircle as shown in Fig. 4.49 and there is no pole enclosed in contour and therefore, we get

$$\int_{-\infty}^{+\infty} \frac{xe^{ix}}{x^2 - a^2}dx = i\pi\Sigma R_0$$

where ΣR_0 is the sum of residues on indented poles. Both poles are simple poles. Hence residue at $z = a$ is

$$R_a = \lim_{z \to a}\frac{(z-a)ze^{iz}}{(z-a)(z+a)} = \frac{e^{ia}}{2}$$

and residue at $z = -a$ is

$$R_{-a} = \lim_{z \to -a}\frac{(z+a)ze^{iz}}{(z-a)(z+a)} = \frac{e^{-ia}}{2}$$

$$\therefore \qquad \Sigma R_0 = R_a + R_{-a} = \frac{1}{2}(e^{ia} + e^{-ia}) = \cos a$$

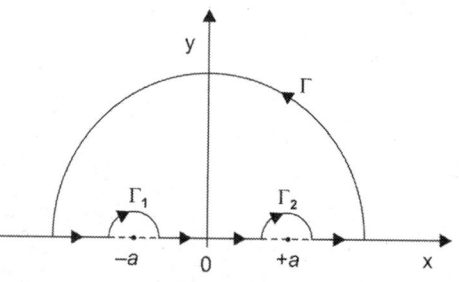

Fig. 6.49

$$\therefore \qquad \int_{-\infty}^{+\infty} \frac{xe^{ix}}{x^2 - a^2}\,dx = i\pi \cos a$$

Equating real and imaginary parts we have

$$\int_{-\infty}^{+\infty} \frac{x \sin x}{x^2 - a^2}\,dx = \pi \cos a$$

and

$$\int_{-\infty}^{+\infty} \frac{x \cos x}{x^2 - a^2}\,dx = 0$$

Example 2. *Show that* $\displaystyle\int_{0}^{+\infty} \frac{x^\alpha}{(1+x)^2}\,dx = \frac{\pi\alpha}{\sin \pi\alpha}, -1 < \alpha < 1$

Solution: Let us choose the integrand $\dfrac{z^\alpha}{(1+z)^2}$ and

solve the integral

$$I = \int_{C} \frac{z^\alpha}{(1+z)^2}\,dz$$

in the complex plane, where the contour C is taken as shown in Fig. 6.50 which is a double circle as the integrand has a double pole at $z = -1$.

In this case, z^α is a branch point and therefore

$$\lim_{|z|\to\infty} zf(z) = \lim_{|z|\to\infty} \frac{z \cdot z^\alpha}{(1+z)^2}$$

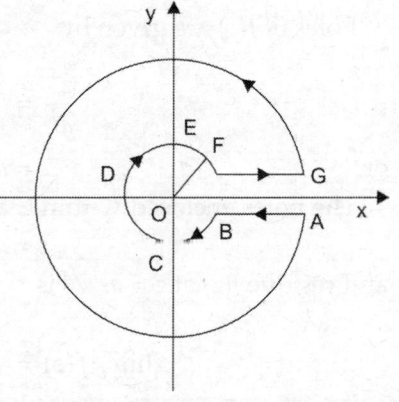

Fig. 6.50

$$= \lim_{|z|\to\infty} \frac{z^{1+\alpha}}{(1+z)^2} = \lim_{|z|\to\infty} \frac{z^{1+\alpha}}{1+z^2+2z}$$

$$= \lim_{|z|\to\infty} \frac{1}{z^{-1-\alpha} + z^{1-\alpha} + 2z^{-\alpha}} = 0 \qquad\qquad [\because\ -1 < \alpha < 1]$$

We have, therefore

$$\int_{0}^{\infty} \frac{x^\alpha}{(1+x)^2}\,dx = \frac{2\pi i \,\Sigma R}{1 - e^{2\pi i \alpha}}$$

where ΣR is the sum of residue of $\dfrac{z^\alpha}{(1+z)^2}$ at the poles enclosed in the contour of

Fig. 6.50.

The function has a double pole at $z = -1$. Hence residue at this pole is

$$\lim_{z\to -1} \frac{\dfrac{d}{dz}(z^\alpha)}{1!} = \lim_{z\to -1}(\alpha z^{\alpha-1}) = \alpha(-1)^{\alpha-1} = -\alpha e^{\pi i \alpha}$$

$$\therefore \qquad \int_{0}^{\infty} \frac{x^\alpha}{(1+x)^2}\,dx = -\frac{2\pi i \alpha e^{\pi i \alpha}}{1 - e^{2\pi i \alpha}} = \frac{\pi\alpha}{\dfrac{e^{\pi i \alpha} - e^{-\pi i \alpha}}{2i}} = \frac{\pi\alpha}{\sin \pi\alpha}$$

Example 3. *Show that m > 0, a > 0 then*

$$\int_0^\infty \frac{x\sin mx}{x^4 + a^4}dx = \frac{\pi}{2a^2}e^{-(ma/\sqrt{2})}\sin\frac{ma}{\sqrt{2}}$$

Solution: Let us consider the function $f(z) = \dfrac{ze^{imz}}{z^4 + a^4}$ in a complex plane and integrate it round the closed contour as shown in Fig. 6.24, then we have

$$\int_C f(z)dz = \int_\Gamma \frac{ze^{imz}}{z^4 + a^4}dz + \int_{-R}^{+R} \frac{xe^{imx}}{x^4 + a^4}dx = 2\pi i\,\Sigma R \qquad (1)$$

where ΣR is the residue of $f(z)$ at its poles enclosed within C.

Poles of $f(z)$ are given by

$$\frac{z^4}{a^4} = -1 = e^{i(2n+1)\pi}$$

or $\qquad\qquad\qquad\qquad z = a\,e^{i(2n+1)\pi/4}$, where $n = 0, 1, 2,...$

The poles enclosed within C are

$$z = ae^{i\pi/4}, \text{ and } z = ae^{3i\pi/4} \text{ only}$$

and residue $f(z)$ at $z = ae^{i\pi/4}$ is

$$\lim_{z\to ae^{i\pi/4}} f(z) = \lim_{z\to ae^{i\pi/4}}\left\{(z - ae^{i\pi/4})\frac{ze^{imz}}{z^4 + a^4}\right\}$$

$$= \left(\frac{ze^{imz}}{4z^3}\right)_{z=ae^{i\pi/4}} = \frac{e^{(ima/\sqrt{2})(1+i)}}{4ia^2}$$

Similarly residue at $z = ae^{i3\pi/4}$, i.e.

$$= \frac{e^{-(ima/\sqrt{2})(1-i)}}{4ia^2}$$

$\therefore \qquad\qquad\qquad \Sigma R = \dfrac{e^{-(ma/\sqrt{2})}}{4ia^2}\left[e^{ima/\sqrt{2}} - e^{-ima/\sqrt{2}}\right]$

$$= \frac{e^{-ma/\sqrt{2}}}{2a^2}\sin\frac{ma}{\sqrt{2}} \qquad (2)$$

Now $\qquad\qquad \left|\int_\Gamma \dfrac{ze^{imz}}{z^4 + a^4}dz\right| = \left|\int_0^\pi \dfrac{R^2 e^{zi\theta}d\theta\, e^{imRe^{i\theta}}}{R^4 e^{4i\theta} + a^4}\right|$

and this approaches to zero as $R \to \infty$

$\therefore \qquad \displaystyle\int_{-\infty}^{+\infty} \frac{x(\cos mx + i\sin mx)}{x^4 + a^4}dx = 2\pi i\,\frac{e^{-ma/\sqrt{2}}}{2a^2}\sin\frac{ma}{\sqrt{2}}$

Equating real and imaginary parts, we have

$$\int_{-\infty}^{+\infty} \frac{x\cos mx}{x^4 + a^4}dx = 0$$

and

$$\int_{-\infty}^{+\infty} \frac{x\sin mx}{x^4 + a^4}dx = \frac{\pi}{a^2}e^{-ma/\sqrt{2}}\sin\frac{ma}{\sqrt{2}}$$

Now

$$\int_{-\infty}^{+\infty} \frac{x\sin mx}{x^4 + a^4}dx = 2\int_{0}^{\infty} \frac{x\sin mx}{x^4 + a^4}dx = \frac{\pi}{a^2}e^{-ma/\sqrt{2}}\sin\frac{ma}{\sqrt{2}}$$

or

$$\int_{0}^{\infty} \frac{x\sin mx}{x^4 + a^4} = \frac{\pi}{2a^2}e^{-ma/\sqrt{2}}\sin\frac{ma}{\sqrt{2}}$$

Example 4. Solve the integral $\int_{0}^{\infty} \frac{\cosh \alpha x}{\cosh \pi x}dx$

Solution: Consider the integral

$$\int_{C} f(z)dz = \int_{C} \frac{e^{\alpha z}dz}{\cosh \pi z}$$

when $f(z) = \dfrac{e^{\alpha z}}{\cosh \pi z}$ and C is contour taken round the rectangle which consists of

x-axis from $-R$ to $+R$ and y-axis from 0 to 1 (Fig. 6.51).

Poles of $f(z)$ are given by $\cosh \pi z = 0$, i.e.

$$\frac{e^{\pi z} + e^{-\pi z}}{2} = 0$$

or

$$e^{\pi z} = -e^{-\pi z} = -\frac{1}{e^{\pi z}}$$

or

$$e^{2\pi z} = -1 = e^{i(2n+1)\pi}$$

or

$$z = \frac{(2n+1)}{2}i \qquad\qquad \text{[where } n = 0 \pm 1 + 2 \pm 2 + \cdots]$$

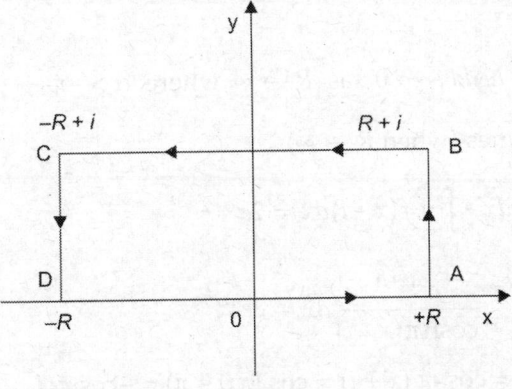

Fig. 6.51

Only the simple pole $z = \dfrac{i}{2}$ lies within the contour.

$\therefore$　Residue of $f(z)$ at $z = \dfrac{i}{2}$ is

$$\left[\frac{e^{\alpha z}}{\dfrac{d}{dz} \cosh \pi z} \right]_{z = \frac{i}{2}} = \left[\frac{e^{i\alpha/2}}{\pi \sinh \pi i/2} \right] = \frac{e^{i\alpha/2}}{\pi i \sin \pi/2} = \frac{e^{i\alpha/2}}{\pi i}$$

Now by Cauchy's residue theorem, we have

$$\int_{-R}^{+R} f(x)dx + \int_0^1 f(R+iy)idy + \int_{+R}^{-R} f(x+i)dx - \int_0^1 f(-R+iy)idy = \frac{2\pi i \, e^{i\alpha/2}}{\pi i} = 2e^{i\alpha/2} \qquad (1)$$

$$\left| \int_0^1 f(R+iy)idy \right| \leq \int_0^1 |f(R+iy)| \, |i| \, |dy|$$

$$\leq \int_0^1 \left| \frac{e^{\alpha(R+iy)}}{\cosh \pi (R+iy)} \right| dy \qquad \text{as } |i| = 1$$

$$\leq \int_0^1 \frac{2e^{\alpha R} \, |e^{\alpha iy}| \, dy}{\left| e^{\pi(R+iy)} + e^{-\pi(R+iy)} \right|}$$

$$\leq \int_0^1 \frac{2e^{\alpha R} dy}{\left| e^{\pi iy} \right| \left(e^{\pi R} + e^{-\pi R} \right)}$$

$$\leq \frac{2e^{\alpha R}}{e^{\pi R} + e^{-\pi R}} \quad \text{which} \to 0 \text{ as } R \to \infty$$

This is satisfied when $\alpha < \pi$

$\therefore$　$\displaystyle \int_0^1 f(R+iy)idy \to 0$ as $R \to \infty$

Similarly

$$\int_0^1 f(-R+iy)idy \to 0 \quad \text{as } R \to \infty \text{ where } \alpha > -\pi$$

Hence Eq. (1) becomes (when $R \to \infty$)

$$\int_{-\infty}^{+\infty} f(x)dx + \int_{+\infty}^{-\infty} f(x+i)dx = 2e^{i\alpha/2}$$

or　$$\int_{-\infty}^{+\infty} \frac{e^{\alpha x}}{\cosh \pi x} dx + \int_{+\infty}^{-\infty} \frac{e^{\alpha(x+i)}}{\cosh \pi (x+i)} dx = 2e^{i\alpha/2}$$

But　$\cosh \pi (x + i) = \cos i\pi (x + i) = \cos (\pi x i - \pi) = -\cos \pi x i$

or　$\cosh \pi (x + i) = -\cosh \pi x$

$$\therefore \qquad \int_{-\infty}^{+\infty} \frac{e^{\alpha x}dx}{\cosh \pi x} - \int_{\infty}^{-\infty} \frac{e^{\alpha(x+i)}}{\cosh \pi x}dx = 2e^{i\alpha/2}$$

or
$$\int_{-\infty}^{+\infty} \frac{e^{\alpha x}dx}{\cosh \pi x} + \int_{-\infty}^{+\infty} \frac{e^{i\alpha}e^{\alpha x}dx}{\cosh \pi x} = 2e^{i\alpha/2}$$

$$(1+e^{i\alpha})\int_{-\infty}^{+\infty} \frac{e^{\alpha x}}{\cosh \pi x}dx = 2e^{i\alpha/2}$$

or
$$\int_{-\infty}^{+\infty} \frac{e^{\alpha x}}{\cosh \pi x}dx = \frac{2e^{i\alpha/2}}{1+e^{i\alpha}} = \frac{2}{e^{i\alpha/2}+e^{-i\alpha/2}} = \sec \alpha/2$$

or
$$\int_{-\infty}^{0} \frac{e^{\alpha x}}{\cosh \pi x}dx + \int_{0}^{\infty} \frac{e^{\alpha x}}{\cosh \pi x}dx = \sec \alpha/2$$

Now if we replace x by $-x$ in the first integral, we get

$$\int_{-\infty}^{0} \frac{e^{\alpha x}}{\cosh \pi x}dx = \int_{0}^{\infty} \frac{e^{-\alpha x}}{\cosh \pi x}dx$$

$$\therefore \qquad \int_{0}^{\infty} \frac{e^{-\alpha x}dx}{\cosh \pi x} + \int_{0}^{\infty} \frac{e^{\alpha x}}{\cosh \pi x}dx = \sec \alpha/2$$

or
$$\int_{0}^{\infty} \frac{e^{\alpha x}+e^{-\alpha x}}{\cosh \pi x} = \sec \alpha/2$$

or
$$2\int_{0}^{\infty} \frac{\cosh x}{\cosh \pi x} = \sec \alpha/2$$

or
$$\int_{0}^{\infty} \frac{\cosh \alpha x}{\cosh \pi x} = \frac{1}{2}\sec \alpha/2$$

SHORT ANSWER QUESTIONS

1. Define a pole of order m.

2. Expand log $(1 + z)$ in Taylor series about $z = 0$

3. Find the residue at each pole of the function $f(z) = \cot z$

Ans. 1

4. Define analytical function of a complex variable.

5. Find the singularities of the function $f(z) = \dfrac{1}{\sin \pi/z}$; what is the nature of singularity at $z = 0$.

Ans. $z = 1, \dfrac{1}{2}, \dfrac{1}{3}, \cdots, 0, z = 0$ is a nonessential singularity.

6. Locate the name of the singularity of $f(z) = \dfrac{\sin z}{z - \pi}$

7. Define a pole and residue of a pole.

8. Define pole and zero of a complex function. Illustrate each with an example.

9. Find the region of convergence of Taylor series of log $(1 + z)$ about $z = 0$.

10. Express Taylor series expansion of function $f(z)$ with centre at z_0.

11. State necessary and sufficient conditions for a function to be analytic.

12. If $f(z)$ is not analytic in a domain, show that $\int_a^b f(z)dz$ depends on path.

13. Expand the function $f(z) = \dfrac{1}{z+1}$ about $z = 1$ in Taylor's series.

Ans. $\dfrac{1}{2}\left(1 + \dfrac{z}{2} + \dfrac{z^2}{4} + \dfrac{z^3}{8} + \cdots\right) - (z^{-1} + z^{-2} + z^{-3} + \cdots 1)$

14. Locate the name and the singularities in the finite z-plane of $f(z) = \dfrac{\log(z + 5i)}{z^3}$

15. Find the kind of singularities of the following.

i. $\dfrac{\cos \pi z}{(z-a)^2}$ at $z = 0$ and $z = \infty$

Ans. Pole of order 2 at $z = a$ and essential singularity.

ii. $\tan\left(\dfrac{1}{z}\right)$ at $z = 0$

Ans. Nonisolated essential singularity.

iii. $\operatorname{cosec} \dfrac{1}{z}$ at $z = 0$

Ans. Nonisolated essential singularity.

iv. $\sin\dfrac{1}{1-z}$ at $z = 1$

Ans. Isolated essential singularity.

16. State Cauchy–Riemann conditions for analyticity of a function of complex variable is cartesian and polar coordinates.

17. Test if the function

i. $\sin z$ and　ii. $\log z$ are analytical

Ans. i. Analytical　ii. Analytical except at $z = 0$

18. If a function $f(z)$ is analytical at z_0, prove that it is continuous.

19. Prove that if a function is analytical on and in side a closed contour, its integral over the function must vanish.

20. State Cauchy residue theorem.

21. Determine whether $f(z) = \dfrac{z}{z+1}, z \neq 1$ satisfies Cauchy–Riemann coordinates.

22. Differentiate between poles and essential singularities of an analytical function.

23. How will you find the residue at a simple pole?

24. State a theorem to find the residue of a function $f(z)$ at infinity.

25. Test if the function i. $\dfrac{1}{z}$ ii. $|z|^2$ are analytic.

Ans. i. Analytic   ii. Not analytic

26. Show that for any analytic function $f = u + iv$, the relation
$$|\Delta u| = |\Delta v| \text{ exists.}$$

27. What do you mean by harmonic function? The real and imaginary parts of a complex function $f(z) = u(x, y) + iv(x, y)$ in a domain have continuous second order partial derivatives and satisfy Laplace equation $\nabla^2 u = \nabla^2 v = 0$. What will be the nature of function $f(z)$?

28. Express Laurent series of analytical function $f(z)$ about centre $z = z_0$.

29. Show that real and imaginary parts of an analytic function $f(z)$ are harmonic.

30. What do you mean by simply-connected and multiply-connected domains?

31. Prove and explain Cauchy integral theorem.

32. Prove and explain Cauchy integral formula.

33. Explain the term 'singular point of an analytical function'.

PROBLEMS

1. Prove that $f(z) = \log z$ has a branch point at $z = 0$.

2. Find branch points and construct branch lines for the functions

(a) $f(z) = \{z/(1-z)\}^{1/2}$ (b) $f(z) = (z^2 - 4)^{1/3}$ (c) $f(z) = \log(z - z^2)$

3. Construct a Riemann surface for the functions (a) $z^{1/2}$ (b) $z^{1/2}(z-1)^{1/2}$ (c) $\left(\dfrac{z+2}{z-2}\right)$

(d) $\log(z + 2)$ (e) $\sin^{-1} z$ (f) $\tan^{-1} z$

4. Show that the Riemann surface of the function $z^{1/2} + z^{1/3}$ has 6 sheets.

5. Find an analytic function $f(z)$ such that
$$\text{Re}\{f'(z)\} = 3x^2 - 4y - 3y^2 \text{ and } f(1 + i) = 0 \qquad \textit{[Ans. } z^3 + 2iz^2 + 6 - 2i\textit{]}$$

6. If u and v are conjugate harmonic functions, prove that
$$dv = \frac{du}{\partial x}dy - \frac{du}{\partial y}dx .$$

7. If u and v are conjugate harmonic functions, prove that
$$\left(\frac{\partial u}{\partial y} - \frac{\partial v}{\partial x}\right) + i\left(\frac{\partial u}{\partial x} + \frac{\partial v}{\partial y}\right)$$

is analytic in R.

8. Express the Cauchy–Riemann equations in terms of the curvilinear coordinates (ξ, η) where $x = e^\xi \cosh \eta$, $y = e^\xi \sinh \eta$.

9. Show that $u(x, y)$ and $v(x, y)$, the real and imaginary parts of an analytic function can have neither a maxima nor a minima in the interior of any region in which the function is analytic.

10. Find the analytic function $F(z) = u(x, y) + iv(x, y)$ if

   (a) $u(x, y) = x^3 - 3xy^2$ (b) $v(x, y) = e^{-y} \sin x$

11. Consider the function $f(z) = (3y^2 - x^3) + i(6xy^2 - 3yx^2)$

   i. Are the Cauchy–Riemann conditions satisfied at each point of the real axis?

   ii. Are the partial derivatives $\dfrac{\partial u}{\partial x}, \dfrac{\partial u}{\partial y}, \dfrac{\partial v}{\partial x}$ and $\dfrac{\partial v}{\partial y}$ continuous at each point of the real axis?

   iii. Is the function $f(z)$ analytic at each point of the real axis?

12. Let $u(x, y)$ be harmonic and have continuous derivatives of order two at least in a region R.

   (a) Show that

$$v(x, y) = \int_{(a,b)}^{(x,y)} \left\{ -\frac{\partial u}{\partial y} dx + \frac{\partial u}{\partial x} dy \right\}$$

   is independent of the path in R joining (a, b) to (x, y).

   (b) Prove that $u + iv$ is an analytic function of $z = x + iy$ in R.

   (c) Prove that v is harmonic in R.

13. Evaluate $\displaystyle\oint_C \frac{z^2 dz}{z^2 + 4}$ where C is the square with vertices at $\pm 2, \pm 2 + 4i$.

14. (a) Show that $\displaystyle\oint_C \frac{dz}{z+1} = 2\pi i$ if C is the circle, $|z| = 2$.

   (b) Use statement (a) to show that

$$\oint_C \frac{(x+1)dx + y\,dy}{(x+1)^2 + y^2} = 0$$

$$\oint_C \frac{(x+1)dy - y\,dy}{(x+1)^2 + y^2} = 2\pi$$

   and verify these results directly.

15. If $t > 0$ and C is any simple closed curve enclosing $z = -1$, prove that

$$\frac{1}{2\pi i}\oint_C \frac{ze^{zt}}{(z+1)^3} dz = \left(t - \frac{t^2}{2} \right) e^{-t}$$

16. Consider the multivalued function $F(z) = \log\dfrac{z+1}{z-1}$. Construct a branch $f_1(z)$ of $F(z)$ which is continuous everywhere except for a cut along the real axis from $z = -1$ to $z = +1$. Specify the branch carefully, i.e. give exact formulas for calculation of the values of $f_1(z)$ for all values of z. Determine the value of

$$\lim_{\varepsilon \to 0} \frac{1}{2}[f_1(x - i\varepsilon) + f_1(x + i\varepsilon)]$$

17. Show that among the following complex series of constants

(a) $\displaystyle\sum_{n=1}^{\infty} \frac{i^n}{n\sqrt{n}}$

(b) $\displaystyle\sum_{n=1}^{\infty} e^{-n(1+i)}$

(c) $\displaystyle\sum_{n=2}^{\infty} \frac{i^n}{\log n}$

(d) $\displaystyle\sum_{n=1}^{\infty} \frac{(1+i)}{n}$

The series (a) and (b) converge absolutely, the series (c) converges but not absolutely, and the series (d) diverges.

18. If $F(z) = \dfrac{3z-3}{(2z-1)(z-2)}$, find a Laurent series of $F(z)$ about $z = 1$ having convergence of $\dfrac{1}{2} < |z-1| < 1$

[*Ans.* $-\dfrac{1}{8}(z-1)^{-4} + \dfrac{1}{4}(z-1)^{-3} - \dfrac{1}{2}(z-1)^{-2} + (z-1)^{-1} - 1 - (z-1) - (z-1)^2 \cdots$]

19. Let $G(z) = (\tan^{-1} z)/z^4$ (a) Expand $G(z)$ in Laurent series. (b) Determine the region of convergence of the series in (a). (c) Evaluate $\displaystyle\oint_C G(z)dz$, where C is a square

with vertices at $2 \pm 2i$, $-2 \pm 2i$. $\left[\textbf{Ans. } (a) \dfrac{1}{z^3} - \dfrac{1}{3z} + \dfrac{z}{5} - \dfrac{z^3}{7} + \cdots (b) |z| > 0 \ (c) -\dfrac{1}{3} \right]$

20. For each of the functions $ze^{1/z^2}, (\sin^2 z)/z, 1/z(z-4)$ which have singularities at $z = 0$ (a) Give a Laurent expansion about $z = 0$ and determine the region of convergence (b) State in each case whether $z = 0$ is a removable singularity, essential singularity or a pole (c) Evaluate the integral of the function about the circle $|z| = 2$.

[*Ans.* (a) $z + z^{-1} + \dfrac{z^{-3}}{2!} + \dfrac{z^{-5}}{3!} + \cdots, |z| > 0$ (b) essential singularity

(a) $2z - \dfrac{2z^3}{3} + \dfrac{4z^5}{45} - \cdots |z| \geq 0$ (b) removable singularity

(a) $\dfrac{1}{16} - \dfrac{z}{64} + \dfrac{z^2}{256} + \cdots 0 < |z|$ (b) pole (c) $2\pi i, 0, \pi i/2$]

21. Show that $u(r, \theta) = \dfrac{R^2 - r^2}{2\pi} \displaystyle\int_0^{2\pi} \dfrac{u(\phi)d\phi}{R^2 - 2rR\cos(\theta - \phi) + r^2}$

$= \dfrac{a_0}{2} + \displaystyle\sum_{n=1}^{\infty} \left(\dfrac{r}{R}\right)^n \{a_n \cos n\theta + b_n \sin n\theta\}$

where $\quad a_n = \dfrac{1}{\pi} \displaystyle\int_0^{2\pi} u(\phi)\cos n\phi\, d\phi, \quad b_n = \dfrac{1}{\pi} \displaystyle\int_0^{2\pi} u(\phi)\sin n\phi\, d\phi$

22. Expand each of the following functions in a Laurent series about $z = 0$, naming the type of singularity in each case.

(a) $(1 - \cos z)/z$　(b) e^{z^2}/z^3　(c) $z^{-1}\cosh z^{-1}$　(d) $z^2 e^{-z^4}$　(e) $z\sinh\sqrt{z}$

[Ans. (a) $\dfrac{z}{2!} - \dfrac{z^3}{4!} + \dfrac{z^5}{6!}\cdots$, removable singularity.

(b) $\dfrac{1}{z^3} + \dfrac{1}{z} + \dfrac{z}{2!} + \dfrac{z^3}{3!} + \dfrac{z^5}{4!} + \dfrac{z^7}{5!} + \cdots$, pole of order 2.

(c) $\dfrac{1}{z} - \dfrac{1}{2!z^3} + \dfrac{1}{4!z^5}\cdots$, essential singularity.

(d) $z^2 - z^6 + \dfrac{z^{10}}{2!} + \dfrac{z^{14}}{3!}\cdots$, ordinary point.

(e) $z^{3/2} + \dfrac{z^{5/2}}{3!} + \dfrac{z^{7/2}}{5!} + \dfrac{z^{9/2}}{7!} + \cdots$, branch point]

23. Prove that Cauchy' theorem and integral formulas can be obtained as special cases of the residue theorem.

24. For each of the following functions determine the poles and the residue at the poles.

i. $\dfrac{2z-1}{z^2 - z - 2}$　ii. $\dfrac{(z+1)^2}{(z-1)^2}$　iii. $\dfrac{\sin z}{z^2}$　iv. $\sec hz$　v. $\cot z$

[Ans. i. $z = -1, 2$, Residue $\dfrac{1}{3}, \dfrac{5}{3}$　ii. $z = 1$, Res $= 4$　iii. $z = 0$, Res 1

iv. $z = \dfrac{1}{2}(2K+1)\pi i$, Res. $(-1)^{k+1}i$, where $K = 0, \pm 1, \pm 2\ldots$

v. $z = K\pi i$ Res. 1, where $K = 0, \pm 1, \pm 2\ldots$]

25. Find the residue of $F(z) = \dfrac{\cot z \cdot \coth z}{z^3}$ at $z = 0$.　　[Ans. $-7/45$]

26. Evaluate $\oint_C \dfrac{2 + 3\sin \pi z}{z(z-1)^2}dz$, where C is square having vertices at $3 + 3i$, $3 - 3i$, $-3 + 3i$, $-3 - 3i$.　　[Ans. $-6\pi i$]

27. Show that $\displaystyle\int_0^\infty \dfrac{\cos 2\pi x}{x^4 + x^2 + 1}dx = -\dfrac{\pi}{2\sqrt{3}}e^{-\pi\sqrt{3}}$

28. Prove that $\displaystyle\int_0^\infty \dfrac{\sin^2 x}{x^2}dx = \dfrac{\pi}{2}$

29. Prove that the sum of the residues of the function $\dfrac{2z^5 - 4z^2 + 5}{3z^6 - 8z + 10}$ at all the poles is 2/3.

30. If n is a positive integer, prove that $\displaystyle\int_0^{2\pi} e^{\cos\theta}\cos(n\theta-\sin\theta)d\theta = \frac{\pi}{n!}$.

31. Prove that $\displaystyle\int_0^\infty \frac{\sin ax}{e^{2\pi x}-1}dx = \frac{1}{4}\coth\frac{a}{2} - \frac{1}{2a}$

[*Hint:* Integrate $\dfrac{e^{iaz}}{e^{2\pi z}-1}$ around a rectangle with vertices at $0, R, R+i, i$ and let $R\to\infty$]

32. Evaluate the following definite integrals.

i. $\displaystyle\int_{-\infty}^{+\infty} e^{i\lambda x}\frac{\sinh ax}{\sinh \pi x}dx$ if $-\pi < a < +\pi$ $\left[\text{Ans. } \dfrac{\sin a}{\cos a + \cosh \lambda}\right]$

ii. $\displaystyle\int_0^\infty \frac{x\sin x}{x^2+1}dx$ [*Ans.* $\pi/2e$]

iii. $\displaystyle\int_{-\infty}^{+\infty} \frac{\cos x}{(x^2+a^2)(x^2+b^2)}dx$ $[a\ne b]$ $\left[\text{Ans. } \dfrac{\pi}{a^2-b^2}\left(\dfrac{e^{-b}}{b}-\dfrac{e^{-a}}{a}\right)\right]$

iv. $\displaystyle\int_0^{2\pi} \cos^n\theta\, d\theta$ $\left[\text{Ans. } \dfrac{n!\pi}{2^{n-1}(n/2!)^2}\; n,\text{ is an even integer}\right]$

v. $\displaystyle\int_{-\infty}^{+\infty} \frac{\cosh x}{\cosh \pi x}dx$ $\left[\text{Ans. } \sec\dfrac{a}{2}(|a|<\pi)\right]$

vi. $\displaystyle\int_{-\infty}^{+\infty} \frac{e^x\cdot x^2}{1+e^{2x}}dx$ [*Ans.* $\pi^3/8$]

[*Hint:* use the rectangle with the sides $y=0, y=\pi, x=-R$ and $x=R$]

vii. $\displaystyle\int_0^{2\pi} \frac{(1+2\cos\theta)^n\cos n\theta}{3+2\cos\theta}d\theta$ $\left[\text{Ans. } \dfrac{2\pi}{\sqrt5}(3-\sqrt5)^n\, n>0\right]$

viii. $\displaystyle\int_0^1 \frac{dx}{\sqrt[3]{x^2-x}}$ $\left[\text{Ans. } \dfrac{2\pi}{\sqrt3}\right]$

ix. $\displaystyle\int_0^\infty \frac{x^{2m}}{x^{2m}+1}dx$ [*Ans.* $\pi/2n \sin\{(2m+1)/2n\}\pi$]

x. $\displaystyle\int_0^\infty \frac{\sin \pi x}{x(1-x^2)}dx$ [*Ans.* π]

33. i. If r is real, prove that $\displaystyle\int_0^\pi \log(1-2r\cos\theta+r^2)d\theta$

$$= \begin{cases} 0 & \text{if } |r|\le 1 \\ \pi\log r^2 & \text{if } |r|>1 \end{cases}$$

ii. Use the result in (*i*) to evaluate $\displaystyle\int_0^{\pi/2} \log\sin\theta\cdot d\theta$ $\left[\text{Ans. } -\dfrac{1}{2}\pi\log 2\right]$

34. Show that the function $u(x,y) = \dfrac{\sin 2x}{\cosh 2y + \cos 2x}$ can serve as a real part of some analytic function $f(z)$. Evaluate $v(x, y) = Imf(z)$. Express $f(z)$ explicitly in terms if z.

35. Determine a transformation which rotates the ellipse $x^2 + xy + y^2 = 5$ so that the major and minor axes are parallel to the coordinate axes. What are lengths of major and minor axes?

36. Let $u = u(x, y)$, $v = v(x, y)$ be transformation of points of the xy plane on to points of the uv plane.

    i.   Show that in order that the transformation preserves angles it is necessary and sufficient that

$$\left(\frac{\partial u}{\partial x}\right)^2 + \left(\frac{\partial v}{\partial x}\right)^2 = \left(\frac{\partial u}{\partial y}\right)^2 + \left(\frac{\partial v}{\partial y}\right)^2$$

    and $\qquad \dfrac{\partial u}{\partial x}\dfrac{\partial u}{\partial y} + \dfrac{\partial v}{\partial x}\dfrac{\partial v}{\partial y} = 0$

    ii.   Deduce from statement (i) that we must have either

    (a) $\dfrac{\partial u}{\partial x} = \dfrac{\partial v}{\partial y}, \dfrac{\partial u}{\partial y} = -\dfrac{\partial v}{\partial x}$ or   (b) $\dfrac{\partial u}{\partial x} = -\dfrac{\partial v}{\partial y}, \dfrac{\partial u}{\partial y} = \dfrac{\partial v}{\partial x}$

    Thus conclude that $u + iv$ must be analytic function of $x + iy$.

37. Suppose the mapping function $w = f(z)$ has Taylor series expansion

$$w = f(z) = f(a) + f^{(1)}(a)(z - a) + \cdots + \frac{f^{(n)}(a)}{n!}(z - a)^n + \cdots$$

Show that $f^{(k)}(a) = 0$ for $k = 0, 1,..., n - 1$ while $f^{(n)} \neq 0$ then angles in the z-plane with vertical at $z = a$ are multiplied by n in the w-plane.

38. A circular cylinder obstacle of radius a rests at the bottom of a channel of fluid which at distances far from the obstacle flows with velocity v_0.

    i.   Prove that the complex potential is given by $\Omega(z) = \pi a v_0 \coth (\pi a/z)$

    ii.   Show that the speed at the top of the cylinder is $\dfrac{1}{4}\pi^2 v_0$ and compare with that for a circular obstacle in the middle of a fluid.

    iii.   Show that the difference in pressure between top and bottom of the cylinder is $\sigma\pi^4 v_0^9 /32$

39. Show that the image of the circle $(x - 1)^2 + y^2 = 1$ under the mapping $w = z^2$ is a cardioid $\rho = 2 (1 - \cos \phi)$, where $w = \rho e^{i\phi}$. Sketch the cardioid and locate the point where the invariance of angles breaks down.

[*Hint*: Express the equation of the circle in polar coordinates in the z-plane].

40. The complex transformation $w = be^{z-a}$, where a and b are complex constants, maps the isoceles triangle T with the vertices at $z_1 = b, z_2 = b + 1, z_3 = b + 1 + i$ into some curvilinear triangle T' in the w-plane. Illustrate the invariance of angles property of this mapping by explicitly calculating the interior angles of T' to be $90°$, $45°$ and $45°$.

41. (a) State and prove Cauchy's integral formula.

    (b) $\displaystyle\int_{-\infty}^{+\infty} \frac{\cos x}{x^2 + a^2}\,dx = \frac{\pi}{2a}e^{-a}$

42. (a) State Jordan' Lemma and hence show that

$$\int_{-\infty}^{+\infty} \frac{\cos x}{x^2 + a^2}\,dx = \frac{\pi e^{-a}}{2a}, \text{ where } a > 0$$

    (b) Explain the meaning of branch points and branch cuts.

    (c) Show that

$$\int_0^\infty \frac{x^{n-1}}{1+x}\,dx = \frac{\pi}{\sin n\pi}, \text{ where } 0 < n < 1$$

43. (a) Derive Cauchy–Riemann conditions for the existence of unique derivative of an arbitrary function of a complex variable.

    (b) Show that $\displaystyle\int_0^\infty \frac{x^{n-1}}{1+x}\,dx = \frac{\pi}{\sin n\pi}$, where $0 < n < 1$

    (c) Evaluate the integral $\displaystyle\int_0^\infty \frac{dx}{x^2 + a^2}$

44. (a) State and prove cauchy integral formula.

    (b) Evaluate (i) $\displaystyle\int_0^\infty \frac{dx}{x^4 + a^4}$ (ii) $\displaystyle\int_0^\infty \frac{\log(1+x^2)}{1+x^2}\,dx$ (iii) $\displaystyle\int_0^\infty \frac{\cos 2ax - \cos 2bx}{x^2}\,dx$

45. (a) Given $v(x, y) = -\cos x \sinh y$, find $u(x, y)$ and $w(z)$.

    (b) By integrating $\dfrac{e^{iz^2}}{z}$ around a suitable contour, show that

$$\int_0^\infty \frac{\sin x^2}{x^2}\,dx = \pi/4$$

    (c) Make a Laurent's series expansion of

$$w(z) = \frac{1}{z(z-1)}$$

    about the point $z = 0$.

    (d) Distinguish between poles and essential singular points.

46. Discuss the difficulties encountered in dealing with multivalued functions. Show how these can be removed with the help of Riemann surfaces. Explain the construction of Riemann surface for $\sqrt{z}$ and $\sqrt{z^2 - a^2}$.

MULTIPLE CHOICE QUESTIONS

1. If z_1 and z_2 are two complex numbers, then the correct relation is
   (a) $|z_1 + z_2| = |z_1| + |z_2|$
   (b) $|z_1 + z_2| \geq |z_1| - |z_2|$
   (c) $|z_1 + z_2| \leq |z_1| + |z_2|$
   (d) $|z_1 + z_2| < |z_1| + |z_2|$

2. If z_1 and z_2 are two complex numbers, then correct relation is
   (a) $|z_1 - z_2| = |z_1| + |z_2|$
   (b) $|z_1 - z_2| \geq |z_1| - |z_2|$
   (c) $|z_1 - z_2| \leq |z_1| + |z_2|$
   (d) $|z_1 - z_2| < |z_1| + |z_2|$

3. Logarithm of a complex number $z = (r, \theta)$ is a
   (a) multivalued function
   (b) singlevalued function
   (c) can be a multivalued function and can be singlevalued function
   (d) data is insufficient

4. For complex function $\arg z + \arg \bar{z}$ is (n is an integer including zero)
   (a) $2\pi n$
   (b) πn
   (c) $(2n + 1)\pi$
   (d) $(2n + 1)\,\pi/2$

5. If z_1 and z_2 are two complex numbers, then $|z_1 - z_2|^2 + |z_1 + z_2|^2$ has the values
   (a) $4|z_1||z_2|$
   (b) $4|z_1|^2|z_2|^2$
   (c) $2|z_1|^2 + 2|z_2|^2$
   (d) $2|z_1|^2 - 2|z_2|^2$

6. The two curves u and v are constant in a complex plane intersect at angles
   (a) $\dfrac{\pi}{4}$
   (b) $\dfrac{\pi}{2}$
   (c) $\dfrac{3\pi}{4}$
   (d) zero

7. If $u(x, y) = x^2 - y^2$ is the real part of an analytic function
   (a) $2xy$
   (b) $x^2 + y^2$
   (c) $4xy$
   (d) $2(x^2 + y^2)$

8. The value of integral $\int_C \bar{z}\,dz$, where C is upper half of the circle $|z| = 1$, from $z = -1$ to $z = +1$ is
   (a) $i\pi$
   (b) $-i\pi$
   (c) i
   (d) $-i$

9. The residue of the function $\dfrac{z^2}{z^2 + a^2}$ at $z = ia$ is
   (a) $\dfrac{ia}{2}$
   (b) $-\dfrac{ia}{2}$
   (c) $\dfrac{ia^2}{2}$
   (d) $-\dfrac{ia^2}{2}$

10. The residue of the function $\dfrac{\cos z}{z}$ at $z = 0$ is

(a) i (b) $-i$

(c) 1 (d) -1

11. The necessary condition for function $f(z)$ to be analytic is $[f(z) = u(x, y) + iv(x, y)]$

(a) $U_x = V_x,\ U_y = -V_y$ (b) $U_x = -V_v,\ U_y = V_x$

(c) $U_x = V_y,\ U_y = -V_x$ (d) $U_x = V_y,\ U_y = V_x$

12. If $f(z)$ is analytic and has derivatives, then

(a) $\dfrac{df}{dz} = \dfrac{\partial f}{\partial y}$ (b) $\dfrac{df}{dz} = \dfrac{\partial f}{\partial x} + i\dfrac{\partial f}{\partial y}$

(c) $\dfrac{df}{dz} = \dfrac{\partial f}{\partial x}$ (d) $\dfrac{df}{dz} = \dfrac{\partial f}{\partial x} - i\dfrac{\partial f}{\partial y}$

13. If $f(z) = u + iv$ is an analytic function, then

(a) $\dfrac{\partial^2 u}{\partial x^2} + \dfrac{\partial^2 v}{\partial y^2} = 0$ (b) $\dfrac{\partial^2 u}{\partial y^2} + \dfrac{\partial^2 v}{\partial x^2} = 0$

(c) $\dfrac{\partial^2 u}{\partial x^2} + \dfrac{\partial^2 u}{\partial y^2} = 0$ (d) $\dfrac{\partial^2 u}{\partial x^2} + 1\dfrac{\partial^2 v}{\partial x^2} = 0$

14. If $f(z)$ is analytic and has a pole of order m at $z = a$ then $\dfrac{1}{f(z)}$ has

(a) zero of order m at $z = a$ (b) zero of order m at $z = \dfrac{1}{a}$

(c) zero of order 1 at $z = a$ (d) zero of order 1 at $z = \dfrac{1}{a}$

15. The residue of $\dfrac{z}{(z-a)(z-b)}$ at infinity is

(a) -1 (b) $+1$

(c) $-i$ (d) $+i$

16. The analytic function $f(z)$ whose real part is $x^2 - y^2$ is

(a) z (b) z^2

(c) $|z|^2$ (d) z^{-2}

17. The function $f(z) = \dfrac{1}{z^2 - 1}$ in the contour C given by $x^2 + y^2 = 4$ has

(a) no simple pole   (b) a simple pole at $z = 1$

(c) two simple poles at $z = \pm 1$ (d) two simple poles at $z = \pm i$

18. The value of $\oint_C \dfrac{dz}{z - 3}$ if C is the circle $|z - 2| = 5$ is

(a) πi (b) $2\pi i$

(c) 0 (d) 2π

19. The residue of the function $f(z) = \dfrac{z^2}{z^2 + 4}$ at $z = 2i$ is

    (a) $e^{i\pi/2}$ (b) $e^{i\pi}$

    (c) $e^{3i\pi/2}$ (d) $e^{-i\pi/2}$

20. The value of magnitude of the integral $\displaystyle\int_C \frac{dz}{z}$, where C is $|z| = r$ is equal to

    (a) $2\pi r$ (b) 2π

    (c) π (d) $\log r$

21. The function $f(z) = |z|^2$ with $z = x + iy$ is

    (a) differentiable for all values of x and y

    (b) differentiable for only positive values of x and y

    (c) differentiable at $x = 0$, $y = 0$ only

    (d) differentiable at $x = 1$ and $y = 1$ only

22. The function $f(z) = z^3$ with $z = x + iy$ is analytic

    (a) in entire z-plane

    (b) for positive values of x and y only

    (c) at $x = 0$, $y = 0$ only

    (d) on the line $y = 1$ only

23. If $f(z)$ is an analytic function of the complex variables z, which satisfies the following conditions:

    i. It is analytic in the upper half plane except at finite number of poles

    ii. $f(z_0 \to 0)$ uniformly as $|z| \to \infty$ for $0 < \arg z < \pi$

    iii. $\displaystyle\lim_{R \to \infty}\int_C^R e^{imz} f(z)dz = 0$, where C is a semicircle of radius R and centre at origin.

    This is called

    (a) Jordan's Lemma   (b) Laurent's theorem

    (c) Cauchy's theorem   (d) Goursat's Lemma

24. If a function $f(z)$ has poles only in the finite part of the z-plane then the function is said to be

    (a) Laurent's function   (b) Lioville's function

    (c) residue function   (d) meromorphic function

25. The Cauchy integral theorem states that if $f(z)\, dz$ is analytic in a simply connected domain D; then $\displaystyle\int_C f(z) = 0$ on every simply closed path C in domain d the condition of analytic in this theorem is

    (a) necessary   (b) sufficient

    (c) necessary and sufficient   (d) arbitrary

26. The complex function $f(z) = e^{-\frac{1}{(z-1)^2}}$ has a point $z = 1$

(a) a pole of order 1   (b) a pole of order 2

(c) an isolated essential singularity   (d) a nonisolated essential singularity

27. For two given function $u = x^2 - y^2$ and $v = \dfrac{y}{z^2 + y^2}$, which of the following is

(a) both functions u and v are harmonic

(b) the function u is harmonic but v is not harmonic

(c) the function v is harmonic but u is not harmonic

(d) both functions u and v are not harmonic

28. The necessary condition for function $f(z) = u + iv$ to be analytical at all points in the region R are

(a) $\dfrac{\partial u}{\partial x} = \dfrac{\partial v}{\partial y}$ and $\dfrac{\partial u}{\partial y} = \dfrac{\partial v}{\partial x}$ provided they exist

(b) $|z^2| = x^2 + y^2$ only

(c) u and v are harmonic functions

(d) $\dfrac{\partial u}{\partial u} = -\dfrac{\partial v}{\partial y}, \dfrac{\partial u}{\partial y} = \dfrac{\partial v}{\partial x}$ provided they exist

29. If a function $f(z)$ has a simple pole of order n at $z = z_0$ then the function $f''(z)/f'(z)$ has

(a) no pole   (b) a simple pole

(c) pole of order $(n-1)$ (d) an essential singularity

30. If C is a circle $|z - 1| = 3$ in the complex plain then $\displaystyle\int_C \dfrac{\cos z}{z - \pi} dz$ is equal to

(a) πi (b) $-\pi i$

(c) $2\pi i$ (d) zero

31. A function which is analytic at all points of the z-plane and finite at infinity

(a) must have singularity   (b) must be zero

(c) must be a constant   (d) can not exist

32. The value of the integral $\displaystyle\int_{-\infty}^{+\infty} \dfrac{\cos mx}{x} dx$ is

(a) 0   (b) 1

(c) $\dfrac{\pi}{2}$ (d) π

33. Which of the following integral have value different from others

(a) $\displaystyle\int_0^\infty \dfrac{\sin x}{x} dx$ (b) $\displaystyle\int_0^\infty \dfrac{\sin mx}{m} dx$

(c) $\displaystyle\int_0^\infty \dfrac{1 - \cos x}{x^2} dx$ (d) $\displaystyle\int_0^\infty \dfrac{\log x}{(1 + x^2)^2} dx$

34. Value of $\oint \dfrac{z^2}{9z^2+4}dz$, where C $|z+1| = 1$ is

(a) $2\pi i$

(b) $-2\pi i$

(c) $\dfrac{\pi}{2i}$

(d) none of these

35. The function $\dfrac{1}{(z-1)^{1/2}}$

(a) is analytic function in the region $|z| < 2$

(b) has a pole at $z = 1$

(c) has a branch point at $z = 1$

(d) has an essential singularity at $z = 1$

36. The value of integral $I = \dfrac{1}{2\pi i}\oint_z \dfrac{dz}{z-3}$, where C is a circle $|z| = 1$ is

(a) 1

(b) 0.5

(c) 2

(d) $e^{|z|}$

37. The real and imaginary parts of function $f(z) = u + iv = \sin z$

(a) $u = \sin x$, $v = \cos x$

(b) $u = \sin x \cos hy$, $v = \cos x \sin hy$

(c) $u = \sin x \cos y$, $v = \cos x \sin y$

(d) $u = e^x$, $v = e^y$

38. The residue of $\dfrac{\sin z}{(1-z^4)}$ at $z = 1$ is

(a) $\dfrac{1}{4}\sinh$

(b) $4 \sinh$

(c) $\sin i$

(d) zero

39. The residue of $\dfrac{z^2}{z^2+a^2}$ at $z = ia$ is

(a) $\dfrac{ia}{2}$

(b) $-\dfrac{ia}{2}$

(c) ia

(d) $-ia$

40. The residue of $\dfrac{z}{(z-a)(z-b)}$ at infinity is

(a) $+1$

(b) -1

(c) 0

(d) $\sqrt{ab}$

41. The residue of $\dfrac{\cos z}{z}$ at $z = 0$ is

(a) $+1$

(b) -1

(c) 0

(d) $\pi/2$

42. The analytic function $f(z)$ of which the real part is $e^x \cos y$ is

(a) e^z

(b) e^{iz}

(c) e^{-iz}

(d) $e^{|z|}$

43. If $f(z)$ is analytic within and on closed contour C and z_0 is any point within C, then

(a) $f(a) = 2\pi i \int \dfrac{f(z)}{z-a} dz$

(b) $f(a) = \dfrac{1}{2\pi i} \int \dfrac{f(z)}{z+a} dz$

(c) $f(a) = 2\pi i \int \dfrac{f(z)}{z+a} dz$

(d) $f(a) = \dfrac{1}{2\pi i} \int \dfrac{f(z)}{z-a} dz$

44. The value of integral $\int \tan z\, dz$, where C is $|z| = 2$ is

(a) 0

(b) πi

(c) $2\pi i$

(d) $\pi/2$

45. If $f(z)$ is continuous in a simply connected domain D and $\int C f(z) dz = 0$ for every closed path in D then $f(z)$ is

(a) constant in D

(b) zero in D

(c) analytic in D

(d) a polynomial in D

46. If $I = \oint_C dz \log(z)$, where C is the unit circle taken anticlockwise and $\log(z)$ is the principle branch of the logarithm function which one of the following is correct.

(a) $I = 0$ by residue theorem

(b) I is not defined as $\log(z)$ has a branch cut

(c) $I \neq 0$

(d) $\oint_C dz \log(z^2) = 2I$

47. The value of $\int_{-1}^{1} \pi(z+1) dz$ is

(a) 0

(b) $2\pi i$

(c) $-2\pi i$

(d) $(-1 + 2i)\pi$ **[GATE]**

48. The value of integral $I = \int_c \dfrac{e^z dz}{z^2 - 32 + 2}$, where C is the circle $|z| = \dfrac{3}{2}$ is

(a) $2\pi i e$

(b) $\pi i e$

(c) $-2\pi i e$

(d) $-\pi i e$ **[GATE]**

49. The value of integral $\oint_C \dfrac{e^z \sin z}{z^2} dz$, where C is the contour of unit circle $|z-2| = 1$ is

(a) $2\pi i$

(b) $4\pi i$

(c) πi

(d) 0   **[GATE]**

50. For complex function $f(z) = \dfrac{e^{\sqrt{z}} - e^{-\sqrt{z}}}{\sin \sqrt{z}}$, which of the following statement is correct?

(a) $z = 0$ is a branch point

(b) $z = 0$ is a pole of order one

(c) $z = 0$ is removable singularity

(d) $z = 0$ is an essential singularity

ANSWERS

1. (c)	2. (b)	3. (a)	4. (a)	5. (c)	6. (b)	7. (a)	8. (b)
9. (a)	10. (c)	11. (c)	12. (c)	13. (c)	14. (a)	15. (a)	16. (b)
17. (c)	18. (b)	19. (a)	20. (b)	21. (c)	22. (a)	23. (a)	24. (d)
25. (a)	26. (b)	27. (a)	28. (a)	29. (b)	30. (c)	31. (c)	32. (a)
33. (d)	34. (d)	35. (c)	36. (d)	37. (b)	38. (a)	39. (a)	40. (b)
41. (a)	42. (a)	43. (d)	44. (c)	45. (c)	46. (a)	47. (b)	48. (c)
49. (a)	50. (c)						

7

Differential Equations and Polynomials

7.1 INTRODUCTION

The analysis of any type of linear system generally leads to a mathematical model in the form of differential equation, in particular, to those of second order. Equations which involve dependent, independent variables and derivatives of the dependent variables with respect to the independent variable are called *differential equations*. *Ordinary differential equations* contain total derivatives (one independent variable) and *partial differential equations* contain partial derivatives with respect to two or more independent variables.

The most general form of a differential equation[*] is

$$a_0(x)\frac{d^n y}{dx^n} + a_1(x)\frac{d^{n-1}y}{dx^{n-1}} + \cdots + a_{n-1}(x)\frac{dy}{dx} + a_n(x)y = f(x) \qquad (7.1)$$

where $a_0(x), a_1(x),..., a_n(x), f(x)$ and $y(x)$ are the functions of the independent variable x. If the term $f(x) = 0$, the equation is said to be *homogeneous*, otherwise, it is said to be *inhomogeneous* and $f(x)$ is called an *inhomogeneous term*.

The *order* of a differential equation is the order of *the highest derivative appearing in the equation*. Equation (7.1) is of order n. If $n = 1$, we have the linear equation of *first order* and if $n = 2$ it is linear equation of *second order*, etc.

The *degree* of differential equation is the *power of the highest derivative after the equation has been rationalised*, i.e. after fractional powers of all derivatives have been removed. Thus, the equation

$$\frac{d^2 y}{dx^2} + \left(\frac{dy}{dx}\right)^2 + xy = 3$$

is of the *second order* and *first degree*, while

$$\frac{d^2 y}{dx^2} + \sqrt{\frac{dy}{dx}} + xy = 0$$

is of the *second order* and *second degree*.

A differential equation is defined as *linear if it does not contain the square or any higher power of dependent variable or any of its derivative or the product of the two*.

---

[*] The partial differential equations and their examples will be dealt later.

The solution of an equation of nth order involves, in principle, carrying out n integrations. Since each of them introduces one arbitrary constant, the final expression for the dependent variable will contain n arbitrary constants. However, a solution in which one or more of these constants are given specific values, for instantce the value zero, will also satisfy the differential equation. In view of this consideration, two types of solutions of an ordinary differential equation of nth order may be distinguished:

i. The *complete* or *general* solution contains its full complement of n independent* arbitrary constants.

ii. *Particular* solutions can be obtained from general solution by fixing one or more variables as constants.

A large percentage of the differential equations which occur in applied problems are linear and of first or second order. There are a few field equations in physics which are of higher order than the second (for example the equations for the transverse motions of stiff plates) but the equations which are important enough to be considered in this book are either of first or second order.

7.2 GENERAL METHOD OF SOLVING CERTAIN DIFFERENTIAL EQUATIONS OF HIGHER ORDER

It seems appropriate, however, to discuss first a few special types of differential equations which can be solved by elementary means. While the theory given in this section is applicable to equations of any order, emphasis will be placed solely on second order equations because of their prominence in mathematical physics.

Linear Equations with Constant Coefficients

If the various coefficients $a_n(x)$, $n = 1, 2,... n$, of Eq. (7.1) are constant, we may write this equation in the form

$$\frac{d^n y}{dx^n} + a_1 \frac{d^{n-1}}{dx^{n-1}} + \cdots + a_{n-1} \frac{dy}{dx} + a_n y = f(x) \tag{7.2}$$

provided $a_0 = 1$.

In discussing this type of equation, it becomes convenient to introduce a new notation, we write $D^{**} = \dfrac{d}{dx}$. This differential equation (Eq. 7.2) under consideration in its

---

* Arbitrary constants are said to be independent if two or more of them cannot be replaced by equivalent single one. Thus, the constants c_1 and c_2 in the functions $ax + c_1$ and $c_1 e^x + c_2 e^x$ are not independent because these functions may be written $ax + c$ and ce^x respectively.

This distinction is elementary. A more adequate analysis would focus attention upon independent *solutions* of the differential equation rather independent constants. Solutions are independent when the so-called Wronskian determinant fails to vanish.

** A symbol such as D, which is meaningless unless applied to a function of x, and therefore not a mathematical quantity in the usual sense, bears the name 'operator'. In the present connection D may be regarded as nothing more than an abbreviation. In quantum mechanics, it will be found that operators such as D are entities of considerable significance which give rise to an operator algebra, quite different in many respect from ordinary algebra.

most general form may be written as:

$$D^n y + a_1 D^{n-1} y + \cdots + a_n y = f(x) \tag{7.3}$$

This may be written in the form

$$(D^n + a_1 D^{n-1} + \cdots + a_n) y = f(x) \tag{7.4}$$

where the significance of the term in parentheses of the left-hand member is that it constitutes an operator that when operating on $y(x)$ leads to the left-hand member of Eq. (7.3).

If $f(x)$ in Eq. (7.4) is placed equal to zero, we obtain the equation

$$L_n(D)y = (D^n + a_1 D^{n-1} + \cdots + a_n) y = 0 \tag{7.5}$$

This is called the reduced equation.

The general solution, Eq. (7.4) consists of the sum of two parts y_c and y_p. y_c is the solution of the reduced equation and is called the complementary function which satisfies the equation

$$L_n(D)y_c = (D^n + a_1 D^{n-1} + \cdots + a_n) y_c = 0 \tag{7.6}$$

whereas y_p is called particular integral and satisfies the equation

$$L_n(D)y_p = (D^n + a_1 D^{n-1} + \cdots + a_n) y_p = f(x) \tag{7.7}$$

Thus, the general solution of a linear differential equation with constant coefficients is the sum of a particular integral y_p and the complementary function y_c. Thus, we obtain

$$y = y_c + y_p \tag{7.8}$$

In order to find y_c, first determine the roots of Eq. (7.5), this is known as auxiliary equation. If these roots are denoted by α_i, the complementary function is

$$y_c = \sum_i c_i e^{\alpha_i x} \tag{7.9}$$

where c is an arbitrary constant. Thus the complementary function has n independent arbitrary constants.

If two of the roots of the auxiliary equation are equal, i.e. $\alpha_1 = \alpha_2$, the complementary function of Eq. (7.9) will contain the part $(c_1 + c_2)e^{\alpha_1 x}$ or $ce^{\alpha_1 x}$, i.e. one arbitrary constant is lost and the solution no longer is complete.

To remove this fault, the part of the solution $(c_1 + c_2)e^{\alpha_1 x}$ must be replaced by $(c_1 x + c_2)e^{\alpha_1 x}$. An extension of this argument leads to the general result. If α is a p-fold root of the auxiliary equation, the complementary function is

$$y_c = c_1 e^{\alpha_1 x} + c_2 e^{\alpha_2 x} + \cdots + c_i (1 + b_1 x + b_2 x^2 + \cdots + b_{p-1} x^{p-1}) e^{\alpha_1 x} + \cdots \tag{7.10}$$

A particular integral of a linear differential $F(D)y = f(x)$ with constant coefficients is given by

$$y = \frac{1}{F(D)} f(x)$$

For certain forms of $f(x)$, the particular integral can be directly written as follows:

i. If $f(x)$ is of the form $e^{\alpha x}$;

$$y = \frac{1}{F(D)} e^{\alpha x} = \frac{1}{F(\alpha)} e^{\alpha x}, \; F(\alpha) \neq 0$$

ii. If $f(x)$ is of the form $\sin(ax + b)$ or $\cos(ax + b)$;

$$y = \frac{1}{F(D^2)} \sin(ax+b) = \frac{1}{F(-a^2)} \sin(ax+b), F(-a^2) \neq 0$$

$$y = \frac{1}{F(D^2)} \cos(ax+b) = \frac{1}{F(-a^2)} \cos(ax+b), F(-a^2) \neq 0$$

iii. If $f(x)$ is of the form x^m;

$$y = \frac{1}{F(D)} x^m = (a_0 + a_1 D + a_2 D^2 + \cdots + a_m D^m) x^m, a_0 \neq 0$$

obtained by expanding $y = \dfrac{1}{F(D)}$ in ascending powers of D and suppressing all terms beyond D^m; since $D^n x^m = 0$, when $n > m$.

iv. If $f(x)$ is of the form $e^{\alpha x} V(x)$

$$y = \frac{1}{F(D)} e^{\alpha x} V(x) = e^{\alpha x} \cdot \frac{1}{F(D+\alpha)} V(x)$$

v. If $f(x)$ is of the form $x V(x)$

$$y = \frac{1}{F(D)} x V = x \frac{1}{F(D)} V(x) - \frac{F'(D)}{\{F(D)\}^2} V(x)$$

Example 1. *Solve*

$$(D^3 - 5D^2 + 8D - 4) y = e^{2x} + 2e^x + 3e^{-x}$$

Solution: The complementary function is

$$y = c_1 e^x + c_2 e^{2x} + c_3 x e^{2x}$$

and particular integral is

$$y = \frac{1}{(D-1)(D-2)^2} e^{2x} + \frac{2}{(D-1)(D-2)^2} e^x + \frac{3}{(D-1)(D-2)^2} e^{-x}$$

In 1st and 2nd term,

$F(a) = F(2) = 0$ and $F(a) = F(1) = 0$

and the short method does not apply. However, we write

$$y = \frac{1}{(D-2)^2} \left\{ \frac{1}{(D-1)} e^{2x} \right\} + \frac{2}{(D-1)} \left\{ \frac{1}{(D-2)^2} e^x \right\} + \frac{3}{(D-1)(D-2)^2} e^{-x}$$

$$= \frac{1}{(D-2)^2} e^{2x} + \frac{2}{D-1} e^x + \frac{3}{(-2)(-3)^2} e^{-x}$$

$$= e^{2x} \iint (dx)^2 + 2e^x \int dx - \frac{1}{x} e^{-x}$$

$$= \frac{1}{2}x^2 e^{2x} + 2xe^x - \frac{1}{6}e^{-x}$$

Hence, the complete solution is

$$y = c_1 e^x + c_2 e^{2x} + c_3 xe^{2x} + \frac{1}{2}x^2 e^{2x} + 2xe^x - \frac{1}{6}e^{-x}$$

Example 2. *Solve*

$$(D^3 + D^2 + D + 1)\, y = \sin 2x + \cos 3x$$

Solution: The complementary function is

$$y = c_1 \cos x + c_2 \sin x + c_3 e^{-x}$$

The particular integral is

$$y = \frac{1}{(D^2 + 1)(D + 1)}\{\sin 2x + \cos 3x\}$$

$$= \frac{1}{(D^2 + 1)(D + 1)}\sin 2x + \frac{1}{(D^2 + 1)(D + 1)}\cos 3x$$

Hence the operator is not of the form $\dfrac{1}{F(D^2)}$. However, the method used is

$$y = \frac{(D-1)}{(D^2 + 1)(D^2 - 1)}\sin 2x + \frac{(D-1)}{(D^2 + 1)(D^2 - 1)}\cos 3x$$

$$= \frac{(D-1)}{(-4+1)(-4-1)}\sin 2x - \frac{(D-1)}{(-9+1)(-9-1)}\cos 3x$$

$$= \frac{1}{15}(D-1)\sin 2x + \frac{1}{80}(D-1)\cos 3x$$

or
$$y = \frac{1}{15}(2\cos 2x - \sin 2x) - \frac{1}{80}(3\sin 3x + \cos 3x)$$

The complete solution is

$$y = c_1 \cos x + c_2 \sin x + c_3 e^{-x} + \frac{1}{15}(2\cos 2x - \sin 2x) - \frac{1}{80}(3\sin 3x + \cos 3x)$$

Example 3. *Solve* $(D^2 + 4)\, y = \cos 2x + \cos 4x$

Solution: The complementary function is

$$y = c_1 \cos 2x + c_2 \sin 2x$$

The particular integral is

$$y = \frac{1}{(D^2 + 4)}(\cos 2x + \cos 4x) = \frac{1}{(D^2 + 4)}\cos 2x + \frac{1}{(D^2 + 4)}\cos 4x$$

The concise method suggested cannot be used to evaluate $\dfrac{1}{D^2 + 4}\cos 2x$ since $f(-a^2) = 0$. As such the following procedure may be used.

Consider $\dfrac{1}{(D^2+4)}\cos(2+h)x = \dfrac{1}{-(2+h)^2+4}\cos(2+h)x = -\dfrac{1}{4h-h^2}\cos(2+h)x$

Using Taylor's theorem, we have

$$\frac{1}{(D^2+4)}\cos(2+h)x = -\frac{1}{4h+h^2}\left\{\cos 2x - hx\sin 2x - \frac{1}{2}(hx)^2\cos 2x + \cdots\right\}$$

The first term $\cos 2x$ is part of the complementary function and need not be considered here. Hence, a particular integral is

$$\frac{1}{(D^2+4)}\cos(2+h)x = \frac{1}{h+4}\left\{x\sin 2x + \frac{1}{2}hx^2\cos 2x + \cdots\right\}$$

Taking limit as $h \to 0$, we obtain

$$\frac{1}{D^2+4}\cos 2x = \frac{1}{4}x\sin 2x$$

Since

$$\frac{1}{D^2+4}\cos 4x = -\frac{1}{12}\cos 4x$$

The complete solution is

$$y = c_1\cos 2x + c_2\sin 2x + \frac{1}{4}x\sin 2x - \frac{1}{12}\cos 4x$$

Example 4. *Solve*

$$(D^2 - 4D + 3)y = 2xe^{3x} + 3e^x\cos 2x$$

Solution: The complementary function is

$$y = c_1e^x + c_2e^{3x}$$

The particular integral is

$$y = \frac{1}{D^2-4D+3}(2xe^{3x} + 3e^x\cos 2x)$$

$$= 2\frac{1}{D^2-4D+3}xe^{3x} + 3\frac{1}{D^2-4D+3}e^x\cos 2x$$

$$= 2e^{3x}\frac{1}{D^2+2D}x + 3e^x\frac{1}{D^2-2D}\cos 2x$$

$$= 2e^{3x}\frac{1}{2D}\cdot\frac{1}{(1+D/2)}x + 3e^x\frac{1}{-4-2D}\cos 2x$$

$$= e^{3x}\cdot\frac{1}{D}\cdot\left(1+\frac{D}{2}\right)^{-1}x + 3e^x\frac{1}{(-2)}\frac{D-2}{D^2-4}\cos 2x$$

$$= e^{3x}\frac{1}{D}(1-D/2)x + \frac{3e^x}{16}(D-2)\cos 2x$$

$$= e^{3x}\frac{1}{D}(x-1/2) + \frac{3}{16}e^x(-2\sin 2x - 2\cos 2x)$$

$$= \frac{1}{2}e^{3x}(x^2-x) - \frac{3}{8}e^x(\sin 2x + \cos 2x)$$

Hence the complete solution is

$$y = c_1 e^x + c_2 e^{3x} + \frac{1}{2} e^{3x} (x^2 - x) - \frac{3}{8} e^x \{\sin 2x + \cos 2x\}$$

Example 5. *Solve*

$$(D^2 - 1) y = x^2 \sin 3x$$

Solution: The complementary function is

$$y = c_1 e^x + c_2 e^{-x}$$

Particular integral is

$$y = \frac{1}{D^2 - 1} x^2 \sin 3x = x \frac{1}{D^2 - 1} x \sin 3x - \frac{2D}{(D^2 - 1)^2} (x \sin 3x)$$

or

$$y = x^2 \frac{1}{D^2 - 1} \sin 3x - x \frac{2D}{(D^2 - 1)^2} \sin 3x$$

$$-2D \left\{ x \frac{1}{D^4 - 2D^2 + 1} \sin 3x - \frac{4D^3 - 4D}{(D^4 - 2D^2 + 1)^2} \sin 3x \right\}$$

$$= x^2 \frac{1}{D^2 - 1} \sin 3x - x \frac{2D}{(D^2 - 1)^2} \sin 3x$$

$$-2D \left\{ x \frac{1}{(D^2 - 1)^2} \sin 3x \right\} + \frac{8D^2}{(D^2 - 1)^3} \sin 3x$$

$$= -\frac{1}{10} x^2 \sin 3x - \frac{3}{50} x \cos 3x - \frac{1}{50} D(x \sin 3x) + \frac{9}{125} \sin 3x$$

$$= -\frac{1}{10} x^2 \sin 3x - \frac{3}{50} x \cos 3x - \frac{3}{50} x \cos 3x - \frac{1}{50} \sin 3x + \frac{9}{125} \sin 3x$$

or

$$y = \frac{13 - 25x^2}{250} \sin 3x - \frac{3}{25} x \cos 3x$$

Hence the complete solution is

$$y = c_1 e^x + c_2 e^{-x} + \frac{13 - 25x^2}{250} \sin 3x - \frac{3}{25} x \cos 3x$$

Linear Equations with Variable Coefficients

The Cauchy linear equation

$$a_0 x^n \frac{d^n y}{dx^n} + a_1 x^{n-1} \frac{d^{n-1}}{dx^{n-1}} + \cdots + a_{n-1} x \frac{dy}{dx} + a_n y = F(x) \tag{7.11}$$

in which $a_0, a_1, ..., a_n$ are constants, and the Legendre linear equation

$$a_0 (bx + c)^n \frac{d^n y}{dx^n} + a_1 (bx + c)^{n-1} \frac{d^{n-1} y}{dx^{n-1}} + \cdots + a_{n-1} (bx + c) \frac{dy}{dx} + a_n y = F(x) \tag{7.12}$$

of which Eq. (7.11) is the special case ($b = 1$ and $c = 0$), may be reduced to linear equations with constant coefficients by properly chosen transformations of the independent variables.

Let $x = e^z$, then if D is defined by $D = \dfrac{d}{dz}$

In Eq. (7.11), the Cauchy linear equation, becomes

$$Dy = \frac{dy}{dx} = \frac{dy}{dz}\frac{dz}{dx} = \frac{1}{x}\frac{dy}{dz} \quad \text{and } xDy = Dy$$

$$D^2y = \frac{d}{dx}\left(\frac{1}{x}\frac{dy}{dz}\right) = \frac{1}{x^2}\left(\frac{d^2y}{dz^2} - \frac{dy}{dz}\right) \quad \text{and } X^2D^2y = D(D-1)y$$

....

$$x^r D^r y = D(D-1)(D-2)...(D-r+1)y$$

After making these replacements Eq. (7.11) becomes

$$[\{a_0\,D(D-1)(D-2)...(D-n+1)\} + \{a_1\,D(D-1)(D-2)...(D-n+2)\}$$
$$+\cdots+ a_{n-1}\,D + a_n]\,y = F(e^z) \tag{7.13}$$

a linear equation with constant coefficients.

In Eq. (7.12) the Legendre linear equation,
let $bx + c = e^z$, then

$$Dy = \frac{dy}{dz}\cdot\frac{dz}{dx} = \frac{b}{(bx+c)}\frac{dy}{dz} \quad \text{and } (bx+c)\,Dy = bDy$$

$$D^2y = \frac{b^2}{(bx+c)^2}\left(\frac{d^2y}{dz^2} - \frac{dy}{dz}\right) \quad \text{and } (bx+c)^2\,D^2y = b^2\,D(D-1)$$

....

$$(bx+c)^r\,D^r y = b^r D(D-1)...(D-r+1)y$$

After making these replacements, Eq. (7.12) becomes

$$\{a_0 b^n\,D(D-1)(D-2)...(D-n+1) + a_1 b^{n-1}D(D-1)(D-2)$$

$$...(D-n+2)+\cdots+ a_{n-1}bD + a_n\}\,y = F\left(\frac{e^z-c}{b}\right) \tag{7.14}$$

a linear equation with constant coefficients.

Example 6. *Solve*

$$(x^3 D^3 + 2xD - 2)\,y = x^2 \log x + 3x$$

Solution: The transformation $x = e^z$ reduces the equation to

$$\{D(D-1)(D-2) + 2(D-2)\}\,y = (D-1)(D^2 - 2D + 2)\,y = ze^{2z} + 3e^z$$

The complementary function is

$$y = c_1 e^z + e^z(c_2 \cos z + c_3 \sin z)$$

The particular integral is

$$y = \frac{1}{D^3 - 3D^2 + 4D - 2}(ze^{2z} + 3e^z)$$

$$= e^{2z} \frac{1}{(D+2)^3 - 3(D+2)^2 + 4(D+2) - 2}z + 3\frac{1}{(D-1)(D^2 - 2D + 2)}e^z$$

$$= e^{2z} \frac{1}{D^3 - 3D^2 + 4D + 2}z + 3\frac{1}{(D-1)(1)}e^z$$

$$= e^{2z}\left(\frac{1}{2} - D\right)z + 3e^z \int dz = e^{2z}\left(\frac{1}{2}z - 1\right) + 3ze^z$$

and the solution is

$$y = c_1 e^z + e^z(c_2 \cos z + c_3 \sin z) + e^{2z}\left(\frac{1}{2}z - 1\right) + 3ze^z$$

$$= c_1 x + x(c_2 \cos\log x + c_3 \sin\log x) + \frac{1}{2}x^2(\log x - 2) + 3x\log x$$

Example 7. *Solve*

$$\{3x + 2)^2 D^2 + 3(3x + 2) D - 36\} y = 3x^2 + 4x + 1$$

Solution: The transformation $3x + 2 = e^z$ reduces the equation to

$$\{9D(D-1) + 9D - 36\}y = 9(D^2 - 4)y = \frac{1}{3}(9x^2 + 12x + 3)$$

or

$$9(D^2 - 4) y = \frac{1}{3}\{(3x+2)^2 - 1\} = \frac{1}{3}(e^{2z} - 1)$$

or

$$(D^2 - 4)y = \frac{1}{27}(e^{2z} - 1)$$

Hence the complete solution is

$$y = c_1 e^{2z} + c_2 e^{-2z} + \frac{1}{27}\left\{\frac{1}{D^2 - 4}e^{2z} - \frac{1}{D^2 - 4}e^{0z}\right\}$$

$$= c_1 e^{2z} + c_2 e^{-2z} + \frac{1}{108}(ze^{2z} + 1)$$

or

$$y = c_1(3x+2)^2 + c_2(3x+2)^{-2} + \frac{1}{108}\{(3x+2)^2 \log(3x+2) + 1\}$$

7.3 LINEAR EQUATIONS WITH VARIABLE COEFFICIENTS—EQUATIONS OF THE SECOND ORDER

Linear second order differential equations with variable coefficients are the most frequently encountered differential equations in applied mathematics and physical problems next to those with constant coefficients. We shall give the brief outline of the common methods of solution for such equations.

A linear differential equation of the second order has the form

$$\frac{d^2y}{dx^2} + R(x)\frac{dy}{dx} + S(x)y = Q(x) \tag{7.15}$$

If the coefficients R and S are constants, the equation can be solved by the methods of the preceding sections, otherwise, no general method is known. In this section certain procedures are given which at times, will yield a solution.

Change of Dependent Variable

Under the transformation

$$y = uv, \text{ where } u = u(x) \text{ and } v = v(x)$$

$$\frac{dy}{dx} = u\frac{dv}{dx} + v\frac{du}{dx} \text{ and } \frac{d^2y}{dx^2} = u\frac{d^2v}{dx^2} + 2\frac{dv}{dx}\frac{du}{dx} + v\frac{d^2u}{dx^2}$$

Equation (7.15) becomes

$$\frac{d^2v}{dx^2}R_1(x)\frac{dv}{dx} + S_1(x) = Q_1(x) \tag{7.16}$$

where

$$R_1(x) = \frac{2}{u}\frac{du}{dx} + R(x),\ Q_1(x) = \frac{Q(x)}{u},$$

and

$$S_1(x) = \frac{1}{u}\left\{\frac{d^2u}{dx^2} + R(x),\frac{du}{dx} + S(x)u\right\}$$

(a) If u is a particular integral of

$$\frac{d^2y}{dx^2} + R(x)\frac{dy}{dx} + S(x)y = 0$$

Then $S_1(x) = 0$ and Eq. (7.16) becomes

$$\frac{d^2v}{dx^2} + R_1(x)\frac{dv}{dx} = Q_1(x) \tag{7.17}$$

On substitution

$$\frac{dv}{dx} = P \text{ and } \frac{d^2v}{dx^2} = \frac{dP}{dx}$$

reduces Eq. (7.17) to linear equation of the first order

$$\frac{dP}{dx} + R_1(x)P = Q_1(x) \tag{7.18}$$

(b) If v is chosen so that

$$R_1(x) = \frac{2}{u}\frac{du}{dx} + R(x) = 0$$

or

$$\frac{du}{u} = -\frac{1}{2}R(x)dx$$

i.e.

$$u = e^{-\frac{1}{2}\int R(x)dx}$$

Now

$$\frac{du}{dx} = -\frac{1}{2}uR(x)$$

and
$$\frac{d^2u}{dx^2} = -\frac{1}{2}R(x)\frac{du}{dx} - \frac{1}{2}\frac{dR}{dx}u$$

so that
$$S_1(x) = S(x) + \frac{R(x)}{u}\frac{du}{dx} + \frac{1}{u}\frac{d^2u}{dx^2}$$

$$= S(x) + \frac{R(x)}{2u}\frac{du}{dx} - \frac{1}{2}\frac{dR}{dx}$$

$$= S - \frac{1}{4}R^2 - \frac{1}{2}\frac{dR}{dx}$$

and
$$Q_1 = Q/u.$$

If $S_1(x) = S - \frac{1}{4}R^2 - \frac{1}{2}\frac{dR}{dx} = A$, a constant, thus Eq. (7.16) becomes

$$\frac{d^2v}{dx^2} + Av = Q/u$$

a linear equation with constant coefficients.

If
$$S_1(x) = A/x^2, \text{ Eq. (7.16) becomes}$$

$$x^2\frac{d^2v}{dx^2} + Av = \frac{Qx^2}{u}$$

the Cauchy's equation and substitution $x = e^z$ will reduce it to one with constant coefficients.

Change of Independent Variable

Let the transformation be $z = f(x)$. Then

$$\frac{dy}{dx} = \frac{dy}{dz} \cdot \frac{dz}{dx}$$

and
$$\frac{d^2y}{dx^2} = \frac{d^2y}{dz^2}\left(\frac{dz}{dx}\right)^2 + \frac{dy}{dz}\frac{d^2z}{dx^2}$$

and Eq. (7.15) becomes

$$\frac{d^2y}{dz^2}\left(\frac{dz}{dx}\right)^2 + \left(\frac{d^2z}{dx^2} + R\frac{dz}{dx}\right)\frac{dy}{dz} + Sy = Q$$

or
$$\frac{d^2y}{dz^2} + \frac{\frac{d^2z}{dx^2} + R\frac{dz}{dx}}{\left(\frac{dz}{dx}\right)^2}\frac{dy}{dz} + \frac{Sy}{\left(\frac{dz}{dx}\right)^2} = \frac{Q}{\left(\frac{dz}{dx}\right)^2} \qquad (7.19)$$

or
$$\frac{dz}{dx} = \sqrt{\frac{\pm S}{a^2}}$$

where a^2 being a positive constant (one may consistently choose $a^2 = 1$).

If now $\dfrac{\dfrac{d^2z}{dx^2} + R\dfrac{dz}{dx}}{\left(\dfrac{dz}{dx}\right)^2} = A$ (a constant)

then Eq (7.19) becomes

$$\frac{d^2y}{dz^2} + A\frac{dy}{dz} \pm a^2 y = \frac{Q}{\left(\dfrac{dz}{dx}\right)^2}$$

a linear equation with constant coefficients.

Operational Factoring

It may be possible to separate the left member of

$$\{P(x)D^2 + R(x)D + S(x)\}y = Q(x)$$

into two linear operators $F_1(D)$ and $F_2(D)$ so that

$$\{F_1(D) \cdot F_2(D)\}y = F_1(D)\,\{F_2(D)y\}^*$$
$$= \{P(x)D_2 + R(x)D + S(x)\}\,y = Q(x) \qquad (7.20)$$

Then setting $F_2(D)y = v$, Eq. (7.20) becomes

$F_1(D)v = Q(x)$, a linear equation of order one.

Example 1. Solve

$$x^2(x+1)\frac{d^2y}{dx^2} - x(2+4x+x^2)\frac{dy}{dx} + (2+4x+x^2)y = -x^4 - 2x^3$$

Solution: The particular integral of the equation

$$x^2(x+1)\frac{d^2y}{dx^2} - x(2+4x+x^2)\frac{dy}{dx} + (2+4x+x^2)y = 0 \text{ is } y = x \text{ becomes}$$

$$R + Sx = -\frac{x(2+4x+x^2)}{x^2(x+1)} + x\frac{2+4x+x^2}{x^2(x+1)} = 0$$

The transformation $y = xv$ gives

$$\frac{dy}{dx} = x\frac{dv}{dx} + v, \quad \frac{d^2y}{dx^2} = x\frac{d^2v}{dx^2} + 2\frac{dv}{dx}$$

and reduces the given equation to

$$x^2(x+1)\left\{\frac{xd^2v}{dx^2} + 2\frac{dv}{dx}\right\} - x\{2+4x+x^2\}\left\{\frac{xdv}{dx} + v\right\}$$

$$+ (2+4x+x^2)xv = -x^4 - 2x^3$$

or

$$\frac{d^2v}{dx^2} - \frac{x-2}{x+1}\frac{dv}{dx} = -\frac{x+2}{x+1}$$

---

* The factors are not commutative if D is treated as an operator. When D is treated as a variable rather than an operator, the factors are commutative.

Putting $\dfrac{dv}{dx} = P$, the equation becomes

$$\frac{dP}{dx} - \frac{x+2}{x+1}P = -\frac{x+2}{x+1}$$

for which

$$e^{-\int\left(1+\frac{1}{x+1}\right)dx} = \frac{e^{-x}}{x+1}$$

is an integrating factor. Then

$$\frac{e^{-x}}{x+1}P = -\int \frac{(x+2)e^{-x}}{(x+1)^2}dx = \frac{e^{-x}}{x+1} + c_1$$

or

$$P = \frac{dv}{dx} = 1 + c_1(1+x)e^x$$

and

$$v = \frac{y}{x} = x + c_1 x e^x + c_2$$

or

$$y = x^2 + c_1 x^2 e^x + c_2 x$$

Example 2. *Solve*

$$\frac{d^2y}{dx^2} - 2\tan x \frac{dy}{dx} + 3y = 2\sec x.$$

Solution: It is seen that $y = \sin x$ is a particular integral of

$$(D^2 - 2\tan xD + 3)y = 0$$

*The transformation $y = v \sin x$ reduces the given equation to

$$\sin x \frac{d^2v}{dx^2} + 2\left(\cos x - \frac{\sin^2 x}{\cos x}\right)\frac{dv}{dx} = 2\sec x$$

or

$$\frac{d^2v}{dx^2} + 2(\cot x - \tan x)\frac{dv}{dx} = 4\,\text{cosec}\,2x$$

The substitution $\dfrac{dv}{dx} = P$ reduces the equation to

$$\frac{dP}{dx} + 2(\cot x - \tan x)P = 4\,\text{cosec}\,2x$$

for which the integrating factor is $\sin^2 2x$.

---

*For the equation $(D^2 + RD + S)\, y = 0$: Particular integral is

(a) $y = x$ if $R + Sx = 0$

(b) $y = e^x$ if $1 + R + Sx = 0$

(c) $y = e^{-x}$ if $1 - R + Sx = 0$

(d) $y = e^{mx}$ if $m^2 + Rm + S = 0$

Then
$$P \sin^2 2x = \int 4\sin 2x \cdot dx = -2\cos 2x + c_1$$

or
$$P = \frac{dv}{dx} = -\frac{2\cos 2x}{\sin^2 2x} + c_1 \cosec^2 2x$$

or
$$v = \frac{y}{\sin x} = -2\int \cosec 2x \cdot \cot 2x\, dx + c_1 \int \cosec^2 2x\, dx$$

$$= \cosec 2x - \frac{c_1}{2}\cot 2x + c_2$$

or
$$y = \frac{1}{2}\sec x + c_1\left(\cos x - \frac{1}{2}\sec x\right) + c_2 \sin x$$

Example 3. Solve

$$\frac{d^2y}{dx^2} - \frac{2}{x}\frac{dy}{dx} + \left(1 + \frac{2}{x^2}\right)y = xe^x$$

Solution: Here $R = -\dfrac{2}{x}$

and
$$S = 1 + \frac{2}{x^2}$$

Hence
$$S - \frac{1}{4}R^2 - \frac{1}{2}\frac{dR}{dx} = 1$$

and
$$u = e^{-\frac{1}{2}\int R\, dx}\, e^{\int \frac{dx}{x}} = x$$

The transformation $y = uv = xv$ gives

$$\frac{dy}{dx} = x\frac{dv}{dx} + v$$

and
$$\frac{d^2y}{dx^2} = x\frac{d^2v}{dx^2} + 2\frac{dv}{dx}$$

and reduces the equation to

$$\frac{d^2v}{dx^2} + v = e^x$$

a linear equation with constant coefficients whose complete solution is

$$v = \frac{y}{x} = c_1 \cos x + c_2 \sin x + \frac{1}{D^2 + 1}e^x$$

$$= c_1 \cos x + c_2 \sin x + \frac{1}{2}e^x$$

Thus
$$y = c_1 x \cos x + c_2 x \sin x + \frac{1}{2}xe^x$$

Example 4. *Solve*

$$\frac{d^2y}{dx^2} - (1 + 4e^x)\frac{dy}{dx} + 3e^{2x}y = e^{2(x+e^x)}$$

Solution: On putting

$$\frac{dz}{dx} = \sqrt{\frac{S}{a^2}} = \sqrt{\frac{3e^{2x}}{3}}^* = e^x$$

$$\frac{\dfrac{d^2z}{dx^2} + R\left(\dfrac{dz}{dx}\right)}{\left(\dfrac{dz}{dx}\right)^2} = \frac{e^x - (1 + 4e^x)e^x}{(e^x)^2} = -4 = A$$

The introduction of $z = e^x$ as new independent variable leads to

$$\frac{d^2y}{dz^2} + A\frac{dy}{dz} + a^2y = \frac{Q}{\left(\dfrac{dz}{dx}\right)^2}$$

or

$$\frac{d^2y}{dz^2} - 4\frac{dy}{dz} + 3y = \frac{e^{2(x+e^x)}}{e^{2x}} = e^{2e^x} = e^{2z}$$

whose complete solution is

$$y = c_1 e^z + c_2 e^{3z} + \frac{1}{D^2 - 4D + 3}e^{2z}$$

$$= c_1 e^z + c_2 e^{3z} - e^{2z}$$

Replacing z by e^x we have

$$y = c_1 e^{e^x} + c_2 e^{3e^x} - e^{2e^x}$$

Example 5. *Solve* $[xD^2 + (1 - x)D - 2(1 + x)]y = e^{-x}(1 - 6x)$

Solution: The equation may be written as

$$[xD + (1 + x)][D - 2]y = e^{-x}(1 - 6x)$$

Putting $(D - 2)y = v$, we have

$$\{xD + (1 + x)\}v = e^{-x}(1 - 6x)$$

or

$$\left(D + \frac{1}{x} + 1\right)v = e^{-x}\left(\frac{1}{x} - 6\right)$$

The integrating factor is xe^x so that

$$xe^x v = \int(1 - 6x)dx = x - 3x^2 + c_1$$

---

* Here the choice of $a^2 = 3$ is one of convenience only. Taking $a^2 = 1$ will yield $A = -\dfrac{4}{\sqrt{3}}$ and the transformation $z = \sqrt{3}\, e^x$ will yield the same solution.

or $$(D-2)y = v = (1-3x)e^{-x} + \frac{c_1 e^{-x}}{x}$$

Here e^{-2x} is an integrating factor, so that

$$ye^{-2x} = \int \left\{ (1-3x)e^{-3x} + \int \frac{c_1 e^{-3x}}{x} \right\} dx$$

$$= xe^{-3x} + c_1 \int \frac{e^{-3x}}{x} dx + c_2$$

or $$y = xe^{-x} + c_1 e^{2x} \int \frac{e^{-3x}}{x} dx + c_2 e^{2x}$$

Example 6. Solve

$$\frac{dy}{dx} + \frac{2}{x} y + \frac{1}{2} x^3 y^2 = \frac{1}{2x}$$

Solution: The Riccati equation* reduced to linear equation of second order by substitution

$$y = \frac{1}{Qu} \frac{du}{dx} = \frac{2}{x^3 u} \frac{du}{dx}$$

or $$\frac{d^2 u}{dx^2} + \left(\frac{2}{x} - \frac{\frac{3}{2} x^2}{\frac{x^3}{2}} \right) \frac{du}{dx} - \frac{1}{2x} \cdot \frac{x^3}{2} u = 0$$

or $$\frac{d^2 u}{dx^2} - \frac{1}{x} \frac{du}{dx} - \frac{1}{4} x^2 u = 0 \qquad (1)$$

Now substitute $$\frac{dz}{dx} = \sqrt{-\frac{S}{1}} = \sqrt{\frac{1}{4} x^2 / \frac{1}{4}} = x$$

reduces Eq. (1) to

$$\frac{d^2 u}{dz^2} - \frac{1}{4} u = 0$$

---

* The Riccati equation $\frac{dy}{dx} + yP(x) + y^2 Q(x) = R(x)$, $Q(x) \neq 0$ is reduced to

$$\frac{d^2 u}{dx^2} + \left(P - \frac{1}{Q} \frac{dQ}{dx} \right) \frac{du}{dx} - RQu = 0$$

by substituting $y = \frac{1}{Qu} \frac{du}{dx}$ and $\frac{dy}{dx} = \frac{1}{Qu} \frac{d^2 u}{dx^2} - \frac{1}{Qu^2} \left(\frac{du}{dx} \right)^2 - \frac{1}{Q^2} \frac{1}{u} \frac{dQ}{dx} \frac{du}{dx}$

whose solution is

$$u = c_1 e^{\frac{1}{2}z} + c_2 e^{-\frac{1}{2}z}$$

Then

$$y = \frac{1}{Qu} \cdot \frac{du}{dx} = \frac{2}{x^3} \cdot \frac{\frac{1}{2}\left(c_1 e^{\frac{1}{2}z} - c_2 e^{-\frac{1}{2}z}\right)}{c_1 e^{\frac{1}{2}z} + c_2 e^{-\frac{1}{2}z}} \cdot x$$

or

$$y = \frac{1}{x^2} \frac{c_1 e^{\frac{1}{4}x^2} - c_2 e^{-\frac{1}{4}x^2}}{c_1 e^{\frac{1}{4}x^2} + c_2 e^{-\frac{1}{4}x^2}} \qquad \qquad \because z = \int x\, dx = \frac{x^2}{2}$$

Example 7. *Solve* $\dfrac{dy}{dx}(\tan x + 3\cos x)y + y^2 \cos^2 x = -2$

Solution: The substitution

$$y = \frac{1}{Qu}\frac{du}{dx} = \frac{\sec^2 x}{u} \cdot \frac{du}{dx}$$

reduces the equation to

$$\frac{d^2u}{dx^2}(\tan x - 3\cos x)\frac{du}{dx} + 2u\cos^2 x = 0 \qquad (1)$$

Now on further substitution $\dfrac{dz}{dx} = \sqrt{\dfrac{2\cos^2 x}{2}} = \cos x$

or $z = \sin x$ and reduces Eq. (1) to

$$\frac{d^2u}{dz^2} - 3\frac{du}{dz} + 2u = 0$$

whose solution is $u = c_1 e^z + c_2 e^{2z}$

Then

$$y = \frac{1}{Qu}\frac{du}{dz} = \frac{\sec^2 x(c_1 e^z + 2c_2 e^{2z})}{c_1 e^z + c_2 e^{2z}}\cos x$$

or

$$y = \frac{\sec x(e^{\sin x} + 2Ke^{2\sin x})}{(e^{\sin x} + Ke^{2\sin x})}$$

where

$$K = \frac{c_2}{c_1}$$

7.4. SERIES SOLUTIONS OF LINEAR DIFFERENTIAL EQUATION OF SECOND ORDER

The solutions of differential equations with constant coefficients can be expressed in terms of elementary function. This is usually no longer true when the coefficients are functions of independent variables. In this case, the solutions lead to transcendental functions, which can be expressed either in terms of infinite series or definite integrals.

The method of series integration of the solution of a differential equation enables to represent one variable in a series of ascending or descending powers of another variable.

Consider the general homogeneous differential equation of second order

$$P_0(x) Y'' + P_1(x) Y' + P_2(x) Y = 0 \tag{7.21}$$

or

$$Y'' + R_1(x) Y' + R_2(x) Y = 0 \tag{7.72}$$

in which $P_i(x)$ are polynomials ($i = 0, 1, 2$).

The point $x = x_0$ in Eq. (7.22) is called an *ordinary point* of differential equation if both of the functions $R_1(x)$ and $R_2(x)$ are analytic[*] at $x = x_0$. If either (or both) of these functions is not analytic at x_0, then x_0 is called *singular point* of the differential equation.

When $x = x_0$ is a *singular point* of differential equation, the procedure of preceding sections will not yield a complete solution.

If the differential Eq. (7.22) possesses a *singular point* at $x = x_0$, then a convergent development of the solution in power series about the point $x = x_0$ having a finite number of terms is possible when and only when $(x - x_0) R_1(x)$ and $(x - x_0)^2 R_2(x)$ remains finite.

If the functions $(x - x_0) R_1(x)$ and $(x - x_0)^2 R_2(x)$ are both analytic at x_0 then x_0 is called a *regular singular point, nonessential singular point* or *removable singular point*. If either (or both) of these functions defined by these products is not analytic at x_0, then x_0 is called an *irregular singular point, essential singularity* or *nonremovable singularity*.

Therefore, we have

i. If the differential equation has no singularity then we are sure to solve it by method of series integration.

ii. If the differential equation has a singularity then this may or may not be solved by the method of series integration.

For nonessential singularity this method is applicable.

For essential singularity this method is not applicable. When $x = x_0$ is a regular singular point of Eqs (7.21) and (7.22) there always exists a series solution of the form

$$Y = \sum_{n=0}^{\infty} A_n (x - x_0)^{m+n} \tag{7.23}$$

with $A_0 \neq 0$. We shall proceed to determine m and A, so that Eq. (7.23) satisfies Eqs (7.21) and (7.22) has the form $f(m) A_0$. The equation

$$f(m) = 0 \tag{7.24}$$

is called the *indicial equation*.

The second order linear homogeneous differential equations with ordinary points or regular singular point x_0 can be solved by the method of series integration and the solution is given as a series developed about the point x_0. But there is still a question to answer that whether the two solutions obtained by this method always exist and are

---

[*] A function $f(x)$ is said to be analytic at $x = x_0$, if its Taylor series about x_0, i.e. $\displaystyle\sum_{n=0}^{\infty} \frac{f^{(n)}(x_0)}{n!} (x - x_0)^n$ exists and converges to $f(x)$ for all x in some open interval including x_0 (*refer* to Chapter 6—Complex Variables).

these solutions independent of each other? The answer is *No*. The existence of two solutions and their linear independence is determined from the roots of the indicial equations. We can classify the roots of the indicial equations into following three classes.

i. The roots are distinct and do not differ by an integer.

The complete solution is then given by

$$Y = A\bar{y}\big|_{m=m_1} + B\bar{y}\big|_{m=m_2} \tag{7.25}$$

where A and B are the constants.

ii. The roots are equal.

The complete solution is then obtained as

$$Y = A\bar{y}\big|_{m=m_1} + B\frac{\partial \bar{y}}{\partial m}\bigg|_{m=m_2} \tag{7.26}$$

iii. The roots are distinct and differ by an integer.

When the two roots $m_1 < m_2$ of the indicial equation differ by an integer, the greater of the two roots m_2 will always yield solution while the smaller root m_1 may or may not. In the latter case, we set $A_0 = B_0(m_2 - m_1)$ and obtain the complete solution as

$$Y = A\bar{y}\big|_{m=m_1} + B\frac{\partial \bar{y}}{\partial m}\bigg|_{m=m_1}$$

Example 1. *Solve in series* $2x^2 y'' - xy' + (x^2 + 1) y = 0$.

Solution: Here $x = 0$ is a regular singular point, hence the series solution method is applicable. Putting

$$y = \sum_{n=0}^{\infty} A_n x^{m+n}$$

We have

$$y' = \sum_{n=0}^{\infty} (m+n) A_n x^{m+n-1}$$

and

$$y'' = \sum_{n=0}^{\infty} (m+n)(m+n-1) A_n x^{m+n-2},$$

on substitution gives

$$\sum_{n=0}^{\infty} 2(m+n)(m+n-1) A_n x^{m+n} - \sum_{n=0}^{\infty} (m+n) A_n x^{m+n} + \sum_{n=0}^{\infty} A_n x^{m+n+2} + \sum_{n=0}^{\infty} A_n x^{m+n} = 0 \tag{1}$$

Equating the coefficient of x^{m+n} equal to zero, we have $A_2, A_3,...$ etc. satisfying the recursion formula

$$A_n = \frac{1}{(m+n)(2m+2n-3)+1} A_{n-2} \quad n \geq 2 \tag{2}$$

The lowest power of x is m. Hence indicial equation is

$$(m-1)(2m-1) = 0 \tag{3}$$

The roots of the indicial equation are $m = \dfrac{1}{2}, 1$ and for either value, the first term ($n = 0$) of Eq. (1)

$$(m-1)(2m-1)A_0 x^m$$

vanishes. Since, however, neither of these values of m will cause the second term ($n = 1$) of Eq. (1)

$$m(2m+1)A_1 x^{m+1}$$

to vanish, we take $A_1 = 0$. Using Eq. (2) it follows that

$$A_1 = A_3 = A_5 = A_7 = \ldots = 0. \text{ Thus}$$

$$\bar{y} = A_0 x^m \left\{ 1 - \frac{1}{(m+2)(2m+1)+1} x^2 + \frac{1}{[(m+2)(2m+1)+1][(m+4)(2m+5)+1]} x^4 \ldots \right\}$$

satisfies

$$2x^2 y'' - xy' + (x^2 + 1) y = 0$$

When $m = \dfrac{1}{2}$ and $A_0 = 1$, we have

$$y_1 = \sqrt{x} \left(1 - \frac{x^2}{6} + \frac{x^4}{168} - \frac{x^6}{11088} + \cdots \right)$$

and when $m = 1$ and $A_0 = 1$, we have

$$y_2 = x \left(1 - \frac{x^2}{10} + \frac{x^4}{360} - \frac{x^6}{28080} + \cdots \right)$$

The complete solution then is

$$Y = Ay_1 + By_2$$

$$A\sqrt{x} \left(1 - \frac{x^2}{6} + \frac{x^4}{168} - \frac{x^6}{11088} + \cdots \right) + Bx \left(1 - \frac{x^2}{10} + \frac{x^4}{360} - \frac{x^6}{28080} + \cdots \right)$$

Since $x = 0$ is the only finite singular point, the series converges for all finite values of x.

Example 2. Solve in series form $x\dfrac{d^2 y}{dx^2} + \dfrac{dy}{dx} + x^2 y = 0$

Solution: Put $y = \displaystyle\sum_{n=0}^{\infty} A_n x^{m+n}$

$\therefore$ $$\frac{dy}{dx} = \sum_{n=0}^{\infty} (m+n) A_n x^{m+n-1}$$

and $$\frac{d^2 y}{dx^2} = \sum_{n=0}^{\infty} (m+n)(m+n-1) A_n x^{m+n-2}$$

Substituting for $\dfrac{d^2 y}{dx^2}, \dfrac{dy}{dx}$ and y in differential equation, we obtain

$$\sum_{n=0}^{\infty} (m+n)(m+n-1) A_n x^{m+n-1} + \sum_{n=0}^{\infty} (m+n) A_n x^{m+n-1} + \sum_{n=0}^{\infty} A_n x^{m+n+2} = 0$$

Equating the coefficient of lowest power of x to zero we get indicial equation. The indicial equation is $(n = 0)$

$$A_0 m^2 = 0$$

The roots of the indicial equation are equal.

The coefficients of A satisfy the recursion formula

$$(m + n)(m + n - 1) A_n + (m + n) A_n + A_{n-3} = 0$$

i.e.

$$A_n = -\frac{1}{(m+n)^2} A_{n-3} \cdot n \geq 3$$

Hence we take $A_0 = 1$ and $A_1 = A_2 = 0$

Then $\quad A_1 = A_4 = A_7 = A_{10} = ... = 0$

and $\quad A_2 = A_5 = A_8 = A_{11} = ... = 0$

$$\bar{y} = x^m \left\{ 1 - \frac{1}{(m+3)^2} x^3 + \frac{1}{(m+3)^2(m+6)^2} x^6 - \frac{1}{(m+3)^2(m+6)^2(m+9)^2} x^9 + \cdots \right\}$$

The roots of the indicial equation are $m = 0$. Hence these correspond but one series solution satisfying differential equation with $m = 0$. However,

$$y_1 = \bar{y}\Big|_{m=0} \quad \text{and} \quad y_2 = \frac{\partial \bar{y}}{\partial m}\Big|_{m=0}$$

are solutions of the given differential equation.

$$\frac{\partial \bar{y}}{\partial m} = \bar{y}\log x + 2x^m \left[\frac{1}{(m+3)^3} x^3 - \left\{ \frac{1}{(m+3)^3(m+6)^2} + \frac{1}{(m+3)^2(m+6)^3} \right\} x^6 \right.$$

$$\left. + \left\{ \frac{1}{(m+3)^3(m+6)^2(m+9)^2} + \frac{1}{(m+3)^2(m+6)^3(m+9)^2} + \frac{1}{(m+3)^2(m+6)^2(m+9)^3} \right\} x^9 \cdots \right]$$

Using the root $m = 0$ of the indicial equation

$$y_1 = \bar{y}_1\Big|_{m=0} = 1 - \frac{1}{3^2} x^3 + \frac{x^6}{3^4(2!)^2} - \frac{x^9}{3^6(3!)^2} + \cdots$$

$$y_2 = \frac{\partial \bar{y}}{\partial m}\Big|_{m=0} = y_1 \log x + 2\left[\frac{1}{3^3} x^3 - \frac{x^6}{3^5(2!)^2}\left(1 + \frac{1}{2}\right) + \frac{x^9}{3^7(3!)^2}\left(1 + \frac{1}{2} + \frac{1}{3}\right) \right] \cdots$$

The complete solution

$$y = Ay_1 + By_2$$

$$= (A + B\log x)\left[1 - \frac{1}{3^2} x^3 + \frac{1}{3^4(2!)^2} x^6 - \frac{1}{3^6(3!)^2} x^9 + \cdots \right]$$

$$+ 2B\left[\frac{1}{3^3} x^3 - \frac{1}{3^5(2!)^2}\left(1 + \frac{1}{2}\right) x^6 + \frac{1}{3^7(3!)^2}\left(1 + \frac{1}{2} + \frac{1}{3}\right) x^9 \cdots \right]$$

The series converges for all finite values of $x \neq 0$.

Example 3. *Solve in series*

$$x\frac{d^2y}{dx^2} - 3\frac{dy}{dx} + xy = 0$$

Solution: Put $y = \sum_{n=0}^{\infty} A_n x^{m+n}$

$\therefore$
$$\frac{dy}{dx} = \sum_{n=0}^{\infty} (m+n) A_n x^{m+n-1}$$

and
$$\frac{d^2y}{dx^2} = \sum_{n=0}^{\infty} (m+n)(m+n-1) A_n x^{m+n-2}$$

Substitution in differential equation yields

$$\sum_{n=0}^{\infty} (m+n)(m+n-1) A_n x^{m+n-1} - 3 \sum_{n=0}^{\infty} (m+n) A_n x^{m+n-1} + \sum_{n=0}^{\infty} A_n x^{m+n+1} = 0$$

The indicial equation is

$$m(m-4) A_0 = 0$$

and its roots are $m = 0$ and 4.

Equating the coefficient of x^{m+n-1} equal to zero we get recursion formula

$$A_n = -\frac{1}{(m+n)(m+n-4)} A_{n-2} [n \geq 2]$$

It is clear that this relation yields finite values when $m = 4$, the larger of the roots, but when $m = 0$, $A_4 = \infty$. Here $m = 0$ does not yield solution.

Hence, we replace A_0 by $B_0(m - 0) = B_0 m$ and adjusting $A_1 = 0$, we have the series

$$\bar{y} = A_0 x^m \left[1 - \frac{1}{(m+2)(m-2)} x^2 + \frac{1}{m(m+4)(m+2)(m-2)} x^4 \right.$$

$$\left. - \frac{1}{(m-2)(m+2)^2 m(m+4)(m+6)} x^6 + \cdots \right]$$

$$= B_0 x^4 \left[m - \frac{m}{(m+2)(m-2)} x^2 + \frac{1}{(m+4)(m+2)(m-2)} x^4 \right.$$

$$\left. - \frac{1}{(m-2)(m+2)^2 (m+4)(m+6)} x^6 + \cdots \right]$$

satisfies the equation

$$x\frac{d^2\bar{y}}{dx^2} - 3\frac{d\bar{y}}{dx} + x\bar{y} = (m-4)mA_0 x^{m-1} = (m-4)m^2 B_0 x^{m-1}$$

Since the right hand members contain the factor m^2, it follows that $\bar{y}$ and $\frac{\partial \bar{y}}{\partial m}$, with $m = 0$ are solutions of the given differential equation.

$$\frac{\partial \bar{y}}{\partial m} = \bar{y}\log x + B_0 x^m \left[1 + \frac{m^2+4}{[(m-2)(m+2)]^2} x^2 - \frac{x^4}{(m-2)(m+2)(m+4)} \right.$$

$$\left\{ \frac{1}{(m-2)} + \frac{1}{(m+2)} + \frac{1}{(m+4)} \right\} + \frac{1}{(m-2)(m+2)^2(m+4)^2(m+6)(m+8)}$$

$$\left\{ \frac{1}{(m-2)} + \frac{2}{(m+2)} + \frac{1}{(m+4)} + \frac{1}{(m+6)} \right\} x^6 - \frac{1}{(m-2)(m+2)^2(m+4)^2(m+6)(m+8)}$$

$$\left. \left\{ \frac{1}{(m-2)} + \frac{2}{(m+2)} + \frac{2}{(m+4)} + \frac{1}{(m+6)} + \frac{1}{(m+8)} \right\} x^6 + \cdots \right]$$

Using the root with $m = 0$ and $B_0 = 1$, we obtain

$$y_1 = \bar{y}\big|_{m=0} = -\frac{1}{2\cdot2\cdot4} x^4 + \frac{1}{2\cdot2^2\cdot4\cdot6} x^6 - \frac{1}{2\cdot2^2\cdot4^2\cdot6\cdot8} x^8 + \cdots$$

and

$$y_2 = \frac{\partial \bar{y}}{\partial m}\bigg|_{m=0} = y_1 \log x + 1 + \frac{1}{2^2} x^2 + \frac{1}{2^5}\cdot\frac{1}{2!} x^4 - \frac{1}{2^6 3! 1!}\left(1 + \frac{1}{2} + \frac{1}{3}\right) x^6$$

$$+ \frac{1}{2^8 4! 2!}\left[\left(1 + \frac{1}{2} + \frac{1}{3} + \frac{1}{4}\right)\frac{1}{2}\right] x^8$$

$$- \frac{1}{2^{10}}\cdot\frac{1}{5!}\cdot\frac{1}{3!}\cdot\frac{1}{3!}\left[\left(1 + \frac{1}{2} + \frac{1}{3} + \frac{1}{4} + \frac{1}{5}\right)\left(\frac{1}{2} + \frac{1}{3}\right)\right] x^{10} + \cdots$$

The complete solution is

$$y = Ay_1 + By_2 = (A + B\log x)\left[-\frac{1}{2^3 2!} x^4 + \frac{1}{2^5 3! 1!} x^6 - \frac{1}{2^7 4! 2!} x^8 + \cdots \right]$$

$$+ B\left[1 + \frac{1}{2^2} x^2 + \frac{1}{2^5 2!} x^4 - \frac{1}{2^6 3! 1!}\left(1 + \frac{1}{2} + \frac{1}{3}\right) x^6 \right.$$

$$\left. + \frac{1}{2^8 4! 2!}\left[\left(1 + \frac{1}{2} + \frac{1}{3} + \frac{1}{4}\right)\frac{1}{2} x^8 \cdots\right] \right.$$

The series converges for all finite values of $x \ne 0$.

Example 4. Solve $(x^2 - x)\dfrac{d^2y}{dx^2} + 3\dfrac{dy}{dx} - 2y = x + \dfrac{3}{x^2}$ near $x = 0$.

Solution: Taking $y = \displaystyle\sum_{n=0}^{\infty} A_n x^{m+n}$

and substituting for y, $\dfrac{dy}{dx}$ and $\dfrac{d^2y}{dx^2}$, the recursion formula is

$$A_n = \frac{m+n-3}{m+n-4} A_{n-1}$$

and the complementary function is

$$\bar{y} = A_0 x^m \left(1 + \frac{m-2}{m-3}x + \frac{m-1}{m-3}x^2 + \frac{m}{m-3}x^3 + \frac{m+1}{m-3}x^4 + \cdots \right)$$

The indicial equation for complementary function is

$$m(4-m)A_0 = 0$$

giving $m = 0$ or 4. Setting $A_0 = 1$ we have for $m = 0$

$$y_1 = 1 + \frac{2x}{3} + \frac{x^2}{3} - \frac{x^4}{3} - \frac{2x^5}{3} - \frac{3x^6}{3} - \frac{4x^7}{3} \cdots$$

and for $m = 4$

$$y_2 = x^4(1 + 2x + 3x^2 + 4x^3 + 5x^4 + \cdots) = \frac{x^4}{(1-x)^2}$$

and

$$y_1 = 1 + \frac{2x}{3} + \frac{x^2}{3} - \frac{x^4}{3}$$

The complementary function is

$$y = A(x^2 + 2x + 3) + \frac{Bx^4}{(1-x)^2}$$

In finding a particular integral, we consider each of the terms of the right member of the given differential equation separately. Setting right member equal to x, that is

$$m(4-m)A_0 x^{m-1} = x \text{ identically.}$$

We have $m = 2$ and $A_0 = \frac{1}{4}$ for $m = 2$, the recursion formula is

$$A_n = \frac{n-1}{n-2}A_{n-1}$$

Thus

$$A_1 = A_2 = A_3 = \ldots = 0$$

The particular integral corresponding to the term x is $\frac{x^2}{4}$

Again, setting the right member equal to $\frac{3}{x^2}$ that is

$$m(4-m)A_0 x^{m-1} = \frac{3}{x^2} \text{ identically.}$$

We have $m = -1$ and $A_0 = -\frac{3}{5}$ for $m = -1$

$$A_n = \frac{n-4}{n-5}A_{n-1}$$

Thus

$$A_1 = \frac{3}{4}A_0, A_2 = \frac{1}{2}A_0, A_3 = \frac{1}{4}A_0$$

$$A_4 = A_5 = \ldots = 0$$

The particular integral corresponding to the term $\dfrac{3}{x^2}$ is

$$-\frac{3}{5}x^{-1}\left(1+\frac{3}{4}x+\frac{1}{2}x^2+\frac{1}{4}x^3\right)$$

The required complete solution is

$$y = A(x^2+2x+3)+\frac{Bx^4}{(1-x)^2}-\frac{3}{5x}-\frac{9}{20}-\frac{3}{10}x+\frac{1}{10}x^2$$

$$= C(x^2+2x+3)+\frac{Bx^4}{(1-x)^2}+\frac{1}{4}x^2-\frac{3}{5x}$$

7.5. THE LEGENDRE EQUATION

The differential equation

$$(1-x^2)\frac{d^2y}{dx^2}-2x\frac{dy}{dx}+n(n+1)y = 0 \tag{7.27}$$

is known in literature as *Legendre's differential equation of degree n,* and arises very frequently in various branches of applied mathematics, specially in the process of obtaining solutions of Laplace's equation in spherical coordinates and hence is of great importance in mathematical applications to Physics and Engineering.

The differential equation of Legendre's type can be solved in series of ascending or descending powers of x, but the solutions in descending powers of x is of more physical importance. Since there is no singularity at $x = 0$, the solution of the equation can be obtained in the form of series developed about $x = 0$.

The solution of the equation in the series form containing descending powers of x is given by

$$y = \sum_{p=0}^{\infty}a_px^{m-p} \tag{7.28}$$

$$\frac{dy}{dx} = \sum_{p=0}^{\infty}a_p(m-p)x^{m-p-1}$$

and

$$\frac{d^2y}{dx^2} = \sum_{p=0}^{\infty}a_p(m-p)(m-p-1)x^{m-p-2}$$

Substituting the values of $\dfrac{dy}{dx}$ and $\dfrac{d^2y}{dx^2}$ in Eq. (7.27), we get

$$\sum_{p=0}^{\infty}\left[(1-x^2)(m-p)(m-p-1)x^{m-p-2}-2x(m-p)x^{m-p-1}+n(n+1)x^{m-p}\right]a_p = 0$$

or $$\sum_{p=0}^{\infty}\left[(m-p)(m-p-1)x^{m-p-2}+\{n(n+1)-2(m-p)-(m-p)(m-p-1)\}x^{m-p}\right]a_p = 0$$

or $$\sum_{p=0}^{\infty}\left[(m-p)(m-p-1)x^{m-p-2}+\{n(n+1)-(m-p)(m-p+1)\}x^{m-p}\right]a_p = 0$$

This relation is an identity and therefore the coefficients of various powers of x are zero.

Equating the coefficients of the highest power of $x(= x^m)$ to zero, we get (put $p = 0$)

$$a_0\{n(n+1) - m(m+1)\} = 0 \qquad (7.29)$$

which gives $m = n$ or $m = -(n+1)$, since $a_0 \neq 0$

Now equating the coefficients of x^{m-1} to zero (put $p = 1$), we get

$$a_1\{n(n+1) - m(m+1)\} = 0 \qquad (7.29a)$$

Using the results of Eq. (7.29), i.e. $m = n$ or $m = -(n+1)$

$$n(n+1) - m(m-1) \neq 0$$

Hence $a_1 = 0$.

Now let us equate the coefficients of x^{m-r} to zero, which gives

$$(m-r+2)(m-r+1)a_{r-2} + \{n(n+1) - (m-r)(m-r+1)\}a_r = 0$$

$$a_r = -\frac{(m-r+2)(m-r+1)}{n(n+1) - (m-r)(m-r+1)}a_{r-2} \qquad (7.30)$$

Thus for $m = n$, the recurrence relation between the coefficients is given by

$$a_r = -\frac{(n-r+2)(n-r+1)}{n^2+n-(n-r)(n-r+1)}a_{r-2}$$

$$= -\frac{(n-r+2)(n-r+1)}{r(2n-r+1)}a_{r-2} \qquad (7.31)$$

and putting $m = -(n+1)$ in Eq. (7.30) we get

$$a_r = -\frac{(-n-r+1)(-n-r)}{n^2+n-\{-n-r-1\}\{-n-r\}}a_{r-2} = \frac{(n+r-1)(n+r)}{r(2n+r+1)}a_{r-2}$$

or

$$a_r = \frac{(n+r-1)(n+r)}{r(2n+r+1)}a_{r-2} \qquad (7.32)$$

Case 1: When $m = n$. By putting $r = 2, 3,...$ in Eq. (7.31), we have

$$a_2 = -\frac{n(n-1)}{2(2n-1)}a_0$$

$$a_3 = -\frac{n(n-1)(n-2)}{3(2n-2)}a_1 = 0 \text{ (since } a_1 = 0)$$

Thus $a_3 = a_5 = a_7 =...= 0$, i.e. all the odd coefficients are zero and

$$a_4 = -\frac{(n-1)(n-3)}{4(2n-3)}a_2 = \frac{n(n-1)(n-2)(n-3)}{2 \cdot 4 \cdot (2n-1)(2n-3)}a_0$$

In this way all the even coefficients can be determined and in general

$$a_{2r} = (-1)^r \frac{n(n-1)(n-2)\cdots(n-2r+2)(n-2r+1)}{2 \cdot 4 \cdot 2r(2n-1)(2n-3)\cdots(2n-2r-1)(2n-2r+1)}a_0$$

Hence the series solution for $m = n$ is

$$y = a_0 x^n \left[1 - \frac{n(n-1)}{2(2n-1)} x^{-2} + \frac{n(n-1)(n-2)(n-3)}{2 \cdot 4 \cdot (2n-1)(2n-3)} x^{-4} + \cdots \right.$$

$$\left. + (-1)^r \frac{n(n-1)(n-2)\cdots(n-2r+1)}{2 \cdot 4 \cdots 2r(2n-1)(2n-3)\cdots(2n-2r+1)} x^{-2r} + \cdots \right] \qquad (7.33)$$

Case 2: When $K = -n - 1$. We have for $r = 2, 3\ldots$ from Eq. (7.32)

$$a_2 = \frac{(n+1)(n+2)}{2(2n+3)} a_0$$

Since a_3 will contain a_1 and hence is zero and such all the odd coefficients $a_5, a_7, a_9, \ldots$ etc. are zero

and

$$a_{2r} = \frac{(n+1)(n+2)\cdots(n+2r)}{2 \cdot 4 \cdots 2r(2n+3)(2n+5)(2n+2r+1)} a_0$$

and therefore the series solution for $m = -n - 1$ is given by

$$Y = a_0 x^{-(n+1)} \left[1 + \frac{(n+1)(n+2)}{2(2n+3)} x^{-2} + \frac{(n+1)(n+2)(n+3)(n+4)}{2 \cdot 4 \cdot (2n+3)(2n+5)} x^{-4} + \cdots \right.$$

$$\left. + \frac{(n+1)(n+2)\cdots(n+2r)}{2 \cdot 4 \cdots 2r(2n+3)(2n+5)\cdots(2n+2r+1)} x^{-2r} + \cdots \right] \qquad (7.34)$$

The solutions given by Eqs (7.33) and (7.34) are independent and hence their sum represents the general solution of Legendre's equation. Whether the solutions are of any interests depends on their convergence properties. The convergence can be checked by the ratio test, i.e.

$$\lim_{j \to \infty} \frac{|y_{j+2}|}{|y_j|} = \frac{1}{x^2}$$

Hence these solutions are convergent for $|x| > 1$. Thus these equations are of physical interest for the values of x except for $-1 < x < 1$. In this interval the solutions are not convergent.

To find convergent solutions in this interval we can solve Legendre's differential equation by developing the series solution in the ascending powers of x. For this let us assume the solution

$$y = \sum_{p=0}^{\infty} a_p x^{m+p}$$

Differentiation and substitution in the Legendre's Eq. (7.27) give

$$\sum_p a_p (m+p)(m+p-1) x^{m+p-2} - \sum_p a_p [(m+p-n)(m+p+n+1)] x^{m+p} = 0$$

Equating the coefficients of lowest power of x, we get

$$a_0 m (m - 1) = 0 \quad \text{which gives } m = 0 \text{ or } m = +1$$

Equating the coefficients of x^{m-1} to zero we get

$$a_1(m+1)m = 0 \quad \text{which gives } a_1 = 0 \text{ for } m = 1,$$

but for $m = 0$, a_1 may or may not be zero.

Equating coefficients of x^{m+r}, we get

$$a_{r+2} = \frac{(m+r-n)(m+r+n+1)}{(m+r+2)(m+r+1)}a_r \tag{7.35}$$

For $m = 0$, first choice of indicial constant, the recurrence relation is

$$a_{r+2} = \frac{(r-n)(r+n+1)}{(r+2)(r+1)}a_r$$

which gives

$$a_2 = -\frac{n(n+1)}{2!}a_0, a_4 = \frac{(2-n)(n+3)}{3\cdot 4}a_2 = \frac{n(n-2)(n+1)(n+3)}{4!}a_0$$

and in general a_{2r} is given by

$$a_{2r} = (-1)^r \frac{n(n-2)\cdots(n-2r+2)(n+1)(n+3)\cdots(n+2r-1)}{(2r)!}a_0$$

Similarly,

$$a_3 = -\frac{(n-1)(n+2)}{2\cdot 3}a_1 = -\frac{(n-1)(n+2)}{3!}a_1$$

$$a_5 = +\frac{(n-1)(n-3)(n+2)(n+4)}{5!}a_1$$

and

$$a_{2r+1} = (-1)^r \frac{(n-1)(n-3)\cdots(n-2r+1)(n+2)(n+4)\cdots(n+2r)}{(2r+1)!}a_1$$

and hence the solution of Legendre's equation is

$$y = a_0\left[1 - \frac{n(n+1)}{2!}x^2 + \frac{n(n-2)(n+1)(n+3)}{4!}x^4 + \cdots\right.$$

$$+(-1)^r \frac{n(n-2)(n-2r+2)(n+1)(n+3)\cdots(n+2r-1)}{(2r)!}x^{2r} + \cdots\right]$$

$$+ a_1x\left[1 - \frac{(n-1)(n+2)}{3!}x^2 + \cdots + (-1)^r \frac{(n-1)(n-3)\cdots(n-2r+1)(n+2)\cdots(n+2r)}{(2r+1)!}x^{2r}\cdots\right]$$

$$\tag{7.36}$$

For second choice of indicial constant $m = 1$, the recurrence relation becomes

$$a_{r+2} = \frac{(r+1-n)(r+n+2)}{(r+3)(r+2)}a_r$$

and

$$a_{2r} = (-1)^r \frac{(n-1)(n-3)\cdots(n-2r+1)(n+2)\cdots(n+2r)}{(2r+1)!}a_0$$

and $a_3 = a_5 = a_7 = \ldots = 0$. Since $a_1 = 0$.

Therefore, solution of Legendre's differential equation becomes

$$y = a_0 x \left[1 - \frac{(n-1)(n+2)}{3!} x^2 + \frac{(n-1)(n-3)(n+2)(n+4)}{5!} x^4 \cdots \right.$$

$$\left. + (-1)^r \frac{(n-1)(n-3)\cdots(n-2r+1)(n+2)(n+4)\cdots(n+2r)}{(2r+1)!} x^{2r} \cdots \right] \qquad (7.36a)$$

Comparing the Eqs (3.36a) and (7.36) we observe that the second part of Eq. (7.36) with coefficients constant a_1 is same as Eq. (7.36a). It seems that a part of solution of Eq. (7.36) is the solution of Eq. (7.36a).

Hence the second part of Eq. (7.36) is not a solution at all and hence, we can set $a_1 = 0$ for $m = 0$. Hence the Eq. (7.36) reduces to

$$y = a_0 \left[1 - \frac{n(n+1)}{2!} x^2 + \frac{n(n-2)(n+1)(n+3)x^4}{4!} + \cdots \right.$$

$$\left. + (-1)^r \frac{n(n-2)\cdots(n-2r+2)(n+1)(n+3)\cdots(n+2r-1)}{(2r)!} x^{2r} + \cdots \right] \qquad ...(7.36b)$$

The solutions of Eqs (7.36a) and (7.36b) can be proved by ratio test as representing convergent series for $-1 < x < 1$. If an infinite series, as the solution of a given differential equation is reduced into a finite series the solution is called polynomial. Thus the solution of Eq. (7.36b) reduces to polynomial, when n is an even positive or an odd negative integer (or zero).

(a) Let n be even and positive, i.e. $n = 2r$ then Eq. (7.36b) reduces to

$$y = a_0 \left[1 - \frac{n(n+1)}{2!} x^2 + \cdots + (-1)^{n/2} \frac{n(n-2)\cdots 4 \cdot 2(n-1)(n+3)\cdots(2n-1)}{n!} x^n \right]$$

On the other hand, under these conditions Eq. (7.33) reduces to

$$y = a_0 x^n \left[1 - \frac{n(n-1)}{(2n-1)} x^2 + \cdots + (-1)^{n/2} \frac{n!}{n(n-2)\cdots 2(n+1)(n+3)\cdots(2n-1)} x^{-n} \right]$$

The two solutions are identical if multiplied by a constant factor

$$(-1)^{n/2} \frac{n(n-2)\cdots 4 \cdot 2(n+1)(n+3)\cdots(2n-1)}{n!}$$

Thus for n being an even positive integer the solution of Eq. (7.36b) coincides to Eq. (7.33).

(b) For n an odd negative integer, the solution of Eq. (7.36b) is identical to Eq. (7.34). The solution [Eq. (7.36a)] reduces to polynomial when n is an odd positive for an even negative integer.

(c) For n an odd positive integer solution of Eq. (7.36a) gives

$$y = a_0 \left[x - \frac{(n-1)(n+2)}{3!} x^3 + \cdots + (-1)^{n/2} \frac{(n-1)(n-3)\cdots 2(n+2)\cdots(2n-1)}{n!} x^n \right]$$

$$(7.36c)$$

While under this condition Eq. (7.33) reduces to

$$y = a_0 x^n \left[1 - \frac{n\ (n-1)}{2\ (2n-1)} x^{-2} + \cdots + (-1)^{(n-1)/2} \frac{n!\, x^{-n+1}}{2\cdot 4\cdots(n-1)(n+2)\cdots(2n-1)} \right]$$

which when divided by coefficient of its last term becomes identical with Eq. (7.36c).

(d) If n is an integer (negative) solutions Eqs (7.36a) and (7.34) become identical.

The solutions of Eqs (7.33) and (7.34) are of great importance for integral values of n. For n a positive integer, the solution of Eq. (7.33) is a polynomial but Eq. (7.34) is an infinite series and the general solution is a linear combination of them. For n a negative integer Eq. (7.33) is an infinite series and Eq. (7.34) is a polynomial.

For $2n$ being equal to some positive odd integer the solution of Eq. (7.33) degenerates into Eq. (7.34). This can be seen by putting $2n = 1 - 1$, for which the denominator in the coefficient of x^{n-2l} and in every subsequent term in Eq. (7.33) vanishes. To remove this difficulty arising due to this fact the entire series may be multiplied by $2(l - r)$ which will reduce all terms of order higher than $(n - 2l)$ to zero while the other terms remain finite. Hence the series begins with the term containing $x^{n-2l} = x^{-n-1}$ and is identical with Eq. (7.34). Thus in this case we get only one solution and this is an infinite series.

In a similar manner it can be shown that for n equal to an odd negative integer the solution of Eq. (7.34) degenerates to Eq. (7.33). The functions of interest are those which remain finite for all values of x for $-1 \le x \le 1$. Such functions exist only when n is a positive or a negative integer for which the consideration is limited to solutions Eqs (7.33) and (7.34) because the others reduce to these solutions. Secondly, the inspection shows that solution of Eq. (7.34) with n replaced by $-(n + 1)$ is identical with solution of Eq. (7.33). Therefore, we may further restrict our consideration to positive values of n (including zero) and retain of Eq. (7.33) as a significant solution.

For the useful polynomial for n equal to positive integer, the constant a_0 in solution Eq. (7.33) is chosen as

$$a_0 = \frac{1\cdot 3\cdot 5\cdots(2n-1)}{n!}$$

Then the solution

$$y = P_n(x) = \frac{1\cdot 3\cdot 5\cdots(2n-1)}{n!} x^n \left[1 - \frac{n\ (n-1)}{2\ (2n-1)} x^{-2} + \cdots \right.$$

$$\left. +(-1)^r \frac{(n-2r+1)\cdots(n-1)n}{2\cdot 4\cdots 2r(2n-2r+1)\cdots(2n-1)} x^{-2r} + \cdots \right] \tag{7.37}$$

is called Legendre's polynomial or Legendre's coefficient or zonal harmonics of first kind. The last term of the series in the bracket is

$$(-1)^{n/2} \frac{n!}{2^{n/2}\left(\dfrac{n}{2}!\right)(2n-1)(2n-3)\cdots(n+1)} x^{-n}$$

or

$$(-1)^{n-1/2} \frac{n!}{2^{n-1/2}\left(\dfrac{n-1}{2}!\right)(2n-1)(2n-3)\cdots(n+2)} x^{-(n-2)}$$

according as n is even or odd.

For n being the negative integer the constant a_0 in solution of Eq. (7.34) is chosen as

$$a_0 = \frac{n!}{1 \cdot 3 \cdot 5 \cdots (2n+1)}$$

and corresponding polynomial is

$$y = Q_n(x) = \frac{n! x^{-n-1}}{1 \cdot 3 \cdot 5 \cdots (2n+1)} \left[1 + \frac{(n+1)(n+2)}{2(2n+3)} x^{-2} + \cdots \right.$$

$$\left. + \frac{(n+1)(n+2)\cdots(n+2r)}{2 \cdot 4 \cdots (2r)(2n+3)\cdots(2n+2r+1)} x^{-2r} \right] \tag{7.38}$$

and is defined as Legendre's polynomial of second kind.

The most general solution of Legendre's differential equation is

$$y = A P_n(x) + B Q_n(x) \tag{7.39}$$

where A and B are arbitrary constants.

For n, a positive integer, the polynomial $P_n(x)$ has the expansion as in Eq. (7.37) ending with term containing x^0 or the term containing x accordingly n is even or odd integer. The expansion can also be written as

$$P_n(x) = \sum_{r=0}^{N} (-1)^r \frac{1 \cdot 3 \cdot 5 \cdots (2n-1)}{n!} \times \frac{(n-2r+1)(n-2r+2)\cdots n}{2 \cdot 4 \cdots (2r)(2n-2r+1)\cdots(2n-1)} x^{n-2r}$$

$$= \sum_{r=0}^{N} (-1)^r \frac{1 \cdot 3 \cdot 5 \cdots (2n-2r-1)}{2^r r!(n-2r)!} x^{n-2r}$$

$$= \sum_{r=0}^{N} (-1)^r \frac{1 \cdot 2 \cdot 3 \cdots (2n-2r)}{2^r r!(n-2r)! 2 \cdot 4 \cdots (2n-2r)} x^{n-2r}$$

$$= \sum_{r=0}^{N} (-1)^r \frac{(2n-2r)!}{2^n r!(n-2r)!(n-r)!} x^{n-2r} \tag{7.40}$$

where $N = \dfrac{n}{2}$ for n even and $N = \dfrac{(n-1)}{2}$ for n odd.

From this expansion, the values of $P_n(x)$ for $n = 0, 1, 2$, etc. can be calculated.

$$\left. \begin{aligned} n &= 0 \quad P_0(x) = 1 \\ n &= 1 \quad P_1(x) = x \\ n &= 2 \quad P_2(x) = 1/2(3x^2 - 1) \\ n &= 3 \quad P_3(x) = 1/2(5x^3 - 3x) \\ n &= 4 \quad P_4(x) = 1/8(35x^4 - 30x^2 + 3x) \\ n &= 5 \quad P_5(x) = 1/8(63x^5 - 70x^3 + 15x) \end{aligned} \right\} \tag{7.41}$$

Legendre polynomial defined by Eq. (7.40) occurs very frequently in large number of problems in physics, engineering and applied mathematics. It is used to find the solution of Laplace and Helmholtz equations in spherical coordinates and hence has its direct applications in the study of potential and thermal distribution in spherical symmetrical bodies. It is also used to express the distribution function of angular correlation between two γ-ray photons emitted in successive transitions of a nuclei and also in the study of scattering problems in the special case of radial incidence.

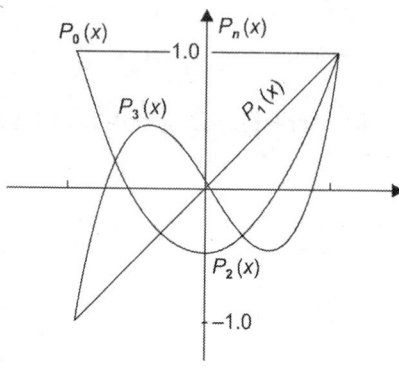

Fig. 7.1

• The graphical representation of some of the Legendre polynomials is given in Fig. 7.1.

The Generating Function of $P_n(x)$

The Legendre polynomial $P_n(x)$ is the coefficient of z^n in the expansion of $(1 - 2xz + z^2)^{-1/2}$ in ascending powers of x, i.e.

$$(1 - 2xz + z^2)^{-1/2} = \sum_{n=0}^{\infty} P_n(x)z^n \tag{7.42}$$

These may be verified for the lower powers of n by expanding Eq. (7.42) by binomial theorem. To prove it for the general term we write

$$(1 - 2xz + z^2)^{-1/2} = [1 - (2xz - z^2)]^{-1/2}$$

$$= 1 + \frac{1}{2}(2xz - z^2) + \frac{\left(\frac{1}{2}\right)\left(\frac{3}{2}\right)}{2}(2xz - z^2)^2 + \cdots$$

$$+ \frac{1 \cdot 3 \cdots (2k-5)}{2^{k-2}(k-2)!}(2xz - z^2)^{k-2} + \frac{1 \cdot 3 \cdots (2k-3)}{2^{k-1}(k-1)!}(2xz - z^2)^{k-1}$$

$$+ \frac{1 \cdot 3 \cdots (2k-1)}{2^k k!}(2xz - z^2)^k + \cdots$$

or $$(1 - 2xz + z^2)^{-1/2} = \sum_{n=0}^{\infty} + \frac{1 \cdot 3 \cdot 5 \cdots (2n-1)}{2^n n!}(2xz - z^2)^n \tag{7.43}$$

But $(2xz - z^2)^n$ on expansion gives

$$(2xz - z^2)^n = \sum_{k=0}^{\infty}(-1)^k \frac{n! z^n}{k!(n-k)!}(2x)^{n-k} z^k$$

Thus Eq. (7.43) becomes

$$(1 - 2xz + z^2)^{-1/2} = \sum_{n=0}^{\infty}\sum_{k=0}^{\infty}(-1)^k \frac{1 \cdot 3 \cdot 5 \cdots (2n-1)n!}{2^n n!(n-k)!k!}(2x)^{n-k} z^{n+k}$$

or $$= \sum_{n=0}^{\infty}\sum_{k=0}^{\infty}(-1)^k \frac{n!(2n)!}{2^{2n}(n!)^2 k!(n-k)!}(2x)^{n-k} z^{n+k} \tag{7.44}$$

This result is obtained by multiplying the numerator and denominator by $2.4...(2n) = 2^n \times n!$

Rearranging the order of summation, Eq. (7.44) can be written as

$$(1-2xz+z^2)^{-1/2} = \sum_{n=0}^{\infty}\sum_{k=0}^{n/2}(-1)^k \frac{(2n-2k)!(2x)^{n-2k}}{2^{2n-2k}k!(n-k)!(n-2k)!}z^n$$

$$(1-2xz+z^2)^{-1/2} = \sum_{n=0}^{\infty}\left[\sum_{k=0}^{[n/2]}(-1)^k \frac{(2n-2k)!}{2^n k!(n-k)!(n-2k)!}x^{n-2k}\right]z^n \tag{7.44a}$$

where $[n/2] = N$ for n even and $(n-1)/2$ for n odd. The series in bracket in this equation is identical to that on the right hand side of Eq. (7.40).

Thus we have

$$(1-2xz+z^2)^{-1/2} = \sum_{n=0}^{\infty}P_n(x)z^n$$

Hence by expanding the function $(1-2xz+z^2)^{-1/2}$ in the binomial expansion in ascending powers of z, we obtain the Legendre polynomials of different orders as the coefficients of corresponding power of z.

Thus the function $(1-2xz+z^2)^{-1/2}$ is defined as the generating function of $P_n(x)$. We can draw the following useful and direct conclusion from Eq. (7.44).

(i) $P_n(-1) = (-1)^n$.

The result directly follows from Eq. (7.44a)

Let $x = -1$

$$(1-2xz+z^2)^{-1/2} = \sum_{n=0}^{\infty}P_n(-1)z^n$$

or

$$\sum_{n=0}^{\infty}P_n(-1)z^n = (1+z)^{-1} = \sum_{n=0}^{\infty}(-1)^n z^n$$

Equating the coefficients of z^n we have

$$P_n(-1) = (-1)^n.$$

(ii) $P_n(-x) = (-1)^n P_n(x)$

The generating function $(1-2xz+z^2)^{-1/2}$ remains unchanged if we replace x by $(-x)$ and z by $(-z)$. Thus we have

$$(1-2xz+z^2)^{-1/2} = \sum_{n=0}^{\infty}P_n(x)z^n$$

and

$$(1-2xz+z^2)^{-1/2} = \left\{1-2(-x)(-z)+(-z)^2\right\}^{-1/2} = \sum_{n=0}^{\infty}P_n(-x)(-z)^n$$

Therefore

$$\sum_{n=0}^{\infty}P_n(x)z^n = \sum_{n=0}^{\infty}(-1)^n P_n(-x)z^n$$

Equating the coefficients of z^n, we have

$$P_n(-x) = (-1)^n P_n(x).$$

Here if we put $x = 1$, we get

$$P_n(-1) = (-1)^n P_n(1) \quad \text{But} \quad P_n(1) = 1$$

Hence $\qquad\qquad P_n(-1) = (-1)^n$ gives the results in statement (i)

(iii) $P_n (\cos \theta)$ can be expressed as a series consisting of cosines of even or odd integer multiples of θ.

Let us put $x = \cos\theta = \dfrac{e^{i\theta} + e^{-i\theta}}{2}$. We have

$$(1 - 2\cos\theta z + z^2)^{-1/2} = \sum_{n=0}^{\infty} P_n(\cos\theta) z^n$$

or

$$\sum_{n=0}^{\infty} P_n(\cos\theta) z^n = (1 - z e^{i\theta})^{-1/2} (1 - z e^{-i\theta})^{-1/2}$$

On expanding the right hand side binomially, we have

$$\sum_{n=0}^{\infty} P_n(\cos\theta) z^n = \left[1 + \frac{1}{2} z e^{i\theta} + \frac{1 \cdot 3}{2 \cdot 4} e^{2i\theta} + \frac{1 \cdot 3 \cdot 5}{2 \cdot 4 \cdot 6} z^3 e^{3i\theta} + \cdots + \frac{1 \cdot 3 \cdots (2n-1)}{2 \cdot 4 \cdot 6 \cdots 2n} z^n e^{-ni\theta} + \cdots \right]$$

$$\times \left[1 + \frac{1}{2} z e^{-i\theta} + \frac{1 \cdot 3}{2 \cdot 4} z^2 e^{-2i\theta} + \frac{1 \cdot 3 \cdot 5}{2 \cdot 4 \cdot 6} z^3 e^{-3i\theta} + \cdots \frac{1 \cdot 3 \cdots (2n-1)}{2 \cdot 4 \cdot 6 \cdots 2n} z^n e^{-ni\theta} + \cdots \right]$$

Equating the coefficients of z^n on both sides, we have

$$P_n(\cos \theta) = \frac{1 \cdot 3 \cdots (2n-1)}{2 \cdot 4 \cdot 6 \cdots 2n} \left[\left(e^{in\theta} + e^{-in\theta} \right) + \frac{2n}{2(2n-1)} \left\{ e^{i(n-2)\theta} + e^{-i(n-2)\theta} \right\} + \cdots \right.$$

$$+ \frac{2n(2n-2)}{(2n-1)(2n-3)} \cdot \frac{1}{2} \cdot \frac{3}{4} \left\{ e^{i(n-4)\theta} + e^{-i(n-4)\theta} \right\} + \cdots$$

$$\left. + \frac{1 \cdot 3 \cdots (2r-1) \cdots 2n(2n-2) \cdots (2n-2r+2)}{2 \cdot 4 \cdot 6 \cdots 2r(2n-1)(2n-3) \cdots (2n-2r+1)} \left\{ e^{i(n-2r)\theta} + e^{-i(n-2r)\theta} \right\} + \cdots \right]$$

or $\quad P_n(\cos \theta) = 2 \dfrac{1 \cdot 3 \cdot 5 \cdots (2n-1)}{2 \cdot 4 \cdots 2n} \left[\cos n\theta + \dfrac{n}{2n-1} \cos(n-2)\theta \right.$

$$+ \frac{1 \cdot n(n-1)}{2!(2n-1)(2n-3)} \cos(n-4)\theta$$

$$\left. + \frac{1 \cdot 3 \cdots (2r-1) \cdots n(n-1) \cdots (n-r+1)}{r!(2n-1)(2n-3) \cdots (2n-2r+1)} \times \cos(n-2r)\theta + \cdots \right] \qquad (7.45)$$

Rodrigues' Formula

An important formula for $P_n(x)$ may be deduced directly from Legendre's differential equation.

Let $\qquad\qquad\qquad y = (x^2 - 1)^n$

$$\therefore \qquad \frac{dy}{dx} = n(x^2 - 1)^{n-1} \cdot (2x)$$

Multiplying both sides by $(x^2 - 1)$, we get

$$(x^2 - 1)\frac{dy}{dx} = n(x^2 - 1)^n \cdot (2x) = 2nxy$$

Differentiating it $(n + 1)$ times using Leibnitz's theorem, we get

$$(x^2 - 1)\frac{d^{n+2}y}{dx^{n+2}} + {}^{n+1}c_1 \frac{d^{n+1}y}{dx^{n+1}}(2x) + {}^{n+1}c_2 \frac{d^n y}{dx^n}(2) = 2n\left[x\frac{d^{n+1}y}{dx^{n+1}} + {}^{n+1}c_1 \frac{d^n y}{dx^n} \cdot 1 \right]$$

or $\qquad (x^2 - 1)\frac{d^{n+2}y}{dx^{n+2}} + 2(n+1)x\frac{d^{n+1}y}{dx^{n+1}} + n(n+1)\frac{d^n y}{dx^n} = 2nx\frac{d^{n+1}y}{dx^{n+1}} + 2n(n+1)\frac{d^n y}{dx^n}$

or $\qquad (x^2 - 1)\frac{d^{n+2}y}{dx^{n+2}} + 2x\frac{d^{n+1}y}{dx^{n+1}} - n(n+1)\frac{d^n y}{dx^n} = 0 \qquad (7.46)$

Putting $V_n = \dfrac{d^n y}{dx^n}$ in Eq. (7.46) we have

$$(x^2 - 1)\frac{d^2 V_n}{dx^2} + 2x\frac{dV_n}{dx} - n(n+1)V_n = 0 \qquad (7.47)$$

This is Legendre's equation. Hence V_n satisfies Legendre's equation. But since

$$V_n = \frac{d^n y}{dx^n} = \frac{d^n}{dx^n}(x^2 - 1)^n \qquad (7.48)$$

V_n is a polynomial of degree n and since Legendre's equation has one and only one distinct solution that of form $P_n(x)$, it follows that $P_n(x)$ is a constant multiple of V_n. Hence we have

$$P_n(x) = c\frac{d^n}{dx^n}(x^2 - 1)^n \qquad (7.49)$$

To determine the value of constant c, we merely consider the highest power of x on each side of equation, that is

$$\frac{2n!}{2^n (n!)^2}x^n = c\frac{d^n(x^{2n})}{dx^n} = c\frac{(2n!)}{n!}x^n$$

Hence $\qquad\qquad c = \dfrac{1}{2^n n!} \qquad (7.50)$

Substituting this value of c into Eq. (7.49) we obtain

$$P_n(x) = \frac{1}{2^n n!}\frac{d^n}{dx^n}(x^2 - 1)^n \qquad (7.51)$$

This is Rodrigue's formula for the Legendre's polynomials.

Orthogonality of Functions

The word orthogonality originally comes from vector analysis, where two vectors u and v are said to be orthogonal if their scalar product becomes zero,

i.e. $\qquad\qquad u \cdot v = u_x v_x + u_y v_y + u_z v_z = 0$

This definition of orthogonality of vectors when generalised for n-dimensional space in which a vector is specified by n components, the condition takes the form

$$\sum_{i=1}^{n} u_i v_i = 0$$

The continuum analogy of this condition can be written as

$$\int_a^b u(x)v(x)dx = 0 \tag{7.51a}$$

where x is a continuous variable. This is orthogonality condition for $u(x)$ and $v(x)$ in the range $a \le x \le b$. In case the function $u(x)$ and $v(x)$ are complex, the definition of orthogonality is generalised as

$$\int_a^b u^*(x)v(x)dx = 0 \tag{7.51b}$$

where $u^*(x)$ is complex conjugate of $u(x)$.

Equation (7.51b) is taken as the basic definition of orthogonality of $u(x)$ and $v(x)$ in the interval (a, b). Now if we have a whole set of function $f_n(x)$ where $n = 1, 2,...,$ etc. and

$$\int_a^b f_n^*(x)f_m(x)dx = \begin{cases} 0 & \text{if } m \ne n \\ \text{const} \ne 0 & \text{if } m = n \end{cases} \tag{7.51c}$$

We call the functions $f_n(x)$ a set of orthogonal functions. For example

$$\int_{-\pi}^{+\pi} \sin nx \cdot \sin mx\, dx = \begin{cases} 0 & \text{if } m \ne n \\ \pi & \text{if } m = n \end{cases}$$

and hence $\sin nx$ form a set of orthogonal functions in the interval $(-\pi$ to $+\pi)$.

Some functions are not orthogonal but can be made orthogonal by introducing a factor $g(x)$ called as weight function, such that

$$\int_a^b f_n^*(x)f_m(x)g(x)dx = \begin{cases} 0 & \text{if } m \ne n \\ \text{const} \ne 0 & \text{if } m = n \end{cases}$$

Now the function $f_n(x)$ and $f_m(x)$ are said to be orthogonal with respect to $g(x)$ over interval (a, b).

In case of Eq. (7.51c) for $m = n$ if the integral has the value

$$\int_a^b f_n^*(x)f_n(x)dx = \text{constant } c^2 \text{ (say)}$$

then

$$\int_a^b \left\{ \frac{1}{c} f_n^*(x) \right\}\left\{ \frac{1}{c} f_n(x) \right\} dx = 1$$

The functions are said to have been normalised to unity. A set of normalised orthogonal functions are called orthonormal. Thus the functions $\phi(x)$ are said to form an orthonormal set of functions in the interval (a, b) if

$$\int_a^b \phi_n(x)\phi_m(x)dx = \begin{cases} 0 & \text{if } m \ne n \\ 1 & \text{if } m = n \end{cases} = \delta_{mn} \tag{7.51d}$$

where δ_{mn} is Kronecker delta symbol and has value 1 for $m = n$ and zero for $m \ne n$.

There are differential equations whose solutions are orthogonal set. Differential equations with this property are called *self-adjoint differential equations*.

Let us consider a general homogeneous second order differential equation

$$Ly(x) = a_0(x)\, y'' + a_1(x)\, y' + a_2(x)\, y = 0$$

where the coefficients $a_0(x)$, $a_1(x)$, $a_2(x)$ are the real functions of x and first two derivatives of $a_0(x)$ as well as first derivative of $a_1(x)$ are continuous in the region of interest $a \le x \le b$.

The above equation is self-adjoint. As such its solutions will be orthogonal set of functions, if

$$L = \bar{L}$$

where $\bar{L}$ is adjoint operator for L given by

$$\bar{L}y(x) = \frac{d^2}{dx^2}[a_0 y] - \frac{d}{dx}[a_1 y] + a_2 y$$

$$= a_0 \frac{dy^2}{dx^2} + \left(2\frac{da_0}{dx} - a_1\right)\frac{dy}{dx} + \left(\frac{d^2 a_0}{dx^2} - \frac{da_1}{dx} + a_2\right)y$$

Now equating $Ly(x) = \bar{L}y(x)$, we get

$$\frac{du_0(x)}{dx} = a_1(x),$$

i.e. if the condition for the coefficients of a differential equation is satisfied, its solutions form the orthogonal set of functions.

Orthogonality of Legendre's Polynomials

For Legendre's differential equation

$$a_0(x) = 1 - x^2 \text{ and } a_1(x) = -2x$$

and therefore the condition $\dfrac{d}{dx}a_0(x) = a_1(x)$ is satisfied. Hence the equation is self-adjoint differential equation and hence its solution $[P_n(x)]$ form the orthogonal set of functions.

Legendre polynomials are a set of orthogonal functions in the interval $(-1, 1)$ like the trigonometric functions $\cos mx$ and $\sin mx$.

$$\int_{-1}^{+1} P_m(x)P_n(x)\,dx = \frac{2}{2n+1}\delta_{mn} \tag{7.52}$$

where $\delta_{mn} = \begin{cases} 0 & \text{if } m \ne n \\ 1 & \text{if } m = n \end{cases}$ is Kronecker delta symbol[*].

Legendre polynomials $P_n(x)$ satisfy the Legendre's differential Eq. (7.27). Thus we have

$$(1-x)^2 \frac{d^2 P_n(x)}{dx^2} - 2x\frac{dP_n(x)}{dx} + n(n+1)P_n(x) = 0$$

---

[*] Detailed properties of this symbol have been discussed in Chapter 3—Tensor Analysis.

or
$$\frac{d}{dx}\left[(1-x^2)\frac{dP_n(x)}{dx}\right]+n(n+1)P_n(x)=0 \qquad (7.53)$$

Similarly
$$\frac{d}{dx}\left[(1-x^2)\frac{dP_m(x)}{dx}\right]+m(m+1)P_m(x)=0 \qquad (7.53a)$$

Multiplying Eq. (7.53) by $P_m(x)$ and Eq. (7.53a) by $P_n(x)$ and then subtracting we have

$$P_m(x)\frac{d}{dx}\left[(1-x)^2\frac{dP_n(x)}{dx}\right]+n(n+1)P_n(x)P_m(x)-P_n(x)\frac{d}{dx}\left[(1-x^2)\frac{dP_m(x)}{dx}\right]$$
$$-m(m+1)\,P_n(x)\,P_m(x)=0$$

or
$$\frac{d}{dx}\left[(1-x^2)\left\{P_m(x)\frac{dP_n(x)}{dx}-P_n(x)\frac{dP_m(x)}{dx}\right\}\right]+(n(n+1)-m(m+1))P_n(x)\,P_m(x)=0$$

Integrating this equation w.r.t. x from -1 to $+1$ we have

$$(1-x^2)\left\{P_m(x)\frac{dP_n(x)}{dx}-P_n(x)\frac{dP_m(x)}{dx}\right\}\Bigg|_{-1}^{+1}+[n(n+1)-m(m+1)]\int_{-1}^{+1}P_n(x)P_m(x)dx=0$$

Because of factor $(1-x^2)$ first term is zero. Thus we have

$$(n-m)(n+m+1)\int_{-1}^{+1}P_n(x)P_m(x)dx=0$$

Thus either $n=m$ or $\int_{-1}^{+1}P_n(x)P_m(x)dx=0$

as $(n+m+1)$ can not be zero because m and n are positive integers. This establishes

$$\int_{-1}^{+1}P_n(x)P_m(x)dx=0 \quad \text{when } m\neq n \qquad (7.54)$$

To calculate integral when $m=n$ we square both sides of Eq. (7.44) and obtain

$$(1-2xz+z^2)^{-1}=\left[\sum_{n=0}^{\infty}P_n(x)z^n\right]^2$$

$$=\sum_{n=0}^{\infty}\{P_n(x)\}^2\,z^{2n}+2\sum_{\substack{n=0\\m=0}}^{\infty}z^{m+n}P_m(x)P_n(x)$$

We now integrate both sides of this equation with respect to x over the interval $(-1, 1)$ and observe that the product terms on the right vanish in view of the orthogonality property [Eq. (7.54)]. We thus obtain

$$\int_{-1}^{+1}\frac{dx}{(1-2xz+z^2)}=\sum_{n=0}^{\infty}z^{2n}\int_{-1}^{+1}\{P_n(x)\}^2\,dx$$

But the integral on the left has the value

$$\int_{-1}^{+1}\frac{dx}{(1-2xz+z^2)}=\frac{1}{z}\{\log(1+z)-\log(1-z)\}$$

$$= 2\left\{1 + \frac{z^2}{3} + \frac{z^4}{5} + \cdots + \frac{z^{2n}}{2n+1} + \cdots\right\}$$

$$= 2\sum_{n=0}^{\infty} \frac{z^{2n}}{2n+1}$$

Thus we have

$$\sum_{n=0}^{\infty} z^{2n} \int_{-1}^{+1} \{P_n(x)\}^2 \, dx = 2\sum_{n=0}^{\infty} \frac{z^{2n}}{2n+1}$$

Equating the coefficients of z^{2n} on both sides we have

$$\int_{-1}^{+1} \{P_n(x)^2 \, dx\} = \frac{2}{2n+1} \tag{7.54a}$$

Combining Eqs (7.54) and (7.54a) we get Eq. (7.52).

Recurrence Relations for $P_n(x)$

These are the relations among the polynomials of different orders and their derivatives.

1. Differentiating Eq. (7.44) with respect to z we have

$$\frac{d}{dz}[1 - 2xz + z^2]^{-1/2} = \frac{d}{dz}\sum_{n=0}^{\infty} P_n(x)z^n = \sum_{n=0}^{\infty} nP_n(x)z^{n-1}$$

or

$$\frac{x-z}{(1 - 2xz + z^2)^{3/2}} = \sum_{n=0}^{\infty} nP_n(x)z^{n-1}$$

or

$$(x - z)(1 - 2xz + z^2)^{-1/2} = (1 - 2xz + z^2)\sum_{n=0}^{\infty} nP_n(x)z^{n-1}$$

or

$$(x - z)\sum_{n=0}^{\infty} P_n(x)z^n = (1 - 2xz + z^2)\sum_{n=0}^{\infty} nP_n(x)z^{n-1}$$

Equating the coefficients of z^n on both sides we have

$$xP_n(x) - P_{n-1}(x) = (n+1)P_{n+1}(x) - 2xnP_n(x) + (n-1)P_{n-1}(x)$$

or

$$(2n+1)xP_n(x) = (n+1)P_{n+1}(x) + nP_{n-1}(x) \tag{7.55}$$

2. Differentiating Eq. (7.44) w.r.t. x we have

$$\frac{z}{(1 - 2xz + z^2)^{3/2}} = \sum_{n=0}^{\infty} \frac{dP_n(x)}{dx} z^n$$

or

$$(1 - 2xz + z^2)\sum_{n=0}^{\infty} P_n'(x)z^n = z(1 - 2xz + z^2)^{-1/2} = z\sum_{n=0}^{\infty} P_n(x)z^n$$

where

$$P_n'(x) = \frac{dP_n(x)}{dx}$$

Equating the coefficients of z^n on both sides of this equation we have

$$P'_n(x) - 2xP'_{n-1}(x) + P'_{n-2}(x) = P_{n-1}(x)$$

Replacing n by $(n + 1)$, we have

$$P'_{n+1}(x) + P'_{n-1}(x) = 2xP'_n(x) + P_n(x) \qquad (7.56)$$

3. Differentiating Eq. (7.55) with respect to x we have

$$(2n+1)P_n(x) + (2n+1)xP'_n(x) = (n+1)P'_{n+1}(x) + nP'_{n-1}(x) \qquad (7.57)$$

Putting Eq. (7.57) + (2n + 1) into Eq. (7.56), yields

$$P'_{n+1}(x) - P'_{n-1}(x) = (2n + 1)\, P_n(x) \qquad (7.58)$$

and now replacing n by $(n - 1)$, we have

$$P'_n(x) = (2n-1)P_{n-1}(x) + P'_{n-2}(x)$$

$$= (2n-1)P_{n-1}(x) + \{2(n-2)-1\}\, P_{n-2-1}(x) + P'_{n-4}(x)$$

$$= (2n-1)P_{n-1}(x) + (2n-5)P_{n-3}(x) + P'_{n-4}(x)$$

or $$P'_n(x) = (2n-1)P_{n-1}(x) + (2n-5)P_{n-3}(x) + (2n-9)P_{n-5}(z) + \cdots \qquad (7.59)$$

i.e. ending with $3P_1(x)$ or $P_0(x)$ according as n is even or odd. The expansion of Eq. (7.59) is an important as well as useful expansion known as *Christoffel's expansion*.

4. Equation (7.57) when added n times to Eq. (7.56), yields

$$P'_{n+1}(x) = (n+1)P_n(x) + xP'_n(x) \qquad (7.60)$$

on replacing n into Eq. (7.60) by $(n - 1)$, we have

$$nP_{n-1}(x) + xP'_{n-1}(x) = P'_n(x)$$

on $$nP_{n-1}(x) = P'_n(x) - xP'_{n-1}(x) \qquad (7.61)$$

Again Eq. (7.57) when added to $(n + 1)$ times to Eq. (7.56), yields

$$P'_{n-1}(x) = xP'_n(x) - nP_n(x) \qquad (7.62)$$

or $$nP_n(x) = x\frac{dP_n(x)}{dx} - \frac{dP_{n-1}(x)}{dx} \qquad (7.62a)$$

Multiplying Eq. (7.62) by x and subtracting from Eq. (7.61), we have

$$n[P_{n-1}(x) - xP_n(x)] = (1-x)^2\frac{dP_n(x)}{dx}$$

or $$(1-x^2)\frac{dP_n(x)}{dx} = n\left[P_{n-1}^{(x)} - xP_n(x)\right] \qquad (7.63)$$

Equation (7.55) on rearranging the terms yields

$$(n+1)xP_n(x) - (n+1)P_{n+1}(x) = nP_{n-1}(x) - nxP_n(x)$$

$$= n\{P_{n-1}(x) - xP_n(x)\}$$

$$= (1-x^2)\frac{dP_n(x)}{dx} \qquad \text{(using Eq. 7.63)}$$

or $\qquad (1-x^2)\dfrac{dP_n(x)}{dx} = (n+1)\{xP_n(x) - P_{n+1}(x)\}$ $\qquad\qquad$ (7.64)

and replacing n by $-(n + 1)$ we have

or $\qquad (1-x^2)P'_{-(n+1)}(x) = -n\{xP_{-(n+1)}(x) - P_{-n}(x)\}$

$\qquad\qquad\qquad\qquad = n\{P_{-n}(x) - xP_{-n(n+1)}(x)\}$ $\qquad\qquad$ (7.65)

5. Multiplying Eq. (7.63) by $(n + 1)$ and Eq. (7.64) by n and adding yields

$$(2n+1)(1-x^2)\frac{dP_n(x)}{dx} = n(n+1)[P_{n-1}(x) - P_{n+1}(x)] \qquad\qquad (7.66)$$

Comparison of Eqs (7.63) and (7.65) gives

$$P_{-(n+1)}(x) = P_n(x) \qquad\qquad (7.66a)$$

Equations (7.55) and (7.66) are called *recurrence relations* and by combining these it is possible to get some more relations which may be useful in certain calculations.

Integral Representation for Legendre Polynomials

Rodrigue's formula as in Eq. (7.51) provides means of developing an integral representation of $P_n(x)$. Let us consider a function $f(z) = (z^2 - 1)^n$, the value of which at any point $z = x$ is given by Cauchy integral formula[*].

$$f(x) = \frac{1}{2\pi i}\oint \frac{f(z)}{z-x}dz$$

where the closed contour encloses the point $z = x$.

or $\qquad (x^2 - 1)^n = \dfrac{1}{2\pi i}\oint \dfrac{f(z)}{z-x}dz$

Thus we have

$$(x^2 - 1)^n = \frac{1}{2\pi i}\oint \frac{(z^2-1)^n}{(z-x)}dz$$

Differentiating this equation n times w.r.t. x and multiplying by $\dfrac{1}{2^n n!}$, we get

$$\frac{1}{2^n n!}\frac{d^n}{dx^n}(x^2-1)^n = \frac{2^{-n}}{2\pi i}\oint \frac{(z^2-1)^n}{(z-x)^{n+1}}dz$$

or using Eq. (7.51) we have, $\qquad\qquad\qquad\qquad\qquad$ [since LHS = $P_n(x)$]

$\therefore \qquad\qquad P_n(x) = \dfrac{1}{2\pi i}\dfrac{1}{2^n}\oint \dfrac{(z^2-1)^n}{(z-x)^{n+1}}dz$ $\qquad\qquad$ (7.67)

This is the *Schlaefi integral* for Legendre's polynomial. Choosing the contour as a circle of radius $(x^2 - 1)^{1/2}$ and centre at x, we may have

$$z - x = (x^2 - 1)^{1/2}e^{i\phi}$$

or $\qquad\qquad\qquad z = x + (x^2 - 1)^{1/2}e^{i\phi}.$

---

[*] Discussed in Chapter 6—*Functions of Complex Variables.*

Making this substitution in Eq. (7.67) we have

$$P_n(x) = \frac{1}{2\pi i} \int_0^{2\pi} \frac{2^n (x^2-1)^{n/2} e^{in\phi} \left\{ x + \sqrt{x^2-1} \cos\phi \right\}^n \cdot i\sqrt{x^2-1} e^{i\phi} \cdot d\phi}{2^n \left\{ \sqrt{x^2-1} e^{i\phi} \right\}^{n+1}}$$

$$= \frac{1}{2\pi} \int_0^{2\pi} \left\{ x + \sqrt{x^2-1} \cos\phi \right\}^n d\phi$$

Since integrand is even function of ϕ, hence we get

$$P_n(x) = \frac{1}{\pi} \int_0^\pi \left\{ x + \sqrt{x^2-1} \cos\phi \right\}^n d\phi \qquad (7.68)$$

Equation (7.68) is called *Laplace's first integral representation of $P_n(x)$.*
If we replace n by $-(n+1)$ in this equation we have

$$P_{-n-1}(x) = \frac{1}{\pi} \int_0^\pi \left\{ x + \sqrt{x^2-1} \cos\phi \right\}^{-n-1} d\phi \qquad (7.68a)$$

But Eq. (7.66a) gives $P_{-(n+1)}(x) = P_n(x)$
Now we have

$$P_n(x) = \frac{1}{\pi} \int_0^\pi \frac{d\phi}{\left\{ x + \sqrt{x^2-1} \cos\phi \right\}^{n+1}} \qquad (7.68b)$$

This is called Laplace's second integral representation for $P_n(x)$. In this equation let us put

$$x + \sqrt{x^2-1} \cos\phi = z$$

$$\therefore \qquad d\phi = \frac{dz}{i(1-2xz+z^2)^{1/2}}$$

Now Eq. (7.68b)

$$P_n(x) = \frac{1}{2\pi i} \oint \frac{z^{-n-1}}{(1-2xz+z^2)^{1/2}} dz \qquad (7.68c)$$

where the closed contour encloses the origin.

This is another integral representation for $P_n(x)$. This equation gives $P_n(x) = R$, where

R is the residue of function $\dfrac{z^{-n-1}}{(1-2xz+z^2)^{1/2}}$ at $z = 0$ or R is the coefficient of z^{-1}

in Laurent's expansion of $\dfrac{z^{-n-1}}{(1-2xz+z^2)^{1/2}}$. Thus R is the coefficient of z^n in the

expansion of $(1 - xz + z^2)^{-1/2}$. Hence we conclude that

$$(1 - 2xz + z^2)^{-1/2} = \sum_{n=0}^\infty P_n(x) z^n$$

which is generating function.

Thus, we have developed generating function for Legendre's polynomial starting with Schlaefi integral.

In case n is not an integer, the point $z = 0$ in integral of Eq. (7.68c) is the branch point and the path of integration is loop around the singular point $z = x \pm \sqrt{x^2 - 1}$ clockwise.

Expansion of an Arbitrary Function in a Series of Legendre Polynomials

If $F(x)$ is sectionally continuous in the interval $(-1, 1)$ and also its derivative $F'(x)$ is sectionally continuous in every interval interior to $(-1, 1)$, it may be shown that $F(x)$ may be expanded in a series of the form

$$F(x) = \sum_{n=0}^{\infty} a_n P_n(x) \tag{7.69}$$

To obtain the general coefficient a_n, we multiply both sides of Eq. (7.69) by $P_n(x)$ and integrate over the interval $(-1, 1)$.

We obtain

$$\int_{-1}^{+1} F(x) P_n(x) dx = a_n \int_{-1}^{+1} [P_n(x)]^2 dx = \frac{2a_n}{2n+1} \tag{7.70}$$

using orthogonality of Legendre polynomials, [Eq. (7.54a)]. The general coefficient of expansion of Eq. (7.69).

$$a_n = \frac{2n+1}{2} \int_{-1}^{+1} F(x) P_n(x) dx \tag{7.71}$$

The expansion of Eq. (7.69) is similar to an expansion of an arbitrary function into a Fourier series.

7.6. ASSOCIATED LEGENDRE POLYNOMIALS

In the solution of some potential problems it is convenient to use certain polynomials closely related to the Legendre polynomials. We shall discuss them briefly in this section.

We consider the equation

$$(1-x^2)\frac{d^2P}{dx^2} - 2x\frac{dP}{dx} + \left\{ n(n+1) - \frac{m^2}{1-x} \right\} P = 0 \tag{7.72}$$

where m in an integer, n a non-negative integer and $|m| \le n$. Since Eq. (7.72) does not depend on the sign of m, it will be sufficient to assume that $0 \le m \le n$.

The substitution

$$P(x) = \text{const } (1 - x^2)^{m/2} u(x)$$
$$= c(1 - x^2)^{m/2} u(x) \tag{7.73}$$

reduces Eq. (7.72) to the form

$$(1-x^2)\frac{d^2u}{dx^2} - 2(m+1)x\frac{du}{dx} + (n-m)(n+m+1) u = 0 \tag{7.74}$$

$$0 \le m \le n.$$

For $m = 0$ this is exactly the Legendre Eq. (7.27), one of whose solution is Legendre polynomial $P_n(x)$.

$$(1-x^2)\frac{d^2P_n}{dx^2} - 2x\frac{dP_n}{dx} + n(n+1)P_n = 0$$

Differentiating the Legendre equation m times, one obtains

$$(1-x^2)\frac{d^2}{dx^2}\left[\frac{d^m P_n}{dx^m}\right] - 2(m+1)x\frac{d}{dx}\left[\frac{d^m P_n}{dx^m}\right] + (n-m)(n+m+1)\frac{d^m P_n}{dx^m} = 0$$

which has just the form of Eq. (7.74). Hence

$$u(x) = c\frac{d^m P_n(x)}{dx^m} \tag{7.75}$$

is a solution of Eq. (7.74). Combining Eqs (7.73) and (7.75), we define the *associated Legendre function* $P_n^m(x)$ as

$$P_n^m(x) = (1-x^2)^{m/2}\frac{d^m P_n(x)}{dx^m}\quad 0\le m\le n \tag{7.76}$$

Here normalisation constant is taken as 1. It is obvious that the associated Legendre functions reduce to the Legendre polynomials for $m = 0$

$$P_n^0(x) = P_n(x)$$

We also notice that if $m > n$ we have

$$P_n^m(x) = 0 \tag{7.77}$$

Using Rodrigue's formula for $P_n(x)$, we obtain the corresponding Rodrigue's formula for $P_n^m(x)$ as

$$P_n^m(x) = \frac{(1-x^2)^{m/2}}{2^n n!}\frac{d^{n+m}}{dx^{n+m}}(x^2-1)^n \quad (0\le m\le n) \tag{7.78}$$

Note: Although the Rodrigue's formula is designed to be used for non-negative values of m, it yields meaningful function even if m is negative so long as $|m|\le n$. In fact many authors define the functions

$$P_n^{-m}(x) = \frac{(1-x^2)^{-m/2}}{2^n n!}\frac{d^{n-m}}{dx^{n-m}}(x^2-1)^n \quad (m>0, m\le n) \tag{7.79}$$

These functions happen to be, however, multiples of functions $P_n^m(x)$, that is one can establish the formula

$$P_n^{-m}(x) = (-1)^m\frac{(n-m)!}{(n+m)!}P_n^m(x) \quad (|m|\le n) \tag{7.80}$$

Associated Legendre functions with negative m are useful in defining spherical harmonics but they are not indispensable. Some authors do not use them at all, while others define them arbitrarily, e.g. by

$$P_n^{-m}(x) = (-1)^m P_n^m \quad \text{or} \quad P_n^{-m} = P_n^m.$$

Unless stated otherwise, we shall assume that (in this book) in symbol $P_n^m(x)$, m is non-negative.

Orthogonality of Associated Legendre Polynomials

Associated Legendre functions with the same index m but with different index n are orthogonal to each other, i.e.

$$\int_{-1}^{+1} P_n^m(x)P_{n'}^m(x)dx = \frac{2}{2n+1}\frac{(n+m)!}{(n-m)!}\delta_{n'n} \tag{7.81}$$

The normalisation integral can be derived as follows. Taking $n = n'$.

$$\int_{-1}^{+1}\left[P_n^m(x)\right]^2 dx = \int_{-1}^{+1}(1-x^2)^m \frac{d^m P_n(x)}{dx^m} \cdot \frac{d^m P_n(x)}{dx^m}dx$$

Integrating by parts, we have

$$\int_{-1}^{+1}\left[P_n^m(x)\right]^2 dx = \int_{-1}^{+1}\frac{d^{m-1}P_n(x)}{dx^{m-1}}\frac{d}{dx}\left[(1-x^2)^m \frac{d^m P_n(x)}{dx^m}\right]dx \qquad (7.82)$$

We observe that

$$\frac{d}{dx}\left[(1-x^2)^m \frac{d^m P_n(x)}{dx^m}\right] = (1-x^2)^m \frac{d^{m+1}P_n(x)}{dx^{m+1}} - 2mx(1-x^2)^{m-1}\frac{d^m P_n(x)}{dx^m} \qquad (7.83)$$

Let us put

$$\frac{d^{m-1}P_n(x)}{dx^{m-1}} = U_{m-1}(x)$$

Now $U_{m-1}(x)$ satisfies the differential equation

$$(1-x^2)U''_{m-1} - 2mxU'_{m-1} + [n(n+1)-(m-1)m]U_{m-1} = 0$$

Equation (7.83) yields

$$\frac{d}{dx}\left[(1-x^2)^m \frac{d^m P_n}{dx^m}\right] = (1-x^2)^{m-1}\left[(1-x^2)U''_{m-1} - 2mxU'_{m-1}\right]$$

$$= -(1-x^2)^{m-1}\left[n(n+1) - m(m-1)\right]U_{m-1}$$

$$= -(1-x^2)^{m-1}(n+m)(n-m+1)\frac{d^{m-1}P_n}{dx^{m-1}} \qquad (7.84)$$

Substitution of Eqs (7.84) and (7.82) yields

$$\int_{-1}^{+1}\left[P_n^m(x)\right]^2 dx = (n+m)(n-m+1)\int_{-1}^{+1}\left[P_n^{m-1}(x)\right]^2 \qquad (7.85)$$

Applying this formula m times, we obtain

$$\int_{-1}^{+1}\left[P_n^m(x)\right]^2 dx = \frac{(n+m)!}{(n-m)!}\int_{-1}^{+1}\left[P_n(x)\right]^2 dx$$

So that*

$$\int_{-1}^{+1}\left[P_n^m(x)\right]^2 dx = \frac{2}{2n+1}\frac{(n+m)!}{(n-m)!}$$

Hence $\qquad \int_{-1}^{+1}P_n^m(x)P_{n'}^m(x)dx = \frac{2}{2n+1}\frac{(n+m)!}{(n-m)!}\delta_{nn'}$

---

* If index $n \neq n'$ in the end we will get integral

$$\int_{-1}^{+1}P_n(x)P_{n'}(x)dx$$

which is zero. Then Eq. (7.54) becomes $\qquad$ (if $n = n'$)

$$\int_{-1}^{+1}\left[P_n(x)\right]^2 dx = \frac{2}{2n+1}$$

Recurrence Relations for $P_n^m(x)$

We had the following relations for Legendre polynomials, i.e. Eqs (7.55), (7.60) and (7.62).

$$(2n + 1)\, xP_n(x) = (n + 1)\, P_{n+1}(x) + nP_{n-1}(x) \tag{7.85a}$$

$$P'_{n+1}(x) = (n + 1)\, P_n(x) + xP'_n(x) \tag{7.85b}$$

$$xP_n(x) + P'_{n-1}(x) = xP'_n(x) \tag{7.85c}$$

Differentiating Eq. (7.85a) m times and multiplying by $(1 - x^2)^{m/2}$, we get

$$(2n+1)xP_n^m(x)+(2n+1)m\sqrt{1-x^2}\,P_n^{m-1}(x) = (n+1)P_{n+1}^m(x)+nP_{n-1}^m(x) \tag{7.86}$$

Differentiating Eqs (7.85b) and (7.85c) $(m - 1)$ times and multiplying by $(1 - x^2)^{m/2}$, we have

$$P_{n+1}^m(x) = (n+m)\sqrt{1-x^2}\,P_n^{m-1}(x)+xP_n^m(x) \tag{7.86a}$$

$$(n-m+1)\sqrt{1-x^2}\,P_n^{m-1}(x)+ P_{n-1}^m(x) = xP_n^m \tag{7.86b}$$

Eliminating $\sqrt{1-x^2}\,P_n^{m-1}$ from Eqs (7.86) and (7.86b) obtain pure recurrence relation in x.

$$(n-m+1)P_{n+1}^m(x)-(2n+1)xP_n^m(x)+(n+m)P_{n-1}^m(x) = 0 \tag{7.87}$$

A second such pure recurrence relation in m can also be obtained by using differential Eq. (7.74) and relation Eq. (7.75).

$$(1-x^2)u''(m+1)xu'\{n(n+1)-m(m+1)\}u = 0$$

where

$$u = \frac{d^m}{dx^m}P_n(x)$$

replacing m by $m - 1$ and multiplying by $(1 - x^2)^{m/2}$ we get

$$\sqrt{1-x^2}\,P_n^{m+1} - 2mxP_n^m +(n+m)(n-m+1)\sqrt{1-x^2}\,P_n^{m-1} = 0 \tag{7.88}$$

Again we have

$$P_n^m = (1-x^2)^{m/2}\frac{d^m}{dx^m}P_n$$

Differentiating and multiplying by $(1 - x^2)$, we have

$$(1-x^2)\frac{dP_n^m}{dx} = \sqrt{1-x^2}\,P_n^{m+1} - mxP_n^m \quad (m\text{-raising}) \tag{7.89}$$

From Eqs (7.88) and (7.89), we have

$$(1-x^2)\frac{dP_n^m}{dx} = mxP_n^m -(n+m)(n-m+1)\sqrt{1-x^2}\,P_n^{m-1} \quad (m\text{-lowering}) \tag{7.90}$$

From Eqs (7.90) and (7.86b), we have

$$(1-x^2)\frac{dP_n^m}{dx} = (n+m)P_{n-1}^m - nxP_n^m \quad (n\text{-lowering}) \tag{7.91}$$

Finally Eqs (7.91) and (7.87) yield

$$(1-x^2)\frac{dP_n^m}{dx} = (n+1)xP_n^m -(n-m+1)P_n^m \quad \text{(}n\text{-raising)} \qquad \text{...(7.92)}$$

Equations (7.89), (7.90), (7.91) and (7.92) are also called *Ladder operations*.

Example 1. *Prove that*

$$\int_{-1}^{+1}\left\{\frac{d}{dx}P_n(x)\right\}^2 dx = n(n+1) \qquad \qquad \text{[Jodhpur 1997]}$$

Solution: From Christoffel's expansion we have Eq. (7.59)

$$P_n'(x) = (2n-1)P_{n-1}(x)+(2n-5)P_{n-3}(x)+(2n-9)P_{n-5}(x)+\cdots$$

ending with $3P_1(x)$ or $P_0(x)$ according as n is even or odd.

Using orthogonality property of Legendre polynomials we get

$$\int_{-1}^{+1}\left\{\frac{d}{dx}P_n(x)\right\}^2 dx = (2n-1)^2 \cdot\frac{2}{2(n-1)+1}+(2n-5)^2\cdot\frac{2}{2(n-3)+1}+\cdots+3^2\cdot\frac{2}{2+1}$$

or $\dfrac{2}{2.0+1}$ according as n is even or odd.

Thus $\displaystyle\int_{-1}^{+1}\left\{\frac{d}{dx}P_n(x)\right\}^2 dx = 2\{(2n-1)+(2n-5)+(2n-9)\cdots+3\}$ [when n is even]

$$= 2\{(2n-1)+(2n-5)+\cdots+1\} \qquad \text{[when } n \text{ is odd]}$$

$$= n(n+1) \text{ for } n \text{ even or odd integer.}$$

Example 2. *Show that*

$$\int_{-1}^{+1}(1-x^2)\left\{\frac{dP_n(x)}{dx}\right\}^2 dx = \frac{2n(n+1)}{2n+1}$$

Solution: We have Eq. (7.66)

$$(1-x^2)\frac{dP_n(x)}{dx} = \frac{n(n+1)}{2n+1}\{P_{n-1}(x)-P_{n+1}(x)\} \qquad \qquad (1)$$

and Eq. (7.59) gives

$$\frac{dP_n(x)}{dx} = (2n-1)P_{n-1}(x)+(2n-5)P_{n-3}(x)+\cdots \qquad \qquad (2)$$

ending with $3P_1(x)$ or $P_0(x)$

Multiplying Eqs (1) and (2) and integrating within the limits -1 to $+1$ and using orthogonality property of Legendre's polynomials we have

$$\int_{-1}^{+1}(1-x^2)\left\{\frac{dP_n(x)}{dx}\right\}^2 dx = \frac{n(n+1)(2n-1)}{2n+1}\int_{-1}^{+1}\{P_{n-1}(x)\}^2 dx$$

$$= \frac{n(n+1)(2n-1)}{(2n+1)} \frac{2}{2(n-1)+1}$$

$$= \frac{2n(n+1)}{(2n+1)}$$

which is the required result.

Example 3. *If* (u, ϕ, z) *and* (r, θ, ϕ) *be the cylindrical and spherical coordinates of the same point and if* $x = \cos \theta$, *prove that*

$$P_n(x) = (-1)^n \frac{r^{n+1}}{n!} \frac{\partial^n}{\partial z^n}\left(\frac{1}{r}\right)$$

Solution: In the given coordinate system

$$r^2 = u^2 + z^2$$

or
$$\frac{1}{r} = (u^2 + z^2)^{-1/2} = \phi(u, z) \tag{1}$$

which is a function of u and z.

By Taylor's theorem we may have

$$\phi(u, z - k) = \phi(u, z) - k\frac{\partial}{\partial z}\phi(u, z) + \frac{k^2}{2!}\frac{\partial^2}{\partial z^2}\phi(u, z) + \cdots + (-1)^n \frac{k^n}{n!}\frac{\partial^n}{\partial z^n}\phi(u, z) + \cdots$$

$$= \sum_{n=0}^{\infty}(-1)^n \frac{k^n}{n!}\frac{\partial^n}{\partial z^n}\phi(u, z) \tag{2}$$

But by Eq. (1), we have

$$\phi(u, z - k) = \left\{u^2 + (z - k)^2\right\}^{-1/2}$$

$$= \left\{u^2 + z^2 - 2zk + k^2\right\}^{-1/2}$$

$$= \left\{r^2 - 2rk\cos\theta + k^2\right\}^{-1/2} \qquad \text{[since } z = r\cos\theta\text{]}$$

or
$$\phi(u, z - k) = r^{-1}\left\{1 - 2\frac{k}{r}\cos\theta + \left(\frac{k}{r}\right)^2\right\}^{-1/2}$$

$$= \frac{1}{r}\sum_{n=0}^{\infty}\left(\frac{k}{r}\right)^n P_n(\cos\theta) \tag{3}$$

Thus from Eqs (2) and (3), we have

$$\sum_{n=0}^{\infty}(-1)^n \frac{k^n}{n!}\frac{\partial^n}{\partial z^n}\phi(u, z) = \frac{1}{r}\sum_{n=0}^{\infty}\left(\frac{k}{r}\right)^n P_n(\cos\theta)$$

Equating the coefficients of k^n on both the sides, we have

$$P_n(\cos\theta) = (-1)^n \frac{r^{n+1}}{n!}\frac{\partial^n}{\partial z^n}\phi(u, z)$$

$$= (-1)^n \frac{r^{n+1}}{n!} \frac{\partial^n}{\partial z^n}\left(\frac{1}{r}\right)$$

Since $\cos\theta = x$ hence

$$P_n(x) = (-1)^n \frac{r^{n+1}}{n!} \frac{\partial^n}{\partial z^n}\left(\frac{1}{r}\right)$$

Hence the result.

Example 4. *If n is a positive integer, prove that*

$$\int_{-1}^{+1} P_n(x)(1-2xz+z^2)^{-1/2}dx = \frac{2z^n}{2n+1}$$

and hence, making use of Rodrigue's formula, deduce that

$$\int_{-1}^{+1} (1-x^2)^n(1-2xz+z^2)^{-n-1/2}dx = \frac{2^{2n+1}(n!)^2}{(2n+1)!}$$

where $P_n(x)$ are Legendre's polynomials.

Solution: From generating function of $P_n(x)$ we have

$$(1-2xz+z^2)^{-1/2} = \sum_n z^n P_n(x)$$

$$\therefore \quad \int_{-1}^{+1} P_n(x)(1-2xz+z^2)^{-1/2}dx = \int_{-1}^{+1} P_n(x)\sum_r z^r P_r(x)dx$$

$$= \sum_r z^r \int_{-1}^{+1} P_n(x)P_r(x)dx$$

$$= \sum_r z^r \frac{2}{2r+1}\delta_{n,r} \qquad [\textit{from} \text{ orthogonality property}]$$

$$= \frac{2z^n}{2n+1}$$

From Rodrigue's formula, we have

$$P_n(x) = \frac{1}{2^n n!}\frac{d^n}{dx^n}(x^2-1)^n$$

$$\therefore \quad \int_{-1}^{+1} \frac{1}{2^n n!}\left\{\frac{d^n}{dx^n}(x^2-1)^n\right\}(1-2xz+z^2)^{-1/2}dx = \frac{2z^n}{2n+1}$$

or

$$\int_{-1}^{+1}\left\{\frac{d^n}{dx^n}(x^2-1)^n\right\}\{1-2xz+z^2\}^{-1/2}dx = \frac{2^n n!2z^n}{2n+1}$$

Integrating by parts, we have

$$\left[(1-2xz+z^2)^{-1/2}\frac{d^{n-1}}{dx^{n-1}}(x^2-1)^n\right]_{-1}^{+1} - \int_{-1}^{+1}-\frac{1}{2}(1-2xz+z^2)^{-3/2}(-2z)\frac{d^{n-1}}{dx^{n-1}}(x^2-1)^n dx$$

$$= \frac{2^{n+1}n!z^n}{2n+1}$$

or
$$(-1)^1 z \int_{-1}^{+1} (1 - 2xz + z^2)^{-3/2} \frac{d^{n-1}}{dx^{n-1}} (x^2 - 1)^n dx = \frac{2^{n+1} n! z^n}{2n + 1}$$

Integrating again by parts, we get

$$(-1)^2 z^2 \int_{-1}^{+1} 1 \cdot 3 \cdot (1 - 2xz + z^2)^{-5/2} \frac{d^{n-2}}{dx^{n-2}} (x^2 - 1)^n dx = \frac{2^{n+1} n! z^n}{(2n + 1)}$$

Continuing this process of integration by parts n-times, we have

$$(-1)^n z^n \int_{-1}^{+1} 1 \cdot 3 \cdot 5 \cdots (2n - 1)(1 - 2xz + z^2)^{-\frac{(2n+1)}{2}} (x^2 - 1)^n dx = \frac{2^{n+1} \cdot n! z^n}{(2n + 1)}$$

or
$$(-1)^{2n} z^n \int_{-1}^{+1} (1 - x^2)^n (1 - 2xz + z^2)^{-\frac{(2n+1)}{2}} dx = \frac{2^{n+1} \cdot n! z^n}{(2n + 1)!} 2^n n!$$

or
$$\int_{-1}^{+1} (1 - x^2)^n (1 - 2xz + z^2)^{-n-1/2} dx = \frac{2^{2n+1} \cdot (n!)^2}{(2n + 1)!}$$

Hence the result.

Example 5. *Prove that*

$$P_n\left(-\frac{1}{2}\right) = P_0\left(-\frac{1}{2}\right) P_{2n}\left(\frac{1}{2}\right) + P_1\left(-\frac{1}{2}\right) P_{2n-1}\left(\frac{1}{2}\right) + \cdots + P_{2n}\left(-\frac{1}{2}\right) P_0\left(\frac{1}{2}\right)$$

Solution: We have

$$(1 - 2zx + z^2)^{-1/2} = \sum_{n=0}^{\infty} z^n P_n(x)$$

Putting $x = \frac{1}{2}$ and $-\frac{1}{2}$ respectively in the above result, we get

$$(1 - z + z^2)^{-1/2} = \sum z^n P_n\left(\frac{1}{2}\right) \tag{1}$$

and
$$(1 - z + z^2)^{-1/2} = \sum z^n P_n\left(-\frac{1}{2}\right) \tag{2}$$

Again putting z^2 for z in Eq. (2) we get

$$(1 + z^2 + z^4)^{-1/2} = \sum z^{2n} P_n\left(-\frac{1}{2}\right) \tag{3}$$

But
$$(1 + z^2 + z^4) = (1 + z + z^2)(1 - z + z^2)$$

or
$$(1 - z + z^2)^{-1/2} (1 + z + z^2)^{-1/2} = \sum z^{2n} P_n\left(-\frac{1}{2}\right)$$

$$= \sum z^r P_r\left(\frac{1}{2}\right) z^m P_m\left(-\frac{1}{2}\right)$$

Equating coefficients of z^{2n} on both sides, we have

$$P_n\left(-\frac{1}{2}\right) = P_0\left(-\frac{1}{2}\right) P_{2n}\left(\frac{1}{2}\right) + P_1\left(-\frac{1}{2}\right) P_{2n-1}\left(\frac{1}{2}\right) + \cdots + P_{2n}\left(-\frac{1}{2}\right) P_0\left(\frac{1}{2}\right).$$

Example 6. *Find the sum of the* $(n + 1)$ *terms of series*

$$\sum_{n=0}^{\infty} (2m+1)P_m(x)P_m(y)$$

Solution: From recurrence relation [Eq. (7.55)], we have

$$(2n + 1)xP_n(x) = (n + 1)\, P_{n+1}(x) + nP_{n-1}(x) \tag{1}$$

and $\qquad (2n + 1)\, yP_n(y) = (n + 1)\, P_{n+1}(y) + nP_{n-1}(y) \tag{2}$

Multiply Eq. (1) by $P_n(y)$ and Eq. (2) by $P_n(x)$ and subtracting, we get

$$(2n+1)(x-y)P_n(x)P_n(y) = \left[(n+1)\{P_n(y)P_{n+1}(x) - P_n(x)P_{n+1}(y)\}\right.$$
$$\left. + n\{P_n(y)P_{n-1}(x) - P_n(x)P_{n-1}(y)\}\right]$$

Putting $n = 1, 2, 3...$ etc. and adding we have

$$\sum_{n=1}^{\infty} (2n+1)(x-y)P_n(x)P_n(y) = (n+1)\left[P_{n+1}(x)P_n(y) - P_{n+1}(y)P_n(x)\right]$$

$$-1\left[P_1(x)P_0(y) - P_1(y)P_0(x)\right]$$

But $\qquad\qquad P_1(x) = x, P_1(y) - y$
$$P_0(x) = 1, P_0(y) = 1$$

$\therefore \qquad P_1(x)\, P_0(y) - P_1(y)\, P_0(x) = (x - y)$
$$= (x - y)\, P_0(x)\, P_0(y)$$

This term when taken to L.H.S gives

$$\sum_{n=0}^{n} (2n+1)(x-y)P_n(x)P_n(y) = (n+1)\left[P_{n+1}(x)P_n(y) - P_{n+1}(y)P_n(x)\right]$$

which is the required sum of $(n + 1)$ terms.

This result is sometimes called as *Christoffel's summation formula.*

7.7. BESSEL'S DIFFERENTIAL EQUATION

Let us consider the linear differential equation

$$x^2 \frac{d^2y}{dx^2} + x\frac{dy}{dx} + (x^2 - n^2)y = 0 \tag{7.93}$$

where n is a constant.

This equation is known in the literature as *Bessel's differential equation*. The solutions of which are defined as *Bessel's functions*.

Equation (7.93) has a nonessential singularity at the point $x = 0$ and therefore its solution can be obtained as a power series developed about this point. Thus, we can assume its solution in the form of the following ascending series

$$Y = \sum_{k=0}^{\infty} a_k x^{m+k} \tag{7.94}$$

Substitution of Eq. (7.94) in Eq. (7.93), yields

$$\sum_{k=0}^{\infty} a_k(m+k)(m+k-1)x^{m+k} + \sum_{k=0}^{\infty} a_k(m+k)x^{m+k} + \sum_{k=0}^{\infty} a_k x^{m+k+2} - \sum_{k=0}^{\infty} n^2 a_k x^{m+k} = 0$$

or

$$\sum_{k=0}^{\infty} a_k\left\{(m+k)^2 - n^2\right\} x^{m+k} + \sum_{n=0}^{\infty} a_k x^{m+k+2} = 0 \qquad (7.95)$$

The indicial equation is

$$a_0(m^2 - n^2) = 0$$

and its roots are

$$m = \pm n \qquad (7.96)$$

Equating the coefficients of x^{m+1} equal to zero, we have

$$a_1\left\{(1+m)^2 - n^2\right\} = 0$$

or $\qquad a_1(m+n+1)\,(m-n+1) = 0$

For either choice of m value given by Eq. (7.96) neither $(m + n +1)$ nor $(m - n + 1)$ vanishes (with an exception for $m = \pm n = -1/2$. Thus, we must have $a_1 = 0$.

Equating the coefficients of x^{m+k+1} in Eq. (7.95) equal to zero we have

$$a_{k+2} = -\frac{a_k}{(m+k+2)^2 - n^2}$$

$$= -\frac{a_k}{(m+k+n+2)(m+k-n+2)} \qquad (7.97)$$

Case 1. When $m = n$, the recurrence relation [Eq. (7.97)] for the coefficients a_k becomes

$$a_{k+2} = \frac{a_k}{(2n+k+2)(k+2)} \qquad (7.98)$$

Substitution in Eq. (7.94) gives

$$y = a_0 x^n\left[1 - \frac{x^2}{2(2n+2)} + \frac{x^4}{2.4(2n+2)(2n+4)} + \cdots + \frac{(-1)^r x^{2r}}{2^{2r} r!(n+1)(n+2)\cdots(n+r)} + \cdots\right]$$

Since $\qquad a_1 = 0, a_1 = a_3 = a_5 = \ldots = \ldots = 0$

or

$$y = a_0 \sum_{r=0}^{\infty} \frac{(-1)^r x^{n+2r}}{2^{2r} r!(n+1)(n+2)\cdots(n+r)} \qquad (7.99)$$

All the coefficients given by the following equation are finite except when n is a negative integer.

$$a_{2r} = \frac{(-1)^r a_0}{2^{2r} r!(n+1)(n+2)\cdots(n+r)}$$

Excluding this case, let us choose

$$a_0 = \frac{1}{2^n \Gamma(n+1)}$$

or for n integer $\qquad a_0 = \frac{1}{2^n n!}$

Then we have

$$y = J_n(x) = \frac{1}{2^n \Gamma(n+1)} \sum_{r=0}^{\infty} (-1)^r \frac{x^{n+2r}}{2^{2r} r!(n+1)(n+2)\cdots(n+r)}$$

$$= \sum_{r=0}^{\infty} \frac{(-1)^r}{r!\Gamma(n+r+1)} \left(\frac{x}{2}\right)^{n+2r} \tag{7.100}$$

This series converges for any finite value of x and represents the function $J_n(x)$ which is defined as *Bessel's function* of first kind of order n.

Case 2. When $m = -n$, the recurrence relation for n noninteger becomes

$$a_{k+2} = -\frac{a_k}{(k+2)(-2n+k+2)} \tag{7.101}$$

and the second solution is obtained by replacing n by $-n$ in Eq. (7.100). Thus we have

$$J_{-n}(x) = \sum_{r=0}^{\infty} (-1)^r \frac{1}{r!\Gamma(-n+r+1)} \left(\frac{x}{2}\right)^{-n+2r} \tag{7.102}$$

and the complete solution of Bessel's differential equation [Eq. (7.93)] for noninteger n is

$$y = AJ_n(x) + BJ_{-n}(x) \tag{7.103}$$

For n integer, the two roots of indicial [Eq. (7.96)], differ by an integer and as discussed in Section (7.4) two independent solutions of the differential equation cannot be obtained by this method of series solution.

From recurrence relation [Eq. (7.101)], we have for n an integer when $k = 2(n-1)$ the coefficient a_{k+2} becomes indeterminate and we have no series solution for second choice of m. Moreover, for n integer the two solutions given by Eqs (7.100) and (7.102) are no more independent. Because the denominator of first n terms $r = 0, 1, 2,..., (n-1)$, contain the factor involving gamma functions of zero or of negative integer which yields zero, i.e.

$$\frac{1}{\Gamma(-n+r+1)} = 0 \quad \text{for} \quad r = 0, 1, 2,...$$

Thus for n integer, the series Eq. (7.102) becomes

$$J_{(-n)}(x) = \sum_{r=n}^{\infty} (-1)^r \frac{1}{r!\Gamma(r-n+1)} \left(\frac{x}{2}\right)^{-n+2r}$$

Putting $r = n + s$ where s is a $+ve$ integer, we have

$$J_{-n}(x) = \sum_{s=0}^{\infty} (-1)^{n+s} \frac{1}{(n+s)!\Gamma(s+1)} \cdot \left(\frac{x}{2}\right)^{n+2s}$$

$$= (-1)^n \sum_{s=0}^{\infty} (-1)^s \frac{1}{s!\Gamma(n+s+1)} \left(\frac{x}{2}\right)^{n+2s} = (-1)^n J_n(x) \tag{7.104}$$

which shows that $J_n(x)$ has either odd or even symmetry.

$J_n(x)$ is an even function if n is an even integer, or an odd function if n is an odd integer. For nonintegral n such simple symmetry is not possessed by $J_n(x)$. This equation also shows that for n integer we have no longer linear dependent solutions and the general solution given by Eq. (7.103) is also a more general solution since

$$y = AJ_n(x) + BJ_{-n}(x) = \{A + (-1)^n B\} J_n(x) = CJ_n(x).$$

We therefore do not have general solution for Bessel's equation for n integer by the method of series substitution. Thus an independent second solution must be found for this case. Second solution of Bessel's equation for integer n yields Bessel's function of second kind.

Now consider the function

$$Y_n(x) = \frac{1}{\sin n\pi}\{\cos n\pi J_n(x) - J_{-n}(x)\} \tag{7.105}$$

Now if n is not an integer, the function is dependent of $J_n(x)$ and since it is a linear combination of $J_n(x)$ and $J_{-n}(x)$ it is a solution of Bessel's differential equation of order n. If now n is an integer, and because of the relation [Eq. (7.104)], we have

$$Y_n(x) = \frac{0}{0}$$

Hence, when n is an integer, we define $Y_n(x)$ to be

$$Y_r(x) = \lim_{r \to 0} \frac{J_r(x)\cos r\pi - J_{-r}(x)}{\sin r\pi}$$

With this definition of $Y_r(x)$ we have, on carrying out the limiting process,

$$\frac{\pi}{2}Y_0(x) = J_0(x)\left(\log\frac{x}{2} + \gamma\right) + \left(\frac{x}{2}\right)^2 - \frac{\left(1+\frac{1}{2}\right)\left(\frac{x}{2}\right)^4}{(2!)^2} + \left(1+\frac{1}{2}+\frac{1}{3}\right)\frac{\left(\frac{x}{2}\right)^6}{(3!)^2} \tag{7.106}$$

where γ is Euler's constant defined by

$$\gamma = \lim_{n \to \infty}\left(1 + \frac{1}{2} + \frac{1}{3} + \cdots + \frac{1}{n} - \log n\right) = 0.5772157 \tag{7.107}$$

Also, when n is any positive integer, we have

$$\pi Y_n(x) = 2J_n(x)\left(\log\frac{x}{2} + \gamma\right) - \sum_{r=0}^{\infty}\frac{(-1)^r\left(\frac{x}{2}\right)^{n+2r}}{r!(n+r)!}\left(\sum_{m=1}^{n+r}m^{-1} + \sum_{m=1}^{r}m^{-1}\right)$$

$$-\sum_{r=0}^{n-1}\left(\frac{x}{2}\right)^{-n+2r}\frac{(n-r+1)!}{r!} \tag{7.108}$$

where for $r = 0$ instead of

$$\sum_{m=1}^{n+r}m^{-1} + \sum_{m=1}^{r}m^{-1}$$

we write

$$\sum_{m=1}^{n}m^{-1}$$

The presence of the logarithmic term in the function shows that these functions are infinite at $x = 0$.

The general solution of Bessel's differential equation may now be written in the form
$$y = C_1 J_n(x) + C_2 Y_n(x) \tag{7.109}$$
where C_1 and C_2 are arbitrary constants.

The function $Y_n(x)$ given by Eq. (7.105) is called the *Newmann function* for order n.

Bassel's Function of the Third Kind (Hankel Function)
These are defined as
$$H_n^{(1)}(x) = J_n(x) + i Y_n(x) \tag{7.110}$$
and
$$H_n^{(2)}(x) = J_n(x) - i Y_n(x) \tag{7.111}$$

Then $J_n(x)\, Y_n(x)$ in terms of $H_n^{(1)}(x)$ and $H_n^{(2)}(x)$ are

$$\left. \begin{array}{l} J_n(x) = \dfrac{1}{2}\left\{ H_n^{(1)}(x) + H_n^{(2)}(x) \right\} \\[12pt] Y_n(x) = -\dfrac{1}{2} i \left\{ H_n^{(1)}(x) - H_n^{(2)}(x) \right\} \end{array} \right] \tag{7.112}$$

Generating Function of $J_n(x)$
$J_n(x)$, the Bessel function of first kind of order n is the coefficient of t^n in the expansion
of the function $e^{\frac{x}{2}\left(t - \frac{1}{t}\right)}$.

or
$$e^{\frac{x}{2}\left(t - \frac{1}{t}\right)} = \sum_{n=-\infty}^{+\infty} J_n(x) t^n \tag{7.113}$$

Expanding the exponentials in Maclaurin series, we have

$$e^{\frac{x}{2}\left(t - \frac{1}{t}\right)} = e^{\frac{1}{2}xt - \frac{x}{2t}}$$

$$= \left[1 + \frac{xt}{2} + \frac{x^2 t^2}{2^2 2!} + \frac{x^3 t^3}{2^3 3!} + \cdots + \frac{x^n t^n}{2^n n!} + \cdots \right] \times \left[1 - \frac{x}{2t} + \frac{x^2}{2^2 2! t^2} \cdots (-1)^n \frac{x^n}{2^n n! t^n} + \cdots \right]$$

In this product the terms free of t are

$$1 - \left(\frac{x}{2}\right)^2 + \frac{1}{(2!)^2}\left(\frac{x}{2}\right)^4 - \frac{1}{(3!)^2}\left(\frac{x}{2}\right)^6 \cdots + (-1)^n \frac{1}{(n!)^2}\left(\frac{x}{2}\right)^{2n} + \cdots = J_0(x)$$

and the coefficient of t^n is

$$\frac{x^n}{2^n n!} - \frac{x^{n+1}}{2^{n+1}(n+1)!} \cdot \frac{x}{2} + \frac{x^{n+2}}{2^{n+2}(n+2)!}\left(\frac{x}{2}\right)^2 \frac{1}{2!} - \frac{x^{n+3}}{2^{n+3}(n+3)!}\left(\frac{x}{2}\right)^2 \frac{1}{3!} + \cdots$$

$$= \frac{1}{n!}\left(\frac{x}{2}\right)^n - \frac{1}{1!(n+1)!}\left(\frac{x}{2}\right)^{n+2} + \frac{1}{2!(n+2)!}\left(\frac{x}{2}\right)^{n+4} \cdots (-1)^r \frac{1}{r!(n+r)!}\left(\frac{x}{2}\right)^{n+2r} + \cdots$$

$$= \sum_{r=0}^{\infty} (-1)^r \frac{1}{r!(n+r)!}\left(\frac{x}{2}\right)^{n+2r} = J_n(x)$$

and the coefficient of $\dfrac{1}{t^n}$ is

$$(-1)^n \left\{ \frac{x^n}{2^n n!} - \frac{x^{n+1}}{2^{n+1}(n+1)!} \cdot \frac{x}{2} + \frac{x^{n+2}}{2^{n+2}(n+2)!} \left(\frac{x}{2}\right)^2 \frac{1}{2!} \cdots + (-1)^r \frac{x^{n+r}}{2^{n+r}(n+r)!} \left(\frac{x}{2}\right)^r \frac{1}{r!} + \cdots \right\}$$

$$= (-1)^n \sum_{r=0}^{\infty} (-1)^r \frac{1}{(n+r)!r!} \left(\frac{x}{2}\right)^{n+2r} = (-1)^n J_n(x) = J_{-n}(x)$$

which gives the result

$$e^{\frac{x}{2}\left(t-\frac{1}{t}\right)} = \sum_{n=-\infty}^{+\infty} J_n(x) t^n$$

Jacobi Series: Equation (7.113) helps us to evaluate the Bessel's functions of different order (for n integer). The generating function also helps us to express trigonometric functions as expansions containing Bessel's functions. To do this let us put $t = e^{i\phi}$, then we have

$$\exp\left\{ix \frac{(e^{i\phi} - e^{-i\phi})}{2i}\right\} = \sum_{n=-\infty}^{+\infty} J_n(x) e^{ni\phi}$$

or

$$e^{ix\sin\phi} = J_0(x) + \left\{J_1(x)(e^{i\phi}) + J_{-1}(x)(e^{-i\phi})\right\} + \left\{J_2(x)(e^{2i\phi}) + J_{-2}(x)(e^{-2i\phi})\right\}$$

$$+ \left\{J_3(x)(e^{3i\phi}) + J_{-3}(x)(e^{-3i\phi})\right\} + \cdots + \left\{J_n(x)(e^{in\phi}) + J_{-n}(x)(e^{-in\phi})\right\} + \cdots$$

and using Eq. (1.104), we have

$$\cos(x\sin\phi) = J_0(x) + 2J_2(x)\cos 2\phi + 2J_4(x)\cos 4\phi + \cdots \tag{7.114}$$

and

$$\sin(x\cos\phi) = 2J_1(x)\sin\phi + 2J_3(x)\sin 3\phi + \cdots \tag{7.115}$$

But replacing ϕ by $\left(\dfrac{\pi}{2} - \phi\right)$ the series expansion of $\cos(x\cos\phi)$ and $\sin(x\cos\phi)$ can be obtained from Eqs (7.114) and (7.115). Thus we have

$$\cos(x\cos\phi) = J_0(x) - 2J_2(x)\cos 2\phi + 2J_4(x)\cos 4\phi + \cdots \tag{7.116}$$

and

$$\sin(x\cos\phi) = 2J_1(x)\cos\phi - 2J_3(x)\cos 3\phi + 2J_5(x)\cos 5\phi + \cdots \tag{7.117}$$

Equations (7.114), (7.115), (7.116) and (7.117) are called *Jacobi series*.

Graphical Representation of Bessel's Functions

Putting $x = 0$ in Eq. (7.113), we have

$$e^0 = 1 = \sum_{n=0}^{\infty} J_n(0) t^n$$

Equating the coefficients of t^n we have $J_0(0) = 1$ and $J_n(0) = 0$ for $n \neq 0$. It shows that all the Bessel's functions of integral order excepting $J_0(0)$ are zero. For $x \to \infty$ it can be shown that

$$J_n(x) = \sqrt{\frac{2}{\pi x}} \cos\left(x - \frac{n\pi}{2} + \frac{\pi}{4}\right)$$

Plot of some of the Bessel's functions as a function of x are shown in Fig. 7.2.

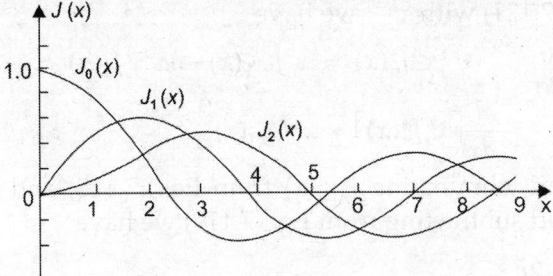

Fig. 7.2

Recurrence Relations for $J_n(x)$

Differentiating Eq. (7.113) partially w.r.t. t, we have

$$\frac{1}{2}x\left(1+\frac{1}{t^2}\right)e^{x/2(t-1/t)} = \sum_{n=-\infty}^{+\infty} nJ_n(x)t^{n-1}$$

or

$$\frac{1}{2}x\left(1+\frac{1}{t^2}\right)\sum_{n=-\infty}^{+\infty} J_n(x)t^n = \sum_{n=-\infty}^{+\infty} nJ_n t^{n-1}$$

Equating coefficients of t^{n-1} on both the sides, we have

$$\frac{1}{2}xJ_{n-1}(x)+\frac{1}{2}xJ_{n+1}(x) = nJ_n(x)$$

or

$$2nJ_n(x) = x\{J_{n-1}(x)+J_{n+1}(x)\} \qquad (7.118)$$

Again differentiating Eq. (7.113) partially w.r.t. x, we get

$$\frac{1}{2}(t-1/t)e^{x/2(t-1/t)} = \sum_{n=-\infty}^{+\infty} J_n'(x)t^n$$

or

$$(t-1/t)\sum_{u=-\infty}^{+\infty} J_n(x)t^n = 2\sum_{n=-\infty}^{+\infty} J_n'(x)t^n$$

Equating the coefficients of t^n on both the sides, we have

$$J_{n-1}(x) - J_{n+1}(x) = 2J_n'(x) \qquad (7.119)$$

If we put $n = 0$, Eq. (7.119) becomes

$$J_{-1}(x) - J_{+1}(x) = 2J_0'(x)$$

or

$$-2J_{+1}(x) = 2J_0'(x) \qquad \text{[since } J_{+1}(x) = -J_{-1}(x)]$$

or

$$J_0'(x) = -J_1(x) \qquad (7.120)$$

Eliminating $J_{n+1}(x)$ between Eqs (7.118) and (7.119) by dividing Eq. (7.118) with x and adding with Eq. (7.119), we have

$$\frac{2n}{x}J_n(x)+2J_n'(x) = 2J_{n-1}(x)$$

or

$$nJ_n(x)+xJ_n'(x) = xJ_{n-1}(x)$$

or

$$xJ_n'(x) = xJ_{n-1}(x) - nJ_n(x) \qquad (7.121)$$

Multiplying Eq. (7.121) with x^{n-1} we have

$$x^n J_n'(x) = x^n J_{n-1}(x) - nx^{n-1} J_n(x)$$

$$\frac{d}{dx}\left[x^n J_n(x)\right] = x^n J_{n-1}(x) \tag{7.122}$$

On the other hand, eliminating $J_{n-1}(x)$ from Eqs (7.118) and (7.119) by dividing Eq. (7.118) with x and subtracting from Eq. (7.119) we have

$$\frac{2n}{x} J_n(x) - 2 J_n'(x) = 2 J_{n+1}(x)$$

or

$$x J_n(x) - n J_n(x) = -x J_{n+1}(x) \tag{1.123}$$

and multiplying by x^{-n-1}, Eq. (7.123) becomes

$$x^{-n} J_n'(x) - nx^{-n-1} J_n(x) = x^{-n} J_{n+1}(x)$$

or

$$\frac{d}{dx}\left[x^{-n} J_n(x)\right] = -x^{-n} J_{n+1}(x) \tag{7.124}$$

The relations from Eqs (7.118) to (7.124) are known as *recurrence relation for Bessel's function*.

Integral Representation of $J_n(x)$

The series Eqs (7.114) and (7.115) known as Jacobi series can be used to derive the integral formulae for $J_n(x)$. If we multiply Eq. (7.114) with $\cos n\phi$ and Eq. (7.115) with $\sin n\phi$ and integrate with respect to ϕ from 0 to π, we get

$$\int_0^\pi \cos(x\sin\phi)\cos n\phi\, d\phi = \begin{cases} 0 \text{ for } n \text{ odd}^* \\ \pi J_n(x) \text{for } n \text{ even or zero} \end{cases} \tag{7.125}$$

and $\displaystyle\int_0^\pi \sin(x\sin\phi)\sin n\phi\, d\phi = \begin{cases} \pi J_n(x) \text{for } n \text{ odd}^* \\ 0 \text{ for } n \text{ even} \end{cases} \tag{7.126}$

Adding Eqs (7.125) and (7.126) we have

$$J_n(x) = \frac{1}{\pi}\int_0^\pi \{\cos(x\sin\phi)\cos n\phi + \sin(x\sin\phi)\sin n\phi\}\, d\phi$$

$$= \frac{1}{\pi}\int_0^\pi \cos(n\phi - x\sin\phi)\, d\phi \tag{7.127}$$

which is the integral representation of $J_n(x)$ for n an odd or even integer. For $n = 0$, this integral reduces to

$$J_0(x) = \frac{1}{\pi}\int_0^\pi \cos(x\sin\phi)\, d\phi \tag{7.128}$$

---

* Here we have used the result

$$\int_0^\pi \cos n\phi\cos m\phi = \frac{\pi}{2}\delta_{mn} \text{ and } \int_0^\pi \sin n\phi\sin m\phi\, d\phi = \frac{\pi}{2}\delta_{mn}$$

Another integral representing $J_n(x)$ may be given by

$$J_n(x) = \frac{1}{\sqrt{\pi}\,\Gamma(n+1/2)}\left(\frac{x}{2}\right)^n \int_0^\pi \cos(x\sin\phi)\cos^{2n}\phi\,d\phi \qquad (7.129)$$

This can be proved by expanding $\cos(x\sin\phi)$ in powers of $x\sin\phi$ as follows.

$$\cos(x\sin\phi) = \sum_{r=0}^{\infty}(-1)^r \frac{x^{2r}}{(2r)!}\sin^{2r}\phi$$

Equation (7.129) becomes

$$J_n(x) = \frac{1}{\sqrt{\pi}}\frac{1}{\Gamma(n+1/2)}(x/2)^n\int_0^\pi \sum_{r=0}^{\infty}(-1)^r\frac{x^{2r}}{(2r)!}\sin^{2r}\phi\cos^{2n}\phi\,d\phi$$

$$= \frac{1}{\sqrt{\pi}}\frac{2}{\Gamma(n+1/2)}\sum_{r=0}^{\infty}\frac{x^{n+2r}}{2^n}\cdot\frac{(-1)^r}{(2r)!}\int_0^\pi \sin^{2r}\phi\cos^{2n}\phi\,d\phi$$

$$= \frac{1}{\sqrt{\pi}}\frac{2}{\Gamma(n+1/2)}\sum_{r=0}^{\infty}(-1)^r\frac{x^{n+2r}}{2^n(2r)!}\frac{\left|\dfrac{(2r+1)}{2}\right|\left|\dfrac{(2n+1)}{2}\right|}{2\Gamma(n+r+1)}$$

$$= \sum_{r=0}^{\infty}(-1)^r\frac{\dfrac{2r-1}{2}\cdot\dfrac{2r-3}{2}\cdots 1/2\cdot 2^{2r}}{\Gamma(n+r+1)(2r)!}\left(\frac{x}{2}\right)^{n+2r*}$$

$$= \sum_{r=0}^{\infty}(-1)^r(x/2)^{n+2r}\cdot 2^r\frac{1}{\Gamma(n+r+1)2r(2r-2)\cdots 4.2}$$

$$= \sum_{r=0}^{\infty}(-1)^r(x/2)^{n+2r}\frac{1}{r!\Gamma(n+r+1)} = J_n(x)$$

Hence $J_n(x)$ may be represented by the integral

$$J_n(x) = \frac{1}{\sqrt{\pi}}\frac{1}{\Gamma(n+1/2)}(x/2)^n\int_0^\pi \cos(x\sin\phi)\cos^{2n}\phi\,d\phi$$

Orthogonality of Bessel's Function

If α and β are the roots of the equation $J_n(\mu) = 0$, then the condition of orthogonality of Bessel's function over the interval $(0, 1)$ is

$$\int_0^1 J_n(\alpha x)J_n(\beta x)x\,dx = 0 \quad \text{for } \alpha \neq \beta$$

---

*$\Gamma n = (n-1)\,\Gamma(n-1)$

and $\Gamma\dfrac{1}{2} = \sqrt{\pi}$

and the condition of normalisation is

$$\int_0^1 [J_n(\alpha x)]^2 \, x \, dx = \frac{1}{2} J_{n+1}^2(\alpha)$$

i.e.

$$\int_0^1 J_n(\alpha x) J_n(\beta x) x \, dx = \frac{1}{2} J_{n+1}(\alpha) J_{n+1}(\beta) \delta_{\alpha\beta} \qquad (7.130)$$

Here $\delta_{\alpha\beta}$ is Kronecker delta.

Proof: We know that $J_n(x)$ is the solution of Bessel's equation

$$\frac{d^2y}{dx^2} + \frac{1}{x}\frac{dy}{dx} + \left(1 - \frac{n^2}{x^2}\right)y = 0$$

Let us replace x by αx and y by u in Bessel's equation, we have

$$\frac{d^2u}{d(\alpha x)^2} + \frac{1}{\alpha x}\frac{du}{d(\alpha x)} + \left(1 - \frac{n^2}{\alpha^2 x^2}\right)u = 0$$

or

$$\frac{1}{\alpha^2}\frac{d^2u}{dx^2} + \frac{1}{\alpha^2 x}\frac{du}{dx} + \left(1 - \frac{n^2}{\alpha^2 x^2}\right)u = 0$$

or

$$x^2\frac{d^2u}{dx^2} + x\frac{du}{dx} + (\alpha^2 x^2 - n^2)u = 0 \qquad (7.130a)$$

must have $J_n(\alpha x)$ as its solution.

Similarly, if we substitute βx for x and v for y in Bessel's equation we have

$$x^2\frac{d^2v}{dx^2} + x\frac{dv}{dx} + (\beta^2 x^2 - n^2)v = 0 \qquad (7.130b)$$

and $J_n(\beta x)$ must be its solution.

(i) $\dfrac{v}{x}$ (ii) $\dfrac{u}{x}$ yields

$$x\left[v\frac{d^2u}{dx^2} - u\frac{d^2v}{dx^2}\right] + \left[v\frac{du}{dx} - u\frac{dv}{dx}\right] + (\alpha^2 - \beta^2)xuv = 0$$

or

$$\frac{d}{dx}\left[x\left(v\frac{du}{dx} - u\frac{dv}{dx}\right)\right] + (\alpha^2 - \beta^2)xuv = 0$$

Since $J_n(\alpha x)$ and $J_n(\beta x)$ are the solutions of Eqs (7.130a) and (7.130b), we have

$$u = J_n(\alpha x) \quad \text{and} \quad v = J_n(\beta x)$$

Therefore,

$$\frac{d}{dx}\left[\left\{xJ_n(\beta x)\frac{dJ_n(\alpha x)}{dx} - xJ_n(\alpha x)\frac{dJ_n(\beta x)}{dx}\right\}\right] + (\alpha^2 - \beta^2)\,xJ_n(\alpha x)\,J_n(\beta x) = 0$$

Integrating w.r.t x between the limits $x = 0$ to $x = 1$ we get

$$\left\{xJ_n(\beta x)\frac{dJ_n(\alpha x)}{dx} - xJ_n(\alpha x)\frac{dJ_n(\beta x)}{dx}\right\}\bigg|_0^1 + \int_0^1 (\alpha^2 - \beta^2)\,xJ_n(\alpha x)J_n(\beta x)\,dx = 0$$

Case 1. If α and β are different roots of $J_n(\mu) = 0$, then $J_n(\alpha) = 0$ and $J_n(\beta) = 0$. The first term vanishes for both the limits. Hence, we have

$$\int_0^1 (\alpha^2 - \beta^2) x J_n(\alpha x) J_n(\beta x) dx = 0$$

and as $\alpha \neq \beta$, we have

$$\int_0^1 x J_n(\alpha x) J_n(\beta x) dx = 0 \qquad (7.131)$$

Case 2. If $\alpha = \beta$

$$\int_0^1 x J_n(\alpha x) J_n \alpha(x) dx = \lim_{x \to \alpha} \frac{\left[x J_n(\beta_x) \dfrac{d}{dx} J_n(\alpha x) - J_n(\alpha x) \dfrac{d}{dx} J_n(\beta_x) \right]_0^1}{\beta^2 - \alpha^2}$$

To evaluate this we let $J_n(\alpha) = 0$, but let $\beta \to \alpha$ as a limit. We have R.H.S

$$= \lim_{\beta \to \alpha} \frac{\left[x J_n(\beta x) \dfrac{d}{dx} J_n(\alpha x) \right]_0^1}{\beta^2 - \alpha^2} \qquad (7.131a)$$

Recurrence relation [Eq. (7.124)] for $J_n(x)$ is

$$\frac{d}{dx}\left[x^{-n} J_n(x) \right] = x^{-n} J_{n+1}(x)$$

or $\quad x^{-n} \dfrac{d}{dx} J_n(x) - n x^{-n-1} J_n(x) = -x^{-n} J_{n+1}(x)$

or $\quad x \dfrac{d}{dx} J_n(x) = n J_n(x) - x J_{n+1}(x)$

Replacing x by (αx) we get

$$\alpha x \frac{d}{d(\alpha x)} J_n(\alpha x) = n J_n(\alpha x) - \alpha x J_{n+1}(\alpha x)$$

$$x \frac{d}{dx} J_n(\alpha x) = n J_n(\alpha x) - \alpha x J_{n+1}(\alpha x) \qquad (7.131b)$$

Substituting in Eq. (7.131a), we have

$$\lim_{\beta \to \alpha} \frac{\left[J_n(\beta x)\{n J_n(\alpha x) - \alpha x J_{n+1}(\alpha x)\} \right]_0^1}{\beta^2 - \alpha^2}$$

$$= \lim_{\beta \to \alpha} \frac{\left[-\alpha x J_n(\beta x) J_{n+1}(\alpha x) \right]_0^1}{\beta^2 - \alpha^2} \qquad [since\ J_n(\alpha x) = 0]$$

$$= \lim_{\beta \to \alpha} \frac{-\alpha J_n(\beta) J_{n+1}(\alpha)}{\beta^2 - \alpha^2}$$

$$= \lim_{\beta \to \alpha} \frac{-\dfrac{\partial}{\partial \beta}[\alpha J_n(\beta) J_{n+1}(\alpha)]}{\dfrac{\partial}{\partial \beta}(\beta^2 - \alpha^2)} = \frac{-\alpha J_{n+1}(\alpha) \dfrac{\partial J_n(\beta)}{\partial \beta}}{2\beta} \qquad (7.131c)$$

Replacing x by β in recurrence relation of Eq. (7.123), we have

$$\beta \frac{d}{d\beta} J_n(\beta) = n J_n(\beta) - \beta J_{n+1}(\beta)$$

or

$$\frac{d}{d\beta} J_n(\beta) = \frac{1}{\beta}\{n J_n(\beta) - \beta J_{n+1}(\beta)\}$$

Substitution in Eq. (7.131b) yields

$$\lim_{\beta \to \alpha} -\frac{\alpha}{2\beta} \frac{1}{\beta}\{n J_n(\beta) - \beta J_{n+1}(\beta)\} J_{n+1}(\alpha)$$

$$= -\frac{\alpha}{2\alpha^2}\{n J_n(\alpha) - \alpha J_{n-1}(\alpha)\} J_{n+1}(\alpha)$$

$$= \frac{1}{2} J_{n+1}^2(\alpha) \qquad\qquad [since\ J_n(\alpha) = 0]$$

Hence, we have

$$\int_0^1 x\{J_n(\alpha x)\}^2\, dx = \frac{1}{2} J_{n+1}^2(\alpha) \qquad\qquad (7.132)$$

or   combining Eqs (7.131) and (7.132), we have

$$\int_0^1 x J_n(\alpha x) J_n(\beta x)\, dx = \frac{1}{2} J_{n+1}^2(\alpha)\delta_{\alpha\beta}$$

We are now in a position to expand the arbitrary function $F(x)$ in the interval from $x = 0$ to $x = 1$ in the series of the form

$$F(x) = \sum_{s=1}^{\infty} c_s J_n(\alpha_s x) \qquad\qquad (7.133)$$

where α_s are the successive positive roots of

$$J_n(\alpha) = 0$$

To obtain the general coefficient c_k of this expansion, we multiply both sides of Eq. (7.133) with $x J_n(\alpha_k x)$ and integrate from $x = 0$ to $x = 1$. We have

$$\int_0^1 x J_n(\alpha_k x) F(x)\, dx = c_k \int_0^1 x J_n^2(\alpha_k x)\, dx$$

$$= c_k \frac{1}{2} J_{n+1}^2(\alpha_k)$$

or

$$c_k = \frac{2}{J_{n+1}^2(\alpha_k)} \int_0^1 x J_n(\alpha_k x) F(x)\, dx \qquad\qquad (7.134)$$

This expansion is analogous to the expansion of an arbitrary function in a Fourier series.

Expressions for $J_n(x)$ when n is Half an Odd Integer

i. Values of $J_{\pm\frac{1}{2}}(x)$

For nonintegral values of n, we have $J_n(x)$ from Eq. (7.100).

Expanding Eq. (7.100), we have

$$J_n(x) = \sum_{r=0}^{\infty} \frac{(-1)^r}{r!\Gamma(n+r+1)}\left(\frac{x}{2}\right)^{n+2r}$$

$$= \frac{x^n}{2^n \Gamma(n+1)}\left[1 - \frac{x^2}{2\cdot 2(n+1)} + \frac{x^4}{2\cdot 4\cdot 2^2(n+1)(n+2)} \cdots\right]$$

Putting $n = \dfrac{1}{2}$ this equation yields

$$J_{+1/2}(x) = \left(\frac{x}{2}\right)^{1/2} \frac{1}{\Gamma 3/2}\left[1 - \frac{x^2}{2\cdot 2\cdot 3/2} + \frac{x^4}{2\cdot 4\cdot 2^2\cdot 3/2\cdot 5/2} \cdots\right]$$

$$= \left(\frac{x}{2}\right)^{1/2} \frac{1}{1/2\sqrt{\pi}}\left[1 - \frac{x^2}{3!} + \frac{x^4}{5!} \cdots\right]$$

$$= \left(\frac{2x}{\pi}\right)^{1/2} \frac{1}{x}\left[x - \frac{x^3}{3!} + \frac{x^5}{5!} \cdots\right]$$

$$J_{+1/2}(x) = \sqrt{\frac{2}{\pi x}}\sin x \qquad\qquad (7.135)$$

and putting $n = -\dfrac{1}{2}$ we have

$$J_{-1/2}(x) = \left(\frac{x}{2}\right)^{-1/2} \frac{1}{\Gamma_{1/2}}\left[1 - \frac{x^2}{2!} + \frac{x^4}{4!} \cdots\right]$$

$$J_{-1/2}(x) = \sqrt{\frac{2}{\pi x}}\cos x \qquad\qquad (7.136)$$

ii. Values of $J_{\pm 3/2}(x)$

Substituting $n = \dfrac{1}{2}$ in recurrence relation [Eq. (7.118)], we get

$$J_{1/2}(x) = x\{J_{-1/2}(x) + J_{3/2}(x)\}$$

or
$$J_{3/2}(x) = \frac{1}{x}J_{1/2}(x) - J_{-1/2}(x)$$

Substituting the values of $J_{1/2}(x)$ and $J_{-1/2}(x)$ from Eqs (7.135) and (7.136), we have

$$J_{3/2}(x) = \frac{1}{x}\cdot\sqrt{\frac{2}{\pi x}}\sin x\cdot - \sqrt{\frac{2}{\pi x}}\cos x$$

$$= \sqrt{\frac{2}{\pi x}}\left\{\frac{\sin x}{x} - \cos x\right\} \qquad\qquad (7.137)$$

Similarly substituting $n = -\dfrac{1}{2}$ in Eq. (7.118), we have

$$-J_{-1/2}(x) = x\{J_{1/2}(x) + J_{-3/2}(x)\}$$

or
$$J_{-3/2}(x) = -\frac{1}{x} \cdot J_{-1/2}(x) - J_{1/2}(x)$$

$$= -\frac{1}{x} \cdot \sqrt{\frac{2}{\pi x}} \cos x - \sqrt{\frac{2}{\pi x}} \sin x$$

$$= -\sqrt{\frac{2}{\pi x}} \left\{ \frac{\cos x}{x} + \sin x \right\} \qquad (7.138)$$

Example 1. *Prove that*

$$J_n(x) = (-2)^n x^n \frac{d^n}{d(x^2)^n} J_0(x)$$

Solution: Bessel's equation of zeroth order is

$$\frac{d^2 y}{dx^2} + \frac{1}{x}\frac{dy}{dx} + y = 0 \qquad (1)$$

The solution is $J_0(x)$ [for $n = 0$]

Put $x^2 = t$ $\therefore \quad \dfrac{dt}{dx} = 2x = 2\sqrt{t}$

So that $\dfrac{dy}{dx} = \dfrac{dy}{dt}\dfrac{dt}{dx} = 2\sqrt{t}\dfrac{dy}{dt}$

and $\dfrac{d^2 y}{dx^2} = \dfrac{d}{dx}\left(\dfrac{dy}{dx}\right) = \dfrac{d}{dx}\left(2\sqrt{t}\cdot\dfrac{dy}{dt}\right) = \dfrac{d}{dt}\left(2\sqrt{t}\dfrac{dy}{dt}\right)\dfrac{dt}{dx}$

$$= 2\cdot 2\sqrt{t}\left\{\sqrt{t}\dfrac{d^2 y}{dt^2} + \dfrac{1}{2}\cdot\dfrac{1}{\sqrt{t}}\dfrac{dy}{dt}\right\}$$

or $\dfrac{d^2 y}{dx^2} = 4t\cdot\dfrac{d^2 y}{dt^2} + 2\dfrac{dy}{dt}$

Substituting the values of $\dfrac{dy}{dx}$ and $\dfrac{d^2 y}{dx^2}$ in Eq. (1) we have

$$4t\frac{d^2 y}{dt^2} + 2\frac{dy}{dt} + \frac{1}{\sqrt{t}}2\sqrt{t}\frac{dy}{dt} + y = 0$$

$$4t\frac{d^2 y}{dt^2} + 4\frac{dy}{dt} + y = 0$$

Differentiating above equation n times w.r.t t by Leibnitz theorem

$$4\left\{t\frac{d^{n+2} y}{dt^{n+2}} + {}^n c_1 \cdot 1 \cdot \frac{d^{n+1} y}{dt^{n+1}}\right\} + 4\frac{d^{n+1} y}{dt^{n+1}} + \frac{d^n y}{dt^n} = 0$$

or
$$4t\frac{d^{n+2}y}{dt^{n+2}} + 4(n+1)\frac{d^{n+1}y}{dt^{n+1}} + \frac{d^n y}{dt^n} = 0$$

Let us put $\phi = \dfrac{d^n y}{dt^n} = \dfrac{d^n J_0}{dt^n}$

we have
$$4t\frac{d^2\phi}{dt^2} + 4(n+1)\frac{d\phi}{dt} + \phi = 0 \tag{2}$$

As $J_n(x)$ is the solution of Bessel's equation of nth order

$$\frac{d^2y}{dx^2} + \frac{1}{x}\frac{dy}{dx} + \left(1 - \frac{n^2}{x^2}\right)y = 0 \tag{3}$$

Substituting $y = x^n z$, we have

$$\frac{dy}{dx} = x^n\frac{dz}{dx} + nx^{n-1}z$$

and
$$\frac{d^2y}{dx^2} = x^n\frac{d^2z}{dx^2} + nx^{n-1}\frac{dz}{dx} + nx^{n-1}\frac{dz}{dx} + n(n-1)x^{n-2}z$$

$$= x^n\frac{d^2z}{dx^2} + 2nx^{n-1}\frac{dz}{dx} + n(n-1)x^{n-2}z$$

Substitution in Eq. (3) yields

$$x^n\frac{d^2z}{dx^2} + 2nx^{n-1}\frac{dz}{dx} + n(n-1)x^{n-2}z + \frac{1}{x}\left\{x^n\frac{dz}{dx} + nx^{n-1}z\right\} + \left(1 - \frac{n^2}{x^2}\right)x^n z = 0$$

or
$$x^n\frac{d^2z}{dx^2} + (2n+1)x^{n-1}\frac{dz}{dx} + x^n z = 0$$

or
$$\frac{d^2z}{dx^2} + \frac{(2n+1)}{x}\frac{dz}{dx} + z = 0 \tag{4}$$

Put $x^2 = t$ in Eq. (4), it is reduced to

$$4t\frac{d^2z}{dt^2} + 4(n+1)\frac{dz}{dt} + z = 0 \tag{5}$$

Comparing Eqs (2) and (5) we get

$$z = \phi\frac{d^n J_0(x)}{dt^n} = \frac{d^n J_0(x)}{d(x^2)^n} \tag{6}$$

But
$$y = x^n z$$

Hence
$$J_n(x) = cx^n\frac{d^n J_0(x)}{d(x^2)^n} \tag{7}$$

where c is a constant to be determined.

As
$$J_0(x) = \sum_{r=0}^{\infty}\frac{(-1)^r x^{2r}}{(2^r r!)^2}$$

and

$$\frac{d^n J_0(x)}{d(x^2)^n} = \frac{d^n}{d(x^2)^n} \sum_{r=0}^{\infty} \frac{(-1)^r x^{2r}}{(2^r r!)^2}$$

$$= \frac{d^n}{d(x^2)^n} \sum_{r=0}^{\infty} \frac{(-1)^{n+r} (x^2)^{n+r}}{\{2^{n+r}(n+r)!\}^2}$$

Since all the terms in which index of x is less than $2n$ will vanish on differentiating n times with respect to x^2.

or

$$\frac{d^n J_0(x)}{d(x^2)^n} = \sum_{r=0}^{\infty} \frac{(-1)^{n+r}(n+r)(n+r-1)\cdots(r+1)(x^2)^r}{2^{2n+2r}\{(n+r)!\}^2}$$

$$= \sum_{r=0}^{\infty} (-1)^{n+r} \frac{(n+r)!}{r!} \frac{x^{2r}}{2^{2n+2r}(n+r!)^2}$$

$$= \sum_{r=0}^{\infty} \frac{(-1)^{n+r} x^{2r}}{2^{2n+2r}(n+r)! r!}$$

$\therefore$

$$J_n(x) = cx^n \frac{d^n J_0(x)}{d(x^2)^n} = cx^n \sum_{r=0}^{\infty} \frac{(-1)^{n+r} x^{2r}}{2^{2n+2r}(n+r)! r!}$$

$$= c \sum_{r=0}^{\infty} \frac{(-1)^{n+r} (x/2)^{n+2r}}{2^n (n+r)! r!}$$

$$= \frac{c(-1)^n}{2^n} \sum_{r=0}^{\infty} \frac{(-1)^r}{(n+r)! r!} \left(\frac{x}{2}\right)^{n+2r}$$

$$= cJ_n(x) \frac{(-1)^n}{2^n}$$

or

$$c = \frac{2^n}{(-1)^n} = (-2)^n$$

Hence

$$J_n(x) = (-2)^n x^n \frac{d^n}{d(x^2)^n} J_0(x)$$

Hence the result.

Example 2. If $n > -1$ show that

$$\int_0^x x^{-n} J_{n+1}(x) dx = \frac{1}{2^n n!} - x^{-n} J_n(x)$$

Solution: From recurrence relation [Eq. (7.124)], we have

$$\frac{d}{dx}\left[x^{-n} J_n(x)\right] = -x^{-n} J_{n+1}(x)$$

Integrating between the limits 0 and x we get

$$\int_0^x x^{-n} J_{n+1}(x)\,dx = \left[-x^{-n} J_n(x)\right]_0^x$$

$$= -x^{-n} J_n(x) + \lim_{x \to 0}\left\{x^{-n} J_n(x)\right\}$$

$$= -x^{-n} J_n(x) + \lim_{x \to 0} \frac{\dfrac{d^n}{dx^n} J_n(x)}{n!}$$

$$= -x^{-n} J_n(x) + \frac{1}{n!}\frac{n!}{2^n \Gamma(n+1)}$$

$$= -x^{-n} J_n(x) + \frac{1}{2^n n!}$$

Hence the result.

Example 3. *Prove that*

i. $J_n J'_{-n} - J'_n J_{-n} = -\dfrac{2\sin n\pi}{\pi x}$

ii. $\dfrac{d}{dx}\left(\dfrac{J_{-n}}{J_n}\right) = -\dfrac{2\sin n\pi}{\pi x J_n^2}$

Solution: i. We know that J_n and J_{-n} are the solutions of Bessel's equation.

$$\frac{d^2 y}{dx^2} + \frac{1}{x}\frac{dy}{dx} + \left(1 - \frac{n^2}{x^2}\right)y = 0$$

or $\qquad J''_n + \dfrac{1}{x}J'_n + \left(1 - \dfrac{n^2}{x^2}\right)J_n = 0 \qquad\qquad (1)$

and $\qquad J''_{-n} + \dfrac{1}{x}J'_{-n} + \left(1 - \dfrac{n^2}{x^2}\right)J_{-n} = 0 \qquad\qquad (2)$

Multiplying (1) by J_{-n} and (2) J_{+n} and subtracting we have

$$(J_{-n}J''_n - J_n J''_{-n}) + \frac{1}{x}(J_{-n}J'_n - J_n J'_{-n}) = 0 \qquad\qquad (3)$$

Now let $\qquad u = J_{-n}J'_n - J_n J'_{-n}$

$\therefore \qquad u' = J_{-n}J''_n + J'_{-n}J'_n - J_n J''_{-n} - J'_n J'_{-n} = J_{-n}J''_n - J_n J''_{-n}$

Hence Eq. (3) can be written as

$$u' + \frac{1}{x}u = 0$$

or $\qquad\qquad u = c/x$, where c is arbitrary constant

i.e. $\qquad\qquad J_{-n}J'_n - J_n J'_{-n} = \dfrac{c}{x} \qquad\qquad (4)$

To find the value of c, compare the coefficients of lowest degree terms on both sides of Eq. (4) bearing in mind that

$$J_n(x) = \sum_{r=0}^{\infty} \frac{(-1)^r}{r!\Gamma(n+r+1)}\left(\frac{x}{2}\right)^{n+2r}$$

$$J_n = \frac{1}{2^n \Gamma(n+1)} x^n$$

and

$$J_{-n} = \frac{1}{2^{-n}\Gamma(-n+1)} x^{-n}$$

The lowest term is obtained by putting $r = 0$

$$\therefore \qquad J_n' = \frac{nx^{n-1}}{2^n\Gamma(n+1)}, J_{-n}' = \frac{1}{2^{-n}\Gamma(-n+1)}(-n)x^{-n-1}$$

Therefore the lowest term in $J_{-n}J_{+n}' - J_n J_{-n}'$ is

$$\frac{1}{2^{-n}\Gamma(-n+1)} x^{-n} \frac{1}{2^n \Gamma(n+1)} nx^{n-1} - \frac{1}{2^n\Gamma(n+1)} x^n \frac{1}{2^{-n}\Gamma(-n+1)}(-n)x^{-n-1}$$

$$= -\frac{n}{n\Gamma n\Gamma(1-n)} \cdot \frac{1}{x}(1+1) = \frac{2\sin n\pi}{x\pi} \quad ^*$$

Hence comparing coefficients of $1/x$ in Eq. (4) we get

$$c = \frac{2\sin n\pi}{\pi}$$

Hence, we have $J_{-n}J_n' - J_n J_{-n}' = \frac{c}{x} = \frac{2\sin n\pi}{\pi x}$

ii. $\dfrac{d}{dx}\left(\dfrac{J_{-n}}{J_n}\right) = \dfrac{J_n J_{-n}' - J_{-n}J_n'}{J_n^2} = \dfrac{2\sin n\pi}{\pi x J_n^2}$

Hence the result.

Example 4. *Prove that*

i. $[J_0(x)]^2 + 2[J_1(x)]^2 + 2[J_2(x)]^2 + \cdots = 1$

ii. $\dfrac{d}{dx}[xJ_n J_{n+1}] = x\left[J_n^2 - J_{n+1}^2\right]$

and hence deduce that

$$x = 2J_0 J_1 + 6J_1 J_2 + \cdots + 2(2n+1)J_n J_{n+1} + \cdots$$

Solution: i. We know that

$$\cos(x\sin\phi) = J_0(x) + 2J_2(x)\cos 2\phi + 2J_4(x)\cos 4\phi + \cdots$$

and

$$\sin(x\sin\phi) = 2J_1(x)\sin\phi + 2J_3(x)\sin 3\phi + 2J_5(x)\sin 5\phi + \cdots$$

$^*\Gamma(n+1)\,n\Gamma n$

and $\Gamma n\Gamma(1-n) = \pi/\sin n\pi$

Now squaring and integrating w.r.t. ϕ between limits 0 to π, we have

$$\int_0^\pi \cos^2(x\sin\phi)d\phi = [J_0(x)]^2\,\pi + 2[J_2(x)]^2\,\pi + 2[J_4(x)]^2\,\pi + \cdots \qquad (1)$$

$$\int_0^\pi \sin^2(x\sin\phi)d\phi = 2[J_1(x)]^2\,\pi + 2[J_3(x)]^2\,\pi + 2[J_5(x)]^2\,\pi + \cdots \qquad (2)$$

Note that

$$\int_0^\pi 2\cos^2 n\phi\,d\phi = \int_0^\pi (1+\cos 2n\phi)d\phi = \pi$$

and

$$\int_0^\pi 2\sin^2 n\phi\,d\phi = \int_0^\pi (1-\cos 2n\phi)d\phi = \pi$$

and

$$\int_0^\pi \cos n\phi \sin m\phi\,d\phi = \int_0^\pi \sin n\phi \sin m\phi\,d\phi = 0$$

Adding Eqs (1) and (2), we get

$$\pi\left[\{J_0(x)\}^2 + 2\{J_1(x)\}^2 + 2\{J_2(x)\}^2 + \cdots\right] = \int_0^\pi \left\{\cos^2(x\sin\phi) + \sin^2(x\sin\phi)\right\}d\phi = \int_0^\pi d\phi = \pi$$

$$\therefore \ \{J_0(x)\}^2 + 2\{J_1(x)\}^2 + 2\{J_2(x)\}^2 + \cdots = 1$$

ii. We have

$$\frac{d}{dx}[xJ_nJ_{n+1}] = J_nJ_{n+1} + x\{J_nJ'_{n+1} + J_nJ'_{n+1}\}$$

$$= J_nJ_{n+1} + J_n(xJ'_{n+1}) + (xJ'_n)J_{n+1}$$

From recurrence formula Eqs (7.123) and (7.121), we have

$$xJ'_n = nJ_n - xJ_{n+1} \qquad (3)$$

$$xJ'_n = -nJ_n + xJ_{n-1}$$

Replacing n by $(n + 1)$ in Eq. (7.121), we have

$$xJ'_{n+1} = -(n + 1)\,J_{n+1} + xJ_n \qquad (4)$$

$$\therefore \quad \frac{d}{dx}[xJ_nJ_{n+1}] = J_nJ_{n+1} + J_n(xJ'_{n+1}) + (xJ'_n)J_{n+1}$$

$$= J_nJ_{n+1} + J_n\{-(n+1)J_{n+1} + xJ_n\} + J_{n+1}\{nJ_n - xJ_{n+1}\}$$

$$= x\left(J_n^2 - J_{n+1}^2\right)$$

Now put $n = 0, 1, 2,...$ respectively and adding after multiplying by 1, 3, 5,..., we get

$$\frac{d}{dx}\left[x(J_0J_1 + 3J_1J_2 + 5J_2J_3 + \cdots)\right] = x\left(J_0^2 - J_1^2\right) + 3x\left(J_1^2 - J_2^2\right) + 5x\left(J_2^2 - J_3^2\right) + \cdots$$

$$= x\{J_0^2 + 2J_1^2 + 2J_2^2 + 2J_3^2 + \cdots\}$$

$$= x \qquad \text{[by part Eq. (1) above]}$$

Integrating both sides between limits 0 to x, we have

$$x\{J_0J_1 + 3J_1J_2 + 5J_2J_3 + \cdots\} = \frac{x^2}{2}$$

or

$$x = 2J_0J_1 + 6J_1J_2 + 10J_2J_3 + \cdots + 2(2n+1)J_nJ_{n+1} + \cdots$$

Example 5. *Starting from general expansion for $J_n(x)$ prove that*

$$\left(\frac{\pi x}{2}\right)J_{3/2}(x) = \frac{\sin x}{x} - \cos x$$

Solution: We have

$$J_n(x) = \sum_{s=0}^{\infty}(-1)^s\frac{1}{s!(n+s)!}\left(\frac{x}{2}\right)^{n+2s}$$

$$= \frac{x^n}{2^n n!} - \frac{x^{n+2}}{2^{n+2}(n+1)!1!} + \frac{x^{n+4}}{2^{n+4}2!(n+2)!} - \cdots$$

$$= \frac{x^n}{2^n n!}\left[1 - \frac{x^2}{2^2 \cdot 1 \cdot (n+1)} + \frac{x^4}{2^4 \cdot 1 \cdot 2 \cdot (n+1)(n+2)} + \cdots\right]$$

$$= \frac{x^n}{2^n\overline{|n+1}}\left[1 - \frac{x^2}{2(2n+2)} + \frac{x^4}{2 \cdot 4 \cdot (2n+2)(2n+4)} + \cdots\right]$$

Putting $n = 3/2$, we have

$$J_{3/2}(x) = \frac{x^{3/2}}{2^{3/2}\overline{|5/2}}\left[1 - \frac{x^2}{2 \cdot 5} + \frac{x^4}{2 \cdot 4 \cdot 5 \cdot 7} - \frac{x^6}{2 \cdot 4 \cdot 6 \cdot 5 \cdot 7 \cdot 9} + \cdots\right]$$

$$= \frac{x^{3/2}}{2^{3/2} \cdot \frac{3}{2} \cdot \frac{1}{2}\sqrt{\pi}}\left[1 - \frac{x^2}{2 \cdot 5} + \frac{x^4}{2 \cdot 4 \cdot 5 \cdot 7} - \frac{x^6}{2 \cdot 4 \cdot 6 \cdot 5 \cdot 7 \cdot 9} + \cdots\right]$$

or

$$\sqrt{\frac{\pi x}{2}}J_{3/2}(x) = \frac{1}{3}\left[x^2 - \frac{x^4}{2 \cdot 5} + \frac{x^6}{2 \cdot 4 \cdot 5 \cdot 7} - \frac{x^8}{2 \cdot 4 \cdot 6 \cdot 5 \cdot 7 \cdot 9} + \cdots\right]$$

$$= \frac{x^2}{3} - \frac{x^4}{2 \cdot 3 \cdot 5} + \frac{x^6}{2 \cdot 3 \cdot 4 \cdot 5 \cdot 7} - \frac{x^8}{2 \cdot 3 \cdot 4 \cdot 6 \cdot 5 \cdot 7 \cdot 9} + \cdots$$

$$= \frac{2x^2}{3!} - \frac{4x^4}{5!} + \frac{6x^6}{7!} - \frac{8x^8}{9!} + \cdots$$

$$= \frac{(3-1)x^2}{3!} - \frac{(5-1)x^4}{5!} + \frac{(7-1)x^6}{7!} - \frac{(9-1)x^8}{9!} + \cdots$$

or

$$\sqrt{\frac{\pi x}{2}}J_{3/2}(x) = \frac{1}{x}\left[x - \frac{x^3}{3!} + \frac{x^5}{5!} - \frac{x^7}{7!} + \cdots\right] - \left[1 - \frac{x^2}{2!} + \frac{x^4}{4!} - \frac{x^6}{6!} + \cdots\right]$$

$$= \frac{\sin x}{x} - \cos x$$

$$J_{3/2}(x) = \sqrt{\frac{2}{\pi x}}\left(\frac{\sin x}{x} - \cos x\right)$$

Example 6. *Derive Bessel's equation from that of Legendre.*

Solution: Legendre's equation is

$$(1-x^2)\frac{d^2y}{dx^2}-(2x)\frac{dy}{dx}+n(n+1)y = 0$$

Differentiating N times and using Leibnitz's theorem, we have

$$(1-x^2)\frac{d^{N+2}y}{dx^{N+2}}+N(-2x)\frac{d^{N+1}y}{dx^{N+1}}+\frac{N(N-1)}{1\cdot 2}(-2)\frac{d^Ny}{dx^N}$$

$$-2x\frac{d^{N+1}y}{dx^{N+1}}+N(-2)\frac{d^Ny}{dx^N}+n(n+1)\frac{d^Ny}{dx^N}=0$$

i.e. $\qquad (1-x^2)\frac{d^{N+2}y}{dx^{N+2}}-2(N+1)x\frac{d^{N+1}y}{dx^{N+1}}+\{n(n+1)-N(N+1)\}\frac{d^Ny}{dx^N}=0$

Now if we put $z=\dfrac{d^Ny}{dx^N}$, then we have

$$(1-x^2)\frac{d^2z}{dx^2}-2(N+1)x\frac{dz}{dx}+\{n(n+1)-N(N+1)\} = 0 \qquad (1)$$

Now let us put $Y=(1-x^2)^{N/2}z$, i.e. $z=Y(1-x^2)^{-N/2}$

So that $\qquad \dfrac{dz}{dx} = (1-x^2)^{-N/2}\dfrac{dY}{dx}+Y(-N/2)(1-x^2)^{-N/2-1}(-2x)$

$$= (1-x^2)^{-N/2}\frac{dY}{dx}+Nx(1-x^2)^{-N/2-1}Y$$

and $\qquad \dfrac{d^2z}{dx^2} = (1-x^2)^{-N/2}\dfrac{d^2Y}{dx^2}+2Nx(1-x^2)^{-N/2-1}\dfrac{dY}{dx}$

$$+N\{(1-x^2)^{-N/2-1}+x(1-x^2)^{-N/2-2}(-N/2-1)(-2x)\}Y$$

or $\qquad \dfrac{d^2z}{dx^2} = (1-x^2)^{-N/2}\dfrac{d^2Y}{dx^2}+2Nx(1-x^2)^{-N/2-1}\dfrac{dY}{dx}$

$$+N\{1+(N+1)x^2\}(1-x^2)^{-N/2-2}Y$$

Substituting in Eq. (1) and dividing throughout with $(1-x^2)^{-N/2}$, we get

$$(1-x^2)\frac{d^2Y}{dx^2}+\{2Nx-2(N+1)x\}\frac{dY}{dx}$$

$$+\left[n(n+1)-N(N+1)-\frac{2Nx^2(N+1)}{1-x^2}+\frac{N\{1+(N+1)x^2\}}{1-x^2}\right]Y =0$$

or $\qquad (1-x^2)\dfrac{d^2Y}{dx^2}-2x\cdot\dfrac{dY}{dx}+\left[n(n+1)-\dfrac{N^2}{1-x^2}\right]Y = 0 \qquad (2)$

Now in order to change independent variable if we put

$$X = n\sqrt{1-x^2}, \text{ i.e. } 1-x^2 = \frac{X^2}{n^2}.$$

which gives

$$\frac{dY}{dx} = \frac{dY}{dX}\cdot\frac{dX}{dx} = \frac{-nx}{\sqrt{1-x^2}}\frac{dY}{dX}$$

and

$$\frac{d^2Y}{dx^2} = \frac{d}{dx}\left\{\frac{dY}{dx}\right\} = \frac{d}{dx}\left[-\frac{nx}{\sqrt{1-x^2}}\cdot\frac{dY}{dX}\right]$$

or

$$\frac{d^2Y}{dx^2} = -n\left[\frac{\sqrt{1-x^2}+\dfrac{x^2}{\sqrt{1-x^2}}}{1-x^2}\frac{dY}{dX}+\frac{x}{\sqrt{1-x^2}}\frac{d^2Y}{dX^2}\frac{dX}{dx}\right]$$

$$= -n\left[\frac{1}{(1-x^2)^{3/2}}\frac{dY}{dX}-\frac{nx^2}{1-x^2}\frac{d^2Y}{dX^2}\right]$$

On substitution in Eq. (2) gives

$$n^2x^2\frac{d^2Y}{dX^2}+\left[\frac{-n}{\sqrt{1-x^2}}+\frac{2nx^2}{\sqrt{1-x^2}}\right]\frac{dY}{dX}+\left\{n(n+1)-\frac{N^2}{1-x^2}\right\}Y = 0$$

i.e.

$$\left(1-\frac{X^2}{n^2}\right)\frac{d^2Y}{dX^2}+\frac{(2x^2-1)}{X}\frac{dY}{dX}+\left\{1+\frac{1}{n}-\frac{N^2}{X^2}\right\}Y = 0 \quad [\because X^2 = n^2(1-x^2)]$$

or

$$\left(1-\frac{X^2}{n^2}\right)\frac{d^2Y}{dX^2}+\left(1-\frac{2X^2}{n^2}\right)\frac{1}{X}\frac{dY}{dX}+\left\{1+\frac{1}{n}-\frac{N^2}{X^2}\right\}Y = 0$$

Now taking the limits when $n \to \infty$, we have

$$\frac{d^2Y}{dX^2}+\frac{1}{X}\frac{dY}{dX}+\left(1-\frac{N^2}{X^2}\right)Y = 0,$$

which is Bessel's Equation.

7.8. HYPERGEOMETRIC (GAUSS) DIFFERENTIAL EQUATION

The differential equation

$$x(1-x)\frac{d^2y}{dx^2}+\left\{c-(a+b+1)x\right\}\frac{dy}{dx}-aby = 0, \tag{7.139}$$

with a, b, c as constants, is known as Gauss hypergeometric equation or simply the Gauss equation or hypergeometric equation. The solution of hypergeometric equation is called the hypergeometric function denoted by $F(a, b, c, x)$. Dividing Eq. (7.139) with $(x^2 - x)$, we have

$$\frac{d^2y}{dx^2}+X_1\frac{dy}{dx}+X_2y = 0 \tag{7.140}$$

where
$$X_1 = \frac{(a+b+1)x - c}{x^2 - x}$$

and
$$X_2 = \frac{ab}{x^2 - x}$$

It is obvious that $X_1 \to \infty$ for $x = 0$, 1, and ∞.

and
$$X_2 \to \infty \text{ for } x = 0 \text{ and } 1$$

Therefore $x = 0$, 1 and ∞ are called singularities of Eq. (7.139). Thus, we first consider the series solution of this equation relatively valid near the regular point at the origin.

We assume the solution in the form

$$y = \sum_{m=0}^{\infty} a_m x^{m+n} \qquad (7.141)$$

where n is a constant, so that

$$y' = \sum_{m=0}^{\infty} a_m (m+n) x^{m+n-1}$$

and
$$y'' = \sum_{m=0}^{\infty} a_m (m+n)(m+n-1) x^{m+n-2}$$

Substituting these values in Eq. (7.139) we get

$$x(1-x) \sum_{m=0}^{\infty} a_m (m+n)(m+n-1) x^{m+n-2} + [c - (a+b+1)x] \sum_{m=0}^{\infty} a_m (m+n) x^{m+n-1}$$

$$- ab \cdot \sum_{m=0}^{\infty} a_m x^{m+n} = 0$$

or
$$\sum_{m=0}^{\infty} a_m \{ (m+n)^2 + (m+n)(a+b) + ab \} x^{m+n} - \sum_{m=0}^{\infty} a_m (m+n)(m+n+c-1) x^{m+n-1} = 0$$

Since this equation is an identity, the coefficient of various powers of x must be equal to zero. The indicial equation is obtained by equating the coefficient of the lowest power of x, i.e. of x^{n-1} equal to zero. We have

$$a_0 n(n+c-1) = 0$$

Since $a_0 \neq 0$ being the coefficient of the first term of the series, we must have

$$n = 0 \quad \text{and} \quad n = 1 - c \qquad (7.142)$$

Equating to zero the coefficient of next higher power of x, i.e. x^n, we get

$$a_0 [n^2 + n(a+b) + ab] - a_1 (1+n)(1+n+c-1) = 0$$

or
$$a_0 \{ n^2 + n(a+b) + ab \} - a_1 (1+n)(n+c) = 0$$

$$\therefore \qquad a_1 = \frac{(n+a)(n+b)}{(1+n)(n+c)} a_0 \qquad (7.143)$$

Again equating the coefficient of x^{m+n} equal to zero we get recurrence relation between the coefficients a_m.

$$a_{m+1} = \frac{(m+n+a)(m+n+b)}{(m+n+1)(m+n+c)} a_m \qquad (7.144)$$

Case 1. When $n = 0$

We have

$$y = \sum_{m=0}^{\infty} a_m x^m$$

and

$$a_{m+1} = \frac{(m+a)(m+b)}{(m+1)(m+c)} a_m$$

Substituting $m = 0, 1, 2,...$ etc., we will get $a_1, a_2, a_3...$ so that

$$a_1 = \frac{a \cdot b}{1 \cdot c} a_0, a_2 = \frac{(1+a)(1+b)}{2 \cdot (1+c)} a_1$$

$$a_2 = \frac{a(1+a) \cdot b(1+b)}{2!c \cdot (1+c)} a_0$$

$$a_3 = \frac{a(a+1)(a+2)b(b+1)(b+2)}{3!c \cdot (c+1)(c+2)} a_0$$

and so on. The general coefficient a_m will be

$$a_m = \frac{a(a+1)(a+2)\cdots(a+m-1)b(b+1)(b+2)\cdots(b+m-1)}{m!c \cdot (c+1)(c+2)\cdots(c+m-1)} a_0$$

$$= \frac{(a)_m (b)_m}{m!(c)_m} a_0$$

where

$$(P)_m = P(P+1)(P+2)\cdots(P+m-1) \text{ and } P = a, b, c$$

Hence

$$y = a_0 \left[1 + \frac{a \cdot b}{1 \cdot c} x + \frac{a(a+1)b(b+1)}{2!c(c+1)} x^2 + \cdots + \frac{(a)_m (b)_m x^m}{m!(c)_m} + \cdots \right]$$

or

$$y = a_0 \sum_{m=0}^{\infty} \frac{(a)_m (b)_m}{m!(c)_m} x^m \qquad (7.145)$$

is the particular solution of given equation and the series is known as *hypergeometric series*. For $a = 1$ and $b = c$

$$y = a_0 \sum_{m=0}^{\infty} x^m$$

is an ordinary geometric series. If $a_0 = 1$, the series of Eq. (7.145) becomes

$$y = \sum_{m=0}^{\infty} \frac{(a)_m (b)_m}{m!(c)_m} x^m \qquad (7.146)$$

Special case of which is a geometric series, called the hypergeometric series.

Convergence of this Series

mth term of the series is

$$Y_m = \frac{(a)_m (b)_m}{m!(c)_m} x^m$$

the $(m + 1)$th term is

$$Y_{m+1} = \frac{(a)_{m+1}(b)_{m+1}}{(m+1)!(c)_{m+1}} x^{m+1}$$

So that

$$\frac{Y_{m+1}}{Y_m} = \frac{(a)_{m+1}(b)_{m+1}}{(m+1)!(c)_{m+1}} x^{m+1} \cdot \frac{m!(c)_m}{(a)_m(b)_m} x^{-m}$$

$$= \frac{(a+m)(b+m)}{(m+1)(c+m)} x$$

$$\therefore \quad \lim_{m \to \infty} \left| \frac{Y_{m+1}}{Y_m} \right| = \lim_{m \to \infty} \frac{\left(1 + \dfrac{a}{m}\right)\left(1 + \dfrac{b}{m}\right)}{\left(1 + \dfrac{1}{m}\right)\left(1 + \dfrac{c}{m}\right)} |x| = |x|$$

This series is convergent if

$$\lim_{m \to \infty} \left| \frac{Y_{m+1}}{Y_m} \right| < 1 \quad \text{or} \quad |x| < 1$$

Obviously the radius of convergence is unity. The series

$$y = \sum_{m=0}^{\infty} \frac{(a)_m(b)_m}{m!(c)_m} x^m$$

for $x < 1$, the above relation is a hypergeometric series and the solution is called hypergeometric function, denoted by $F(a, b, c, x)$.

Thus, the solution of the hypergeometric equation for $n = 0$ is

$$y = a_0 F(a, b, c, x) \tag{7.147}$$

Case 2. When $n = 1 - c$

The solution of hypergeometric differential equation for this value of n is given by

$$y = \sum_{m=0}^{\infty} \frac{(1-c+m+a)(1-c+m+b)}{(1-c+m+1)(1-c+m+c)} a_m$$

$$= \frac{(a'+m)(b'+m)}{(c'+m)(m+1)} a_m$$

where $a' = a - c + 1$, $b' = 1 - c + b$ and $c' = 2 - c$

Hence for $n = 1 - c$, we have

$$a_1 = \frac{a'b'}{c' \cdot 1} a_0, \quad a_2 = \frac{a'(a'+1)b'(b'+1)}{c'(c'+1) \cdot 2!} a_0$$

and

$$a_m = \frac{(a')_m(b')_m}{m!(c')_m} a_0$$

The solution then becomes

$$y = a_0 x^{1-c} \left\{ 1 + \frac{a'b'}{1 \cdot c'} x + \cdots + \frac{(a')_m(b')_m}{m!(c')_m} + \cdots \right\}$$

$$= a_0 x^{1-c} \sum_{m=0}^{\infty} \frac{(a')_m(b')_m}{m!(c')_m} x^m$$

$$= a_0 x^{1-c} F(a', b', c', x)$$

$$= a_0 x^{1-c} F\{a - c + 1, b - c + 1, 2 - c, x\} \tag{7.148}$$

Thus, there are two independent (particular) solutions [Eqs (7.147) and (7.148)] of hypergeometric equation in the form of the series developed about $x = 0$. Here c should not be one or a negative integer because two solutions will be identical for these cases.

If $c = 1$, then

$$F\{a - c + 1, b - c + 1, 2 - c, x\} = F(a, b, 1, x)$$

and $$F(a, b, c, x) = F(a, b, 1, x)$$

Hence both the series are identical. When c is $a - ve$ integer say $-p$, then

$$(-p)_m = (-p)(-p+1)\cdots 0 \cdot 1 \cdots (-p+m-1) = 0$$

thereby $a_m = \infty$ and thus the solution of the type $F(a, b, c, x)$ cannot be obtained. Thus for $c \neq 1, -1, -2...$ the two solutions for $n = 0$ and $n = 1 - c$ are linearly independent and so the general solution of hypergeometric equation can be written as

$$y = AF(a, b, c, x) + Bx^{1-c} F(a - c + 1, b - c + 1, 2 - c, x) \tag{7.149}$$

where A and B are arbitrary constants.

(a) For singularity at $x = 1$, the solution is obtained by developing the series about $x = 1$. For this let us put $1 - x = t$.

So that $$\frac{dt}{dx} = -1$$

$$\frac{dy}{dx} = \frac{dy}{dt} \cdot \frac{dt}{dx} = -\frac{dy}{dt}$$

and $$\frac{d^2y}{dx^2} = \frac{d}{dx}\left(\frac{dy}{dx}\right) = \frac{d}{dx}\left(-\frac{dy}{dt}\right)$$

$$= -\frac{d^2y}{dt^2} \cdot \frac{dt}{dx} = \frac{d^2y}{dt^2}$$

Thus Eq. (7.139) is reduced to

$$t(1-t)\frac{d^2y}{dt^2} + \{c - (a+b+1)(1-t)\}\left(-\frac{dy}{dt}\right) - aby = 0$$

$$t(1-t)\frac{d^2y}{dt^2} + \{(a+b-c+1) + (a+b+1)t\}\frac{dy}{dt} - aby = 0$$

This equation is similar to Eq. (7.139) except that c is replaced by $c' = a + b - c + 1$ and x by $t = 1 - x$.

Hence by similar procedure the roots of indicial equation obtained are 0 and $1 - c' = c - a - b$ and correspondingly the solution is

$$y = AF(a, b, c't) + Bt^{1-c'} F(a - c' + 1, b - c' + 1, 2 - c', t)$$

or $$y = AF\{a, b, (a+b-c+1), (1-x)\} + B(1-x)^{c-b-a}$$

$$xF\{(c-b), (c-a), (c-a-b+1), (1-x)\} \tag{7.149a}$$

(b) When singularity occurs at $x = \infty$, the solution is obtained by developing the series about $x = \infty$, for this we substitute

$$x = \frac{1}{t}, \quad \text{i.e.} \quad t = \frac{1}{x}$$

or

$$\frac{dt}{dx} = -\frac{1}{x^2} = -t^2$$

or

$$\frac{dy}{dx} = \frac{dy}{dt} \cdot \frac{dt}{dx} = -t^2 \frac{dy}{dt}$$

and

$$\frac{d^2y}{dx^2} = \frac{d}{dx}\left(\frac{dy}{dx}\right) = \frac{d}{dt}\left(\frac{dy}{dx}\right)\frac{dt}{dx} = \frac{d}{dt}\left(-t^2 \frac{dy}{dt}\right)(-t^2) = t^2\left\{2t \cdot \frac{dy}{dt} + t^2 \frac{d^2y}{dt^2}\right\}$$

Substituting in Eq. (7.139), we get

$$t^2(1-t)\frac{d^2y}{dt^2} + \left\{2t(1-t) - (a+b+1)t + ct^2\right\}\frac{dy}{dt} + aby = 0 \qquad (7.150)$$

Let the series solution of Eq. (7.150) be

$$y = \sum_{m=0}^{\infty} a_m t^{m+n} \qquad (7.151)$$

Then

$$\frac{dy}{dt} = \sum_{m=0}^{\infty} a_m(m+n)t^{m+n-1}$$

and

$$\frac{d^2y}{dt^2} = \sum_{m=0}^{\infty} a_m(m+n)(m+n-1)t^{m+n-2}$$

Substituting in Eq. (7.150) yields

$$t^2(1-t)\sum_m a_m(m+n)(m+n-1)t^{m+n-2} + \left\{2t(1-t) - (a+b+1)t + ct^2\right\}$$

$$\times \sum_m a_m(m+n)t^{m+n-1} + ab\sum_m a_m t^{m+n} = 0$$

or

$$\sum_{m=0}^{\infty}\left\{(m+n)(m+n-1) + (2-a-b-1)(m+n) + ab\right\}a_m t^{m+n}$$

$$-\sum_{m=0}^{\infty}\left\{(m+n)(m+n-1) + (2-c)(m+n)\right\}a_m t^{m+n+1} = 0$$

This equation is an identity, and therefore coefficients of various powers of t must be zero, the indicial equation then is obtained by equating coefficients of lowest power of t, i.e. t^n equal to zero. Hence we have

$$a_0\left\{n(n-1) + (2-a-b-1)(n) + ab\right\} = 0$$

or

$$a_0(n-a)(n-b) = 0,$$

i.e. roots of indicial equation are $n = a$ or $n = b$

and

$$a_{m+1} = \frac{(m+n)(m+n+1-c)}{\left\{(m+n+1)(m+n+1-a-b) + ab\right\}} \qquad (7.152)$$

Case 1. When $n = a$

$$a_{m+1} = \frac{(m+a)(m+a+1-c)}{(m+a+1)(m-b+1)+ab}a_m$$

$$= \frac{(a+m)(a-c+m+1)}{(m+1)(a-b+m+1)}a_m \qquad (7.153)$$

Substituting $m = 0, 1, 2, 3$, etc.

$$a_1 = \frac{a(a-c+1)}{1(a-b+1)}a_0$$

$$a_2 = \frac{(a+1)(a-c+2)}{2(a-b+2)}a_1 = \frac{a(a+1)(a-c+1)(a-c+2)}{2.1(a-b+1)(a-b+2)}a_0$$

$$a_m = \frac{(a)_m(a-c+1)_m}{m!(a-b+1)_m}a_0$$

Hence
$$y = t^a \sum_{m=0}^{\infty} \frac{(a)_m(a-c+1)_m}{m!(a-b+1)_m}a_0 t^m$$

$$= a_0 t^a F\{a,(a-c+1),(a-b+1),t\}$$

As
$$t = \frac{1}{x} \text{ we have}$$

$$y = a_0 x^{-a} F\left\{a,(a-c+1),(a-b+1),\frac{1}{x}\right\}$$

Case 2. When $n = b$. We have series solution

$$y = a_0 x^{-b} F\left\{b,(b-c+1),(b-a+1),\frac{1}{x}\right\}$$

Hence the general equation as a series developed about $x = \infty$ is

$$y = Ax^{-a} F\left\{a,(a-c+1)(a-b+1),\frac{1}{x}\right\} + Bx^{-b}\left\{b,(b-c+1),(b-a+1),\frac{1}{x}\right\} \qquad (7.154)$$

Elementary Properties of Hypergeometric Functions

(i) Symmetry Property

Hypergeometric function is symmetrical with respect to parameters a and b, i.e. it does not change if the parameters a and b are interchanged.

We have

$$F(a, b, c, x) = \sum_{m=0}^{\infty} \frac{(a)_m(b)_m x^m}{(c)_m m!}$$

$$= \sum_{m=0}^{\infty} \frac{(b)_m(a)_m}{(c)_m m!} x^m = F(b, a, c, x)$$

i.e.
$$F\begin{bmatrix} a & b \\ & & x \\ c \end{bmatrix} = F\begin{bmatrix} b & a \\ & & x \\ c \end{bmatrix}$$

Symbolically hypergeometric function $F(a, b, c, x)$ is denoted by symbol $F\begin{bmatrix} a & b \\ & & x \\ c \end{bmatrix}$

and is called *Riemann F-function.*

(ii) Differentiation of Hypergeometric Function

We have

$$F(a, b, c, x) = \sum_{m=0}^{\infty} \frac{(a)_m (b)_m}{(c)_m m!} x^m$$

$$\therefore \quad \frac{dF}{dx}(a,b,c,x) = \sum_{m=1}^{\infty} \frac{(a)_m (b)_m}{(c)_m (m-1)!} x^{m-1}$$

Let us put $m - 1 = n$

$$\therefore \quad \frac{dF}{dx}(a,b,c,x) = \sum_{n=0}^{\infty} \frac{(a)_{n+1} (b)_{n+1}}{(c)_{n+1} n!} x^n$$

$$= \sum_{n=0}^{\infty} \frac{a(a+1)_n \, b(b+1)_n}{c(c+1)_n \, n!} x^n$$

$$= \frac{ab}{c} F\{(a+1),(b+1),(c+1)x\}^* \qquad (7.155)$$

Similarly
$$\frac{d^2 F}{dx^2} = \frac{ab}{c} \frac{(a+1)(b+1)}{c+1} F\{(a+2)\cdot(b+2),(c+2),x\}$$

Repeating r times we have

$$\frac{d^r F}{dx^r} = \frac{(a)_r (b)_r}{(c)_r} F\{(a+r),(b+r)(c+r),x\} \qquad (7.156)$$

If $x = 0$

$$F(a, b, c, 0) = \lim_{x \to 0} \sum_{m=0}^{\infty} \frac{(a)_m (b)_m}{(c)_m m!} x^m = 1$$

Similarly $F\{(a+r),(b+r),(c+r)0\} = 1$

$$\therefore \quad \frac{d}{dx} F(a,b,c,x)_{x=0} = \frac{ab}{c}$$

$^*(a)_{m+1} = a(a+1)(a+2)\cdots(a+m)$
$\quad = a\{(a+1)(a+2)\cdots(a+1+m-1)\}$
$\quad = a(a+1)_m$

(iii) Integral Representation

We have

$$F(a, b, c, x) = \sum_{m=0}^{\infty} \frac{(a)_m (b)_m}{(c)_m m!} x^m$$

$$= \sum_{m=0}^{\infty} \frac{\Gamma(a+m)}{\Gamma(a)} \frac{\Gamma(b+m)}{\Gamma(b)} \frac{\Gamma(c)}{\Gamma(c+m)} \frac{x^m}{m!}$$

since

$$(a)_m = \frac{\Gamma(a+m)}{\Gamma(a)}$$

Multiplying and dividing by $\Gamma(c - b)$, we have

$$F(a, b, c, x) = \frac{\Gamma(c)}{\Gamma(a)\Gamma(b)\Gamma(c-b)} \sum_{m=0}^{\infty} \frac{\Gamma(a+m)\Gamma(b+m)\Gamma(c-b)}{\Gamma(c+m)} \frac{x^m}{m!}$$

Now β-function is defined as

$$\beta(r, n) = \int_0^1 t^{r-1}(1-t)^{n-1} dt = \frac{\Gamma(r)\Gamma(n)}{\Gamma(r+n)}$$

where $r > 0$ and $n > 0$

Hence the hypergeometric function $F(a, b, c, x)$ in terms of beta function can be written as

$$F(a, b, c, x) = \frac{\Gamma(c)}{\Gamma(a)\Gamma(b)\Gamma(c-b)} \sum_{m=0}^{\infty} \Gamma(a+m)\beta\{(b+m),(c-b)\} \frac{x^m}{m!}$$

$$= \frac{\Gamma(c)}{\Gamma(a)\Gamma(b)\Gamma(c-b)} \sum_{m=0}^{\infty} \Gamma(a+m) \int_0^1 t^{b+m-1}(1-t)^{c-b-1} \frac{x^m}{m!} dt$$

$$= \frac{1}{\beta\{b,(c-b)\}} \sum_{m=0}^{\infty} (a)_m \int_0^1 t^{b-1}(1-t)^{c-b-1} \frac{(xt)^m}{m!} dt$$

$$= \frac{1}{\beta\{b,(c-b)\}} \int_0^1 t^{b-1}(1-t)^{c-b-1} \sum_{m=0}^{\infty} \frac{(a)_m (xt)^m}{m!} dt$$

But

$$\sum_{m=0}^{\infty} \frac{(a)_m (xt)^m}{m!} = 1 + a\frac{xt}{1!} + a(a+1)\frac{(xt)^2}{2!} + \cdots$$

$$= (1 - xt)^{-a}$$

∴

$$F(a, b, c, x) = \frac{1}{\beta\{b,(c-b)\}} \int_0^1 t^{b-1}(1-t)^{c-b-1}(1-xt)^{-a} dt \qquad (7.157)$$

This formula is very useful.

(a) This is used to derive the value of hypergeometric functions with unit arguments

$$F(a, b, c; 1) = \frac{1}{\beta\{b,(c-b)\}} \int_0^1 (1-t)^{c-b-1} t^{b-1}(1-t)^{-a} dt$$

$$= \frac{1}{B\{b,(c-b)\}} \int_0^1 t^{b-1}(1-t)^{c-b-a-1}\,dt$$

$$= \frac{B\{b,(c-a-b)\}}{B\{b,(c-b)\}} \qquad\qquad (7.157a)$$

Replacing beta function by gamma function and using the relation $\beta(m,n)=\dfrac{\overline{|m}\,\overline{|n}}{\overline{|m+n}}$

Eq. (7.157a) becomes

$$F(a,b,c;\,1) = \frac{\overline{|c}\;\overline{|c-a-b}}{\overline{|(c-a)}\,\overline{|(c-b)}} \qquad\qquad (7.157b)$$

This formula is known as *Gauss formula*. If we put $a = -n$ (a negative integer) then

$$\frac{\overline{|(c-a-b)}}{\overline{|(c-b)}} = \frac{\overline{|(c-b+n)}}{\overline{|(c-b)}} = \frac{1\cdot 2\cdots(c-b+n-1)}{1\cdot 2\cdots(c-b-1)}$$

$$= (c-b)(c-b+1)\cdots(c-b+n-1) = (c-b)_n$$

and

$$\frac{\overline{|(c-a)}}{\overline{|c}} = \frac{\overline{|(c-n)}}{\overline{|c}} = \frac{1\cdot 2\cdots(c-1)c\cdots(c+n-1)}{1\cdot 2\cdots(c-1)}$$

$$= c(c+1)\cdots(c+n-1) = (c)_n$$

so

$$F(a,b,c;\,1) = F(-n,b,c;1) = \frac{\overline{|c}\;\overline{|c-b+n}}{\overline{|c+n}\,\overline{|c-b}}$$

$$= \frac{(c-b)_n}{(c)_n} \qquad\qquad (7.157c)$$

This formula is known as *Vandermonde's theorem* in mathematics.

(b) If we put $x = -1$ and $a = 1 + b - c$, then we have

$$F(a,b,b-a+1;\,-1) = \frac{\Gamma(1+b-a)}{\Gamma b\,\Gamma(1-a)} \int_0^1 (1-t^2)^{-a}t^{b-1}\,dt$$

$$= \frac{1}{2}\frac{\Gamma(1+b-a)}{\Gamma b\,\Gamma(1-a)} \int_0^1 (1-\xi)^{-a}\xi^{(b-1)/2}\,\frac{d\xi}{\sqrt{\xi}}$$

where $\xi = t^2$

or

$$F(a,b,b-a+1;\,-1) = \frac{\Gamma(1+b-a)}{\Gamma b\,\Gamma(1-a)}\cdot\frac{1}{2}B\left(\frac{1}{2}b,1-a\right)$$

$$= \frac{\Gamma(1+b-a)}{\Gamma b\,\Gamma(1-a)}\cdot\frac{1}{2}\frac{\Gamma\left(\frac{1}{2}b\right)\Gamma(1-a)}{\Gamma\left(1-a+\frac{1}{2}b\right)}$$

$$= \frac{\Gamma(1+b-a)}{\Gamma(1+b)} \cdot \frac{\Gamma\left(1+\frac{1}{2}b\right)}{\Gamma\left(1+\frac{1}{2}b-a\right)}$$

so $\qquad F(a, b, b-a+1; -1) = \dfrac{\Gamma(1+b-a)\Gamma\left(1+\frac{1}{2}b\right)}{\Gamma(1+b)\Gamma\left(1+\frac{1}{2}b-a\right)}$

This is known as *Kummer's theorem.*

(iv) Representation of Various Functions in Terms of Hypergeometric Functions
(a) Legendre Polynomials
From Rodrigue's formula of Legendre polynomials, we have

$$P_n(x) = \frac{1}{2^n n!} \frac{d^n}{dx^n}(x^2 - 1)^n$$

$$= \frac{1}{n!} \frac{d^n}{dx^n}\left[(x-1)^n \left\{\frac{1}{2}(x+1)\right\}^n\right]$$

$$= \frac{1}{n!} \frac{d^n}{dx^n}\left[(x-1)^n \left\{1 - \frac{1}{2}(1-x)\right\}^n\right]$$

$$= \frac{(-1)^n}{n!} \frac{d^n}{dx^n}\left[(1-x)^n \left\{1 - n \cdot \frac{1}{2}(1-x) + \frac{n(n-1)}{2!}\frac{(1-x)^2}{4} - \frac{n(n-1)(n-2)}{3!}\frac{(1-x)^3}{8} + \cdots\right\}\right]$$

$$= \frac{(-1)^n}{n!} \frac{d^n}{dx^n}\left[(1-x)^n - \frac{n}{2}(1-x)^{n+1} + \frac{n(n-1)}{2!2^2}(1-x)^{n+2} - \frac{n(n-1)(n-2)}{3!2^3}(1-x)^{n+3} + \cdots\right]$$

$$= \frac{(-1)^n}{n!}\left[(-1)^n n! - \frac{n}{2}(-1)^n \frac{(n+1)!}{1!}(1-x) + \frac{n(n-1)}{2!}\frac{(-1)^n}{2^2}\frac{(n+2)!}{2!}(1-x)^2 + \cdots\right]$$

$$= 1 + \frac{(-n)(n+1)}{1 \cdot 1!}\left(\frac{1-x}{2}\right) + \frac{(-n)(-n+1)(n+1)(n+2)}{2 \cdot 1 \cdot 2!}\left(1 - \frac{x}{2}\right)^2 + \cdots$$

$$\therefore \qquad P_n(x) = F\left(-n, n+1, 1, \frac{1-x}{2}\right) \qquad\qquad (7.158)$$

Substituting $x = \cos\theta$ in Eq. (7.158) we have

$$P_n(\cos\theta) = F\left(-n, n+1, 1, \sin^2\frac{\theta}{2}\right) \qquad\qquad (7.158a)$$

Substituting $x = -\cos\theta$ we have

$$P_n(-\cos\theta) = F\left(-n, n+1, 1, \cos^2\frac{\theta}{2}\right)$$

or $\qquad (-1)^n P_n(\cos\theta) = F\left(-n, n+1, 1, \cos^2\frac{\theta}{2}\right) \qquad$ [*since* $P_n(-x) = (-1)^n P_n(x)$]

$$[P_n(\cos\theta)] = (-1)^n F\left(-n, n+1, 1, \cos^2\frac{\theta}{2}\right)$$

$$= (-1)^n F\left(n+1, -n, 1, \cos^2\frac{\theta}{2}\right) \qquad (7.158b)$$

since in hypergeometric series a and b can be interchanged.

(b) Elliptical Integrals

The elliptical integrals of the first kind and second kind are defined as

$$K(x) = \int_0^{\pi/2} \frac{d\phi}{[1-x^2\sin^2\phi]^{1/2}} \qquad \text{[1st kind]}$$

$$E(x) = \int_0^{\pi/2} [1-x^2\sin\phi]^{1/2} d\phi \qquad \text{[2nd kind]}$$

Both of them can be represented in terms of hypergeometric functions as seen here.

$$K(x) = \int_0^{\pi/2} (1-x^2\sin^2\phi)^{-1/2} d\phi$$

$$= \int_0^{\pi/2} \sum_r \left(\frac{1}{2}\right)\left(\frac{1}{2}+1\right)\cdots\left(\frac{1}{2}+r-1\right) \frac{x^{2r}\sin^{2r}\phi}{r!} d\phi \qquad (7.159)$$

$$= \sum_r \frac{\left(\frac{1}{2}\right)_r x^{2r}}{r!} \int_0^{\pi/2} \sin^{2r}\phi\, d\phi$$

$$= \frac{\pi}{2} \sum_r \frac{\left(\frac{1}{2}\right)_r \left(\frac{1}{2}\right)_r}{(1)_r r!} (x^2)^r = \frac{\pi}{2} F\left(\frac{1}{2}, \frac{1}{2}, 1, x^2\right) \qquad (7.159a)$$

Since
$$\int_0^{\pi/2} \sin^{2r}\phi\, d\phi = \frac{\Gamma\left(r+\frac{1}{2}\right)}{\Gamma(r+1)} \frac{\sqrt{\pi}}{2} = \frac{\left(\frac{1}{2}\right)_r}{r!} \frac{\pi}{2} = \frac{\left(\frac{1}{2}\right)_r}{(1)_r} \frac{\pi}{2}$$

Similarly
$$E(x) = \frac{\pi}{2} F\left(-\frac{1}{2}, \frac{1}{2}, 1, x^2\right) \qquad (7.159b)$$

Hence both the elliptical functions can be represented in terms of hypergeometric functions.

(c) Elementary Functions

In hypergeometric functions $F(a, b, c; x)$ if a or b is a negative integer, then the series stops after certain terms and hence reduces to polynomial.

For example if $b = 0$ then

$$F(a, b, c; x) = F(a, 0, c; x) = 1 + \frac{ab}{c}x + \cdots = 1$$

and
$$F(a, -2, c; x) = 1 - \frac{2ax}{c} + \frac{a(a+1)}{c(c+1)}x^2$$

Thus hypergeometric function for $b = -2$ reduces to a simple algebraic function (elementary function).

The linear transformation of hypergeometric functions give us

$$F(a, b, c; x) = (1-x)^{-a} F\left\{a, c-b, c; \frac{x}{x-1}\right\}$$

and

$$F(a, b, c; x) = (1-x)^{-b} F\left\{c-a, b, c; \frac{x}{x-1}\right\}$$

Equating R.H.S we have

$$F\left(a, c-b, c; \frac{x}{x-1}\right) = \left(1 - \frac{x}{x-1}\right)^{b-c} F\left\{c-a, c-b, c, \frac{x/(x-1)}{\left[\left\{\frac{x}{x-1}\right\}-1\right]}\right\}$$

$$= (x-1)^{c-b} (-1)^{c-b} F(c-a, c-b, c; x)$$

$$= (1-x)^{c-b} F(c-a, c-b, c; x)$$

Hence

$$F(a, b, c; x) = (1-x)^{-a} F\left\{a, c-b, c; \frac{x}{x-1}\right\}$$

$$= (1-x)^{c-b-a} F\{c, -a, c-b, c; x\} \tag{7.159c}$$

RHS of (7.159c) is polynomial for $c = b$, i.e.

$$(1-x)^{-a} = F(a, b, b; x)$$

$$\therefore \qquad F(a, b, b; x) = \text{algebraic function } (1-x)^{-a} \tag{7.159d}$$

Let $b = 1$ and $a = -n$ then

$$(1-x)^n = F(-n, 1, 1; x) \tag{7.159e}$$

If

$$a = \frac{1}{2} \text{ then } (1-x)^{-1/2} = F\left(\frac{1}{2}, 1, 1; x\right) \tag{7.159f}$$

If x is replaced by $(1 - x)$ and $a = -n$, $b = 1$, then

$$x^n = F(-n, 1, 1; 1-x) \tag{7.159g}$$

Hence, the algebraic functions like x^n, $(1-x)^n$, $(1-x)^{-1/2}$ and $(1-x)^{-a}$ can be represented in terms of hypergeometric functions.

7.9. CONFLUENT HYPERGEOMETRIC FUNCTIONS

The hypergeometric differential equation is given by

$$(x^2 - x)\frac{d^2y}{dx^2} + \{(1+a+b)x - c\}\frac{dy}{dx} + aby = 0$$

If we replace x by $\dfrac{x}{b}$ then we have

$$x\left(1 - \frac{x}{b}\right)\frac{d^2y}{dx^2} + \left\{c - \left(1 + \frac{1+a}{b}\right)x\right\}\frac{dy}{dx} - ay = 0 \tag{7.160}$$

Solution of which is represented by the function

$$F\left(a,b,c,\frac{x}{b}\right)$$

If we make $b \to \infty$ then Eq. (7.160) reduces to

$$x\frac{d^2y}{dx^2}+(c-x)\frac{dy}{dx}-ay = 0 \tag{7.161}$$

The equation is called the *confluent hypergeometric differential equation*, solution of which is represented by

$$\lim_{b\to\infty}F\left(a,b,c,\frac{x}{b}\right) = \lim_{b\to\infty}\sum_{m=0}^{\infty}\frac{(a)_m(b)_m x^m}{(c)_m m!b^m}$$

but

$$\lim_{b\to\infty}\frac{(b)_m}{b^m} = 1$$

Hence

$$\lim_{b\to\infty}\left(a,b,c,\frac{x}{b}\right) = \sum_{m=0}^{\infty}\frac{(a)_m x^m}{(c)_m m!} = F(a,c,x) \tag{7.162}$$

This function is the solution of confluent differential equation and hence is called *confluent hypergeometric function*. Solution of this differential equation may also be obtained directly by the series integration method.

Hence the complete solution of confluent hypergeometric equation is given by

$$y = AF(a;c;x)+Bx^{1-c}F\{(a-c+1),(2-c),x\}$$

Here A and B are arbitrary constants and c is neither zero nor a negative integer.

Convergence of Confluent Hypergeometric Function

The mth term of series representing confluent hypergeometric function is

$$U_m = \frac{(a)_m x^m}{(c)_m m!}$$

and $(m + 1)$th term is

$$U_{m+1} = \frac{(a)_{m+1} x^{m+1}}{(c)_{m+1}(m+1)!}$$

So that

$$\lim_{m\to\infty}\left|\frac{U_{m+1}}{U_m}\right| = \lim_{m\to\infty}\left|\frac{(a+m)x}{(c+m)(m+1)}\right| = 0$$

Hence $\left|\dfrac{U_{m+1}}{U_m}\right| < 1$ for every value of x, thus the series is always convergent.

Integral Representation of Confluent Hypergeometric Function

We have

$$\frac{(a)_k}{(c)_k} = \frac{\overline{|c|}\overline{(a+k)}}{\overline{|a|}\overline{(c+k)}}=\frac{\overline{|c}}{\overline{|a|}\overline{(c-a)}}\frac{\overline{|a+k}\overline{|c-a}}{\overline{|c+k}}$$

But
$$\frac{\overline{|m|n}}{\overline{|m+n}} = \beta(m,n) = \int_0^1 t^{m-1}(1-t)^{n-1} dt$$

∴
$$\frac{\Gamma(a+k)\Gamma(c-a)}{\Gamma(c+k)} = \beta(a+k, c-a) = \int_0^1 t^{a+k-1}(1-t)^{c-a-1} dt$$

Hence
$$\frac{(a)_k}{(c)_k} = \frac{\Gamma c}{\Gamma a \Gamma(c-a)}\int_0^1 t^{a+k-1}(1-t)^{c-a-1} dt$$

$$= \frac{1}{(a,c-a)}\int_0^1 t^{a+k-1}(1-t)^{c-a-1} dt$$

So
$$F(a, c; x) = \sum_{k=0}^{\infty}\frac{(a)_k x^k}{(c)_k k!} = \sum_{k=0}^{\infty}\frac{x^k}{k!}\frac{1}{\beta(a,c-a)}\int_0^1 t^{a+k-1}(1-t)^{c-a-1} dt$$

$$= \frac{1}{\beta(a,c-a)}\int_0^t t^{a-1}(1-t)^{c-a-1} dt \sum_{k=0}^{\infty}\frac{(xt)^k}{k!}$$

But
$$\sum_{k=0}^{\infty}\frac{(xt)^k}{k!} = 1 + \frac{xt}{1!} + \frac{x^2 t^2}{2!} + \cdots \infty = e^{xt}$$

Hence
$$F(a, c; x) = \frac{\Gamma c}{\Gamma a \Gamma c - a}\int_0^1 e^{xt} t^{a-1}(1-t)^{c-a-1} dt \qquad (7.162a)$$

This is integral formula representing the confluent hypergeometric functions. Here if we put $t = 1 - s$.

then
$$F(a, c; x) = \frac{\Gamma c}{\Gamma a \Gamma c - a}\int_0^1 e^{x(1-s)}(1-s)^{a-1} s^{c-a-1} ds$$

$$= \frac{\Gamma c e^x}{\Gamma a \Gamma c - a}\int_0^1 e^{-xs} s^{c-a-1}(1-s)^{a-1} ds$$

But we know that

$$\int_0^1 e^{xt} t^{a-1}(1-t)^{c-a-1} dt \text{ can be written as}$$

$$\frac{\Gamma a \Gamma(c-a)}{\Gamma c} F(a,c;x)$$

Hence $\int_0^1 e^{-xs} s^{c-a-1}(1-s)^{a-1} ds = \dfrac{\Gamma c - a \Gamma a}{\Gamma c} F(c-a, c; -x)$

or
$$F(a, c; x) = e^x \cdot F(c-a; c; -x) \qquad (7.162b)$$

In general, if we put $x = 0$ in Eq. (1) then

$$F(a; c; 0) = \frac{\Gamma c}{\Gamma a \Gamma(c-a)}\int_0^1 t^{(a-1)}(1-t)^{c-a-1} dt$$

$$= \frac{\Gamma c}{\Gamma a \Gamma(c-a)}\beta(a, c-a) = 1$$

Example 1. *Express associated Legendre polynomial $P_n^m(x)$ in terms of hypergeometric function.*

Solution: From Eq. (7.158) we have

$$P_n(x) = F\left(-n, n+1, 1, \frac{1-x}{2}\right)$$

Then

$$P_n^m(x) = (1-x^2)^{m/2} \frac{d^m}{dx^m} P_n(x)$$

$$= (1-x^2)^{m/2} \frac{d^m}{dx^m} F\left(-n, n+1, 1, \frac{1-x}{2}\right)$$

$$= (1-x^2)^{m/2} \frac{d^m}{d\left(\frac{1-x}{2}\right)^m} F\left(-n, n+1, 1, \frac{1-x}{2}\right) \left(-\frac{1}{2}\right)^m$$

$$= (-1)^m (1-x^2)^{m/2} \frac{1}{2^m} \frac{d^m}{d\left(\frac{1-x}{2}\right)^m} F\left(-n, n+1, 1, \frac{1-x}{2}\right)$$

$$= (-1)^m (1-x^2)^{m/2} \cdot \frac{1}{2^m} \cdot \frac{(-n)_m (n+1)_m}{(1)_m} \times F\left(-n+m, m+n+1, m+1, \frac{1-x}{2}\right)$$

[using Eq. (7.156)*]*

or

$$P_n^m(x) = \frac{(n+m)!}{(n-m)!} \frac{(1-x^2)^{m/2}}{2^m m!} F\left(-n+m, m+n+1, m+1, \frac{1-x}{2}\right)$$

Example 2. *Express Bessel's function in terms of confluent hypergeometric function and hence derive the integral formula for Bessel's function.*

Solution: Bessel's equation is given as

$$\frac{d^2 y}{dx^2} + \frac{1}{x}\frac{dy}{dx} + \left(1 - \frac{n^2}{x^2}\right) y = 0$$

Let us substitute $y = z^n e^{-zi} F$

We have

$$zF'' + \{(2n+1) - 2iz\} F' - i(2n+1)F = 0 \tag{1}$$

which is confluent hypergeometric equation for

$$a = n - \frac{1}{2}, c = 2n+1, x = 2iz$$

Therefore its solution is $F\left[\left(n+\frac{1}{2}\right), 2n+1, 2iz\right]$

But

$$y = J_n(z) cz^n e^{-zi} F\left(n+\frac{1}{2}, 2n+1, 2iz\right).$$

Substituting integral form of confluent hypergeometric function, we have

$$J_n(z) = \frac{1}{2^n \Gamma(n+1)} z^n e^{-zi} \frac{\Gamma(2n+1)}{\Gamma\left(n+\dfrac{1}{2}\right)\Gamma\left(n+\dfrac{1}{2}\right)} \int_0^1 e^{2zit} t^{n-\frac{1}{2}} (1-t)^{n-\frac{1}{2}} dt$$

Here we have set the constant $c = \dfrac{1}{2^n \Gamma(n+1)}$

or

$$J_n(z) = \frac{(z/2)^n}{\Gamma(n+1)} \frac{\Gamma(2n+1)}{\Gamma\left(n+\dfrac{1}{2}\right)\Gamma\left(n+\dfrac{1}{2}\right)} \int_0^1 e^{zi(2t-1)} t^{n-\frac{1}{2}} (1-t)^{n-\frac{1}{2}} dt$$

Let us put $2t = 1 + u$.

$$J_n(z) = \frac{(z/2)^n \Gamma(2n+1)}{\Gamma(n+1)\Gamma\left(n+\dfrac{1}{2}\right)\Gamma\left(n+\dfrac{1}{2}\right)} \cdot \frac{1}{2} \int_{-1}^{+1} e^{ziu} \frac{(1+u)^{n-\frac{1}{2}}}{2^{n-\frac{1}{2}}} \frac{(1-u)^{n-\frac{1}{2}}}{2^{n-\frac{1}{2}}} du$$

$$= \left(\frac{z}{2}\right)^n \frac{\Gamma(2n+1)}{\Gamma(n+1)\Gamma\left(n+\dfrac{1}{2}\right)\Gamma\left(n+\dfrac{1}{2}\right)} \frac{1}{2^{2n}} \int_{-1}^{+1} e^{ziu} (1+u)^{n-\frac{1}{2}} (1-u)^{n-\frac{1}{2}} du$$

From Legendre's duplication formula, we have

$$\Gamma(2n+1) = \frac{2^{2n}}{\sqrt{\pi}} \Gamma\left(n+\frac{1}{2}\right)\Gamma(n+1)$$

∴

$$J_n(z) = \frac{(z/2)^n}{\sqrt{\pi}\,\Gamma\left(n+\dfrac{1}{2}\right)} \int_{-1}^{+1} (1-u^2)^{n-\frac{1}{2}} e^{ziu} du$$

which is the integral formula for $J_n(z)$. Now putting $u = \cos\theta$, the formula becomes

$$J_n(z) = \frac{(z/2)^n}{\sqrt{\pi}\,\Gamma(n+1/2)} \int_0^\pi e^{iz\cos\theta} \cdot \sin^{2n}\theta \cdot d\theta$$

which is well-known integral formula for Bessel's function of first kind.

Example 3. Show that $P_n(\cos\theta) = \cos^{2n}\theta/2\, F(-n, -n, 1, -\tan^2\theta/2)$

Solution: We have from first Laplace integral

$$P_n(\mu) = \frac{1}{\pi} \int_0^\pi \left\{\mu \pm \sqrt{\mu^2 - 1}\cos\phi\right\}^n d\phi$$

∴ $P_n(\cos\theta) = \dfrac{1}{\pi} \displaystyle\int_0^\pi \{\cos\theta + i\sin\theta\cos\phi\}^n d\phi$ (taking + ve sign)

$$= \frac{1}{\pi} \int_0^\pi \left\{\cos^2\theta/2 - \sin^2\theta/2 + 2i\sin\theta/2\cos\theta/2 \left(\frac{e^{i\phi} + e^{-i\phi}}{2}\right)\right\}^n d\phi$$

$$= \frac{1}{\pi}\int_0^\pi \left\{\cos\theta/2(\cos\theta/2 + i\sin\theta/2e^{i\phi} + i\sin\theta/2e^{-i\phi}(\cos\theta/2 + i\sin\theta/2e^{i\phi}))\right\}^n d\phi$$

$$= \frac{1}{\pi}\int_0^\pi \left\{(\cos\theta/2 + i\sin\theta/2e^{i\phi})(\cos\theta/2 + i\sin\theta/2e^{-i\phi})\right\}^n d\phi$$

$$= \frac{1}{\pi}\cos^{2n}\theta/2\int_0^\pi (1 + i\tan\theta/2e^{i\phi})^n(1 + i\tan\theta/2e^{-i\phi})^n d\phi$$

$$= \frac{1}{\pi}\cos^{2n}\theta/2\int_0^\pi \left\{1 + in\tan\theta/2e^{i\phi} - \frac{n(n-1)}{2!}\tan^2\theta/2e^{2i\phi} + \cdots\right\}$$

$$\times\left\{1 + in\tan\theta/2e^{-i\phi} - \frac{n(n-1)}{2!}\tan^2\theta/2e^{-2i\phi} + \cdots\right\}d\phi$$

Here diagonal product will give cosines of multiples of ϕ and when they will be integrated within limits 0 to π will vanish. We have only to consider the column products, i.e.

$$P_n(\cos\theta) = \frac{1}{\pi}\cos^{2n}\theta/2\int_0^\pi \left\{1 - n^2\tan^2\theta/2 + \frac{n^2(n-1)^2}{2!2!}\tan^4\theta/2\right\}d\phi$$

$$= \cos^{2n}\theta/2\left[1 - n^2\tan^2\theta/2 + \frac{n^2(n-1)^2}{2!2!}\tan^4\theta/2\cdots\right]$$

$$= \cos^{2n}\theta/2\left[1 + \frac{(-n)(-n)}{1\cdot 1}(-\tan^2\theta/2)\right.$$

$$\left. + \frac{(-n)(-n+1)(-n)(-n+1)}{1\cdot 2\cdot 1\cdot(1+1)}\times(-\tan^2\theta/2)^2\cdots\right]$$

$$= \cos^{2n}\theta/2\, F(-n,-n,1,-\tan^2\theta/2).$$

Example 4. *Show that Legendre's equation*

$$(1-\mu^2)\frac{d^2y}{d\mu^2} - 2\mu\frac{dy}{d\mu} + n(n+1)y = 0$$

changes into the hypergeometric form.

$$x(1-x)\frac{d^2y}{dx^2} + \{c - (a+b+1)x\}\frac{dy}{dx} - aby = 0.$$

Hence show by comparison that its complete primitive is

$$AF\left(-\frac{n}{2},\frac{n+1}{2},\frac{1}{2},\mu^2\right) + B\mu F\left(\frac{n+2}{2}, -\frac{(n-1)}{2},\frac{3}{2},\mu^2\right)$$

Solution: Let us substitute $\mu^2 = x$ in the given equation

$$\therefore \qquad \frac{dy}{d\mu} = 2\mu\frac{dy}{dx} \text{ and } \frac{d^2y}{d\mu^2} = 2\frac{dy}{dx} + 4\mu^2\frac{d^2y}{dx^2}$$

The given Legendre's equation becomes

$$x(1-x)\frac{d^2y}{dx^2} + \left(\frac{1}{2} - \frac{3}{2}x\right)\frac{dy}{dx} + \frac{n(n+1)}{4}y = 0 \tag{1}$$

The complete primitive of

$$x(1-x)\frac{d^2y}{dx^2} + \{c - (a+b+1)x\}\frac{dy}{dx} - aby = 0 \tag{2}$$

is $AF(a, b, c, x) + Bx^{1-c} F\{(a+1-c)(b+1-c), (2-c)x\}$

Comparing Eqs (1) and (2)

$$c = \frac{1}{2}, a+b+1 = \frac{3}{2} \quad \text{and} \quad ab = \frac{n(n+1)}{4}$$

Solving for a and b, we have $a = \dfrac{n+1}{2}$ and $b = -n/2$.

Hence the complete primitive is

$$AF\left(\frac{n+1}{2}, -\frac{n}{2}, \frac{1}{2}, \mu^2\right) + B\mu F\left(\frac{n+2}{2}, -\frac{(n-1)}{2}, \frac{3}{2}, \mu^2\right)$$

Example 5. *Prove that the contour integral*

$$I = \int_c \frac{e^\lambda \lambda^{a-b}}{(\lambda - x)^a} d\lambda,$$

where c is a contour such that real $(\lambda) = -\infty$ *at both its ends, satisfy confluent hypergeometric differential equation*

Solution:
$$I = \int_c \frac{e^\lambda \lambda^{a-b}}{(\lambda - x)^a} d\lambda$$

$$\frac{dI}{dx} = \int_c \frac{e^\lambda \lambda^{a-b} a}{(\lambda - x)^{a+1}} d\lambda$$

and
$$\frac{d^2 I}{dx^2} = \int_c \frac{e^\lambda \lambda^{a-b} a(a+1)}{(\lambda - x)^{a+2}} d\lambda$$

$$\therefore \quad x\frac{d^2 I}{dx^2} + (b-x)\frac{dI}{dx} - aI = \int_c e^\lambda \lambda^{a-b}\left\{\frac{a(a+1)x}{(\lambda-x)^{a+2}} + \frac{a(b-x)}{(\lambda-x)^{a+1}} - \frac{a}{(\lambda-x)^a}\right\} d\lambda \tag{1}$$

Now $\dfrac{a(a+1)x}{(\lambda-x)^{a+2}} + \dfrac{a(b-x)}{(\lambda-x)^{a+1}} - \dfrac{a}{(\lambda-x)^a}^* = \dfrac{a(a+1)\lambda}{(\lambda-x)^{a+2}} - \dfrac{a(a+1)}{(\lambda-x)^{a+1}} - \dfrac{a\lambda}{(\lambda-x)^{a+1}} + \dfrac{ab}{(\lambda-x)^{a+1}}$ \tag{2}

$$* \quad \frac{a(a+1)x}{(\lambda-x)^{1+2}} + \frac{a(b-x)}{(\lambda-x)^{a+1}} - \frac{a}{(\lambda-x)^a} = \frac{a(a+1)\{\lambda-(\lambda-x)\}}{(\lambda-x)^{a+2}} + \frac{ab}{(\lambda-x)^{a+1}} - \frac{a\{\lambda-(\lambda-x)\}}{(\lambda-x)^{a+1}} - \frac{a}{(\lambda-x)^a}$$

$$= \frac{a(a+1)\lambda}{(\lambda-x)^{a+2}} - \frac{a(a+1)}{(\lambda-x)^{a+1}} + \frac{ab}{(\lambda-x)^{a+1}} - \frac{a\lambda}{(\lambda-x)^{a+1}}$$

$$= \frac{a(a+1)\lambda}{(\lambda-x)^{a+2}} - \frac{a(a+1)}{(\lambda-x)^{a+1}} - \frac{a\lambda}{(\lambda-x)^{a+1}} + \frac{ab}{(\lambda-x)^{a+1}}$$

Integrating by parts, we have

$$\int_c e^\lambda \lambda^{a-b+1} \frac{a(a+1)}{(\lambda-x)^{a+2}} d\lambda = -\left[\frac{e^\lambda a\lambda^{a-b+1}}{(\lambda-x)^{a+1}}\right]_c + \int_c \frac{ae^\lambda}{(\lambda-x)^{a+1}}\left\{\lambda^{a-b+1}+(a-b+1)\lambda^{a-b}\right\}d\lambda \qquad (3)$$

R.H.S. of Eq. (1) in light of Eq. (2) is

$$= \int_c \frac{e^\lambda \lambda^{a-b+1} a(a+1)}{(\lambda-x)^{a+2}} d\lambda - \int_c \frac{e^\lambda \lambda^{a-b} a(a+1)}{(\lambda-x)^{a+1}} d\lambda - \int_c \frac{e^\lambda \lambda^{a-b+1} a}{(\lambda-x)^{a+1}} d\lambda + \int_c \frac{abe^\lambda \lambda^{a-b}}{(\lambda-x)^{a+1}} d\lambda$$

Substitution from Eq. (3) gives R.H.S of Eq. (1)

$$= -\left[\frac{e^\lambda a\lambda^{a-b+1}}{(\lambda-x)^{a+1}}\right]_c + \int_c \frac{ae^\lambda \lambda^{a-b+1}}{(\lambda-x)^{a+1}} d\lambda + \int_c \frac{e^\lambda \lambda^{a-b} a(a+1)}{(\lambda-x)^{a+1}} d\lambda$$

$$- \int_c \frac{abe^\lambda \lambda^{a-b}}{(\lambda-x)^{a+1}} d\lambda - \int_c \frac{a(a+1)e^\lambda \lambda^{a-b}}{(\lambda-x)^{a+1}} d\lambda - \int_c \frac{ae^\lambda \lambda^{a-b+1}}{(\lambda-x)^{a+1}} d\lambda + \int_c \frac{abe^\lambda \lambda^{a-b}}{(\lambda-x)^{a+1}} d\lambda$$

$$= -\left[\frac{e^\lambda a\lambda^{a-b+1}}{(\lambda-x)^{a+1}}\right]_c \to 0 \text{ since at the end of } c\text{-real } (\lambda) = -\infty.$$

Thus Eq. (1) becomes

$$x\frac{d^2 I}{dx^2} + (b-x)\frac{dI}{dx} - aI = 0$$

which is confluent hypergeometric equation.

Example 6. *Express Laguerre polynomial in terms of confluent hypergeometric function.*

Solution: The Leguerre polynomials are defined as

$$L_n(x) = \sum_{r=0}^{n} (-1)^r \frac{(n!)^2}{(n-r)!(r!)^2} x^r$$

But

$$\frac{(-1)^r}{(n-r)!} = (-1)^r \frac{n(n-1)\cdots(n-r+1)}{n!}$$

$$= \frac{(-n)(-n+1)(-n+2)\cdots(-n+r-1)}{n!} = \frac{(-n)_r}{n!}$$

$$\therefore \qquad L_n(x) = \sum_{r=0}^{n} \frac{(-n)_r (n!)^2 x^r}{(n!)(r!)^2}$$

$$= n! \sum_{r=0}^{n} \frac{(-n)_r x^r}{(1)_r r!} \qquad [\text{since } r! = (1)_r]$$

$$= n!\, F(-n, 1, x).$$

Example 7. *Express error functions in terms of confluent hypergeometric function.*

Solution: Error function is defined as

$$\text{erf}(x) = \frac{2}{\sqrt{\pi}} \int_0^x e^{-t^2} dt = \frac{2}{\sqrt{\pi}} \int_0^x \sum_{k=0}^{\infty} (-1)^k \frac{t^{2k}}{k!} dt$$

$$= \frac{2}{\sqrt{\pi}} \sum_{k=0}^{\infty} \frac{(-1)^k}{k!} \int_0^x t^{2k} dt$$

$$= \frac{2}{\sqrt{\pi}} \sum_{k=0}^{\infty} \frac{(-1)^k}{k!} \frac{x^{2k+1}}{2k+1} = \frac{2}{\sqrt{\pi}} x \sum_{k=0}^{\infty} \frac{(-1)^k}{k!} \frac{x^{2k}}{(2k+1)}$$

$$= \frac{2}{\sqrt{\pi}} x \sum_{k=0}^{\infty} \frac{(-x^2)^k}{(2k+1)k!}$$

But

$$(2k+1) = \frac{(2k+1)(2k-1)(2k-3)\cdots 1}{(2k-1)(2k-3)\cdots 1}$$

$$= \frac{\left(\dfrac{1}{2}\right)\left(\dfrac{3}{2}\right)\cdots \dfrac{(2k-1)}{2}(2k+1)}{\dfrac{(2k-1)}{2} \cdot \dfrac{(2k-3)}{2}\cdots \left(\dfrac{1}{2}\right)}$$

$$= \frac{\dfrac{3}{2}\cdots \dfrac{2k-1}{2}, \dfrac{2k+1}{2}}{\left(\dfrac{1}{2}\right)\cdots \left(\dfrac{2k-1}{2}\right)} = \frac{\left(\dfrac{3}{2}\right)_k}{\left(\dfrac{1}{2}\right)_k}$$

Hence the error function $\text{erf}(x)$ is given by

$$\text{erf}(x) = \frac{2}{\sqrt{\pi}} x \sum_{k=0}^{\infty} \frac{\left(\dfrac{1}{2}\right)_k (-x)^{2k}}{\left(\dfrac{3}{2}\right)_k k!}$$

$$= \frac{2}{\sqrt{\pi}} xF\left(\frac{1}{2}, \frac{3}{2}; -x^2\right)$$

Example 8. *Represent Fresnel's integrals of 1st kind and 2nd kind in confluent hypergeometric functions.*

Solution: Fresnel's integrals are represented as follows.

$$c(x) = \frac{1}{2\sqrt{2}}\left[e^{\pi i/4}\phi\left(\frac{\sqrt{\pi}}{\sqrt{2}} xe^{-\pi i/4}\right) + e^{-\pi i/4}\phi\left(\frac{\sqrt{\pi}}{\sqrt{2}} xe^{\pi i/4}\right)\right] \qquad (1)$$

and

$$s(x) = \frac{1}{2i\sqrt{2}}\left[e^{\pi i/4}\phi\left(\frac{\sqrt{\pi}}{\sqrt{2}} xe^{-\pi i/4}\right) - e^{-\pi i/4}\phi\left\{\left(\frac{\sqrt{\pi}}{\sqrt{2}}\right) xe^{\pi i/4}\right\}\right] \qquad (2)$$

where φ functions are probability functions.

or
$$\phi(z) = \frac{2}{\sqrt{\pi}} \int_0^z e^{-t^2}\, dt$$

$$\therefore \quad \phi\left\{\sqrt{\frac{\pi}{2}} \cdot x \cdot e^{-\pi i/4}\right\} = \frac{2}{\sqrt{\pi}} \int_0^{\sqrt{\frac{\pi}{2}} x e^{-\pi i/4}} e^{-t^2}\, dt$$

or
$$\phi\left[\left\{\sqrt{\frac{\pi}{2}} x e^{-\pi i/4}\right\}\right] = \sqrt{2} x e^{-\pi i/4} \sum_{k=0}^{\infty} \frac{\left(\frac{1}{2}\right)_k \left(\frac{1}{2}\pi i x^2\right)^k}{\left(\frac{3}{2}\right)_k k!}$$

$$= \sqrt{2}\, x e^{-\pi i/4} F\left(\frac{1}{2}, \frac{3}{2}; \frac{\pi i x^2}{2}\right)$$

Similarly
$$\phi\left\{\frac{\pi}{2} x e^{i\pi/4}\right\} = \sqrt{2}\, x e^{\pi i/4} F\left(\frac{1}{2}, \frac{3}{2}; -\frac{\pi i x^2}{2}\right)$$

Hence
$$c(x) = \frac{x}{2}\left[F\left(\frac{1}{2}, \frac{3}{2}; \frac{\pi i x^2}{2}\right) + F\left(\frac{1}{2}, \frac{3}{2}; \frac{-\pi i x^2}{2}\right)\right] \tag{3}$$

$$s(x) = \frac{x}{2i}\left[F\left(\frac{1}{2}, \frac{3}{2}; \frac{\pi i x^2}{2}\right) - F\left(\frac{1}{2}, \frac{3}{2}; \frac{-\pi i x^2}{2}\right)\right] \tag{4}$$

Hence the Fresnel's integrals of first and second kinds can be represented in terms of confluent hypergeometric functions.

7.10. HERMITE DIFFERENTIAL EQUATION

The differential equation

$$\frac{d^2y}{dx^2} - 2x\frac{dy}{dx} + 2py = 0 \tag{7.163}$$

where p is a constant, is called *Hermite differential equation*. There is no singularity in the finite plane of the differential equation and hence it can be solved by the method of series integration. The series solution of Hermite equation may be expressed as

$$y = \sum_{m=0}^{\infty} a_m x^{m+n} \tag{7.164}$$

On differentiation and substitution in Eq. (7.163) gives

$$\sum_m a_m(m+n)(m+n-1)x^{m+n-2} - \sum_m 2a_m(m+n-p)x^{m+n} = 0 \tag{7.165}$$

The indicial equation is obtained by equating the coefficient of lower power of x equal to zero. The indicial equation is

$$a_0 n(n-1) = 0 \tag{7.166}$$

As $a_0 \neq 0$ being the coefficient of first term, therefore we must have either

$$n = 0 \quad \text{or} \quad n = 1 \tag{7.167}$$

Equating the coefficient x^{n-1} equal to zero, we have

$$a_1 n(n+1) = 0 \tag{7.168}$$

Since $n + 1 \neq 0$ for any value of n given by Eq. (7.167) hence Eq. (7.168) implies that either

$$n = 0 \quad \text{or} \quad a_1 = 0 \quad \text{or both are zero.}$$

The recurrence relation between the coefficient is obtained by equating coefficient of $x^{m+n} = 0$. Thus we have

$$a_{m+2}(m+2+n)\,(m+n+1) = 2(m+n-p)a_m$$

or

$$a_{m+2} = \frac{2(m+n-p)}{(m+n+2)(m+n+1)} a_m \tag{7.169}$$

Since we have two possible values of n, now there arise two cases.

Case 1. When $n = 0$ we have from Eq. (7.169)

$$a_{m+2} = \frac{2m - 2p}{(m+2)(m+1)} a_m \tag{7.170}$$

This gives

$$a_2 = -\frac{2p}{2!} a_0$$

and

$$a_4 = \frac{4-2p}{4\cdot 3}\cdot a_2 = \frac{4-2p}{4\cdot 3}\cdot\frac{(-2p)}{2!} a_0 = -\frac{4-2p}{4!}2p\cdot a_0$$

$$= \frac{(-2)^2 p(p-2)}{4!}\cdot a_0$$

Thus

$$a_{2m} = (-2)^m \frac{p(p-2)\cdots(p-2m+1)}{(2m)!} a_0$$

and similarly odd coefficients are given by

$$a_{2m+1} = \frac{(-2)^m (p-1)(p-3)\cdots(p-2m+1)}{(2m+1)!} a_1$$

$$\therefore \quad y = \sum_{m=0}^{\infty} a_m x^m = a_0 + a_1 x + a_2 x^2 + a_3 x^3 + \cdots$$

$$= a_0\left[1 - \frac{2p}{2!}x^2 + \frac{2^2 p(p-2)}{4!}x^4 - \cdots + \frac{(-2)^m p(p-2)\cdots(p-2m+2)}{(2m)!} + \cdots\right]$$

$$+ a_1\left[x - \frac{2(p-1)}{3!}x^3 + \cdots + \frac{(-2)^m (p-1)(p-3)\cdots(p-2m\cdots+1)}{(2m+1)!}x^{2m+1} + \cdots\right] \tag{7.171}$$

Case 2: When $n = 1$, it will yield the solution which is 2nd part of Eq. (7.171), i.e.

$$a_1\left[x - \frac{2(p-1)}{3!}x^3 + \cdots + \frac{(-2)^m (p-1)(p-3)\cdots(p-2m+1)}{(2m+1)!}x^{2m+1} + \cdots\right]$$

Hence, we conclude that in Eq. (7.171), $a_1 = 0$ and the two separate solutions of Hermite equation are

$$y_1 = a_0 \sum_{m=0}^{\infty} \frac{(-2)^m p(p-2)\cdots(p-2m+2)x^{2m}}{(2m)!} \qquad (7.172)$$

and

$$y_2 = a_1 \sum_{m=1}^{\infty} \frac{(-2)^m (p-1)(p-3)\cdots(p-2m+1)}{(2m+1)!}x^{2m+1} \qquad (7.173)$$

and the general solution of Hermite equation is

$$y = Ay_1 + By_2\ldots \qquad (7.174)$$

where A and B are arbitrary constants.

Hermite Polynomials

When p is an even integer and $a_0 = (-1)^{p/2} \dfrac{p!}{(p/2)!}$, obviously the series Eq. (7.172)

terminates at the $\left(\dfrac{p}{2}+1\right)$th term. This value of y is known as the Hermite polynomial

of order p, for p even, i.e.

$$H_p(x) = (2x)^p - \frac{p(p-1)}{1!}(2x)^{p-2} + \frac{p(p-1)(p-2)(p-3)}{2!}(2x)^{p-4}\cdots$$

$$+(-1)^p \frac{p(p-1)\cdots(p-2r+1)(2x)^{p-2r}\cdots}{r!} +(-1)^{p/2}\frac{p!}{(p/2)!} \qquad (7.175)$$

When p is an odd integer and $a_1 = (-1)^{p-1/2}\dfrac{(p+1)!}{(p+1)/2!}$, the series Eq. (7.173) termi-

nates at $\left(\dfrac{p-1}{2}\right)$th term. This value of y is known as the Hermite polynomial of order

p for p odd, i.e.

$$H_p(x) = \left\{(2x)^p - \frac{p(p-1)}{1!}(2x)^{p-2} + \cdots + (-1)^r \frac{p(p-1)\cdots(p-2r+1)}{r!}(2x)^{p-2r}\right.$$

$$\left. +\cdots+(-1)^{p-1/2}\frac{(p+1)!}{(p+1)/2!}x\right\} \qquad (7.176)$$

Thus, we have Hermite polynomials of degree p, p being a positive integer.

$$H_p(x) = \sum_{r=0}^{m}(-1)^r \frac{p!}{r!(p-2r)!}(2x)^{p-2r}$$

where

$$m = \begin{cases} p/2 \text{ if } p \text{ is even} \\ \dfrac{1}{2}(p-1) \text{ if } p \text{ odd} \end{cases}$$

Generating Function of Hermite Polynomials

The generating function for Hermite polynomials is

$$G(x, t) = e^{-t^2+2tx} = \sum_{p=0}^{\infty} \frac{H_p(x)}{p!} t^p \tag{7.177}$$

We have
$$G(x, t) = e^{2tx} \cdot e^{-t^2} = \sum_{r=1}^{\infty} \frac{(2tx)^r}{r!} \sum_{s} \frac{(-t^2)^s}{s!}$$

$$= \sum_{r,s} (-1)^s \frac{(2x)^r (t)^{r+2s}}{r!s!}$$

The coefficient of t^p (for fixed value of s) on R.H.S is obtained by putting $r + 2s = p$, i.e. $r = p - 2s$ and is given by

$$(-1)^s \frac{(2x)^{p-2s}}{(p-2s)!s!}$$

The total coefficient of t^p is obtained by summing the overall allowed values of s and since $r = p - 2s$

Thus if p is even, s goes from 0 to $p/2$ and if p is odd s goes from 0 to $\frac{p-1}{2}$.

$$\therefore \text{ Coefficient of } t^p = \sum_{s=0}^{p/2} (1)^s \frac{(2x)^{p-2s}}{(p-2s)!s!} = \frac{H_p(x)}{p!}$$

Thus the function $e^{-t^2+2tx} = e^{\{x^2-(t-x)^2\}}$ generates all the Hermite polynomials and hence it is called the *generating function of Hermite polynomials*.

Recurrence Formulae for Hermite Polynomials

We have

$$e^{2xt-t^2} = \sum_{n=0}^{\infty} \frac{H_n(x)}{n!} t^n \tag{7.178}$$

i. Differentiating w.r.t. x we have

$$2te^{2xt-t^2} = 2\sum_{h=0}^{\infty} \frac{H_n(x)}{n!} t^{n+1} = \sum_{n=0}^{\infty} \frac{H_n'(x)t^n}{n!}$$

Equating the coefficients of t^n on both sides we have

$$\frac{2H_{n-1}(x)}{(n-1)!} = \frac{H_n'(x)}{n!} \quad \text{or } H_n'(x) = 2nH_{n-1}(x) \tag{7.179}$$

ii. Differentiating Eq. (7.178) w.r.t. t we have

$$\sum_{n=1}^{\infty} \frac{H_n(x)}{(n-1)!} t^{n-1} = (2x - 2t)e^{2xt-t^2}$$

$$= 2x \sum_{n=0}^{\infty} \frac{H_n(x)}{n!} t^n - 2t \sum_{n=0}^{\infty} \frac{H_n(x)}{n!} t^n$$

or $$2x\sum_{n=0}^{\infty}\frac{H_n(x)}{n!}t^n = 2\sum_{n=0}^{\infty}\frac{H_n(x)}{n!}t^{n+1} + \sum_{n=1}^{\infty}\frac{H_n(x)}{(n-1)!}t^{n-1}$$

Equating the coefficients of t^n on both the sides, we have

$$2x\frac{H_n(x)}{n!} = 2\frac{H_{n-1}(x)}{(n-1)!} + \frac{H_{n+1}(x)}{n!}$$

or $$2xH_n(x) = 2nH_{n-1}(x) + H_{n+1}(x) \qquad (7.180)$$

iii. Subtracting Eq. (7.179) from Eq. (1.80), we have

$$H_n'(x) = 2xH_n(x) - H_{n+1}(x) \qquad (7.181)$$

iv. Since $H_n(x)$ is the solution of Hermite differential equation

$$Y'' - 2xY' + 2nY = 0$$

$\therefore$ $$H_n''(x) - 2xH_n'(x) + 2nH_n(x) = 0 \qquad (7.182)$$

Relations Eqs (7.172)–(7.182) are known as *recurrence relations for Hermite polynomials.*

Rodrigue's Formula for Hermite Polynomials

From generating function we have

$$e^{x^2-(t-x)^2} = \sum_{n=0}^{\infty}\frac{H_n(x)}{n!}t^n$$

Differentiating n times w.r.t. t, we have

$$e^{x^2}\frac{d^n}{dt^n}e^{-(t-x)^2} = H_n(x) + H_{n+1}(x)t + \cdots \qquad \ldots(7.183)$$

But we have

$$\frac{d}{dt}e^{-(t-x)^2} = -2(t-x)e^{-(t-x)^2}$$

and $$\frac{d}{dx}e^{-(t-x)^2} = 2(t-x)e^{-(t-x)^2}$$

$\therefore$ $$\lim_{t\to 0}\frac{d}{dt}e^{-(t-x)^2} = 2xe^{-(t-x)^1}\lim_{t\to 0}(-1) = \frac{d}{dx}e^{-(t-x)^2}$$

Similarly $$\lim_{t\to 0}\frac{d^2}{dt^2}e^{-(t-x)^2} = (-1)^2\lim_{t\to 0}\frac{d^2}{dx^2}e^{-(t-x)^2}$$

In general $$\lim_{t\to 0}\frac{d^n}{dt^n}e^{-(t-x)^2} = (-1)^n\lim_{t\to 0}\frac{d^n}{dx^n}e^{-(t-x)^2}$$

Taking limit as $t \to 0$ we have from Eq. (7.183)

$$\lim_{t\to 0}e^{x^2}\frac{d^n}{dt^n}e^{-(t-x)^2} = (-1)^n e^{x^2}\lim_{t\to 0}\frac{d^n}{dx^n}e^{-(t-x)^2} = H_n(x)$$

or
$$(-1)^n e^{x^2} \frac{d^n}{dx^n} e^{-x^2} = H_n(x)$$

or
$$H_n(x) = (-1)^n e^{x^2} \frac{d^n}{dx^n} (e^{-x^2}) \qquad (7.184)$$

which is Rodrigue's differential form of Hermite polynomial.

Integral form of Hermite Polynomial

Let us assume

$$Y_n = \frac{1}{2\pi i} \oint t^{-n-1} e^{\{x^2 - (t-x)^2\}} dt \qquad (7.185)$$

where the contour is taken around a circle having centre at origin.

Differentiating Eq. (7.185) w.r.t x we have

$$\frac{dY_n}{dx} = \frac{1}{2\pi i} \oint 2t^{-n} e^{\{x^2 - (t-x)^2\}} dt$$

and
$$\frac{d^2Y_n}{dx^2} = \frac{1}{2\pi i} \oint 4t^{-n+1} e^{\{x^2 - (t-x)^2\}} dt$$

and
$$Y_n'' - 2xY_n' + 2nY_n = \frac{1}{2\pi i} \oint 4t^{-n+1} e^{\{x^2 - (t-x)^2\}} dt$$

$$- \frac{2x}{2\pi i} \oint 2t^{-n} e^{\{x^2 - (t-x)^2\}} dt + \frac{2n}{2\pi i} \oint t^{-n-1} e^{\{x^2 - (t-x)^2\}} dt$$

$$= \frac{1}{2\pi i} \oint (4t^2 - 4xt + 2n) t^{-1-n} e^{\{x^2 - (t-x)^2\}} dt$$

$$= \frac{-2}{2\pi i} \oint \frac{d}{dt} \left[t^{-n} e^{\{x^2 - (t-x)^2\}} \right] dt$$

The value of integral can be solved by contour integration method and comes out to be zero.

∴
$$Y_n'' - 2xY_n' + 2nY_n = 0$$

which is the Hermite equation, and hence the value of Y_n given by Eq. (7.185) is also a solution of Hermite equation.

So, we have
$$H_n(x) = cY_n(x) \qquad (7.186)$$

where c is a constant.

Putting $x = 0$ in Eq. (7.175), we have

$$H_n(0) = (-1)^{n/2} \frac{n!}{(n/2)!} \qquad (7.187)$$

and from Eq. (7.185), we have

$$Y_n(0) = \frac{1}{2\pi i} \oint t^{-n-1} e^{-t^2} dt = \frac{(-1)^{n/2}}{(n/2)!} \qquad (7.188)$$

Comparison between Eqs (7.186), (7.187) and (7.188) gives

$$c = n!$$

Hence

$$H_n(x) = \frac{n!}{2\pi i} \oint t^{-n-1} e^{\{x^2 - (t-x)^2\}} dt$$

Orthogonality of Hermite Polynomial

The orthogonal property of Hermite polynomials is

$$\int_{-\infty}^{+\infty} e^{-x^2} H_n(x) H_m(x) dx = 2^n n! \sqrt{\pi} \, \delta_{mn} \tag{7.189}$$

where δ_{mn} is Kronecker delta such that $\delta_{mn} = 1$ for $m = n$ and 0 for $m \neq n$.

Since $H_n(x)$ is the solution of Hermite equation of order n, we have

$$H_n''(x) - 2x H_n'(x) + 2n H_n(x) = 0$$

Multiplying throughout by e^{-x^2}, we have

$$e^{-x^2} H_n''(x) - 2x e^{-x^2} H_n'(x) + 2n e^{-x^2} H_n(x) = 0$$

or

$$\frac{d}{dx}\left[e^{-x^2} H_n'(x)\right] + 2n e^{-x^2} H_n(x) = 0 \tag{7.189a}$$

Similarly, we have

$$\frac{d}{dx}\left[e^{-x^2} H_m'(x)\right] + 2m e^{-x^2} H_m(x) = 0 \tag{7.189b}$$

Multiplying Eq. (7.189a) by $H_m(x)$ and Eq. (7.189b) by $H_n(x)$ and subtracting we have

$$2(n-m)e^{-x^2} H_m(x) H_n(x) = H_n(x)\frac{d}{dx}\left[e^{-x^2} H_m'(x)\right] - H_m(x)\frac{d}{dx}\left[e^{-x^2} H_n'(x)\right]$$

$$= \frac{d}{dx}\left[e^{-x^2}\left\{H_n(x) H_m'(x) - H_n'(x) H_m(x)\right\}\right]$$

Integrating w.r.t. x between limits $-\infty$ to $+\infty$ we have

$$2(n-m)\int_{-\infty}^{+\infty} e^{-x^2} H_m(x) H_n(x) dx = \left[e^{-x^2}\left\{H_n(x) H_m'(x) - H_n'(x) H_m(x)\right\}\right]_{-\infty}^{+\infty}$$

$$= 0 \qquad\qquad [\text{since } e^{-x^2} = 0 \text{ for } x = \pm\infty]$$

Thus for $m \neq n$ we have

$$\int_{-\infty}^{+\infty} e^{-x^2} H_m(x) H_n(x) dx = 0 \tag{7.190}$$

From generating function of Hermite polynomials, we have

$$e^{2xt - t^2} = \sum_{n=0}^{\infty} \frac{H_n(x)}{n!} t^n$$

and

$$e^{2xh - h^2} = \sum_{m=0}^{\infty} \frac{H_m(x)}{m!} h^m$$

Multiplying together by e^{-x^2}, we have

$$e^{-x^2}e^{2xt-t^2}\cdot e^{2xh-h^2} = \sum_{n=0}^{\infty}\sum_{m=0}^{\infty}\frac{H_n(x)t^n}{n!}\frac{H_m(x)h^m}{m!}e^{-x^2}$$

and integrating with respect to x from $-\infty$ to $+\infty$, we have

$$\int_{-\infty}^{+\infty}e^{-x^2}e^{2xt-t^2}e^{2xh-h^2}dx = \sum_{n=0}^{\infty}t^nh^n\int_{-\infty}^{+\infty}\frac{e^{-x^2}[H_n(x)]^2}{(n!)^2}dx$$

$$+\sum_{\substack{n=0\\m\neq n}}^{\infty}\sum_{m=0}^{\infty}t^nh^m\int_{-\infty}^{+\infty}\frac{e^{-x^2}[H_n(x)H_m(x)]}{m!n!}dx$$

The last term on RHS vanishes by virtue of Eq. (7.190). Therefore, we have

$$\sum_{n=0}^{\infty}\frac{(th)^n}{(n!)^2}\int_{-\infty}^{+\infty}e^{-x^2}[H_n(x)]^2dx = \int_{-\infty}^{+\infty}e^{(-x^2+2xt-t^2+2xh-h^2)}dx$$

$$= \int_{-\infty}^{+\infty}e^{-(x-t-h)^2}e^{2ht}dx = \sqrt{\pi}\,e^{2ht}$$

$$\sum_{n=0}^{\infty}\frac{(th)^n}{(n!)^2}\int_{-\infty}^{+\infty}e^{-x^2}[H_n(x)]^2 = \sqrt{\pi}\sum_{n=0}^{\infty}\frac{2^n(th)^n}{n!}$$

Equating the coefficients of $(th)^n$ on both sides, we have

$$\int_{-\infty}^{+\infty}e^{-x^2}[H_n(x)]^2dx = n!\sqrt{\pi}\,2^n \qquad (7.191)$$

Combining Eqs (7.190) and (7.191), we have

$$\int_{-\infty}^{+\infty}e^{-x^2}H_n(x)H_m(x)dx = 2^nn!\sqrt{\pi}\,\delta_{mn},$$

and the function is defined by

$$y = e^{-x^2/2}H_n(x)$$

is called *Hermite (orthogonal) function*.

___

Example 1. *Prove that for $m < n$*

$$\frac{d^m}{dx^m}H_n(x) = \frac{2^mn!}{(n-m)!}H_{n-m}(x)$$

Solution: We have

$$\sum_{n=0}^{\infty}\frac{H_n(x)}{n!}t^n = e^{-t^2+2tx}$$

Differentiating partially w.r.t. x, m times, we have

$$\sum_{n=0}^{\infty}\frac{t^n}{n!}\frac{d^m}{dx^m}H_n(x) = \frac{d^m}{dx^m}\{e^{-t^2+2tx}\} = (2t)^me^{-t^2+2tx}$$

$$= (2t)^m\sum_{n=0}^{\infty}\frac{H_n(x)}{n!}t^n = 2^m\sum_{n=0}^{\infty}\frac{H_n(x)}{n!}t^{n+m}$$

Equating the coefficients of t^n (by putting $n = n - m$ in R.H.S., we have

$$\frac{1}{n!}\frac{d^m}{dx^m}H_n(x) = \frac{2^m H_{n-m}(x)}{(n-m)!}$$

or

$$\frac{d^m}{dx^m}H_n(x) = \frac{2^m \cdot n!}{(n-m)!}H_{n-m}(x)$$

Example 2. If $y_n = e^{-x^2/2}H_n(x)$, i.e. Hermite orthogonal function and

$$I_{mn} = \int y_m(x)y_n(x)dx$$

then prove that $\qquad I_{mn} = 2nI_{(n-1)(n-1)}.$

Solution. From recurrence Eq. (7.180) we have

$$2xH_n(x) = 2nH_{n-1}(x) + H_{n+1}(x)$$

Multiplying both sides by $e^{-x^2/2}$ we have

$$2xy_n(x) = 2ny_{n-1}(x) + y_{n+1}(x)$$

Multiplying this equation by $y_{n-1}(x)$ and integrating with respect to x from $-\infty$ to $+\infty$ we have

$$\int_{-\infty}^{+\infty} 2xy_n(x)y_{n-1}(x)dx = 2n\int_{-\infty}^{+\infty} y_{n-1}(x)y_{n-1}(x)dx + \int_{-\infty}^{+\infty} y_{n+1}(x)y_{n-1}(x)dx$$

$$= 2nI_{(n-1),(n-1)} \qquad (1)$$

Since $\qquad I_{(n+1)(n-1)} = 0$

From Rodrigue's formula we have

$$H_n(x) = (-1)^n e^{x^2}\frac{d^n}{dx^n}(e^{-x^2})$$

or

$$y_n(x) = (-1)^n e^{x^2/2}\frac{d^n}{dx^n}(e^{-x^2}) \qquad (2)$$

Substituting this value of $y_n(x)$ in Eq. (1), we have

$$\int_{-\infty}^{+\infty} 2x(-1)^n e^{x^2/2}\frac{d^n}{dx^n}(e^{-x^2})(-1)^{n-1}e^{x^2/2}\frac{d^{n-1}}{dx^{n-1}}(e^{-x^2})dx = 2nI_{n-1,n-1}$$

or

$$-\int_{-\infty}^{+\infty} 2xe^{x^2}\frac{d^n}{dx^n}(e^{-x^2})\frac{d^{n-1}}{dx^{n-1}}(e^{-x^2})dx = 2nI_{n-1,n-1}$$

Integrating L.H.S by parts taking $(2xe^{x^2})$ as second function

and $\dfrac{d^n}{dx^n}(e^{-x^2})\dfrac{d^{n-1}}{dx^{n-2}}(e^{-x^2})$ as first function

we have

$$\left[-e^{x^2}\frac{d^n}{dx^n}(e^{-x^2})\cdot\frac{d^{n-1}}{dx^{n-1}}(e^{-x^2})\right]_{-\infty}^{+\infty}+\int_{-\infty}^{+\infty}e^{+x^2}\frac{d}{dx}\left[\frac{d^n}{dx^n}(e^{-x^2})\cdot\frac{d^{n-1}}{dx^{n-1}}(e^{-x^2})\right]dx$$

$$=\int_{-\infty}^{+\infty}e^{x^2}\left[\frac{d^{n+1}}{dx^{n+1}}(e^{-x^2})\cdot\frac{d^{n-1}}{dx^{n-1}}(e^{-x^2})+\frac{d^n}{dx^n}(e^{-x^2})\cdot\frac{d^n}{dx^n}(e^{-x^2})\right]dx$$

$$=\int_{-\infty}^{+\infty}e^{x^2}\frac{d^{n+1}}{dx^{n+1}}(e^{-x^2})\cdot\frac{d^{n-1}}{dx^{n-1}}(e^{-x^2})dx+\int_{-\infty}^{+\infty}e^{x^2}\frac{d^n}{dx^n}(e^{-x^2})\cdot\frac{d^n}{dx^n}(e^{-x^2})dx$$

$$=\int_{-\infty}^{+\infty}(-1)^{n+1}y_{n+1}(x)(-1)^{n-1}y_{n-1}(x)dx+\int_{-\infty}^{+\infty}(-1)^n y_n(x)(-1)^n y_n(x)dx$$

By using Eq. (2).

$\therefore$ L.H.S $= I_{n+1,\,n-1} + I_{n;\,n}$

R.H.S $= 2nI_{(n-1),\,(n-1)}$

or $I_{nn} = 2nI_{n-1,\,n-1}$

Since $I_{(n+1)(n-1)} = 0$

Hence the result.

Example 3. Prove that $H_n(-x) = (-1)^n H_n(x)$

Solution: We have

$$H_n(x) = \sum_{r=0}^{n/2}\frac{(-1)^r n!(2x)^{n-2r}}{(n-2r)!r!}$$

and

$$H_n(-x) = \sum_{r=0}^{n/2}\frac{(-1)^r n!(-2x)^{n-2r}}{(n-2r)!r!}$$

$$=\sum_{r=0}^{n/2}\frac{(-1)^r(-1)^{n-2r} n!(2x)^{n-2r}}{(n-2r)!r!}$$

$$=(-1)^n\sum_{r=0}^{n/2}\frac{(-1)^r n!(2x)^{n-2r}}{(n-2r)!r!}=(-1)^n H_n(x)$$

7.11. LAGUERRE'S DIFFERENTIAL EQUATION

The differential equation

$$xy'' + (1-x)y' + ny = 0 \tag{7.192}$$

where n is constant, is known as Laguerre's differential equation. This equation has removable singularity at $x = 0$ hence its solution can be found using series integration method. Let the series solution of above equation be

$$y = \sum_{r=0}^{\infty}a_r x^{m+r} \tag{7.193}$$

Substitution of Eq. (7.193) in Eq. (7.192) yields

$$\sum_{r=0}^{\infty} a_r \left[(m+r)^2 x^{m+r-1} - (m+r-n) x^{m+r} \right] = 0 \qquad (7.194)$$

The indicial equation then is

$$a_0 m^2 = 0$$

giving $m = 0$. Since a_0 is taken as arbitrary constant.

Equating the coefficient x^{m+r} equal to zero, we have

$$a_{r+1}(m+r+1)^2 = (m+r-n) a_r$$

or

$$a_{r+1} = \frac{m+r-n}{(m+r+1)^2} a_r$$

and since $m = 0$

$$a_{r+1} = \frac{r-n}{(r+1)^2} a_r$$

This is recurrence relation for the coefficients a_r.

Substituting $r = 0, 1, 2, 3$ etc. we get all the coefficients as:

$$a_1 = -na_0 = (-1)^1 na_0 = \frac{(-1)^1}{(1!)^2} na_0$$

$$a_2 = \frac{1-n}{2^2} a_1 = \frac{(-n)(1-n)}{(2)^2} a_0 = \frac{n(n-1)}{(2!)^2} a_0 = \frac{(-1)^2 n(n-1)}{(2!)^2} a_0$$

$$a_3 = \frac{(2-n)}{3^2} a_2 = \frac{(2-n)}{3^2} \frac{(-1)^2 n(n-1)}{(2!)^2} a_0$$

$$= \frac{(-1)^3 n(n-1)(n-2)}{(3!)^2} a_0$$

$$\vdots \qquad \vdots \qquad \vdots$$

$$a_r = \frac{(-1)^r n(n-1)(n-2)\cdots(n-r+1)}{(r!)^2} a_0$$

$$\therefore \quad y = \sum_{r=0}^{\infty} a_r x^r = a_0 \left[1 - nx + \frac{n(n-1)}{(2!)^2} x^2 + \cdots + \frac{(-1)^r n(n-1)\cdots(n-r+1)}{(r!)^2} x^r + \cdots \right] \qquad (7.195)$$

If n is a positive integer, the series terminates when $r = n + 1$ and if we put $a_0 = n!^*$, then solution y representing Eq. (7.195) becomes Laguerre polynomial $L_n(x)$.

$$L_n(x) = (-1)^n \left[x^n - \frac{x^2}{1!} x^{n-1} + \frac{x^2(n-1)^2}{2!} x^{n-2} + \cdots + (-1)^n n! \right]$$

---

[*] Some author use $a_0 = 1$ and Leguerre polynomial is

$$L_n(x) = \sum_{r=0}^{n} \frac{(-1)^r n!}{(r!)^2 (n-r)!} x^r$$

The two expressions merely differ by a constant factor.

$$= \sum_{r=0}^{n} \frac{(-1)^r (n!)^2}{(r!)^2 (n-r)!} x^r \tag{7.196}$$

Thus a Laguerre's polynomial is the solution of Laguerre differential equation.

Generating Function for Laguerre Polynomials

The generating function for Laguerre polynomials is

$$f(x, t) = \frac{e^{-xt/(1-t)}}{1-t} = \sum_{n=0}^{\infty} \frac{L_n(x)t^n}{n!} \quad \text{for } |t| < 1 \tag{7.197}$$

We have

$$(1-t)^{-1} e^{-xt/(1-t)} = \sum_{r=0}^{\infty} \frac{(-1)^r \cdot t^r x^r}{(1-t)^{r+1}} \frac{1}{r!}$$

$$= \sum_{r=0}^{\infty} \frac{(-1)^r}{r!} t^r x^r (1-t)^{-(r+1)}$$

$$= \sum_{r=0}^{\infty} \frac{(-1)^r}{r!} x^r t^r \sum_{s=0}^{\infty} \frac{(r+1)(r+2)\cdots(r+s)}{s!} t^s$$

$$= \sum_{r=0}^{\infty} \frac{(-1)^r}{r!} x^r t^r \sum_{s=0}^{\infty} \frac{(r+s)!}{r!s!} t^s$$

$$= \sum_{r=0}^{\infty} \frac{(-1)^r (r+s)!}{(r!)^2 s!} x^r t^{s+r} \tag{7.198}$$

The coefficient of t^n (for fixed value of r) on RHS is obtained by putting $r + s = n$, i.e. $s = n - r$ and is given by

$$(-1)^r \frac{n!}{(r!)^2 (n-r)!} x^r$$

The net coefficient of t^n is obtained by summing over all allowed values of r.
As $s = n - r$ and $s \geq 0$ $\therefore n - r \geq 0$ or $r \leq n$.
Hence net coefficient of t^n on RHS of Eq. (7.198) is

$$\sum_{r=0}^{\infty} \frac{(-1)^r n!}{(r!)^2 (n-r)} x^r = \frac{L_n(x)}{n!}$$

Hence we write

$$\frac{e^{-xt/(1-t)}}{1-t} = \sum_{n=0}^{\infty} \frac{L_n(x)}{n!} t^n$$

Thus the function $\dfrac{e^{-xt/(1-t)}}{1-t}$ generates all Laguerre's polynomials and hence it is called the *generating function of Laguerre polynomials*.

Rodrigue's Differential Formula for Laguerre Polynomials

Differentiating the generating function Eq. (7.197) n times w.r.t. t, we have

$$\frac{d^n}{dt^n}\frac{e^{-xt/1-t}}{1-t} = \frac{d^n}{dt^n}\left\{(1-t)^{-1}e^{\left(1-\frac{1}{1-t}\right)x}\right\}$$

$$= e^x\frac{d^n}{dt^n}\left\{(1-t)^{-1}e^{-x/1-t}\right\}$$

$$= \frac{d^n}{dt^n}\sum_{n=0}^{\infty}\frac{L_n(x)t^n}{n!}$$

or

$$e^x\frac{d^n}{dt^n}\left\{(1-t)^{-1}e^{-x/1-t}\right\} = L_n(x)+L_{n+1}(x)t+\cdots \qquad (7.199)$$

But

$$\frac{\partial}{\partial t}\left\{(1-t)^{-1}e^{-x/1-t}\right\} = \frac{1-x-t}{(1-t)^3}e^{-x/1-t}$$

$$\lim_{t\to 0}\frac{\partial}{\partial t}\left\{(1-t)^{-1}e^{-x/1-t}\right\} = (1-x)e^{-x}\frac{\partial}{\partial x}(xe^{-x})$$

Similarly

$$\lim_{t\to 0}\frac{\partial^2}{\partial t^2}\left\{(1-t)^{-1}e^{-x/1-t}\right\} = \frac{\partial^2}{\partial x^2}(x^2e^{-x})$$

and

$$\lim_{t\to 0}\frac{\partial^n}{\partial t^n}\left\{(1-t)^{-1}e^{-x/1-t}\right\} = \frac{\partial^n}{\partial x^n}(x^ne^{-x}) \qquad (7.200)$$

Taking the limits as $t\to 0$, Eq. (7.199) in view of Eq. (7.200) became

$$L_n(x) = e^x\frac{d^n}{dx^n}(x^ne^{-x}) \qquad (7.201)$$

which is Rodrigue's representation of Laguerre's polynomial.

Integral Formula of Laguerre Polynomials

Let us assume the integral

$$y_n = \frac{1}{2\pi i}\oint\frac{t^{-n-1}}{1-t}e^{-xt/(1-t)}dt \qquad (7.202)$$

So that

$$y_n' = \frac{1}{2\pi i}\oint\frac{t^{-n}}{(1-t)^2}e^{-xt/(1-t)}dt$$

and

$$y_n'' = \frac{1}{2\pi i}\oint\frac{t^{-n+1}}{(1-t)^3}e^{-xt/(1-t)}dt$$

Laguerre's polynomial is the solution of the differential equation

$$xy'' + (1-x)y' + ny = 0$$

Substituting the values of y_n, y_n' and y_n'', we have L.H.S.

$$\frac{1}{2\pi i}\oint\left[\frac{xt^2}{(1-t)^2}-\frac{(1-x)t}{(1-t)}+n\right]\frac{t^{-n-1}}{1-t}e^{-xt/(1-t)}dt = \frac{1}{2\pi i}\oint\frac{d}{dt}\left[\frac{t^{-n}}{1-t}e^{-xt/(1-t)}\right]dt$$

which is zero since the quantity in brackets takes same values at the initial and final points of the contour.

i.e. $\qquad xy_n'' + (1-x)y_n' + ny_n = 0.$

Hence y_n represents the solution of Laguerre's equation and hence we have

$$L_n(x) = cy_n(x)$$

where c is a constant. If we put $x = 0$, then we have

$$L_n(0) = n!$$

and $\qquad\qquad y_n(0) = \dfrac{1}{2\pi i}\displaystyle\int \dfrac{t^{-n-1}}{1-t}\,dt = 1^*$

i.e. $\qquad\qquad c = n!$

$\therefore \qquad\qquad L_n(x) = \dfrac{n!}{2\pi i}\displaystyle\oint \dfrac{t^{-n-1}}{1-t}\,e^{-xt/(1-t)}\,dt$

Recurrence Formula for Laguerre Polynomials

i. Differentiating the generating function w.r.t. t we have

$$\dfrac{(1-x-t)}{(1-t)^2}e^{-xt/(1-t)} = \sum_{n=1}^{\infty}\dfrac{L_n(x)t^{n-1}}{(n-1)!}$$

or $\qquad (1-x-t)\displaystyle\sum_{n=0}^{\infty}\dfrac{L_n(x)t^n}{n!} = (1-t)^2\sum_{n=1}^{\infty}\dfrac{L_n(x)t^{n-1}}{(n-1)!}$

Equating the coefficients of t^n, we have

$$(1-x)\dfrac{L_n(x)}{n!} - \dfrac{L_{n-1}(x)}{(n-1)!} = \dfrac{L_{n+1}(x)}{n!} - 2\dfrac{L_n(x)}{(n-1)!} + \dfrac{L_{n-1}(x)}{(n-2)!}$$

Multiplying throughout with $n!$ and rearranging the terms we have

$$(1 + 2n - x)\,L_n(x) = n^2 L_{n-1}(x) + L_{n+1}(x) \qquad\qquad (7.203)$$

ii. Differentiating the generating function w.r.t. x we have

$$-(1-t)^{-1}\dfrac{t}{1-t}e^{-xt/(1-t)} = \sum_{n=0}^{\infty}\dfrac{L_n'(x)}{n!}t^n$$

$$-\sum_{n=0}^{\infty}\dfrac{L_n(x)}{n!}t^n = (1-t)\sum_{n=0}^{\infty}\dfrac{L_n'(x)}{n!}t^n$$

Equating the coefficients of t^n we have

$$-\dfrac{L_{n-1}(x)}{(n-1)!} = \dfrac{L_n'(x)}{n!} - \dfrac{L_{n-1}'(x)}{(n-1)!}$$

or $\qquad\qquad L_n'(x) = nL_{n-1}'(x) - nL_{n-1}(x)$

$$L_n'(x) = n\{L_{n-1}'(x) - L_{n-1}(x)\} \qquad\qquad (7.204)$$

---

* The integral is evaluated by the method of contour integration as in Chapter 6—*Complex Variables*.

iii. Differentiating Eq. (7.203) w.r.t. x we have

$$(1+2n-x)L_n'(x)-L_n(x) = n^2 L_{n-1}'(x)+L_{n+1}'(x) \tag{7.205}$$

Replacing n by $n+1$ in Eq. (7.204) we have

$$L_{n+1}'(x) = (n+1)L_n'(x)-(n+1)L_n(x) \tag{7.206}$$

Substituting the values of $L_{n+1}'(x)$ and $L_{n-1}'(x)$ from Eqs (7.206), (7.204) in (7.205) we have

$$(1+2n-x)L_n'(x)-L_n(x) = n^2 L_{n-1}(x)+nL_n'(x)+(n+1)L_n'(x)-(n+1)L_n(x)$$

or

$$xL_n(x) = nL_n(x) - n^2 L_{n-1}(x) \tag{7.207}$$

iv. We have the generating function as

$$\frac{1}{1-t}e^{-xt/1-t} = \sum_{n=0}^{\infty}\frac{L_n(x)}{n!}t^n$$

Differentiating w.r.t. x we have

$$\sum_{n=0}^{\infty}\frac{L_n'(x)}{n!}t^n = \left(-\frac{t}{1-t}\right)\frac{1}{1-t}e^{-xt/1-t} = -t(1-t)^{-1}\sum_{r=0}^{\infty}\frac{L_r(x)t^r}{r!}$$

$$= (-t)(1-t)^{-1}\sum_{r=0}^{\infty}\frac{L_r(x)}{r!}t^r = -t\sum_{s=0}^{\infty}t^s\sum_{r=0}^{\infty}\frac{L_r(x)}{r!}t^r$$

or

$$\sum_{n=0}^{\infty}\frac{L_n'(x)t^n}{n!} = -\sum_{r,s=0}^{\infty}\frac{L_r(x)t^{r+s+1}}{r!}$$

For the coefficient of t^n and for fixed value of r, we have

$$r+s+1 = n$$

The net coefficient of t^n on RSH is obtained by summing over all allowed values of r.

$$s = n-r-1 \text{ and } s \geq 0$$

$\therefore$

$$n-r-1 \geq 0 \text{ or } r \leq n-1$$

$\therefore$ Equating coefficients of t^n, we have

$$\frac{L_n'(x)}{n!} = -\sum_{r=0}^{n-1}\frac{L_r(x)}{r!}$$

or

$$L_n'(x) = n!\sum_{r=0}^{n-1}\frac{L_r(x)}{r!} \tag{7.208}$$

Equations (7.203), (7.204), (7.207) and (7.208) are called *recurrence relations for Laguerre polynomials*.

Orthogonal Property of Laguerre Polynomials

The Laguerre polynomials do not form an orthogonal set themselves because the Laguerre's differential equation is not self-adjoint. However, the related set of functions

$$\phi_n(x) = \frac{1}{n!}e^{-x/2}L_n(x) \tag{7.209}$$

form an orthogonal set for the interval $0 \le x \le \infty$, i.e.

$$\int_0^\infty \phi_m(x)\phi_n(x)dx = \delta_{mn} \qquad (7.210)$$

To prove it, we have

$$L_n(x) = e^x \frac{d^n}{dx^n}(x^n e^{-x})$$

or

$$e^{-x} x^m L_n(x) = x^m \frac{d^n}{dx^n}(x^n e^{-x})$$

On integrating, we have

$$\int_0^\infty e^{-x} x^m L_n(x)dx = \int_0^\infty x^m \frac{d^n}{dx^n}(x^n e^{-x})dx$$

$$= x^m \frac{d^{n-1}(x^n e^{-x})}{dx^{n-1}}\Big|_0^\infty - \int_0^\infty mx^{m-1} \frac{d^{n-1}}{dx^{n-1}}(x^n e^{-x})dx$$

On substituting the limits, the first part vanishes. And now integrating m times we have

$$\int_0^\infty e^{-x} x^m \frac{d^n}{dx^n}(x^n e^{-x})dx = (-1)^m m! \int_0^\infty \frac{d^{n-m}}{dx^{n-m}}(x^n e^{-x})dx$$

$$= 0 \qquad [\text{If } n > m] \qquad (7.211)$$

Similarly,

$$\int_0^\infty e^{-x} x^n \frac{d^n}{dx^n}(x^m e^{-x})dx = 0 \qquad [\text{If } n < m] \qquad (7.212)$$

But $L_n(x)$ is polynomial of degree n in x and $L_m(x)$ is polynomial of degree m in x. Hence,

$$\int_0^\infty e^{-x} L_m(x)L_n(x)dx = 0, \quad \text{for } m > n \text{ and for } m < n$$

For $m = n$

$$\int_0^\infty e^{-x} \{L_n(x)\}^2 dx = \int_0^\infty \begin{bmatrix} (-1)^n e^{-x} x^n + e^{-x} \times \text{functions involving} \\ \text{powers } x \text{ other than } n \end{bmatrix} L_n(x)dx \qquad (7.213)$$

From Rodrigue's formula, we have

$$e^{-x} L_n(x) = \frac{d^n}{dx^n}(x^n e^{-x}) = (-1)^n x^n e^{-x} + \text{functions involving powers of } x \text{ other than } n.$$

But $\int_0^\infty e^{-x} x^m L_n(x)dx = 0$. From Eqs (7.211) and (2.212).

Hence all the terms except first term on RHS of Eq. (7.213) are zero. Hence

$$\int_0^\infty e^{-x} \{L_n(x)\}^2 dx = (-1)^n \int_0^\infty e^{-x} x^n L_n(x)dx$$

But

$$e^{-x} L_n(x) = \frac{d^n}{dx^n}(x^n e^{-x})$$

We have

$$\int_0^\infty e^{-x} \{L_n(x)\}^2 \, dx = (-1)^n \int_0^\infty x^n \frac{d^n}{dx^n}(x^n e^{-x}) \, dx$$

Integrating by parts we have

$$\int_0^\infty e^{-x} \{L_n(x)\}^2 \, dx = (-1)^n x^n \frac{d^{n-1}}{dx^{n-1}}(x^n e^{-x}) \Big|_0^\infty$$

$$+ (-1)(-1)^n \int_0^\infty n x^{n-1} \frac{d^{n-1}}{dx^n}(x^n e^{-x}) \, dx$$

First term vanishes for both limits.

$$\therefore \qquad \int_0^\infty e^{-x} \{L_n(x)\}^2 \, dx = (-1)^n (-1) \int_0^\infty n x^{n-1} \frac{d^{n-1}}{dx^{n-1}}(x^n e^{-x}) \, dx$$

$$= (-1)^n (-1)^n n! \int_0^\infty x^n e^{-x} \, dx$$

$$= (n!)^2.$$

or $$\int_0^\infty \frac{e^{-x/2}}{n!} L_n(x) \frac{e^{-x/2}}{n!} L_n(x) \, dx = 1 \qquad (7.214)$$

Combining Eqs (7.209), (7.211), (7.212) and (7.214) we have

$$\int_0^\infty \phi_m(x) \phi_n(x) \, dx = \delta_{mn}$$

7.12. ASSOCIATED LAGUERRE DIFFERENTIAL EQUATION AND ASSOCIATED LAGUERRE POLYNOMIALS

The differential equation

$$x \frac{d^2 y}{dx^2} + (p + 1 - x) \frac{dy}{dx} + ny = 0 \qquad (7.215)$$

is called the associated Laguerre differential equation. If u is the solution for Laguerre equation of order $(n + p)$, then we may substitute u for y in Laguerre equation of order $(n + p)$, and we have

$$x \frac{d^2 u}{dx^2} + (1 - x) \frac{du}{dx} + (n + p)u = 0 \qquad (7.216)$$

Differentiating Eq. (7.216) p times with respect to x by Leibnitz's theorem, we have

$$\left[x \frac{d^{p+2} u}{dx^{p+2}} + p \cdot \frac{d^{p+1} u}{dx^{p+1}} \right] + \left[(1 - x) \frac{d^{p+1} u}{dx^{p+1}} - p \cdot \frac{d^p u}{dx^p} \right] + (n + p) \frac{d^p u}{dx^p} = 0$$

or $$x \frac{d^{p+2} u}{dx^{p+2}} + (p + 1 - x) \frac{d^{p+1} u}{dx^{p+1}} + n \frac{d^p u}{dx^p} = 0 \qquad (7.217)$$

which shows that $y = \dfrac{d^p u}{dx^p}$ satisfies associated Laguerre equation.

And if we denote the solution of Laguerre equation of order $(n + p)$ by L_{n+p} then the associated Laguerre polynomial is defined by

$$L_n^p(x) = (-1)^p \frac{d^p}{dx^p} \left[L_{n+p}(x) \right] \tag{7.218}$$

We have
$$L_n(x) = \sum_{r=0}^{n} \frac{(-1)^r (n!)^2}{(n-r)!(r!)^2} x^r$$

$\therefore$
$$L_{n+p}(x) = \sum_{r=0}^{n+p} \frac{(-1)^r \{(n+p)!\}^2}{(n+p-r)!(r!)^2} x^r$$

and differentiating p times with respect to x we have

$$\frac{d^p}{dx^p}\left[L_{n+p}(x) \right] = \frac{d^p}{dx^p} \sum_{r=0}^{n+p} \frac{(-1)^r \{(n+p)!\}^2}{(n+p-r)!(r!)^2} x^r$$

Since
$$\frac{d^p x^r}{dx^p} = 0 \quad \text{for } r < p$$

$\therefore$
$$\frac{d^p}{dx^p}\{L_{n+p}(x)\} = \sum_{r=p}^{n+p} \frac{(-1)^r \{(n+p)!\}^2}{(n+p-r)!(r!)^2} r\{(r-1)(r-2)\cdots(r-p+1)\} x^{r-p}$$

$$= \sum_{r=p}^{n+p} \frac{(-1)^r \{(n+p)!\}^2}{(n+p-r)!(r!)^2} \frac{r!}{(r-p)!} x^{r-p}$$

$$= \sum_{r=p}^{n+p} \frac{(-1)^r \{(n+p)!\}^2}{(n+p-r)!r!} \frac{1}{(r-p)!} x^{r-p}$$

Let $r - p = s$

$$\frac{d^p}{dx^p}\{L_{n+p}(x)\} = \sum_{s=0}^{n} (-1)^{p+s} \frac{\{(n+p)!\}^2}{(n-s)!(p+s)!s!} x^s = (-1)^p \sum_{s=0}^{n} (-1)^s \frac{\{(n+p)!\}^2}{(n-s)!(p+s)!s!} x^s$$

But
$$L_n^p(x) = (-1)^p \frac{d^p}{dx^p}\{L_{n+p}(x)\} = \sum_{s=0}^{n} (-1)^s \frac{\{(n+p)!\}^2}{(n-s)!(p+s)!s!} x^s \tag{7.219}$$

This equation represents the associated Laguerre polynomial in series form.

Generating Function for Associated Laguerre Polynomials

If $L_n(x)$ is the Laguerre polynomial of degree n, then we have

$$\frac{1}{1-t} e^{-xt/1-t} = \sum_{n=0}^{\infty} \frac{L_n(x)t^n}{n!}$$

Differentiating partially with respect to x, p times, we have

$$\frac{1}{1-t} \frac{d^p}{dx^p}\{e^{-xt/1-t}\} = \frac{d^p}{dx^p} \sum_{n=0}^{\infty} \frac{L_n(x)t^n}{n!}$$

Since $\dfrac{d^p}{dx^p} L_n(x) = 0$ for $0 \le n \le p$, hence putting $n = m + p$, we have

$$\frac{1}{1-t}\frac{d^p}{dx^p}\{e^{-xt/1-t}\} = \frac{d^p}{dx^p}\sum_{m=0}^{\infty}\frac{L_{m+p}(x)}{(m+p)!}t^{m+p}$$

$$= \sum_{m=0}^{\infty}\frac{t^{m+p}}{(m+p)!}\frac{d^p}{dx^p}L_{m+p}(x)$$

or

$$\frac{(-1)^p t^p}{(1-t)^{p+1}}e^{-xt/1-t} = \sum_{m=0}^{\infty}\frac{t^{m+p}}{(m+p)!}\frac{d^p}{dx^p}L_{m+p}(x)$$

or

$$\frac{e^{-xt/1-t}}{(1-t)^{p+1}} = \sum_{m=0}^{\infty}\frac{(-1)^p \cdot t^m}{(m+p)!}\frac{d^p}{dx^p}L_{m+p}(x)$$

$$= \sum_{m=0}^{\infty}\frac{L_m^p(x)t^m}{(m+p)!} \tag{7.220}$$

It is obvious that the function

$$f(x, t) = \frac{e^{-xt/1-t}}{(1-t)^{p+1}} \tag{7.221}$$

generates all the associated Laguerre's polynomials and hence it is called the *generating function of associated Laguerre polynomials*.

Rodrigues' Representation of Associated Laguerre Polynomials

$$L_n^p(x) = \frac{(n+p)!}{n!}e^x x^{-p}\frac{d^n}{dx^n}(e^{-x}x^{n+p}) \tag{7.222}$$

Differentiating the function $e^{-x} x^{n+p}$, n times with respect to x using Leibnitz theorem, we have

$$\frac{d^n}{dx^n}\{e^{-x}x^{n+p}\} = \left\{ x^{n+p}(-1)^n e^{-x} + n(n+p)x^{n+p-1}(-1)^{n-1}e^{-x}\right.$$

$$+ \frac{n(n-1)(n+p)(n+p-1)x^{n+p-2}}{2!}(-1)^{n-2}e^{-x} + \cdots$$

$$\left. + (-1)\frac{n(n+p)!}{(p+1)!}e^{-x}x^{p+1} + e^{-x}\frac{(n+p)!}{p!}x^p \right\}$$

$$= e^{-x}x^p n!\left\{ \frac{(-1)^n}{n!}x^n + \frac{(-1)^{n-1}(n+p)}{(n-1)!}x^{n-1}\right.$$

$$\left. + (-1)^{n-2}\frac{(n+p)(n+p-1)}{(n-2)!2!}x^{n-2} + \cdots + \frac{(n+p)!}{p!n!} \right\}$$

$$= e^{-x}x^p n! \left\{ \frac{(-1)^n \cdot (n+p)!}{(n-n)!(n+p)!n!}x^n \right.$$

$$+ \frac{(-1)^{n-1}(n+p)!x^{n-1}}{\{n-(n-1)\}!\{n+p-1\}!(n-1)!} + \cdots + \left. \frac{(n+p)!}{n!p!0!} \right\}$$

$$= e^{-x}x^p n! \sum_{r=0}^{\infty} \frac{(-1)^r \cdot (n+p)!}{(n-r)!(p+r)!r!}x^r$$

$$= \frac{e^{-x}x^p \cdot n!}{(n+p)!} \sum_{r=0}^{\infty} \frac{(-1)^r \cdot \{(n+p)!\}^2}{(n-r)!(p+r)!r!}x^r$$

$$= \frac{e^{-x}x^p \cdot n!}{(n+p)!} L_n^p(x) \text{ from Eq. (7.219)}$$

or $$L_n^p(x) = \frac{(n+p)!}{n!}e^x x^{-p} \frac{d^n}{dx^n}\left\{e^{-x}x^{n+p}\right\}$$

The is called *Rodrigues' formula for associated Laguerre polynomials.*

Orthogonal Property of Associated Laguerre Polynomials

The associated Laguerre's polynomials do not form an orthogonal system. However, the related set of functions

$$\phi_n(x) = \frac{e^{-x/2}x^{p/2}L_n^p(x)}{(n+p)!} \tag{7.223}$$

form an orthogonal set for the interval $0 \le x \le \infty$, i.e.

$$\int_0^\infty \frac{e^{-x}x^p L_n^p(x)L_m^p(x)}{(n+p)!(m+p)!}dx = \frac{(n+p)!}{n!}\delta_{mn} \tag{7.224}$$

To prove, we have generating function

$$\sum_{m=0}^{\infty} \frac{L_m^p(x)t_1^m}{(m+p)!} = \frac{1}{(1-t_1)^{p+1}}e^{-t_1x/1-t_1}$$

and $$\sum_{n=0}^{\infty} \frac{L_n^p(x)t_2^n}{(n+p)!} = \frac{1}{(1-t_2)^{p+1}}e^{-t_2x/(1-t_2)}$$

Multiplying, we have

$$\sum_{n,m=0}^{\infty} \frac{L_m^p(x)L_n^p(x)t_1^m t_2^n}{(m+p)!(n+p)!} = \{(1-t_1)(1-t_2)\}^{-(p+1)}e^{-t_1x/1-t_2}e^{-t_2x/1-t_2}$$

Multiplying both sides by $e^{-x}x^p$ and integrating w.r.t. to x between limits 0 and ∞, we have

$$I = \sum_{n,m=0}^{\infty} t_1^m t_2^n \int_0^\infty \frac{e^{-x}x^p L_m^p(x)L_n^p(x)}{(m+p)!(n+p)!}dx$$

$$= \int_0^\infty \{(1-t_1)(1-t_2)\}^{-(p+1)} x^p e^{-x} e^{-t_1 x/(1-t_1)} e^{-t_2 x/(1-t_2)} dx$$

$$= \{(1-t_1(1-t_2)\}^{-(p+1)} \int_0^\infty x^p e^{-x(1-t_1 t_2)/(1-t_1)(1-t_2)^*} dx$$

$$= \{(1-t_1)(1-t_2)\}^{-(p+1)} \left\{ \frac{1-t_1 t_2}{(1-t_1)(1-t_2)} \right\}^{-(p+1)} p!$$

$$= p!(1-t_1 t_2)^{-(p+1)}$$

$$= p! \left[1 + \frac{(p+1)}{1!} t_1 t_2 + \frac{(p+1)(p+2)}{2!} t_1^r t_2^2 + \cdots + \frac{(p+1)(p+2)\cdots(p+r)}{r!} t_1^r t_2^r + \cdots \right]$$

$$= p! \sum_{r=0}^\infty \frac{(p+1)(p+2)\cdots(p+r)}{r!} t_1^r t_2^r$$

Equating the coefficients of $t_1^m t_2^n$, we observe that this coefficient is zero on R.H.S, i.e.

$$\int_0^\infty \frac{e^{-x} x^p L_m^p(x) L_n^p(x) dx}{(m+p)!(n+p)!} = 0 \quad \text{for } m \neq n$$

For $m = n$ we have

$$\int_0^\infty \frac{e^{-x} x^p L_m^p(x) L_n^p(x) dx}{\{(n+p)!\}^2} = \frac{(p+n)!}{n!}$$

Hence $\quad \int_0^\infty \phi_n(x) \cdot \phi_m(x) dx = \frac{(p+n)!}{n!} \delta_{mn}$

Recurrence Formulae for Associated Laguerre Polynomials

i. We have

$$L_n^p(x) = \sum_{r=0}^n \frac{(-1)^r \{(n+p)!\}^2}{(n-r)!(p+r)!r!} x^r$$

$$= \sum_{r=0}^{n-1} \frac{(-1)^r \{(n+p)!\}^2}{(n-r)!(p+r)!r!} x^r + \frac{(-1)^n \cdot (n+p)!}{n!} x^n$$

$$= \sum_{r=0}^{n-1} \frac{(-1)^r \{(n+p-1)!\}^2 (n+p)^2}{(n-r-1)!(p+r-1)!r!} \cdot \frac{1}{(n-r)(p+r)} x^r + \frac{(-1)^n (n+p)!}{n!} x^n$$

But $\quad \dfrac{n+p}{(n-r)(p+r)} = \dfrac{1}{p+r} + \dfrac{1}{n-r}$

$$^* \int_0^\infty x^p e^{-\alpha x} dx = \alpha^{-(p+1)} p!$$

$$\therefore \qquad L_n^p(x) = \sum_{r=0}^{n-1} \frac{(-1)^r \left\{(n+p-1)!\right\}^2 (n+p)}{(n-r-1)!(p+r-1)!r!} \times \frac{x^r}{(p+r)}$$

$$+ \sum_{r=0}^{n-1} \frac{(-1)^r \left\{(n+p-1)!\right\}^2 (n+p)}{(n-r-1)!(p+r-1)!} \cdot \frac{x^r}{(n-r)} + (-1)^n \frac{(n+p)!}{n!} x^n$$

$$= \sum_{r=0}^{n-1} \frac{(-1)^r \left\{(r+p-1)!\right\}^2 (n+p)}{(n-r-1)!(p+r)!r!} x^r + \sum_{r=0}^{n} \frac{(-1)^r \left\{(n+p-1)!\right\}^2 (n+p)x^r}{(n-r-1)!(p+r-1)!r!(n-r)}$$

$$= \sum_{r=0}^{n-1} \frac{(-1)^r \left[\left\{(n-1)+p!\right\}\right]^2 (n+p)}{\left\{(n-1)-r\right\}!(p+r)!r!} x^r + \sum_{r=0}^{n} \frac{(-1)^r \left\{(n+p-1)!\right\}^2 (n+p)}{(n-r)!(p+r-1)!r!} x^r$$

$$= (n+p)L_{n-1}^p(x) + (n+p)L_n^{p-1}(x)$$

or $\qquad L_n^p(x) = (n+p)L_{n-1}^p(x) + (n+p)L_n^{p-1}(x)$ \hfill (7.225)

ii. Recurrence relation for Laguerre polynomial, Eq. (7.207) gives

$$xL_n'(x) = nL_n(x) - n^2 L_{n-1}(x)$$

Replacing n by $n + p$, we have

$$xL_{n+p}'(x) = (n+p)L_{n+p}(x) - (n+p)^2 L_{n+p-1}(x)$$

Differentiating both sides of above equation p times w.r.t. x by Leibnitz theorem, we have

$$x\frac{d^p}{dx^p}L_{n+p}'(x) p\frac{d^{p-1}}{dx^{p-1}}L_{n+p}'(x) = (n+p)\frac{d^p}{dx^p}L_{n+p}(x) - (n+p)^2 \frac{d^p}{dx^p}L_{n+p-1}(x)$$

or $\qquad x\frac{d^p}{dx^p}L_{n+p}'(x) = n\frac{d^p}{dx^p}L_{n+p}(x) - (n+p)^2 \frac{d^p}{dx^p}L_{n+p-1}(x)$

$$\left(\text{since } \frac{d^{p-1}}{dx^{p-1}}L_{n+p}'(x) = \frac{d^p}{dx^p}L_{n+p}(x)\right)$$

Multiplying throughout with $(-1)^p$ and using

$$L_n^p(x) = (-1)^p \frac{d^p}{dx^p}L_{n+p}(x)$$

We have $\qquad xL_n'^p(x) = nL_n^p(x) - (n+p)^2 L_{n-1}^p(x)$

iii. We have

$$L_n^p(x) = \sum_{r=0}^{n} \frac{(-1)^r \left\{(n+p)!\right\}^2}{(n-r)!(p+r)!r!} x^r$$

Differentiating w.r.t. to x we have

$$L'^p_n(x) = \sum_{r=1}^{n} \frac{(-1)^r \{(n+p)!\}^2}{(n-r)!(p+r)!(r-1)!} x^{r-1}$$

Put $r - 1 = s$ or $r = 1 + s$

$$L'^p_n(x) = \sum_{s=0}^{n-1} \frac{(-1)^{s+1} \{(n-1+p+1)!\}^2}{\{(n-1)-s\}!\{(p+1)+s\}!s!} x^s$$

$$= -1 \sum_{s=0}^{n-1} \frac{(-1)^s \left[\{(n-1)+(p+1)\}!\right]^2}{\{(n-1)-s\}!\{(p+1)+s\}!s!} x^s$$

Hence $\qquad L'^p_n(x) = -1 L^{p+1}_{n-1}(x)$ \hfill (7.226)

iv. From recurrence relation Eq. (7.203) for Laguerre polynomials,
we have $L_{n+1}(x) = (2n + 1 - x) L_n(x) - n^2 L_{n-1}(x)$
Replacing n by $n + p$, we have

$$L_{n+p+1}(x) = (2n+2p+1-x)L_{n+p}(x) - (n+p)^2 L_{n+p-1}(x)$$

Differentiating both sides p times with respect to x.
We have

$$\frac{d^p}{dx^p} L_{n+p+1}(x) = (2n+2p+1-x)\frac{d^p}{dx^p} L_{n+p}(x)$$

$$-p\frac{d^{p-1}}{dx^{p-1}} L_{n+p}(x) - (n+p)^2 \frac{d^p}{dx^p} L_{n+p-1}(x)$$

Multiplying throughout with $(-1)^p$ and using the relation

$$L^p_n(x) = (-1)^p \frac{d^p}{dx^p} L_{n+p}(x)$$

We have $\qquad L^p_{n+1}(x) = (2n+2p+1-x)L^p_n(x) + pL^{p-1}_{n+1}(x) - (n+p)^2 L^p_{n-1}(x)$ \hfill (7.226a)

Recurrence relation Eq. (7.225) on replacing n by $n + 1$.

$$L^p_{n+1}(x) = (n+p+1)L^p_n(x) + (n+p+1)L^{p-1}_{n+1}(x) \qquad (7.226b)$$

Eliminating $L^{p-1}_{n+1}(x)$ between Eqs (7.226a) and (7.226b), we have

$$(n+1)L^p_{n+1}(x) = (n+1+p)(2n+p+1-x)L^p_n(x) - (n+p)^2(n+p+1)L^p_{n-1}(x)$$

or $\qquad \dfrac{(n+1)}{n+1+p} L^p_{n+1}(x) = (2n+p+1-x)L^p_n(x) - (n+p)^2 L^p_{n-1}(x)$ \hfill (7.227)

Equations (7.219), (7.220), (7.221) and (7.222) are called the *recurrence relations for associated Laguerre polynomials*.

SHORT ANSWER QUESTIONS

1. What do you mean by order and degree of a differential equation? Explain with two examples.

2. Solve the differential equation $\dfrac{d^2y}{dx^2} + y = \sec x \tan x$

3. Prove that $P_n(-x) = (-1)^n P_n(x)$

4. State orthogonal properties of Legendre polynomials. Hence, find the value of

   i. $\displaystyle\int_{-1}^{+1} [P_4(x)]^2\, dx$ ii. $\displaystyle\int_{-1}^{+1} P_3(x)P_4(x)\, dx$

5. Prove that

   i. $\cos(x\cos\phi) = J_0(x) - 2\cos 2\phi\, J_2(x) + 2\cos 4\phi\, J_4(x)$ and

   ii. $\sin(x\sin\phi) = 2\cos\phi\, J_1(x) - 2\cos 3\phi\, J_3(x) + 2\cos\phi\, J_5(x)$

6. Give the orthogonality relation connecting the Hermite polynomials.

7. Show that $J_n(-x) = (-1)^n J_n(x)$

8. Where are the singular points of the Bessel differential equation,
   $$x^2y'' + xy' + (x^2 - m^2)\, y = 0 \text{ located.}$$

9. Show that $H_{2n}(0) = (-1)^n \dfrac{(2n)!}{n!}$

10. State orthogonal properties of Laguerre polynomials.

11. What do you mean by hypergeometric functions?

12. Write Laguerre's differential equation. Write its solution.

13. Is $J_m(0) = 0$ for all nonnegative integer of m? Justify your answer.

14. Obtain the integral representation of $J_n(x)$.

15. Give Bessel's differential equation. What are its independent solutions. Write generating function for Bessel's function $J_n(x)$.

16. Evaluate $P_n(0)$ when n is odd and even.

17. Prove that $P'_n(1) = \dfrac{1}{2}n(n+1)$.

18. Show that $P_n(1)$ and $P_n(-1) = (-1)^n P(1)$.

19. What do you mean by ordinary regular singularity?

20. Write the Wronskian's second order differential equation. What is its physical significance.

21. When a power series is restricted to a finite number of terms, what is the name of the resulting series? Explain with an example.

22. Discuss whether $x = 0$ is a ordinary regular singular point or irregular singular point of differential equation
    $$x^2y'' + \left(x^2 + \frac{7}{81}\right)y = 0$$

23. Show that $P_n(-x) = (-1)^n P_n(x)$. Hence prove that

    i. $P_{2n}(-x) = P_{2m}(x)$ and   ii. $P_{2m+1}(-x) = -P_{2m+1}(x)$

24. Prove that i. $P_{2n}(0) = (-1)^n \dfrac{(2n)!}{2^{2n}(n!)^2}$ ii. $P_{2n+1}(0) = 0$

25. Why is the region of solutions of Legendre's differential equation is restricted in the range $-1 \le x \le 1$?

26. Prove that $J_{y_2}(x) = \sqrt{\dfrac{2}{\pi x}} \sin x.$

27. Find the numerical value of the quantity $J_4(3)$ using the recurrence relation.

28. Show that $x^2 = \dfrac{1}{2} - H_0(x) + \dfrac{1}{4}H_2(x)$

29. Write Rodrigue's formula for Laguerre polynomials. Hence, find $L_n(x)$, $L_1(x)$ and $L_2(x)$.

30. Show that $L_1(x) = L_1'(x) - \dfrac{1}{2}L_2'(x).$

31. Prove that $H_n(-x) = (-1)^n H_n(x), \, n \ge 0.$

32. Prove that $J_0'(x) = -J_1(x).$

33. Evaluate $J_{-1/2}(x).$

34. Prove that $\dfrac{2}{3}P_2(x) + \dfrac{1}{3}P_0(x) = x^2.$

35. Prove that $P_n(x) = (-1)^n P_n(-x).$

36. Write Rodrigue's formula for Legendre polynomials.

37. Find the solution of differential equation

$$xy'' + (1 - 2x)y' + (n-1)x = 0.$$

38. How will you find the solution of a differential equation by the method of separation of variables.

39. Write Legendre's differential equation. Write its two independent solutions.

40. Show that Legendre's equation is self-adjoint.

41. Write generating function by Legendre's polynomials and prove it.

42. Show that $xP_{n-1}' - P_n'(x) = (n+1)P_{n+1}(x).$

43. Evaluate $J_{3/2}(x)$ and $J_{-3/2}(x).$

44. Show that $\displaystyle\int_0^\infty x^n J_{n-1}(x)\,dx = x^n J_n(x).$

45. Show that i. $H_0(x) = 1$ ii. $H_3(x) = 8x^2 - 12x.$

    iii. $H_n(0) = (-1)^n \dfrac{n!}{\left(\dfrac{n}{2}\right)!}$, if n is an even integer and zero if n is an add integer.

46. Find the value of $L_1(x).$

1. If $y = \dfrac{\sin n\theta}{\cos\theta/2}, x = \sin(\theta/2);$

prove that $(1-x^2)\dfrac{d^2y}{dx^2} - 3x\dfrac{dy}{dx} + (4n^2-1)y = 0$

and hence $y = 2n\left[x + \dfrac{4(1-n^2)}{3!}x^3 + \dfrac{4^2(1-n^2)(2^2-n^2)}{5!}x^5 + \cdots\right]$

and $\dfrac{\cos n\theta}{\cos\theta/2} = 1 - \dfrac{1-4n^2}{2!}x^2 + \dfrac{(1-4n^2)(3^2-4n^2)}{4!}x^4 + \cdots$

2. Obtain the general solution of the hypergeometric equation

$$x(x-1)\dfrac{d^2F}{dx^2} + [(a+b+1)x - c]\dfrac{dF}{dx} + abF = 0$$

about a singular point $x = 0$.

3. Show that

$$e^{x(t-1/t)/2} = \sum_{n=-\infty}^{+\infty} J_n(x)t^n$$

and hence deduce the relation

$$J_n(x) = \dfrac{1}{\pi}\int_0^\pi \cos(n\theta - x\sin\theta)d\theta$$

where n is any integer.

4. Show that Legendre polynomials can be expressed as

$$P_n(x) = \dfrac{1}{2^n n!}\dfrac{d^n}{dx^n}(x^2-1)^n$$

5. Find the solution of the following equation

$$\dfrac{d^2y}{dx^2} + p(x)\dfrac{dy}{dx} + q(x)y = 0$$

   i. When $p(x)$ and $q(x)$ have no singular points

   ii. When the equation has only singular point

   iii. When the equation has an irregular singular point

6. (a) Prove that $\displaystyle\int_{-1}^{+1}[P_n(x)]^2\,dx = \dfrac{2}{2n+1}$

(b) Prove that $\dfrac{d}{dx}\left[x^{-n}J_n(x)\right] = x^{-n}J_{n+1}(x)$

(c) Write the generating function for Hermite polynomials and obtain the first three Hermite polynomials from it.

7. (a) For the Legendre's polynomials, prove the following recurrence relation

$$(l+1)P_{l+1}(x) - (2l+1)xP_l(x) + lP_{l-1}(x) = 0$$

(b) If n is a positive integer, prove that $J_n(x)$ is the coefficient of x^n in the expansion of $e^{x(z-1/z)/2}$ in ascending and descending powers of z.

(c) Discuss the orthogonality properties of Laguerre polynomials.

8. Show that;

   i. For Bessel functions

     (a) $xJ_n'(x) + nJ_n(x) - xJ_{n-1}(x) = 0$

     (b) $J_{n+1}(x) = \dfrac{n}{2}J_n(x) - J_n'(x)$

   ii. For Hermite polynomials

     $H_{n+1}(x) = 2xH_n(x) - 2nH_{n-1}(x)$

   iii. $\displaystyle\int_{-1}^{+1} P_l^m(x)P_{l'}^m(x)\,dx = \dfrac{2}{(2l+1)}\dfrac{(l+m)!}{(l-m)!}\delta_{l,l'}$

9. Prove that if $x = \sin\theta/2$ and $y = \cos n\theta$

$$(1 + x^2)\, y'' - xy' + 4n^2 y = 0,$$

hence prove that

$$\cos n\theta = 1 - \frac{n^2}{2!}(2\sin\theta/2)^2 + \frac{n^2(n^2-1^2)}{4!}(2\sin\theta/2)^4 + \cdots$$

and $\dfrac{\sin n\theta}{\cos\theta/2} = 2\sin\dfrac{1}{2}\theta - \dfrac{n(n^2-1)}{3!}(2\sin\theta/2)^3 + \cdots$

10. Write Legendre differential equation and solve it by series integration method and prove that if $P_n(x)$ is a solution of the equation, then

$$(2n + 1)xP_n = (n + 1)P_{n+1} + nP_{n-1}$$

11. Write the differential equation obeyed by Legendre polynomials. Show that

   i. $P_n(x) = \dfrac{1}{2^n n!}\dfrac{d^n}{dx^n}(x^2 - 1)^n$ ii. $P_n(1) = 1$

12. Prove the following.

   i. $nP_n(x) = (2n - 1)xP_{n-1}(x) - (n-1)P_{n-2}(x)$

   ii. $(2n + 1)P_n(x) = \dfrac{dP_{n+1}(x)}{dx} - \dfrac{dP_{n-1}(x)}{dx}$

   iii. $\displaystyle\int_{-1}^{+1} xP_n(x)P_{n-1}(x)\,dx = \dfrac{2n}{4n^2 - 1}$

13. Show that;

   i. $\displaystyle\int_{-1}^{+1} P_0(x)\,dx = 2$

   ii. $\dfrac{1 - z^2}{(1 - 2xz + z^2)^{3/2}} = \displaystyle\sum_{n=0}^{\infty}(2n + 1)P_n(x)z^n$

14. Prove that;

i. $\int_{-1}^{+1} P_n(x)P_m(x)dx = \dfrac{2}{2m+1}\delta_{m,n}$

where $\delta_{m,n}$ is Kronecker delta symbol.

ii. $(1-2tx+t^2)^{-1/2} = \displaystyle\sum_{n=0}^{\infty} t^n P_n(x)$

where $P_n(x)$ is Legendre polynomial.

15. Solve completely the given Bessel's equation

$$x^2\dfrac{d^2y}{dx^2} + x\dfrac{dy}{dx} + \left(x^2 - \dfrac{1}{4}\right)y = 0,$$

prove that the two solutions can be written in the form

$$\dfrac{\sin x}{\sqrt{x}} \text{ and } \dfrac{\cos x}{\sqrt{x}}$$

16. Show that

$$\int_{-1}^{+1} x^m P_n(x)dx = 0 \text{ for } m < n$$

$$= \dfrac{m(m-1)(m-2)\cdots(m-n+2)}{(m+n+1)(m+n-1)\cdots(m-n+3)} \text{ for } m > n-1$$

when n is a +ve integer.

17. Define associated Legendre polynomials and prove their orthogonality condition.

If $P_n^m(x)$ are the associated Legendre polynomials, show that

$$P_n^{-m}(x) = (-1)^m \dfrac{(n-m)!}{(n+m)!} P_n^m(x)$$

18. The associated Legendre polynomial $P_n^m(x)$ satisfies self-adjoint differential equation

$$(1-x^2)P_n'''^m(x) - 2xP_n'^m(x) + \left[n(n+1) - \dfrac{m^2}{1-x^2}\right]P_n^m(x) = 0$$

From the differential equation for $P_n^m(x)$ and $P_n^k(x)$, show that

i. $\int_{-1}^{+1} P_n^m(x)P_n^k(x)\dfrac{dx}{1-x^2} = 0 \text{ for } k \ne m$

$$= \dfrac{(n+m)!}{m(n-m)!} \text{ for } k = m$$

ii. $\int_{-1}^{+1} P_p^m(x)P_q^m(x)dx = \dfrac{2}{2q+1}\dfrac{(q+m)!}{(q-m)!}\delta_{p\cdot q}$

19. Prove that a general solution of differential equation

$$x^2 y'' + xy' + (x^2 - n^2)y = 0$$

for all values of n is

$$y(x) = c_1 J_n(x) + c_2 y_n(x)$$

where $J_n(x)$ and $y_n(x)$ are the Bessel's functions of the first and second kind respectively, c_1 and c_2 being constants.

20. For Bessel's function of 1st kind, prove that

i. $J_{-n}(x) = (-1)^n J_n(x)$

ii. $J_{n-1}(x) - J_{n+1}(x) = 2J_n'(x)$

iii. $J_0(x) = -J_1(x)$

iv. $\dfrac{2nJ_n(x)}{x} = J_{n-1}(x) + J_{n+1}(x)$

v. $\dfrac{d}{dx}[x^n J_n(ax)] = ax^n J_{n-1}(x)$

vi. $J_n(x+y) = \displaystyle\sum_{r=-\infty}^{+\infty} J_r(x) J_{n-r}(y)$. For integral values of n.

vii. $J_{5/2}(x) = \sqrt{\dfrac{2}{\pi x}}\left[(3-x^2)\dfrac{\sin x}{x^2} - \dfrac{3\cos^2 x}{x}\right]$

21. If a and b are different roots of $J_n(\alpha) = 0$, prove that

$$\int_0^1 x J_n(ax) J_n(bx)\,dx = \frac{1}{2}J_{n+1}^2(a)\delta_{a,b}$$

22. Show that;

i. $\displaystyle\int_0^\infty e^{-ax} J_0(bx)\,dx = \dfrac{1}{\sqrt{(a^2+b^2)}}; \; a,b>0$

ii. $\displaystyle\int_0^x x^{n+1} J_n(x)\,dx = x^{n+1} J_{n+1}(x)$

iii. $\displaystyle\int_0^\infty J_0(bx)\,dx = \dfrac{1}{b}$

iv. $J_n(x) = \dfrac{1}{2\pi}\displaystyle\int_0^{2\pi} \cos(x\sin\theta - n\theta)\,d\theta$

v. $\dfrac{d}{dx}\left(\dfrac{J_{-n}(x)}{J_n(x)}\right) = -\dfrac{2\sin n\pi}{\pi x J_n^2}$

23. Prove that;

i. $J_0(x) + 2J_2(x) + 2J_4(x) + \cdots = 1$

ii. $J_0(x) - 2J_2(x) + 2J_4(x) + \cdots = \cos x$

iii. $J_1(x)+3J_3(x)+5J_5(x)+\cdots=x/2$

iv. $2^2 J_2(x)-4^2 J_4(x)+6^2 J_6(x)+\cdots=x\sin x/2$

v. $1^2 J_1(x)-3^2 J_3(x)+5^2 J_5(x)+\cdots=x\sin x/2$

vi. $[J_0(x)]^2+2[J_1(x)]^2+2[J_2(x)]^2+\cdots=1$

24. Prove that

    i. $4J_n''(x)=J_{n-2}(x)-2J_n(x)+J_{n+2}(x)$

    ii. $\dfrac{x}{2}J_n(x)=(n+1)J_{n+1}(x)-(n+3)J_{n+2}(x)+(n+5)J_{n+3}(x)\cdots$

    iii. $J_2-J_0=2J_0''$

    iv. $J_0''=x^{-1}J_0'$

    v. $J_2+3J_0'+4J''=0$

    vi. $J_n N_n'-J_n' N_n=\dfrac{2}{\pi x}$

    vii. $J_n N_{n+1}-J_{n+1}N_n=-\dfrac{2}{\pi x}$

    where N_n is Neumann function.

25. Obtain the polynomial solutions of differential equation
$$y''-2xy'+2xy=0$$
for integral values of n and obtain the first four polynomials.

26. For Hermite polynomials, show that

    i. $2nH_{n-1}(x)=H_n'(x)$

    ii. $2xH_n(x)-H_{n-1}(x)=H_n'(x)$

    iii. $2xH_n(x)-2nH_{n-1}(x)=H_{n+1}(x)$

    vi. $|H_{n+1}(x)|\le|H_n(x)|$

27. Prove the orthogonality relation for Hermite polynomial in the form
$$\int_{-\infty}^{+\infty}e^{-x^2}H_m(x)H_n(x)dx=2^n n!\pi^{1/2}\delta_{m,n}$$

28. Show that

    i. $\displaystyle\int_{-\infty}^{+\infty}H_n(x)e^{-x^2/2}dx=\begin{cases}\dfrac{2\pi n!}{(x/2)!} & \text{for } n \text{ even}\\[2mm] 0 & \text{for } n \text{ odd}\end{cases}$

    ii. $\displaystyle\int_{-\infty}^{+\infty}xH_n(x)e^{-x^2/2}dx=\begin{cases}0 \text{ for } n \text{ even}\\[2mm] 2\pi\dfrac{(n+1)!}{\left(n+\dfrac{1}{2}\right)!} & n \text{ odd}\end{cases}$

29. With $Q_n = e^{-x^2/2}H_n(x)/(2^n n!\pi^{1/2})^{1/2}$, show that

i. $\dfrac{1}{\sqrt{2}}\left(x+\dfrac{d}{dx}\right)Q_n(x) = \sqrt{n}\,Q_{n-1}(x)$

ii. $\dfrac{1}{\sqrt{2}}\left(x-\dfrac{d}{dx}\right)Q_n(x) = (n+1)^{1/2}Q_{n-1}(x)$

30. What are Laguerre polynomials? Derive orthogonality relation

$$\int_0^\infty e^{-x}L_m(x)L_n(x)dx = \delta_{mn}$$

31. From generating function derive Rodrigue's representation

$$L_n^k(x) = \frac{e^x x^{-k}}{n!}\frac{d^n}{dx^n}(e^{-x}x^{n+k})$$

32. If $|x| < 1$ and $\left|\dfrac{x}{1-x}\right| < 1$ then show that

$$F(a, b, c, x) = (1-x)^a F\left(a,c-b,c,\frac{x}{1-x}\right)$$

33. Show that

i. $P_n(x) = F\left(-n,n+1,1,\dfrac{1-x}{2}\right)$

ii. $P_n(\cos\theta) = \cos^{2n}\theta/2\,F(-n,-n+1-\tan^2\theta/2)$

$\qquad = F(-n,n+1,1,\sin^2\theta)$

$\qquad = \cos^n\theta F\left(-\dfrac{n}{2},\dfrac{n+1}{2},1,-\tan^2\theta/2\right)$

iii. $P_{2n}(x) = (-1)^n\dfrac{(2n)!}{2^{2n}n!n!}F(-n,n+1/2,1/2,x^2)$

iv. $P_{2n+1}(x) = (-1)^n\dfrac{(2n+1)!}{2^{2n}n!n!}xF(-n,n+3/2,3/2,\,x^2)$

34. Prove the

i. $xF(a+1,c+1,x) = c[F(a+1,c,x)-F(a,c,x)]$

ii. $aF(a+1,c+1,x) = (a-c)F(a,c+1,x)+cF(a,c,x)$

iii. $(a+x)F(a+1,c+1,x) = (a-c)F(a,c+1,x)+cF(a+1,c,x)$

iv. $acF(a+1,c,x) = c(a+x)F(a,c,x)-x(c-a)F(a,c+1,x)$

v. $(c-a)xF(a,c+1,x) = c(x+c-1)F(a,c,x)+c(1-c)F(a,c-1,x)$

35. Verify that the Legendre function of the second kind $Q_n(x)$ is given by

$$Q_n(x) = \frac{\pi^{1/2}n!}{\left(n+\dfrac{1}{2}\right)!(2x)^{n+1}}F\left(\frac{n}{2}+\frac{1}{2},\frac{n}{2}+1,\frac{n}{2}+\frac{3}{2},x^{-2}\right)$$

36. Obtain the regular solution of the differential equation

$$\frac{d^2Y}{dx^2} + (E - X^2)Y = 0$$

making the substitution $Y(x) = e^{-x^2/2}F(x)$, where $F(x)$ is some series solution and E is a constant.

Discuss the orthogonality property of polynomials so obtained.

MULTIPLE CHOICE QUESTIONS

1. Legendre differential equation has singular points
(a) $(0, \infty)$
(b) $(-\infty, \infty)$
(c) $(-1, 1)$
(d) none of these

2. The value of $P_n(x)$ is

(a) $\frac{d^n}{dx^n}(x^{-2} - 1)^n$

(b) $\frac{1}{2^n n!}\frac{d^n}{dx^n}(x^2 - 1)^n$

(c) $(-1)^n x^n \frac{d}{d(x^2)^n}P_0(x)$

(d) $\frac{2^n}{n!}\frac{d^{n-1}}{dx^{n-1}}(x-1)^{2n}$

3. If $P_n(x)$ is Legendre polynomial of order n then $P'_n(-x)$ is equal to
(a) $(-1)^{n+1}P'_n(x)$
(b) $(-1)^n P'_n(x)$
(c) $(-1)^n P_n(x)$
(d) $P''_n(x)$

4. If $J_n(x)$ is Bessel's function of first kind of order n then the value of integral
$\int_0^{\pi}[J_{-2}(x) - J_2(x)]dx$ is equal to
(a) 0
(b) –2
(c) 2
(d) 1

5. The value of Hermite polynomial $H_1(n)$ is
(a) zero
(b) one
(c) x
(d) $2x$

6. $P_n(-x)$ has the value
(a) $P_n(x)$
(b) $(-1)^n P_n(x)$
(c) $-P_n(x)$
(d) 0

7. The value of $P_n(1)$ is
(a) 0
(b) 1
(c) –1
(d) $(-1)^n$

8. If $P_n(x)$ is Legendre polynomial of order n, the value of $3x^2 + 3x + 1$ can be expressed as
(a) $2P_2 + 3P_1$
(b) $4P_2 + 2P_1 + P_0$
(c) $3P_1 + 3P_2 + P_0$
(d) $2P_2 + 3P_1 + 2P_0$

9. The value of integral $\int xJ_0(x)dx$ is equal to
(a) $xJ_1(x) - J_0(x)$
(b) $xJ_1(x)$
(c) $J_1(x)$
(d) $x^2J_n(x)$

10. The value of $\int_{-\infty}^{+\infty} e^{-x}[H_n(x)]^2 dx$ is equal to

(a) $2^n n!\sqrt{\pi}$

(b) $\dfrac{2^n n!}{\sqrt{\pi}}$

(c) 1

(d) $\dfrac{\sqrt{\pi}}{2^n n!}$

11. The value of $\int_0^\infty e^{-x}[L_n(x)]^2 dx$ is

(a) 0

(b) 1

(c) $n!$

(d) $\dfrac{1}{2^n n!}$

12. The value of $J_{1/2}(\pi/2)$ is

(a) 0

(b) 1

(c) $\dfrac{\pi}{2}$

(d) $\dfrac{2}{\pi}$

13. In terms of Legendre polynomials $x^2 + x$ is equal to

(a) $P_0 + P_1 + 2P_2$

(b) $\dfrac{1}{3}P_0 + P_1 + \dfrac{1}{3}P_2$

(c) $\dfrac{1}{3}P_0 + P_1 + \dfrac{2}{3}P_2$

(d) $\dfrac{1}{3}P_0 + \dfrac{2}{3}P_2$

14. A second order linear differential equation of the term $x^2\dfrac{d^2y}{dx^2} + x\dfrac{dy}{dx} + (x^2 - a^2)y = 0$, where n is constant is called
   (a) Bessel's differential equation   (b) Legendre's differential equation
   (c) Languerre's differential equation (d) Poisons' equation

15. The value of $\int_{-1}^{+1} P_0(x)dx$ is

(a) 0

(b) x

(c) 1

(d) 2

16. The modified Bessel's functions $J_n(x)$ and $K_n(x)$ are
   (a) oscillatory in nature   (b) exponential in nature
   (c) linear in nature   (d) constant in nature

17. If P_n is Legendre's polynomial, then $\int_{-1}^{+1}[P_n(x)]^2 dx$ is

(a) 1

(b) $\dfrac{1}{2}(2n-1)$

(c) $\dfrac{2}{2n+1}$

(d) $\dfrac{n(n+1)}{2}$

18. The value of $J_{-1/2}(\pi/2)$ is

(a) 0

(b) 1

(c) $\dfrac{\pi}{2}$

(d) $\dfrac{2}{\pi}$

19. Which of the following is not equal to 1

(a) $P_0(x)$

(b) $J_0(x)$

(c) $H_0(x)$

(d) $L_0(x)$

20. The value of $F(1, 2, x)$ is

(a) 1

(b) –1

(c) $\dfrac{1}{2}$

(d) $\dfrac{3}{2}$

21. The value of $J_{1/2}^2(x) + J_{-1/2}^2(x)$ is

(a) 0

(b) $\dfrac{2}{\pi x}$

(c) $\sqrt{\dfrac{2}{\pi x}}\, e^{ix}$

(d) 1

22. If $P_n(x)$ is Legendre polynomial, then the value of $\displaystyle\int_{-1}^{+1} P_n(x) P_m(x)\,dx$ $(m \neq n)$ is equal to

(a) 0

(b) 1

(c) $\dfrac{2}{(2n+1)(2m+1)}$

(d) $\dfrac{m!n!}{2(m+n)!}$

23. The value of $\dfrac{2}{3} P_2(x) + \dfrac{1}{3} P_0(x)$ is

(a) x

(b) x^2

(c) x^3

(d) $x^2 + 2/3$

24. The orthogonality condition for Legendre's polynomial $\displaystyle\int_{-1}^{+1} P_m(x) P_n(x)\,dx = 0$ is true for

(a) $m = n$

(b) $m \neq n$

(c) $m = n = 0$

(d) none of these

25. For integral value of n, $J_n(-x)$ is

(a) $(-1)^n J_n(x)$

(b) $J_n(x)$

(c) 0

(d) $(-1)^n \pi J_n(x)$

26. The associated Legendre polynomial is given by

(a) $P_n^m(x) = \dfrac{1}{2^n n!\, dx^n}\dfrac{d^n}{}(x^2 - 1)^n$

(b) $P_n^m(x) = (1 - x^2)^{m/2} \dfrac{d^m}{dx^m} P_n(x)$

(c) $P_n^m(x) = \dfrac{1}{2^n m!}\dfrac{d^n}{dx^n}(x^2 - 1)^n$

(d) $P_n^m(x) = \dfrac{1}{2^{n+m} m!n!}\dfrac{d^{m+n}}{dx^{m+n}}(x^2 - 1)^{m+n}$

27. The Legendre polynomial $P_n(x)$ has
    (a) a real zero between 0 and 1
    (b) n zeros of which only 1 is between −1 and +1
    (c) $(2n − 1)$ real zeros between −1 and +1
    (d) no real zero between 0 and 1

28. The value of $\dfrac{J_{1/2}(x)}{J_{-1/2}(x)}$ is

    (a) 1   (b) $\tan x$
    (c) $\cot x$ (d) $\tanh (x)$

29. Which function represents $F(1, 1, 2, x)$?

    (a) $\log (1 - x)$ (b) $\dfrac{1}{x}\log(1-x)$

    (c) $-\dfrac{1}{x}\log(1+x)$ (d) $-\dfrac{1}{x}\log(1-x)$

30. The independent solutions of the equation $\dfrac{d^2y}{dx^2} - 3\dfrac{dy}{dx} + 2y = 0$ are

    (a) e^{2x} and e^{-x} (b) e^{2x} and e^x

    (c) $\dfrac{1}{x}$ and x^2 (d) $\sin 2x$ and $\cos x$

31. The value $P_3(x)$ is

    (a) $\dfrac{1}{2}(3x^2 - 1)$ (b) $3x$

    (c) $\dfrac{1}{2}(5x^3 - 3x)$ (d) $\dfrac{1}{2}(5x^3 - 1)$

32. The value of $J_n(x)$ is

    (a) $\dfrac{d^n}{dx^n}(x^2 - 1)^n$ (b) $\dfrac{1}{2^n n!}\dfrac{d^n}{dx^n}(x^2 - 1)^n$

    (c) $(-2)^n x^n \dfrac{d}{d(x^2)^n}J_0(x)$ (d) $\dfrac{2^n}{n!}\dfrac{d^{n-1}}{dx^{n-1}}(x-1)^{2n}$

33. The value of Bessel's function $J_0(0)$ is
    (a) zero   (b) 1
    (c) 2   (d) none of these

34. The associated Legendre equation $P_m^n(x) = 0$, if
    (a) $m = n$ (b) $m = n = 0$
    (c) $m > n$ (d) $m < n$

35. The value of $H'_{2n}(0)$ is

(a) $(-1)^n 2^m \left(\dfrac{1}{2}\right)^n$

(b) $(-1)^n 2^{2n+1} \left(\dfrac{3}{2}\right)^n$

(c) 0

(d) -1

36. The value of integral $\displaystyle\int_{-1}^{+1} x^2 P_3(x)dx$ is equal to

(a) $\dfrac{2}{5}$

(b) $\dfrac{2}{7}$

(c) $\dfrac{1}{2^2}\dfrac{1}{31}$

(d) 0

37. The Rodrigue's formula for $P_n(x)$, the Legendre polynomial of degree n is expressed as $P_n(x) = K \dfrac{d^n}{dx^n}\left|(x^2 - 1)^n\right|$. Then the value of K is

(a) $\dfrac{n!}{2^n}$

(b) $\dfrac{2^n}{n!}$

(c) $\dfrac{1}{2^n n!}$

(d) $\dfrac{1}{2^n (n!)^2}$

38. The value of integral $\displaystyle\int_0^\infty J_0(\beta x)dx$ is

(a) 0

(b) 1

(c) $\dfrac{1}{\beta}$

(d) β

39. The value of integral $\displaystyle\int_0^\infty e^{-\alpha x} J_0(x)dx$ is

(a) 1

(b) $\dfrac{1}{\sqrt{1+\alpha^2}}$

(c) $\sqrt{1+\alpha^2}$

(d) $\dfrac{1-\alpha}{1+\alpha}$

40. Which of the following is not correct?

(a) $x^2 - 1$

(b) $2x^2 - 1$

(c) $4x^2 - 2$

(d) $\dfrac{1}{2}(3x^2 - 1)$

41. Which of the following is not correct?

(a) $H_{2n}(0) = (-1)^n \dfrac{2n!}{n!}$

(b) $H_{2n+1}(0) = 0$

(c) $H'_{2n}(0) = 0$

(d) $H'_{2n+1}(0) = 0$

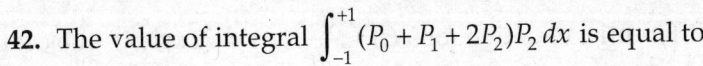

42. The value of integral $\int_{-1}^{+1}(P_0 + P_1 + 2P_2)P_2\, dx$ is equal to

   (a) $\dfrac{2}{3}$

   (b) $\dfrac{4}{5}$

   (c) 2^{-1}

   (d) 0

43. $J_0(x)$ and $J_1(x)$ are Bessel's function, then $J_1'(x)$ is

   (a) $J_0(x) - \dfrac{1}{x}J_1(x)$

   (b) $J_0(x) + \dfrac{1}{x}J_1(x)$

   (c) $-J_0(x)$

   (d) $J_0(x) - \dfrac{1}{x^2}J_1(x)$

44. The generating function for Bessel's polynomials is

   (a) $e^{x(z-z^{-1})/2}$

   (b) $e^{x^2/2}$

   (c) $e^{-x(z-z^{-1})/2}$

   (d) $(1 - 2xz + z^2)^{-1/2}$

45. If the solution of Legendre differential equation as infinite series is reduced to finite series, then the solution is called

   (a) polynomial

   (b) binomial

   (c) trinomial

   (d) Green's function

46. The value of $J_{-1/2}(x)$ is

   (a) $\sqrt{\dfrac{2}{\pi x}}$

   (b) $\dfrac{1}{\sqrt{\pi x}}\cos x$

   (c) $\sqrt{\dfrac{2}{\pi x}}\sin x$

   (d) $\sqrt{\dfrac{2}{\pi x}}\cos x$

47. The value of Bessel's function $J_{1/2}(x)$ is

   (a) $\sqrt{\dfrac{1}{\pi x}}\sin x$

   (b) $\sqrt{\dfrac{1}{\pi x}}\cos x$

   (c) $\sqrt{\dfrac{2}{\pi x}}\sin x$

   (d) $\sqrt{\dfrac{2}{\pi x}}\cos x$

48. Consider Bessel equation $v = 0$, $\dfrac{d^2y}{dx^2} + \dfrac{1}{x}\dfrac{dy}{dx} + y = 0$ which one of the statement is correct.

   (a) equation has regular singular point at $x = 0$ and $x = \infty$.

   (b) equation has two linearly in dependent solution that are entirely different.

   (c) equation has an entire solution and a second linearly independent singular at $z = 0$.

   (d) limit $x \to \infty$ taken along x-axis exists for both the linearly independent solutions

   [GATE]

49. The solution of the differential equation for $Y(t) = \dfrac{d^2Y}{dt^2} - Y = 2\cosh(t)$, subject to the initial conditions $Y(0) = 0$, $\left.\dfrac{dY}{dt}\right|_{t=0} = 0$ is

(a) $\dfrac{1}{2}\cosh t + t\sinh(t)$

(b) $-\sinh(t) + t\cosh(t)$

(c) $t\cosh t$

(d) $t\sinh t$

50. Given the recurrence relation for the Legendre polynomials
$$(2n + 1)\, xP_n(x) = (n + 1)\, P_{n+1}(x) + nP_{n-1}(x)$$
which of the following integrals has a nonzero value?

(a) $\displaystyle\int_{-1}^{+1} x^2 P_n(x) P_{n+1}(x)\,dx$

(b) $\displaystyle\int_{-1}^{+1} xP_n(x) P_{n+1}(x)\,dx$

(c) $\displaystyle\int_{-1}^{+1} x[P_n(x)]^2\,dx$

(d) $\displaystyle\int_{-1}^{+1} x^2 P_n(x) P_{n+2}(x)\,dx$ **[GATE]**

ANSWERS

1. (a)	2. (b)	3. (a)	4. (a)	5. (d)	6. (b)	7. (b)	8. (d)
9. (b)	10. (a)	11. (b)	12. (d)	13. (c)	14. (a)	15. (d)	16. (b)
17. (c)	18. (a)	19. (b)	20. (c)	21. (b)	22. (a)	23. (b)	24. (b)
25. (a)	26. (b)	27. (a)	28. (b)	29. (d)	30. (b)	31. (c)	32. (c)
33. (a)	34. (d)	35. (c)	36. (d)	37. (c)	38. (c)	39. (b)	40. (c)
41. (d)	42. (b)	43. (a)	44. (a)	45. (a)	46. (d)	47. (c)	48. (c)
49. (d)	50. (a)						

Integral Transforms

8.1. INTRODUCTION

In the theory of operational calculus, a mathematical technique that has proved to be a powerful method for solving differential equations and general system analysis is the *integral transform*. A linear integral transformation of the function $F(t)$ is represented by the following equation:

$$f(s) = T\{F(t)\} = \int_a^b F(t)K(x,t)dt \tag{8.1}$$

where $f(s)$ is called integral transform of $F(t)$ w.r.t. the Kernel $K(s, t)$ which denotes some prescribed function of the variable t and a parameter s. Integral transforms are used in a variety of applications, for example:

   i. Solution of integral equations of the convolution or faultung type

  ii. Solution of ordinary differential equations and partial differential equations

 iii. The theory of prime numbers

 iv. Evaluation of definite integrals

  v. Derivation of asymptotic, power and Fourier series

 vi. Summation of power and Fourier series

vii. The solution of nonlinear ordinary differential equations

Different kinds of integral transforms are obtained depending on the Kernel $K(s, t)$ used and the range of integration, e.g.

  i. $f(s) = \int_0^\infty e^{-st} \cdot F(t)dt$ (8.2)

     is known as Laplace transform.

 ii. $f(s) = \int_0^\infty F(t)t\,J_n(st)dt$ (8.3)

     is known as Hankel transform (Fourier–Bessel transform).

     Here $J_n(st)$ is Bessel's function of order n.

iii. $f(s) = \int_0^\infty F(t)t^{s-1}dt$ (8.4)

     is known as Mellin transform.

iv. $f(s) = \int_0^\infty F(t)e^{-ist}dt$ (8.5)

is known as Fourier transform.

In this chapter, we will introduce the integral transform of Laplace and develop the methods of their application.

8.2. LAPLACE TRANSFORM AND ITS PROPERTIES

Given any function $F(t)$, we can perform the integration

$$f(s) = \int_0^\infty e^{-st}F(t)dt$$ (8.6)

provided that the integral exists. It can be shown that this integral does exist if $F(t)$ is sectionally continuous and of exponential order as $t \to \infty$, i.e. if $F(t)$ satisfies these criteria then there exists some domain of s for which the integral exists[*]. The function $f(s)$ so constructed is called Laplace transform of $F(t)$. Simply by selecting various functions $F(t)$ and carrying out the indicated integration, we can construct a table of transform pairs, function $F(t)$ and their transform $f(s)$ (Table 8.1).

Also we can develop some general properties of the pairs, $F(t)$ and $f(s)$ directly from the definition.

Let us introduce the notation used in this book. If a function of t is indicated in terms of capital letters such as $F(t)$, $G(t)$, $Y(t)$ etc. The Laplace transform of the function is denoted by the corresponding lower case letters, i.e. $f(s)$, $g(s)$, $y(s)$, etc.

Thus $$f(s) = L\{F(t)\} = \int_0^\infty e^{-st}F(t)dt$$

Here L is often called Laplace transformation operator and

$$F(t) = L^{-1}\{f(s)\}$$

Here L^{-1} is called inverse Laplace transform.

i. Linearity property

If α_1 and α_2 are any constants and $f_1(s)$ and $f_2(s)$ are Laplace transforms of $F_1(t)$ and $F_2(t)$ respectively, then it follows from the definition that

$$L\{\alpha_1 F_1(t) + \alpha_2 F_2(t)\} = \alpha_1 L\{F_1(t)\} + \alpha_2 L\{F_2(t)\}$$
$$= \alpha_1 f_1(s) + \alpha_2 f_2(s)$$ (8.7)

ii. Translation or shifting property

(a) If $$L\{F(t)\} = f(s)$$

then $$L\{e^{at}F(t)\} = f(s-a)$$ (8.8)

Proof: $$L\{e^{at}F(t)\} = \int_0^\infty e^{at}e^{-st}F(t)dt$$

$$= \int_0^\infty e^{-(s-a)t}F(t)dt = f(s-a)$$

---

[*] *Sectionally continuous* means the domain of t can be subdivided into intervals in each of which $F(t)$ is continuous. Exponential order means that there exists some value, p, such that $\lim_{t\to\infty} |F(t)|e^{-pt}$ (real part $p > 0$) is bounded.

(b) If $L\{F(t)\} = f(s)$ and $G(t) = \begin{cases} F(t-a) & t > 0 \\ 0 & t < a \end{cases}$

Proof: $L\{G(t)\} = \displaystyle\int_0^\infty e^{-st}G(t)dt$

$$= \int_0^a 0 \cdot e^{-st}dt + \int_a^\infty e^{-st}F(t-a)dt$$

$$= \int_a^\infty e^{-st}F(t-a)dt$$

Put $t - a = P$ $\therefore$ $dt = dP$

$\therefore$ $\displaystyle\int_a^\infty e^{-st}F(t-a)dt = \int_a^\infty e^{-s(P+a)}F(P)dP$

$$= e^{-sa}\int_0^\infty e^{-sP}F(P)dP = e^{-sa}f(s)$$

or $L\{G(t)\} = e^{-sa}f(s)$ (8.9)

Hence the result.

iii. Change of scale property

If $L\{F(t)\} = f(s)$, then $L\{F(at)\} = \dfrac{1}{a}f(s/a)$

Proof: $L\{F(at)\} = \displaystyle\int_0^\infty e^{-st}F(at)dt$

Put $at = P$ $\therefore$ $t = \dfrac{P}{a}$ and $dt = \dfrac{1}{a}dP$

$\therefore$ $\displaystyle\int_0^\infty e^{-sP/a}F(P)\dfrac{dP}{a} = \dfrac{1}{a}\int_0^\infty e^{-sP/a}F(P)dP = \dfrac{1}{a}f\{s/a\}$

or $L\{F(at)\} = \dfrac{1}{a}f\{s/a\}$ (8.10)

iv. Laplace transform of derivatives

(a) If $L\{F(t)\} = f(s)$, then

$L\{F'(t)\} = sf(s) - F(0)$

provided $F(t)$ is of exponential order as $t \to \infty$ and sectionally continuous (and finite) at $t = 0$.

Proof: $L\{F'(t)\} = \displaystyle\int_0^\infty e^{-st}F'(t)dt$

$$= e^{-st}F(t)\big|_0^\infty - \int_0^\infty e^{-st}(-s)F(t)dt \qquad \text{(integrating by parts)}$$

$$= -F(0) + sf(s) = sf(s) - F(0) \qquad (8.11)$$

Since $F(t)$ is of exponential order as $t \to \infty$ and sectionally continuous at $t = 0$.

(b) $L\{F''(t)\} = s^2 f(s) - sF(0) - F'(0)$

Proof: $L\{F''(t)\} = \int_0^\infty e^{-st} F''(t) dt$

$= e^{-st} F'(t)\big|_0^\infty + \int_0^\infty s \cdot e^{-st} F'(t) dt$ (integrating by parts)

$= -F'(0) + s\{sf(s) - F(0)\}$ [*using* Eq. (8.11)]

$\therefore \qquad L\{F''(t)\} = s^2 f(s) - sF(0) - F'(0)$ (8.12)

and in general

$L\{F^{(n)}(t)\} = s^n f(s) - s^{n-1} F(0) - s^{n-2} F^{(1)}(0) \cdots$

$\cdots - sF^{(n-2)}(0) - F^{(n-1)}(0)$ (8.13)

v. Laplace transform of integrals

If $\qquad L\{F(t)\} = f(s)$ then

$$L\left\{ \int_0^t F(u) du \right\} = \frac{f(s)}{s}$$

Proof: Let $\qquad G(t) = \int_0^t F(u) du$ then $G(0) = 0$

and $\qquad G'(t) = F(t)$

Now $\qquad L \cdot \{G'(t)\} = sL\{G(t)\} - G(0) = sL\{G(t)\}$

$\therefore \qquad L\{G(t)\} = L \cdot \left\{ \int_0^t F(u) du \right\} = \frac{1}{s} L\{G'(t)\} = L\{F(t)\} \frac{1}{s}$

or $\qquad L\left\{ \int_0^t F(u) du \right\} = \frac{f(s)}{s}$ (8.14)

vi. Derivatives of Laplace transform

If $\qquad L\{F(t)\} = f(s)$ then

$$L\{t^n F(t)\} = (-1)^n \frac{d^n}{ds^n} f(s)$$ [where $n = 1, 2, 3...$]

Proof: We have

$$f(s) = \int_0^\infty e^{-st} F(t) dt$$

Then by Leibnitz's rule for differentiating under the integral sign

$$\frac{df(s)}{ds} = \frac{d}{ds} \int_0^\infty e^{-st} F(t) dt = \int_0^\infty \frac{\partial}{\partial s} e^{-st} F(t) dt$$

$$= -\int_0^\infty e^{-st} \{tF(t)\} dt = (-1)L\{tF(t)\}$$ (8.15)

which proves the theorem for $n = 1$.

To establish the theorem in general, we use mathematical induction. Assume the theorem to be true for $n = p$, i.e.

$$\int_0^\infty e^{-st} \{t^p F(t)\} dt = (-1)^p \frac{d^p f(s)}{ds^p}$$

then $\dfrac{d}{ds}\displaystyle\int_0^\infty e^{-st}\{t^p F(t)\}dt \ = \ (-1)^p \dfrac{d^{p+1}f(s)}{ds^{p+1}}$

Then by Leibnitz's rule for differentiating under integral sign gives

$$\int_0^\infty \frac{\partial e^{-st}}{\partial s}\{t^p F(t)\}dt \ = \ -\int_0^\infty e^{-st}\{t^{p+1}F(t)\}dt = (-1)^p \frac{d^{p+1}(s)}{ds^{p+1}}$$

or $\displaystyle\int_0^\infty e^{-st}\{t^{p+1}F(t)\}dt \ = \ (-1)^{p+1}\dfrac{d^{p+1}(s)}{ds^{p+1}}$ (8.16)

It follows that if the theorem holds for $n = P$ then theorem holds for $n = P + 1$. But by Eq. (8.15) the theorem is true for $n = 1$.

Hence it is true for $n = 1 + 1 = 2$ and $n = 2 + 1 = 3$ etc. and thus for all positive values of n.

vii. Integration of Laplace transform

If $L\{F(t)\} = f(s)$ then

$$\int_s^\infty f(s)ds \ = \ L\left\{\frac{F(t)}{t}\right\}$$

Proof: We have $f(s) = \displaystyle\int_0^\infty e^{-st}F(t)dt$

Integrating w.r.t. s within the limits from s to ∞ we have

$$\int_s^\infty f(s)ds \ = \ \int_s^\infty ds\int_0^\infty e^{-st}F(t)dt$$

s and t being independent variables the order of integration can be interchanged. Thus, we have

$$\int_s^\infty f(s)ds \ = \ \int_0^\infty dt\int_s^\infty e^{-st}F(t)ds$$

or $\displaystyle\int_s^\infty f(s)ds \ = \ \int_0^\infty dt\left[e^{-st}\dfrac{F(t)}{(-t)}\right]_s^\infty$

$$= \ \int_0^\infty e^{-st}\left\{\frac{F(t)}{t}\right\}dt$$

or $\displaystyle\int_s^\infty f(s)ds \ = \ L\left\{\dfrac{F(t)}{t}\right\}$ (8.17)

viii. Initial and final value theorems

If $L\{F(t)\} = f(s)$

Then

(a) $\displaystyle\lim_{t\to 0} F(t) = \lim_{s\to\infty} sf(s)$ (Initial value theorem)

(b) $\displaystyle\lim_{t\to\infty} F(t) = \lim_{s\to 0} sf(s)$ (Final value theorem)

Proof: We have from Eq. (8.11)

$$L\{F'(t)\} \ = \ \int_0^\infty e^{-st}F'(t)dt = sf(s) - F(0)$$ (8.18)

Taking the limit as $s \to \infty$, we have

$$\lim_{s \to \infty} \int_0^\infty e^{-st} F'(t) dt = 0 = \lim_{s \to \infty} sf(s) - F(0)$$

or

$$F(0) = \lim_{s \to \infty} sf(s)$$

But

$$\lim_{t \to 0} F(t) = F(0)$$

$\therefore$

$$\lim_{t \to 0} F(t) = \lim_{s \to \infty} sf(s) \tag{8.19}$$

On the other hand, taking the limit as $s \to 0$ in Eq. (8.18), we have LHS is

$$\lim_{s \to 0} \int_0^\infty e^{-st} F'(t) dt = \int_0^\infty F'(t) dt = \lim_{t \to \infty} \{F(t) - F(0)\}$$

and RHS is

$$\lim_{s \to 0} sf(s) - F(0)$$

$\therefore$

$$\lim_{t \to \infty} \{F(t) - F(0)\} = \lim_{s \to 0} sf(s) - F(0)$$

or

$$\lim_{t \to \infty} F(t) = \lim_{s \to 0} sf(s) \tag{8.20}$$

ix. Periodic functions

If $F(t)$ is a periodic function of period $T > 0$ then

$$LF\{(t)\} = \frac{\int_0^T e^{-st} F(t) dt}{1 - e^{-sT}}$$

Proof: We have

$$LF\{(t)\} = \int_0^\infty e^{-st} F(t) dt$$

or

$$LF\{(t)\} = \int_0^T e^{-st} F(t) dt + \int_T^{2T} e^{-st} F(t) dt + \int_{2T}^{3T} e^{-st} F(t) dt + \cdots$$

In the second integral, put

$$t = u + T$$

and in third integral $t = u + 2T$ etc. Then

$$L\{F(t)\} = \int_0^T e^{-su} F(u) du + \int_0^T e^{-s(u+T)} F(u+T) du + \int_0^T e^{-(u+2T)s} F(u+2T) du + \cdots$$

Since $F(t)$ is a periodic function of period T, we have

$$F(u) = F(u + T) = F(u + 2T) = F(u + 3T) = \cdots$$

$\therefore$

$$L\{F(t)\} = \int_0^T e^{-su} F(u) du + e^{-st} \int_0^T e^{-su} F(u) du + e^{-2sT} \int_0^T e^{-su} F(u) du + \cdots$$

$$= \left(1 + e^{-st} + e^{-2sT} + \cdots\right) \int_0^T e^{-su} F(u) du$$

or

$$L\{F(t)\} = \frac{\int_0^T e^{-su} F(u) du}{1 - e^{-ST}} \tag{8.21}$$

8.3. METHODS OF FINDING LAPLACE TRANSFORMS OF SOME SPECIAL FUNCTIONS

Series Expansion Method

In this method, the function is first expanded in the form of power series and then Laplace transform of each term is obtained. For example, let the function be $\sin \sqrt{t}$.

We have

$$\sin \sqrt{t} = \sqrt{t} - \frac{(\sqrt{t})^3}{3!} + \frac{(\sqrt{t})^5}{5!} - \frac{(\sqrt{t})^7}{7!} + \cdots$$

$$= t^{1/2} - \frac{t^{3/2}}{3!} + \frac{t^{5/2}}{5!} - \frac{t^{7/2}}{7!} + \cdots$$

Then Laplace transform is

$$L\{\sin \sqrt{t}\} = L\{t^{1/2}\} - L\left\{\frac{t^{3/2}}{3!}\right\} + L\left\{\frac{t^{5/2}}{5!}\right\} - L\left\{\frac{t^{7/2}}{7!}\right\} + \cdots$$

But by definition of Laplace transforms, we have

$$L\{t^n\} = \int_0^\infty e^{-st} t^n dt = \frac{n!}{s^{n+1}} \quad \text{if } n \text{ is an integer}$$

$$= \frac{\Gamma(n+1)}{s^{n+1}} \quad \text{if } n \text{ is a noninteger}$$

$$L\{\sin \sqrt{t}\} = \frac{\Gamma 3/2}{s^{3/2}} - \frac{\Gamma 5/2}{3! s^{5/2}} + \frac{\Gamma 7/2}{5! s^{7/2}} - \frac{\Gamma 9/2}{7! s^{9/2}} + \cdots \qquad \text{(integrating by parts)}$$

$$= \frac{\sqrt{\pi}}{2s^{3/2}} \left\{1 - \left(\frac{1}{2^2 s}\right) + \frac{1}{2!}\left(\frac{1}{2^2 s}\right)^2 - \frac{1}{3!}\left(\frac{1}{2^2 s}\right)^2 + \cdots\right\} = \frac{\sqrt{\pi}}{2s^{3/2}} e^{-1/4s} \qquad (8.22)$$

Differential Equation Method

In this method, first a differential equation satisfied by function $F(t)$ is obtained and the Laplace transform of function $F(t)$ is determined. For example, the function $F(t)$ whose Laplace transform is to be determined be $\sin \sqrt{t}$. Thus

$$F(t) = \sin \sqrt{t}$$

or

$$F'(t) = \frac{\cos \sqrt{t}}{2\sqrt{t}}$$

and

$$F''(t) = \frac{1}{2}\left\{\frac{\cos \sqrt{t}}{-2(t)^{3/2}} - \frac{1}{2t}\sin(\sqrt{t})\right\} = \frac{1}{2t}\left[-\frac{\cos \sqrt{t}}{2\sqrt{t}} - \frac{\sin \sqrt{t}}{2}\right]$$

or

$$4t\, F''(t) + 2F'(t) + F(t) = 0 \qquad (8.23)$$

The differential Eq. (8.23) is satisfied by the function $F(t)$. Taking the Laplace transformation, and using Eqs (8.15), (8.11) and (8.12), we have

$$L\{4tF''(t)\} = 4L\{tF''(t)\} = -4\frac{d}{ds}[L\{F''(t)\}]$$

$$= -4\frac{d}{ds}\{s^2 f(s) - sF(0) - F'(0)\}$$

$$= -4s^2 f'(s) - 8sf(s) + 4F(0)$$

Similarly, $L\{2F'(t)\} = 2sf(s) - 2F(0)$ and $L\{F(t)\} = f(s)$

Substitution in Eq. (8.23) yields

$$4s^2 f'(s) + (6s - 1)f(s) = 0$$

or

$$\frac{f'(s)}{f(s)} = -\frac{6s-1}{4s^2}$$

Integrating w.r.t. *s* we have

$$\log f(s) = \int \frac{1-6s}{4s^2}\,ds = \int\left(\frac{1}{4s^2} - \frac{3}{2s}\right)ds$$

$$= -\frac{1}{4s} - \frac{3}{2}\log s + \log c$$

where $\log c$ is the constant of integration.

$$f(s) = \frac{ce^{-\frac{1}{4s}}}{s^{3/2}} \tag{8.24}$$

For large *s*, Eq. (8.24) becomes

$$f(s) = \frac{c}{s^{3/2}}.$$

For small *t* we have $\sin\sqrt{t} = \sqrt{t}$, and

$$L\{\sqrt{t}\} = \frac{\Gamma 3/2}{s^{3/2}}$$

$$= \frac{\sqrt{\pi}}{2s^{3/2}} \qquad \left[using\ L\{t^n\} = \frac{\Gamma(n+1)}{s^{n+1}}\right]$$

Equating the limiting values we have

$$c = \frac{\sqrt{\pi}}{2}$$

Substitution in Eq. (8.24) gives

$$f(s) = \frac{\sqrt{\pi}}{2s^{3/2}}e^{-\frac{1}{4s}}$$

which is the same as Eq. (8.22).

Example 1. *Find the Laplace transform of*

$$F(t) = \int_0^t \frac{\sin u}{u}\,du$$

Solution: **Method 1.**

Let

$$F(t) = \int_0^t \frac{\sin u}{u}\,du$$

Then $$F(0) = 0 \quad \text{and} \quad F'(t) = \frac{\sin t}{t}$$

Taking Laplace transforms we have

$$L\{tF'(t)\} = L\{\sin t\} = \frac{1}{s^2 + 1}$$

or $$-\frac{d}{ds} L\{F'(t)\} = \frac{1}{s^2 + 1} \qquad\qquad [using \text{ Eq. (8.15)}]$$

or $$-\frac{d}{ds}\{sf(s) - F(0)\} = \frac{1}{s^2 + 1} \qquad\qquad [using \text{ Eq. (8.11)}]$$

or $$\frac{d}{ds}\{sf(s)\} = -\frac{1}{s^2 + 1}$$

Integrating we have

$$sf(s) = -\tan^{-1} s + c$$

Using the initial value theorem

$$\lim_{s\to\infty} sf(s) = \lim_{t\to 0} F(t) = F(0) = 0$$

so that $$c = \frac{\pi}{2}$$

Thus $$sf(s) = \frac{\pi}{2} - \tan^{-1} s = \tan^{-1}\frac{1}{s}$$

or $$f(s) = \frac{1}{s}\tan^{-1}\frac{1}{s}$$

Method 2. Using infinite series, we have

$$\int_0^t \frac{\sin u}{u}\,du = \int_0^t \frac{1}{u}\left(u - \frac{u^3}{3!} + \frac{u^5}{5!} - \frac{u^7}{7!} + \cdots\right)du$$

$$= t - \frac{t^3}{3.3!} + \frac{t^5}{5.5!} - \frac{t^7}{7.7!} + \cdots$$

$$\therefore \quad L\cdot\left\{\int_0^t \frac{\sin u}{u}\,du\right\} = L\left\{t - \frac{t^3}{3.3!} + \frac{t^5}{5.5!} - \frac{t^7}{7.7!} + \cdots\right\}$$

$$= \frac{1}{s^2} - \frac{3!}{3.3!s^4} + \frac{5!}{5.5!s^5} - \frac{7!}{7.7!s^7} + \cdots$$

$$= \frac{1}{s}\left[\frac{\left(\frac{1}{s}\right)}{1} - \frac{\left(\frac{1}{s}\right)^3}{3} + \frac{\left(\frac{1}{s}\right)^5}{5} - \frac{\left(\frac{1}{s}\right)^7}{7} + \cdots\right]$$

$$= \frac{1}{s}\tan^{-1}\frac{1}{s}$$

using the series

$$\tan^{-1} x = x - \frac{x^3}{3} + \frac{x^5}{5} - \frac{x^7}{7} + \cdots \qquad\qquad [|x| < 1]$$

Example 2. *Prove that*

(i) $L\{C_i(t)\} = L\left\{\displaystyle\int_t^\infty \frac{\cos u}{u}\,du\right\} = \dfrac{\log(s^2+1)}{2s}$

(ii) $L\{E_i(t)\} = L\left\{\displaystyle\int_t^\infty \frac{e^{-u}}{u}\,du\right\} = \dfrac{\log(s+1)}{s}$

(iii) $L\{e_r f(\sqrt t)\} = L\left\{\dfrac{2}{\sqrt\pi}\displaystyle\int_0^{\sqrt t} e^{-u^2}\,du\right\} = \dfrac{1}{s\sqrt{s+1}}$

Solution: (i) Let $F(t) = \displaystyle\int_t^\infty \frac{\cos u}{u}\,du$

so that $\qquad\qquad\qquad F'(t) = -\dfrac{\cos t}{t}$ and $\quad tF'(t) = -\cos t$

Taking the Laplace transform, we have

$$L\{tF'(t)\} = -\frac{d}{ds}\{LF'(t)\}$$

$$= -\frac{d}{ds}\{sf(s) - F(0)\} = -L\{\cos t\}$$

or $\qquad\qquad \dfrac{d}{ds}\{sf(s)\} = L\{\cos t\} = \dfrac{s}{s^2+1}$

Then by integration,

$$sf(s) = \frac{1}{2}\log(s^2+1) + c$$

Using final value theorem, we have

$$\lim_{s\to 0} sf(s) = \lim_{t\to\infty} F(t) = 0$$

$$c = 0$$

Thus $\qquad\qquad sf(s) = \dfrac{1}{2s}\log(s^2+1)$

or $\qquad\qquad\quad f(s) = \dfrac{1}{2s}\log(s^2+1)$

(ii) Let $\qquad\qquad F(t) = \displaystyle\int_t^\infty \frac{e^{-u}}{u}\,du.$ Then $tF'(t) = -e^{-t}$

Taking Laplace transform,

$$L\{tF'(t)\} = -\frac{d}{ds}[L\{F'(t)\}]$$

$$= -\frac{d}{ds}\{sf(s) - F(0)\} = -L\{e^{-t}\}$$

or $\qquad\qquad \dfrac{d}{ds}\{sf(s)\} = \dfrac{1}{s+1}$

Integrating we have

$$sf(s) = \log(s+1) + c$$

Applying final value theorem

$$\lim_{s \to 0} sf(s) = \lim_{t \to \infty} F(t) = 0$$

or $c = 0$

or $f(s) = \dfrac{\log(s+1)}{s}$

(iii) $L\left\{\dfrac{2}{\sqrt{\pi}} \displaystyle\int_0^{\sqrt{t}} e^{-u^2} du\right\} = L\left\{\dfrac{2}{\sqrt{\pi}} \displaystyle\int_0^{\sqrt{t}} \left(1 - u^2 + \dfrac{u^4}{2!} - \dfrac{u^6}{3!} \cdots\right) du\right\}$

$$= L\left\{\dfrac{2}{\sqrt{\pi}}\left(\sqrt{t} - \dfrac{t^{3/2}}{3} + \dfrac{t^{5/2}}{5.2!} - \dfrac{t^{7/2}}{7.3!} + \cdots\right)\right\}$$

$$= \dfrac{2}{\sqrt{\pi}}\left\{\dfrac{\Gamma 3/2}{s^{3/2}} - \dfrac{\Gamma 5/2}{3s^{5/2}} + \dfrac{\Gamma 7/2}{5.2! s^{7/2}} - \dfrac{\Gamma 9/2}{7.3! s^{9/2}} + \cdots\right\}$$

$$= \dfrac{1}{s^{3/2}}\left[1 - \dfrac{1}{2.s} + \dfrac{1.3}{2.4 s^2} - \dfrac{1.3.5}{2.4.6 s^3} + \cdots\right]$$

$$= \dfrac{1}{s^{3/2}}\left(1 + \dfrac{1}{s}\right)^{-1/2} = \dfrac{1}{s\sqrt{s+1}}$$

Example 3. $L\{\log t\} = \dfrac{\Gamma'(1) - \log s}{s} = -\dfrac{\gamma + \log s}{s}$

where $\gamma = 0.5772156\ldots$ is Euler's constant.

Solution: We have

$$\Gamma(r) = \int_0^\infty u^{r-1} e^{-u} du$$

Differentiating w.r.t. r we have

$$\Gamma'(r) = \int_0^\infty u^{r-1} e^{-u} \log u\, du$$

or $\Gamma'(1) = \displaystyle\int_0^\infty \log u\, e^{-u} du$

Let $u = st,\ s > 0$

Then $\Gamma'(1) = s\displaystyle\int_0^\infty e^{-st}\{\log s + \log t\} dt$

Now $L\{\log t\} = \displaystyle\int_0^\infty e^{-st} \log t\, dt$

$$= \dfrac{\Gamma'(1)}{s} - \int_0^\infty e^{-st} \log s \cdot dt$$

$$= \dfrac{\Gamma'(1)}{s} - \dfrac{\log s}{s} = -\dfrac{\gamma + \log s}{s}^{*}$$

Hence the result.

---

* In Chapter 4—*Factorial and Related Functions*, we have proved that $\Gamma'(1) = -\gamma$.

Example 4. *Evaluate*

(i) $\int_0^\infty te^{-2t}\cos t \cdot dt = \dfrac{3}{25}$

(ii) $\int_0^\infty \dfrac{e^{-\sqrt{2}t}\sinh t \cdot \sin t}{t}dt = \dfrac{\pi}{8}$

(iii) $\int_0^\infty e^{-t}erf(\sqrt{t})dt = \dfrac{1}{\sqrt{2}}$

(iv) $\int_0^\infty J_0(t)dt = 1$

Solution: (i) $\qquad L \cdot \{t\cos t\} = \int_0^\infty e^{-st} \cdot t\cos t\, dt = -\dfrac{d}{ds}L\{\cos t\}$

$$= -\dfrac{d}{ds}\left(\dfrac{s}{s^2+1}\right) = \dfrac{s^2-1}{(s^2+1)^2}$$

Putting $s = 2$ we have

$$\int_0^\infty e^{-2t}t\cos t\, dt = \dfrac{(2)^2-1}{\{(2)^2+1\}^2} = \dfrac{3}{25}$$

(ii) $\qquad \int_0^\infty \dfrac{e^{-\sqrt{2}t}\sinh t\sin t}{t}dt = \int_0^\infty \dfrac{e^{-(\sqrt{2}-1)t}\sin t}{2t}dt - \int_0^\infty \dfrac{e^{-(\sqrt{2}+1)t}\sin t}{2t}dt$

Since $\qquad\qquad\qquad \sinh t = \dfrac{e^{+t}-e^{-t}}{2}$,

Now $\qquad\qquad L\left\{\dfrac{\sin t}{t}\right\} = \int_s^\infty L\{\sin t\}ds$ $\qquad\qquad\qquad$ [*using* Eq. (8.17)]

$$= \int_s^\infty \dfrac{ds}{s^2+1}[\tan^{-1}s]_s^\infty = \dfrac{\pi}{2} - \tan^{-1}s$$

Hence $\qquad\qquad L\left\{\dfrac{\sin t}{t}\right\} = \int_s^\infty \dfrac{e^{-st}\cdot\sin t}{t}dt = \dfrac{\pi}{2} - \tan^{-1}s$

Putting $s = (\sqrt{2}-1)$, we have

$$\int_0^\infty \dfrac{e^{-(\sqrt{2}-1)t}\sin t}{2t}dt = \dfrac{1}{2}\left\{\dfrac{\pi}{2} - \tan^{-1}(\sqrt{2}-1)\right\} \qquad\qquad (1)$$

and putting $s = (\sqrt{2}+1)$, we have

$$\int_0^\infty \dfrac{e^{-(\sqrt{2}+1)t}\sin t}{2t}dt = \dfrac{1}{2}\left\{\dfrac{\pi}{2} - \tan^{-1}(\sqrt{2}+1)\right\}$$

or $\qquad \int_0^\infty \dfrac{e^{-\sqrt{2}t}\sinh t\sin t}{t}dt = \int_0^\infty \dfrac{e^{-(\sqrt{2}-1)t}\sin t}{2t}dt - \int_0^\infty \dfrac{e^{-(\sqrt{2}+1)t}\sin t}{2t}dt$

$$= \dfrac{1}{2}\left\{\dfrac{\pi}{2} - \tan^{-1}(\sqrt{2}-1)\right\} - \dfrac{1}{2}\left\{\dfrac{\pi}{2} - \tan^{-1}(\sqrt{2}+1)\right\}$$

$$= \dfrac{1}{2}\left\{\tan^{-1}(\sqrt{2}+1) - \tan^{-1}(\sqrt{2}-1)\right\}$$

$$= \dfrac{1}{2}\tan^{-1}1 = \dfrac{1}{2}(\pi/4) = \pi/8$$

(iii) From Example 2, part (iii) we have

$$L\{erf\sqrt{t}\} = \int_0^\infty e^{-st} erf(\sqrt{t})dt = \frac{1}{s\sqrt{s+1}}$$

Putting $s = 1$, we have

$$\int_0^\infty e^{-t} erf(\sqrt{t})dt = \frac{1}{\sqrt{1+1}} = \frac{1}{\sqrt{2}}.$$

(iv) **Method 1.** We have defined Bessel's function of order n as

$$J_n(t) = \frac{t^n}{2^n \Gamma(n+1)}\left\{1 - \frac{t^2}{2(2n+2)} + \frac{t^4}{2\cdot4(2n+2)(2n+4)}\cdots\right\}$$

Putting $n = 0$

$$J_0(t) = 1 - \frac{t^2}{2^2} + \frac{t^4}{2^2\cdot4^2} - \frac{t^6}{2^2\cdot4^2\cdot6^2} + \cdots$$

Then $$L\{J_0(t)\} = \int_0^\infty e^{-st} J_0(t)dt$$

$$= L\left[1 - \frac{t^2}{2^2} + \frac{t^4}{2^2\cdot4^2} - \frac{t^6}{2^2\cdot4^2\cdot6^2} + \cdots\right]$$

$$= \frac{1}{s} - \frac{1}{2^2}\cdot\frac{2!}{s^3} + \frac{1}{2^2\cdot4^2}\cdot\frac{4!}{s^5} - \frac{1}{2^2\cdot4^2\cdot6^2}\cdot\frac{6!}{s^7} + \cdots$$

$$= \frac{1}{s}\left\{1 - \frac{1}{2}\left(\frac{1}{s^2}\right) + \frac{1\cdot3}{2\cdot4}\left(\frac{1}{s^4}\right) + \frac{1\cdot3\cdot5}{2\cdot4\cdot6}\left(\frac{1}{s^6}\right) + \cdots\right\}$$

$$= \frac{1}{s}\left\{\left(1 + \frac{1}{s^2}\right)^{-1/2}\right\} = \frac{1}{\sqrt{s^2+1}}$$

Putting $s = 0$, we have

$$\int_0^\infty J_0(t)dt = 1$$

Method 2. Using differential equations.

$J_n(t)$—the Bessel's function of order n satisfies the Bessel's differential equation

$$t^2\frac{d^2 J_n(t)}{dt^2} + \frac{tdJ_n(t)}{dt} + (t^2 - n^2)J_n(t) = 0$$

Putting $n = 0$, we have $J_0(t)$ satisfies the differential equation

$$tJ_0''(t) + J_0'(t) + tJ_0(t) = 0$$

Taking the Laplace transform of both sides we have

$$L\{tJ_0''(t)\} + L\{J_0'(t)\} + L\{tJ_0(t)\} = 0$$

or $$-\frac{d}{ds}L\{J_0''(t)\} + L\{J_0'(t)\} - \frac{d}{ds}L\{J_0(t)\} = 0$$

If $L\{J_0(t)\} = y$ we have

$$-\frac{d}{ds}\{s^2 y - sJ_0(0) - J_0'(0)\} + \{sy - J_0(0)\} - \frac{dy}{ds} = 0$$

But $J_0' = 0$ and $J_0(0) = 1$

or
$$\frac{dy}{ds} = -\frac{sy}{s^2 + 1}$$

Thus
$$y = \frac{c}{\sqrt{s^2 + 1}}$$

Using initial value theorem, we have

$$\lim_{t \to 0} F(t) = \lim_{s \to \infty} sf(s)$$

$$\lim_{s \to \infty} sy = \lim_{s \to \infty} \frac{cs}{\sqrt{s^2 + 1}} = c \lim_{t \to \infty} J_0(t) = 1$$

or
$$c = 1$$

$\therefore$
$$y = \frac{1}{\sqrt{s^2 + 1}}$$

But
$$y = L\{J_0(t)\} = \int_0^\infty e^{-st} J_0(t) dt = \frac{1}{\sqrt{s^2 + 1}}$$

Putting $s = 0$, we have

$$\int_0^\infty J_0(t) dt = 1$$

Hence the result.

Example 5. *Find the Lapalce transforms of*

(i) $\{t^{1/2} + t^{-1/2}\}$ (ii) $\dfrac{e^{-2t}}{\sqrt{t}}$ (iii) $t^{7/2} e^{3t}$

(iv) $\{e^{-t} \sin^2 t\}$ (v) $\displaystyle\int_0^t \frac{1 - e^{-u}}{u} du$ (vi) $\displaystyle\int_0^t (u^2 - u + e^{-u}) du$

Solution: (i) $L\{t^{1/2} + t^{-1/2}\} = L\{t^{1/2}\} + L\{t^{-1/2}\}$

$$= \frac{\left\lfloor\dfrac{3}{2}\right.}{s^{3/2}} + \frac{\left\lfloor\dfrac{1}{2}\right.}{s^{1/2}}$$

$$= \frac{1}{2}\frac{\sqrt{\pi}}{s^{3/2}} + \frac{\sqrt{\pi}}{s^{1/2}} = \frac{\sqrt{\pi}}{2\sqrt{s}}\left\{\frac{1}{s} + 2\right\}$$

$$= \frac{1}{2}\sqrt{\frac{\pi}{s}}\left\{\frac{2 + s}{s}\right\}$$

(ii)
$$L\left\{\frac{1}{\sqrt{t}}\right\} = L\{t^{-1/2}\} = \frac{\left\lfloor\dfrac{1}{2}\right.}{s^{1/2}} = \frac{\sqrt{\pi}}{s^{1/2}}$$

We have if

$$L\{F(t)\} = f(s)$$

Then

$$L\{e^{at}F(t)\} = f(s-a) \qquad\qquad [using \text{ (Eq. 8.8)}]$$

$$\therefore \qquad L\{e^{-2t}t^{-1/2}\} = \sqrt{\dfrac{\pi}{(s+2)}}$$

(iii)

$$L\{t^{7/2}\} = \dfrac{\left|\dfrac{9}{2}\right.}{s^{9/2}} = \dfrac{\dfrac{7}{2}\cdot\dfrac{5}{2}\cdot\dfrac{3}{2}\cdot\dfrac{1}{2}\sqrt{\pi}}{s^{9/2}}$$

$$= \dfrac{105}{16}\dfrac{\sqrt{\pi}}{s^{9/2}}$$

$$\therefore \qquad L\{e^{3t}\cdot t^{7/2}\} = \dfrac{105}{16}\dfrac{\sqrt{\pi}}{(s-3)^{9/2}}$$

(iv)

$$L\{e^{-t}\sin^2 t\} = L\left\{e^{-t}\dfrac{(1-2\cos 2t)}{2}\right\}$$

$$= L\left\{\dfrac{e^{-t}}{2}\right\} - L\left\{\dfrac{e^{-t}\cdot\cos 2t}{2}\right\}$$

Now

$$L\{1\} = \dfrac{1}{s} \qquad \therefore \quad L\{e^{-t}\} = \dfrac{1}{s+1}$$

$$L\{\cos 2t\} = \dfrac{s}{s^2+4} \qquad \therefore \quad L\{e^{-t}\cos 2t\} = \dfrac{s+1}{(s+1)^2+4}$$

$$\therefore \qquad L\{e^{-t}\sin^2 t\} = \dfrac{1}{2}\left\{\dfrac{1}{s+1} - \dfrac{s+1}{s^2+2s+5}\right\}$$

$$= \dfrac{2}{(s+1)(s^2+2s+5)}$$

(v) Let

$$F(t) = \int_0^t \dfrac{1-e^{-u}}{u}\,du \qquad \therefore \quad F(0) = 0$$

and

$$F'(t) = \dfrac{1-e^{-t}}{t}$$

$$\therefore \qquad L\{F'(t)\} = sL\{F(t)\} - F(0) = sL\{F(t)\}$$

But

$$L\{F'(t)\} = L\left\{\dfrac{1}{t}\right\} - L\left\{\dfrac{e^{-t}}{t}\right\} = \int_s^\infty \dfrac{1}{s}\,ds - \int_s^\infty \dfrac{ds}{1+s}$$

$$L\{F'(t)\} = -\log s + \log(1+s) = \log\left(1+\dfrac{1}{s}\right)$$

$$\therefore \qquad sL\{F(t)\} = \log\left(1+\dfrac{1}{s}\right)$$

or

$$L\cdot\{F(t)\} = L\int_0^t \dfrac{1-e^{-u}}{u}\,du = \dfrac{1}{s}\log\left(1+\dfrac{1}{s}\right)$$

(vi) Let
$$F(t) = \left\{ \int_0^t (u^2 - u + e^{-u}) du \right\} \qquad \qquad \therefore \ F(0) = 0$$

and
$$F'(t) = (t^2 - t + e^{-t})$$

$$\therefore \qquad L\{F'(t)\} = sL\{F(t)\} - F(0) = L\{t^2 - t + e^{-t}\}$$

or
$$sL\{F(t)\} = L\{t^2\} - L\{t\} + L\{e^{-t}\}$$

$$= \frac{2!}{s^3} - \frac{1}{s} + \frac{1}{s+1} = \frac{2}{s^3} - \frac{1}{s} + \frac{1}{s+1}$$

$$\therefore \qquad L\{F(t)\} = L\int_0^t (u^2 - u + e^{-u}) du$$

$$= \frac{2}{s^4} - \frac{1}{s^2} + \frac{1}{s(s+1)} = \frac{(2 + 2s - s^2)}{s^4(1+s)}$$

8.4. THE INVERSE LAPLACE TRANSFORM

If Laplace transform of function $F(t)$ is $f(s)$, i.e. $L\{F(t)\} = F(s)$, then $F(t)$ is called an inverse Laplace transform of $f(s)$, i.e.

$$F(t) = L^{-1}\{f(s)\} \qquad \qquad (8.25)$$

where L^{-1} is called the inverse Laplace transformation operator.

It can be shown that the inverse Laplace transform is not unique in the strict sense of the concept of uniqueness. To show this let us define a null function[*] $N(t)$ such that for all $t > 0$

$$\int_0^t N(u) du = 0 \qquad \qquad (8.26)$$

The Laplace transform of this function is zero, i.e.
$$L\{N(t)\} = 0$$

Hence
$$L\{F(t) + N(t)\} = L\{F(t)\} + L\{N(t)\}$$
$$= L\{F(t)\} = f(s)$$

Consequently
$$L^{-1}\{f(s)\} = F(t)$$

and also
$$L^{-1}\{f(s)\} = F(t) + N(t)$$

From this it follows that we can have two different functions with the same Laplace transforms.

If we allow null functions, we see that the inverse Laplace transform is not unique. It is unique, however, if we disallow null functions [this does not, in general, arise in case of physical interest]. We can state this observation in terms of a theorem known as Lerch's theorem.

Lerch's theorem: If we restrict ourselves to function $F(t)$ which are sectionally continuous in every finite interval $0 \le t \le a$ and of exponential order for $t > a$, then the inverse Laplace transform of $f(s)$, i.e. $L^{-1}\{f(s)\} = F(t)$ is unique. We shall always assume such uniqueness unless stated otherwise.

---

[*] In general, any function which is zero at all but a countable set of points, i.e. a set of points which can be put into one-to-one correspondence with the natural numbers 1, 2, 3... is a null function.

S.No.	$\{F(t)\}$	$f(s) = L\{F(t)\}$

Table 8.1: Laplace transforms of some elementary functions

S.No.	$\{F(t)\}$	$f(s) = L\{F(t)\}$
1.	1	$\dfrac{1}{s}; s>0$
2.	t	$\dfrac{1}{s^2}; s>0$
3.	t^n (n is an integer > 0)	$\dfrac{n!}{s^{n+1}}; s>0$
4.	$\dfrac{1}{\sqrt{t}}$	$\sqrt{\dfrac{\pi}{s}}; s>0$
5.	$\sqrt{t}$	$\dfrac{1}{2}\dfrac{\sqrt{\pi}}{s^{3/2}}; s>0$
6.	$t^{n-1/2}$ (n is an integer $n>0$)	$\dfrac{\sqrt{\pi}}{2^n}\dfrac{1\cdot3\cdot5\cdots(2n-1)}{s^{n+1/2}}; s>0$
7.	e^{at}	$\dfrac{1}{s-a}; s>0$
8.	$\sin at$	$\dfrac{a}{s^2+a^2}; s>0$
9.	$\cos at$	$\dfrac{s}{s^2+a^2}; s>0$
10.	$\sinh at$	$\dfrac{a}{s^2-a^2}; s>\lvert a\rvert$
11.	$\cosh at$	$\dfrac{s}{s^2-a^2}; s>\lvert a\rvert$
12.	$J_0(at)$	$\dfrac{1}{\sqrt{s^2+a^2}}$
13.	e^{iat}	$\dfrac{s+ia}{s^2+a^2}$
14.	$\delta(t)$ (Dirac delta function)	1
15.	$\log t$	$\dfrac{\Gamma'(1)-\log s}{s}$ or $-\dfrac{\gamma+\log s}{s}$ (γ is Euler's constant)

8.5. CONVOLUTION OR FALTUNG THEOREM

If $L\{F(t)\} = f(s)$ and $L\{G(t)\} = g(s)$, then

$$L^{-1}\{f(s)g(s)\} = \int_0^t F(u)G(t-u)\,du = F^*G^1$$

The theorem is proved if we can prove that

$$L\left\{\int_0^t F(u)G(t-u)\,du\right\} = f(s)\,g(s) \tag{8.27}$$

---

[*] The integral $\int_0^t F_1(u)F_2(t-u)\,du$ is often denoted by $F_1 * F_2$ and called the convolution or Faltung of F_1 and F_2.

To show this we have

$$L\left\{\int_0^t F(u)G(t-u)du\right\} = \int_{t=0}^\infty e^{-st}dt\int_{u=0}^t F(u)G(t-u)du$$

$$= \int_{t=0}^\infty\int_{u=0}^t e^{-st}F(u)G(t-u)du\,dt$$

$$= \lim_{M\to\infty}\int_{t=0}^M\int_{u=0}^t e^{-st}F(u)G(t-u)du\,dt \qquad (8.28)$$

The upper limits in integrals are held finite in order to avoid complications when changing the limits. The region of integration of Eq. (8.28) in the tu plane is shown shaded in Fig. 8.1(a).

The shaded region R_{tu} of the tu plane is transformed into the shaded region R_{uv} of the uv plane shown in Fig. 8.1(b). Then by the theorem on transformation of multiple integrals we have

$$\int_{t=0}^M\int_{u=0}^t e^{-st}F(u)G(t-u)du\,dt = \int_{u=0}^M\int_{v=0}^t e^{-s(u+v)}F(u)G(v)\left|\frac{\partial(u,t)}{\partial(u,v)}\right|du\,dv$$

where the Jacobian transformation is

$$J = \frac{\partial(u,t)}{\partial(u,v)} = \begin{vmatrix}\dfrac{\partial u}{\partial u} & \dfrac{\partial u}{\partial v}\\[2mm] \dfrac{\partial t}{\partial u} & \dfrac{\partial t}{\partial v}\\[2mm] \dfrac{\partial u}{\partial u} & \dfrac{\partial v}{\partial v}\end{vmatrix} = \begin{vmatrix}1 & 0\\ 1 & 1\end{vmatrix} = 1$$

Then we have

$$\lim_{M\to\infty}\int_{t=0}^M\int_{u=0}^t e^{-st}F(u)G(t-u)du\,dt = \lim_{M\to\infty}\int_{u=0}^M\int_{v=0}^M e^{-s(u+v)}F(u)\,G(v)\,du\,dv$$

$$= \int_0^\infty e^{-su}F(u)du\int_0^\infty e^{-sv}G(v)dv$$

$$= f(s)\,g(s)$$

or

$$L\left\{\int_0^t F(u)G(t-u)du\right\} = f(s)\,g(s)$$

which establishes the theorem.

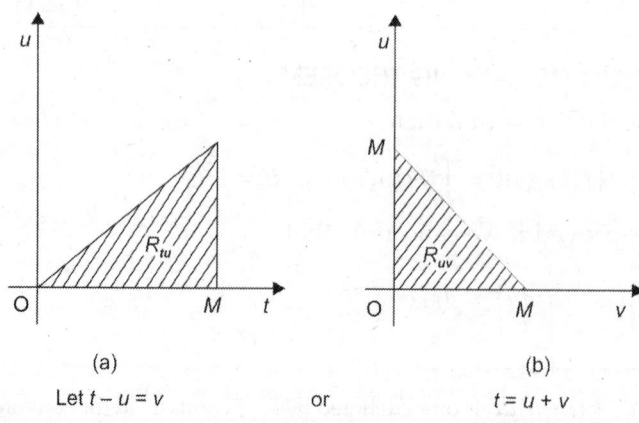

(a)   or   (b)

Let $t-u=v$ $t=u+v$

Fig. 8.1

8.6. HEAVYSIDE EXPANSION FORMULA (THEOREM)

If $P(s)$ and $Q(s)$ are polynomials such that $P(s)$ has degree less than that of $Q(s)$ and $Q(s)$ has n distinct zeros[*] α_k, $k = 1, 2, 3...$, then

$$L^{-1}\left\{\frac{P(s)}{Q(s)}\right\} = \sum_{k=1}^{n} \frac{P(\alpha_k)}{Q'(\alpha_k)}e^{\alpha kt} \qquad \qquad ...(8.29)$$

Since $Q(s)$ is a polynomial with n distinct zeros, $\alpha_1, \alpha_2,..., \alpha_n$, we can write

$$\frac{P(s)}{Q(s)} = \frac{A_1}{s-\alpha_1} + \frac{A_2}{s-\alpha_2} + \cdots + \frac{A_k}{s-\alpha_k} + \cdots + \frac{A_n}{s-\alpha_n}$$

Multiplying both sides by $s - \alpha_k$ and taking the limit as $s \to \alpha_k$ we have

$$\lim_{s\to\alpha_k} \frac{P(s)}{Q(s)}(s-a_k) = A_k$$

or

$$A_k = \lim_{s\to\alpha_k} P(s) \lim_{s\to\alpha_k} \frac{s-\alpha_k}{Q(s)} = P(\alpha_k)\lim_{s\to\alpha_k}\frac{1}{Q'(s)}$$

$$= \frac{P(\alpha_k)}{Q'(\alpha_k)}$$

Thus

$$\frac{P(s)}{Q(s)} = \frac{P(\alpha_1)}{Q'(\alpha_1)}\cdot\frac{1}{s-\alpha_1} + \frac{P(\alpha_2)}{Q'(\alpha_2)}\cdot\frac{1}{s-\alpha_2} + \cdots + \frac{P(\alpha_k)}{Q'(\alpha_k)}\frac{1}{s-\alpha_k} + \cdots n \text{ terms}$$

Then taking the inverse Laplace transform we have

$$L^{-1}\left\{\frac{P(s)}{Q(s)}\right\} = \frac{P(\alpha_1)}{Q'(\alpha_1)}e^{\alpha_1 t} + \cdots + \frac{P(\alpha_k)}{Q'(\alpha_k)}e^{\alpha_k t} + \cdots + \frac{P(\alpha_n)}{Q'(\alpha_n)}e^{\alpha_n t}$$

$$= \sum_{k=1}^{n} \frac{P(\alpha_k)}{Q'(\alpha_k)}e^{\alpha_k t}$$

8.7. METHODS OF FINDING INVERSE LAPLACE TRANSFORMS

1. **Partial fraction method:** Any rational function $\dfrac{P(s)}{Q(s)}$ where $P(s)$ and $Q(s)$ are polynomials, with the degree of $P(s)$ less than that of $Q(s)$ can be written as the sum of rational functions called partial fractions having the form $\dfrac{A}{(as+b)^r}$, $\dfrac{As+B}{(as^2+bs+c)^r}$, where $r = 1, 2, 3,....$ By finding the inverse Laplace transform of each of the partial fractions, we can find $L^{-1}\left\{\dfrac{P(s)}{Q(s)}\right\}$.

2. **Series method:** If $f(s)$ has a series expansion in inverse powers of s given by

$$f(s) = \frac{a_0}{s} + \frac{a_1}{s^2} + \frac{a_2}{s^3} + \cdots$$

---

[*] see Chapter 6—*Complex Variables.*

Then under suitable conditions we can find inverse Laplace transform term-by-term as

$$F(t) = L^{-1}\{f(s)\}$$

$$= a_0 + a_1 t + \frac{a_2 t^2}{2!} + \frac{a_3 t^3}{3!} + \cdots$$

3. Method of differential equations

4. Differentiation with respect to parameter

5. The complex inversion formula: It provides a direct means for obtaining the inverse Laplace transform of a given function. If $L\{F(t)\} = f(s)$ then $L^{-1}\{f(s)\}$ is given by

$$F(t) = \frac{1}{2\pi i} \int_{\alpha-i\infty}^{\alpha+i\infty} e^{st} f(s) ds; \ t > 0 \qquad (8.30)$$

and
$$F(t) = 0 \quad \text{for } t < 0.$$

This result is called the *complex inversion integral* or formula. It is also known as *Bromwich's integral formula.*

To prove this, we start with the Fourier integral formula which is written as

$$\phi(t) = \frac{1}{2\pi} \int_{-\infty}^{+\infty} e^{i\lambda t} d\lambda \int_{-\infty}^{+\infty} \phi(u) e^{-i\lambda u} du \qquad (8.31)$$

Let
$$\phi(t) = e^{-\alpha t} F(t) \quad \text{for } t > 0$$
$$= 0 \quad \text{for } t < 0$$

$$\therefore \qquad e^{-\alpha t} F(t) = \frac{1}{2\pi} \int_{-\infty}^{+\infty} e^{i\lambda t} d\lambda \int_{-\infty}^{+\infty} e^{-\alpha u} F(u) e^{-i\lambda u} du$$

$$= \frac{1}{2\pi} \int_{-\infty}^{+\infty} e^{i\lambda t} d\lambda \int_{-\infty}^{+\infty} e^{-(\alpha+i\lambda)u} F(u) du$$

$$= \frac{1}{2\pi} \int_{-\infty}^{+\infty} e^{i\lambda t} f(\alpha + i\lambda) d\lambda$$

Let $\alpha + i\lambda = s$ $\qquad \therefore \qquad d\lambda = \dfrac{ds}{i}$

$$\therefore \qquad e^{-\alpha t} F(t) = \frac{1}{2\pi i} \int_{\alpha-i\infty}^{\alpha+i\infty} e^{(s-\alpha)t} f(s) ds$$

$$= \frac{e^{-\alpha t}}{2\pi i} \int_{\alpha-i\infty}^{\alpha+i\infty} e^{st} f(s) ds$$

or
$$F(t) = \frac{1}{2\pi i} \int_{\alpha-i\infty}^{\alpha+i\infty} e^{st} f(s) ds$$

$$= \sum \text{residues of } e^{st} f(s) \text{ at poles of } f(s)$$

By using the residue theorem.

Example 1. *Find inverse Laplace transforms of*

(i) $\dfrac{1}{(s+1)(s^2+1)}$

(ii) $\dfrac{s^2}{(s^2+4)^2}$

(iii) $\dfrac{1}{s^2(s+1)^2}$

(iv) $\dfrac{s}{(s^2+a^2)^2}$

Solution: (i) **Method 1.** We can write $\dfrac{1}{(s+1)(s^2+1)} = \dfrac{1}{s+1} \cdot \dfrac{1}{s^2+1}$

Since $L^{-1}\left\{\dfrac{1}{s+1}\right\} = e^{-t}$ and $L^{-1}\left\{\dfrac{1}{s^2+1}\right\} = \sin t$

Then using convolution theorem, we have

$$L^{-1}\left\{\dfrac{1}{s+1} \cdot \dfrac{1}{s^2+1}\right\} = \int_0^t e^{-u} \sin(t-u)\,du$$

On integration, we have

$$L^{-1}\left\{\dfrac{1}{s+1} \cdot \dfrac{1}{s^2+1}\right\} = \dfrac{1}{2}\{\sin t \quad \cos t \mid e^{-t}\}$$

Method 2. Breaking the function in partial fractions, we have

$$\dfrac{1}{(s+1)(s^2+1)} = \dfrac{A}{s+1} + \dfrac{Bs+C}{(s^2+1)}$$

Multiplying throughout with $s + 1$ and putting $s = -1$ yields

$$A = \dfrac{1}{2}$$

On further simplification yields $B = -\dfrac{1}{2}$ and $c = \dfrac{1}{2}^{*}$

Hence we have

$$\dfrac{1}{(s+1)(s^2+1)} = \dfrac{1}{2}\left\{\dfrac{1}{s+1} + \dfrac{1-s}{s^2+1}\right\}$$

$$= \dfrac{1}{2}\left\{\dfrac{1}{s+1} - \dfrac{s}{s^2+1} + \dfrac{1}{s^2+1}\right\}$$

---

* $\dfrac{1}{(s+1)(s^2+1)} = \dfrac{1}{2}\dfrac{1}{(s+1)} + \dfrac{Bs+c}{(s^2+1)} = \dfrac{\dfrac{1}{2}(s^2+1)+(s+1)(Bs+c)}{(s+1)(s^2+1)}$

$$= \dfrac{s^2\left(\dfrac{1}{2}+B\right)+(B+c)s+\left(c+\dfrac{1}{2}\right)}{(s+1)(s^2+1)}$$

Equating the coefficients of s, s^2 on both sides, we have

$$B = -\dfrac{1}{2} \text{ and } C = +\dfrac{1}{2}$$

$$\therefore \quad L^{-1}\left\{\frac{1}{(s+1)(s^2+1)}\right\} = \frac{1}{2}L^{-1}\left\{\frac{1}{s+1}\right\} - \frac{1}{2}L^{-1}\left\{\frac{s}{s^2+1}\right\} + \frac{1}{2}L^{-1}\left\{\frac{1}{s^2+1}\right\}$$

$$= \frac{1}{2}e^{-t} - \frac{1}{2}\cos t + \frac{1}{2}\sin t = \frac{1}{2}\{e^{-t} - \cos t + \sin t\}$$

(ii)
$$\frac{s^2}{(s^2+4)^2} = \frac{s}{(s^2+4)} \times \frac{s}{(s^2+4)}$$

Since $L^{-1}\left\{\dfrac{s}{s^2+4}\right\} = \cos 2t$

Hence using convolution theorem we have

$$L^{-1}\left\{\frac{s^2}{(s^2+4)^2}\right\} = L^{-1}\left\{\frac{s}{s^2+4} \cdot \frac{s}{s^2+4}\right\} = \int_0^t \cos 2u \cos 2(t-u) du$$

$$= \int_0^t \cos 2u\{\cos 2t \cos 2u + \sin 2t \sin 2u\} du$$

$$= \int_0^t \cos 2t \cdot \cos^2 2u\, du + \int_0^t \sin 2t \cos 2u \cdot \sin 2u\, du$$

$$= \frac{\cos 2t}{2}\int_0^t (1 + \cos 4u) du + \frac{\sin 2t}{2}\int_0^t \sin 4u\, du$$

$$= \frac{\cos 2t}{2}\left[u + \frac{\sin 4u}{4}\right]_0^t + \frac{\sin 2t}{2}\left[\frac{-\cos 4u}{4}\right]_0^t$$

$$= \frac{\cos 2t}{2}\left[t + \frac{\sin 4t}{4}\right] + \frac{\sin 2t}{2}\left[\frac{1}{4} - \frac{\cos 4t}{4}\right]$$

$$= \frac{1}{2}t\cos 2t + \frac{1}{8}(\cos 2t \sin 4t - \sin 2t \cos 4t) + \frac{1}{8}\sin 2t$$

$$= \frac{1}{2}t\cos 2t + \frac{1}{8}\sin 2t + \frac{1}{8}\sin 2t\,.$$

$$= \frac{1}{2}t\cos 2t + \frac{1}{4}\sin 2t$$

(iii)
$$\frac{1}{s^2(s+1)^2} = \frac{1}{s^2} \cdot \frac{1}{(s+1)^2}$$

and
$$L^{-1}\left\{\frac{1}{s^2}\right\} = t \quad \text{and} \quad L^{-1}\left\{\frac{1}{(s+1)^2}\right\} = te^{-1}$$

$$\therefore \quad L^{-1}\left\{\frac{1}{s^2} \cdot \frac{1}{(s+1^2)}\right\} = \int_0^t ue^{-u}(t-u) du \qquad \text{(using convolution theorem)}$$

$$= t\int_0^t ue^{-u} du - \int_0^t u^2 e^{-u} du$$

$$= t\left[\left\{\frac{ue^{-u}}{(-1)}\right\}_0^t - \int_0^t \frac{e^{-u}}{(-1)} du\right] - \left\{\frac{u^2 e^{-u}}{(-1)}\right\}_0^t + \int_0^t \frac{2ue^{-u}}{(-1)} du$$

$$= t(-te^{-t}) + t\int_0^t e^{-u}\,du + t^2 e^{-t} - 2\int_0^t ue^{-u}\,du$$

$$= t\left\{\frac{e^{-u}}{(-1)}\right\}_0^t - 2\left\{\frac{ue^{-u}}{(-1)}\right\}_0^t + 2\int_0^t \frac{e^{-u}}{(-1)}\,du$$

$$= t(1-e^{-t}) + 2te^{-t} - 2\left\{\frac{e^{-u}}{-1}\right\}_0^t$$

$$= t(1-e^{-t}) + 2\,te^{-t} + 2\,(e^{-t}-1)$$

$$= t(1+e^{-1}) + 2\,(e^{-t}-1)$$

(iv)
$$\frac{s}{(s^2+a^2)^2} = \frac{1}{a}\frac{s}{s^2+a^2}\cdot\frac{a}{s^2+a^2}$$

$$L^{-1}\left\{\frac{s}{(s^2+a^2)}\right\} = \cos at \quad \text{and} \quad L^{-1}\left\{\frac{a}{s^2+a^2}\right\} = \sin at$$

and using convolution theorem, we have

$$L^{-1}\left\{\frac{s}{(s^2+a^2)^2}\right\} = \frac{1}{a}\int_0^t \cos au \sin a(t-u)\,du$$

$$= \frac{1}{a}\int_0^t \cos au\,(\sin at \cos au - \cos at \sin au)\,du$$

$$= \frac{1}{a}\sin at\int_0^t \cos^2 au\,du - \frac{1}{a}\cos at\int_0^t \cos au \sin au\,du$$

$$= \frac{1}{2a}\sin at\int_0^t (\cos 2au + 1)\,du - \frac{1}{2a}\cos at\int_0^t \sin 2au\,du$$

$$= \frac{t\sin at}{2a}$$

Example 2. *Find inverse Laplace transform of*

(i) $e^{-\sqrt{s}}$

(ii) $\dfrac{e^{-x\sqrt{s}}}{\sqrt{s}}$

(iii) $\log\dfrac{s^2}{s^2+4}$

(iv) $\dfrac{e^{-s}(1-e^{-s})}{s(s^2+1)}$

Solution: (i) **Method 1.** Let $y(s) = e^{-\sqrt{s}}$

Then
$$\frac{dy}{ds} = -\frac{e^{-\sqrt{s}}}{2s^{1/2}} \quad \text{and} \quad \frac{d^2y}{ds^2} = \frac{e^{-\sqrt{s}}}{4s} + \frac{e^{-\sqrt{s}}}{4s^{3/2}}$$

Thus
$$4s\frac{d^2y}{ds^2} + 2\frac{dy}{ds} - y = 0 \tag{1}$$

If
$$L\{Y(t)\} = y(s) \text{ then}$$

$$\frac{d^2y}{ds^2} = L\{t^2 Y(t)\}$$

or
$$s\frac{d^2y}{ds^2} = L\left\{\frac{d}{dt}t^2Y(t)\right\} = L\{t^2Y'(t) + 2tY(t)\} \tag{2}$$

and
$$\frac{dy}{ds} = L\{-tY(t)\} \tag{3}$$

Using Eqs (2) and (3) one yields
$$4L\{t^2Y'(t) + 2tY(t)\} - 2L\{tY(t)\} - L\{Y(t)\} = 0°$$

or
$$\frac{dY(t)}{Y(t)} + \frac{(6t-1)}{4t^2}dt = 0$$

or
$$\log Y(t) + \frac{3}{2}\log t + \frac{1}{4t} = c_1$$

or
$$Y(t) = \frac{c}{t^{3/2}}e^{-1/4t}$$

or
$$tY(t) = \frac{c}{t^{1/2}}e^{-1/4t}$$

and
$$L\{tY(t)\} = -\frac{d}{ds}L\{Y(t)\}$$

$$= -\frac{d}{ds}(e^{-\sqrt{s}}) = \frac{e^{-\sqrt{s}}}{2\sqrt{s}}$$

For large value of t,
$$tY(t) \sim \frac{c}{t^{1/2}}$$

and
$$L\{tY(t)\} \sim L\left\{\frac{c}{t^{1/2}}\right\} = \frac{c\sqrt{\pi}}{\sqrt{s}}$$

For small value of s,
$$\frac{e^{-\sqrt{s}}}{2\sqrt{s}} \sim \frac{1}{2\sqrt{s}}$$

Hence using final value theorem, we have
$$c = \frac{1}{2\sqrt{\pi}}$$

or
$$L^{-1}\{e^{-\sqrt{s}}\} = Y(t) = \frac{1}{2\sqrt{\pi}}\frac{e^{-1/4t}}{t^{3/2}}$$

Method 2. Using infinite series, we have
$$L^{-1}\{e^{-\sqrt{s}}\} = L^{-1}\left\{1 - s^{1/2} + \frac{s}{2!} - \frac{s^{3/2}}{3!} + \frac{s^2}{4!} - \frac{s^{5/2}}{5!} + \cdots\right\}$$

$$= L^{-1}\{1\} - L^{-1}\{s^{1/2}\} + L^{-1}\left(\frac{s}{2!}\right) - L^{-1}\left(\frac{s^{3/2}}{3!}\right) + \cdots$$

Using the result

$$L^{-1}\{s^{p+1/2}\} = \frac{t^{-p-3/2}}{\Gamma(-p-1/2)} \quad \text{and} \quad L^{-1}\{s^p\} = 0$$

Hence we have

$$L^{-1}\{e^{-\sqrt{s}}\} = \frac{t^{-3/2}}{2\sqrt{\pi}} - \left(\frac{1}{2}\right)\left(\frac{3}{2}\right)\frac{t^{-5/2}}{3!\sqrt{\pi}} + \left(\frac{1}{2}\right)\left(\frac{3}{2}\right)\left(\frac{5}{2}\right)\frac{t^{-7/2}}{5!\sqrt{\pi}} + \cdots$$

Since $\quad \Gamma\left(-P-\dfrac{1}{2}\right) = \dfrac{(-1)^{p+1}\sqrt{\pi}}{\sqrt{\pi}}\left(\dfrac{2}{1}\right)\left(\dfrac{2}{3}\right)\left(\dfrac{2}{5}\right)\cdots\left(\dfrac{2}{2P+1}\right)$

or $\quad L^{-1}\{e^{-\sqrt{s}}\} = \dfrac{1}{2\sqrt{\pi}}\cdot\dfrac{1}{t^{3/2}}\left\{1-\left(\dfrac{1}{2^2t}\right)+\left(\dfrac{1}{2^2t}\right)^2\cdot\dfrac{1}{2!}-\left(\dfrac{1}{2^2t}\right)^3\cdot\dfrac{1}{3!}+\cdots\right\}$

$$= \frac{1}{2\sqrt{\pi}}\frac{e^{-1/4t}}{t^{3/2}}$$

(ii) $\quad \dfrac{e^{-x\sqrt{s}}}{\sqrt{s}} = \dfrac{1}{\sqrt{s}}\left\{1-x\sqrt{s}+\dfrac{x^2s}{2!}-\dfrac{x^3s^{3/2}}{3!}+\dfrac{x^4s^2}{4!}-\dfrac{x^5s^{5/2}}{5!}+\dfrac{x^6s^3}{6!}\cdots\right\}$

$\therefore\quad L^{-1}\left\{\dfrac{e^{-x\sqrt{s}}}{\sqrt{s}}\right\} = L^{-1}\left\{s^{-1/2}-x+\dfrac{x^2s^{1/2}}{2!}-\dfrac{x^3\cdot s^2}{3!}+\dfrac{x^4s^{3/2}}{4!}-\dfrac{x^5s^2}{5!}+\dfrac{x^6\cdot s^{5/2}}{6!}-\cdots\right\}$

Using the result

$$L^{-1}\{s^{p+1/2}\} = \frac{t^{-p-3/2}}{\Gamma\left(-P-\dfrac{1}{2}\right)}$$

$$= \frac{(-1)^{p+1}}{\sqrt{\pi}}\left(\frac{1}{2}\right)\left(\frac{3}{2}\right)\left(\frac{5}{2}\right)\cdots\left(\frac{2P+1}{2}\right)t^{-(p+3/2)}$$

and $\quad L^{-1}\{s^p\} = 0$

We have

$$L^{-1}\left\{\frac{e^{-x\sqrt{s}}}{\sqrt{s}}\right\} = \frac{1}{\sqrt{\pi t}} - \frac{1}{\sqrt{\pi t}}\left(\frac{x^2}{4t}\right) + \left(\frac{x^2}{4t}\right)^2\frac{1}{2!\sqrt{\pi t}} + \left(\frac{x^2}{4t}\right)^3\cdot\frac{1}{3!}\frac{2}{\sqrt{\pi t}} + \cdots$$

$$= \frac{1}{\sqrt{\pi t}}\left\{1-\left(\frac{x^2}{4t}\right)+\left(\frac{x^2}{4t}\right)^2\cdot\frac{1}{2!}-\left(\frac{x^2}{4t}\right)^3\cdot\frac{1}{3!}+\cdots\right\}$$

$$= \frac{1}{\sqrt{\pi t}}e^{-x^2/4t}$$

(iii) $\quad f(s) = \log\dfrac{s^2}{s^2+4} = 2\log s - \log(s^2+4)$

or $\quad -f'(s) = -\dfrac{2}{s}+\dfrac{2s}{s^2+4}$

or $\qquad L^{-1}\{-f'(s)\} = -2 + 2\cos 2t$

If $L\{F(t)\}$ is $f(s)$ then

$$L\{tF(t)\} = -\frac{df_{(s)}}{ds}$$

or $\qquad tF(t) = -2 + 2\cos 2t$

or $\qquad F(t) = L^{-1}\{f(s)\} = -\frac{2}{t}(1 - \cos 2t)$

(iv) $\qquad f(s) = \dfrac{e^{-s}(1 - e^{-s})}{s(s^2 + 1)} = \dfrac{e^{-s}}{s} - \dfrac{e^{-s} \cdot s}{s^2 + 1} - \dfrac{e^{-2s}}{s} + \dfrac{e^{-2s} \cdot s}{s^2 + 1}$

or $\qquad L^{-1}\{f(s)\} = L^{-1}\left\{\dfrac{e^{-s}}{s}\right\} - L^{-1}\left\{\dfrac{e^{-s}s}{s^2 + 1}\right\} - L^{-1}\left\{\dfrac{e^{-2s}}{s}\right\} + L^{-1}\left\{\dfrac{e^{-2s}s}{s^2 + 1}\right\}$

$$L^{-1}\left\{\dfrac{e^{-s}}{s}\right\} = 1 \quad \text{and} \quad L^{-1}\left\{\dfrac{e^{-2s}}{s}\right\} = 1 \quad \text{since } L^{-1}\{s^p\} = 0$$

$$L^{-1}\left\{\dfrac{e^{-s}s}{s^2 + 1}\right\} = \cos(t - 1) \quad \text{and} \quad L^{-1}\left\{\dfrac{e^{-2s}s}{s^2 + 1}\right\} = \cos(t - 2)^*$$

Hence, $\qquad L^{-1}\{f(s)\} = \{1 - \cos(t - 1)\} - \{1 - \cos(t - 2)\}$

8.8. APPLICATIONS OF LAPLACE TRANSFORMS

Laplace transforms are used to

i. solve differential equations:

The method provides the most convenient means for solving the differential equations of electrical network and mechanical oscillation. The chief advantage of this method is that it is very direct and easy on tedious evaluations of arbitrary constants.

ii. solve some complicated definite integrals.

iii. solve integral and differential equation.

---

*If $\qquad L^{-1}\{f(s)\} = F(t)$ then $L^{-1}\left\{e^{-as}f(s)\right\} = \begin{cases} F(t - a) & t > a \\ 0 & t < a \end{cases}$

Proof: $\qquad f(s) = \displaystyle\int_0^\infty e^{-st}F(t)dt \quad \text{or} \quad e^{-as}f(s) = \int_0^\infty e^{-as}e^{-st}F(t)dt$

or $\qquad e^{-as}f(s) = \displaystyle\int_0^\infty e^{-(a+t)s}F(t)dt = \int_a^\infty e^{-su}F(u - a)du \qquad [\because u = t + a]$

$$= \int_0^a e^{-su}(0)du + \int_a^\infty e^{-su}F(u - a)du = \int_0^\infty e^{-su}F(u - a)du$$

or $\qquad L^{-1}\left\{e^{-as}f(s)\right\} = F(t - a) \qquad\qquad [\text{for } t > a].$

Example 1. *Solve the differential equation*

$$\frac{d^2Y}{dt^2} + 2\frac{dY}{dt} + 5Y = e^{-t}\sin t$$

such that $Y(0) = 0$ and $\left.\frac{dY}{dt}\right|_{t=0} = 1$

Solution: Given

$$\frac{d^2Y}{dt^2} + 2\frac{dY}{dt} + 5Y = e^{-t}\sin t$$

Taking Laplace transforms, we have

$$L\left\{\frac{d^2Y}{dt^2}\right\} + 2L\left\{\frac{dY}{dt}\right\} + 5L\{Y\} = L\{e^{-t}\sin t\}$$

$$\left\{s^2y - sY(0) - Y'(0)\right\} + 2\left\{sy - Y(0)\right\} + 5y$$

$$= \frac{1}{(s+1)^2 + 1} = \frac{1}{s^2 + 2s + 2}$$

where y is $L\{Y(t)\}$

or $\quad\quad (s^2 + 2s + 5)y = 1 + \dfrac{1}{s^2 + 2s + 2}$

or $\quad\quad y = \dfrac{1}{s^2 + 2s + 5} + \dfrac{1}{(s^2 + 2s + 2)(s^2 + 2s + 5)}$

$$= \frac{1}{s^2 + 2s + 5} + \frac{1}{3}\left\{\frac{1}{s^2 + 2s + 2} - \frac{1}{s^2 + 2s + 5}\right\}$$

$$= \frac{2}{3} \cdot \frac{1}{s^2 + 2s + 5} + \frac{1}{3} \cdot \frac{1}{s^2 + 2s + 2}$$

$$= \frac{1}{3} \cdot \frac{2}{(s+1)^2 + (2)^2} + \frac{1}{3} \cdot \frac{1}{(s+1)^2 + (1)^2}$$

or $\quad Y(t) = L^{-1}\{y\} = \dfrac{1}{3}L^{-1}\left\{\dfrac{2}{(s+1)^2 + (2)^2}\right\} + \dfrac{1}{3}L^{-1}\left\{\dfrac{1}{(s+1)^2 + (1)^2}\right\}$

$$= \frac{1}{3} \cdot e^{-t}\sin 2t + \frac{1}{3} \cdot e^{-t}\sin t$$

$$= \frac{1}{3}e^{-t}\{\sin t + \sin 2t\}$$

Example 2. *A particle of mass m moves along the x-axis and is attracted towards origin O with a force numerically equal to Kx, K > 0. A damping force given by $\beta \cdot \dfrac{dx}{dt}, \beta > 0$ also acts. Discuss the motion, treating all cases, assuming that $x(0) = x_0$ and $x'(0) = v_0$.*

Solution: The equation of motion is

$$m\frac{d^2X}{dt^2} = -KX - \beta\frac{dX}{dt}$$

or
$$\frac{d^2X}{dt^2} + 2\alpha\frac{dX}{dt} + \omega^2 X = 0 \tag{1}$$

where
$$\alpha = \frac{\beta}{2m} \quad \text{and} \quad \omega^2 = \frac{K}{m}$$

Taking the Laplace transforms we have

$$\{s^2 x - sX(0) - X'(0)\} + 2\alpha\{sx - X(0)\} + \omega^2 x = 0$$

where x is Laplace transform of $X(t)$

or
$$(s^2 + 2\alpha s + \omega^2)x = sx_0 + (2\alpha x_0 + v_0)$$

or
$$x = \frac{sx_0 + \alpha x_0}{s^2 + 2\alpha s + \omega^2} + \frac{\alpha x_0 + v_0}{s^2 + 2\alpha s + \omega^2}$$

Case 1. $\omega^2 - \alpha^2 > 0$

$$X(t) = L^{-1}\{x\} = L^{-1}\left\{\frac{(s+\alpha)x_0}{(s+\alpha)^2 + (\omega^2 - \alpha^2)}\right\} + \frac{v_0 + \alpha x_0}{\sqrt{\omega^2 - \alpha^2}}\left\{\frac{\sqrt{\omega^2 - \alpha^2}}{(s+\alpha)^2 + \omega^2 - \alpha^2}\right\}$$

$$= x_0 e^{-\alpha t}\cos\sqrt{\omega^2 - \alpha^2}\,t + \frac{v_0 + \alpha x_0}{\sqrt{\omega^2 - \alpha^2}}e^{-\alpha t}\sin\sqrt{\omega^2 - \alpha^2}\,t$$

The motion is called damped oscillatory motion. The particle oscillates about O. The amplitude of motion becomes smaller with each swing.

The period of oscillation is $\dfrac{2\pi}{\sqrt{\omega^2 - \alpha^2}}$

Case 2. $\omega^2 - \alpha^2 = 0$

$$X(t) = L^{-1}\{x\} = L^{-1}\left\{\frac{x_0}{s+\alpha}\right\} + \frac{v_0 + \alpha x_0}{(s+\alpha)^2}$$

$$= x_0 e^{-\alpha t} + (v_0 + \alpha x_0)te^{-\alpha t}$$

$$= e^{-\alpha t}\{x_0 + (v_0 + \alpha x_0)t\}$$

Here the particle does not oscillate about O. Instead it approaches O gradually but never reaches it. The motion is called critically damped motion, since any decrease in the damping constant β, would produce oscillation.

Case 3. $\omega^2 - \alpha^2 < 0$

$$X(t) = L^{-1}\{x\} = L^{-1}\left\{\frac{(s+\alpha)x_0}{(s+\alpha)^2 - (\alpha^2 - \omega^2)}\right\} + L^{-1}\left\{\frac{v_0 + \alpha x_0}{(s+\alpha)^2 - (\alpha^2 - \omega^2)}\right\}$$

$$= x_0 e^{-\alpha t}\cosh\sqrt{\alpha^2 - \omega^2}\,t + \frac{v_0 + \alpha x_0}{\sqrt{\alpha^2 - \omega^2}}e^{-\alpha t}\sin\sqrt{\alpha^2 - \omega^2}\,t$$

The motion is called overdamped motion and is nonoscillatory.

Example 3. *An inductor L, a resistor R and a capacitor C are connected in series with an emf E. At t = 0, the charge on the capacitor and current in the circuit are zero. Find the current and charge at any time t > 0.*

Solution: Let Q and I be the instantaneous charge and current respectively at time t.

We have

$$L\frac{dI}{dt} + RI + \frac{Q}{C} = E$$

or $\qquad \dfrac{d^2Q}{dt^2} + \dfrac{R}{L}\dfrac{dQ}{dt} + \dfrac{Q}{LC} = \dfrac{E}{L}$

Taking Laplace transform and let $q = L\{Q(t)\}$, we have

$$\{s^2 q - sQ(0) - Q'(0)\} + \frac{R}{L}\{sq - Q(0)\} + \frac{1}{LC}q = \frac{E}{L}\frac{1}{s}$$

But $Q(0) = 0$

and $\quad Q'(0) = 0$

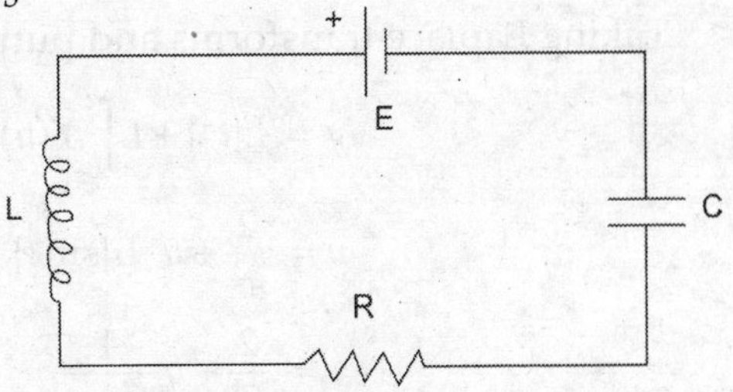

or $\quad \left(s^2 + \dfrac{R}{L}s + \dfrac{1}{LC}\right)q = \dfrac{E/L}{s}$

or $\qquad q = \dfrac{E}{s\left(s^2 + \dfrac{R}{L}s + \dfrac{1}{LC}\right)}$

Fig. 8.2

$$= \frac{EC}{s} - \frac{EC\left(s + \dfrac{R}{2L}\right)}{\left(s + \dfrac{R}{2L}\right)^2 + \left(\dfrac{1}{LC} - \dfrac{R^2}{4L^2}\right)} - \frac{\dfrac{ECR}{2L}}{\left(s + \dfrac{R}{2L}\right)^2 + \left(\dfrac{1}{LC} - \dfrac{R^2}{4L^2}\right)}$$

or $\quad Q(t) = L^{-1}\{q\} = EC - ECe^{-\frac{R}{2L}t}\cos\sqrt{\dfrac{1}{LC} - \dfrac{R^2}{4L^2}}\, t - \dfrac{ECR}{2L\sqrt{\dfrac{1}{LC} - \dfrac{R^2}{4L^2}}}e^{-\frac{R}{2L}t}\sin\sqrt{\dfrac{1}{LC} - \dfrac{R^2}{4L^2}}\, t$

or $\qquad Q(t) = EC - ECe^{-\frac{R}{2L}t}\cos\omega t - \dfrac{ECR}{2L\omega}e^{-\frac{R}{2L}t}\sin\omega t$

where $\qquad \omega = \sqrt{\dfrac{1}{LC} - \dfrac{R^2}{4L^2}}$

and current $\qquad I = \dfrac{dQ(t)}{dt} = \dfrac{ECR}{2L}e^{-\frac{R}{2L}t}\cos\omega t$

$$+ EC\omega e^{-\frac{R}{2L}t}\sin\omega t + \frac{ECR}{4L^2\omega}e^{-\frac{R}{2L}t}\sin\omega t - \frac{ECR}{2L}e^{-\frac{R}{2L}t}\cos\omega t$$

$$= \frac{EC}{\omega}\left(\omega^2 + \frac{R^2}{4L^2}\right)e^{-\frac{R}{2L}t}\sin\omega t$$

$$= \frac{E}{\omega L}e^{-\frac{R}{2L}t}\sin\omega t$$

[**Note:** Solve the above problem, if an alternating emf $E = E_0 \sin pt$ is applied]

Example 4. *Solve the integral equations*

(i) $Y(t) = t^2 + \int_0^t Y(u)\sin(t-u)\,du$

(ii) $\int_0^t \dfrac{Y(u)}{\sqrt{t-u}}\,du = 1 + t + t^2$

Solution: (i) $Y(t) = t^2 + \int_0^t Y(u)\sin(t-u)\,du$

Taking Laplace transforms and putting $L\{Y(t)\} = y$, we have

$$y = L\{t^2\} + L\int_0^t Y(u)\sin(t-u)\,du$$

$$y = \frac{2}{s^3} + y\cdot L\{\sin t\} \qquad\qquad (using \text{ convolution theorem})$$

$$= \frac{2}{s^3} + y\frac{1}{s^2+1}$$

or $\qquad y\left(1 - \dfrac{1}{s^2+1}\right) = \dfrac{2}{s^3}$

or $\qquad y = \dfrac{2(s^2+1)}{s^5} = \dfrac{2}{s^3} + \dfrac{2}{s^5}$

$\therefore \qquad Y(t) = L^{-1}\{y\} = 2L^{-1}\left(\dfrac{1}{s^3}\right) + 2L^{-1}\left(\dfrac{1}{s^5}\right)$

$$= 2\cdot\frac{t^2}{2!} + 2\cdot\frac{t^4}{4!} = t^2 + \frac{t^4}{12}$$

(ii) $\qquad \int_0^t \dfrac{Y(u)}{\sqrt{t-u}}\,du = 1 + t + t^2$

Taking Laplace transform and $L\{Y(t)\} = y$, we have

$$L\{Y(t)\}\, L\{t^{-1/2}\} = L\{1\} + L\{t\} + L\{t^2\} \qquad (using \text{ convolution theorem})$$

$$y\cdot\frac{\Gamma 1/2}{s^{1/2}} = \frac{1}{s} + \frac{1}{s^2} + \frac{2}{s^3}$$

or $\qquad y = \dfrac{1}{|1/2}\left[\dfrac{1}{s^{1/2}} + \dfrac{1}{s^{3/2}} + \dfrac{2}{s^{5/2}}\right]$

or $\qquad Y(t) = L^{-1}\{y\} = \dfrac{1}{\Gamma\dfrac{1}{2}}L^{-1}\left\{\dfrac{1}{s^{1/2}} + \dfrac{1}{s^{3/2}} + \dfrac{2}{s^{5/2}}\right\}$

$$= \frac{1}{\Gamma\dfrac{1}{2}}\left[\frac{t^{-1/2}}{\Gamma\dfrac{1}{2}} + \frac{t^{1/2}}{\Gamma\dfrac{3}{2}} + \frac{2t^{3/2}}{\Gamma\dfrac{5}{2}}\right] = \frac{1}{\sqrt{\pi}}\left[\frac{t^{-1/2}}{\sqrt{\pi}} + \frac{t^{1/2}}{\dfrac{1}{2}\sqrt{\pi}} + \frac{2t^{3/2}}{\dfrac{3}{2}\cdot\dfrac{1}{2}\sqrt{\pi}}\right]$$

$$= \frac{1}{\pi}\left[t^{-1/2} + 2t^{1/2} + \frac{8}{3}t^{3/2}\right] = \frac{t^{-1/2}}{3\pi}[3 + 6t + 8t^2]$$

Example 5. *Solve the Bessel's equation of zero order*

$$tY''(t) + Y'(t) + tY(t) = 0$$

under the condition that $Y(0) = I$ and $Y'(0) = 0$ and its derivatives have transforms.

Solution: Taking the Laplace transforms, we have

$$L\{tY''(t)\} + L\{Y'(t)\} + L\{tY(t)\} = 0$$

or

$$-\frac{d}{ds}L\{Y''(t)\} + L\{Y'(t)\} - \frac{d}{ds}L\{Y(t)\} = 0$$

$$-\frac{d}{ds}\{s^2 y - sY(0) - Y'(0)\} + \{sy - Y(0)\} - \frac{dy}{ds} = 0$$

where

$$L\{Y(t)\} = y$$

or

$$-2sy - s^2\frac{dy}{ds} + 1 + sy - 1 - \frac{dy}{ds} = 0$$

or

$$(1 + s^2)\frac{dy}{ds} = -sy$$

or

$$\frac{dy}{y} = -\frac{s}{1+s^2}ds$$

or

$$y = \frac{c}{\sqrt{1+s^2}}$$

where c is constant of integration.

or

$$y = \frac{c}{s}\left(1 + \frac{1}{s^2}\right)^{-1/2}$$

$$= \frac{c}{s}\left(1 - \frac{1}{2}\cdot\frac{1}{s^2} + \frac{1\cdot 3}{2^2\cdot 3!}\cdot\frac{1}{s^4} + \cdots\right) s > 1$$

$$= c\sum_{n=0}^{\infty}\frac{(-1)^n(2n)!}{(2^n\cdot n!)^2 s^{2n+1}}$$

or

$$Y(t) = L^{-1}\{y\} = c\sum_{n=0}^{\infty}\frac{(-1)^n\cdot(2n)!}{(2^n n!)^2}\cdot\frac{t^{2n}}{(2n)!}$$

$$= c\sum_{n=0}^{\infty}\frac{(-1)^n t^{2n}}{(2^n\cdot n!)^2}$$

But $Y(0) = 1$ we have $c = 1$

or

$$Y(t) = \sum_{n=0}^{\infty}\frac{(-1)^n t^{2n}}{(2^n\cdot n!)^2}$$

which is Bessel's function of zero order, i.e.

$$Y(t) = J_0\cdot(t)$$

Example 6. *Solve the following differential equation*

$$\frac{d^2X}{dt^2} + \lambda^2 X = 0$$

Solution: Taking the Laplace transform of the given equation, we have

$$L\left\{\frac{d^2X}{dt^2}\right\} + \lambda^2 L\{X\} = 0$$

or $s^2x - sX(0) - X'(0) + \lambda^2 x = 0$

where $L\{X(t)\} = x$ and putting $X(0) = c_1$ and $X'(0) = c_2$

We have

$$x = \frac{c_1 s + c_2}{s^2 + \lambda^2} = \frac{c_1 s + c_2}{(s + i\lambda)(s - i\lambda)}$$

$$= \frac{c_1/2 + ic_2/2\lambda}{(s + i\lambda)} + \frac{c_1/2 - ic_2/2\lambda}{(s - i\lambda)}$$

Applying Laplace's inverse theorem[*] we get

$$X(t) = L^{-1}\{x\} = \frac{A_1}{2\pi i}\int_{\alpha - i\infty}^{\alpha + i\infty}\frac{e^{st}}{s + i\lambda}ds + \frac{A_2}{2\pi i}\int_{\alpha - i\infty}^{\alpha + i\infty}\frac{e^{st}}{s - i\lambda}ds$$

where $A_1 = \dfrac{c_1}{2} + \dfrac{ic_2}{2\lambda}$ and $A_2 = \dfrac{c_1}{2} - \dfrac{ic_2}{2\lambda}$

Let $I_1 = \int_{\alpha - i\infty}^{\alpha + i\infty}\frac{e^{st}}{s + i\lambda}ds$

The complex function $\dfrac{e^{st}}{s + i\lambda}$ has a simple pole at $s = -i\lambda$ and $I_1 = 2\pi i$ (residue at $s = i\lambda$)[**].

Residue at $(s = -i\lambda) = \lim\limits_{s \to -i\lambda}\frac{(s + i\lambda)e^{st}}{(s + i\lambda)} = e^{-i\lambda}$

∴ $I_1 = 2\pi i e^{-i\lambda t}$

---

[*] Laplace inverse theorem states that if $F(t)$ is a continuous, and of exponential order and has derivative and $L\{F(t)\} = f(s)$ then

$$F(t) = L^{-1}\{f(s)\} = \frac{1}{2\pi i}\lim_{\beta \to \infty}\int_{\alpha - i\beta}^{\alpha + i\beta} e^{st} f(s)\, ds \quad t \geq 0$$

$$= \frac{1}{2\pi i}\int_{\alpha - i\alpha}^{\alpha + i\alpha} e^{st} f(s)\, ds$$

where the integral on right hand side is Bromwich integral.
[**] Cauchy's residue theorem discussed in Chapter 6—*Complex Variables*

Similarly
$$I_2 = \int_{\alpha-i\infty}^{\alpha+i\infty} \frac{e^{st}}{s-i\lambda} ds = 2\pi i \, e^{i\lambda t}$$

$\therefore$
$$X(t) = A_1 e^{-i\lambda t} + A_2 e^{i\lambda t}$$

8.9. THE FOURIER TRANSFORM

The Fourier transform of $F(x)$ is defined as

$$F\{F(x)\} = \int_{-\infty}^{+\infty} e^{-i\lambda x} \cdot F(x)dx = f(\lambda) \tag{8.32}$$

The function $f(\lambda)$ is called the Fourier transform of $F(x)$. The function $F(x)$ is the inverse Fourier transform of $f(\lambda)$ and is written as

$$F(x) = F^{-1}\{f(\lambda)\} = \frac{1}{2\pi}\int_{-\infty}^{+\infty} e^{i\lambda x} f(\lambda)d\lambda \tag{8.33}$$

We also call Eq. (8.33) an inverse formula corresponding to Eq. (8.32).

Note: The constants preceding the integral sign can be any constants whose product is $\frac{1}{2\pi}$. If taken as $\frac{1}{\sqrt{2\pi}}$, we obtain the so called symmetric form.

8.10. FOURIER SINE AND COSINE TRANSFORM

i. The (infinite) Fourier sine transform of $F(x)$, $0 < x < \infty$ is defined as

$$f_s(\lambda) = \int_0^\infty F(x)\sin \lambda x \, dx \tag{8.34}$$

The function $F(x)$ is then called the inverse Fourier sine transform of $f_s(\lambda)$ and is given by

$$F(x) = \frac{2}{\pi}\int_0^\infty f_s(\lambda)\sin \lambda x \, d\lambda \tag{8.35}$$

ii. The (infinite) Fourier cosine transform of $F(x)$, $0 < x < \infty$, is defined as

$$f_c(\lambda) = \int_0^\infty F(x)\cos \lambda x \, dx \tag{8.36}$$

The function $F(x)$ is then called the inverse Fourier cosine transform of $f_c(\lambda)$ and is given by

$$F(x) = \frac{2}{\pi}\int_0^\infty f_c(\lambda)\cos \lambda x \, d\lambda \tag{8.37}$$

iii. The finite Fourier sine transform of $F(x)$, $0 < x < 1$, is defined as

$$f_s(n) = \int_0^t F(x)\sin\frac{n\pi x}{l}dx \tag{8.38}$$

where n is an integer. The function $F(x)$ is then called the inverse finite Fourier sine transform of $f_s(n)$ and is given by

$$F(x) = \frac{2}{l}\sum_l^\infty f_s(n)\sin\frac{n\pi x}{l} \tag{8.39}$$

iv. The finite Fourier cosine transform of $F(x)$, $0 < x < 1$, is defined as

$$f_c(n) = \int_0^l F(x)\cos\frac{n\pi x}{l}dx \tag{8.40}$$

where n is an integer. The function $F(x)$ is then called the inverse finite Fourier cosine transform of $f_c(n)$ and is given by

$$F(x) = \frac{1}{l}f_c(0) + \frac{2}{l}\sum_{n=1}^{\infty} f_c(n)\cos\frac{n\pi x}{l} \tag{8.41}$$

8.11. FOURIER SINE AND COSINE TRANSFORM OF DERIVATIVES

If u is some function of x and t, then finite Fourier sine and cosine transform of $\dfrac{\partial u}{\partial x}$

and $\dfrac{\partial^2 u}{\partial x^2}$ for $0 < x\ 1$ and $t > 0$ is obtained as follows:

i. By definition, the Fourier sine transform of $\dfrac{\partial u}{\partial x}$ is equal to

$$F_s\left\{\frac{\partial u}{\partial x}\right\} = \int_0^l \frac{\partial u}{\partial x}\sin\frac{n\pi x}{l}dx$$

On integrating by parts, we have

$$F_s\left\{\frac{\partial u}{\partial x}\right\} = \int_0^l \frac{\partial u}{\partial x}\sin\frac{n\pi x}{l}dx$$

$$= u(x,t)\sin\frac{n\pi x}{l}\Big|_0^l - \frac{n\pi}{l}\int_0^l u(x,t)\cos\frac{n\pi x}{l}dx$$

$$= -\frac{n\pi}{l}F_c\{u(x,t)\} \tag{8.42}$$

ii. The finite Fourier cosine transform of $\dfrac{\partial u}{\partial x}$ is equal to

$$F_c\left\{\frac{\partial u}{\partial x}\right\} = \int_0^l \frac{\partial u}{\partial x}\cos\frac{n\pi x}{l}dx$$

$$= u(x,t)\cos\frac{n\pi x}{l}\Big|_0^l + \frac{n\pi}{l}\int_0^l u(x,t)\sin\frac{n\pi x}{l}dx$$

$$= \frac{\pi n}{l}F_s\{u(x,t)\} + u(l,t)\cos n\pi - u(0,t) \tag{8.43}$$

iii. Replacing $u(x, t)$ by $\dfrac{\partial u}{\partial x}$ in Eq. (8.42) we have

$$F_s\left\{\frac{\partial^2 u}{\partial x^2}\right\} = -\frac{n\pi}{l}F_c\left\{\frac{\partial u}{\partial x}\right\}$$

and substituting the value of $F_c\left\{\dfrac{\partial u}{\partial x}\right\}$ from Eq. (8.43), we have

$$F_s\left\{\frac{\partial^2 u}{\partial x^2}\right\} = -\frac{n\pi}{l}\left[\frac{n\pi}{l}\cdot F_s\{u(x,t\} + u(l,t)\cos n\pi - u(0,t)\right]$$

$$= -\frac{n^2\pi^2}{l^2}F_s\{u(x,t)\} + \frac{n\pi}{l}\{u(0,t) - u(l,t)\cos n\pi\} \qquad (8.44)$$

iv. Replacing $u(x, t)$ by $\dfrac{\partial u}{\partial x}$ in Eq. (8.43), we have

$$F_c\left\{\frac{\partial^2 u}{\partial x^2}\right\} = \frac{n\pi}{l}F_x\left\{\frac{\partial u}{\partial x}\right\} + u_x(l,t)\cos n\pi - u_x(0,t)$$

and substituting the value of $F_x\left(\dfrac{\partial u}{\partial x}\right)$ from Eq. (8.40), we have

$$F_c\left\{\frac{\partial^2 u}{\partial x^2}\right\} = -\frac{n^2\pi^2}{l^2}F_c\{u_x(x,t)\} - \{u_x(0,t) - u_x(l,t)\cos n\pi\} \qquad (8.45)$$

where u_x denotes the partial derivative with respect to x.

8.12. THE CONVOLUTION OR FALTUNG THEOREM FOR FOURIER TRANSFORM

The convolution of two functions $F(x)$ and $G(x)$ where $-\infty < x < +\infty$ is defined as

$$F\{F(x)*G(x)\} = F\left[\int_{-\infty}^{+\infty} F(u)G(x-u)du\right] \qquad (8.46)$$

The convolution theorem for Fourier transforms says that the Fourier transform of the convolution of $F(x)$ and $G(x)$ is the product of the Fourier transform of $F(x)$ and $G(x)$, i.e.

$$F\{F(x)*G(x)\} = F\{F(x)\}\ F\{G(x)\}, \qquad (8.47)$$

and to prove this theorem let us take Fourier transform of $F(x)*G(x)$. We have

$$F\{F(x)*G(x)\} = \int_{-\infty}^{+\infty} e^{-i\lambda x}dx\int_{-\infty}^{+\infty} F(u)G(x-u)du$$

Let $x - u = P$ or $dx = dP$

$$\therefore \qquad F\{F(x)*G(x)\} = \int_{-\infty}^{+\infty}\int_{-\infty}^{+\infty} e^{-i\lambda(P+u)}\cdot F(u)G(P)dudP$$

$$= \int_{-\infty}^{+\infty} e^{-i\lambda P}\cdot G(P)dP\int_{-\infty}^{+\infty} e^{-i\lambda u}F(u)du$$

$$= F\{F(x)\}\cdot F\{G(x)\} = f(\lambda)g(\lambda) \qquad (8.48)$$

8.13. SOME APPLICATIONS OF FOURIER TRANSFORMS

(i) The Fourier transforms are used to solve differential equations arising in boundary value problems of physics and mechanics. These are evident from the following examples.

Example 1. *Use finite Fourier transforms to solve*

$$\frac{\partial u}{\partial t} = \frac{\partial^2 u}{\partial x^2}$$

$$u(0, t) = 0, \quad u(4, t) = 0, \ u(x, 0) = 2x, \text{ when } 0 < x < 4, t > 0.$$

Solution: Take the finite Fourier sine transform (with $l = 4$) of both sides of partial differential equation. We obtain

$$F_s\left\{\frac{\partial u}{\partial t}\right\} = F_s\left\{\frac{\partial^2 u}{\partial x^2}\right\}$$

or

$$\int_0^4 \frac{\partial u}{\partial t}\sin\frac{n\pi x}{4}dx = -\frac{n^2\pi^2}{16}F_s\{u(x,t)\} + \frac{n\pi}{4}\{u(0,t) - u(4,t)\cos n\pi\}$$

Writing $u(n, t) = F_s\{u(x, t)\}$ and using the condition

$$u(0, t) = 0, \quad u(4, t) = 0. \text{ We have}$$

$$\frac{d}{dt}\int_0^4 u(x,t)\sin\frac{n\pi x}{4}dx = \frac{du(n,t)}{dt} - \frac{n^2\pi^2}{16}u(n,t)$$

or

$$u(n, t) = ce^{-\frac{n^2\pi^2}{16}t} \qquad (1)$$

where c is the constant of integration.

Taking the Fourier sine transform of the condition

$$u(x, 0) = 2x, \text{ we have}$$

$$F_s\{u(x, 0)\} = u(n, 0) = F_s\{2x\}$$

or

$$u(n, 0) = \int_0^4 \sin\frac{n\pi x}{4}\cdot 2x\, dx = -\frac{32}{n\pi}\cos n\pi \qquad (2)$$

From Eqs (1) and (2), $c = -\frac{32}{n\pi}\cos n\pi$

or

$$u(n, t) = -\frac{32}{n\pi}\cos n\pi\, e^{-\frac{n^2\pi^2}{16}t}$$

or

$$u(x, t) = F_s^{-1}\{u(n,t)\}$$

$$= \frac{2}{4}\sum_{n=1}^{\infty}\frac{(-32)}{n\pi}\cos n\pi\, e^{-\frac{n^2\pi^2}{16}t}\cdot\sin\frac{n\pi}{4}x$$

$$= -\frac{16}{\pi}\sum_{n=1}^{\infty}\frac{\cos n\pi}{n}\sin\frac{n\pi x}{4}e^{-\frac{n^2\pi^2}{16}t}$$

Example 2. *A flexible string of length π is slightly stretched between points $x = 0$ and $x = \pi$ on the x-axis, its ends fixed at these points. When set into small transverse vibrations, the displacement $y(x, t)$ from the x-axis of any point x at time t is given by*

$$\frac{\partial^2 y}{\partial t^2} = a^2\frac{\partial^2 y}{\partial x^2}$$

Using finite Fourier transforms, find a solution of this equation satisfying the conditions $y(0, t) = y(\pi, t) = 0$, $y_1(x, 0) = 0$

$$y(x, 0) = 0.1 \sin x + 0.01 \sin 4x, \text{ for } 0 < x < \pi, t > 0$$

Solution: i. Take the finite Fourier sine transform with $l = \pi$ of both sides of partial differential equation. We have,

$$F_s\left\{\frac{\partial^2 y}{\partial t^2}\right\} = a^2 F_s\left\{\frac{\partial^2 y}{\partial x^2}\right\}$$

or
$$\int_0^\pi \frac{\partial^2 y}{\partial t^2} \sin nx\, dx = a^2 F_s\left\{\frac{\partial^2 y}{\partial x^2}(x,t)\right\}$$

Let $y(n, t) = F_s\{y(x, t)\}$ and using the condition

$$y(0, t) = y(\pi, t) = 0.$$

We have
$$\frac{d^2 y(n,t)}{dt^2} = -a^2 n^2 y(n, t)$$

or
$$y(n, t) = A \sin ant + B \cos ant \qquad (1)$$

where A and B are arbitrary constants of integration.

Using the condition of $y_t(x, 0) = 0$ and taking Fourier sine transform we have

$$\frac{\partial y}{\partial t}(n,0) = 0$$

giving
$$A = 0$$

∴
$$y(n, t) = B \cos nat \qquad (2)$$

Taking the Fourier sine transform of the condition

$$y(x, 0) = 0.1 \sin x + 0.01 \sin 4x$$

We have
$$F_s\{y(x, 0)\} = 0.1\, F_s\{\sin x\} + 0.01\, F_s\{\sin 4x\}$$

or
$$y(n, 0) = 0.1\int_0^\pi \sin x \sin nx\, dx + 0.01\int_0^\pi \sin 4x \sin nx\, dx$$

or
$$y(n, 0) = \begin{cases} 0.1\dfrac{\pi}{2} & \text{for } n = 1 \\[2mm] 0.01\dfrac{\pi}{2} & \text{for } n = 4 \\[2mm] 0 \text{ for all other value of } x \end{cases}$$

or
$$B = 0.1\frac{\pi}{2} \quad \text{for } n = 1$$

$$= 0.01\frac{\pi}{2} \quad \text{for } n = 4$$

$$= 0 \text{ for all other values of } n.$$

Taking the inverse Fourier transform of Eq. (2), we have

$$y(x, t) = \frac{2}{\pi}\sum_{n=1}^{\infty} B \cos ant \sin nx$$

$$= 0.1 \cos at \sin x + 0.01 \cos 4 at \sin 4x.$$

ii. The Fourier transforms are also used to evaluate integrals.

Example 3. *Show that*

$$\int \frac{\cos \lambda x}{1+\lambda^2} d\lambda = \frac{\pi}{2} e^{-x} \text{ for } x \geq 0$$

Solution: Let $F(x) = e^{-x}$.

The Fourier integral theorem gives

$$F(x) = \frac{2}{\pi} \int_0^\infty \cos \lambda x \, d\lambda \int_0^\infty F(x) \cos \lambda x \, dx$$

Then

$$\frac{2}{\pi} \int_0^\infty \cos \lambda x \, d\lambda \int_0^\infty e^{-x} \cos \lambda x \, dx = e^{-x}$$

But

$$\int_0^\infty e^{-x} \cos \lambda x \, dx = \frac{1}{1+\lambda^2}, \text{ we have}$$

$$\frac{2}{\pi} \int_0^\infty \frac{\cos \lambda x}{1+\lambda^2} d\lambda = e^{-x}$$

or

$$\int_0^\infty \frac{\cos \lambda x}{1+\lambda^2} d\lambda = \frac{\pi}{2} e^{-x}$$

SHORT ANSWER QUESTIONS

1. Define the term 'integral transform'.
2. What do you mean by Laplace transform.
3. State and explain any two properties of Laplace transform.
4. Find Laplace transform of Dirac delta functions.
5. State and explain change of scale property of Laplace transform.
6. State and prove linearity theorem of Fourier's transform.
7. Find Laplace transform of error function.
8. State and explain convolution theorem of Laplace transform.
9. State convolution theorem for inverse Laplace transform.
10. State and explain linearity property of Laplace transform.
11. Find the Laplace transform of e^{at}.
12. State and explain linearity property of Laplace transform.
13. Define Fourier integral transform of a function.
14. What are infinite Fourier sine and cosine transforms?
15. State and explain change of scale property of Fourier transform.
16. State and explain shifting property of Fourier transform.
17. State and explain Parseval's convolution theorem regarding Fourier transform.
18. What do you mean by Fourier transform of the derivative of a function.
19. Show that $F.T\left(\dfrac{d^n f}{dt^n}\right) = (i\alpha)^n FT[f(t)]$

20. What do you mean by derivative of Laplace transform.

21. Find the Laplace transform of periodic function $f(t + T) = f(t)$.

22. Express Laplace transform of f^n in terms of gamma functions.

23. What do you mean by inverse Laplace transform?

24. State Fourier—Mellin theorem.

PROBLEMS

1. (a) If $L\{F(t)\} = f(x)$, where L stands for Laplace transform show that

$$L^{-1}\{e^{ks} f(s)\} = F(t + k) \text{ for } k > 0,$$

provided that $F(t) = 0$ for $0 < t < k$.

   (b) Define Fourier transform and inverse Fourier transform. Compute the Fourier transform of a triangular pulse given by

$$f(t) = at \text{ for } 0 < t \leq T$$
$$= 0 \text{ for } t > T.$$

2. (a) Find the Laplace transform of $\dfrac{t^2}{2!} e^{-at}$

   (b) State and prove Faltung theorem for Fourier transforms.

   (c) Using Laplace transforms, solve the equation

$$\frac{d^2 x}{dt^2} + w^2 x = 0$$

Subject to the conditions

$$x = x_0 \text{ at } t = 0, \ \frac{dx}{dt} = 0 \text{ at } t = 0$$

3. (a) Derive Fourier integral formula mentioning its condition of validity and hence give the inversion formula for Fourier transform.

   (b) The Laplace transform of function $f(t)$ is defined as

$$L\{F(t)\} = \int_0^\infty F(t)e^{-st} dt$$

Show that

   i. $L\{t^n\} = \dfrac{n!}{s^{n+1}}, n \geq 0$

   ii. $L\{e^{-at} \sin wt\} = \dfrac{w}{(s+a)^2 + w^2}$

4. Define Fourier transform and inverse Fourier transform. Compute the Fourier transform of a rectangular pulse given by

$$f(t) = 1 \text{ for } |t| < 1$$
$$= 0 \text{ for } |t| > 1$$

5. Evaluate the following integrals operationally:

a. $\int_0^\infty \frac{\sin tx}{x^{1/2}}dx \left(\frac{\pi}{2t}\right)^{1/2}$ $t>0$

b. $\int_{-\infty}^\infty \frac{x\sin tx}{a^2+x^2}dx = \pi e^{-at}$ $a>0$

c. $\int_0^\infty \frac{e^{-tx^2}}{(1+x^2)}dx = \frac{\pi}{2}e^t erf\, t^{1/2}$

d. $\int_0^\infty t^3 e^{-t} \sin t\, dt = 0$

e. $\int_0^\infty \frac{e^{-t}\sin t}{t}dt = \frac{\pi}{4}$

f. $\int_0^\infty ue^{-u^2} J_0(au)du\frac{1}{2}e^{-a^2/4}$

g. $\int_0^\infty te^{-t}E_i(t)dt - \log 2 - \frac{1}{2}$

h. $\int_0^\infty ue^{-u^2} erf\, u\cdot du = \frac{\sqrt{2}}{4}$

6. Find the Laplace transforms of

a. $L\{\sin t^2\} = \sum_{n=1}^\infty \frac{(-1)^{n-1}(4n-2)!}{(2n-1)!s^{2n+1}}$

b. $L\{\sin^6 t\} = \frac{6!}{s(s^2+4)(s^2+16)(s^2+36)}$

and hence generalize the formula for $L\{\sin^{2n} t\}$

c. $L\{\cos \log t\delta(t-\pi) = -e^{-\pi s} \log \pi$

d. $L\{e^{-t}\sin^2 t\} = \frac{3}{(s+1)(s^2+2s+5)}$

e. $L\left\{\int_0^t \frac{1-e^{-u}}{u}du\right\} = \frac{1}{s}\log\left(1+\frac{1}{s}\right)$

f. $L\{e^{2t}s_i(t)\} = \frac{\tan^{-1}(s-2)}{(s-2)}$

g. $L\{ts_i(t)\} = \frac{\tan^{-1} s}{s^2} - \frac{1}{s(s^2+1)}$

h. $L\{\cosh at \cos at\} = \frac{s^2}{s^4+4a^4}$

i. $L\{\sinh at \cos at\} = \frac{as^2+2a^3}{s^4+4a^4}$

j. $L\{\sinh at \cos at\} = \frac{as^2-2a^3}{s^4+4a^4}$

k. $L\{\sinh at \sin at\} = \frac{2a^2s}{s^4+4a^4}$

7. If $L\{F(t)\} = f(s)$ show that

$$L\left\{\int_0^\infty \frac{t^u F(u)}{\Gamma(u+1)}du\right\} = \frac{f(\log s)}{s\log s}$$

8. If $L_n(t)$, $n = 0, 1, 2...$ are the Laguerre polynomials, prove that

$$\sum_{n=1}^{\infty} \frac{L_n(t)}{n!} = e^{J_0(2\sqrt{t})},$$

where
$$L_n(t) = e^t \frac{d^n}{dt^n}(t^n e^{-t})$$

9. Find inverse Laplace transform of the following.

(a) $L^{-1}\left\{\dfrac{1}{s}\log\dfrac{(s^2+a^2)}{s^2+b^2}\right\} = \displaystyle\int_0^t \dfrac{\cos au - \cos bu}{u}\, du$

(b) $L^{-1}\left\{\dfrac{1}{(s^2+1)^2}\right\} = \dfrac{1}{8}[(3-t^2)\sin t - 3t\cos t]$

(c) $L^{-1}\left\{\dfrac{s}{(s^2+4)^3}\right\} = \dfrac{1}{64}t(\sin 2t - 2t\cos 2t)$

(d) $L^{-1}\left\{\dfrac{1}{(8s-27)^{1/3}}\right\} = t^{-2/5} e^{27t/8} 2\Gamma(1/3)$

(e) $L^{-1}\left\{\dfrac{1}{\sqrt{s^2-4s+20}}\right\} = e^{2t}J_3(4t)$

(f) $L^{-1}\dfrac{8s+20}{\sqrt{s^2-12s+32}} = 2e^{6t}(4\cosh 2t + 7\sinh 2t) = 11e^{8t} - 3e^{4t}$

(g) $L^{-1}\left\{\dfrac{1}{1+\sqrt{s}}\right\} = \dfrac{t^{-1/2}}{\sqrt{\pi}} - e^t \operatorname{erfc}(\sqrt{t})$

(h) $L^{-1}\left\{\dfrac{1}{s+e^{-s}}\right\} = \displaystyle\sum_{n=0}^{\infty} \dfrac{(-1)^n(t-n)^n}{n!}$

(i) $L^{-1}\left\{e^{-3s-2\sqrt{s}}\right\} = \dfrac{1}{\sqrt{\pi(t-3)^3}} e^{-1/(t-3)}$

(j) $L^{-1}\left\{\dfrac{s^2}{s^2+u^2}\right\} = \dfrac{1}{2}t\cos 2t + \dfrac{1}{4}\sin 2t$

(k) $L^{-1}\left\{\dfrac{1}{(s^3+1)^3}\right\} = \dfrac{1}{8}(3-t^2)\sin t - 3t\cos t$

10. Solve the differential equation: $\dfrac{\partial u}{\partial t} = 2\dfrac{\partial^2 u}{\partial x^2}$, $u(0,t) = 0, u(5,t) = 0, u(x,0) = 10\sin 4\pi x$

[*Ans.* $u(x,t) = 10e^{-32\pi^2 t}\sin 4\pi x$]

11. Solve

$$\frac{\partial^2 y}{\partial t^2} = 16\frac{\partial^2 y}{\partial x^2}, y_x(0,t) = 0, \ y(3,t) = 0, \ y(x,0) = 0$$

and $y_t(x, 0) = 12\cos \pi x + 16\cos 3\pi x - 8\cos 5\pi x$

     [*Ans.* $y(x, t) = 12\cos \pi x \sin 4\pi t + 16\cos 3\pi x \sin 12\pi t - 8\cos 5\pi x \sin 20\pi t$]

12. Solve the following integral equations.

(a) $y(t) = t + 2\int_0^t \cos(t-u)y(u)du$ [$y(t) = t + 2 + 2\,(t-1)e^t$]

(b) $\int_0^t y(u)\sin(t-u)du = 2y(t) + t - 2$ [$y(t) = 1$]

(c) $\int_0^t y(u)\cos(t-u)du = y'(t);$ if $y(0) = 1$ $\left[y(t) = 1 + \frac{1}{2}t^2\right]$

(d) $\int_0^t y''(u)y'(t-u)du = y'(t) - y(t)$ if $y(0) = y'(0) = 0; \ y(t) \neq 0$

13. Find the infinite Fourier sine and cosine transformations of

     (a) $F(x) = x^2$ $(0 < x < 1)$

$$\left[\textbf{Ans.}\ (i)\ \frac{2l^3}{n^3\pi^3}(\cos n\pi - 1) - \frac{l^3}{n\pi}\cos n\pi \text{ if } n = 1,2,3\cdots \text{ and } \frac{l^3}{3} \text{ if } n = 0 \right.$$

$$(ii)\ \frac{2l^3}{n^2\pi^2}(\cos n\pi - 1)\Big]$$

     (b) $F(x) = \frac{1}{2}(\pi - x)x$ $(0 < x < \pi)$

$$\left[\textbf{Ans.}\ (i)\ \frac{1-\cos n\pi}{n^3}\quad (ii)\ -\frac{\pi}{2n^2}(1+\cos n\pi)\right]$$

     (c) $F(x) = 1$ $(0 < x < 1)$

$$\left[\textbf{Ans.}\ (i)\ \frac{l(1-\cos n\pi)}{n\pi}\quad (ii)\ 0 \text{ if } n = 1,2,...;1 \text{ if } n = 0\right]$$

14. If $Fc\{F(x)\} = \dfrac{6\left(\dfrac{\sin n\pi}{3} - \cos n\pi\right)}{(2n+1)\pi}$ for $n = 1, 2, 3...$ and $\dfrac{2}{\pi}$ for $n = 0$. Find $F(x)$ where $0 < x < 4$.

$$\left[\textbf{Ans.}\ \frac{1}{2\pi} + \frac{3}{\pi}\sum_{n=1}^{\infty}\left(\frac{\dfrac{\sin n\pi}{2} - \cos n\pi}{2n+1}\right)\cos\frac{n\pi x}{4}\right]$$

15. If $f(n) = \dfrac{\cos\left(\dfrac{2n\pi}{3}\right)}{(2n+1)^2}$. Find (a) $F_s^{-1}\{f(n)\}$ and (b) $F_c^{-1}\{f(n)\}$ if $0 < n < 1$

$$\left[\textbf{Ans.}\ (a)\ 2\sum_{n=1}^{\infty}\frac{\cos(2n\pi/3)}{(2n+1)^2}\sin n\pi x \quad (b)\ 1 + 2\sum_{n=1}^{\infty}\frac{\cos(2n\pi/3)}{(2n+1)^2}\cos n\pi x\right]$$

16. Find the Fourier sine transform of $e^{-x}, x \geq 0$ and show that

$$\int_0^\infty \frac{x \sin mx}{x^2 + 1} dx = \frac{\pi}{2} e^{-m}; \quad m > 0$$

Explain from the viewpoint of Fourier integral theorem; why the result does not hold for $m = 0$.   [**Ans.** $\lambda/(1 + \lambda^2)$]

17. Solve $\dfrac{\partial u}{\partial t} = \dfrac{\partial^2 u}{\partial x^2}, 0 < x < 6, t > 0$ subject to conditions

$$u(0,t) = 0, u(6,t) = 0, u = (x,0) = \begin{cases} 1 & 0 < x < 3 \\ 0 & 3 < x < 6 \end{cases}$$

$$u(x,t) = \sum_{n=1}^\infty \left\{ \frac{1 - \cos n\pi/3}{n\pi} \right\} e^{-n^2\pi^2 t/36} \sin \frac{n\pi x}{6}$$

18. A flexible string of length p is tightly stretched between $x = 0$ and $x = p$ on the x-axis, its ends fixed at these points when set into small vibrations, the displacement $y(x, t)$ from the x-axis of any point x at time t is given by

$$\frac{\partial^2 y}{\partial t^2} = a^2 \frac{\partial^2 y}{\partial x^2}$$

using finite Fourier transform, find the solution of this equation which satisfies the conditions $y(0, t) = 0$, $y(\pi, t) = 0$, $y(x, 0)$, $y(x, 0) = 0.1 \sin x + 0.01 \sin 4x$, $y_t(x, 0) = 0$ for $0 < x < \pi, t > 4$.

[**Ans.** $y(x, t) = 0.1 \sin x \cos at + 0.01 \sin 4x \cos 4 at$]

19. Establish the following formulas:

(a) $F_s\{e^{-x} \cos x\} = \sqrt{\dfrac{2}{\pi}} \dfrac{\lambda^3}{\lambda^4 + 4}$

(b) $F_s\{F(x)\} = \sqrt{\dfrac{2}{\pi}} \dfrac{\sin \lambda x}{1 - \lambda^2}$; where $F(x) = \begin{cases} \sin x & \text{for } 0 \leq x \leq \pi \\ 0 & \text{for } x > \pi \end{cases}$

(c) $F_s\{xe^{-ax}\} = \sqrt{\dfrac{2}{\pi}} \dfrac{a^2 - \lambda^2}{(a^2 + \lambda^2)^2}$

(d) $F_c\{x^{\alpha-1}\} = \sqrt{\dfrac{2}{\pi}} \Gamma(\alpha) \dfrac{\cos \alpha\pi/2}{\lambda^\alpha} (0 < \alpha < 1)$

20. Show that $L\{\cos Kt\} = \dfrac{s}{s^2 + K^2}$ and $L\{\sin Kt\} = \dfrac{K}{s^2 + K^2}$

MULTIPLE CHOICE QUESTIONS

1. Which of the following represents Fourier transform?

(a) $g(\lambda) = \displaystyle\int_0^\infty f(t) e^{-i\lambda t} dt$

(b) $g(\lambda) = \displaystyle\int_0^\infty f(t) e^{-\lambda t} dt$

(c) $g(\lambda) = \displaystyle\int_0^\infty f(t) e J_n(\lambda t) dt$

(d) $g(\lambda) = \displaystyle\int_0^\infty f(t) t^{\lambda-1} dt$

2. Laplace transform of $t^{-1/2}$ is

(a) $\dfrac{\sqrt{\pi}}{S}$

(b) $\sqrt{\dfrac{\pi}{S}}$

(c) $\dfrac{S}{\sqrt{\pi}}$

(d) $\dfrac{1}{\sqrt{\pi S}}$

3. Laplace transform of $t^n \, e^{at}$ is

(a) $\dfrac{1}{(s+a)^n}$

(b) $\dfrac{1}{(s+a)^{n+1}}$

(c) $\dfrac{n!}{(s+a)^n}$

(d) $\dfrac{n!}{(s+a)^{n+1}}$

4. Inverse Laplace transform of $f(s) = \dfrac{k}{(s^2 + k^2)}$ is

(a) $\sin kt$

(b) $\cos kt$

(c) $\sinh kt$

(d) none of these

5. Inverse Laplace transform of $f(s) = \dfrac{a}{s^2 - a^2}$ is

(a) $\sin(at)$

(b) $\cos(at)$

(c) $\sinh(at)$

(d) $\cosh(at)$

6. Laplace transform of $e^{at} \cos \omega t$ is

(a) $\dfrac{s-a}{(s-a)^2 + w^2}$

(b) $\dfrac{s+a}{(s+a)^2 + w^2}$

(c) $\dfrac{s}{(s-a)^2 + w^2}$

(d) $\dfrac{s+a}{(s-a)^2 + w^2}$

7. Laplace transforms of $J_0(t)$ is

(a) $\dfrac{1}{s}$

(b) $\dfrac{1}{\sqrt{s^2 + 1}}$

(c) $\dfrac{1}{s^2 + 1}$

(d) $\dfrac{1}{\sqrt{1 - s^2}}$

8. Which of the following represents Laplace's transform?

(a) $g(s) = \displaystyle\int_0^\infty f(t) e^{-ist} dt$

(b) $g(s) = \displaystyle\int_0^\infty f(t) e^{-st} dt$

(c) $g(s) = \displaystyle\int_0^\infty f(t) J_n(st) dt$

(d) $g(s) = \displaystyle\int_0^\infty f(t) e^{s-t} dt$

9. Laplace transform of $\dfrac{\sin at}{t}$ is

(a) $\tan^{-1}\left(\dfrac{s}{a}\right)$

(b) $\sin^{-1}\left(\dfrac{s}{a}\right)$

(c) $\sec^{-1}\left(\dfrac{s}{a}\right)$

(d) does not exist

10. Laplace transform of $\sin at$ is

(a) $\dfrac{a}{s^2 + a^2}$

(b) $\dfrac{s}{s^2 + a^2}$

(c) $\dfrac{a}{s^2 - a^2}$

(d) $\dfrac{s}{s^2 - a^2}$

11. Laplace transform of function $f(t) = e^{-t} \cos 2t$ is

(a) $\dfrac{s^2 + 2s + 5}{(s^2 + 2s - 3)^2}$

(b) $\dfrac{s^2 - 2s + 5}{(s^2 + 2s - 3)^2}$

(c) $\dfrac{4s + 4}{(s^2 - 2s - 3)^2}$

(d) $\dfrac{4s - 1}{(s^2 + 2s - 3)^2}$

12. Laplace transform of $\cos at$ is

(a) $\dfrac{a}{s^2 + a^2}$

(b) $\dfrac{s}{s^2 + a^2}$

(c) $\dfrac{a}{s^2 - a^2}$

(d) $\dfrac{s}{s^2 - a^2}$

13. Inverse Laplace transform of $\dfrac{1}{s^2 + a^2}$ is

(a) $\sin at$

(b) $\sin h\, at$

(c) $\cos at$

(d) $\cos h\, at$

14. Laplace transform of $\dfrac{\cos at}{t}$ is

(a) $\tan^{-1}\left(\dfrac{s}{a}\right)$

(b) $\cot^{-1}\left(\dfrac{s}{a}\right)$

(c) $\sec^{-1}\left(\dfrac{s}{a}\right)$

(d) does not exist

15. The Fourier transform of Gaussian function e^{-x^2} is

(a) $\dfrac{1}{\sqrt{2}} e^{-\omega^2/4}$

(b) $\dfrac{1}{\sqrt{2\pi}} e^{-\omega^2/2}$

(c) $\dfrac{1}{\sqrt{2\pi}} e^{-\omega^2}$

(d) $\sqrt{\dfrac{2}{\pi}}\, e^{-\omega^2/2}$

16. Fourier transform of $\dfrac{df}{dt}$ is $\left[Ft\left(\dfrac{df}{dt}\right) \right]$ is

(a) $\dfrac{\omega}{\sqrt{2\pi}} \displaystyle\int_{-\infty}^{+\infty} \dfrac{df}{dt} e^{-i\omega t} \cdot dt$

(b) $\sqrt{\dfrac{\omega}{2\pi}} \displaystyle\int_{-\infty}^{+\infty} \dfrac{df}{dt} e^{-i\omega t} \cdot dt$

(c) $\dfrac{i\omega}{\sqrt{2\pi}} \displaystyle\int_{-\infty}^{+\infty} \dfrac{df}{dt} e^{-i\omega t} \cdot dt$

(d) $\dfrac{1}{i\omega\sqrt{2\pi}} \displaystyle\int_{-\infty}^{+\infty} \dfrac{df}{dt} e^{-i\omega t} \cdot dt$

17. Laplace transform of $t + t^2$ is

(a) $\dfrac{1}{s} + \dfrac{1}{s^2}$

(b) $\dfrac{1}{s^2} + \dfrac{1}{s^3}$

(c) $\dfrac{1}{s^2} + \dfrac{2}{s^3}$

(d) $\dfrac{2}{s^2} + \dfrac{3}{s^3}$

18. Laplace transform of periodic function $f(t) = f(t + T)$ $t > 0$ is

(a) $\dfrac{1}{T}\displaystyle\int_0^T e^{-st} f(t)\,dt$

(b) $\dfrac{2}{T}\displaystyle\int_0^T e^{-st} f(t)\,dt$

(c) $\dfrac{1}{1 - e^{-sT}}\displaystyle\int_0^T e^{-st} f(t)\,dt$

(d) $\dfrac{(1 - e^{-sT})}{T}\displaystyle\int_0^T e^{-st} f(t)\,dt$

19. Laplace transform of exponential integral function $\displaystyle\int_t^\infty \dfrac{e^{-x}}{x}\,dx$ is

(a) $\dfrac{1 - e^{-s}}{s}$

(b) $-\dfrac{e^{-s}}{s^2}$

(c) $s \log (s + 1)$

(d) $\dfrac{\log(s + 1)}{s}$

20. Laplace transform of function $F(t) = 1$ is

(a) $\dfrac{1}{s}$ for values of s

(b) $\dfrac{1}{s}$ for $s > 0$

(c) $\dfrac{1}{s}$ for $s < 0$

(d) $\dfrac{1}{s^2}$ for $s > 0$

21. Laplace transform of e^{at} is

(a) $\dfrac{1}{s - a}$

(b) $\dfrac{1}{s + a}$

(c) $\dfrac{2}{s - a}$

(d) $\dfrac{2}{s + a}$

22. Laplace transform of $t^2 e^{-at}$ is

(a) $\dfrac{2}{s + a}$

(b) $\dfrac{2}{(s + a)^2}$

(c) $\dfrac{2}{(s + e)^2}$

(d) none of these

23. The Laplace transform of $F(t) = t$ $(t > 0)$ is

(a) $\dfrac{1}{s}$

(b) $\dfrac{1}{s^2}$

(c) s^2

(d) $\dfrac{1}{1 + s^2}$

24. Laplace transform of function $f(t) = e^{-at} \sin 3t$ is

(a) $\dfrac{3}{s^2 + q}$

(b) $(s - a)^2 + 9$

(c) $\dfrac{3}{(s+a)^2} + s$

(d) $\dfrac{9}{(s+3)^2 + s}$

25. The Fourier sine transform of function $f(x) = e^{-ax}$ is

(a) $\dfrac{s}{a^2 + s^2}$

(b) $\dfrac{a}{a^2 + s^2}$

(c) $\dfrac{s \sin ax}{\sqrt{a^2 + x^2}}$

(d) $\dfrac{a \sin ax}{\sqrt{a^2 + x^2}}$

26. The Fourier cosine transform of function $f(x) = e^{-ax}$ is

(a) $\dfrac{s}{a^2 + s^2}$

(b) $\dfrac{a}{a^2 + s^2}$

(c) $\dfrac{s \cos ax}{\sqrt{a^2 + s^2}}$

(d) $\dfrac{a \cos ax}{\sqrt{a^2 + s^2}}$

27. Fourier transform of function $f(t) = e^{-t+1}$ is

(a) zero

(b) $\dfrac{2}{1+s}$

(b) $\dfrac{1}{(1+s^2)}$

(d) none of these

28. If $F(w) = \sqrt{\dfrac{2}{\pi}} \int_0^\infty f(x) \sin(\omega x) dx$, then

(a) $F(\omega)$ is Fourier sine transform of $f(x)$

(b) $F(\omega)$ is Fourier cosine transform of $f(x)$

(c) $F(\omega)$ is Laplace sine transform of $f(x)$

(d) $F(\omega)$ is Laplace cosine transform of $f(x)$

29. Fourier transform of the function $e^{-|x|}$ is

(a) $\dfrac{2}{1+s^2}$

(b) $\dfrac{2}{(1+s)^3}$

(c) $\dfrac{2}{s^2 + s}$

(d) $\dfrac{2}{s^2 + 2s + 1}$

30. Inverse Laplace transform of $\dfrac{1}{s(s^2 + 1)}$ is

(a) $(1 - \cos t)$

(b) $(1 + \cos t)$

(c) $(1 - \sin t)$

(d) $(1 + \sin t)$

31. If the Fourier transform $F[\delta(x - a)] = \exp[-2\pi va]$ then $P^{-1}(\cos 2\pi av)$ will correspond to

(a) $\delta(x - a) - \delta(x + a)$ (b) a constant

(c) $\dfrac{1}{2}[\delta(x - a) + i\delta(x + a)]$ (d) $\dfrac{1}{2}[\delta(x - a) + \delta(x + a)]$ **[GATE]**

32. If $f(x) = \begin{cases} 0 & \text{for } x < 3 \\ x - 3 & \text{for } x \geq 3 \end{cases}$ then the Laplace transform of $f(x)$ is

(a) $s^{-2} e^{3s}$ (b) $s^2 e^{-3s}$

(b) s^{-2} (d) $s^{-2} e^{-3s}$ **[GATE]**

ANSWERS

1. (a)	2. (b)	3. (d)	4. (b)	5. (c)	6. (a)	7. (b)	8. (b)
9. (b)	10. (a)	11. (a)	12. (b)	13. (b)	14. (d)	15. (a)	16. (c)
17. (c)	18. (c)	19. (d)	20. (a)	21. (a)	22. (c)	23. (b)	24. (c)
25. (a)	26. (b)	27. (c)	28. (a)	29. (a)	30. (a)	31. (d)	32. (d)

9

Partial Differential Equations

9.1 INTRODUCTION

The mathematical formulation of a physical problem frequently takes the form of a partial differential equation together with subsidiary conditions which the solution must satisfy. These *boundary conditions* are dictated by the physical nature of the problem (the requisite number and boundary conditions that must be given in order that a solution of the problem exists, is purely a mathematical question).

In this text, we will not concern ourselves with proving that the solution of a boundary value problem, as described above, is unique. Instead, we will assume that a solution of the partial differential equation, obtained by any means which satisfies all the boundary conditions is the unique solution. Thus, throughout the text, we forego mathematical rigor in order to devote our attention to methods of solution.

9.2. PARTIAL DIFFERENTIAL EQUATIONS IN PHYSICS

The most commonly occurring partial differential equations in physical problems are linear. A linear partial differential equation is one having the properties that if $\phi_1(x, y, z, t)$ and $\phi_2(x, y, z, t)$ are two distinct solutions then $\phi_1 + \phi_2$ is the solution, and also $A\phi_1$ or $B\phi_2$ is a solution where A or B is constant.

Typical partial differential equations used in study of physics are:

i. Laplace's equation

$$\nabla^2\phi = \left(\frac{\partial^2\phi}{\partial x^2} + \frac{\partial^2\phi}{\partial y^2} + \frac{\partial^2\phi}{\partial z^2}\right) = 0 \tag{9.1}$$

ii. Three dimensional wave equation

$$\nabla^2\phi = \frac{1}{c^2}\frac{\partial^2\phi}{\partial t^2} \tag{9.2}$$

where c is wave velocity.

iii. Poisson's equation

$$\nabla^2\phi = \frac{\rho}{\varepsilon} \tag{9.3}$$

where ρ is the function of position coordinates and ε is dielectric constant of the medium.

4. Diffusion equation or heat flow equation

$$\nabla^2\phi = \frac{1}{k}\frac{\partial\phi}{\partial t} \tag{9.4}$$

where ϕ is temperature and $k = \dfrac{\text{Thermal conductivity}}{\text{Speicfic heat} \times \text{density}}$

5. Schrödinger equation

$$-\frac{\hbar^2}{2m}\nabla^2\phi + V\phi = E\phi = i\hbar\frac{\partial\phi}{\partial t} \tag{9.5}$$

6. Hamilton–Jacobi equation

$$\nabla\phi \cdot \nabla\phi + V = \frac{\partial\phi}{\partial t} \tag{9.6}$$

7. Stress equation

$$\nabla^4\phi = 0 \tag{9.7}$$

All of these except the Hamilton–Jacobi equation are linear. Furthermore, all of these except the Poisson and Hamilton–Jacobi are homogeneous. That is, the dependent function ϕ, occurs to the same power in every term of the equation.

General techniques for solving these partial differential equations are the method of separation of variables, integral solutions employing Green's function, method of integral transform, and method of numerical calculations. In this chapter, we will be mainly concerned with the method of separation of variables. In this method, the partial differential equation splits into ordinary differential equations which may be solved by the methods described in the Chapter on *Differential Equations and Polynomials*.

9.3. LAPLACE'S EQUATION IN THREE DIMENSIONS

Rectangular Coordinates

The Laplace's equation in Cartesian coordinates is expressed as

$$\nabla^2\phi = \frac{\partial^2\phi}{\partial x^2} + \frac{\partial^2\phi}{\partial y^2} + \frac{\partial^2\phi}{\partial z^2} = 0$$

We begin to solve this equation by the method of separation of variables. We put

$$\phi = X(x)\,Y(y)\,Z(z) \tag{9.8}$$

and obtain

$$\frac{1}{X}\cdot\frac{d^2X}{dx^2} + \frac{1}{Y}\frac{d^2Y}{dy^2} + \frac{1}{Z}\frac{d^2Z}{dz^2} = 0 \tag{9.9}$$

Each one of these terms must separately equal to a constant and the sum of these constants (which we write as k_x^2, k_y^2, k_z^2) must vanish. Thus

$$\phi_{k_x k_y k_z} = e^{k_x x}e^{k_y y}e^{k_z z} = e^{(k_x x + k_y y + k_z z)} \tag{9.10}$$

where

$$k_x^2 + k_y^2 + k_z^2 = 0 \tag{9.11}$$

If k_x, k_y or k_z is zero, the corresponding factor in Eq. (9.10) must be replaced by $(a_1 x + a_2)$, etc.

A more general solution would be

$$\phi = \sum_{k_x k_y k_z} c_{k_x k_y k_z} e^{(k_x x + k_y y + k_z z)} \tag{9.12}$$

In this connection, it is sometimes convenient to regard k_x, k_y, k_z formally as the components of a vector k. Equation (9.10) may then be written as

$$\phi_k = c(k) e^{k \cdot r} \quad |k| = 0 \tag{9.13}$$

Cylindrical Coordinates (r, θ, z)

The Laplace's equation expressed in terms of the cylindrical coordinates is

$$\nabla^2 \phi = \frac{1}{r}\frac{\partial}{\partial r}\left(r\frac{\partial \phi}{\partial r}\right) + \frac{1}{r^2}\frac{\partial^2 \phi}{\partial \theta^2} + \frac{\partial^2 \phi}{\partial z^2} \tag{9.14}$$

In order to obtain solution of this equation, we put

$$\phi = R(r)\, Z(z)\, \theta(\theta) \tag{9.15}$$

Substitution in Eq. (9.14) and dividing with ϕ will result in

$$\frac{1}{R}\frac{d^2 R}{dr^2} + \frac{1}{r}\cdot\frac{dR}{dr}\cdot\frac{1}{R} + \frac{1}{r^2\theta}\frac{d^2\theta}{d\theta^2} + \frac{1}{Z}\frac{d^2 Z}{dz^2} = 0$$

Clearly the last term on left must be constant and let us put it equal to $-m^2$. Then

$$Z(z) = c_1\, e^{\pm imz} \tag{9.16}$$

The remaining equation

$$\frac{1}{R}\frac{d^2 R}{dr^2} + \frac{1}{R}\cdot\frac{1}{r}\frac{dR}{dr} + \frac{1}{r^2\theta}\frac{d^2\theta}{d\theta^2} + \frac{1}{Z}\frac{d^2 Z}{dz^2} = 0$$

When multiplied by r^2 separates again into two equations

$$\frac{1}{\theta}\cdot\frac{d^2\theta}{d\theta^2} = -l^2 \quad \text{or} \quad \frac{d^2\theta}{d\theta^2} + l^2\theta = 0 \tag{9.17}$$

and

$$r^2\frac{d^2 R}{dr^2} + r\frac{dR}{dr} - (m^2 r^2 + l^2)R = 0 \tag{9.18}$$

Equation (9.17) has the solution

$$\theta = c_2 e^{\pm il\theta} \tag{9.19}$$

while the Eq. (9.18) turns into Bessel's equation when the substitution $imr = x$ is made for it, then reads:

$$x^2\frac{d^2 R}{dr^2} + x\frac{dR}{dx} + (x^2 - l^2)R = 0 \tag{9.20}$$

The solution of this equation is

$$R(x) = J_l(x) = J_l(irm) \tag{9.21}$$

Combining these results we have

$$\phi = c_{ml} J_l(irm) e^{\pm i(mz \pm l\theta)} \tag{9.22}$$

when $l = 0$, $\theta = a_1\theta + a_2$; hence we obtain another solution of lesser generality than Eq. (9.22) becomes

$$\phi_{m,0} = c_{m,0} J_0(irm) e^{\pm imz}(a_1\theta + a_2) \tag{9.22a}$$

On the other hand, when $m = 0$, $Z = b_1z + b_2$ instead of Eq. (9.16). The equation for $R(r)$ takes the form

$$r^2 \frac{d^2R}{dr^2} + r\frac{dR}{dr} - l^2R = 0$$

whose solution then is

$$R(r) = r^{\pm l}$$

Hence
$$\phi_{0l} = r^{\pm l}(b_1z + b_2)e^{\pm il\theta} \tag{9.22b}$$

Finally, when both l and m are zero, the solution may be seen to takes the form
$$\phi_{00} = (a_1 \log r + a_2)(b_1z + b_2)(c_1\theta + c_2) \tag{9.22c}$$

The most general function satisfying Laplace's equation is a superposition of solutions of Eqs (9.22), (9.22a), (9.22b), and (9.22c).

Spherical Polar Coordinates (r, θ, ψ)

Laplace's equation when transformed to polar coordinates reads

$$\nabla^2\phi = \frac{1}{r^2}\frac{\partial}{\partial r}\left(r^2\frac{\partial\phi}{\partial r}\right) + \frac{1}{r^2 \sin\theta}\cdot\frac{\partial}{\partial\theta}\left(\sin\theta\frac{\partial\phi}{\partial\theta}\right) + \frac{1}{r^2 \sin^2\theta}\frac{\partial^2\phi}{\partial\psi^2} = 0 \tag{9.23}$$

In order to solve this equation using separation of variables method we put $\phi = R(r)\,\theta(\theta)\,\xi(\psi)$

Substituting in Eq. (9.23) and dividing with ϕ, we have

$$\frac{1}{R}\frac{1}{r^2}\frac{\partial}{\partial r}\left(r^2\frac{\partial R}{\partial r}\right) + \frac{1}{r^2 \sin\theta}\frac{1}{\theta}\frac{\partial}{\partial\theta}\left(\sin\theta\cdot\frac{\partial\theta}{\partial\theta}\right) + \frac{1}{r^2 \sin^2\theta}\frac{1}{\xi}\frac{\partial^2\xi}{\partial\psi^2} = 0$$

Multiplying by $r^2 \sin^2\theta$ will isolate the term $\dfrac{\partial^2\xi}{\partial\psi^2}$ as the only one depending on ψ from the remainder of the equation. If, therefore, we put it equal to $-m^2$ so that
$$\xi = ce^{\pm imy} \tag{9.24}$$

Equation (9.23) takes the form

$$\frac{\sin^2\theta}{R}\cdot\frac{d}{dr}\left(r^2\frac{dR}{dr}\right) + \frac{\sin\theta}{\theta}\cdot\frac{d}{d\theta}\left(\sin\theta\frac{d\theta}{d\theta}\right) - m^2 = 0 \tag{9.25}$$

When this is divided throughout by $\sin^2\theta$, the terms involving r are clearly separated from those involving θ. Hence, we have

$$\frac{1}{\theta\sin\theta}\cdot\frac{d}{d\theta}\left(\sin\theta\cdot\frac{d\theta}{d\theta}\right) - \frac{m^2}{\sin^2\theta} + c = 0 \tag{9.26}$$

and
$$\frac{1}{R}\frac{d}{dr}\left(r^2\frac{dR}{dr}\right) - c = 0 \tag{9.27}$$

where c denotes the same constant in both equations. It will prove convenient to write this constant in the form $c = l\,(l + 1)$. Let us now make the substitution $\sin\theta = x$ in Eq. (9.26), we get

$$(1-x^2)\frac{d^2\theta}{dx^2} - 2x\frac{d\theta}{dx} + \left[l(l+1) - \frac{m^2}{1-x^2}\right]\theta = 0 \tag{9.27a}$$

This is the associated Legendre equation. Its solutions are

$$\theta = P_l^m(x) \text{ and } Q_l^m(x) \tag{9.28}$$

the associated Legendre functions.

The function $R(r)$ is now easily obtained by solving Eq. (9.27) which now becomes

$$r^2 \frac{d^2R}{dr^2} - 2r\frac{dR}{dr} - l(l+1)R = 0 \tag{9.29}$$

Its solution obviously is

$$R(r) = a_1 r^l + a_2 r^{-(l+1)} \tag{9.30}$$

In view of Eqs (9.24), (9.28) and (9.30), we conclude that a solution of Laplace's equation in polar coordinates has the form

$$\phi_{lm} = \left\{ a_1 r^l + a_2 r^{-(l+1)} \right\} P_l^m(\cos\theta) e^{\pm im\psi} \tag{9.31}$$

and the general solution will be a superposition of any number of such functions.

9.4. WAVE EQUATION IN THREE DIMENSIONS

Rectangular Coordinates

To give a concise definition of a wave in physical descriptive terms is an easy matter. Mathematically, it is defined as the condition of physical quantity ϕ, which satisfies the differential equation

$$c^2 \nabla^2 \phi = \frac{\partial^2 \phi}{\partial t^2} \tag{9.32}$$

where c is called the phase velocity of the wave. In general, c may be a function of space coordinates (wave travelling in a nonhomogeneous medium). This general case is of special interest in quantum or wave mechanics and in certain branches of optics and will not be dealt here.

In present section, c will be considered constant, that is independent of space and time. We try the form of solution

$$\phi = X(x)\, Y(y)\, Z(z)\, T(t) \tag{9.33}$$

Substitution in wave Eq. (9.32) and dividing by $XYZT$, we have

$$\frac{1}{X}\frac{d^2X}{dx^2} + \frac{1}{Y}\frac{d^2Y}{dy^2} + \frac{1}{Z}\frac{d^2Z}{dz^2} = \frac{1}{c^2 T}\frac{d^2T}{dt^2} \tag{9.34}$$

Each term is function of only one of the independent variables.

Hence if the equation is to hold for all values of x, y, z, t, each term must be constant and every expression of the form

$$\phi = A e^{i(lx + my + nz - \omega t)} \tag{9.35}$$

where A, l, m, n, ω are constants and

$$l^2 + m^2 + n^2 = \frac{\omega^2}{c^2} \tag{9.36}$$

is a solution, so is any sum of expressions of this form. The complex exponential can be replaced by cosines and sines.

Cylindrical Coordinates (r, θ, z)

The wave equation expressed in terms of cylindrical coordinates is

$$\frac{1}{r}\frac{\partial}{\partial r}\left(r\frac{\partial \phi}{\partial r}\right)+\frac{1}{r^2}\frac{\partial^2 \phi}{\partial \theta^2}+\frac{\partial^2 \phi}{\partial z^2}=\frac{1}{c^2}\frac{\partial^2 \phi}{\partial t^2} \qquad (9.37)$$

Using the *separation of variables method* for its solution, we have

$$\phi(r,\,\phi,\,z,\,t) = R(r),\,\theta(\theta),\,Z(z)\,T(t) \text{ gives}$$

$$\frac{1}{R}\cdot\frac{1}{r}\frac{d}{dr}\left(r\frac{dR}{dr}\right)+\frac{1}{r^2}\cdot\frac{1}{\theta}\frac{d^2\theta}{d\theta^2}+\frac{1}{Z}\frac{d^2Z}{dz^2}=\frac{1}{c^2T}\frac{d^2T}{dt^2} \qquad (9.38)$$

Each side of this equation must be equal to the constant which we shall call $-k^2$.

Hence,
$$\frac{d^2T}{dt^2}=-c^2k^2T^2=-\omega^2T$$

or
$$T(t) = Ae^{\pm i\omega t} \qquad (9.39)$$

With this substitution, Eq. (9.38) becomes

$$\frac{1}{R}\cdot\frac{1}{r}\frac{d}{dr}\left(r\frac{dR}{dr}\right)+\frac{1}{r^2}\cdot\frac{1}{\theta}\frac{d^2\theta}{d\theta^2}+\frac{1}{Z}\frac{d^2Z}{dz^2}+k^2=0 \qquad (9.40)$$

The term $\dfrac{d^2Z}{\partial z^2}$ depends only on z and variation in r and θ does not change this term as such must be constant.

Putting

$$\frac{1}{Z}\frac{d^2Z}{dz^2}=-m^2$$

has the solution

$$Z = Be^{\pm imz} \qquad (9.41)$$

Equation (9.40) now on multiplying with r^2 becomes

$$\frac{1}{R}r\frac{d}{dr}\left(r\frac{dR}{dr}\right)+\frac{1}{\theta}\frac{d^2\theta}{d\theta^2}+(k^2-m^2)r^2=0 \qquad (9.42)$$

The term $\dfrac{1}{\theta}\dfrac{d^2\theta}{d\theta^2}$ is independent of r and hence must equal to a constant. Putting this equal to $-n^2$ we have

$$\frac{d^2\theta}{d\theta^2}=-n^2\theta \qquad (9.43)$$

which has the solution

$$\theta = ce^{\pm in\theta} \qquad (9.44)$$

The remaining R in Eq. (9.42) in becomes

$$r^2\frac{d^2R}{dr^2}+r\frac{dR}{dr}+\{(k^2-m^2)r^2-n^2\}R=0 \qquad (9.45)$$

and putting $\sqrt{(k^2-m^2)}\,r=x$, Eq. (9.45) reduces to Bessel's equation

$$x^2 \frac{d^2 R}{dx^2} + x \frac{dR}{dx} + (x^2 - n^2)R = 0$$

and its solution is Bessel's function, i.e.

$$R(x) = J_n(x) = J_n\left\{\sqrt{(k^2 - m^2)}\, r\right\} \tag{9.46}$$

Combining these results we have

$$\phi = A_{kmn} J_n\left\{\sqrt{(k^2 - m^2)}\, r\right\} e^{\pm i(mz \pm n\theta \pm \omega t)} \tag{9.47}$$

When $m = 0$, $Z = (a_1 z + a_2)$
and we obtain another solution

$$\phi_{k0n} = J_n(kr)(a_1 z + a_2)e^{\pm i(n\theta \pm \omega t)} \tag{9.48}$$

On the other hand when $n = 0$

$$\theta = (b_1 \theta + b_2)$$

and we obtain other solution.

$$\phi_{km0} = J_n\left\{\sqrt{(k^2 - m^2)}\, r\right\}(b_1 \theta + b_2)e^{\pm i(mz \pm \omega t)} \tag{9.49}$$

and finally when both $m = n = 0$, the solution takes the form

$$\phi_{k00} = J_0(kr)(a_1 z + a_2)(b_1 \theta + b_2)e^{\pm i\omega t} \tag{9.50}$$

The most general function satisfying wave equation is a superposition of solutions Eqs (9.47), (9.48), (9.49) and (9.50).

Spherical Polar Coordinates (r, θ, ψ)

Wave equation when transformed to spherical polar coordinates reads

$$\nabla^2 \phi = \frac{1}{r^2} \frac{\partial}{\partial r}\left(r^2 \frac{\partial \phi}{\partial r}\right) + \frac{1}{r^2 \sin\theta} \cdot \frac{\partial}{\partial \theta}\left(\sin\theta \frac{\partial \phi}{\partial \theta}\right) + \frac{1}{r^2 \sin\theta} \cdot \frac{\partial^2 \phi}{\partial \psi^2}$$

$$= \frac{1}{c^2} \frac{\partial^2 \phi}{\partial t^2} \tag{9.51}$$

We begin by solving the wave equation in spherical polar coordinates by taking the solution of the form

$$\phi(X, t) = X(x)\,(T/t) \tag{9.52}$$

where $X(x)$ is the space function.

Substituting this trial solution into partial differential equation [Eq. (9.51)] and dividing by XT gives

$$\frac{\nabla^2 X}{X} = \frac{1}{c^2} \cdot \frac{1}{T} \frac{d^2 T}{dt^2}$$

The left side is a function of x only, the right side is a function of T only. They must therefore be constants equal to (say) $-k^2$ ($-k^2$ or k is called a separation constant). Thus, the equation separates into two, the time dependent being

$$\frac{d^2 T}{dt^2} = -\omega^2 T, \quad \text{where } \omega = kc.$$

The solution is

$$T = A_1 \sin \omega t + A_2 \cos \omega t \quad \text{or} \quad T = Ae^{\pm i\omega t} \qquad (9.53)$$

Now let us turn to the second equation, involving the space function $X(x)$. It is also called the Helmholtz equation

$$\nabla^2 X + k^2 X = 0$$

or, in spherical coordinates

$$\left[\frac{1}{r^2} \frac{\partial}{\partial r}\left(r^2 \frac{\partial}{\partial r}\right) + \frac{1}{r^2 \sin\theta} \cdot \frac{\partial}{\partial\theta}\left(\sin\theta \frac{\partial}{\partial\theta}\right) + \frac{1}{r^2 \sin^2\theta} \cdot \frac{\partial^2}{\partial\psi^2}\right] X + k^2 X = 0$$

Let $X = R(r)\,\Phi(\theta)\,\xi(\psi)$. Substituting and dividing throughout by $R\theta\phi$ gives

$$\frac{1}{r^2}\cdot\frac{1}{R}\frac{d}{dr}\left(r + \frac{dR}{dr}\right) + \frac{1}{r^2\Phi\sin\theta}\frac{d}{d\theta}\left(\sin\theta\frac{d\Phi}{d\theta}\right) + \frac{1}{r^2\xi\sin^2\theta}\frac{d^2\xi}{d\psi^2} + k^2 = 0$$

If we multiply throughout by $r^2 \sin^2\theta$, third term would depend only on ψ, while the rest would depend only on r and θ. Therefore

$$\frac{1}{\xi}\frac{d^2\xi}{d\psi^2} = \text{const} = -m^2$$

or

$$\frac{d^2\xi}{d\psi^2} + m^2\xi = 0$$

With solutions

$$\xi = B_1 \sin m\psi + B_2 \cos m\psi \quad \text{or} \quad \xi = Be^{\pm im\psi} \qquad (9.54)$$

The r, θ equation becomes

$$\frac{1}{r^2}\cdot\frac{1}{R}\frac{d}{dr}\left(r^2\frac{dR}{dr}\right) + \left(\frac{1}{r^2\Phi\sin\theta}\frac{d}{d\theta}\right)\left(\sin\theta\frac{d\Phi}{d\theta}\right) - \frac{m^2}{r^2\sin^2\theta} + k^2 = 0$$

If we multiply throughout by r^2, the first and fourth term depend only on r, while the second and third depend only on θ. Thus

$$\frac{1}{\sin\theta}\frac{d}{d\theta}\left(\sin\theta\frac{d\Phi}{d\theta}\right) - \frac{m^2}{\sin^2\theta}\cdot\Phi = \text{const}\times\Phi = -l(l+1)\Phi \qquad (9.55)$$

and

$$\frac{1}{r^2}\frac{d}{dr}\left(r^2\frac{dR}{dr}\right) + \left[k^2 - \frac{l(l+1)}{r^2}\right]R = 0 \qquad (9.56)$$

If we put $\cos\theta = x$, Eq. (9.55) becomes

$$(1-x^2)\frac{d^2\Phi}{dx^2} - 2x\frac{d\Phi}{dx} + \left[l(l+1) - \frac{m^2}{1+x^2}\right]\Phi = 0 \qquad (9.57)$$

Equation (9.57) is the associated Legendre equation. Its solutions are associated Legendre polynomials

$$\Phi = P_l^m(x) \quad \text{and} \quad \Phi = Q_l^m(x) \qquad (9.58)$$

The radial Eq. (9.56) with the change of dependent variable $R = \dfrac{u}{\sqrt{r}}$ becomes

$$\frac{d^2u}{dr^2} + \frac{1}{r}\frac{du}{dr} + \left[k^2 - \frac{\left(l+\frac{1}{2}\right)^2}{r^2}\right]u = 0$$

which is just the Bessel's equation and its solution is Bessel's polynomials, i.e.

$$R = \frac{J_{l+1/2}(kr)}{\sqrt{r}} \quad \text{and} \quad \frac{Y_{l+1/2}(kr)}{\sqrt{r}} \tag{5.59}$$

It is convenient to define spherical Bessel's functions by

$$J_l(x) = \sqrt{\frac{\pi}{2}}J_{l+1/2}(x) \quad \text{and} \quad n_l(x) = \sqrt{\frac{\pi}{2x}}Y_{l+1/2}(x)$$

Hence the solution of radial equation can be written as

$$R(r) = J_l(k_r) \text{ and } n_l(k_r) \tag{9.60}$$

Thus, we have found the following solution

$$\nabla^2\phi - \frac{1}{c^2}\frac{\partial^2\phi}{\partial t^2} = 0$$

$$\Rightarrow \qquad \phi = \begin{Bmatrix} e^{+i\omega t} \\ e^{-\omega i} \end{Bmatrix} \begin{Bmatrix} e^{+im\phi} \\ e^{-im\phi} \end{Bmatrix} \begin{bmatrix} P_l^m(\cos\theta) \\ Q_l^m(\cos\theta) \end{bmatrix} \begin{Bmatrix} J_1(kr) \\ n_1(kr) \end{Bmatrix} \tag{9.61}$$

where each bracket represents a linear combination of the two functions inside. Any linear combination of such solutions is again a solution because of the linearity of the original differential equation.

9.5. EXAMPLES: SOLUTIONS OF LAPLACE'S EQUATION

Line Charge between Two Earthed Conducting Plates

Consider a pair of parallel grounded conducting plates at $y = 0$ and $y = a$, as shown in (Fig. 9.1) with a line charge parallel to the z-axis at point $(0, d)$. We seek a solution valid between the plates, assuming there are no other charges. The problem is thus two-dimensional, one for which rectangular coordinates are appropriate. Let us assume that

$$\phi(x, y) = X(x)\ Y(y)$$

where X is function of x alone and Y is a function of y. Except for the point $(0, d)$, the equation to be satisfied is

$$\nabla^2\phi = Y\frac{d^2X}{dx^2} + X\frac{d^2Y}{dy^2} = 0$$

or $\qquad \dfrac{1}{X}\dfrac{d^2X}{dx^2} + \dfrac{1}{Y}\dfrac{d^2Y}{dy^2} = 0$

Since x and y can vary independently, both terms must be independent of either variable and we can write

$$\frac{1}{X}\cdot\frac{d^2X}{dx^2} = -\frac{1}{Y}\cdot\frac{d^2Y}{dy^2} = k \tag{9.62}$$

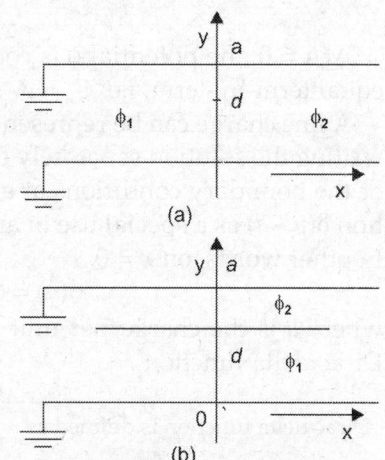

Fig. 9.1: Showing two ways of dividing the space between capacitor plates containing the charge at $x = 0$, $y = d$.

The constant k is called the separation parameter. If there are no restrictions on k, the product of the general solutions of the ordinary differential equations is a general solution of the two-dimensional Laplace equation. The boundary conditions of the physical problem, however, will limit both the nature of the solutions and values of the separation parameters. The solution we seek is a sum of allowed product solutions, with coefficients determined so that the boundary conditions are exactly satisfied.

There remains a choice in the sign of the separation constant k and thus in the nature of the corresponding solutions. Let us first assume that k of Eq. (9.62) is positive, $k = k^2$, so that the ordinary differential equations become

$$\frac{d^2Y}{dy^2} + k^2 Y = 0$$

$$\frac{d^2X}{dx^2} - k^2 X = 0$$

having general solutions

$$\left. \begin{array}{l} Y = A\sin ky + B\cos ky \\ X = ce^{kx} + De^{-kx} \end{array} \right\} \tag{9.63}$$

The boundary conditions to be satisfied are

$$\phi = 0 \text{ at } y = 0, \text{ and } y = a, \text{ and } x = \pm\infty$$

The potential may be made to vanish at the plates simply by setting $B = 0$ and limiting k to the values $n\pi/a$, where n is an integer. The conditions at plus and minus infinity along x, however, cannot be satisfied simultaneously by either term of X, so that we must write the solutions separately for regions of positive and negative x;

$$\left. \begin{array}{l} \displaystyle \mathop{\phi}_{-\infty<x<0} = \sum_{n=1}^{\infty} c_n e^{n\pi x/a} \sin\frac{n\pi y}{a} \\ \\ \displaystyle \mathop{\phi}_{0<x<\infty} = \sum_{n=1}^{\infty} A_n e^{-n\pi x/a} \sin\frac{n\pi y}{a} \end{array} \right] \tag{9.64}$$

At $x = 0$, the potential ϕ is continuous, so that the coefficients in the two series are equal term-by-term, i.e. $C_n = A_n$.

A line charge can be represented by a two-dimensional δ-function. But since we are writing the solution separately for regions 1 and 2, it is possible to use our knowledge of the boundary conditions at a surface charge and employ a one-dimensional function $\delta(y - d)$ as a special use of an arbitrary charge distribution along the surface $x = 0$. In other words, on $x = 0$.

$$\sigma(y) = q\delta(y - d) \tag{9.65}$$

where q is the charge per unit length perpendicular to the xy-plane and $\delta(y - d)$ is Dirac delta function[*].

---

[*] Dirac delta function is defined as

$$\int_0^a \delta(y - d)\,dy = 1$$

and has the property

$$\int_0^a F(y)\delta(y - d)\,dy = F(d), \quad y \le d \le a$$

Now taking the flux from the line charge into account the potential must then satisfy the conditions

$$\frac{\sigma(y)}{\varepsilon_0} = \frac{q\delta(y-d)}{\varepsilon_0} = \left[\frac{\partial\phi_1}{\partial x} - \frac{\partial\phi_2}{\partial x}\right]_{x=0}$$

$$= \sum_{n=1}^{\infty} A_n \frac{2n\pi}{a} \sin\frac{n\pi y}{a} \tag{9.66}$$

The Fourier coefficients in Eq. (9.66) are determined in the usual way by multiplying both sides by $\sin\dfrac{m\pi y}{a}$ and integrating from 0 to a. All terms of series on the right will vanish except that for which $m = n$. Therefore,

$$\frac{q}{\varepsilon_0}\sin\frac{n\pi d}{a} = \frac{2n\pi}{a} \cdot A_n \frac{a}{2} \quad \text{or} \quad A_n = \frac{q}{\varepsilon_0 n\pi} \cdot \sin\frac{n\pi d}{a} \tag{9.67}$$

The entire solution is then

$$\phi_1 = \frac{q}{\varepsilon_0\pi} \sum_{n=1}^{\infty} \frac{1}{n}\sin\frac{n\pi d}{a} e^{n\pi x/a}\sin\frac{n\pi y}{a}$$

$$\phi_2 = \frac{q}{\varepsilon_0\pi} \sum_{n=1}^{\infty} \frac{1}{n}\sin\frac{n\pi d}{a} e^{-n\pi x/a}\sin\frac{n\pi y}{a} \tag{9.68}$$

Equation (9.68) gives the complete solution of the problem.

It is instructive to note that the same potential may look quite different with the opposite choice of sign for the separation parameter in Eq. (9.62). If we put $k = -k^2$ the solution for $X(x)$ is just $\cos kx$ since the potential is obviously an even function of x but no limitations are imposed on k.

No single function Y will vanish on the two conducting plates and we must again divide the region into two parts this time by the plane $y = d$ (Fig. 9.1b). It can be easily verified that for any k the two solutions which vanish at $y = 0$ and $y = a$ and are continuous at $y = d$ are

$$Y_1 = \frac{\sinh ky}{\sinh kd}, \quad 0 < y < d$$

$$Y_2 = \frac{\sinh k(a-y)}{\sinh k(a-d)}, \quad d < y < a \tag{9.69}$$

The potentials are integrals over k, with coefficients which we may call $A(k)$

$$\phi_1 = \int_{-\infty}^{+\infty} A(k)\cos kx \frac{\sinh ky}{\sinh kd}dk$$

$$\phi_2 = \int_{-\infty}^{+\infty} A(k)\cos kx \frac{\sinh k(a-y)}{\sinh k(a-d)}dk \tag{9.70}$$

The charge density on the plane $y = d$ is now a function of x and the condition in the potentials is

$$\frac{q\delta(x)}{\varepsilon_0} = \left[\frac{\partial\phi_1}{\partial y} - \frac{\partial\phi_2}{\partial y}\right]_{y=d}$$

$$= \int_{-\infty}^{+\infty} A(k)\cos k(x)\left[\frac{\cosh kd}{\sinh kd} + \frac{\cosh k(a-d)}{\sinh k(a-d)}\right]k\,dk$$

$$= \int_{-\infty}^{+\infty} A(k)\frac{\cos kx \sinh ka}{\sinh kd \sinh k(a-d)}k\,dk$$

$$= A(k)\frac{2\pi\delta(x)\sinh ka}{\sinh kd \sinh k(a-d)}k^{*}$$

or
$$A(k) = \frac{q}{2\pi\varepsilon_0}\frac{\sinh kd \sinh k(a-d)}{k \sinh ka} \tag{9.71}$$

and the potential is given explicitly by

$$\phi_1 = \frac{q}{2\pi\varepsilon_0}\int_{-\infty}^{+\infty}\frac{\sinh k(a-d)\cos kx \sinh ky}{k \sinh ka}dk$$

$$\phi_2 = \frac{q}{2\pi\varepsilon_0}\int_{-\infty}^{+\infty}\frac{\sinh kd \cos kx \sinh k(a-y)}{k \sinh ka}dk \tag{9.72}$$

The Potential of Dielectric Sphere in Uniform Field

Let us consider a dielectric sphere of inductive capacity k in the presence of a uniform field whose force lines are parallel to the x-axis as shown in Fig. 9.2. The lines of electric displacement are shown.

Since the field at infinity is uniform, the potential is given by
$$\phi_\infty = -E_0 x = -E_0 r \cos\theta = -E_0 r P_1(\cos\theta) \tag{9.73}$$

From the solution of Laplace equation in spherical polar coordinates, Eqs (9.28) and (9.30) can be written as

$$\phi_1 = \sum_{n=0}^{\infty} A_n r^n P_n(\cos\theta)$$

$$\phi_2 = \sum_{n=0}^{\infty} B_n r^{-n-1} P_n(\cos\theta) - E_0 r P_1(\cos\theta) \tag{9.74}$$

for the potentials inside and outside the sphere. The boundary conditions, $\phi_1 = \phi_2$

and $k\dfrac{\partial\phi_1}{\partial r} = \dfrac{\partial\phi_2}{\partial r}$ at $r = a$ must hold for all

values of the angle θ. We therefore evaluate the constants A_n and B_n by equating the coefficients of $P_n(\cos\theta)$ for same n, and find that

$$A_0 = B_0 = 0 = A_n = B_n \text{ for } n \text{ greater than } 1$$

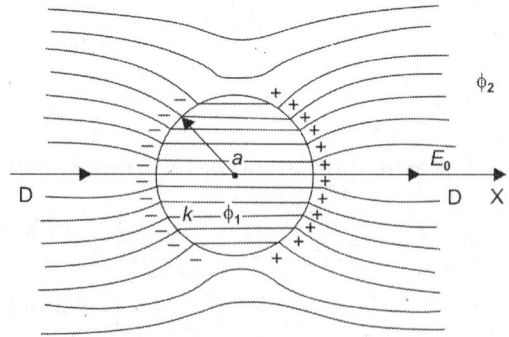

Fig. 9.2: Dielectric sphere in a uniform field

$$^*\int_{-\infty}^{+\infty}\cos kx\,dk = 2\pi\delta(x),$$

as it possesses all the properties that of a Dirac delta function and 2π is the normalising constant.

$$A_1 = -\frac{3E_0}{k+2} \quad \text{and} \quad B_1 = \frac{(k-1)E_0 a^3}{k+2}^*$$

The potentials are therefore

$$\left.\begin{array}{l} \phi_1 = -\dfrac{3E_0 r}{k+2}\cos\theta \\[3mm] \phi_2 = \dfrac{(k-1)\,E_0 a^3\cos\theta}{(k+2)}\dfrac{1}{r^2} - E_0 r\cos\theta \end{array}\right\} \tag{9.75}$$

Note that the field E inside the sphere is uniform, but is smaller than the field at infinity by the ratio $3/k + 2$. It is seen that induced field of the sphere in the region outside the sphere is that of a dipole whose moment is

$$P = 4\pi\varepsilon_0 a^3\left(\frac{k-1}{k+2}\right)E_0 \tag{9.76}$$

Let a quantity L known as the depolarisation factor for a dielectric body, be defined as

$$\frac{L}{\varepsilon_0} = \frac{|E_0| - E\,\text{inside}|}{|\rho\,\text{inside}|}$$

Remembering that

$$\rho = \varepsilon_0 (k-1)\,E$$

gives

$$\frac{L}{\varepsilon_0} = \frac{E_0 - \dfrac{3E_0}{k+2}}{\varepsilon_0 (k-1)\dfrac{3E_0}{k+2}} = \frac{1}{3\varepsilon_0}$$

or

$$L = \frac{1}{3}$$

Thus for a sphere $L = \dfrac{1}{3}$, while for a thin rod oriented parallel to the field $L \approx 0$ and for a thin disc oriented normal to the field $L = 1$. Thus, the electric field within a dielectric body in a uniform field is always smaller than the field at a large distance, while the dielectric displacement is always larger.

The Potential of a Dielectric Sphere and a Point Charge

Problems involving point charges and boundaries which have spherical symmetry can be solved in terms of the functions of Section 9.3. Let us consider a point charge

---

$^*\phi_1 = \phi_2$ at $r = a$ gives for $n = 1$

$$A_1 a = B_1 a^{-2} - E_0 a \tag{a}$$

$k\dfrac{\partial\phi_1}{\partial r} = \dfrac{\partial\phi_2}{\partial r}$ at $r = a$ gives for $n = 1$

$$kA_1 = -2B_1 a^{-3} - E_0 \tag{b}$$

Solving Eqs (a) and (b) gives A_1 and B_1

and a dielectric sphere of radius a, as shown in Fig. 9.3 with r_0 being the distance from the centre of the sphere to the point charge. We may obtain the potential in the radial and angular functions obtained in Eq. (9.31) with separation parameter m equal to zero. We shall need three expressions for ϕ.

$$
\left.
\begin{aligned}
\phi_1^* &= \sum_{n=0}^{\infty} A_n r^n P_n(\cos\theta) \\[2mm]
\scriptstyle 0<r<a \\[2mm]
\phi_2 &= \frac{q}{4\pi\varepsilon_0 r_0} \sum_{n=1}^{\infty} \left(\frac{r}{r_0}\right)^n P_n(\cos\theta) + \sum_{n=0}^{\infty} B_n r^{-(n+1)} P_n(\cos\theta) \\[2mm]
\scriptstyle 0<r<r_0 \\[2mm]
\phi_3 &= \frac{q}{4\pi\varepsilon_0 r} \sum_{n=1}^{\infty} \left(\frac{r_0}{r}\right)^n P_n(\cos\theta) + \sum_{n=0}^{\infty} B_n r^{-(n+1)} P_n(\cos\theta) \\[2mm]
\scriptstyle r_0<r<\infty
\end{aligned}
\right\}
\tag{9.77}
$$

The boundary conditions are

$$\phi_1 = \phi_2 \quad \text{at } r = a$$

and

$$k\frac{\partial\phi_1}{\partial r} = \frac{\partial\phi_2}{\partial r} \quad \text{at } r = a$$

Equating the coefficients of $P_n(\cos\theta)$, the boundary conditions yield

$$A_n a^n = \frac{qa^n}{4\pi\varepsilon_0 r_0^{n+1}} + B_n a^{-(n+1)}$$

$$k\cdot A_n n a^{n-1} = \frac{qna^{n-1}}{4\pi\varepsilon_0 r_0^{n+1}} - (n+1)B_n a^{-(n+2)}$$

Fig. 9.3: Dielectric sphere and point charge

Solving for A_n and B_n, we have

$$A_n = \frac{q(2n+1)}{4\pi\varepsilon_0 r_0^{n+1}\{(n+1)+kn\}}$$

and

$$\left. \begin{aligned} \\ B_n &= \frac{qn(1-k)a^{(2n+1)}}{\{(n+1)+kn\}4\pi\varepsilon_0 r_0^{n+1}} \end{aligned} \right\} \tag{9.78}$$

The required potentials are obtained when these values of A_n and B_n are substituted in Eq. (9.77).

If V_n a homogeneous function of (x, y, z) of degree n is a solution of Laplace's equation, then $\dfrac{V_n}{r^{2n+1}}$ is also a solution of Laplace equation.

$$^*R = (r^2 + r_0^2 - 2rr_0\cos\theta)^{1/2}$$

or

$$\frac{1}{R} = \frac{1}{r_0}\left[\left(\frac{r}{r_0}\right)^n + 1 - 2\left(\frac{r}{r_0}\right)\cos\theta\right]^{-1/2} = \frac{1}{r_0}\sum_{n=0}^{\infty}\left(\frac{r}{r_0}\right)^n P_n(\cos\theta) \quad \text{for } r < r_0$$

$$= \frac{1}{r}\left[1 + \left(\frac{r_0}{r}\right)^2 - 2\left(\frac{r_0}{r}\right)\cos\theta\right]^{-1/2} = \frac{1}{r}\sum_{n=0}^{\infty}\left(\frac{r_0}{r}\right)^n P_n(\cos\theta) \quad \text{for } r_0 < r$$

The solution of Laplace equation is Eq. (9.31)

$$\phi = \left(Ar^n + \frac{A}{r^{n+1}} \right) S_n$$

where S_n is function of n, θ and ψ, i.e. $S_n = S_n(n, \theta, \psi)$ and called a surface harmonic of degree n and by the addition of such solutions, the most general solution of Laplace's equation may be obtained. Thus

$$\phi = \sum \left(A_n r^n + \frac{B_n}{r^{n+1}} \right) S_n$$

Any solution of Laplace's equation is said to be a spherical harmonic. A solution which is homogeneous in x, y, z (in polar coordinates r, θ and ψ) and of dimensions n is said to be a spherical harmonic of degree n. Accordingly it must be of the form $Ar^n S_n$.

Now V_n is spherical harmonic of degree n then

$$V_n = Ar^n S_n$$

Now,

$$\frac{V_n}{r^{2n+1}} = \frac{AS_n}{r^{n+1}}$$

i.e. $\dfrac{V_n}{r^{2n+1}}$ is a spherical harmonic of degree $(n + 1)$ in r and hence must be the solution of Laplace equation.

Point Charge in Grounded Conducting Cylinder

Because of azimuthal symmetry, the entire potential is independent of θ (cylindrical coordinates r, θ, z) and the radial solutions are confined to $J_0(imr)$ or $J_0(kr)$ [$im = k$].

Thus the desired solutions have the form

$$\phi_1 = \sum_{n=1}^{\infty} A_n e^{k_n z} J_0(k_n r) \quad \text{for } z < 0$$

$$\phi_2 = \sum_{n=1}^{\infty} A_n e^{-k_n z} J_0(k_n r) \quad \text{for } z > 0$$

(9.79)

where k_n is such that at the boundary of the cylinder $r = r_0$, $J_0(k_n r_0) = 0$. The flux conditions at the plane $z = 0$ can be stated by means of a two dimensional δ-function defined by the relation

$$2\pi \int_0^{r_0} \delta(r) r \, dr = 1 \tag{9.80}$$

so that

$$\frac{q\delta(r)}{\varepsilon_0} = \left[\frac{\partial \phi_1}{\partial z} - \frac{\partial \phi_2}{\partial z} \right]_{z=0}$$

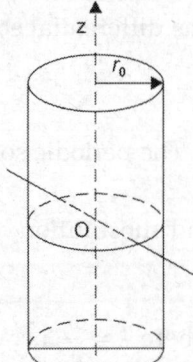

Fig. 9.4: Point charge inside a grounded conducting cylinder

$$= 2 \sum_{n=0}^{\infty} k_n A_n J_0(k_n r) \tag{9.81}$$

If we multiply both sides of Eq. (9.81) by $J_0(k'_n r) \, r \, dr$ and integrate over r, we obtain using orthogonality properties of Bessel's functions as[*]

[*]If a and b are different roots of $J_n(\mu\rho) = 0$ then

$$\int_0^\rho J_n(ar) J_n(br) r \, dr = 0 \quad \text{or} \quad \int_0^\infty F(x) \delta(x - a) dx = F(a)$$

$$\frac{qJ_0(0)}{2\pi\varepsilon_0} = 2k_n A_n \int_0^{r_0} [J_0(k_n r)]^2 r dr$$

or $$A_n = \frac{q/\varepsilon_0}{4\pi k_n} \cdot \frac{J_0(0)}{\int_0^{r_0} [J_0(k_n r)]^2 r dr} \qquad (9.82)$$

If we take into account the boundary condition at r_0, i.e. $J_0(k_n r_0) = 0$ then from orthogonal property of Bessel's functions, we have

$$\int_0^{r_0} [J_0(k_n r)]^2 r dr = \frac{r_0^2}{2}[J_1(k_n r_0)]^2 \qquad (9.83)$$

Also $J_0(0) = 1$ and using Eqs (9.83) in (9.82) we have

$$A_n = \frac{\frac{q}{\varepsilon_0}}{2\pi k_n} \cdot \frac{1}{r_0^2 [J_1(k_n r)]^2}$$

and $$\phi_1 = \frac{q}{2\pi r_0^2 \varepsilon_0} \sum_{n=1}^{\infty} e^{k_n z} \frac{J_0(k_n r)}{k_n \{J_1(k_n r)\}^2} \qquad \text{for } z < 0$$

and $$\phi_2 = \frac{q}{2\pi r_0^2 \varepsilon_0} \sum_{n=1}^{\infty} e^{-k_n z} \frac{J_0(k_n r)}{k_n \{J_1(k_n r)\}^2} \qquad \text{for } z > 0$$

9.6. EXAMPLES: SOLUTIONS OF WAVE EQUATION

Vibrations of Circular Membrane (Round Drum Head)

The differential equation describing small oscillations is

$$\nabla^2 u = \frac{1}{c^2}\frac{\partial^2 u}{\partial t^2} \qquad (9.83a)$$

The periodic solution describing normal modes of the drum will have the form
$$u(x, t) = u(x)\, e^{-i\omega t} \qquad (9.83b)$$
and substitution of Eqs (9.83a) in (9.83b) yields
$$\nabla^2 u + k^2 u = 0$$

where $k = \dfrac{\omega}{c}$, i.e. wave number.

In two-dimensional polar coordinates

$$\nabla^2 = \frac{1}{r}\cdot\frac{\partial}{\partial r}\left(r\frac{\partial}{\partial r}\right) + \frac{1}{r^2}\frac{\partial}{\partial \theta^2}$$

Then, we have

$$\frac{1}{r}\frac{\partial}{\partial r}\left(r\frac{\partial u}{\partial r}\right) + \frac{1}{r^2}\frac{\partial^2 u}{\partial \theta^2} + k^2 u = 0$$

and its solution of using Eq. (9.48) is
$$u = J_n(kr)\, e^{-in\theta}$$
Here the function $Y_n(kr)$ has been eliminated because it becomes infinite at $r \to 0$.

The requirement that our solution be single valued means that n must be an integer. If the membrane is clamped at its outer edge, $r = R$, we must have

$$u = 0 \quad \text{at } r = R$$

Therefore

$$J_n(kR) = 0$$

The zeros of Bessel's functions are tabulated in various books[*]. Some of the zeros are (approximately) as follows

$$\left.\begin{array}{ll} J_0(x) = 0; & x \approx 2.40, 5.52, 8.65 \cdots \\ J_1(x) = 0; & x \approx 3.83, 7.02, 10.17 \cdots \\ J_2(x) = 0; & x \approx 5.14, 8.42, 11.62 \cdots \end{array}\right\} \tag{9.84}$$

Thus the lowest modes of our membrane are

(a) $K = \dfrac{2.40}{R}, \omega = 2.40\dfrac{c}{R}$ and $u \propto J_0\left(2.45\dfrac{r}{R}\right)$

(b) $K = \dfrac{3.83}{R}, \omega = 3.83\dfrac{c}{R}$ and $u \propto J_1\left(3.83\dfrac{r}{R}\right)\left\{\begin{array}{l}\cos\theta \\ \sin\theta\end{array}\right\}$ (9.85)

(c) $K = \dfrac{5.14}{R}, \omega = 5.14\dfrac{c}{R}$ and $u \propto J_2\left(5.14\dfrac{r}{R}\right)\left\{\begin{array}{l}\cos 2\theta \\ \sin 2\theta\end{array}\right\}$

(d) $K = \dfrac{5.52}{R}, \omega = 5.52\dfrac{c}{R}$ and $u \propto J_0\left(5.52\dfrac{r}{R}\right)$

Figure 9.5 indicates the nodes, i.e. places where u is always zero. Note that there are two independent modes belonging to (b) and (c) frequencies. This is an example of degeneracy.

Vibrations of Rectangular Membrane

In this case, it is assumed that the displacement at the boundary is zero so that

$$u = 0 \quad \text{for} \quad \left\{\begin{array}{ll} x = 0 & \text{and} \quad x = a \\ y = 0 & \text{and} \quad y = b \end{array}\right.$$

Now the two dimensional wave equation in Cartesian coordinates becomes

$$\frac{1}{X}\frac{d^2X}{dx^2} + \frac{1}{Y}\frac{d^2Y}{dy^2} + \frac{\omega^2}{c^2} = 0$$

or

$$\frac{1}{X}\frac{d^2X}{dx^2} + \frac{\omega^2}{c^2} = -\frac{1}{Y}\frac{d^2Y}{dy^2} = k^2$$

Thus

$$\frac{d^2Y}{dy^2} + k^2 Y = 0$$

and

$$\frac{d^2X}{dx^2} + \left(\frac{\omega^2}{c^2} - k^2\right)X = 0$$

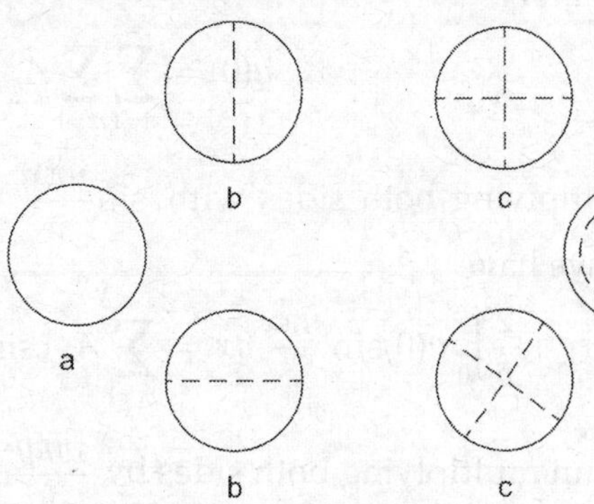

Fig. 9.5

[*]Abramowitz M. and Stegun I. *Handbook of Mathematical Functions*, Dower: New York (1965).

and their solution can be written as

$$Y = B\sin(ky + \delta)$$

$$\left. X = A\sin\left[\sqrt{\frac{\omega^2}{c^2} - k^2}\, x + \delta'\right]\right\}$$

(9.86)

with B, A, δ and δ' are constants.

Now the boundary condition $n = 0$ at $x = 0$ and $x = a$ yields

$$\delta' = 0 \quad \text{and} \quad \sqrt{\frac{\omega^2}{c^2} - k^2} \cdot a = n\pi$$

with n as integer.

So

$$X = \sum_n A_n \sin\frac{n\pi x}{a}$$

and the boundary condition $u = 0$, at $y = 0$ and $y = b$ yields

$$\delta = 0 \text{ and } kb = m\pi$$

where m is an integer. Thus gives

$$Y = \sum_m B_m \sin\frac{m\pi y}{b}$$

The two boundary conditions yield

$$\omega = \pi c \sqrt{\frac{n^2}{a^2} + \frac{m^2}{b^2}}$$

(9.87)

Hence the complete solution is

$$u = \sum_{n=1}^{\infty} A_n \sin\frac{n\pi x}{a} \sum_{m=1}^{\infty} B_m \sin\frac{m\pi y}{b}(\cos\omega t + i\sin\omega t)$$

or

$$u = \sum_{n=1}^{\infty}\sum_{m=1}^{\infty} [A_{mn}\cos\omega t + B_{mn}\sin\omega t]\sin\frac{n\pi x}{a}\cdot\sin\frac{m\pi y}{b} \quad (9.88)$$

where A_{mn} and B_{mn} are new constants which may be determined from initial displacement $u(0)$ and initial velocity $\dot{u}(0)$. We thus have

$$u(0) = \sum_{n=1}^{\infty}\sum_{m=1}^{\infty} A_{mn}\sin\frac{n\pi x}{a}\cdot\sin\frac{m\pi y}{b}$$

Multiplying both sides with $\sin\dfrac{n\pi x}{a}$ and then integrating within limits $x = 0$ to $x = a$, we have

$$\frac{2}{a}\int_0^a u(0)\sin\frac{n\pi x}{a}dx = \sum_{m=1}^{\infty} A_{mn}\sin\frac{m\pi y}{b}$$

Again multiplying both sides by $\dfrac{m\pi y}{b}$ and then integrating within the limits $y = 0$ to $y = b$, we have

$$\frac{4}{ab}\int_0^a\int_0^b u(0)\sin\frac{n\pi x}{a}\cdot\sin\frac{m\pi y}{b}dx\,dy = A_{mn}$$

(9.89)

Similarly, we have

$$\frac{4}{ab\omega} \int_0^a \int_0^b \dot{u}(0) \sin\frac{n\pi x}{a} \sin\frac{m\pi y}{b} dx\, dy = B_{mn} \tag{9.90}$$

Thus after determining A_{mn} and B_{mn}, we may get the solution of the problem in the form of Eq. (9.88) with value of ω given by Eq. (9.87).

Acoustic Radiations from Breathing Sphere

(a) In this case, the sphere performs volume oscillations without distortion. It is characterised by the two boundary conditions.

$$P\Big|_{r=r_0} = P_0 e^{-i\omega t}$$

$$P\Big|_{r\to\infty} = f(r,\theta) e^{i(Kr-\omega t)}$$

Conditions (i) states that at the surface of sphere $r = r_0$, all points shall be in phase and (ii) implies that at infinity, the wave shall be an outgoing one. We limit ourselves to monochromatic waves (pure tones), so that there is only one value of K or ω. Using the spherical polar coordinates, the condition (i) gives that at $r = r_0$ there must not be functional dependence either on θ or ψ, i.e. both l and m are zero. Hence Eq. (9.61) reduces to

$$P = c \{a_1 J_0(Kr) + a_2 N_0(Kr)\} e^{-i\omega t} \tag{9.91}$$

where c, a_1 and a_2 are constants.

But

$$J_0(kr) = \sqrt{\frac{\pi}{2kr}} J_{1/2}(kr)$$

and

$$N_0(kr) = \sqrt{\frac{\pi}{2kr}} Y_{1/2}(kr) = \sqrt{\frac{\pi}{2kr}} J_{-1/2}(kr)$$

Hence Eq. (9.90) becomes

$$P = c\{a_1 J_{1/2}(kr) + a_2 J_{-1/2}(kr)\} \sqrt{\frac{\pi}{2kr}} e^{-i\omega t}$$

$$= c\left\{a_1 \frac{\sin(kr)}{kr} - a_2 \frac{\cos(kr)}{kr}\right\} e^{-i\omega t}{}^*$$

In order to satisfy condition (ii), we put

$$a_1 = i, a_2 = -1$$

obtaining

$$P = \frac{c}{kr} e^{i(Kr-\omega t)}$$

as our final result.

(b) When the sphere of the preceding example vibrates, not with spherical symmetry, but in such a way that condition (i) reads

$$P\Big|_{r=r_0} = \text{const} = \cos\theta\, e^{-i\omega t},$$

$^* J_{1/2}(x) = \sqrt{\frac{2}{\pi x}} \sin x$ and $J_{-1/2}(x) = \sqrt{\frac{2}{x\pi}} \cos x$

it is said to emit *dipole waves*. Condition (*ii*) remaining unchanged for all the functions composing spherical harmonic*. Only P_1^0 (cos θ) is a cosine function. Therefore *l* must be 1. Hence Eq. (9.61) now reduces to

$$P = c\{a_1 J_1(kr) + a_2 N_1(kr)\} e^{-i\omega t} \cdot \cos\theta$$

But

$$J_1(kr) = \sqrt{\frac{\pi}{2kr}} J_{3/2}(kr)$$

and

$$N_1(kr) = \sqrt{\frac{\pi}{2kr}} \cdot Y_{3/2}(kr) = \sqrt{\frac{\pi}{2kr}} J_{-3/2}(kr)$$

so that

$$P = c\cos\theta \left[a_1 \left\{ \frac{\sin kr}{(kr)^2} - \frac{\cos kr}{kr} \right\} + a_2 \left\{ -\frac{\sin kr}{kr} - \frac{\cos kr}{(kr)^2} \right\} \right] e^{-i\omega t}$$

If this expression is to satisfy the condition (*ii*) it is necessary to choose

$$a_1 = -1 \text{ and } a_2 = -i$$

So that

$$P = c\cos\theta \left\{ \frac{1}{kr} + \frac{i}{(kr)^2} \right\} e^{i(kr - \omega t)}$$

The constant *c* may be complex. If it is written

$$c = c_1 + ic_2$$

The real part of *P* which alone is of interest, will be

$$\text{Re } P = \left(\frac{c_1}{kr} - \frac{c_2}{k^2 r^2} \right) \cos\theta \cdot \cos(kr - \omega t) - \left(\frac{c_1}{k^2 r^2} + \frac{c_2}{kr} \right) \cos\theta \cdot \sin(kr - \omega t)$$

For small values of *r*

$$\text{Re } P = -\frac{\cos\theta}{k^2 r^2} [c_2 \cos(kr - \omega t) + c_1 \sin(kr - \omega t)]$$

For large values of *r*

$$\text{Re } P = \frac{\cos\theta}{rk} [c_2 \cos(kr - \omega t) + c_1 \sin(kr - \omega t)]$$

If $c_1 = 0$, the disturbance is of the form cos (*kr* – ω*t*) near the surface of the sphere, but of the sine form at infinity. If $c_2 = 0$, the reverse is true. There occurs, therefore, a curious change of phase as the wave moves out.

9.7. SOLUTION OF HEAT FLOW EQUATION

Equation (9.4) is known as heat flow equation or the diffusion equation, where φ is the temperature and *k* is a constant characteristic of the material through which the heat is flowing. The current density vector in electromagnetic field, the potential and the current propagating along an electrical cable, the excess hydrostatics pressure at any point in consolidation of soil, and the concentration of material in homogeneous material, also satisfy this equation.

---

* A sum of the form

or $Y_l(\theta, \psi) = \sum\limits_{m=l}^{l} c_m P_l^m (\cos\theta) e^{im\psi}$

with arbitrary coefficient c_m is called spherical harmonic.

Now, let us consider a bar of length l and a uniform section, the diameter of which is small in comparison to length with its surface impervious to heat so that there is no radiation from the sides. If one end of the bar be taken as origin and distance along the bar by x, then we have one dimensional heat flow equation as

$$k\frac{\partial^2 \phi}{\partial x^2} = \frac{\partial \phi}{\partial t} \qquad (9.92)$$

with k as constant.

Here two different boundary conditions are possible.

i. Ends of the bar kept at constant temperature (zero), i.e.

$\phi = 0$ when $x = 0$ and $x = l$ for all values of t

$\phi = F(x)$ for $t = 0$, $\phi \neq \infty$ for $t = \infty$

ii. Ends of the bar are impervious to heat, i.e.

$\dfrac{\partial \phi}{\partial x} = 0$ at $x = 0$ and $x = l$ for all values of t

$\phi = F(x)$ for $t = 0$, $\phi \neq \infty$ for $t = \infty$.

We will solve Eq. (9.92) using the method of separation of variables and try as solution of the form

$$\phi(x, t) = e^{ml}u(x) \qquad (9.93)$$

where m is a constant and $u(x)$ is a function of x to be determined. Substitution in Eq. (9.92) yields

$$\frac{d^2 u}{dx^2} = \frac{m}{k}u = -a^2 u \qquad (9.94)$$

where $$a^2 = -\frac{m}{k}$$

and the general solution of Eq. (9.94) is

$$u = A \sin ax + B \cos ax \qquad (9.94a)$$

The boundary conditions (i) yield

$$B = 0 \quad \text{and} \quad a = \frac{r\pi}{l} \text{ with } r = 0, 1, 2, 3...$$

Corresponding to each value of r we have a solution of differential equation [Eq. (9.94)] of the form

$$U_r = A_r \sin\frac{r\pi x}{l}$$

with A_r, the arbitrary constant and the possible values of constant m may be written as

$$m_r = -k\frac{r^2\pi^2}{l^2} \qquad (9.95)$$

To each value of r there corresponds a solution of the differential Eq. (9.92) of the form

$$\phi_r = A_r \sin\frac{r\pi x}{l} \cdot e^{m_r t} = A_r \sin\frac{r\pi x}{l} \cdot e^{-k\frac{r^2\pi^2 t}{l^2}}$$

that satisfies the boundary conditions. By summing over all values of r, we construct the general solution

$$\phi = \sum_{r=1}^{\infty} A_r e^{-k\frac{r^2\pi^2 t}{l^2}} \cdot \sin\frac{r\pi x}{l} \tag{9.96}$$

Using the condition $\phi = F(x)$ for $t = 0$ the arbitrary constant A_r can be evaluated. Thus we obtain

$$A_r = \frac{2}{l}\int_0^l F(x)\sin\frac{r\pi x}{l}\cdot dx \tag{9.97}$$

Equation (9.96) with the values of A_r given by Eq. (9.97) gives the solution of the problem.

iii. If instead of the ends of the bar being kept at temperature zero, they are impervious to heat, then the statement of the problem is defined by case (ii). In this case we have from Eq. (9.94a)

$$\frac{du}{dx} = Aa \cos ax - Ba \sin ax$$

and this is to vanish at $x = 0$ and $x = l$, we must have

$$A = 0 \quad \text{and} \quad \sin al = 0$$

or $\qquad a = \dfrac{r\pi}{l}, \quad r = 0, 1, 2...$

and as in case (i), we obtain the general solution

$$\phi = B_0 + \sum_{r=1}^{\infty} B_r e^{-k\frac{r^2\pi^2}{l^2}t} \cdot \cos\frac{r\pi x}{l} \tag{9.98}$$

with $\qquad B_0 = \dfrac{1}{l}\int_0^l F(x)dx$

and $\qquad B_r = \dfrac{2}{l}\int_0^l F(x)\cos\dfrac{r\pi x}{l}dx$

It is interesting to note that when $t = \infty$, we have

$$\phi = B_0 = \frac{1}{l}\int_0^l F(x)dx$$

the average initial temperature of the bar. This result might of course have been inferred directly from the fact that no heat leaves the bar.

iv. If the temperature on one face ($x = 0$) of the bar is given as a sinusoidal function of the time, the boundary conditions for Eq. (9.92) are

$$\phi = \phi_0 \sin \omega t \quad \text{for } x = 0, \quad \phi \neq \infty \text{ at } x = \infty$$
$$= \text{Im} \cdot (V_0 e^{i\omega t})$$

where Im stands for *imaginary part of the solution*

For the solution of Eq. (9.92) let us assume
$$\phi(x, t) = \text{Im} [u(x) e^{i\omega t}]$$

Discarding Im symbol and substituting in Eq. (9.92), we obtain

$$i\omega u = k\frac{\partial^2 u}{\partial x^2} \quad \text{or} \quad \frac{\partial u^2}{\partial x^2} = a^2 u \tag{9.99}$$

where

$$a^2 = \frac{i\omega}{k} = \frac{\omega}{k}e^{i\left(\frac{\pi}{2}+2n\pi\right)} \qquad\qquad n = 0, 1, 2, 3\ldots$$

or

$$a = \sqrt{\frac{\omega}{k}}\,e^{i\left(\frac{\pi}{4}+n\pi\right)}$$

and the positive root may be written in the form

$$a = \sqrt{\frac{\omega}{k}}\,e^{\frac{i\pi}{4}} = \sqrt{\frac{\omega}{2k}}(1+i) \tag{9.100}$$

Solution of Eq. (9.99) may be written in the form

$$u = Ae^{ax} + Be^{-ax} \tag{9.101}$$

where A and B are arbitrary constants. Now at $x = \infty$, the temperature is finite and hence we must have $A = 0$. At $x = 0$, $t = 0$, we have

$$u = B = V_0$$

Hence the solution is

$$\phi(x, t) = \text{Im}\,[V_0\,e^{-ax}\,e^{i\omega t}]$$

$$= \text{Im}\left[V_0 e^{-\sqrt{\frac{\omega}{2k}}x}\,e^{i\left(\omega t - \sqrt{\frac{\omega}{2k}}x\right)}\right] = V_0 e^{-\sqrt{\frac{\omega}{2k}}x}\sin\left\{\omega t - \sqrt{\frac{\omega}{2k}}x\right\} \tag{9.102}$$

The negative roots lead to the same solution.

We thus see that the temperature decreases in magnitude and changes in phase as we proceed farther into solid. This result explains why the daily variation of the temperature of the Earth's surface cannot be traced to a depth of 3 or 4 ft, while the annual variation cannot be traced beyond a depth of 60 or 70 ft. The greater the frequency of the variation of the surface temperature, the more rapidly is the amplitude of the variation attenuated. This phenomenon has its counterpart in electromagnetic theory and is known in the electrical literature as the *skin effect*.

Three-Dimensional Heat Flow

Let us find the temperature within a cube of side L, initially at temperature $\phi = 0$, which at time $t = 0$ is immersed in a heat bath at temperature $\phi = \phi_0$. For this we must solve the equation

$$\nabla^2\phi = \frac{1}{k}\frac{\partial\phi}{\partial t} \tag{9.103}$$

Assuming the t solution of the form

$$T \propto e^{-\lambda t}$$

Then

$$\nabla^2\phi + \frac{\lambda}{k}\phi = 0$$

or

$$\frac{\partial^2\phi}{\partial x^2} + \frac{\partial^2\phi}{\partial y^2} + \frac{\partial^2\phi}{\partial z^2} = -\frac{\lambda}{k}\phi$$

and separation of variables gives

$$\phi \propto e^{i\alpha x} \cdot e^{i\beta y} \cdot e^{i\gamma z}$$

with
$$\alpha^2 + \beta^2 + \gamma^2 = \frac{\lambda}{k}$$

Now choosing the origin at one corner of the cube, the condition is

$$\phi = \phi_0 \text{ for } x = 0 \text{ and } x = L$$
$$\phi = \phi_0 \text{ for } y = 0 \text{ and } y = L$$
$$\phi = \phi_0 \text{ for } z = 0 \text{ and } z = L$$

Note that this is an inhomogeneous boundary condition, although a very simple one. A particular solution of Eq. (9.103) is $\phi = \phi_0$, and the complementary function must satisfy the corresponding homogeneous boundary conditions $\phi = 0$ on the surface. Thus we write

$$\phi = \phi_0 \sum_{lmn} C_{lmn} \sin\frac{l\pi x}{L} \sin\frac{m\pi y}{L} \sin\frac{n\pi z}{L} e^{-\lambda_{lmn}t} \qquad (9.104)$$

where
$$\lambda_{lmn} = k\frac{\pi^2}{L^2}\{l^2 + m^2 + n^2\}$$

To determine the constants C_{lmn}, we have the condition that $\phi = 0$ when $t = 0$. Therefore

$$\sum_{lmn} C_{lmn} \sin\frac{l\pi x}{L} \sin\frac{m\pi y}{L} \sin\frac{m\pi z}{L} = -\phi_0$$

Multiplying by $\sin\frac{l'\pi x}{L} \sin\frac{m'\pi y}{L} \sin\frac{n'\pi z}{L}$ and integrating over the whole cube, the result is

$$C_{lmn} = \begin{cases} -\dfrac{64\phi_0}{\pi^3 lmn} & \text{if } l, m, n \text{ are odd} \\ 0 & \text{if } l, m, n \text{ are even} \end{cases}$$

Thus for $t > 0$

$$\phi = \phi_0\left\{1 - \frac{64}{\pi^3}\sum_{lmn\,\text{odd}}\frac{1}{lmn}\sin\frac{l\pi x}{L}\sin\frac{m\pi y}{L}\sin\frac{n\pi z}{L}\right\} \times \exp\left[-(l^2 + m^2 + n^2)k\frac{\pi^2 t}{L^2}\right] \qquad (9.105)$$

For $t \geq \dfrac{L^2}{k}$, only the first term in the sum matters

i.e,
$$l = m = n = 1$$

and
$$\phi = \phi_0\left[1 - \frac{64}{\pi^3}\cdot\sin\frac{\pi x}{L}\sin\frac{\pi y}{L}\sin\frac{\pi z}{L}\exp\left(-\frac{3k\pi^2 t}{L^2}\right)\right] \qquad (9.105a)$$

Our solution is not useful for small t.

Two-Dimensional Heat Conduction

Let us consider a thin rectangular plate whose surface is impervious to heat flow has at $t = 0$ an arbitrary distribution of temperature. Its four edges are kept at zero temperature (Fig. 9.6). It is required to determine the subsequent temperature of the plate as t increases.

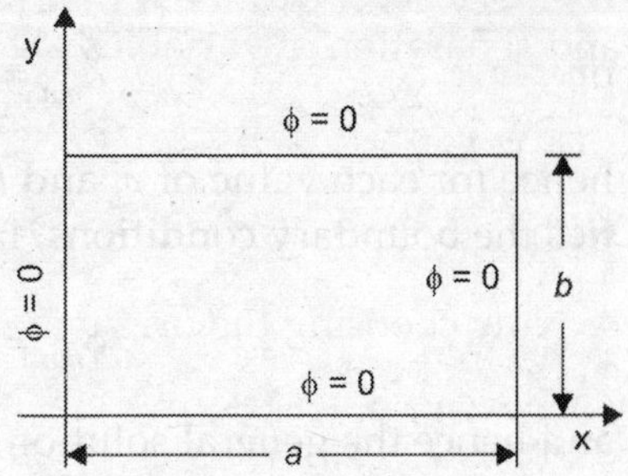

Let the plate extends from $x = 0$ to $x = a$ and from $y = 0$ to $y = b$. Expressing the problem mathematically we must solve the equation

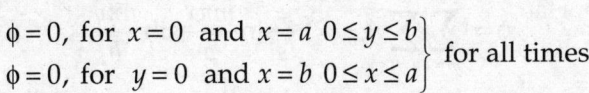

$$\frac{\partial \phi}{\partial t} = k\left(\frac{\partial^2 \phi}{\partial x^2} + \frac{\partial^2 \phi}{\partial y^2}\right) \qquad (9.106)$$

Fig. 9.6

Subject to the conditions

$$\left.\begin{array}{l} \phi = 0, \text{ for } x = 0 \text{ and } x = a \ 0 \leq y \leq b \\ \phi = 0, \text{ for } y = 0 \text{ and } x = b \ 0 \leq x \leq a \end{array}\right\} \text{ for all times}$$

and $\phi = \phi(x, y)$ for $0 \leq x \leq a, 0 \leq y \leq b$ at $t = 0$
and $\phi = 0$ at $t = \infty$.

Let us assume the solution of Eq. (9.106) of the form

$$\phi = e^{-\Box t} F_1(x) F_2(y)$$

and its substitution in Eq. (9.106) of the form

$$\frac{1}{F_1}\frac{d^2 F_1}{dx^2} + \frac{d^2 F_2}{dy^2}\cdot\frac{1}{F_2} - -\frac{\lambda}{k}$$

and this may be written as

$$\frac{1}{F_1}\frac{d^2 F_1}{dx^2} + \frac{\lambda}{k} = -\frac{d^2 F_2}{dy^2}\cdot\frac{1}{F_2} = c^2 \qquad (9.107)$$

The left-hand member of Eq. (9.107) is a function of x only and the right-hand member of Eq. (9.107) is a function of y only and hence both are equal to constants which we call c^2. Separating into two equations, we have

$$\frac{d^2 F_1}{dx^2} + \left(\frac{\lambda}{k} - c^2\right) F_1 = 0$$

and

$$\frac{d^2 F_2}{dy^2} + c^2 F_2 = 0$$

These equations have the solutions

$$F_1 = A_1 \sin px + B_1 \cos px \quad \text{with } p^2 = \left(\frac{\lambda}{k} - c^2\right)$$

and $F_2 = A_2 \sin cy + B_2 \cos cy$

where A's and B's are arbitrary constants. Now to satisfy boundary conditions, it is obvious that $\phi = 0$ for $x = 0$ and $y = 0$ must yield $B_1 = B_2 = 0$. Also $\phi = 0$ for $x = a$ and $y = b$ yields

$$\sin pa = 0 \quad \text{or} \quad p = \frac{m\pi}{a} \qquad m = 0, 1, 2\ldots$$

$$\sin cb = 0 \quad \text{or} \quad c = \frac{\pi n}{b} \qquad n = 0, 1, 2\ldots$$

But $p^2 = \frac{\lambda}{k} - c^2$

or
$$\lambda_{mn} = k(p^2 + c^2) = k\left[\left(\frac{m\pi}{a}\right)^2 + \left(\frac{n\pi}{b}\right)^2\right]$$

hence for each value of m and n we find a particular solution of Eq. (9.106) that satisfies the boundary conditions. Hence the solution is of the form

$$\phi_{mn} = A_{mn}e^{-\lambda_{mn}t}\sin\frac{m\pi x}{a}\cdot\sin\frac{n\pi y}{b}$$

and hence the general solution is

$$\phi = \sum_{m=1}^{\infty}\sum_{n=1}^{\infty}A_{mn}e^{-\lambda_{mn}t}\sin\frac{m\pi x}{a}\sin\frac{n\pi y}{b} \tag{9.108}$$

where A_{mn} are arbitrary constants that must be determined from the initial condition $\phi = \phi(x, y)$ at $t = 0$,

At $t = 0$

$$\phi(x, y) = \sum_{m=1}^{\infty}\sum_{n=1}^{\infty}A_{mn}\sin\frac{m\pi x}{a}\cdot\sin\frac{n\pi y}{b}$$

and multiplying both sides with $\sin\dfrac{m'\pi x}{a}\cdot\sin\dfrac{n'\pi x}{b}$ (m' and n' are integers) and

integrating from $x = 0$ to $x = a$ and $y = 0$ to $y = b$, we obtain

$$A_{m'n'} = \frac{4}{ab}\int_{x=0}^{x=a}\int_{y=0}^{y=b}\phi(x,y)\sin\frac{m'\pi x}{a}\cdot\sin\frac{n'\pi x}{b}dxdy$$

which determines the arbitrary constants of the general solution Eq. (9.108).

9.8. GENERAL THEOREMS

Here are some general theorems useful in the solution of certain heat flow and diffusion problems.

Theorem 1: If $u_1(x, t)$, $u_2(y, t)$ and $u_3(z, t)$ are the solutions of three linear heat flow equations

$$\frac{\partial u_1}{\partial t} = k\frac{\partial^2 u_1}{\partial x^2},\ \frac{\partial u_2}{\partial t} = k\frac{\partial^2 u_2}{\partial y^2},\ \frac{\partial u_3}{\partial t} = k\frac{\partial^2 u_3}{\partial z^2}$$

then $u = u_1\,u_2\,u_3$ is necessarily a solution of the three-dimensional heat flow equation.

$$\frac{\partial u}{\partial t} = k\left(\frac{\partial^2 u}{\partial x^2} + \frac{\partial^2 u}{\partial y^2} + \frac{\partial^2 u}{\partial z^2}\right)$$

The proof of this follows directly by substituting $u = u_1\,u_2\,u_3$ in the three-dimensional heat flow equation.

Then we have

$$\frac{\partial}{\partial t}(u_1u_2u_3) = k\left[\frac{\partial^2(u_1u_2u_3)}{\partial x^2} + \frac{\partial^2(u_1u_2u_3)}{\partial y^2} + \frac{\partial^2(u_1u_2u_3)}{\partial z^2}\right]$$

or $\quad u_2 u_3 \dfrac{\partial u_1}{\partial t} + u_3 u_1 \dfrac{\partial u_2}{\partial t} + u_1 u_2 \dfrac{\partial u_3}{\partial t} = k\left[u_2 u_3 \dfrac{\partial^2 u_1}{\partial x^2} + u_1 u_3 \dfrac{\partial^2 u_2}{\partial y^2} + u_1 u_2 \dfrac{\partial^2 u_3}{\partial z^2} \right]$

and substitution of three linear heat flow equations we obtain an identity, which proves the theorem.

Theorem 2: If $u_1\ (r, t)$ is a solution of the symmetric two-dimensional equation

$$\frac{\partial u_1}{\partial t} = k\left(\frac{\partial^2 u_1}{\partial r^2} + \frac{1}{r}\frac{\partial u_1}{\partial r} \right)$$

and $u_2(z, t)$ is a solution of the linear heat flow equation

$$\frac{\partial u_2}{\partial t} = k\frac{\partial^2 u_2}{\partial z^2}$$

then $u = u_1 u_2$ is necessarily a solution of the three-dimensional heat flow equation

$$\frac{\partial u}{\partial t} = k\left(\frac{\partial^2 u}{\partial r^2} + \frac{1}{r}\frac{\partial u}{\partial r} + \frac{\partial^2 u}{\partial z^2} \right)$$

The proof is again identical to that of theorem I.

9.9. GREEN'S FUNCTIONS

Green's functions provide a general method for solving differential equation satisfying certain boundary conditions in the form of an integral equation.

We will consider one-dimensional case for simplicity.

Let us consider inhomogeneous Sturm–Liouville equation

$$LY(x) = \frac{d}{dx}\left[P(x)\frac{dY(x)}{dx} \right] - q(x)Y(x) = -f(x) \qquad (9.109)$$

where $Y(x)$ is function of x and $L = \left[P(x)\dfrac{d}{dx} \right] - q(x)$ is the self-adjoint differential operator. The function $Y(x)$ satisfies certain boundary conditions at the end point a and b of the specified interval (a, b). Now we proceed to define a strange and arbitrary function $G(x, t)$ called Green's function over the range (a, b).

i. The interval (a, b) is divided by a parameter t and let us denote $G(x)$ by $G_1(x)$ for $a \le x \le t$ and $G(x)$ by $G_2(x)$ for $t < x \le b$.

ii. The function $G_1(x)$ and $G_2(x)$ each satisfy the homogeneous Sturm–Liouville's equation

$$\left. \begin{array}{l} LG_1(x) = 0 \ \text{for}\ a \le x < 1 \\ LG_2(x) = 0 \ \ \text{for}\ \ t \le x < b \end{array} \right\} \qquad (9.110)$$

iii. At $x = a$, $G_1(x)$ satisfies the boundary conditions imposed on $Y(x)$. At $x = b$, $G_2(x)$ satisfies the boundary conditions imposed on $Y(x)$. For convenience in renormalising the boundary conditions are taken to be homogeneous, i.e.

at $x = a$, $\qquad\qquad Y(a) = 0, \qquad\quad Y'(a) = 0$

at $x = b$, $\qquad\qquad Y(b) = 0, \qquad\quad Y'(b) = 0$

or $\qquad\quad \alpha_1 Y(a) + \beta_1 Y'(a) = 0$

and $\qquad\quad \alpha_2 Y(b) + \beta_2 Y'(b) = 0$

where α and β are constants.

iv. $G(x)$ is continuous, i.e.
$$G_1(t) = G_2(t)$$

v. $G'(x)$ and $G''(x)$ are continuous at every point within the range of x except at $x = t$, where it is discontinuous so that

$$G'(t + 0) - G'(t - 0) = -\frac{1}{P(t)}$$

To find the solution of Eq. (9.109), we consider $u_1(x)$ and $u_2(x)$ as the solutions of the homogeneous of Sturm–Liouville's equation such that $u_1(x)$ and $u_2(x)$ satisfy boundary conditions at $x = a$ and $x = b$ respectively.

Then we may take

$$G(x, t) = \begin{cases} c_1 u_1(x) & \text{for } x < t \\ c_2 u_2(x) & \text{for } x > t \end{cases} \tag{9.111}$$

The continuity condition at $x = t$ requires

$$c_1 u_1(t) = c_2 u_2(t) \quad \text{or} \quad c_1 u_1(t) - c_2 u_2(t) = 0 \tag{9.112}$$

The discontinuity condition of the first derivatives

$$c_2 u_2'(t) - c_1 u_1'(t) = -\frac{1}{P(t)} \tag{9.113}$$

The two solutions $u_1(x)$ and $u_2(x)$ are independent if the Wronskian determinant

$$\begin{vmatrix} u_1(t) & u_2(t) \\ u_1'(t) & u_2'(t) \end{vmatrix} = 0, \quad \text{or} \quad u_1(t)u_2'(t) - u_2(t)u_1'(t) = 0 \tag{9.114}$$

Let us assume that $u_1(x)$ and $u_2(x)$ are linearly independent since no Green's function exists when $u_1(x)$ and $u_2(x)$ are linearly dependent. For independent $u_1(x)$ and $u_2(x)$, we have the Wronskian

$$u_1(t)u_2'(t) - u_2(t)u_1'(t) = \frac{A}{P(t)} \tag{9.115}$$

where A is a constant. This equation is sometimes called *Abel's formula*. Comparing Eqs (9.113) and (9.115), we get

$$c_1 = -\frac{u_2(t)}{A} \quad \text{and} \quad c_2 = -\frac{u_1(t)}{A} \tag{9.116}$$

With these values of c_1 and c_2, Eq. (9.113) is obviously satisfied. Substituting the values of c_1 and c_2, the Green's function from Eq. (9.111) is written as

$$G(x, t) = \begin{cases} -\dfrac{1}{A} u_1(x) u_2(t) & \text{for } x < t \\ -\dfrac{1}{A} u_1(t) u_2(x) & \text{for } x > t \end{cases} \tag{9.117}$$

From this, the symmetry property of Green's function is obvious, i.e.
$$G(t, x) = G(x, t)$$

The physical interpretation of this property is, a cause at t yields the same effect on x as a cause at x produces at t. We have constructed the Green's function.

Now it remains to show that the solution of Eq. (9.109) in terms of Green's function is

$$Y(x) = \int_a^b G(x,t)f(t)dt \tag{9.118}$$

This can be shown by direct substitution. Substituting $G(x, t)$ from Eqs (9.117) in (9.118), we get

$$Y(x) = -\frac{1}{A}\int_a^x u_2(x)u_1(t)f(t)dt - \frac{1}{A}\int_x^b u_1(x)u_2(t)f(t)dt \tag{9.119}$$

Differentiating we get

$$Y'(x) = -\frac{1}{A}\int_a^x u_2'(x)u_1(t)f(t)dt - \frac{1}{A}\int_x^b u_1'(x)u_2(t)f(t)dt \tag{9.120}$$

and
$$Y''(x) = -\frac{1}{A}\int_a^x u_2''(x)u_1(t)f(t)dt - \frac{1}{A}\int_x^b u_1''(x)u_2(t)f(t)dt$$

$$-\frac{1}{A}\big[u_1(x)u_2'(x) - u_2(x)u_1'(x)\big]f(x)$$

$$= -\frac{u_2''(x)}{A}\int_a^x u_1(t)f(t)dt - \frac{u_1''(x)}{A}\int_x^b u_2(t)f(t)dt - \frac{f(x)}{P(x)} \tag{9.121}$$

Using Eqs (9.113) and (9.116), and
substituting Y, Y' and Y'' in Eq. (9.109), we get

$$LY(x) = P(x)\frac{d^2Y(x)}{dx^2} + \frac{dP(x)}{dx}\frac{dY}{dx} - q(x)Y(x)$$

$$= -\frac{1}{A}\int_a^x u_1(t)f(t)dt\left\{P(x)\frac{d^2u_2(x)}{dx^2} + \frac{dP(x)}{dx}\cdot\frac{du_2(x)}{dx} - q(x)u_2(x)\right\}$$

$$= -\frac{1}{A}\int_x^b u_2(t)f(t)dt\left\{P(x)\frac{d^2u_1(x)}{dx^2} + \frac{dP(x)}{dx}\cdot\frac{du_1(x)}{dx} - q(x)u_1(x)\right\} - f(x) \tag{9.122}$$

Since $u_1(x)$ and $u_2(x)$ are considered as the solutions of the homogeneous Sturm–Liouville's equation, i.e.

$$Lu_1(x) = P(x)\frac{d^2u_1(x)}{dx^2} + \frac{dP(x)}{dx}\frac{du_1(x)}{dx} - q(x)u_1(x) = 0$$

and
$$Lu_2(x) = P(x)\frac{d^2u_2(x)}{dx^2} + \frac{dP(x)}{dx}\frac{du_2(x)}{dx} - q(x)u_2(x) = 0$$

Hence Eq. (9.122) gives

$$LY(x) = P(x)\frac{d^2Y(x)}{dx^2} + \frac{dP(x)}{dx}\frac{dY(x)}{dx} - q(x)Y(x) = -f(x)$$

Thus Eq. (9.109) is satisfied.

9.10. SOLUTION OF POISSON'S EQUATION USING GREEN'S FUNCTIONS

In presence of electric charges the electrostatic potential ϕ satisfies the Poisson's equation

$$\nabla^2\phi = -\frac{\rho}{\varepsilon} \text{ (MKS units)} \tag{9.123}$$

where ρ is the charge density and ε is the permittivity of the medium.

If the charges are point charges q_i, then the general solution of Poisson's equation is the superposition of single charge solutions given by

$$\phi = \frac{1}{4\pi\varepsilon_i}\sum_i \frac{q_i}{r_i} \tag{9.124}$$

If in place of discrete point charges the system consists of continuous charge distribution with charge density ρ, then Eq. (9.124) becomes

$$\phi(r) = \frac{1}{4\pi\varepsilon}\int \frac{\rho dv}{r} \tag{9.125}$$

If the charge is away from the origin at $r = r_1$ (say) then electrostatic potential at position r_2 is given by

$$\phi(r_2) = \frac{1}{4\pi\varepsilon}\int \frac{\rho(r_1)dr}{|r_2 - r_1|} \tag{9.126}$$

If ϕ is the potential corresponding to given charge distribution and therefore satisfies Poisson's equation [Eq. (9.123)] and $G(r_1, r_2)$ be the Green's function required to satisfy Poisson's equation with a point source at the point r_1. Then we have

$$\nabla^2 G(r_1, r_2) = -\delta(r_1 - r_2) \tag{9.127}$$

Physically $G(r_1, r_2)$ is the potential at r_2 corresponding to a unit source at r_1.

By Green's theorem in usual form

$$\iiint_V (\phi\nabla^2 G - G\nabla^2\phi)dV = \iint_S (\phi\nabla G - G\nabla\phi)dS$$

Assuming that the integral falls off faster than r^{-2}, we may simplify our problem by taking the volume so large that the surface integral vanishes, leaving

$$\iiint_V (\phi\nabla^2 G - G\nabla^2\phi)dV = 0$$

i.e.

$$\iiint_V \phi\nabla^2 G dV = \iiint_V G\nabla^2\phi dV \tag{9.128}$$

Using Eqs (9.123), (9.127) and (9.128), we have

$$-\iiint_V \phi(r_1)\delta(r_2 - r_1)dV = \iiint \frac{G(r_1, r_2)\rho(r_1)}{\varepsilon}dV$$

Using the properties of Dirac delta functions we have

$$\phi(r_2) = \frac{1}{\varepsilon}\iiint G(r_1, r_2)\rho(r_1)dV \tag{9.129}$$

Gauss law gives

$$\iiint \nabla^2\left(\frac{1}{r}\right)dV = \begin{cases} 0 & \text{if volume excludes the origin} \\ -4\pi & \text{if volume includes origin} \end{cases}$$

Using Dirac delta functions this result can be written as

$$\nabla^2\left(\frac{1}{4\pi r}\right) = -\delta(r)$$

or in case where the electrostatic charge is shifted from origin to position $r = r_1$ and $r_{12} = r_2 - r_1$, we have

$$\nabla^2\left(\frac{1}{4\pi r_{12}}\right) = -\delta(r_2 - r_1) \tag{9.130}$$

Comparing Eq. (9.130) with (9.127) we have

$$G(r_1, r_2) = \frac{1}{4\pi r_{12}} = \frac{1}{4\pi(r_2 - r_1)}$$

Substitution of Eq. (9.129), the solution of Poisson's equation becomes

$$\phi(r_2) = \frac{1}{4\pi\varepsilon}\iiint_V \frac{\rho(r_1)}{|r_2 - r_1|}dV \tag{9.131}$$

which is in complete agreement with Eq. (9.126).

Example 1. *A long cylinder is made of two halves, the upper half is at the temperature ϕ_1 while the lower half is at the temperature ϕ_2. Find the distribution of the temperature inside the cylinder.*

Solution: Since the temperature is not the function of time, hence the differential equation of the problem is Laplace equation

$$\nabla^2\phi = 0$$

Here we have a long cylinder with its axis along the z-axis. So z-axis has no effect on the distribution of the temperature because of symmetry about it.

Hence considering the two-dimensional Laplace equation in cylindrical coordinates, its general solution is given by Eq. (9.22) as

$$\phi = a_0 + \sum_{l=1}^{\infty}[a_l\cos l\theta + b_l\sin l\theta]r^l + \sum_{l=1}^{\infty}(c_l\cos l\theta + d_l\sin l\theta)r^{-l}$$

But at the centre $r = 0$, the temperature is finite, i.e. ϕ = finite hence

$$c_l = d_l = 0$$

Thus for finite ϕ at $r = 0$

$$\phi = a_0 + \sum_{l=1}^{\infty}[a_l\cos l\theta + b_l\sin l\theta]r^l$$

Now suppose $\phi = f(\theta)$, at $r = R$, then

$$f(\theta) = a_0 + \sum_{l=1}^{\infty}[a_l\cos l\theta + b_l\sin l\theta]R^l \tag{1}$$

Integrating within limits from 0 to 2π, we have

$$a_0 = \int_0^{2\pi} f(\theta)d\theta \tag{2}$$

Multiplying Eq. (1) by $\cos l\theta$ and then integrating within the limits 0 to 2π, we have

$$a_l = \frac{1}{R^l} \cdot \frac{1}{\pi} \int_0^{2\pi} f(\theta) \cos l\theta \cdot d\theta \qquad (3)$$

Similarly

$$b_l = \frac{1}{R^l} \cdot \frac{1}{\pi} \int_0^{2\pi} f(\theta) \sin l\theta \cdot d\theta \qquad (4)$$

But we have $f(\theta) = \phi_1$ for $\pi > \theta > 0$ (upper half)

and $f(\theta) = \phi_2$ for $2\pi > \theta > \pi$ (lower half)

Substituting these values of $f(\theta)$ in Eqs (2), (3) and (4), we have

$$a_0 = \frac{1}{2\pi} \int_0^{\pi} \phi_1 d\theta + \frac{1}{2\pi} \int_{\pi}^{2\pi} \phi_2 d\theta = \frac{\phi_1 + \phi_2}{2}$$

$$a_l = \frac{1}{R^l} \cdot \frac{1}{\pi} \int_0^{\pi} \phi_1 \cos l\theta \cdot d\theta + \frac{1}{R^l} \cdot \frac{1}{\pi} \int_{\pi}^{2\pi} \phi_2 \cos l\theta\, d\theta = 0$$

and

$$b_l = \frac{1}{R^l} \cdot \frac{1}{\pi} \int_0^{\pi} \phi_1 \sin l\theta \cdot d\theta + \frac{1}{R^l} \cdot \frac{1}{\pi} \int_{\pi}^{2\pi} \phi_2 \sin l\theta \cdot d\theta$$

$$= \frac{1}{\pi R^l} \frac{\phi_1 - \phi_2}{l} \cdot (1 - \cos l\pi)$$

$$= \frac{2}{\pi R^l} \frac{(\phi_1 - \phi_2)}{l} \left.\begin{array}{c}\\\end{array}\right\} \text{ for } l \text{ odd.}$$

$$= 0 \qquad\qquad \text{ for } n \text{ even}$$

Hence temperature distribution is given by

$$\phi = \frac{\phi_1 + \phi_2}{2} + \frac{2(\phi_1 - \phi_2)}{\pi} \sum_{l\,\text{odd}} \frac{r^l}{lR^l} \sin l\theta$$

Solution 2. *Find out the permanent temperature within a solid sphere of radius unity when one half of the surface of the sphere is kept at constant temperature 0 and the other half of the surface is at 1°C.*

Solution: The given conditions are

(i) $\phi = 1$ for $0 < \theta < \pi/2$ (ii) $\phi = 0$ for $\pi/2 < \theta < \pi$

We consider the distribution for upper half, i.e for $\pi > \theta > 0$ and the distribution for lower half will be found symmetrically about the z-axis. Hence taking the two-dimensional Laplace equation in spherical polar coordinates, we have

$$r^2 \frac{\partial^2 \phi}{\partial r^2} + 2r \frac{\partial \phi}{\partial r} + \frac{1}{\sin\theta} \frac{\partial}{\partial\theta}\left(\sin\theta \cdot \frac{\partial \phi}{\partial\theta}\right) = 0$$

The general solution of this equation can be written using Eq. (9.31) with $m = 0$ as

$$\phi = \sum_{l=0}^{\infty}\left(a_l r^l + \frac{b_l}{r^{l+1}}\right) P_l(\cos\theta)$$

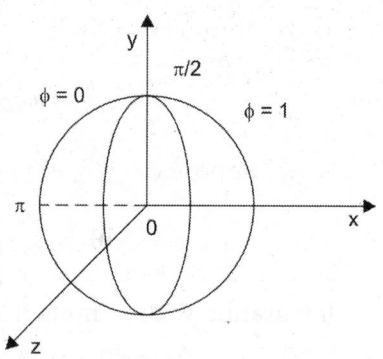

Fig. 9.7

But at the centre $r = 0$ the temperature is finite, i.e. $b_l = 0$ and hence we have

$$\phi = \sum_{l=0}^{\infty} a_l r^l P_l(\cos\theta) \qquad (1)$$

On the surface of sphere $r = 1$. Hence for surface

$$\phi = \sum_{l=0}^{\infty} a_l P_l(\cos\theta)$$

$$= \sum_{l=0}^{\infty} a_l P_l(\mu), \quad \text{where } \mu = \cos\theta$$

Multiplying the equation by $P_l(\mu)$ and integrating within limits from $\mu = -1$ to $\mu = +1$, we have using orthogonal properties of Legendre polynomials,

$$a_l = \frac{2l+1}{2} \int_{-1}^{+1} \phi P_l(\mu) d\mu$$

$$= \frac{2l+1}{2} \int_{\pi}^{0} \phi P_l(\cos\theta) d(\cos\theta) = \frac{2l+1}{2} \int_{0}^{\pi} \phi P_l(\cos\theta)\sin\theta\, d\theta$$

$$= \frac{2l+1}{2} \int_{0}^{\pi/2} \phi P_l(\cos 0)\sin 0\, d0 + \int_{\pi/2}^{\pi} \phi P_l(\cos 0)\sin 0\, d0$$

But $\phi = 0$ for $\pi/2 < \theta < \pi$ and $\phi = 1$ for $\pi/2 > \theta > 0$

So $$a_l = \frac{2l+1}{2} \int_{0}^{\pi/2} P_l(\cos\theta)\sin\theta \cdot d\theta = \frac{2l+1}{2} \int_{0}^{l} P_l(x) dx$$

Therefore, we have

$$a_0 = \frac{1}{2} \int_{0}^{1} dx = \frac{1}{2} \qquad\qquad \text{for } P_0(x) = 1$$

$$a_1 = \frac{3}{2} \int_{0}^{1} P_1(x) dx = \frac{3}{2} \int_{0}^{1} x dx = \frac{3}{4} \qquad \text{for } P_1(x) = x$$

$$a_2 = \frac{5}{2} \int_{0}^{1} P_2(x) dx = \frac{5}{2} \int_{0}^{1} \frac{3x^2-1}{2} dx = 0 \quad \text{for } P_2(x) = \frac{3x^2-1}{2}$$

$$a_3 = \frac{7}{2} \int_{0}^{1} P_3(x) dx = \frac{7}{2} \int_{0}^{1} \frac{(5x^3-3x)}{2} dx \quad \text{for } P_3(x) = \frac{5x^3-3x}{2}$$

$$= -\frac{7}{16} \quad \text{and so on.}$$

Thus we have

$$\phi = \sum_{l=0}^{\infty} a_l r^l P_l(\mu)$$

$$= \frac{1}{2} + \frac{3}{4} r P_1(\cos\theta) - \frac{7}{16} r^3 P_3(\cos\theta) + \cdots$$

which is the required solution.

Example 3. *If the surface of the sphere of radius a is kept at a fixed distribution of electric potential of the form u = F(θ), then find the potential inside and outside the spherical surface assuming that the space inside and outside the sphere to be free of charges.*

Solution: Since the surface of the sphere of radius a is kept at a fixed distribution of electric potential, the boundary conditions are

$$\phi = F(\theta) \text{ at } r = a \tag{1}$$

and $\phi \to 0$ as $r \to \infty$, i.e. potential is zero at infinity $\qquad$ (2)

Case 1. Outside the sphere

Since the space outside the sphere is free of charges hence potential function ϕ must satisfy Laplace equation

$$\nabla^2\phi = 0$$

Because of spherical symmetry the general solution of Eq. (9.31) will be independent of ψ and hence

$$\phi = \sum_{n=0}^{\infty}(A_n r^n + B_n r^{-n-1})P_n(\cos\theta)$$

The boundary condition $\phi \to 0$ at $r \to \infty$ gives $A_n = 0$. Hence

$$\phi = \sum_{n=0}^{\infty}\frac{B_n}{r^{n+1}}P_n(\cos\theta)$$

Using boundary condition $\phi = F(\theta)$ at $r = a$ we have

$$F(\theta) = \sum_{n=0}^{\infty}\frac{B_n}{a^{n+1}}P_n(\cos\theta)$$

Multiplying both sides by P_l (cos θ) and integrating w.r.t (cos θ), we have

$$\int_{-1}^{+1}F(\theta)\cdot P_l(\cos\theta)d(\cos\theta) = \sum_{n=0}^{\infty}\frac{B_n}{a^{n+1}}\int_{-1}^{+1}P_n(\cos\theta)P_l(\cos\theta)d(\cos\theta)$$

$$= \frac{B_l}{a^{l+1}}\cdot\frac{2}{2l+1}$$

or $\qquad B_l = a^{l+1}\cdot\dfrac{2l+1}{2}\displaystyle\int_{-1}^{+1}F(\theta)P_l(\cos\theta)d(\cos\theta)$

$$= a^{l+1}\cdot\frac{2l+1}{2}\int_{0}^{\pi}F(\theta)P_l(\cos\theta)\sin\theta\cdot d\theta$$

Thus $\qquad \phi = \displaystyle\sum_{n=0}^{\infty}\frac{a^{n+1}}{r^{n+1}}\cdot\frac{2n+1}{2}P_n(\cos\theta)\int_{0}^{\pi}F(\theta)P_n(\cos\theta)\sin\theta\cdot d\theta$

Case 2. Inside the sphere

Since $\phi \neq 0$ at $r = 0$, hence $B_n = 0$. Hence the solution of Laplace equation for $r < a$ is

$$\phi = \sum_{n=0}^{\infty}A_n r^n P_n(\cos\theta)$$

and $\qquad F(\theta) = \displaystyle\sum_{n=0}^{\infty}A_n a^n P_n(\cos\theta) \qquad\qquad$ since $\phi = F(\theta)$ at $r = a$.

Multiplying both sides with $P_l(\cos\theta)$ and integrating w.r.t. $\cos\theta$, we have

$$\int_{-1}^{+1} F(\theta)P_l(\cos\theta)d(\cos\theta) = \sum_{n=0}^{\infty} A_n a_n \int_{-1}^{+1} P_n(\cos\theta)\cdot P_l(\cos\theta)d(\cos\theta)$$

$$= A_l a^l \cdot \frac{2}{2l+1}$$

or

$$A_l = \frac{2l+1}{2a^l}\int_{-1}^{+1} F(\theta)P_l(\cos\theta)d(\cos\theta)$$

$$= \frac{2l+1}{2a^l}\int_0^{\pi} F(\theta)P_l(\cos\theta)\sin\theta\cdot d\theta$$

Hence

$$\phi = \sum_{n=0}^{\infty}\frac{2n+1}{2a^n}r^n P_n(\cos\theta)\int_0^{\pi} F(\theta)P_n(\cos\theta)\sin\theta\cdot d\theta$$

Example 4. *Determine the electrostatic potential at any point due to a uniformly charged circular ring of small cross-section lying in xy plane and carrying total charge q.*

Solution: Let O be the centre of the ring of the radius a lying in xy plane. The axis of symmetry of the ring is z-axis and therefore the potential is independent of y. As there is no charge in the space interior and exterior to ring, Laplace's equation $\nabla^2\phi = 0$ is satisfied and therefore the general solution may be expressed as

$$\phi = \sum_{n=0}^{\infty}\left(A_n r^n + \frac{B_n}{r^{n+1}}\right)P_n(\cos\theta) \qquad (1)$$

Along the symmetry axis, i.e. z-axis, $\theta = 0$ and $r = z$ hence potential at any point Q (since each point of the ring is at the same distance $\sqrt{a^2+z^2}$ from point Q) is,

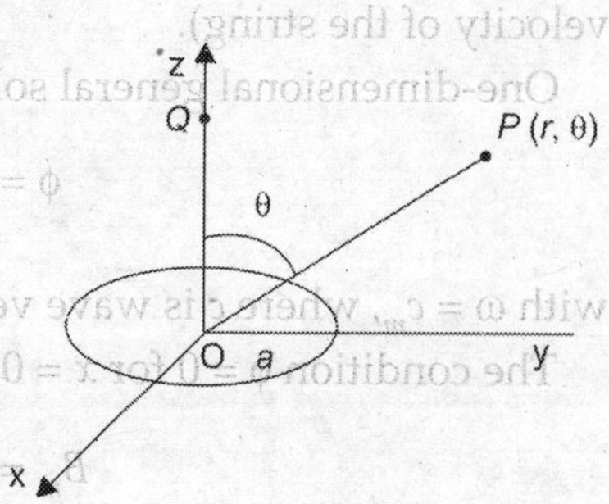

$$\phi = \frac{1}{4\pi\varepsilon_0}\frac{q}{\sqrt{a^2+z^2}}$$

For $z < a$, we have

$$\phi = \frac{1}{4\pi\varepsilon_0}\cdot\frac{q}{a}\left(1+\frac{z^2}{a^2}\right)^{-1/2}$$

$$= \frac{1}{4\pi\varepsilon_0}\cdot\frac{q}{a}\left[1-\frac{z^2}{2a^2}+\frac{1.3}{2.4}\frac{z^4}{a^4}+\cdots\right]$$

$$= \frac{1}{4\pi\varepsilon_0}\cdot\frac{q}{a}\sum_{s=0}^{\infty}(-1)^s\cdot\frac{2s!}{2^{2s}(s!)^2}\left(\frac{z}{a}\right)^{2s} \qquad (2)$$

For $z > a$

$$\phi = \frac{1}{4\pi\varepsilon_0}\cdot\frac{q}{z}\left(1+\frac{a^2}{z^2}\right)^{-1/2}$$

Fig. 9.8

$$= \frac{1}{4\pi\varepsilon_0} \cdot \frac{q}{z} \left[1 - \frac{a^2}{2z^2} + \frac{1.3a^4}{2.4z^4} + \cdots \right]$$

$$= \frac{1}{4\pi\varepsilon_0} \cdot \frac{q}{z} \sum_{s=0}^{\infty} (-1)^s \frac{2s!}{2^{2s}(s!)^2} \left(\frac{a}{z} \right)^{2s} \qquad (3)$$

Comparing Eqs (1), (2) and (3), we have $B_n = 0$ for $z < a$ and A_n are coefficients of Eq. (2). While $A_n = 0$ for $z > a$ and B_n are the coefficients of Eq. (3). Thus we have

$$\phi = \frac{1}{4\pi\varepsilon_0} \cdot \frac{q}{a} \left[P_0 \cos(\theta) - \frac{1}{2} \frac{r^2}{a^2} P_2(\cos\theta) + \frac{1}{2} \cdot \frac{3}{4} \frac{r^4}{a^4} P_4(\cos\theta) + \cdots \right] \text{ for } r < a$$

and $$\phi = \frac{1}{4\pi\varepsilon_0} \cdot \frac{q}{a} \left[\frac{a}{r} P_0 \cos(\theta) - \frac{1}{2} \cdot \frac{a^3}{r^3} P_2(\cos\theta) + \frac{1}{2} \cdot \frac{3}{4} \frac{a^s}{r^5} P_4(\cos\theta) + \cdots \right] \text{ for } r > a.$$

Example 5. *If a string is plucked at its mid-point by the displacement h, find the displacement at any point.*

Solution: Initial displacement position of the string is as shown in Fig. 9.9. It is obvious from the figure that

$$\phi(x) = \frac{2hx}{l} \qquad 0 \leq x \leq \frac{l}{2}$$

$$\phi(x) = \frac{2h(l-x)}{l} \qquad \frac{l}{2} < x < l \Bigg\} \text{ at } t = 0$$

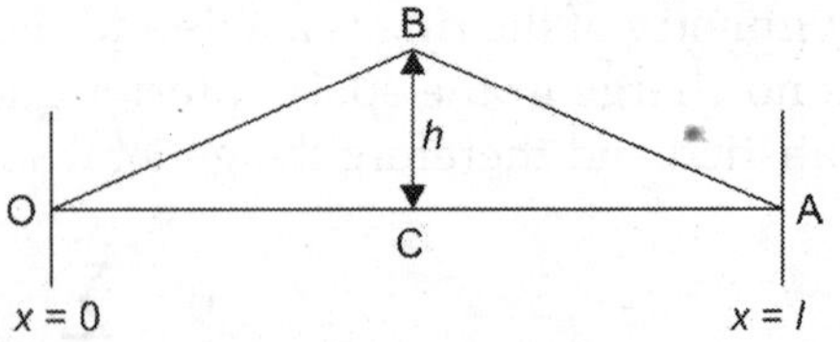

and $\phi' = 0$ for all x at $t = 0$. (ϕ' is transverse velocity of the string).

Fig. 9.9

One-dimensional general solution of wave equation can be written as [Eq. (9.35)].

$$\phi = \sum_m (A_m \sin mx + B_m \cos mx)(c_m \sin \omega t + D_m \cos \omega t)$$

with $\omega = c_m$, where c is wave velocity.

The condition $\phi = 0$ for $x = 0$ and $x = l$ gives

$$B_m = 0 \quad \text{and } m = \frac{n\pi}{l}$$

and $\phi' = 0$ at $t = 0$ gives $c_m = 0$. Hence general solution can be written as

$$\phi = \sum_n a_n \sin \frac{n\pi}{l} x \cdot \cos \frac{n\pi ct}{l}$$

Multiplying with $\sin \frac{n'\pi x}{l}$ and integrating within the limits 0 to l for $t = 0$, we have

$$a_n = \frac{2}{l} \int_0^l \phi(x) \sin \frac{n\pi x}{l} dx$$

$$= \frac{2}{l} \int_0^{l/2} \frac{2hx}{l} \cdot \sin \frac{n\pi x}{l} dx + \frac{2}{l} \int_{l/2}^l \frac{2h(l-x)}{l} \times \sin \frac{n\pi x}{l} dx$$

$$= \frac{4h}{n^2\pi^2}[1-(-1)^n]\sin\frac{n\pi}{2}$$

$$= \begin{cases} 0 & \text{if } n \text{ is even} \\ \dfrac{8h}{n^2\pi^2}\sin\dfrac{n\pi}{2} & \text{if } n \text{ is odd} \end{cases}$$

$$\phi = \frac{8h}{\pi^2}\sum_{n=1}^{\infty} \frac{\sin\dfrac{(2n-1)\pi x}{l}}{(2n-1)^2}\cdot\cos\frac{\pi(2n-1)ct}{l}$$

$$= \frac{8h}{\pi^2}\sum_{n=1}^{\infty} \frac{1}{(2n-1)^2}\cdot\sin\frac{(2n-1)\pi x}{l}\cos\frac{\pi(2n-1)ct}{l}$$

Example 6. *Derive the general expression of displacement of string on a small portion of which a transverse force is applied.*

Solution: The situation described in this problem suggests an idealisation, a force applied at a point on the string. Mathematically speaking this idealisation corresponds to a consideration of the limiting case of a force.

$$F(x) = \begin{cases} 0, & x < \xi - \Delta/2 \\ F, & \xi - \Delta/2 < x < \xi + \Delta/2 \\ 0, & x > \xi + \Delta/2 \end{cases} \qquad (1)$$

where length Δ of the portion of string acted on by the force is allowed to go to zero.

In terms of applied force the equation of motion for the string is

$$\frac{\partial^2\phi}{\partial x^2} = -\frac{F(x)}{T} = f(x) \qquad (2)$$

where T is tension along the string.

Expressing force given by Eq. (1) by Dirac delta functions $\delta(x)$

$$\delta(x) = \begin{cases} 0 & x < -\Delta/2 \\ 1/\Delta & -\Delta/2 < x < +\Delta/2 \\ 0 & x > \Delta/2 \end{cases} \qquad (3)$$

In order to solve the Eq. (2) for a force concentrated at point $x = \xi$, we first work out the solution of

$$\frac{d^2\phi}{dx^2} = -\delta(x - \xi) \qquad (4)$$

The solution satisfies the homogeneous equation $\dfrac{\partial^2\phi}{\partial x^2} = 0$ at all points for which $x \neq \xi$. To obtain its behavior at $x = \xi$ we integrate both sides from $x = \xi - \varepsilon$ to $x = \xi + \varepsilon$ where ε is a vanishingly small quantity. We have[*]

$$\frac{d\phi}{dx} = -1$$

---

[*] Using the property of Dirac delta functions we have $\int \delta(x)dx = 1$

We see that the solution must have a unit change of slope at $x = \xi$. If the supports of the string are rigid, the shape of the string of length l for a force $F = T$ concentrated at $x = \xi$ must be

$$\phi = G(x/\xi) = \begin{cases} x(l-\xi)/l & \text{for } 0 < x < \xi \\ \xi(l-x)/l & \text{for } \xi < x < l \end{cases}$$

as shown in Fig. 9.10.

Here the function $G(x/\xi)$ is the Green's function for Eq. (ii) for point $x = \xi$. We see that the solution for string with force concentrated at $x = \xi$ is $\dfrac{F}{T} G(x/\varepsilon)$ and the solution for force F_1 concentrated at ξ_1 and F_2 at ξ_2 is

$$\frac{F_1}{T} G\left(\frac{x}{\xi_1}\right) + \frac{F_2}{T} G\left(\frac{x}{\xi_2}\right)$$

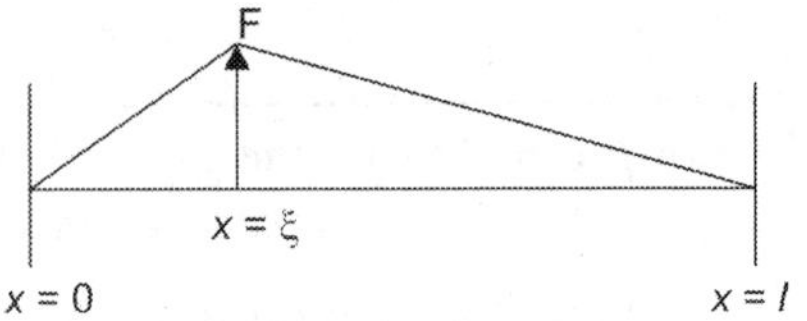

Going from sum to integral and using property of Dirac delta function

$$\int F(x)\delta(x-a)dx = F(a)$$

Fig. 9.10

We see that the steady state shape of a string under tension T between rigid supports a distance l apart, under the action of a steady transverse force $F(x)$ is

$$\phi = \int_0^l \frac{F(\xi)}{T} G\left(\frac{x}{\xi}\right) d\xi$$

Thus Green's function, which is the solution for a concentrated force at $x = \varepsilon$, can be used to obtain a solution of Poisson's equation for a force of arbitrary form, distributed along the string.

Example 7. *An infinitely long uncharged conducting cylinder of circular cross-section is placed in a uniform electric field E_0 in the direction of x-axis, with its axis at right angles to the lines of force. Find the component of the field induced by the presence of cylinder.*

Solution: A conducting cylinder with its axis perpendicular to a uniform electric field causes distortion in that field. For example, a very long fine copper wire stretched midway between two large plates of a parallel-plate capacitor will cause distortions in the field near the wire, like those shown in (Fig. 9.11).

The solutions of $\nabla^2\phi = 0$ obtained in this section can be used to calculate the field around such a cylinder. The uniform field shown in Fig. 9.11a before the cylinder is introduced can be produced by one infinite conducting plate at $x = -x$ with the potential $E_0 x$ and another at $x = -x$ with potential $E_0 x$, and another at $x = x$ with potential $-E_0 x$ where $x \geq a$. The potential of the uniform field is then

$$V = -E_0 x = -E_0 r \cos\theta$$

in the cylindrical coordinates of (Fig. 9.11b). With the cylinder in place, $\nabla^2\phi = 0$ in the space between the plates and the cylinder.

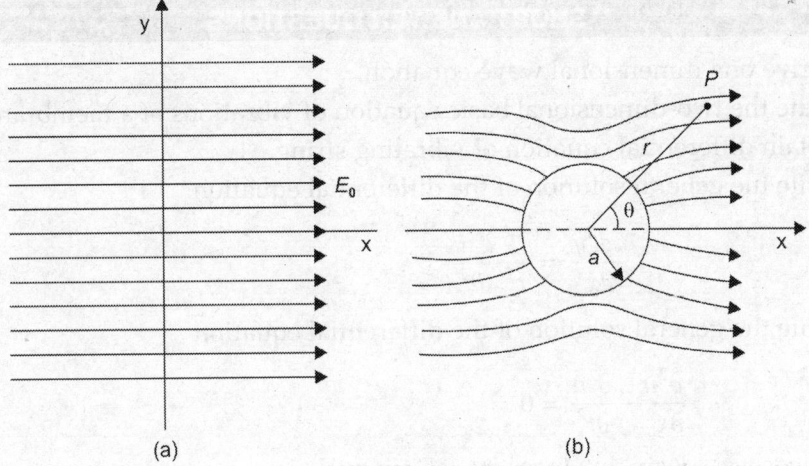

(a)   (b)

Fig. 9.11

If the cylinder axis coincides with the z-axis, the boundary conditions are

$$\phi = 0 \text{ at } r = a \text{ and } \phi = -E_0 r \cos\theta \text{ as } r \geq 0$$

Since $\cos\theta$ is the only function of θ in these boundary conditions the solution may be assumed to be

$$\phi = Ar\cos\theta + \frac{B\cos\theta}{r}$$

Substitution of the boundary conditions at $r \geq a$ gives

$$\phi = Ar\cos\theta = -E_0 r \cos\theta \text{ or } A = -E_0$$

and substitution of the boundary conditions at $r = a$ gives

$$\phi = -E_0 a\cos\theta + \frac{B\cos\theta}{a} = 0 \text{ or } B = E_0 a^2$$

Hence the solution is

$$\phi = -E_0 r\cos\theta + E_0 a^2 \frac{\cos\theta}{r} \tag{1}$$

The field is

$$E = -\nabla\phi = \left((E_0\cos\theta) + \frac{E_0 a^2}{r}\cos\theta\right)a_r + \left(-E_0\sin\theta + \frac{E_0 a^2}{r^2}\sin\theta\right)a_\theta$$

which may be written in the form

$$E = E_0 i + \frac{E_0 a^2}{r^2}(a_r\cos\theta + a_\theta\sin\theta) \tag{2}$$

The first term is just the uniform field that exists in the absence of the cylinder, while the second term is the field due to the cylinder. Figure 9.11b shows the field lines obtained from Eq. (2). It is interesting to note that on the surface of the cylinder where $r = a$, the field is $2E_0$ at $\theta = 0$ or π and zero at $\theta = \pi/2$ or $3\pi/2$. The latter result should have been expected, since the tangential component of E must be zero.

SHORT ANSWER QUESTIONS

1. Derive one-dimensional wave equation.
2. Write the two-dimensional basic equation of vibrations of a membrane.
3. Obtain differential equation of vibrating string.
4. Write the general solution of the differential equation

$$\frac{d^2y}{dx^2} = \frac{1}{c^2}\frac{d^2y}{dt^2}$$

5. Write the general solution of the differential equation

$$\frac{\partial^2 u}{\partial x^2} - \frac{\partial^2 u}{\partial t^2} = 0$$

Ans. $[(A \cos mx + B \sin mx)(c \cos mt + D \sin mt)]$

6. Find the solution of heat flow equation

$$\nabla^2 u = \frac{1}{h^2}\frac{\partial u}{\partial t}$$

7. What is heat flow equation? Express in cartesian, cylindrical and spherical coordinates.
8. Write Laplace's and Poisson's equation in Cartesian coordinates.
9. Write two dimensional heat flow equation under steady state.
10. What do you mean by surface harmonic and surface zonal harmonic?
11. Obtain the differential equation of a hanging chain.
12. If $\psi_1(x, t)$, $\psi_2(y, t)$ and $\psi_3(z, t)$ are solutions of three linear heat flow equations in $x, y,$ and z variables respectively, prove that $\psi = \psi_1 \psi_2 \psi_3$ is necessarily a solution of three-dimensional heat flow equation.
13. Show that the general solution of one-dimensional wave equation can be represented by

$$Y(x, t) = \frac{1}{2}[f(x+ct) + f(x-ct)]$$

14. The sides of a brick are kept at zero temperature and its internal temperature is given arbitrarily at $t = 0$ to be $f(x, y, z)$. Find the subsequent temperature.
15. State any two frequently used partial differential equations in physical problems.
16. It is possible to use the methods of electrical circuit theory to solve problems in one-dimensional heat conduction, why?
17. Write down the classical wave equation in spherical polar coordinates (expressing all symbols). Briefly outline the method of separation of variables.
18. Write the telegraph equations for potential and current.
19. Illustrate the use of variable separation technique to split a partial differential equation of n variables in n ordinary differential equations.
20. Solve one-dimensional heat flow equation.
21. Derive heat coordination equation.

22. Write Helmholtz equation. State one physical situation where it is used.
23. Define diffusivity of a substance.
24. Write two-dimensional Laplace's equation in cylindrical coordinates (r, θ).

PROBLEMS

1. (a) Write the Laplace equation in cylindrical coordinates and discuss the problem of a grounded conducting cylinder.

   (b) Express div A in spherical polar coordinates.

2. Express the Maxwell's equation

$$\nabla \times E + \frac{1}{c}\frac{\partial B}{\partial t} = 0$$

   componentwise in cylindrical polar coordinate.

3. (a) Obtain the solution of Laplace equation in cylindrical coordinates

   (b) A point charge is placed at the origin of coordinates inside a grounded conducting cylinder. Find the potential at a point (i) inside (ii) outside the cylinder.

4. (a) A uniformly stretched thin circular membrane of radius a is fixed along its boundary. It is given an initial displacement $u_0(r)$ at the centre and then released from rest. Derive expressions for the displacement at any time t, and the frequency of vibration.

   (b) Obtain a solution of one-dimensional wave equation in Cartesian coordinates which represents two waves moving along positive and negative directions of x-axis.

5. (a) Find the solution of wave equation in spherical polar coordinates.

   (b) If U_n a homogeneous function of (x, y, z) of degree n, is a solution of Laplace's equation (i.e. a spherical harmonics), show that $r^{-2n-2} V_n$ is also a solution of Laplace equation.

   (c) A dielectric sphere is placed in a uniform electric field. Find the potential at a point external to the sphere.

6. (a) Obtain the solution of the wave equation in spherical polar coordinates.

   (b) A line charge is placed in space between two parallel grounded infinite plates. The line charge is parallel to the plates. Derive an expression for the potential in the entire region of space between the two plates.

7. Show that the potential due to conducting disc of radius a carrying a charge q is

$$\phi(r, z) = \frac{q}{4\pi\varepsilon_0 a}\int_0^\infty e^{-k|z|}J_0(kr)\frac{\sin ka}{k}\,dk$$

   in cylindrical coordinates.

8. An infinite circular cylindrical sheet of radius a is divided longitudinally into quarters which are raised to potentials $V, 0, -V, 0$ respectively. Show that the potential inside the cylinder is given by

$$\phi = \frac{V}{\pi}\left\{\tan^{-1}\left(\frac{2ay}{a^2 - r^2}\right) + \tan^{-1}\frac{2ax}{a^2 - r^2}\right\}$$

What is the potential outside?

9. Consider the motion of a stretched string, whose displacement $y(x, t)$ obeys the differential equation

$$\frac{1}{c^2}\frac{\partial^2 y}{\partial t^2} = \frac{\partial^2 y}{\partial x^2}$$

The boundary conditions are

i. The ends of the string are fixed $y(0, t) = y(L, t) = 0$
ii. The shape of the string is given at $t = 0$, $y(x, 0) = f(x)$
iii. The shape of the string is given at another time T, $y(x, T) = y(x)$.

This is an example of the applications of a Dirichlet boundary condition on a closed boundary (in the $x - t$ plane) to a hyperbolic differential equation.

Does the problem have a solution? Discuss separately the cases.

(a) $\dfrac{cT}{L}$ = integer   (b) $\dfrac{cT}{L}$ = rational number   (c) $\dfrac{cT}{L}$ = irrational number

10. The temperature in an infinite cylindrical rod of radius a satisfies the condition

i. $\nabla^2 T = \alpha^{-1}\dfrac{\partial T}{\partial t}$ $\left(\alpha = \dfrac{k}{c\rho}\right)$

ii. $T = 0$ at $t = 0$
iii. $T = T_0 \cos\phi$ at $\rho = a$
    Find $T(\rho, \phi, t)$ for $t > 0$.

11. An infinite long plane and uniform plate is bounded by two parallel edges and an edge at right angles to them. The breadth is π, and the end is maintained at temperature ϕ_0 at all points and the edge at temperature zero. Show that the steady state temperature is given by

$$\phi = \frac{4\phi_0}{\pi}\left(e^{-y}\sin x + \frac{1}{3}e^{-3y}\sin 3x + \cdots\right) = \frac{2\phi_0}{\pi}\tan^{-1}\frac{\sin x}{\sinh y}$$

12. A rectangular plate has sides of length a and b maintained at temperature 0 and 1 respectively. Find the steady temperature at any point and show that at the centre of the rectangle its value is

$$\frac{4}{\pi}\sum_{n=0}^{\infty}\frac{(-1)^n}{2n+1}\sec h(2n+1)\frac{\pi a}{2b}$$

13. The differential equation governing the displacement of a viscous damped string is

$$\frac{\partial^2 y}{\partial t^2} = c^2\frac{\partial^2 y}{\partial x^2} - 2k\frac{\partial y}{\partial t}$$

Find the general solution of this equation when the string has an initial displacement $y = y_0(x)$ and an initial velocity

$$\frac{\partial y}{\partial t} = v_0(x) \text{ at } t = 0$$

14. A bar length l is heated so that its two ends are at temperature zero. If initially the temperature is given by $u = cx(l-x)/l^2$, find the temperature at time t.

15. A rectangular plate bounded by the lines $x = 0$, $y = 0$, $x = a$, $y = b$, has an initial distribution of temperature given by

$$v = A\sin\frac{\pi x}{a}\sin\frac{\pi y}{b}$$

The edges are kept at zero temperature, and the plane faces are impermeable to heat. Find the temperature at any point and time, and show that very close to any corner of the plate the lines of equal temperature and flow of heat are orthogonal system of rectangular hyperbolas.

Show that the heat lost by plate across the edges up to time t is

$$\frac{4mAab}{\pi^2}\left\{1 - e^{-h^2\left(\frac{1}{a^2}+\frac{1}{b^2}\right)\pi^2 t}\right\}$$

where m is the thermal capacity of the plate per unit area.

16. A bar of uniform cross section is covered with impermeable varnish and extends from the point $x = 0$ to infinity. The bar being throughout at temperature zero, the extremity is brought at time $t = 0$ to temperature ϕ_0 and kept at this temperature. Find the distribution of temperature in the bar at any subsequent time, and verify that the solution gives the obvious solution at $t = \infty$.

17. A string of length l is composed of two separate strings each of different materials fastened together at point a. The segment of string from 0 to the point a is of one material and the segment from a to l is that of the other. Formulate the equation describing this problem; along with the appropriate boundary conditions, and develop the solution in general terms assuming that the initial position is given by $f(x)$ for $0 < x < l$. The string has no initial velocity.

    [**Hint:** Assume that the string as a whole is at the same frequency]

18. Find the solution for the vibrations of a string whose initial velocity is zero and initial distribution is

    (a) $f(x) = A(\sin x - \sin 2x)$ $0 \le x \le \pi$

    (b) $f(x) = 0.1\, x(\pi - x)$ $0 \le x \le \pi$

    as shown in Fig. 9.12.

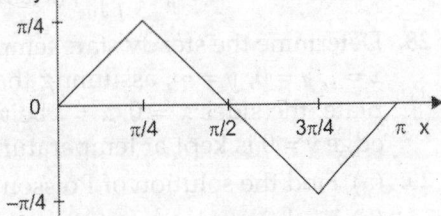

Fig. 9.12

19. Transform two-dimensional Laplace equation into polar coordinates and then solve it to obtain the expressions for circular harmonics.

20. Explain any two-dimensional physical problem illustrating the solution of Laplace's equation by the method of separation of variables.

21. (a) Discuss the importance of Laplace equation and express it in

    (i) cylindrical coordinate system

    (ii) spherical coordinate system.

    (b) Obtain the general solution of Laplace's equation in polar coordinates.

22. (a) Derive the solution of Laplace's equation for a problem having Azimuth symmetry.

(b) A circular ring has charge q per unit length on its circumference. Find the electric potential at all points not lying on the ring.

23. By expressing $\nabla^2 v$ in spherical polar coordinates, solve completely the equation $\nabla^2 v = 0$ in terms of (r, θ, ϕ). Give physical interpretation of two simplest solutions.

24. Obtain solution of three-dimensional Laplace equation in cylindrical coordinates which is finite on the axis.

25. (a) Solve Laplace's equation in spherical polar coordinates.

(b) Hence write down the values of first four spherical harmonics.

(c) Discuss the properties of spherical harmonics.

26. (a) Discuss the general solution of Laplace's equation in spherical polar coordinates.

(b) Prove that the distribution of temperature in the steady state of circular plate of radius R which has one half the circumference at temperature T_1 and other half at T_2 is given by

$$T = \frac{T_1 + T_2}{2} + \frac{2(T_1 - T_2)}{\pi} \sum_{n=1}^{\infty} \left(\frac{r}{R}\right)^{2n-1} \frac{\sin(2n-1)\theta}{2n-1}$$

27. (a) Obtain one-dimensional heat flow equation along a bar in the form of

$$\frac{\partial u}{\partial t} = h^2 \frac{\partial^2 u}{\partial x^2}$$

(b) Find its solution by the method of separation of variables in the form of

$$U = \sum_{n=1}^{\infty} A_n e^{-t\left(\frac{n\pi h}{l}\right)^2} \sin\frac{n\pi x}{l} t,$$

where $\qquad A_n = \frac{2}{l}\int_0^l f(x)\sin\frac{n\pi x}{l} dx$

28. Determine the steady state temperature in thin plate bounded by the lines $x = 0$, $x = l$, $y = 0$, $y = \infty$, assuming that heat cannot escape from either surface of the plate, the sides $x = 0$, $x = l$ being kept at zero temperature and also the lower edge $y = 0$ is kept at temperature $F(x)$ and the edge $y = \infty$ at zero temperature.

29. (a) Find the solution of Poisson's equation using Green's function technique.

(b) Find the solution of the equation $\nabla^2 \phi = Ar + B$ inside a sphere $r < a$, if the boundary $\phi_{r=a} = 0$ is satisfied on the sphere.

30. The temperature u at any point x of a rod at time t satisfies the differential equation

$$\frac{\partial u}{\partial t} = h^2 \frac{\partial^2 u}{\partial x^2}$$

If initially, i.e. at $t = 0$ the temperature at any point x of the rod is $f(x)$, prove that the temperature $u(x, t)$ will be

$$u(x, t) = \frac{1}{\sqrt{\pi}} \int_{-\infty}^{+\infty} f(x + 2ht^{1/2}\omega)e^{-\omega^2} d\omega$$

31. Find the steady state temperature distribution in a sector of a circular plate of radius 10 and angle $\pi/4$ if the temperature is maintained at $0°$ along the radii and at $100°$ along the curved edge.

$$\left[\textbf{Ans. } \phi = \frac{400}{\pi} \sum_{\text{odd}} \frac{1}{n} \left(\frac{r}{20} \right)^{4n} \sin\theta \right]$$

32. A uniform conducting sphere of radius a and the thermometric conductivity h^2 is initially at temperature 0. Heat is supplied uniformly throughout the sphere in such a manner that the temperature would rise at a rate P if there was conduction. The outside is maintained at temperature 0. Show that at any subsequent time the temperature at any point is

$$\phi = \frac{1}{6}\frac{P}{h^2}(a^2 - r^2) + \sum_n \frac{2P}{h^2 r}(-1)^n \sin\frac{n\pi r}{a} \exp(-b^2 h^2 \pi^2 t / a)$$

If r is small, show that an approximation to $-\dfrac{\partial\phi}{\partial r}$ at a point outside is $P\left(\dfrac{2}{h}\sqrt{\dfrac{t}{\tau}} - \dfrac{t}{a} \right)$

33. Show that the allowed angular frequencies of a vibrating circular membrane are determined by the equation

$$J_n\left[a\sqrt{\frac{m\omega^2}{T}} \right] = 0, \quad n = 0, 1, 2...$$

where a is the radius of the circular membrane, m the mass per unit area of the membrane and T is the tension of unit length.

34. A string of length l has zero initial velocity and the initial displacement given as

$$\phi(0) = \begin{cases} \dfrac{4h}{l}x & \text{for } 0 < x\dfrac{l}{4} \\ \dfrac{2h(l-x)}{l} & \text{for } \dfrac{l}{4} < x < \dfrac{l}{2} \\ 0 & \text{for } \dfrac{l}{2} < x < l \end{cases}$$

Find the displacement as function of x and t.

35. Obtain the solution for a vibrating string obeying the equation

$$\frac{\partial^2 u}{\partial x^2} = \frac{1}{c^2}\frac{\partial^2 u}{\partial t^2} \quad -l < x < l$$

with the boundary conditions

$$u(-l, t) = u(l, t) = 0 \text{ for all } t.$$

$$u(x, 0) = \begin{cases} k(l-x) & 0 < x < l \\ k(l+x) & -l < x < 0 \end{cases}$$

36. Solve the partial differential equation for the vibration of a square elastic membrane of side L, fixed at edges

$$\frac{\partial^2\phi}{\partial x^2}+\frac{\partial^2\phi}{\partial y^2}=\frac{1}{v^2}\frac{\partial^2\phi}{\partial t^2}$$

where v is the velocity of elastic wave and $\phi(x, y, t)$ is the displacement of the point (x, y) at time t normal to the plane of the membrane. Obtain the lowest frequency of vibration if $v = 10{,}000$ m/sec and $L = \sqrt{2}$ meters. **[Delhi 1994]**

MULTIPLE CHOICE QUESTIONS

1. Which of the following is the wave equation?

   (a) $\dfrac{\partial^2\phi}{\partial x^2}=c^2\dfrac{\partial^2\phi}{\partial t^2}$

   (b) $\dfrac{\partial^2\phi}{\partial x^2}=\dfrac{1}{c^2}\dfrac{\partial^2\phi}{\partial t^2}$

   (c) $\dfrac{\partial\phi}{\partial t}=c^2\dfrac{\partial^2\phi}{\partial x^2}$

   (d) $\nabla^2\phi+k^2\phi=0$

2. The equation of vibrations of hanging chain is

   (a) $\dfrac{\partial^2u}{\partial x^2}+\dfrac{\partial^2u}{\partial y^2}=\sqrt{\dfrac{g}{x}}\dfrac{\partial^2u}{\partial t^2}$

   (b) $x\dfrac{\partial^2u}{\partial x^2}+y\dfrac{\partial^2u}{\partial y^2}=\sqrt{\dfrac{g}{x}}\dfrac{\partial^2u}{\partial t^2}$

   (b) $x\dfrac{\partial^2u}{\partial x^2}+y\dfrac{\partial^2u}{\partial y^2}=\dfrac{1}{g}\dfrac{\partial^2u}{\partial t^2}$

   (d) $\dfrac{\partial^2u}{\partial x^2}+\dfrac{\partial u}{\partial x}=\dfrac{x}{g}\dfrac{\partial^2u}{\partial t^2}$

3. Laplace equation may be used in the study of
   (a) steady state temperature in a region containing source of heat
   (b) steady state temperature in a region containing no heat source
   (c) variable state temperature in a region with no heat source
   (d) variable state temperature in a region containing heat source

4. In Laplace and Helmholtz equations which is common in separated differential equations?
   (a) equations in r only
   (b) equation in r and θ only
   (c) equations in θ and ϕ only
   (d) equations in r, θ and ϕ

5. Heat flow equation is
   (a) $\nabla^2\phi = 1$
   (b) $\nabla^2\phi = 0$
   (c) $\nabla^2\phi = \dfrac{1}{h^2}\dfrac{\partial\phi}{\partial t}$
   (d) $\nabla^2\phi = \dfrac{1}{c^2}\dfrac{\partial^2\phi}{\partial t^2}$

6. d'Alembert's solution of one-dimensional liberating string is
   (a) $f(x + ct)$
   (b) $f(x - ct)$
   (c) $f_1(x - ct) + f_2(x - ct)$
   (d) $f_1(x + ct) \times f_2(x - ct)$

7. The normal modes of a vibrating string are ($p = 1, 2, 3...$)

   (a) $\dfrac{1}{Pl}\sqrt{\dfrac{T}{m}}$

   (b) $\dfrac{P}{2l}\sqrt{\dfrac{T}{m}}$

   (c) $\dfrac{2l}{P}\sqrt{\dfrac{T}{m}}$

   (d) $\dfrac{2P}{l}\sqrt{\dfrac{T}{m}}$

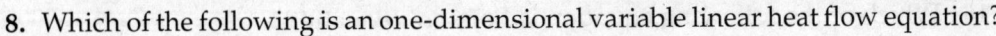
8. Which of the following is an one-dimensional variable linear heat flow equation?

(a) $\dfrac{\partial^2 u}{\partial x^2} = \dfrac{1}{c^2} \dfrac{\partial^2 u}{\partial t^2}$

(b) $\dfrac{\partial^2 u}{\partial x^2} = \dfrac{1}{h^2} \dfrac{\partial u}{\partial t}$

(c) $\dfrac{\partial^2 u}{\partial x^2} = 0$

(d) $\dfrac{\partial^2 u}{\partial x^2} + k^2 u = 0$

9. The partial differential equation $\dfrac{\partial^2 u}{\partial t^2} - c^2 \dfrac{\partial^2 u}{\partial x^2} = 0$ is the

(a) wave equation
(b) heat equation
(c) Laplace equation
(d) Helmholtz equation

10. The solution of equation of hanging chain will involve
(a) Legender's polynomials
(b) Bessel's functions
(c) Hermite polynomials
(d) Leguerre's polynomials

11. The equation of wave propagation in a gas is

(a) $\nabla^2 \phi = \dfrac{\gamma P}{\rho} \dfrac{\partial^2 \phi}{\partial t^2}$

(b) $\nabla^2 \phi = \dfrac{\rho}{\gamma P} \dfrac{\partial^2 \phi}{\partial t^2}$

(c) $\nabla^2 \phi = \sqrt{\dfrac{\gamma P}{\rho}} \dfrac{\partial^2 \phi}{\partial t^2}$

(d) $\nabla^2 \phi = \sqrt{\dfrac{\rho}{\gamma P}} \dfrac{\partial^2 \phi}{\partial t^2}$

12. Laplace's differential equation is
(a) $\nabla^2 \phi = 0$
(b) $\nabla^2 \phi = \rho$
(c) $\nabla^2 \phi = -1$
(d) $\nabla^2 \phi = +1$

13. The displacement function $Y(x, t)$ for a string fixed at both ends and having initial displacement $Y(x, 0) = 0.2 \sin \dfrac{\pi x}{2}$ is $Y(x, t) = 0.2 \sin Px \cos Pvt$. The value of P must be

(a) π

(b) $\dfrac{\pi}{2}$

(c) $\dfrac{2}{\pi}$

(d) depends on point of plucking

14. The solution of Laplace's equation in spherical polar coordinates will involve
(a) Legender's polynomials
(b) Bessel's functions
(c) Hermite polynomials
(d) Leguerre's polynomials

15. If gravitational force is taken into account, the equation of motion of horizontal vibrating string will be

(a) $\dfrac{\partial^2 y}{\partial t^2} = c^2 \dfrac{\partial^2 y}{\partial x^2} + g$

(b) $\dfrac{\partial^2 y}{\partial t^2} = c^2 \dfrac{\partial^2 y}{\partial x^2} - g$

(c) $\dfrac{\partial^2 y}{\partial x^2} = \dfrac{1}{c^2} \dfrac{\partial^2 y}{\partial x^2} + g$

(d) $\dfrac{\partial^2 y}{\partial t^2} = \dfrac{1}{c^2} \dfrac{\partial^2 y}{\partial x^2} - g$

16. Which of the following equation is Poisson's equation?

(a) $\nabla^2 u = 0$

(b) $\nabla^2 u = -\dfrac{\rho}{\varepsilon_0}$

(c) $\rho \nabla^2 u = \varepsilon_0$

(d) $\nabla^2 u = \dfrac{\partial^2 u}{\partial t^2} = \dfrac{\rho}{\varepsilon_0}$

17. In heat flow equation $\nabla^2 u = \dfrac{1}{h^2}\dfrac{\partial u}{\partial t}$ the quantity h is called

(a) Planck's constant

(b) conductivity

(c) heat flow constant

(d) diffusivity

18. If X is the function of x only, and Y is function of y only and Z is the function z only, then solution of Laplace's equation $\dfrac{\partial^2 \phi}{\partial x^2} + \dfrac{\partial^2 \phi}{\partial y^2} + \dfrac{\partial^2 \phi}{\partial z^2} = 0$ may be expressed as

(a) $u = X + Y + Z$

(b) $u = XYZ$

(c) $u = \dfrac{XYZ}{X+Y+Z}$

(d) $u = \dfrac{X+Y+Z}{XYZ}$

19. The basic equation of vibration of membrane in two dimension is

(a) $\dfrac{\partial^2 u}{\partial x^2} + \dfrac{\partial^2 u}{\partial y^2} = 0$

(b) $\dfrac{\partial^2 u}{\partial x^2} + \dfrac{\partial^2 u}{\partial y^2} = \dfrac{\rho}{\varepsilon_0}$

(c) $\dfrac{\partial^2 u}{\partial x^2} + \dfrac{\partial^2 u}{\partial y^2} = \dfrac{1}{c^2}\dfrac{\partial^2 u}{\partial t^2}$

(d) $\dfrac{\partial^2 u}{\partial x^2} + \dfrac{\partial^2 u}{\partial y^2} = \dfrac{1}{c^2}\dfrac{\partial^2 u}{\partial t^2}$

20. The equation of motion of a vibrating string is

(a) $\dfrac{\partial^2 y}{\partial t^2} = \dfrac{T}{\sigma}\dfrac{\partial^2 y}{\partial x^2}$

(b) $\dfrac{\partial^2 y}{\partial x^2} = \dfrac{T}{\sigma}\dfrac{\partial^2 y}{\partial t^2}$

(c) $\dfrac{\partial^2 y}{\partial x^2} = \sqrt{\dfrac{T}{\sigma}\dfrac{\partial^2 y}{\partial t^2}}$

(d) $\dfrac{\partial^2 y}{\partial x^2} = \dfrac{T}{\rho}\dfrac{\partial^2 y}{\partial t^2}$

21. The telegraph equation is

(a) $\dfrac{\partial^2 Y}{\partial x^2} = Rc\dfrac{\partial I}{\partial t}, \dfrac{\partial I}{\partial x^2} = Rc\dfrac{\partial V}{\partial t}$

(b) $\dfrac{\partial^2 V}{\partial x^2} = Lc\dfrac{\partial^2 V}{\partial t^2}, \dfrac{\partial^2 I}{\partial x^2} = Lc\dfrac{\partial^2 I}{\partial t^2}$

(c) $\dfrac{\partial^2 V}{\partial x^2} = \dfrac{L}{R}\dfrac{\partial I}{\partial t}, \dfrac{\partial^2 I}{\partial x^2} = \dfrac{L}{R}\dfrac{\partial V}{\partial t}$

(d) $\dfrac{\partial^2 V}{\partial x^2} = \dfrac{1}{Lc}\dfrac{\partial^2 V}{\partial t^2}, \dfrac{\partial^2 I}{\partial t^2} = \dfrac{1}{Lc}\dfrac{\partial^2 I}{\partial t^2}$

22. The wave equation for electromagnetic potential A and ϕ (when current density J and charge density ρ are zero) are

(a) $\nabla^2 A = \dfrac{\mu}{\varepsilon}\dfrac{\partial^2 A}{\partial t^2}, \nabla^2 \phi = \dfrac{\mu}{\varepsilon}\dfrac{\partial^2 \phi}{\partial t^2}$

(b) $\nabla^2 A = \dfrac{1}{\mu\varepsilon}\dfrac{\partial^2 A}{\partial t^2}, \nabla^2 \phi = \dfrac{1}{\mu\varepsilon}\dfrac{\partial^2 \phi}{\partial t^2}$

(c) $\nabla^2 A = \mu\varepsilon\dfrac{\partial^2 A}{\partial t^2}, \nabla^2 \phi = \mu\varepsilon\dfrac{\partial^2 \phi}{\partial t^2}$

(d) $\nabla^2 A = \sqrt{\dfrac{\mu}{\varepsilon}\dfrac{\partial^2 A}{\partial t^2}}, \nabla^2 \phi = \sqrt{\dfrac{\mu}{\varepsilon}\dfrac{\partial^2 \phi}{\partial t^2}}$

23. The relation between phase velocity u and group velocity V_g is

(a) $V_g = u - \lambda \dfrac{du}{d\lambda}$

(b) $u = V_g - \lambda \dfrac{dV_g}{d\lambda}$

(c) $V_g = 2\lambda \dfrac{du}{d\lambda}$

(d) $u = 2\lambda \dfrac{dV_g}{d\lambda}$

24. Which of the following waves can not be propagated along the waveguide?

(a) TE waves

(b) TM waves

(c) TEM waves

(d) radio waves

25. The differential equation $x\dfrac{du}{dx} + y\dfrac{du}{dy} = 0$, by method of separation of variables is

(c = constant)

(a) $e^x y^{-x}$

(b) $\left(\dfrac{x}{y}\right)^c$

(c) $(xy)^c$

(d) $e^{(x+y)/2}$

26. A string fixed at both ends is plucked at its mid point by the displacement h. The displacement at distance from the fixed end at time t is given by

$$y(x, t) = A\left(\sin\frac{\pi x}{l}\cos\frac{\pi vt}{l} - \frac{1}{9}\sin\frac{3\pi x}{l}\cos\frac{3\pi vt}{l}\cdots\right)$$ the value of A is

(a) h

(b) $\dfrac{2h}{\pi^2}$

(b) $\dfrac{4h}{\pi^2}$

(d) $\dfrac{8h}{\pi^2}$

ANSWERS

1. (a)	2. (c)	3. (b)	4. (c)	5. (c)	6. (c)	7. (b)	8. (b)
9. (a)	10. (b)	11. (b)	12. (a)	13. (b)	14. (a)	15. (b)	16. (b)
17. (c)	18. (b)	19. (c)	20. (a)	21. (a)	22. (c)	23. (a)	24. (c)
25. (b)	26. (d)						

10

Linear Vector Spaces

10.1. INTRODUCTION

We come across several systems in chemical engineering, which can be described by linear equations. Such equations are typical and they arise in modelling many systems in chemical engineering as well as other disciplines. We have already developed analytical methods of solutions to linear equations which arise in modelling of systems. We have developed these methods in a general setting and applied it to a wide class of linear equations, e.g. algebraic equations, ordinary differential equations, and partial differential equations. This demonstrates the universal nature and the common basis of the traditional methods used in solving these systems.

The universal technique is based on concepts from linear algebra and their generalisation. In previous chapters, we restricted ourselves to the problems with finite degrees of freedom. These are governed by linear algebraic equations and linear ordinary differential equations. These arise in the steady state modelling and the transient analysis of lumped systems, respectively. The spatial dependence of the variables in such systems is neglected, but the time dependence is retained. Examples of such systems include distillation columns, stirred tank reactors, and so on. The number of dependent variables in all these systems is finite. Such systems also occur in the solution of partial differential equations using numerical techniques as finite differences, etc. The dependent variables describe a physical system and hence are restricted to be necessarily real throughout this book. In this section, we will investigate equations of the form

$$\frac{du}{dt} = Lu + b \tag{10.1}$$

where u, b are vectors each with n coordinates and L is an $n \times n$ matrix.

Our universal method of solution of these different systems is rooted deeply in concepts of vector spaces and their generalisation. These are developed in the rest of this chapter. In Chapter 2, we have discussed the matrix as a linear operator. The properties and definitions of the linear operator are introduced in the general setting. This enables their ready extension to infinite dimensional operators which occur when dealing with partial differential equations (*refer to* Chapter 9). A word on system dimension is in order at this stage. The dimension or order of a physical system is used to denote the number of dependent variables which interact with each other and determine system behaviour. It does not represent the geometrical dimension, i.e. the number of spatial coordinates. The dimension of a vector space, on the other hand, is used to denote the number of coordinates of each element belonging to that space.

10.2. VECTORS

Students of physics like to think of vectors as a directed line segment, i.e. quantities which have magnitude as well as direction. Mathematics students visualise vectors as an ordered *n*-tuple of numbers written as a column or a row. These are different but equivalent ways of representing the same quantity. The former is called *geometric representation* and the latter *algebraic representation*. Each has its own advantages. The former helps in visualising and understanding the different concepts like distance, length, etc. in a vector space. The extension of these concepts and their generalisation to higher dimensional spaces is possible only because of the latter. We will discuss both these methods of representing a vector.

Consider a plane with two perpendicular axes Ox, Oy (Fig. 10.1a). The origin O is the point of intersection of the two axes. A vector **OP** in this plane is represented by a directed line starting from O. Its magnitude and direction are given by the length and direction of the line segment OP. This is the geometric representation of a vector in the plane which is a two-dimensional space.

The same vector can also be represented by the coordinates of the point $P(x_1, y_1)$. The two coordinates (x_1, y_1) represent the distance of P from O measured parallel to the two axes Ox, Oy, respectively. A vector in this space is an ordered pair (2-tuple) of numbers written as a column $\begin{bmatrix} x_1 \\ y_1 \end{bmatrix}$ or a row $[x_1, y_1]$. This is the algebraic representation of the vector **OP**.

Consider three mutually perpendicular axes Ox, Oy, Oz in the three-dimensional space we live in (Fig. 10.1b). A point P in this space is represented by its coordinates (x_1, y_1, z_1) or an ordered 3-tuple. The coordinates measure the distance of P from O measured parallel to the three axes. The ordering is important as the first coordinate always denotes the distance parallel to the Ox axis, the second, the distance parallel to the Oy axis, and so on. This is the algebraic representation of the vector. The geometrical representation here again is the directed line segment OP.

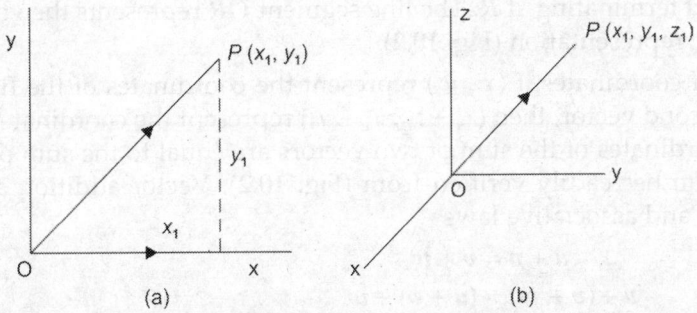

Fig. 10.1: Vector space illustration of geometric and algebraic representations of vectors: (a) Two-dimensional space (b) Three-dimensional space

We have discussed the geometric and algebraic representations of vectors so far, for two-dimensional vector spaces. We next describe the algebraic properties in detail. This necessitates the introduction of many formal mathematical concepts. A detailed formal discussion of these concepts can be found in Murdoch (1995) and Kreyszig (1982). These definitions are important since they allow us to easily extend the methods and techniques developed in the context of finite dimensional vector spaces to infinite dimensional systems or partial differential equations. The theory we develop, allows us to discuss developments in general terms, irrespective of whether the operator L in Eq. (10.1) is a matrix or a differential operator.

10.3. VECTOR SPACES

The notion of vector space will now be formally defined. This is based on the concept of closure.

Closure: A set of numbers is said to be closed under an operation when any two elements of the set subject to the operation yields a third element belonging to the same set. For example, the set of integers is closed under the operation of addition, as the sum of two integers yields a third integer. Closure under other operations as subtraction and multiplication is defined in a similar way. A set of numbers is closed under division if for every a, b, belonging to the set with $b \neq 0$, a/b is a member of the set. The set of all integers 1 (positive, negative and zero) is not closed under division. The integers 2, 3 give rise to 2/3 or 3/2 which are both nonintegers and do not belong to set 1. They are, however, closed under addition.

Field: A field is a set of numbers closed under addition, subtraction, multiplication and division. The set of integers is not closed under division. It is hence not a field. The set of real (complex) numbers forms a field and is denoted by $\mathbb{R}(\mathbb{C})$.

Scalar multiplication: A scalar k multiplying a vector u yields another vector whose length is increased (or decreased) by a factor of $|k|$. If k is positive (negative), the direction of the vector is unchanged (reversed) in the geometric representation. In the algebraic representation, each coordinate of the new vector is k times the corresponding previous coordinate.

Vector addition: Two vectors OP and OQ can be added in the geometric representation using the parallelogram law. The first vector is drawn starting at the origin O. The second vector is drawn parallel to itself, starting from the terminating point of the first vector (P) and terminating at R. The line segment OR represents the vectorial sum in the geometric representation (Fig. 10.2).

In terms of coordinates if (x_1, y_1) represent the coordinates of the first vector and (x_2, y_2) the second vector, then $(x_1 + x_2, y_1 + y_2)$ represent the coordinates of the vector sum. The coordinates of the sum of two vectors are equal to the sum of their coordinates. This can be readily verified from (Fig. 10.2). Vector addition also obeys the commutative and associative laws

$$u + v = v + u$$
$$u + (v + w) = (u + v) + w$$

The algebraic representation of a vector as an n-tuple allows the easy verification of these laws. We have now established all the concepts required to define vector space.

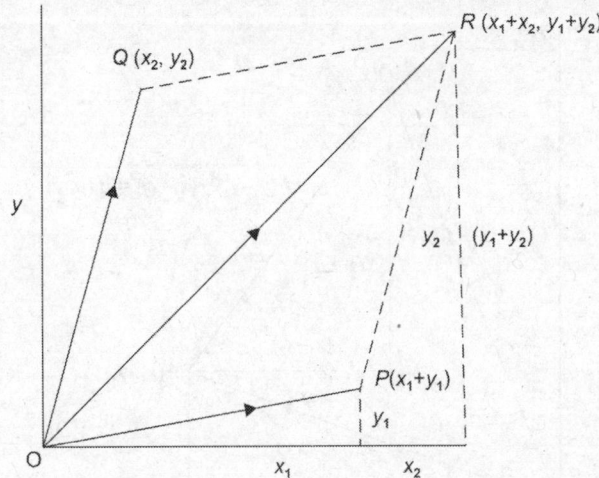

Fig. 10.2: Geometric representation of vector addition in two-dimensional space

A vector space or a linear space has two kinds of elements:

1. A field $\mathbb{F}$ which consists of scalars. These could be real or complex, i.e. $\mathbb{F}$ can be $\mathbb{R}$ or $\mathbb{C}$.

2. A set of vectors $\mathbb{V}$, which is closed under the operations of scalar multiplication and vector addition.

The implies that, for $c_1, c_2 \in \mathbb{F}$ and $u^1, u^2 \in \mathbb{V}$, the vectors $c_1 u^1, c_2 u^2$ and $c_1 u^1 + c_2 u^2$ belong to $\mathbb{V}$. This is true for all $c^i \in \mathbb{F}$ and $u^i \in \mathbb{V}$. In this section, we use the first few letters of the alphabet b, c to denote scalars in $\mathbb{F}$ and the last few alphabets u, v to denote vectors in $\mathbb{V}$. The superscript 'i' in the vector indicates the ith vector.

n-dimensional real (complex) vectors are said to belong to the vector space by $\mathbb{R}^n(\mathbb{C}^n)$. Every vector in $\mathbb{R}^n(\mathbb{C}^n)$ has n-coordinates, each of which is a real (complex) number. More specifically, the vectors in the plane belong to $\mathbb{R}^2$. For the present context we take the dimension of a vector space to be synonymous with the number of coordinates of a vector in a vector space. The formal definition of dimension of a vector space is based on the algebraic concept of linear dependence and will be introduced later.

10.4. METRICS, NORMS AND INNER PRODUCTS

Concepts like distance between vectors, length of a vector, angle between vectors will now be explained in two- and three-dimensional spaces using the geometric representation [Murdoch (1995)]. They will be extended to higher dimensional spaces using the algebraic representation.

Distance between points (vectors): Let $u(x_1, y_1)$, $v(x_2, y_2)$ be two vectors in $\mathbb{R}^2$ as shown in Fig. 10.3. Each vector has its starting point at the origin O. The distance between the vectors denoted by $d(u, v)$ is defined *classically as*

$$d(u, v) = +\sqrt{(x_1 - x_2)^2 + (y_1 - y_2)^2}$$

(10.2)

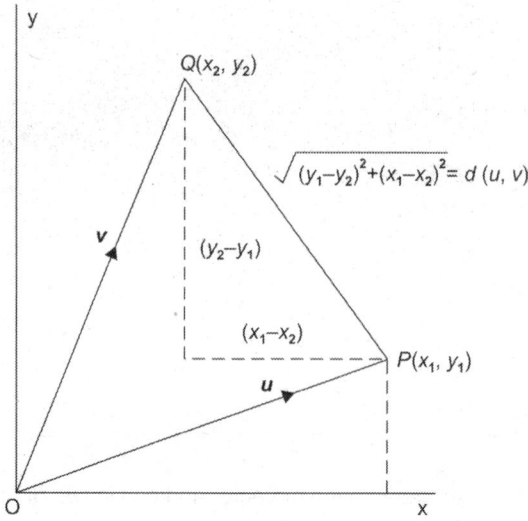

Fig. 10.3: Geometric representation of distance or metric between vectors, $d(u, v)$

This definition can be extended directly to a three-dimensional space $\mathbb{R}^3$ as

$$d(u, v) = +\sqrt{(x_1 - x_2)^2 + (y_1 - y_2)^2 + (z_1 - z_2)^2} \qquad (10.2a)$$

$d(u, v)$ is a scalar and is also called the metric. The definition in Eqs (10.2) and (10.2a) is called the Euclidean metric. It represents the geometric distance between the terminal points of the vectors.

Length: The geometric (Euclidean) length of a vector u is the distance of its terminal point from the origin. This is also called the norm of the vector and is denoted by $\|u\|$. Since the origin has zero as its coordinates, we have

$$\|u\| = \sqrt{x_1^2 + y_1^2} \qquad \text{in } \mathbb{R}^2 \qquad (10.3)$$

$$= \sqrt{x_1^2 + y_1^2 + z_1^2} \qquad \text{in } \mathbb{R}^3 \qquad (10.3a)$$

This is identical to the length of a line in Euclidean geometry and thus Eq. (10.3) is called the Euclidean norm.

Angle: The angle between two vectors (u and v) is indirectly given by the inner-product between them. This is a measure of the projection of a vector $u(x_1, y_1)$ in the direction of the vector $v(x_2, y_2)$ and is denoted by $\langle u, v \rangle$. For vectors in a plane, this is defined as

$$\langle u, v \rangle = x_1 x_2 + y_1 y_2 \qquad (10.4)$$

The inner-product is a scalar. Equation (10.4) is analogous to (in fact, a scalar multiple of) the dot product between vectors in physics. The inner-product directly yields the cosine of the angle θ between the two vectors as given by

$$\cos \theta = \langle u, v \rangle / \|u\| \|v\| \qquad (10.4a)$$

The length of a vector, the distance and the angle between two vectors have been defined using the algebraic representation (i.e. their coordinates). The geometric representation allows us to visualise these in the lower dimensional spaces $\mathbb{R}^2$, $\mathbb{R}^3$. These concepts can be extended to higher dimensions easily, using the algebraic representation as column vectors.

For notational convenience, from now on, we represent the ith coordinate of the row vector u by u_i and of v by v_i, i.e.

$$u = (u_1, u_2,...,u_n), \quad v = (v_1, v_2,...,v_n)$$

This differs from the notation used so far, where x represents the first coordinate, y the second coordinate, etc. The change in the notation, to which we will adhere for the rest of this chapter, is to introduce mathematical elegance and facilitate generalisation. We do not restrict ourselves to a maximum dimension of 26, which is, what we would have to if we had used the English alphabets to represent the different coordinates. A further word on notation is necessary before we continue any further. A subscript i denotes the ith coordinate and a superscript j the jth vector. So x^j is the jth vector, whereas x_i is a scalar representing the ith coordinate of the vector x. Capital letters will be used to denote operators, i.e. marix operators, differential operators or boundary operators.

We digress from this notation when we talk of maps. While dealing with maps and the Newton–Raphson method, the subscript 'n' is used to refer to the nth iterate of the map, and not the nth coordinate of a vector. Most of our discussion will centre on one-dimensional maps. Hence there is no vector. This change in notation is necessitated by the need for consistency with literature. Also, by dimension of a system, we mean the number of independent state variables needed to describe the system completely. A finite-dimensional system is modelled by a finite number of ordinary differential equations or algebraic equations. An infinite-dimensional system, on the other hand, is modelled by an infinite number of equations. These occur while solving boundary value problems. We would like to emphasise that by *infinite dimensional*, we do not mean spatially unbounded or extending to infinity in space. Our systems can be spatially bounded and still be infinite dimensional. Here the system state is determined by specifying the values of a variable in an interval (or an infinity of points). These variables interact with each other to determine system behaviour.

The ordered n-tuple, i.e. a vector can be written as a row vector or a column vector. A distinct identity is maintained for a row vector as opposed to a column vector. This is necessary to carry out algebraic operations on the vectors. Transposing a column vector, i.e. writing the column as a row while preserving the order of the coordinates, yields a row vector. Similarly, a row vector can be transposed to yield a column vector. The transpose operation is denoted by a superscript t as

$$\begin{pmatrix} x_1 \\ x_2 \\ \vdots \\ x_n \end{pmatrix}^t = (x_1, x_2,...,x_n) \tag{10.5}$$

and

$$(x_1, x_2,...,x_n)^t = \begin{pmatrix} x_1 \\ x_2 \\ \vdots \\ x_n \end{pmatrix} \tag{10.5a}$$

10.4.1. Metric Space

The Euclidean distance defined in Eq. (10.2) can be extended to higher dimensional spaces. The distance $d(u, v)$ between two vectors (u, v) in $\mathbb{R}^n$ is the positive scalar defined as

$$d(u, v) = \left[\left(\sum_{i=1}^{n} (u_i - v_i)^2 \right) \right]^{1/2} \tag{10.6}$$

In higher dimensional spaces, we lose the geometric significance of the Euclidean distance as we cannot visualise the space. Equation (10.6a) is a general and abstract representation of the distance between u, v. The loss of the physical significance of distance prompts us to seek other permissible definitions in $\mathbb{R}^n$. These can be allowed as long as they satisfy certain axioms. The axioms ensure that the definitions do not violate the basic properties which we expect a function such as distance to satisfy. The definitions are a representative measure of the distance and will not correspond to the Euclidean (geometrical) distance between vectors. These should also be valid for lower, i.e. two- and three-dimensional spaces.

A metric $d(u, v)$ is defined so as to satisfy the following axioms:

D1: $d(u, v) \geq 0$

D2: $d(u, v) = 0$ implies $u = v$

D3: $d(u, w) \leq d(u, v) + d(v, w)$

D4: $d(u, v) = d(v, u)$

These axioms are mathematical statements of the properties that the distance between two vectors must possess. D1 states that the distance between two points cannot be negative. D2 states, if the distance between two points is zero, then the two points are identical. D3 is the mathematical representation of the triangle inequality, i.e. two sides of a triangle are always greater than the third side (Fig. 10.4). The last axiom says that the distance between u and v is the same as that between v and u (refer to Stakgold, 1979).

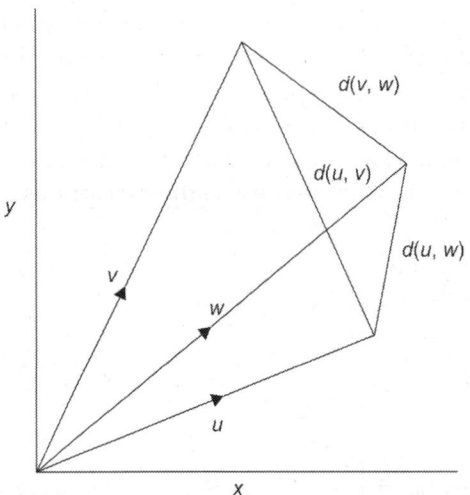

Fig. 10.4: Illustration of the triangle inequality in a two-dimensional space

A metric space consists of a set of elements where a metric is defined. This does not have to be a linear space or a vector space, as the axioms ($D1$–$D4$) do not involve concepts of scalar multiplication or vector addition. The Euclidean metric defined on $\mathbb{R}^n$ in Eq. (10.6) is denoted by $d_2(u, v)$. Euclidean metric defined on $\mathbb{R}^n$ are

$$d_1(u, v) = \sum_{i=1}^{n} |u_i - v_i| \qquad (10.6a)$$

$$d_\infty(u, v) = \max_{i \in 1, n} |u_i - v_i| \qquad (10.6b)$$

Example 10.1. *Determine if* $d(u,v) = \sum_{i=1}^{4} (u_i - v_i)$ *is a suitable metric on* $\mathbb{R}^n$.

Solution: $d(u,v) = \sum_{i=1}^{4} (u_i - v_i)$ must satisfy the four axioms for all vectors in $\mathbb{R}^n$. It is easy to demonstrate that the axioms are violated by taking an example. Let $u = (0, 1, 0, 0)^t$ and $v = (0, 2, 0, 0)^t$.

$$d(u, v) = -1 < 0, \text{ for } u \neq v$$

The first axiom is violated, hence this is not an appropriate metric.

Example 10.2. *Compute* $d_1(u, v)$, $d(u, v)$, $d_\infty(u, v)$ *for*

$$u = (1, 2, 3, 4)^t, \ v = (3, 5, 2, 1)^t$$

Solution:

$$d_1(u, v) = \sum_{i=1}^{4} |u_i - v_i| = |1-3| + |2-5| + |3-2| + |4-1| = 9$$

$$d_2(u, v) = \left(\sum_{i=1}^{4} (u_i - v_i)^2 \right)^{1/2} = \sqrt{23}$$

$$d_\infty(u, v) = \max_{i \in 1, 4} |u_i - v_i| = 3$$

For the same two vectors (u, v), the three metrics yield different values for distance. They are a measure of the separation, the difference between u and v. All three matrics, however, become identically zero when u equals v as they must in order to satisfy axiom $D2$.

The concept of a metric is very important as it generates the notion of convergence. In many problems the solution is obtained numerically through an iterative process. The iterates generate a sequence of vectors. The sequence is said to converge to the solution if the metric between two successive iterates is less than a desired tolerance. The tolerance determines the accuracy desired. The metric employed to check convergence can be any of the above valid definitions.

10.4.2. Normed Linear Space

The norm of a vector is an abstract generalisation of its length. A normed linear space is a linear space in which a norm is defined. The norm is a non-negative number denoted by $\|u\|$.

The axioms have to satisfy a norm are

    N1: $\|u\| > 0$ for $u \neq 0$

    N2: $\|u\| = 0$ if and only if $u = 0$

    N3: $\|\alpha u\| = |\alpha|\,\|u\|$

    N4: $\|u + v\| \leq \|u\| + \|v\|$

These axioms are mathematical statements of the properties we expect a norm to possess. The first two axioms state that the length of a nonzero vector is always positive. The only vector that can have a zero norm is the zero vector and only the zero vector has zero norm. The third axiom tells us that the length of a scalar multiple of a vector u is the scalar multiple times $\|u\|$. The absolute value on the scalar is necessary to satisfy the first axiom. The fourth axiom is the triangle inequality which we saw earlier in D4. The third and fourth axioms involve the concepts of scalar multiplication and vector addition. A norm has to be hence necessarily defined on a linear space. This ensures that αu, $u + v$, occurring in axioms N3, N4 belong to the same vector space as u, v.

A metric is generated by a norm on a linear space. This is obtained from the relation

$$d(u, v) = \|u - v\| \tag{10.7}$$

The metric so generated satisfies the norms D1–D4. Thus, a normed linear space is necessarily a metric space.

A norm like the metric can be defined on a linear space in many ways. The various definitions have the significance of a length. They are not equal to the Euclidean or geometrical length. The Euclidean norm in $\mathbb{R}^n$ is given by

$$\|u\|_2 = \left(\sum_{i=1}^{n} u_i^2 \right)^{1/2} \tag{10.8}$$

The norm generates the Euclidean metric via Eq. (10.7). It has the significance of the geometric length of a vector in $\mathbb{R}^2$, $\mathbb{R}^3$.

Other valid definitions of the norm in $\mathbb{R}^n$ are

$$\|u\|_1 = \sum_{i=1}^{n} |u_i| \tag{10.8a}$$

$$\|u\|_\infty = \max_{i \in 1, n} |u_i| \tag{10.8b}$$

Norms [Eqs (10.8a) and (10.8b)] generate the metrics defined in Eqs (10.6a) and (10.6b) through Eq. (10.7).

Example 10.3. *Determine if $\|u\| = \max\limits_{i \in 1, n}$ is a valid definition of a norm in $\mathbb{R}^n$.*

Solution: We consider an element u in $\mathbb{R}^n$ $(-2, -3, -3... -3)$. This has the last $n - 1$ coordinates as -3, and the first coordinate as -2.

$$\|u\| = -2 < 0$$

This violates axiom N1 and hence is not a valid norm.

10.4.3. Inner-Product Space

The inner-product between two vectors u, v is a measure of the angle between them. It represents the projection of one vector in the direction of the second vector. This is denoted as $\langle u,v \rangle$ and is defined on a real linear space so as to satisfy the following axioms:

I1: $\langle u,v \rangle > 0$ for $u \neq 0$
$\langle u,u \rangle = 0$ for $u = 0$

I2: $\langle u,v \rangle = \langle v,u \rangle$

I3: $\langle u,\alpha v \rangle = \langle \alpha u,v \rangle = \alpha \langle u,v \rangle$

I4: $\langle u,v+w \rangle = \langle u,v \rangle + \langle u,w \rangle$

The inner-product is a scalar quantity. A suitable candidate satisfying the above axioms in $\mathbb{R}^n$ is

$$\langle u,v \rangle = \sum_{i=1}^{n} u_i v_i \tag{10.9}$$

The axioms involve scalar multiplication and vector addition. So the inner-product like the norm can be defined only on a linear space. An inner-product generates a norm of the form

$$\|u\| = \langle u,u \rangle^{1/2} \tag{10.10}$$

This norm satisfies all axioms N1–N4. A metric can be generated using this norm from

$$d(u, v) = \|u - v\| = \langle u - v, u - v \rangle^{1/2} \tag{10.10a}$$

An inner-product space is necessarily a normed linear space and a metric space. A normed linear space may not be an inner-product space. This indicates a hierarchy in the different concepts of angle, length and distance. The concept of an angle generates the notion of length which in turn defines distance. If distance is defined in a space, it does not imply length or angle is defined in it. The axioms I1–I4 and the definition [Eq. (10.9)] are valid only for real vector spaces. The field F is the set of real numbers and the vectors have only real numbers as their coordinates. The reason for this can be best understood from the example below.

Example 10.4. $u = [2i, 2i]^t$ is a nonzero element of the two-dimensional complex vector space.

Solution: Using Eq. (10.9), we obtain

$$\langle u,u \rangle = 2i \cdot 2i + 2i \cdot 2i = -8$$

This violates axiom I1. The inner-product as defined in Eq. (10.9) is invalid. This is overcome by redefining the inner-product more generally in a complex vector space as

$$\langle u,v \rangle = \sum_{i=1}^{n} \bar{u}_i v_i \tag{1}$$

where the bar over u denotes complex conjugate. Using this, we have

$$\langle u,u \rangle = 8$$

The axioms an inner-product must satisfy in a complex vector space are modified as

IC1: $\langle u,u \rangle > 0$ for $u \neq 0$

$\quad\quad \langle u,u \rangle = 0$ for $u = 0$

IC2: $\langle u,v \rangle = \overline{\langle v,u \rangle}$

IC3: $\langle u,\alpha v \rangle = \alpha \langle u,v \rangle$

$\quad\quad \langle \alpha u,v \rangle = \bar{\alpha} \langle u,v \rangle$

IC4: $\langle u,v+w \rangle = \langle u,v \rangle + \langle u,w \rangle$

The inner-product defined in Eq. (10.9) is similar to the dot product arising in vector algebra and physics. It is a measure of the cosine of the angle between two vectors.

Two vectors u, v are said to be orthogonal when $\langle u,v \rangle = 0$. This is an important concept as it is often convenient to deal with an orthogonal set of vectors. Two vectors are said to be *orthonormal*, if they are orthogonal to each other and each of them possesses a unit norm. In an orthogonal set of vectors. Each vector is orthogonal to every other member of that set.

10.5. LINEAR DEPENDENCE AND DIMENSION

10.5.1. Linear Dependence

Two vectors u^1, u^2 from a vector space are linearly dependent if constants c^1, c^2 one or both are nonzero exist, such that

$$c_1 u^1 + c_2 u^2 = 0 \quad\quad\quad (10.11)$$

For nonzero u^1, u^2 a nonzero c_1 implies c_2 is also nonzero and we have $u^1 = -c_2 u^2 / c_1$. The linear dependence of two nonzero vectors implies one is a scalar multiple of another. The vectors are independent if the relation [Eq. (10.11)] is satisfied only for $c_1 = c_2 = 0$. We now extend this definition to a set of r vectors.

A linear relation for r vectors $\{u^1, u^2,...,u^r\}$ is an equation of the form

$$\sum_{i=1}^{r} c_i u^i = 0 \quad\quad\quad (10.11a)$$

where the c_i's are scalars. A linear relation with c_i's identically zero is called the *trivial relation*. A nontrivial relation has some nonzero c_i's. A vector set $\{u^i\}$ is dependent if it satisfies a nontrivial relation. A set of vectors which satisfies only the trivial relation and no nontrivial relation is independent.

For a set of r dependent vectors there is at least one nonzero c_i such that Eq. (10.11a) is satisfied. Without any loss of generality we assume this to be c_1. This allows u^1 to be written as a linear combination of the other vectors as

$$u^1 = \sum_{i=2}^{r} \frac{c_i u^i}{c_1} \qu\quad\quad\quad (10.11b)$$

The linear relation [Eq. (10.11a)] for vectors with n coordinates each, yields n equations in r unknowns $c_1, c_2,...,c_r$. The vector u^i is denoted as $(u_1^i, u_2^i,...,u_n^i)^t$. Here u_j^i represents the jth coordinate of vector u^i. The unknowns c_i are obtained by equating each coordinate of the vector sum in Eq. (10.12b) to zero. This yields

$$\sum_{j=1}^{r} u_i^j c_j = 0 \quad \text{for} \quad i = 1,..., n \qu\quad\quad (10.12)$$

The coordinates u_i^j are known and the unknowns are the c_i's. The set is independent if the only solution to the system [Eq. (10.12)] is $c_i = 0$ for all i. It is dependent if a
nontrivial solution, such that at least one $c_i \neq 0$ satisfies Eq. (10.12).

The linear relation [Eq. (10.11a)], as we have seen, yields n equations in r unknowns.
If $r > n$ the c_i's can be determined, assigning nonzero values to $r - n$ of them and
evaluating the rest from Eq. (10.12). Therefore, a set with $r > n$ is always dependent.
For $r \leq n$, we have to solve for the c_i's and evaluate the nature of the set.

Example 10.5. *Is the set* $(2\ 1\ 3)^t$, $(4\ 1\ 5)^t$, $(2\ 0\ 2)^t$ *a dependent set?*

Solution: The linear relation $\sum_{i=1}^{3} c_i u^i = 0$ results in

$$2c_1 + 4c_2 + 2c_3 = 0$$
$$c_1 + c_2 = 0$$
$$3c_1 + 5c_2 + 2c_3 = 0$$

A possible solution to this system is $c_1 = c_2 = c_3 = 0$. This trivial solution is always a
condition, as the equations are homogeneous. We have to determine if this system can
admit a nonzero solution. We have $c_1 = -c_2$ from the equation.

$$2c_2 + 2c_3 = 0,$$

when we eliminate c_1 from them. Assigning $c_2 = -1$, a nonzero arbitrary number is
obtained, i.e. $c_1 = 1$, $c_3 = 1$ as a set of nonzero solutions. We conclude that the three
vectors form a dependent set. There is an infinite number of such nonzero solutions.
These can be obtained by assigning different values arbitrarily to c_2.

Example 10.6. *Show* $(2\ 1\ 3)^t$, $(4\ 1\ 5)^t$, *and* $(2\ 0\ 2)^t$ *form a linearly independent set.*

Solution: Equating coordinates $\sum_{i=1}^{3} c_i u^i = 0$, we get

$$2c_1 + 4c_2 + 2c_3 = 0$$
$$c_1 + c_2 = 0$$
$$3c_1 + 5c_2 + 2c_3 = 0$$

This is reduced to

(a) $c_1 = -c_2$
(b) $2c_2 + 2c_3 = 0$
(c) $2c_2 + 4c_2 = 0$

This system of equations has only the zero solution and the given set is an independent set.

Example 10.7. *Find whether the following sets are dependent or independent:*

(a) $(2\ 0\ 4)^t$ $(2\ 0\ 8)^t$ $(2\ 1\ 3)^t$ $(4\ 1\ 5)^t$
(b) $(2\ 1\ 4)^t$ $(6\ 3\ 12)^t$
(c) $(2\ 1\ 4)^t$ $(3\ 3\ 12)^t$

Solution: (a) We obtain three equations in the four unknown c_i's:

$$2c_1 + 2c_2 + 2c_3 + 4c_4 = 0$$
$$c_3 + c_4 = 0$$
$$4c_1 + 8c_2 + 3c_3 + 5c_4 = 0$$

We assign $c_3 = 1$ arbitrarily, and solve for c_1, c_2, c_4 to yield $c_1 = 3/2$, $c_2 = -1/2$. $c_4 = -1$. This is a dependent set, as we expect, since the number of vectors exceeds the number of coordinates of each vector $(r > n)$.

(b) The linear relation yields

$$2c_1 + 6c_2 = 0$$
$$c_1 + 3c_2 = 0$$
$$4c_1 + 12c_2 = 0$$

An admissible solution is $c_1 = -3$, $c_2 = 1$. Therefore, this is a dependent set.

(c) Equating the coordinates of the vectors to zero yields

$$2c_1 + 3c_2 = 0$$
$$c_1 + 3c_2 = 0$$
$$4c_1 + 12c_2 = 0$$

The only solution here is the trivial solution and the given two vectors are independent. The concept of linear dependence plays a vital role in the solution of linear algebraic equation. This also formally defines the dimension of a vector space.

10.5.2. Dimension of Vector Space

So far we have identified the dimension of a real vector space as being equal to the number of coordinates of the vector in it. Thus in $\mathbb{R}^3$ each vector has three coordinates and we said that the vectors belong to a three-dimensional space. We will now formally define the dimension of a vector space. This is based on the algebraic concept of linear independence (*refer to* Stakgold). This approach also generates the notions of a subspace and a basis.

Dimension: The dimension of a vector space $\mathbb{V}$ is equal to the maximum number of linearly independent vectors in $\mathbb{V}$.

Theorem 1: The dimension of $\mathbb{R}^n$ is n.

Proof: A set of r vectors with $r > n$ in $\mathbb{R}^n$ forms a linearly dependent set as seen earlier. The dimension $\mathbb{R}^n \leq n$. If at least one set of n independent vectors in $\mathbb{R}^n$ can be obtained, this will be the maximum number and the dimension $\mathbb{R}^n$ will be n. Let e^i denote the vector with ith coordinate 1 and all other coordinates zero. So e^3 is $(0, 0, 1, 0, 0,...,0)^t$. The linear relation

$$\sum_{i=1}^{n} c_i e^i = 0, \tag{10.13}$$

implies $[c_1, c_2,...,c_n]^t = 0$. The only solution to the linear relation is $c_i = 0$ for all i.

The set $(e^1, e^2,...,e^n)$ is hence linearly independent. N is the maximum number of linearly independent vectors in $\mathbb{R}^n$. Consequently the dimension of $\mathbb{R}^n$ is n. If the dimension of a space is n and does not imply that the space is $\mathbb{R}^n$. The elements can belong to $\mathbb{R}^p$, where $p > n$. This gives rise to the notion of subspace.

10.5.3. Subspace

The set $\{u^1,...,u^m\}$ is called a generating set of a vector space $\mathbb{V}$ if every vector u in $\mathbb{V}$ can be expressed as a linear combination of this set. $\mathbb{V}$ consists of all scalar multiples and linear combinations of the members of the generating set. If each u^i has n coordinates and the number of linearly independent vectors in the set $\{u^i\}$ is less than n, the space $\mathbb{V}$ is a subspace of $\mathbb{R}^n$. We explain this concept of a subspace in detail with illustrations.

Every vector in $\mathbb{R}^2$ can be written as a linear combination of two independent vectors u^1, u^2. These generate the space $\mathbb{R}^2 \cdot u^1$ by itself generates many vectors which are scalar multiples of u^1. These vectors lie on the directed line segment u^1. They form a one-dimensional subspace in the two-dimensional space generated by u^1, u^2 (Fig. 10.5). Every vector in this subspace is specified by two coordinates. Similarly, all scalar multiples of u^2 generate another one-dimensional subspace.

The three-dimensional space $\mathbb{R}^3$ is generated by u^1, u^2, u^3. u^1 generates a one-dimensional subspace, u^2, u^3. A two-dimensional subspace consists of all scalar multiples of u^2, u^3 and their linear combinations. Similarly, linear combinations of u^1, u^2 and u^1, u^3 generate at least two other two-dimensional subspaces of $\mathbb{R}^3$.

Example 10.8. *Vectors* $(1, 0)^t$ $(1, 1)^t$ *generate* $\mathbb{R}^2$.

Solution: Vectors $(2, 2)^t$, $(1, 1)^t$ generate a one-dimensional subspace in $\mathbb{R}^n$. The subspace consists of all multiples of $(1, 1)^t$ (Fig. 10.5).

Theorem 2: Let the m vectors $\{u^1, u^2,...,u^m\}$ be a linearly independent generating set of vector space $\mathbb{V}$. A set of r vectors, $r > m$ is linearly dependent in $\mathbb{V}$.

Proof: Let $(v^1, v^2,...,v^r)$ be a set or r vectors in $\mathbb{V}$. Each of the v^r's can be written as a linear combination of the u's as the u^i's are a generating set. So

$$v^j = \sum_{i=1}^m a_{ji} u^i \text{ for } j = 1...r \tag{1}$$

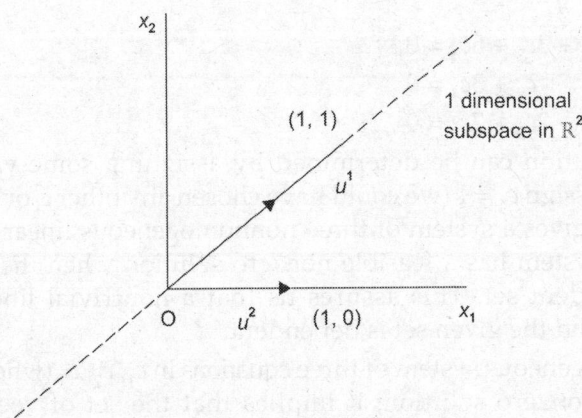

Fig. 10.5: A one-dimensional subspace in $\mathbb{R}^2$, consisting of all scalar multiples along u^1 (indicated by dashed line)

The c_i's in the linear relation

$$\sum_{i=1}^{r} c_i v^i = 0 \tag{2}$$

decide if the set $\{v^1, v^2,...,v^r\}$ is either dependent or independent. Expressing each v^i in Eq. (10.15b) using Eq. (10.15a), we obtain

$$\sum_{i=1}^{r} c_i \sum_{k=1}^{m} a_{ik} u^k = 0 \tag{3}$$

or

$$\sum_{k=1}^{m} \left(\sum_{i=1}^{r} c_i a_{ik} \right) u^k = 0, \text{ changing the order of the summations.}$$

As the u^k's form an independent set, this relation among u^k's hold if and only if

$$\sum_{i=1}^{r} c_i a_{ik} = 0 \quad \text{for } k = 1... m \tag{4}$$

This is a system of m equations in the r unknown c_i's ($r > m$). This admits a nontrivial solution $(c_1, c_2,...,c_r)$ as seen earlier, which renders $\{v_i\}$ a dependent set from Eq. (2.15b). Since $\{v^i\}$ was chosen arbitrarily, the theorem holds for any set of r vectors ($r > m$).

Example 10.9. *Consider the set of four vectors in* $\mathbb{R}^3$

$$(1\ 2\ 3)^t \quad (2\ 1\ 3)^t \quad (3\ 1\ 2)^t \quad (5\ 3\ 7)^t$$

Solution: The dimension of $\mathbb{R}^3$ is 3. Hence we have at most three vectors which form an independent set. The given set of vectors must therefore be necessarily dependent. Consider

$$\sum_{i=1}^{4} c_i u^i = 0$$

This yields a system of three equations (one for each coordinate) in the four unknowns c_i, $i = 1, 4$.

$$c_1 + 2c_2 + 3c_3 + 5c_4 = 0$$
$$2c_1 + c_2 + c_3 + 3c_4 = 0$$
$$3c_1 + 3c_2 + 2c_3 + 7c_4 = 0$$

A nonzero solution can be determined by assigning some value to one of the variables. Let us assign $c_4 = 1$ (we could have chosen any other c_i or assigned any other value). This then gives a system of three nonhomogeneous linear equations in three unknowns. The system has a feasible nonzero solution when the first three vectors form an independent set. This assures us that a nontrivial linear relation exists between the u^i's and the given set is dependent.

If the nonhomogeneous system of three equations in c_1, c_2, c_3 (when we assign $c_4 = 1$) has no feasible nonzero solution, it implies that the set of vectors u^1, u^2, u^3 are dependent. Here we can assign c_4 the value zero and seek a nonzero solution for c_1, c_2, c_3. This will again yield the result such that the set of vectors is dependent. We conclude that a set of p vectors in $\mathbb{R}^n$, where $p > n$, is necessarily dependent.

10.5.4. An Application of the Concept of Linear Dependence

The concept of linear dependence plays a vital role in sensor placement in a chemical plant. It enables us to determine the variables which need to be measured, which will allow us to determine the state of the plant completely.

The system consisting of five equations in nine unknowns can be written as

$$
\begin{bmatrix}
-1 & -1 & 1 & 0 & 0 & 0 & 0 & 0 & 0 \\
0 & 0 & 1 & -1 & -1 & 0 & 0 & 0 & 0 \\
0 & 0 & 0 & 1 & 0 & -1 & -1 & 0 & 0 \\
0 & 1 & 0 & 0 & -1 & 0 & 0 & 1 & 0 \\
0 & 0 & 0 & 0 & 0 & 1 & 0 & 1 & -1
\end{bmatrix}
\begin{bmatrix}
F_1 \\ F_2 \\ F_3 \\ F_4 \\ F_5 \\ F_6 \\ F_7 \\ F_8 \\ F_9
\end{bmatrix} = 0
\tag{10.14}
$$

The five rows of the matrix A are vectors in $\mathbb{R}^9$. These five vectors are linearly independent. They span a five-dimensional subspace in the nine-dimensional space $\mathbb{R}^9$. One has to specify four variables to determine all the others uniquely. This however cannot be any four variables as we will see now.

The system Eq. (10.14) has nine flow-rates $\{F_1, F_2,..., F_9\}$. The mass balances across each unit yields a system of five equations, viz.

$$
\begin{aligned}
F_3 - F_2 - F_1 &= 0 \\
F_3 - F_4 - F_5 &= 0 \\
F_4 - F_7 - F_6 &= 0 \\
F_2 + F_8 - F_6 &= 0 \\
F_8 - F_9 + F_6 &= 0
\end{aligned}
\tag{10.14a}
$$

When four of the flow-rates are measured and specified, the remaining five can be predicted from Eqs (10.14) and (10.14a). Specifying F_1, F_5, F_6, F_9. We can recast the equations in vectorial form as

$$
\begin{bmatrix}
-1 & 1 & 0 & 0 & 0 \\
0 & 1 & -1 & 0 & 0 \\
0 & 0 & 1 & -1 & 0 \\
1 & 0 & 0 & 0 & 0 \\
0 & 0 & 0 & 0 & 0
\end{bmatrix}
\begin{bmatrix}
F_2 \\ F_3 \\ F_4 \\ F_7 \\ F_8
\end{bmatrix} =
\begin{bmatrix}
F_1 \\ F_5 \\ F_6 \\ F_5 \\ F_9 - F_6
\end{bmatrix}
\tag{10.14b}
$$

or $Au = b$.

It can be verified that the five rows of A are linearly independent and we can solve for $[F_2, F_3, F_4, F_7, F_8]$. Thus, the state of the entire plant can be uniquely determined experimentally by measuring the four variables $[F_1, F_5, F_6, F_9]$.

Consider now a situation where $[F_1, F_2, F_5, F_8]$ are the four flow-rates measured.

The system of Eq. (10.14b) for the unknowns can be recast as

$$\begin{bmatrix} 1 & 0 & 0 & 0 & 0 \\ 1 & -1 & 0 & 0 & 0 \\ 0 & 1 & -1 & -1 & 0 \\ 0 & 0 & 0 & 0 & 0 \\ 0 & 0 & 1 & 0 & -1 \end{bmatrix} \begin{bmatrix} F_3 \\ F_4 \\ F_6 \\ F_7 \\ F_9 \end{bmatrix} = \begin{bmatrix} F_1 + F_2 \\ F_5 \\ 0 \\ F_5 - F_2 - F_8 \\ -F_8 \end{bmatrix} \qquad (10.14c)$$

The five rows of matrix A are linearly dependent. This can be seen directly, as the fourth row vector is the zero vector (every set containing the zero vector is dependent). This is system of four independent equations in five unknowns. The state of the plant cannot therefore be obtained uniquely by specifying F_1, F_2, F_5, F_8.

As a second example, consider the splitter–mixer network in Fig. 10.6. The system is characterised by five flow-rates $\{F_1, F_2,...,F_5\}$. The mass balance across each unit can be written to yield the three equations

$$F_1 - F_2 - F_3 = 0$$
$$F_2 - F_4 = 0$$
$$F_3 + F_4 - F_5 = 0$$

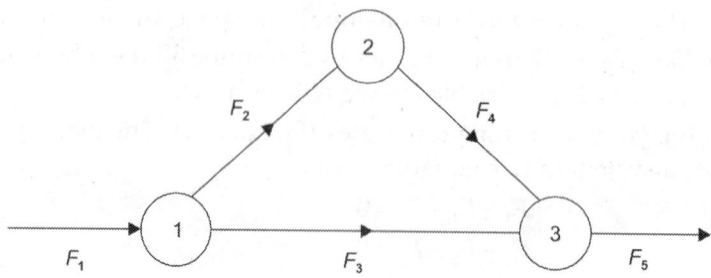

Fig. 10.6: A splitter–mixer network

The mass balance across units 1, 2, yield

$$F_1 = F_4 + F_3$$

and across units 2, 3, result in

$$F_2 + F_3 = F_5$$

It would appear that we now have a system of five equations for the F_i's. This can be cast in vectorial form as

$$\begin{bmatrix} 1 & -1 & -1 & 0 & 0 \\ 0 & 1 & 0 & -1 & 0 \\ 0 & 0 & 1 & 1 & -1 \\ 1 & 0 & -1 & -1 & 0 \\ 0 & 1 & 1 & 0 & -1 \end{bmatrix} \begin{bmatrix} F_1 \\ F_2 \\ F_3 \\ F_4 \\ F_5 \end{bmatrix} = 0 \qquad (10.15)$$

The rows of the matrix are vectors in $\mathbb{R}^5$. The five row vectors $(r^1, r^2,...,r^5)$ are however, not independent. Clearly,

$$r^4 - r^1 - r^2 = 0, \qquad r^5 - r^2 - r^3 = 0$$

The five row vectors in Eq. (10.16) span a three-dimensional subspace in $\mathbb{R}^5$. Generating more equations, without including any additional information, does not enable us to obtain all five F_i's. One has to specify two of the F_i's (such that the resulting three equations are independent), to estimate the remaining F_i's.

10.5.5. Basis

Definition: A linearly independent generating set of a vector space $\mathbb{V}$ is called a basis of $\mathbb{V}$.

Theorem 3*: Consider $\mathbb{R}^n$ a vector space of dimension n. Every basis of $\mathbb{R}^n$ contains exactly n vectors and these are linearly independent. Conversely, a set of n linearly vectors forms a basis of $\mathbb{R}^n$.

Theorem 4: Every vector in a vector space can be expressed in one and only one way as a linear combination of its basis.

Proof: Let $\{e^i\}_{i=1,n}$ be a basis set for an n-dimensional space $\mathbb{R}^n$. A vector u in $\mathbb{R}^n$ can be expressed as a linear combination of $\{e^i\}$ as

$$u = \sum_{i=1}^n c_i e^i \tag{10.16}$$

If this representation is nonunique, there is another set of scalars b_i such that

$$u = \sum_{i=1}^n b_i e^i \tag{10.16a}$$

Subtracting, we get

$$0 = \sum_{i=1}^n (c_i - b_i) e^i \tag{10.16b}$$

Since e^i is a basis, it is a linearly independent set, and Eq. (10.16b) is satisfied only for $c_i = b_i$.

An arbitrary vector u in $\mathbb{R}^n$ can be expressed in terms of a basis $\{u^i\}_{i=1,n}$ uniquely as

$$u = \sum_{i=1}^n c_i u^i \tag{10.17}$$

Equating the n coordinates on both sides, we obtain n equations in n unknown c_i's. The determination of c_i's in the representation [Eq. (10.17)] is cumbersome and tedious. The c_i's now do not have the significance of distance measured parallel to a basis vector as the basis set may not be orthogonal.

Consider now a situation where the basis is orthogonal or orthonormal. Here every vector in the set is orthogonal to every other vector of the set. Consider an orthogonal basis $\{e^i\}_{i=1,n}$. An arbitrary u can be expressed as a linear combination of this basis as

$$u = \sum_{i=1}^n b_i e^i \tag{10.18}$$

---

* we *refer to* the interested reader, any standard text of linear algebra for proof to this theorem.

The b_i's can be found taking the inner-product of both sides with e^j. Using the definition of the real inner-product, this yields

$$\langle u, e^j \rangle = \sum_{i=1}^{n} \langle b_i, e^i, e^j \rangle = b_j \|e^j\|^2$$

as $\langle e^i, e^j \rangle = 0$ for $i \neq j$ (since they are orthogonal) or

$$b_j = \langle u, e^j \rangle / \|e^j\|^2 \qquad (10.19)$$

Example 10.10. Express the vector $(8\ 5\ 11)^t$ in terms of the basis $(1\ 2\ 3)^t$, $(2\ 1\ 3)^t$, $(3\ 1\ 2)^t$. We seek

$$u = \sum_{i=1}^{3} c_i u^j$$

Solution: The coefficients c_i are obtained by solving
$$c_1 + 2c_2 + 3c_3 = 8$$
$$2c_1 + c_2 + c_3 = 5$$
$$3c_1 + 3c_2 + 2c_3 = 11$$

This system has the solution $c_1 = c_3 = 1$, $c_2 = 2$. Every vector in $\mathbb{R}^3$ can be similarly represented uniquely in terms of this basis. The coefficients c_i are uniquely obtained by solving a set of linear equations as we have shown. This is an inefficient and a tedious exercise, especially for higher-dimensional systems. We illustrate with an example how this problem is can be solved when the basis is orthogonal.

Example 10.11. The set $u^1 = (1\ 2\ 3)^t$, $u^2 = (3\ -3\ 1)^t$, $u^3 = (-11\ -8\ 9)^t$ forms an orthogonal basis in $\mathbb{R}^3$. Represent the vector $b(4\ 5\ 6)^t$ in terms of this basis.

Solution: We write $b = \sum_{i=1}^{3} c_i u^i$. Taking the inner-products of both sides with u^1, we have

$$\langle u^1, b \rangle = \langle u^1, c_1 u^1 \rangle + \langle u^1, c_2 u^2 \rangle + \langle u^1, c_3 u^3 \rangle$$

$$= \langle u^1, u^2 \rangle = \langle u^1, u^3 \rangle = 0$$

as the basis set is orthogonal. We now obtain

$$c_1 = \frac{\langle u^1, b \rangle}{\langle u^1, u^1 \rangle} = \frac{32}{14}$$

Similarly,

$$c_2 = \frac{\langle u^2, b \rangle}{\langle u^2, u^2 \rangle} = \frac{3}{19}$$

$$c_3 = \frac{\langle u^3, b \rangle}{\langle u^3, u^3 \rangle} = -\frac{15}{133}$$

Example 10.12. Express the vector $u(7, 3, 9)^t$ in $\mathbb{R}^3$ as a linear combination of the following basis sets:

i. $u^1(1, 1, 1)^t$, $u^2(2, 1, 3)^t$, $u^3(3, 0, 4)^t$

ii. $u^1(1, 1, 1)^t$, $u^2(1, 0, -1)^t$, $u^3(1, -2, 1)^t$

Solution: (i) The basis set (a) is not an orthogonal basis as $\langle u^i, u^j \rangle = 0$ for $i \neq j$ is not satisfied. We write

$$u = \sum_{i=1}^{3} c_i u^t$$

The scalars c_i are obtained by equating the coordinates on both sides. This yields

$$c_1 + 2c_2 + 3c_3 = 7$$
$$c_1 + c_2 = 3$$
$$2c_1 + 3c_2 + 4c_3 = 9$$

Solving $c_1 = 2, c_2 = c_3 = 1$.

(ii) The basis set (b) is orthogonal as $\langle u^i, u^j \rangle = 0$ for $i \neq j$. The inner-product we work on is

$$\langle u, v \rangle = \sum_{i=1}^{n} u_i v_i$$

We represent

$$u = \sum_{i=1}^{3} b_i u^i$$

The b_j's are obtained by taking the inner-product with u^j on both sides. This yields

$$b_j = \frac{\langle u, u^j \rangle}{\langle u^j, u^j \rangle}$$

Substituting for u, u^j, we obtain $b_1 = 19/3$, $b_2 = -1$, $b_3 = 5/3$. Each coefficient c_i in the representation [Eq. (10.15)] can be obtained in a relatively elegant manner, when the basis set is orthogonal. This is particularly effective for higher dimensional systems. Here, each coefficient is determined independently, i.e. b_i is independent of b_j for $i \neq j$. This is an important feature, when we work in infinite dimensional spaces.

The coordinates of a vector represented in terms of a basis $\{u^i\}$ are the coefficients of the linear combination of the basis set in which it is expressed. The vector u in Eq. (10.18) in the basis $\{u^i\}$ has coordinates $(c_1, c_2,...,c_n)^t$ and in the basis $\{e^i\}$ as in Eq. (10.19) has coordinates $(b_1, b_2,...,b_n)^t$.

Normally, it is preferred to work in an orthogonal basis than in a general linearly independent basis. We now discuss a technique which can convert a linearly independent basis to an orthonormal basis.

10.6. GRAM–SCHMIDT ORTHONORMALISATION

The representation of a vector in terms of an orthogonal basis is very elegant as we have just seen. The Gram–Schmidt orthonormalisation technique enables us to obtain an orthonormal set from a linearly independent set. We will explain this method geometrically in $\mathbb{R}^2$ and generalize it to $\mathbb{R}^n$ for large n using the algebraic concepts developed so far.

Let u^1, u^2 be two independent vectors in $\mathbb{R}^2$ as shown in (Fig. 10.7a). Therefore, the vectors are not collinear. e^1 is the unit vector along u^1. OA is the projection of u^2 on u^1. Its magnitude, i.e. length, is given by $\langle u^2, e^1 \rangle$. The line segment AB, i.e. vector g^2, is orthogonal to u^1 and clearly is the vectorial difference between u^2 and OA or $\langle u^2, e^1 \rangle e^1$. That is,

$$g^2 = u^2 - \langle u^2, e^1 \rangle e^1 \tag{10.20}$$

The unit vector in direction of g^2 is

$$e^2 = g^2 / \|g^2\| \tag{10.20a}$$

A second set of orthogonal vectors f^1, f^2 can be constructed from u^1, u^2 by reordering the vectors as u^2, u^1. This is illustrated in Fig. 10.7b. In an n-dimensional space from an independent basis it is possible to generate n orthonormal basis, by changing the order of the vectors.

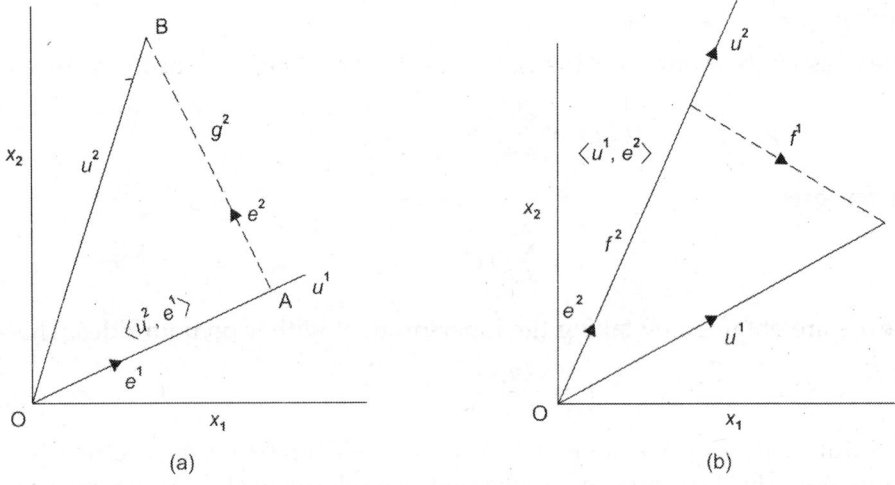

Fig. 10.7: Geometric illustration of Gram–Schmidt orthonormalisation in $\mathbb{R}^2$
(a) u^1 is chosen as the first vector; (b) u^2 is chosen as the first vector

We extend the above geometric method to an n-dimensional vector space $\mathbb{V}$, algebraically with the basis $\{u^1, u^2, ..., u^n\}$. The first

$$e^1 = u^1 / \|u^1\| \tag{10.21}$$

e^1 clearly has unit norm. e^2 is found such that it is orthogonal to e^1 and has unit norm. We construct g^2 as we did earlier, by subtracting from u^2, its projection on e^1. We normalise g^2 to obtain e^2.

$$g^2 = u^2 - \langle u^2, e^1 \rangle e^1 \tag{10.22}$$

$$e^2 = g^2 / \|g^2\| \tag{10.22a}$$

g^2 is clearly not the zero vector, since it is a nontrivial combination of two independent vectors, g^2 is orthogonal to e^1 as

$$\langle g^2, e^1 \rangle = \langle u^2, e^1 \rangle - \langle u^2, e^1 \rangle \langle e^1, e^1 \rangle = 0$$

A nonzero g^3 is obtained by subtracting from u^3, its projections on e^1 and e^2. This yields

$$g^3 = u^3 - \langle u^3, e^1 \rangle e^1 - \langle u^3, e^2 \rangle e^2 \qquad (10.23)$$

The unit vector e^3 is defined as

$$e^3 = g^3 / \|g^3\| \qquad (10.23a)$$

Clearly, $\langle g^3, e^1 \rangle = \langle g^3, e^2 \rangle = 0$. This construction process is continued for all n vectors. The general relationship to obtain g^k, e^k is

$$g^k = u^k \sum_{i=1}^{k-1} \langle u^k, e^i \rangle e^i \text{ for } k = 1,..., n \qquad (10.24)$$

$$e^k = g^k / \|g^k\| \qquad (10.24a)$$

Example 10.13. *Generate an orthonormal set from the linearly independent set* $(2, 0, 1)^t$, $(2, 1, 3)^t$, $(4, 1, 2)^t$ *in* R^3.

We use the inner-product definition

$$\langle u, v \rangle = \sum_{i=1}^{n} u_i v_i$$

$$g^1 = u^1 = (2, 0, 1)^t$$

This yields

$$e^1 = \frac{1}{\sqrt{5}} (2,0,1)^t$$

$$g^2 = u^2 - \langle u^2, e^1 \rangle e^1$$

$$= (2,1,3)^t - \frac{7}{\sqrt{5}} \frac{1}{\sqrt{5}} (2,0,1)^t$$

$$= \frac{1}{\sqrt{5}} (-4,5,8)^t$$

$$e^2 = \frac{1}{\sqrt{105}} (-4,5,8)^t$$

$$g^3 = u^3 - \langle u^3, e^1 \rangle e^1 - \langle u^3, e^2 \rangle e^2$$

$$= (4,1,2)^t - \frac{10}{\sqrt{5}} \frac{1}{\sqrt{5}} (2,0,1)^t - \frac{1}{21} \frac{1}{\sqrt{105}} (-4,5,8)^t$$

$$= \frac{4}{21} (1,4,-2)^t$$

$$e^3 = \frac{1}{\sqrt{21}} (1,4,-2)^t$$

The set

$$\{e^i\},\ e^1 = \frac{1}{\sqrt{5}}(2,0,1)^t,\ e^2 = \frac{1}{\sqrt{105}}(-4,5,8)^t, e^3 = \frac{1}{\sqrt{21}}(1,4,-2)^t$$

is an orthonormal basis for R^2. Here,

$$\langle e^i, e^j \rangle = \delta_{ij}$$

where δ_{ij}, the Kronecker delta function, is defined as

$$\delta_{ij} = 1 \quad \text{for } i = j$$
$$= 0 \quad \text{for } i \neq j$$

More details of the method of Gram–Schmidt orthogonalisation can be found in Noble and Daniel (1977), Stakgold (1979).

PROBLEMS

1. State which of the following statements are true or false with reasons and/or examples.

   i. A metric space is necessarily a vector space.

   ii. An inner-product space is necessarily a vector-space.

   iii. A normed vector space is necessarily an inner-product space.

   iv. A normed vector space is necessarily a metric space.

   v. Every vector space is a metric space.

   vi. A set of n linearly independent vectors is a basis for $\mathbb{R}^n$.

   vii. If V is a subspace of W, then (a) the number of coordinates in a vector in V is less than W, and (b) the number of linearly independent vectors in V is less than those in W.

   viii. A basis in a vector space has only one-orthonormal basis.

   ix. Every vector space has only an orthogonal basis.

2. Consider the set of nonnegative integers I. Is this set a vector space? Is it a metric space?

3. Verify that the three definitions (2.8a)–(2.8c) satisfy the axioms of a norm. Calculate all three norms for u, v, w defined as
   $$u = [1 \ 4 \ 6]^t, \quad v = [2 \ 1 \ 3]^t, \quad w = [1 \ -2 \ 0]^t$$
   Find $d_\infty(u, v), d_1(u, w), d_2(v, w)$,

4. Show that the three norms in Eq. (10.8) generate the metric d_1, d_2, d_∞ respectively.

5. Show that for vectors $u, v \in \mathbb{C}^n, \langle u, v \rangle = \Sigma \bar{u}_i v_i$ is a valid definition of an inner-product.

6. Consider $u, v \in \mathbb{R}^n$. Is $d(u, v) = |x_2 - y_2|$ a valid definition of a metric in $\mathbb{R}^4$?

7. Show that $\langle u + v, u - v \rangle = \|u\|^2 - \|v\|^2$.

8. Is $\begin{bmatrix} 1 \\ 2 \\ 3 \end{bmatrix}, \begin{bmatrix} 2 \\ 3 \\ 1 \end{bmatrix}, \begin{bmatrix} 3 \\ 1 \\ 2 \end{bmatrix}$ a basis in $\mathbb{R}^3$? Why?

9. Is $\begin{pmatrix} 1 \\ 2 \end{pmatrix}\begin{pmatrix} 3 \\ 1 \end{pmatrix}$ a basis in $\mathbb{R}^2$? Is this set orthogonal? Obtain two different orthonormal bases from this. How many orthonormal bases can you have in $\mathbb{R}^2$?

10. Check if the following sets are dependent or independent:

   i. $\begin{pmatrix} 2 \\ 1 \\ 3 \end{pmatrix}\begin{pmatrix} 4 \\ 1 \\ 5 \end{pmatrix}\begin{pmatrix} 2 \\ 2 \\ 4 \end{pmatrix}$,   ii. $\begin{pmatrix} 2 \\ 1 \\ 3 \end{pmatrix}\begin{pmatrix} 4 \\ 1 \\ 5 \end{pmatrix}\begin{pmatrix} 2 \\ 0 \\ 2 \end{pmatrix}$,   iii. $\begin{pmatrix} 2 \\ 1 \\ 3 \end{pmatrix}\begin{pmatrix} 4 \\ 2 \\ 6 \end{pmatrix}$

11. i. Consider $(2\ 1\ 3)^t, (4\ 1\ 5)^t, (2\ 2\ 4)^t$. Is this a basis in $\mathbb{R}^3$? Is this a generating set in $\mathbb{R}^3$? Is this set orthogonal? From this set obtain an orthonormal set.

   ii. The vectors $[3\ 2\ 3]^t\ [2\ -3\ 0]^t$ belong to $\mathbb{R}^3$. Make an orthogonal basis for $\mathbb{R}^3$ out of this set.

12. From the basis $[1\ 2\ 3]^t\ [2\ 1\ 3]^t\ [4\ 3\ 6]^t$, generate an orthonormal basis.

13. Are the following sets dependent:

   i. $[1, 2, 3]^t\ [4\ 5\ 6]^t\ [10\ 11\ 12]^t$
   ii. $[1\ 2\ 3]^t\ [4\ 8\ 9]^t$

14. Find a basis for the vector space generated by:

   $(1\ 3\ 7)^t,\ (2\ -1\ 0)^t,\ (1\ 10\ -21)^t,\ (4\ -3\ 2)^t,$

   What is the dimension of the space? Does the set $(1\ -1\ 1)^t, (1\ 1\ -3)^t, (1\ 2\ 5)^t$ span the same space?

15. Find the dimension of the vector space spanned by:

   $(1\ -1\ 1\ 2)^t,\ [2\ -3\ -3\ 2]^t,\ [-1\ 2\ 3\ 1]^t,\ [1\ 1\ 1\ 7]^t$

16. Obtain an orthonormal set spanning the same subspace as

   $u^1 = [1\ 1\ 1\ -1]^t,\ u^2 = [2\ -1\ -1\ 1]^t,\ u^3 = [-1\ 2\ 2\ 1]^t$

17. Consider the sets:

   i. $[1\ 2\ 3]^t,\ [4\ 5\ 6]^t,\ [7\ 8\ 9]^t,\ [2\ 3\ 4]^t$
   ii. $[1\ 2\ 3]^t,\ [2\ 1\ 3]^t,\ [3\ 1\ 2]^t,\ [5\ 3\ 7]^t$

   Are these dependent? Obtain a basis that spans the space generated by these vectors.

18. Compute the norm of the vector
   $x = (1 + i, 1 - i)^t$

19. Consider a set of reversible elements, i.e. any molecular reaction as shown in Fig. 10.8. Determine the algebraic equation governing the equilibrium concentration of A_1, A_2, A_3. Is this uniquely determined? Determine the number of concentrations does one need to specify to determine the equilibrium state uniquely? Write the equations in vectorial form. How many independent rows and columns does the matrix have? All rate constants are in S^{-1}.

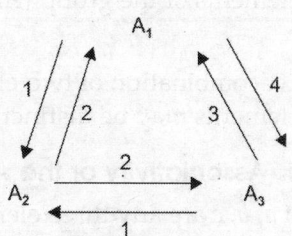

Fig. 10.8: Reaction scheme

Group Theory

11.1. INTRODUCTION

Group theory, as an abstract mathematical theory was developed a long before it found applications in physics, with the development of quantum mechanics, in the first quarter of twentieth century. It was realised that this abstract mathematical theory can be of considerable use in analysing the eigen states of a physical system. Similarly classification of solids in terms of their crystal structures occupies a prominent position in solid state physics. The full symmetry of a crystal lattice comprises all operations which transforms the crystal lattice into itself. This includes translations, rotations, inversions, time reversal and combinations, etc. The important point about the set of these symmetry operations of the crystal is that it forms an abstract group, clearly the key to the mathematical structure of the crystal lattice is provided by the group theory.

In this chapter, we first develop the mathematical theory and then will come to its applications in physics.

11.2. GROUP

A set of distinct elements which can be combined together according to certain mathematical axioms is called a group (G). The process of combination of elements of a group through a mathematical axiom is called group multiplication which must satisfy the following four properties.

i. Closure or Closure Axiom

If A and B are any two elements in group $G(A, B, C,...$ etc.) then the product of any two elements of the group (AB) is a unique element which also belongs to that group, e.g.

$$AB \in G \quad \text{or} \quad BA \in G, \tag{11.1}$$

i.e. combination of two elements by a group operation gives a third element. The two elements may be distinct or same.

ii. Associativity or the Associative Axiom

If a, b, c are any three elements of a group G, then

$$A(BC) = (AB)C = (CA)B \tag{11.2}$$

This is true for all elements in the set

In the integer group, $G(0, 1, 2, 3...)$, for example

$$(0 + 1) + 2 = 0 + (1 + 2)$$

iii. Existence of the Identity Element (The Identity Axiom):

There is an element I in a group G such that $AI = A$, where A is any element in G, and I is called the identity element such that

$$\left. \begin{array}{l} AI = IA = A \\ BI = IB = B \\ CI = IC = C \end{array} \right\} \qquad (11.3)$$

In the integer group $G(0, 1, 2,...)$, zero is the identity element (I) and in triangle rotation group I will be a no rotation element.

iv. The Existence of Inverse (Inverse Axiom):

This property means that corresponding to every element A of group G, there exists an element (denoted by A^{-1}) in G such that

$$AA^{-1} = I \qquad (11.4)$$

A^{-1} is then called the inverse of the element A.

In the integer group G (...–3, –2, –1, 0, 1, 2, 3...), the inverse of an element is negative and in the triangle group the inverse is the relation that brings the triangle back to its original position after the elements rotation operation (i.e. rotation of 2π).

From the above axioms we can prove that $IA = A$ and $A^{-1}A = I$

By definition of identity element, we have

$$I1 = I$$

Now multiplying the relation $AA^{-1} = I$ by I, we have   (from the left side)

$$IAA^{-1} = I \quad \text{or} \quad (IA)A^{-1} = I$$

Thus $IA = A$

Similarly multiplying $AA^{-1} = I$ by A from right hand side we obtain

$$AA^{-1}A = IA$$

or $A(A^{-1}A) = A \quad \text{and} \quad A^{-1} \cdot A = I$

In general, group multiplication is not cumutative. A cumutative group is known as an *abelian group*.

The number of elements in a group is called its order. If n is finite, i.e. a group consists of finite number of elements, it is called *finite group* and if it contains infinite number of elements, it is called *infinite group*.

A group consisting of a single element A and its powers A^2, A^3, A^4,... $= I$ is called a cyclic group of order n. If n is the smallest positive number integer for which $A^n = I$ then n (the order of cyclic group) is also known as the order of the element A.

Some Examples of Groups

Example 1. Four numbers 1, i, $-i$ and -1 form a group of order 4 under ordinary multiplication. Here $i = \sqrt{-1}$. The multiplication table is shown in Table 11.1.

Table 11.1: Multiplication table

	1	i	-1	$-i$
1	1	i	-1	$-i$
i	i	-1	$-i$	1
-1	-1	$-i$	1	i
$-i$	$-i$	1	i	-1

Table 11.1 shows that multiplication of any two elements gives one of the element. Thus the closure property is satisfied, thus multiplication is associative.

Element 1 is the identity element. Also $i(-i) = 1$ and $(-1)(-1) = 1$ showing that $-i$ is the inverse of the element i and vice versa, where as -1 is its own inverse.

If the element i of the group be represented by A then the element of the group can be represented by A, A^2, A^3, $A^4 = I$. This then represents a cyclic group (by definition) of order 4. A cyclic group of order n is denoted by C.

Example 2. A set of all positive and negative integrals including zero form a group under addition. The addition of two such integers gives another integer which is a member of this group. Such addition is associative. The identity element is zero (0) because $N + 0 = N$. For every integer N, $-N$ which belongs to the set is its inverse $N + (-N) = 0$.

The same set of positive and negative integers, however do not form a group under ordinary multiplication. The product of two such integers is another integer. The multiplication is associative. The integer is the identity element. But for a given integer N, the inverse would be N^{-1} which is not an integer if $N \neq 1$.

Example 3. The following four matrices form a group under matrix multiplication.

$$I = \begin{pmatrix} 1 & 0 \\ 0 & 1 \end{pmatrix}, \ A = \begin{pmatrix} 0 & 1 \\ -1 & 0 \end{pmatrix}, \ B = \begin{pmatrix} -1 & 0 \\ 0 & -1 \end{pmatrix}, \ C = \begin{pmatrix} 0 & -1 \\ 1 & 0 \end{pmatrix}$$

We can verify by direct multiplication of any two of these matrices, which gives one of these 4 matrices. Matrix multiplication is associative. The unit matrix is $\begin{pmatrix} 1 & 0 \\ 0 & 1 \end{pmatrix}$, i.e. identity element. Also $AC = CA = I$ and $B^2 = I$, indicating that A and C are the inverse of each other and B is its own inverse.

Example 4. A set of all $n \times n$ unitary matrices form a group under matrix multiplication.

If U_1 and U_2 are two $n \times n$ unitary matrices, then $U = U_1 U_2$ is also a $n \times n$ unitary matrix.

$$U^\dagger U = (U_1 U_2)^\dagger U_1 U_2 = U_2^\dagger (U_1^\dagger, U_1) U_2 = U_2^\dagger U_2 = 1$$

This multiplication is associative. The $n \times n$ unit matrix which is also a unitary matrix is the identity element. Also for unitary matrix U, its adjoint U^+ is the inverse. But U^+ is also a unitary matrix.

$$(U^+)^+U^+ = UU^+ = 1 \quad \text{and} \quad U^+(U^+)^+ = U^+U = 1$$

Example 5. The set of all symmetry transformations which leave a physical system invariant forms a group.

Group multiplication in this case is two successive transformations. If each transformation leaves the system invariant, the product will also leave the system invariant. The group product thus is a symmetry transformation. The successive applications of these transformation is associative. The identity is the transformation which leaves the system undisturbed. Finally for every symmetry transformations there is the inverse operation (for example the inverse of transformation $\pi/2$ rotation anticlockwise is transformation $\pi/2$ rotation clockwise). The inverse operation will also leave the physical system invariant and is thus a symmetry operation.

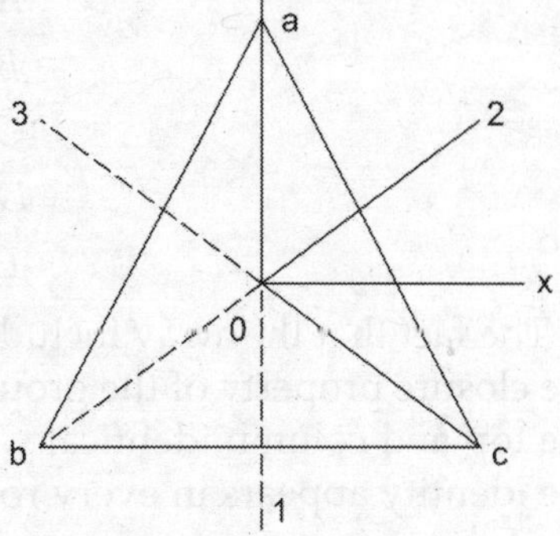

In order to illustrate the multiplication in two consecutive transformations, we take an specific example of symmetry transformations of an equilateral triangle (Fig. 11.1).

The various symmetry operations with their corresponding symbols are:

    I: no transformation

    A: anticlockwise rotation by $\dfrac{2\pi}{3}$ about z-axis

    B: anticlockwise rotation by $\dfrac{4\pi}{3}$ about z-axis

    C: reflection at axis 1

    D: reflection at axis 2

    E: reflection at axis 3

Fig. 11.1

In this case, operations I, A, B, C, D, E form a group. We can easily show that a combination of any two of these operations is equivalent to one of the above operations, let us find the effect of operations CA on the equilateral triangle (Fig. 11.2).

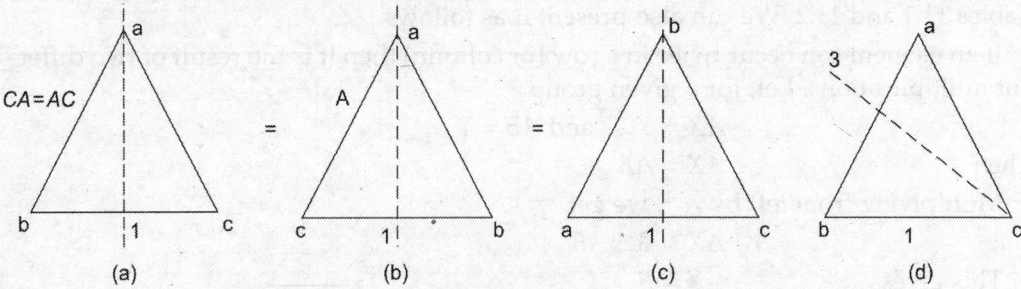

$CA = AC$ A

(a)   (b)   (c)   (d)

Fig. 11.2

C : reflection at axis 1; Transforms (a) into (b).

A : anticlockwise rotation about z-axis by $2\pi/3$. This transforms (b) into (c). But this is equivalent to E thus $CA = E$.

Hence the transformation (I, A, B, C, D, E) form a group, this group is known as the group dihedral, D_3.

The rules of multiplication for a finite group can be consistently expressed by a multiplication table. The multiplication table for the symmetry group of an equilateral triangle is given in Table 11.2.

Table 11.2: Multiplication table

	I	A	B	C	D	E
I	I	A	B	C	D	E
A	A	B	I	E	C	D
B	B	I	A	D	E	C
C	C	D	E	I	A	B
D	D	E	C	B	I	A
E	E	C	D	A	B	I

The fact that the array includes only the elements listed at the side and top verifies the closure property of the group. The identity element produces a row identical with the top and column identical with the side. The existence of an inverse signifies that the identity appears in every row and column.

Example 6. The permutation of n identical objects form a group. This group is denoted by S_n. Each permutation can be considered a transformation on the system. Two such successive operations are equivalent to an other permutation.

We can workout multiplication table of the group, we find that the table is same as for the dihedral group D_3.

The group S_n plays a very important role in physics. For a system of n identical particles the Hamiltonian remains unchanged under a permutation of the particles. This fact can be utilised to classify the energy eigen-states of the system.

11.3. REARRANGEMENT LEMMA

This lemma states that in a group multiplication table, each element of group occurs only once in a row or a column, i.e. a group multiplication table is merely a rearrangement of elements of the group. This is true in general and can be seen by examining Tables 11.1 and 11.2. We can also present it as follows.

If an element can occur twice in a row (or column) then it is the result of two different multiplications. Let, for a given group

$$AX = Y \quad \text{and } AB = Y$$

Then $$AX = AB$$

Multiplying from left by A^{-1}, we get

$$A^{-1} AX = A^{-1} AB$$

This gives $$X = B$$

This is a contradiction because all the elements are supposed to be distinct.

11.4. SOME DEFINITIONS

Subgroup

A subset of elements of a group (G) form a group under same group multiplication as that of G is called a subgroup (H) of G.

For every group there are tow trivial subgroups (i) the identity element I itself forms a subgroup (ii) the whole group G also can be considered as subgroup of itself. A subgroup (H) of group (G) is said to be a proper subgroup if $H \neq G$.

Examples of subgroup H

(a) For a group of 4 numbers $1, i, -i, -1$ $(i = \sqrt{-1})$, the numbers $(1, -1)$ form a subgroup.

(b) All even numbers, positive and negative including zero form a subgroup of all integers under addition.

(c) For a group of four matrices

$$I = \begin{pmatrix} 1 & 0 \\ 0 & 1 \end{pmatrix}, A \begin{pmatrix} 0 & 1 \\ -1 & 0 \end{pmatrix}, B = \begin{pmatrix} -1 & 0 \\ 0 & -1 \end{pmatrix}, C = \begin{pmatrix} 0 & -1 \\ 1 & 0 \end{pmatrix}$$

the matrices I, B form a subgroup under matrix multiplication.

(d) For a set of $n \times n$ unitary matrices forming a group under matrix multiplication, a set of $n \times n$ unitary matrices with determinant $+1$ form a subgroup.

(e) If we consider a group of symmetry operations for equilateral triangle (Fig. 11.1) such as transformation (I, A, B, C, D, E), then proper subgroups are

$$H_1 = (I, A, B), \quad H_2 = (I, C), \quad H_3 = (I, D), \quad H_4 = (I, I), \quad H_5 = (I).$$

Conjugate Elements

In group G if A and X are two elements such that XAX^{-1} also belongs to G then the elements A and XAX^{-1} are said to be conjugate of each other.

A collection of elements of a group G which are conjugate to each other is called a *class*.

In example (a) above, each element of the group is conjugate to only itself and forms a class with one element. To test this let is consider the conjugate of 1.

$$(1)^{-1} = 1, \quad (-1) i (-1)^{-1} = 1, \quad (-i) 1 (-i)^{-1} = 1$$

Thus 1 is conjugate to only itself and forms a class with one element, similarly it can be shown that i is conjugate to only itself.

This is a general characteristics of all abelian groups. In every group, the identity element I forms a class by itself. It is due to the fact that the only element conjugate to I is I.

$$XIX^{-1} = XX^{-1} = I \quad \text{for any } X \text{ in } G$$

The dihedral group, D_3 which we have defined with the transformation group of equilateral triangle is no abelion, we can easily show that only conjugate element to A is element B. Thus A and B form a class of group D_3. Similarly, we can show that the elements $C, D,$ and E are conjugate to one another and hence form a class. Thus, the classes of group D_3 are $I, (A, B), (C, D, E)$.

It is worthwhile to observe that A and B are both rotations whereas C, D, E are reflections about some axes.

Some Results Regarding Classes

(i) No element can be common between two distinct classes

This result be proved by using mathematical induction, i.e. we will assume a common element in two classes and then show that it is not possible.

Let A be a common element between two distinct classes of a group G.

$$C_1 = (A, B, C...)$$

and
$$C_2 = (A', B', C'...)$$

Since A and B are conjugate to each other then

$$A = XBX^{-1}$$

when X is some element of G.

Similarly
$$A = YB'Y^{-1}$$

Since A and B' are conjugate to each other and Y is some other element of G, then

$$XBX^{-1} = YB'Y^{-1}$$

or
$$(Y^{-1} X) B(X^{-1}Y) = B'$$

or
$$(Y^{-1}X) B(Y^{-1}X)^{-1} = B'$$

But $Y^{-1}X$ belongs to G, then the last equation implies that B and B' are conjugate to each other. Thus B' should also belong to c_1 and B to c_2 proceeding in the same way. We conclude that the members of c_1 and c_2 are identical which contradicts the assumption that c_1 and c_2 are distinct classes. Thus, A which is a common element between two distinct classes c_1 and c_2 is not possible.

(ii) A group consists of entire classes

A group consists of entire classes follows from the result and no element can be common between two distinct classes (a result we proved above).

An identity element forms a class by itself. Let us represent this class by c_1. We then choose any element of the group and find elements conjugate to it. They are put together to form class c_2. If this does not exhaust all the elements of the group, we start with one of the remaining elements and find the elements conjugate to it. Those elements which are distinct from the elements of c_1 and c_2 form class c_3. In this way, we continue to form distinct classes c_1, c_2, c_3,...,c_n till all the elements are exhausted and we have obtained n classes, in this way group G is broken into n classes.

Coset: Let $H = (H_1, H_2...)$ be a subgroup of group G and X is an element of G not contained is H, then multiplying H with X from left side we get

$$XH = X(H_1, H_2...) = (XH_1, XH_2...) \qquad (11.4a)$$

XH is called left coset of H.

Similarly a right coset is defined as the set of elements

$$HY = (H_1, H_2...), Y = (HY, H_2Y,...) \qquad (11.5)$$

Here Y is also an element of G not contained in H, i.e. $Y \in G$.

It should be noted that a coset is not a subgroup of G. It is due to the fact that identity element (I) does not exist in XH or HY, in place of element I there will be either

$$IH_i = H_i \quad \text{or} \quad H_iI = H_i$$

where H_i is ith element of subgroup H.

It can be very easily proved that two left cosets of a subgroup H either coincide completely or have no elements common is them.

Proof: Consider two left cosets

$$XH = (XH_1, XH_2...)$$

and

$$YH = (YH_1, YH_2...)$$

Let the two elements XH_1 and YH_1 in them are common. This means $XH_1 = YH_1$, this gives

$$Y^{-1}X = H_1H^{-1}$$

This shows that $Y^{-1}X$ is a member of the subgroup H, therefore if we multiply all the elements of H by $Y^{-1}X$ from left we get back H (rearrangement lemma), i.e.

$$Y^{-1}XH = H \quad \text{or} \quad XH = YH$$

This means that the two must coincide completely. The other possibility is that no element is common between the cosets.

Theorem relating group (G) and its subgroup (H)

This theorem states that the order of finite group (G) must be an integral multiple of the order of any of its subgroup (H).

To prove this theorem, let H be a subgroup of order of the group G. Let us form a left coset XH, where $X \in G$ but is not contained in H, by this process we get h elements belonging to G which are distinct from the elements of H. If this does not exhaust, all the elements of G, we repeat the exercise by forming a new coset. Everytime this is done new different elements of G are obtained. The procedure is repeated till all the elements of G are exhausted. The cosets can not have any element in common. Hence, we conclude that number of elements of G is just a multiple of the number of elements of the subgroup H.

Let

$$H = (H_1, H_2,...,H_n)$$

be a subgroup of G. We now construct a set

$$XHX^{-1} = (XH_1X^{-1}, XH_2X^{-1},..., XH_nX^{-1})$$
$$= H'_1, H'_2,...,H'_n$$

This new set so formed is also a subgroup of G. We can prove this as follows:

if $H_iH_i = H_k$ then

$$H'_iH'_i = (XH_iX^{-1})(XH_iH^{-1}) = XH_iH_iX^{-1} = XH_kX^{-1} = H'_k$$

Then the sets of elements $(H_1, H_2,...,H_n)$ and $(H'_1, H'_2,...,H'_n)$ will have the same multiplication table (Table 11.2). If one set forms a group, so will the other.

Subgroup $(XH_1X^{-1}, XH_2X^{-1}... XH_nX^{-1})$ is called a conjugate subgroup of H in G.

Invariant Subgroups and Factor Group

A subgroup H is called an invariant subgroup, if all its conjugate subgroups in G are identical with H.

The whole group G and the identity element I are two trivial examples of invariant subgroup.

Similarly in group $G = (1, i, -i, 1)$, subgroup $(1, -1)$ is an invariant subgroups.

For the dihedral group D_2 (rotation of an equilateral triangle), the elements (I, A, B) form an invariant subgroup.

All the left and right sets of an invariant subgroup are identical. This property of an invariant subgroup can be verified as follows.

We know

$$XHX^{-1} = (XH_1X^{-1}, XH_2X^{-1},...,XH_nX^{-1}) = H$$

This gives

$$XH = HX \tag{11.6}$$

This proves the property mentioned above.

For an invariant subgroup, the sets $(H_1, H_2,...,H_n)$ and $(XH_1X^{-1}, XH_2X^{-1},...,XH_nX^{-1})$ are identical except for a possible rearrangement, this means that for every H_i, $i = 1, 2, 3,...,n)$ we can find H_j, $j = 1, 2, 3... n$ such that

$$XH_iX^{-1} = H_j \tag{11.7}$$

This then implies that for every member H_i in this subgroup, all its conjugate elements are also in the subgroup. This means that an invariant subgroup consists of entire classes. As an example of this we see that an invariant subgroup (I, A, B) of D_3 consists of classes I and (A, B).

An invariant subgroup H together with its cosets forms a group under certain group multiplication rules. We will find now this rule.

Let us take cosets

$$K_1 = XH \quad \text{and} \quad K_2 = YH.$$

When we multiply them, all the elements of the first are multiplied by all the elements of the second and repetition if any in this is omitted. It is found that the product of these cosets is another coset of H.

$$K_1K_2 = (HX)(YH)$$
$$= (XH_1, XH_2,...,XH_n)(YH_1, YH_2,...,YH_n)$$
$$= XH_1YH_1, XH_2YH_1,...,XH_nYH_1, XH_1YH_2,...,XH_n$$
$$\quad YH_2,...,XH_1YH_n, XH_2YH_n,...,XH_nYH_n$$

or

$$K_1K_2 = X(H_1Y H_2Y,...,H_nY) H_1 X(H_1Y_1H_2Y,...,H_nY)$$
$$\quad H_2,...,X(H_1Y_1H_2Y,...,H_nY) H_n$$
$$= XK'_2H \tag{11.8}$$

where

$$K'_2 = (H_1Y, H_2Y,...,H_n^{-1}) = HY \tag{11.9}$$

Since H is an invariant subgroup

$$K'_2 = HY = YH$$

Thus

$$K_1K'_2 = XY HH \tag{11.10}$$

HH means multiplying every element of H with all its elements. If repetitions are omitted this will yield H. Thus

$$K_1K_2 = XYH \tag{11.11}$$

Right hand side of Eq. (11.11) is another coset of H. This proves the closure property of H. This multiplication is also associative. The subgroup H plays the role of the identity element.

$$K_1H = XHH = XH = K_1 \tag{11.12}$$

and
$$HK_1 = HXH = XHH = XH = K_1 \tag{11.13}$$

The coset $K_3 = X^{-1}H$ is the inverse of K_1

and
$$K_1K_3 = XX^{-1}H = H \tag{11.14}$$

The invariant subgroup H and cosets K_i ($i = 1, 2, 3,...$ etc.) satisfy all the group axioms under the coset multiplication rule given, therefore form a group. This group is denoted by G/H and is called factor group or quotient group. The order of this group g/n, where n is order of group G.

11.5. HOMOMORPHISM AND ISOMORPHISM

Isomorphisms

Consider the groups of same order n, i.e.

$$G = (G_1, G_2, G_3,..., G_n)$$

and
$$G' = (G'_1, G'_2, G'_3,..., G'_n)$$

If there is one to one correspondence between the elements of these groups, i.e. $G_i \rightarrow G'_i$ such that

$$G_i G'_i = G_K \text{ implies } G'_i G'_j = G'_k$$

and vice versa, i.e.

$$G'_i G'_j = G'_k \text{ implies } G_i G_j = G_k$$

Such groups are called *isomorphic to each other*.

For two isomorphic groups, their order should be same, i.e. they must have the same number of elements and their multiplication table must also be same. The groups may be entirely same dihedral group D_3 and the permeation group S_3 are isomorphic groups.

Similarly group $(1, i, -i, -1)$ is isomorphic to cyclic group $(A, A^2, A^3, A^4 = I)$.

Homomorphism

If there is group multiplication correspondence between elements of two groups (not necessarily one to one correspondence) such that $G_i G_j = G_k$ implies $G'_i G'_j = G'_k$. The two groups are said to be homomorphic groups. This means the mapping of several elements of group G is one element of G' such that multiplication is preserved. The inverse is also true. Isomorphism is a special case of homomorphism.

To understand homomorphism, consider the mapping of elements of group $(1, i, -i, -1)$ to those of group $(A, A^2 = I)$

Here $(1, -1) \rightarrow I$ and $(i, -i) \rightarrow A$.

The product $(-1)(i)$ should correspond to IA and infect $-i$ corresponds A.

It can be easily verified that multiplication is preserved for all other products.

The mapping of symmetry group D_3 of the equilateral triangle to the group $(1, -1)$ is homomorphic. If we take $(I, A, B) \rightarrow 1$ and $(C, D, E) \rightarrow -1$.

Since $A \rightarrow 1, C \rightarrow 1$ then $E = AC$ should correspond to (1) (–1) = –1. Which is indeed the case here. Similarly all the other products are preserved. Thus group D_3 is homomorphic to the group (1, –1).

A trivial example of homomorphic is the mapping of all the elements of G to the identity element I.

11.6. REPRESENTATION OF GROUPS

The most important concept in the applications of group theory to the solution of the classification and related problems in solid state physics and other branches is that of the representation of groups.

If there is homomorphic mapping of group $G = (G_i)$ to a set of operators $D(G_i)$ on a vector space V, then the set of operators $D(G_i)$ is said to form a representation of the group in vector space V.

We know that operators on a vector space can be represented by matrices. In order to simplify matters here in group representation the operators are represented by matrices rather than operators themselves. Hence, in the text of this chapter, we will adapt following definition for group representation.

By group representation $G = (G_i)$, we mean a homomorphic mapping of G to a set of $n \times n$ nonsingular matrices $D(G_i)$. This means for every element G_i, we have $(n \times n)$ nonsingular matrix $D(G_i)$. If

$$G_i G_j = G_k \quad \text{then} \quad D(G_i) D(G_j) = D(G_k)$$

Then identity element I corresponds to the $n \times n$ unit matrix. If $G_k = I$, the identity element, then G_i and G_k are inverse of each other, then

$$D(G_i) D(G_j) = 1$$

and matrices $D(G_i)$ and $D(G_j)$ are inverse of each other and they have to be non-singular. If the mapping is isomorphic then the representation is called a faithful representation. The dimensions of the matrices is called dimensions of the representation.

Some examples of group representation are given here.

Example 1. Consider the cyclic group $C_4 = (A, A^2, A^3, A^4 = I)$ and the set of matrices

$$D(A) = \begin{pmatrix} 0 & 1 \\ -1 & 0 \end{pmatrix}, D(A^2) = \begin{pmatrix} -1 & 0 \\ 0 & -1 \end{pmatrix}, D(A^3) = \begin{pmatrix} 0 & -1 \\ 1 & 0 \end{pmatrix}$$

and

$$D(I) = \begin{pmatrix} 1 & 0 \\ 0 & 1 \end{pmatrix}$$

Solution: These matrices form a group under matrix multiplication and there is an isomorphism between the groups and C_4. The correspondence between elements of the two groups is clear from the labelling of the matrices. This set of matrices form a two dimension faithful representation of the group C_4. But this is not the only representation of C_4. We can have also the following one-dimensional representation.

$$D(I) = 1, \quad D(A) = -1, \quad D(A^2) = 1, \quad D(A^3) = -1$$

Another one-dimensional representation is given by

$$D(I) = 1, \quad D(A) = -1, \quad D(A^2) = 1, \quad D(A^3) = 1$$

In this way, there are a large number of representations of a group. Hence to distinguish them a superscript such as $D^{(v)}(A)$, $D^{(u)}(A)$... etc is used. In order to distinguish above two attributes, (one-dimensional), it can be written as $D^{(1)}$ and $D^{(2)}$ respectively. Two additional one-dimensional representations of group C_4 can be

$$D^{(3)}(I) = 1, \quad D^{(3)}(A) = i, \quad D^{(3)}(A^2) = -1, \quad D^{(3)}(A^3) = -i$$

and
$$D^{(4)}(I) = 1, \quad D^{(4)}(A) = -i \quad D^{(4)}(A^2) = -1 \quad D^{(4)}(A^3) = i$$

The above discussion clearly signifies that the representation has nothing to do with the order of the group. We see that group C_4 is of order 4 but it has one and two dimensional representations. Higher dimensional representations of C_4 are also possible.

Example 2. Representation of the triangle (dihedral group D_3)

Solution: Let us construct a (2×2) matrix that have the same multiplication table as given for the triangle group. Let under rotation through an angle θ, a point $P(x, y)$ goes to $Q(x', y')$ as shown in Fig. 11.3. Then relation between P and Q is given by

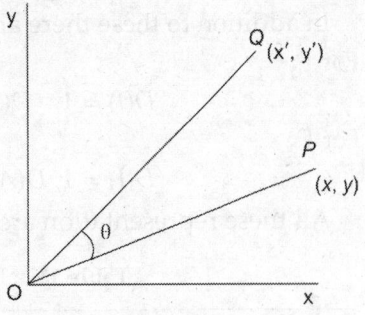

Fig. 11.3

$$\begin{pmatrix} x' \\ y' \end{pmatrix} = \begin{pmatrix} \cos\theta & -\sin\theta \\ \sin\theta & \cos\theta \end{pmatrix} \begin{pmatrix} x \\ y \end{pmatrix} \qquad (11.15)$$

The operations I, A and B correspond to rotation $\theta = 0$, $\dfrac{2\pi}{3}$ and $\dfrac{4\pi}{3}$ respectively.

Thus we can represent them by matrix,

$$D(I) = \begin{pmatrix} 1 & 0 \\ 0 & 1 \end{pmatrix}, \; D(A) = \begin{pmatrix} -1/2 & \sqrt{3}/2 \\ \sqrt{3}/2 & -1/2 \end{pmatrix}, \; D(B) = \begin{pmatrix} -1/2 & -\sqrt{3}/2 \\ -\sqrt{3}/2 & -1/2 \end{pmatrix} \quad ...(11.16)$$

Similarly for reflections, the coordinates of the vertices a, b, c relative to the centroid are (Fig. 11.2).

$$a\left(0, \frac{2}{\sqrt{3}}\right), \; b\left(-1, -\frac{1}{\sqrt{3}}\right), \; \text{and } c\left(1, -\frac{1}{\sqrt{3}}\right)$$

Under reflection (C) $(b \to c)$, under reflection (D) $(c \to a)$ and under (E) $(a \to b)$. We observe that

$$\begin{pmatrix} +1 \\ -1/\sqrt{3} \end{pmatrix} = \begin{pmatrix} -1 & 0 \\ 0 & 1 \end{pmatrix} \begin{pmatrix} -1 \\ -1/\sqrt{3} \end{pmatrix}$$

$$\begin{pmatrix} 0 \\ 2/\sqrt{3} \end{pmatrix} = \begin{pmatrix} -1/2 & -\sqrt{3}/2 \\ \sqrt{3}/2 & -1/2 \end{pmatrix} \begin{pmatrix} 1 \\ -1/3 \end{pmatrix}$$

$$\begin{pmatrix} -1 \\ -1/\sqrt{3} \end{pmatrix} = \begin{pmatrix} 1/2 & -\sqrt{3}/2 \\ -\sqrt{3}/2 & -1/2 \end{pmatrix} \begin{pmatrix} 0 \\ 2/\sqrt{3} \end{pmatrix}$$

We obtain

$$D(C) = \begin{pmatrix} -1 & 0 \\ 0 & 1 \end{pmatrix}, \quad D(D) = \begin{pmatrix} -1/2 & \sqrt{3}/2 \\ \sqrt{3}/2 & -1/2 \end{pmatrix}$$

$$D(E) = \begin{pmatrix} 1/2 & -\sqrt{3}/2 \\ -\sqrt{3}/2 & -1/2 \end{pmatrix}$$

(11.17)

We can verify easily that these six (2 × 2) matrices constitute a representation of the group of equivalent triangle. This representation is usually denoted by Γ_3.

In addition to these there are two more one-dimensional representations Γ_1 and Γ_2. For Γ_1.

$$D(I) = 1, \quad D(A) = 1, \quad D(B) = 1, D(C) = 1, \quad D(D) = 1, \text{ and } D(E) = 1$$

For Γ_2

$$D(I) = 1, D(A) = 1, D(B) = 1 \; D(C) = -1, \quad D(D) = -1, \text{ and } D(E) - 1$$

All these representation are summarised in Table 11.3.

Table 11.3: Representation of the dihedral group (D_3)

	I	A	B	C	D	E
Γ_1	1	1	1	1	1	1
Γ_2	1	1	1	-1	-1	-1
$\Gamma_3/3$	$\begin{pmatrix} 1 & 0 \\ 0 & 1 \end{pmatrix}$	$\begin{pmatrix} -1/2 & -\sqrt{3}/2 \\ \sqrt{3}/2 & -1/2 \end{pmatrix}$	$\begin{pmatrix} -1/2 & \sqrt{3}/2 \\ -\sqrt{3}/2 & -1/2 \end{pmatrix}$	$\begin{pmatrix} -1 & 0 \\ 0 & +1 \end{pmatrix}$	$\begin{pmatrix} 1/2 & \sqrt{3}/2 \\ \sqrt{3}/2 & -1/2 \end{pmatrix}$	$\begin{pmatrix} 1/2 & -\sqrt{3}/2 \\ -\sqrt{3}/2 & -1/2 \end{pmatrix}$

11.7. REDUCIBLE AND IRREDUCIBLE REPRESENTATION

Two representations $D(G_i)$ and $D'(G_i)$ are said to be equivalent if $D'(G_i)$ can be written as

$$D'(G_i) = D(u)^{-1} D(G_i) D(u)$$

(11.18)

Hence u is the element of the group. $D(u)$ is called similarity transformation and is a nonsingular matrix.

Representation D and D' related by Eq. (11.18) are said to be equivalent. This means conjugate elements represent the same physical quantity in different coordinate systems are related to one another by a unitary or similarity transformations. In other words, the equivalent representations are physically indistinguishable.

Elements belonging to the same conjugate class have equivalent representation and one can not relate the elements to one another belonging to different conjugate classes by applying a unitary transformation. Such representation are called inequivalent. Clearly, the number of inequivalent representations of a group will be the same as the number of conjugate classes in the group.

If two representations $D^{(u)}(G_i)$ and $D^{(v)}(G_i)$ of a group which may or may not be equivalent, then the set of matrices of the form

$$D(G_i) = \begin{pmatrix} D^{(u)}(G_i) & 0 \\ 0 & D^{(v)}(G_i) \end{pmatrix}$$

(11.19)

is also a representation of group.

A representation is said to be reducible if all the matrices in the representation can be brought to the above block diagonal by the same similarity transformation. If no similarity transformation can bring all the matrices to the block diagonal from their representation is said to be irreducible.

If the representing matrix can be reduced to block diagonal form Eq. (11.19), the representation is said to be fully irreducible.

In Eq. (11.19), if representations $D^{(u)}(G_i)$ and $D^{(v)}(G_i)$ are irreducible, then the representation $D(G_i)$ is said to be decomposed into irreducible representation $D^{(u)}(G_i)$ and $D^{(v)}(G_i)$.

To test the reducibility property of a representation $D(G_i)$, we find a similarity transformation P such that on being applied to every matrix of $D(G_i)$ reduces each matrix to the form

$$P^{-1}D(G_i)P = \begin{pmatrix} D^{(u)}(G_i) & 0 & 0 \\ 0 & D^{(v)}(G_i) & 0 \\ 0 & 0 & D^{(\alpha)}(G_i) \end{pmatrix} \tag{11.20}$$

i.e. $D(G_i)$ consists of zero everywhere except for submatrices along the diagonal. If such reduction is possible, the representation is said to be reducible, otherwise the representation is called irreducible.

Clearly irreducible representation can not be equivalent unless they are identical (Eq. 11.19) can be expressed in the form of a direct sum

$$D = D' + D'' \tag{11.21}$$

For example to understand what we have read above, we take the two-dimensional representation of C_4.

$$D(A) = \begin{pmatrix} 0 & 1 \\ -1 & 0 \end{pmatrix}, \quad D(A^2) = \begin{pmatrix} -1 & 0 \\ 0 & -1 \end{pmatrix}$$

$$D(A^3) = \begin{pmatrix} 0 & -1 \\ 1 & 0 \end{pmatrix}, \quad D(I) = \begin{pmatrix} 1 & 0 \\ 0 & 1 \end{pmatrix}$$

Let us transform all these matrices by similarity transformation $P^{-1}D(G_i)P$, where

$$P = \frac{1}{\sqrt{2}} \begin{pmatrix} 1 & 1 \\ i & -i \end{pmatrix} \text{ then}$$

$$D'(A) = P^{-1}D(A)P$$

$$= \frac{1}{\sqrt{2}} \begin{pmatrix} 1 & -i \\ i & i \end{pmatrix} \begin{pmatrix} 0 & 1 \\ -1 & 0 \end{pmatrix} \frac{1}{\sqrt{2}} \begin{pmatrix} 1 & 1 \\ i & -i \end{pmatrix}$$

$$= \begin{pmatrix} i & 0 \\ 0 & -i \end{pmatrix}$$

Similarly $\quad D'(A^2) = \begin{pmatrix} -i & 0 \\ 0 & -i \end{pmatrix}, D'(A^3) = \begin{pmatrix} -i & 0 \\ 0 & i \end{pmatrix}$ and $D'(I) = \begin{pmatrix} i & 0 \\ 0 & i \end{pmatrix}$

Obviously
$$D'(A) = \begin{pmatrix} D^{(3)}(A) & 0 \\ 0 & D^{(4)}(A) \end{pmatrix}, \quad D'(A^2)\begin{pmatrix} D^{(3)}(A^2) & 0 \\ 0 & D^{(4)}(A^2) \end{pmatrix}$$

$$D'(A^3) = \begin{pmatrix} D^{(3)}(A^3) & 0 \\ 0 & D^{(4)}(A^3) \end{pmatrix}, \quad D'(I)\begin{pmatrix} D^{(3)}(A^3) & 0 \\ 0 & D^{(4)}(A^3) \end{pmatrix}$$

Thus the given two-dimensional representation of C_4 is reducible and can be decomposed into one-dimensional representation $D^{(3)}$ and $D^{(4)}$. This can be symbolically written as
$$D' = D^{(3)} + D^{(4)}$$

11.8 THE DIRECT PRODUCT REPRESENTATION

We have seen that a new representation $D'(X)$ of higher dimensions can be formed by adding two given irreducible representations $D^{(\mu)}(X)$ and $D^{(\nu)}(X)$ of a group in the following way

$$D'(X) = \begin{pmatrix} D^{(\mu)}(X) & 0 \\ 0 & D^{(\nu)}(X) \end{pmatrix} \quad D'(X) = \begin{pmatrix} D^{(\mu)}(X) & 0 \\ 0 & D^{(\nu)}(X) \end{pmatrix}$$

The irreducible representation can be repeated as many times as we wish.

We can also form a higher dimensional representation by constructing the direct product of matrices $D^{(\mu)}(X)$ and $D^{(\nu)}(X)$. The meaning of constructing the direct product of matrices we mean that a matrix $D'(X)$ is obtained when each element is obtained by multiplying each element of $D^{(\mu)}(X)$ with every element of $D^{(\nu)}(X)$ such that

$$D^{\mu}_{lm}(X)D^{\nu}_{pq} = D_{lp,\,mq}(X) \tag{11.21a}$$

Here if matrices $D^{(\mu)}(X)$ and $D^{(\nu)}(X)$ are of dimensions n_μ and n_ν respectively then matrix $D(X)$ is of dimensions $n_\mu n_\nu$. It can be easily verified that the matrix $D(X)$ form a representation of the given group. To prove this, let us consider the product

$$\sum_{lm} D^{(\mu)}_{lm}(X)D^{(\mu)}_{mn}(Y)\sum_q D^{(\nu)}_{pq}(X)D^{(\nu)}_{qr}(Y)$$

$$= \left[D^{(\mu)}(X)D^{(\mu)}(Y) \right]_{ln}\left[D^{(\nu)}(X)D^{(\nu)}(Y) \right]_{pr}$$

$$= \left[D^{(\mu)}(XY) \right]_{ln}\left[D^{(\nu)}(XY) \right]_{pr}$$

$$= D_{lp,\,nr}(XY) \tag{11.22}$$

This product is

$$= \sum_m \sum_q D^{(\mu)}_{lm}(X)D^{(\nu)}_{pq}(X)D^{(\mu)}_{mn}(Y)D^{(\nu)}_{qr}(Y)$$

$$= \sum_m \sum_q D_{lp,\,mq}(X)D_{mq,\,nr}(Y) \tag{11.23}$$

From Eqs (11.22) and (11.23)

$$[D(X)D(Y)]_{lp,nr} = D_{lp,nr}(XY)$$

or $$D(X)\,D(Y) = D(XY) \tag{11.24}$$

Hence the matrices $D(X)$ form a representation of the group. A direct product representation is in general reducible.

The regular representation:

If the group multiplication table is rearranged such that the identity element I occurs along the diagonal, then this is also the representation of a finite group.

We construct matrices $D(X)$ by substituting unity in the table where the element X occurs and putting zero at all other places. Such matrices form a representation of the group.

To prove this, let top row of the multiplication table be labelled as $G_1 = I, G_2, G_3...G_n$

	$G_1 = I$ G_2 G_3 $\cdots$ G_n
G_1^{-1}	
G_2^{-1}	
G_3^{-1}	
$\vdots$	
G_n^{-1}	

Since identity element occurs along the diagonal, the left column must be $G_1^{-1}, G_2^{-1}, G_3^{-1}, \cdots, G_n^{-1}$. By definition, the matrix $D(G)$ is such that

$$\left.\begin{array}{l} D_{ij}(G_k) = 0 \ \ \text{if} \ \ G_i^{-1}G_j \neq G_k \\ \qquad\quad = 1 \ \ \text{if} \ \ G_i^{-1}G_j = G_k \end{array}\right\} \tag{11.25}$$

Now $$[D(G_k)D(G_h)]_{ij} = \sum_i D_{il}(G_k)D_j(G_n) \tag{11.26}$$

In each row or column of the matrices, only one element is unity others are zero (rearrangement lemma). Therefore in the sum over l above, only one term can possibly contribute. If this term corresponds to $l = m$, we have

$$[D(G_k)D(G_h)]_{ij} = D_{im}(G_k)D_{mj}(G_h) \tag{11.27}$$

But $D_{im}(G_k)$ is nonzero and is equal to unity only if

$$G_i^{-1}G_m = G_k \ \ \text{or} \ \ G_m = G_iG_k.$$

Similarly $$D_{mj}(G_h) = 1 \ \text{only if} \ G_m^{-1}G_i = G_h \ \text{or} \ G_m = G_jG_h^{-1}$$

otherwise $D_{mj}(G_h) = 0$. Thus

$$[D(G_k)D(G_h)] = 1 \ \text{only if} \ G_iG_k = G_jG_h^{-1} \ \text{if} \ G_i^{-1}G_j = G_kG_h,$$

otherwise it is equal to zero. Therefore by definition

$$[D(G_k)D(G_h)]_{ij} = [D(G_kG_h)]_{ij} \tag{11.28}$$

Hence $D(G_k) D(G_h) = D(G_k G_h)$, proving that the matrices $D(G_k)$ form a representation. This representation is known as the regular representation. The regular representation is reducible.

11.9. ORTHOGONALITY THEOREM AND SCHUR'S LEMMAS

Orthogonality theorem answers the important questions like, whether a given representation is reducible or not? If the representation is reducible then how to break it into irreducible representations? To prove orthogonality theorem, we will first see Schur's lemmas. There are two schur's lemmas.

Schur's lemma 1: If a matrix A commutes with all the matrices, $D(G_i)$ of an irreducible representation of group G then A is a constant multiple of the unit matrix.

Schur's lemma 2: (*a*) If D and D' be two irreducible representations of same dimensions of a group G and A is matrix such that

$$D(G_i) A = AD'(G_i) \tag{11.29}$$

for all G_i then either D and D' are equivalent or $A = 0$

(*b*) In the above if D and D' are of different dimensions then $A = 0$

Proof: Lemma 1: We know that every representation of a finite group is equivalent to a unitary representation. Hence we can write the matrix $D(G_i)$ is also unitary. For such representation

$$D^+(G_i) = D^{-1}(G_i) = D(G_i^{-1})$$

We also assume the matrix A to be Hermitian.

Since matrices A_1 and $D(G_i)$ commute, they can have simultaneous eigen vectors, A being a $(n \times n)$ Hermitian matrix has n eigen vectors which can be chosen to be orthogonal.

Let
$$\phi_i = \begin{pmatrix} u_{1i} \\ u_{2i} \\ \vdots \\ u_{ni} \end{pmatrix}$$

denote the normalized eigen vectors of A belonging to the eigen values a_i, $i = 1, 2, 3,...,n$ and let us for a unitary matrix

$$u = \begin{pmatrix} u_{11} & u_{12} & \cdots & u_{1n} \\ u_{21} & u_{22} & \cdots & u_{2n} \\ \vdots & & & \\ u_{n1} & u_{n2} & \cdots & u_{nn} \end{pmatrix} = \begin{pmatrix} \phi_1 & \phi_2 & \cdots & \phi_n \end{pmatrix}$$

Then
$$Au = (\alpha_1\phi_1\alpha_2\phi_2 + \cdots + \alpha_n\phi_n)$$

and
$$u^+Au = u^{-1}Au$$

$$= \begin{pmatrix} \alpha_1 & 0 & 0 & \cdots & 0 \\ 0 & \alpha_2 & 0 & \cdots & 0 \\ 0 & & & & \\ \vdots & & & & \\ 0 & 0 & 0 & \cdots & \alpha_n \end{pmatrix}$$

The similarity transformation vAU brings A to the diagonal form, since ϕ_i $(i=1,2,3,...,n)$ are simultaneous eigen vectors of $D(G_i)$, the transformation $u^{-1}D(G_i)u$ also diagonalizes $D(G_i)$. This holds for all $G_i \in G$. This means that the representation is reducible which is a contradiction. Therefore A must be a multiple of the unit matrix[*].

Proof: Lemma 2(a): $D(G_i)A = AD'(G_i)$ for all $G_i \in G$

Taking Hermitian conjugate of both sides we have

$$A^+D^+(G_i) = D'^+(G_i)A^+$$

or

$$A^+D(G_i^{-1}) = D'(G_i^{-1})A^+$$

or

$$A+D(G_i^{-1})A = D'(G_i^{-1})A^+A \tag{11.30}$$

Since G_i^{-1} is an element of G it follows from Eq. (11.29)

$$D(G_i^{-1})A = AD'(G_i^{-1}) \tag{11.31}$$

Substituting in Eq. (11.30), we get

$$A^+AD'(G_i^{-1}) = D'(G_i^{-1})A^+A \tag{11.32}$$

A^+A commutes with all the matrices $D'(G_i^{-1})$ of the irreducible representation D'. Hence by lemma 1

$$A^+A = \lambda 1 \tag{11.33}$$

If D and D' both are of same dimension n, A is a $(n \times n)$ matrix. Then $\det(A^+A) = \lambda^n$ or $|\det A|^n = \lambda^n$.

If $\lambda \neq 0$ $\det A \neq 0$ and A^{-1} exists. Then from Eq. (11.29)

$$D(G_i) = AD'(G_i)A^{-1} \tag{11.34}$$

i.e. two representations are equivalent.

If $\lambda = 0$ $A^+A = 0$. Equating ith element from both sides, we get

$$\sum_k A_{ki}^* A_{ki} = 0 \quad \text{or} \quad \sum_k |(A_{ki})|^2 = 0$$

This implies that $A_{ki} = 0$ for a given i and $k = 1, 2, 3,...,n$. Since this is true for any i, we have $A = 0$

Proof: Lemma 2(b): Now let us consider two irreducible representation of different dimensions. Their dimensions be n and n' respectively and $n < n'$. A be a $(n \times n')$ matrix. Let us construct a matrix A' $(n' \times n')$ by adding $(n' - n)$ rows with zeros to A.

Then $A'^+A' = A^+A = \lambda I$

or $|\det A'|^2 = \lambda^n \tag{11.35}$

But $\det A' = 0$

Therefore $A^+A' = 0$

If $A' = 0$ implies $A = 0$ [from Lemma 2(a)]

---

[*]Result is taken from Chapter 11—*Linear Vector Spaces*

In all our above discussion, we have assumed A to be Hermitian, if it is not we can define two Hermitian matrices.

$$A_1 = \frac{A + A^+}{2}$$

and
$$A_2 = \frac{A - A^+}{2i}$$

Now if A commutes with $D(G_i)$ for all $G_i \in G$, so does A'.

or
$$AD(G_i) = D(G_i) A$$
$$D^+(G_i) A^+ = A^+ D^+(G_i)$$
or
$$D(G_i^{-1}) A^+ = A^+ D(G_i^{-1}) \tag{11.36}$$

Hence A_1 and A_2 also commute with $D(G_i)$ for all $G_i \in G$, A_1 and A_2 being Hermitian and we can proceed as above to prove the Lemma.

Orthogonality Theorem

If $D^{(\mu)}$ and $D^{(\nu)}$ are two irreducible representations of a group G of order n; then orthogonality condition is

$$\sum_{k=1}^{n} D_{il}^{(\mu)}(G_k) D_{mj}^{(\nu)}(G_k^{-1}) = \frac{n}{n_\mu} \delta_{\mu\nu} \delta_{ij} \delta_{ml} \tag{11.37}$$

where n_μ is the dimension of the representation $D^{(\mu)}$ and dimensions (D^ν) are n_ν. Let us consider the matrix

$$A = \sum_{k=1}^{n} D^{(\mu)}(G_i) \times D^{(\nu)}(G_k^{-1}) \tag{11.38}$$

where matrix X_n is $n_\mu \times n_\nu$.

$$D^{(\mu)}(G_i) A = \sum_{k=1}^{n} D^{(\mu)}(G_i) D^{(\mu)}(G_k) \times D^{(\nu)}(G_k^{-1}) D^{(\nu)}(G_i^{-1}) D^{(\nu)}(G_i)$$

$$= \sum_{j=1}^{n} D^{(\mu)}(G_i) \times D^{(\nu)}(G_j^{-1}) D^{(\nu)}(G_i) \qquad \text{[where } G_i G_k = G_i]$$

$$= AD^{(\nu)}(G_i) \tag{11.39}$$

This gives $A = 0$ \hfill [*using* Lemma 2(*b*)]

Till now we have taken matrix X as an arbitrary matrix. Let us take matrix X such that its lmth element is unity and rest of the elements are zero. This can be written as

$$X_{pq} = \delta_{lp} \delta_{mq}$$

Now equating ljth element of A to zero, then from Eq. (11.38), we have

$$\sum_{k=1}^{n} D_{il}^{(\mu)}(G_k) D_{mj}^{(\nu)}(G_k^{-1}) = 0 \tag{11.40}$$

If we put $\mu = \nu$ in Eq. (11.39), we have
$$D^{(\mu)}(G_i) A = AD^{(\mu)}(G_i) \tag{11.41}$$

Since $A = \lambda 1$ (using lemma 1), then taking ij^{th} element of A we get from Eq. (11.38)

$$\lambda \delta_{ij} = \sum_{k=1}^{n} D_{ij}^{(\mu)}(G_k) D_{ml}^{(\mu)}(G_k^{-1}) \tag{11.42}$$

putting $i = j$ and summing over i, we get

$$\lambda n_\mu = \sum_{k=1}^{n} \sum_{i=1}^{n} D_{il}^{(\mu)}(G_k) D_{mj}^{(\mu)}(G_k^{-1})$$

or $$\lambda n_\mu = \sum_{k=1}^{n} \left[D^{(\mu)}(G_k^{-1}) D^{(\mu)}(G_k) \right]_{ml} \equiv n \delta_{ml}$$

or $$\lambda = \frac{n \delta_{ml}}{n_\mu} \tag{11.43}$$

Substituting in Eq. (11.42)

$$\sum_{k=1}^{n} D_{il}^{(\mu)}(G_i) D_{mj}^{(\mu)}(G_k^{-1}) = \frac{n}{n_\mu} \delta_{ij} \delta_{lm} \tag{11.44}$$

Equations (11.40) and (11.44) can be combined to get

$$\sum_{k=1}^{n} D_{il}^{(\mu)}(G_k) D_{mj}^{(v)}(G_k^{-1}) = \frac{n}{n_\mu} \delta_{\mu v} \delta_{ij} \delta_{ml} \tag{11.45}$$

If the representation is unitary, then $D_{mj}^{(v)}(G_k^{-1}) = D_{jm}^{*(v)}(G_k)$ then Eq. (11.45) reduces to

$$\sum_{k=1}^{n} D_{il}^{(\mu)}(G_k) D_{jm}^{*(v)}(G_k) = \frac{n}{n_\mu} \delta_{\mu v} \delta_{ij} \delta_{ml} \tag{11.46}$$

For fixed values of μ and i, $D_{il}^{(\mu)}(G_k)$ can be considered a vector of n components $D_{1l}^{(\mu)}(G_1), D_{2l}^{(\mu)} ... D_{nl}^{(\mu)}(G_n)$.

For representation of $D^{(\mu)}$ there are n_μ^2 vectors in the set ($i = 1, 2, 3,...,n_\mu,$ $l = 1, 2,...,n_\mu$) which are orthogonal to one another. But the total number of mutually orthogonal vectors is less than the dimensions of the space. Therefore

$$\sum_{\mu} n_\mu^2 \leq n \tag{11.47}$$

But while discussing reducible and irreducible representations, we have shown that equality does exist, i.e.

$$\sum_{\mu} n_\mu^2 = n \tag{11.47a}$$

11.10. CHARACTER OF A REPRESENTATION AND CHARACTER TABLE

Character of the representation $D^{(\mu)}(G_i)$ is the sum of the diagonal elements of $D^{(\mu)}(G_i)$. Mathematically it is represented as character

$$X^{(\mu)}(G_i) = \sum_{k} D_{kk}^{(\mu)}(G_i) \text{ for } G_i \in G \tag{11.48}$$

or $$X^{(\mu)}(G_i) = T_r D^{(\mu)}(G_i)$$

For two equivalent representations, their characters are same, if
$$D^{(\mu)}(G_i) = P^{-1} D^{(\mu)}(G_i) P$$

Then
$$T_r D^{(\mu)'}(G_i) = T_r \left[P^{-1} D^{(\mu)}(G_i) P \right] = T_r [PP^{-1} D^{(\mu)}(G_i)]$$
$$= T_r D^{(\mu)}(G_i) \tag{11.49}$$

Here $D^{(\mu)'}(G_i)$ and $D^{(\mu)}(G_i)$ are two equivalent traces and hence their characters are also same as Eq. (11.49).

The characters of the representations identify the equivalent representations, we also infer that in a given representation, characters of group elements belonging to the same class are the same. To prove this let G_i and G_j belong to the same class such that

$$G_i = G_k \, G_i \, G_k^{-1}$$

Then
$$D(G_i) = D(G_k)D(G_j)D(G_k^{-1}) = D(G_k)D(G_j)D^{-1}(G_k)$$

and
$$X(G_i) = T_r \, D(G_i) = T_r \, D(G_k) \, D(G_j) \, D^{-1}(G_k)$$

or
$$X(G_i) = T_r D(G_j) \tag{11.50}$$

In Eq. (11.46), let us put $i = l$ and $m = j$ and sum over i and j we get

$$\sum_{k=1}^{n} X^{(\mu)}(G_k) X^{(\nu)*}(G_k) = n\delta_{\mu\nu} \tag{11.51}$$

If in the group G, there is a class of n_i elements then each element in this class will make the same contribution to the sum, thus Eq. (11.51) can be written as

$$\sum_{i} X_i^{(\mu)} X_i^{(\mu)*} n_i = n\delta_{\mu\nu} \tag{11.52}$$

where sum is the left hand side runs over all the classes.

If n_k is the number of classes of the group, then $n_i^{1/2} X_i^{(\mu)}$ can be considered as n_k components of a vector. The number of such vectors is the number of nonequivalent irreducible representations N of the group. Equation (11.52) indicates that these vectors are mutually orthogonal in space and must be less than or equal to the dimensions of the space, i.e. $N \le n_k$. But in fact equality actually exists, i.e.

$$N = n_k \tag{11.53}$$

i.e. the number of nonequivalent irreducible representations of a group = the number of classes in the group, $n_i^{1/2} X_0^{(\mu)}$ forms n_k mutually orthogonal vectors in a n_k dimensional space. Thus they form a complete set and vector X_i (which is a character corresponding to class i, in the space can be expressed as a linear combination of them)

$$X_i = \sum_{\mu=1}^{N} C_\mu X_i^{(\mu)} n_i^{1/2} \tag{11.54}$$

where C_μ are the expansion coefficients. Multiplying by $X_i^{(\nu)*} n_i^{1/2}$ and summing over i, we get

$$C_\nu = \frac{1}{n} \sum_{i=1}^{n_k} X_i X_i^{(\nu)*} n_i^{1/2} \tag{11.55}$$

---

* It is a convension to indicate the number of a class by putting the number in front of the class.

Substituting from Eq. (11.55) in to Eq. (11.54), we get

$$X_i = \sum_\mu \sum_j X_j X_i^{(\mu)} X_j^{(\mu)*} \frac{n_i^{1/2} n_j^{1/2}}{n} \tag{11.56}$$

Hence

$$\sum_\mu X_i^{(\mu)} X_j^{(\mu)*} = \frac{n}{n_i^{1/2} n_j^{1/2}} \delta_{ij}$$

or

$$\sum_\mu X_i^{(\mu)} X_j^{(\mu)*} n_i = n_n \delta_{ij} \tag{11.57}$$

This is second orthogonality relation satisfied by the characters of an irreducible representation.

Character Table

A character table is a table in which the characters corresponding to different classes for all the possible irreducible representations of a finite group are tabulated. The various irreducible representations are indicated in a column in the left. The classes with the elements in bracket are written on top[#]. The character table can be prepared without a knowledge of the irreducible representations of group.

Theorem: It is defined as all irreducible representations of abelian group are one-dimensional helps in constructing the character tables for various groups. Hence we will prove this theorem before illustrating a few examples of the character table.

Consider an irreducible representation $D(G_i)$, $G_i \in G$ of an abelian group G. Let G_k be a given element of G since the group in abelian, then group product of elements commutes, i.e.

$$D(G_i) \, D(G_k) = D(G_k) \, D(G_i) \text{ for all } G_i \in G$$

Schur's lemma 1 gives that $D(G_i)$ must be multiple of a unit matrix, i.e

$$D(G_k) = \lambda_k 1$$

This also applies to all the matrices $D(G_k)$, $G_k \in G$. This means representation $D(G_i)$ is equivalent to the one-dimensional representation $G_i \to \lambda_i$

To construct a character table we will adopt a method given as under:

1. The number of equivalent irreducible representations of the group is equal to the number of classes. The dimensions of their representations are obtained from the Eq. (11.47)

$$\sum_n n_\mu^2 = n$$

   where n is the order of the group. If the group is abelian all the irreducible representations are one-dimension.

2. Identity element I is a class by itself. It is written as first class in the top row. In an irreducible representation $D^{(\mu)}$ whose dimensionality is n_μ, I is represented by $(n_\mu \times n_\mu)$ unit matrix. Therefore

$$X^{(\mu)}(I) = n_\mu$$

   The first column in the table can then be easily written down.

[#]It is a convension to indicate the number of a class by putting the number in front of the class.

3. Every group has one-dimensional representation in which every element is represented by unity. This gives us a row in the character table in which each character is unity.

4. Other rows can be constructed from the orthogonality among various rows expressed by Eq. (11.52), i.e.

$$\sum_i X_i^{(\mu)} X_i^{(\nu)*} n_i = n\delta_{\mu\nu}$$

In case $\mu = \nu$, we get

$$\sum_i |X_i^{(\mu)}|^2 n_i = n$$

5. Orthogonality among various columns of the table is given by Eq. (11.57).

$$\sum_i X_i^{(\mu)} X_i^{(\mu)*} n_i = n\delta_{ij}$$

.can be used to fill the blanks in the character table. Usually these rules are sufficient to complete the table.

Example 1: To construct character table of the group C_4. Since the given group is abelian as such all its irreducible representations are of one-dimension.

From the relation $\sum_\mu n_\mu^2 = 4$ with $n_\mu = 1$ for all μ, this means there are four irreducible representations. Thus, we have a 4×4 table with all the elements of first row and first column are unity.

Table 11.4: Character table of group C_4

	$C_1(I)$	$C_2(A)$	$C_3(A^2)$	$C_4(A^3)$
$D^{(1)}$	1	1	1	1
$D^{(2)}$	1	i	-1	$-i$
$D^{(3)}$	1	-1	1	-1
$D^{(4)}$	1	$-i$	-1	i

Since $A^4 = I$, $\quad D^{(\mu)}(A^4) = [D^{(\mu)}(A)]^4 = D^{(\mu)}(I)$

The representations are one-dimensional

$$D^{(\mu)}(A) = X^{(\mu)}(A)$$

Hence $\qquad [X^{(\mu)}(A)]^4 = X^{(\mu)}(I) = 1$

or $\qquad X^{(\mu)}(A) = \exp\left(i\frac{2n\pi}{4}\right), \, n = 0,1,2,3,...$

$n = 0$ corresponds to $D^{(1)}$, $n = 1$ corresponds $D^{(2)}$

$n = 3$ corresponds to $D^{(3)}$, and $n = 3$ corresponds $D^{(4)}$

This gives the seccond column of the table.

The rest of the characters can be obtained by using orthogonality of columns and rows.

Since the group is abelian, the simplest way to obtain them is to use the result

$$X^{(\mu)}(A^n) = D^{(\mu)}(A^n) = [D^{(\mu)}(A)]^n = [X^{(\mu)}(A)]^n$$

Here orthogonality relations [Eqs (11.52) and (11.57)] are satisfied.

Example 2. To construct the character table of dihedral group D_3 consisting of the three classes $I, (AB), (CDE)$.

Table 11.5: Character table of dihedral group D_3

	$C_1(I)$	$2C_2(A,B)$	$3C_3(C,D,E)$
$D^{(1)}$	1	1	1
$D^{(2)}$	1	1	−1
$D^{(3)}$	2	−1	0

There are in all three irreducible representations, let their dimensions be n_1, n_2 and n_3 then

$$n_1^2 + n_2^2 + n_3^2 = 6$$

This equation is satisfied if we take $n_1 = 1$, $n_2 = 1$ and $n_3 = 2$. This shows that there are two one-dimensional ($D^{(1)}$ and $D^{(2)}$) and one two-dimensional ($D^{(3)}$) irreducible representations. The first column in Table 11.5 is thus easily obtained. The first row is 1, 1, 1.

In one-dimensional representation $X(C^2) = 1$ since $C^2 = I$

and $\qquad X(C^2) = D(C^2) = [D(C)]^2 = X(C^2) = 1$

Thus $\qquad X(C) = \pm 1$

But $\qquad X^{(1)}(C) = \pm 1$

$\therefore \qquad X^{(2)}(C) = -1$

Using orthogonality relation [Eq. (11.52)], we get

$$1 + 2X^{(2)}C_2 - 3 = 0 \quad \text{or} \quad X^{(2)}C_2 = 1$$

The rest of the characters $X^{(3)}C_2$ and $X^{(3)}C^3$ can be obtained using orthogonality relation [Eq. (11.52)].

$$2 + 2X^{(3)}C_2 + 3X^{(3)}C_3 = 0$$

and $\qquad 2 + 2X^{(3)}C_2 - 3X^{(3)}C_3 = 0$

Solving we get $X^{(3)}C_3 = 0$ and $X^{(3)}C_2 = -1$.

11.11. DECOMPOSING A REDUCIBLE REPRESENTATION INTO IRREDUCIBLE ONES

In this section, we will discuss:

(a) How to know whether a representation is reducible or not.

(b) If it is reducible, then how to decompose it into a number of irreducible representations, we will answer these questions using orthogonality relation [Eq. 11.51].

Consider a representation $D(G_i)$ with character $X(G_i)$. If the representation is reducible then it can be written in terms of irreducible representation $D^{(\mu)}(G_i)$ as

$$D(G_i) = \sum_\mu a_\mu D^{(\mu)}(G_i) \tag{11.58}$$

Here coefficient a_μ gives how many times the irreducible representation $D^{(\mu)}(G_i)$ occurs in $D(G_i)$. a_μ are thus a positive integer or zero.

Taking the trace of both sides of Eq. (11.59), we get

$$X(G_i) = T_r D(G_i) = \sum_\mu a_\mu X^{(\mu)}(G_i) \tag{11.59}$$

$$\because \qquad T_r D^{(\mu)}(G_i) = X(G_i)$$

Multiplying both sides of Eq. (11.59) with $X^{(\mu)*}(G_i)$ and summing over all of i, we obtain

$$\sum_i X(G_i) X^{(\mu)*}(G_i) = \sum_\mu a_\mu \sum_i X^{(\mu)}(G_i) X^{(\mu)*}(G_i) = \sum_\mu a_\mu n_k \delta_{\mu\nu} \qquad [\textit{from Eq. (11.51)}]$$

or

$$a_\mu = \frac{1}{n} \sum_i X(G_i) X^{(\mu)*}(G_i) \tag{11.60}$$

If all a_μ's are zero except one, the representation is said to be irreducible, otherwise it is reducible.

Another criterion of testing irreducibility of a representation $D(G_i)$ is to observe that irreducible $D(G_i) \equiv D^{(\nu)}(G_i)$ then

$$a_\mu = \delta_{\mu\nu}$$

and

$$X(G_i) = X^{(\nu)}(G_i) \tag{11.61}$$

Then

$$\sum_i X(G_i) X^*(G_i) = \sum_i X^{(\nu)}(G_i) X^{(\nu)*}(G_i) = n$$

Hence if representation is irreducible then

$$\sum_i |X(G_i)|^2 = n \tag{11.62}$$

As an example, two-dimensional representation of C_4 group have character

$$X(A) = 0, \quad X(A^2) = -2, \; X(A^3) = 0, \quad X(I) = 2 \text{ and}$$

$$\sum_i |X(G_i)|^2 = 2 \neq 4$$

This result shows that the representation is reducible since group C_4 is an abelian, as such all its irreducible representations are one-dimensional. This means the given two-dimensional representation must be reducible.

Using Eq. (11.60) we find $a_1 = a_2 = 0$ and $a_3 = a_4 = 1$

Thus

$$D = D^{(3)} + D^{(4)}$$

The above discussion gives the utility and importance of a character table.

The characters of the irreducible representation of a group are called primitive or simple characters. The character of a reducible representation is called a compound character.

Theorem[#]: The regular representation contains each irreducible representation as many times as its dimensions. We will prove it now using Eq. (11.60).

Let regular representation of a group be represented as

$$D^{reg}(G_i) = \sum_\mu D^{(\mu)}(G_i), \text{ then}$$

$$a_\mu = -\sum_i X^{reg}(G_i) X^{(\mu)*}(G_i) \tag{11.63}$$

From definition of the regular representation, the characters $X^{reg}(G_i)$ are given by

$$\begin{cases} X^{reg}(G_i) = 0 & \text{if } G_i \neq I \\ \quad\quad = n & \text{if } G_i = I \end{cases}$$

Substituting the result in Eq. (11.63) above, we get

$$a_\mu = \frac{1}{n^n} X^{(\mu)*}(I) = l_\mu$$

where l_μ is the dimension of the irreducible representation $D^{(\mu)}(G_i)$.

$$\therefore \quad\quad D^{reg}(G_i) = \sum_\mu l_\mu D^{(\mu)}(G_i)$$

This shows that $D^{reg}(G_i)$ contains the irreducible representation $D^{(\mu)}(G_i)$, l_μ times.

11.12. CONSTRUCTION OF REPRESENTATION

Now we will take up the problem of how to construct representations of a group and the connection of group theory to physics? The two problems are related to each other.

In physical applications of group theory, the group elements are the symmetry operations on space coordinates. The result of a symmetry operation R on space coordinates can be taken as a set of transformations. If $[D(R)]_{ij}$ is a matrix corresponding to symmetry operation R, then a set of transformation can be taken as,

$$X'_i = \sum_{i=1}^{3} [D(R)]_{ij} X_j \tag{11.63a}$$

These matrices form a representation of a group. To check this let us consider two successive transformations.

$$X' \rightarrow RX, \quad X'' \rightarrow SX'$$

and

$$X'' = \sum_j D(S)_{ij} X'_j$$

$$= \sum_j \sum_k [D(s)]_{ij} [D(R)]_{ik} X_k$$

or

$$X'' = \sum_k [D(S)D(R)]_{ik} X_k \tag{11.64}$$

But

$$X_\mu = \sum_k D(SR)_{ik} X_k$$

$$\therefore \quad\quad D(SR) = D(S)\, D(R) \tag{11.65}$$

[#]We have stated earlier also.

This shows that matrices form a representation of the group. Thus a representation can be constructed by considering the effects of symmetry operators on the space coordinates and finding the matrix (R). To illustrate the above discussion we refer to Example 2 in Section 11.6

In order to construct other representations, the method adopted above is to be generalized. In above procedure, the symmetry operators are operating on the space coordinates (x, y, z) but in a generalized method we define operators which operate on a space function $\phi(x)$ [$\phi(x)$ stands for $\phi(x, y, z)$]. These operators O_R which are related to the symmetry transformation operators R are defined as

$$O_R \phi(x) = \phi(R^{-1}x)$$

It can then be proved that these operators O_R form a group G' which is isomorphic to the original group G of symmetry transformations.

To prove this, let us consider two successive transformations S and R In function space $\phi(x)$, these transformations give rise to

$$O_S O_R \phi(x) = O_S \psi(x), \quad \text{where } \psi(x) = O_R \phi(x)$$
$$= \psi(S^{-1}x)$$
$$= \phi(R^{-1}S^{-1}x)$$

Then
$$O_S O_R \phi(x) = \phi[(SR)^{-1}x]$$
$$= O_{SR}\,\phi(x) \tag{11.66}$$

Hence the result.

Since G and G' are isomorphic then if representation of G' is constructed we can easily get representation of G.

Construction of representation of G' transformation: Consider a set of functions $\phi(x)$ and transformation O_R operate on them. Here R runs over all group elements. In this way a new set of functions is generated. We then find a basis set such that the new set of functions so formed are expressed. Let us call these basis functions as $\phi_1(x)$, $\phi_2(x),...,\phi_n(x)$. Then the functions $O_R\phi_p(x)$ can be expressed as a linear combination of the basis set, i.e.

$$O_R\phi_p(x) = \sum_{n=1}^{n} D_{\nu\mu}(R)\phi_\nu(x) \tag{11.67}$$

This way the matrices $D_{\nu\mu}(R)$ generate a representation of the group G', we prove this as follows.

We know
$$O_R O_S \phi_p(x) = O_S \sum_{n=1}^{n} D_{\nu\mu}(R)\phi_\nu(x)$$

$$= \sum_{x=1}^{n} D_{\nu\mu}(R) \sum_{\lambda=1}^{n} D_{\lambda\nu}(S)\phi_\lambda(x) = \sum_{\lambda=1}^{n} [D(s)DCR]_{\lambda\mu}\phi_\lambda(x) \tag{11.68}$$

But
$$O_R O_S \phi_p(x) = O_{SR}\,\phi_p(x) = \sum_{\lambda=1}^{\lambda\mu} D(SR)_{\lambda\mu}\,\phi_\lambda(x) \tag{11.68a}$$

Hence $D(SR) = D(S)\,D(R)$. The matrices $D(R)$ form a representation of G' and therefore also of G.

11.13. REPRESENTATION OF GROUPS AND QUANTUM MECHANICS

A given energy quantum mechanical system is described by a Schrödinger equation

$$H\psi = E\psi \tag{11.69}$$

where H is the Hamiltonian of the system and E total energy. Let $O_R, O_S,..$etc. is a set of all operators which commute with the Hamiltonian. This set of operators form a group. In this group, the identity element (I) is the unit operator which belongs to this set. The product of two operators which commutes with the Hamiltonian also commute with H. The inverse of operator O_R also commutes with H, i.e.

$$[O_R O_S^{-1} H] = 0$$

or $\qquad [O_R H] O_S^{-1} + O_R [O_S^{-1} H] = 0 \tag{11.70}$

But $\qquad\qquad [O_R H] = 0 \quad \therefore \quad [O_S^{-1} H] = 0$

The group consisting of all operators which commute with H is known as symmetry group of the Hamiltonian (or of the Schrödinger equation). Operating O_R from left on the Schrödinger equation we get

$$O_R H\phi = EO_R\phi$$

or $\qquad\qquad HO_R\phi = EO_R\phi \tag{11.71}$

Thus $O_R\phi$ is an eigen function of H corresponding to the same energy as the eigen function ϕ. If an eigen function is given then a set of other degenerate eigen functions can be generated by operating on it by all the symmetry operators O_R. If all the degenerate eigen functions can be generated in this way, the degeneracy is called *normal*. We call degeneracy as accidental if there are other degenerated states which can not be obtained in this way. In fact accidental degeneracy is a wave phenomenon[*], in our further discussion, we will assume that the system under investigation has normal degeneracy.

We consider a state with n-fold degeneracy. Since a linear combination of degenerate eigen functions is also eigen functions, hence we have a freedom in choosing the eigen functions is case of normal degeneracy.

Let $\psi_1, \psi_2,...,\psi_n$ be n linearly independent degenerate eigen functions corresponding to an energy state we choose these to be orthonormal operating on any number ψ_P of the set by O_R we get another degenerate eigen function. We can express the operation $O_R\psi_P$ as a linear combination of a set, i.e.

$$O_R\psi_P = \sum_{v=1}^{n} D_{vP}(R)\psi_v \tag{11.72}$$

We have discussed that matrices $D(R)$ form a representation of the symmetry group of the Hamiltonian. This representation is irreducible. This is due to the fact that we have assumed only normal degeneracy (i.e. there is no accidental degeneracy), i.e. all the n degenerate states are generated by the operation of the set of operators O_R on any one of them. The dimensions of the matrices $D(R)$ are thus $(n \times n)$ and it is not

---

[*]The degeneracy of 2S and 2P levels of the hydrogen atom is a well known example of accidental degeneracy.

possible to have lower dimensional matrices corresponding to these transformations. This leads to the following important results.

i. If a level has n-fold degeneracy, the degenerate eigen functions give rise to an n-dimensional irreducible representation of the symmetry group of the Hamiltonian.

ii. Using the knowledge of degeneracy of the energy levels of a system, they can be classified according to the irreducible representations of the symmetry group.

iii. The irreducible representations of the symmetry group of the Hamiltonian can be used to infer the possible degeneracy in the energy levels.

iv. If the symmetry group is abelian, all the irreducible representations are one-dimensional which means that there is no degeneracy.

v. For symmetry group D_1—the group which has 2 one-dimensional and 1 two-dimensional irreducible representations the levels should be either double degenerate or there should be no degeneracy. For such a system, there is no level which has four-fold degeneracy.

11.14. LIE GROUPS AND LIE ALGEBRAS

Let us consider a group of transformations

$$X' = T_n X = X + n \tag{11.73}$$

where $n = 0, \pm 1, \pm 2 + \cdots$. The identify element of the group corresponds to $n = 0$

The product of two transformations T_m and T_n is T_{m+n}.

Then

$$T_m T_n X = T_m(X + n) = X + n + m = T_{m+n} X \tag{11.74}$$

The inverse of transformation T_n is T^{-n}. All the properties of the group can be expressed is terms of a single parameter n. To find the product $T_m T_n$ means we have to find an integer I which is a function of m and n, i.e. $I = f(m, n)$ such that

$$T_I = T_m T_n$$

In above (particular) example $f(m, n) = m + n$. The inverse of the T_n is an integer $\bar{n}$ such that $f(n, \bar{n}) = n + \bar{n} = 0$. This group is an example of one parameter group of infinite order. In the above example also, the parameter takes only discrete values. But if we consider the group or rotation about z-axis, the group is a one parameter continuous group of infinite order. This group is denoted by $SO(2)$. In this group the angle of rotation in the product of two rotations $R_2(\theta_1)$ and $R_2(\theta_2)$ is the rotation $R_2(\theta_3)$ such that $\theta_3 = f(\theta_1, \theta_2) = \theta_1 + \theta_2$. Where θ_3 is the continuous function of the parameters θ_1 and θ_2. In general for a continuous group, the product of $G(a)$ and $G(b)$ is $G(c)$ such that $C = f(a, b)$ is a continuous function of the parameter has to be replaced by a set of x-parameters.

In a continuous group if $f(a, b) = c$ is an analytic function of a and b, and $G(\bar{a})$ is the inverse of $G(a)$, then $\bar{a}$ is also an analytic function of a. This group is called a *lie group*. The group of rotations about the z-axis is an example of one parameter lie group. In both the examples discussed above the parameters corresponding to the identity element have zero value.

For generating a lie group, the main requirement is that the identity element has the value zero. This can always be easily managed. Here, we take a simple example of generating a lie group in which we shall assume that the identity element has zero values of all the r-parameters a_i of a r-parameter group.

Consider an infinitesimal rotation around the z-axis through an angle $d\theta$. Under this transformation the coordinate (x, y) of a point in the plane XY change to $(X'Y')$ such that

$$\left.\begin{array}{l} X' = x + dx = x\cos d\theta + y\sin d\theta \approx x + yd\theta \\ Y' = y + dy = -x\sin d\theta + y\cos d\theta \approx -xd\theta + y \end{array}\right\} \tag{11.75}$$

The change $d\phi$ under infinitesimal transformation is given by

$$d\phi = \frac{\partial\phi}{\partial x}\frac{dx}{d\theta} + \frac{\partial\phi}{\partial y}\frac{dy}{d\theta} \tag{11.76}$$

Using Eq. (11.75) we get

$$d\phi = d\theta\left(y\frac{\partial}{\partial x} - x\frac{\partial}{\partial y}\right), \quad \phi = (-id\theta)X\phi \tag{11.77}$$

where $X = \dfrac{1}{i}\left(x\dfrac{\partial}{\partial y} - y\dfrac{\partial}{\partial x}\right)$ is called the generator of the lie group.

In general for the r-parameter lie group of transformations

$$X_r = f(x_1, x_2,...,x_n, a_1, a_2,...,a_r)$$

In an n-dimensional space, the changed ϕ in a function under an infinitesimal transformation is given by

$$d\phi = \sum_{k=1}^{r}(-ida_k)X_k\phi \tag{11.78}$$

Here X_k $(k = 1, 2,...,r)$ are known as generators of the x-parameter lie group. From Eq. (11.78), we get

$$X_k\phi = -\frac{1}{i}\frac{\partial\phi}{\partial a_k} \tag{11.79}$$

For $SO(2)$ group if $R_z(\theta)$ denotes the operator in two-dimensional space (x, y) representing rotation through an angle θ about z-axis then

$$d\phi = R_z(d\theta)\,\phi - \phi \tag{11.80}$$

and substituting from Eq. (11.77), we get

$$R_z(d\theta)\,\phi = (1 - id\theta X)\phi$$

or

$$R_z(d\theta) = 1 - id\theta X \tag{11.81}$$

Equation (11.81) is an alternative definition of the generators. In general for lie group of r-parameter if the group elements are denoted by $R(a_1, a_2,...,a_r)$

$$R(da_1, da_2,...da_r) = 1 - i\sum_{k=1}^{r}da_k X_k \tag{11.82}$$

and

$$R_z(\theta + d\theta) = R_z(\theta)R_z(d\theta) = R_z(\theta) - id\theta R_z(\theta)X$$

or
$$R_z(\theta + d\theta) = R_z(\theta) + d\theta \frac{R_z}{d\theta}$$

Comparing these two equations, we get

$$\frac{R_z}{d\theta} = -iR_z(\theta)X \tag{11.83}$$

Solution of this equation is

$$R_z(\theta) = e^{-iX\theta} \tag{11.84}$$

under the condition $R_z(0) = 1$

We have called X as generator because it generates all rotations about the z-axis using the operator $e^{iX\theta}$.

For the group SO(2), the generator

$$X = -\frac{1}{i}\left(x\frac{\partial}{\partial y} - y\frac{\partial}{\partial x}\right)$$

is component of angular momentum in z-direction (J_z) in units of $\frac{h}{2\pi}(\bar{h})$

$\therefore$
$$R_z(\theta) = \exp(-i J_z \theta) \tag{11.85}$$

The generators X_i of a lie group have the property that the commutator of generators X_i and X_j is a linear combination of the generators, i.e.

$$[X_i, X_j] = \sum_k C_{ij}^k X_k \tag{11.86}$$

The constants C_{ij}^k are known as structure constants of the lie group and the set of commutation rules [Eq. (11.86)] is known as *lie algebra*.

11.15. THREE–DIMENSIONAL ROTATION GROUP SO(3)

A group denoted $SO(3)$ is formed by a set of all rotations in three-dimensional space. The property of this group leaves the length of a vector $\gamma = (x, y, z)$ invariant after transformation. In general, the group of all orthogonal transformations in three-dimensional space which leaves the length of the vector γ invariant is denoted by $O(3) \cdot SO(3)$ is a subgroup of $O(3)$.

In a three-dimensional space, let an orthogonal transformation matrix represented by A, then

$$A\tilde{A} = 1 \tag{11.87}$$

or
$$\det A \cdot \det \tilde{A} = 1$$

Since $\det \tilde{A} = \det A$, we have
$$(\det A)^2 = 1 \quad \text{or} \quad \det A = \pm 1$$

A pure rotation corresponds to $\det A = +1$

Thus a group of all orthogonal transformations in three-dimensional a space with determinant $+1$ is defined as the $SO(3)$ group. On the other hand, orthogonal transformations with determinant -1 involves either pure inversion or a combination of inversion and rotation.

To construct a rotational group, the parameters for an arbitrary rotation in three-dimension space are taken in Euler angles α, β, γ, where α is the angle of rotation about z-axis, β is the angle rotation about y-axis and γ is the angle of rotation about z-axis. The rotations are shown in Fig. 11.4.

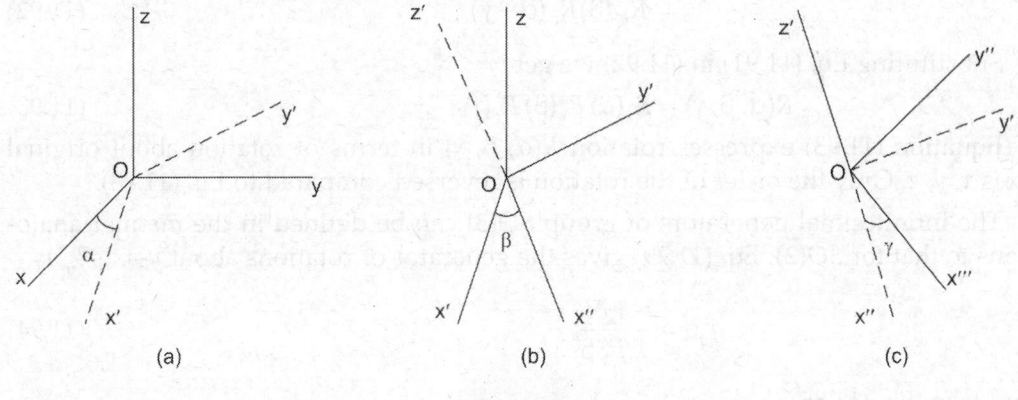

(a)   (b)   (c)

Fig. 11.4

All these rotations are anticlockwise. An arbitrary rotation $R(\alpha, \beta, \gamma)$ characterised by the Euler angles α, β, γ can be written as

$$R(\alpha, \beta, \gamma) = R_{z'}(\gamma)R_{y'}(\beta)R_x(\alpha) \tag{11.88}$$

For convenience, we will take rotation $R(\alpha, \beta, \gamma)$ about the original axis x, y, z instead of the new positions of axes y' and z'.

Let us consider axes A, B and C. A and B are at angle θ to each other and C is perpendicular to both A and B (Fig. 11.5). Let us first rotate A about C by an angle θ (anticlockwise). This coincides A with B. Then we rotate about B by an angle ϕ, and finally we rotate about C by an angle θ (clockwise). This rotation will bring A back to its original position. This whole operation is equivalent to rotation through angle ϕ about axis A. We can write all this transformations by the relation

$$R_B(\phi) = R_C^{-1}(\theta)R_B(\phi)R_c(\theta) \tag{11.89}$$

This gives an important result that rotation about different axes by the same angle belong to the same class.

Using Eq. (11.89) for Fig. 11.4a, we can write

$$R_z(\gamma) = R_{y'}^{-1}(\beta)R_{z'}(\gamma)R_{y'}(\beta)$$

or $$R_{z'}(\gamma) = R_{y'}(\beta)R_z(\gamma)R_{y'}^{-1}(\beta) \tag{11.90}$$

From Fig. 11.4a, we get

$$R_y(\beta) = R_z^{-1}(\alpha)R_{y'}(\beta)R_z(\alpha)$$

or $$R_{y'}(\beta) = R_z(\alpha)R_y(\beta)R_z^{-1}(\alpha) \tag{11.91}$$

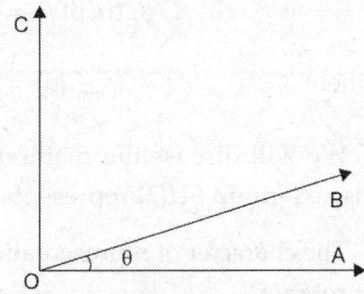

Fig. 11.5

Equations (11.88) and (11.90) give

$$R(\alpha, \beta, \gamma) = \left[R_{y'}(\beta) R_z(\gamma) R_{y'}^{-1}(\beta) \right] R_{y'}(\beta) R_z(\alpha)$$

$$= R_{y'}(\beta) R_z(\gamma) R_z(\alpha)$$

$$= R_{y'}(\beta) R_z(\alpha + \gamma) \tag{11.92}$$

Substituting Eq. (11.91) in (11.92) we get

$$R(\alpha, \beta, \gamma) = R_z(\alpha) R_y(\beta) R_z(\gamma) \tag{11.93}$$

Equation (11.93) expresses rotation $R(\alpha, \beta, \gamma)$ in terms of rotation about original axes x, y, z. Only the order of the rotation is reversed compared to Eq. (11.88).

The infinitesimal generators of group $SO(3)$ can be defined in the manner analogous to that for $SO(2)$. Eq. (11.79) gives the generator of rotations about y-axis J_y as

$$J_y \phi = -\frac{1}{i} \frac{\partial \phi}{\partial \beta} \tag{11.94}$$

and from Eq. (11.85).

$$R_z(\beta) = \exp(-i\beta J_y) \tag{11.95}$$

The irreducible representation $D^{(J)}_{m'm}(\alpha, \beta, \gamma)$ of SO(3) can be obtained by operating $R(\alpha, \beta, \gamma)$ on the set of eigen functions ψ_m^J of J^2 and J_z. Hence

$$J^2 \psi_m^J = f(J+1) \psi_m^J \tag{11.96}$$

and

$$J_z \psi_m^J = m \psi_m^J$$

Thus

$$R(\alpha, \beta, \gamma) \psi_m^J = \sum_m \psi_{m'}^J \cdot D^{(J)}_{m'm}(\alpha, \beta, \gamma) \tag{11.97}$$

where

$$D^{(J)}_{m'm}(\alpha, \beta, \gamma) = \left[\psi_{m'}^{(J)} R(\alpha, \beta, \gamma) \psi_m^J \right] \tag{11.98}$$

But

$$R(\alpha, \beta, \gamma) = R_z(\alpha) R_y(\beta) R_z(\gamma)$$

$$= \exp(-i\alpha J_z) \exp(-i\beta J_y) \exp(-i\gamma J_z) \tag{11.99}$$

Substituting Eq. (11.99) in (11.98) and using Eq. (11.96), we get

$$D^{(J)}_{m'm}(\alpha, \beta, \gamma) = e^{-i\alpha m'} \left(\psi_{m'}^J e^{-i\beta J_y} \psi_m^J \right) e^{-i\gamma m} = e^{i\alpha m'} d^J_{m'm}(\beta) e^{-i\gamma m} \tag{11.100}$$

where

$$d^J_{mm'}(\beta) = \left[\psi_{m'}^J \exp(-i\beta J_y) \psi_m^J \right] \tag{11.101}$$

We will discuss the method to evaluate $d^J_{mm'}(\beta)$ in next section, where we will discuss group $SU(2)$ representation.

The character of representation $D^{(J)}_{m'm}(\alpha, \beta, \gamma)$ can be obtained by using the property of rotation.

The rotation by the same angle about different axes belong to the same class. Hence the character depends only on the angle of rotation and not on the axis about which the rotation takes place.

The matrix corresponding to the rotation by an angle α about z-axis is

$$F_{m'm}^{(J)}(\alpha,0,0) = e^{-i\alpha m}\delta_{mm'}$$

The character corresponding to this class is

$$X^J(\alpha) = \sum_{m=-j}^{+J} e^{-i\alpha m} = \frac{\sin\left(J+\dfrac{1}{2}\right)\alpha}{\sin\alpha/2} \qquad (11.102)$$

11.16. SPECIAL UNITARY GROUPS SU(2) AND SU(3)

In an n-dimensional vector space, a set of all unitary transformations form a group $u(n)$ called unitary group. This group is isomorphic to the group of all unitary matrices which form a group under matrix multiplication. $SU(2)$ is the group of all (2×2) unitary matrices with determinant $+1$ (special unitary group). Similarly a set of all (3×3) unitary matrices with determinant $+1$ form the group $SU(3)$ under matrix with multiplication. In general, $SU(n)$ the special unitary group is a subgroup of $u(n)$.

A general (2×2) complex matrix

$$u = \begin{pmatrix} p & r \\ q & s \end{pmatrix}$$

can be specified by eight real parameters. Since u is unitary, we have

$$u^+u = uu^+ = 1$$

The gives the following conditions:

$$pp^* + rr^* = 1 \qquad (11.103)$$
$$qq^+ + ss^* = 1 \qquad (11.104)$$
$$pq^* + rs^* = 0 \qquad (11.105)$$
$$p^*q + r^*s = 0 \qquad (11.106)$$

Also $\det u = ps - rq = 1 \qquad (11.107)$

Equation (11.106) gives

$$\frac{s}{q} = -\frac{p^*}{r^*} \qquad (11.108)$$

Substituting Eq. (11.108) in (11.107), we get

$$-(pp^* + rr^*)\frac{q}{r} = 1$$

But $pp^* + rr^* = 1$ \hfill [*from* Eq. (11.103)

$$\therefore \qquad q = -r^* \qquad (11.109)$$

and from Eq. (11.108) $\qquad s = p^*$

The number of real parameters needed to specify an element of the group is thus 3 which is the same as for the group $SO(3)$. The group $SU(2)$ and $SU(3)$ are closely related. There is a powerful homomorphism between $SU(2)$ and $SO(3)$.

The Homomorphism between SU(2) and SO(3)

Consider a vector $\gamma = (x, y, z)$ is an ordinary three-dimension space and a (2×2) matrix.

$$A = \begin{pmatrix} z & x-iy \\ x+iy & -z \end{pmatrix} \tag{11.110}$$

Let us make a unitary transformation uAu^{-1}, where

$$u = \begin{pmatrix} p & q \\ -q^* & p^* \end{pmatrix} \text{ such that } pp^* + qq^* = 1$$

By this transformation we get another matrix A'

$\therefore$ $$A' = uAu^{-1} \tag{11.111}$$

A is Hermitian and traceless. In a unitary transformation these properties are preserved. Therefore A' is also Hermitian and traceless. We can write

$$A' = \begin{pmatrix} z' & x'-iy' \\ x'+iy' & -z' \end{pmatrix} \tag{11.112}$$

Using Eq. (11.111), we can express x', y', z' in terms of x, y, z. Now

$$\det A' = \det (uAu^{-1}) = \det A$$

or $$-(x'^2 + y'^2 + z'^2) = -(x^2 + y^2 + z^2) \tag{11.113}$$

Thus associated with a 2×2 unitary matrix u, there is a (3×3) matrix $R(u)$ which transforms vector $\gamma = (x, y, z)$ to vector $\gamma' = (x', y', z')$ such that the length of the vector is unchanged. $R(u)$ is therefore an orthogonal matrix and determinant of $R(u)$ is either $+1$ or -1. If

$$u = \begin{pmatrix} 1 & 0 \\ 0 & 1 \end{pmatrix}, \quad A' = A \text{ and } R(u) = \begin{pmatrix} 1 & 0 & 0 \\ 0 & 1 & 0 \\ 0 & 0 & 1 \end{pmatrix}$$

having det $R(u) = +1$

Since all such matrices are to be generated by a continuous variation of the parameters p and q, the determinant of $R(u)$ can not change discontinuously to -1. Determinant of $R(u)$ therefore is $+1$ and $R(u)$ is an element of $SO(3)$.

Now we will prove that all rotations, some elements of $SU(2)$ can be associated. As earlier stated an arbitrary rotation can be specified by the Euler angles (α, β, γ). Using [Eqs (11.111) and (11.112)], we can easily prove

1. Matrix $u = \begin{pmatrix} e^{i\alpha/2} & 0 \\ 0 & e^{-i\alpha/2} \end{pmatrix}$ corresponds to rotation α about z-axis.

2. Matrix $u = \begin{pmatrix} \cos\beta/2 & \sin\beta/2 \\ -\sin\beta/2 & \cos\beta/2 \end{pmatrix}$ corresponds to a rotation of angle β about the

y-axis therefore the unitary matrix

$$u(\alpha, \beta, \gamma) = \begin{pmatrix} e^{i\gamma/2} & 0 \\ 0 & e^{-i\gamma/2} \end{pmatrix} \begin{pmatrix} \cos\beta/2 & \sin\beta/2 \\ -\sin\beta/2 & \cos\beta/2 \end{pmatrix} \begin{pmatrix} e^{+i\alpha/2} & 0 \\ 0 & e^{-i\alpha/2} \end{pmatrix}$$

corresponds to $R(\alpha, \beta, \gamma)$. Direct multiplication gives

$$u(\alpha, \beta, \gamma) = \begin{pmatrix} \cos\beta/2\, e^{i(\alpha+\gamma)/2} & \sin\beta/2\, e^{i(\gamma-\alpha)/2} \\ -\sin\beta/2\, e^{-i(\gamma-\alpha)/2} & \cos\beta/2\, e^{-i(\alpha+\gamma)/2} \end{pmatrix}$$

$$\rightarrow R(\alpha, \beta, \gamma) \tag{11.114}$$

Equation (11.114) shows that two unitary matrices $u(\alpha, \beta, \gamma)$ and $u(\alpha, \beta + 2\pi, \gamma)$ though they are not identical due to presence of half angles give same rotation $R(\alpha, \beta, \gamma)$ because rotation $R(\alpha, \beta, \gamma)$ and $R(\alpha, \beta + 2\pi, \gamma)$ are identical. Thus there is two to one correspondence between the matrices $u(\alpha, \beta, \gamma)$ and $R(\alpha, \beta, \gamma)$. The two successive unitary transformations $u(\alpha_1, \beta_1, \gamma_1)\, u(\alpha_2, \beta_2, \gamma_2)$ correspond to $R(\alpha_1, \beta_1, \gamma_1)\, R(\alpha_2, \beta_2, \gamma_2)$. This gives complete homomorphism between $SU(3)$ and $SO(3)$.

11.17. THE IRREDUCIBLE REPRESENTATION OF SU(2)

Consider a vector in a complex two-dimensional vector space. Let us denote it by

$$\begin{pmatrix} \alpha_1 \\ \alpha_2 \end{pmatrix}$$

Under $SU(2)$ transformation it gets transformed to

$$\begin{pmatrix} \alpha_1' \\ \alpha_2' \end{pmatrix}$$

The most general form of the matrix of transformation is

$$\begin{pmatrix} p & q \\ -q^* & p^* \end{pmatrix} \text{ with } pp^* + qq^* = 1$$

Then
$$\begin{pmatrix} \alpha_1' \\ \alpha_2' \end{pmatrix} = \begin{pmatrix} p & q \\ -q^* & p^* \end{pmatrix} \begin{pmatrix} \alpha_1 \\ \alpha_2 \end{pmatrix} \tag{11.114a}$$

Then
$$\left. \begin{array}{l} \alpha_1' = p\alpha_1 + q\alpha_2 \\ \alpha_2' = -q^*\alpha_1 + p^*\alpha_2 \end{array} \right\} \tag{11.114b}$$

We now define a set of functions

$$f_m^{(j)} = \frac{\alpha_1^{j+m}\, \alpha_2^{j-m}}{[(j+m)!(j-m)!]^{1/2}} \tag{11.114c}$$

where j either a positive integer (including zero) or a half integral (i.e. half of an odd integer). For a given j, m can take $(2j + 1)$ values as

$$-j, -(j-1), -(j-2),..., j - 1j.$$

It is now easy to prove that as a result of transformation (11.114a), the $(2j + 1)$ function [Eq. (11.115)] for a given value j, get transformed into linear combinations of themselves, thus

$$f_m^{(j)} \to f_m^{(j)'} = \frac{(\alpha_1')^{j+m}(\alpha_2')^{j-m}}{[(j+m)!(j-m)!]^{1/2}}$$

$$= \frac{(p\alpha_1 + q\alpha_2)^{j+m}(-q^*\alpha_1 + p^*\alpha_2)^{j-m}}{[(j+m)!(j-m)!]^{1/2}} \tag{11.114d}$$

Expanding $(p\alpha_1 + q\alpha_2)^{j+m}$ and $(-q^*\alpha_1 + p^*\alpha_2)^{j-m}$ using binomial theorem, we get

$$f_m^{(j)'} = \sum_{k=0} \sum_{l=0} \frac{[(j+m)!(j-m)!]^{1/2} (\beta\alpha_1)^{j+m-k}}{k!(k+m-k)! \, l \,!(j-m-l)!}$$

$$= (q\alpha_2)^k (-q^*\alpha_1)^{l-m-l}(p^*\alpha_2)^l \tag{11.114e}$$

The upper limits of the summation should be such that neither $(j + m - k)$ nor $(j - m - k)$ become negative.

If we define $m' = j - k - l$, (11.114d) reduces to

$$f_m^{(j)'} = \sum_{m=-j}^{+j} \frac{\alpha_1^{j+m'} \alpha_2^{j-m'}}{[(j+m')!(j-m')!]^{1/2}}$$

or

$$f_m^{(j)'} = \sum_{k=0} \frac{[(j+m)!(j-m)!(j+m')!(j-m')!]}{k!(j+m-k)!(j-m'-k)!(m'-m+k)!}$$

$$= (p)^{j+m-k}(p^*)^{j-m'-k}(q)^k(-q^*)^{m'-m+k}$$

$$= \sum_{m'=-j}^{+j} f_{m'}^{(j)} D_{m'm}^{(j)}(p,q) \tag{11.115}$$

where

$$D_{m'm}^{(j)}(p,q) = \sum_{k=0} \frac{[(j+m)!(j-m)!(j+m')!(j-m')!]^{1/2}}{k!(j+m-k)!(j-m'-k)!(m'-m+k)!}$$

$$= (p)^{j-m-k}(p^*)^{j-m'-k}(q)^k(-q^*)^{m'-m+k} \tag{11.116}$$

The matrices $D_{m'm}^j(p,q)$ therefore form a $(2j + 1)$ dimensional representation of $SU(2)$. This representation is irreducible.

To prove irreducibility of this representation let us consider a $(2j + 1) \times (2j + 1)$ matrix A which commutes with all the matrices $D^{(j)}(p, q)$, i.e.

$$AD^{(j)}(p, q) = D^{(j)}(p, q)A \tag{11.117}$$

The equation should be true for all values of p and q. If $q = 0$ Eq. (11.114b) reduces to

$$\alpha'_1 = p\alpha_1 \quad \text{and} \quad \alpha'_2 = p^*\alpha_2$$

and

$$|p|^2 + |q|^2 = |p|^2 = 1. \text{ Hence we can write}$$

$$p = e^{-i\alpha/2}$$

Equation (11.114c) gives

$$f_m^{(j)} = f_m^{(j)'} = e^{i\alpha m} f_m^{(j)} \tag{11.118}$$

Comparing it with Eq. (11.116), we get

$$D_{m'm}^{(j)}(p,0) = e^{i\alpha m}\delta_{mm'} \tag{11.119}$$

Equating rs^{th} element of Eq. (11.117a) on both sides, we get

$$\sum_k A_{rk}\delta_{ks}\, e^{i\alpha s} = \sum_k \delta_{rk} e^{i\alpha r} A_{kq}$$

or

$$A_{rs}e^{i\alpha s} = A_{rs}e^{i\alpha r} \tag{11.120}$$

This is true for all r and s, only if A is diagonal, i.e. $A_{rs} = A_r\delta_{rs}$

Now we choose another matrix $D^{(j)}(p, q)$ with $q \neq 0$ since A commutes with all matrices $D^{(j)}(p, q)$ and if we equate rs^{th} elements of Eq. (11.117a), we get

$$A_r D_{rs}^{(j)}(pq) = D_{rs}^{(j)}(p,q) A_s \tag{11.121}$$

Since all the elements of $D^{(j)}(p, q)$ are not equal to zero, we must have $A_r = A_s$ for all r and s. Thus, A is a multiple of the unit matrix.

Here we have proved that matrix A which commutes with all matrices $D^{(j)}(p, q)$ is a multiple of unit matrix. Now using converse of Schur's lemma we can conclude that the matrices $D^{(j)}(p, q)$ form an irreducible representation of $SU(2)$ of dimension $(2j + 1)$.

11.18. REPRESENTATION OF SO(3) FROM THOSE OF SU(2)

We have seen homomorphism between $SU(2)$ and $SO(3)$, as such the irreducible representations of $SU(2)$ are also the irreducible representations of $SO(3)$. We have also proved that unitary matrix [Eq. (11.114)].

$$u(\alpha, \beta, \gamma) = \begin{pmatrix} \cos\beta/2\, e^{i(\alpha+\gamma)/2} & \sin\beta/2\, e^{i(\gamma-\alpha)/2} \\ -\sin\beta/2\, e^{-i(\gamma-\alpha)/2} & \cos\beta/2\, e^{-i(\alpha+\gamma)/2} \end{pmatrix}$$

corresponds to rotation of the axes by Euler angles α, β, γ. Informing representations, we adopted the convention that the operators operate on functions instead of on the coordinate axes. The convention is equivalent to reverse rotation, i.e.

$$R^{-1}(\alpha, \beta, \gamma) = R(-\gamma, -\beta, -\alpha)$$

and the corresponding unitary matrix can be written as

$$u(-\gamma, -\beta, -\alpha) = \begin{pmatrix} \cos\beta/2\, e^{-i(\alpha+\gamma)/2} & -\sin\beta/2\, e^{-i(\alpha-\gamma)/2} \\ -\sin\beta/2\, e^{i(\alpha-\gamma)/2} & \cos\beta/2\, e^{i(\alpha+\gamma)/2} \end{pmatrix}$$

Comparing with general form of (2×2) unitary matrix with determinant 1, we get

$$p = \cos\beta/2 \, e^{-i(\alpha-\gamma)/2} \quad \text{and} \quad q = -\sin\beta/2 \, e^{-i(\alpha-\gamma)/2}$$

and Eq. (11.117) gives a $(2j + 1)$ dimensional irreducible representation of $SO(3)$

$$D^j_{m,m'}(\alpha,\beta,\gamma) = e^{-im'\alpha} e^{-im\gamma}$$

$$= \sum \frac{[(j+m)!(j-m)!(j-m')!(j+m')!]^{1/2}}{k!(j+m-k)!(j-m'+k)!(m'-m+k)!} \times (\cos\beta/2)^{2j+m-m'-2k}(\sin\beta/2)^{m'-m+2k}$$

$$\tag{11.122}$$

If we compare this equation with Eq. (11.100) we can find expression for $d^{(j)}_{m'm}(\beta)$ is

$$D^{(J)}_{m'm}(\alpha,\beta+2\pi,\gamma) = (-1)^{2J} D^{(J)}_{m'm}(\alpha,\beta,\gamma) \tag{11.123}$$

Here if j has only integral values as in case of orbital angular momentum,

then $\qquad D^{(J)}_{m'm}(\alpha,\beta+2\pi,\gamma) = D^J_{m'm}(\alpha,\beta,\gamma) \tag{11.124}$

and the representation is single valued. But if j is half integral, then

$$D^{(J)}_{m'm}(\alpha,\beta+2\pi,\gamma) = -D^J_{m'm}(\alpha,\beta,\gamma) \tag{11.125}$$

and $\qquad D^{(J)}_{m'm}(\alpha,\beta+2\pi,\gamma) = D^{(J)}_{m'm}(\alpha,\beta,\gamma) \tag{11.126}$

In this case the representation is double valued.

11.19. GENERATORS OF UNITARY GROUPS

We need $2n^2$ real parameters to specify a unitary matrix $u(n)$. The unitary condition gives n^2 representations. This reduces the number of independent real parameters to n^2. Hence there are n^2 generators of $u(n)$. If H is Hermitian then any unitary matrix u can be expressed as

$$u = \exp(iH)$$

and a general $n \times n$ matrix can be written as

$$u(a_1, a_2, a_3, ..., a_{n^2}) = \exp\left[-i\sum_{r=1}^{n^2} a_r H_r\right] \tag{11.126a}$$

where a_r are n^2 real parameters and H_r are n^2 linearly independent Hermitian matrices of order $(n \times n)$. If we consider infinitesimal parameters then we have

$$u(da_1, da_2, ..., da_{n^2}) = 1 - i\sum_{r=1}^{n^2} da_r H_r \tag{11.126b}$$

Comparing Equation (11.126b) with Eq. (11.82) we can identify n^2 Hermitian matrices H_r as the generators of $u(n)$. We can get generators of $SU(n)$ by imposing the condition that

$$\det u = +1.$$

Now from Eq. (11.126a)

$$\det u(a_1, a_2, a_3, ..., a_{n^2}) = \exp\left[-i \operatorname{trace} \sum_{r=1}^{n^2} a_r H_r\right]$$

$$= \exp\left[-i \sum_{r=1}^{n^2} a_r \operatorname{trace} H_r\right] \qquad (11.126c)$$

The number of independent parameters are reduced to $(n^2 - 1)$ by imposing the condition that $\det u(a_1, a_2, a_3, ..., a_{n^2}) = +1$. If we put $a_{n^2} = 0$, then

$$\sum_{r=1}^{n^2-1} a_r \operatorname{trace} H_r = 0 \qquad (11.126d)$$

Since, we can choose parameters a_r such that only one of them is nonzero and the rest are zeros, it follows that

$$\operatorname{trace} H_r = 0 \quad \text{for} \quad r = 1, 2, 3, ..., (n^2 - 1)$$

Hence we can take any set of traces, linearly independent $(n \times n)$ Hermitian matrices, as the generators of $SU(n)$. The generators are not unique. To get the generators of $U(n)$ we have to add to the set of $H_r[r = 1, 2, 3, ..., (n^2 - 1)]$, the $(n \times n)$ unit matrix, for the group $SU(2)$ the convenient choice of the generators is

$$H_1 = \begin{pmatrix} 0 & 1 \\ 1 & 0 \end{pmatrix}, H_2 = \begin{pmatrix} 0 & -i \\ i & 0 \end{pmatrix}, H_3 = \begin{pmatrix} 1 & 0 \\ 0 & -1 \end{pmatrix} \qquad (11.127)$$

These are Pauli's spin matrices and are frequently used in quantum mechanics.

11.20. GROUP SU(3)

A special unitary group $SU(3)$ is a set of (3×3) unitary matrices with determinant $+1$ which forms a group under matrix multiplication.

Number of generators for this group $= 3^2 - 1 = 8$

The following traceless, (3×3) Hermitian matrices is a convenient choice

$$\lambda_1 = \begin{pmatrix} 0 & 1 & 0 \\ 1 & 0 & 0 \\ 0 & 0 & 0 \end{pmatrix}, \lambda_2 = \begin{pmatrix} 0 & i & 0 \\ i & 0 & 0 \\ 0 & 0 & 0 \end{pmatrix}, \lambda_3 = \begin{pmatrix} 1 & 0 & 0 \\ 0 & -1 & 0 \\ 0 & 0 & 0 \end{pmatrix}$$

$$\lambda_4 = \begin{pmatrix} 0 & 0 & 1 \\ 0 & 0 & 0 \\ 1 & 0 & 0 \end{pmatrix}, \lambda_5 = \begin{pmatrix} 0 & 0 & -i \\ 0 & 0 & 0 \\ i & 0 & 0 \end{pmatrix}, \lambda_6 = \begin{pmatrix} 0 & 0 & 0 \\ 0 & 0 & 1 \\ 0 & 1 & 0 \end{pmatrix}$$

$$\lambda_7 = \begin{pmatrix} 0 & 0 & 0 \\ 0 & 0 & -i \\ 0 & i & 0 \end{pmatrix}, \lambda_8 = \frac{1}{\sqrt{3}} \begin{pmatrix} 1 & 0 & 0 \\ 0 & 1 & 0 \\ 0 & 0 & -2 \end{pmatrix} \qquad (11.128)$$

We can obtain commutation relations satisfied by the generators by direct matrix multiplication. They have the form

$$[\lambda_i, \lambda_j] = 2i \sum_k C_{ij}^k \lambda_k \tag{11.129}$$

C_{ij}^k are called structure constants and are characteristics of the group and are independent of the particular choice of the generators.

Some of the structure constants for the group $SU(3)$ are given here

$$C_{12}^3 = 1, \; C_{14}^7 = \frac{1}{2}, \; C_{15}^6 = -\frac{1}{2}, \; C_{24}^6 = \frac{1}{2}, \; C_{23}^7 = \frac{1}{2}, \; C_{34}^5 = \frac{1}{2}$$

$$C_{36}^7 = -\frac{1}{2}, \; C_{45}^8 = \frac{\sqrt{3}}{2}, \; C_{67}^8 = \frac{\sqrt{3}}{2} \tag{11.130}$$

We can use the property

$$C_{ij}^k = -C_{ji}^k$$

to obtain other nonzero structure constants. This property can be proved using Eq. (11.129). The rest of the constants are zero.

The rank of lie group is defined as the number of mutually commuting generators such operators can be constructed from the generators. For the group $SO(3)$, the generators are J_x, J_y, X_z. No two of them commute and the rank of lie group is one. It has operator

$$J^2 = J_x^2 + J_y^2 + J_z^2$$

This is called Casimir operator. In fact Casimir operator gives rank of the group, because it is a general result that Casimir operators are equal to the rank of the group.

The eigen values of j^2 for group $SO(3)$ and $j(j + 1)$, where j is a positive integer or a half-integer. The irreducible representation of $SO(3)$ group can be labelled by i, in the same way as the eigen values of Casimir operator are labelled, i.e.

$$\Lambda = \sum_{i=1}^{8} \lambda_i^2 \tag{11.131}$$

$SU(3)$ group has rank 2 and can be characterised by two non-negative integers (m, n). Group $SU(3)$ has another Casimir operator which is formed of trilinear combination of the generators. The irreducible representations of $SU(3)$ group can be labelled by (m, n). The dimension of the representation is given by

$$d = \frac{(1+m)(1+n)(2+m+n)}{2} \tag{11.132}$$

The lowest dimension is 1 corresponding to $m = 0$, $n = 0$. $(1, 0)$ and $(0, 1)$ are 2 three-dimensional representations. It is a convention that the representations are denoted by their dimensions. In this notation $(1, 0) = 3$, $(0, 1) = 3$, $(2, 0) = 6$, $(1, 1) = 8$, $(3, 0) = 10$ etc.

11.21. SOME APPLICATIONS OF GROUP THEORY IN PHYSICS

Classification of States

We have already discussed the use of group theory in quantum mechanics in Section 11.13, it is shown that the eigen functions of an n-fold degenerate level form an n-dimensional irreducible representation of the symmetry group of the Hamiltonian assuming there is no accidental degeneracies. We can easily infer the possible degeneracies of the states from the knowledge of the symmetry group of the Hamiltonian.

To elaborate this we take the case of the classification of the energy eigen functions of the hydrogen atom. The Hamiltonian is rotationally invariant and the eigen states can be classified according to the irreducible representation of $SO(3)$. If we ignore spin, these representations can be characterised by integer l, the dimensions of the corresponding representation being $(2l + 1)$. The wave functions can then be labelled as ψ_{lm}, where m can take $(2l + 1)$ values $(-l, -l + 1, -l + 2,...., l -1)$. The eigen functions of any system having spherical symmetry can be labelled in the above manner.

We can also take the system whose symmetry group is the group proper rotation which leaves the cube invariant (the group is called O group). The group consists of 24 elements and has five classes hence there are five irreducible representations. They are usually labelled as A_1, A_2, I, T_1, T_2. Here A_1 and A_2 are one-dimensional, I is two-dimensional and T_1, T_2 are three-dimensional representations of the group. An eigen state of the system, if it is doubly degenerated, then the wave functions for these states must belong to the two-dimensional irreducible representation of group O. Then we can choose two orthogonal functions $\psi_1^{(E)}$ and $\psi_2^{(E)}$ to label the states. A state (level) with no degeneracy should belong to either A_1 or A_2 and the corresponding eigen functions can be then labelled as $\psi^{(A_1)}$ and $\psi^{(A_2)}$. In general, the wave function belonging to irreducible representation Γ would be denoted by $\psi_\mu^{(\Gamma)}$. A group element G operating on $\psi_\mu^{(\Gamma)}$ will give a linear combination of all the degenerate eigen functions belonging to the irreducible representation Γ, i.e.

$$G\psi_\mu^{(\Gamma)} = \sum_\Gamma \psi_{\mu'}^{(\Gamma)} D_{\mu\mu'}^{(\Gamma)}(G)$$

The wave function $\psi_\mu^{(\Gamma)}$ is said to transform according to the uth row of the irreducible representation.

Classification of Elementary Particles

In nuclear and elementary particle physics, groups $SU(2)$ and $SU(3)$ have wide applications. The irreducible representations of $SU(2)$ and $SU(3)$ are used to classify the elementary particles in various multiplets. The proton (p) and neutron (n) have almost the same mass and are identical so far as nuclear forces are concerned. The small difference in their mass is due to the electromagnetic interaction because proton has charge where as neutron is a neutral particle. If we neglect this difference in mass, the proton and neutron can be treated as two degenerate energy states forming an irreducible representation of the group $SU(2)$. A transformation from a proton to a neutron or vice versa can be thought of as a rotation in hypothetical space called the *iso-spin* in space. Just as rotation in ordinary three-dimensional space preserves the length r of the vector r, the rotation in iso-spin space conserves a quantity called iso-spin (I). The proton and the neutron differ in their z-component of iso-spin (I_z). According to this picture they form an iso-spin doublet with $I = 1/2$ and $I_z = +1/2$ and $-1/2$.

Higher dimensional representations of the $SU(2)$ group corresponding to iso-spin can be formed by taking the direct product of the (2×2) representations. This representation is reducible and breaks up into an irreducible representation of three-dimensional and a one-dimensional representation (this is equivalent to saying that coupling two angular momenta each equal to $1/2$, one get angular momenta 1 and 0). In mathematical notation this can be written as

$$2 \oplus 2 = 3 \oplus 1$$

Another example is of pions. Three pions $\pi^{\pm}$ and π^0 correspond to three representations. They have almost equal mass and are identical so far as strong interaction is concerned. Thus, they also form an iso-spin triplet with $I = 1$ and $I_z = \pm 1, 0$.

The necessity of additional quantum number called strangeness was felt during the study of production and decay of some particles like Λ^0, K^0 etc. Hence another particle λ was added to iso-spin doublet ($n-p$) λ is an iso-spin singlet ($I = 0$) and has strangeness $S = -1$. From this it was predicted that these particles would have the same rest mass if they had only strong interaction. This degenerate triplet then forms the basis for a (3×3) representation of the group $SU(3)$. As in case of group $SU(2)$ higher dimensional representations of $SU(3)$ can be constructed by considering direct products of (3×3) representations and breaking them into irreducible. This in mathematical notation can be written as

$$3 \oplus 3 = 6 \oplus 3$$

The attempt to group together the known elementary particles into multiplets of 6 and 3 failed.

Unlike $SU(2)$ group, the group $SU(3)$ has two inequivalent ($3 \times \bar{3}$) irreducible representations $3 \equiv (1, 0)$ and $\bar{3} \equiv (0, 1)$. Taking the direct product of 3, and $\bar{3}$ and breaking it into irreducible representations, we get

$$3 \otimes 3 = 8 \oplus 1$$

Gell–Mann postulated that the eight mesons $\pi^{\pm}$, π^0, $k^{\pm}$, k^0, R^0 and η^0 belong to 8 representations of $SU(3)$. Similarly by considering the direct product

$$3 \otimes 3 \otimes 3 = 10 \oplus 8 \oplus 8 + 1$$

According to Gell–Mann eight hyperons, n, p, $\Sigma^{\pm}$, Σ^0, E^-, E^0, Λ also belong to 8 representations while the lowest representation of SU(3) correspond to particles actually observed in nature like $p-n$. The 3-dimensional representation of $SU(3)$ corresponds to particles (decomposed by p', n', λ) which have not been observed in nature so far. Gell–Mann proposed that the lowest dimensional representation of $SU(3)$ which is observed in nature in an octet. This is the famous 'eight fold way' of Gell–Mann, the 10 representation of baryons is also observed in nature. To this set of representation, the particles are Δ^-, Δ^0, Δ^+, Δ^{++}, Σ^{*0}, Σ^{*-}, Σ^{*+}, Σ^{*-}, E^{*0}, Q^{-1}

In all the above multiplets (of baryons), the baryon number, intrinsic spin and parity are the same for the members of the multiplets. The only difference is in their wasses, which is the consequence of the breaking of the $SU(3)$ symmetry in nature.

Matrix Elements and Solution Rules

In quantum mechanics, we often come across problems to evaluate matrix elements of an operator V between two states of a system. For example, in perturbation theory (time dependent or time independent), operator V is perturbation and states involved are those of the unperturbed system. The states can be classified according to the irreducible representations of the symmetry group G of the unperturbed Hamiltonian and can be denoted by $\phi_m^{(\mu)}$. In general it will be a sum of terms $V_\rho^{(\lambda)}$ transforming according to the irreducible representations $\Gamma^{(\lambda)}$ of the same symmetry group G, using group theory we can easily infer whether a matrix element of the form $\left(\phi_m^{(\mu)}, v_\rho^{(\lambda)}\phi_n^{(v)}\right)$ vanishes or not in case of time dependent perturbation this shows that the transition is allowed or not and thus gives solution rules.

Since $V_\rho^{(\lambda)}$ and $\phi_n^{(v)}$ transform according to the irreducible representations $\Gamma^{(\lambda)}$ and $\Gamma^{(v)}$, the product $V_\rho^{(\lambda)} \times \phi_n^{(v)}$ transforms according to the direct product representation $\Gamma^{(\lambda)} \times \Gamma^{(v)}$. In general this representation is reducible and can be decomposed into various irreducible representations of group G, i.e.

$$\Gamma^{(\lambda)} \times \Gamma^{(v)} = \sum_p a_p \Gamma^{(p)}$$

Since the basic functions belonging to different irreducible representations of a group are orthogonal, the matrix element $(\phi_m^{(\mu)}, V_p^{(\lambda)}, \phi_n^{(v)})$ will vanish unless the representation $\Gamma^{(\mu)}$ occurs in the decomposition of $\Gamma^{(\lambda)} \times \Gamma^{(v)}$.

To understand all this, we take the example of the selection rules for electric dipole transitions if the symmetry group of the system is O the electric dipole moment operator is qr and under rotation through an angle Q about the z-axis, its components transform

$$\begin{pmatrix} qx' \\ qy' \\ qz' \end{pmatrix} = \begin{pmatrix} \cos\theta & -\sin\theta & 0 \\ \sin\theta & \cos\theta & 0 \\ 0 & 0 & 1 \end{pmatrix} \begin{pmatrix} qx \\ qy \\ qz \end{pmatrix}$$

The character corresponding to rotation θ of the representation according to which qr transforms is therefore

$$X(\theta) = 1 + 2 \cos\theta$$

If we call this representation Γ, we get the following characters of the elements of O in the following representation (Table 11.6).

Thus, we have established the irreducible representation of O group according to which operator $V = qr$ transforms, and we can now form the various products.

Table 11.6				
I	$8C_3$	$3C_2 = 3C_4^2$	$6C_2$	$6C_4$
Γ 3	0	-1	-1	1

Splitting or Energy Levels

When a perturbation is added, the symmetry group of the Hamiltonian changes. For example in case of Zeeman effect, the atom is placed in magnetic field and the symmetry of the total Hamiltonian is reduced. The Hamiltonian is still invariant under rotation about an axis coinciding with the direction of magnetic field. The Hamiltonian's symmetry group now becomes $SO(2)$ which is a subgroup of $SO(3)$. $SO(2)$ group is an abelian group and has only one-dimensional representations. The irreducible representation of $SO(3)$ characterised by l can be decomposed into $(2l + 1)$ irreducible representations of group $SO(2)$. A $(2l + 1)$ fold degenerate level ψ_m^l will therefore split into $(2l + 1)$ different levels.

If the atom is in crystal form, its symmetry group is also changed. If the atom is placed at the centre of a cubic crystal, the symmetry group of the Hamiltonian changes from $SO(3)$ to O. The complete symmetry group of a cube is the group O_n which is the direct product of O and the inversion group (I, i). But so far as splitting of energy levels is concerned the inclusion of inversion does not give any new information. The group O has two-fold, three-fold and four-fold axes of rotation (an n-fold axis of rotation C_n is a symmetry axis of rotation by an angle $2\pi/n$). The character table for the group is given below (Table 11.7).

Table 11.7					
	I	$8C_3$	$3C_2 = 3C_4^2$	$6C_2$	$6C_4$
Λ_1	1	1	1	1	1
Λ_2	1	1	1	-1	-1
I	2	-1	2	0	0
T_1	3	0	-1	-1	1
T_2	3	0	-1	1	-1

A $(2l + 1)$ fold degenerate state belonging to irreducible representation $D^{(l)}$ of $SO(3)$ group splits in field. To prove this, we will first calculate the characters of the elements of O in the representation $D^{(l)}$. To do this we use the result of Eq. (11.102).

$$X^{(l)}(\alpha) = \frac{\sin\left(l + \frac{1}{2}\right)\alpha}{\sin \alpha/2}$$

The characters for a few values of l are given in Table 11.8.

Table 11.8					
	I	$8C_3$	$3C_2 = 3C_4^2$	$6C_2$	$6C_4$
$D^{(0)}$	1	1	1	1	1
$D^{(1)}$	1	0	-1	-1	1
$D^{(2)}$	3	-1	1	1	-1
$D^{(3)}$	7	1	-1	-1	-1

An irreducible representation $D^{(l)}$ of group $SO(3)$ will in general not be an irreducible representation of O and $D^{(l)}$, can be decomposed into irreducible representation of O using Eq. (11.60) which gives the following results:

$$D^{(0)} = \Lambda_1, \ D^{(1)} = T_1, \ D^{(2)} = I + T_2, D^3 = \Lambda_2 + T_1 + T_2$$

$D(2)$ is a five dimensional representation. Since there are no five dimensional irreducible representations of O we can easily conclude that $D(2)$ should be decomposed into lower dimensional representations. Thus, a level characterised by $l = 2$ should split into two levels. All levels with $l > 2$ should also split. The levels corresponding to $l = 0$ and $l = 1$, however belong to irreducible representations of O and therefore do not split.

We can form a number of products which are given below.

$$T_1 \times \Lambda_1 = T_1, \quad T_1 \times \Lambda_2 = T_2, \quad T_1 \times I = T_1 + T_2$$
$$T_1 \times T_1 = \Lambda_1 + I + T_1 + T_2, \ T_1 \times T_2 = \Lambda_2 + I + T_1 + T_2$$

These give us allowed transitions. They are,

$$\Lambda_1 \leftrightarrow T_1, \ \Lambda_1 \leftrightarrow T_2, \ I \leftrightarrow T_1, T_2 \ T_1 \leftrightarrow T_2$$

Allowed transitions form solution rules.

PROBLEMS

1. By considering the symmetry transformation on the coordinates X_t construct the irreducible representation of D_4.
2. For the group D_3, the symmetry group of the equilateral triangle, construct the regular representation. Is the representation irreducible or reducible? If it is reducible, decompose it into irreducible representations.
3. A proper subgroup H of G is a subgroup which is identical with G. Prove the following.
   If H is a proper subgroup of G and H consists of complete classes, then it is an invariant subgroup.
4. Show that there can be two nonisomorphic groups of order 4.
5. Show that an element of a group G constitutes a class by itself, if and only if it commutes with all the elements of G.
6. Prove that the following sets form groups under the given rules of group multiplication.
   i. all $m \times n$ matrices under matrix multiplication
   ii. the n roots of $Z^n = 1$ under ordinary multiplication
   iii. the set of numbers $(1, 2,...,n-1)$, where n is a prime number, under multiplication modulo n (in multiplication modulo n, the product is to be divided with n and the remainder is to be taken as the result of multiplication).
7. Prove that a finite group whose order is a prime number must be a cyclic group.
8. Show that the set of elements which are inverse of the elements of a class of a group also forms a class.

9. Show that Lorentz transformations in the x-direction

$$x' = \frac{x - vt}{\sqrt{1 - v^2/c^2}}$$

and

$$t' = \frac{t - vx/c^2}{\sqrt{1 - v^2/c^2}}$$

form a group considering infinitesimal transformations. Find the generator.

10. By taking direct product of the irreducible representations of D_3, construct new representations.

11. Show that the transformations $X' = ax + b$, $a \neq 0$ form a lie group. Find the generators.

12. Consider the symmetry group of transformation of a square (this is a dihedral group D_4).

   (a) Write down the symmetry operations and workout the group multiplication table.

   (b) Identify all the subgroups.

   (c) Identify the classes.

13. Consider a class C of a group. Each element of C is used to generate a cyclic group. Prove that the order is same for each element of the class.

14. Let C_i and C_j be two classes in a group G. We define the two classes as follows: each element of C_i is multiplied by each element of C_j keeping each element as many times as it occurs:

   (a) Prove that $C_i\, C_j$ consists of entire classes. This can be stated mathematically as

$$C_i C_j = \sum_k C_{ijk} C_k$$

   where C_{ijk} are constants.

   (b) Prove that $C_i C_j = C_j C_i$ even if the group is not abelian (This implies $C_{ijk} = C_{jik}$)

   (c) Form the products $C_1 C_2$, $C_1 C_3$, $C_2 C_3$ for the three classes of the group D_3 and find the constants C_{ijk}.

15. (a) If the matrices $D(R)$ form a representation of the group G, prove that the complex conjugate matrices $D(R)$ also form a representation of G.

   (b) Prove that neither the matrices $D^{-1}(R)$ nor the matrices $D^+(R)$ form representations of G.

16. Let H be a subgroup of G, prove that a coset XH of H in G is not a subgroup of G by proving that the identity element I can not be XH.

17. (a) Determine the number and the dimensions of the inequivalent irreducible representations of D_4, the symmetry group of a square.

   (b) Construct the character table of D_4.

SHORT ANSWER QUESTIONS

1. What is the matrix representation of a group?

   Ans. A group of nonsingular matrices which is homomorphic to the original group is a matrix representation. The order of the matrices is the dimensionality of the representation.

2. What is an abelian group?

   Ans. A group where multiplication rule commutes also.

3. What is isomorphism and homomorphism in group theory?

   Ans. Two groups are said to be isomorphic if there is a one to one correspondence between their elements. The two groups are homomorphic if a many to one correspondence exists.

4. What is an order of a group?

   Ans. The number of elements in the group is the order of the group.

5. What do you mean by an invariant group?

   Ans. A subgroup composed of complete classes is called invariant subgroup.

6. What is a subgorup?

   Ans. A subset of elements in a group satisfying the group postulates is called a subgroup.

7. What do you mean by conjugate elements and classes?

   Ans. If A and X belong to a group and if $B = X^{-1} AX$, then A and B are conjugate to each other. The set of all elements of a group conjugate to A is called the conjugate class of A.

8. What do you mean by 'group multiplication table'?

   Ans. A table indicating the results of pairs of group elements. Each group element occurs only once in each row or column in the table (rearrangement theorem).

9. What is a cyclic group?

   Ans. A group is said to be cyclic if it is composed of $I, A, A^2, ..., A^{n-1}$ where $A^n = I$.

Numerical Methods

12.1. INTRODUCTION

There are many physical problems which can be reduced to the solution of differential equations, linear or nonlinear, ordinary or partial which are difficult to solve analytically. Numerical methods can help to find the solutions at various points under the given initial boundary conditions. The important methods of solving ordinary differential equations numerically are as follows:

   i. Picard's method of successive approximations

  ii. Euler's method

 iii. Modified Euler's method

 iv. Talor's series method

  v. Milne's method

 vi. Runge's method

 vii. Runge–Kutta method

viii. Adam–Bashforth method

12.2. PICARD'S METHOD OF SUCCESSIVE APPROXIMATIONS

Let us consider the differential equation

$$\frac{dy}{dx} = f(x, y), \text{ with } y = y_0 \text{ at } x = x_0 \tag{12.1}$$

Equation (12.1) may be expressed as

$$\int_{y_0}^{y} dy = \int_{x_0}^{x} f(x,y)dx \implies y = y_0 + \int_{x_0}^{x} f(x,y)dx \tag{12.2}$$

As a first approximation, replace of y by y_0 in $f(x, y)$ to give

$$y_1 = y_0 + \int_{x_0}^{x} f(x,y_0)dx \tag{12.3}$$

Similarly, successive approximations, viz. second, third, fourth,..., etc.

$$y_2 = y_0 + \int_{x_0}^{x} f(x, y_1) dx \qquad (12.4)$$

$$y_3 = y_0 + \int_{x_0}^{x} f(x, y_2) dx \qquad (12.5)$$

---- ---- ---- ---- ---- ---- ---- ---

$$y_{n+1} = y_0 + \int_{x_0}^{x} f(x, y_n) dx$$

The process is repeated till one obtains two successive approximations become nearly equal.

Example 1. *Use Piccard's method to approximate y when x = 0.2, given the differential equation*

$$\frac{dy}{dx} = x - y \text{ with } y_0 = 1 \text{ at } x_0 = 0$$

Solution: We have $\dfrac{dy}{dx} = x - y$ with $y_0 = 1$ at $x_0 = 0$

$\therefore$

$$y_1 = y_0 + \int_{x_0}^{x} f(x, y_0) dx = 1 + \int_{x_0}^{x} (x-1) dx = \frac{x^2}{2} - x + 1$$

Now, if $x = 0.2$,

$$y_1 = \frac{(0.2)^2}{2} - 0.2 + 1 = 0.82$$

$$y_2 = y_0 + \int_{x_0}^{x} f(x, y_1) dx$$

$$= 1 + \int_{x_0}^{x} \left[x - \left(\frac{x^2}{2} - x + 1 \right) \right] dx$$

$$= -\frac{x^3}{3} + x^2 - x + 1$$

Now, putting $x = 0.2$ again

$$y_2 = -\frac{(0.2)^3}{3} + (0.2)^2 - (0.2) + 1 = 0.8373$$

Further,

$$y_3 = y_0 + \int_{x_0}^{x} f(x, y_2) dx$$

$$= 1 + \int_{0}^{x} \left[x - \left(\frac{x^3}{3} + x^2 - x + 1 \right) \right] dx$$

$$= \frac{x^4}{24} - \frac{x^3}{3} + x^2 - x + 1 \qquad (1)$$

Putting $x = 0.2$ in Eq. (1), one obtains

$$y_3 = \frac{(0.2)^4}{24} - \frac{(0.2)^3}{3} + (0.2)^2 - 0.2 + 1 = 0.8374$$

Further,

$$y_4 = y_0 + \int_{x_0}^{x} f(x, y_3) dx = 1 + \int_0^x \left[x - \left(\frac{x^4}{24} - \frac{x^3}{3} + x^2 - x + 1 \right) \right] dx$$

$$= -\frac{x^5}{120} + \frac{x^4}{12} - \frac{x^3}{3} - x + 1 \qquad (2)$$

Putting $x = 0.2$ in the Eq. (2), we obtain

$$y_4 = \frac{(0.2)^5}{120} + \frac{(0.2)^4}{12} - \frac{(0.2)^3}{3} - (0.2)^2 - 0.2 + 1 = 0.83746$$

Thus, we see that $y_3 \cong y_4$

Hence, the required solution is $y = 0.83746$

Example 2. *Find the solution of* $\dfrac{dy}{dx} = 1 + xy$ *by Picard's method passing through* (0, 1), *correct to three places of decimal for* $0 < x < .5$.

Solution: Given $x = x_0 = 0$ and $y = y_0 = 1$,

First approximation $y^{(1)} = y_0 + \int_{x_0=0}^{x} (1 + xy_0) dx = 1 + \int_0^x (1 + x) dx$

$$= \left(1 + x + \frac{x^2}{2} \right) \qquad (1)$$

So

$$y^{(2)} = y_0 + \int_0^x \left(1 + xy^{(1)} \right) dx = 1 + \int_0^x \left\{ 1 + x \left(1 + x + \frac{x^2}{2} \right) \right\} dx$$

$$= 1 + \left\{ x + \frac{x^2}{2} + \frac{x^3}{3} + \frac{x^4}{8} \right\} \qquad (2)$$

Similarly

$$y^{(3)} = 1 + \int_0^x \left[1 + x \left(1 + x + \frac{x^2}{2} + \frac{x^3}{3} + \frac{x^4}{8} \right) \right] dx$$

$$= 1 + \int_0^x \left\{ x + \left(1 + x + x^2 + \frac{x^3}{6} \right) \right\} dx = 1 + x + x^2 + \frac{x^3}{3} + \frac{x^4}{24} \qquad (4)$$

for $x = 0.1$	$y^{(1)}$	$y^{(2)}$	$y^{(3)}$
	1.105	1.11016	1.1103
for $x = 0.2$	$y^{(1)}$	$y^{(2)}$	$y^{(3)}$
			1.2427

Exact value: We have $\dfrac{dy}{dx} - y = x$, equation is linear

with $\text{IF} = e^{-\int dx} = e^{-x}$, $\text{PI} = -x - 1$,   so $y = ce^{-x} - x - 1$

$1 = c - 0 - 1 \Rightarrow c = 2$,   so $y = 2e^{-x} - x - 1$

12.3. EULER'S METHOD

The basic approximation in Euler's method is approximating the curve by its tangent and turning it after every interval h in the direction of the tangent at the corresponding points on the curve (Fig. 12.1). From $\dfrac{dy}{dx} = f(x,y), y_1 \approx y_0 + h\left(\dfrac{dy}{dx}\right)$ at x_0, y_0, and so on.

Let $\dfrac{dy}{dx} = f(x,y)$ be the given equation whose curve passes through $P(x_0, y_0)$. Let the value of y be required for $x = x_0 + l = x_0 + nh$, where $x_0 + h$, $x_0 + 2h,...,x_0 + nh$ are n point $P_1, P_2,...,P_n$, of subdivision between P and Q $(x = x_0 + l)$ at P_0 y for curve $\left\{\dfrac{dy}{dx} = f(x,y)\right\}$

coincides with tangent at P_0 whose equation is $y - y_0 = \dfrac{dy}{dx}(x - x_0)$. This is solution curve at P_0.

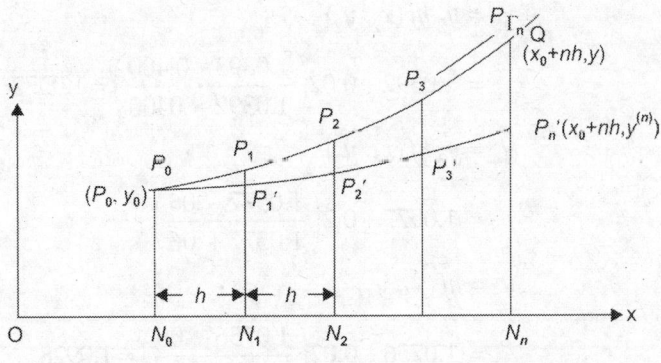

Fig. 12.1

At P_1 and P_1', $x = x_0 + h$, P_1 is point on the actual curve and P_1' is point on the solution curve.

At P_2 and P_2', $x = x_0 + 2h$, P_2 is on the actual curve and P_2' is on the solution curve, where $P_1' P_2'$ is paralled to tangent at P_1. Similary $P_2' P_3'$ is parallel to the tangent at P_2.

In this way the ordinate of the actual curve is $y = P_n N_n$

Ordinate of solution is $P_n' N_n$

Deviation between actual curve and Euler's solution curve $P_n P_n'$, as it goes on changing and in some cases do not converge to the actual solution. It is of little importance.

Example 1. Solve by Euler's method $\dfrac{dy}{dx} = \dfrac{y - x}{y + x}$, given that the curve passes through (0, 1). Find y for $x = 0.1$.

Solution: The starting point (x_0, y_0) is $= (0, 1)$, so $x_0 = 0$ and $y_0 = 1$
Taking only one interval between $x = 0$ and $x = 0.1$

The theoretical solution is

$$y = y_0 + \frac{dy}{dx}(x - x_0)$$

or

$$y_1 = 1 + \frac{1-0}{1+0}(x-0) = 1 + 1(0.1) = 1.1 \tag{1}$$

Dividing the interval from $x_0 = 0$ to $x_n = 0.10$, in 5 parts at $x_1 = 0.02$, $x_2 = 0.04$, $x_3 = 0.06$, $x_4 = 0.08$ and $x_5 = 0.10$

$$y_1 = y_0 + h\frac{dy}{dx} = 1 + 0.02\left(\frac{1-0}{1+0}\right) = 1.02 \tag{2}$$

$$y_2 = y_1 + hf(x_1, y)$$

$$= 1.02 + 0.2\left(\frac{1.02 - 0.2}{1.02 + .02}\right) = 1.0392 \tag{3}$$

Now,

$$y_3 = y_2\, hf(x_2, y_2)$$

$$= 1.0392 + 0.02\left(\frac{1.0393 - 0.400}{1.0392 + .0400}\right) = 1.0577 \tag{4}$$

Now

$$y_4 = y_3\, hf(x_3, y_3)$$

$$= 1.0577 + 0.2\left(\frac{1.0577 - .06}{1.0577 + .06}\right)$$

Now

$$y_5 = hf(x_4, y_4)$$

$$= 1.0756 + 0.02\left(\frac{1.0756 - .08}{1.0756 + .08}\right) = 1.0928$$

Hence for $x = 0.1$, $y = 1.0928$.

12.4. MODIFIED EULER'S METHOD

Let C be the actual curve defined by $\frac{dy}{dx} = f(x,y)$ and passing through $P_0(x_0, y_0)$ (Fig. 12.2).

Let $P_1, P_2,...$ be points on this curve with abscissa $x_0 + h$, $x_0 + 2h,...$

In Euler's method, we approximated the curve by the tangents at P_0, whose ordinate at $x = x_0 + h$ was

$$P_1 N_1 = y_1^{(1)} = y_0 + f(x_0, y_0) \cdot (x_0 + h - x_0)$$

$$= y_0 + hf(x_0, y_0) \tag{12.6}$$

In modified Euler's method, we approximate the curve by the line $P_0 P_1''$, passing through $P(x_0, y_0)$ and whose

slope

$$m_0 = \frac{f(x_0, y_0) + f(x_0, y_1^{(1)})}{2} \tag{12.7}$$

So at P_1''

$$y_1 = y_0 + h\frac{f(x_0, y_0) + f(x_0 + h, y_1^{(1)})}{2} \tag{12.8}$$

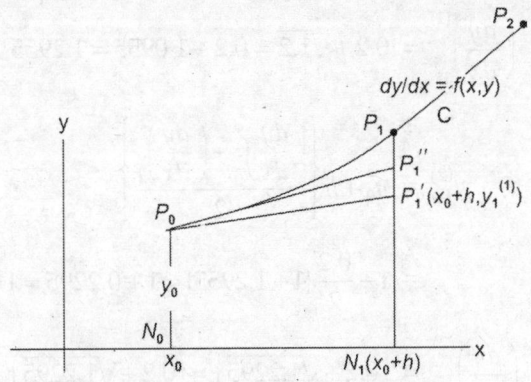

Fig. 12.2

Those slope at $(x_0 + h)$ is taken by the average of slope of the curve at x_0 and $(x_0 + h)$, proceeding in this way, we get successively

$$Y_{n+1} = y_n + \frac{h}{2}\left\{ f(x_n, y_n) + f\left(x_{x+1}, y_{n+1}^{(1)}\right)\right\} \tag{12.9}$$

where

$$\int_{y_n}^{y_{n+1}} \frac{dy}{dx} dx = \int_{x_0}^{x_{n+1}} f(x, y) dx$$

or

$$y_{n+1} - y_n = \frac{h}{2}\left(y_n^1 + y_{n+1}^1\right) \text{ (employing the trapezoidal rule)}$$

$$= \frac{h}{2}\left[f(x_n, y_n) + f\left(x_{n+1}, y_{n+1}^{(1)}\right)\right] \tag{12.10}$$

which is same as Eq. (12.10) and

$$y_{n+1}^{(1)} = y_n + hf(x_n, y_n) \tag{12.11}$$

Thus the modified Euler's method is represented by Eqs (12.5) and (12.6). Employing the trapezoidal rule, the whole series (the Taylor's expansion), is approximated only by its first two terms. So local truncation error is $-h^3 y'''(x_n + \theta h)\ 0 < \theta < 1$. Due to less error, compared with the main Euler's method, the modified method gives a better approximation y_n instead of y_n^1 in Euler's method. Euler's and modified Euler's methods may be viewed as predicator and corrector formulas.

Example 1. *Using Euler' modified method, solve the equation* $\dfrac{dy}{dx} = x + |\sqrt{y}| = f(x,y)$, *given that* $y = 1$, *when* $x = 0$. *Solve it for* $0 \le x \le 0.6$ *in steps of 0.2.*

Solution: We have

$$x_0 = 0, \quad y_0 = 1, \quad \text{and} \quad h = 0.2$$

at P_1' ordinate

$$= y_1^{(1)} = y_0 + hf(x_0, y_0)$$

$$= 1 + 0.2[0 + \sqrt{1}] = 1.2 \tag{1}$$

Now $\dfrac{dy}{dx}$ at $P_1' = (x + \sqrt{y})$ at $x = 0.2,\ y = y^{(1)} = 1.2$

$$\left(\frac{dy}{dx}\right)_1^{(1)} = 0.2 + \sqrt{1.2} = 0.2 + 1.0955 = 1.2955$$

$$y_1^{(2)} = y_0 + h\left\{\frac{\left(\frac{dy}{dx}\right)_0 + \left(\frac{dy}{dx}\right)_1}{2}\right\}$$

$$= 1 + \frac{0.2}{2}\{1 + 1.2955\} = 1 + 0.2295 = 1.2295$$

$$\left(\frac{dy}{dx}\right)_1^{(2)} = \{0.2 + \sqrt{1.2295}\} = \left(0.2 + \sqrt{1.2295}\right)$$

$$= (0.2 + 1.1089) = 1.3089$$

$$y_1^{(3)} = y_0 + \frac{h}{2}\left[\left(\frac{dy}{dx}\right)_{0,0} + \left(\frac{dy}{dx}\right)_1^{(2)}\right] = 1 + \frac{0.2}{2}[1 + 1.3089] = 1.2309$$

$$\left(\frac{dy}{dx}\right)_1^{(3)} = f(x_1, y_1^{(3)})$$

$$= 0.2 + \sqrt{1.2309} = 0.2 + 1.10995 = 1.3095$$

$$y_1^{(4)} = y_0 + \frac{h}{2}\left[\left(\frac{dy}{dx}\right)_0 + \left(\frac{dy}{dx}\right)_1^{(3)}\right] = 1 + \frac{0.2}{2}[1 + 1.3095] = 1.2309$$

As $y_1^{(4)} = y_1^{(3)}$, So we stop here and take $y_1 = 1.2309$.

Calculation for y_2 – At $x = x_2 = 0 + 2(0.2) = 0.4$

$$y_2^{(1)} = y_1 + h(x_1, y_1)$$

$$= 1.2309 + 0.2[0.2 + \sqrt{1.2309}]$$

$$= 1.2309 + 0.2[0.2 + 1.1095] = 1.2309 + 0.2619 = 1.4928$$

$$y_2^{(2)} = y_1 + \frac{h}{2}\left[\left(\frac{dy}{dx}\right)_1 + \left(\frac{dy}{dx}\right)_2^{(1)}\right]$$

$$= 1.2309 + 0.1\left[(x_1 + \sqrt{y_1}) + (x_2 + \sqrt{y_2^{(1)}})\right]$$

$$= 1.2309 + 0.1\left[0.2 + \sqrt{1.2309} + 0.4 + \sqrt{1.4928}\right]$$

$$= 1.2309 + 0.1(2.9313) = 1.5253$$

$$= 1.2309 + 0.1\left[1.3095 + 0.4 + \sqrt{1.5240}\right] = 1.5253$$

$$= 1.2309 + [0.2 + 1.1095 + 0.4 + 1.2218]$$

$$y_2^{(2)} = 1.2309 + 0.1(2.9313) = 1.5240$$

$$y_2^{(3)} = y_1 + \frac{h}{2}\left[\left(\frac{dy}{dx}\right)_1 + \left(\frac{dy}{dx}\right)_2^{(2)}\right]$$

$$= 1.2309 + 0.1\left[1.3095 + x_2\sqrt{y_2^{(2)}}\right]$$

$$= 1.2309 + 0.1\left[1.3095 + 0.4 + \sqrt{1.5240}\right] = 1.5253$$

$$y_2^{(4)} = y_1 + \frac{h}{2}\left[\left(\frac{dy}{dx}\right)_1 + \left(\frac{dy}{dx}\right)_2^{(3)}\right]$$

$$= 1.2309 + 0.1\left[1.3095 + 0.4 + \sqrt{1.5253}\right]$$

$$= 1.2309 + 0.1[1.3095 + 0.4 + 1.2350] = 1.5253$$

As $y_2^{(3)} = y_2^{(4)}$, so we stop and take $y_2 = 1.5253$

Calculation for y_3 at $x = x_3 = 0.6$

$$x_2 = 0.4 y_2 = 1.5253$$

$$y_3^{(1)} = y_2\, hf(x_2 + y_2)$$

$$= 1.5253 + 0.2\left[0.4 + \sqrt{1.5253}\right]$$

$$= 1.5253 + 0.2[0.4 + 1.2350] = 1.5253 + 0.3270 = 1.8523$$

$$y_3^{(2)} = y_2 + \frac{h}{2}\left[\left(\frac{dy}{dx}\right)_2 + \left(\frac{dy}{dx}\right)_3^{(1)}\right]$$

$$= 1.5253 + 0.1\left[x_2 + \sqrt{y_2} + x_3 + \sqrt{y_3^{(1)}}\right]$$

$$= 1.5253 + 0.1[0.4 + 1.2350 + 0.6 + 1.3610] = 1.8849$$

$$y_3^{(3)} = y_2 + \frac{h}{2}\left[\left(\frac{dy}{dx}\right)_2 + \left(\frac{dy}{dx}\right)_3^{(2)}\right]$$

$$= 1.5253 + 0.1\left[0.4 + 1.2350 + 0.6 + \sqrt{1.8840}\right]$$

$$= 1.5253 + 0.1[1.0 + 1.2350 + 1.3729] = 1.8861$$

$$y_3^{(4)} = y_2 + \frac{h}{2}\left[\left(\frac{dy}{dx}\right)_2 + \left(\frac{dy}{dx}\right)_3^{(3)}\right]$$

$$= 1.5253 + 0.1\left[0.4 + 1.2350 + 0.6 + \sqrt{1.8861}\right]$$

$$= 1.5253 + 0.1[1 + 1.2350 + 1.3734] = 1.8861$$

Now as $y_3^{(3)} = y_3^{(4)}$, so we stop and take $y^3 = 1.8861$.

12.5. TAYLOR'S SERIES METHOD

Given the differential equation $\dfrac{dy}{dx} = f(x,y)$, being satisfied at $x = x_0$ and $y = y_0$ to solve the equation.

We have the solution $y = y_0$ at $x = x_0$ for a point $x = x_0 + h$

Let the solution be $y(x_0 + h)$ but by Taylor's expansion

$$y(x_0 + h) = y(x_0) + \frac{h}{\underline{|1}} y'(x_0) + \frac{h^2}{\underline{|2}} y''(x_0) + \cdots \tag{12.12}$$

as

$$\frac{dy}{dx} = f(x, y) \tag{12.13}$$

at

$$y = y_0 \qquad \frac{dy}{dx} = f(x, y_0) \tag{12.14}$$

By differentiating Eq. (12.14) any number of times, we can get $y(x_0 + h)$

Having known x and y at $P_1(x_0 + h)$, $y(x_0 + h)$, we can get value of y for other values like $(x_0 + 2h)$, etc. of x.

Example 1. Solve $\dfrac{dy}{dx} = x + y$, by Taylor's series method numerically given at $x = 1, y = 0$, find value upto $x = 1.2$ with $h = 0.1$. Compare the result with the explicit solution.

Solution: $\dfrac{dy}{dx} = x + y \qquad \Rightarrow \qquad \left(\dfrac{dy}{dx}\right)_{\text{at } P_0(1,0)} = x + 0 = 1$

$\dfrac{d^2y}{dx^2} = 1 + \dfrac{dy}{dx} \qquad \Rightarrow \qquad \left(\dfrac{d^2y}{dx^2}\right)_{\text{at } P_0} = 1 + 1 = 2$

$\dfrac{d^3y}{dx^3} = 0 + \dfrac{d^2y}{dx^2} \qquad \Rightarrow \qquad y'''(1, 0) = y''(1, 0) = 2$

For $h = 0.1$

$$y(1 + h) = y(1) + \frac{h}{\underline{|1}} y'(1) + \frac{h^2}{\underline{|2}} y''(1) + \frac{h^3}{\underline{|3}} y'''(1)$$

$$= 0 + 0.1(1) + \frac{(0.1)^2}{2}(2) + \frac{(0.1)^3}{6}(2) + \frac{(0.1)^4}{24} \times 2$$

$$= 0.1 + 0.01 + 0.0003333 + 0.000008$$

$$= 0.111033847 = 0.110 \text{ (approx.)}$$

For $x_1 = 1.1$ with $y_1 = 0.11$

$$\frac{dy}{dx} = x + y = 1.21$$

$$\frac{d^2y}{dx^2} = 1 + \frac{dy}{dx} = 2.21, \quad \frac{d^2y}{dx^2} = \frac{d^3y}{dx^3} = 2.21$$

So
$$y(1.1 + 0.1) = y'(1.1) + \frac{(0.1)}{\underline{1}} y''(1.1) + \frac{(0.1)^2}{\underline{2}} y'''(1.1) + \cdots$$

$$= 0.11 + 0.1(1.21) + \frac{.01}{2}(2.21) + \cdots = 0.232 \text{ (approx.)}$$

Analytically $\qquad y = -x - 1 + 2e^{x-1}$ at $x = 1.2 = -1.2 - 1 + 2 (1.221) = 0.252$

12.6. MILNE'S PREDICTOR–CORRECTOR METHOD

Newton's forward interpolation formula

Let $y = f(x)$ be a function of x which assumes the values $f(a), f(a + h), f(a + 2h),..., f(a + nh)$ for $(n + 1)$ equidistant values $a, a + h, a + 2h,..., a + nh$ of the independent variable x. Then the Newton's forward interpolation formula is

$$f(a + h\mu) = f(a) + \mu\Delta f(a) + \frac{\mu(\mu - 1)}{2!} \Delta^2 f(a) + \cdots + \frac{\mu(\mu - 1)(\mu - 2)\cdots(\mu - n - 1)}{n!} \Delta^n f(a)$$

$$\left[\text{where} \quad \mu = \frac{x - a}{h} \right]$$

This formula is useful for interpolating the value of $f(x)$ near the beginning of the set of values given. h is called off differencing while Δ is forward difference operator.

Newton's Backward interpolation formula

$$f(a + nh + \mu b) = f(a + nh) + \mu\nabla f (a + nh)$$

$$+ \frac{\mu(\mu + 1)}{1!} \nabla^2 f(a + nh) + \cdots + \frac{\mu(\mu + 1)\cdots\mu + \overline{n - 1}}{n!} \nabla^n f(a + nh)$$

This formula is useful when the value of $f(x)$ is required near the end of the table.

Interpolation and Extrapolation

Interpolation: It is the technique of obtaining a value of the function for any intermediate value of the independent variable provided the values of the function corresponding to that variable are given in a tabular form.

Extrapolation: It is the technique of computing the value of the function outside the range of given values.

Example: Use following data and explain the terms interpolation and extrapolation.

x	0	4	8	12	16	20
$f(x)$	5	7	10	13	17	22

The technique of finding $f(2), f(3), f(5), f(10), f(14)$, etc. is the interpolation while that of finding $f(-1), f(21), f(24)$, etc. is the extrapolation.

Following assumptions are made for interpolation:

    i. The given function should be either in an increasing order or in a decreasing order, without any sudden jumps.

    ii. If the given function is a polynomial, then it can be represented as a polynomial with a good degree of approximation.

For interpolation, the simplest technique is the graphical method. In this technique, we plot $y(x)$ against x on suitable scale on y- and x-axes. Then, one has to draw a smooth curve through the points and read from the graph, the value of y, corresponding to the required value of x. One can obtain the same result by using the interpolation formulae.

To solve the equation $\dfrac{dy}{dx} = f(x, y)$ satisfying the condition $x = x_0,\ y = y_0$ we proceed as follows:

Newton's forward interpolation formula applied to $y' = f(x\ y)$ is

$$y' = y_0' + u\Delta y_0' + \frac{u(u-1)}{2}\Delta^2 y_0' + \frac{u(u-1)(u-2)}{6}\Delta^3 y_0' + \cdots \tag{12.15}$$

where $\qquad x = x_0 + uh \quad \text{or} \quad u = \dfrac{x - x_0}{h}$

Integrating Eq. (12.15) w.r.t. x over x_0 to $x_0 + 4h \quad$ or $\quad y = 0$ to $u = 4$

$$\int_{x_0}^{x_0+4h} y'\, dx = \int_0^4 \left\{ y_0' + u\Delta y_0' + \frac{u^2 - u}{2}\Delta^2 y_0' \right\} h\, dy$$

So

$$y_4 - y_0 = h\left[uy_0' + \frac{u^2}{2}\Delta y_0' + \left(\frac{u^3}{3} - \frac{u^2}{2} \right)\frac{1}{2}\Delta^2 y_0' \cdots \right]_0^4$$

$$= h\left[4y_0' + \frac{4^2}{2}\Delta y_0' + \left(\frac{20}{3}\Delta^2 y_0' + \frac{8}{3}\Delta^3 y_0' + \frac{28}{90}\Delta^4 y_0' \right) \right]$$

$$= h\left[4y_0' + 8(y_1' - y_0') + \frac{20}{3} + (y_2' - 2y_1' + y_0') + \frac{8}{3}(y_3' - 3y_2' + 3y_1' - y_0') \right]$$

So

$$y_4 = y_0 + \frac{4h}{3}[2y_1' - y_2' + y_3'] + \frac{28}{90}h\Delta^4 y_0 \tag{12.16}$$

This formula is Milne's extrapolation (predictor) formula. It is used to predict the value of y_4, if the values of y_0, y_1, y_2, y_3 is known.

To obtain the corrector formula, we integrate Eq. (12.15) over $x = x_0$ to $x_2 = x + 2h$ or from $u = 0$ to $u = 2$ and get

$$y_2 = y_0 + h\left[2y_0' + 2\Delta y_0' + \frac{1}{3}\Delta^2 y_0' - \frac{1}{90}\Delta^4 y_0' \right] \quad \text{as } \Delta \equiv (E - 1)$$

We have

$$y_2 = y_0 + h\left[2y_0' + 2(y_1' - y_0') + \frac{1}{3}(y_2' - 2y_1' + y_0') - \frac{1}{90}(y_4' - 4y_3' + 6y_2' - 4y_1' + y_0') \right]$$

$$y_2 = y_0 + \frac{h}{3}(y_0' + 4y_1' + y_2') - \frac{h}{90}\Delta^4 y_0' \tag{12.17}$$

This is Milne's corrector formula. The value of y_4 obtained from Eqs (12.16) and (12.17) can be put as

$$y_{n+1} = y_{n-1} + \frac{4h}{3}[2y_{n-2} - y_{n-1} + y'_n] \tag{12.18}$$

and

$$y_{n+1}^{(1)} = y_{n-1} + \frac{h}{3}(y'_{n-1} + 4y'_n + y'_{n+1}) \tag{12.19}$$

Error in Eq. (12.18) is of order $h\frac{28}{90}\Delta^4 y'_0$ and that in Eq. (12.19) it is of the order

$-\frac{h}{90}\Delta^4 y'_0$. These errors are of opposite signs and very small in magnitude.

Example 1. *Tabulate, by Milne's corrector formula, the numerical solution of* $\frac{dy}{dx} = x+y$,

given when $x = x_0 = 0$, $y = y_0 = 1$. *Obtain y for x = 0.20 and for x = 0.30.*

Solution: As the values of y for $x = 0.50, 0.10, 0.15$ and 0.20 should be known for this method, we calculated these values by using picard's method.

Picard's method

We start with $\frac{dy}{dx} = f(x,y_0) = x+1$

so

$$\int_0^1 dy = \left[\frac{x^2}{2} + x\right]_0^x$$

$$y_1 - y_0 = x + \frac{x^2}{2} \text{ and } y_0 = 1$$

$$y^{(1)} = 1 + x + \frac{x^2}{2} \text{ as } y_0 = 1$$

Second approximation is

$$\frac{dy}{dx} = x + y^{(1)} = x + 1 + x + \frac{x^2}{2}$$

so

$$\int_{y_0}^{y^{(1)}} dy = \int_0^x \left(x + 1 + x + \frac{x^2}{2}\right) dx, \quad y^{(2)} = 1 + x + x^2 + \frac{x^3}{6}$$

Now, we have

x	0	0.5	0.10	0.15
	y_0	y_1	y_2	y_3
y	1.000	1.0525	1.1103	1.1736
	y'_0	y'_1	y'_2	y'_3
$\frac{dy}{dx} =$	1.000	1.025	1.2103	1.3236

$$y = \frac{dy}{dx} = x + y$$

By Milne's predictor formula, we have

$$y_4 = y_0 + \frac{4h}{3}(2y_1' - y_2' + y_3')$$

$$= 1 + \frac{4(0.05)}{3}[2.205 - 1.2103 + 2.6472]$$

$$= 1.2448 \text{ (predicted value)} \tag{1}$$

It is corrected by the corrector formula

$$y_{n+1^{(1)}} = y_{n-1} + \frac{h}{3}(y_{n-1}' + 4y_n' + y_{n+1}') \tag{2}$$

Putting $n = 3$,

$$y_4^{(1)} = y_2 + \frac{h}{3}[y_2' + 4y_3' + y_4']$$

$$= 1.1103 + \frac{0.05}{3}[1.2103 + 5.2944 + 1.4428] = 1.2428 \tag{3}$$

By Eqs (1) and (2), the predicted and corrected values are same.

Taking $\qquad y_4 = y_{.20} = 1.2428,$

$$y_4' = x_4 + y_4 = 0.2 + 1.2428 = 1.4428 \tag{4}$$

Putting $n = 4$, from the predictor formula,

$$y_{n+1} = y_{n-3} + \frac{4h}{3}(2y_{n-2}' - y_{n-1}' + 2y_n') \tag{5}$$

$$y_5 = y_1 + \frac{4h}{3}[2y_2' - y_3' - 2y_4']$$

$$= 1.0525 + \frac{4(0.5)}{4}[2.4206 - 1.3236 + 2.8856] = 1.3180 \tag{6}$$

so $\qquad y_5^{(1)} = x_5 + y_5 = 0.25 + 1.3180 = 1.5680 \tag{7}$

Using corrector formula Eq. (2) for $n = 4$, we have

$$y_5^{(1)} = y_3 + \frac{h}{3}[y_3' + 4y_4' + y_5']$$

$$= 1.1735 + \frac{.05}{3}[1.3236 + 5.7712 + 1.5680] = 1.3180 \tag{8}$$

Again the predicted and corrected values of y_5 as given by Eqs (6) and (8) are equal.

Taking $y_5 = 1.3180$ and $y_5^1 = x_5 + y_5 = 0.25 + 1.3180 = 1.5680$ and putting $n = 5$ in Eq. (5), the predicted value

$$y_6 = y_2 + \frac{4h}{3}[2y_3' - y_4' + 2y_5']$$

$$= 1.1103 + \frac{4 \times .05}{3}[2.6472 - 1.4428 + 3.1360] = 1.3997 \tag{9}$$

Using corrector formula [Eq. (2)], taking $n = 5$, corrected value is

$$y_6^{(1)} = y_4 + \frac{h}{3}[y_4' + 4y_5' + y_6']$$

$$= 1.248 + \frac{0.05}{3}[1.428 + 6.2720 + 1.6997]$$

$$= 1.248 + \frac{0.05}{3}[1.428 + 6.2720 + 1.6997]$$

$$y_6^{(1)} = 1.248 + \frac{0.5}{3} \times 9.41459 = 1.3997 = y_{.30} \tag{10}$$

It is same as the predicted value given by Eq. (7)

Hence the solution is

x	0.20	0.25	0.30
y	1.4428	1.3180	1.3997

12.7. RUNGE–KUTTA METHOD

The methods of Runge and Runge–Kutta are most widely used methods, having less calculations and suitability for computation of higher order derivatives become more complicated.

Given $\dfrac{dy}{dx} = f(x, y)$ with condition that at $x = x_0$, $y = y_0$

To solve the equation, we have

$$\frac{dy}{dx} = \frac{\delta y}{\delta x} \text{ (approx)} = f(x, y)$$

So $\delta y = \delta x f(x, y)$

This idea is used in defining 4 increments k_1, k_2, k_3 and k_4 in y corresponding to $\delta x = h$

or

$$k_1 = hf(x_0, y_0) \tag{12.20}$$

$$k_2 = hf\left(x_0 + \frac{h}{2}, y_0 + \frac{k_1}{2}\right) \tag{12.21}$$

$$k_3 = hf\left(x_0 + \frac{h}{2}, y_0 + \frac{k_2}{2}\right) \tag{12.22}$$

$$k_4 = hf(x_0 + h, y_0 + k_3) \tag{12.23}$$

Finally, we define $k = \dfrac{1}{6}(k_1 + 2k_2 + 2k_3 + k_4) \tag{12.24}$

The value of y at $x = x_0 + h$ is $y = y_0 + k \tag{12.25}$
where k is given by Eq. (12.24)

This is known as Runge–Kutta method of fourth order.

Example 1. *Apply fourth order Runge–Kutta method to find approximate value of y at x = 0.2, given* $\dfrac{dy}{dx} = x + y$ *and for* $x = x_0 = 0$, $y = y_0 = 1$.

$$\frac{dy}{dx} = f(x, y) = x + y$$

Solution: Here $x_0 = 0$, $y_0 = 1$, $h = 0.2$, $f(x_0, y_0) = 0 + 1 = 1$

$$k_1 = hf(x_0, y_0) = 0.2(1) = 0.2 \tag{1}$$

$$k_2 = hf\left(x_0 + \frac{h}{2}, y_0 + \frac{k_1}{2}\right)$$

$$= 0.2\left[0 + \frac{0.2}{2} + 1 + \frac{0.2}{2}\right] = 0.24 \tag{2}$$

$$k_3 = hf\left(x_0 + \frac{h}{2}, y_0 + \frac{k_2}{2}\right)$$

$$= 0.2\left[0 + 0.1 + 1 + \frac{0.24}{2}\right] = 0.244 \tag{3}$$

$$k_4 = hf(x_0 + h_1 y_0 + k_3)$$
$$= 0.2[0 + 0.2 + 1 + 0.244] = 0.2888 \tag{4}$$

Now

$$k = \frac{1}{6}[k_1 + 2k_2 + 2k_3 + k_4]$$

$$= \frac{1}{6}[0.2 + 0.48 + 0.488 + 0.2888] = 0.2428$$

Hence

$$y = y_0 + k = 1.2428$$

Example 2. *Apply Runge–Kutta method to find an approximate value of y for x = 0.2 in steps of 0.1 if* $\dfrac{dy}{dx} = x + y^2$, *given* $y = 1$, *when* $x = 0$

Solution: Here we have to take $h = 0.1$ and carry out calculations in two steps: first finding $y(0.1)$ and then $y(0.2)$.

Step 1: We have $x_0 = 0$, $y_0 = 1$, $h = 0.1$, $f(x, y) = x + y^2$

$$k_1 = hf(x_0, y_0) = 0.1(0 + 1^2) = 0.1 \tag{1}$$

$$k_2 = hf\left(x_0 + \frac{h}{2}, y_0 + \frac{k_1}{2}\right)$$

$$= 0.1\left[0 + 0.05 + \left(1 + \frac{0.1}{2}\right)^2\right] = 0.1152 \tag{2}$$

$$k_3 = 0.1\left[0.05 + \left(1 + \frac{0.1152}{2}\right)^2\right] = 0.1168 \tag{3}$$

$$k_4 = h\left[0.1+(1.1168)^2\right]=0.1347 \tag{4}$$

Hence $\quad k = \dfrac{1}{6}[k_1 + 2k_2 + 2k_3 + k_4] = \dfrac{1}{6}[0.1+0.2304+0.2336+0.1347]=0.1165$

$$y = y_0 + k = 1 + 0.1165 = 1.1165 \tag{5}$$

Step 2: For $x = x_1 = 0.1$, $y = y_1 = 1.1165$, $h = 0.1$

$$k_1 = hf(x_1, y_1) = 0.1\left[0.1+(1.1165)^2\right]=0.1347$$

$$k_2 = hf\left(x_1+\frac{h}{2},y_1+\frac{k_1}{2}\right)=0.1\left[0.15+\frac{(1.0673+0.1165)^2}{(1.1838)^2}\right]=0.1551$$

$$k_3 = hf\left(x_1+\frac{h}{2},y_1+\frac{k_2}{2}\right)=0.1\left[0.15+(1.1940)^2\right]=0.1576$$

$$k_4 = hf(x_1+h,y_1+k_3)=0.1\left[0.2+(1.2741)^2\right]$$
$$= 0.1[0.2 + 1.6230] = 0.1823$$

$$k = \frac{1}{6}[k_1+2k_2+2k_3+k_4]=0.1570$$

Hence $\qquad y(0.2) = y_1 + k = 1.1165 + 0.1570 = 1.2735$

MISCELLANEOUS EXAMPLES

Example 1. *Apply Picard's method to find the solution of* $\dfrac{dy}{dx}=y-x$, *with* $x = 0, y = 2$ *up to the third order approximation. Here* x_0, $y_0 = 2, f(x_0, y_0) = y - x$.

Solution: Starting with

$$\frac{dy}{dx} = f(x_1, y_0) = 2 - x \tag{1}$$

so $\qquad \displaystyle\int_2^{y_1} dy = \int_0^x (2-x)\,dx$

or $\qquad y_1 = 2+2x-\dfrac{x^2}{2} \tag{2}$

Now $\qquad \dfrac{dy}{dx} = f(x, y) = 2x-\dfrac{x^2}{2}-x$

so $\qquad \displaystyle\int_2^y dy = \left(\dfrac{x^2}{2}-\dfrac{x^3}{6}\right)$

or $\qquad y = y_2 = 2+\dfrac{x^2}{2}-\dfrac{x^3}{6} \tag{3}$

Now $\qquad \dfrac{dy}{dx} = f(x,y_2)=\left(2+\dfrac{x^2}{2}-\dfrac{x^3}{6}-x\right)$

Integrating,
$$\int_2^y dy = \int_0^x \left(2 + \frac{x^2}{2} - \frac{x^3}{6} - x\right) dx$$

or
$$y - 2 = 2x + \frac{x^3}{6} - \frac{x^4}{24} - \frac{x^2}{2}$$

so
$$y = y_3 \left(2 + 2x + \frac{x^3}{6} - \frac{x^4}{24} - \frac{x^2}{2}\right) \tag{4}$$

Example 2. *Use Picard's method, to solve* $\dfrac{dy}{dx} = x + y^2$, *given* $y = 0$, *when* $x = 0$, *we have* $x_0 = y_0 = 0$ *and* $f(x, y) = x + y^2$.

Solution:
$$\frac{dy}{dx} = f(x, y_0) = x + 0 = x \tag{1}$$

so
$$\int_0^y dy = \int_0^x x\, dx$$

or
$$y = y_1 = \frac{x^2}{2} \tag{2}$$

Now taking
$$\frac{dy}{dx} = f(x, y_1) = x + \frac{x^4}{4}$$

Integrating,
$$\int_0^y dy = y = \int_0^x \left(x + \frac{x^4}{4}\right) = \frac{x^2}{2} + \frac{x^5}{20} \tag{3}$$

Now taking
$$\frac{dy}{dx} = f(x, y_2) = x + \left(\frac{x^2}{2} + \frac{x^5}{20}\right)^2$$

$$y = y_3 = \int_0^x f\left(x + \frac{x^4}{4} + \frac{2x^7}{40} + \frac{x^{10}}{400}\right) dx$$

$$y_3 = \frac{x^2}{2} + \frac{x^5}{20} + \frac{x^8}{160} + \frac{x^{11}}{4400}$$

Example 3. *Use Milne's method to calculate* $y(0.8)$ *and* $y(1.0)$ *for* $\dfrac{dy}{dx} = 1 + y^2$, *given*

x	$x_0 = 0$	0.2	0.4	0.6
y	$y_0 = 0$	$y_1 = 0.2027$	$y_2 = 0.4228$	$y_3 = 0.6841$
we find y_1	1	1.04108	1.17875	1.46778

Solution: To obtain solution of
$$\frac{dy}{dx} = f(x, y) = 1 + y^2 \tag{1}$$

With $x_0 = 0$, $y_0 = 0$, h = interval for $x = 0.2 = h$

Milne's predictor formula is

$$y_{n+1} = y_{n-3} + \frac{4h}{3}(2y'_{n-2} - y'_{n-1} + 2y'_n) \tag{2}$$

and the correcter formula is

$$y_{n+1} = y_{n-1} + \frac{h}{3}(y'_{n-1} + 4y'_n + y'_{n+1}) \tag{3}$$

Taking $n = 3$ in Eq. (2)

$$y_4 = y_0 + \frac{4h}{3}[2y'_1 - y'_2 + 2y'_3] \tag{4}$$

$$= 0 + \frac{8}{30}[2(1.04108 - 1.1787 + 2(1.4678)] = 1.0237 \text{ (predicted value)}$$

$$y_4 = 1.0237$$

$$y'_4 = 1 + y^2_4 = 1 + (1.0237)^2 = 2.0480$$

Corrected value for $n = 3$

$$y_4^{(1)} = y_2 + \frac{h}{3}[y'_2 + 4y'_3 + y'_4]$$

$$= 0.4228 + \frac{0.2}{3}[1.1787 + 4(1.4678) + 2.0480]$$

$$y(0.8) = 0.4228 + 0.6065 = 1.0293$$

For $y(1.0)$

x	y_1	y_2	y_3	y_4
	0.2	0.4	0.6	0.8
y	0.2027	0.4228	0.6841	1.0293
$y'_1 = 1 + y^2$	1.0411	1.1787	1.4678	2.0594

Predictor value

$$y_5 = y_1 \frac{4h}{3}[2y'_2 - y'_3 + 2y'_4]$$

$$= 0.2027 + \frac{0.8}{3}[2 \times 1.1787 - 1.4678 + 2(2.0594]$$

$$= 0.2027 + \frac{8}{30}[2.3574 - 1.4678 + 4.1188] = 1.5382$$

$$y'_5 = 1 + y^2_5 = 1 + (1.5382)^2 = 3.3660$$

Corrected value

$$y_5^{(1)} = y_3 + \frac{0.2}{3}[y'_3 + 4y'_4 + y'_5]$$

$$= 0.6841 + \frac{2}{30}[1.467 + 4 \times 2.0594 + 3.3360]$$

$$= 0.6841 + \frac{2}{30}[1.4678 + 8.2376 + 3.3360]$$

$$= 0.6841 + 0.8714 = 1.5555$$

Example 4. *Using Picard's method obtain solution of* $\frac{dy}{dx} = x + x^4y$ *given* $y(0) = 3$. *Find* $y(0.1), y(0.2), y(0.3)$.

Solution: $f(x, y) = x + x^4y$, where $x_0 = 0$, $y_0 = 3$

so $f(x, y_0) = x + 3x^4$

$\therefore$

$$y_1 = y_0 + \int_0^x (x + 3x^4)dx = 3 + \frac{x^2}{2} + \frac{3x^5}{5} \tag{2}$$

Now $f(x, y) = x + x^4\left(3 + \frac{x^2}{2} + \frac{3}{5}x^5\right)$

So

$$y_2 = y_0 + \int_0^x \left(x + 3x^4 + \frac{1}{2}x^6 + \frac{3}{5}x^9\right)dx$$

$$= 3 + \frac{x^2}{2} + \frac{3x^5}{5} + \frac{x^7}{14} + \frac{3x^{10}}{50}$$

Similarly

$$y_3 = 3 + \frac{x^2}{2} + \frac{3}{5}x^5 + \frac{1}{14}x^7 + \frac{3}{50}x^{10} + \frac{x^{12}}{168} + \frac{x^{15}}{250}$$

$$y(0.1) = 3 + 0.005 = 3.005 \text{ (approx.)}$$

$$y(0.2) = 3 + \frac{(0.2)^2}{2} + \frac{3}{5}(0.2)^5 + \cdots = 3.0202$$

$$y(0.3) = 3 + \frac{(0.3)^2}{2} + \frac{3}{5}(0.3)^5 + \cdots = 3.0464$$

Example 5. *Use modified Euler's method with one step to find y at* $x = 0.1$ *where* $\frac{dy}{dx} = x^2 + y$, *given* $y(0) = 0.94$.

Solution: $\frac{dy}{dx} = (x^2 + y)$

$$y_{(1)}^{(1)} = y_0 + hf(x_0, y_0) = 0.94 + 0.1 [0^2 + 0.94] = 1.034 \tag{1}$$

$$y_1 = y_0 + \frac{h}{2}[f(x_0, y_0) + f(x_1, y_1^{(1)})]$$

$$= 0.94 + .05 [(0 + 0.94) + (0.1)2 + 1.034]$$

$$= 1.0392$$

Example 6. Use Milne's method to find the solution of $\dfrac{dy}{dx} = x - y^2$ at $x = 0.8$, given

x	0	0.2	0.4	0.6
y	0	0.2	0.795	0.1762

Solution: Solving, we obtain

x	0	0.2	0.4	0.6
$y' = x - y^2$	$y'_0 = 0$	$y'_1 = 0.1996$	$y'_2 = 0.3937$	$y'_3 = 0.5689$

Using predictor formula

$$y_4 = y_0 + \frac{4h}{3}[2y'_1 - y'_2 + 2y'_3]$$

$$= 0 + \frac{0.8}{3}[2(0.1996) - (0.3937) + 2(0.5689)]$$

$$= \frac{0.8}{3} \times 1.1433 = 0.3049$$

Hence $\qquad y'_4 = x_4 - y_4^2 - 0.8 - (0.3049)^2 - 0.7072$

By corrector formula

$$y_4^{(1)} = y_2 + \frac{h}{3}[y'_2 + 4y'_3 + y'_4]$$

$$= 0.0795 + \frac{2}{30}[0.3937 + 4(0.5689 + 0.7072)]$$

$$= 0.795 + \frac{2}{30} \times 3.3765 = 0.0795 + 0.2251 = 0.3046$$

Example 7. Use Milne's Predictor–corrector method to find $y(0.4)$, given $\dfrac{dy}{dx} = 2e^x - y$

x	0	0.1	0.2	0.3
y	2	2.01	2.04	2.09

Solution:

x	0	0.2	0.4	0.6
$y' = 2e^x - y$	0	0.2003	0.4028	0.6097

By predictor formula

$$y_4 = y_0 + \frac{4h}{3}[2y'_1 - y'_2 + 2y'_3]$$

$$y'_4 = 2e^4 - y = 0.8213$$

Using Milne's predicted method, we have

$$y_4^{(c)} = y_2 + \frac{h}{3}[y_2' + 4y_3' + y_4']$$

$$= 2.04 + \frac{1}{30}[0.4028 + 4(0.6097) + 0.8213] = 2.1621$$

Hence $y(0.4) = 2.1621$

Example 8. *Using Runge–Kutta method find y when x = 1.2 in steps of 0.1 given that*

$$\frac{dy}{dx} = x^2 + y^2 \text{ and } y^{(1)} = 1.5$$

Solution: Here $f(x, y) = x^2 + y^2$, $h = 0.1$

For $y(1.1)$, we have $x_0 = 1$, $y_0 = 1.5$

$$k_1 = hf(x_0, y_0) = 0.1[1^2 + (1.5)] = 0.325$$

$$k_2 = hf\left(x_0 + h, y_0 + \frac{k_1}{2}\right) = 0.1[(0.105)^2 + (1.6625)^2] = 0.3866$$

$$k_3 = hf\left(x_0 + h, y_0 + \frac{k_2}{2}\right)$$

$$= 0.1[(1.05)^2 + (1.6933)] = 0.3969$$

$$k_4 = hf[(x_0 + h, y_0 + k_3)] = 0.1[(1.1)^2 + (1.8969)^2] = 0.4808$$

$\therefore$ $$y_1 = y_0 + \frac{1}{6}(k_1 + 2k_2 + 2k_3 + k_4)$$

$$= 1.5 + \frac{1}{6}(0.325 + 0.3866 + 0.3969 + 0.4808) = 1.8954$$

For $y(1.2)$, we have $x_1 = 1.1$, $y_1 = 1.8954$

$$k_1 = 0.1\left[(1.1)^2 + (1.8954)^2\right] = 0.4802$$

$$k_2 = (0.1)\left[(1.15)^2 + (2.1355)^2\right] = 0.5882$$

$$k_3 = (0.1)\left[(1.15)^2 + (2.1895)^2\right] = 0.6116$$

$$k_4 = (0.1)\left[(1.2)^2 + (2.5070)^2\right] = 0.7725$$

$$y(1.2) = 1.8954 + \frac{1}{6}[0.4802 + 1.764 + 1.2232 + 0.7725] = 2.5041$$

Example 9. *Use Runge–Kutta 4th order method to solve* $\frac{dy}{dx} = -2xy^2$, *given* $y(0) = 1$, *with* $h = 0.2$ *for* $x = 0.2$ *and* 0.4

Solution: $f(x, y) = -2xy^2$, $x_0 = 0$, $y_0 = 1$
$$f(x_0, y_0) = 0$$

i. For
$$h = 0.2, x = 0.2$$
$$k_1 = hf(x_0, y_0) = 0 \tag{1}$$

$$k_2 = hf\left(x_0 + \frac{h}{2}, y_0 + \frac{k_1}{2}\right)$$

$$= 0.2f(0 + 0.1, 1 + 0) = 0.2f(0.1, 1)$$
$$k_2 = 0.2(-2)(0.1)1^2 = -.04 \tag{2}$$

$$k_3 = hf\left(x_0 + \frac{h}{2}, y_0 + \frac{k_2}{2}\right)$$

$$= 0.2f(0.1, 1 - 0.02) = 0.2f(0.1, 0.98)$$
$$= (0.2)(-2)(0.1)(0.98)^2 = -0.384 \tag{3}$$
$$k_4 = hf(x_0 + h, y_0 + k_3)$$
$$= 0.2f(0.2, 1 - 0.0284) = 0.2f(0.2 - 0.9616)$$
$$= (0.2)(-2)(0.2)(-0.9616)2 = -0.0740 \tag{4}$$

$$k = \frac{1}{6}[k_1 + 2k_2 + 2k_3 + k_4] = -0.03846$$

$$y(0.2) = y_0 + k$$
$$= 1 - 0.03846 = 0.96154 \tag{5}$$

ii. To find y taking two steps of $h = 0.2$ at $x = 0.4$.

We will start with initial value as the value at $x = 0.2$

So now $x_1 = 0.2$ and $y_1 =$ the value of y obtained above $= 0.96154$
$$k_1 = hf(k_1, y_1)$$

$$= 0.2(-2)x_1 y_1^2$$
$$= -4(0.2)(0.96154)^2 = -0.08 \times 0.924567 = -0.0739648$$

$$k_2 = hf\left(x_1 + \frac{h}{2}, y_1 + \frac{k_1}{2}\right)$$

$$= (0.2)(-2)(0.3)\left(0.96154 + \frac{0.0739}{2}\right)^2$$

$$\approx -0.10257$$

$$k_3 = hf\left(x_1 + \frac{h}{2}, y_1 + \frac{k_2}{2}\right)$$

$$\approx 0.09942$$
$$k_4 = hf(x_1 + h, y_1 + k_3)$$
$$= -(0.2)2(0.4)(0.96154 - 0.9942)^2$$
$$\approx 0.118917$$

$$k = \frac{1}{6}[k_1 + 2k_2 + 2k_3 + k_4] \approx -0.09948$$

$\therefore$

$$y(0.2) = 0.96154$$
$$y(0.4) = 0.96154 - 0.09948 = 0.86205$$

NUMERICAL SOLUTION OF PARTIAL DIFFERENTIAL EQUATIONS

All the problems of engineering and physics reduce to the solution of differential equations both ordinary or partial. The solution has to be obtained satisfying the given boundary conditions. Theoretical solution of these equations is possible only in some very particular simple cases. So help is taken in solving these equations by numerical methods.

One can take up the solution of partial differential equations by the method of *finite differences*, which is most commonly used.

Method of finite differences: In this method, the partial derivatives occurring in the main equation and the given boundary conditions are replaced by their finite difference approximations. Thus, the problem changes into a linear difference equation, which is solved by iterative procedure numerically with the help of computers.

Finite difference approximations to derivatives: Consider a region R of the xy-plane around the point (x, y) at which the solution of our problems is required. Let the region be divided into a network of small rectangles of sides $\Delta x = h$ and $\Delta y = k$ by drawing lines parallel to the coordinate axes separated by distance h and k. The points of intersection of these lines are called *mesh points* (lattice points or grid points). Now using Taylor's series expansion, we have

$$u(x + h, y + k) = u(x,y) + \frac{1}{\lfloor 1}\left(h\frac{\partial u}{\partial x} + k\frac{\partial u}{\partial y}\right) + \cdots$$

$$\frac{1}{\lfloor 2}\left(h^2\frac{\partial^2 u}{\partial x^2} + 2hk\frac{\partial^2 u}{\partial x \partial y} + k^2\frac{\partial^2 u}{\partial y^2}\right) + \cdots \tag{12.26}$$

and

$$u(x + h, y - k) = u(x,y) - \frac{1}{\lfloor 1}\left(h\frac{\partial u}{\partial x} + k\frac{\partial u}{\partial y}\right) + \cdots$$

$$\frac{1}{\lfloor 2}\left(h^2\frac{\partial^2 u}{\partial x^2} + 2hk\frac{\partial^2 u}{\partial x \partial y} + k^2\frac{\partial^2 u}{\partial y^2}\right) + \cdots \tag{12.27}$$

Taking $k = 0$, (keeping y constant), the finite difference approximations for $\frac{\partial u}{\partial x}$ are

$$\frac{\partial u}{\partial x} = \frac{u(x+h,y) - u(x,y)}{h} + 0(h) \tag{12.28}$$

$0(h)$ means terms of the order of h or terms containing h and its higher powers,

or

$$\frac{\partial u}{\partial x} = \frac{u(x,y) - u(x - hy)}{h} + 0(h) \tag{12.29}$$

Taking average of Eqs (12.8) and (12.9)

$$\frac{\partial u}{\partial x} = \frac{u(x,+h,y) - u(x - h,y)}{2h} + 0(h^2) \tag{12.30}$$

and

$$\frac{\partial^2 u}{\partial x^2} = \frac{u(x - h,y) - 2u(x,y) + u(x + h,y)}{h^2} + 0(h^2) \tag{12.31}$$

The given region (say the rectangle) is divided into smaller parts of $\delta x = h$ and $\delta y = k$, with origin at the centre of rectangle (Fig. 12.3), we have

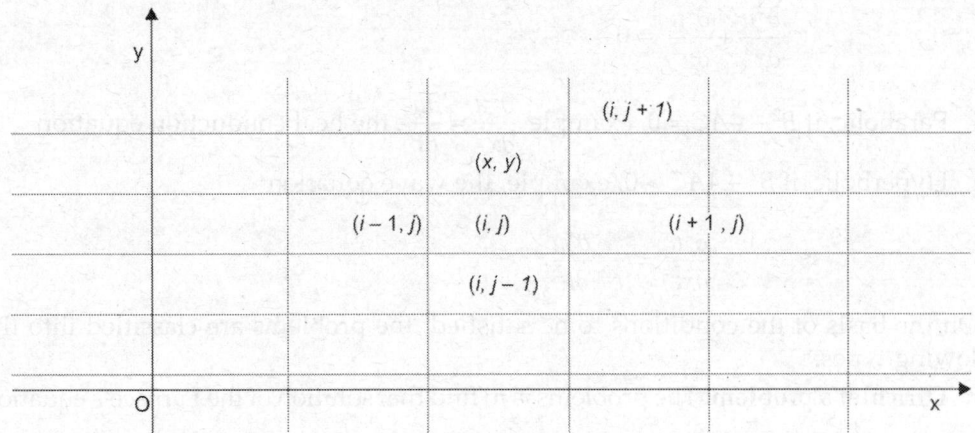

Fig. 12.3

Writing $u(x, y)\, u_{i,j}$, the above approximations become

$$u_x = \frac{u_{i+1,j} - u_{ij}}{h} + O(h) \tag{12.32}$$

$$= \frac{u_{i,j} - u_{i-1,j}}{h} + O(h) \tag{12.33}$$

$$= \frac{u_{i+1,j} - u_{i-1,j}}{2h} + O(h^2) \tag{12.34}$$

and $$u_{xx} = \frac{u_{i-1,j} - 2u_{i,j} + u_{i+1,j}}{h^2} + O(h^2) \tag{12.35}$$

Similary $$u_y = \frac{u_{i,j+1} - u_{ij}}{k} + O(k) \tag{21.36}$$

$$= \frac{u_{i,j} - u_{i,j-1}}{k} + O(k^2) \tag{12.36a}$$

$$= \frac{u_{i,j+1} - u_{i,j-1}}{2k} + O(k^2) \tag{12.36b}$$

and $$u_{yy} = \frac{u_{i,j-1} - 2u_{i,j+1}}{2k^2} + O(k^2) \tag{12.36c}$$

Replacing the partial derivatives by their different values using Eqs (12.32) and (12.39), we get the finite difference analogues of the given equation.

General second order linear partial differential equations: A general linear partial differential equations of order 2 in two independent variables is of the form

$$A(x,y)\frac{\partial^2 u}{\partial x^2} + B(x,y)\frac{\partial u}{\partial y \partial x} + C(x,y)\frac{\partial^2 u}{\partial y^2} + F\left(x,y,u\frac{\partial u}{\partial x}, v\frac{\partial v}{\partial y}\right) = 0 \tag{12.37}$$

The equation is classified and is called

   i. Elliptic, if $B^2 - 4AC < O$, example Laplace's equation

$$\frac{\partial^2 u}{\partial x^2} + \frac{\partial^2 u}{\partial y^2} = 0$$

  ii. Parabolic, if $B^2 - 4AC = 0$, example $\dfrac{\partial^2 u}{\partial x^2} = \dfrac{\partial u}{\partial t}$, the heat conduction equation

 iii. Hyperbolic, if $B^2 - 4AC > 0$, example, the wave equation

$$\frac{\partial^2 u}{\partial x^2} = \frac{1}{c^2}\frac{\partial^2 u}{\partial t^2}$$

On the basis of the conditions to be satisfied, the problems are classified into the following types:

   i. **Dirichlet's problem:** The problems is to find that solution of the Laplace's equation

$$\frac{\partial^2 u}{\partial x^2} + \frac{\partial^2 u}{\partial y^2} = 0 \text{ in a region } R \text{ and which is equal to a cotinuous function } f(x, y) \text{ on}$$

the boundary C of the region R. So $u(x, y) =$ given function $f(x, y)$ on C.

  ii. **Cauchy's problem:** The problem is to solve the wave equation $\dfrac{\partial^2 u}{\partial t^2} - \dfrac{\partial^2 u}{\partial x^2} = 0$

for $t > 0$ given the form of the string and its velocity at $t = 0$, represented by

     (a) $u(x, 0) = f(x)$ and (b) $\left(\dfrac{\partial u}{\partial t}\right)_{t=0} = g(x)$, where f and $g(x)$ are known functions.

 iii. Solving the heat condition equation $\dfrac{\partial^2 u}{\partial x^2} = \dfrac{\partial u}{\partial t}$, for $t > 0$ given the initial heat distribution $u(x, 0) = f(x)$.

Solution of Laplace Equation

The equation

$$\frac{\partial^2 u}{\partial x^2} + \frac{\partial^2 u}{\partial y^2} = 0 \tag{12.37a}$$

is to be solved for a region R bounded by a curve C on which $u = f(x, y)$ is given.

Taking $\Delta x = h = k = \Delta y$, let as explained above the region R be divided into a network of square meshes and the difference approximation analogues of the given equation becomes

$$\{u_{i-1,j} - 2u_{i,j} + u_{j+1,j}\} + \{u_{i,j-1} - 2u_{i,j} + u_{i,j+1}\} = 0 \tag{12.37b}$$

or

$$u_{ij} = \frac{1}{4}[u_{i-1,j} + u_{i+1,j} + u_{i,j-1} + u_{i,j+1}] \tag{12.38}$$

This shows that the value of u at the point (i, j)... is the average of its value at four neighbouring points on its left, right, above, and below (Fig. 12.4). It is called the *standard 5 points formula.*

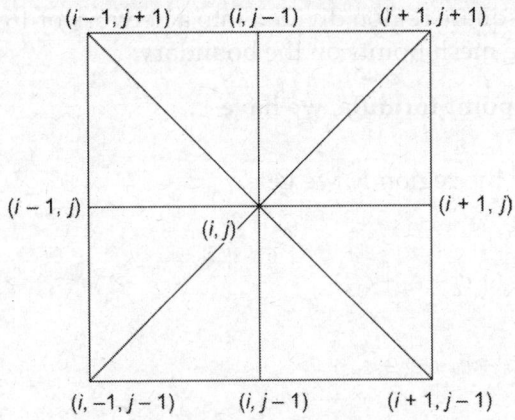

Fig. 12.4: Diagonal 5 points formula

The formula

$$u_{ij} = \frac{1}{4}\left[(u_{i-1,j-1} + u_{i+1,j+i}) + (u_{i+1,j-1} + u_{i-1,j+1})\right] \qquad (12.38a)$$

where the value of u at a mesh points (i, j) is the average of its values at the four neighbouring diagonal mesh points. It is called the diagonal 5 points formula. The justification of diagonal 5 points formula lies in the fact that the Laplace equation

$\dfrac{\partial^2 u}{\partial x^2} + \dfrac{\partial^2 u}{\partial y^2} = 0$ remains invariant in form if the axes be rotated through 45°.

Nine Point Formula

Based on the value of u at nine points including the point (i, j) the formula

$$u_{ij} = \frac{1}{20}\left[(u_{i-1,j-1} + u_{i+1,j+1}) + (u_{i-1,j+1} + u_{i+1,j-1})\right]$$

$$+ \frac{1}{5}\left[u_{i-1,j} + u_{i+1,j} + u_{i,j-1} + u_{i,j+1}\right]$$

has error of the order of $0(h^6)$ and is known as the 9 points formula (Fig. 12.5).

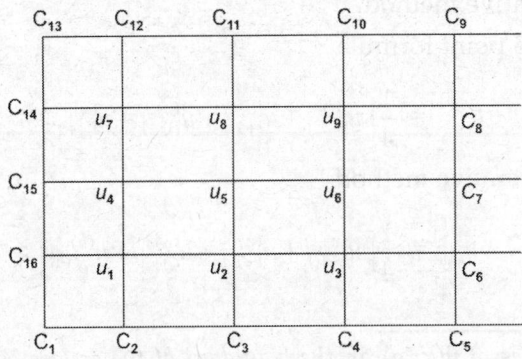

Fig. 12.5: Nine points formula

Let the region R be square region divided into a network of 16 meshes let $c_1, c_2, ..., c_{16}$ be the value of u at the mesh points on the boundary.

Using the diagonal 5 point formula, we have

$$\frac{\partial^2 u}{\partial x^2} + \frac{\partial^2 u}{\partial y^2} = 0 \text{ in the region } R, \text{ we get}$$

$$u_5 = \frac{1}{4}(c_1 + c_5 + c_9 + c_{13})$$

Also

$$u_1 = \frac{1}{4}(c_1 + c_3 + c_5 + c_{15})$$

$$u_3 = \frac{1}{4}(c_3 + c_5 + c_7 + u_{15})$$

$$u_9 = \frac{1}{4}(u_5 + c_7 + c_9 + c_{11})$$

$$u_7 = \frac{1}{4}(c_{15} + u_5 + c_{11} + c_{13})$$

Using standard 5 points formula

$$u_2 = \frac{1}{4}(c_1 + u_3 + c_3 + c_5)$$

$$u_4 = \frac{1}{4}(c_{15} + u_5 + u_1 + u_7)$$

$$u_6 = \frac{1}{4}(u_5 + c_7 + u_3 + u_9)$$

$$u_8 = \frac{1}{4}(u_7 + u_9 + u_5 + c_{11})$$

When all $u_i's$ ($i = 1, 2, ..., 9$) are calculated, their accuracy is improved by the following iterative methods.

i. Point-Jacabi-iterative method

Using standard 5 point formula

$$u_{i,j}^{n+1} = \frac{1}{4}\left[u_{i-1,j}^{(n)} + u_{i+1,j}^{(n)} + u_{i,j-1}^{(n)} + u_{i,j+1}^{(n)} \right]$$

ii. Gauss–Seidel interative method

$$u_{i,j}^{n+1} = \frac{1}{4}\left[u_{i-1,j}^{(n+1)} + u_{i+1,j}^{(n)} + u_{i,j-1}^{(n+1)} + u_{i,j+1}^{(n)} \right]$$

Exampe 1. *Given values of u(x, y) on the boundary of the square as shown in the figure, evaluate u(x, y) satisfying $\nabla^2 u = 0$, at the pivotal point shown in Fig. 12.6.*

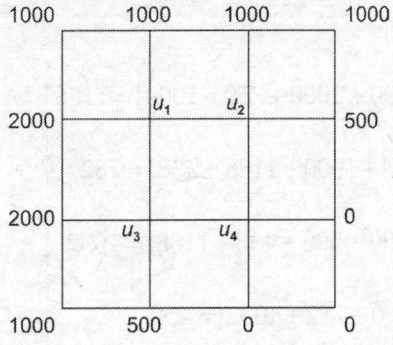

Fig. 12.6

Solution: To get the initial values of u_1, u_2, u_3, u_4, let us start with the assumption $u_4 = 0$. Then

$$\left. \begin{array}{l} u_1^{(1)} = \dfrac{1}{4}(2000 + 1000 + 1000 + 0) = 100 \text{ (diag. formula)} \\[3mm] u_2^{(1)} = \dfrac{1}{4}(u_1 + 500 + u_4 = 0 + 1000) = 625 \text{ (std. formula)} \\[3mm] u_3^{(1)} = \dfrac{1}{4}(2000 + 0 + 500 + 1000) = 875 \text{ (std. formula)} \\[3mm] u_4^{(1)} = \dfrac{1}{4}[875 + 0 + 0 + 625] = 375 \text{ (std. formula)} \end{array} \right\} \tag{2}$$

On second iteration using the formula, we have

$$\left. \begin{array}{l} u_1^{(n+1)} = \dfrac{1}{4}[1000 + 2000 + u_2^{(n)} + u_3^{(n)}] \\[3mm] u_2^{(n+1)} = \dfrac{1}{4}[500 + 1000 + u_1^{(n+1)} + u_4^{(n)}] \\[3mm] u_3^{(n+1)} = \dfrac{1}{4}[2000 + 500 + u_4^{(n)} + u_1^{n+1}] \\[3mm] u_4^{(n+1)} = \dfrac{1}{4}[0 + 0 + u_2^{(n+1)} + u_3^{(n+1)}] \end{array} \right\} \tag{3}$$

Using $n = 1$

$$\left. \begin{array}{l} u_1^{(2)} = \dfrac{1}{4}[1000 + 2000 + 625 + 875 = 1125 \\[3mm] u_2^{(2)} = \dfrac{1}{4}[500 + 1000 + 1125 + 375] = 750 \\[3mm] u_3^{(2)} = \dfrac{1}{4}[2000 + 500 + 1125 + 375] = 1000 \\[3mm] u_4^{(2)} = \dfrac{1}{4}[0 + 0 + 750 + 1000] = 438 \end{array} \right\} \tag{4}$$

Taking $n = 2$ in Eq. (3)

$$\left.\begin{aligned}
u_1^{(3)} &= \frac{1}{4}[1000 + 2000 + 750 + 1000] = 1188 \\[6pt]
u_2^{(3)} &= \frac{1}{4}[500 + 1000 + 1188 + 438] = 782 \\[6pt]
u_3^{(3)} &= \frac{1}{4}[2000 + 500 + 438 + 1188] = 1032 \\[6pt]
u_4^{(3)} &= \frac{1}{4}[0 + 0 + 782 + 1032] = 454
\end{aligned}\right\} \qquad (5)$$

Taking $n = 3$ in Eq. (4)

$$\left.\begin{aligned}
u_1^{(4)} &= \frac{1}{4}[1000 + 2000 + 782 + 1032] = 1,204 \\[6pt]
u_2^{(4)} &= \frac{1}{4}[500 + 1000 + 1204 + 454] = 789 \\[6pt]
u_3^{(4)} &= \frac{1}{4}[2000 + 500 + 1204 + 454] = 1,040 \\[6pt]
u_4^{(4)} &= \frac{1}{4}[0 + 0 + 789 + 1040] = 458
\end{aligned}\right\} \qquad (6)$$

Similarly

$$\left.\begin{aligned}
u_1^{(5)} &= 1207 \\
u_2^{(5)} &= 791 \\
u_3^{(5)} &= 1041 \\
u_4^{(5)} &= 458
\end{aligned}\right\} \qquad (7)$$

$$\left.\begin{aligned}
u_1^{(6)} &= 1208 \\
u_2^{(6)} &= 791 \\
u_3^{(6)} &= 1041 \\
u_4^{(6)} &= 458.25
\end{aligned}\right\} \qquad (8)$$

As there is no significant difference between the fourth, fifth and sixth iteration values, the solution is $u_1 = 1208$, $u_2 = 792$, $u_3 = 1042$, $u_4 = 458$.

Example 2. *Solve* $\dfrac{\partial^2 u}{\partial x^2} + \dfrac{\partial^2 u}{\partial y^2} = 0$ *by Jacobi's method (Fig. 12.7).*

Solution: With the help of given value (starting assumption),

let $\qquad u_1^{(0)} = 0,\ u_2^{(0)} = 0,\ u_3^{(0)} = 1,\ u_4^{(0)} = 1$

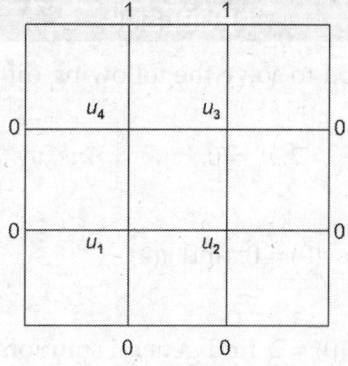

Fig. 12.7

Using standard formula

$$u_1^{(1)} = \frac{1}{4}[0+0+u_2^{(0)}+u_4^{(0)}] = \frac{0+0+0+1}{4} = 0.25$$

Its iteration

$$u_2^{(1)} = \frac{1}{4}[0+0+u_1^{(0)}+u_3^{(0)}] = \frac{0+0+0+1}{4} = 0.25$$

$$u_3^{(1)} = \frac{1}{4}[1+u_2^{(0)}+u_4^{(0)}+0] = \frac{0+0+1+0}{4} = 0.50$$

$$u_4^{(1)} = \frac{1}{4}[0+1+u_1^{(0)}+u_3^{(0)}] = 0.50$$

Using Jacobi's method

$$u_{i,j}^{n+1} = \frac{1}{4}\left[u_{i-1,j}^{(n)}+u_{i+1,j}^{(n)}+u_{i,j-1}^{(n)}+u_{i,j+1}^{(n)}\right]$$

2nd iterations

$$u_1^{(2)} = \frac{1}{4}(0+0+0.25+0.50) = 0.1875$$

$$u_2^{(2)} = \frac{1}{4}(0+0+0.25+0.50) = 0.1875$$

$$u_3^{(2)} = \frac{1}{4}(0+1+0.25+0.50) = 0.4375$$

$$u_4^{(2)} = \frac{1}{4}(0+1+0.25+0.50) = 0.4375$$

3rd iteration $u_1^{(3)} = u_2^{(3)} = 0.15625;\ u_3^{(3)} = u_4^{(3)} = 0.40625$

4th iteration $u_1^{(4)} = u_2^{(4)} = 0.14062;\ u_3^{(4)} = u_4^{(4)} = 0.39062$

5th iteration $u_1^{(5)} = u_2^{(5)} = 0.13281;\ u_3^{(5)} = u_4^{(5)} = 0.38281$

6th iteration $u_1^{(6)} = u_2^{(6)} = 0.12891;\ u_3^{(6)} = u_4^{(6)} = 0.37891$

PROBLEMS

1. Employ Picard's method to solve the following differential equations.

   i. $\dfrac{dy}{dx} = 1 + xy$, with $x_0 = 2, y_0 = 0$.
   $$\left[\text{Ans. } y_3 = \frac{x^4}{12} - \frac{5x^3}{6} + \frac{7x^2}{2} - \frac{35x}{6} + \frac{73}{12}\right]$$

   ii. $\dfrac{dy}{dx} = x^2 + y^2$, given $y(0) = 0$, find $y(4)$.　　　　　　　　　[*Ans.* 0.0214]

   iii. $\dfrac{dy}{dx} = y - x$, given $y(0) = 2$, find general solution upto x^4.

   $$\left[\text{Ans. } y = 2 + 2x + \frac{x^2}{2} + \frac{x^3}{6} - \frac{x^4}{24}\right]$$

2. Given $\dfrac{dy}{dx} = 1 + y^2$ with boundary conditions $x_0 = 0$, $y_0 = 0$. Solve by Picards

   method.
   $$\left[\text{Ans. } y = x + \frac{x^3}{3} + \frac{2x^5}{15}\right]$$

3. Use Taylor' series method to find $y(0.5)$ taking $h = 0.5$ when $\dfrac{dy}{dx} = x^2 + y^2$ and $y(0) = 1$. Find y correct to three places of decimal.　　　　　[*Ans.* $y = 1.0525$]

4. Solve $\dfrac{dy}{dx} = \dfrac{1}{x+y}$ for $x = 0.5$ to $x = 2$, $h = 0.5$ using Runge–Kutta method with

   $x_0 = 0$,

   [*Ans.* $y_{0.5} = 1.3571, y_1 = 1.5837, y_{1.5} = 1.7565, y_2 = 1.8950$]

5. Use Runge–Kutta method to solve the following.

   i. $\dfrac{dy}{dx} = xy$ and $x = 1.4$ given $x_0 = 1, y_0 = 2$ (take $h = 0.2$)　　[*Ans.* $y_{1.4} = 2.99948$]

   ii. $\dfrac{dy}{dx} = 1 + y^2$, given $y(0) = 0$, in steps of $h = 0.2$ find $y(0.2), y(0.4), y(0.6)$

   [*Ans.* 0.2027, 0.4228, 0.6841]

   iii. $\dfrac{dy}{dx} = x^2 + y^2$, given $y(0) = 1$, find $y(0.2)$　　　　　　　　[*Ans.* 1.2735]

6. Solve $\dfrac{dy}{dx} = x - y^2$ by Runge–Kutta method when $y(0) = 1$, Find value of y for

   $x = 0.2$　　　　　　　　　　　　　　　　　　　　　　　　　　[*Ans.* 0.76972]

7. Solve by Milne's method $\dfrac{dy}{dx} = 1 + xy^2, y(0) = 1$, for $x = 0.4, 0.5$, using Milne's, method given

x	0.1	0.2	0.3
y	1.105	1.223	1.355

$$\left[Ans.\ y_4^{(1)} = 1.538 \right]$$

8. Use Milne's method to solve $\dfrac{dy}{dx} = x + y$, when $y(0) = 1$, Find $y(0.2)$ and $y(0.3)$

[*Ans.* 1.2428, $y(0.3) = 1.3997$]

9. Solve by Milne's method $\dfrac{dy}{dx} = 2e^x - y$ at $x = 0.4, 0.5$, given

x	0	0.1	0.2	0.3
y	2	2.01	2.04	2.09

[*Ans.* $y_4 = 2.162,\ y_5 = 2.256$]

10. Solve $\dfrac{dy}{dx} = x + y$, where at $x = 0$, $y = 1$ by Runge–Kutta method from $x = 0$ to $x = 0.4$ with $h = 0.1$ [*Ans.* $y_1 = 1.1103,\ y_2 = 1.2428,\ y_3 = 1.5836$]

11. Use Taylor's series method for the solution of $\dfrac{dy}{dx} = 2y + 3e^x$ with $x_0 = 0, y_0 = 0$.

[*Ans.* $y(1) = 13.91, y(1.1) = 17.87$]

12. Solve $\dfrac{dy}{dx} = x - y^2$ given $y(0) = 1$ taking $h = 0.2$. Find solution for $x = 0.2, 0.4$ and 0.6.

[*Ans.* $y(.2) = 0.8512, y(.4) = 0.8798, y(.6) = 0.7620$]

13. Using Euler's modified method solve the following.

i. $\dfrac{dy}{dx} = \log(x + y)$, given $y(0) = 1$. Find y for $x = 0.2$ [*Ans.* 1.0490]

ii. $\dfrac{dy}{dx} = x^2 + y$, given $y(0) = 1$. Find $y(0.02), y(0.04), y(0.06)$

[*Ans.* 1.0202, 1.0408, 1.0619]

iii. $\dfrac{dy}{dx} = y^2 - \dfrac{y}{x}$, given $y(1) = 1$, in steps of 0.1 determine $y(1.1)$ to $y(1.6)$

[*Ans.* 1.0046, 1.0186, 1.0420, 1.0753, 1.1241, 1.1837]

14. Solve by Euler's method $\dfrac{dy}{dx} = 1 - y$, given $y_0 = 0$ for $0 \leq x \leq 0.2$ taking $h = 0.1$

[*Ans.* $y_2 = 0.181418$]

15. Solve $\dfrac{dy}{dx} = xy$, by Euler's method when $y(0) = 1$. Find $y(0.4)$ taking size of step

$h = 0.1$ [*Ans.* 1.0611]

16. Solve $\dfrac{dy}{dx} = \sqrt{x+y}$, given $y(0.4) = 0.41$ by Runge–Kutta method taking $h = 0.2$

            [*Ans.* 0.8490]

17. Solve by Taylor's series method $\dfrac{dy}{dx} = x - y^2$ with $y(0) = 1$ [*Ans.* 0.9138]

Solve by Taylor's series method $\dfrac{dy}{dx} = x + y^2$, given $y(0) = 1$ [*Ans.* 0.1266]

18. Use modified Euler's method to solve $\dfrac{dy}{dx} = x + y$, given $y(0) = 1$, taking $h = 0.2$

(step size) obtain value of y at $x = 1$.

19. Using Runge–Kutta method of fourth order and solve

$$\frac{dy}{dx} = \frac{y^2 - x^2}{y^2 + x^2} \text{ with } y(0) = 1$$

at $x = 0.02, 0.4$ [*Ans.* 1.196; 1.3752]

20. Using central difference approximation, solve $\nabla^2 u = 0$ at the nodal points of the square grid with boundary values as indicated in the Fig. 12.8

[*Ans.* $u_1 = 26.66$, $u_2 = 33.33$, $u_3 = 43.33$, $u_4 = 46.66$]

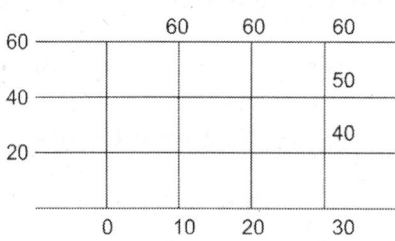

Fig. 12.8

21. Using central difference approximation, solve $\nabla^2 u = 0$ for the grid system as indicated in Fig. 12.9.   [*Ans.* $u_1 = 0.99$, $u_2 = 1.49$, $u_3 = 0.49$]

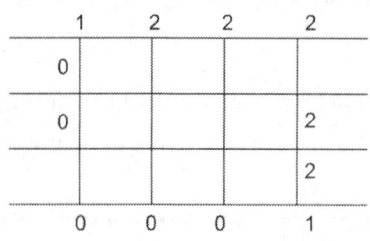

Fig. 12.9

MULTIPLE CHOICE QUESTIONS

1. The square root of a number N is to be obtained by applying the Newton–Raphson iteration to the equation $x^2 - N = 0$. If i denotes the iteration index, the correct iterative scheme will be

   (a) $x_{i+1} = \dfrac{1}{2}\left(x_i + \dfrac{N}{x_i}\right)$

   (b) $x_{i+1} = \dfrac{1}{2}\left(x_i^2 + \dfrac{N}{x_i^2}\right)$

   (c) $x_{i+1} = \dfrac{1}{2}\left(x_i + \dfrac{N^2}{x_i}\right)$

   (d) $x_{i+1} = \dfrac{1}{2}\left(x_i - \dfrac{N^2}{x_i}\right)$

2. Gauss–Seidel iterative method can be used for solving a set of
   (a) linear differential equation only
   (b) linear algebraic equations only
   (c) both linear and nonlinear algebraic equations
   (d) both linear and nonlinear differential equations

3. The convergence of the bisection method is
   (a) cubic
   (b) quadratic
   (c) linear
   (d) none of these

4. For the differential equation $\dfrac{dy}{dx} = x - y^2$, given that

x	0	0.2	0.4	0.6
y	0	0.02	0.0795	0.1762

   Using predictor–correction method, the y at next value of x is

   (a) 0.5114
   (b) 0.4648
   (c) 0.3046
   (d) 0.2498

5. Which one of the following is correct?
   (a) Bisection method is used for iteration
   (b) Regula–Falsi method is direct method
   (c) Secant method is direct method
   (d) Newton–Raphson method is not iterative method

6. If $n = 3$, $a_0 = 1$, $a_1 = 0$, $a_2 = -1$, $a_3 = -11$, then the root of the equation between 2 and 3 by Regula–Falsi method is
   (a) 2.0
   (b) 2.09
   (c) 2.9
   (d) 2.2

7. If $e^0 = 1$, $e^1 = 2.72$, $e^2 = 7.39$, $e^3 = 20.09$ and $e^4 = 54.60$, then by Simpson's $\dfrac{1}{3}$rd rule value of $\displaystyle\int_0^4 e^x dx$ is
   (a) 52.78
   (b) 53.87
   (c) 5.278
   (d) 5.387

8. Using bisection method, the negative root $x^3 - 4x + 9 = 0$ correct to three decimal places is
   (a) –2.506
   (b) –2.706
   (c) –2.406
   (d) none of these

9. A real root of the equation $x - \cos x = 0$ by the method false position correct to four decimal places is
   (a) 0.7391
   (b) 0.7439
   (c) 0.7347
   (d) none of these

10. Using Newton–Raphson method, a root correct to three decimal places of the equation $e^x = 1 + 2x$ is
    (a) 1.256
    (b) 1.255
    (c) 1.286
    (d) none of these

11. Match the items in columns I and II using the codes given below the columns.

Column I	Column II
(P) Gauss–Seidel method	(1) Interpolation
(Q) Forward Newton method	(2) Nonlinear differential equation
(R) Runge–Kutta method	(3) Numerical integration
(S) Trapeoidal rule	(4) Linear algebraic equation

    Codes

	P	Q	R	S
(a)	1	2	3	4
(b)	2	3	4	1
(c)	3	4	2	1
(d)	4	1	2	3

12. Newton–Raphson method is used to compute a root of the equation $x^3 - 13 = 0$ with 3.5 as the initial value. The approximation after one iteration is
    (a) 3.575
    (b) 3.677
    (c) 3.667
    (d) 3.607

13. Using Newton–Raphson method, a root correct to three decimal places of the equation $\sin x = 1 - x$ is
    (a) 0.5111
    (b) 0.500
    (c) 0.555
    (d) none of these

14. A root of the equation $x^3 - x - 11 = 0$ correct to four decimals using bisection method is
    (a) 2.3737
    (b) 2.3838
    (c) 2.3736
    (d) none of these

15. Using bisection method the negative value of $x^3 - x + 11 = 0$ is
    (a) –2.3736
    (c) –2.3838
    (c) –2.3737
    (d) none of these

16. Which of the following statements applies to the bisection method used for finding roots of functions?
    (a) Converges within a few iteration
    (b) Guaranteed to work for all continuous functions
    (c) It is faster than the Newton–Raphson method
    (d) Requires that there will be no error in determining the sign of the function

17. The Newton–Raphson method is used to find the root of the equation $x^2 - 2 = 0$, if iterations are started from -1, the iterations will be
    (a) converged to -1
    (b) converged to $\sqrt{2}$
    (c) converged to $-\sqrt{2}$
    (d) not converged

18. Newton–Raphson iteration formula for finding $\sqrt[3]{C}$, where $C > 0$ is
    (a) $x_{n+1} = \dfrac{2x_n^2 + \sqrt[3]{C}}{3x_n^2}$
    (b) $x_{n+1} = \dfrac{2x_n^3 - \sqrt[3]{C}}{3x_n^2}$
    (c) $x_{n+1} = \dfrac{2x_n^2 + C}{3x_n^2}$
    (d) $x_{n+1} = \dfrac{2x_n^2 - C}{3x_n^2}$

19. The iteration formula to find square root of a positive real number b using the Newton–Raphson method is
    (a) $x_{k+1} = \dfrac{3(x_k + b)}{2x_k}$
    (b) $x_{k+1} = \dfrac{(x_k^2 + b)}{2x_k}$
    (c) $x_{k+1} = \dfrac{x_k - 2x_k}{(x_k^2 + b)}$
    (d) none of these

20. Match the following and choose the correct combination:

Group I	Group II
(E) Newton–Raphson method	(1) Solving nonlinear equation
(F) Runge–Kutta method	(2) Solving linear simultaneous equations
(G) Simpson's Rule	(3) Solving ordinary differential equations
(H) Gauss elimination	(4) Numerical integration
	(5) Interpolation
	(6) Calculation of eigen values

    Codes:
    (a) E-6, F-1, G-5, H-3
    (b) E-1, F-6, G-4, H-3
    (c) E-1, F-3, G-4, H-2
    (d) E-5, F-3, G-4, H-1

21. Given $a > 0$, we wish to calculate its reciprocal value $\dfrac{1}{a}$ by using Newton–Raphson method for $f(x) = 0$. The Newton–Raphson algorithm for the function will be
    (a) $x_{k+1} = \dfrac{1}{2}\left(x_k + \dfrac{a}{x_k}\right)$
    (b) $x_{k+1} = \left(x_k + \dfrac{a}{2}x_k^2\right)$
    (c) $x_{k+1} = 2x_k - ax_k^2$
    (d) $x_{k+1} = x_k - \dfrac{a}{2}x_k^2$

22. A numerical solution of the equation $f(x) = x + \sqrt{x} - 3 = 0$ can be obtained using Newton–Raphson method. If the starting value of $x = 2$ for ten iterations, the value of x that is to be used in the next step is
    (a) 0.306
    (b) 0.739
    (c) 1.694
    (d) 2.306

23. A real root of equation $x^3 + x^2 - 1 = 0$ by iteration (method of successive approximation) method is
    (a) 0.7548765
    (b) 0.7548756
    (c) 0.7548776
    (d) 0.7548764

24. A real root of equation $\cos x = 3x - 1$ correct to seven decimal places by the method of successive approximation is
    (a) 0.6071016
    (b) 0.6071015
    (c) 0.60711516
    (d) none of these

25. Using trapezoidal rule and the table given below, $\int_4^{5.2} \ln x \, dx$ will be

x	4	4.2	4.4	4.6	4.8	5	5.2
$\ln x$	1.39	1.44	1.48	1.53	1.57	1.61	1.65

    (a) 1.8277
    (b) 1.9284
    (c) 1.6424
    (d) 0.98795

26. Using trapezoidal rule, taking 10 equal intervals $\int_0^{\pi/12} \sin x \, dx$, will be
    (a) 1.902
    (b) 1.941
    (c) 1.888
    (d) 0.99795

27. If $y_1 = 4$, $y_3 = 12$, $y_4 = 19$ and $y_x = 7$, then x will be
    (a) 1.42
    (b) 1.68
    (c) 1.86
    (d) 1.98

28. The table below gives values of a function $F(x)$ obtained for values of x at intervals 0.25.

x	0	0.25	0.5	0.75	1.00
$f(x)$	1	0.9412	0.8	0.64	0.50

The value of the integral of the function between the limit 0 to 1 using Simpson's rule is
    (a) 0.7854
    (b) 2.3562
    (c) 3.1416
    (d) 7.5000

29. In the solution of the following set of linear equations by Gauss elimination using partial pivoting $5x + y + 2z = 34$; $4y - 3z = 12$ and $10x - 2y + z = -4$; the pivots for elimination of x and y are
    (a) 10 and 4
    (b) 10 and 2
    (c) 5 and 4
    (d) 5 and –4

30. The equation $x^3 + 4x - 9 = 0$, needs to be numerically solved using the Newton–Raphson method is

(a) $x_{k+1} = \left(\dfrac{2x_k^3 + 9}{3x_k^2 + 4} \right) \cdot 8$

(b) $x_{k+1} = \dfrac{3x_k^2 + 4}{2x_k^2 + 9}$

(c) $x_{k+1} = x_k - 3x_k^2 + 4$

(d) $x_{k+1} = \dfrac{4x_k^2 + 3}{9x_k^2 + 2}$

SOLUTIONS

1. (a) Given, $x^2 - N = 0$

$$f(x) = x^2 - N$$
$$f'(x) = 2x$$

$$f_{x+i} = \frac{f(x_i)}{f'(x_i)} = x_i \frac{x_i^2 - N}{2x_i} = \frac{1}{2}\left[x_i + \frac{N}{x_i} \right]$$

2. (b) Linear algebraic equation
3. (c) Linear, because the order of convergence of bisection method is $1/2$.
4. (c) We have, $x = 0, 0.2, 0.4, 0.6$

$$f(x) = x - y^2$$

We get, $f_1(x) = 0.1996, f_2(x) = 0.3937, f_3(x) = 0.5689$

Using predictor formula

$$y_4^P = y_0 + \frac{4}{3}h(2f_1 - f_2 + 2f_3) \qquad \text{(here } h = 0.2)$$

$$= 0 + \frac{0.8}{3}[2(0.1996) - (0.3937) + 2(0.5689)] = 0.3049$$

$$y_4^C = y_2 + \frac{h}{3}\left(f_2 - 4f_3 + f_4^P \right)$$

$$f_4^{(P)} = f(x_4, y_4^P) = f(0.8, 0.3049)$$

$$y_4^{(C)} = 0.0795 + \frac{0.02}{3}[0.3937 + 4(0.5689) + 0.7070] = 0.3046$$

5. (a) Bisection method, Regula–Falsi method, Secant method, Newton–Raphson methods are iterative methods.

6. (b) If $n = 3, a_0 = 1, a_1 = 0, a_2 = -2, a_3 = -5$, then the root of the equation between 1.75 and 2.5 by Regula–Falsi method is obtained from the equation $x^3 - 2x - 5 = 0$

$$x_1 = \frac{bf(a) - af(b)}{f(a) - f(b)}$$

where $a = 1.75, \ b = 2.5$

$$x_1 = 2.0189$$

and $f(x_1) = f(2.0189) = -0.8901 < 0$

Repeating the process, we get $= 2.094$

7. (b) Here, $h = 1$

By Simpson rule,

$$\int_0^4 e^x dx = \frac{h}{3}[(e^0 + e^4) + 4(e^1 + e^3) + 2e^2]$$

$$\int_0^4 e^x dx = \frac{1}{3}[(55.60) + (91.24) + (14.78)]$$

8. (a)

9. (a)

10. (a)

11. (d) Gauss–Seidel method—linear algebraic equation; Newton's forward interpolation; Runge–Kutta method—nonlinear differential equation; trapezoidal rule—numerical integration

12. (d) $f(x) = x^2 - 13 = 0$. $f'(x) = 2x$

Initial value approximation

$$x_0 = 3.5$$

If x_1 be the approximation after one iteration. By Newton–Raphson method

$$x_1 = x_0 - \frac{f(x_0)}{f'(x_0)} = 3.5 - \frac{(3.5)^2 - 13}{2 \times 3.5} = 3.607$$

13. (a)

14. (c)

15. (a)

16. (b) Bisection method is used for finding the roots of continuous functions.

17. (c)

18. (c)

19. (b)

20. (b)

21. (c) Given, $a > 0$ and $x = \frac{1}{a}$ or $a = \frac{1}{x}$

Let $f(x) = \frac{1}{x} - a = 0$

Here, we wish to find the reciprocal of $\left(\frac{1}{a}\right)$

$$f'(x) = -\frac{1}{x^2}$$

Now, by Newton–Raphson method

$$x_{k+1} = x_k - \frac{f(x_k)}{f'(x_k)} = -\frac{\left(\frac{1}{x_k} - 1\right)}{\left(-\frac{1}{x_k^2}\right)} = x_k + (x_k - ax_k^2)$$

$$x_{k+1} = 2x_k - ax_k^2$$

22. $x_{n+1} = x_n - \dfrac{f(x_n)}{f'(x_n)}$, given $f_{(x)} \equiv x + \sqrt{x} - 3 = 0$

$$f'(x) = 1 + \frac{1}{2\sqrt{x} - 3} \quad \text{or} \quad f(2) = (2 + \sqrt{2} - 3) = \sqrt{2} - 1$$

$$f'(2) = \frac{2\sqrt{2} + 1}{2\sqrt{2}}$$

or $$x_{n+1} = 2 - \frac{\sqrt{2} - 1}{\frac{2\sqrt{2} + 1}{2\sqrt{2}}} = 1.694 \qquad\qquad (\because x_0 = 2)$$

23. (c) A necessary condition

24. (a)

25. (a) By trapezoidal rule, we have

$$\int_{x_0}^{x_0 + nh} f(x)\,dx = \frac{h}{2}[(y_0 + y_n) + 2(y_1 + y_2 + y_3 + \cdots + y_{n-1})] \qquad \left(\because h = \frac{5.2 - 4}{6}\right)$$

$$= \frac{0.2}{2}[(1.39 + 1.65) + 2(1.44 + 1.48 + 1.53 + 1.57 + 1.61)] = 1.82764$$

25. (d) Divide $\left[0, \dfrac{\pi}{2}\right]$ into ten equal parts each of width $h = \dfrac{\pi}{20}$. Then, the value of $y = \sin x$ are given in the table below.

x	0	$\dfrac{\pi}{20}$	$\dfrac{2\pi}{20}$	$\dfrac{3\pi}{20}$	$\dfrac{4\pi}{20}$	$\dfrac{5\pi}{20}$
$\sin x$	0	0.15643	0.30902	0.45399	0.5877	0.70711

x	$\dfrac{6\pi}{20}$	$\dfrac{7\pi}{20}$	$\dfrac{8\pi}{20}$	$\dfrac{9\pi}{20}$	$\dfrac{10\pi}{20}$
$\sin x$	0.80902	0.89101	0.95106	0.9876	1

Using trapezoidal rule,

$$\int_0^{\pi/2} \sin x\,dx = \frac{h}{2}[(y_0 + y_{10}) + 2(y_1 + y_2 + y_3 + y_4 + y_5 + y_6 + y_7 + y_8 + y_9)]$$

$$= \frac{\pi}{40}[12.70624] = 0.99795$$

27. (c) $$x = \frac{(-5)(-12)}{(-8)(-15)}(1) + \frac{(3)(-12)}{(8)(-7)}(3) + \frac{(3)(-5)}{(15)(7)}(4) = \frac{1}{2} + \frac{27}{14} - \frac{4}{7} = 1.86$$

28. (a) Given, $h = 0.25$

x	0	0.25	0.5	0.75	1.00
$f(x)$	1	0.9412	0.8	0.64	0.50

By Simpson's $\dfrac{1}{3}$ rd rule

$$\int_{x_0}^{x_0+nh} F(x)\,dx = \dfrac{0.25}{3}[(1+0.50)+2(0.8)+4(0.9412+0.64)]$$

$$= \dfrac{0.25}{3}[(1+0.50)+2(0.8)+4(0.9412+0.64)]$$

$$= \dfrac{0.25}{3}(9.4248) = 0.7854.$$

29. (a) The given system of equations is

$$5x + y + 2z = 34 \tag{1}$$
$$4y - 3z = 12 \tag{2}$$
$$10x - 2y + z = -4 \tag{3}$$

Now, we follow the Gauss elimination method. First we eliminate the term x in Eqs (2) and (3) with the help of Eq. (1) but Eq. (2) does not have any x term, so we only eliminate x in Eq. (3) and coefficient of which is called our first pivot.

I pivot:

$$10x - 2y + z = -4$$
$$2(34 - y - 2z) - 2y + z = -4$$
$$68 - 2y - 4z - 2y + z = -4$$

II pivot:

$$4y - 3z = 12 \tag{4}$$

Now, we eliminate the term y in Eqs (2) and (4) and the coefficient of y is called our second pivot.

$$4y - 3z = 12$$
$$4y + 3z = -72$$
$$\overline{---}$$
$$6z = 60$$
$$z = 10$$

So, system of equations becomes

$$5x + y + 2z = 34$$
$$4y - 3z = 12$$
$$\overline{z = 10}$$
$$\text{First pivot} = 10$$
$$\text{Second pivot} = 4$$

30. (a) Let $\quad f(x) = x^3 + 4x - 9 = 0$
$$f'(x) = 3x^2 + 4 = 0$$

Now, Newton–Raphson method gives

$$x_{k+1} = x_k - \dfrac{f(x_k)}{f'(x_k)} \quad \text{or} \quad x_k - \dfrac{x_k^3 + 4x_k - 9}{3x_k^2 + 4}$$

$$x_{k+1} = \dfrac{3x_k^3 + 4x_k - x_k^3 - 4x_k + 9}{3x_k^2 + 4} \quad \text{or} \quad \left(\dfrac{2x_k^3 + 9}{3x_k^2 + 4}\right)$$

13

Probability and Statistics

A. PROBABILITY

13.1. BASIC DEFINITIONS

1. **Deterministic experiments:** Experiments which have a single outcome are called *deterministic*. Example. Pump air in a ballon and it will expand.

2. **Random experiment:** If an experiment may result in any one of its various outcomes, then it is called a *random experiment.* Example: Tossing a coin has two outcomes. It may result in head or tail. Rolling a die has six outcomes as any one of its six faces may come uppermost. So a random experiment done in an unbiased manner and whose result is based on chance factor.

3. **Event:** Each one of the several outcomes of a random experiment is called an *elementary event.* If an event includes several elementary events, it is called a compound event. In rolling a cubical die, with faces marked from 1 to 6, the face marked 5, coming uppermost is an elementary event but an even number coming uppermost in a compound event, as it is made of three elementary events, viz. (2, 4, 6), coming uppermost.

4. **Sample space:** The set of all possible outcomes of a random experiment is called its *sample space.*

    Sample space $\qquad S = (H, T)$, in toss of a coin.

    $\qquad\qquad\qquad S = \{1, 2, 3, 4, 5, 6\}$, in rolling a die.

    If 50 tickets marked 1 to 50 are included in a lottery then sample space

    $$S = \{1, 2, 3,... 49, 50\}$$

5. **Trial:** Performing a random experiment is called a trial. Example: Tossing a coin; drawing a card from a pack of 52 cards.

6. **Exhaustive cases:** If an event $B = B_1 \cup B_2... \cup B_n$ then $B_1, B_2,...,B_n$ are its exhaustive cases, where $B_i \cap B_j = \phi$, $i \neq j$, i, $j = 1, 2,..., n$. Thus, exhaustive case of an event is the set of various mutually inclusive outcomes in which the event B may result in a trial.

7. **Mutually exclusive:** If two events be such that one precludes the happening of the other, they are called *mutually exclusive.* Thus, in toss of a coin, head and tail both cannot come together, so they are mutually exclusive.

If $E_1, E_2,..., E_n$ be the set of mutually exclusive exhaustive cases into which an event E may result, then

Sample space of $E = E_1 U E_2,..., U E_n.$

8. **Probability:** If an event A can happen in m ways and fails to happen in n ways in a random experiment, then

$P(A)$ = Probability of happening of $A = \dfrac{m}{m+n}$

$P(A\text{-not}) \equiv P(\overline{A})$ = Probability of not happening of $A = \dfrac{n}{m+n}$

Thus, probability is a numerical fraction

$$0 \le P(A) \le 1,$$

$$P(A) + P(\overline{A}) = \dfrac{m}{m+n} + \dfrac{n}{m+n} = 1$$

We should note that if probability of an event be zero, it does not mean that the event will never happen. May be, that it happens at the first trial and then may not happen at all. Similarly, if the probability of an event be one, it does not mean that the event will always happen. It may fail even at the first trial and then happen always.

For example, suppose one has only one ticket in a lottery of 10 lac tickets, though his chance of winning is almost zero, yet if his number comes in the draw, he wins and the event has happened at the first trial inspite of its probability being almost zero.

Theoretical and Experimental Probability

In toss of a coin (theoretically), a priori probability $P(H) = P(T) = \dfrac{1}{2}$. But on actual tossing experimentally if a coin when tossed 1000 times, head results 502 times and tail 498 times and if this be the average of several such trials then experimental probability or a posteriori probability of head is 0.502.

Equally likely cases: If two events, in an experiment, are equally probable, they are called equally likely. Thus $P(H) = P(T)$, in toss of a coin, head and tail are said to be equally likely to occur.

Independent events: If two or more events can happen simultaneously then two such events are said to be independent if the happening or nonhappening of one of them does not affect the happening or nonhappening of the other. We will see later that for such events $P(AB) = P(A) \cdot P(B)$, where $P(AB)$ denotes the probability of the simultaneous occurrence of A and B. Example: Let a coin be tossed and a die be rolled then the probability of head in the toss and face marked 5 coming uppermost is $\dfrac{1}{2} \times \dfrac{1}{6}$ as sample space $S = \{H1, H2, H3, H4, H5, H6, T1, T2, T3, T4, T5, T6\}$.

As all the 12 outcomes are equally likely and probability of each is $\dfrac{1}{12}$.

Hence $P(H5) = \dfrac{1}{12}$. Also $P(H) \cdot P(5) = \dfrac{1}{2} \times \dfrac{1}{6} = \dfrac{1}{12}$

So the events are independent.

Conditional Probability

The probability of happening of an event A when another event B is known to have already happened is called *conditional probability* of A and denoted by $P(A/B)$.

Odds in favour of an event A

It is defined to be $= \dfrac{P(A)}{P(\bar{A})}$ $\bar{A} \equiv$ not-A

Odds against an event

It is defined to be $= \dfrac{P(\bar{A})}{P(A)}$

If the probability of an event $A \equiv \dfrac{P(A)}{P(\bar{A})} = \dfrac{2}{3}$ then $P(\bar{A}) = 1 - \dfrac{2}{5} = \dfrac{3}{5}$

Odds in favour of $A = \dfrac{P(A)}{P(\bar{A})} = \dfrac{2}{3}$ and odds against $A = \dfrac{P(\bar{A})}{P(A)} = \dfrac{3}{2}$

Note: The standard notations will be adopted without defining them again and again.

Example 1. *A die is rolled in an unbiased way. Find the probability of getting (i) an even number (ii) a multiple of 3 and even also (iii) a multiple of 3 or an even number.*

Solution: Sample space $= S\,\{1, 2, 3, 4, 5, 6\}$. All events are equally likely, so

$$P(1) = P(2) = P(3) = P(4) = P(5) = P(6) = \dfrac{1}{6}$$

i. Getting of an even number $\equiv$ event A, then

$$P(A) = \text{prob of getting 2 or 4 or 6}$$

$$= \text{prob }(2) + \text{prob }(4) + \text{prob }(6) = \dfrac{1}{6} + \dfrac{1}{6} + \dfrac{1}{6} = \dfrac{3}{6}.$$

ii. Getting of numbers which are both even and also multiple of $3 \equiv$ event C. Such number is only no. 6

So, probability of $(B) = \dfrac{1}{6}$

iii. Getting the number which is either even or multiple of $3 \equiv$ event C. Such numbers are 2, 4, 6, 3.

So $P(C) = \dfrac{4}{6}$

Example 2. *A bag contains 5 white and 4 black balls, three balls are drawn at random, find the chance that (i) they are all white (ii) no white (ii) only one white.*

Solution: Total number of balls $= 9$

No. of ways in which three balls can be taken out of $9 = {}^9C_3$.

No. of ways that all are white, means all the three are drawn from 5 white balls, the possible ways of doing so = $^5C_3$.

So prob. of getting all white = $\dfrac{\text{No. of favourable ways}}{\text{Total number of ways}} = \dfrac{^5C_3}{^9C_3} = \dfrac{5}{42}$

Prob. of getting no white = Prob. of getting all black = $\dfrac{^4C_3}{^9C_3} = \dfrac{1}{21}$

Prob. (1 W and 2 B) = $\dfrac{^5C_3 \times ^4C_3}{^9C_3} = \dfrac{5}{14}$

Example 3. *Write the sample space for the sum of points in a single throw with two dice with their respective probabilities.*

Solution: Minimum number on each die is 1 and maximum is 6

So minimum sum = 1 + 1 = 2 and maximum = 6 + 6 = 12

Sample space = {2, 3, 4, 5, 6, 7, 8, 9, 10, 11, 12}. Sum = 7 can be formed when

Point on one die	Point on other	No. of ways
1	+ 6	
2	+ 5	
3	+ 4	
4	+ 3	= 6
5	+ 2	
6	+ 1	

Similarly calculations can be done for other numbers also and we find:

Sum	2	3	4	5	6	7	8	9	10	11	12
Number of favorable ways	1	2	3	4	5	6	5	4	3	2	1
									Total = 6 × 6 = 36		
Respective probabilities	$\dfrac{1}{36}$	$\dfrac{2}{36}$	$\dfrac{3}{36}$	$\dfrac{4}{36}$	$\dfrac{5}{36}$	$\dfrac{6}{36}$	$\dfrac{5}{36}$	$\dfrac{4}{36}$	$\dfrac{3}{36}$	$\dfrac{2}{36}$	$\dfrac{1}{36}$

Example 4. *Find the chance of having the sum greater than 7 in a single throw with two dice.*

Solution: Sum of number on the dice with their respective number of favourable cases is as shown below.

Sum	2	3	4	5	6	7	8	9	10	11	12
Number of favourable ways for the above sum	1	2	3	4	5	6	5	4	3	2	1
										Total = 36	

Probability (sum = 8, 9, or 10 or 11 or 12)

$$= \frac{\text{Sum of their favourable cases}}{\text{Total number of cases}} = \frac{5+4+3+2+1}{36} = \frac{15}{36}$$

Example 5. *The dice are thrown simultaneously. Find the probability that the sum of the number on the two dice is 10.*

Solution: The experiment of throwing two dice can result in $6 \times 6 = 36$ ways. Now we will determine the favourable number of cases.

The number of one die is given by exponents of the function

$$f(x) = x + x^2 + x^3 + x^4 + x^5 + x^6$$

$$= x\frac{(1-x^6)}{1-x} \quad \text{(sum of a GP)}$$

The two dice are same, therefore the sum of the number on two dice are given by the coefficients of x in the function.

$$[f(x)]^2 = \left[x\frac{(1-x^6)}{1-x} \right]^2 = x^2(1-x^6)^2(1-x)^{-2}$$

$$= x^2[1 - 2x^6 + x^{12}](1-x)^{-2}$$

The coefficient of x^{10} in the above function will give the favourable number of cases, i.e. sum on two faces is ten. This is given by

$$^{-2}C_8(-1)^8 - 2\,^{-2}C_2 = {}^{8+2-1}C_8 - 2\,^{2+2-1}C_2$$

$$= {}^9C_8 - 2\,^3C_2$$

$$= 9 - 6 = 3$$

∴ The required probability is

$$\frac{3}{36} = \frac{1}{12}$$

Example 6. *A has one third shares in a lottery in which there are 3 prizes and 6 blanks; B has, share in the lottery in which there is one prize and 2 blanks: show that A's chance of success to B is 16 is to 7.*

Solution: In 3 draws made by A, the possibilities with their respective probabilities are as shown below.

A	No prize drawn	1 prize + 2 blanks	2 prizes + 1 blank	3 prizes
Probabilites	$\dfrac{^6C_3}{^9C_3}$	$\dfrac{^6C_2 \times\,^3C_1}{^9C_3}$	$\dfrac{^6C_1 \times\,^3C_2}{^9C_3}$	$\dfrac{^3C_3}{^9C_3}$
	$\dfrac{6\times5\times4}{6\times84}$	$\dfrac{15\times3}{84}$	$\dfrac{18}{84}$	$\dfrac{1}{84}$

$P(A)$ = Probability of A getting at least one prize

$$= \frac{45+18+1}{84} = \frac{64}{84} = \frac{16}{21}$$

Similarly $P(B)$ = B's chance of success

$$= \frac{^1C_1}{^3C_1} = \frac{1}{3}$$

$$\frac{P(A)}{P(B)} = \frac{16}{21} \times \frac{3}{1} = \frac{16}{7}$$

Example 7. *In hand at whist what is the chance that, the 4 kings are held by a specified player.*

Solution: Let the players be A, B, C, and D. Then A is to have 4 kings and 9 other cards out of 13 cards dealt to him

Its probability $= \dfrac{^4C_4 \times {}^{48}C_9}{^{52}C_{13}} = \dfrac{1 \times \lfloor 48 \; \lfloor 39 \; \lfloor 13}{\lfloor 9 \; \lfloor 39 \quad \lfloor 52}$

$$= \frac{10 \times 11 \times 12 \times 13}{52 \times 51 \times 50 \times 49} = \frac{11}{4165}$$

Example 8. *In shuffling a pack of 52 cards, four are accidently dropped, find the chance that the missing cards be one from each suit.*

Solution: If 52 cards are there and 13 cards of each suit (heart, diamond, spade and club). We have to find the probability of taking 4 cards from 52 cards, one card of each suit and for that the required probability is

$$P = \frac{(^{13}C_1)(^{13}C_1)(^{13}C_1)(^{13}C_1)}{^{52}C_4} = \frac{13 \times (13)^3 \times 24}{52 \times 51 \times 50 \times 49} = \frac{2197}{20825}$$

Example 9. *In a single throw with two dice, find the chance of getting a doublet.*

Solution: Doublet means getting either one of $(1, 1), (2, 2), (3, 3), (4, 4), (5, 5), (6, 6)$.

So required probability $= \dfrac{\text{No. of favourable cases}}{\text{Total number of cases}} = \dfrac{6}{36} = \dfrac{1}{6}$

Example 10. *In how many ways can 4 letters be posted if there are 5 letter boxes available. What is the chance that all are posted in one box.*

Solution: For each letter the number of ways of posting = 5

So, total number of ways of posting 4 letters = $5 \times 5 \times 5 \times 5$ = 625.

Now there are 5 ways in which all the letters are put in one of the boxes, viz. in box B_1 or B_2 or B_3 or B_4 or B_5. So required probability $= \dfrac{5}{625}$

Example 11. *A box contains 100 bulbs out of which 20 are defective, if 10 bulbs be taken for inspection, find the chance that:*

   i. all are defective

  ii. none is defective

 iii. at least one is defective

Solution: Number of ways of choosing 10 balbs out of $100 = {}^{100}C_{10}$

$$\text{Probability (all are defective)} = \frac{{}^{20}C_{10}\,(\text{all chosen from 20 defectives})}{{}^{100}C_{10}}$$

$$\text{Prob. (all are good)} = \frac{{}^{80}C_{10}}{{}^{100}C_{10}}$$

Probability at least one is defective

$$= 1 - \text{Probability that all are good}$$

$$= 1 - \frac{{}^{80}C_{10}}{{}^{100}C_{10}}$$

Example 12. *Thirteen persons take their seats at a round table, show that it is 5 : 1 against two particular persons sitting together.*

Solution: Number of ways in which 12 persons may sit

$$= \lfloor 13 - 1 = \lfloor 12$$

Let two persons A_1 and A_2 be replaced by C, then C and 11 others = total 12

They can arrange in $\lfloor 12 - 1 = \lfloor 11$ ways.

A_1 and A_2 can arrange as $A_1 A_2$ and $A_2 A_1$ in 2 ways.

So probability that A_1 and A_2 are togather $= \dfrac{2\lfloor 11}{\lfloor 12} = \dfrac{1}{6}$

Probability that they are not near to each other $= 1 - \dfrac{1}{6} - \dfrac{5}{6}$

Odds against they being together $= \dfrac{P(\bar C)}{P(C)} = \dfrac{1}{6} \div \dfrac{6}{5} = \dfrac{5}{1} = 5:1.$

Example 13. *A and B are two independent witnesses in a case. The probability that A speaks truth is x and the probability that B speaks truth is y. In narrating a certain event if they agree, show that the event actually happened is* $\dfrac{xy}{1 - x - y + 2xy}$

Solution: When they agree, then both are speaking truth or both are speaking untruth.

The probability of this is $xy + (1 - x)(1 - y)$

Then probability that both say truth $= \dfrac{xy}{xy + (1 - x)(1 - y)}$

Example 14: *A, B, C toss a coin in succession, on the understanding that one to get head would win. Find their respective chances of winning.*

Solution: *Probability of A:* Probability of head at 1st, 4th, 6th, 10th chances $p + q.q.q + q^6 P...$ It is infinite GP, with common ratio $= q^3$.

So
$$P(A) = \frac{p}{1-q^3} = \frac{\frac{1}{2}}{1-\frac{1}{8}} = \frac{4}{7}$$

Probability of B: If at 1st throw A does not get head, then B would get the chance B will have the 2nd, 5th, 8th,... chances.

So
$$P(B) = qp + q^4p + q^7p... = \frac{qp}{1-q^3}$$

$$= \frac{\frac{1}{2} \times \frac{1}{2}}{1-\frac{1}{8}} = \frac{2}{7}$$

If chances of C comes, he will be getting head at 3rd, 6th, 9th,...

So
$$P(C) = q^2p + q^5p+\cdots \frac{a^2p}{1-q^3} = \frac{\frac{1}{8}}{1-\frac{1}{8}} = \frac{1}{7}$$

Thus if A gets the first chance, B second and C third and then further chances in the same way till at last some one gets head, then

$$P(A) = \frac{4}{7}; P(B)\frac{2}{7}; P(C) = \frac{1}{7}$$

13.2. LAW OF ADDITION OF PROBABILITIES

For two events A and B, which are parts of the same sample space S
$$P(A \cup B) = P(A) + P(B) - P(AB)$$
also written as $P(A + B) = P(A) + P(B) - P(AB)$, in usual standard notations.

Let A and B be events made up of n and m elementary events out of which x elementary event x are common (Fig. 13.1).

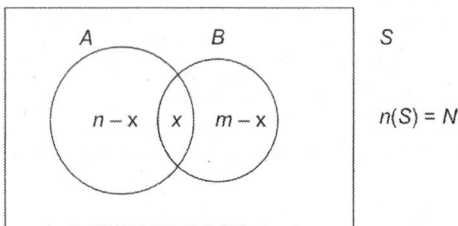

Fig. 13.1

The $n(A) \equiv$ number of elements in $A = n$

$n(B) \equiv$ number of elements in $B = m$

$n(AB) \equiv$ number of elements common to both A and $B = x$

Number of elements in

$$A \cup B = n(A \cup B) \equiv (n - x) + x + (m - x) = n + m - x$$

So $P(A \cup B) = \dfrac{n + m - x}{N} = \dfrac{n}{N} + \dfrac{m}{n} - \dfrac{x}{n} = P(A) + P(B) - P(AB)$ (13.1)

If $A \cap B = \phi \Rightarrow$ number of elements in $A \cap B = 0$

or $x = 0$

Then above theorem reduces to

$$P(A \cup B) = P(A) + P(B),$$ (13.2)

when A and B are disjoint.

For three events A, B and C

$$P(A + B + C) = P(A) + P(B) + P(C) - \{P(AB) + P(BC) + P(CA)\} + P(ABC)$$

Law of Multiplication of Probabilities

For two events A and B

$$P(AB) = P\left(\frac{A}{B}\right) \cdot P(B) = P\left(\frac{B}{A}\right) \cdot P(A),$$

where $P\left(\dfrac{A}{B}\right)$ is the conditional probability of $A \equiv$ probability of happening of A,

assuming B has already happened, and $P\left(\dfrac{B}{A}\right)$ is the conditional probability of B = the

probability of B, when it is known that A has already happened.

Proof: Let the events not be independent and have something in common

Number of elements in A:

$$n(A) = x + y$$
$$n(B) = y + z$$
$$n(A \cap B) = y$$

So $P(A) = \dfrac{x + y}{N}$, $P(B) = \dfrac{y + z}{N}$, $P(AB) = \dfrac{y}{N}$ (13.3)

$$P\left(\frac{A}{B}\right) = \text{Probability of } A, \text{ when the number of elements in the}$$

sample space reduces to $n(B) = y + z$

$$= \frac{y}{y + z}$$ (13.4)

$$P\left(\frac{B}{A}\right) = \text{Probability of } B, \text{ when } A \text{ has happened or the number}$$

of elements in the new sample space reduces to $n(A) = x = y$, so

$$P\left(\frac{B}{A}\right) = \frac{y}{x+y} \tag{13.5}$$

Now

$$P(AB) = \frac{y}{N}$$

$$= \frac{y}{x+y} \cdot \frac{x+y}{N} \quad \text{or} \quad = \frac{y}{y+z} \cdot \frac{y+z}{N}$$

So

$$P(AB) = P\left(\frac{B}{A}\right) \cdot P(A) \quad \text{or} \quad P\left(\frac{A}{B}\right) \cdot P(B) \tag{13.6}$$

13.3. EVENTS INDEPENDENT

If A and B are not part of the same event, they are independent, when the happening of A does not dependent on the happening or nonhappening of B, then

$$P\left(\frac{A}{B}\right) = P(A) \tag{13.7}$$

and similarly

$$P\left(\frac{B}{A}\right) = P(B) \tag{13.8}$$

Putting the value of $P\left(\frac{A}{B}\right)$ and $P\left(\frac{B}{A}\right)$ in Eq. (13.6), we get

$$P(AB) = P\left(\frac{A}{B}\right) P(B) = P\left(\frac{B}{A}\right) P(A)$$

So

$$= P(A) P(B) = P(B) P(A)$$

Hence if events A and B be independent, then

$$P(AB) = P(A) \times P(B) \tag{13.9}$$

13.4. PROBABILITY OF HAPPENING OF AT LEAST ONE OF THE SEVERAL INDEPENDENT EVENTS

If $A_1, A_2,..., A_n$ be n independent events with probabilities $p_1, p_2,...,p_n$ then probability of at least one of the several events is given by

$$P(A_1 + A_2 + \cdots + A_n) = 1 - P(\bar{A}_1 \cdot \bar{A}_2, \cdots, \bar{A}_n)$$

and if $A_1, A_2,...,A_n$ be independent then

$$P(\bar{A}_1 \cdot \bar{A}_2, \cdots, \bar{A}_n) = P(\bar{A}_1)P(\bar{A}_1), \cdots, P(\bar{A}_n)$$

So

$$P(A_1 + A_2 + \cdots + A_n) = 1 - (1 - P_1)(1 - P_2),...,(1 - P_n) \tag{13.10}$$

Proof: We give the proof in a simple case of two events
We know

$$A \cup \bar{A} = S \quad \text{(The sample space)}$$
$$B \cup \bar{B} = S$$

$$(13.11)$$

so

$$(A \cup \bar{A}) \cap (B \cup \bar{B}) = S \cap S = S$$

or

$$\{A \cap B + A \cap \bar{B} + \bar{A} \cap B\} = S$$

as

$$\{A \cap B + A \cap \bar{B} + \bar{A} \cap B\} \cap (\bar{A} \cap \bar{B}) = \phi$$

We have $\left[\{A \cap B + A \cap \bar{B} + \bar{A} \cap B\} + (\bar{A} \cap \bar{B}) \right] = P(S) = 1$

But

$$P(A \cap B + A \cap \bar{B} + \bar{A} \cap B) \qquad (13.12)$$

taking probability of at least one of the events A, B, we have

$$P(A + B) + P(\bar{A}\,\bar{B}) = 1$$

or

$$P(A + B) = 1 - P(\bar{A}\,\bar{B})$$

If A and B be independent

$$P(A + B) = 1 - P(\bar{A})P(\bar{B})$$
$$= 1 - (1 - P_1)(1 - P_2) \qquad (13.13)$$

Generalizing the above result, we get the result Eq. (13.10)

13.5. THEOREMS OF TOTAL PROBABILITY OF COMPOUND EVENTS

Theorem 1: For an event A with sample space S,

$$P(A) = \sum_{i=1}^{n} P(A_i) P\left(\frac{A}{A_i}\right),$$

where $A_i,\ i = 1, 2, ..., n$ are pairwise disjoint
events then $A_1 \cup A_2, ..., \cup A_n = S$
and $A_i \cap A_j = \phi \quad$ for $\ i \neq j$

Proof: As $A = A \cap S = A \cap \{A_1 \cup A_2, ..., \cup A_n\}$

$$A = (AA_1) \cap (AA_2), ..., \cap (AA_n)$$

As these are disjoint

$$P(A) = P(A_1 A) + P(AA_2) + \cdots + P(AA_n)$$

$$= P\left(\frac{A}{A_1}\right) P(A_1) + P\left(\frac{A}{A_2}\right) P(A_2) + \cdots + P\left(\frac{A}{A_n}\right) P(A_n)$$

$$= P\left(\frac{A}{A_1}\right) P(A_1) + \cdots + P\left(\frac{A}{A_n}\right) P(A_n) = \sum_{i=1}^{n} P\left(\frac{A}{A_i}\right) P(A_i) \quad (13.14)$$

In particular $P(A) = P(AB) + P(A\bar{B})$ as $B \cup \bar{B} = S$ and $B \cap \bar{B} = \phi$.

Theorem 2: Baye's Theorem

If an event A can happen with any of the event $A_1, A_2, ..., A_n$ which are pairwise disjoints and such that $A_1 \cap A_2, ..., \cap A_n$ are the whole sample spaces then the conditional probability

$$P\left(\frac{A_i}{A}\right) = \frac{P(A_i A)}{P(A)} = \frac{P\left(\frac{A}{A_i}\right)P(A_i)}{P(A)} \tag{13.15}$$

or writing value of $P(A)$ in Eq. (13.15), we have

$$P\left(\frac{A_i}{A}\right) = \frac{P\left(\frac{A}{A_i}\right)P(A_i)}{\sum_{i=1}^{n} P\left(\frac{A}{A_i}\right)P(A_i)} \tag{13.16}$$

Proof: As a sample case, let

$$A_1 \cup A_2 \cup A_3 = S \tag{13.17}$$

then

$$P(A) = P(AS) = P\{A(A_1 + A_2 + A_3)\}$$

$$= P(AA_1) + P(AA_2) + P(AA_3) \tag{13.18}$$

Assuming A_1, A_2, A_3 to be pairwise disjoint:

Now

$$P\left(\frac{A_i}{A}\right) = \frac{P(AA_i)}{P(A)} = \frac{P\left(\frac{A}{A_i}\right)P(A_i)}{P(AS)}$$

$$= \frac{P\left(\frac{A}{A_i}\right)P(A_i)}{P\left(A\sum_{i=1}^{3} A_i\right)}$$

$$= \frac{P\left(\frac{A}{A_i}\right)P(A_i)}{P(AA_1) + P(AA_2) + P(AA_3)}$$

$$= \frac{P\left(\frac{A}{A_i}\right)P(A_i)}{P\left(\frac{A}{A_1}\right)P(A_1) + P\left(\frac{A}{A_2}\right)P(A_2) + P\left(\frac{A}{A_3}\right)P(A_3)}$$

Note 1: The probabilities $P(A_1), P(A_2), ..., P(A_n)$ are all termed 'a priori probabilities' because there are probabilities before we have any additional information about the experiment.

Note 2: The probabilities $P\left(\frac{A}{A_1}\right), P\left(\frac{A}{A_2}\right), ..., P\left(\frac{A}{A_n}\right)$, are all called 'likelyhoods' because they indicate how likely the event A under consideration is to occur, given each 'a prior' probability.

Note 3: The probabilities $P\left(\dfrac{A}{A_1}\right), P\left(\dfrac{A}{A_2}\right), ..., P\left(\dfrac{A}{A_n}\right)$, are all called *'a posteriori probabilities'* because they are determined after the results of the experiment (that A has occurred) is known.

13.6. MATHEMATICAL EXPRESSION

If X is a random variate which assume values $x_1, x_2,..., x_n$ with frequencies $f_1, f_2,...f_n$, then mean value of X is commonly known as *expectation* and is denoted by $E(X)$. If $p(x_i)$ is the probability of x taking the value x_i which is given by

$$p(x_i) = \frac{f_i}{\displaystyle\sum_{i=1}^{n} f_i}$$

and expected value of X is then defined by

$$E(X) = \sum_{i=1}^{n} x_i p(x_i) \tag{13.19}$$

In general, mathematical expectation of any function $\phi(x)$ is given by

$$E[\phi(x)] = \sum_{i=1}^{n} \phi(x_i) p(x_i) \tag{13.20}$$

Notes:

1. If $\phi(x_i) = x_i$, Eq. (13.20) gives the mean value of x, as it reduces to Eq. (13.19) mean is denoted by μ.

2. If $\phi(x_i) = [x_i - \mu]^r$ then Eq. (13.20) reduces to

$$E[f(x_i)] = \sum_{i=1}^{n} (x_i - \mu)^r p(x_i)$$

which is called the rth moment about mean and is denoted by μ_r.

3. If in particular $r = 2$, then $\phi(x_i - \mu)^2 = \displaystyle\sum_{i=1}^{n} (x_i - \mu)^2 p(x_i)$ is called the variance of x and is denoted by σ^2.

The positive square root of variance σ is called *standard deviation*.

Theorem 1: For two random variates x and y,
$$E(x + y) = E(x) + E(y)$$

Proof: Let x and y be two random variates where x assumes m values $x_1, x_2,...,x_m$ and y assumes the n values $y_1, y_2,...,y_n$ then

$$E(x + y) = \sum\sum \frac{1}{NM} f_{ij}(x_i + y_j),$$

where f_{ij} is the frequency for $x = x_i, y = y_j$

$$= \sum_i \left(\sum_i \frac{f_{ij}}{N} \right) \frac{1}{M} x_i + \sum_j \left(\sum_j \frac{f_{ij}}{N} \right) y_j$$

$$= \sum_i p_{i0} x_i + \sum_j p_{0j} y_j$$

$$= E(x) + E(y)$$

Theorem 2: If x and y be two random variates which are independent then

$$E(xy) = E(x) \, E(y)$$

Proof: For independent variates

$$f_{ij} = (f_{i0}) \, (f_{0j}),$$

So

$$E(xy) = \sum_i \sum_j f_{ij} x_i y_j$$

$$\sum_i \left(\sum_i f_{ij} y_j \right) x_i = \sum_i f_{i0} x_i \sum_j f_{ij} f_j,$$

$$= E(x) \cdot E(y)$$

Example 1. *In a city there are three factories A, B, C which produce 30%, 50% and 20% of an item, respectively. The defective items produced by them are 5%, 2% and 7% (an item purchased from the market is found to be defective). Find the probability that it was produced by the factory C.*

Solution: We have

$$P(A) = \frac{3}{10}, P(B) = \frac{5}{10}, P(C) = \frac{2}{10}$$

Let D denotes that the item purchased is defective

Also given is $P\left(\dfrac{\text{Defective}}{A} \right) = \dfrac{5}{100}, P\left(\dfrac{D}{B} \right) = \dfrac{2}{100}, P\left(\dfrac{D}{C} \right) = \dfrac{7}{100}$

Now $P\left(\dfrac{C}{D} \right) =$ Probability that the defective items were produced by C

$$= \frac{P\left(\dfrac{D}{C} \right) P(C)}{P\left(\dfrac{D}{A} \right) P(A) + P\left(\dfrac{D}{B} \right) P(B) + P\left(\dfrac{D}{C} \right) P(C)}$$

$$= \frac{\dfrac{7}{100} \times \dfrac{2}{100}}{\dfrac{5}{100} \times \dfrac{3}{10} + \dfrac{2}{100} \times \dfrac{5}{10} + \dfrac{7}{100} \times \dfrac{2}{10}} = \frac{14}{15 + 10 + 14} = \frac{14}{39}$$

Similarly, probability that it was produced by A

$$= \frac{P\left(\dfrac{D}{A}\right)P(A)}{P(D)} = \frac{15}{39}$$

Probability that the defective item was produced by B

$$P\left(\frac{D}{B}\right) = \frac{P(DB)}{P(B)} = \frac{P\left(\dfrac{D}{B}\right)P(B)}{P(D)} = \frac{10}{39} .$$

Example 2. *The probability of happening of an event A is* $\dfrac{1}{2}$ *and that of B is* $\dfrac{1}{3}$. *If probability of the happening of at least one of them be* $\dfrac{2}{3}$, *show that A and B are independent.*

Solution: Given is

$$P(A) = \frac{1}{2}, \ P(B) = \frac{1}{3}, \ \text{and} \ P(A + B) = \frac{2}{3},$$

or $\qquad P(A + B) = P(A) + P(B) - P(AB),$

putting the given values, we have

$$\frac{2}{3} - \frac{1}{2} - \frac{1}{3} = -P(AB),$$

so $\qquad P(AB) = -\left\{\dfrac{4-3-2}{6}\right\} = \dfrac{1}{6}$

and $\qquad P(A), P(B) = \dfrac{1}{2} \times \dfrac{1}{3} = \dfrac{1}{6}$

So, we have $\qquad P(AB) = \dfrac{1}{6} = P(A) \cdot P(B),$

which means that the events A and B are independent.

Example 3. *Probability that A would solve a problem is* $\dfrac{1}{3}$ *and for B it is* $\dfrac{1}{4}$. *If they both try, find the probability that the problem will be solved.*

Solution: A and B are separate individuals, so they are individuals which implies
$$P(AB) = P(A) \cdot P(B)$$

Now that problem is solved means that either one of them or they both are able to solve it for which the probability
$$P(A + B) = P(A) + P(B) - P(AB)$$

$$= \frac{1}{3} + \frac{1}{4} - \frac{1}{3} \times \frac{1}{4} = \frac{6}{12} = \frac{1}{2}$$

Example 4. *There are three urns containing 2 white and 3 black balls; 3 white and 2 black and 3 white and 1 black balls respectively. If each urn is equally likely to be chosen, find the probability that a ball drawn from one of them is white.*

Solution:
$$P\left(\frac{W}{U_1}\right) = \frac{2}{5}, P\left(\frac{W}{U_2}\right) = \frac{3}{5}, P\left(\frac{W}{U_3}\right) = \frac{3}{4}$$

$$P(U_1) = P(U_2) = P(U_3) = \frac{1}{3}$$

Sample space of urns $\quad S = \{U_1, U_2, U_3\}, U_1 U U_2 U U_3 = S$ and
$$W \cap S = W$$

So
$$
\begin{aligned}
P(W) &= P\{W(U_1 + U_2 + U_3)\} \\
&= P(WU_1) + P(WU_2) + P(WU_3) \\
&= P\left(\frac{W}{U_1}\right)P(U_1) + P\left(\frac{W}{U_2}\right)P(U_2) + P\left(\frac{W}{U_3}\right)P(U_3) \\
&= \frac{2}{5} \times \frac{1}{3} \times \frac{3}{5} \times \frac{1}{3} \times \frac{3}{4} \times \frac{1}{3} = \frac{7}{12}
\end{aligned}
$$

Example 5. *If p be the probability for a man aged x years to die in next one year, find the probability that out of n men $A_1, A_2,...,A_n$, each aged x years, A_1 will be the first to die.*

Solution: We know $P(A_1 + A_2 + \cdots + A_n) + P(\bar{A}_1\bar{A}_2,...,\bar{A}_n) = 1$

Probability that at least one out of $A_1, A_2,..., A_n$ dies
$$
\begin{aligned}
&= P(A_1 + A_2 + \cdots + A_n) \\
&= 1 - P(\bar{A}_1\bar{A}_2,...,\bar{A}_n), \text{ as } A_1, A_2,..., A_n \text{ are independent} \\
&= 1 - (1-p)(1-p)...\, n \text{ factors } \{1 - (1-p^n)\}
\end{aligned}
$$

Probability that A_1 dies first $= \left(\dfrac{1_{C_1}}{n_{C_1}}\right) \times \{1 - (1-p)^n\}$

$$= \frac{1}{n}\{1 - (1-p)^n\}$$

Example 6. *A and B throw a die for a prize of Rs. 11.00, which is to be won by the player who first throws six. Find their expectations if the first chance is given to A.*

Solution: Probability of getting 6 at first chance $= \dfrac{1}{6} = p$

Probability of 6 coming at second throw $= pq$, where $q = 1 - p = \dfrac{5}{6}$

Probability of 6 coming at 3rd throw $= q^2 p$ and so on. Now till 6 comes, A will have the 1st, 3rd, 5th,... chance.

So $\qquad P(A) = p + q^2p + q^4p$, infinite GP with common ratio $= q^2$

$$= \frac{p}{1-q^2} = \frac{\dfrac{1}{0.6}}{1-\dfrac{25}{36}} = \frac{6}{11}$$

$P(B)$ = Probability of 6 at 2nd, 4th, 6th throws

$$= qp + q^3p + \cdots = \frac{qp}{1-q^2} = \frac{5}{11}$$

Expected value of A = Probability of A's success × value of prize

$$= \frac{6}{11} \times 11 = ₹6.00$$

Similarly $\qquad E(B) = \frac{5}{11} \times 11 = ₹5.00$

Example 7. *Two cards are drawn from a pack of 52 cards, what is the probability that either both are red or both are kings?*

Solution: Number of ways of drawing 2 cards from 52 cards $= {}^{52}C_2$

$$= \frac{52 \times 51}{1 \times 2} = 26 \times 51$$

Probability that both are red

$$= P(A) = \frac{{}^{26}C_2}{{}^{52}C_2} = \frac{13 \times 25}{26 \times 51} = \frac{25}{102}$$

Probability that both are kings

$$= P(B) = \frac{{}^{4}C_2}{{}^{52}C_2} = \frac{6}{26 \times 51} = \frac{1}{13 \times 17}$$

Now $\qquad P(A + B) = P(A) + P(B) - P(AB)$

So $\qquad P(A + B) = \frac{25}{102} + \frac{1}{13 \times 17} - \frac{2C_2}{52C_2}$

$$= \frac{25 \times 13 + 6 - 1}{13 \times 17 \times 6} = \frac{330}{13 \times 17 \times 6} = \frac{55}{221}$$

Example 8. *A box contains 20 bolts and 30 nuts. Half of the bolts and nuts are rusted. If two items are drawn at random, what in the probability that either both are rusted or both are bolts.*

Solution: Let both rusted nuts be denoted by R, then $P(R) = \dfrac{{}^{25}C_2}{{}^{50}C_2}$

Both drawn are bolts be denoted by B, then $P(B) = \dfrac{^{20}C_2}{^{50}C_2}$

Probability that they are rusted bolts $= P(RB) = \dfrac{^{10}C_2}{^{50}C_2}$

so
$$P(R + B) = P(R) + P(B) - P(RB)$$

$$= \frac{^{25}C_2 + {}^{20}C_2 - {}^{10}C_2}{^{50}C_2} = \frac{25 \times 12 + 10 \times 19 - 45}{5 \times 5 \times 7 \times 7} = \frac{89}{245}$$

Example 9. *A bag has 4 white and 3 black and a second bag has 3 white and 5 black balls. A ball is drawn from the first bag and without noting its colour, is put into the second bag. Then a ball is drawn from the second bag and is found to be white. Find its probability.*

Solution: Let E = the event that ball drawn from bag is white then $P(E) = \dfrac{4}{7}$ and is black then $P(E) = \dfrac{3}{7}$.

Probability of a white ball from bag B_2, when E has happened

$$= P\left(\frac{W}{E}\right) = \frac{3+1}{8+1} = \frac{4}{9}$$

$$P\left(\frac{W}{\overline{E}}\right) = \frac{3}{8+1} = \frac{3}{9}$$

$$P(W) = P\left(\frac{W}{E}\right)P(E) + P\left(\frac{W}{\overline{E}}\right)P(\overline{E})$$

$$= \frac{4}{9} \times \frac{4}{7} + \frac{3}{9} \times \frac{3}{7} = \frac{25}{63}$$

Example 10. *If $P(A) = \dfrac{3}{8}$, $P(B) = \dfrac{1}{2}$, $P(A \cap B) = \dfrac{1}{4}$, find $P\left(\dfrac{A}{B}\right)$ and $P\left(\dfrac{B}{A}\right)$*

Solution:
$$P(\overline{A}) = 1 - \frac{3}{8} = \frac{5}{8}, \quad P(\overline{B}) = 1 - \frac{1}{2} = \frac{1}{2}$$

$$P(A + B) = P(A) + P(B) - P(AB) = \frac{3+4-2}{8} = \frac{5}{8}$$

$$P(\overline{A}\,\overline{B}) = 1 - P(A + B) = 1 - \frac{5}{8} = \frac{3}{8}$$

Now
$$P\left(\frac{\overline{A}}{\overline{B}}\right) = \frac{P(\overline{A}\,\overline{B})}{P(\overline{B})} = \frac{3}{8} \times \frac{2}{1} = \frac{3}{4}$$

$$P\left(\frac{\overline{B}}{\overline{A}}\right) = \frac{P(\overline{B}\,\overline{A})}{P(\overline{A})} = \frac{3}{8} \times \frac{8}{5} = \frac{3}{5}$$

Example 11. *A speaks truth 4 out of 5 times. A die is rolled. He reports that it is 6. What is the chance that there is six actually.*

Solution: Let A be the event that A speaks truth

then
$$P(A) = \frac{4}{5}, \quad P(\bar{A}) = 1 - \frac{4}{5} = \frac{1}{5}$$

B be the event that 6 occurs on the die

then
$$P(B) = \frac{1}{6}, \quad P(\bar{B}) = \frac{5}{6}$$

6 has actually occurred means B has happened

A reports six means A speaks truth, meaning A has happened, then

$$P\left(\frac{B}{A}\right) = \frac{P(BA)}{P(A)} = \frac{P\left(\frac{A}{B}\right)P(B)}{P\left(\frac{A}{B}\right)P(B) + P\left(\frac{A}{B}\right)P(\bar{B})}$$

$$= \frac{P\left(\frac{A \text{ reports } 6}{6 \text{ has occurred}}\right) \times (\text{Probability of } 6)}{P\left(\frac{A \text{ speaks truth}}{6 \text{ has occurred}}\right)\text{Prob}(6) + P\left(\frac{A \text{ reports } 6}{6 \text{ has not occurred}}\right)P(6\text{not})}$$

$$= \frac{P(A\text{speaks truth}) \times \frac{1}{6}}{\frac{4}{5} \times \frac{1}{6} + (A\text{speaks false}) \times \frac{5}{6}} = \frac{\frac{4}{5} \times \frac{1}{6}}{\frac{4}{5} \times \frac{1}{6} + \frac{1}{5} \times \frac{5}{6}} = \frac{4}{5}$$

Example 12. *A letter is known to have come either from TATANAGAR or CALCUTTA. On the envelope only two consecutive letters TA are visible. What is the chance that the letter has come from (i) Calcutta (ii) Tatanagar.*

Solution: Word Calcutta has consecutive pairs of letters in its spelling, CA, AL, LC, CU, UT, TT, TA = 7. One pair is TA.

So probability of $\left(\dfrac{TA}{\text{Calcutta}}\right) = \dfrac{1}{7}$

TATANAGAR has the pairs $TA, AT, TA, AN\ NA, AG, GA, AR = 8$

There are two pairs of TA

So probability (TA) from $TATANAGAR = \dfrac{2}{8} = \dfrac{1}{4}$

Probability (letter coming from Calcutta) = $P(\text{Cal}) = \dfrac{1}{7}$

$P(\text{Cal}) = \dfrac{1}{2}$ = Probability (letter coming from Tatanagar)

$$P\left(\frac{\text{Calcutta}}{(TA)\text{readable}}\right) = \frac{P[(\text{Calcutta})(TA)]}{P(TA)}$$

$$= \frac{P\left(\dfrac{TA}{\text{Cal}}\right)P(\text{Cal})}{P\left(\dfrac{TA}{\text{Cal}}\right)P(\text{Cal}) + P\left(\dfrac{TA}{\text{Cal}}\right)P(\text{Cal})}$$

$$= \frac{\dfrac{1}{7} \times \dfrac{1}{2}}{\dfrac{1}{7} \times \dfrac{1}{2} + \dfrac{1}{4} \times \dfrac{1}{2}} = \frac{4}{11}$$

Similarly $\qquad P\left(\dfrac{\text{Tatanagar}}{TA\,\text{readable}}\right) = \dfrac{7}{11}$

Example 13. *Two players A and B of equal skill play a set of games. They leave off playing when a wants 3 points and B needs 2 points to win. If the stake is of Rs. 16.00, find how to divide it in A and B.*

Solution: *A* can win in the following ways ($A \equiv A$ wins, $B \equiv B$ wins)

$A\ A\ A$ (A wins only in three games), its probability $= \dfrac{1}{8}$

$B\ A\ A\ A$, its probability $= \dfrac{1}{16}$

$A\ B\ A\ A$, its probability $= \dfrac{1}{16}$

$A\ A\ B\ A$, its probability $= \dfrac{1}{16}$

$P(B) = P(BB) + P(ABB) + P(BAAB) + P(BAB) + P(ABAB) + P(AABB)$; total $= \dfrac{5}{16}$

So $\qquad\qquad P(A) = \dfrac{5}{16}$ and $P(B) = 1 - \dfrac{5}{16} = \dfrac{11}{16}$

$$P(B) = \dfrac{1}{4} + \dfrac{2}{8} + 3 \times \dfrac{1}{16} = \dfrac{11}{16}$$

$$A\text{'s share} = \dfrac{5}{16} \times 16 = ₹\ 5.00$$

$$B\text{'s share} = \dfrac{11}{16} \times 16 = ₹\ 11.00$$

Example 14. *In an examination with multiple type questions with four choices, a candidate answers either by his skill or by guess or by copying. The probability of making correct guess is* $\frac{1}{3}$ *and that of copying is* $\frac{1}{6}$. *The probability of correct answer by copying is* $\frac{1}{8}$. *Find the probability that he knew the answer to the question; given that he answered it correctly.*

Solution: Let us denote correct answering by $\equiv C$

answering by ability by $= A$

answering by guess work by $= G$

answering by copying by $= U$ (unfair means)

Given $\qquad P(G) = \frac{1}{3}, P(UM) = \frac{1}{6}, P(A) = 1 - \frac{1}{3} - \frac{1}{6} = \frac{3}{6}$

$$P\left(\frac{C}{UM}\right) = \frac{1}{8}$$

options by guess work 25% may be correct

$$P\left(\frac{C}{G}\right) = \frac{1}{4}, = P\left(\frac{C}{A}\right) = 1$$

Probability $\left(\dfrac{A}{\text{correct answer} = C}\right) = \dfrac{P(AC)}{P(C)}$

$$= \frac{P\left(\dfrac{C}{A}\right)P(A)}{P\left(\dfrac{C}{A}\right)P(A) + P\left(\dfrac{C}{G}\right)P(G) + P\left(\dfrac{C}{UM}\right)P(UM)}$$

$$= \frac{1 \times \dfrac{3}{6}}{1 \times \dfrac{3}{6} + \dfrac{1}{4} \times \dfrac{1}{3} + \dfrac{1}{8} \times \dfrac{1}{6}} = \frac{3 \times 8}{24 + 4 + 1} = \frac{24}{29}$$

Example 15. *In a bolt factory, machine M_1, M_2 and M_3 manufacture 25%, 35%, and 40% of the total production respectively. Of their output 5%, 4% and 2% are defective bolts. A bolt is drawn at random from the product and is found to be defective. What is the probability that it was manufactured by machine M_2.*

Solution: Let A_1, A_2 and A_3 denote the events that a bolt selected at random is manufactured by the machine M_1, M_2 and M_3 respectively and let A denote the event of its being defective. Then

$$P(A_1) = 0.25, \quad P(A_2) = 0.35, \quad \text{and} \quad P(A_3) = 0.40.$$

The probability of drawing a defective bolt manufactured by the machine M_1 is $P(A/A_1) = 0.05$. Similary $P(A/A_2) = 0.04$ and $P(A/A_3) = 0.02$

By Baye's theorem, we have

$$P(A_2/2) = \frac{P(A_2) \cdot P(A/A_2)}{P(A_1) \cdot P(A/A_1) + P(A_2)P(A/A_2) + P(A_3) \cdot P(A/A_3)}$$

$$P(A_2/A) = \frac{0.35 \times 0.04}{0.25 \times 0.3 + 0.35 \times 0.04 + .4 \times 0.02}$$

$$= \frac{0.041}{0.345} = 0.41$$

13.7. THEORETICAL PROBABILITY DISTRIBUTIONS

This can be classified under following two groups:

1. **Observed probability distribution:** The observed probability distribution is based on actual observations and experimentation.

2. **Theoretical probability distribution:** If certain hypothesis is assumed, it is sometimes possible to derive mathematically what the probability distribution of certain sample space should be. Such distributions are called theoretical probability distributions. For example, binomial, poisson, normal probability distribution.

Definitions

1. **Random variable:** If the value of a variable can not be exactly predicted in advance but depends upon some chance factor, then the variable is called a random variable. A random variable is also called *stochastic variable.*

2. **Discrete random variable:** It is a variable which can assume only isolated values. For examples.

    i. The number of students appearing in a class test consisting of 40 students is a discrete random variable as it can not assume values other than 0, 1, 2,...,40.

    ii. The number of defective screws in a lot of 10 screws is a discrete random variable as it can assume values 0, 1, 2,...,10 only.

3. **Continuous random variable:** A variable which can assume any value within an interval, i.e. all values of a continuous scale. For example

    i. The weight of a group of individuals.

    ii. The height of a groups of individuals.

Discrete probability distribution: Let a random variable X assume values $x_1, x_2,...,x_n$ with probabilities $p_1, p_2,...,p_n$ respectively, where $P(X = x_i) = p_i \geq 0$ for each x_i

and $p_1 + p_2 + \cdots + p_n = \sum_{i=1}^{n} p_i = 1$. Then

$$X: \quad x_1, x_2,...,x_n$$

$$P(X): p_1, p_2,...,p_n$$

is called the discrete probability distribution for X.

4. **Continuous probability distribution:** Let x be a random variate which assume value in the interval. $(-\infty, \infty)$, we define probability in the interval (a, b) by the integral.

$$P(a, \leq x \leq b) = \int_a^b f(x) dx \tag{13.21}$$

$f(x)$ is called probability density function which satisfies

    i. $f(x) \geq 0$ for all x in $(-\infty, \infty)$

    ii. $\int_{-\infty}^{\infty} f(x)dx = 1$

So the variate x with its probability given by Eq. (13.21) is a continuous variate and its distribution is called the continuous probability distribution.

For example: Normal probability distribution, exponential distribution etc.

13.7.1. Binomial Distribution

If we make a series of n trials, each resulting into a success or a failure = (nonsuccess) and if the probability p of success remains same at each trial, then the probabilities of 0, 1, 2,...,n successes are the successive items in the binomial expansion of $(q + p)^n$. So the probability of r successes and $(n - r)$ failures in n trials is the $(r + 1)$th term $T_{r+1} = {}^nC_r\, q^{n-r}\, p^r$, where $q = 1 - p$. This probability distribution is called the *binomial distribution* and represented by

$$P(r) = {}^nC_r p^r q^{n-r}\,;\ r = 0, 1, 2,...,n$$

The total probability is one

The variate values with their respective probabilities in the binomial distribution are

Number of successes	$x = 0$	1	2	...	r	...	n
Corresponding probabilities	q^n	${}^nC_1 q^{n-1}p$	${}^nC_2 q^{n-2}p^2$		${}^nC_r q^{n-r}p^r$		${}^nC_n p^r$

sum of probabilities

$$= q^n + {}^nC_1 q^{n-1}p + \cdots + {}^nC_n p^n$$
$$= (q + p)^n \ \text{ but } q + p = 1$$
$$= 1^n = 1$$

Note: n and p occurring in the binomial distribution are called parameters.

In a Binomial distribution, we have

    i. n, the number of trial is finite,

    ii. each trial has only two possible outcomes usually called success or failure,

    iii. all the trials are independent,

    iv. p (or q) is constant throughout for all the trials.

Mean of the binomial distribution

We have

$$\bar{x} = \frac{f_1 x_1 + f_2 x_2 + \cdots + f_n x_n}{N}$$

$$= p_1 x_1 + p_2 x_2 + \cdots + p_n x_n,$$

where relative frequency $\dfrac{f_r}{N} = p_r$

In binomial distribution x assumes the $(n + 1)$ values 0, 1, 2,...,n

thus
$$\bar{x} = \sum_{r=0}^{n} \left({}^n C_r q^{n-r} p^r \right) r$$

$$= \sum_{r=0}^{n} \frac{(n)!}{(r)!(n-r)!} q^{n-r} p^r$$

$$= \sum_{r=0}^{n} \frac{(n)!}{(r-1)!\{(n-1)-(r-1)\}!} q^{[(n-1)(r-1)]} (p^{r-1} p)$$

Let $r - 1 = s$, then $\sum \dfrac{n(n-1)!}{s!\{(n-1-s)!\}} \cdot q^{(h-1)s} - sp^s \cdot p$

$$= np \sum_{s=0}^{n-1} {}^{(n-1)} C_s q^{(h-1)s} p^s$$

Mean $\bar{x}$ (x = binomial variate)

Mean $= np (q + p)^{n-1} = np(1)^{n-1} = np$ $\hspace{2cm}$ (13.22)

Variance of the Binomial Distribution

Variance of x is defined as

$$x = \sum p_i (x_i - \bar{x})^2$$

If x be a binomial variate, then
$$= \sum p_i \left\{ x_i^2 - 2\bar{x} x_i + (\bar{x})^2 \right\}$$

$$= \sum p_i x_i^2 - 2\bar{x} \sum p_i x_i + (\bar{x})^2 \sum p_i$$

$$= \sum p_i x_i^2 - 2\bar{x}\bar{x} + (\bar{x})^2 \times 1$$

$$= \sum p_i x_i^2 - (\bar{x})^2 \hspace{2cm} (13.23)$$

Now
$$\sum_{i=0}^{n} p_i x_i^2$$

$$\sum p_i x_i^2 = {}^n C_i q^{n-i} p^i (x_i = i)^2$$

So
$$= \sum_{i=0}^{n} \frac{(n)!}{(i)!(n-i)!} i^2 q^{n-i} p^i$$

Now
$$\frac{i^2}{(i)!} = \frac{i}{(i-1)!} = \frac{i-1+1}{(i-1)} = \frac{1}{(i-2)!} + \frac{1}{(i-1)}$$

$$\sum p_i x_i^2 = \sum_{i=2}^{n} \frac{(n)! \, q^{n-i} p^i}{(i-2)(n-i)!} + \sum_{i=0}^{n} \frac{(n)!}{(i-1)!(n-1)!} q^{n-1} p^i$$

Let $i - 2 = r$, [Eq. (13.21)] and $i - 1 = s$ in 2nd sum

$$= \sum_{i=2}^{n} \frac{n(n-1)(n-2)!}{(r)!(n-2)!-r} q^{(n-2)-r} p^r p^2 + \sum_{i=2}^{n} \frac{n(n-1)!}{(s)!(n-1)!-s} q^{n-1-s} p^s p^1$$

$$= n(n-1)p^2(q+p)^{n-2} + np(q+p)^{n-1}$$

$$= \{n(n-1)p^2 + np\} \text{ as } q+p = 1 \tag{13.24}$$

Putting the value in Eq. (13.23)

Variance $\quad x = \sigma^2 = n(n-1)p^2 + np - (np)^2$

(x = binomial variate)

Variance $\quad \sigma^2 = np\{np - p + 1 - np\} = np(1-p) = npq$ $\tag{13.25}$

Standard deviation (binomial)

$$\sigma = \sqrt{\text{variance}} = \sqrt{npq} \tag{13.26}$$

13.7.2. Recurrence Formula for the Binomial Distribution

In the binomial probability distribution of n trial the number of r success and $n - r$ failure is given by

$$P(r) = {}^nC_r \, p^r q^{n-r}$$

where p is the probability of successes and $q = 1 - p$ is the probability of failure of a single trial.

Similarly $\qquad P(r+1) = {}^nC_{r+1} \, p^{r+1} q^{n-r-1}$

$$\therefore \qquad \frac{P(r+1)}{P(r)} = \frac{\dfrac{(n)! p^{r+1} \cdot q^{n-r-1}}{(r+1)(n-r-1)}}{\dfrac{(n)! p^r \cdot q^{n-r}}{(r)!(n-r)!}} = \frac{n-r}{r+1} \cdot \frac{p}{q}$$

$$\Rightarrow \qquad P(r+1) = \frac{n-r}{r+1} \cdot \frac{p}{q} \cdot P(r)$$

which is the required recurrence formula. This formula is used to find the probabilities, for any value of r if the same for one value of r is known.

Most Probable Number Mode of the Binomial Distribution

A random variable X, having the values 0, 1, 2,... r,... n, with binomial probabilities $P(r)$; $r = 0, 1,... n$, is said to be most probable if the probability $P(r)$ is maximum, i.e. if

$$P(r-1) \le p(r) \ge P(r+1) \text{ for all } r.$$

$$\Rightarrow \qquad {}^nC_{r-1} \, p^{r-1} q^{n-r+1} \le {}^nC_r \, p^r q^{n-r} \ge {}^nC_{r+1} \, p^{r+1} q^{n-r-1}$$

$$\Rightarrow \qquad \frac{(n)!}{(r-1)!(n-r+1)} p^{r-1} q^{n-r+1} \le \frac{(n)!}{(r)!(n-r)!} p^r q^{n-r} \ge \frac{(n)!}{(r+1)!(n-r)} p^{r+1} q^{n-r-1}$$

$$\Rightarrow \qquad \frac{q^2}{(n-r+1)(n-r)} \le \frac{pq}{r\cdot(n-r)} \ge \frac{p^2}{r(r+1)}$$

$$\text{(I)} \qquad\qquad \text{(II)} \qquad \text{(III)}$$

Taking statements of (I) and (II) terms

$$\frac{q^2}{(n-r+1)(n-r)} \le \frac{pq}{r(n-r)}$$

$$qr \le p(n-r+1) \tag{13.27}$$

$$= r \le p(n+1)$$

Taking statements of (II) and (III), we get

$$q(r+1) \ge p(n-r)$$

$$r \ge pn - q = p(n+1) - 1 \tag{13.28}$$

Combining Eqs (13.27) and (13.28), we get

$$p(n+1) - 1 \le r \le p(n+1) \tag{13.29}$$

So $X = r$ is most probable number if it satisfy inequality Eq. (13.29).

13.7.3. Application of Binomial Distribution

The binomial probability is applied in solving the problems concerning:

i. Finding the number of successes in a experiment consisting of n trial, finding the number of defective articles manufactured by a machine.

ii. Estimation of reliability of a system. From the observed data, mean and standard deviation are found and then by using these parameters in binomial distribution theoretical data are obtained. These theoretical and observed data are compared for estimating the reliability of the system.

iii. Number of rounds fired from a gun hitting a target.

iv. Radar detection

Example 1. *During war, 1 ship out of 9 was sunk on an average in making a certain voyage. What was the probability that exactly 3 out of a convoy of 6 ships would arrive safely?*

Solution: The probability of a ship arriving safely

$$p = 1 - \frac{1}{9} = \frac{8}{9} \qquad\qquad \because \ q = \frac{1}{9}$$

∴ The probability that exactly 3 ships arrive safely $= P(r = 3)$

$$= {}^6C_3 (p)^3 (q)^3$$

$$= \frac{6 \times 5 \times 4}{3 \times 2 \times 1} \cdot \left(\frac{8}{9}\right)^3 \left(\frac{1}{9}\right)^3$$

$$= 20 \times \frac{8 \times 8 \times 8}{9^6} = \frac{10240}{9^6}$$

Example 2. *If a coin is tossed 5 times, write the probability distribution for the number of heads obtained.*

Solution: As the probability of success (obtaining head) remains same at every toss. The variate values will be from 0 to 5 (integral) with probabilities as the successive terms in the expansion of $(q + p)^5$, where $q = p = \dfrac{1}{2}$ so

Variate values	0	1	2	3	4	5
Corresponding probabilities	q^5	$^5C_1 q^4 p$	$^5C_2 q^3 p^2$	$^5C_3 q^2 p^3$	$^5C_4 q p^4$	p^5
	$= \dfrac{1}{32}$	$= \dfrac{5}{32}$	$= \dfrac{10}{32}$	$= \dfrac{5}{32}$	$= \dfrac{1}{32}$	

Example 3. *Out of 3 defective and 7 good bulbs, three bulbs are drawn at a time. Write the probability distribution of the defective bulbs drawn. Find the mean of the distribution.*

Solution: Number of defective bulbs drawn may be 0, 1, 2, 3 with respective probabilities.

No. of defectives drawn	0	1	2	3
Corresponding probabilities	$\dfrac{^7C_3}{^{10}C_3}$	$\dfrac{^7C_2 \times ^3C_1}{^{10}C_3}$	$\dfrac{^7C_1 \times ^3C_2}{^{10}C_3}$	$\dfrac{^3C_3}{^{10}C_3}$
	$= \dfrac{35}{120}$	$= \dfrac{63}{120}$	$= \dfrac{21}{120}$	$= \dfrac{1}{120}$

$$\text{Mean } \bar{x} = \sum p_i x_i = 0 + \frac{63}{120} \times 1 + \frac{21}{120} \times 2 + \frac{1}{120} \times 3 = \frac{108}{120}$$

Example 4. *Assuming that on an average, one telephone out of 10 is busy. Six telephone numbers are randomly selected and called. Find the probability that 4 out of them would be busy.*

Solution: Probability p for a line to be busy is $p = 0.1$

So $\qquad\qquad\qquad q = 1 - p = 0.9$

Required probability

$$= \frac{6 \times 5}{2}(0.1)^4 (0.9)^2 = 1.215 \times 10^{-3}$$

Example 5. *If the probability of success in trials with constant probability P be $= \dfrac{1}{10}$, find the number of trials necessary so that the probability of getting at least one success is greater than $\dfrac{1}{2}$.*

Solution: Let each trial be denoted by event A_i, then

$$P(A_1 + A_2 + \cdots + A_n) + P(\bar{A}_1 \bar{A}_2 \cdots \bar{A}_n) = 1$$

So probability of getting at least one success in n trials

$$= P(A_1 + A_2 + \cdots + A_n)$$

$$= 1 - P(\bar{A}_1 \bar{A}_2 \cdots \bar{A}_n) \text{, as each of the trials are independent}$$

$$= 1 - P(\bar{A}_1) P(\bar{A}_2) \cdots P(\bar{A}_n)$$

$$= 1 - (1 - p)(1 - p)\ldots n \text{ factors}$$

$$= 1 - (1 - p)^n, \text{ we want}$$

$$1 - (1 - p)^n > \frac{1}{2} \quad \text{or} \quad \left(1 - \frac{1}{2}\right) > (1 - 0.1)^n$$

or
$$\frac{1}{2} > (0.9)^n$$

or
$$\log_{10} \frac{1}{2} > n \log_{10} \frac{9}{10}$$

or
$$-\log 2 > -n \log \frac{10}{9}$$

or
$$n > \frac{\log 2}{\log 10 - 2\log 3}$$

$$= \frac{0.3010}{1 - 2 \times 4771} = 6.5, \ n \text{ is an integral}$$

so
$$n = 7$$

Example 6. *The probability that a bulb produced by a factory will fuse before 100 days of use is 0.05. Find the probability that out of 5 such bulbs (i) one or (ii) not more than one or (iii) more than one or (iv) at least one will fuse before 100 days of use.*

Solution: With $p = 0.05$, $q = 1 - p = 0.95$, $x = 5$, the probability distribution of number of bulbs fused is

Number of bulbs fused	0	1	2	3	4	5
Corresponding probabilities	q^5	$5q^4 p$	$10q^3 p^2$	$10 q^2 p^3$	$5q p^4$	p^5

i. Probability of none fused
$$= q^5 = (0.95)^5 \tag{1}$$

ii. Probability of not more than one fused
$$= q^5 + 5q^4 p$$
$$= (0.95)4 (0.95 + 5 \times 0.5)$$
$$= (1.2) (0.95)^4 \tag{2}$$

iii. P (more than one fused)
$$= 1 - P(0) - P(1) = 1 - (0.12) (0.95)^4 \tag{3}$$

iv. P (at least one will fuse)
$$= 1 - P(0) = 1 - q^5 = 1 - (0.95)^5 \tag{4}$$

Example 7. *The mean and variance of a binomial distribution are 4 and* $\dfrac{4}{3}$ *respectively.*
Find $P(x \geq 1)$.

Solution: Given

$$np = 4, \quad npq = \frac{4}{3}$$

so

$$q = \frac{1}{3}, \quad p = 1 - \frac{1}{3} = \frac{2}{3}$$

$$p(x \geq 1) = 1 - q^n = 1 - \frac{1}{3^n} = \left(1 - \frac{1}{3^6}\right)$$

Example 8. *If 10% of the pens manufactured by a company are defective, find the probability that a box of 12 pens contains (i) two defectives (ii) at least two defectives.*

Solution: $P(\text{not good}) = q = 0.1$ $P(\text{good}) = p = 0.9, \quad n = 12$

  i. Probability (2 defectives)
$$^{12}C_2 (0.9)^{10} (0.1)^2 = 0.2301$$

  ii. Probability (at least 2 defectives)
$$= 1 - P(0) - P(1)$$
$$= 1 - (0.9)^{12} - 12(0.9)^{11} (0.1) = 0.341$$

Example 9. *Probability that a man aged 60 years will remain alive till 70 is 0.65. What is the probability that out of 10 such men, at least 7 would be alive at 70.*

Solution: $P = 0.65, \quad q = 0.35, \quad n = 10$

7	8	9	10
$^{10}C_7 (0.65)^7 (0.35)^3$	$^{10}C_8 (0.65)^8 (0.35)^2$	$^{10}C_9 (0.65)^9 (0.35)$	$^{10}C_{10} (0.65)^{10}$

Required probability
$$(x \geq 7) = P(7) + P(8) + P(9) + P(10)$$
$$= (0.65)^7 \left\{ 120(0.35)^3 + 45(0.65)(0.35)^2 + 10(0.65)^2 (0.35) + (0.65)^3 \right\} = 0.515$$

13.8. Poisson's Distribution

In a series of n trials with probability of success p remaining constant at each trial, the probability of r successes is given by

$$P(x - r) = \frac{(n)!}{(r)!(n-r)!} p^r (1-p)^{n-r}$$

If the probability p of success at an individual trial be very small tending to zero and the number of trials n be very large tending to infinity such that the product np remains finite equal to m.

Then

$$P(r) = \frac{n(n-1)\cdots(n-r+1)p^r}{(r)!} \frac{(1-p)^n}{(1-p)^r}$$

$$\lim_{\substack{n \to \infty \\ np \to m \\ p \to 0}} = \frac{np(np-p)\cdots(np-rp+p)}{(r)!} \frac{\left(1 - \dfrac{m}{n}\right)^n}{\left(1 - \dfrac{m}{n}\right)^r}$$

so

$$P(r) = \frac{m^r}{(r)!} e^{-m}$$

This limiting form of binomial distribution with above probability is called Poisson's distribution. Here the variate x assumes discrete values $x = r$, $r = 0, 1, 2,...,\infty$.

Total probability $= \displaystyle\sum_{r=0}^{\infty} P(x = r)$

$$= e^{-m}\left\{1 + \frac{m}{(1)!} + \frac{m^2}{(2)!} + \cdots\right\} = e^{-m} e^m = 1 \tag{13.30}$$

Mean of Poisson's Distribution

$$\text{Mean } \bar{x} = \sum_{i=0}^{\infty} p_i x_i, \text{ where } x_i = 0, 1, 2,...$$

$$= \sum_{0}^{\infty} e^{-m} \frac{m^i}{(i)!} i = m e^{-m} \sum_{1}^{\infty} \frac{m^{i-1}}{(i-1)!}$$

$$\bar{x} = m e^{-m} e^m = m \tag{13.31}$$

Variance of Poisson's Distribution

The variance is given as

$$\sigma^2 = \sum p_i (x_i - \bar{x})^2$$

$$= \sum p_i x_i^2 - 2\bar{x} \sum p_i x_i + (\bar{x})^2 \sum p_i$$

$$= \sum p_i x_i^2 - 2\bar{x}\bar{x} + (\bar{x})^2 = \sum p_i x_i^2 - (\bar{x})^2 \tag{13.32}$$

Now

$$p(r) = e^{-m} \cdot \frac{m^r}{(r)!} r = 0,1,2,...$$

$$\sum_{0}^{\infty} p_i r_i^2 = \sum_{r=0}^{\infty} e^{-m} \frac{m^r}{(r)!} \cdot r^2 = \sum_{0}^{\infty} e^{-m} m^r \left(\frac{r-1+1}{r-1}\right)$$

$$= m^2 e^{-m} \sum_{2}^{\infty} \frac{m^{r-2}}{(r-2)!} + m e^{-m} \sum_{r=1}^{\infty} \frac{m^{r-1}}{(r-1)!}$$

$$= m^2 e^{-m} e^m + m e^{-m} e^m = m^2 + m$$

Hence from Eq. (13.13) variance (Poisson distribution)

$$= \sum p_i x_i^2 - (\bar{x})^2$$

$$= m^2 + m - (m)^2 = m \qquad (13.33)$$

Thus, in the Poisson's distribution

$$P(r) = \frac{e^{-m} m^r}{(r)!}$$

Mean $\bar{x}$ = variance $\quad x = \sigma^2(x) = m \qquad (13.34)$

Standard deviation $\quad \sigma = \sqrt{m} \qquad (13.35)$

Note: The mean and variance of the Poisson distributions are same.

Some examples of Poisson's Variates
  i. Number of deaths occurring in a country due to a rare disease
 ii. Number of faulty cells coming in a telephone exchange
iii. Number of typing mistakes in a number of typed pages
 iv. Number of air accidents in a year
  v. Number of cars passing a point in a specified period
 vi. Number of defective pieces in a pack of 100 or more.

13.8.1. Recurrence Formula for the Poisson Distribution

For the poisson distribution of r successes, we have

$$P(r) = e^{-m} \cdot \frac{m^r}{(r)!}$$

where m is the mean of successes.

Also $\qquad P(r + 1) = e^{-m} \cdot \frac{m^{r+1}}{(r+1)!}$

$\therefore \qquad \dfrac{P(r+1)}{P(r)} = \dfrac{m}{r+1}$

$\Rightarrow \qquad p(r + 1) = \dfrac{m}{r+1} \cdot P(r)$

This is called the recurrence formula for Poisson's distribution which gives the probabilities of various values of r if one value for particular value of r is known.

13.8.2. Applications of Poisson Distribution

The distribution is applied to problems concerning
  i. Demand pattern for certain spare parts.
 ii. Arrival pattern of 'defective vehicles in a workshop' or patients in a hospital.
iii. Number of fragments from a shell hitting a target.
 iv. Number of defective articles in a consignment of large articles.

13.8.3. Mode of Poisson Distribution

The mode of a Poisson's variate is that value of the variate r for which the probability is maximum, i.e. r is a mode if

$$P(r-1) \le P(r) \ge p(r+1)$$

$$\Rightarrow \qquad e^{-m} \cdot \frac{m^{r-1}}{(r-1)!} \le e^{-m} \cdot \frac{m^r}{(r)!} \ge e^{-m} \cdot \frac{m^{r-1}}{(r+1)!}$$

$$\Rightarrow \qquad 1 \le \frac{m}{r} \ge \frac{m^2}{(r+1)r}$$

$$\Rightarrow \qquad r \le m \text{ and } r+1 \ge m$$

$$\therefore \qquad m-1 \le r \le m$$

Thus if m is not an integer, the mode is the integral value between m and $m-1$ and if m is an integer then m and $m-1$ are the modes so any one of them can be taken as mode. The distribution in this case is bimodal.

Example 1. *If the variance of the Poisson distribution is 2, find the probabilities for $r = 1, 2, 3, 4$ from the recurrence relation of the Poisson distribution.*

Solution: The parameter m of the Poisson distribution is same as variance:

$$\therefore \qquad m = 2$$

$$\therefore \qquad P(r) = e^{-2} \cdot \frac{2^r}{(r)!}$$

$$P(r+1) = e^{-2} \cdot \frac{2^{r+1}}{(r+1)!} \text{ or } = \frac{2}{r+1} p(r)$$

which is the recurrence relation.

Putting $r = 0, 1, 2, 3$ and 4, we get

$$P(1) = 2.P(0) = 2e^{-2} = 2 \times 0.1353 = 0.2706$$

$$P(2) = \frac{2}{2} P(1) = 0.2706$$

$$P(3) = \frac{2}{3} P(2) = 0.1804$$

$$P(4) = \frac{1}{2} P(3) = 0.0902$$

Example 2: *A car hire firm has two cars, which it hires out every day. The number of demands each day are distributed following the Poisson distribution with mean $m = 1.5$. Calculate the proportion of days on which (i) neither car is used (ii) the proportion of days on which some demand is refused.*

$$\text{(use } e^{-1.5} = 0.2231)$$

Solution: i. Probability of zero demand = probability (no car is used)

$$= e^{-m} \frac{m^0}{(0)!} = e^{-1.5} = 0.2231$$

ii. Probability (as there are only 2 cars, if demand $x \geq 3$) some demands will be refused.

$$= \text{Probability } (x \geq 3) = 1 - e^m \left\{ 1 + \frac{m}{(1)!} + \frac{m^2}{(2)!} \right\}$$

$$= 1 - 0.2231 (1 + 1.5 + 1.125) = 0.1913$$

Example 3. *If 2% electric bulbs manufactured by a company be defective, find the probability of (i) 0 (ii) 1 (iii) 2 (iv) 3 defectives in a lot of 100 bulbs using Poisson's distribution. ($e^{-2} = 0.1353$).*

Solution: $P = \dfrac{2}{100}$, $n = 100$, so $np = m = 2$

i. $P(x = 0) = \dfrac{e^{-m} m^0}{(0)!} = 0.1353$

ii. $P(x = 1) = e^{-m} \dfrac{m^1}{(1)!} = 2(0.1353) = 0.2706$

iii. $P(x - 2) - e^{-m} \dfrac{m^2}{(2)!} = 2(0.1353) = 0.2706$

iv. $P(x = 3) = e^{-m} \dfrac{m^3}{6} = (1.33)(0.1353) = 0.18$

Example 4. *If X be a Poisson variate with $3P(X = 2) = 2P(X = 1)$, evaluate $P(X = 0)$.*

Solution:

$$3e^{-m} \frac{m^2}{2} = 2e^{-m} \frac{m}{1}, \text{ so } m = \frac{4}{3}$$

$$P(X = 0) = e^{-\frac{4}{3}} \frac{m^0}{0!} = e^{-\frac{4}{3}}$$

Example 5. *Between 2 to 4 pm, the average number of phone calls per minute coming into the switch board of a company is 2.5. Find the probability that during one minute. There will be (i) 0 (ii) (1) (iii) 2 (iv) 3 (v) 4 or fewer (vi) more than 6 phone calls (use $e^{-2.5} = 0.08208$).*

Solution: Average = mean = $m = 2.5$

i. $P(x = 0) = e^{-m} = 0.08208$

ii. $P(x = 1) = e^{-m} \dfrac{m}{1} = 2.5 \times 0.08208 = 0.2050$

iii. $P(x = 2) = e^{-2.5} \dfrac{(2.5)^2}{2} = 3.125(.0825) = 0.2565$

iv. $P(x = 3) = e^{-2.5} \dfrac{(2.5)^3}{6} = \dfrac{15.375}{6}(0.0825) = 0.2138$

v. $P(x \leq 4) = P(0) + P(1) + P(2) + P(3) + P(4) = 0.8911$

vi. $P(x \geq 4) = 1 - \{P(0) + P(1) + P(2) + P(3) + P(4) + P(5) + P(6)\} = 0.0142$

Example 6. *A manufacture of screws knows that 4% of his product is defective. If he sells the screws in boxes of 100 and guarantees that not more than 5 screws will be defective, what is the probability that a box will fail to meet the guarantee?*

Solution: Here $P = \dfrac{4}{100};\ n = 100$ so $m = np = 4$

The box fails in guarantee, when number of defectives are 6 or more

$$P(X \geq 6) = 1 - [P(0) + P(1) + P(2) + P(3) + P(4) + P(5)]$$

$$= 1 - e^{-4}\left[1 + \frac{4}{1} + \frac{4^2}{2} + \frac{4^3}{6} + \frac{4^4}{24} + \frac{4^5}{120}\right]$$

Example 7. *A large number of observations on a given solution which contained bacteria were made by taking samples of 1 cc each and noting down the number of bacteria present in each sample. Assuming the poisson distribution and given that 10% of the samples contained no bacteria, find the average number of bacteria per cc.*

Solution: Let average number of bacteria per *cc* be *m*.

According to Poisson distribution, probability of *r* bacteria is given by

$$P(r) = e^{-m} \cdot \frac{m^r}{r!}$$

But $\quad\quad\quad\quad\quad\quad P(0) = 0.1 \text{ (given)}$

$\therefore \quad\quad\quad\quad\quad\quad\quad 0.1 = e^{-m}$

$\Rightarrow \quad\quad\quad\quad\quad\quad\quad e^m = 10$

$\quad\quad\quad\quad\quad\quad\quad\quad m = \log 10$

$\Rightarrow \quad\quad\quad\quad\quad\quad\quad m = 2.3026$

Example 8. *A manufacturer of cotter pins knows that 5% of his products are defective. If he sells cotter pins in boxes of 100 and guarantees that not more than 10 pins will be defective, what is the probability that a box will fail to meet the guaranteed quality $[e^{-5} = 0.0006738]$?*

Solution: We have $m = np = 100 \times \dfrac{5}{100} = 5$

$\therefore$ Required probability of failing to meet the guarantee is the probability of getting 11 or more defective pins.

$$= 1 - [P(0) + P(1) + \cdots + P(10)]$$

$$= 1 - \sum_{r=0}^{10} e^{-5} \frac{5^r}{r!} = 1 - (0.00674) \times \sum_{r=0}^{10} \frac{5^r}{r!}$$

13.9. NORMAL PROBABILITY DISTRIBUTION

So far we have discussed two discrete distributions, namely binomial and Poisson, with the help of which we are able to enumerate the probability of success or failure occurring in a fixed number of independent trials. However, in practice we come across a number of biological, economic, industrial measurements where the variables are continuous in nature, and as such can be adequately described only by a continuous probability distribution. One of the most important continuous probability distributions in the entire field of statistics is the normal probability distribution.

The graphical shape of the normal distribution, called the normal curve, is the bell-shaped smooth symmetrical curve as shown in the Fig. 13.2 having the density function defined by

$$f(x) = \frac{1}{\sigma\sqrt{2\pi}} e^{-\frac{1}{2}\left(\frac{x-\mu}{\sigma}\right)^2} \qquad [-\infty < x < \infty] \qquad (13.36)$$

where μ = mean of the normal distribution

σ = standard deviation of the distribution

π = 3.14159 (approx.)

e = 2.718 (approx.)

μ and σ are also known as parameters of the normal distribution.

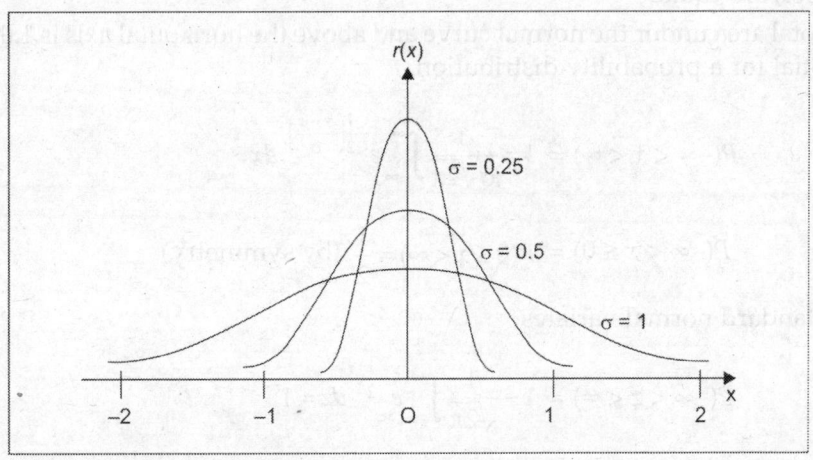

Fig. 13.2: For density function (1) with $m = 0$, for various values of σ.

Definition

Normal probability distribution: The probability distribution with density function given by Eq. (13.36) is called normal distribution or the Gaussian distribution, x is called the normal variate with mean μ and standard deviation σ and is denoted as

$$x : N (\mu, \sigma)$$

Standard normal variate: If x is a normal variate with mean μ and standard deviation σ, then

$$z = \frac{x - \mu}{\sigma}$$

is called standard normal variate. It has 0 mean and 1 as standard deviation. The density function is obtained from Eq. (13.36) by putting $\mu = 0$ and $\sigma = 1$,

i.e.
$$f(z) = \frac{1}{\sqrt{2\pi}} e^{-\frac{1}{2}z^2} \quad [-\infty < z < \infty] \tag{13.37}$$

13.9.1. Properties of Normal Distribution or the Normal Curve

Some of the important properties of normal distribution or curve may be listed as under Figs 13.3 and 13.4:

1. The curve is symmetrical about the vertical axis through the mean, i.e. if we fold the curve along with this vertical axes, the two halves of the curve coincide.

2. The mean, median and mode of the distribution are indentical, i.e.
$$\text{Mean} = \text{Mode} = \text{Median}$$

3. The curve is asymptotic to x-axis, i.e. it becomes closer and closer to x-axis but never actually touches it.

4. The height of the curve declines symmetrically in either direction from the maximum point. Therefore, the ordinates for values of $x = \mu \pm k$, where k is a real number, are equal.

5. The total area under the normal curve and above the horizontal axis is 1.0 which is essential for a probability distribution.

So
$$P(-\infty < x < \infty) = 1 = \frac{1}{\sigma\sqrt{2\pi}} \int_{-\infty}^{\infty} e^{-\frac{1}{2}\left(\frac{x-\mu}{\sigma}\right)^2} dx \tag{13.38}$$

and
$$P(-\infty < x \le 0) = P(0 \le x < \infty) = \frac{1}{2} \quad \text{(by symmetry)}$$

For standard normal variates

$$P(-\infty < z < \infty) = 1 - \frac{1}{\sqrt{2\pi}} \int_{-\infty}^{\infty} e^{-\frac{1}{2}z^2} dz = 1 \tag{13.39}$$

$$P(z \le \infty) = P(z \ge 0) = \frac{1}{\sqrt{2\pi}} \int_{0}^{\infty} e^{-\frac{1}{2}z^2} dz = \frac{1}{2} \tag{13.40}$$

6. Out of the most important property of the normal curve is the area property (probability in the given interval). Since the shape of the curve is completely determined by its parameters m and s so the probability in the certain interval depends on these parameters. Some important probabilies are given as

    i. $P(\mu - \sigma < x < \mu + \sigma) = 0.6827$ (13.41)

    ii. $P(\mu - 2\sigma < x < \mu + 2\sigma) = 0.9545$ (13.42)

    iii. $P(\mu - 3\sigma < x < \mu + 3\sigma) = 0.9973$ (13.43)

Thus, practically, we can say that whole area under the normal curve lies within $\mu \pm 3\sigma$ limits, which are also called 3σ limits.

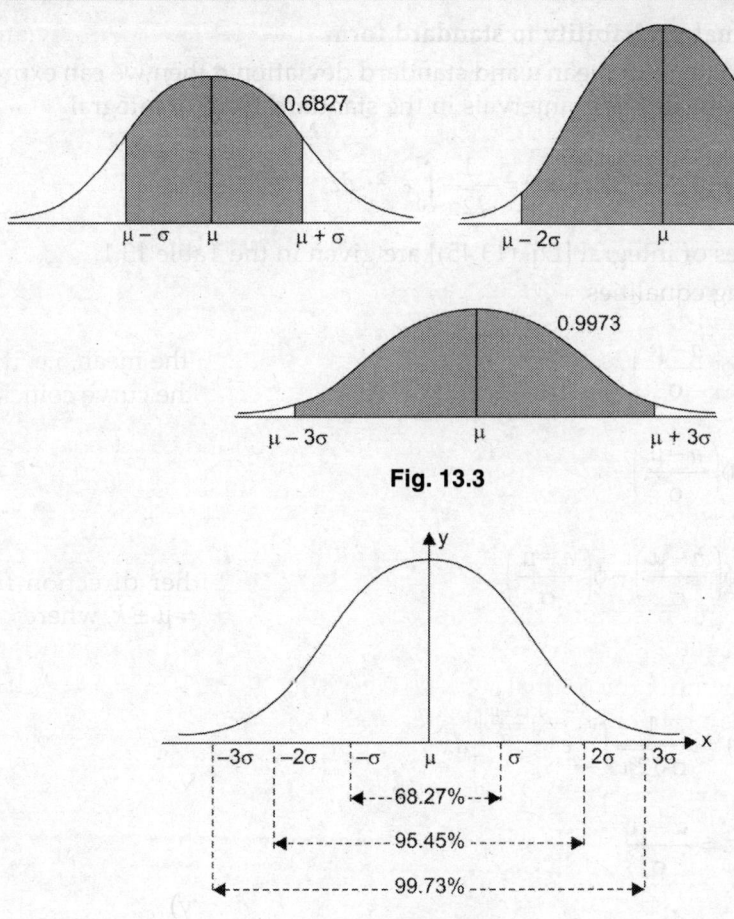

Fig. 13.3

Fig. 13.4

13.9.2. Normal Probabilities for Intervals

Let x be a normal variate with mean μ and standard deviation σ with density function given by Eq. (13.36) then the probability that x lies in the interval $a \leq x \leq b$ is defined by

$$P(a \leq x \leq b) = \int_a^b f(x)\,dx = \frac{1}{\sigma\sqrt{2\pi}} \int_a^b e^{\frac{-(x-\mu)^2}{2\sigma^2}}\,dx \tag{13.44}$$

The above integral determines the area under the normal curve bounded by two ordinates at $x = a$ and $x = b$.

Thus, the integral [Eq. (13.44)] is not easy to findout but can be expressed in terms of the integral

$$\phi(t) = \frac{1}{\sqrt{2\pi}} \int_0^t e^{-\frac{1}{2}z^2}\,dz \tag{13.44a}$$

The values of $\phi(t)$ are given in Table 13.1 for various values of t. But $\phi(t)$ is equal to $P(0 < z < t)$, probability of standard normal variate in the interval $(0, t)$.

Reduction of normal probability to standard form

If x is a normal variate with mean μ and standard deviation σ then we can express the probability in various different intervals in the standard form of integral

$$P(0 \le z \le t) = \phi(t) = \frac{1}{\sqrt{2\pi}} \int_0^t e^{-\frac{1}{2}z^2} dz \qquad (13.45)$$

For $t > 0$, the values of integral [Eq. (13.45)] are given in the Table 13.1.
Prove the following equalities.

a. $P(x \le a) = \frac{1}{2} + \phi\left(\frac{a-\mu}{\sigma}\right)$ \qquad\qquad (13.46)

b. $P(x \ge a) = \frac{1}{2} - \phi\left(\frac{a-\mu}{\sigma}\right)$ \qquad\qquad (13.47)

c. $P(a \le x \le b) = \phi\left(\frac{b-\mu}{\sigma}\right) - \phi\left(\frac{a-\mu}{\sigma}\right)$ \qquad (13.48)

d. $\phi(-a) = -\phi(a)$ \qquad\qquad (13.49)

Proof: a. $P(x \le a) = \frac{1}{\sigma\sqrt{2\pi}} \int_{-\infty}^{a} e^{-\frac{1}{2}\left(\frac{x-\mu}{\sigma}\right)^2} dx$

Put $\qquad\qquad z = \dfrac{x-\mu}{\sigma}$

or $\qquad\qquad dz = \dfrac{dx}{\sigma}$

$$P(x \le a) = \frac{1}{2\sqrt{\pi}} \int_{-\infty}^{a-\mu/\sigma} e^{-\frac{1}{2}z^2} dz$$

$$= \frac{1}{\sqrt{2\pi}} \int_{-\infty}^{0} e^{-\frac{1}{2}z^2} dz + \int_{0}^{a-\mu/\sigma} e^{-\frac{1}{2}z^2} dz$$

$$= \frac{1}{2} + \phi\left(\frac{a-\mu}{\sigma}\right) \qquad [(using \; Eq. \; (13.40)]$$

b. $\qquad\qquad P(x \ge a) = \frac{1}{\sigma\sqrt{2\pi}} \int_{a}^{\infty} e^{-\frac{1}{2}\left(\frac{x-\mu}{\sigma}\right)} dx$

Put $\qquad\qquad z = \dfrac{x-\mu}{\sigma}$

$$= \frac{1}{\sqrt{2\pi}} \int_{a-\mu/\sigma}^{\infty} e^{-\frac{1}{2}z^2} dz$$

$$= \frac{1}{\sqrt{2\pi}} \int_0^\infty e^{-\frac{1}{2}z^2} dz - \frac{1}{\sqrt{2\pi}} \int_0^{a-\mu/\sigma} e^{-\frac{1}{2}z^2} dz$$

$$= \frac{1}{2} - \phi\left(\frac{a-\mu}{\sigma}\right)$$

c. $\quad P(a \le x \le b) = \frac{1}{\sigma\sqrt{2\pi}} \int_a^b e^{-\frac{1}{2}\left(\frac{x-\mu}{\sigma}\right)} dx$

Put $\quad z = \frac{x-\mu}{\sigma}$

$$= \frac{1}{\sqrt{2\pi}} \int_{a-\mu/\sigma}^{b-\mu/\sigma} e^{-\frac{1}{2}z^2} dz$$

$$= \frac{1}{\sqrt{2\pi}} \int_0^{b-\mu/\sigma} e^{-\frac{1}{2}z^2} dz - \frac{1}{\sqrt{2\pi}} \int_0^{a-\mu/\sigma} e^{-\frac{1}{2}z^2} dz$$

$$= \phi\left(\frac{b-\mu}{\sigma}\right) - \phi\left(\frac{a-\mu}{\sigma}\right)$$

d. $\quad \phi(-a) = \frac{1}{\sqrt{2\pi}} \int_0^{-a} e^{-\frac{1}{2}z^2} dz$

Put $\quad z = -u$
$\quad\quad dz = -du$

$$\phi(-a) = \frac{-1}{\sqrt{2\pi}} \int_0^a e^{-\frac{1}{2}u^2} du = -\phi(a)$$

Thus, from the above result, we can find out the probabilities of a normal variate in any interval by converting into the function $\phi(t)$ the values of which are given in the table for positive values of t. If t is negative then by using result (d) we can also find the value of $\phi(t)$ as

$$\phi(t) = -\phi(-t)$$

t negative $\Rightarrow$ $(-t)$ is positive.

$\phi(t)$ is equal to $P(0 < z < t)$, where z is standard normal variate.

Example 1. *For* $-\infty < x < \infty$, *and probability density*

$$f(x) = \frac{1}{\sigma\sqrt{2\pi}} e^{-\frac{(x-\mu)^2}{2\sigma^2}}$$

Show that the total probability is 1.

Solution: Total probability

$$I = \int_{-\infty}^\infty f(x)\,dx = \frac{1}{\sigma\sqrt{2\pi}} \int_{-\infty}^\infty e^{-\left(\frac{x-\mu}{\sqrt{2}\sigma}\right)^2} dx$$

Let $\qquad\qquad\dfrac{x-\mu}{\sigma\sqrt{2}} = y$ so $dx = \sigma\sqrt{2}\,dy$

$$I = \frac{\sigma\sqrt{2}}{\sigma\sqrt{2\pi}}\int_{-\infty}^{\infty} e^{-y^2}\,dy$$

$$= \frac{1}{\sqrt{\pi}}\int_{-\infty}^{\infty} e^{-y^2}\,dy = \frac{1}{\sqrt{\pi}}2\int_{0}^{\infty} e^{-y^2}\,dy \qquad\qquad \text{As } f(-y) = f(y)$$

Let $\qquad\qquad y^2 = u \quad \text{or} \quad y = \sqrt{u}$

$$I = \frac{2}{\sqrt{\pi}}\int_{0}^{\infty} \frac{1}{2}e^{-u}u^{1/2-1}\,du = \frac{1}{\sqrt{\pi}}\Gamma(1/2) = 1$$

by Gamma integral $\displaystyle\int_{0}^{\infty} e^{-x}x^{n-1}\,dx = \Gamma n$. Hence proved.

Example 2. *Show that the mean of the normal distribution with probability density*

$$f(x) = \frac{1}{\sigma\sqrt{2\pi}}e^{-\frac{(x-\mu)^2}{2\sigma^2}}\ [-\infty < x < \infty]$$

is μ and variance is σ^2.

Solution: i. $\quad \bar{x} = \displaystyle\int_{-\infty}^{\infty} xf(x)\,dx$

$$= \frac{1}{\sigma\sqrt{2\pi}}\int_{-\infty}^{\infty} xe^{-\left(\frac{x-\mu}{\sqrt{2}\sigma}\right)^2}dx, \qquad \text{let } \frac{x-\mu}{\sigma\sqrt{2}} = y$$

$$= \frac{\sigma\sqrt{2}}{\sigma\sqrt{2}\sqrt{\pi}}\int_{-\infty}^{\infty} (\mu + \sigma\sqrt{2}y)e^{-y^2}\,dy$$

$$= \frac{1}{\sqrt{\pi}}\left[\mu\underbrace{\int_{-\infty}^{\infty} e^{-y^2}\,dy}_{\text{even function}} + \sigma\sqrt{2}\underbrace{\int_{-\infty}^{\infty} ye^{-y^2}\,dy}_{\text{odd function}}\right]$$

$$= \frac{1}{\sqrt{\pi}}\left[\mu\,2\int_{0}^{\infty} e^{-y^2}\,dy + 0\right]$$

$$\bar{x} = \mu\frac{2}{\sqrt{\pi}}\frac{\sqrt{\pi}}{2} \quad (\text{see Example 1}) = \mu$$

ii. $\qquad \text{Variance} = \displaystyle\int_{-\infty}^{\infty} (x-\mu)^2 f(x)\,dx$

$$= \frac{1}{\sigma\sqrt{2\pi}}\int_{-\infty}^{\infty} (x-\mu)^2 e^{-\left(\frac{x-\mu}{\sqrt{2}\sigma}\right)^2}dx, \qquad \text{let } \frac{x-\mu}{\sigma\sqrt{2}} = y$$

$$= \frac{1}{\sigma\sqrt{2}\sqrt{\pi}}\sigma\sqrt{2}\int_{-\infty}^{\infty}2\sigma^2 y^2 e^{-y^2}dy = \frac{2\sigma^2}{\sqrt{\pi}}\cdot 2\int_{0}^{\infty}y^2 e^{-y^2}dy$$

Let $\qquad y^2 = u \qquad$ so $\quad dy = \dfrac{1}{2\sqrt{u}}du$

$\therefore \quad$ Variance $= \dfrac{4\sigma^2}{\sqrt{\pi}}\displaystyle\int_{0}^{\infty}\dfrac{u^2 e^{-u}}{2\cdot(\sqrt{u})}du$

$$= \frac{2\sigma^2}{\sqrt{\pi}}\int_{0}^{\infty}u^{\frac{3}{2}-1}e^{-u}du$$

$$= \frac{2\sigma^2}{\sqrt{\pi}}\times\Gamma\left(\frac{3}{2}\right) = \frac{2}{\sqrt{\pi}}\sigma^2\times\frac{1}{2}\times\sqrt{\pi} = \sigma^2$$

Example 3. *Find the mean deviation from the mean of a normal variate x.*

Solution: Let x be normal variate with mean m and variance s^2. Then mean deviation from the mean is defined by

$$\int_{-\infty}^{\infty}|x-\mu|f(x)dx = \frac{1}{\sigma\sqrt{2\pi}}\int_{-\infty}^{\infty}|(x-\mu)|e^{-\frac{1}{2}\left(\frac{x-\mu}{\sigma}\right)^2}dx \qquad \left[\text{put } z = \frac{x-\mu}{\sigma}\right]$$

$$= \frac{\sigma}{\sqrt{2\pi}}\int_{-\infty}^{\infty}|z|e^{-\frac{z^2}{2}}dz$$

$$= \frac{\sigma}{\sqrt{2\pi}}\int_{-\infty}^{0}-ze^{-\frac{z^2}{2}}dz + \int_{0}^{\infty}ze^{-\frac{z^2}{2}}dz$$

$$= \frac{2\sigma}{\sqrt{2\pi}}\int_{0}^{\infty}ze^{-\frac{z^2}{2}}dz \qquad \left[\text{put } \frac{z^2}{2} = t\right]$$

$$= \frac{2\sigma}{\sqrt{2\pi}}\int_{0}^{\infty}e^{-t}dt = \frac{2\sigma}{\sqrt{2\pi}}\times[-e^{-t}]_{0}^{\infty}$$

$$= \sqrt{\frac{2}{\pi}}\cdot\sigma = \frac{4}{5}\sigma$$

Example 4. *Find the area under the standard normal curve for (–1.25 < z + 0.6)*

Solution: Required area = Area for (–1.25 < z < 0) + area for (0 < z < +0.6)

$\qquad\qquad$ = Area for (0 < z < 1.25) + area for (0 < z < + 0.6)

$\qquad\qquad$ = ϕ(1.25) + ϕ(0.6)

$\qquad\qquad$ = 0.3944 + 0.2257 = 0.6201 $\quad$ (*refer to* Table 13.1)

Example 5. _Find the area under the normal curve for (i) z < –18 (ii) z > 0.5 (iii) z < 2.3 (iv) z > –2.25._

Solution: i.

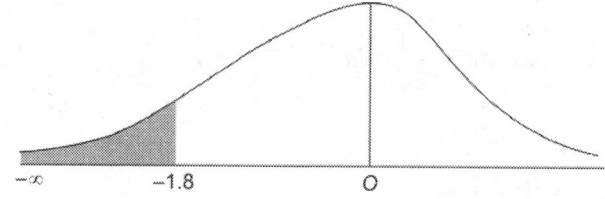

Fig. 13.5

Required area

$$= \text{Area } (-\infty < z < 0) - \text{area } (-1.8 < z < 0)$$
$$= 0.5 \text{ Area } (0 < z < 1.8) = 0.5 - \phi(1.8)$$
$$= 0.5 - .4641 = 0.0359$$

ii.

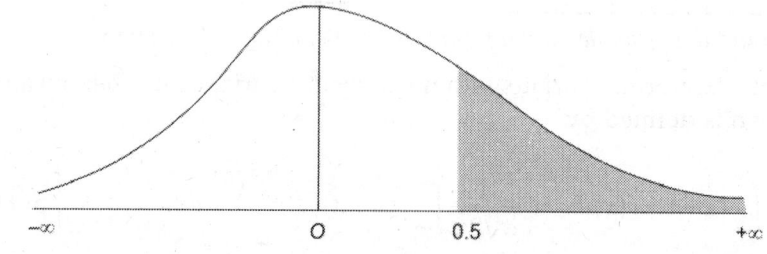

Fig. 13.6

Required area

$$= \text{Area } (0.5 < z < +\infty)$$
$$= \text{Area } (0 < z < +\infty) - \text{area } (0 < z < 0.5)$$
$$= 0.5 - 0.1915 = 0.3085$$

iii.

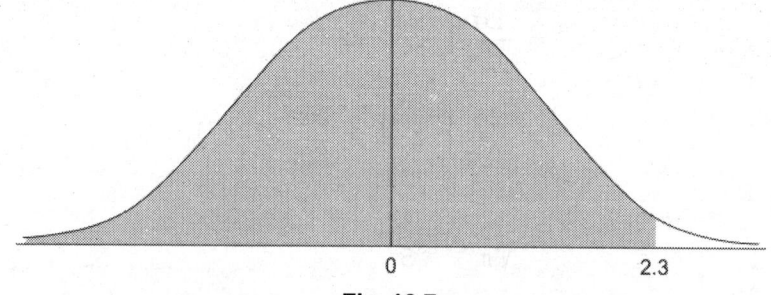

Fig. 13.7

Required area

$$= \text{Area } (z < 2.3) = \text{Area } (-\infty < z < 0) + \text{area } (0 < z < 2.3)$$
$$= 0.5 + \phi(2.3)$$
$$= 0.5 + 0.4893 = 0.9893$$

iv.

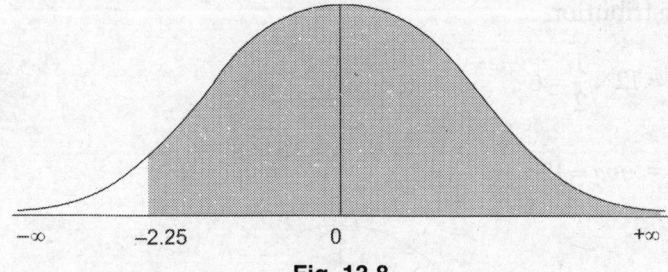

Fig. 13.8

Required area

$$= \text{Area } (z > -2.25) = \text{Area } (-2.25 < z < 0) + \text{area } (0 < z < +\infty)$$

$$\phi(2.25) + \frac{1}{2}$$

$$= \text{Area } (0 < z < +2.25) + 0.5 = 0.4878 + 0.5 = 0.9878$$

Example 6. *Assuming the mean height of soldiers to be m = 68.22 inches with variance s = 10.8 (inches)². Find how many soldiers in a regiment of 1000 would you expect to be above 6 feet tall (given area under the normal curve for A(0 < z < 0.35) = 0.1368 and for A(0 < z < 1.15) = 0.3746.*

Solution:
$$z = \frac{x - \mu}{\sigma} = \frac{72 - 68.22}{\sqrt{10.8}} = \frac{3.78}{3.286} = 1.15$$

$$P(z > 1.15) = 0.5 - P(0 < z < 1.15) = 0.5 - \phi(1.15)$$
$$= 0.5 - 0.3746 = 0.1254$$

Hence the number of soldiers in $N = 1000$
$$= Np = 1000 \times 0.1254 = 125.4$$

So 125 soldiers will be more than 6 ft tall.

Example 7. *A coin is tossed 12 times. Find the probability, both exactly and by fitting a normal distribution of getting (i) 4 heads (ii) at most 4 heads.*

Solution: By binomial

$$p = q = \frac{1}{2}, \ n = 12$$

$$P(4 \text{ heads}) = {}^{12}C_4 q^{12-4} p^4 = \frac{11 \times 5 \times 9}{2^{12}} = 0.121$$

$$P(\text{no of heads} \leq 4) = P(0) + P(1) + P(2) + P(3) + P(4)$$

$$= \frac{1}{2^{12}} \left[{}^{12}C_0 + {}^{12}C_1 + {}^{12}C_2 + {}^{12}C_3 + {}^{12}C_4 \right]$$

$$= \frac{1 + 12 + 66 + 220 + 495}{2^{12}} = \frac{794}{2^{12}} = 0.194$$

By normal distribution

Mean $\mu = np = 12 \times \dfrac{1}{2} = 6$

Variance $\sigma^2 = npq = 12 \times \dfrac{1}{2} \times \dfrac{1}{2} = 3$

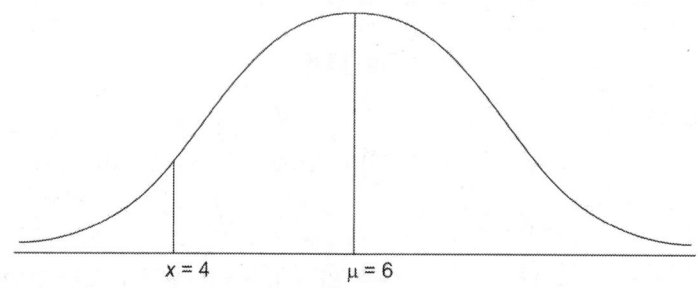

$x = 4$ $\mu = 6$

Fig. 13.9

For $n \le 4$

$$z_1 = \frac{x - \mu}{\sigma} = \frac{4.5 - 6}{\sqrt{3}} = -0.5 \times \sqrt{3} = -0.5 \times 1.773 = -0.866$$

Probability $(-\infty < z < -0.866) = 0.5 - P(0 < z < 0.866)$

Probability number of heads $\le 4 = 0.5 - 0.3065 = 0.1935$

Probability of 4 heads by normal distribution

$$= P(3.5 < x < 4.5)$$

$$= P\left(\frac{3.5 - 6}{\sqrt{3}} < z < \frac{4.5 - 6}{\sqrt{3}} \right) = P(-1.231 < z < -0.7385)$$

$$= P(0 < z < 1.231) - P(0 < z < 0.7385) = \phi(1.231) - \phi(0.73)$$

$$= 0.3907 - 0.2700 = 0.120$$

Example 8. *Marks obtained by a number of students are found to be distributed normally with mean 65 and variance 25. If 3 students are taken randomly, find the probability that 2 of them have marks above 70 (given area under the normal curve from z = 0 to z = 1 is 0.34).*

Solution: Given $\sigma = \sqrt{25} = 5$

Mean $\mu = 65, x = 70$

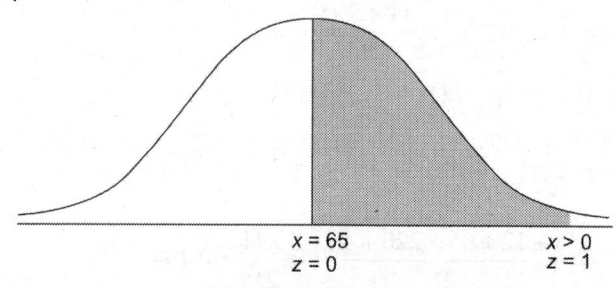

$x = 65$ $x > 0$
$z = 0$ $z = 1$

Fig. 13.10

So $\qquad \dfrac{x - \mu}{\sigma} = z_1$

$$z = \dfrac{70 - 65}{5} = 1$$

$$P(x > 70) = P(z > 1) = 0.5 - P(0 < 2 < 1)$$

$$= \dfrac{1}{2} - \phi(1) = 0.500 - 0.34 = 0.16$$

So $\qquad p = 0.16$

$\qquad\qquad q = 0.84$

Probability that 2 out of 3 have marks above 70

$$= {}^3C_2 p^2 q = 3(0.16)^2(0.84) = 0.0645$$

Example 9. *In an examination the mean marks were 78 with standard deviation 10. Determine the grades of the students whose marks were 93, 78 and 62 respectively. Find the marks of the students whose grades were –0.6 and 1.2 respectively, assuming normal distribution.*

Solution: By grades here, we mean the z values.

i. When $x_1 = 93$, $\qquad z_1 = \dfrac{93 - 78}{10} = 1.5$

ii. When $x_2 = 78$, $\qquad z_2 = \dfrac{78 - 78}{10} = 0$

iii. When $x_3 = 62$, $\qquad z_3 = \dfrac{62 - 78}{10} = -1.6$

iv. When $x_4 = -0.6$, $-0.6 = \dfrac{x_4 - 78}{10} \Rightarrow x_4 \square 72$

v. When $z_5 = 1.2$, $\qquad 1.2 = \dfrac{x_5 - 78}{10} \Rightarrow x_5 = 90$

Example 10. *In an examination the mean marks were 60, with standard deviation 5. If the marks be distributed normally find the percentage of students getting (i) more than 60 (ii) less than 56 (iii) whose marks lie between 45 and 65. If the top 5% of the students are to be given grade A, what are the minimum marks to receive grade A.*

Solution: i. $z_1 = \dfrac{x_1 - \mu}{\sigma} = \dfrac{60 - 60}{\sigma} = 0$

$$P(x > 60) \Rightarrow 1 - P(-\infty < z < 0) = 1 - 0.5 = 0.5$$
$$= 50\% \text{ received above } 60\%$$

ii. $z = \dfrac{56 - 60}{5} = -0.8$

$$P(x < 56) = P(z < -0.80) = 0.5 - P(0 < z < 0.8) = 0.288 = 0.212$$

So 21.2% students obtained marks less than 56%.

iii. $P(45 < z < 65) = P\left(\dfrac{45-60}{5} < z < \dfrac{65-60}{5}\right)$

$= P(-3 < z < +1)$

$= P(0 < z < 3) + P(0 < z < 1) = \phi(3) + \phi(1)$

$= 0.4987 + 0.3413 = 0.8400$　　(*refer to* Table 13.1)

So 84% students got marks between 45 and 65.

iv. If $P(-\infty < z < z') = 0.95$

then $P(0 < z < z') = 0.95 - 0.50 = 0.45$

From Table 13.1: $P(0 < z < 1.65) = 0.4505$

So $z' = 1.65 \quad \Rightarrow \quad x' - 60 = 5(1.65)$

or $x' = 68.25$

So minimum marks for A grade = 69.

If 10% of the students fail, then the minimum passing marks

$$P(-\infty < z < -z_0) = 0.1$$

or 　　　　　　$$P(0 < z < z_0) = 0.5 - 0.1 = 4$$

From Table 13.1 $P(0 < z < 1.29) = 0.4015$

so 　　　　　　　　　　$$z_0 = 1.29$$

$$\frac{x_0 - 60}{5} = -1.29 \quad \Rightarrow \quad x_0 = 60 - 6.45 = 53.45$$

Hence the pass marks were 54.

MISCELLANEOUS EXAMPLES

Example 1: *A die is cast until 6 appears. What is the probability that it must be cast more than 5 times?*

Solution: Here, probability of getting 6 is $p = \dfrac{1}{6}$

Then, 　　　　　　　　　$q = \dfrac{5}{6}$

If X is the number of tosses required for the first success, then

$$P(X = x) = q^{x-1}\, p \text{ for } x = 1, 2, 3,...$$

Required probability = $P(X \le 5)$

$$= 1 - P(x \le 5)$$

$$= 1 - \sum_{x=1}^{5} \left(\frac{5}{6}\right)^{x-1} \cdot \left(\frac{1}{6}\right)$$

$$= 1 - \frac{1}{6}\left\{1 + \left(\frac{5}{6}\right) + \left(\frac{5}{6}\right)^2 + \left(\frac{5}{6}\right)^3 + \left(\frac{5}{6}\right)^4\right\}$$

$$= \left(\frac{5}{6}\right)^5$$

Example 2. *In 256 sets of 12 tosses of a coin, in how many cases one can expect 8 heads and 4 tails.*

Solution: P (head) $= \dfrac{1}{2}$ and P (tail) $= \dfrac{1}{2}$

By binomial distribution, probability of 8 heads and 4 tails in 12 trials is

$$P(X = 8) = {}^{12}C_8 \left(\frac{1}{2}\right)^8 \left(\frac{1}{2}\right)^4$$

$$= \frac{12!}{8!\,4!} \cdot \frac{1}{2^{12}} = \frac{495}{4096}$$

The expected number of such cases in 256 sets

$$= 256 \times P(x = 8) = 256 \times \frac{495}{4096} = 30.9 \approx 31$$

Example 3. *The melting point X of a certain specimen be assumed to be a continuous random variable which is uniformly distributed over the interval [110, 120]. Find the density function of X, mean of X, variance of X and $P(112 \le x \le 115)$.*

Solution: Here, $a = 110$, $b = 120$

$$f(x) = \begin{cases} \dfrac{1}{120 - 110}, & 110 \le x \le 120 \\ 0, & \text{otherwise} \end{cases}$$

$$= \begin{cases} \dfrac{1}{10}, & 110 \le x \le 120 \\ 0, & \text{otherwise} \end{cases}$$

Mean of $\quad X = \dfrac{a+b}{2} = 115$

Variance of $X = \dfrac{(b-a)^2}{12} = \dfrac{(120-110)^2}{12} = \dfrac{25}{3}$

$$P(112 \le x \le 115) = \int_{112}^{115} f(x)\,dx$$

$$= \int_{112}^{115} \frac{1}{10}\,dx = \frac{3}{10}$$

Example 4. *Find median of the following data:*

Cost	10–20	20–30	30–40	40–50	50–60
Items	4	5	3	6	3

Solution:

Cost	Item (f)	Cumulative frequency (cf)
10–20	4	4
20–30	5	9 = C
30–40	3 = f	12
40–50	6	18
50–60	3	21

Here, $\sum f_i = 21 = N, C = 9, h = 10$

Hence $\dfrac{21}{2} = 10.5$

The median class is 30–40

$$\text{Median} = L + \frac{\left(\dfrac{N}{2} - C\right)}{f} \times h = 30 + \frac{10}{3}(10.5 - 9) = 35$$

Example 5. *The two regression equations of the variables x and y are x = 19.13 – 0.87y and y = 11.64 – 0.50x. Find (i) mean of x (ii) mean of y (iii) the correlation coefficient between x and y.*

Solution: Since, mean of x and mean of y lie on the two regression lines, we have

$$\bar{x} = 19.13 - 0.87\bar{y} \tag{1}$$

$$\bar{y} = 11.64 - 0.50\bar{x} \tag{2}$$

Multiplying Eq. (2) by 0.87 and subtracting from Eq. (1), we have

$$[1 - (0.87)(0.50)]\bar{x} = 19.13 - (11.64)(0.87)$$

$$0.57\bar{x} = 9.00$$

$$\bar{x} = 15.79$$

$$\bar{y} = 11.64 - (0.50)(15.79) = 3.74$$

∴ Regression coefficient of y on x is –0.50 and that of x on y is –0.87

Now, since the coefficient of correlation is the geometric mean between the two regression coefficients

$$r = \sqrt{(-0.50)(-0.87)} = \sqrt{0.43} = -0.66$$

[negative sign is taken since both the regression coefficients are negative].

Example 6. *If the probability of a bad reaction from a certain injection is 0.001, determine the chance that out of 2000 individuals more than two will get a bad reaction.*

Solution: It follows Poisson distribution, as the probability of occurrence is very small.

Mean $m = np = 2000(0.0001) = 2$

Probability that more than 2 will get a bad reaction = 1 – [probability that no one gets a bad reaction + probability that two get bad reaction]

$$= 1 - \left[e^{-m} + \frac{m^1 e^{-m}}{1!} + \frac{m^2 e^{-m}}{2!} \right]$$

$$= 1 - \left[\frac{1}{e^2} + \frac{2}{e^2} + \frac{2}{e^2} \right] = 1 - \frac{5}{e^2} = 0.32$$

Example 7. *Ten participants in a contest are ranked by two judges as follows*

x	1	6	5	10	3	2	4	9	7	8
y	6	4	9	8	1	2	3	10	5	7

Calculate the rank correlation coefficient ρ.

Solution: $d_i = x_i - y_i$, then

$$d_i = 5, 2, -4, 2, 2, 0, 1, -1, 2, 1$$

$$\sum d_i^2 = 25 + 4 + 16 + 4 + 4 + 0 + 1 + 1 + 4 + 1 = 60$$

$$\rho = 1 - \frac{6 \sum d_i^2}{n^3 - n}$$

$$= 1 - \frac{6 \times 60}{990} = 0.6 \text{ (nearly)}$$

Example 8. *What is the chance that a leap year selected at random will contain 53 Sundays?*

Solution: A leap year consists 366 days, so there are 52 full weeks (and hence 52 Sundays) and two extra days. These two days can be (i) Monday, Tuesday (ii) Tuesday, Wednesday (iii) Wednesday, Thursday (iv) Thursday, Friday (v) Friday, Saturday (vi) Saturday, Sunday (vii) Sunday, Monday.

∴ Required probability = $\frac{2}{7}$.

Example 9. *Find the chance of throwing (a) four (b) an even number with an ordinary six faced die.*

Solution: There are six possible ways in which the die can fall and of these there is only one way of throwing 4.

a. Thus, the required chance = $\frac{1}{6}$

b. There are six possible ways in which the die can fall and of these there are only three ways of getting, 2, 4 and 6

Thus, the required chance = $\frac{3}{6} = \frac{1}{2}$.

Example 10. *If four coins are tossed, find the chance that there should be two heads and two tails.*

Solution: p = Probability of a head = $\dfrac{1}{2}$

$\qquad q$ = Probability of a tail = $1 - p = \dfrac{1}{2}$

∴ Probability of getting 2 heads (and 2 tails)

When four coins are tossed

$$= p(2) = {}^4C_2\, p^2 q^{4-2} = 6 \times \left(\frac{1}{2}\right)^2 \times \left(\frac{1}{2}\right)^2 = \frac{3}{8}$$

Example 11: *An incomplete frequency distribution is given as below:*

Variable	10–20	20–30	30–40	40–50	50–60	60–70	70–80
Frequency	12	30	?	65	?	25	18

Given that the total frequency is 229 and median is 46, find the missing frequencies.

Solution:

Cf	12	42	$42 + f_1$	$107 + f_1$	$107 + f_1 + f_2$	$132 + f_1 + f_2$	$150 + f_1 + f_2$

Let f_1 and f_2 be the missing frequencies of the 30–40 and 50–60 respectively.

Since, the median lies in the class 40–50

$$46 = 40 + \frac{\dfrac{229}{2} - (12 - 30 + f_1)}{65} \times 10$$

$\qquad f_1 = 33.5$ which can be taken as 34

$\qquad f_2 = 229 - (12 + 30 + 34 + 65 + 25 + 18) = 45$

Example 12. *Two cards are drawn in succession from a pack of 52 cards. Find the chance that the first is a king and the second a queen, if the first card is (i) replaced (ii) not replaced.*

Solution: i. The probability of drawing a king = $\dfrac{4}{52} = \dfrac{1}{13}$

If the card is replaced, the pack will again have 52

Cards of that the probability of drawing a queen is $\dfrac{1}{13}$

The two events being independent, the probability of drawing both cards in succession

$$= \frac{1}{13} \times \frac{1}{13} = \frac{1}{169}$$

ii. The probability of drawing a king $= \dfrac{1}{13}$

If the card is not replaced, the pack will have 51 cards,

so that the chance of drawing a queen is $\dfrac{4}{51}$

Hence, the probability of drawing both cards

$$= \dfrac{1}{13} \times \dfrac{4}{51} = \dfrac{4}{663}$$

Example 13. *There are three bags: first containing 1 white, 2 red, 3 green balls; second 2 white, 3 red, 1 green balls. Two balls are drawn from a bag chosen at random. These are found to be one white and one red. Find the probability that the ball so drawn came from the second bag?*

Solution: Let B_1, B_2, B_3 certain to the first, second, third bags chosen and A: the two balls are white and red.

Now, $\qquad P(B_1) = P(B_2) = P(B_3) = \dfrac{1}{3}$

$$P\left(\dfrac{A}{B_1}\right) = P$$

(a white and red ball are drawn from both bags)

$$= \dfrac{(^1C_1 \times {}^2C_1)}{{}^6C_2} = \dfrac{2}{15}$$

Similarly, $\qquad P\left(\dfrac{A}{B_2}\right) = \dfrac{(^2C_1 \times {}^3C_1)}{{}^6C_2} = \dfrac{2}{5}$

$$P\left(\dfrac{A}{B_3}\right) = \dfrac{(^3C_1 \times {}^1C_1)}{{}^6C_2} = \dfrac{1}{5}$$

By Baye's theorem

$$P\left(\dfrac{B_2}{A}\right) = \dfrac{P(B_2)P\left(\dfrac{A}{B_2}\right)}{P(B_1)P\left(\dfrac{A}{B_1}\right) + P(B_2)P\left(\dfrac{A}{B_2}\right) + P(B_3)P\left(\dfrac{A}{B_3}\right)}$$

$$= \dfrac{\dfrac{1}{3} \times \dfrac{2}{5}}{\dfrac{1}{3} \times \dfrac{2}{15} + \dfrac{1}{3} \times \dfrac{2}{5} + \dfrac{1}{3} \times \dfrac{1}{5}} = \dfrac{6}{11}$$

Example 14. Find the total number of possible events that can occur for an experiment.

Solution: Let sample space S consists of n sample points. Then, total number of possible events

$$= \text{total number of subsets of } S$$

$$= |P(S)|$$

where, $P(S)$ is power set of $S = 2^n$.

PROBLEMS

1. One student is selected from a class with 4 boys and 6 girls, find the chance for this student to be a girl.
$$\left[Ans. \; \frac{6}{10} \right]$$

2. Find the chance that a card selected from a pack of 52 cards is an ace, king, queen or a Jack.
$$\left[Ans. \; \frac{4}{13} \right]$$

3. What is the chance of having 53 Sundays in a year when the year is (i) 2003, (ii) 2004.
$$\left[Ans. \; \frac{1}{7}, \frac{2}{7} \right]$$

4. A number is chosen from natural numbers from 1 to 100. What is the chance that it is neither divisible by 3 nor by 5 nor by both.

Hint: $\left(1 - \dfrac{33 + 20 - (\text{multiples of } 15)6}{100} \right) = 100 - \dfrac{47}{100}$
$$\left[Ans. \; \frac{53}{100} \right]$$

5. Three cards are drawn from a pack of 52, what is the probability that all three are kings.
$$\left[Ans. \; \frac{1}{5525} \right]$$

6. A speaks truth in 60% cases and B in 90% what is the probability of (i) both agreeing in narrating a case (ii) both contradicting.   [*Ans.* (i) 0.58 (ii) 0.42]

7. The probability of A passing an examination is 0.6 and B passing it is 0.7, what is the chance that (i) at least one of them pass (ii) only one of them pass.

[*Ans.* (i) 0.88 (ii) 0.46]

8. Two persons throw a die alternately till one who gets 6 would win. Find their respective chance, if A gets the first chance.
$$\left[Ans. \; (i) \frac{6}{11} \; (ii) \frac{5}{11} \right]$$

9. In a hand at whist (game of playing cards in which 4 persons play with 13 cards distributed to each player), what is probability that king and queen of trump suit is held by a single player.
$$\left[Ans. \; \frac{1}{17} \right]$$

10. The odds against a certain event A are as 5 to 2, and the odds in favor of another event B which is independent of A are as 6 to 5. Find the chance of at least one of the them to happen.

$$\left[Ans. \frac{64}{77} \right]$$

11. If on an average 1 vessel in every 10 is wrecked, find the chance that out of 5 vessels expected to arrive at least 4 will come safely.

$$[Ans.\ p^5 + 5p^4 q = (q)^4\ [0.9 + 5(0.1)] = (0.9)^4 \times 14]$$

12. In a single throw with two dies, A would win when his sum of points on two dice is 6 and B would win when his sum is 7. If A is given the start, show that the chance of A and B are as 30 : 31.

13. A manufacturer supplies quarter horse-power motors in lots of 25. A buyer, before taking a lot, tests at random a sample of 5 motors and accepts the lot if they are all good, otherwise he rejects the lot. Find the probability that

  i. he will accept a lot containing 5 defective motors.

  ii. he will reject a lot containing only one defective motor.

$$[Ans.\ (i)\ 0.292\ (ii)\ 0.2]$$

14. Each coefficient in the equation $ax^2 + bx + c = 0$ is determined by throwing an ordinary die. Find the probability that the given equation will have real roots.

$$\left[Ans. \frac{43}{216} \right]$$

15. If four squares are chosen at random on a chess-board, find the chance that they should be in a diagonal line.

$$\left[Ans. \frac{364}{^{64}C_4} = \frac{91}{158844} \right]$$

16. A bag contains 10 white, 9 black, 8 red and 3 blue balls. Balls are drawn only by one at random from the bag until 2 blue balls are obtained at the 11th draw. Find the probability of drawing 2 blue balls upto 11th draw.

$$\left[Ans. \frac{19}{406} \right]$$

17. A bag has 10 white and 15 black balls. Two balls are drawn in succession. Find the probability that the first be white and second black.

$$\left[Ans. \frac{1}{4} \right]$$

18. If $P(A) = \frac{1}{2}$, $P(B) = \frac{1}{3}$ and $P(A + B) = \frac{2}{3}$, show that the events are independent.

19. A has purchased 5 tickets of a lottery of 100 tickets, while his brother B has purchased 7 tickets, if they purchase 2 tickets more sharing equally, find the chance that (i) the prize comes in their house (ii) if it is known than B has won the prize, what is the chance that it will be shared with A.

$$\left[Ans.\ (i)\ \frac{7}{50},\ (ii)\ \frac{2}{9} \right]$$

20. In a single throw with 2 dice, what is the chance of getting a number greater than 7.

$$\left[Ans. \frac{5}{12} \right]$$

21. A bag contains 5 white, 7 red and 8 black balls. If four balls are drawn in secession one by one without replacement, what is the chance that all are white.

$$\left[Ans.\frac{10}{969} \right]$$

22. There are three boxes containing 1 white, 2 red and 3 black balls; 2 white, 3 red and 1 black ball; 3 white, 1 red and 2 black balls respectively. A box is chosen at random and from it two balls are drawn at random. The two balls are one red and one white. Find the probability that these come from: (i) the first box (ii) the second box (iii) the third box.

$$\left[Ans.\ (i)\ \frac{2}{11}\ \ (ii)\ \frac{6}{11}\ \ (iii)\ \frac{3}{11} \right]$$

23. A die is rolled twice and the sum of the numbers appearing on them in found to be 7. What is the conditional probability that 2 has appeared at least once.

$$\left[Ans.\ \frac{1}{3} \right]$$

24. Three groups of children contain 3 girls and 1 boy; 2 girls and 2 boys; and 1 girl and 3 boys respectively. One child is selected at random from each group. Show that the chance that the three selected consists of 1 girl and 2 boys is $\frac{13}{32}$.

25. One hundred indentical coins each with probability of head = p, $0 < p < 1$ are tossed; if probability of 50 heads be equal to probability of 51 heads, show that $p = \frac{51}{100}$.

26. If $P(A) = \frac{1}{3}$, $P(B) = \frac{3}{4}$ and A and B be independent. Find:

   (i) $P(A \cup B)$

   (ii) $P\left(\frac{A}{A \cup B} \right)$

   (iii) $P\left(\frac{B}{A \cup B} \right)$

$$\left[Ans.\ (i)\ \frac{5}{6}\ \ (ii)\ \frac{2}{5}\ \ (iii)\ \frac{9}{10} \right]$$

27. Three urns A, B, C contains 6 red and 4 black; 2 red and 6 black: 1 red and 8 black balls respectively. A ball drawn from one of the urns is found to be red, find the probability that it was from urn A.

$$\left[Ans.\frac{108}{173} \right]$$

28. An insurance company insured 100 scooter drivers, 200 cars drivers and 300 truck drivers. The probabilities of accident by them are 0.01, 0.3 and 0.15 respectively. If an accident occurred, find the probability that it was done by a scooter driver.

$$\left[Ans.\frac{1}{52} \right]$$

29. There are 3 true coins and 1 false coin with 'head' on both sides. A coin is chosen at random and tossed 4 times. If 'head' occurs all the four times, what is the probability that the false coin has been chosen and used?

$$\left[Ans.\frac{16}{19} \right]$$

30. For a certain binary communication channel, the probability that a transmitted '0' is received as a '0' is 0.95 and the probability that a transmitted '1' is received as '1' is 0.90. If the probability that a '0' is transmitted is 0.4, find the probability that

    i. a '1' is received
    ii. a '1' was transmitted given that a '1' was received

$$\left[\textbf{Ans. } (i)\ 0.56\ (ii)\ \frac{27}{28} \right]$$

31. 10% of the bolts produced by a machine are defective. Find the chance that out of 10 bolts chosen at random (i) none (ii) 2 or (iii) at most 2 are defective.

$$\left[\textbf{Ans.} (i)\ (0.9)^{10}\ (ii)\ 45(0.9)^8(0.1)^2\ (iii)\ (0.9)^{10}\left(1+\frac{10}{9}+\frac{5}{9}\right) \right]$$

32. A cubical die is thrown in sets of 8. The occurrence of 5 or 6 is called a success. In what proportion of sets do you expect three successes.

$$\left[\textbf{Ans.} \frac{56 \times 32}{3^8} \times 100 = 27.3\% \right]$$

33. If the probability of a defective bulb be 0.1, find the mean and standard deviation of defectives in a lot of 500 bulbs. $\left[\textbf{Ans. } \bar{x} = 50,\ \sigma = 6.7 \right]$

34. If the mean of a binomial distribution be 3 and variance $\sigma^2 = \dfrac{3}{2}$. Find the probability of at least 4 successes.

$$\left[\textbf{Ans. } \frac{15+6+1}{64} = \frac{11}{32} \right]$$

35. A binomial random variate satisfies the condition $\dfrac{P(X=5)}{P(X=2)} = \dfrac{2}{135}$ when $n = 6$.

Find p, $\bar{x}$ and σ for the distribution. $\left[\textbf{Ans. } p = 0.25,\ \bar{x} = 1.5,\ \sigma = 1.06 \right]$

36. Find the binomial distribution with mean 10 and variance 6.

$$\left[\textbf{Ans. } p = 0.4,\ n = 25 \right]$$

37. In a hurdle race, a player has to cross 5 hurdles. The probability of clearing each hurdle is $\dfrac{5}{6}$. What is the probability that he will knock down fewer than 2 hurdles.

$$\left[\textbf{Ans. } 2\left(\frac{5}{6}\right)^2 \right]$$

38. What is the least number of times, a pair of dice must be thrown so that the probability of getting the sum = 12 at least once be greater than 0.5.

$$\left[\textbf{Ans. } n > \frac{\log 2}{\log 36 > \log 35} = \frac{0.3010}{0.122} = 24.67 \right]$$

39. The items produced by a company contains 5% defective items. Show that the probability of getting 2 defectives in a sample of 10 is $\dfrac{45 \times 10^8}{20^{10}}$.

40. The probability that a bomb droped from a plane strike the target is 1/5. If six such bombs are droped find the probability that at least two will strike the target.
           [*Ans.* 0.345, 0.246]

41. The incidence of occupational disease in an industry is such that the workers have 20% chance of suffering from it. Find the chance that out of 6 workers, 4 or more will catch the disease.   [*Ans.* 0.01696]

42. If m things are distributed among 'a' men and 'b' women, show that the probability that the number of things received by men is odd.

$$\left[Ans. \ \frac{1}{2}\left(\frac{(b+a)^m - (b-a)^n}{(b+a)^m} \right) \right]$$

43. If the chance that one of the 10 telephone lines is busy at any instant of time is 0.2, what is the chance that
   i. 5 lines are busy at a time
   ii. most probable number of busy lines and their probability
   iii. all the lines are busy at a time   [*Ans.* (i) 0.0258 (ii) 2, 0.0457 (iii) 1.024×10^{-7}]

44. If 2% bolts are defective, find the probability of 5 defectives or less in a lot of 200 bolts ($e^{-4} = 0.0183$).

$$\left[Ans. \ e^{-4}\left(1 + 4 + \frac{4^2}{2!} + \frac{4^3}{3!} + \frac{4^4}{4!} + \frac{4^5}{5!} \right) = 0.785 \right]$$

45. If 3% electric bulbs manufactured by a company are defective, find the probability of 5 defectives in a lot of 100 bulbs ($e^{-3} = 0.0498$).

$$\left[Ans. \ e^{-3}\frac{3^5}{5!} = 0.1008 \right]$$

46. If the probability of an individual out of 2000 individuals suffering from a bad reaction from a drug is 0.001. Find the probability that (i) exactly 3 (ii) more than 2 individuals will suffer by bad reaction (use $e^{-2} = 0.1353$).
         [*Ans.* (i) 0.1804, (ii) 0.3235]

47. A random variate X follows the Poisson's distribution and satisfies
$3P(X = 2) = P(X = 4)$; find (i) $P(X = 0)$ (ii) $P(X = 5)$ [*Ans.* (i) 0.0025 (ii) 1.162]

48. If X be a Poisson variate such that $3P(X = 3) = 4P(X = 4)$, find $P(X = 7)$.

$$\left[Ans. \ \frac{3^7 e^{-3}}{7!} \right]$$

49. Fit a Poisson's distribution to the set of observations.

x	0	1	2	3	4
$F(x) = frequency$	122	60	15	2	1

[*Ans.* mean = 1]

50. Razor blades supplied by a manufacturing company in packets of 10. There is a probability of 1 in 100 blades to be defective. Using Poisson's distribution, calculate the number of packets containing (i) one defective blade (ii) no defective blades and (iii) all defective blades in consignment of 10000 packets.

[*Ans.* (i) 905 (ii) 9048 (iii) 0]

51. Let m and μ^r denote the mean and central rth moment of a Poisson distribution. Prove

$$\mu_{r+1} = r \cdot m \mu_{r-1} + m \frac{d\mu_r}{dm}$$

where

$$\mu_r = \sum_{k=0}^{\infty} (k-m)^r e^{-m} \cdot \frac{m^k}{k!}$$

52. Find the probability that the standard normal variate z to lie between
   i. $0 < z < 1.4$ ii. $-1.3 < z < 0$
   iii. $0.75 < z < 2.34$ iv. $-2 < z < 1.6$

53. In a normal distribution with mean $\mu = 50$ and standard deviation $\sigma = 6$, find $P(38 < x < 59)$. [*Ans.* 0.9104]

54. In a normal distribution with mean $\mu = 50$ and $\sigma = 10$, find x_1 such that $P(-\infty < x < x_1) = 0.16$ and x_2 such that $P(x_2 < x < \infty) = 0.13$ [*Ans.* $x_1 = 40, x_2 = 61.3$]

55. In a large group of men, 5% are under 160 cm height and 40% are between 160 to 170 cm. Assuming the height to be distributed normally, find the mean and standard deviation of the heights. [*Ans.* $\mu = 170$, $\sigma = 6.58$]

56. Fit a normal curve to the following frequency distribution.

x	4	6	8	10	12	14	16	18	20	22	24
$f(x)$	1	7	15	22	35	43	38	20	13	5	1

$$\left[\begin{array}{l} \textbf{\textit{Ans.}}\ \mu = 13.85,\ S.D = 3.85 \\[2mm] f(x) = \dfrac{1}{3.85\sqrt{2\pi}} e^{-\frac{(x-13.85)^2}{29.33}} \end{array} \right]$$

57. 1000 bulbs with mean life 120 days are installed in a factory. If their life follow the normal distribution with $\sigma = 20$ days. Find (i) how many bulbs will fuse before 100 days (ii) how many days after 10% will fuse.

[*Ans.* (i) 159 (ii) 94 days]

58. One thousand candidates in an examination were grouped into 3 classes I, II, and III in descending order of merit. The numbers in the first two classes were 50 and 350 respectively. The highest and lowest marks in class II were 60 and 50 respectively. Assume the distribution to be normal, prove that the average marks is 48.2 and standard deviation is 7.1.

59. The spot speeds at particular location are normally distributed with a mean of 51.7 k/h and a standard deviation of 8.3 k/h. What is the probability that
   i. speed exceeds 65 k/h?
   ii. speed lie between 40 k/h and 70 k/h [*Ans.* (i) 5.48% (ii) 90.7%]

60. The skulls are classified as A, B, C according as length-breadth index under 75, between 75 and 80, find approximately the mean and standard deviation of a series in which A are 58%, B are 38% and C are 4% being given that if

$$\phi(t) = \frac{1}{\sqrt{2\pi}} \int_0^t e^{-z^2/z} dz$$

then $\phi(0.20) = 0.08$ and $\phi(1.75) = 0.46$ [*Ans.* $\mu = 74.4$; $\sigma = 3.2$]

61. The distribution of weekly wages for 500 workers in a factory is approximately normal with the mean and standard deviation of ₹ 75 and ₹ 15. Find the number of workers who receive weekly wages.

(i) less than ₹ 45 (ii) more than ₹ 90.

Given that if $\phi(t) = \frac{1}{\sqrt{2\pi}} \int_0^t e^{-\frac{1}{2}z^2} dz$, then

$$\phi(1) = 0.3413, \quad \phi(2) = 0.4772$$ [*Ans.* (i) 11 (ii) 79]

62. A sample of 100 dry battery cells tested to find the length of life, it was found that mean length $\mu = 12$ hours and standard deviation $\sigma = 3$ hours.

Assuming the data to be normally distributed, what percentage of battery cells are expected to have life

i. more than 15 hours

ii. less than 6 hours

Given that if $\phi(t) = \frac{1}{\sqrt{2\pi}} \int_0^t e^{-\frac{1}{2}z^2} dz$,

$$\phi(1) = 0.3413, \quad \phi(2) = 0.4772$$ [*Ans.* (a) 15.87% (b) 2.28%]

MULTIPLE CHOICE QUESTIONS

1. The standard deviation of spot speed of vehicles in a highway is 8.8 km/h and the mean speed of the vehicles is 33 km/h the coefficient of variation in speed is
   (a) 0.1517   (b) 0.1867
   (c) 0.2666   (c) 0.3646   **[GATE 2007]**

2. There are 25 calculators in a box. Two of them are defective. Suppose 5 calculators are randomly picked for inspections (i.e. each has the same chance of being selected), what is the probability that only one of the defective calculators will be included in the inspection?

   (a) $\dfrac{1}{2}$ (b) $\dfrac{1}{3}$

   (c) $\dfrac{1}{4}$ (d) $\dfrac{1}{5}$ **[GATE 2006]**

3. A hydraulic structure has four gates which operate independently. The probability of failure of each gate is 0.2. Given that gate 1 has failed, the probability that both gates 2 and 3 will also fail is
   (a) 0.240   (b) 0.200
   (c) 0.040   (d) 0.008   **[GATE 2004]**

4. X is a uniformly distributed random variable that takes value between 0 and 1. The value of $E[X^3]$ will be

(a) 0

(b) $\dfrac{1}{8}$

(c) $\dfrac{1}{4}$

(d) $\dfrac{1}{2}$ **[GATE 2008]**

5. If P and Q are two random events, then which of the following is true?

(a) Independence of P and Q implies that probability $(P \cap Q) = 0$

(b) Probability $(P \cup Q) \geq$ probability (P) + probability (Q)

(c) If P and Q are mutually exclusive, then they must be independent

(d) Probability $(P \cap Q) \leq$ probability (P) **[GATE 2005]**

6. A box contains 4 white balls and 3 red balls. In succession, two balls are randomly selected and removed from the box. Given that the first removed ball is white, the probability that the second removed ball is red is

(a) $\dfrac{1}{3}$

(b) $\dfrac{3}{7}$

(c) $\dfrac{1}{2}$

(d) $\dfrac{4}{7}$ **[GATE 2010]**

7. There are two containers, with one containing 4 red and 3 green balls and other containing 3 blue and 4 green balls. One ball is drawn at random from each container. The probability that one of the balls is red and the other is blue will be

(a) $\dfrac{17}{49}$

(b) $\dfrac{9}{49}$

(c) $\dfrac{12}{49}$

(d) $\dfrac{3}{7}$ **[GATE 2011]**

8. n couples are invited to a party with the condition that every husband should be accompanied by his wife. However, a wife need not be accompanied by her husband. The number of different gathering possible at the party is

(a) 2^n

(b) 3^n

(c) $\dfrac{(2n)!}{2^n}$

(d) $\dbinom{2n}{n}$ **[GATE 2003]**

9. A deck of 5 cards (each carrying a distinct number from 1 to 5) is shuffled thoroughly. Two cards are then removed one at a time from the desk. What is the probability that the two cards are selected with the number of the first card being one higher than the number on the second card?

(a) $\dfrac{1}{5}$

(b) $\dfrac{4}{25}$

(b) $\dfrac{1}{4}$

(d) $\dfrac{2}{5}$ **[GATE 2011]**

10. Consider a company that assembles computers. The probability of a faulty assembly of any computer is P. The company therefore, subjects each computer to a testing process. This testing process gives the correct result for any computer with a probability of q. What is the probability of a computer being declared faulty?

(a) $pq + (1 - p)(1 - q)$

(b) $(1 - q)p$

(c) $(1 - p)q$

(d) pq [GATE 2010]

11. An unbalanced dice with 6 faces, number from 1 to 6 is thrown. The probability that the face value is odd is 90% of the probability that the face value is even. The probability of getting any even numbered face is the same. If the probability that the face is even, given that it is greater than 3 is 0.75, which one of the following options is closest to the probability that the face value exceeds 3?

(a) 0.4533

(b) 0.468

(c) 0.485

(d) 0.492 [GATE 2009]

12. Three values of x and y are to be fitted in a straight line in the form $y = a + bx$ by the method of least squares. Given $\Sigma x = 6$, $\Sigma y = 21$, $\Sigma x^2 = 14$ and $\Sigma xy = 46$, the value of a and b are respectively

(a) 2 and 3

(b) 1 and 2

(c) 2 and 1

(d) 3 and 2 [GATE 2008]

13. If probability density function of a random variable X is $f(x) = x^2$ for $-1 \leq x \leq 1$ and $f(x) = 0$ for any other values of x, then the percentage probability $P\left(-\dfrac{1}{3} \leq x \leq \dfrac{1}{3}\right)$ is

(a) 0.247

(b) 2.47

(c) 24.7

(d) 247 [GATE 2008]

14. Aishwarya studies either computer science or mathematics every day. If she studies computer science on a day, then the probability that she studies mathematics the next day is 0.6. If she studies mathematics on a day, then the probability that she studies computer science the next day is 0.4. Given that Aishwarya studies computer science on Monday, what is the probability that she studies computer science on Wednesday?

(a) 0.24

(b) 0.36

(c) 0.4

(d) 0.6 [GATE 2008]

15. For each element in a set of size $2n$, an unbiased coin is tossed. The $2n$ coins are tossed independently. An element is chosen, if the corresponding coin toss were head. The probability that exactly n elements are chosen is

(a) $\dfrac{\dbinom{2n}{n}}{4^n}$

(b) $\dfrac{\dbinom{2n}{n}}{2^n}$

(c) $\dfrac{1}{\dbinom{2n}{n}}$

(d) $\dfrac{1}{2}$ [GATE 2006]

16. Two n bit binary strings S_1 and S_2 are chosen randomly with uniform probability. The probability that the humming distance between these strings (the number of bit positions where two strings differ) is equal to

(a) $\dfrac{{}^nC_d}{2^n}$

(b) $\dfrac{{}^nC_d}{2^d}$

(c) $\dfrac{d}{2^n}$

(d) $\dfrac{1}{2^d}$ **[GATE 2004]**

17. A program consists of two modulus executed sequentially. Let $f_1(t)$ and $f_2(t)$ denote the probability density functions of time taken to execute the two modulus respectively. The probability density function of the overall line taken to execute the program is given by

(a) $f_1(t) + f_2(t)$

(b) $\displaystyle\int_0^t f_1(x) \cdot f_2(x)\,dx$

(c) $\displaystyle\int_0^t f_1(x) f_2(t-x)\,dx$

(d) $\max\{f_1(t), f_2(t)\}$ **[GATE 2003]**

18. A fair coin is tossed 10 times. What is the probability that only the first two tosses will yield heads?

(a) $\left(\dfrac{1}{2}\right)^2$

(b) ${}^{10}C_2\left(\dfrac{1}{2}\right)^2$

(c) $\left(\dfrac{1}{2}\right)^{10}$

(d) ${}^{10}C_2\left(\dfrac{1}{2}\right)^{10}$ **[GATE 2009]**

19. A probability density function is of the form $P(x) = k\,e^{-\alpha|x|}$, $x \in (-\infty, \infty)$. The value of k is

(a) 0.5

(b) 1

(c) 0.5α

(d) α **[GATE 2006]**

20. A fair dice is rolled twice. The probability that an odd number will follow an even number is

(a) $\dfrac{1}{2}$

(b) $\dfrac{1}{6}$

(c) $\dfrac{1}{3}$

(d) $\dfrac{1}{4}$ **[GATE 2005]**

21. A fair coin is tossed independently four times. The probability of the event 'the number of time heads shown up is more than the number of time tails shown up' is

(a) $\dfrac{1}{16}$

(b) $\dfrac{1}{8}$

(c) $\dfrac{1}{4}$

(d) $\dfrac{5}{16}$ **[GATE 2010]**

22. A fair dice is tossed two times. The probability that the second toss results in a value that is higher than the first toss is

   (a) $\dfrac{2}{36}$

   (b) $\dfrac{2}{6}$

   (c) $\dfrac{5}{12}$

   (d) $\dfrac{1}{2}$ **[GATE 2011]**

23. A examination consists of two papers, paper 1, paper 2. The probability of failing in paper 1 is 0.3 and that in paper 2 is 0.2. Given that a student has failed in paper 2, the probability of failing in Paper 1 is 0.6. The probability of a student failing on both the paper is

   (a) 0.5

   (b) 0.18

   (c) 0.12

   (d) 0.06   **[GATE 2007]**

24. The mean of a set of number is $\bar{x}$. If each number is increased by 1 then variance of the new set is

   (a) $\bar{x}$

   (b) $\bar{x} + \lambda$

   (c) $\lambda \bar{x}$

   (d) none of these   **[IES]**

25. If mode of a data is 18 and mean is 24, then median is

   (a) 18

   (b) 24

   (c) 22

   (d) 21   **[IES]**

26. If μ is mean of distribution, then $\Sigma f_i(y_i - \mu)$ is equal to

   (a) MD

   (b) standard deviation

   (c) 0

   (d) none of these   **[IES]**

27. $b_{xy} \times b_{yx}$ is equal to

   (a) $\rho(X, Y)$

   (b) Cov (x, y)

   (c) $\{\rho(x, y)\}^2$

   (d) none of these   **[IES]**

28. If two lines of regression are at right angles, then $\rho(X, Y)$ is equal to

   (a) 1

   (b) –1

   (c) 1 or –1

   (d) 0   **[IES]**

29. Two coins are simultaneously tossed. The probability of two heads appearing simultaneously is

   (a) $\dfrac{1}{8}$

   (b) $\dfrac{1}{6}$

   (c) $\dfrac{1}{4}$

   (d) $\dfrac{1}{2}$ **[GATE 2009]**

30. Which one of the following statements is not true?

   (a) The measure of skewness is dependent upon the amount of dispersion

   (b) In a symmetric distribution, the value of mean, mode and median are the same

   (c) In a positive skewed distribution, mean > median < mode

   (d) In a negative skewed distribution, mode > mean > median   **[GATE 2005]**

31. A box contains 10 screws, 3 of which are defective. Two screws are drawn at random with replacement. The probability that more of the two screws will be defective is
    (a) 100%
    (b) 50%
    (c) 49%
    (d) none of these   **[GATE 2003]**

32. Let $f(x)$ be the continuous probability density function of a random variable X. The probability that $a < X \le b$ is
    (a) $f(b - a)$
    (b) $f(b) - f(a)$
    (c) $\int_a^b f(x)dx$
    (d) $\int_a^b xf(x)dx$ **[IES]**

33. If a fair coin is tossed four times. What is the probability that two heads and two tails will results?
    (a) $\dfrac{3}{8}$
    (b) $\dfrac{1}{2}$
    (c) $\dfrac{5}{8}$
    (d) $\dfrac{3}{4}$ **[GATE 2004]**

34. Assume for simplicity that N people, all born in April (a month of 30 days) are collected in a room. Consider, the event of atleast two people in the room being born on the same date of the month, even in different years, e.g. 1980 and 1985. What is the smallest N. So that probability of this event exceeds 0.5?
    (a) 20
    (b) 7
    (c) 15
    (d) 16   **[GATE 2009]**

35. A loaded dice has following probability distribution of occurrences

Dice value	1	2	3	4	5	6
Probability	$\dfrac{1}{4}$	$\dfrac{1}{8}$	$\dfrac{1}{8}$	$\dfrac{1}{8}$	$\dfrac{1}{8}$	$\dfrac{1}{4}$

If three identical dice as the above are thrown, the probability of occurrence of values 1, 5, and 6 on the three dice is
    (a) same as that of occurrence 3, 4, 5
    (b) same as that of occurrence 1, 2, 5
    (c) $\dfrac{1}{128}$
    (d) $\dfrac{5}{8}$ **[GATE 2007]**

36. A fair coin is tossed three times in succession If the first toss produces a head, then the probability of getting exactly two heads in three tosses is
    (a) $\dfrac{1}{8}$
    (b) $\dfrac{1}{2}$
    (c) $\dfrac{3}{8}$
    (d) $\dfrac{3}{4}$ **[GATE 2005]**

37. If the difference between expectation of the square of random variable and the square of the expectation of the random variable is denoted by R, then
    (a) $R = 0$
    (b) $R < 0$
    (c) $R \ge 0$
    (d) $R > 0$ **[GATE 2011]**

Table 13.1: Area under the standard normal curve $\phi(t) = \dfrac{1}{\sqrt{2\pi}} \displaystyle\int_0^t e^{-\frac{1}{2}z^2}\, dz \ (t > 0)$

$\downarrow t \rightarrow$	00	.01	.02	.03	.04	.05	.06	.07	.08	.09
0	.0000	.0040	.0080	.0120	.0160	.0199	.0239	.0279	.0319	.0359
.1	.0398	.0438	.0478	.0517	.0557	.0596	.0636	.0675	.0714	.0753
.2	.0793	.0832	.0871	.0910	.0948	.0987	.1026	.1064	.1103	.1141
.3	.1179	.1217	.1255	.1293	.1331	.1368	.1406	.1443	.1480	.1517
.4	.1554	.1591	.1628	.1664	.1700	.1736	.1772	.1808	.1844	.1879
.5	.1915	.1950	.1985	.2019	.2054	.2088	.2123	.2157	.2190	.2224
.6	.2257	.2291	.2324	.2357	.2399	.2422	.2454	.2486	.2517	.2549
.7	.2580	.2611	.2642	.2673	.2703	.2734	.2764	.2794	.2823	.2852
.8	.2881	.2910	.2939	.2967	.2995	.3023	.3051	.3078	.3106	.3133
.9	.3159	.3186	.3212	.3238	.3264	.3289	.3315	.3340	.3365	.3389
1.0	.3413	.3438	.3461	.3485	.3508	.3531	.3554	.3577	.3599	.3621
1.1	.3643	.3665	.3686	.3708	.3729	.3749	.3770	.3790	.3810	.3830
1.2	.3849	.3869	.3888	.3907	.3925	.3944	.3962	.3980	.3997	.4015
1.3	.4032	.4049	.4066	.4082	.4099	.4115	.4131	.4147	.4162	.4177
1.4	.4192	.4207	.4222	.4236	.4251	.4265	.4279	.4292	.4306	.4319
1.5	.4332	.4345	.4357	.4370	.4382	.4394	.4406	.4418	.4429	.4441
1.6	.4452	.4463	.4474	.4484	.4495	.4505	.4515	.4525	.4535	.4545
1.7	.4554	.4564	.4573	.4582	.4591	.4599	.4608	.4616	.4625	.4633
1.8	.4641	.4649	.4656	.4664	.4671	.4678	.4686	.4693	.4699	.4706
1.9	.4713	.4719	.4726	.4732	.4738	.4744	.4750	.4756	.4761	.4767
2.0	.4772	.4778	.4783	.4788	.4793	.4798	.4803	.4808	.4812	.4817
2.1	.4821	.4826	.4830	.4834	.4838	.4842	.4846	.4850	.4854	.4857
2.2	.4861	.4864	.4868	.4871	.4875	.4878	.4881	.4884	.4887	.4890
2.3	.4893	.4896	.4898	.4901	.4904	.4906	.4909	.4911	.4913	.4916
2.4	.4918	.4920	.4922	.4925	.4927	.4929	.4931	.4932	.4934	.4936
2.5	.4938	.4940	.4941	.4943	.4945	.4946	.4948	.4949	.4951	.4952
2.6	.4953	.4955	.4956	.4957	.4959	.4960	.4961	.4962	.4963	.4964
2.7	.4965	.4966	.4967	.4968	.4969	.4970	.4971	.4972	.4973	.4974
2.8	.4974	.4975	.4976	.4977	.4977	.4978	.4979	.4979	.4980	.4981
2.9	.4981	.4982	.4982	.4983	.4984	.4984	.4985	.4985	.4986	.4986
3.0	.4987	.4987	.4987	.4988	.4988	.4989	.4989	.4989	.4990	.4990
3.1	.4990	.4991	.4991	.4991	.4992	.4992	.4992	.4992	.4993	.4993
3.2	.4993	.4993	.4994	.4994	.4994	.4994	.4994	.4905	.4995	.4995
3.3	.4995	.4995	.4995	.4996	.4996	.4996	.4996	.4996	.4996	.4997
3.4	.4997	.4997	.4997	.4997	.4997	.4997	.4997	.4997	.4997	.4998
3.5	.4998	.4998	.4998	.4998	.4998	.4998	.4998	.4998	.4998	.4998
3.6	.4998	.4998	.4999	.4999	.4999	.4999	.4999	.4999	.4999	.4999
3.7	.4999	.4999	.4999	.4999	.4999	.4999	.4999	.4999	.4999	.4999
3.8	.4999	.4999	.4999	.4999	.4999	.4999	.4999	.4999	.4999	.4999
3.9	.5000	.5000	.5000	.5000	.5000	.5000	.5000	.5000	.5000	.5000

ANSWERS

1. (c)	2. (b)	3. (c)	4. (c)	5. (d)	6. (c)	7. (c)	8. (b)
9. (d)	10. (a)	11. (b)	12. (d)	13. (b)	14. (c)	15. (a)	16. (a)
17. (c)	18. (c)	19. (c)	20. (d)	21. (d)	22. (c)	23. (c)	24. (b)
25. (c)	26. (c)	27. (c)	28. (d)	29. (c)	30. (d)	31. (d)	32. (c)
33. (a)	34. (b)	35. (c)	36. (b)	37. (c)			

Appendix

GATE 2011

1. Two matrices A and B are said to be similar if $B = P^{-1}AP$ for some invertible matrix P. Which of the following statements is not true?
   (a) Det A = Det B
   (b) Trace of A = Trace of B
   (c) A and B have the same eigen vectors
   (d) A and B have the same eigen values

2. If a force F is derivable from a potential function $v(r)$, where r is the distance from the origin of the coordinate system it follows that

   (a) $\vec{\nabla} \times F = 0$
   (b) $\vec{\nabla} \cdot r = 0$
   (c) $\vec{\nabla} v = 0$
   (d) $\nabla^2 v = 0$

3. A (3×3) matrix has elements such that its trace is 13 and determinant is 36. The eigen values of the matrix are all known to be positive integers. The largest eigen value of the matrix is
   (a) 18
   (b) 12
   (c) 9
   (d) 6

4. The unit vector normal to the surface $x^2 + y^2 - z - 1 = 0$ at the point $P(1, 1, 1)$ is

   (a) $\dfrac{1}{\sqrt{3}}(i + j - k)$
   (b) $\dfrac{1}{\sqrt{6}}(2i + j - k)$

   (c) $\dfrac{1}{\sqrt{6}}(i + 2j - k)$
   (d) $\dfrac{1}{3}(2i + 2j - k)$

5. Consider a cylinder of height h and radius a closed at both ends centered at the origin. Let $ix + jy + kz$ be the position vector and n a unit normal vector to the surface. The surface integral $\int_S r \cdot n \, ds$ over the closed surface of cylinder is
   (a) $2\pi a^2 (a + h)$
   (b) $3\pi a^2 h$
   (c) $2\pi a^2 h$
   (d) zero

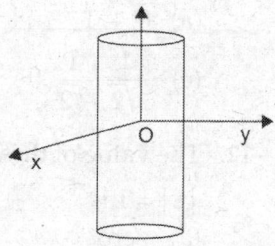

Fig. A.1

6. Solutions to the differential equation $\dfrac{dy}{dx} = -\dfrac{x}{y+1}$ are a family of

   (a) circles with differential radii
   (b) circles with different centres
   (c) straight lines with different slopes
   (d) straight lines with different intercepts on the y-axis

   Common data for Question numbers 7 and 8. Consider a function $f(z) = \dfrac{z\sin z}{(z-\pi)^2}$ of a complex variable z.

7. Which of the following statement is true for the function f(z)?
   (a) f(z) is analytic everywhere in the complex plane
   (b) f(z) has a zero at $z = \pi$
   (c) f(z) has a pole of order 2 at $z = \pi$
   (d) f(z) has a simple pole at $z = \pi$

8. Consider a counterclockwise circular contour $|z| = 1$ about the origin. The integral $\int f(z)dz$ over the contour is
   (a) $-i\pi$ (b) zero
   (c) $i\pi$ (d) $2i\pi$

GATE 2012

9. Identify the correct statement for the vectors $a = 3i + 2j$ and $b = i + 2j$
   (a) the vectors a and b are linearly independent
   (b) the vectors a and b are linearly dependent
   (c) the vectors a and b are orthogonal
   (d) the vectors a and b are normalized

10. The number of independent components of a symmetric tensor A_{ij} with indices $i, j = 1, 2, 3$ is
    (a) 1   (b) 3
    (c) 6   (d) 9

11. The eigen values of the matrix $\begin{pmatrix} 0 & 1 & 0 \\ 1 & 0 & 1 \\ 0 & 1 & 0 \end{pmatrix}$ are

    (a) $0, 1, 1$ (b) $0, -\sqrt{2}, \sqrt{2}$
    (c) $\dfrac{1}{\sqrt{2}}, \dfrac{1}{\sqrt{2}}, 0$ (d) $\sqrt{2}, \sqrt{2}, 0$

12. The value of the integral $\oint_C e^{1/z}dz$ using the contour C of circle with unit radius $|z| = 1$ is
    (a) zero   (b) $1 - 2\pi i$
    (c) $1 + 2\pi i$ (d) $2\pi i$

GATE 2013

13. $f(x)$ is a symmetric periodic function of x, i.e. $f(x) = f(-x)$. Then in general, the Fourier series of the function $f(x)$ will be of the form

(a) $f(x) = \sum_{-\infty}^{+\infty} \pm \left[a_n \cos(nkx) + b_n \sin(nkx) \right]$

(b) $f(x) = a_0 + a_n \cos(nkx)$

(c) $f(x) = \sum_{-\infty}^{+\infty} \pm \left[b_n \sin(nkx) \right]$

(d) $f(x) = a_0 + \sum_{-\infty}^{+\infty} \pm \left[b_n \sin(nkx) \right]$

14. In the most general case which one of the following quantities is not a second order tensor?

(a) stress   (b) strain

(c) moment of inertia   (d) pressure

15. An electron is moving with a velocity of $0.8c$ in the same direction as that of a moving photon. The relative velocity of the electron with respect to photon is

(a) c (b) $-c$

(c) $0.15c$ (d) $-0.15c$

16. Which of the following pairs of the given functions $F(t)$ and its Laplace transform $f(s)$ are not correct?

(a) $F(t) = \delta(t)$, $f(s) = 1$ (singularity at $+ 0$)

(b) $F(t) = 1$, $f(s) = \dfrac{1}{s}$, $(s > 0)$

(c) $F(t) = \sin kt$, $f(s) = \dfrac{s}{s^2 + k^2}$ $(s > 0)$

(d) $F(t) = te^{kt}$, $f(s) = \dfrac{1}{(s-k)^2}$, $(s > k, \ s > 0)$

17. If A and B are constant vectors, then $\nabla(A \cdot B \times r)$ is

(a) AB (b) $A \times B$

(c) r (d) zero

18. $\overline{\left| \left(n + \dfrac{1}{2} \right) \right.}$ is equal to (given $\overline{|(n+1)} = n\overline{|n}$ and $\overline{\left| \left(\dfrac{1}{2} \right) \right.} = \sqrt{\pi}$)

(a) $\dfrac{n!}{2^n} \sqrt{n}$ (b) $\dfrac{2n!}{n!2^n} \sqrt{\pi}$

(c) $\dfrac{2n!}{n!2^{2n}} \sqrt{\pi}$ (d) $\dfrac{n!}{2^{2n}} \sqrt{\pi}$

19. For function $f(z) = \dfrac{16z}{(z+3)(z-1)^2}$, the residue at the pole $z = 1$ is (your answer should be an integer).

20. The degenerate eigen value of the matrix $\begin{pmatrix} 4 & -1 & -1 \\ -1 & 4 & -1 \\ -1 & -1 & 4 \end{pmatrix}$

is_____ (your answer should be an integer).

GATE 2014

21. Unit vector perpendicular to the surface $x^2 + y^2 + z^2 = 3$ at point $(1, 1, 1)$ is

(a) $\frac{1}{\sqrt{3}}(x+y-z)$

(b) $\frac{1}{\sqrt{3}}(x-y-z)$

(c) $\frac{1}{\sqrt{3}}(x-y-z)$

(d) $\frac{1}{\sqrt{3}}(x+y+z)$

22. Which one of the following quantities is invariant under Koventz transformation
    (a) charge density
    (b) charge
    (c) current
    (d) electric field

23. The matrix $\begin{pmatrix} 1 & 1+i \\ 1-i & -1 \end{pmatrix}$ is

(a) orthogonal
(b) symmetric
(c) antisymmetric
(d) unitary

24. The length element ds of an arc is given by $(ds)^2 = 2(dx^1)^2 + (dx^2)^2 + \sqrt{3}\,dx^3 dx^2$. The metric tensor g_{ij} is

(a) $\begin{bmatrix} 2 & \sqrt{3} \\ \sqrt{3} & 1 \end{bmatrix}$

(b) $\begin{bmatrix} 2 & \sqrt{3/2} \\ \sqrt{3/2} & 1 \end{bmatrix}$

(c) $\begin{bmatrix} 2 & 1 \\ \sqrt{3/2} & \sqrt{3/2} \end{bmatrix}$

(d) $\begin{bmatrix} 1 & \sqrt{3/2} \\ \sqrt{3/2} & 2 \end{bmatrix}$

GATE 2015

25. If $f(x) = e^{-x^2}$ and $g(x) = |x|e^{-x^2}$, then
    (a) f and g are differentiable everywhere
    (b) f is differentiable everywhere but g is not
    (c) g is differentiable everywhere but f is not
    (d) g is discontinuous at $x = 0$

26. Consider $w = f(z) = u(x, y) + iv(x, y)$ to be an analytic function in a domain D. Which one of the following options is not correct?
    (a) $u(x, y)$ satisfies Laplace equation in D
    (b) $v(x, y)$ satisfies Laplace equation in D
    (c) $\int_{z_1}^{z_2} f(z)dz$ is dependent on the choice of the contour between z_1 and z_2 in D
    (d) $f(z)$ can be Taylor expanded in D

27. The value of $\int_0^3 t^2 \delta(3t-6)dt$ is_____ (upto one decimal place).
    (a) 1.8
    (b) 2.3
    (c) 2.4
    (d) 1.3

28. Consider a complex function $f(z) = \dfrac{1}{2\left(z+\dfrac{1}{2}\right)\cos(z\pi)}$. Which one of the following statements is correct?

    (a) $f(z)$ has simple poles at $z = 0$ and $z = -\dfrac{1}{2}$

    (b) $f(z)$ has a second order pole at $z = -\dfrac{1}{2}$

    (c) $f(z)$ has infinite number of second order poles

    (d) $f(z)$ has all simple poles

29. A function $y(z)$ satisfies the ordinary differential equation $y'' + \dfrac{1}{z}y' - \dfrac{m^2}{z^2}y = 0$, where $m = 0, 1, 2, 3,....$ Consider the four statements P, Q, R, S as given below:

    P. z^m and z^{-m} are linearly independent solutions for all values of m
    Q. z^m and z^{-m} are linearly independent solutions for all values of $m > 0$
    R. $\ln z$ and 1 are linearly independent solutions for $m = 0$
    S. z^m and $\ln z$ are linearly independent solutions for all values of m

    The correct option for the combination of valid statements is
    (a) $P, R,$ and S
    (b) P and R
    (c) Q and R
    (d) R and S

30. The Heaviside function is defined as
    $H(t) = \begin{cases} +1 & \text{for } t > 0 \\ -1 & \text{for } t < 0 \end{cases}$ and its Fourier transform is given by $-2i/\omega$. The Fourier

    transform of $\dfrac{1}{2}\left[H\left(t+\dfrac{1}{2}\right) - H\left(t-\dfrac{1}{2}\right)\right]$ is

    (a) $\dfrac{\sin\left(\dfrac{\omega}{2}\right)}{\dfrac{\omega}{2}}$
    (b) $\dfrac{\cos\left(\dfrac{\omega}{2}\right)}{\sin\left(\dfrac{\omega}{2}\right)}$

    (c) $\sin\left(\dfrac{\omega}{2}\right)$
    (d) 0

GATE 2016

31. Consider the linear differential equation $\dfrac{dy}{dx} = xy$. If $y = 2$ at $x = 0$, then the value of y at $x = 2$ is given by
    (a) e^{-2}
    (b) $2e^{-2}$
    (c) e^2
    (d) $2e^2$

32. Which of the following is an analytic function of z everywhere in the complex plane?
    (a) z^2
    (b) $(z^*)^2$
    (c) $|z|^2$
    (d) $\sqrt{z}$

33. The direction of ∇f for a scalar field
    $$f(x, y, z) = \frac{1}{2}x^2 - xy + \frac{1}{2}z^2$$
    at a point $P(1, 2, 2)$ is
    (a) $\dfrac{(-j-2k)}{\sqrt{5}}$
    (b) $\dfrac{(-j+2k)}{\sqrt{5}}$
    (c) $\dfrac{(j-2k)}{\sqrt{5}}$
    (d) $\dfrac{(j+k)}{\sqrt{5}}$

34. A periodic function $f(x)$ of period 2π is defined in the interval $(-\pi < z < \pi)$ as
    $$f(x) = \begin{cases} -1 & -\pi < x < 0 \\ 1 & 0 < x < \pi \end{cases}$$

    The appropriate Fourier series expansion for $f(x)$ is
    (a) $f(x) = (4/\pi) [\sin x + (\sin 3x)/3 + (\sin 5x)/5 +\cdots]$
    (b) $f(x) = (4/\pi) [\sin x - (\sin 3x)/3 + (\sin 5x)/5 - \cdots]$
    (c) $f(x) = (4/\pi) [\cos x + (\cos 3x)/3 + (\cos 5x)/5 +\cdots]$
    (d) $f(x) = (4/\pi) [\cos x - (\cos 3x)/3 + (\cos 5x)/5-\cdots]$

GATE 2017

35. The contour integral $\int \dfrac{dz}{1+z^2}$ evaluated along a contour going from $-\infty$ to $+\infty$ along the real axis and closed in the lower half-plane by a half circle is equal to _____ (upto two decimal places).

36. The integral $\int_0^\infty x^2 e^{-x^2} dx$ is equal to _____ (up to two decimal places).

37. The imaginary part of an analytic complex function is $v(x, y) = 2xy + 3y$. The real part of the function is zero at the origin. The value of the real part of the function at $1 + i$ is _____ (upto two decimal places).

38. Let X be a column vector dimension $n > 1$ with at least one nonzero entity. The number of nonzero eigen values of the matrix $M = XX^T$ is
    (a) 0
    (b) n
    (c) 1
    (d) $n-1$

39. Consider the differential equation $\dfrac{dy}{dx} + y\tan(x) = \cos(x)$. If $y(0) = 0$, $y\left(\dfrac{\pi}{3}\right)$ is _____ (upto two decimal places).

SOLUTIONS

1.(c) If A and B are square matrices of the same type and P is invertible then matrices A and $B = P^{-1}AP$, have the same characteristic roots.

If $\qquad B = P^{-1}AP$

then $\quad B - \lambda I = P^{-1}AP - \lambda I$

$$= P^{-1}AP - P^{-1}\lambda IP = P^{-1}(A - \lambda I)P$$

where I is the identity matrix $|B - \lambda I| = 0$ is called the characteristic equation and λ, 0 are the characteristic roots or eigen values.

$$\begin{aligned} |B - \lambda I| &= |P^{-1}(A - \lambda I)P| = |P^{-1}||A - \lambda I||P| \\ &= |A - \lambda I||P^{-1}||P| = |A - \lambda I||PP^{-1}| \\ &= |A - \lambda I||I| = |A - \lambda I| \end{aligned}$$

Thus, the matrices A and $B = P^{-1}AP$ have the same characteristic equations and hence characteristic roots or eigen values since the sum of the eigen values of a matrix is equal to the trace of the matrix and product of the eigen values of matrix is equal to determinant of matrix hence option (c) is incorrect.

2.(a) Since force F is derivable from potential $v(r)$, hence

$$F = \nabla v(r) \quad \text{and} \quad \nabla \times F = -\nabla \times \nabla [v(r)] = 0$$

$\because \nabla$ and $\nabla v(r)$ are parallel to each other and cross product of two parallel vectors is zero

3.(c) We know that for any matrix

  i. The product of eigen values is equal to determinant of that matrix

  ii. The sum of eigen values is equal to trace of that matrix

If λ_1, λ_2 and λ_3 are three eigen values of (3×3) matrix then

$\lambda_1 + \lambda_2 + \lambda_3 = 13 \quad \text{and} \quad \lambda_1 \lambda_2 \lambda_3 = 36$

Hence, the largest eigen value of the matrix is 9.

4.(d) The equation of the surface is $x^2 + y^2 - z - 1 = 0$

Gradient of this function

$$= \nabla(x^2 + y^2 - z - 1)$$

$$= \left(\frac{\partial}{\partial x}i + \frac{\partial}{\partial y}j + \frac{\partial}{\partial z}k \right)(x^2 + y^2 - z - 1)$$

$$= 2xi + 2yj - k$$

$\therefore$ Value of gradient at point $(1, 1, 1)$ is

$$= \frac{2i + 2j - k}{\sqrt{(2)^2 + (2)^2 + 1}} = \frac{2i + 2j - k}{3}$$

5.(b) To calculate surface integral, we require the surface normal n on the top surface of the cylinder $n = k$ and $r \cdot n = r \cdot k = z$. (Fig. A.2) where r is a position vector, $r = xi + yj + zk$

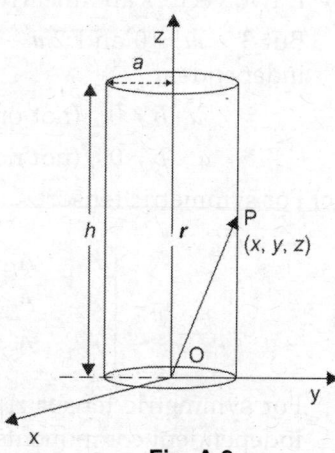

Fig. A.2

Thus $\int_S r \cdot ds = \int r \cdot n\,ds = z \int_S ds$

For top surface $\int ds = \pi a^2$ and $z = h$ $\therefore$ $\int_S r \cdot ds = h\pi a^2$

Position vector $xi + yj$ is normal to the curved surface of the cylinder and the unit vector normal to it is

$$n = \frac{xi + yj}{\sqrt{x^2 + y^2}}$$

and $\quad r \cdot n = \dfrac{x^2 + y^2}{\sqrt{x^2 + y^2}} = \sqrt{x^2 + y^2} = a$

$\therefore$ Surface integral $= \int_S r \cdot ds = \int_s r \cdot n\,ds$

$$= a \int_S ds = a \cdot 2\pi ah = 2\pi ha^2$$

Thus, surface integral $= 2\pi ha^2 + \pi ha^2 = 3\pi ha^2$

6.(a) $\dfrac{dy}{dx} = -\dfrac{x}{y+1}$ $\Rightarrow$ $ydy + dy + dx = 0$

or $\quad \dfrac{x^2}{2} + \dfrac{y^2}{2} + y = C$

or $x^2 + y^2 + 2y = 2C$ or $(x - 0)^2 + (y + 1)^2 = (2C + 1)$
which is a family of circles with different radii.

7.(c) $f(z) = \dfrac{z \sin z}{z - \pi}$ has a pole of order z at $z = \pi$

8.(b) $\oint f(z)dz = 0$

9.(a) If two vectors are linearly dependent then $a + mb = 0$ for some value of m.
But $3 + m = 0$ and $2m + 2 = 0$ do not have any solution, so they are linearly independent.

$\qquad a \cdot b \neq 0$ (not orthogonal)

$\qquad a \times b \neq 0$ (not normalized)

10.(c) For symmetric tensor

$$A_{ij} = \begin{bmatrix} A_{11} & A_{12} & A_{13} \\ A_{21} & A_{22} & A_{23} \\ A_{31} & A_{32} & A_{33} \end{bmatrix}$$

For symmetric tensor $A_{12} = A_{21}$, $A_{23} = A_{32}$ and $A_{13} = A_{31}$. Hence there are 6 independent components.

11.(b) $A = \begin{pmatrix} 0 & 1 & 0 \\ 1 & 0 & 1 \\ 0 & 1 & 0 \end{pmatrix}$

on diagonalizing the matrix A $|A - \lambda I| = 0$

$\therefore$ $\begin{vmatrix} 0 & 1 & 0 \\ 1 & 0 & 1 \\ 0 & 1 & 0 \end{vmatrix} - \lambda \begin{vmatrix} 1 & 0 & 0 \\ 0 & 1 & 0 \\ 0 & 0 & 1 \end{vmatrix} = 0$

or $\begin{vmatrix} -\lambda & 1 & 0 \\ 1 & -\lambda & 1 \\ 0 & 1 & -\lambda \end{vmatrix} = 0$ or $-\lambda(\lambda^2 - 1) + \lambda = 0$

or $\lambda(\lambda^2 - 0) = 0$ or $\lambda = 0, +\sqrt{2}, -\sqrt{2}$

12.(d) Let $I = \oint_C e^{1/z} dz$

$\therefore$ $I = \oint_C \left(1 + \frac{1}{z} + \frac{1}{2z^2} + \cdots\right) dz$

Here singularity of $e^{1/z}$ is $z = 0$ and lies inside the contour $|z| = 1$ (Fig. A.3). Now using Cauchy residue theorem

$\oint_C e^{1/2} dz = 2\pi i \sum R = 2\pi i \lim_{z \to 0} |1| = 2\pi i$

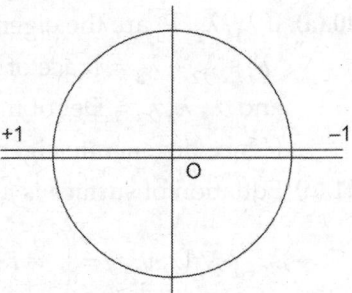

Fig. A.3

13.(b) For a symmetric function coefficients $b_n = 0$. Hence solution is (b)

14.(d) Pressure is not a tensor. All others are tensors of rank 3

15.(b) Speed of light in vacuum is an absolute constant, and independent of the relative motion of the source and observer. In electron frame, the speed of photon is c hence in photon frame electron has velocity $-c$.

16.(c) Laplace transform is given by

$$f(s) = \int_0^\infty f(t) e^{-st} dt$$

Laplace transform of $\sin kl$ is $\dfrac{a}{s^2 + k^2}$ which gives option (c) wrong.

17.(b) $\mathbf{B} \times \mathbf{r} = \begin{vmatrix} i & j & k \\ B_x & B_y & B_z \\ x & y & z \end{vmatrix} = i(zB_y - yB_z) + j(xB_z - zB_x) + k(yB_x - xB_y)$

$\therefore$ $\mathbf{A} \cdot \mathbf{B} \times \mathbf{r} = A_x(zB_y - yB_z) + A_y(xB_z - zB_x) + A_z(yB_x - xB_y)$

$= x(A_yB_z - A_zB_y) + y(A_zB_x - A_xB_z) + z(A_xB_y - A_yB_x)$

$\therefore$ $\nabla[\mathbf{A} \cdot \mathbf{B} \times \mathbf{r}] = i(A_yB_z - A_zB_y) + j(A_zB_x - A_xB_z) + k(A_xB_y - A_yB_x) = \mathbf{A} \times \mathbf{B}$

18.(c) Since $\overline{|h+1} = n\overline{|h}$. Hence

$$\left|h+\frac{1}{2}\right. = \left(n-\frac{1}{2}\right)\left|n-\frac{1}{2}\right. = \left(h-\frac{1}{2}\right)\left(n-\frac{3}{2}\right)\left|h-\frac{3}{2}\right. = \left(h-\frac{1}{2}\right)\left(n-\frac{3}{2}\right)\left(h-\frac{5}{2}\right)\left|h-\frac{5}{2}\right.$$

$$= \left(h-\frac{1}{2}\right)\left(n-\frac{3}{2}\right)\left(n-\frac{5}{2}\right)\cdots\frac{1}{2}\left|\frac{1}{2}\right. = \frac{(2n-1)(2n-3)(2n-5)\cdots1\left|\frac{1}{2}\right.}{2^n}$$

$$= \frac{2n(2n-1)(2n-2)(2n-3)(2n-n)(2n-5)\cdots1\left|\frac{1}{2}\right.}{2^n\cdot2^n\,n(n-1)(n-2)\cdots1} = \frac{2n!}{2^{2n}n!}\sqrt{\pi}$$

19.(a) Residue $R = \lim\limits_{z\to1}(z-1)^2 f(z)$

$$= \lim\limits_{z\to1}\frac{16z}{z+3} = \frac{16}{4} = 4$$

20.(a) If $\lambda_1, \lambda_2, \lambda_3$ are the eigen values of the matrix, then

$\lambda_1 + \lambda_2 + \lambda_3 = $ Trace of matrix $= 4 + 4 + 4 = 12$

and $\lambda_1\lambda_2\lambda_3 = $ Det of matrix $= 50$, its values are 2, 5, 5

Hence degenerate eigen value of the matrix is 5.

21.(d) Equation of surface is $f(x, y, z) = x^2 + y^2 + z^2 - 3$

$$\therefore \quad \nabla f(x, y, z) = \left(\frac{\partial}{\partial x}i + \frac{\partial}{\partial y}j + \frac{\partial}{\partial z}k\right)(x^2 + y^2 + z^2 - 3) = 2xi + 2yj + 2zk$$

and value of gradient at (1, 1, 1) is

$$\nabla f(1, 1, 1) = 2i + 2j + 2k$$

$$\therefore \quad \text{unit vector} = \frac{2i + 2j + 2k}{\sqrt{4+4+4}} = \frac{i + j + k}{\sqrt{3}}$$

22.(b) Charge is invariant under Lorentz transformation.

23.(d) A square matrix is a unitary matrix if its conjugate transpose is equal to its inverse matrix, i.e.

$$A^H = A^{-1}$$

24.(b) The matrix tensor g_{ij} is

$$g_{ij} = \begin{bmatrix} dx & \dfrac{\sqrt{(2)^2-(1)^2}}{2}dx^1dx^2 \\[3mm] \dfrac{\sqrt{(2)^2-(1)^2}}{2}dx^1dx^2 & dx^2 \end{bmatrix}$$

$$\Rightarrow \quad \begin{bmatrix} 2 & \sqrt{\dfrac{3}{2}} \\[3mm] \sqrt{\dfrac{3}{2}} & 1 \end{bmatrix}$$

25.(b) Given, $f(x) = e^{-x^2}$ and $g(x) = |x|e^{-x^2}$

$\qquad f(x) = e^{-x^2}$ for $\forall x \in R$

$\qquad g(x) = xe^{-x^2}$ for $x \geq 0$

$\qquad\qquad = xe^{-x^2}$ for $x < 0$

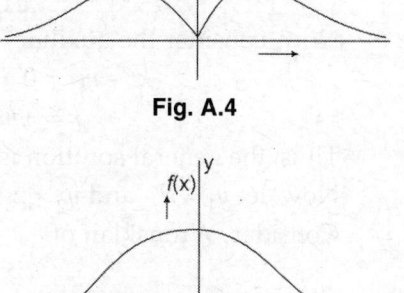

Fig. A.4

Graph of $f(x) = e^{-x^2}$ is shown in Fig. A.4.

Due to presence of $|x|$ in $g(x)$.

Corner (at origin) will be formed which makes $g(x)$ nondifferentiable at those points.

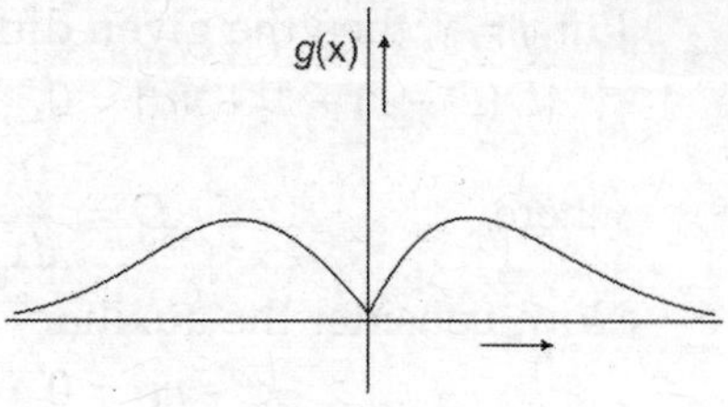

26.(c) Given function $f(z) = u(x, y) + v(x, y)$.

Clearly, $f(x)$ is harmonic function thus $u(x, y)$ and $v(x, y)$ will satisfy Laplace equation.

Fig. A.5

$f(z)$ cannot be Taylor expanded because $u(x, y)$ and $v(x, y)$ are independent functions.

27.(d) Let $I = \int_0^3 t^2 \delta(3t - 6)dt = \int_0^3 t^2 \times \delta\{3(t-2)\} dt$

$\qquad = \dfrac{1}{3} \int_0^3 t^2 \delta(t-2)dt \qquad \left[\because \delta(at-b) = \dfrac{1}{|a|}\delta\left(t - \dfrac{b}{a}\right) \right]$

$\qquad = \dfrac{1}{3}[2]^2 \qquad\qquad \left[\because \int_{-\infty}^{\infty} \delta(t-t_0)\phi(t)\lambda t = \phi(t_0) \right]$

$\qquad = \dfrac{4}{3} = 1.3$

28.(b) Given, $f(z) = \dfrac{1}{2\left(z + \dfrac{1}{2}\right)\cos(z\pi)}$

In the denominator, two terms give zeros at $z = -\dfrac{1}{2}$

$\qquad \cos(z\pi) = \cos\left(-\dfrac{\pi}{2}\right) = 0,$ or $z = -\dfrac{1}{2}$

$\qquad z + \dfrac{1}{2} = -\dfrac{1}{2} + \dfrac{1}{2} = 0$

Thus, $f(z)$ has second order pole at $z = -\dfrac{1}{2}$.

29.(c) Consider the given differential equation

$\qquad y'' + \dfrac{1}{z}y' - \dfrac{m^2}{z^2}y = 0$

$\Rightarrow \qquad z^2 y'' + zy' - m^2 y = 0$

Put $y = x$, then the given differential equation reduces to

$$\{D(D-1) + D - m^2\} = 0,$$

where
$$D = \frac{d}{dx}$$

Now, consider the auxiliary equation

$$k^2 - m^2 = 0 \qquad\qquad \text{(replacing } D \text{ by } k \text{ and } y \text{ by 1)}$$

$$\Rightarrow \qquad\qquad k = \pm m$$

Thus, the general solution is $y = c_1 z^m + c_2 z^{-m}$

Now, let $y_1 = z^m$ and $y_2 = z^{-m}$

Consider, Wronskian of

$$y_1, y_2 = W(y_1, y_2) = \begin{vmatrix} z^m & z^{-m} \\ mz^{m-1} & (-m)z^{m-1} \end{vmatrix}$$

$$= (-m)z^{-1} - mz^{-1}$$

$$= -2mz^{-1} \neq 0 \text{ for } m > 0 \text{ but not for } m = 0$$

$\Rightarrow y_1$ and y_2 are linearly independent solution of the given equation.

Also, for $m = 0$, given differential equation becomes

$$y'' + \frac{1}{z}y' = 0 \qquad\qquad\qquad\qquad (1)$$

Clearly, $y = \ln z$ and $y = 1$ satisfy the above Eq. (1)

Thus, $y = \ln z$ and $y = 1$ are the solution of Eq. (1)

Consider, $\begin{vmatrix} \ln z & 1 \\ \dfrac{1}{z} & 0 \end{vmatrix} = 0 - \dfrac{1}{z} \neq 0$

$\therefore$ $\ln z$ and 1 are linearly independent.

Thus, $\ln z$ and 1 are linearly independent solution of Eq. (1) for $m = 0$.

30.(a) Given, $\qquad\qquad H(w) = -\dfrac{2i}{\omega}$

Now, $\qquad\qquad F_1 = F\left[H\left(t+\dfrac{1}{2}\right)\right] = e^{-j\omega\left(-\frac{1}{2}\right)}H(\omega) = e^{\frac{j\omega}{2}}H(\omega)$

$$F_2 = F\left[H\left(t-\dfrac{1}{2}\right)\right] = e^{-j\omega\left(\frac{1}{2}\right)}H(\omega) = e^{-\frac{j\omega}{2}}H(\omega)$$

Thus, required fourier transform

$$F = \frac{1}{2}[F_1 - F_2] = \frac{1}{2}\left[e^{\frac{j\omega}{2}} - e^{-\frac{j\omega}{2}}\right]H(\omega)$$

$$= \frac{1}{2}\left[2i\sin\frac{\omega}{2}\right]\left[-\frac{2i}{\omega}\right] = \frac{\sin\left(\dfrac{\omega}{2}\right)}{\dfrac{\omega}{2}}$$

31.(d)
$$\frac{dy}{dx} = xy$$

$$\frac{dy}{y} = x\,dx \implies \int \frac{dy}{y} = \int x\,dx$$

$$\implies \qquad \log y = \frac{x^2}{2} + C$$

$$\implies \qquad y = e^{\frac{x^2}{2}+C} = e^{\frac{x^2}{2}} \cdot e^C = K \cdot e^{x^2/2}$$

At $\qquad x = 0,\ y = 2$

$\implies \qquad 2 = K \times e^0$

$\implies \qquad K = 2$

Hence, $\qquad y = 2e^{x^2/2}$

When $\qquad x = 2,\ y = 2 \cdot e^{4/2}$

$\implies \qquad y = 2 \cdot e^2$

32.(a) Let $\qquad z = x + iy$

$\rightarrow \qquad z^2 - (x + iy)^2 = (x^2 - y^2) + 2xyi$

Let $\qquad u = x^2 - y^2$ and $v = 2xy$

$\therefore \qquad \dfrac{\delta u}{\delta x} = 2x$ and $\dfrac{\delta v}{\delta y} = 2x$

Also, $\qquad \dfrac{\delta v}{\delta x} = 2y = \dfrac{\delta u}{\delta y}$

$\implies$ Cauchy–Riemann equation is satisfied.

This function is differentiable also.

Therefore, $f(z) = z^2$ is an analytic function.

33.(b)
$$\nabla f = i\frac{\partial f}{\partial x} + j\frac{\partial f}{\partial y} + k\frac{\partial f}{\partial z}$$

$$= i(x - y) + j(-x) + k(z)$$

At $(1, 1, 2)$; $\nabla f = 0\,i - j + 2k$

$$u = \text{direction} = \frac{-j + 2k}{\sqrt{5}}$$

34.(a) Given that $f(x) = \begin{cases} -1 & -\pi < x < 0 \\ 1 & 0 < x < \pi \end{cases}$

Now, we can write Fourier coefficients as

$$a_0 = \frac{2}{T}\int_{-T/2}^{T/2} f(x)\,dx$$

$$= \frac{2}{2\pi} \int_{-\pi}^{\pi} f(x)dx \qquad\qquad [\because T = 2\pi]$$

$$= \frac{1}{\pi}\left[\int_{-\pi}^{0} f(x)dx + \int_{0}^{\pi} f(x)dx \right]$$

$$= \frac{1}{\pi}\left[\int_{-\pi}^{0} (-1)dx + \int_{0}^{\pi} 1 dx \right]$$

$$= \frac{1}{\pi}\left[-(x)_{-\pi}^{0} + (x)_{0}^{\pi} \right] = \frac{1}{\pi}[-\pi + \pi] = 0$$

$$a_1 = \frac{2}{T} \int_{-T/2}^{T/2} f(x)\cos x\, dx$$

$$= \frac{2}{2\pi} \int_{-\pi}^{\pi} f(x)\cos x\, dx$$

$$= \frac{1}{\pi} \int_{-\pi}^{\pi} f(x)\cos x\, dx$$

$$= \frac{1}{\pi}\left[\int_{-\pi}^{\pi} f(x)\cos x\, dx + \int_{0}^{\pi} f(x)\cos x\, dx \right]$$

$$= \frac{1}{\pi}\left[\int_{-\pi}^{0} (-1)\cos x\, dx + \int_{0}^{\pi} (1)\cos x\, dx \right]$$

$$= \frac{1}{\pi}\left[(-1)(\sin x)_{-\pi}^{0} + (\sin x)_{0}^{\pi} \right]$$

$$= \frac{1}{\pi}\left[(-1)(0 - 0) + 0 \right] = 0$$

$$b_1 = \frac{2}{T} \int_{-T/2}^{T/2} f(x)\sin x\, dx$$

$$= \frac{2}{2\pi} \int_{-\pi}^{\pi} f(x)\sin x\, dx$$

$$= \frac{1}{\pi}\left[\int_{-\pi}^{0} f(x)\sin x\, dx + \int_{0}^{\pi} f(x)\sin x\, dx \right]$$

$$= \frac{1}{\pi}\left[\int_{-\pi}^{0} (-1)\sin x\, dx + \int_{0}^{\pi} \sin x\, dx \right] = \frac{1}{\pi}\left[(\cos x)_{-\pi}^{0} + (-\cos x)_{0}^{\pi} \right]$$

$$b_1 = \frac{1}{\pi}\left[1(-1) - (-1 - 1) \right] = \frac{1}{\pi}[2 + 2] = \frac{4}{\pi}$$

Similarly by solving, we get $a_1 = 0$, $b_2 = 0$ and $a_3 = 0$

$$\therefore \quad b_3 = \frac{1}{\pi}\left[\int_{-\pi}^{0} -\sin 3x\, dx + +\int_{0}^{\pi} \sin 3x\, dx\right]$$

$$= \frac{1}{\pi}\left[\left(\frac{\cos 3x}{3}\right)_{-\pi}^{0} - \left(\frac{\cos 3x}{3}\right)_{0}^{\pi}\right]$$

$$= \frac{1}{\pi}\left[\frac{1}{3}(1+1) - \frac{1}{3}(-1-1)\right] = \frac{1}{\pi}\left[\frac{2}{3} + \frac{2}{3}\right] = \frac{4}{3\pi}$$

Therefore, the appropriate Fourier Series is

$$f(x) = a_0 + a_1 \cos x + b_1 \sin x + a_2 \cos 2x$$
$$+ b_2 \sin 2x + a_3 \cos 3x + b_3 \sin 3x + \cdots$$

$$= 0 + 0 + \frac{4}{\pi}\sin x + 0 + 0 + 0 + \frac{4}{3\pi}\sin 3x + \cdots$$

$$= \frac{4}{\pi}\left[\sin x + \frac{\sin 3x}{3} + \cdots\right]$$

35. $\oint \dfrac{dz}{1+z^2}$ from $-\infty$ to $+\infty$

Using the residue theorem,

$$\oint \frac{dz}{1+z^2} = 2\pi i \sum \text{Residues}$$

$$2\pi i \times \frac{1}{2i} = \pi = 3.14$$

36. $\quad l = \displaystyle\int_{0}^{\infty} x^2 e^{-x^2}\, dx,\left[\text{let } x^2 = u,\ 2x\, dx = du \Rightarrow dx = \frac{du}{2\sqrt{u}}\right]$

$$l = \int_{0}^{\infty} u e^{-u} \frac{du}{2\sqrt{u}} = \frac{1}{2}\int_{0}^{\infty} u^{1/2} e^{-u}\, du$$

$$= \frac{1}{2}\left[\left(\frac{3}{2}\right)\right] = \frac{1}{2} \times \frac{1}{2} \times \sqrt{\pi} = \frac{\sqrt{\pi}}{4} = 0.44$$

37. $\quad z = u(x, y) + iv(x, y)$ and $v(x, y) = 2xy + 3y$

$$\frac{\partial v(x,y)}{\partial y} = 2x \quad \Rightarrow \quad \frac{\partial v(x,y)}{\partial y} = \frac{\partial u(x,y)}{\partial x}$$

$$\therefore \quad \frac{\partial u}{\partial x} = 2x$$

$$u = \int 2x\, dx = \frac{2x^2}{2} + c = x^2 + c$$

$$u(x, 0) \Rightarrow c = 0$$
$$u = x^2$$
$$u(1 + i) = (1 + i)^2 = 1 + i^2 + 2i = 0 + 2i$$
Real part = 0.00

38.(c) $X = \begin{pmatrix} 1 \\ 2 \\ 3 \\ \vdots \\ n \end{pmatrix}, X^T = 1,2,3,\cdots,n$

$M = XX^T$ is a square matrix of order $n \times n$, where I have taken all entries nonzero.

M is a matrix of rank 1 where one eigen value is nonzero (which is true) and all other $n - 1$ eigenvalue are zeroes.

39. $\dfrac{dy}{dx} + y \tan x = \cos x$

Integrating factor $(IF)y = e^{\int \tan x \, dx} = e^{\ln(\sec x)} = \sec x$

$$(IF)y = \int (IF)\cos x \, dx$$

$$y(\sec x) = \int \sec x \cos x \, dx$$

$$y = \frac{x}{\sec x} = x\cos x + C$$

$$y(0) = 0 \quad \Rightarrow \quad C = 0$$

$$y = x\cos x, y\left(\frac{\pi}{3}\right) = \frac{\pi}{3}\cos\left(\frac{\pi}{3}\right)$$

$$= \frac{\pi}{3} \times \frac{1}{2} = \frac{\pi}{6} = 0.52.$$

Suggested Readings

GB Arfken and HJ Weber, *Mathematical Methods for Physicists*, 6th edn, Elsevier, Amsterdam (2005).

S Nair, *Advanced Topics in Applied Mathematics*, Cambridge University Press, Cambridge (2009).

P Blanchard and E Bruning, *Mathematical Methods in Physics*, Birkhauser, Boston (2003).

E Butkov, *Mathematical Physics*, Addison Wesley, Reading, MA (1968).

RV Churchill *et al*, *Complex Variables*, 3rd edn, McGraw Hill, New York (1974).

SM Lea, *Mathematics for Physicists*, Thomson Brooks Cole, Belmont, CA (2004).

J Mathwes and RL Walker, *Mathematical Methods of Physics*, Benjamin, Menlo Park (1964).

PM Morse and H Fressback, *Methods of Theoretical Physics*, McGraw Hill, New York (1953).

M Hamermesh, *Group Theory and Its Application to Physical Problems*, Addison Wesley, Reading, MA (1962).

WH Press *et al*, *Numerical Recipes*, Cambridge University Press, Cambridge (1986).

M Abramowit and IA Stegun, *Handbook of Mathematical Functions*, National Bureau of Standards, Washington DC (1970).

IS Gradshteyn and IM Ryzkhik, *Tables of Integrals, Series and Products*, 4th ed, Academic Press, New York (1965).

W Feller, *An Introduction to Probability Theory and Its Application* (Vols 1 and 2), Wiley, New York (1961, 1966).

AC Aitken, *Determinants and Matrices*, Wiley, New York (1975).

HD Block, *Introduction to Tensor Analysis*, Merrill Columbus, Ohio (1962).

AD Booth, *Numerical Methods*, Plenum, New York (1975).

RD Carmichael, *Introduction to the Theory of Groups of Finite Order*, Dover, New York (1956).

A Cohen, *An Elementary Treatise on Differential Equations*, DC Heath, Boston, Massachusetts (1933). ··

M Hall, *Theory of Groups*, Macmillan, New York (1967).

PG Hoel, *Introduction to Mathematical Statistics*, Wiley, New York (1966).

Index